W9-BAG-303

HOLT

BIOLOGY

Visualizing Life

George B. Johnson

HOLT, RINEHART AND WINSTON

Harcourt Brace & Company

Austin • New York • Orlando • Atlanta • San Francisco • Boston • Dallas • Toronto • London

Acknowledgments

Staff Credits

Executive Editor
Ellen Standafer

Managing Editor
William Wahlgren

Senior Editor
Susan Feldkamp

Project Editors
Carolyn Biegert
Mitchell Leslie
Jennifer Linn
Ann Tarleton

Copyediting
Steve Oelenberger
Laurie Baker

Prepress
Beth Prevelige
Simira Davis
Rose Degollado

Media
Susan Mussey

Manufacturing
Mike Roche
Laura Cuellar

Design
Foca, Inc.

Cover Design and Electronic File Production
Morgan-Cain & Associates

How does this flower attract pollinators? See page 386.

Contributing Writers

David Jaeger
Will C. Wood High School
Vacaville, CA

Nancy Morvillo, Ph.D.
Howard Hughes Medical Institute
Undergraduate Biology Educational Program
State University of New York at Stony Brook
Stony Brook, NY

Karen M. Nein
Science Consultant
Denver, CO

Salvatore Tocci
Science Department
East Hampton High School
East Hampton, NY

Suzanne Weisker
Science Teacher and Department Chair
Will C. Wood High School
Vacaville, CA

Lab Reviewers

George Nassis
Ken Rainis
Geoffrey Smith
WARD'S Natural Science Establishment
Rochester, NY

Ted Parker
Forest Grove, OR

Mark Stallings, Ph.D.
Chair, Science Department
Gilmer High School
Ellijay, GA

Jay A. Young, Ph.D.
Chemical Safety Consultant
Silver Spring, MD

Reviewers

Lois Bergquist, Ph.D.
Professor of Microbiology
Los Angeles Valley College
Van Nuys, CA

Margaret Bradley-Schwartz
Biology Teacher
Lexington High School
Lexington, MA

G. Lynn Carlson, Ph.D.
Senior Lecturer/Lab Manager
University of Wisconsin–Parkside
Kenosha, WI

Pat Carter
Biology/Physics Teacher
Central High School
Evansville, IN

Phillip Creighton, Ph.D.
Dean, School of Science and Technology
Salisbury State University
Salisbury, MD

Mary Pitt Davis
Science Teacher
Long Reach High School
Columbia, MD

Portions of this work were published in previous editions.

Printed in the United States of America

ISBN 0-03-016723-X

13 14 15 16 048 06 05 04 03

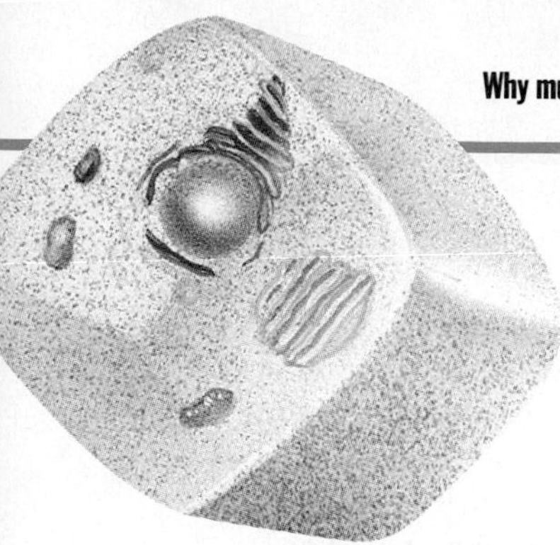

Why must cells be small? See page 43 to find out.

Michael Dole
Science Teacher/Dept. Chairperson
Covina High School
Covina, CA

William J. Ehmann, Ph.D
Chair, Department of Environmental Science
Trinity College of Washington, DC
Washington, DC

David G. Futch, Ph.D.
Associate Professor
Department of Biology
San Diego State University
San Diego, CA

Jim Gardner
Science Department Chair
Central High School
Evansville, IN

Herbert Grossman, Ph.D.
Associate Professor of Botany
Pennsylvania State University
State College, PA

Frank Harrold, Ph.D.
Professor of Anthropology
Department of Sociology and Anthropology
The University of Texas–Arlington
Arlington, TX

Arthur Hulse, Ph.D.
Department of Biology
Indiana University of PA
Indiana, PA

Clifford Keller, Ph.D.
Institute of Neuroscience
University of Oregon
Eugene, OR

Hillar Klandorf, Ph.D.
Associate Professor
Division of Animal and Veterinary Science
West Virginia University
Morgantown, WV

Patty Korn
McLean High School
McLean, VA

Lynne Krenicky
Biology Teacher
Clifton High School
Clifton, NJ

Jo Ann D. Lane
Science Department Chair
St. Ignatius High School
Cleveland, OH

Sheila Logiudice
Biology Instructor
Science/Technology
Lowell High School
Lowell, MA

Dan Millin
Biology Instructor
West Bend East High School
West Bend, WI

Martin Nickels, Ph.D.
Professor of Anthropology
Anthropology Program
Illinois State University
Normal, IL

David O. Norris, Ph.D.
Professor of Biology
E.P.O. Biology
University of Colorado
Boulder, CO

William H. Peltz
Greenwich Academy
Greenwich, CT

Sharon Perlman
Biology Teacher, Environmental Specialist
Dade County Public Schools Center for Environmental Education
Miami, FL

Terence M. Phillips, Ph.D., D.Sc.
Professor of Medicine/Immunochemistry
George Washington University Medical Center
Washington, DC

Joseph B. Schiel, Jr., Ph.D.
Biology Teacher
Artesia High School
Artesia, NM

Richard Storey, Ph.D.
Professor and Chair of Biology
Department of Biology
Colorado College
Colorado Springs, CO

Peter Swanson
Biology Teacher
Quincy High School
Quincy, MA

Mary Kay Thomas
Bowling Green High School
Bowling Green, MO

Donald W. Tuff, Ph.D.
Department of Biology
Southwest Texas State University
San Marcos, TX

Betty H. Smith
Assistant Principal/Curriculum Coordinator
Pineville High School
Pineville, LA

E. Peter Volpe, Ph.D.
Professor of Basic Medical Sciences
Division of Basic Medical Sciences
Mercer University School of Medicine
Macon, GA

To learn how a lizard changes its body temperature, see page 543.

A Message From the Author

You are about to have a lot of fun. You might doubt this because you think science is supposed to be hard and dull, right? It's not. Especially not biology! Biology is the study of the living world—of dinosaurs; of AIDS and acid rain; and of how bees fly, what tigers eat, and why people grow old. Remember when you were young and tried to capture lightning bugs or touch snowflakes—experiencing the sheer wonder of looking at things to see how nature works? That's what biology is, looking at nature and asking questions about it. Each of us was acting as a biologist when we were kids, and there was nothing dull about the excitement we felt then. And biology doesn't have to be dull now. The questions are every bit as interesting as they were, and seeking the answers is every bit as much fun. The secret is to not get bogged down in details. Scientists have learned a lot of information—after all, they have been studying biology for hundreds of years—and wading through this sea of facts can be discouraging. People who find science hard usually get bogged down learning information. So don't do it! Learn ideas instead. At the heart of biology is a set of simple ideas that explain why things work the way they do and how they got that way. Focus on understanding these ideas and you won't lose sight of the questions, and the fun.

Of course, the 35 chapters you see listed on the facing page do contain information that is important to you personally. You need to know how to avoid catching AIDS, why smoking cigarettes will give you cancer, and what you can do to help save the environment. Also, in a broader sense, biology is important to you because you are going to have to live the rest of your life in a world very different from today's. You will live in a rapidly changing world crowded with people, and biology will play a very important role in it.

In writing this book, I have tried to practice what I preach, to focus on ideas rather than information, and to not lose sight of the fun of science. You will notice that there are a lot of pictures in this book; I believe learning is visual and that pictures help. So take it seriously, keep track of the ideas—and have fun!

George B. Johnson, Ph.D.
Washington University
St. Louis, MO

Contents in Brief

Table of Contents

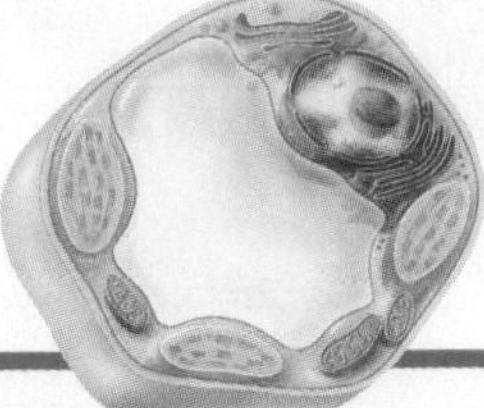

What's inside this plant cell? To find out, see page 54.

Unit One Study of Life

How does this structure turn light energy into chemical energy? See page 88.

Unit Two Continuity of Life

How do bacteria divide? Find the answer on page 104.

What makes the Gulf seaside sparrow unique? See page 184.

Unit Two *continued*

Take a tour of a dinosaur on page 207.

Unit Three The Environment

How can community action transform this vacant lot? See page 304.

Unit Four Diversity of Life

How does a *Paramecium* move? See page 361.

Find out how water travels through a plant on page 396.

CONTENTS

Unit Five Animal Kingdom

Learn the secrets of arthropod success starting on page 450.

What is "double-loop" circulation? To find out, see page 466.

How is this beetle beneficial to humans? See page 508.

Unit Five *continued*

To learn about the biology of this turtle, see page 548.

Unit Six Human Life

The leg bone's connected to the hipbone? Check it out on page 588.

CONTENTS

Unit Six *continued*

What's happening here? See page 637.

A taco tastes great, but now what happens? Learn about digestion starting on page 706.

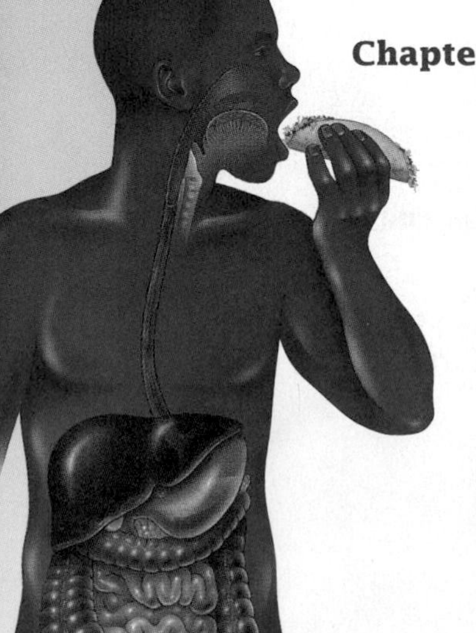

Unit Six *continued*

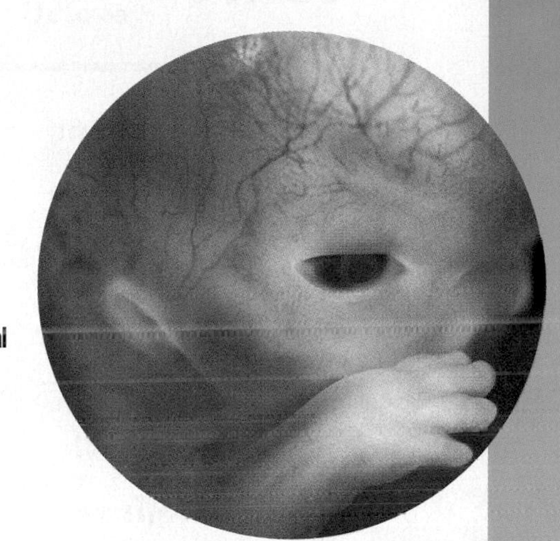

What key events occur during fetal development? See page 733.

Practice your graphing skills starting on page 832.

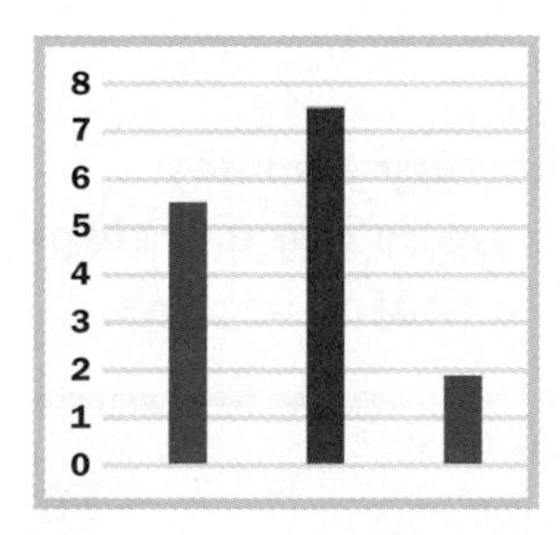

Features

Tours

Take a *Tour* to discover the sizes, habitats, and evolutionary relationships of living things.

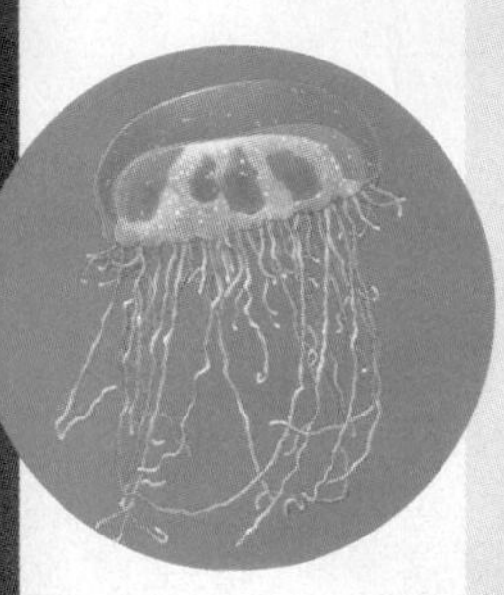

A Closer Look

Discover key structures and learn how they function by taking *A Closer Look*.

Evolutionary Journeys

Travel on an *Evolutionary Journey* through time as you trace the development of each of these major organs.

Career Opportunities

How can your knowledge of biology lead to an exciting job? Find out in *Career Opportunities*.

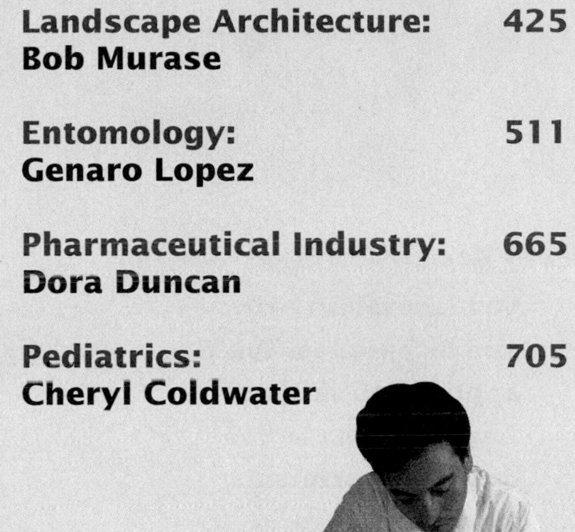

Science, Technology, and Society

Science, Technology, and Society explores the conflicts that can arise between new technologies and the needs of society.

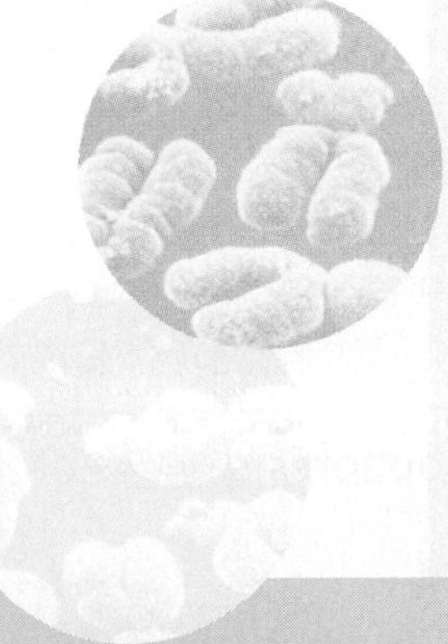

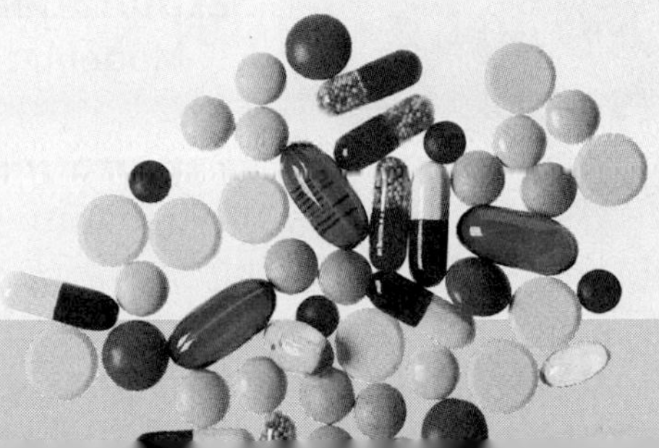

Lab Program

- *Explorations* teach you laboratory skills and techniques. You'll apply these skills in *Investigations* as you design experiments to solve real-world problems.
- Using CD-ROM technology, *Interactive Explorations* allow you to see and control biological phenomena in ways never before possible.

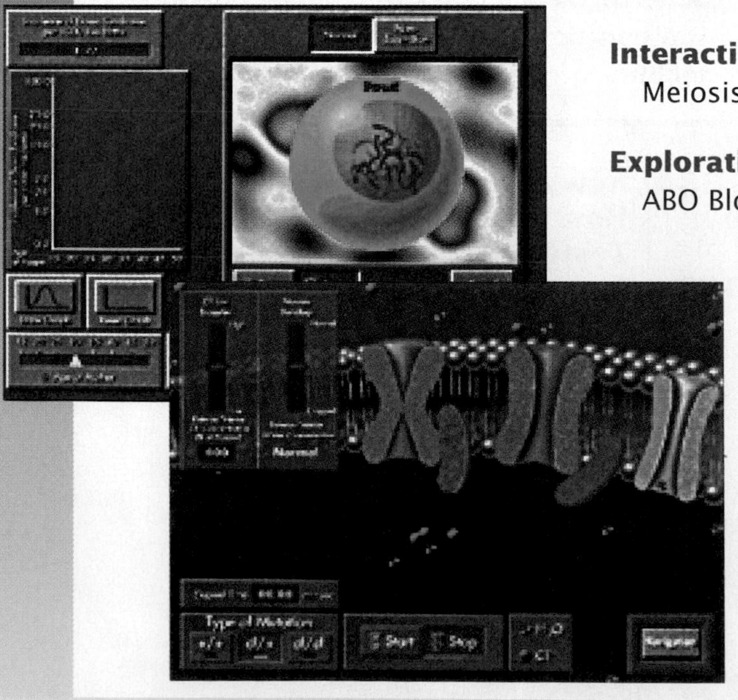

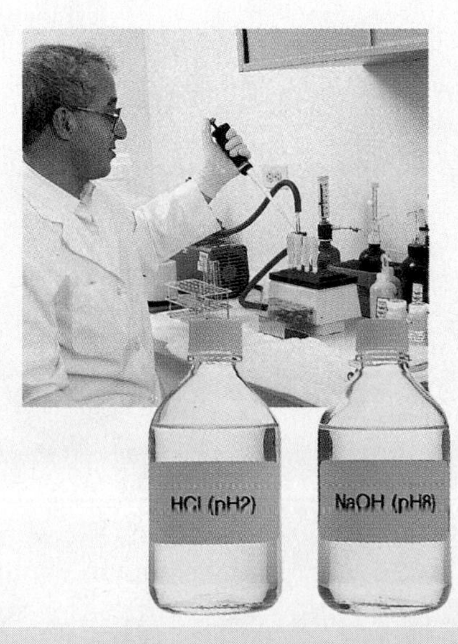

HCl (pH2)
NaOH (pH8)

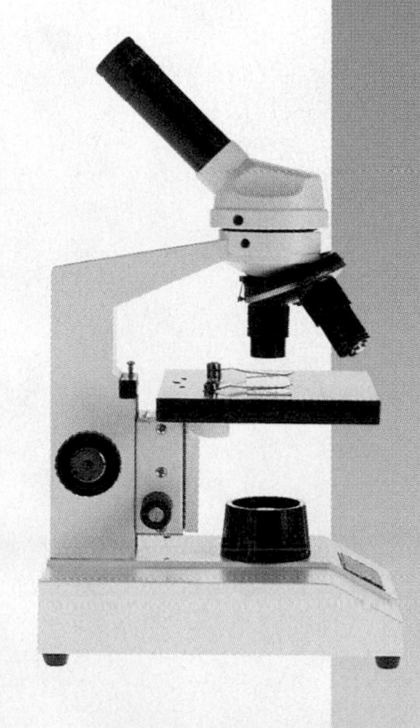

UNIT

One

Study of Life

Cilia lining the human windpipe

GENES AND INHERITANCE

You are a unique combination of traits—hair color, eye color, body shape, and others—controlled by genes inherited from your parents and your ancestors. But some genes passed from parent to child are defective and cause diseases like cystic fibrosis (CF), a lung condition that causes death in nine out of ten patients before age 30.

Making a Diagnosis

Doctors discovered that people with CF have very salty sweat. They also learned that high levels of salt collect in the cells of the lungs, pancreas, and liver of people with this disease. The salt pulls water out of normal mucus, leaving the mucus thick and sticky.

Finding the Cause

The CF gene, like all genes, carries instructions for making proteins. In CF patients, certain protein structures that pump water and salt into and out of cells are built incorrectly. A defective gene carries the wrong instructions for making these proteins and is the cause of this inherited disease.

LOOKING AHEAD

- How could you find out if you are a carrier of a defective gene? See **Science, Technology, and Society: DNA Profiling: Promise or Peril?** pages 38–39.
- Where in a cell are instructions for making proteins? See **A Closer Look: Eukaryotic Cell,** page 53.
- Who helps scientists find answers to the mysteries of cell structure? See **Career Opportunities: Science Technology,** page 55.
- How do protein channels restrict the kind of particles that enter and exit a cell? See pages 69–71.

CHAPTER 1

The Science of Biology

Scientists are just beginning to identify and catalog the great diversity of life found in tropical rain forests, like this one in Costa Rica.

Biology is the study of the living world. Biologist Terry Erwin, at left, collects and studies tropical insects. Biologists also study how a child grows, what dinosaurs were like, and many other topics. People have puzzled about life for thousands of years, but only in the last 150 years have we really begun to understand how the world came to be the way it is.

1-1 Biology Today

OBJECTIVES

1. **List three diseases for which scientists are seeking cures.**
2. **Describe two environmental problems and the ways scientists are trying to solve them.**
3. **Explain three ways you can use biological knowledge to improve your health.**

Studying Life

Science is a way of investigating the world in order to form general rules about what causes things to happen. A scientist is an observer who searches for knowledge to find solutions to problems. Science has changed the world rapidly in modern times, and new and important scientific discoveries, like the one shown in **Figure 1-1**, continue to be made. What scientists have learned about life and how they learned it are the subjects of this text.

You are about to embark on the study of biology, the science of life. **Biology** is the study of living things, including ourselves. While biological scientists, called biologists, continue to learn new things all the time, they now have a pretty clear picture of what living things are like and how they function. Biologists have learned enough, in fact, to begin to make genetic tools that can change the living world in new and important ways—an exciting and sobering prospect. Biology's impact on our daily lives has become so great that every person needs to know about biology in order to make many personal decisions.

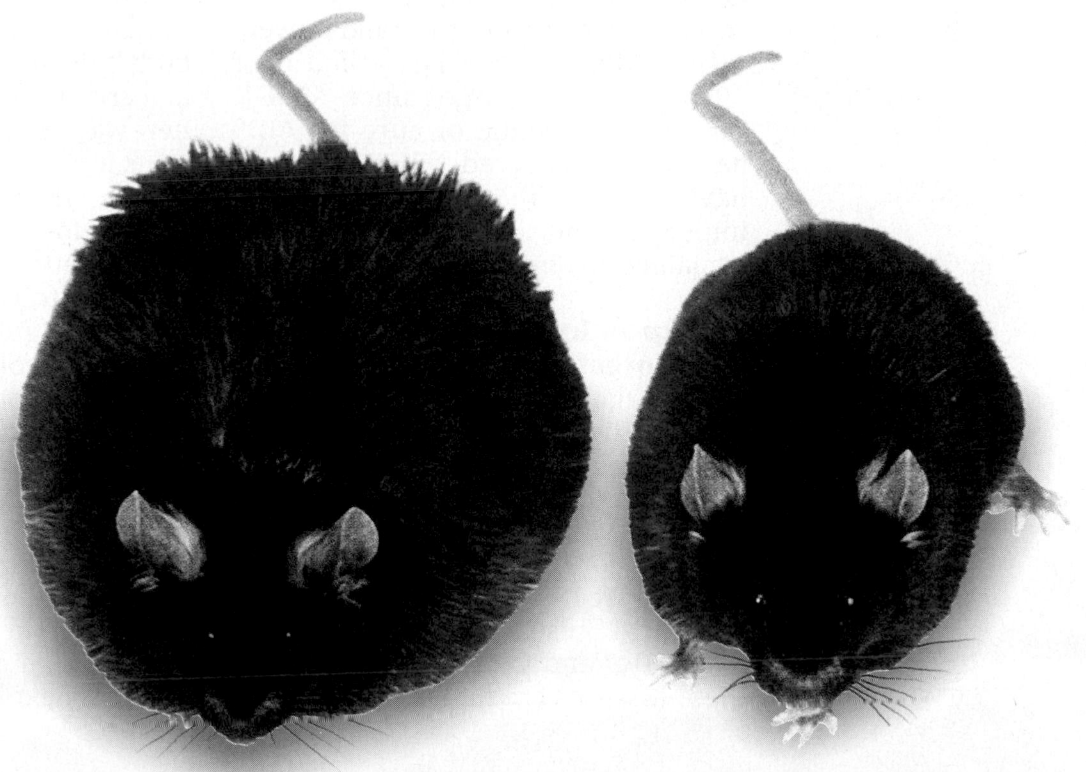

Figure 1-1
Biologist Jeffrey Friedman and his colleagues have discovered a protein that helps the body regulate weight. Mice that produce too little of this protein, like the one on the left, become grossly overweight. Friedman and other scientists are investigating how this protein functions in humans.

"What do I need to know about AIDS?"

Figure 1-2

a **What you learn about AIDS could save your life. There is no vaccine or cure for AIDS, and your knowledge of how the AIDS virus is transmitted is the first step in protecting yourself against infection.**

Biology and Medicine

Perhaps the most direct effect of biology on our lives is in medicine, where scientific advances are improving health and health care every day. For example, stop for a moment and think about the headlines you read in the newspaper this week and the news stories you heard on the radio or TV. Mixed in with the bad news and politics is the news that modern science is exploring ways to cure inherited disorders like cystic fibrosis and muscular dystrophy. Scientists are also searching for ways to treat and prevent AIDS, a disease that has killed over 300,000 Americans since 1981. Though no vaccine or cure for AIDS has been discovered, what scientists have learned about this disease is important for everyone to know, as explained in **Figure 1-2**.

Biological knowledge is used to fight infectious diseases

At the beginning of the twentieth century, the diseases influenza (flu), tuberculosis, and pneumonia were the top three causes of death in the United States. The influenza epidemic of 1918–1919 killed 22 million people worldwide in just 18 months. Now, thanks to biological research that produced antibiotics and vaccines, deaths from these diseases are far less common.

The battle against disease is far from won, however. More than 1 million people are likely to die of malaria this year. Spread by mosquitoes, this disease is common in tropical regions—more than 250 million people suffer from malaria at any one time. Almost every child under the age of five who gets malaria dies. The organism that causes malaria has a very complex life cycle and is therefore difficult to control. Using modern genetic engineering techniques, scientists are trying to design a vaccine that will stimulate the body's defenses to attack the organism at a critical stage of its life cycle. Also, new ways to reduce mosquito populations are being researched.

New strains of bacteria that cause tuberculosis, the fatal lung disease, have recently arisen. These strains are resistant to today's antibiotics. Partly as a result, tuberculosis will kill about 3 million people worldwide this year.

AIDS is another disease that is a subject of intensive research. AIDS is caused by a virus that destroys the body's ability to defend itself from infections. Because the virus changes the characteristics of its surface so quickly, normal vaccines don't work.

No one yet knows the solutions to these problems, but many approaches are being explored.

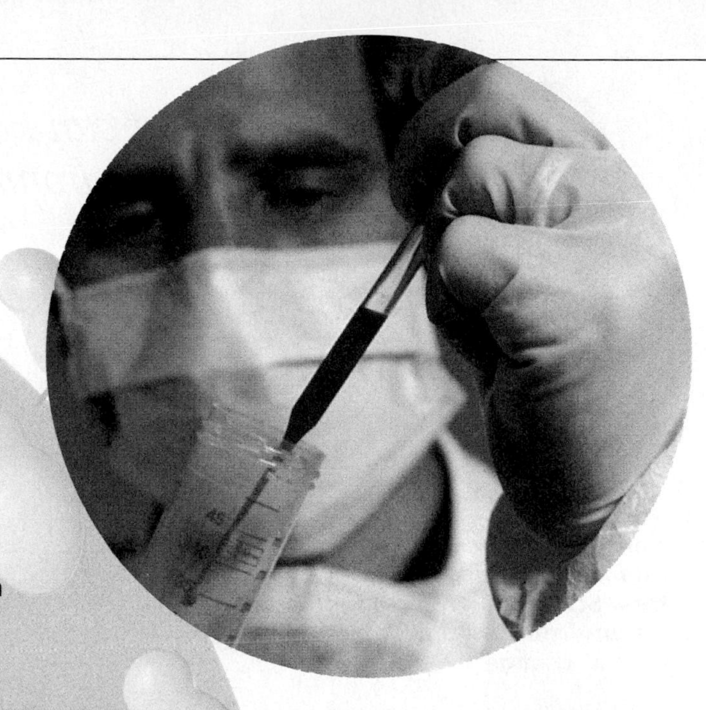

b Researchers, like this one at the Clinical Immunology Lab at East Tennessee State University, use what has been discovered about AIDS when they plan new experiments.

Biological knowledge is used to help cure genetic disorders

Among the most discouraging medical problems are fatal disorders that result from defective genes. **Genes** are the inherited "operating instructions" for the body. Two fatal genetic disorders are cystic fibrosis and muscular dystrophy. Cystic fibrosis affects about 30,000 Americans. Ninety percent of these individuals will die before their 30th birthday. Their lungs become clogged with thick mucus because of a defect in a single gene. Muscular dystrophy kills about 1 in 10,000 humans. Their muscles waste away because of a defect in another gene.

Until recently, such genetic disorders were almost always fatal. But in 1990 biologists used an approach called genetic engineering to transfer copies of a normally functioning gene from a healthy individual into a patient with defective copies of that gene. Three years later, this new therapy was first tried against cystic fibrosis. Transferring normally functioning genes into affected individuals offers the first real hope of curing genetic disorders. A major effort is now underway to identify and sequence every gene in the human body, a program that may open the way to curing many other genetic disorders.

c Politicians must understand the biology of AIDS if they are to make informed decisions about providing care facilities, funding research, and protecting the rights of AIDS victims.

"What can I do to help preserve the environment?"

Figure 1-3
Humans are changing the planet faster and on a greater scale than ever before in history. Scientists are seeking solutions to the environmental problems created by this rapid change.

a In 1989, the *Exxon Valdez* spilled its load of oil on the coast of Alaska. This cleanup crew is using high-pressure sprayers to help remove the oil from the beach.

Biology and the Environment

When you were born, the world contained just over 4 billion people. By the year 2000, the world's population will exceed 6 billion. The exploding human population is placing great stress on the planet by using more energy, consuming more resources, and producing more waste than ever before. One of the greatest challenges we face is finding ways to support so many people without harming the Earth. **Figure 1-3** shows examples of some of the environmental problems we face and describes some of the ways these problems are being addressed.

Biological knowledge is used to help feed a hungry world

One of the most immediate challenges facing today's world is producing enough food to feed its expanding population. Only three kinds of plants (rice, wheat, and corn) provide half of all human energy requirements worldwide. Researchers are presently working to increase the amount of food that can be obtained from a farm without requiring large amounts of fertilizers, pesticides, and expensive, energy-consuming equipment. New varieties of plants are being developed that are more resistant to disease and more tolerant of poor growing conditions. Also, scientists have used genetic engineering to add genes to existing crop plants. These genes increase the productivity of the crop plants by increasing resistance to pests, increasing growth rates, or improving nutritional quality.

Our effect on the environment is often negative

Think for a moment about the materials you use and discard in a day—plastic, paper, glass, and metal. We are beginning to exceed the Earth's capacity to absorb the waste we generate. Releasing pollutants into the atmosphere, for example, can produce acid rain that kills forests and poisons lakes. Industrial chemicals once considered harmless are destroying the ozone in the atmosphere that shields us from the sun's harmful rays. And the burning of gasoline, oil, natural

c The productivity of agricultural land is threatened by the loss of topsoil. Scientists are developing crops that can be grown without tilling the soil to remove weeds, thus reducing soil erosion.

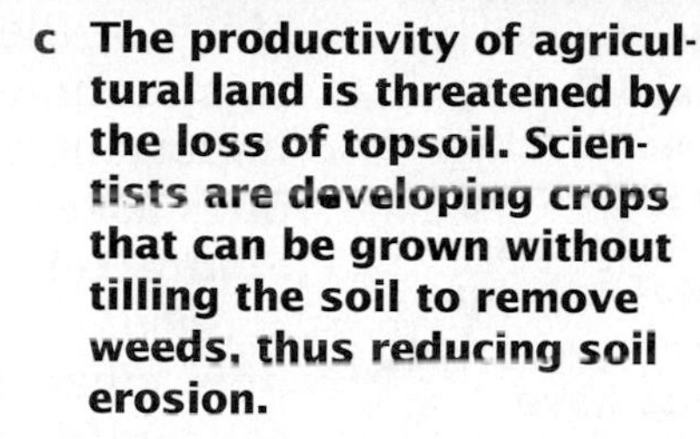

b More kinds of animals and plants live in tropical rain forests than anywhere else on Earth. Scientists like E. O. Wilson of Harvard University are trying to slow the rapid destruction of tropical rain forests.

gas, and coal releases large amounts of carbon dioxide into the atmosphere, causing the Earth's average temperature to rise because carbon dioxide traps the sun's heat.

Preserving a healthy world for your children

An old saying that is attributed to many different sources, from Native Americans to Amish farmers, states that "we do not inherit the world from our parents; we borrow it from our children." To ensure that our children have the resources they need tomorrow, we must stop wasting precious resources, such as topsoil and ground water, that are in limited supply.

Most important, we must not destroy the world's biological diversity. Unfortunately, we are doing just that, and on a very large scale. For example, the world's tropical rain forests are being destroyed at a breathtaking rate, along with much of the world's biological richness. Every second, about 3 hectares (1.3 acres) of tropical rain forest are destroyed—cut for lumber, burned to create fields for growing crops, or cleared to make pastures for grazing cattle. If this practice continues at such an alarming rate, only small patches of rain forest will remain by the middle of the twenty-first century. About half of the world's species live in tropical rain forests, and many of them will disappear as the rain forests are destroyed. Within the next 100 years, more than two-thirds of the species of organisms on Earth may become extinct—about the same percentage of species as disappeared 65 million years ago at the end of the age of the dinosaurs. Biologists and others are working to save as much as possible by creating preserves and by educating the public about the need to save some of the world's biological richness for future generations. However, if we wish to save biological diversity around the world, we all will have to take an active role.

"My uncle died of a heart attack three years ago, at age 45. Will I have heart problems too?"

"My mother had to have a skin cancer removed. Will I get cancer too?"

Figure 1-4
Asking questions is the first step toward finding answers. A knowledge of biology will help answer many of the questions these young adults have.

Biology and You

Much of the information provided in this course will help you answer important personal questions like the ones asked by the students shown in **Figure 1-4**.

Should I smoke? Biologists now know how cigarette smoking causes lung cancer. Chemicals in the smoke enter the cells of your lungs and damage them, and cancer results. Most people who die of lung cancer today are smokers. Smoking is also associated with other lung diseases and with heart disease. One out of three teenagers who take up smoking will die prematurely as a result. You will learn more about the dangers of smoking in Chapters 30 and 32.

How can I avoid developing heart disease? Biologists have learned that diet and exercise have a major influence on whether you are likely to have a heart attack or stroke. You will learn some things you can do to prevent heart disease in Chapter 32.

What are the risks in taking drugs? Most mind-altering drugs are addictive and thus very harmful. This includes the nicotine in cigarettes and the alcohol in wine, beer, and spirits. Biologists have learned much about the physical basis of drug dependency and about the far-reaching effects drugs can have. You will learn about the harmful effects of drugs in Chapter 30.

Will I get cancer? Cancer is the second leading cause of death in the United States. The good news is that scientists have learned that many cancers, including skin cancer, are related to lifestyle and diet and can thus be prevented. You will learn more about cancer in Chapters 6 and 33.

"Many of my friends smoke, and they say smoking is all right as long as you quit when you're older. Is that true?"

"I see other kids using drugs. They say they can stop whenever they want. I wonder if that's true."

SECTION REVIEW

1. **Why is research in the field of genetic engineering considered so important to medicine?**
2. **What environmental problems are being created by the rapidly growing human population?**
3. **List two personal decisions that a knowledge of biology can help you make.**

Science is a way of investigating the world. A scientist is an observer who is driven by the search for new knowledge and who uses methods that differ from those of writers or public officials. Public officials may make conclusions after thinking through a problem. However, scientists make hypotheses after observing a number of specific cases.

1-2 Science Is a Search for Knowledge

OBJECTIVES

1. **Describe the role of testing in verifying hypotheses.**
2. **Describe the importance of controls in testing hypotheses.**
3. **Compare the scientific definition of *theory* with the use of the word in everyday language.**

A Case Study in Science

Perhaps the best way to see how science works is to look at a real case in which science has been used to solve a problem and improve human health. One good example is the deadly disease malaria. **Figure 1-5** shows the global distribution of malaria. In 1941, more than 4,000 Americans died of malaria. In the year 2000, by contrast, fewer than five people are likely to die of malaria in the United States. This disease has been virtually eliminated from the United States and many other countries as a result of work begun by one man. His story is a very real example of how science works. His investigations were simple. They involved careful observation and the formulation of clear questions. Finding the cause of malaria was one of the greatest medical advances of all time.

Observations suggest questions to investigate

In the summer of 1897 an English physician, Ronald Ross, was working in a remote field hospital in Secunderabad, India. Ross set out to discover how malaria was transmitted. Of all tropical diseases, malaria was the greatest killer, taking more than a million lives a year in India alone. Doctors knew that the disease was caused by a microscopic parasite called *Plasmodium*, which could be found in the blood of malaria victims. However, no one was sure how the parasite was transmitted from one person to another. Working alone, Ross discovered how malaria is spread.

Ross observed that patients in the field hospital who did not have malaria were more likely to develop the deadly

Figure 1-5
Malaria kills more than 1 million people per year. The disease is found mainly in tropical regions in Africa, Latin America, and Asia (areas shown in red on the map).

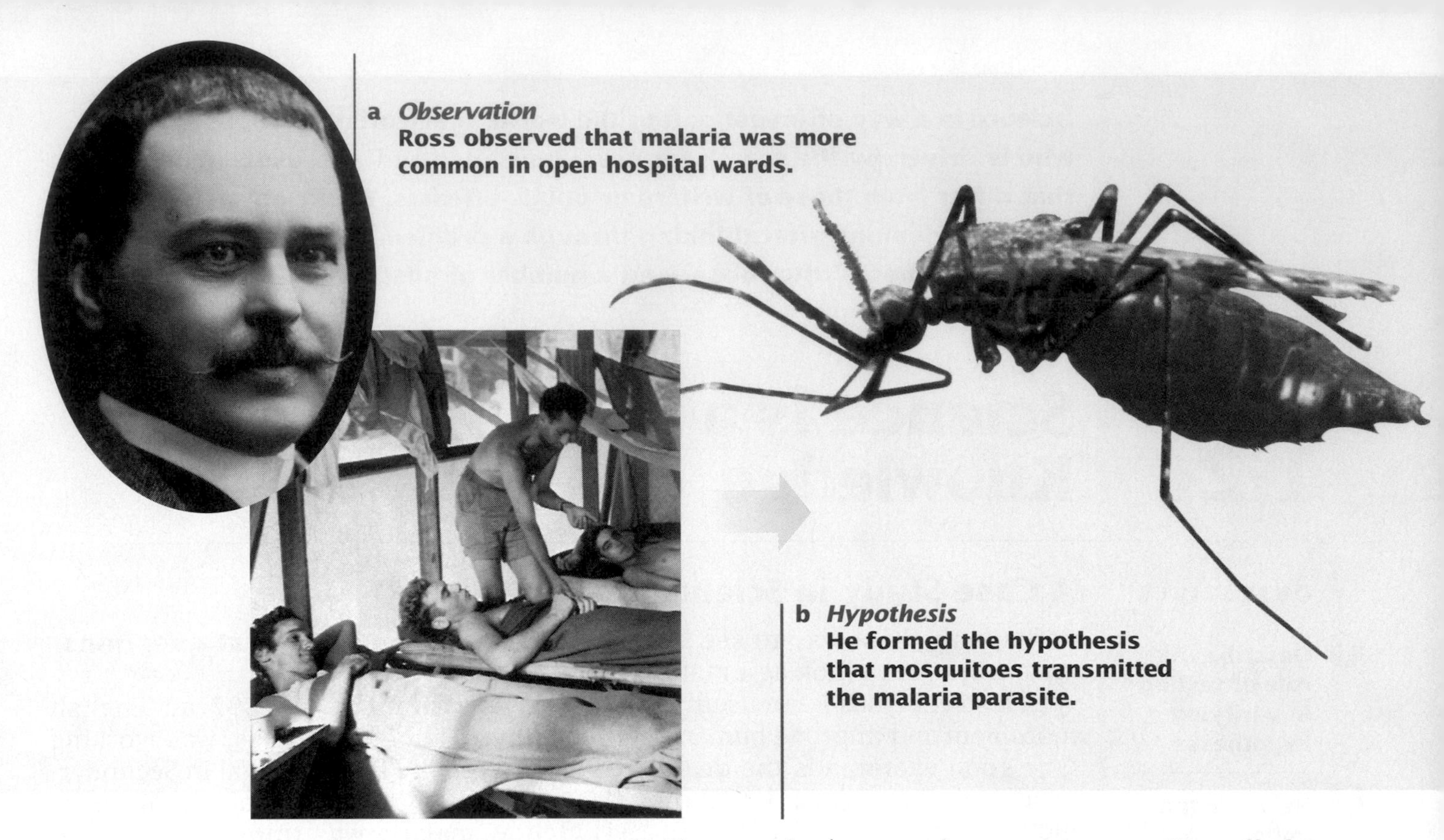

a *Observation*
Ross observed that malaria was more common in open hospital wards.
b *Hypothesis*
He formed the hypothesis that mosquitoes transmitted the malaria parasite.

Figure 1-6 Follow the steps of Ronald Ross's investigation of malaria.

disease in the open wards (those without screens or netting) than in wards with closed windows or screens. Ross wondered why people in open wards were much more likely to get malaria than those in closed wards.

Hypothesis: The basis of further investigation

Observing this pattern, Ross suggested an explanation for why people in open wards, like the one shown in **Figure 1-6a**, were more likely to get malaria. Such an explanation is called a hypothesis. A **hypothesis** is a testable explanation for an observation. Ross proposed that mosquitoes in the open wards might have spread the disease from patients with malaria to patients who did not have the disease. By observing the mosquitoes closely, Ross noted that they were the type known as *Anopheles*. Using this fact, Ross formulated a hypothesis. Ross's hypothesis was that *Anopheles* mosquitoes were spreading malaria from one patient to another.

Predictions: The framework for testing hypotheses

Ross knew that if his hypothesis were correct, he could reasonably expect several consequences. We call these expected consequences predictions. A **prediction** is what you expect to happen if a hypothesis is accurate. Ross predicted that *if* the *Anopheles* mosquitoes were spreading malaria (hypothesis), *then* mosquitoes that had bitten malaria patients and sucked up some of their blood would have picked up the *Plasmodium* parasite (prediction). Ross also predicted that parasites would be alive within the mosquito.

Testing under controlled conditions can verify predictions

The controlled test of a hypothesis is called an **experiment**. Ross carried out a type of experiment known as a control experiment. In a **control experiment**, a group that *has not been* exposed to the variable suspected of causing an effect is compared with one or more groups that *have been* exposed to the variable. The group not exposed to the variable is called a **control group** and serves as the standard for comparison. Since the suspected variable in Ross's experiment was exposure to malaria parasites from the blood of malaria victims, he compared mosquitoes that had fed on malaria patients with those that had fed on uninfected individuals.

c ***Predictions***
He predicted that only mosquitoes that had bitten malaria patients would carry the parasite, the small dark flecks visible below.

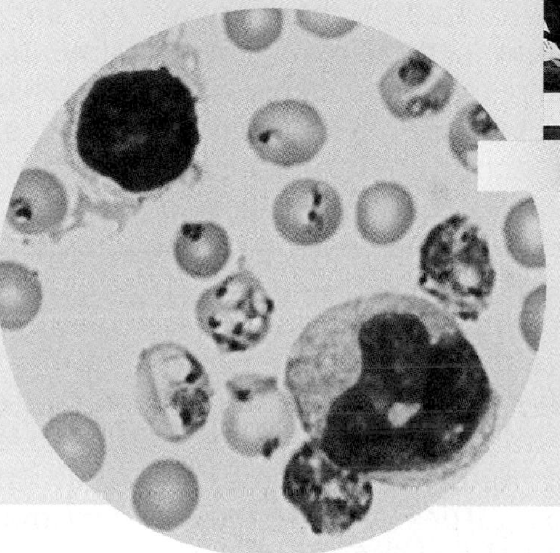

d ***Control experiment***
As a control experiment, Ross checked for the parasite in mosquitoes that had never bitten malaria patients.

e ***Theory***
Ross's findings were subjected to additional tests and were confirmed by other scientists.

Ross first looked for living malaria parasites in *Anopheles* mosquitoes that had fed on the blood of malaria patients. He carefully dissected each mosquito's stomach and found the live parasites. As a control group, Ross used mosquitoes that had not bitten someone with malaria. He checked to see if these mosquitoes also contained the parasites. He knew that if both groups of mosquitoes contained the parasites, then malaria patients could not possibly be the source of the parasites in the mosquitoes, and his hypothesis must be wrong. Gathering newly hatched mosquitoes that had not yet fed, he allowed them to mature and then fed them blood from people who didn't have malaria. When he examined the stomachs of these mosquitoes, he found no parasites. The control group of mosquitoes did not contain malaria parasites, confirming his hypothesis.

Theories: Explanations for observations

Ross reconfirmed and extended his hypothesis with later experiments. For instance, he suspected that the parasites must make their way from the mosquito's stomach to its salivary glands. In this way, the parasites would be transferred with the mosquito's saliva to the next person bitten. Ross showed by careful dissection that the parasites spread through an infected mosquito's body and were indeed present in the salivary glands. Other scientists also conducted experiments that confirmed Ross's hypothesis. A collection of related hypotheses that have been tested and supported is called a theory. A **theory** is a unifying explanation for a broad range of observations. Theories can have a major effect on science when they tie many accepted hypotheses together into a unified concept. Ross's theory that malaria is transmitted from one person to another by *Anopheles* mosquitoes was an important milestone in medicine. The idea that malaria epidemics could be prevented by combating mosquitoes was first put forth in a letter written by Ross to the government of India in 1901. Before the end of that year, American army doctors had eliminated almost all malaria from Havana, Cuba—where malaria had been at an epidemic stage—by greatly reducing the mosquito population. Few advances in the history of medicine have been more dramatic.

Figure 1-7 Cartoonist Gary Larson has a unique view of Ross's theory, but he has made an important biological error. Can you spot it?

"What a day! . . . I must have spread malaria across half the country."

Science requires continued verification of hypotheses

The essence of science is to reject any hypothesis not supported by observations and the results of control experiments. A new hypothesis is examined very closely to see what it predicts, and the predictions are then rigorously tested. If the predictions are not supported by the results of experiments or observations, the hypothesis is rejected. If they are confirmed, the hypothesis is subjected to further testing. One critical aspect of science is that a scientist's work is held up for review by other scientists. The validity of one's hypothesis is questioned by others until similar results are obtained from similar control experiments. This system of checking and rechecking hypotheses helps ensure that scientific reporting is factual and objective.

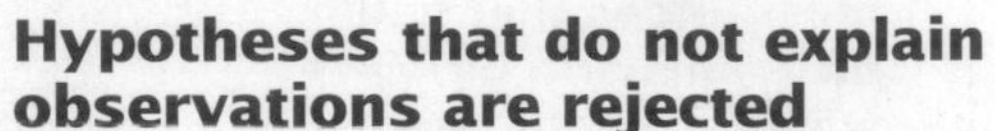

Hypotheses that do not explain observations are rejected

A scientist works by systematically showing that certain hypotheses are invalid, that is, that they are not consistent with the results of experiments. The results of all experiments are used to evaluate alternative hypotheses. An experiment is successful when it shows that one or more of the alternative hypotheses are inconsistent with observations. By conducting experiments, Ross was able to eliminate the hypothesis that mosquitoes *could* transmit the malaria parasite without biting malaria victims. He retained the hypothesis that if mosquitoes did bite malaria victims, then the mosquitoes could transmit the parasite. Scientific progress is often made in the same way a marble statue is made—by chipping away the unwanted bits.

Theories Have Limited Certainty

Theories are the solid foundation of science, that of which we are most certain. But in science there is no absolute certainty, no scientific "truth." The possibility always remains that future evidence will cause a theory to be revised or discarded. A scientist's acceptance of a theory is always provisional. For example, **Figure 1-8** describes how the accepted view of our solar system changed as new evidence became available.

The word *theory* is used differently by scientists than by the general public. To a scientist, a theory represents that of which he or she is most certain. To the general public, the word *theory* implies a lack of knowledge, a guess. How often have you heard someone say "It's only a theory" to imply lack of certainty? As you can imagine, confusion often results. In this text, the word *theory* will always be used in its scientific sense, as a generally accepted scientific principle.

Some theories are so strongly supported that the likelihood of their being rejected in the future is very small. Most of us would be willing to bet that the sun will rise in the east tomorrow, or that if an apple is dropped, it will fall. In physics, the theory of the atom is universally accepted, even though no one had seen an atom until recently. In biology, the theory of evolution by natural selection is so strongly supported by evidence that biologists accept it with as much certainty as physicists accept the theory of the atom. We will examine the theory of evolution in Chapter 10. This theory is very important to biologists because it provides the framework that unifies biology as a science.

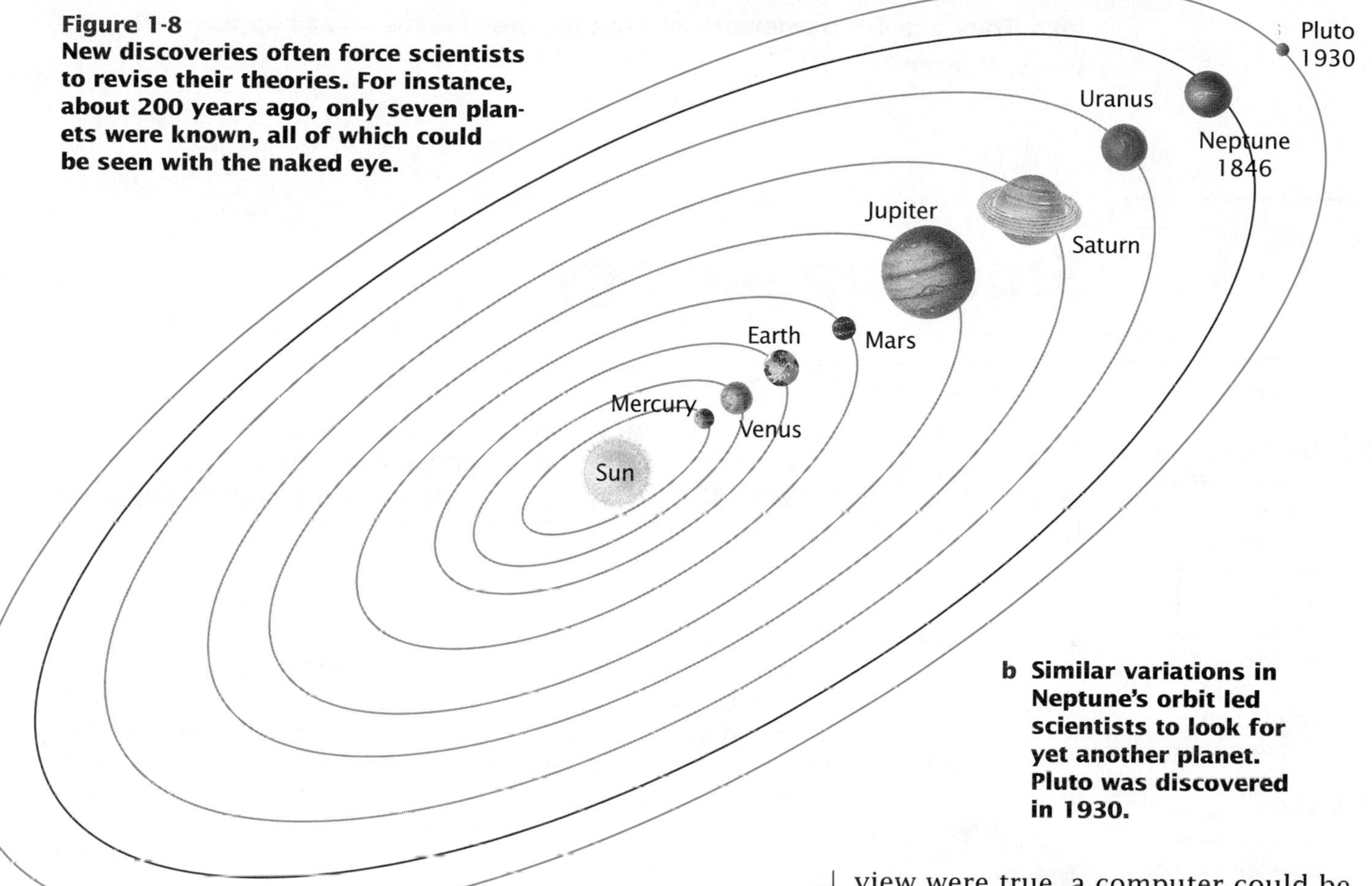

a **A "wobble" in the orbit of Uranus led scientists to suspect that a planet existed beyond Uranus. That planet, discovered in 1846, was Neptune.**

Figure 1-8
New discoveries often force scientists to revise their theories. For instance, about 200 years ago, only seven planets were known, all of which could be seen with the naked eye.

b **Similar variations in Neptune's orbit led scientists to look for yet another planet. Pluto was discovered in 1930.**

Scientific method: the systematic study of a question or problem

It was once fashionable to claim that scientific progress was the result of applying a series of steps called "the scientific method." In this view, science is a sequence of logical "either/or" steps, each step rejecting one of two incompatible alternatives. Trial-and-error testing could inevitably lead one through a maze of uncertainty. If this view were true, a computer could be programmed to be a good scientist. But science is not done this way. If you ask successful scientists how they do their work, you will find that they design experiments with a good idea of the results they will get. Not just any hypothesis is tested—only a hunch or educated guess that is based on all the scientist knows and that allows his or her imagination full play. Because insight and imagination are so important in scientific progress, some scientists are better than others.

SECTION REVIEW

1. **Under what conditions is a hypothesis supported?**
2. **Why was it necessary for Ross to examine the stomachs of mosquitoes that had not fed on blood from malaria patients?**
3. **Would Ross's hypothesis have been supported if the mosquitoes in the control group had contained malaria parasites? Explain your answer.**
4. **Write a sentence using *theory* in its scientific sense.**

From the mountain of information scientists have collected about living things, general principles of particular importance have emerged. You will encounter these principles repeatedly as you explore the ideas in this text. They are the framework of biology, the skeleton that supports all you will learn.

1-3 Studying Biology

OBJECTIVES

1. **List and describe the six major themes of biology.**
2. **Identify one way in which each theme relates to humans.**
3. **Relate evolution to natural selection.**

Themes Unify Ideas

This text uses six unifying principles, or themes, as a framework for your study. These themes are introduced here, and you will see them repeatedly throughout the text. Your goal should be to understand how the many topics you study in biology are examples of these themes.

Theme 1: Cell Structure and Function

All organisms are composed of **cells**, tiny compartments surrounded by membranes. Your body has about 10 trillion cells. Some of these cells like the white blood cells shown in **Figure 1-9**, perform highly specialized functions. The complex chemical processes that occur within cells are much the same in all organisms, and all cells have the same basic structure. A covering called a membrane surrounds the cell and controls what information and materials enter and leave. An internal fluid and an internal framework give shape to the cell and support its other components. A central zone or a nucleus contains the cell's genes, which are coded within long complex molecules called DNA. Many cells contain specialized structures that carry out some of the cell's activities. Cellular activities are often influenced by outside molecules that attach to certain proteins in the cell's membrane. Much of a cell's biology is determined by the nature of the proteins in its membrane. You will learn much more about cellular processes in Chapters 3, 4, 5, and 6.

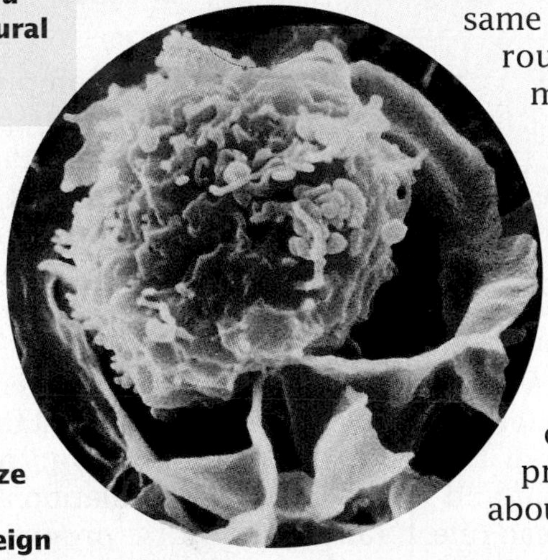

Figure 1-9 White blood cells recognize and destroy invading foreign cells, thus protecting the body from infection.

Theme 2: Stability and Homeostasis

Control is an essential aspect of life. All organisms must maintain a constant internal environment in order to function properly. For example, your body temperature must not vary by more than a few degrees. Your body has many mechanisms for maintaining **homeostasis**, a state of constant internal conditions. Some of these mechanisms involve circulating chemical signals called hormones; other mechanisms involve cells called neurons that carry electrical signals. Systems of neurons, called nervous systems, are capable of producing very complex behaviors, learning, and conscious thought.

Figure 1-10
The similarities between offspring and their parents is evident in this family of cheetahs.

Theme 3: Heredity

In organisms composed of many cells, such as humans, most of the cells are specialized, each performing distinct functions. For example, red blood cells carry oxygen, muscle cells contract, and nerve cells conduct electrical signals. All of the many different kinds of specialized cells, however, are descended from the same fertilized egg cell. As the cells grow and divide, their genes manage an orderly process of change called **development**. The process is controlled by genes and is the same in all humans.

You have probably noticed that children tend to resemble their parents. The transmission of characteristics from parents to their offspring is called **heredity**. Children resemble their parents because instructions for development are passed from parents to offspring. These instructions are in the form of genes. Sometimes "damage" to the genes occurs. These changes, called **mutations**, are usually harmful, but occasionally mutations help an organism to survive. For example, as you will learn in Chapter 10, a mutation in the gene for the blood protein hemoglobin produces resistance to malaria.

Theme 4: Evolution

Biologists have long suspected that life on Earth is the result of evolution. **Evolution** is inherited change in the characteristics of organisms over time. Living things are slowly changing and have apparently been changing since they first appeared on Earth. Charles Darwin proposed that this change is the result of **natural selection**. In natural selection, those organisms with favorable versions of genes are more likely to survive and reproduce. These favorable genes enable an organism to better meet the many challenges presented by its environment. Darwin's theory provides biology with a basis for understanding the diversity of life on Earth. The living things found on Earth today, including humans, resulted from a long history of organisms adjusting to a diverse and changing environment. A **species** is a group of organisms that look similar and can produce fertile offspring in their environment. Natural selection leads to changes in species over time.

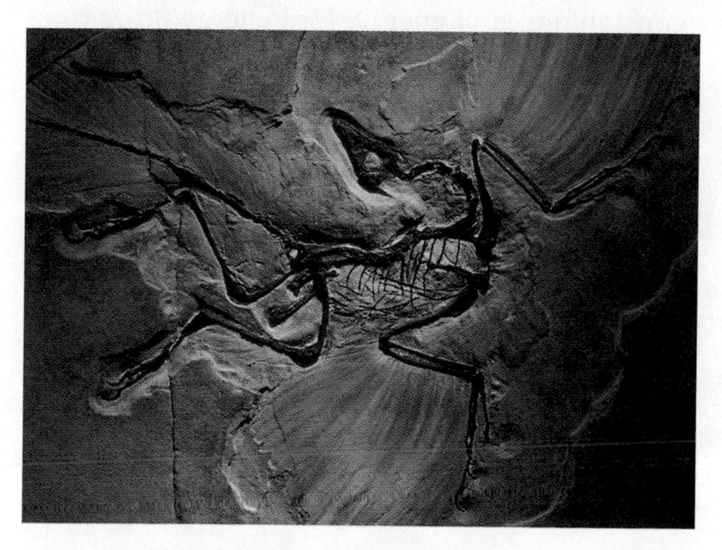

Figure 1-11
The earliest known bird, like its modern relatives, had feathers.

Theme 5: Interdependence

Living things interact with each other and with the nonliving part of their environment in complex ways. Ecology is the study of these interactions. A biological community is a group of interacting organisms. Biological communities are highly structured. The species they contain are interdependent—they interact in ways that influence their survival. For example, owls must capture mice for food. Mice, in turn, must eat plants. Neither owls nor mice can survive without their food supply. The interdependence in biological communities is the result of a long process of evolution in which species have adjusted to each other. The complex web of interactions in a biological community is easily disrupted when the environment is polluted and individual species become extinct, as is happening in much of the world today.

Figure 1-12
All organisms interact with other organisms. For this mouse, its interaction with the owl is fatal.

Theme 6: Matter, Energy, and Organization

Living things are made of the same materials as the rest of the universe—atoms assembled into molecules. All of the physical principles that apply to stars and home computers also apply to you and to every other living creature. Living things differ from nonliving ones only in their degree of organization.

All organisms require energy to carry out life processes. Organisms use energy to grow and to carry out their activities. Without it, life soon stops. Almost all the energy that drives life on Earth is obtained from the sun. Plants capture the energy of sunlight and use it to make complex molecules in a process called photosynthesis. These molecules then serve as the source of fuel for animals that eat them. Maintaining the complexity of living organisms requires a constant input of energy. This energy flows from plants to plant-eating organisms to meat-eating organisms. The availability of energy is a major factor in limiting the size and complexity of biological communities.

SECTION REVIEW

1. **Write a brief summary of the six themes presented in this section.**
2. **List one way you depend on other organisms.**
3. **How is your body temperature related to the theme of *Stability and Homeostasis*?**
4. **Give a brief definition of evolution.**

CHAPTER

1 Highlights

Life is plentiful in this rain forest. A single tree may contain hundreds of species of insects.

	Key Terms	Summary
1-1 Biology Today	science (p. 5) biology (p. 5) gene (p. 7)	• Biologists are combating infectious diseases. Some diseases have been conquered, but AIDS and malaria remain major killers. • Genetic disorders such as cystic fibrosis may be cured through the the transfer of normally functioning genes to affected individuals. • Humans are rapidly changing the environment, and the growing human population is straining the Earth's ability to support it.
1-2 Science Is a Search for Knowledge	hypothesis (p. 12) prediction (p. 12) experiment (p. 12) control experiment (p. 12) control group (p. 12) theory (p. 13)	• Scientific progress is made by posing hypotheses and testing their predictions. Verification of hypotheses is required before they are widely accepted. • Control experiments are important in testing hypotheses. Control groups are used for comparison with experimental groups. • A theory links well-supported hypotheses together in one concept. • All scientific theories can be overturned by new evidence.
1-3 Studying Biology	cell (p. 16) homeostasis (p. 16) development (p. 17) heredity (p. 17) mutation (p. 17) evolution (p. 17) natural selection (p. 17) species (p. 17)	• All organisms are made of cells. • Offspring tend to resemble their parents because genetic information is passed from parents to offspring. • Charles Darwin proposed that living things evolve, or change, through the process of natural selection. • Organisms interact with and depend on other living things. • All living things are composed of atoms and require energy. • Organisms maintain stable internal conditions.

Pollution is one of the many human-caused environmental problems we face today.

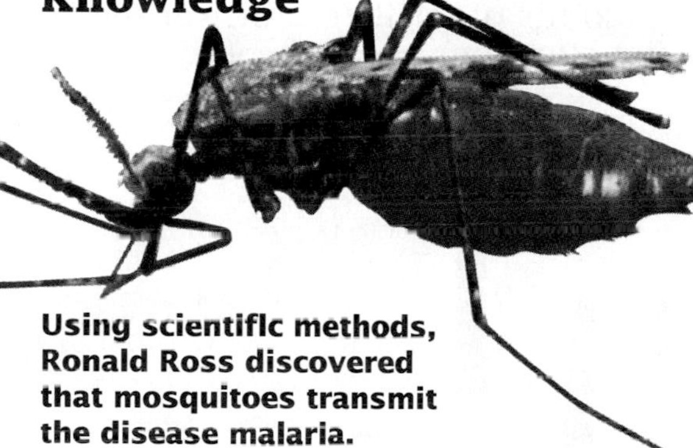

Using scientific methods, Ronald Ross discovered that mosquitoes transmit the disease malaria.

All life is the product of evolution.

Understanding Vocabulary

1. Using each set of words below, write one or more sentences summarizing information learned in this chapter.
   a. hypothesis, prediction, experiment
   b. development, heredity, mutation
   c. evolution, natural selection, species
2. For each pair of terms, explain the differences in their meanings.
   a. science, biology
   b. gene, heredity
   c. malaria, AIDS

Understanding Concepts

3. **Relating Concepts** Copy the unfinished concept map below onto a sheet of paper. Then complete the concept map by writing the correct word or phrase in each oval containing a question mark.

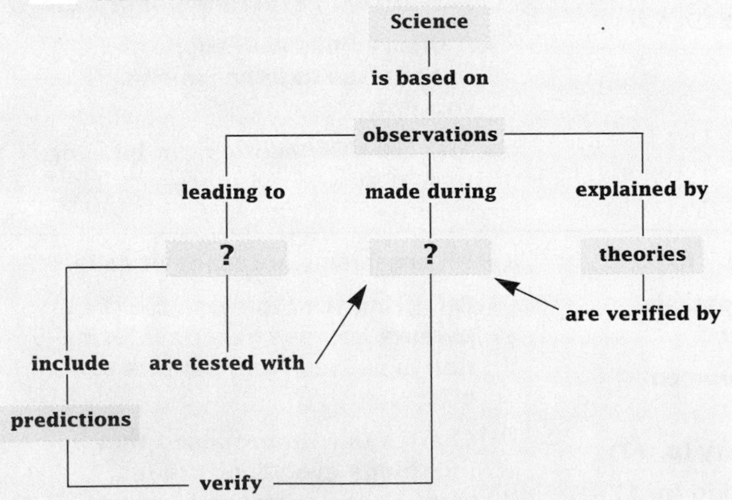

4. **Summarizing Information** Name two genetic disorders. What causes these disorders? Explain the new approach to curing genetic disorders that scientists are developing.
5. **Recognizing Relationships** How is the burning of coal, oil, and natural gas changing the atmosphere?
6. **Recognizing Relationships** Explain why knowledge of the relationships between cigarette smoking and lung cancer and between diet and heart disease are important.
7. **Analyzing Methods** Explain why it is important that the experimental group and the control group differ in only one way in an experiment.
8. **Analyzing Methods** After his initial series of experiments, Ross went on to show that the *Plasmodium* parasites move from the stomach of the mosquito to her salivary glands. Why was it necessary for Ross to make this demonstration?
9. **Analyzing Data** A scientist conducted an experiment to test which kind of fertilizer was the most effective. The graph below shows the data from the experiment. Plants in Group A and Group B received different kinds of fertilizer. Answer the following questions about the graph.

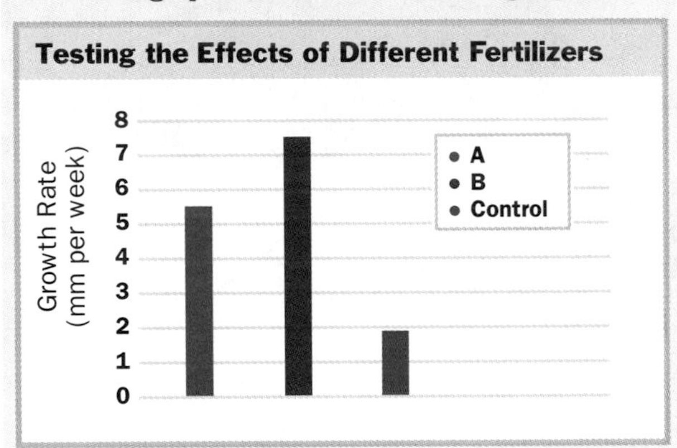

   a. Are both kinds of fertilizer effective? Explain your answer.
   b. Which kind of fertilizer should the control group receive? Explain your answer.
   c. Suppose that the plants in the Group B had been grown in full sunlight while the plants in the control group and in Group A had been grown partially in the shade. How would this influence the interpretation of the results?
10. **Identifying Concepts** Explain why a scientist is unlikely to say, "It's only a theory," when referring to a scientific principle.
11. **Summarizing Information** What is homeostasis? Give an example of homeostasis in your body.

Interpreting Graphics

12. Look at the photograph below of a scientist at work. Based on your examination of the photo, write a hypothesis about what you think the scientist could be investigating. Then write a prediction about what you would expect to occur if the hypothesis is true.

Design an experiment to test the hypothesis you wrote above.

Reviewing Themes

13. ***Interdependence***
You could not survive without other species. Identify at least two ways in which you depend on other species. Also identify two ways in which you affect other species.

14. ***Stability and Homeostasis***
How do sweating, heavy breathing, and thirst after strenuous exercise help maintain homeostasis?

Thinking Critically

15. **Determining Factual Accuracy** The following is an argument for reducing the amount of money the government spends on medical research: "In 1900, three diseases—influenza, tuberculosis, and pneumonia—were the leading causes of death in the United States. Advances in medicine have completely eliminated these diseases from the United States and the rest of the world. Now the leading cause of death for Americans is accidents. Since accidents are preventible, our country can reduce the amount of money we spend on medical research." Identify two factual errors in this argument.

Life/Work Skills

16. **Working Cooperatively** Work in a group to set up a recycling project in your school. Find out ways to recycle materials such as paper, cardboard, aluminum cans, and plastic. Also design a program to reuse items such as books, containers from the cafeteria, packaging materials, and so forth. Explore methods of selling the recycled goods that you cannot use, and find out the processes that those goods go through to emerge as "new" materials.

17. **Communicating** Investigate one of the different fields of biology, such as botany, cell biology, evolutionary biology, microbiology, parasitology, ornithology, zoology, forestry, biochemistry, or bacteriology. Write a report about a scientist who works in one of these fields. What does he or she do? What kind of education does he or she have? Where does the scientist work? What salary ranges can be expected for careers in this field?

Activities and Projects

18. **History Connection** Edward Jenner conducted a daring experiment that led to the eradication of smallpox in the world. What was his hypothesis? What led him to formulate this hypothesis?

19. **Career Connection** Many careers rely on some biological knowledge. Interview people in the following professions and find out what they know about biology and why this knowledge is important to their job: greenhouse or nursery operator, high school athletic coach, swimming pool manager, school nurse, and wildlife illustrator or photographer.

Discovering Through Reading

20. Read the article "On the Brink: Hawaii's Vanishing Species" by Elizabeth Royte in *National Geographic*, September 1995, pages 2–37. Why has Hawaii become the endangered species capital of the United States?

CHAPTER 2

REVIEW
- biological themes (Section 1-3)
- the term *cell* (Glossary)

Discovering Life

Life is abundant on this coral reef. All living things share a set of unique characteristics.

Biology is the study of life. But how can you tell living things from nonliving things? For instance, imagine that you are walking in the woods and you unexpectedly encounter a blob, like the one shown below, lying still on the forest floor. Is the blob alive? How would you decide? Make a list of the things you might do to determine whether the blob is alive.

2-1 What Is Life?

OBJECTIVES

1. **Explain the difficulty in defining life using only visually observable properties.**
2. **Describe five properties shared by all living organisms.**
3. **Relate the properties of life to the biological themes presented in Chapter 1.**

First Guesses at Defining Life

In making your list about the blob shown in **Figure 2-1**, the first thing you would probably look for would be movement.

Movement Almost all animals move around. Squirrels dash along tree branches, sharks knife through the water, and humans ride bicycles and run. However, movement from one place to another is not in itself a sure sign of life. A tree does not move about, for example, but it is alive. A cloud does move about, but it is not alive. Even if you could see the blob move on the ground in front of you, that would not be enough to tell you that the blob is alive.

Sensitivity So what should you do next? One thing you might do is poke the blob to see whether it responds. If you did that, you would be checking its **sensitivity**, its ability to respond to a stimulus (being poked).

Almost all organisms respond to stimuli. Deer flee from sounds that they associate with danger, and plants grow toward light. Air movement, sound, light, and temperature are all stimuli. But not every stimulus produces a visible response. Can you imagine getting a response from kicking a redwood tree? If the blob just sits there after you poke it, and the tree doesn't move after you kick it, that doesn't mean they are not alive. Sensitivity, while a better criterion than movement, is still not a good single characteristic to define life.

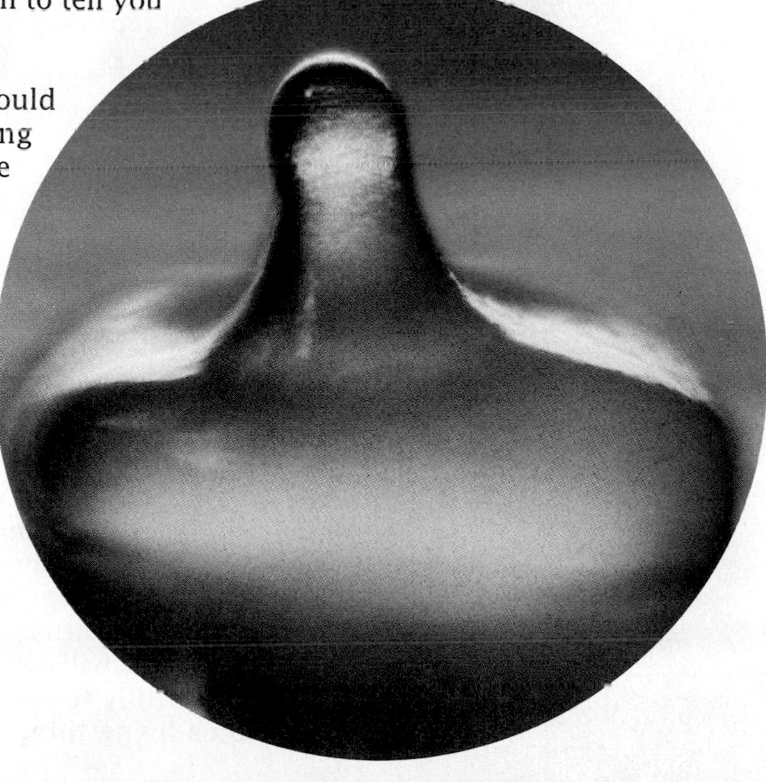

Figure 2-1
If you found this blob on the ground, would you think it was alive? How would you be able to tell? What are the characteristics of living things?

Development The next thing you might do is watch the blob to see if it changes. You would be looking for some signs of development. Development is an orderly progression that leads to greater specialization.

Most organisms exhibit development. For instance, development is the process by which an acorn grows into an oak tree. You too are a product of the processes of development. You started life as a single cell too small to see. As you developed within your mother, some of your cells specialized to become nerve, muscle, and skin cells. But not all living things change markedly as they develop. Bacteria simply grow and reproduce by splitting in half. Nor should all things that undergo orderly, progressive change be considered alive. The lines you see in rocks reflect the progressive, orderly laying down of material over a period of time, but the rock was never alive. The blob may or may not change as you watch it, but either way you cannot be sure if it is alive.

Complexity At this point you might walk up to the blob, pick it up, and examine it more closely to see how complex it is.

All living things are complex. Even the simplest bacterium contains many intricate structures. Complex organization is essential to life. However, complexity is not solely a characteristic of life. A computer is also complex, and so is a television, but they are not alive. So even though complexity is a necessary condition of life, the fact that the blob may seem complex when you examine it does not tell you that the blob is alive.

Death It might occur to you that the blob is the remains of a creature that was once alive but is now dead.

All living things die, but inanimate objects do not. Death is not the same thing as ceasing to function. A car that breaks down does not die. You may say, "The car died," but the now-broken car was never alive. Death is simply the termination of life. Unless one can detect life, death is a meaningless concept.

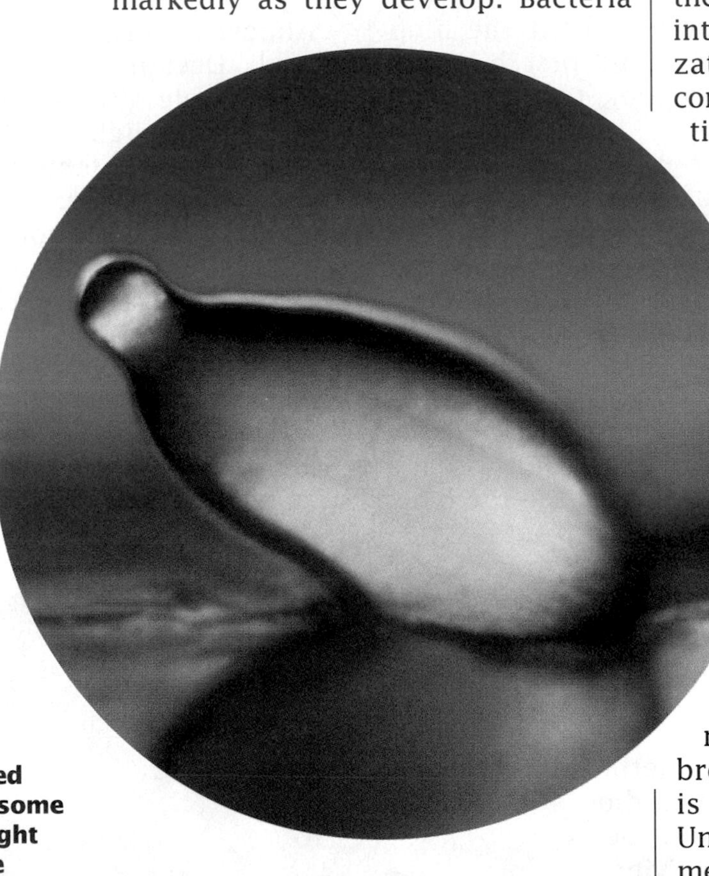

Figure 2-2 If you watched the blob for some time, you might see it change shape.

Life Has Five Characteristic Properties

You can now see that it is not so easy to determine whether something is alive. If you are going to determine whether the blob shown in **Figure 2-2** is alive, you will have to learn a great deal more about it. You will need to examine it more carefully to see how the blob resembles living things. All known organisms share certain general properties. These characteristics have been passed down from the very first organisms that evolved on Earth. It is by these properties that we recognize other living things—these properties define what we mean by *life.* If the blob displays these properties, it is alive.

Sensitivity ***Many living things are sensitive to external stimulation.*** **Slugs move toward light.**

Development ***Many living things grow and change.*** **When a slug reaches a new place, the colony differentiates into a base, a stalk, and a swollen head.**

Complexity ***Many living things are complex.*** **When slime mold amoeba cells exhaust the local supply of bacteria, they aggregate into a mobile colony called a slug, which migrates to another area.**

Death ***All living things die.*** **When nutrients are unavailable or the environment is unfavorable, the slug may die.**

Homeostasis ***All living things maintain homeostasis.*** **Within a slime mold amoeba cell, levels of foods are kept high; if they fall, the cell begins to seek other cells, which it joins to form a slug like the one in Figure 2-2.**

Reproduction ***All living things reproduce.*** **The head of the slug bursts, releasing spores that drift on air currents to new areas.**

Heredity ***All living things contain genetic information in DNA.*** **Each spore contains all the instructions for assembling a new amoeba cell.**

Cellular Organization ***All living things are made of cells.*** **When spores released from a slug encounter moisture, they develop into amoeba cells and begin moving around.**

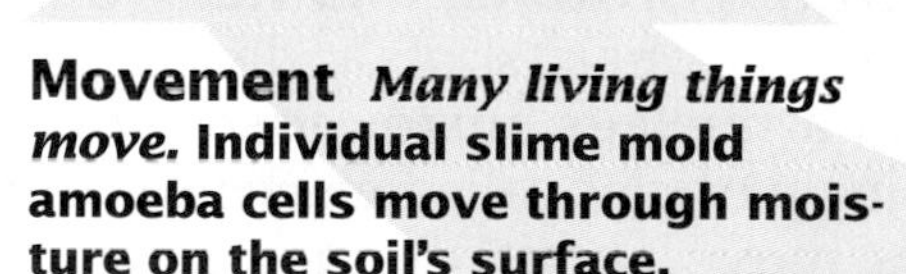

Movement ***Many living things move.*** **Individual slime mold amoeba cells move through moisture on the soil's surface.**

Metabolism ***All living things use energy.*** **Slime molds obtain their energy by ingesting bacteria in the soil.**

Figure 2-3
To be classified as a living thing, the blob must show the five characteristics listed in red type. The blob meets all the requirements for life. The figure also shows a summary of the first guesses about the blob in black type.

All living things are characterized by cellular organization, metabolism, reproduction, homeostasis, and heredity. Look at **Figure 2-3**. It shows how these characteristics apply to the blob, which is a living organism called a cellular slime mold. Living things use energy to grow and move in a process called *metabolism.* All living things must also maintain relatively stable internal conditions through a process called *homeostasis.* All living things contain genetic information. Through *reproduction*, organisms produce new organisms that are genetically similar, or sometimes identical, to themselves. The transmission of genetic information from parent to offspring is called *heredity.*

The blob is alive because it has all five of these characteristics. These five properties define the core of your study of biology. Notice that the characteristics are closely related to the biological themes presented in Chapter 1. For example, the theme Stability and Homeostasis is related to the characteristic of homeostasis because homeostasis is defined as the stability of internal conditions. Keep these relationships in mind as you study biology.

SECTION REVIEW

1. **Why is it insufficient to use movement as a characteristic to define life?**
2. **What are five properties that define life?**
3. **How is the slime mold dependent on other organisms in its surroundings?**
4. **Explain why a computer is not alive.**

The human body is a chemical machine. Everything humans do, from the smashing stroke of a tennis player to the deepest thought of a scientist or poet, can be understood as a chemical process. Because chemistry and biology are closely related, a brief introduction to chemistry will help you better understand how living organisms function.

2-2 Basic Chemistry

OBJECTIVES

1. **Describe the structure of an atom.**
2. **Identify the differences between atoms, elements, ions, and molecules.**
3. **Distinguish between covalent and ionic bonds, and explain how both form.**

Atoms: The Basic Structural Units of Matter

All matter in the universe is composed of tiny particles called atoms. An **atom** is the smallest particle of matter that retains its chemical properties. For example, the smallest particle of carbon that still has all the chemical properties of carbon is a carbon atom.

An **element** is a substance composed of only one type of atom. More than 100 different elements have been discovered. Each element is symbolized by a one- or two-letter abbreviation. For example, iron's symbol is Fe, uranium's is U, and calcium's is Ca. Only 11 elements are common in living things. Over 99 percent of the atoms in your body are either nitrogen, N, oxygen, O, carbon, C, or hydrogen, H.

Atoms are composed of electrons, protons, and neutrons

An atom consists of a dense core called a nucleus surrounded by tiny moving particles called **electrons**. Electrons move about the nucleus. The nucleus contains two kinds of particles, protons and neutrons. Most of the interior of an atom is empty space. If the nucleus of an atom were the size of an apple, the nearest electron could be more than a mile away. **Figure 2-4** shows a highly simplified model of the structure of an atom.

Each proton in the nucleus has a positive charge, and the neutrons have no charge. Each electron has a negative charge. Atoms contain equal numbers of electrons and protons; therefore, atoms have no charge. Electrons move about the nucleus in different energy levels. The farther an electron is from the nucleus, the more energy the electron has.

Figure 2-4
The nucleus of a carbon atom contains six neutrons, shown here in red, and six protons, shown here in blue. Six electrons move around the nucleus, traveling at very high speeds.

Figure 2-5
a Table salt, or sodium chloride, NaCl, is a crystal composed of ions of sodium and chloride held together by ionic bonds.

b At very high magnification, a crystal of salt looks like this.

c The salt crystal is held together by ionic bonds between positively charged sodium ions and negatively charged chloride ions.

d A sodium ion is positively charged because it has lost the single electron from its outer energy level. A chloride ion is negatively charged because it has gained an electron.

Formation of Bonds Stabilizes Atoms

When an atom reacts with other atoms, it gains, loses, or shares electrons. An atom that has gained or lost electrons is called an **ion**. If the ion has more protons than electrons, it is said to be positively charged. If the ion has more electrons than protons, it is said to be negatively charged. Ions with positive charges are electrically attracted to ions with negative charges.

The force holding two atoms or ions together is called a **chemical bond**. All chemical bonds involve interactions between electrons in the energy levels farthest from the nucleus. These outer electrons determine an atom's chemical behavior. In other words, the atom's outermost electrons determine its ability to react with other atoms. Atoms bond to fill their outer energy levels with electrons. A full outer energy level makes an atom chemically stable. Carbon, nitrogen, and oxygen require eight electrons to fill their outer energy levels. Since these atoms all have fewer than eight electrons in their outer energy levels, they form bonds to gain, lose, or share electrons.

An **ionic bond** is the force of attraction between oppositely charged ions. To form an ionic bond, one atom loses one or more electrons so that it has a full outer energy level. The other atom involved in bonding gains one or more electrons so that it too has a full outer energy level. **Figure 2-5** illustrates ionic bonding in table salt.

Molecules result from the formation of covalent bonds

A second type of chemical bond results when two atoms share one or more electrons. This type of chemical bond is called a **covalent bond**. A covalent bond is different from an ionic bond because electrons are *shared*, rather than lost or gained. For example, an atom of oxygen will form covalent bonds with two atoms of hydrogen. The result is a molecule of water.

Figure 2-6
a An atom of oxygen has six electrons in its outer energy level and needs two more electrons to be stable.

b Hydrogen has one electron in its sole energy level, but it needs a total of two to fill this level.

c Water, H_2O, is composed of molecules. Because they share electrons, all three atoms in the molecule have complete outer energy levels.

A **molecule** is a group of atoms held together by covalent bonds. The force that holds these atoms together comes from the sharing of electrons. Most molecules are made of more than two atoms because most atoms must share electrons with more than one other atom to fill their outer energy levels. Oxygen atoms are able to form covalent bonds with two other atoms. Nitrogen atoms can form covalent bonds with up to three other atoms. Carbon atoms can form covalent bonds with as many as four different atoms.

The models in **Figure 2-6** show the shared electrons in a water molecule. Each hydrogen atom shares two electrons with the oxygen atom. A bond in which two electrons are shared is called a single bond. Atoms can also form covalent bonds in which two pairs of electrons are shared—called a double bond—or three pairs of electrons are shared—called a triple bond. Acetic acid, a component of vinegar, is an example of a molecule that contains a double bond. Three models of the structure of acetic acid are shown in **Figure 2-7**.

Figure 2-7
A molecule of acetic acid contains a double bond between one oxygen and one carbon atom. Its formula can be written as $C_2H_4O_2$.

a A simple model of acetic acid shows its double bond.

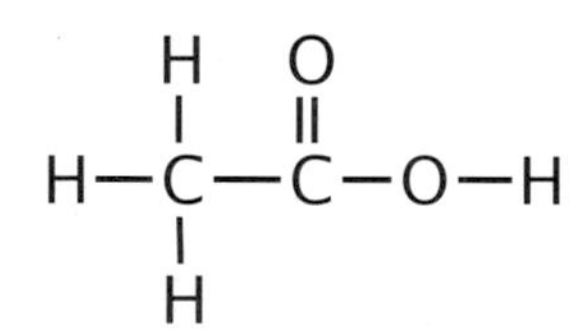

b In the structural formula, a single line between atoms represents a single bond and a pair of lines represents a double bond.

c This space-filling model shows the three-dimensional shape of the molecule.

SECTION REVIEW

1. **Describe the structure of an atom.**
2. **How do atoms differ from ions?**
3. **How are elements related to molecules?**
4. **How do covalent bonds differ from ionic bonds?**

You have just learned how atoms combine to form molecules. Small molecules also join together to form very large and complex molecules that have important functions in living things. There are four types of these large molecules in your body. They are composed almost entirely of carbon, hydrogen, and oxygen. In this section you will learn the basic structure and function of each kind of large molecule.

2-3 Molecules of Life

OBJECTIVES

1. **Define sugars, and describe the process that occurs in the formation of polysaccharides.**
2. **Describe the properties of lipids.**
3. **Explain the factors that affect the three-dimensional structure of proteins.**
4. **Define nucleic acids, and describe their functions.**

Organic Molecules Are Derived From Carbon

Of the 92 different kinds of atoms that occur naturally, carbon is the most closely associated with living things, including the many creatures living on the coral reef shown in **Figure 2-8**. In the last section, you read that carbon has four electrons in its outer energy level and that a carbon atom seeks to fill that energy level by sharing electrons with other atoms. Carbon atoms form long chains that are the backbone of many different kinds of molecules. Molecules with carbon-carbon bonds are called **organic molecules**.

Organic molecules join together to form more complex molecules

Just as atoms can be joined to form molecules, molecules can be linked together to form **macromolecules** *(MAHK roh MAWL uh kyools)*. All organisms are composed of four major classes of macromolecules: carbohydrates, lipids, proteins, and nucleic acids. Each class of macromolecule is composed of long chains of nearly identical subunits, and the different classes of macromolecules are distinguished primarily by the nature of the subunit they contain. The chemical bonds that link the subunit molecules together in macromolecules are the same kind of covalent bonds that hold atoms together in a molecule. The forming and breaking of bonds is called a **chemical reaction**.

Figure 2-8
Diamonds are pure carbon. Carbon-containing molecules are the primary components of living things, including the plants and animals that live on this coral reef.

Carbohydrates Are Energy Sources

A **carbohydrate** *(KAHR boh HEYE drayt)* is an organic macromolecule composed of carbon, hydrogen, and oxygen in a ratio of one carbon atom to two hydrogen atoms to one oxygen atom. Some carbohydrates, such as table sugar, are simple, small molecules. Other carbohydrates, like the starch in potatoes, exist as chains that are hundreds of subunits long. Carbohydrates contain many carbon-hydrogen bonds. They are well suited to be energy sources because their bonds store considerable energy.

Carbohydrates can be either simple or complex molecules

Among the simplest carbohydrates are sugars, small molecules that taste sweet. While sugars may have as few as three carbon atoms, the sugars involved in energy storage, like glucose, have six. These sugars have the formula $C_6H_{12}O_6$. Complex carbohydrates are made by linking individual sugars together to form long chains called polysaccharides. Large polysaccharides are insoluble in water; that is, they do not dissolve in water. They can be deposited in specific storage areas in a cell. This ability to store energy in the form of polysaccharides enables organisms to build up energy reserves.

Starch and glycogen are both complex carbohydrates

Starch is a polysaccharide composed of glucose subunits. Amylose is the simplest kind of starch. It exists as a long, unbranched chain of glucose molecules. Baking or boiling starchy plants such as potatoes and corn breaks these chains into shorter fragments that are soluble and can be taken up by the body.

Humans consume a great deal of carbohydrates; the seeds of rice, wheat, and corn supply about one-half of all the calories used by people. Animals store glucose in the form of long, branched chains called glycogen. **Figure 2-9** shows the structural differences between starch and glycogen.

Cellulose provides structural support

Many organisms use polysaccharides as structural molecules as well as for energy storage. Plants manufacture a polysaccharide called cellulose. Cellulose consists of glucose subunits linked in a way that most animals cannot break down. Cellulose forms the major part of the cell walls of plants. When you eat plants containing cellulose, it passes through your body undigested. This undigested cellulose is one kind of dietary fiber and is an important component of your diet. In contrast, cows and horses are able to digest the cellulose in plants. In their stomach or intestines, these animals have microorganisms that break down cellulose. You lack these microorganisms and would starve on a diet of grass.

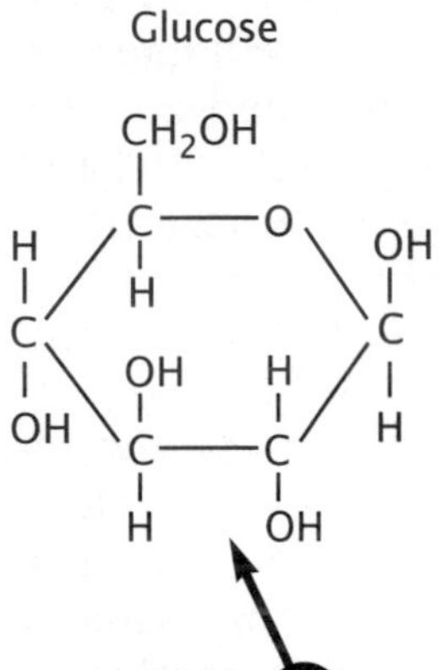

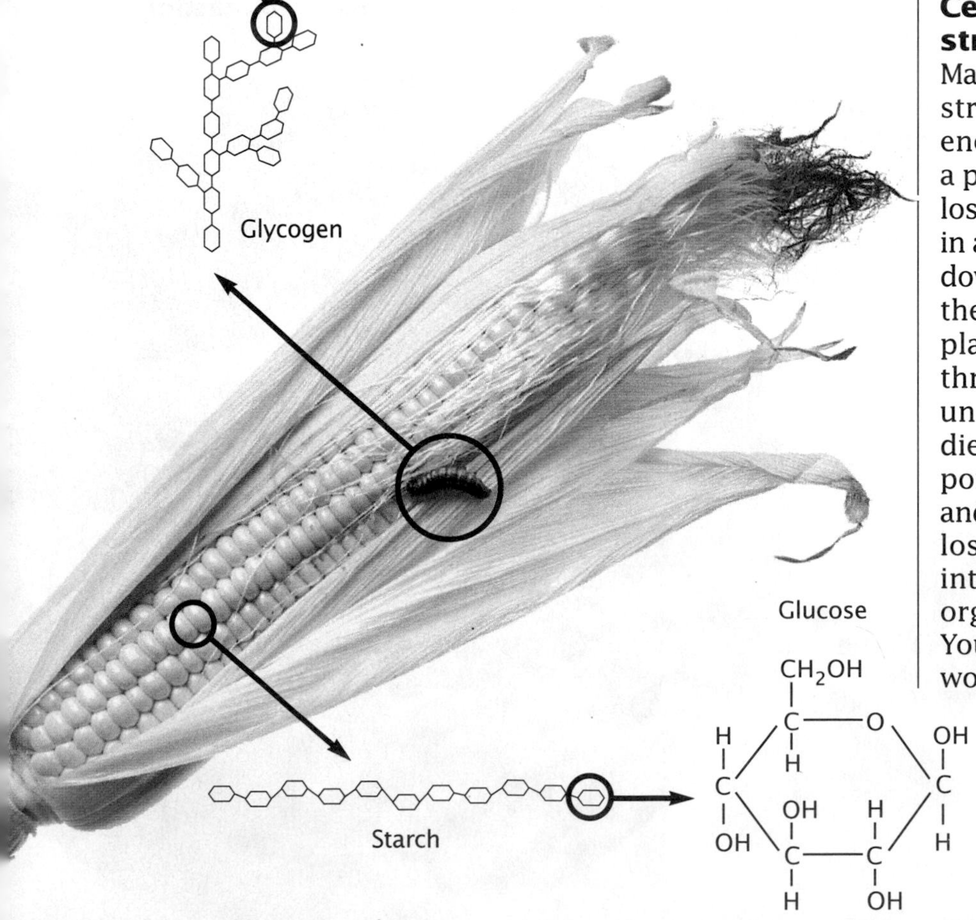

Figure 2-9 Starch and glycogen are macromolecules composed of glucose subunits (symbolized by a six-sided molecule). Starch, the carbohydrate found in corn, is chopped into glucose subunits and converted to glycogen when it is eaten by this worm.

Glucose

Glycogen

Starch

Glucose

Lipids Store Energy

A **lipid** *(LIHP ihd)* is an organic macromolecule that is not soluble (won't dissolve) in water, but is soluble (will dissolve) in oil. The most important lipids are fats, which are energy storage molecules. Fats have more carbon-hydrogen bonds than do carbohydrates. Therefore, a gram of fat contains more energy than a gram of carbohydrate. As you can see in **Figure 2-10**, a fat molecule consists of fatty acids, which are long chains composed mainly of CH_2 units, joined to a molecule of glycerol, the backbone of the fat molecule. There are many kinds of fats, which differ in the kind of fatty acids attached to the glycerol molecule.

In addition to fats, there are many other types of lipids. Among them are steroids *(STEHR oydz)* and waxes. Steroids include the sex hormones and cholesterol. Earwax and beeswax are examples of waxes.

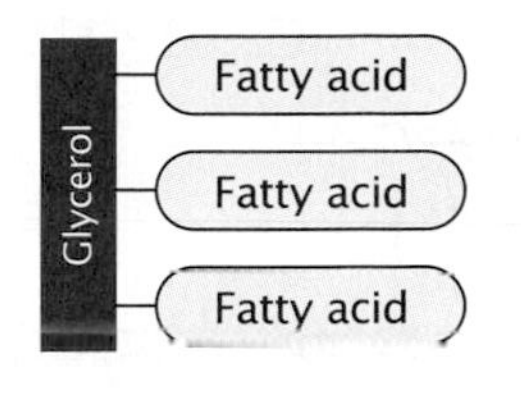

Figure 2-10

a Fats are composed of two kinds of molecules: glycerol and fatty acids. Fats always contain three fatty-acid molecules.

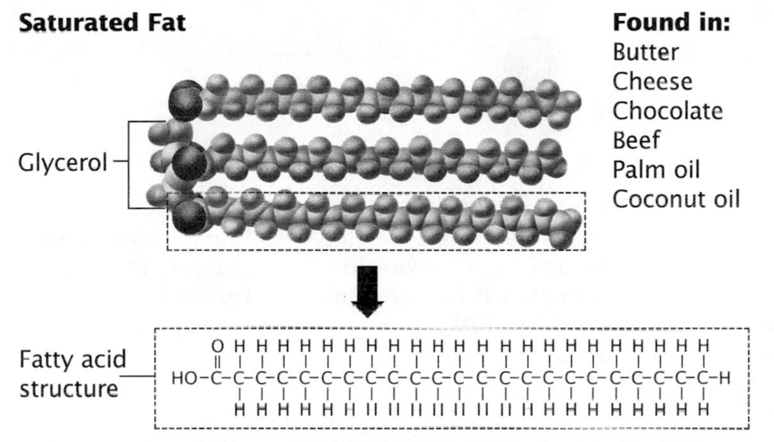

b In saturated fats, the fatty acids contain no double bonds between carbon atoms. In this space-filling model, carbon is indicated by green balls, hydrogen by blue balls, and oxygen by red balls. The same color-coding holds for the next diagram.

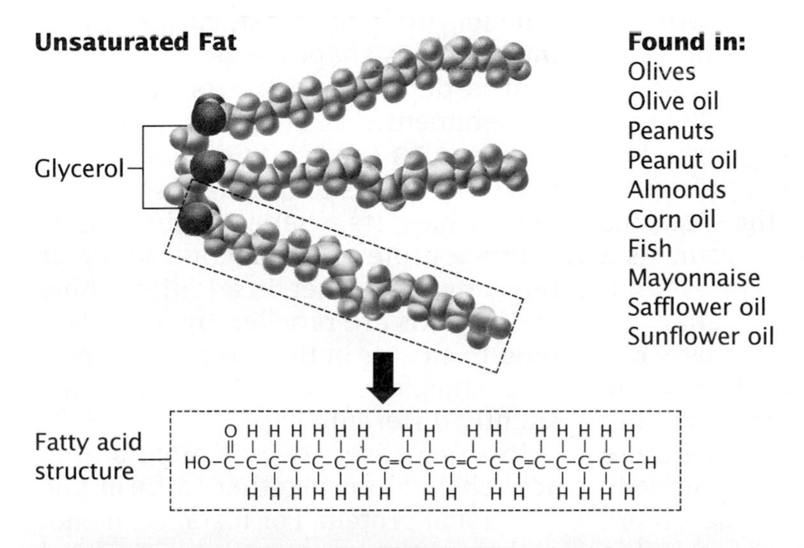

c In unsaturated fats, the fatty acids contain carbon-carbon double bonds. As a result, the fatty-acid chains have kinks in them.

Fats can be saturated or unsaturated

There are two major categories of fats: saturated fats and unsaturated fats. As you can see in **Figure 2-10b**, most carbon atoms in a **saturated fat** are bonded to two hydrogen atoms. The carbon atoms are "saturated" with hydrogen atoms; that is, the carbon atoms are bonded to as many hydrogen atoms as possible. If some carbons within a fatty acid chain are linked by double bonds, kinks occur in the chain. This type of fat is called an **unsaturated fat**. At room temperature, unsaturated fats usually exist as liquids called oils. In contrast, most saturated fats are solid at room temperature. Palm oil and coconut oil are exceptions because they are liquid at room temperature. They are classified as saturated fats because they lack carbon-carbon double bonds. The amount of saturated fat in your diet has an important effect on your health. People who eat diets high in saturated fats are more likely to suffer from heart disease than are people whose diets are low in saturated fats. You will learn more about this topic in Chapter 34.

It is possible to make an oil into a solid fat through a chemical reaction that adds hydrogen to the carbons in the oil. The double bonds become single bonds. This is often done to the oils in peanut butter. The unsaturated peanut fats are converted to saturated fats so that they don't separate as oils while the jar sits on the store shelf.

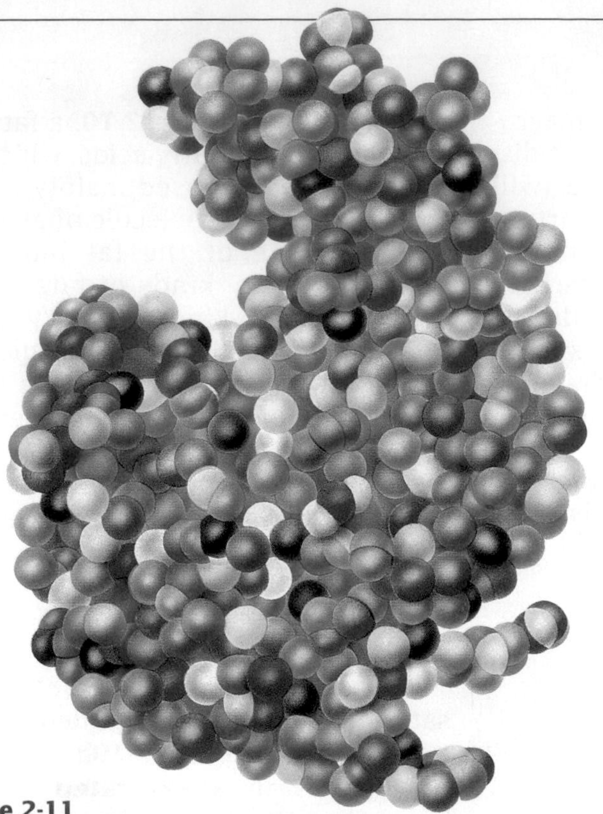

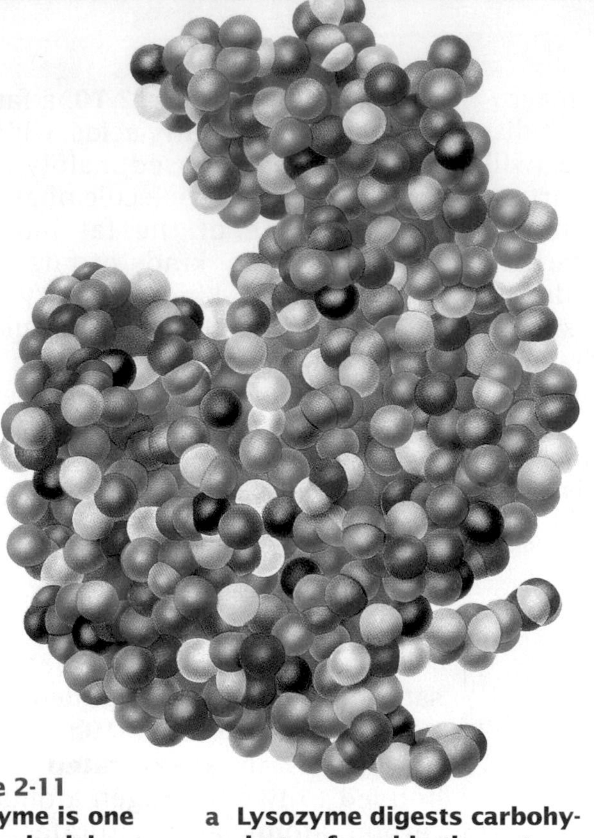

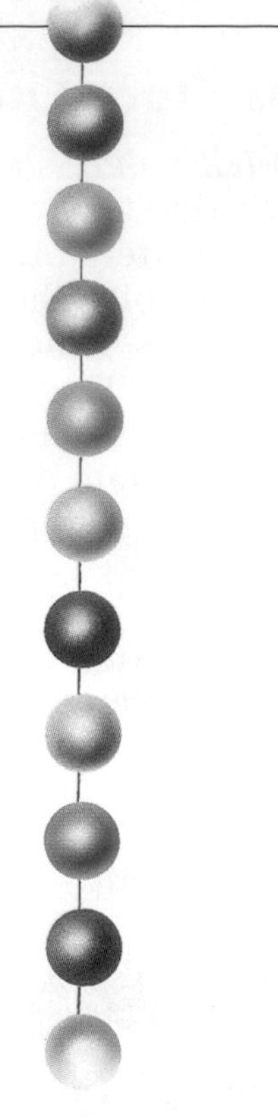

```
    H   H   O
    |   |   ||
H — N — C — C — O — H
        |
    H — C — H
        |
        S
        |
        H
```

Cysteine

```
    H   H   O
    |   |   ||
H — N — C — C — O — H
        |
    H — C — H
    H   |
    |   |
H — C — C — H
    |   |
    H   |
    H — C — H
        |
        H
```

Leucine

Figure 2-11
Lysozyme is one of your body's defensive proteins. It is found in tears, sweat, saliva, mucus, and other body fluids. Lysozyme has a very specific function that is determined by its shape—it destroys bacteria.

a Lysozyme digests carbohydrates found in the outer coating of some kinds of bacteria. These carbohydrates fit into the groove at the upper left of the lysozyme molecule. Lysozyme cannot digest other kinds of molecules, such as proteins or fats, because they do not fit into this groove.

b If you could unfold and stretch out lysozyme, you would find that it is a single chain of 129 amino acids.

c Cysteine and leucine are two of the amino acids found in lysozyme.

Proteins Provide Structure and Increase Reaction Rate

A **protein** *(PROH teen)* is an organic macromolecule composed of long chains of subunits called amino acids. All living things use the same 20 kinds of amino acids to construct proteins. Proteins differ in the number of amino acids they contain and in the sequence in which the amino acids occur. As a consequence, there is an almost endless variety of possible proteins. For example, just consider how many different proteins containing 100 amino acids (a typical length for a protein) could be made by a cell. Since there are 20 amino acids that can potentially occupy each position in the chain, there are many trillions of possible proteins of this length.

The importance of proteins as biological molecules stems from their tendency to bend, twist, curl, and fold into complex shapes. A protein's function is determined by its particular three-dimensional shape, as illustrated in **Figure 2-11**. What determines a protein's shape? Its amino acid sequence determines its shape. Amino acids within a protein interact, attracting or repelling each other. In addition, some amino acids are repelled by water and tend to cluster in the interior of a protein. Thus, each kind of protein has a slightly different shape.

In some cases, two or more amino acid chains join together to form one functional protein. For instance, hemoglobin is the protein in your blood that transports oxygen. A molecule of hemoglobin contains four separate amino acid chains.

In the next few chapters, you will study various types of proteins. Pay particular attention to the shapes of these macromolecules because shape determines a protein's biological function.

Proteins function as structural molecules and enzymes

Proteins often play structural roles in organisms. Cartilage and tendons are made of a protein called collagen, which also gives strength to your skin and bones. A protein called keratin forms the horns of a rhinoceros and the feathers of birds, as well as your own hair.

Proteins play a second very important role in organisms—they act as enzymes. Enzymes increase the rate at which chemical reactions occur, such as those that take place during metabolism. Most of the chemical reactions necessary for growth, movement, and other body activities could not take place without enzymes. You will learn more about enzymes in Chapter 5.

Nucleic Acids Contain Genetic Information

The fourth major class of organic macromolecules is the **nucleic** *(noo CLAY ihk)* **acids**. There are two types of nucleic acids: DNA (deoxyribonucleic acid) and RNA (ribonucleic acid). Subunits of DNA and RNA are called **nucleotides** *(noo CLAY oh teyedz)*. In DNA, these nucleotides form a series of units called genes, which encode information concerning how a given organism will grow and develop. RNA is involved in making working copies of genes. These copies are used in assembling amino acids to make proteins.

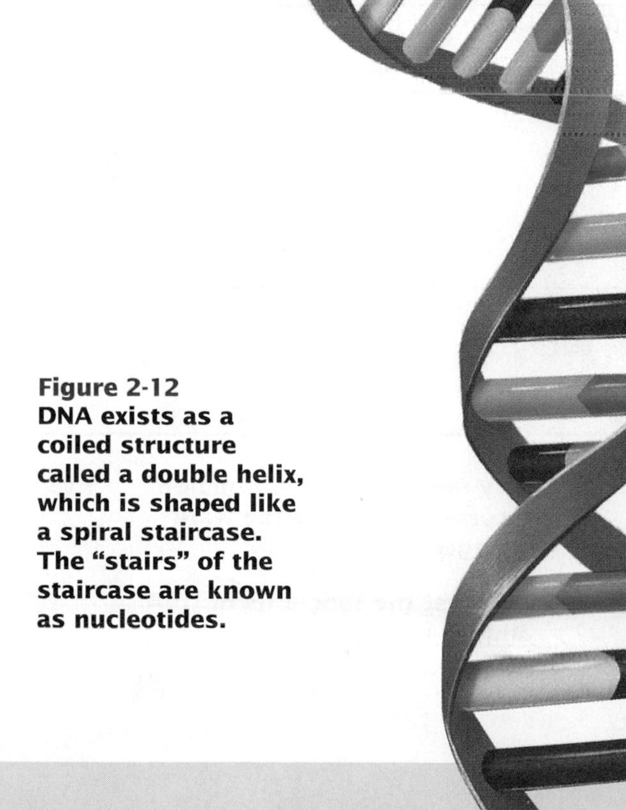

Figure 2-12
DNA exists as a coiled structure called a double helix, which is shaped like a spiral staircase. The "stairs" of the staircase are known as nucleotides.

DNA is a double helix

A DNA molecule consists of two interlocking, coil-shaped strands and resembles a spiral staircase. Chemists refer to this coiling structure as a double helix. **Figure 2-12** shows the structure of a DNA double helix. DNA encodes the sequences of all the cell's proteins, as well as information that determines when each protein is to be produced. In simple cells like bacteria, the DNA exists as a long molecule. But in complex cells like those of your body, the DNA folds and twists into compact bundles associated with proteins. These protein-DNA bundles are called **chromosomes** *(KROHM oh sohmz)*.

RNA helps in the synthesis of proteins

RNA molecules have a variety of shapes, depending on their function in the cell. RNA molecules that are associated with proteins serve as scaffolds for the assembly of all the different proteins in the cell. Other RNA molecules exist in the cell as long, single-stranded threads that carry DNA's message from one part of the cell to another. You will learn more about how DNA and RNA work together to build proteins in Chapter 8.

Macromolecule Summary

The four basic macromolecules (carbohydrates, lipids, proteins, and nucleic acids) have the same fundamental structure. All four are composed of long chains of similar subunits. Using **Table 2-1** you can compare and contrast the structures and functions of these macromolecules.

Table 2-1 Classes of Macromolecules

Class	Typical Subunit	Roles	Examples	
Carbohydrates	Sugar	Store energy	Starch, glycogen	Starch granules
		Structural components	Cellulose, chitin	
Lipids	Fatty acid	Store energy	Body fat	Human fat cells
		Form membranes	Cell membrane	
		Steroid hormones	Testosterone	
Proteins	Amino acid	Serve as enzymes	Lactase	Hair
		Structural components	Hair, cartilage	
		Peptide hormones	Insulin	
Nucleic Acids	Nucleotide	Store genetic information	DNA	Chromosomes
		Direct production of proteins	RNA	

SECTION REVIEW

1. **What is the difference between a sugar and a polysaccharide?**
2. **What is the difference between saturated and unsaturated fats?**
3. **Why might a protein function incorrectly if its shape were changed?**
4. **Contrast the functions of DNA and RNA.**

CHAPTER

2 Highlights

These seeds are the start of new life, but is a seed alive? Use the five properties of life to decide.

	Key Terms	Summary
2-1 What Is Life?	sensitivity (p. 23)	• Some of the most obvious properties of life cannot be used alone to decide whether something is alive because these properties can also appear in nonliving things. • All living things have all five of these characteristics: cellular organization, metabolism, reproduction, homeostasis, and heredity.
2-2 Basic Chemistry	atom (p. 26) element (p. 26) electron (p. 26) ion (p. 27) chemical bond (p. 27) ionic bond (p. 27) covalent bond (p. 27) molecule (p. 28)	• There are more than 100 kinds of atoms. Each kind is known as an element. More than 99 percent of the atoms in your body are nitrogen, oxygen, carbon, or hydrogen. • An ionic bond is the attraction between ions with opposite charges. • A covalent bond is the attraction between atoms that share electrons. Molecules are groups of atoms held together by covalent bonds.
2-3 Molecules of Life	organic molecule (p. 29) macromolecule (p. 29) chemical reaction (p. 29) carbohydrate (p. 30) lipid (p. 31) saturated fat (p. 31) unsaturated fat (p. 31) protein (p. 32) nucleic acid (p. 33) nucleotide (p. 33) chromosome (p. 33)	• Carbon is the most abundant element in living things. Molecules with carbon-carbon bonds are called organic compounds. • Organisms use carbohydrates to store energy and provide structural support. • Lipids are not soluble in water. Fats store more energy than carbohydrates because they have more carbon-hydrogen bonds. • The sequence of amino acids in a particular protein determines its shape and chemical properties. The shape of a particular protein determines its function. • Nucleic acids contain genetic information and direct protein production.

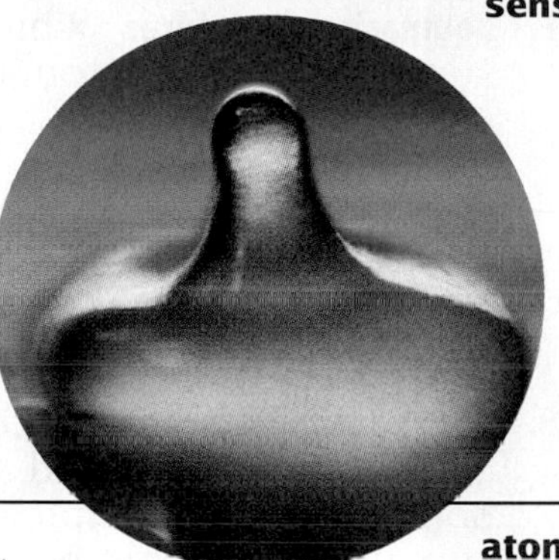

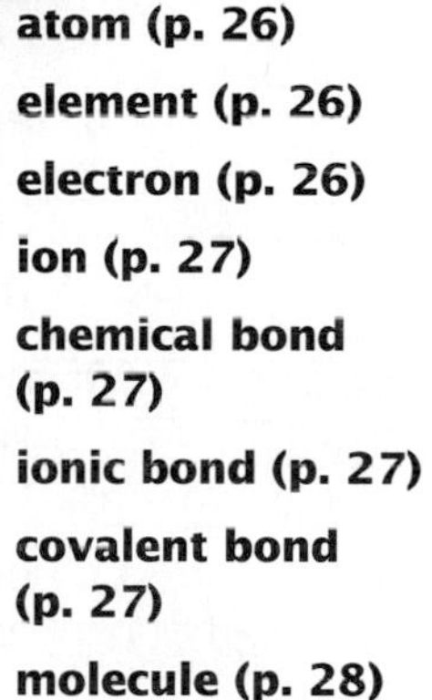

Is this blob alive? How would you go about answering this question?

Atoms are the basic structural units of matter.

Carbon is the primary chemical component of the macromolecules found in living things, including the plants and animals on this coral reef.

CHAPTER 2

review

Understanding Vocabulary

1. Select the term that does not fit with the others and explain why.
   a. metabolism, reproduction, sensitivity, homeostasis, heredity
   b. proton, molecule, neutron, electron
   c. carbohydrate, nucleic acid, water, protein, lipid

2. For each pair of terms, explain the differences in their meanings.
   a. atoms, elements
   b. ions, molecules
   c. covalent bond, ionic bond

Understanding Concepts

3. **Relating Concepts** Copy the unfinished concept map below onto a sheet of paper. Then complete the concept map by writing the correct word or phrase in each oval that contains a question mark.

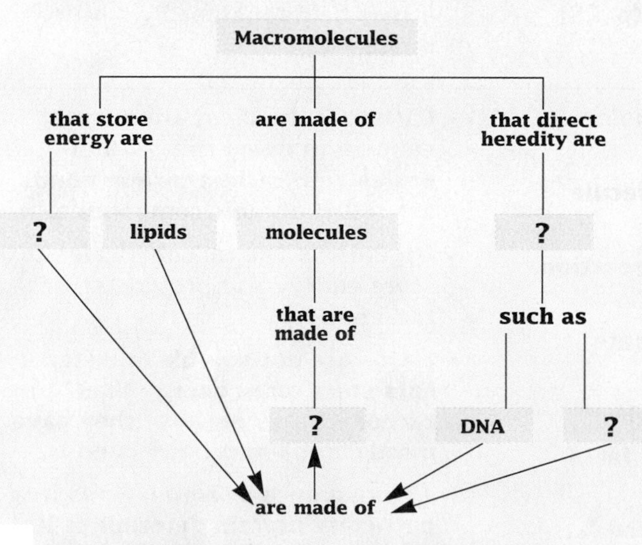

4. **Summarizing Information** Explain why using only visually observable properties is not sufficient to define life.

5. **Inferring Conclusions** Decide which of the following objects are alive: automobile, fossilized dinosaur bone, human intestinal cell, tree. Explain how you made your decisions.

6. **Recognizing Relationships** All living things need energy. How does a slime mold cell obtain energy? How does it respond to shortages of food?

7. **Comparing Structures** Which of the following statements about atoms and ions is true?
   a. An ion always has more electrons than an atom of the same element.
   b. An atom and ion of the same element always have the same number of protons.
   c. Ions are attracted to other ions with the same charge.

8. **Summarizing Information** List two differences between an ionic bond and a covalent bond. Give an example of a molecule that contains a covalent bond.

9. **Recognizing Relationships** Carbohydrate and lipid molecules both store energy. Explain why a lipid contains more energy per gram than a carbohydrate.

10. **Inferring Relationships** Describe two ways in which carbohydrates, proteins, lipids and nucleic acids are similar.

11. **Inferring Conclusions** On a recent Arctic expedition, a biochemist discovered an unknown substance. After performing several experiments, she determined that the substance had the following characteristics: contains carbon, hydrogen, and oxygen and is soluble in oil but not in water. What kind of substance did the biochemist discover? Explain your answer.

12. **Identifying Variables** What factors determine the three-dimensional shape of a protein molecule?

13. **BUILDING ON WHAT YOU HAVE LEARNED** As you learned in Chapter 1, all organisms require energy to carry out life processes. Because they need energy, all living things, including humans, must have ways to store the energy for when they need it. Where do humans get their energy? In what two ways do humans store energy?

Interpreting Graphics

14. Examine the figure shown below.

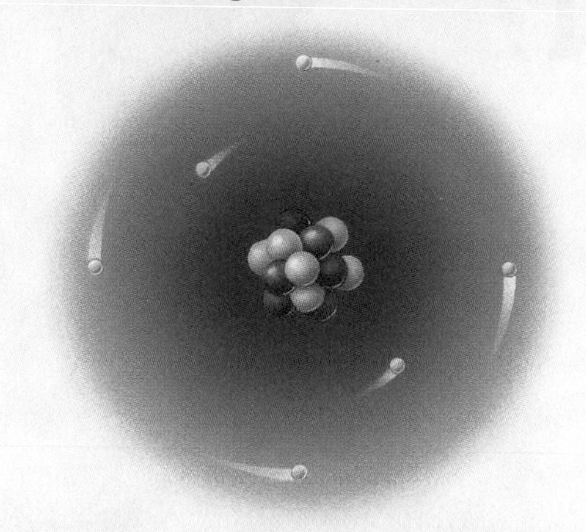

a. What kind of charge does a proton have? a neutron? an electron?
b. Where are the protons and neutrons located?

Reviewing Themes

15. ***Stability and Homeostasis***
Describe the process of homeostasis and explain why it is one of the characteristics of all living things.

16. ***Matter, Energy, and Organization***
What determines the chemical behavior of an atom?

Thinking Critically

17. **Justifying Conclusions** Why do football players and other athletes often consume a diet rich in complex carbohydrates while training?

18. **Recognizing Logical Inconsistencies** "Scientists know that the genetic information in a human cell is carried on the structures called chromosomes, which are found in the nucleus. Proteins are found in chromosomes, so proteins must be the genetic molecule." Explain the logical flaw in this argument.

19. **Justifying Conclusions** Movement is not a characteristic used to define life because some organisms don't move. All organisms eventually die, but death is not a defining characteristic of life either. Explain why death is not a good characteristic for defining life.

Life/Work Skills

20. **Working Cooperatively** Work together with a classmate to research and write a report on kwashiorkor, a severe form of malnutrition that is caused by the lack of dietary protein. What are the symptoms of this condition? Include facts and examples to support your information.

Activities and Projects

21. **Health Connection** Today, many health experts and nutritionists warn people against consuming too much saturated fat. Use library references or information from an on-line database to compare the average daily diet of an adult living in the United States with that of a person living along the Mediterranean Sea (Greece, Italy, France, Portugal, or Spain) or in Japan. How do these diets differ in the amounts and kinds of fish, meat, fruits, vegetables, carbohydrates, and fats consumed daily? What effects do these dietary differences have on health?

22. **Career Connection: Dietitian** Dietitians provide information on proper nutrition. They know the chemical makeup of foods and the effects foods have on the human body. Use a handbook of nutrition to evaluate the nutritional content of the following three typical lunches: (1) cheeseburger, french fries, milk shake; (2) baked potato, sour cream, butter, cheddar cheese, soft drink; (3) pizza with cheese, pepperoni, mushrooms, soft drink.

23. **History Connection** Use references in your local or school library to investigate the research that lead to the discovery of the structure of an atom. Create a time line of the major events and the individuals responsible for the discoveries.

Discovering Through Reading

24. Read the article "The Facts About Fats" in *Consumer Reports*, May 1994, page 308. Describe two ways that excess fat in the diet can be harmful.

Science, Technology, and Society

DNA Profiling: Promise or Peril?

DNA profiling can find the gene responsible for a hereditary disease, reveal family relationships, or link a suspect to the scene of a crime. Yet its benefits must be carefully weighed against potential dangers.

DNA Takes the Witness Stand
Just as a bar code specifies a certain product, the patterns produced by a DNA sample are specific to each individual. A tiny sample of blood, semen, saliva, or bone at a crime scene provides a sufficient amount of DNA for analysis. The patterns obtained can be compared with the patterns of DNA in blood or tissue samples obtained from suspects. Because no two individuals have the same DNA pattern, every person has a unique "DNA fingerprint." This "DNA fingerprint" can be used to establish a link between a suspect and a crime scene.

Forecasting Disease
Using gel electrophoresis, researchers can identify the adult who carries a gene that causes a hereditary disease—such as cystic fibrosis, for example—or a newborn who has the disease. Sometimes the presence of a particular gene does not necessarily guarantee that the individual will have the disease. For example, a recently discovered gene increases an individual's risk of colon cancer. People who have this gene would be wise to have regular checkups, even though it is not certain that they will someday develop colon cancer.

Mapping Family Relationships
The ability of DNA profiling to reveal the base sequences of selected genes makes it an invaluable tool for studying the inheritance of these genes within a family. The characteristic DNA patterns are passed from one generation to the next and thus provide evidence that can be used to determine the identities of family members. In Argentina, DNA profiling has enabled children whose parents were abducted in 1977 by military security forces to be reunited with

their grandparents years later. Biologists studying animal populations have used DNA profiling to measure genetic variety, an important indicator of population diversity. Anthropologists studying ancient civilizations have performed a similar test using DNA from specimens that may be hundreds or even thousands of years old.

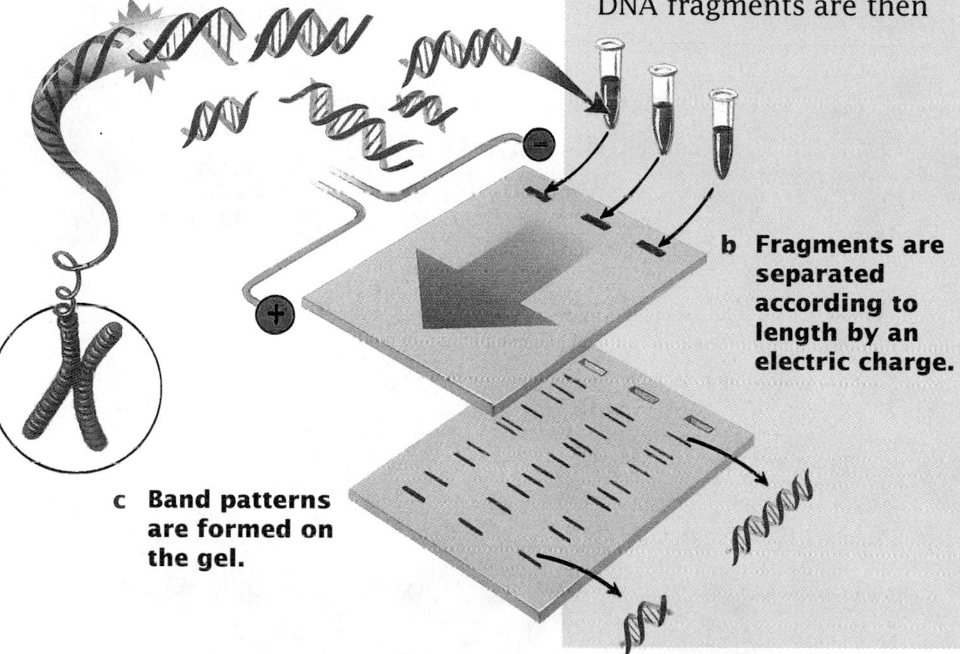

a DNA is fragmented by restriction enzyme treatment.
b Fragments are separated according to length by an electric charge.
c Band patterns are formed on the gel.

Technique

Gel Electrophoresis
Analysis of an individual's DNA begins with the use of restriction enzymes, which cut the DNA at specific base sequences, leaving fragments of various lengths. These fragments vary distinctively from one individual to the next. The DNA fragments are then placed on a bed of semisolid gel. Electrodes at either end of the gel generate an electric current. The DNA, which is negatively charged, moves through the gel toward the positive electrode. Because the smaller DNA fragments move faster than the larger fragments, the separated fragments form a pattern of bands. This pattern, which looks similar to the barcode used by supermarkets, is an individual's DNA profile.

Applications: pinpointing genes responsible for hereditary diseases, confirming family relationships and individual identities, and studying the diversity of genes within populations

Users: medical geneticists, physicians, research biologists, forensic analysts

Analyzing the Issues

Is genetic testing foolproof?

1. Many scientists have warned that DNA profiles can be misinterpreted. Others caution that the possibility of human error must be recognized and urge the adoption of industry-wide quality control standards. Use references available in the library or research an on-line database, and write a report summarizing the ways in which DNA evidence can be altered or misused.

Should genetic testing be mandatory?

2. Read "DNA Profiling," in *National Geographic*, May 1992, pages 112–124. Do you think the police searching for Lynda Mann's killer were justified in requiring DNA profiling of the young men from nearby towns? Would a similar action in the United States violate the Fourth Amendment? Use references available in the library or research an on-line database to find information supporting your viewpoint. Write a report summarizing your conclusions.

Will genetic testing lead to discrimination?

3. Many fear that evidence linking genes with certain diseases opens the door for genetic discrimination. Should employers and insurers have access to an individual's DNA profile? What legislation exists to protect people against genetic discrimination? Write to the following organization for information:

The Council for Responsible Genetics
19 Garden Street
Cambridge, MA 02138

Who will have access to your genetic profile?

4. The Fourth Amendment to the United States Constitution protects citizens against unwarranted search and seizure. Could law enforcement officials order genetic testing of blood or tissue specimens previously collected for another purpose—such as for a medical exam—for comparison if they suspect that an individual has committed a crime? Use references available in the library or research an on-line database to learn more about the role of the Fourth Amendment.

CHAPTER 3

REVIEW
- cells (Section 1-3)
- characteristics of living things (Section 2-1)
- electrons, covalent bonding (Section 2-2)
- lipids, proteins (Section 2-3)

Cells

A cell, such as one of these *Streptococcus* bacteria, is the smallest unit of life.

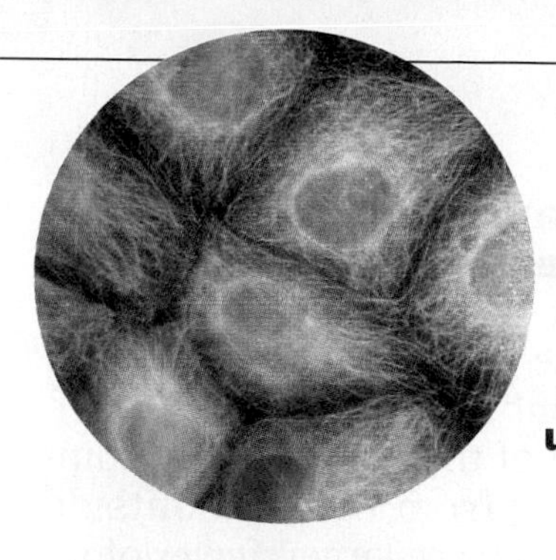

Your body is composed of trillions of cells. As long as these cells work normally, you are generally unaware of them. However, abnormal cell behavior can bring about changes such as the uncontrolled growth of these cancer cells. If we are ever to find cures for this deadly disease and others, we must first look closely at cells to understand how they function.

3-1 At the Boundary of the Cell

OBJECTIVES

1. **Define the function of a cell membrane.**
2. **Explain why smaller cells function more efficiently than larger cells.**
3. **Distinguish between polar and nonpolar molecules.**
4. **Compare the interaction of water molecules with each other and with lipid molecules.**

Cells: The Smallest Vessels of Life

Look at the organisms below and think about the characteristics you share with them. You, the squid, the earthworm, and all other living things are made of tiny compartments called cells. A **cell** is the smallest unit that can carry out all the activities necessary for life. Most microscopic organisms, such as the protist seen in **Figure 3-1**, are made up of a single cell.

In some ways, a cell is like a submarine. A submarine has a tough outer hull that seals the vessel and protects the equipment inside. Similarly, an outer surface called the **cell membrane** shields the delicate internal machinery of a cell. Like a submarine's hull, the cell membrane separates what is inside from what is outside. However, a cell membrane is more complex than the hull of a submarine. It must have "gates" that allow the raw materials needed by the cell to enter and that allow harmful waste products to exit promptly. Without this flow of materials, a cell would not be able to survive. By regulating what goes into and out of a cell, the cell membrane helps to maintain the internal environment of the cell.

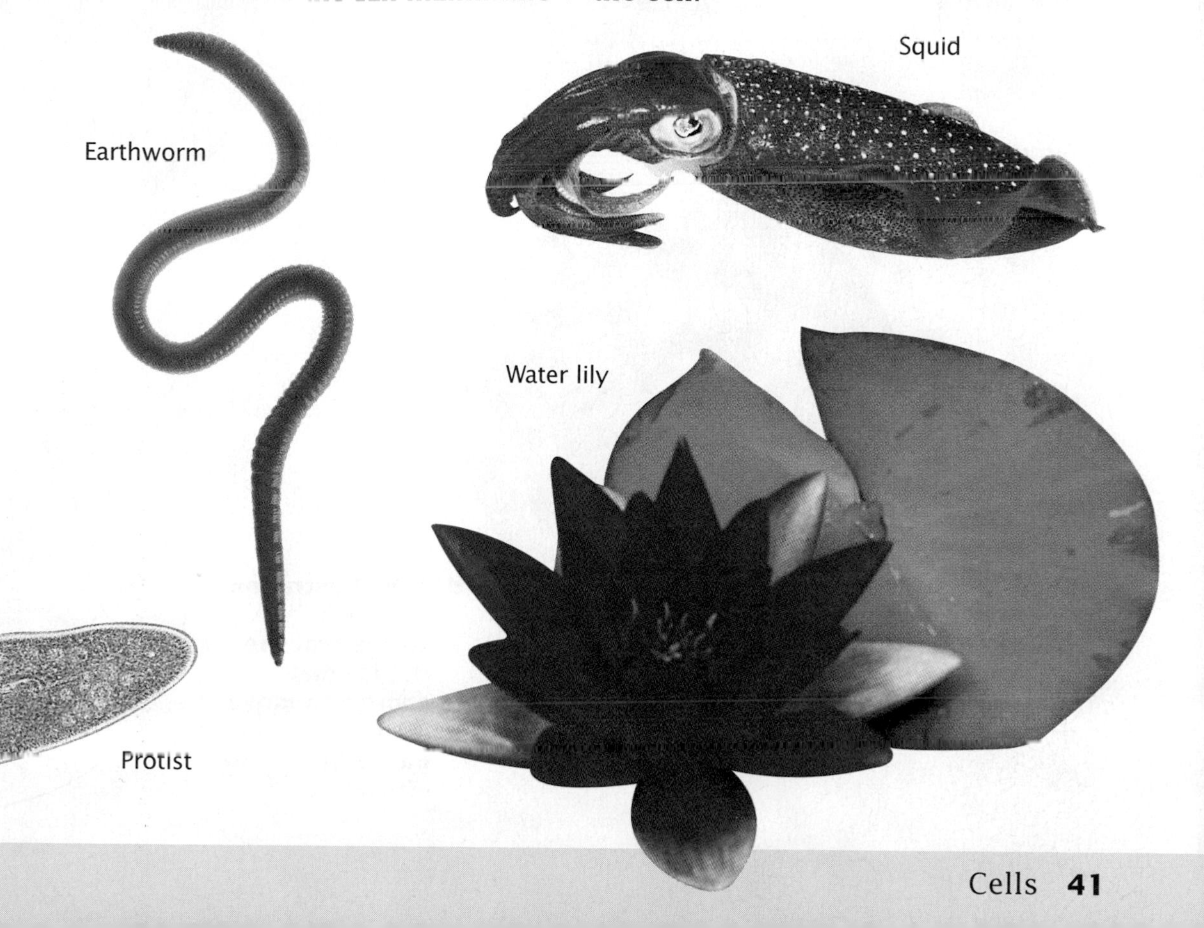

Figure 3-1
Even though these organisms look very different, they share an important characteristic of life—they are made of cells.

What Limits the Size of a Cell?

The most obvious thing that can be said about cells is that they are very small. Although cell sizes vary (for example, the diameter of a human egg cell is about 14 times greater than that of a red blood cell), most cells are so small that you can see them only through a microscope. Study **Figure 3-2**. What factors do you think limit the size of a cell?

Cell parts cannot be too far from the cell membrane

Every bit of food and information needed by the cell must enter through the cell membrane. When cells are small, no part of their complex machinery lies too far from the area outside the cell. If a cell were larger, fewer of its interior structures could be near the cell membrane. That is bad for just the same reason that long supply lines are bad for an army—too many things could go wrong, and responses to information would be too slow. Thus, small cells work more efficiently because their "supply lines" are short.

c . . . and in this illustration you can see that the skin is a complex organ that is made of many different kinds of cells.

Figure 3-2
a The skin of this finger . . .

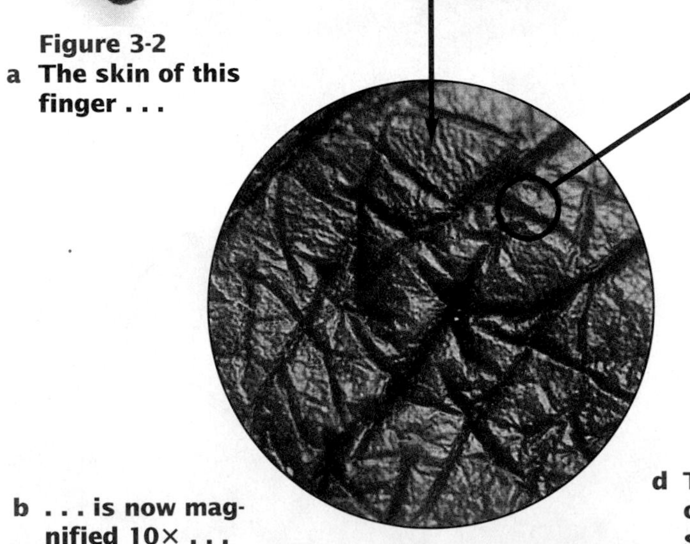

b . . . is now magnified 10× . . .

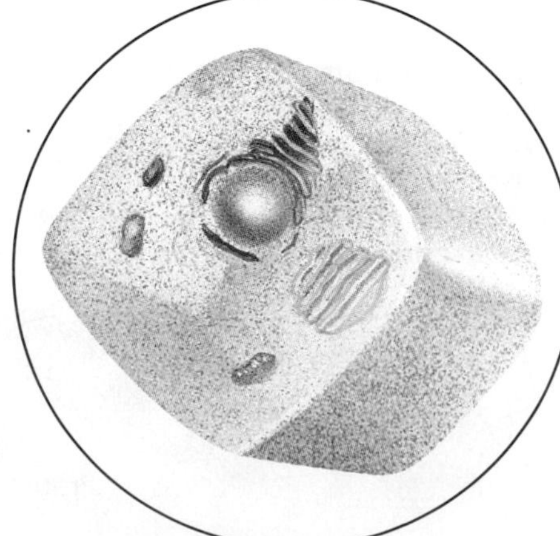

d This illustration of a skin cell shows that the cell is small enough to move materials into and out of its interior efficiently.

The Ratio of Surface Area to Volume in Cells

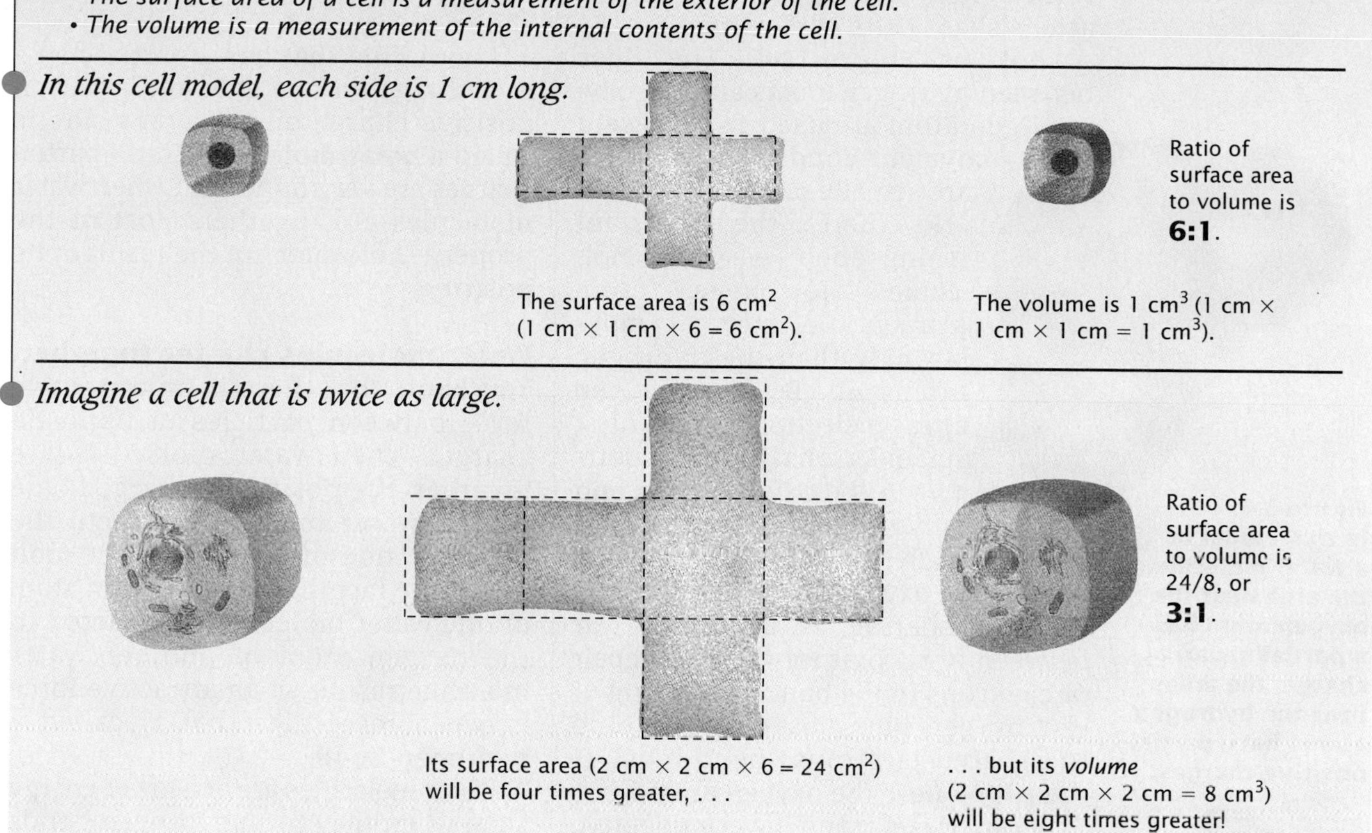

• The surface area of a cell is a measurement of the exterior of the cell.
• The volume is a measurement of the internal contents of the cell.
In this cell model, each side is 1 cm long.
The surface area is 6 cm^2 (1 cm × 1 cm × 6 = 6 cm^2).
The volume is 1 cm^3 (1 cm × 1 cm × 1 cm = 1 cm^3).
Ratio of surface area to volume is **6:1**.
Imagine a cell that is twice as large.
Its surface area (2 cm × 2 cm × 6 = 24 cm^2) will be four times greater, . . .
. . . but its *volume* (2 cm × 2 cm × 2 cm = 8 cm^3) will be eight times greater!
Ratio of surface area to volume is 24/8, or **3:1**.

Figure 3-3
As a cell gets larger, its volume increases faster than its surface area. A cell's surface area must be large enough to meet the needs of its volume.

A cell's volume increases faster than its surface area

As a cell grows, it takes in more food and creates more wastes. Since these substances must pass into and out of the "gates" in the cell membrane, the membrane must be large enough to service the cell's needs. As the cell grows, so does its membrane. But cells cannot grow indefinitely. So what limits cell size?

One factor is the relationship between the surface area and the volume of the cell. As a cell grows, its volume increases at a much faster rate than its surface area. A small cell, such as the one shown in the top row of **Figure 3-3**, has enough surface area to meet its needs. But a cell as large as the one shown in the bottom row might not. The ratio of a cell's surface area to its volume ultimately limits how large that cell can become. Cells cannot grow so large that their surface area becomes too small to take in enough food and remove enough wastes.

Water and the Cell

All cells are surrounded by water. Single-celled creatures swim in small ponds as well as in vast oceans. Even the cells of your body, such as blood cells or skin cells, are surrounded by a thin film of water. Water is present inside the cell too. All of the complex machinery inside the cell perform their functions in water. The cell membrane is shaped by the water found inside and outside of the cell. To understand how water can shape a cell membrane, you must first look closely at the structure of a water molecule.

Water is a polar molecule

The chemical formula for water is H_2O. A water molecule is made of two hydrogen atoms and one oxygen atom that are bonded together. These bonds form when hydrogen and oxygen atoms share pairs of electrons.

Look at **Figure 3-4**. The lines between hydrogen atoms and the oxygen atom are used to represent covalent bonds. These bonds are actually pairs of electrons. Now here's the important thing about a water molecule: the oxygen atom attracts electrons more strongly than the hydrogen atoms do. Because oxygen attracts electrons so strongly, the electrons in the bonds between the oxygen atom and each hydrogen atom are not shared equally. They are more likely to be near the oxygen atom. Think of this unequal sharing as a tug of war between two atoms for the shared pair of electrons in the bond. In this tug of war, oxygen wins most of the time, so the shared electrons spend most of their time near the oxygen atom.

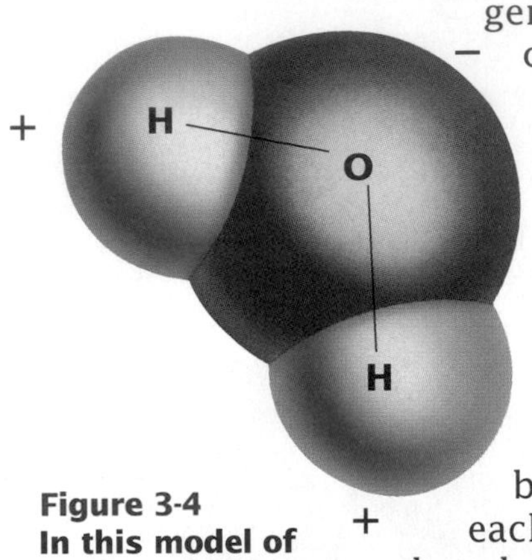

Figure 3-4
In this model of a water molecule, the area near the oxygen atom has a partial negative charge; the areas near the hydrogen atoms have partial positive charges.

Because electrons have a negative charge, the oxygen part of the water molecule is slightly negative. The hydrogen atoms in the water molecule have slightly positive charges because electrons rarely spend their time near the positive hydrogen nuclei. We think of the oxygen side of the molecule as having a partial negative charge and the hydrogen side of the molecule as having a partial positive charge, as shown in **Figure 3-4**.

A molecule that has a partial negative charge on one side and a partial positive charge on the other side is called a **polar molecule**. These partial charges are very important when water molecules are together. Most of the properties of water are the result of its polarity.

Water molecules cluster together

You know that there is an attractive force between particles of opposite charge. When water molecules are together, the positively charged side of one water molecule attracts the negative side of another water molecule. The fact that the hydrogen atom of one water molecule is attracted to the oxygen atom of another water molecule results in an attractive force between molecules that is called a **hydrogen bond**.

Water molecules are at a lower energy state when they form hydrogen bonds with each other. Since all things tend toward lower energy, there is a natural tendency for water molecules to form hydrogen bonds. When forming hydrogen bonds, the water molecules will cluster together, as illustrated in **Figure 3-5**, creating a large number of bonds.

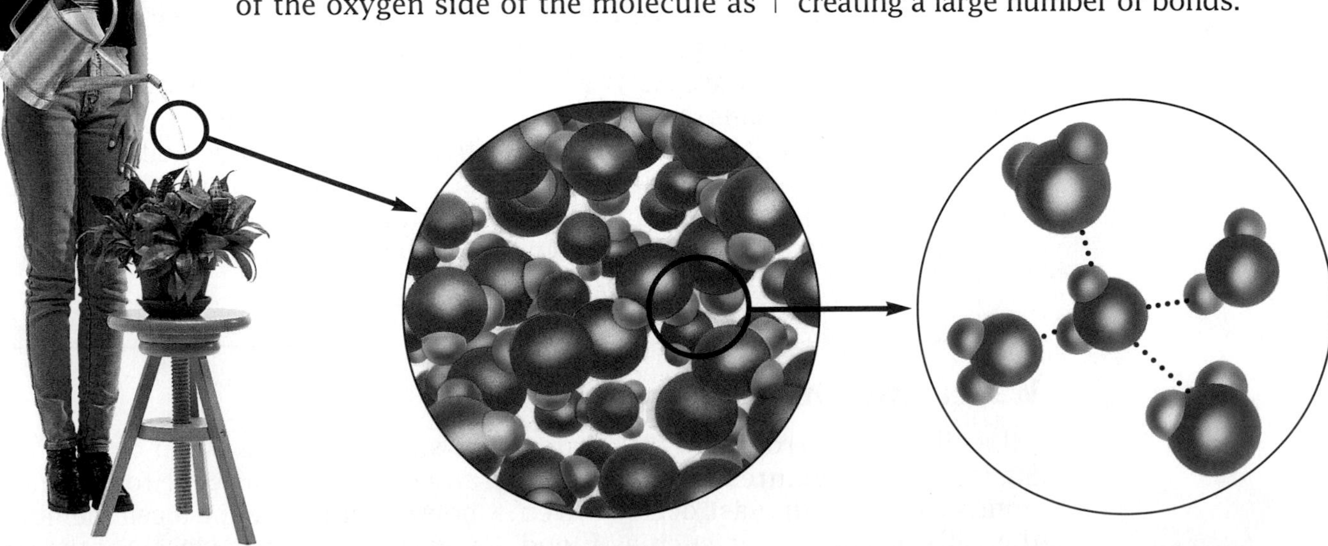

Figure 3-5
a Water in its liquid state is composed of . . .
b . . . many rapidly moving water molecules.
c These water molecules form clusters due to the formation of hydrogen bonds.

Water and the Cell Membrane

Now that you know something about the nature of water molecules, you can look more closely at how water shapes the cell membrane. The basic plan of the cell membrane begins with a sheet of lipids. You learned in Chapter 2 that lipids are molecules such as fats and oils. The interaction between water and lipids shapes the cell membrane.

Figure 3-6 shows what happens when oil is poured into a beaker of water and thoroughly mixed. Soon after mixing, small beads of oil form. Eventually the water and oil separate into two distinct layers. The water and oil won't stay mixed. Why? Water molecules start to cluster together because they form hydrogen bonds with each other. Unlike water, which is polar, lipids are **nonpolar molecules** because they have no negative and positive poles. Polar and nonpolar substances, like oil and water, will separate after being mixed. The oil, which is a nonpolar lipid, is not attracted to the water. Because the water molecules attract one another, the oil is pushed away. In this way, lipids interact only with each other. You can see the lipid and water layers in **Figure 3-6d**.

Figure 3-6
a If you pour oil into a beaker of water . . .

b . . . and stir it thoroughly, . . .

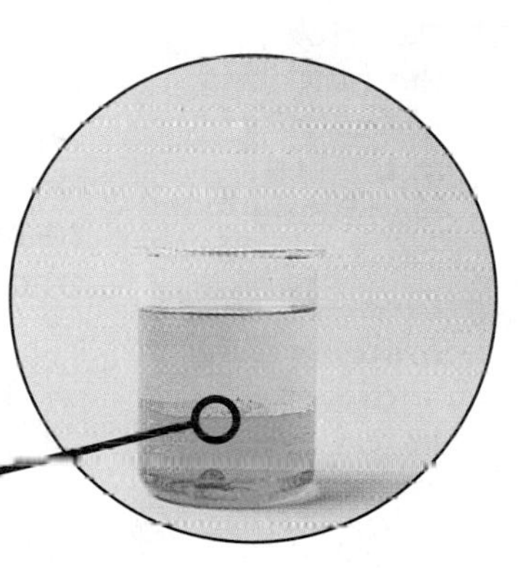

c . . . the oil and water will separate because water is polar and oil is nonpolar.

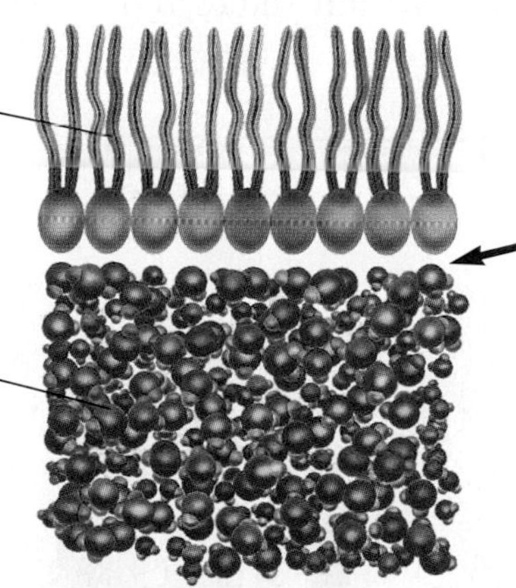

d The force of water molecules pushing a sheet of lipids is the force that shapes a cell's membrane.

SECTION REVIEW

1. **What is the function of a cell membrane?**
2. **Why is it advantageous for a cell to grow only to a certain size and then divide into two smaller cells?**
3. **How does a polar molecule differ from a nonpolar molecule? Give an example of each molecule.**
4. **How do the properties of water help shape a cell membrane?**

Proteins determine most of the functional characteristics of a cell membrane. If a cell's proteins are defective, the cell's function may be impaired, which can have devastating effects. For example, scientists have found that a single faulty gene encodes for a defective protein that causes the deadly disease cystic fibrosis. If scientists can correct the protein, they may be able to cure the disease.

3-2 Membrane Architecture

OBJECTIVES

1. **Illustrate the arrangement of phospholipids in a lipid bilayer.**
2. **Describe two characteristics of a lipid bilayer.**
3. **Describe the functions of proteins in the cell membrane.**
4. **Explain how proteins stay anchored in the cell membrane.**

Structure of the Lipid Bilayer

Recall that the cell membrane is not just an envelope enclosing the contents of a cell. It is more like a security guard controlling the flow of materials into an official building. In order to understand how a cell membrane exercises this control, you must first take a look at its structure.

The basic building block of the cell membrane is a kind of lipid molecule called a **phospholipid**, which has both polar and nonpolar regions. As you can see in **Figure 3-7a**, a phospholipid is an unusual-looking molecule: a short "head" is joined to two long "tails." The head of a phospholipid contains phosphorus and nitrogen, which make it polar. Therefore, the head is attracted to water. The long tails of a phospholipid molecule are nonpolar, however, so water molecules tend to push them away.

Water can interact with the polar heads and repel the nonpolar lipid tails best if the phospholipids are aligned in two layers, as shown in **Figure 3-7b**. This double layer of phospholipids forms a flexible **lipid bilayer**, which is shown in **Figure 3-7c**. Notice that the polar heads of the phospholipids point toward the water inside and outside the cell. The tails are buried in the interior of the membrane, directing themselves away from the water.

Figure 3-7
a **This phospholipid molecule . . .**
b **. . . is part of a lipid bilayer.**
c **The lipid bilayer forms the framework of the cell membrane.**

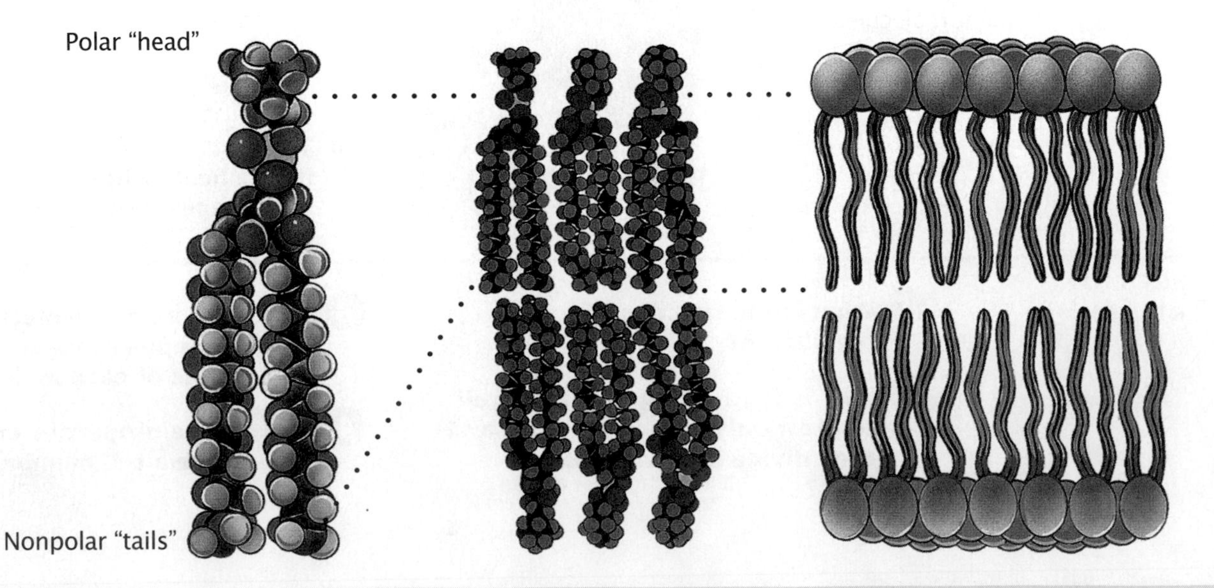

Characteristics of the Lipid Bilayer

Lipid bilayers stop polar molecules

The lipid bilayers found in cell membranes have two important characteristics. One important characteristic is that most polar molecules cannot go across it. Polar molecules are attracted to the water inside the cell or to the water outside. However, they cannot interact with the nonpolar tails of the phospholipids within the lipid bilayer. The result is that the interior part of the cell membrane acts as a barrier to polar molecules. But most food molecules and other substances needed by the cell are polar. So the cell must have a way of allowing these molecules across. If the cell membrane were made only of a lipid bilayer, there would be no way for food and other things to pass into and out of the cell.

The solution is to have passageways through the barrier. The cell membrane has passageways made of proteins, such as those shown in **Figure 3-7d**. By making "gates" that can open and shut in the lipid bilayer, proteins enable the passageways to regulate precisely the substances that go into and out of cells. You will see later that cell membrane proteins have other roles, too.

Lipid bilayers are fluid

A second important characteristic of lipid bilayers is that their phospholipid and protein molecules are not rigidly fixed in place. These molecules move about like rubber life preservers floating on the surface of a swimming pool. The lipid bilayer is fluid.

Because they are not fixed in place, the phospholipid and protein molecules that make up the cell membrane can shift from one region of the cell membrane to another. This is important because cell membranes can be structured to fit the needs of different cell types.

Evolution

All cells have a cell membrane constructed of a lipid bilayer. What does this suggest about evolutionary relationships among living things?

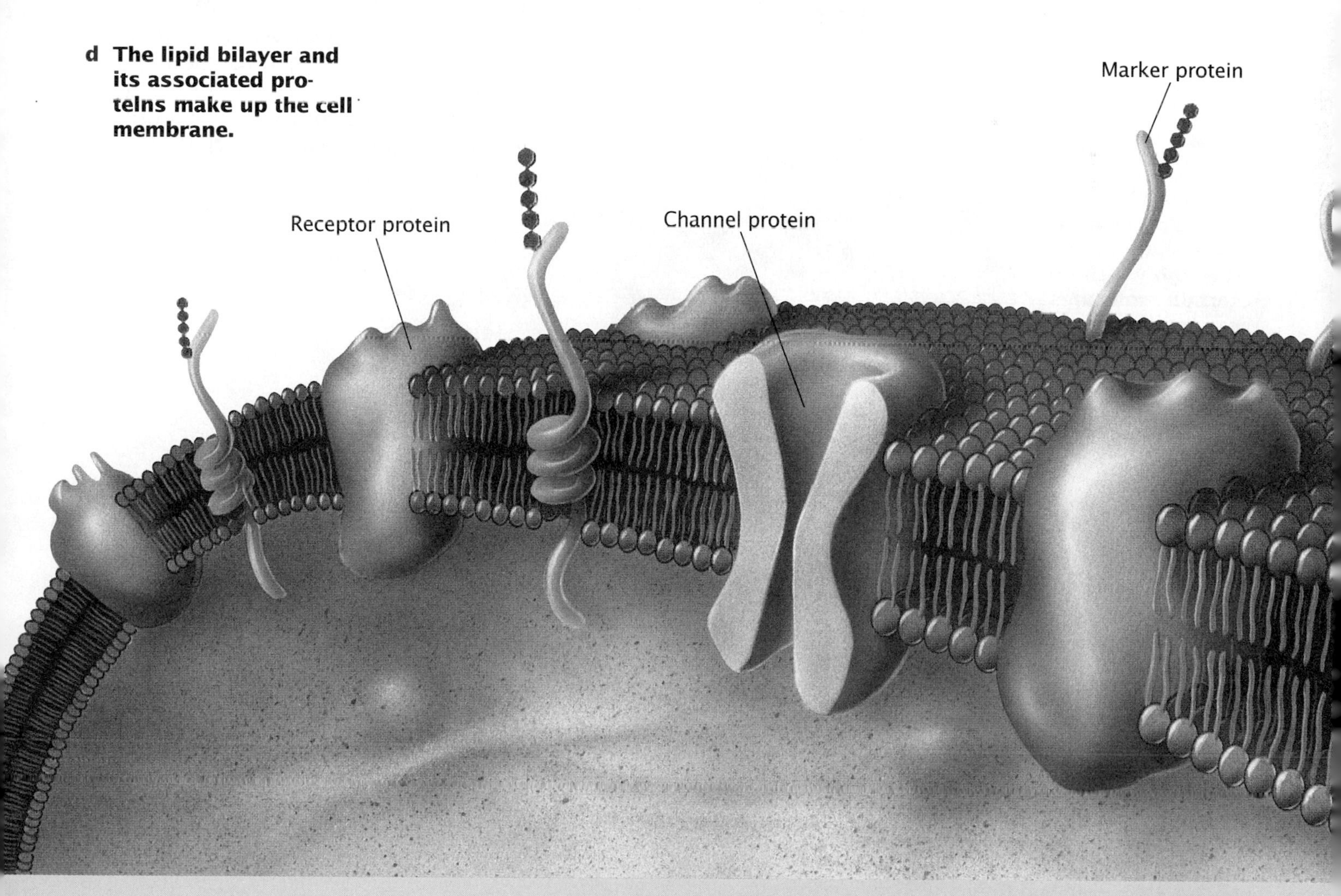

d The lipid bilayer and its associated proteins make up the cell membrane.

Roles of Cell Membrane Proteins

If you were able to peer at the surface of a cell, it would look like a smooth sea of phospholipids interrupted by proteins sticking out from the surface, some like boulders, others like tall trees. Proteins that protrude from the cell membrane may serve as channels, receptors, or markers.

Channels allow some molecules to pass through the membrane

Channels through a cell membrane are formed by doughnut-shaped proteins, as you can see in **Figure 3-8**. Many molecules and ions that are needed by the cell cross the membrane through these passageways. However, these channels are like locked doors—only people with a key can enter. In the same way, each channel will admit only certain molecules.

Receptors transfer information

Receptor proteins in the cell membrane are shaped like boulders, as shown in **Figure 3-9**. Receptors convey information from the external environment into the cell. The part of the receptor that sticks out from the cell membrane has a specific shape. Only molecules with the right shape can fit into these receptors. When a molecule and its receptor are locked together, a change occurs to the part of the receptor that faces inside the cell. This change triggers responses inside the cell.

Markers help identify cells

Cell surface markers are elongated proteins, often with short chains of carbohydrates attached, as you can see in **Figure 3-10**. These proteins are the "name tags" of the cell. Every cell of your body has markers on its surface saying that it is a part of you and nothing else. These markers organized your first body tissues before you were born. As you developed and new tissues and organs were formed, these markers directed your cells to their proper locations. In Chapter 4, you will learn how cell surface markers help a cell interact with its environment.

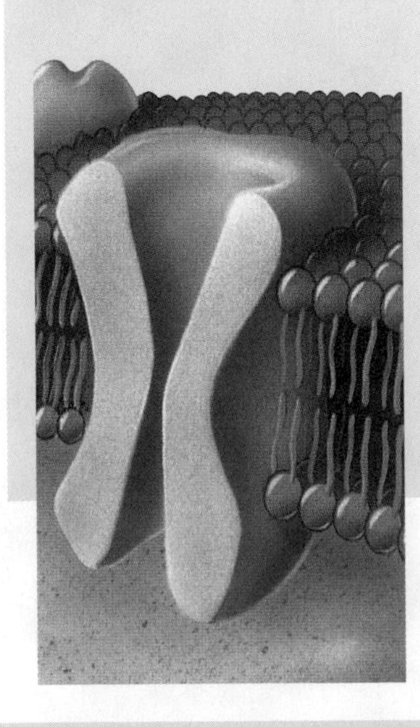

Figure 3-8
Channels
These proteins act as passageways through which only certain molecules can pass.

Figure 3-9
Receptors
These proteins transmit information into the cell by reacting to certain other molecules.

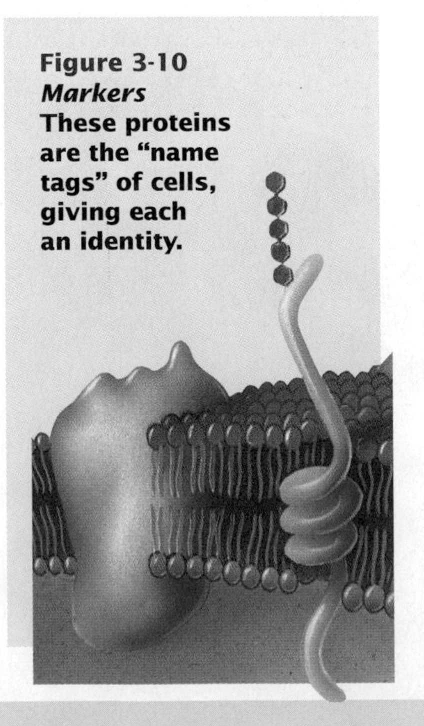

Figure 3-10
Markers
These proteins are the "name tags" of cells, giving each an identity.

Protein Structure and the Cell Membrane

Recall from Chapter 2 that proteins are molecules made of subunits called amino acids. Proteins in the cell membrane can serve as channels, receptors, or markers. How are proteins able to play such different roles? And how do proteins, which can be polar molecules, fit into the inner nonpolar region of the lipid bilayer? To answer these questions, take a closer look at the structure of protein.

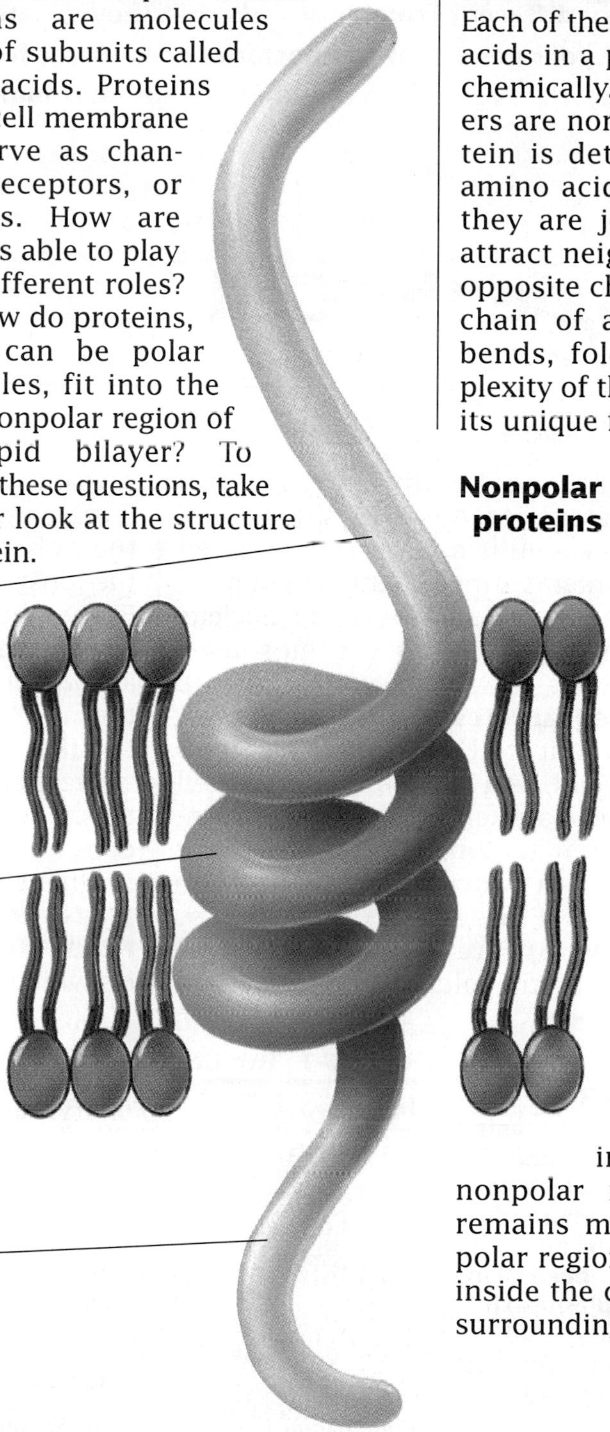

Figure 3-11

a The amino acids in this region of the protein are mostly polar. Therefore, this polar region is compatible with the water outside the cell.

b The amino acids in this region of the protein are mostly nonpolar. Therefore, this region is compatible with the nonpolar area in the center of the bilayer.

c This region is also polar and is compatible with the water inside the cell.

Protein structure is variable

Each of the 20 different kinds of amino acids in a protein is slightly different chemically. Some are polar, while others are nonpolar. The shape of a protein is determined by its particular amino acids and the order in which they are joined. Some amino acids attract neighboring amino acids with opposite charges. As a result, a single chain of amino acids in a protein bends, folds, and twists. The complexity of the shape gives each protein its unique function.

Nonpolar regions anchor proteins in the membrane

A protein that fits into a cell membrane has three basic sections, which are pointed out in **Figure 3-11**. The two end sections contain many polar amino acids that form hydrogen bonds with water. However, the middle section is made of many nonpolar amino acids. This nonpolar coil fits into the nonpolar interior of the lipid bilayer, allowing the protein to float in the membrane like a ship at sea. The protein is anchored into the membrane by this nonpolar region, but the protein remains mobile. The protein's nonpolar region prevents it from sinking inside the cell or floating off into the surrounding water.

SECTION REVIEW

1. **How do phospholipids interact with water to form a lipid bilayer?**
2. **Why is a lipid bilayer said to be fluid? What is the importance of this property?**
3. **Explain three roles proteins play in cell membranes.**
4. **How does a protein stay anchored in a cell membrane?**

Have you ever had a case of strep throat? If so, your cells were under attack by bacteria. Bacteria are single-celled organisms. Bacterial cells are very different from your cells, but they are nevertheless the ancestors of your cells. How do bacterial cells differ from your cells, and how did bacteria evolve into the kind of cells that make up your body?

3-3 Inside the Cell

OBJECTIVES

1. **Distinguish between eukaryotic cells and prokaryotic cells.**
2. **Relate the success of eukaryotes to the presence of organelles.**
3. **Describe a theory that explains the evolution of eukaryotic cells.**
4. **Compare three types of microscopes commonly used by biologists.**

Two Types of Cells

All cells can be divided into two large categories. A eukaryotic *(yoo KAR ee oht ik)* cell is a large, complex cell that contains a membrane-bound compartment called a **nucleus**. The nucleus contains DNA within chromosomes. Organisms with eukaryotic cells are called **eukaryotes** *(yoo KAR ee ohtz)*. A prokaryotic cell, or **prokaryote** *(pro KAR ee oht)*, is a very small, simple cell that lacks a nucleus. Its DNA is a single, circular molecule that is not enclosed in a membrane-bound compartment. Prokaryotic organisms, or prokaryotes, are ancient life-forms. Fossils of the oldest cells on Earth reveal that prokaryotes existed 3.5 billion years ago. Although bacteria are the only living prokaryotes, they are the most numerous of all living things. Every other living organism is a eukaryotic organism, or eukaryote. Compare a few of the general characteristics of these two cell types in **Table 3-1**.

Eukaryotes and prokaryotes share several characteristics. Both kinds of cells have a cell membrane. They also contain a mostly fluid internal environment called the **cytoplasm** *(SYT uh plaz uhm)*. They both also contain **ribosomes** *(RY buh sohmz)*, structures on which proteins are made.

Table 3-1 Two Cell Types

Prokaryote	Eukaryote
No nucleus	Nucleus
No membrane-bound organelles	Many organelles
Most 1–10 μm in size	Many 2–1,000 μm in size
Evolved 3.5 billion years ago	Evolved 1.5 billion years ago
Only bacteria	All other cells

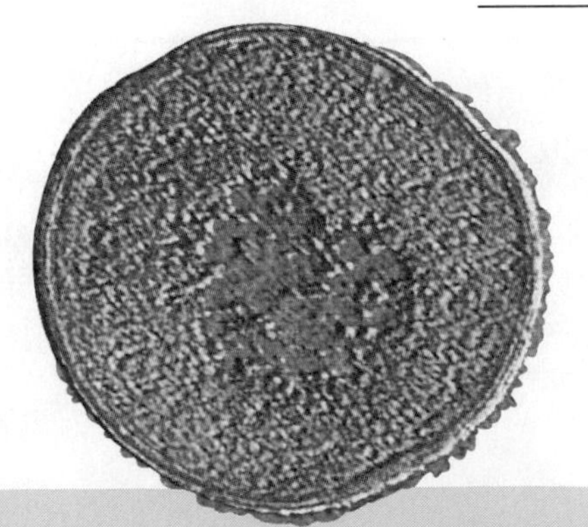

Table 3-1
a This bacterium, *Streptococcus pneumoniae*, is an example of a prokaryotic cell. It has been magnified approximately 40,000 times in this micrograph.

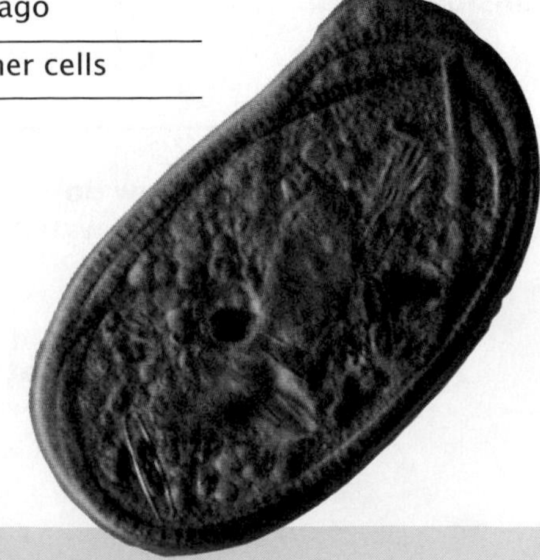

b This protist, *Chilodonella*, is an example of a eukaryotic cell. It has been magnified 150 times in this micrograph.

Eukaryotic Cells Have Compartments

The nucleus is not the only compartment inside a eukaryotic cell. The cytoplasm consists of an aqueous space called the **cytosol** *(SYT uh sohl)* and a variety of structures known as organelles. Each **organelle** is a specialized compartment that carries out one or more specific functions. Some organelles supply energy to other parts of the cell, while some use energy to make molecules. Other organelles package these products for delivery to other sites in the cell or outside the cell. Many organelles, such as the nucleus, are bound by membranes. You can see a variety of organelles in the cell in **Figure 3-12**.

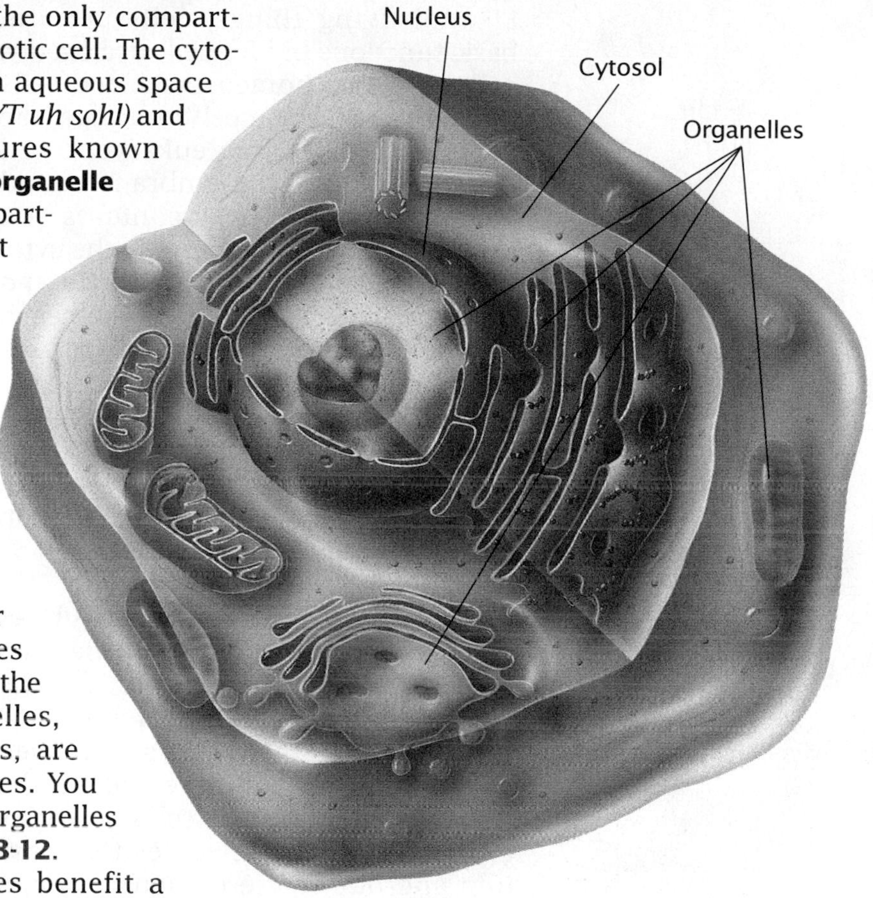

How do organelles benefit a eukaryotic cell? Imagine if you had to attend class in a building without internal walls. How would you get any work done if the gymnasium, cafeteria, library, and office were in the same room? The organelles of a eukaryotic cell are like the many separate rooms in a building. The separate compartments allow specialized activities to be restricted to particular places.

Figure 3-12
One way that a eukaryotic cell differs from a prokaryotic cell is that it contains organelles, specialized compartments that carry out specific functions.

Organelles contribute to the specialization of eukaryotic cells

The organelles in a eukaryotic cell are continuously working, moving, reproducing, and disappearing. Networks of these busy cellular components enable eukaryotic cells to specialize and perform different functions.

Think about the different kinds of cells that make up your body. A muscle cell in your leg must be able to contract quickly. Therefore, it contains a greater number of energy-converting organelles than a bone cell contains. Likewise, the striped, pulsating cells in your heart are very different from the boxlike cells that make up the layers of skin covering your hand. And these skin cells differ greatly from the nerve cells that send electrical signals from your brain throughout your body. Other examples of specialized eukaryotic cells include gland cells, which secrete life-sustaining substances, and sperm cells, which need to be able to move to perform their job.

Keep in mind that all of the cells in your body originated from a single fertilized egg cell. As these cells divided, they changed in slightly different ways, becoming specialized for specific functions. Specialization enabled single-celled eukaryotes to evolve into multicellular organisms capable of surviving in a variety of environments.

Organelles: A Cell's Laborers

Like all living things, cells perform basic functions of life. Cells use energy, cells maintain homeostasis, and cells reproduce. Like the cell's nucleus, the other organelles in a eukaryotic cell are specialized membrane-bound compartments whose contents are separated from other parts of the cytoplasm. Isolated from the hustle and bustle of the rest of the cell, each organelle stands ready to perform a specific task. The coordinated activities of the different organelles enable cells to go about the business of living. *A Closer Look: Eukaryotic Cell*, on page 53, highlights the major organelles in the eukaryotic cell.

The nucleus directs cell activities

The nucleus directs cell activities and serves as the storage center for the cell's DNA. Much like wearing two layers of clothing, the nucleus is encased in a double membrane called the nuclear envelope. The cell's activities are regulated by molecules that pass into and out of the nucleus through the nuclear envelope. How do they get through? Scattered over the surface of the nuclear envelope are shallow pits called nuclear pores, which provide passageways for these molecules.

Cells manufacture and release energy

By now you are aware that the life of a cell is not a restful one. Your cells are always at work. Where do they get the energy to perform all of life's tasks? The energy that drives a cell's activities is converted within organelles called **mitochondria** *(myt uh KAHN dree uh)*. These organelles are specialized to convert energy stored in food. The number of mitochondria in most cells varies. A muscle cell in your heart, which may pump more than 70 times per minute, can contain thousands of mitochondria. A mature red blood cell has none.

The significant differences between you and plants is the source of the food processed by mitochondria for your energy. How do plants provide their mitochondria with food molecules? Plant cells contain **chloroplasts**, organelles that have the amazing ability to make chemical energy in the form of sugars, using air, water, and the energy from sunlight. This process is called photosynthesis. Chloroplasts are found mostly in algae, such as seaweed, and in green plants.

Cells maintain homeostasis

In order for a cell to function properly, supplies must be transported from one part of a cell to another. In a bacterium, materials move about at random. In eukaryotic cells, some of the molecular traffic is directed by an extensive system of membranes called the **endoplasmic reticulum** *(ehn duh PLAZ mihk rih TIHK yuh luhm)*, or ER. Weaving in sheets throughout the cell, the ER creates a series of channels that acts as a highway system through the cytoplasm. This system of internal compartments is a fundamental distinction between eukaryotes and their prokaryote ancestors.

The cell manufactures many proteins and lipids on the surface of the ER. Some of these are used within the cell to replace damaged parts. Other products, such as digestive enzymes and hormones, are exported from the cell. Those proteins and lipids destined for export are enclosed within vesicles that bud off from the ER surface. These molecules are then transported to another organelle called the **Golgi** *(GOHL jee)* **apparatus** (named after Camillo Golgi, the scientist who first described them). The Golgi apparatus is the cell's packaging and distribution center. It puts the finishing touches on molecules and releases them in the membrane-wrapped vesicles. Substances packaged in this way can be sent to different cellular compartments or exported from the cell. Some vesicles fuse with the cell membrane, releasing their contents outside the cell.

Eukaryotic Cell

Eukaryotic cells contain DNA within the nucleus and have membrane-bound organelles. Eukaryotes evolved approximately 1.5 billion years ago.

A Closer Look

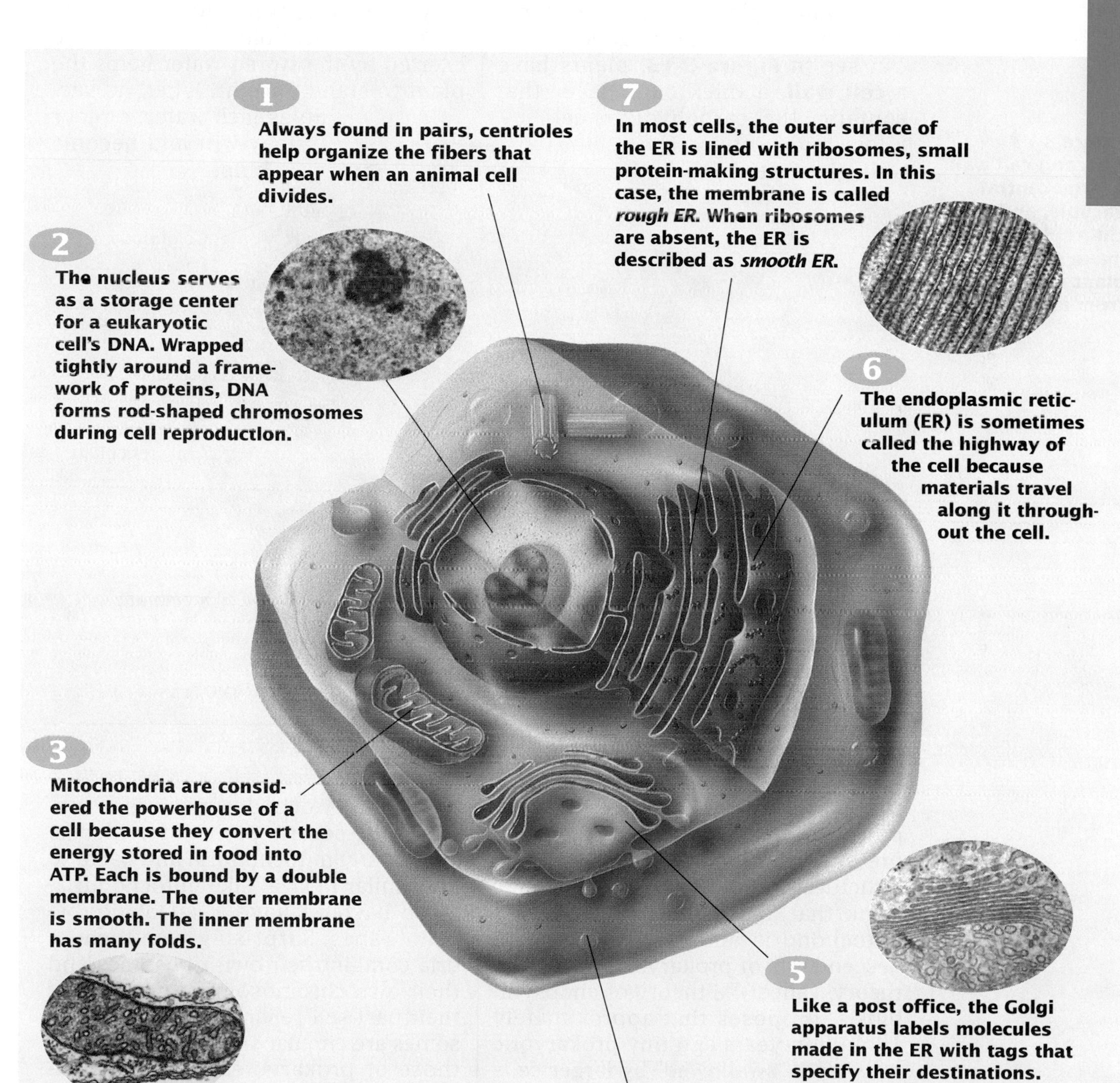

1
Always found in pairs, centrioles help organize the fibers that appear when an animal cell divides.
2
The nucleus serves as a storage center for a eukaryotic cell's DNA. Wrapped tightly around a framework of proteins, DNA forms rod-shaped chromosomes during cell reproduction.
3
Mitochondria are considered the powerhouse of a cell because they convert the energy stored in food into ATP. Each is bound by a double membrane. The outer membrane is smooth. The inner membrane has many folds.
4
Enclosed in membranes, vacuoles are fluid-filled spaces surrounded by membranes. Vacuoles function in digestion, storage, support, and water balance.
5
Like a post office, the Golgi apparatus labels molecules made in the ER with tags that specify their destinations.
6
The endoplasmic reticulum (ER) is sometimes called the highway of the cell because materials travel along it throughout the cell.
7
In most cells, the outer surface of the ER is lined with ribosomes, small protein-making structures. In this case, the membrane is called *rough ER*. When ribosomes are absent, the ER is described as *smooth ER*.

Plant Cells Differ From Animal Cells

Although all eukaryotic cells contain a nucleus and organelles, not all cells contain the same organelles. For example, many green-plant cells contain chloroplasts. Because they face environmental challenges that other organisms don't, plants have additional organelles and cell structures. As you can see in **Figure 3-13**, plants have a **cell wall**, a thick outer layer that contains the carbohydrate cellulose and protein. The cell wall gives strength and rigidity to the cell. Although vacuoles are found in both animal and plant cells, they are often more highly developed in plant cells. Plant cells store waste products, nutrients, and water in large centrally located vacuoles that may occupy between 30 and 90 percent of the cell's volume. The pressure exerted by the stored water helps the plant to stand upright. When its vacuoles lack water, a plant may wilt and become limp.

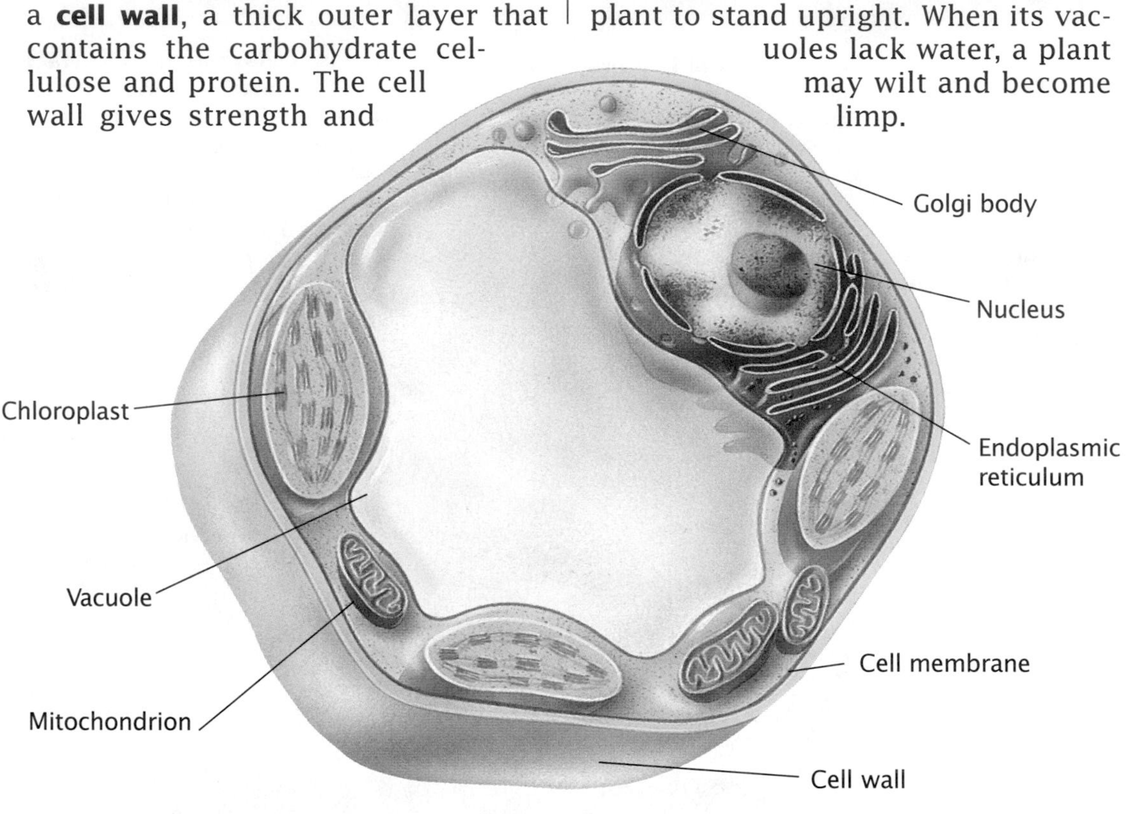

Figure 3-13
A strong cell wall, a large central vacuole, and chloroplasts that make sugar set plant cells apart from animal cells.

The Origin of Eukaryotic Cells

If one of the main differences between prokaryotic and eukaryotic cells is the presence of organelles, how did these structures arise? Most biologists now think that at least two organelles—mitochondria and chloroplasts—are descendants of prokaryotic cells. This theory, called the theory of endosymbiosis, proposes that approximately 1.5 billion years ago tiny prokaryotic cells were "swallowed" by larger cells but were not digested. The prokaryotic "trespassers" remained inside the larger cells, gradually losing their ability to live independently. The two cells formed a partnership—the trespassers became organelles, thus forming the first eukaryotic cells.

The theory of endosymbiosis is supported by several pieces of evidence. First, many mitochondria and bacteria are similar in size. Like mitochondria, some bacteria have a set of double membranes. Surprisingly, mitochondria contain their own ribosomes and their own chromosomes, evidence of their past self-reliance. Their chromosomes are circular strands of DNA, like those of prokaryotes. And the ribosomes in mitochondria are structurally similar to those found in prokaryotes. Also, some organelles divide in a manner similar to bacteria. According to the theory of endosymbiosis, bacteria capable of photosynthesis gave rise to chloroplasts found in plants and algae.

Science Technology

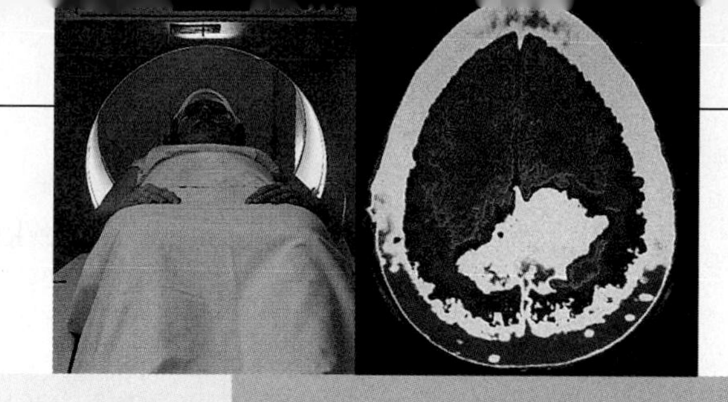

Career Opportunities

What would it be like to . . .

- perform cutting-edge research and investigations?
- have the expertise to work in hospitals, industry, or education?

These are just a few of the opportunities open to people who work in the field of science technology.

Science/Math Career Preparation

High school
- Biology
- Chemistry
- Math

College
- Depends on area of specialization; varies from a 2-year to a 4-year degree

Science Technology: What Is It?

Science technology encompasses the tools, equipment, and procedures used in science and medicine. A science technician is skilled in carrying out specialized procedures and techniques, usually under the supervision of a scientist or doctor. In hospitals, technicians do laboratory work or handle sophisticated diagnostic equipment. In industry, technicians are involved in production of pharmaceutical products, food processing, and manufactured goods.

Employment Outlook

As scientific investigations grow more complex, technicians will be needed to use practical skills, new technologies, and analytical methods.

Career Focus: Phyllis Stout

Technician at a Biological Supplies Company

"I can't remember a time in my life when I wasn't interested in science. I loved my high school biology classes, especially lab work. After I graduated, I began working at Ward's Natural Science, Inc. as a clerk filling orders. Soon, an opportunity to advance came along and for six months I trained to prepare and assemble skeletons. One of my projects involved preparing a bison skeleton for a college in Canada."

What I Do: "A few years later, I trained to work in the microscope-slide department preparing slides in all stages of production. Preparing a microscope slide is meticulous but fulfilling work. After slicing the tissues of a specimen, I stain it using one of many different dyes and techniques. It is then mounted on a glass slide and protected with a coverslip. Last year, I helped prepare over 200,000 slides that were distributed to students all over the world."

To Find Out More

1. For further information, write to:
   American Society for Clinical Laboratory Science
   7910 Woodmont Ave.
   Suite 530
   Bethesda, MD 20814

2. Visit your school counselor to find out which colleges and universities offer programs for science technicians. Review the prerequisites and course requirements. Make a poster showing the range of job opportunities in this field.

3. Find articles that describe a recent advance in medical technology. Prepare a report that describes the technology and the circumstances under which it is used.

Viewing the Cell

The existence of cells became known only after microscopes were developed in the 1600s. Our knowledge of cell structure is the result of better and more powerful microscopes. Each kind of microscope has its own strengths and limitations. Scientists have learned which microscopes can give the most information about the objects they are examining.

Two concepts are important when using a microscope to examine a specimen: magnification and resolution. **Magnification** is the microscope's ability to make an object appear larger. **Resolution** is its capacity to distinguish different objects that are close to one another.

A compound microscope uses two lenses to magnify cells

A microscope that has two lenses is called a compound microscope, shown in **Figure 3-14a**. A typical compound microscope has a light source that shines light up through the specimen, the object being examined. Light passes through the objective lens and then through the eyepiece. The image you see is magnified by both lenses. Total magnification is determined by multiplying the magnification of one lens by that of the other. If your microscope has an eyepiece lens that magnifies 10 times and an objective lens that magnifies 40 times, the specimen appears 400 times larger than it actually is.

When you look at a living cell through a compound microscope, it appears transparent, with slight variations in thickness and density. Contrast is often obtained by slicing the samples of living materials into very thin sections and staining the cells being studied. Since parts of the cell differ in their abilities to soak up various dyes, it is possible to use different stains to make specific cell structures stand out from each other. Looking at a cell this way has disadvantages—only one slice of cell can be seen, and, of course, the cells prepared in this manner die. **Figure 3-14b** shows a micrograph of a *Paramecium* magnified 100 times under a compound microscope. A micrograph is a photograph of an object as seen through a microscope.

Figure 3-14

a In a compound light microscope, light passes through a specimen and is bent and directed through the eyepiece.

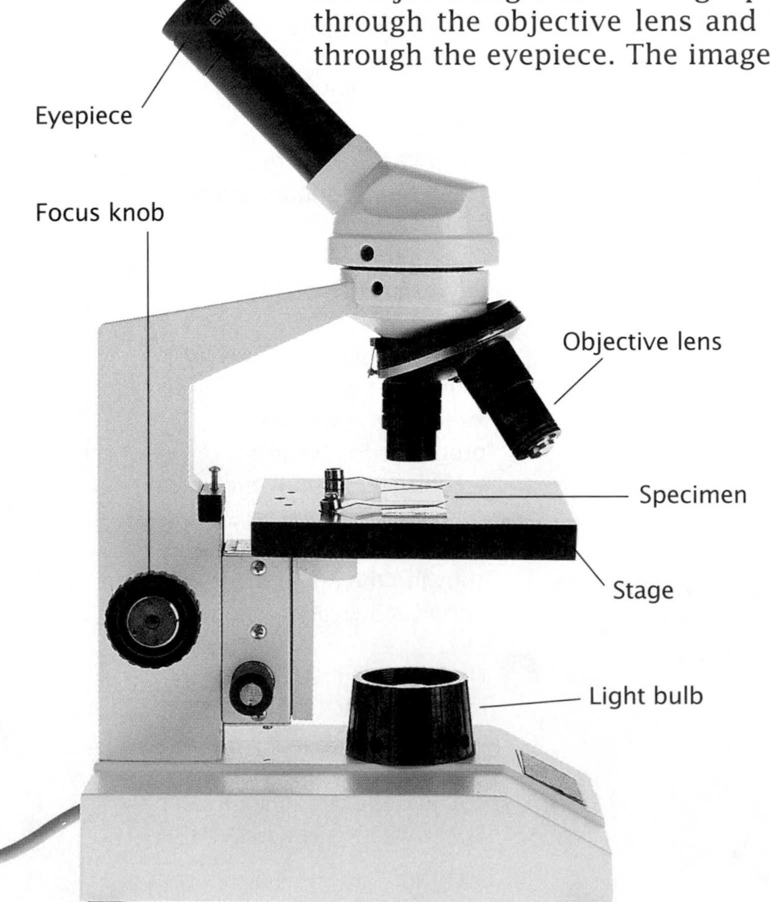

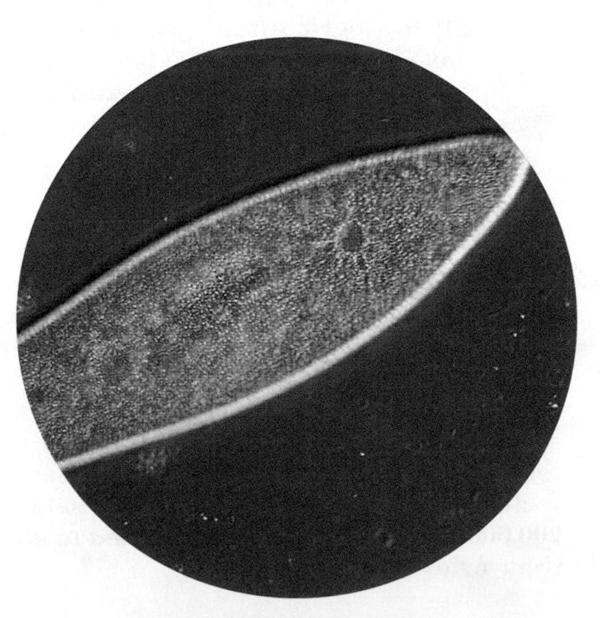

b *Paramecium caudatum*, magnified 100 times, can be studied easily.

Electron microscopes use a beam of electrons to reveal detail

Because light has wavelike characteristics, there is a limit to the size of objects that can be viewed as a sharp, focused image. Practically speaking, a bacterium with a diameter of 0.5 µm is about the smallest living thing that can be distinguished using a good mass-produced light microscope. In the early part of the twentieth century, physicists showed that an accelerated beam of electrons has wavelike properties similar to those of light. This discovery made possible a breakthrough in microscopy, the electron microscope. Electron microscopes show details not visible with light microscopes. However, they are very expensive and hence much less available than light microscopes. Also, specimens must be elaborately prepared before they can be viewed.

The transmission electron microscope (TEM) uses a beam of electrons instead of light to form images, as illustrated in **Figure 3-15a**. The specimen being examined must be sliced into extremely thin sections to allow the electrons to pass through and strike a photographic plate. These sections are treated with stains that block electrons, making details visible. The image produced is called a transmission electron micrograph. It offers very high magnification and resolution, but it does not give information about the three-dimensional shape of the specimen, as you can see in **Figure 3-15b**. Because specimens must be sliced and stained, it is impossible to view living cells with a TEM.

Figure 3-15
a In a transmission electron microscope, electrons pass through a specimen and form an image on a piece of film.

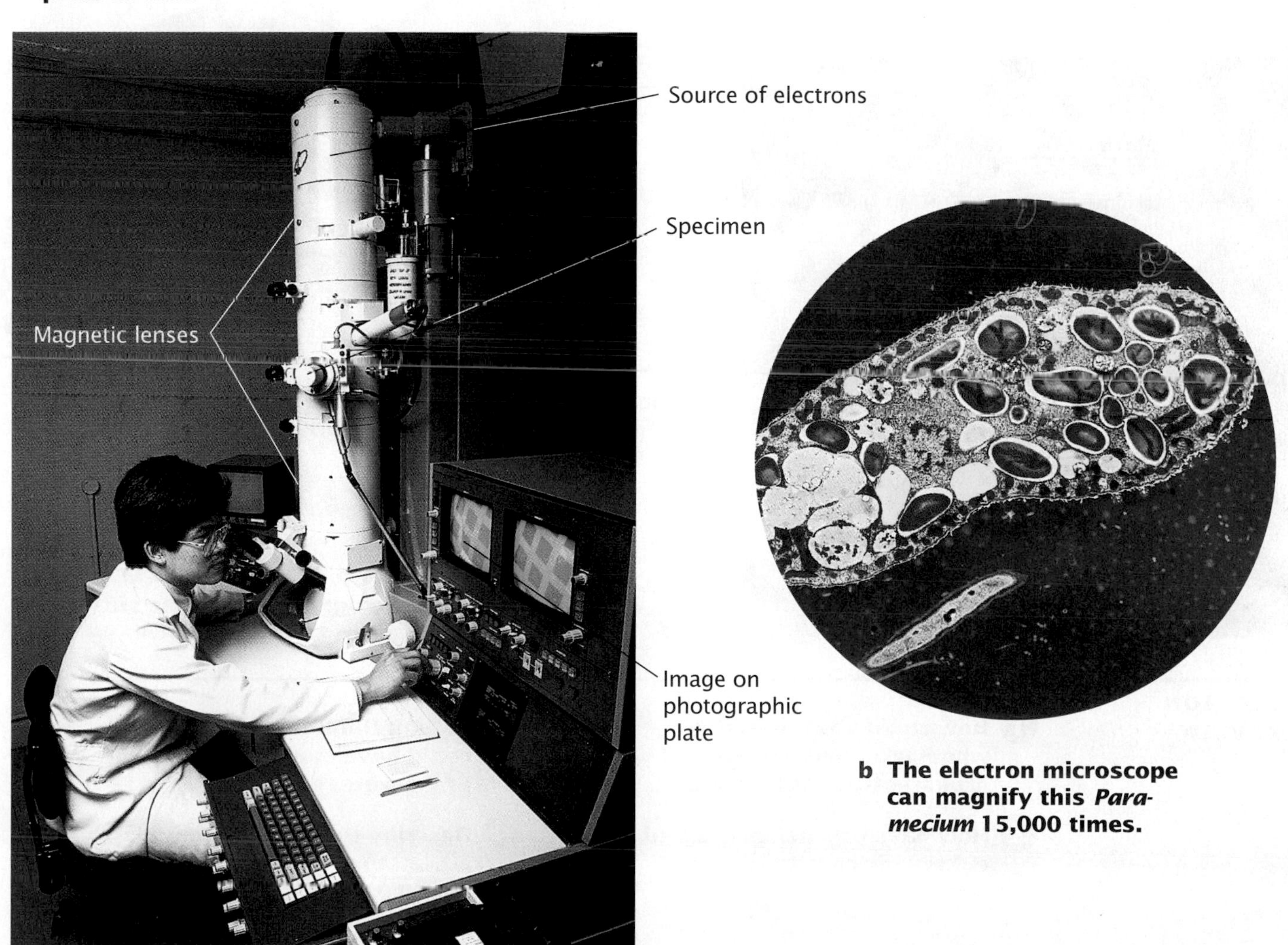

b The electron microscope can magnify this *Paramecium* 15,000 times.

The scanning electron microscope (SEM), shown in **Figure 3-16a**, enables biologists to see the surface of the specimen being studied. Specimens are not sliced, but are placed on a small metal cylinder and coated with a very thin layer of metal. A beam of electrons moves back and forth across the specimen in a manner similar to the electron beam in a television tube. The electrons do not pass through the specimen but instead bounce off its surface, forming an image that can be viewed on a video screen, as seen in **Figure 3-16b**. Because electrons bounce off molecules of gas in the air, the specimen must be viewed in a vacuum chamber. Therefore, living cells cannot be viewed with the SEM.

Micrographs made with electron microscopes are always black and white. This is because the beam of electrons is only one wavelength. The electrons strike film that detects the presence or absence of electrons—light or dark. To enhance cellular detail, color can be added when the film is developed.

A recent development in electron microscopy is the scanning tunneling microscope (STM). It uses an ultrafine, electron-emitting probe that systematically skims the surface of the object, moving up and down to accommodate the contours of the specimen. These movements are used by a computer to reconstruct an image of the specimen's surface. The STM can be used to view living cells.

Figure 3-16
a In a scanning electron microscope, a beam of electrons moves across a specimen and forms an image that is detected and displayed on a video screen.

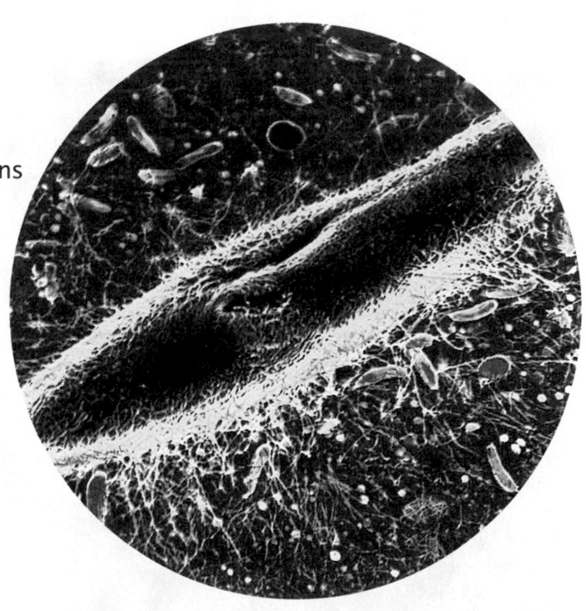

b This color-enhanced SEM micrograph shows the surface of the *Paramecium.*

SECTION REVIEW

1. **How could you determine whether an unidentified cell was prokaryotic or eukaryotic?**
2. **How have organelles enabled eukaryotic cells to become successful?**
3. **Do you think that eukaryotes could have evolved without prokaryotes? Explain.**
4. **Describe the benefits and disadvantages of using an electron microscope to study how substances enter a cell.**

CHAPTER

3 Highlights

In about two years, this boy has grown from a single cell to more than a trillion cells.

	Key Terms	Summary
3-1 At the Boundary of the Cell The cell is the fundamental unit of life. 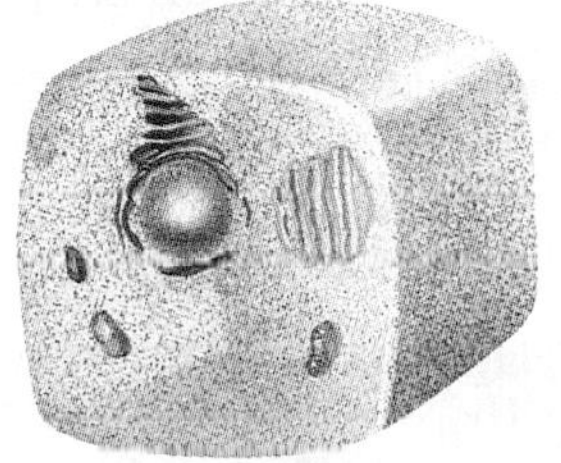	cell (p. 41) cell membrane (p. 41) polar molecule (p. 44) hydrogen bond (p. 44) nonpolar molecule (p. 45)	• A cell is the smallest unit of life. • Smaller cells function more efficiently than larger cells. • Water molecules attract each other and help shape the cell membrane.
3-2 Membrane Architecture The cell membrane is the interface between a cell and its environment. 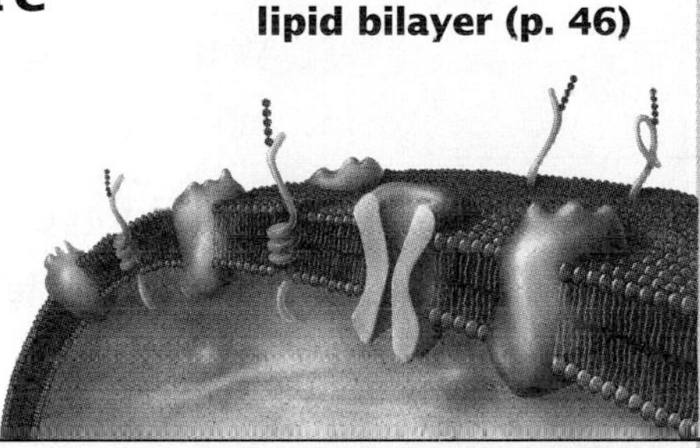	phospholipid (p. 46) lipid bilayer (p. 46)	• The cell membrane is a flexible bilayer made of phospholipids, proteins, and other molecules. • Protein in the cell membrane serve many purposes. Some enable molecules to enter and exit the cell, some enable cells to communicate with their environments, and some identify cells.
3-3 Inside the Cell  A eukaryotic cell contains many membrane-bound compartments called organelles. Each organelle carries out specific functions.	nucleus (p. 50) prokaryote (p. 50) eukaryote (p. 50) cytoplasm (p. 50) ribosome (p. 50) cytosol (p. 51) organelle (p. 51) mitochondrion (p. 52) chloroplast (p. 52) endoplasmic reticulum (p. 52) Golgi apparatus (p. 52) cell wall (p. 54) magnification (p. 56) resolution (p. 56)	• All cells have DNA and cytoplasm. • A eukaryotic cell has a nucleus and other membrane-bound organelles, whereas a prokaryotic cell does not. • Organelles are compartments that carry out life-supporting activities. Their presence enables eukaryotic cells to become specialized for specific functions. • Evidence shows that organelles arose from prokaryotic trespassers in larger cells. • Unlike animal cells, plant cells have cell walls, vacuoles, and chloroplasts. • A compound microscope can magnify a specimen according to the multiplication of the power of its two lenses. • An electron microscope uses beams of electrons to form images magnified up to 200,000 times.

CHAPTER 3

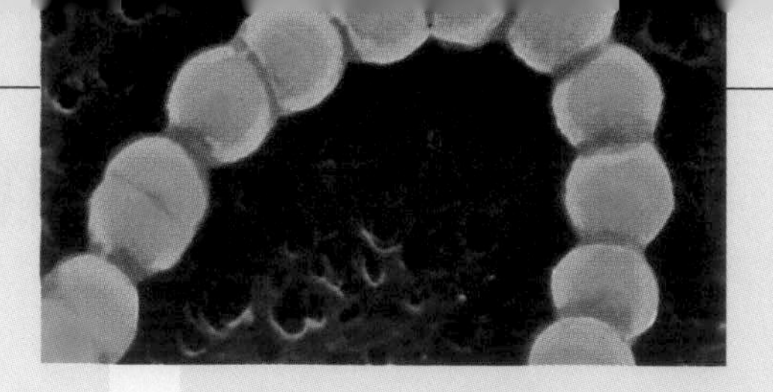

review

Understanding Vocabulary

1. For each pair of terms, explain the differences in their meanings.
   a. cell wall, cell membrane
   b. polar molecule, nonpolar molecule
   c. prokaryote, eukaryote
   d. cytoplasm, cytosol

2. For each set of terms, complete the analogy.
   a. polar molecule : water :: nonpolar molecule : ________
   b. prokaryotic cell : circle of DNA :: eukaryotic cell : ________
   c. protein manufacture : ribosomes :: extract energy stored in food : ________

Understanding Concepts

3. **Relating Concepts** Copy the unfinished concept map below onto a sheet of paper. Then complete the concept map by writing the correct word or phrase in each box containing a question mark.

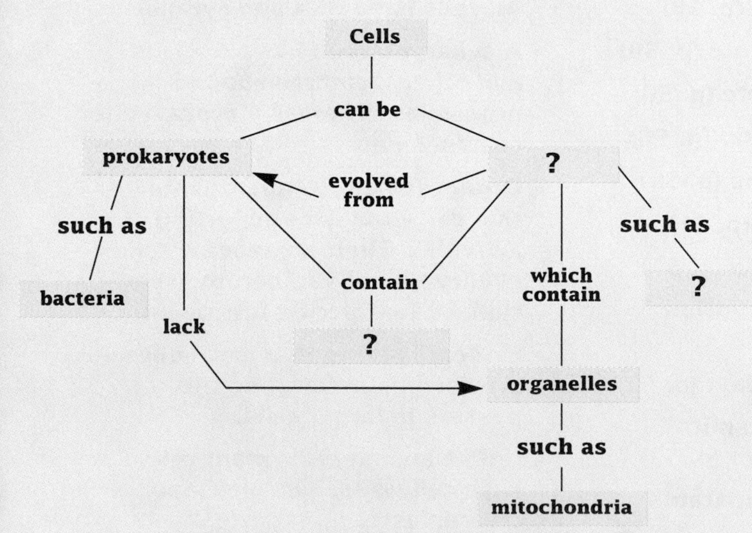

4. **Comparing Functions** In what ways is a cell membrane similar to and different from the hull of a submarine?

5. **Inferring Conclusions** How would a cell be affected if the cell membrane were completely solid and watertight?

6. **Recognizing Relationships** What are the advantages of being composed of many small cells instead of a few large ones?

7. **Inferring Relationships** Compare how water molecules interact with each other with how they interact with lipid molecules. Include an illustration with your answer.

8. **Comparing Structures** Describe how the structure of the lipid bilayer that makes up the cell membrane acts as a barrier to polar molecules.

9. **Summarizing Data** Describe two characteristics of a lipid bilayer. Explain how each characteristic affects the cell membrane.

10. **Comparing Structures** Explain how the structure of a protein enables it to fit in the cell membrane. Include an illustration with your answer.

11. **Inferring Relationships** Explain the following statement: Each cell is a prisoner of its lipid bilayer.

12. **Recognizing Relationships** Describe the functions of three cell membrane protein types. How does the structure of each kind of protein suit its function?

13. **Comparing Structures** In what ways does a bacterium swimming in a pond differ from a cell in your body? How is it similar?

14. **Summarizing Information** Describe two kinds of evidence that suggest eukaryotes evolved from prokaryotes.

15. **Drawing Conclusions** Although cells were first discovered during the 1660s, how would you explain the fact that most of our knowledge about cell structure and function has occurred within the last 50 years?

16. **Drawing Conclusions** Which kind of microscope would you use to study living cells: a compound microscope, a transmission electron microscope, or a scanning electron microscope? Support your answer.

17. **BUILDING ON WHAT YOU HAVE LEARNED** Name five organelles found in cells, and describe how each organelle enables the cell to display the properties of life as they are described in Chapter 2.

Interpreting Graphics

18. In the illustration below, identify the polar heads and the nonpolar tails in the lipid bilayer. How many layers make up the cell membrane? Why are the layers aligned in the manner shown? Describe the pathway that molecules and ions would take to get across the cell membrane.

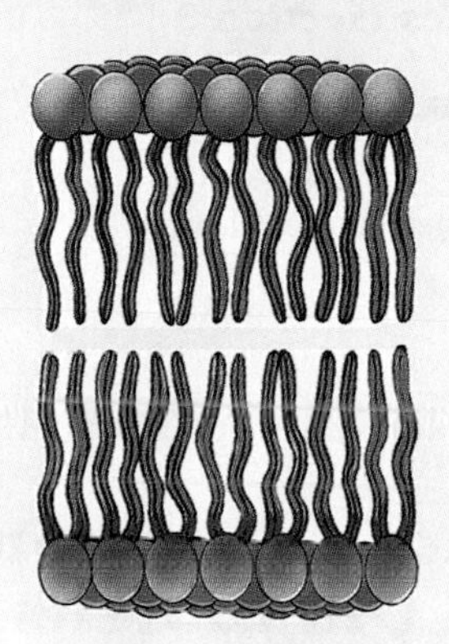

Reviewing Themes

19. ***Cell Structure and Function***
How do chloroplasts and mitochondria function in the capture and release of energy within the cell?

20. ***Evolution***
Describe the theory of endosymbiosis, and list three types of evidence that support it. How does this theory support the theme of evolution?

Thinking Critically

21. **Distinguishing Relevant Information** To determine if an automobile had been involved in a hit-and-run accident, police scientists took samples from matter ground into a smashed fender. They discovered that the material had cell walls and chloroplasts. Had the automobile hit anyone? Explain your answer.

22. **Evaluating Results** A scientist decides to remove one particular kind of organelle from a eukaryotic cell to see if the cell can survive without it. The scientist chooses to remove all mitochondria from the cell. The cell subsequently dies. Explain the probable reason.

23. **Recognizing Logical Inconsistencies** Suppose in the future that cell biologists were able to add any type of organelle or cell structure to any kind of living cell. Would you encourage a scientist to add cell walls to animal cells? Explain your answer.

Life/Work Skills

24. **Finding Information** Many careers require and use knowledge about cell structure and function. Use library references or search an on-line database to research careers such as that of an electron microscopist. What kinds of special equipment and materials do people in these careers use? What kind of educational background and training do jobs in this area require?

25. **Using Technology** Electron microscopes make it possible to see viruses, bacterial cells, and other small organisms in amazing detail. Find references that show details of microorganism structures discovered during the last 20 years. Make a table listing the kinds of organisms and the structures revealed.

Activities and Projects

26. **Unit Focus** Use library references, search an on-line database, or use an electronic encyclopedia to investigate the progress that has been made in gene therapy. Create an illustrated poster depicting at least 25 genetic disorders that doctors are seeking to cure. Include information about research and the technology used to carry out the research.

27. **Health Connection** Work together with some of your classmates to research and prepare a report on the differences between normal human cells and cancer cells. Find out about the differences in cell structure, reproduction rate, and metabolism. Find out how these differences are exploited to create cancer therapy drugs. Make a poster to go along with your report.

28. **Art Connection** Before photographs were commonly used in biology, drawings of cells and organisms were the most accurate way to share information. Describe the benefits and disadvantages of using art instead of a photograph.

Discovering Through Reading

29. Read the article "Crucibles in the Cell" in *Discover*, January 1995, pages 90–91. What new cellular organelles have been discovered and what are their functions? How are they thought to be involved in disease?

CHAPTER 4

REVIEW
- ion (Section 2-2)
- cell membrane structure (Section 3-2)
- characteristics of prokaryotes and eukaryotes (Section 3-3)

The Living Cell

This plant cell is able to move water molecules, food particles and other ions through its membrane.

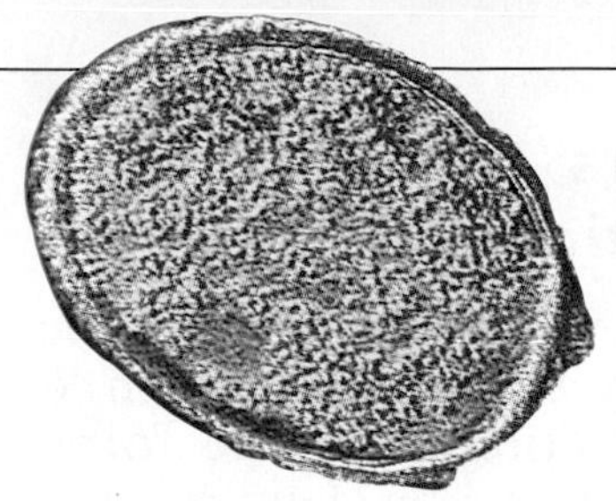

Like all living things, a cell, like this bacterium, needs information about its environment. In the same way that your senses enable you to see, hear, and feel things in your world, proteins in a plasma membrane connect a cell to its surroundings. These proteins are the cell's avenues of communication.

4-1 How Cells Communicate

OBJECTIVES

1. **Recognize the importance of cell communication in multicellular organisms.**
2. **Describe two ways receptor proteins transmit information.**
3. **Explain how an electrical signal affects a voltage-sensitive channel.**
4. **Evaluate the role of cell surface markers in the human body.**

Cell Communication in Multicellular Organisms

A significant difference between your cells and the cells of bacteria and most protists has to do with communication. Your cells are able to "talk" to one another because they have the ability to send and receive information. These "conversations" are essential for all **multicellular organisms** (such as animals and plants), which are made of many cells. They enable all of your cells to coordinate their behavior so that your body functions as a whole rather than as individual cells. The information that a cell receives from its neighbors allows it to determine its position in the body, to adjust its metabolism to suit its particular function, and to grow and divide at the proper time. Without communication, multicellular life would not be possible. **Figure 4-1** shows a stage of development in which communication among cells ensures their correct position in the organism's future shape.

Figure 4-1

a This starfish embryo (magnified 1700 times) is only hours old, but its cells are already communicating with each other to find their proper position in the developing individual.

b This communication ensures that the embryo will grow into its normal adult form shown here approximately 1/3 actual size.

Figure 4-2

a Your body keeps you energized even when you don't have time for a meal. When your blood-sugar levels are low, cells in your pancreas release the hormone glucagon.

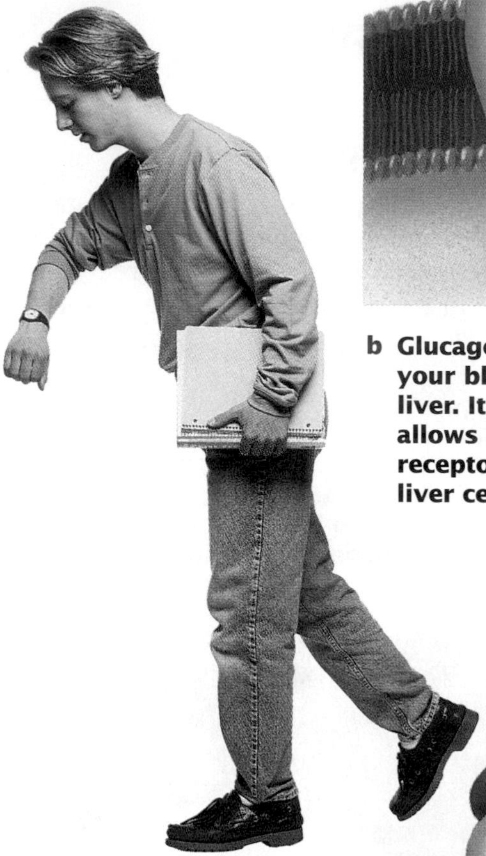

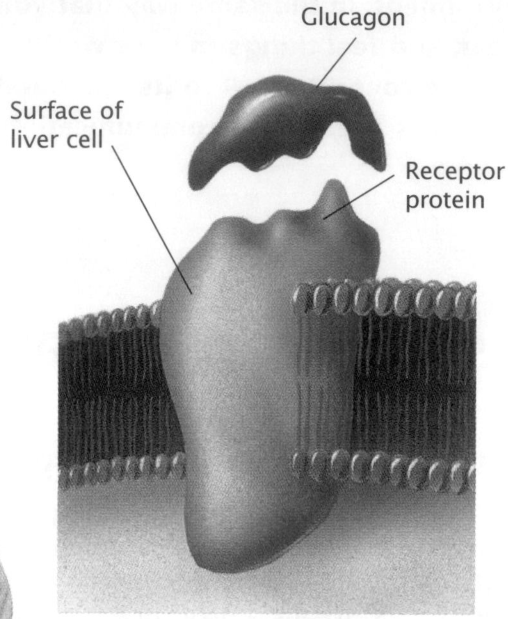

Glucagon

Surface of liver cell

Receptor protein

b Glucagon travels through your bloodstream to your liver. Its unique shape allows it to bind to specific receptors on the surface of liver cells.

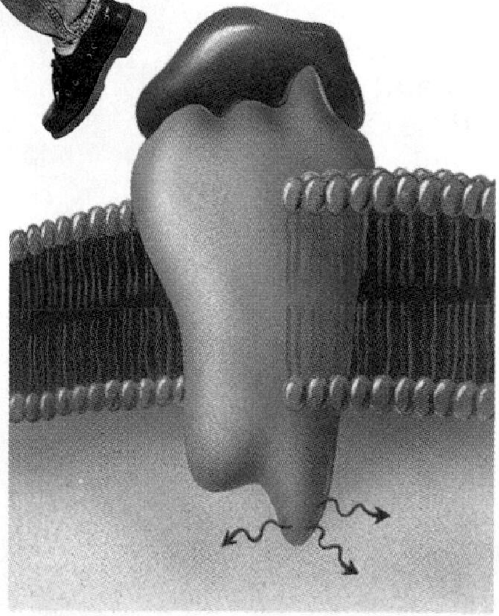

c The binding of glucagon to its receptor activates a second messenger molecule inside the cell. The second messenger relays glucagon's important message to an enzyme inside the cell, which causes blood-sugar levels to rise.

Receptors Transmit Information

Cells in your body are constantly bombarded with news sent by other cells. This information is in the form of chemical signals. The chemical signals that a cell responds to are determined by the cell's **receptor proteins**. Some receptor proteins are located on the cell's surface and function like tiny antennas, transmitting information from the world outside the cell to its interior. Other receptor proteins are located in the cell's interior.

Some receptors work outside the cell

Most receptor proteins extend through the cell's membrane. This enables a chemical signal that can't pass through the membrane to send a message into a cell. For example, a hormone is a chemical signal that acts as a messenger. Many hormones made of proteins cannot penetrate the lipid bilayer of the cell membrane. In order to deliver their messages, these hormones must bind to surface receptor proteins. One such hormone is glucagon. Glucagon is made in your pancreas and released when your blood-sugar levels are low. It travels to your liver through your bloodstream, carrying its urgent message. When glucagon reaches the liver, it causes liver cells to release stored supplies of sugar, which restores blood-sugar levels to normal.

How do receptor proteins deliver glucagon's message? **Figure 4-2** illustrates the action of glucagon. When glucagon reaches the liver, it binds to receptor proteins on the surface of liver cells. This binding causes the receptor proteins to send the message inside the cell. Glucagon's message is transferred to a second messenger, just as one runner passes a baton to the next runner in a relay race. In this case, the second messenger is a compound called cyclic AMP, which triggers an enzyme that causes the release of stored sugar. As a result, blood-sugar levels rise.

Some receptors work inside the cell

Unlike protein hormones, hydrophobic hormones are able to deliver their message by passing through the lipid bilayer of the cell's membrane. These hormones bind to receptor proteins located inside the cell rather than on its surface. Once the hormone and receptor bind, they travel as a unit to the cell's DNA. This unit has the right shape to bind to a particular gene and trigger the production of a new protein. The new protein will then influence chemical activity in the cell and play a vital role in metabolism or development. Examples of hormones that use these receptors are thyroxin, hormones that regulate the body's metabolic rate, and estrogen and testosterone, hormones responsible for the development and maintenance of sexual characteristics.

Some Channels Can Respond to Electricity

Some cells are specialized for conducting information from one cell to another. For example, your nervous system, shown in **Figure 4-3**, is made up of cells that can respond to electrical signals carried by ions. These cells are studded with **voltage-sensitive channels**, proteins that are embedded in the cell membrane. **Figure 4-4** illustrates how voltage-sensitive channels work. Each channel is lined with charged amino acids. At one end of the channel is a gate that is usually closed. However, when ions carrying an electrical signal arrive, the gate flips open and allows the ions to enter the cell. Voltage-sensitive channels are important because they allow electrical signals to travel along nerves as fast as 100 m/s. Without them, your brain and nerves couldn't function, as you'll see in Chapter 30.

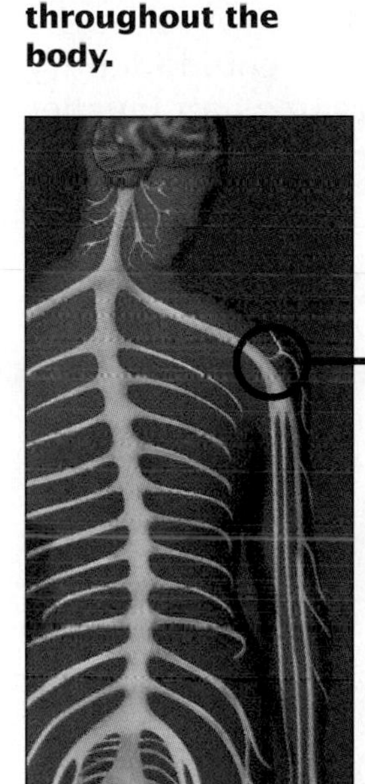

Figure 4-3
Nerve cells called neurons carry electrical signals throughout the body.

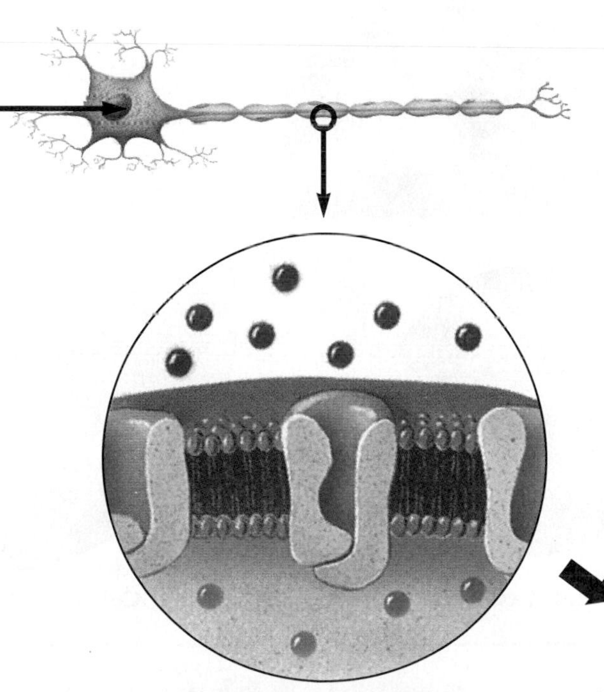

Figure 4-4
a Voltage-sensitive channels are embedded in the cell membrane of a neuron. When the concentration of ions is balanced on both sides of the membrane, the channels are closed.

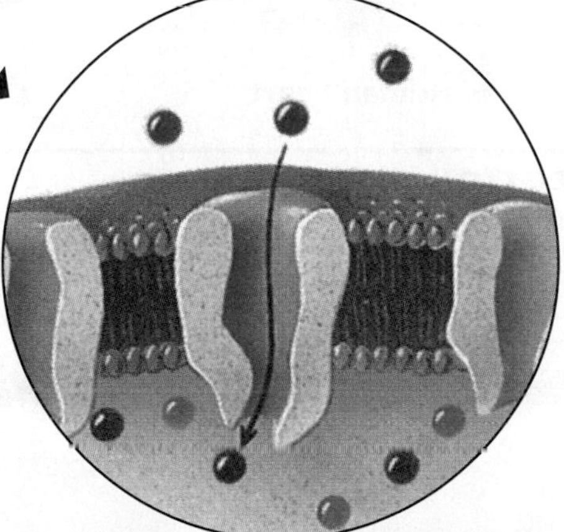

b But when additional ions upset the balance, the voltage-sensitive channels spring open and ions pass through.

Markers Identify Cells

Just as an athlete wears a uniform bearing a name and number, a cell displays specific proteins that identify its "team" and "position." These unique sets of membrane proteins are called **cell surface markers**. Why should a cell spend energy making these markers? Cell surface markers distinguish your cells from other cells, enabling your immune system to recognize and destroy uninvited guests, such as harmful bacteria. Cell surface markers are also used during early development to determine a cell's fate—whether it will become a blood cell, a muscle cell, or a kidney cell.

The uniqueness of an individual's cell surface markers explains why organ donors and recipients must be carefully matched. The doctors who operated on the organ transplant patient shown in **Figure 4-5** had to be sure that the donated heart was a good match before they attempted a transplant. A good match can be defined as two organs that share many of the same identifying proteins on the surface of their cells. As with all proteins, the structure of cell surface markers is determined by genes. If the donor and recipient are related, a successful match is likely. However, cell surface markers are unique, so perfect matches occur only between identical twins. Most organ recipients must therefore take medication to suppress the immune system, which could identify the transplanted organ as an invader and attack it.

Figure 4-5
a This man's heart transplant was successful because the cell surface markers on his cells closely match the cell surface markers on the donor's heart.

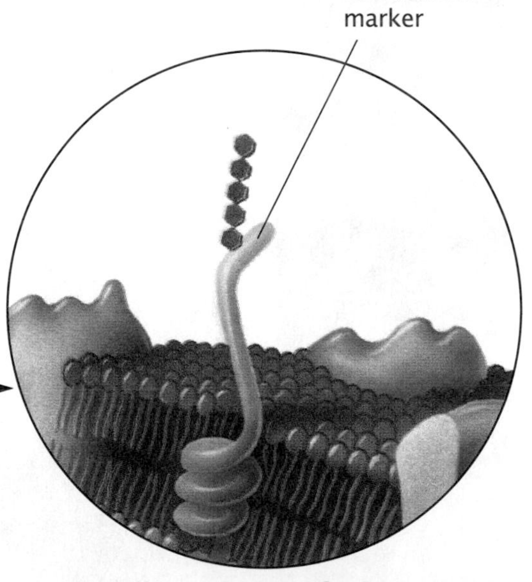

b Human heart

c Heart cells

d Cell membrane of a heart cell

SECTION REVIEW

1. **Why is cellular communication essential for multicellular organisms?**
2. **In what two ways do receptor proteins affect cellular activity?**
3. **How do voltage-sensitive channels respond to electrical signals?**
4. **Why are protein markers important when matching organ donors to recipients?**

On a hot day or after a hard workout, you might reach for a cold sports drink. This type of beverage contains water and ions to replenish what your body lost through perspiration. These ions enter your cells through protein channels. In this section you will take a close look at how cells take in substances and discard wastes.

4-2 Movement of Substances Into and Out of Cells

OBJECTIVES

1. **Describe diffusion and osmosis.**
2. **Distinguish facilitated diffusion from active transport.**
3. **Explain how the sodium-potassium pump transports ions.**
4. **Describe how large substances can enter and exit cells.**

Diffusion and Osmosis

In addition to communicating, cells move water molecules, food particles, and other ions through their membranes. Some molecules, such as water, pass through most membranes freely. Other molecules rely on help from proteins in the plasma membrane.

Diffusion mixes molecules

Look at **Figure 4-6**, which shows a few drops of dye solution that have been placed onto a layer of gelatin in a beaker. What is happening? Within an hour, molecules of dye have begun to move into the gelatin.

The mixing of two substances by the random motion of molecules is called **diffusion**. As shown in **Figure 4-6**, molecules diffuse from a region where their concentration is high (the drops of dye) to a region where their concentration is low (the gelatin). Diffusion stops when the two kinds of molecules are evenly dispersed.

Figure 4-6
a The leisurely pace of diffusion can be demonstrated by placing a few drops of food coloring onto a layer of gelatin.

b In an hour, the food coloring will begin to slowly diffuse.

c Hours later, molecules of food coloring have diffused still further throughout the gelatin.

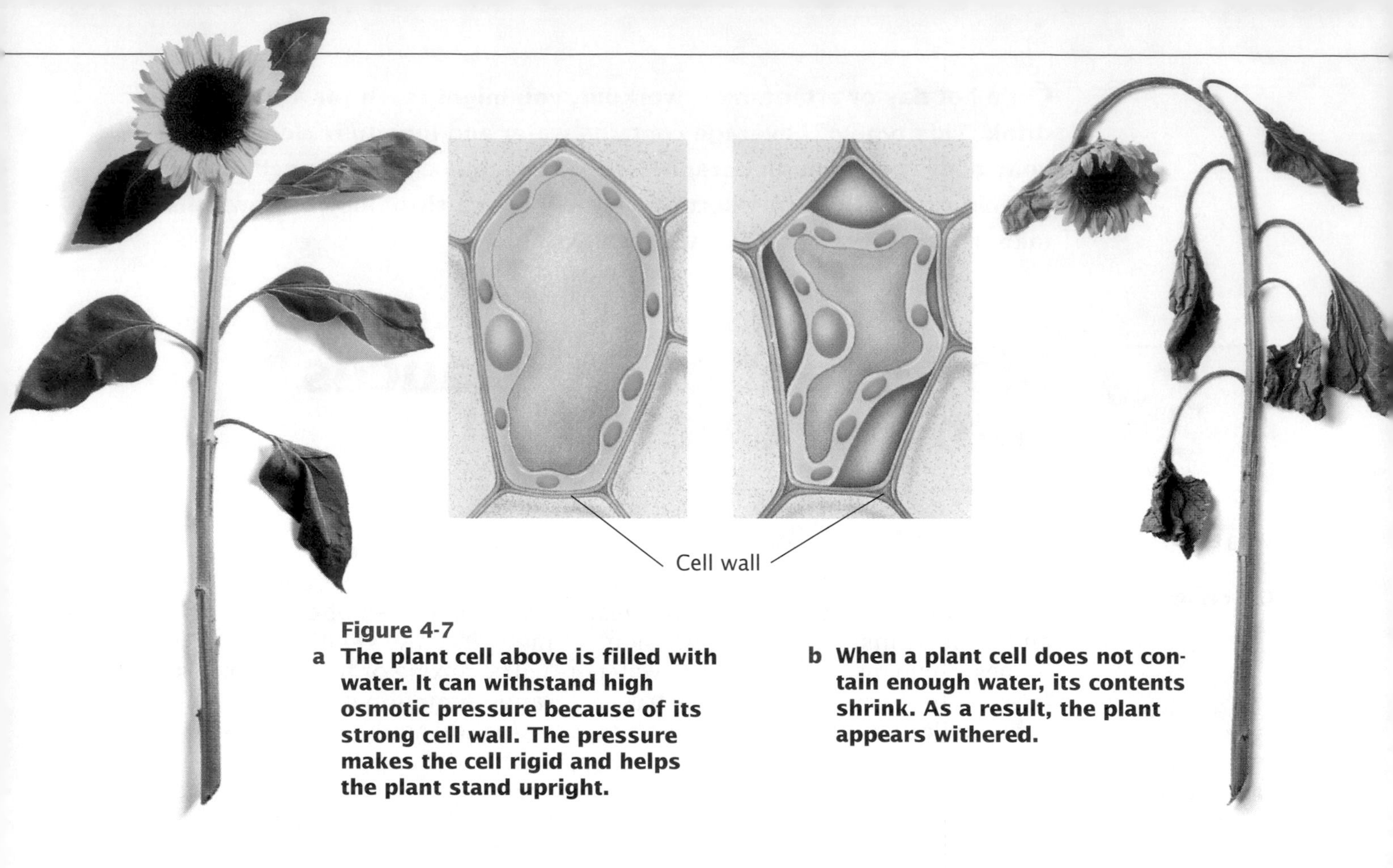

Figure 4-7

a **The plant cell above is filled with water. It can withstand high osmotic pressure because of its strong cell wall. The pressure makes the cell rigid and helps the plant stand upright.**

b **When a plant cell does not contain enough water, its contents shrink. As a result, the plant appears withered.**

Cell Structure and Function

How does osmosis enable a cell to respond to changes within an organism?

Water passes through membranes by osmosis

Water enters and leaves living cells by diffusing across cell membranes. As you learned in Chapter 3, certain molecules can pass through a cell membrane, but others cannot. Because water molecules are very small, they are able to slip freely through the tiny gaps in the lipid bilayer of a cell membrane. Water molecules constantly move back and forth across cell membranes. This free motion of water has a very important consequence—it enables cells to absorb water.

The diffusion of water across a membrane, such as the cell membrane, is called **osmosis** *(ahz MOH sihs)*. As water enters a cell by osmosis, the water molecules move from an area of higher concentration to an area of lower concentration. Cells also contain sugars, proteins, and other substances. Therefore, the concentration of water outside the cell is usually greater than the concentration of water inside the cell. As a result, cells tend to continually gain water by osmosis.

Water in a cell creates osmotic pressure

When water enters a cell, it causes the cell to swell, much as air inflates a balloon. This force creates a pressure called **osmotic** *(ahz MAH tihk)* **pressure**. If osmotic pressure is very high, it can cause a cell to burst. Most cells cannot withstand high osmotic pressure unless their membranes are braced to resist swelling. Many kinds of organisms, such as the plant in **Figure 4-7**, have cell walls to support their cells and prevent them from bursting.

How do organisms without cell walls keep from bursting? Many single-celled organisms, such as paramecia, have organelles that pump excess water out of the cell. In multicellular animals, a balance exists between the concentration of fluids inside and outside the cells. In your body, for example, the concentrations of salts, sugars, and other ions are about the same inside cells as in the fluids surrounding them. Thus, water enters the cells of the body at the same rate that it leaves.

Channels and Pumps Provide Selective Transport

Unlike water molecules, many substances cannot easily pass through the cell membrane. Molecules such as sugars and proteins are often too large to slip through the gaps between the phospholipids. In addition, these molecules are often polar and cannot pass through the nonpolar region of the lipid bilayer. To move into and out of cells, polar molecules and ions use channels and pumps embedded in the cell membrane. These structures are made of protein. Protein channels are shown in **Figure 4-8**.

The transport of specific substances by means of protein channels and pumps is called **selective transport**. This form of transport is said to be selective because channels and pumps enable only certain substances to pass into or out of a cell. By using selective transport, a cell can control the substances that enter and leave. There are two kinds of selective transport: facilitated diffusion and active transport.

Facilitated diffusion is toll-free

The simplest selective transport channels are similar to tunnels. As long as a molecule or ion can fit, it is free to pass through the channel in either direction, as shown in **Figure 4-8**. These channels allow each kind of molecule or ion to diffuse toward the side of the cell where it is least concentrated. Eventually, diffusion balances the concentrations of that molecule or ion on both sides of the membrane. This form of selective transport is called **facilitated diffusion** because the channels assist, or facilitate, the diffusion of substances. The movement of substances is caused by their different concentrations inside and outside the cell. Therefore, a cell does not have to expend energy to obtain these substances. Many ions, such as potassium ions, enter the cell using facilitated diffusion. Potassium ions help maintain the proper balance of electrical charges inside and outside of the cell.

Figure 4-8

a Ahhhh! There is nothing like the taste of a good thirst quencher after a tough workout. Have you ever wondered how the cells of your body replenish nutrients?

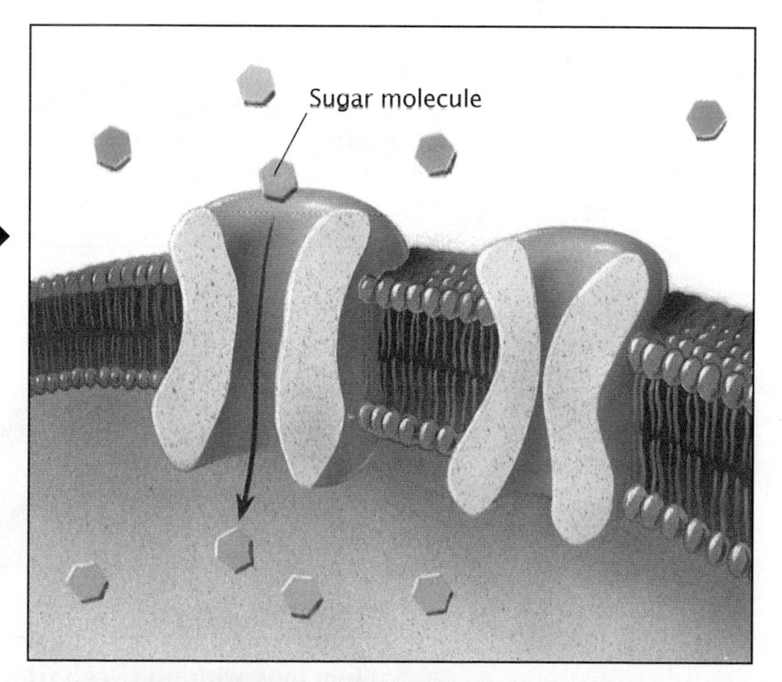

b Your bloodstream carries minerals and sugars to your body's cells. If the molecule has a certain shape, it can freely enter the cell through protein channels.

Active-transport moves molecules in one direction

If all selective transport occurred by means of channels, all kinds of molecules and ions would simply diffuse into or out of a cell. In fact, this is not the case. Unlike channels, selective-transport pumps allow molecules and ions to pass in only one direction. Like turnstiles at a stadium, pumps admit certain molecules and ions into the cell but do not let them out. This strategy enables a cell to stockpile certain substances in far greater concentrations than they occur outside the cell.

In order to operate pumps, a cell must use some of its energy to change the shape of the pump protein. The cell also uses energy to move a substance across the cell membrane to a region of higher concentration. Therefore, the operation of pumps is called **active transport**. Active transport plays a critical role in moving many important materials, such as sugars and amino acids. Almost all of the active transport in animal cells is carried out by only two kinds of pumps: the sodium-potassium pump and proton pumps.

The sodium-potassium pump enables your cells to conduct nerve impulses

Without sodium-potassium pumps, you and most other animals would be immobile lumps. As you can see in **Figure 4-9**, the **sodium-potassium pump** moves sodium ions out of cells and potassium ions into cells. In just 1 second, each channel can transport more than 300 sodium ions out of a cell. This flow of ions is vital for restoring a balance needed to keep a nerve cell functioning. More than one-third of all the energy expended by your body's cells is spent driving the sodium-potassium pump.

Sodium-potassium pumps also help transport sugars and amino acids into your cells. When a few sodium ions remain in the cell, facilitated diffusion channels in the cell membrane allow sodium ions to rush back into the cell. Because so many sodium ions rush back in through these channels, large numbers of partner molecules are pulled through as well, even if they are already present in high concentrations within the cell. The cell-membrane channels that admit sodium ions and their partners are called coupled channels.

A proton pump is the key to cell metabolism

Cells rely on proton pumps to move protons across membranes. Just as the sodium-potassium pump causes a buildup of sodium ions outside the cell, the **proton pump** expels protons, causing large numbers of protons to accumulate on one side of a membrane. The protons then diffuse back into the cell through certain channels.

Figure 4-9
The sodium-potassium pump moves sodium ions out of cells and potassium ions into cells.

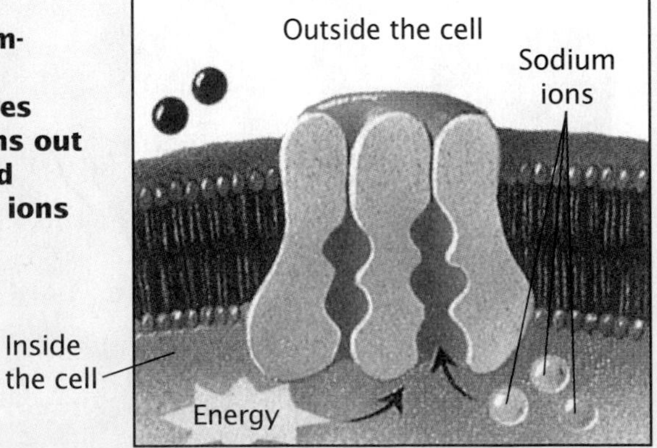

a Sodium ions within the cell fit precisely into receptor sites on the protein channel.

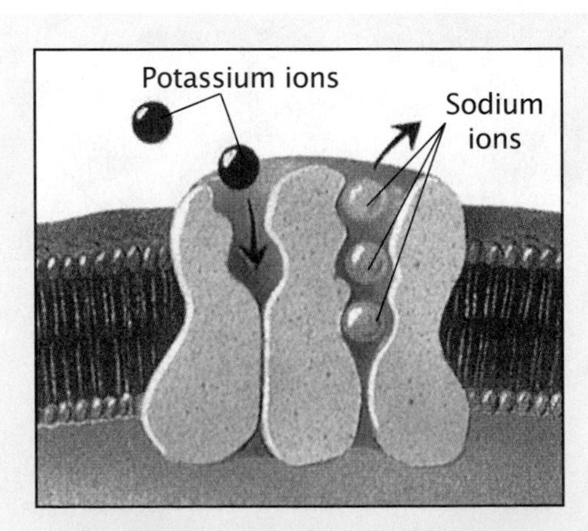

b The channel changes shape, pumping the sodium ions across the membrane. Potassium ions outside the cell move into receptor sites.

Proton pumps are an important mechanism in cell metabolism, as you will learn in Chapter 5. Proton pumps depend on chemical or light energy to function. In green plants, a proton pump enables cells to convert light energy into chemical energy. Another proton pump enables your cells to transform energy obtained from the food you eat into chemical energy you can use.

Defective channel proteins can be lethal

A cell's proteins are built according to genetic instructions that evolved millions of years ago and that have stood the test of time. Any changes in a protein's structure can alter or impair its function and, in turn, threaten the cell's normal activities. The effects can be devastating. Consider, for example, the disorder known as cystic fibrosis. Cystic fibrosis causes the body to produce very thick mucus, which builds up in the lungs, pancreas, and liver. As a result, people with cystic fibrosis have extreme difficulty breathing, and they cannot produce the enzymes needed to completely digest food. The thick mucus also coats the hairlike projections that line air passages and inhibits their sweeping motion. Without the cleansing effect of the hairs, harmful bacteria and debris remain in the lungs and increase the chances of infection.

What causes cystic fibrosis? Research into its cause reads like a detective story. A key clue was that people with cystic fibrosis have very salty sweat, an indication of abnormally high levels of sodium and chloride ions. This salt imbalance is what causes thick mucus to accumulate in a patient's lungs—the high levels of salt in the cells draw water out of the mucus, causing it to thicken.

The level of salt within a cell is determined by the movement of sodium and chloride ions across the cell membrane. These ions pass through protein channels embedded in the cell membrane. Tests of cystic fibrosis patients revealed that their cells are unable to export chloride ions. For some reason their chloride channels do not work.

Searching for the genes that encode chloride channels, researchers eventually isolated the gene that causes cystic fibrosis. Normally, this gene encodes a protein channel that helps move chloride ions out of the cell. In cystic fibrosis patients, the gene encodes for a misshapen protein channel that cannot function properly. In Chapter 9, you will read how this information could yield new methods of treatment and a possible cure for cystic fibrosis patients.

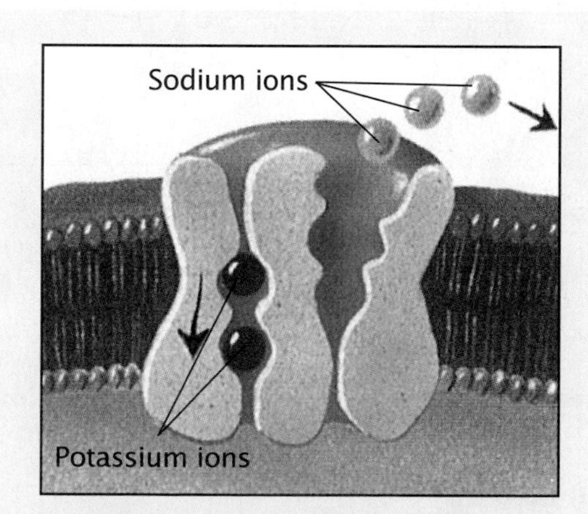

c Sodium ions are released and cannot reenter through this channel. At the same time, potassium ions are pumped across the channel into the cell.

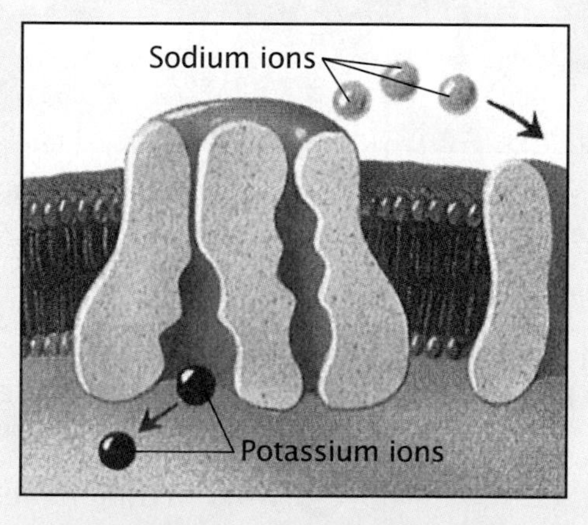

d Potassium ions are released inside the cell. Sodium ions outside the cell, along with sugar molecules, later enter the cell through coupled channels.

Moving Large Particles

Matter, Energy, and Organization

What would happen to a cell if its membrane were solid and impenetrable?

Sometimes particles enter a cell without actually crossing the membrane. For example, some single-celled eukaryotes take in food particles by engulfing them with armlike extensions, as shown in **Figure 4-10**. When the edges of the extensions meet, they fuse together, capturing the particle within a sac inside the cell. The process of bringing particles into a cell using extensions of the membrane is called **endocytosis**. The engulfed material can consist of particles or liquid. In a reverse process, called **exocytosis**, material within sacs inside the cell is discharged from the cell. Exocytosis releases secretions and wastes from the cell.

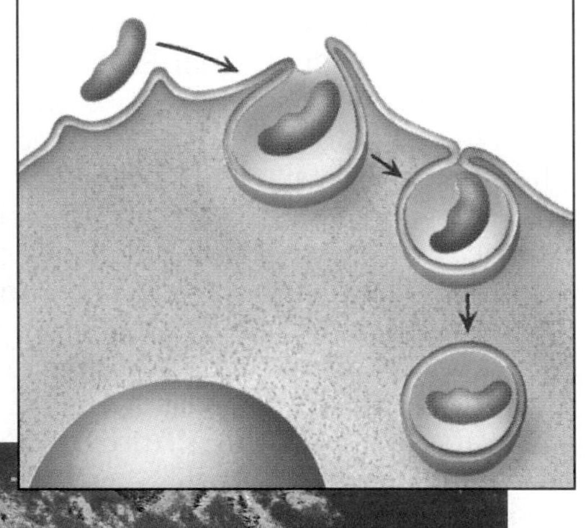

Figure 4-10

a In the process of endocytosis, armlike extensions of the cell reach out to surround and engulf particles that are too large to pass through the plasma membrane.

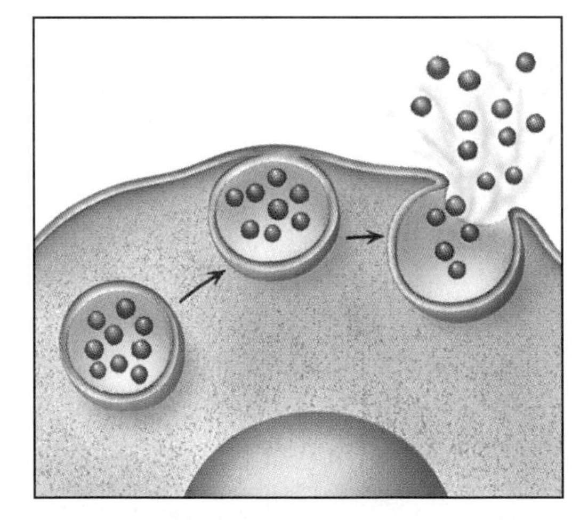

b In the process of exocytosis, a sac of cellular waste inside the cell fuses with the plasma membrane and empties its contents outside the cell.

c In this blood capillary, materials outside the capillary are engulfed and taken in.

SECTION REVIEW

1. **What occurs in the processes of osmosis and diffusion?**
2. **What distinguishes facilitated diffusion from active transport?**
3. **What is the function of the sodium-potassium pump? How does it work?**
4. **How can a cell take in particles too large to cross its membrane?**

CHAPTER

4 Highlights

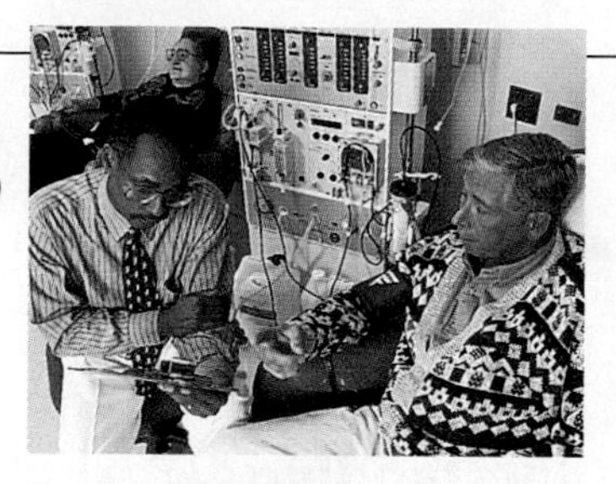

Kidney dialysis uses an artificial membrane to filter toxins out of the blood.

4-1 How Cells Communicate

Chemical signals like glucagon bind to receptor proteins on the cell's surface to send a message to its interior.

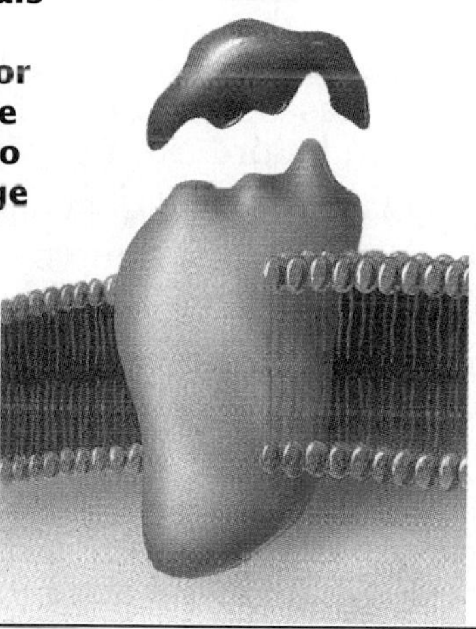

Key Terms

multicellular organism (p. 63)

receptor protein (p. 64)

voltage-sensitive channel (p. 65)

cell surface marker (p. 66)

Summary

- Multicellular organisms could not exist without cell-to-cell communication.
- Proteins embedded in the surface of the cell and within the cell's cytoplasm are capable of transmitting messages carried by other molecules.
- Some proteins are able to respond to electrical currents.
- Other proteins identify a cell as belonging to a specific individual.

4-2 Movement of Substances Into and Out of Cells

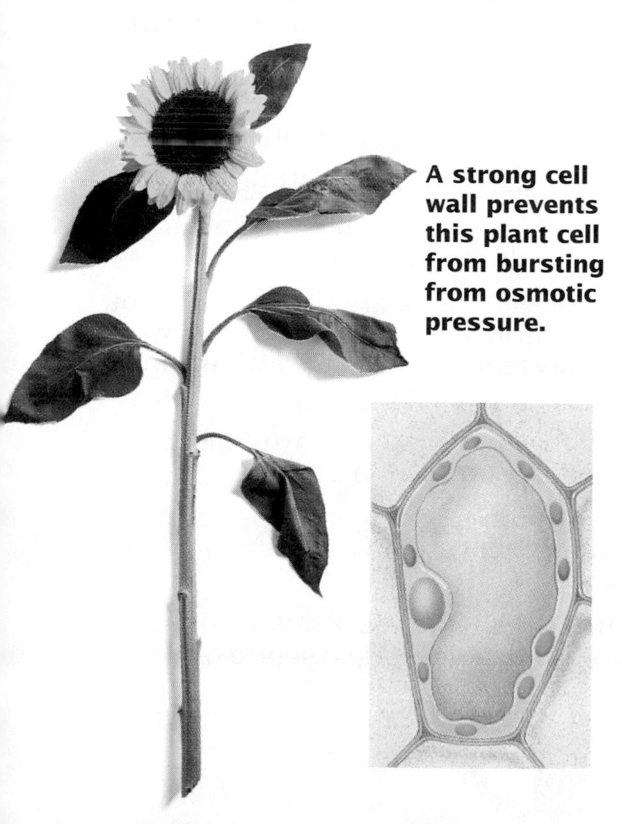

A strong cell wall prevents this plant cell from bursting from osmotic pressure.

Key Terms

diffusion (p. 67)

osmosis (p. 68)

osmotic pressure (p. 68)

selective transport (p. 69)

facilitated diffusion (p. 69)

active transport (p. 70)

sodium-potassium pump (p. 70)

proton pump (p. 70)

endocytosis (p. 72)

exocytosis (p. 72)

Summary

- Some substances enter a cell by diffusion, the movement of molecules from a higher to a lower concentration.
- Water enters and leaves a cell by osmosis, the diffusion of water across a cell membrane.
- Selective transport and facilitated diffusion do not require energy to move substances into cells. Active transport requires energy and can move substances to a region of higher concentration.
- Larger particles are moved into and out of a cell by endocytosis and exocytosis.

CHAPTER 4

review

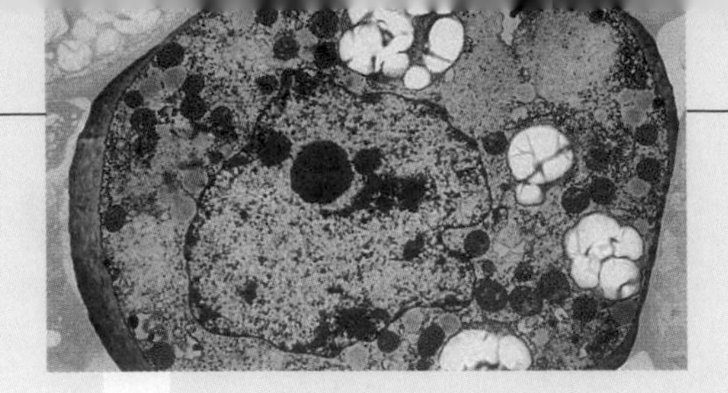

Understanding Vocabulary

1. For each pair of terms, explain the difference in their meanings.
   a. diffusion, osmosis
   b. facilitated diffusion, active transport
   c. endocytosis, exocytosis
2. Using each set of words below, write one or more sentences summarizing information learned in this chapter.
   a. receptor protein, voltage-sensitive channel, cell surface marker
   b. active transport, sodium-potassium pump, proton pump

Understanding Concepts

3. **Relating Concepts** Copy the unfinished concept map below onto a sheet of paper. Then complete the concept map by writing the correct word or phrase in each oval containing a question mark.

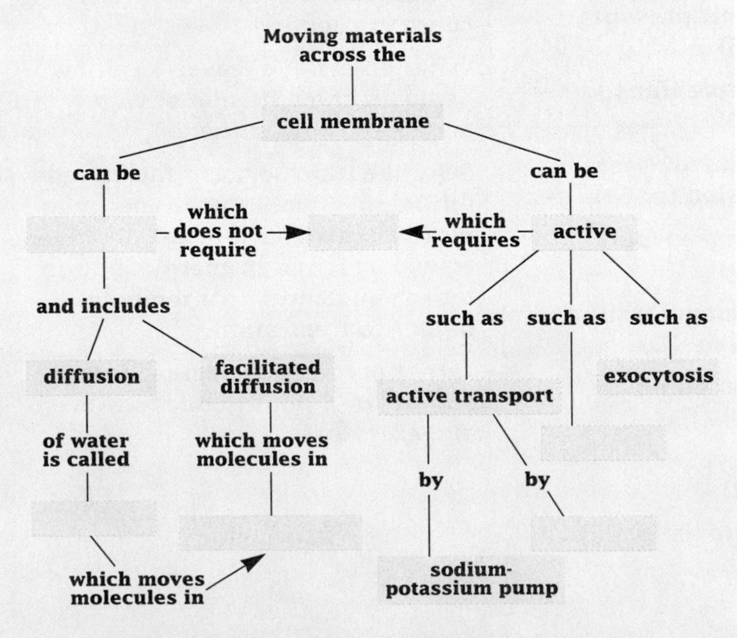

4. **Inferring Relationships** Explain why it is important for cells to communicate with one another.
5. **Summarizing Information** Explain how voltage-sensitive channels work and why they are important.
6. **Comparing Structures and Functions** Where are the two kinds of receptor proteins located? How does each type of receptor protein function?
7. **Predicting Outcomes** Insulin, a hormone that regulates the level of sugar in your blood, reacts with receptor proteins in much the same way as glucagon. Based on what you know about the way glucagon reacts with receptor proteins, describe how insulin sends a message to a cell that it should take in and store extra sugar.
8. **Recognizing Relationships** Explain the importance of the relationship between cellular communication and multicellularity.
9. **Summarizing Information** Describe how cells communicate using both chemical and electrical signals.
10. **Recognizing Relationships** How do cell surface markers affect the success of organ transplants? What can be done to increase the success of organ transplants?
11. **Identifying Functions** What are two active transport mechanisms that enable cells to store certain substances?
12. **Comparing Structures** Compare the methods by which plant and animal cells keep from bursting due to osmotic pressure.
13. **Analyzing Results** People sometimes sprinkle salt on weeds growing through cracks in their sidewalks and driveways. Based on what you know about osmosis, explain why salt would kill the weeds.
14. **Analyzing Methods** Compare the processes of facilitated diffusion and active transport. In what ways are they alike? In what ways are they different?
15. **Identifying Functions** Describe two functions of the sodium-potassium pump.
16. **Comparing Processes** Compare the similarities and differences between diffusion and osmosis.
17. **BUILDING ON WHAT YOU HAVE LEARNED** How does the molecular structure of water enable it to interact with sugars, proteins, and other polar molecules within the cell?

Interpreting Graphics

18. Study the drawing of the sodium- potassium pump shown below. Label the structures in the diagram by matching each letter in the list below to a numbered structure. Describe what happens when a sodium ion moves into a receptor site on the protein channel.

a. sodium ions
b. protein channel
c. lipid bilayer
d. potassium ions

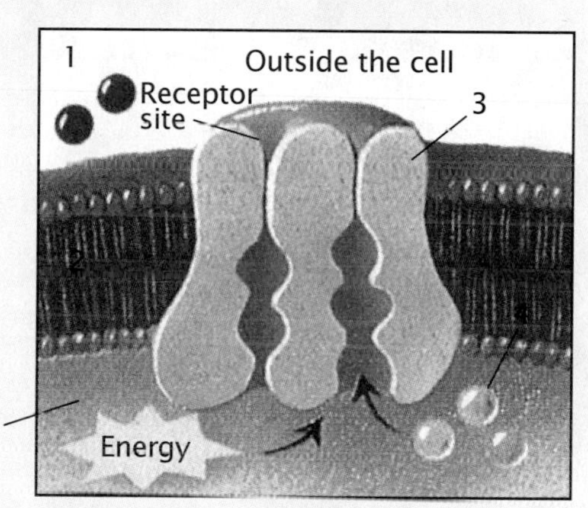

Reviewing Themes

19. ***Cell Structure and Function***
Predict what would happen if red blood cells and plant cells were each placed in containers of distilled water.

20. ***Heredity***
Your friend needs a kidney transplant. Explain why the best source for a donated kidney would be from your friend's identical twin brother.

Thinking Critically

21. **Evaluating Results** A scientist takes skin cells from a frog and a human and places them both in an 0.8 percent salt solution. He was later surprised to find that the frog cells had swollen and burst and that the human cells had shrunk. Based on what you know about osmosis, how would you explain these results?

22. **Evaluating an Argument** How would you respond to someone who says that both osmosis and active transport can occur in a dead cell? Use examples from the chapter to support your response.

Life/Work Skills

23. **Finding Information** When a person's kidneys fail to remove wastes, that person may need to undergo kidney dialysis—the artificial filtering of blood. Consult references in your library to find out how membranes are used in kidney dialysis machines, how frequently dialysis is performed, and what costs are involved. Summarize your findings in a presentation to your class.

24. **Working Cooperatively** Because it involves a defect in cell membrane channel proteins, the disease cystic fibrosis affects the body in many ways. Treatments involve overcoming the various effects of the disease. Work in cooperative teams to research the ways in which cystic fibrosis damages the body. Identify which organs are likeliest to be severely affected by the disease and describe treatments that combat the effects of cystic fibrosis on each organ. Work with the other members of your team to present your findings to your class.

Activities and Projects

25. **Health Connection** Find out how disinfectants, such as alcohol, are able to kill disease-causing bacteria. What part of the bacterial cell is affected by disinfectants? Summarize your findings in a report that includes other methods used to destroy bacteria.

Discovering Through Reading

26. Read the article "The Early Days of Transplantation" by Thomas E. Starzl in *Journal of the American Medical Association*, December 1994, page 1705. What is the single most important case in the history of transplantation? Explain your answer.

27. Read the article "Biology on Ice" by Wallace Ravven in *Discover*, August 1994, pages 36–41. Why would it be beneficial to be able to freeze human organs prior to transplanting them? What damage does freezing and thawing do to an organ? What are two ways to prevent the formation of ice crystals when freezing tissues?

CHAPTER

5

REVIEW
- organic molecules (Section 2-3)
- cell structure (Section 3-3)
- the term *energy*

Energy and Life

Impalas expend a great deal of energy as they bound across wetlands in Botswana.

Just as mechanical engineers study car engines to learn how cars run, scientists study cells to understand how living things work. Like cars, living things are complex machines, full of delicate details, and driven by chemical energy. To understand how your body works, you must look "under the hood" at the chemical "machinery" in your cells.

5-1 Cells and Chemistry

OBJECTIVES

1. **Distinguish chemical reactions that absorb energy from those that release energy.**
2. **Describe the role of activation energy in chemical reactions.**
3. **Describe how an enzyme interacts with its substrate.**
4. **Analyze the role of enzymes in cells.**

Chemical Reactions in Living Things

Fizzing test tubes are not the only places where chemical reactions occur. Chemical reactions within cells drive all the activities associated with life. During chemical reactions, atoms of the reactants (starting materials) are rearranged to form the products (ending materials). Chemical reactions are written in the following form:

$$\text{reactants} \longrightarrow \text{products}$$

In your cells, chemical reactions rearrange the atoms in glucose molecules, making new products and releasing energy. A potato, like the one in **Figure 5-1a**, is an excellent food because it is crammed with starch (long chains of glucose molecules). In plant cells, chemical reactions that absorb energy make glucose and other organic molecules that plants use for energy and growth.

Figure 5-1
a When you eat a potato, chemical reactions in your mouth and small intestine convert starch into glucose. Other reactions break down glucose and release energy that your body uses to do work.

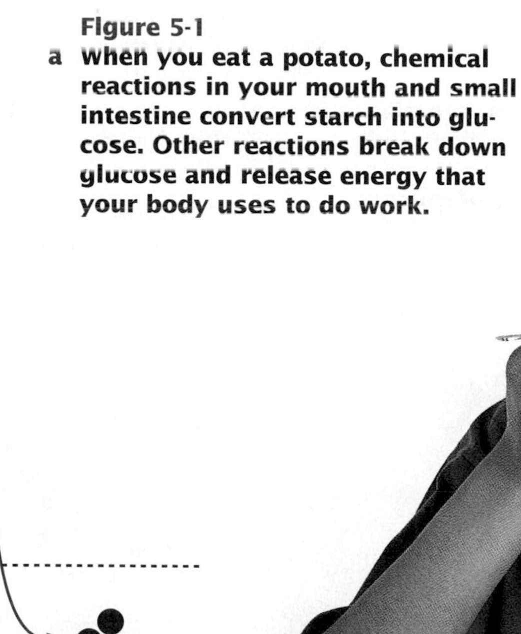

b This graph represents the breakdown of glucose. Notice that energy is released as the reaction's product forms.

Chemical reactions can release or absorb energy

The graph shown in **Figure 5-1b** on the previous page represents a chemical reaction that releases energy. The breakdown of glucose is an example of this type of chemical reaction. Your cells use the energy released during the breakdown of glucose to make proteins and the other organic molecules that make up your body.

Chemical reactions that build macromolecules such as proteins absorb energy. A chemical reaction that absorbs energy is represented by the graph in **Figure 5-2b**. The formation of glucose in a potato plant is an example of a chemical reaction that absorbs energy.

At any moment, thousands of chemical reactions are going on in every cell of an organism's body—a living symphony of chemistry. Collectively, all of the chemical reactions occurring in an organism are called **metabolism** *(muh TAB uh liz uhm).*

Figure 5-2

a This potato plant uses energy from light to form the chemical bonds of glucose.

Chemical reactions need some energy to get started

The heat from a flame ignites the logs in a campfire. The spark from a spark plug causes the gasoline in an engine to ignite. In both of these cases, an input of energy is used to start a chemical reaction. The energy needed to start a chemical reaction is called **activation energy**. Think of a boulder that you want to roll down a hill. To get it rolling downhill, you must first give it a push. Activation energy is simply a chemical push that gets a chemical reaction going.

Even if a chemical reaction releases energy, activation energy must still be supplied before that reaction can occur. For example, the combustion of gasoline provides the energy needed to power an automobile. But only after the key is turned in the ignition will sparks from the spark plugs ignite the gasoline in the engine's cylinders. The sparks provide the activation energy that triggers the burning of gasoline.

Enzymes lower activation energy for chemical reactions

Like engines, cells consume fuel. The burning of gasoline fuel in an engine requires a spark or high temperature to get the reaction going fast enough to power a car. Most chemical reactions that happen in cells also require very high temperatures in order to proceed fast enough to keep a cell alive. Such high temperatures kill most cells.

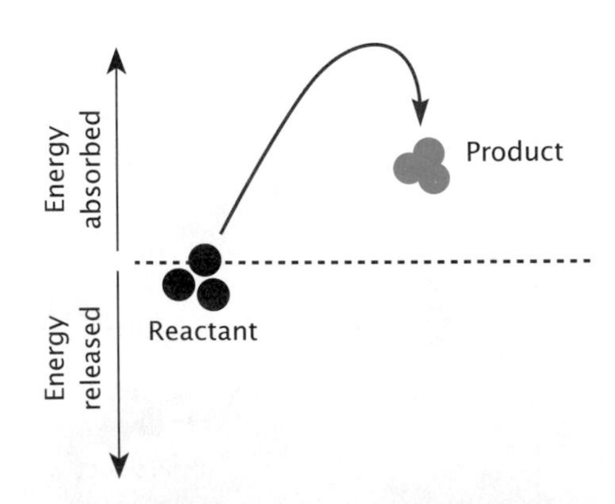

b This graph represents the making of glucose. Notice that energy is absorbed as the reaction's product forms.

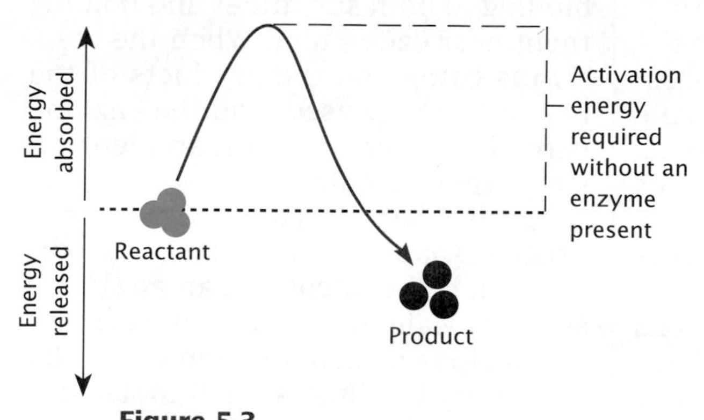

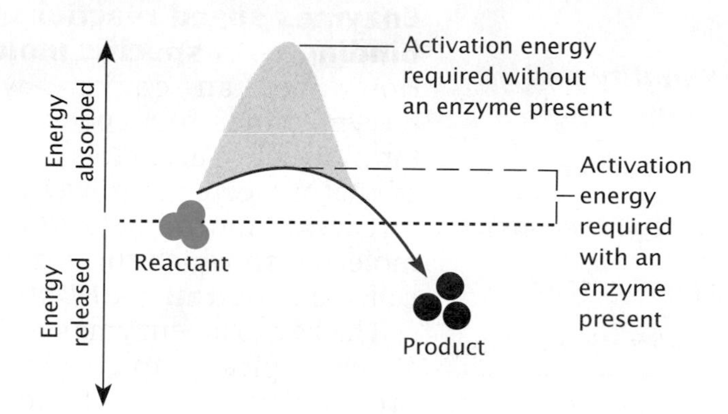

Figure 5-3
a Activation energy must be supplied for most chemical reactions to occur.

b Enzymes lower the amount of activation energy required to start a chemical reaction.

Fortunately, the chemical reactions in cells occur quickly and at relatively low temperatures through the action of enzymes. As you read in Chapter 2, **enzymes** are proteins that can hasten a chemical reaction. How do enzymes increase the speed of a reaction?

A chemical reaction could proceed faster if less energy were needed to get it started. Think about the activation energy needed to roll a heavy boulder over a hill. One way to reduce the amount of energy necessary would be to reduce the hill's size. Once the boulder is at the top of the hill, digging away some of the ground in front of the boulder would reduce the amount of energy needed to send it rolling downhill. In a sense, using an enzyme in a reaction is like lowering the top of the hill. Enzymes cause reactions to occur with less activation energy, as shown in **Figure 5-3**.

Enzymes are biological catalysts. A **catalyst** makes a chemical reaction proceed faster, but it is not used up by the reaction. An enzyme-catalyzed reaction is faster because it has a lower activation energy than does an uncatalyzed reaction.

Actions of Biological Catalysts

To understand the importance of enzymes, consider how your blood picks up carbon dioxide and delivers it to your lungs to be exhaled. Carbon dioxide is a waste product that will poison the body if it is not continuously removed. When carbon dioxide is converted to carbonic acid, it can be safely carried within the bloodstream. However, the chemical reaction that converts carbon dioxide to carbonic acid is very slow. In a given volume of water, only 200 molecules of carbonic acid form in an hour.

Fortunately, an enzyme called carbonic anhydrase is present in the blood. This enzyme increases the rate of this reaction to form 600,000 molecules of carbonic acid every second. The enzyme, shown in **Figure 5-4**, accelerates the reaction rate about 10 million times! Without carbonic anhydrase, your blood would quickly become poisoned with carbon dioxide.

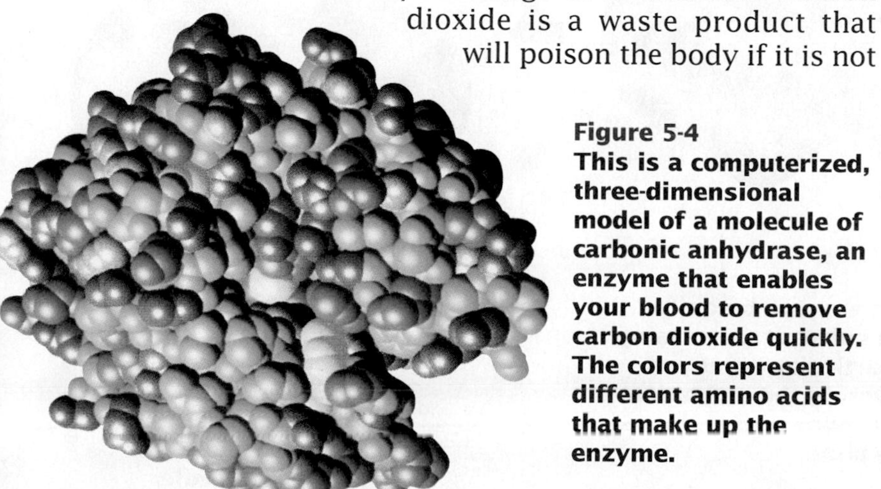

Figure 5-4
This is a computerized, three-dimensional model of a molecule of carbonic anhydrase, an enzyme that enables your blood to remove carbon dioxide quickly. The colors represent different amino acids that make up the enzyme.

Stability and Homeostasis

A conductor controls the music of an orchestra by dictating when each instrument plays. How is this similar to the way your body controls chemical reactions?

Enzymes speed reactions by binding with specific molecules

How does an enzyme work? An enzyme binds to a specific molecule (or molecules) and stresses the bonds of that molecule in a way that makes a reaction more likely to occur. The molecule on which the enzyme acts is called a **substrate** *(SUHB strayt).*

The key to an enzyme's activity is its shape. Typically, an enzyme is a large protein that consists of a folded chain of hundreds of amino acids. Each enzyme has one or more deep folds on its surface. These folds form pockets called **active sites**. As you can see in **Figure 5-5**, an enzyme's substrate will fit into an active site just as your feet fit comfortably into your favorite pair of sneakers.

An enzyme makes a reaction proceed more quickly by lowering activation energy in one of two ways. When a substrate molecule binds to the active site of an enzyme, the molecule is held close to a certain part of the enzyme's surface. The enzyme may put a strain on a particular chemical bond in the substrate molecule and thus make the bond more likely to break. An enzyme may also encourage the formation of a bond between two substrates by binding to both substrates and holding them near each other. When the reaction is complete, the products of the reaction are released, and the enzyme is ready to combine with an identical substrate molecule.

Because an enzyme must have a precise shape in order to work correctly, a cell can control an enzyme's activity by altering the enzyme's shape. The shapes of many enzymes can be altered by binding "signal" molecules to their surfaces. The new shape produced by binding the signal molecule acts to turn the enzyme "on" or "off."

Cells have thousands of different enzymes

A cell contains thousands of different kinds of enzymes, each promoting a different chemical reaction. Not all cells contain the same enzymes. As you read this page, the chemical reactions going on in a nerve cell in your eye are very different from the chemical reactions in one of your red blood cells because the two kinds of cells contain different enzymes. The enzymes active at any one time in a cell determine what happens in that cell, just as traffic lights control the flow of traffic in a city.

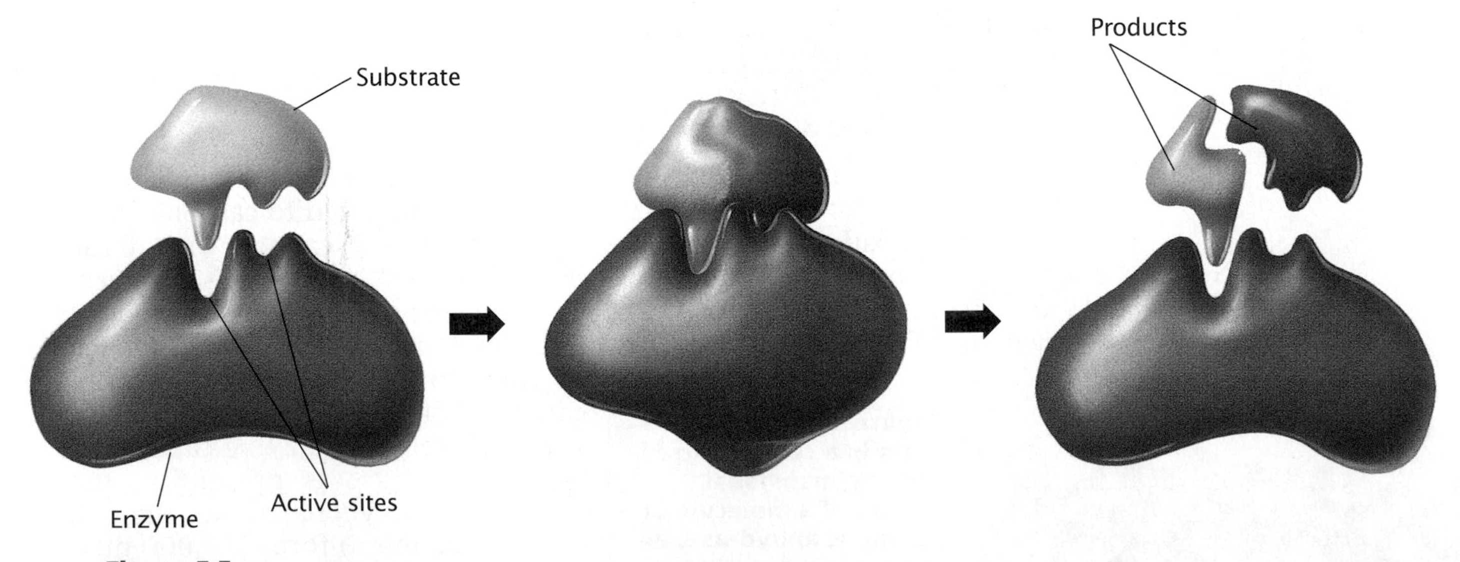

Figure 5-5

a Enzymes lower the activation energy of a reaction. Here an enzyme is about to bind with its substrate.

b The enzyme binds to the substrate, stressing a particular bond and lowering the activation energy needed to break the bond.

c When the products of the reaction are released, the enzyme returns to its original shape and is ready to combine with another substrate molecule.

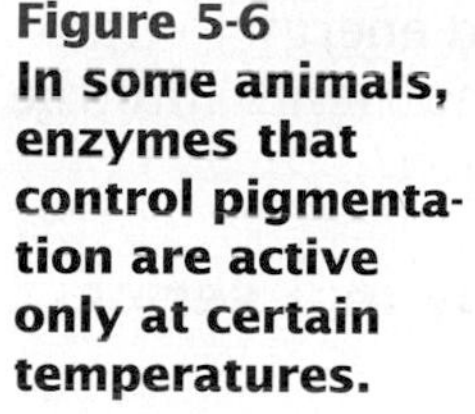

Figure 5-6
In some animals, enzymes that control pigmentation are active only at certain temperatures.

a **The darker parts of the Siamese cat are also the cooler regions of its body. This is because the enzyme catalyzing the formation of the dark pigment is more active in cooler temperatures.**

b **Similarly, the fur of the adult northern seal is dark because the enzyme that catalyzes pigment formation is more active in cold arctic temperatures. The newborn seal is white because the enzyme was less active in the warmer temperatures inside its mother's body.**

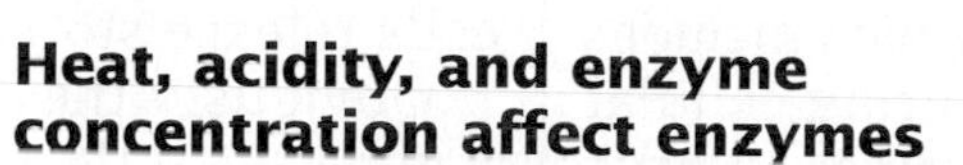

Heat, acidity, and enzyme concentration affect enzymes

Each enzyme functions best within a certain temperature range. When temperatures become too low or too high, reaction rates decrease sharply. For instance, many enzymes in your body shut down when you have a high fever. If the internal body temperature of a human being were to reach 44°C (112°F), many enzymes would be destroyed, and the individual would probably die. For additional examples of how enzymes are affected by temperature, look at **Figure 5-6**. How does temperature affect the pigmentation of the Siamese cat and the northern seal?

Another factor influencing enzyme activity is acidity (the strength of the acids in body fluids). When an organism's acidity is too high or too low, most enzymes cease to function properly. One exception, however, is the enzyme pepsin, which functions in the stomach's highly acidic environment.

Finally, the rate of an enzyme-catalyzed reaction is affected by the concentration of the enzyme and the substrate. The rate of a chemical reaction in the body, for example, can be accelerated by increasing the concentration of the enzyme necessary to catalyze that reaction. This is how your body controls its development.

SECTION REVIEW

1. **How does the outcome of a chemical reaction that absorbs energy differ from one that releases energy?**
2. **What is activation energy, and how does it relate to cell metabolism?**
3. **Explain how an enzyme can increase the rate of a chemical reaction.**
4. **Why do you think it is advantageous for the human body to have many different enzymes?**

If you stopped eating, you would eventually die. Why? Because the food you eat provides the energy needed for life. All of the properties by which we define life—growth, movement, sensitivity, and reproduction—use energy. Just as logs must be continually supplied to a campfire to keep the fire burning, energy must be continually supplied to living cells to keep life going.

5-2 Cells and Energy

OBJECTIVES

1. **Explain how cells use energy.**
2. **Compare the way a fire releases energy from fuel with the way a cell releases energy from fuel.**
3. **Describe the role played by ATP in cells.**
4. **Recognize the path of energy between plants and animals in the living world.**

How Cells Use Energy

Living cells use energy to do all those things that require work. One of the most obvious of these is movement. Some bacteria and protozoans swim about, propelling themselves through fluids by rapidly spinning flagella. Cells also use energy to change their shape. White blood cells, for example, must extend and retract their cell membranes when they engulf invading bacteria in the bloodstream, as shown in **Figure 5-7**. Metabolism, which includes tasks such as manufacturing cellular components and repairing cellular structures, requires great quantities of energy. Cells also need energy to transport food and other nutrients into the cell and to expel wastes.

Cells release energy in a series of small steps

Perhaps you have heard that jogging or some other exercise "burns" a lot of calories. The word *burn* is often used to describe what happens when your cells release stored energy from food.

Obviously the "burning" of food in living cells differs from the burning of logs in a campfire. When logs burn, the energy stored in wood is released in a single reaction as heat and light. But this is not what happens in cells. Instead, energy stored in food molecules is released at each step in a series of enzyme-catalyzed chemical reactions, as shown in **Figure 5-8**. The product of one chemical reaction is a reactant in the next reaction.

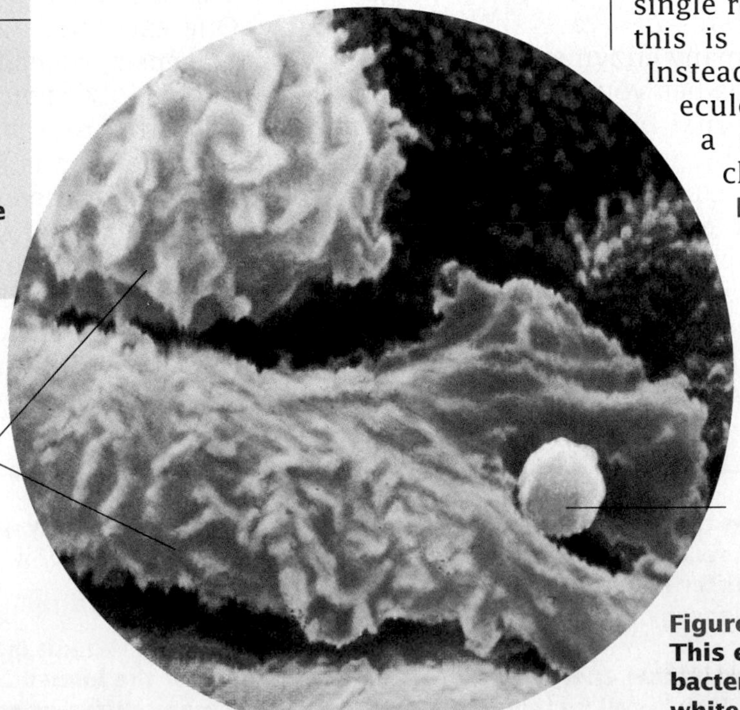

White blood cells

Bacterium

Figure 5-7
This electron micrograph shows a bacterium being engulfed by a human white blood cell. The white blood cell uses energy when changing its shape.

Figure 5-8
Energy is released from starch through a series of enzyme-catalyzed chemical reactions. The product of one reaction becomes the reactant of the next. Energy is released by each reaction.

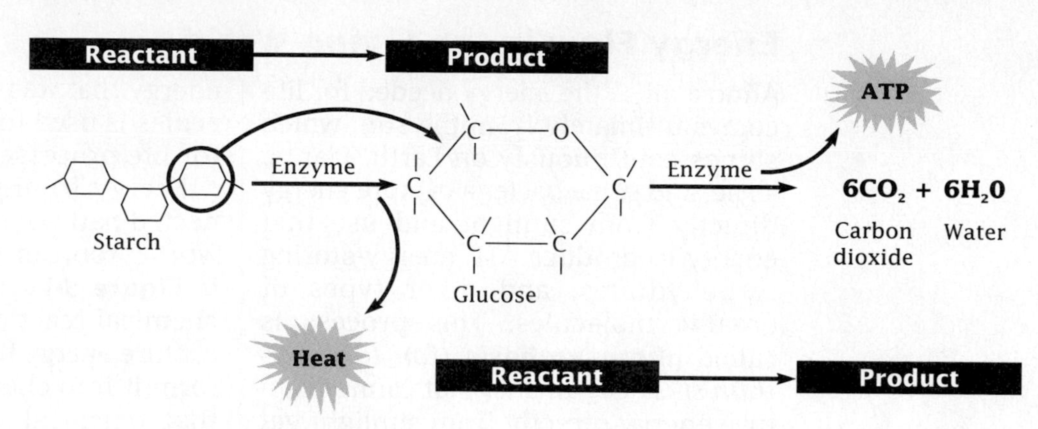

Cells transfer energy from food to a molecule called ATP

When living cells break down molecules, some of the energy released from the molecules is in the form of heat. The rest of the energy is stored temporarily in molecules of adenosine triphosphate, or **ATP**, as shown in **Figure 5-8**. Like the money you use to pay for your needs, ATP is a portable form of "energy currency" inside of cells. ATP molecules deliver energy to wherever it is needed in a cell.

As **Figure 5-9** shows, ATP contains three phosphate groups. The term *triphosphate* means "three phosphate groups." When the outer phosphate group of ATP is removed, ADP (adenosine diphosphate) is formed. The term *diphosphate* means "two phosphate groups." The reaction that forms ADP from ATP releases energy in a way that enables cells to use the energy. The following equation summarizes this reaction:

$$\text{ATP} \xrightarrow{H_2O} \text{ADP} + \text{P} + \text{energy}$$

In cells, almost all chemical reactions that absorb energy require less energy than is released by the removal of a phosphate group from ATP. Therefore, ATP is able to drive most of a cell's activities.

Cells use ATP to drive many chemical reactions

A steady supply of ATP is necessary to ensure that a cell can perform all the tasks essential for life. Making molecules, for example, uses energy. The reactions that build new molecules cannot begin until something supplies the energy needed by the reaction. That something, as you have probably guessed, is ATP.

The energy released from the breakdown of ATP can provide energy for other reactions. When a reaction that absorbs energy in a cell is driven by the conversion of ATP to ADP, it is called a coupled reaction. Coupled reactions occur often during metabolism.

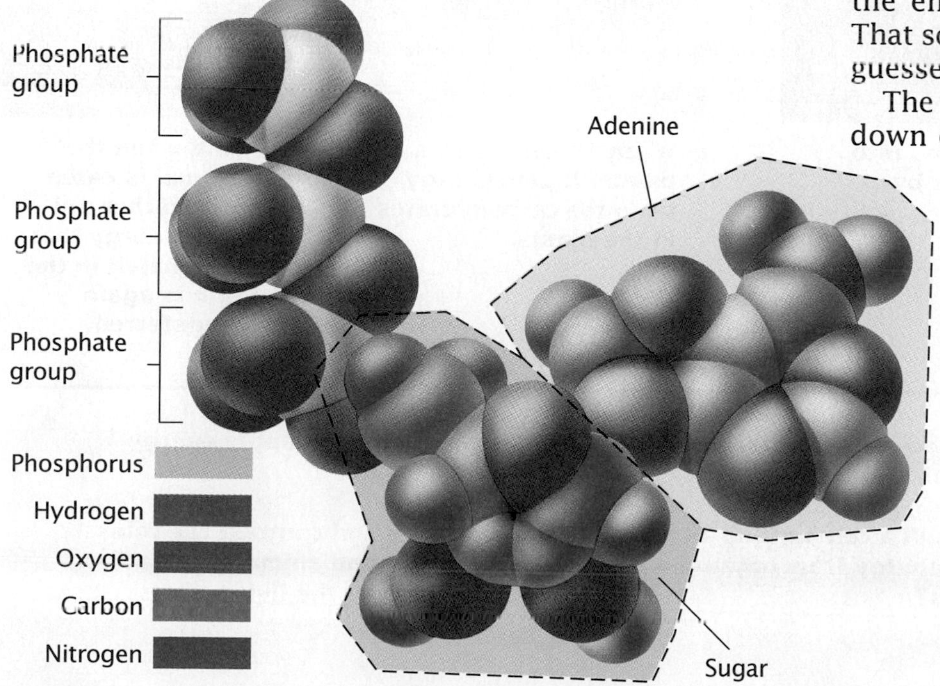

Figure 5-9
An ATP molecule is made of a sugar, adenine, and a chain made of three phosphate groups. Energy is released when the outermost phosphate group is removed.

Energy Flow in the Living World

Almost all of the energy needed for life comes ultimately from the sun, which shines continuously on Earth. Plants, algae, and some bacteria capture energy directly from sunlight and use that energy to produce ATP, energy-storing carbohydrates, and other types of organic molecules. This process is called **photosynthesis** *(foh toh SIHN thuh sihs)*. Organisms that cannot capture energy directly from sunlight get the energy for life by consuming food, which contains carbohydrates and other organic molecules.

All living things use a process called **cellular respiration** to obtain energy from carbohydrates and other organic molecules. During cellular respiration, energy that was stored in organic molecules is used to make the ATP needed for life processes.

Energy from the sun flows in connected pathways throughout the living world. You can see one such pathway in **Figure 5-10**. The energy-absorbing chemical reactions of photosynthesis capture energy from sunlight and transform it into chemical energy. Some of that chemical energy is transferred when one organism eats another. The energy-releasing reactions of cellular respiration enable living things to use that chemical energy to do work. You will learn more about the processes of photosynthesis and cellular respiration in Sections 5-3 and 5-4.

Figure 5-10
a Light energy streaming from the sun . . .

b . . . is converted into carbohydrates by plants.

c When an animal eats plants, it gets energy from the carbohydrates in the plants.

d And when that animal is eaten by another animal, energy that originated in the sun is again transferred.

SECTION REVIEW

1. **What life activities performed by cells require energy?**
2. **Why doesn't a cell burn up as it releases energy from organic molecules?**
3. **How does ATP supply energy for the cell?**
4. **Compare and contrast the roles of plants and animals in the flow of energy in the living world.**

Have you ever eaten beef enchiladas? The beef came from a cow that ate grass. Other parts of the enchiladas came directly from plants. With few exceptions, you end up with plants (or some other photosynthetic organism) if you trace your food back to its origin. Clearly, you depend on plants for food, as do all plant eaters and organisms that eat plant eaters. The energy in that food came from sunlight.

5-3 Photosynthesis

OBJECTIVES

1. **Describe how photosynthetic organisms use energy from light to make organic molecules and oxygen.**
2. **Explain how plant cells use the ATP and NADPH that forms during photosynthesis.**
3. **Analyze the roles of electrons and protons in photosynthesis.**
4. **Summarize the three stages of photosynthesis.**

Harnessing the Sun's Energy

Each day, the light energy that reaches Earth equals the energy released by the explosions of about 1 million atomic bombs. Plants, algae, and some bacteria capture about 1 percent of that energy and convert it to chemical energy through photosynthesis. During photosynthesis, energy captured from sunlight is used to obtain electrons, which carry the energy through a series of chemical reactions.

Photosynthesis has three stages

In the first stage of photosynthesis, energy is captured from light. During the second stage, the energy is used to make ATP and an energy-carrying compound called NADPH. During the third stage, the ATP and NADPH are used to power the manufacture of energy-rich carbohydrates using CO_2 from the air.

This method of photosynthesis evolved among ancient bacteria more than 3 billion years ago. Today, all plants and algae use this type of photosynthesis. During the process, electrons are obtained by removing hydrogen atoms from water (H_2O) molecules, and oxygen (O_2) molecules are released, as you can see in **Figure 5-11**. Earth's atmosphere is now rich in oxygen gas because of the oxygen released by photosynthesis.

Figure 5-11
In bright light, scores of tiny bubbles appear on the leaves of this water plant. The bubbles contain oxygen gas that is produced through the process of photosynthesis.

Stage 1: Capturing Light Energy

Now that you know plants use energy from light for photosynthesis, you might be wondering where the energy is in light. Light is a form of electromagnetic radiation, which is energy transmitted by special waves. When the sun shines on you, your body is bombarded by all types of radiation emitted from the sun.

The amount of energy transmitted by an electromagnetic wave depends on its wavelength—shorter wavelengths have more energy than longer wavelengths. Some wavelengths of electromagnetic radiation, such as those in X rays and ultraviolet light, carry a large amount of energy. Other wavelengths, such as those in radio waves, carry very little energy. Our eyes detect only wavelengths that carry intermediate amounts of energy. These wavelengths make up "visible" light, as shown in **Figure 5-12**. Plants absorb mostly blue and red light and reflect much of what is left of the visible light, which is why most plants look green.

Certain molecules absorb light

How can a leaf or a human eye "choose" which colors of light to absorb? Certain molecules within these structures are able to absorb particular wavelengths of light. Molecules that absorb light are called **pigments**. Your eyes contain a pigment called retinal, which absorbs all wavelengths of visible light. Insect eyes contain a pigment that absorbs wavelengths with more energy than those absorbed by retinal, so insects can see ultraviolet light that humans cannot see.

When a pigment absorbs light, electrons of certain atoms in the pigment molecules are boosted to a higher energy level. Energy transferred from the light to the electrons causes this move. Just as climbing to the next rung of a ladder requires you to raise your foot an exact distance, boosting an electron requires an exact amount of energy. Therefore, a particular kind of pigment can absorb only wavelengths with the appropriate amount of energy.

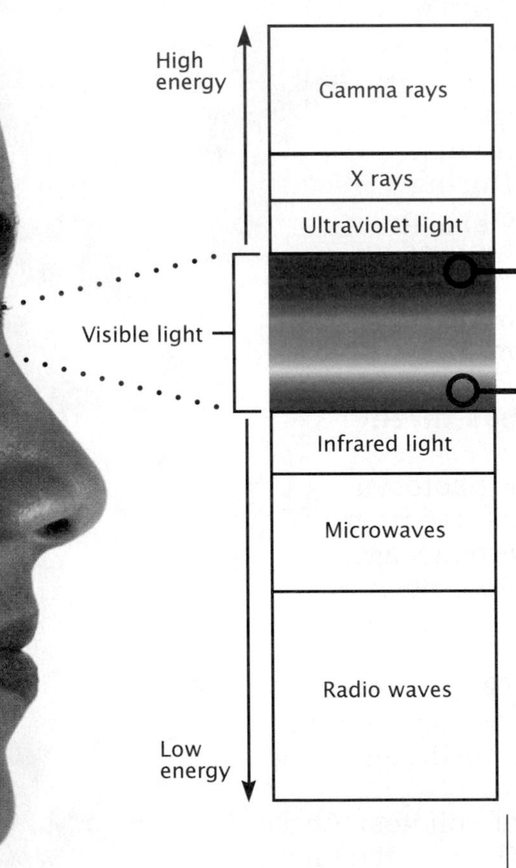

Figure 5-12
The entire range of wavelengths of electromagnetic radiation, from low-energy radio waves to high-energy gamma rays, is called the electromagnetic spectrum. Visible light is the portion of the spectrum that humans can see.

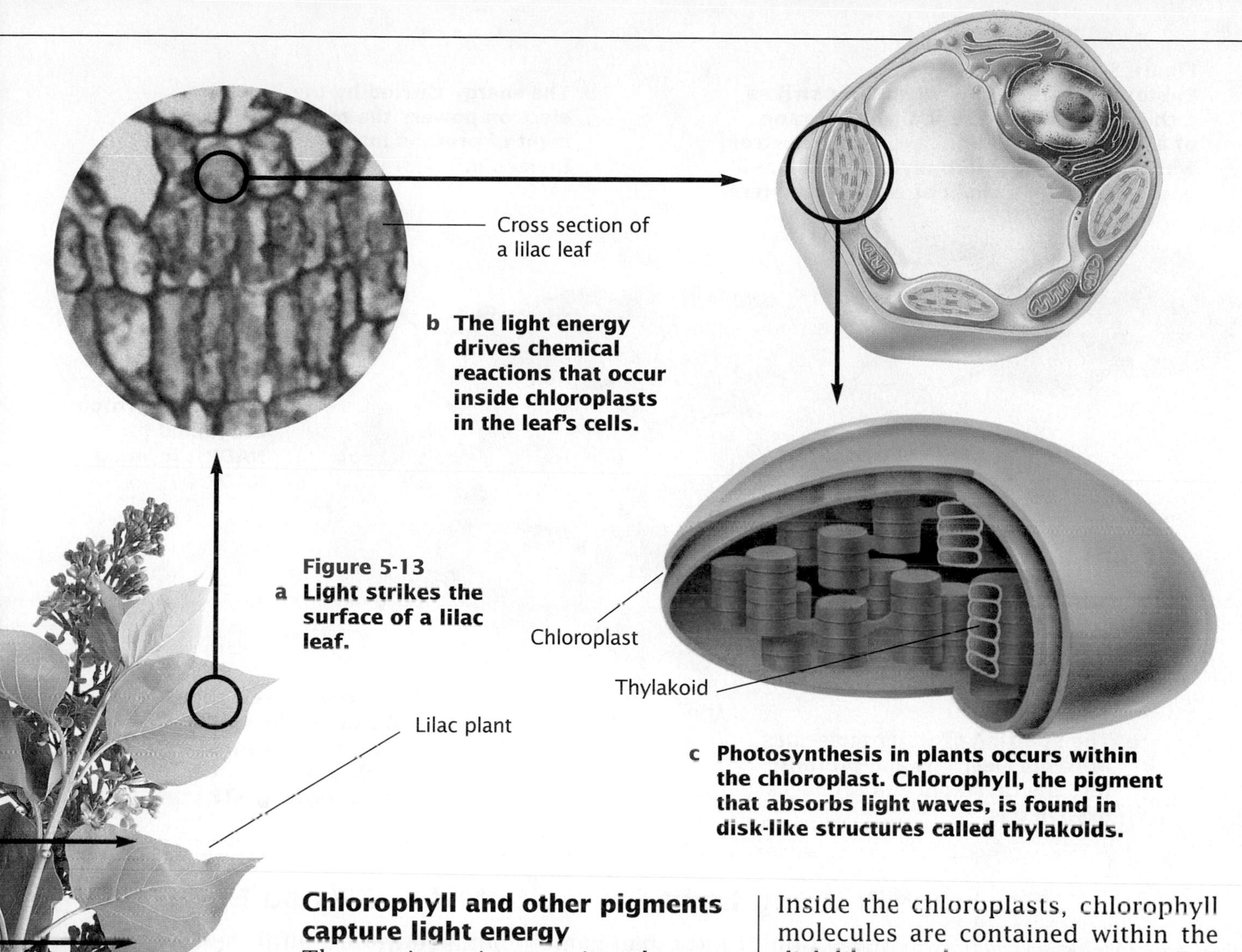

Figure 5-13
a Light strikes the surface of a lilac leaf.
b The light energy drives chemical reactions that occur inside chloroplasts in the leaf's cells.
c Photosynthesis in plants occurs within the chloroplast. Chlorophyll, the pigment that absorbs light waves, is found in disk-like structures called thylakoids.

Chlorophyll and other pigments capture light energy

The major pigment in plants is **chlorophyll**, which absorbs mainly red and blue light and reflects green light. Plants also contain other pigments that help in gathering energy for photosynthesis. These pigments, called carotenoids, absorb slightly different wavelengths of light and reflect colors such as yellow and orange. In most leaves, the green reflected by chlorophyll covers up the colors reflected by other pigments. Chlorophyll, however, breaks down in cold temperatures, enabling the colors of the carotenoids to be seen in the yellow and orange leaves of autumn.

Where are the pigments for photosynthesis located? Look at the structure of the leaf seen in **Figure 5-13**. Chlorophyll and the other pigments that assist in photosynthesis are found within the chloroplasts of plant cells. Inside the chloroplasts, chlorophyll molecules are contained within the disk-like membrane structures called thylakoids.

When light strikes the thylakoids in a chloroplast, electrons within certain chlorophyll molecules are boosted to a higher energy level. These "excited" (boosted) electrons jump from their chlorophyll molecules to a nearby protein molecule in the thylakoid membrane. Each excited electron is then passed through the membrane from one electron-accepting molecule to another, like a ball being passed down a line of people.

Before excited electrons can leave their chlorophyll molecules, the electrons must be replaced by other electrons. Water supplies these electrons. Plants obtain electrons from water by splitting water molecules, H_2O. As water molecules split, chlorophyll takes the electrons from the hydrogen atoms, leaving protons. The remaining oxygen atoms combine to form oxygen gas.

Figure 5-14

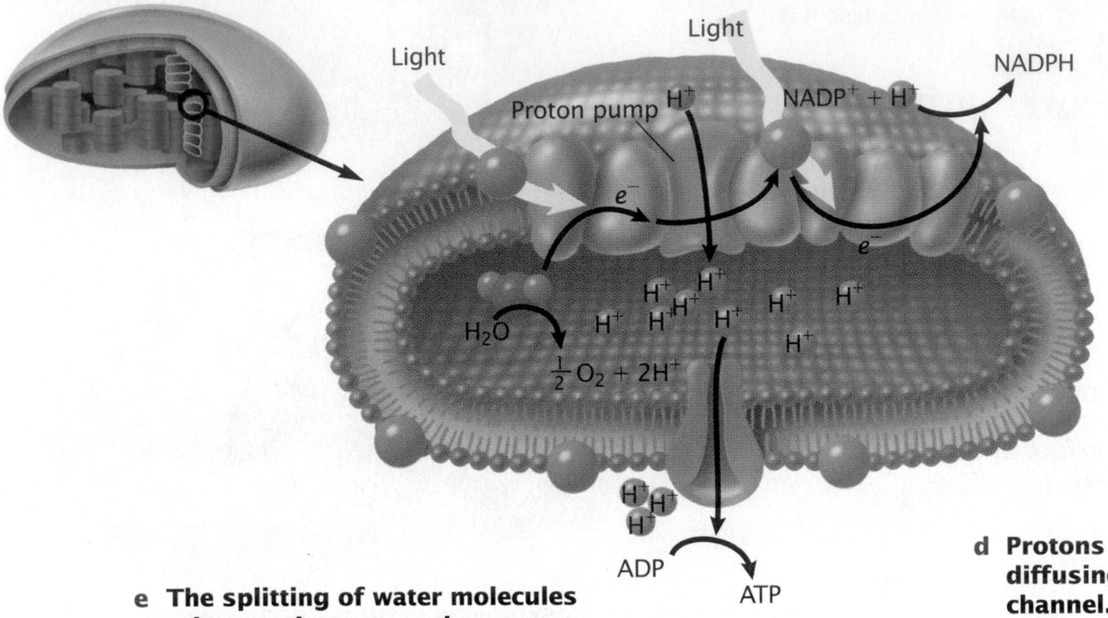

a Photosynthesis begins when light strikes a chlorophyll molecule in the membrane of a thylakoid. The light excites an electron, which leaves the chlorophyll molecule and passes through a chain of electron carriers.

b The energy carried by the electron powers the movement of protons into the thylakoid.

c When light strikes a second chlorophyll molecule, it excites an electron that passes through another chain of electron carriers. The electron and a proton are joined to $NADP^+$, forming NADPH.

d Protons leave a thylakoid by diffusing through a protein channel. The force of their exit adds a phosphate group to ADP, forming ATP.

e The splitting of water molecules releases electrons and protons into the thylakoid and produces oxygen gas.

Stage 2: Using Light Energy to Make ATP and NADPH

Cell Structure and Function

How does the structure of the thylakoid enable plants to capture light energy and transform it into chemical energy?

Just as the action of a football game depends on moving the ball down the field, energy flow in cells depends on passing excited electrons from one carrier to another. This movement of excited electrons transfers the energy contained in those electrons from carrier to carrier as well. Follow the path taken by excited electrons in stage 2 of photosynthesis in **Figure 5-14**. The journey begins when chlorophyll absorbs light and loses an electron, as **Figure 5-14a** indicates. This electron is replaced by an electron from water, as **Figure 5-14e** indicates.

ATP is made when protons flow through a protein channel

Each excited electron from chlorophyll loses some of its energy as it passes through a series of carriers in a thylakoid membrane. This energy is used to pump protons across the membrane and into the interior of the thylakoid, as **Figure 5-14b** indicates. Protons are pumped inward until the thylakoid is "bursting at the seams." These protons are driven by diffusion through the only available exits, protein channels like the one in **Figure 5-14d**. These channels use the force of protons exiting a thylakoid to add a phosphate group to an ADP molecule, making ATP.

Additional light energy is absorbed to make NADPH

After the energy of excited electrons is used to make ATP, the electrons are accepted by other chlorophyll molecules. These molecules absorb more light, which excites some of their electrons. The excited electrons pass through another series of electron carriers. Finally, these electrons are attached to an electron acceptor called $NADP^+$, forming the energy carrier NADPH. These steps are pointed out by **Figure 5-14c**. The ATP and NADPH generated in stage 2 help power stage 3 of photosynthesis—the building of carbohydrates.

Stage 3: Building Carbohydrates

In the final stage of photosynthesis, carbon atoms are captured from carbon dioxide in the air and used to make organic molecules, which store energy. A series of reactions called the Calvin cycle produces organic molecules that can be used to assemble more complex organic molecules, such as the carbohydrates, lipids, and proteins needed for energy and growth. Because this series of reactions also regenerates its starting molecule, it forms a cycle. The Calvin cycle, which is summarized in **Figure 5-15**, is named for Melvin Calvin, the American biochemist who worked out the series of reactions in the cycle. The energy to fuel the Calvin cycle comes from the ATP and NADPH made during stage 2 of photosynthesis.

The many steps in the overall process of photosynthesis are often represented by the following single equation:

$$6CO_2 + 6H_2O \xrightarrow{\text{(light)}} C_6H_{12}O_6 + 6O_2$$

However, this equation does not show how photosynthesis happens. It merely says that 6 carbon dioxide molecules, 6 water molecules, and light are needed to form one 6-carbon organic molecule and 6 molecules of oxygen.

Plants use the organic molecules they produce during photosynthesis for their life processes. For example, sugar made in the leaves of a potato plant can be used to make cellulose for building new cell walls. Some of the sugar is stored as starch in the potato tuber. The plant may later break down the starch to make the ATP needed for energy, as you will see in Section 5-4.

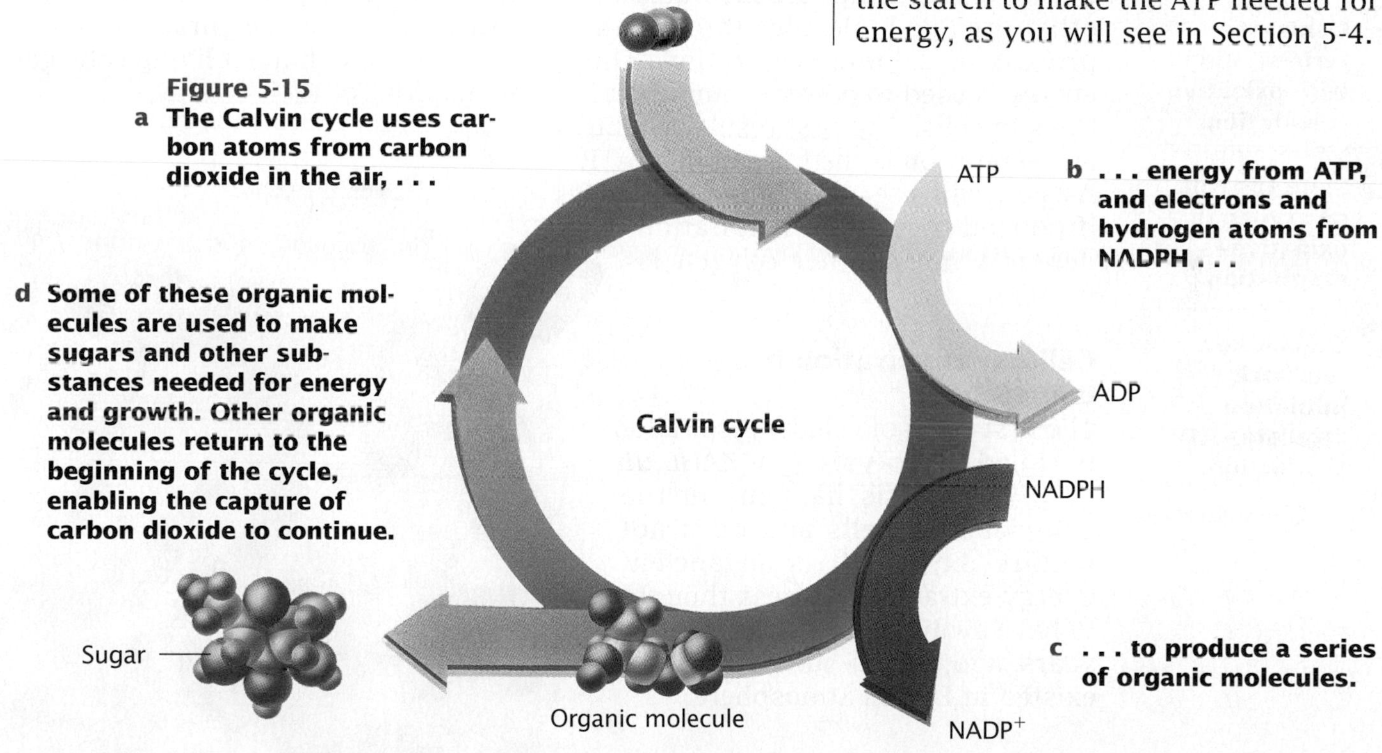

Figure 5-15

a The Calvin cycle uses carbon atoms from carbon dioxide in the air, . . .

b . . . energy from ATP, and electrons and hydrogen atoms from NADPH . . .

c . . . to produce a series of organic molecules.

d Some of these organic molecules are used to make sugars and other substances needed for energy and growth. Other organic molecules return to the beginning of the cycle, enabling the capture of carbon dioxide to continue.

SECTION REVIEW

1. **How is the energy in light captured by photosynthetic organisms?**
2. **What is the general role played by both ATP and NADPH during photosynthesis?**
3. **How are electrons and protons used during the photosynthetic process?**
4. **How does stage 2 of photosynthesis depend on stage 1? How does stage 3 depend on stage 2?**

Although a plant produces carbohydrates only in cells with chloroplasts, all cells of the plant use these carbohydrates for energy. In both plants and animals—indeed, in almost all organisms—the energy for living is obtained by recycling the carbohydrate molecules produced by photosynthesis. The energy in these molecules is released in cellular respiration.

5-4 Cellular Respiration

OBJECTIVES

1. **Explain the role of cellular respiration in making energy available to living things.**
2. **Contrast fermentation with oxidative respiration.**
3. **State the role of oxygen in oxidative respiration.**
4. **Explain how feedback inhibition regulates ATP production.**

Releasing Energy From Organic Molecules

You have seen how energy from the sun is converted into chemical energy by photosynthesis. The chemical energy is stored in organic molecules such as carbohydrates. Living things such as those seen in **Figure 5-16** release energy from carbohydrates and other organic molecules through the process of cellular respiration. This energy is used to power chemical reactions in cells. The first result of cellular respiration is the formation of ATP. As you will see, the amount of ATP produced by cellular respiration depends on whether oxygen is present.

Cellular respiration has two stages

The first stage of cellular respiration is called **glycolysis** *(gly KAHL uh sihs)*. Glycolysis happens in the cytoplasm of cells and does not require oxygen. It is an ancient energy-extracting process thought to have evolved more than 3 billion years ago, when no oxygen gas existed in Earth's atmosphere.

In most living things, a second stage of cellular respiration, called **oxidative respiration**, follows glycolysis. Oxidative respiration, which requires oxygen, happens within mitochondria. It is far more effective than glycolysis at recovering energy from organic molecules. Oxidative respiration is the method by which most living cells get the majority of their energy.

Figure 5-16
The plants that this cow is eating obtained energy from the sun. In turn, the plants provide the cow with the energy it needs to carry out its life activities. Both plants and animals use the process of cellular respiration to release energy stored in organic molecules.

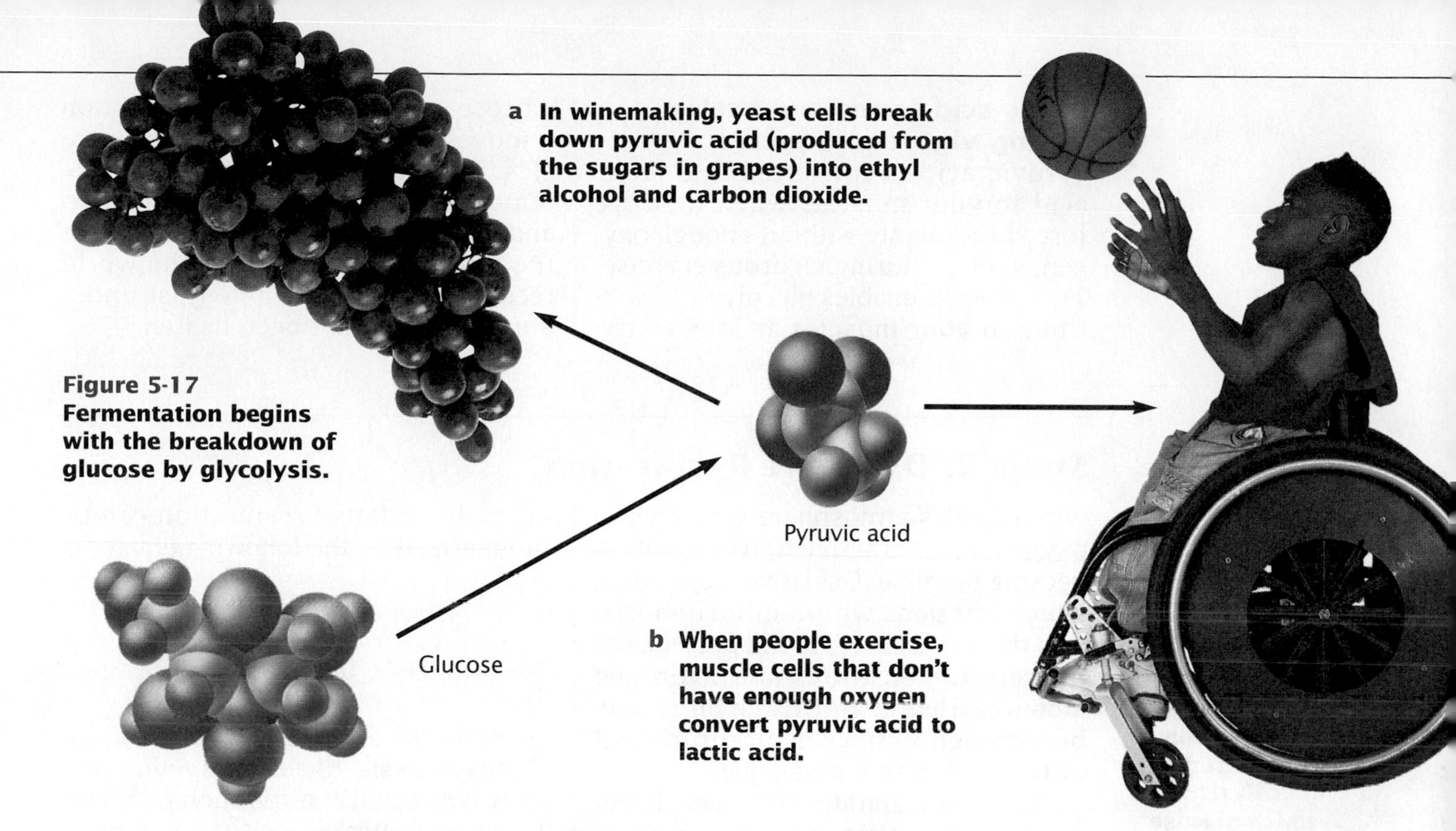

a In winemaking, yeast cells break down pyruvic acid (produced from the sugars in grapes) into ethyl alcohol and carbon dioxide.
Pyruvic acid
Glucose
b When people exercise, muscle cells that don't have enough oxygen convert pyruvic acid to lactic acid.

Figure 5-17
Fermentation begins with the breakdown of glucose by glycolysis.

Stage 1: Glycolysis

Glycolysis is one of the oldest biological processes we know. Ancient bacteria used it to make the ATP needed to drive chemical reactions within their cells. Today, every living thing, including you, still uses glycolysis.

Glycolysis breaks down glucose into smaller molecules

The word *glycolysis* means "the splitting of glucose." In a series of reactions, one molecule of glucose is split into two smaller molecules of a compound called **pyruvic** *(py ROO vihk)* **acid**. Although cells must use some ATP to begin glycolysis, the process produces a little more ATP than it uses. For each molecule of glucose that enters glycolysis, a cell gains two molecules of ATP.

During glycolysis, electrons carried on hydrogen atoms are stripped from glucose molecules. The electrons and hydrogen atoms are donated to an electron acceptor called NAD^+, forming the energy carrier NADH. For glycolysis to continue, however, electrons donated to NAD^+ to form NADH must be donated to some other organic molecule. This process frees the NAD^+ to accept more electrons and hydrogen atoms from glycolysis. Thus, glycolysis is followed by another set of reactions—either fermentation or oxidative respiration. The pyruvic acid molecules made during glycolysis still contain energy, but they cannot be broken down further to release energy if no oxygen is present.

Fermentation occurs in the absence of oxygen

The incomplete breakdown of organic compounds in the absence of oxygen is called **fermentation**. Like glycolysis, fermentation evolved before Earth's atmosphere contained oxygen and does not use oxygen. Today, cells still use fermentation to produce small amounts of ATP when oxygen is not available.

During fermentation, the electrons and hydrogen atoms carried by NADH are attached to pyruvic acid molecules. This process regenerates NAD^+, which enables glycolysis to continue. In some organisms, pyruvic acid is broken down into carbon dioxide and ethyl alcohol (ethanol), which is the alcohol in beer and wine. In other organisms, pyruvic acid is converted to lactic acid. These two types of fermentation are summarized in **Figure 5-17**.

Lactic acid forms in muscles during vigorous exercise

Pyruvic acid is converted into lactic acid in your muscles when they are forced to operate without enough oxygen, such as during vigorous exercise. This process enables glycolysis to continue in your muscles as long as the glucose supply lasts. Blood circulation removes excess lactic acid from muscles. But when lactic acid cannot be removed quickly enough, it builds up and your muscles get tired. Because of the buildup of lactic acid, the world record for running a mile (just under 4 minutes) has not been beaten.

Stage 2: Oxidative Respiration

When Earth's atmosphere grew rich in oxygen, an alternative to fermentation became possible. Oxidative respiration, which happens within mitochondria, was this alternative. In the presence of oxygen, the electrons and hydrogen atoms carried by NADH can ultimately be attached to oxygen atoms instead of to pyruvic acid molecules.

Oxidative respiration involves chemical reactions that make much more ATP than is made by fermentation. **Figure 5-18** compares the results of fermentation and oxidative respiration. The complete breakdown of glucose through oxidative respiration can be summarized by the following equation:

$$C_6H_{12}O_6 + 6O_2 \longrightarrow 6CO_2 + 6H_2O + 36ATP$$

As with the summary equation for photosynthesis, this equation does not show how cellular respiration happens. It merely indicates that the complete breakdown of a glucose molecule uses 6 oxygen atoms and forms 6 carbon dioxide molecules, 6 water molecules, and as many as 36 ATP molecules.

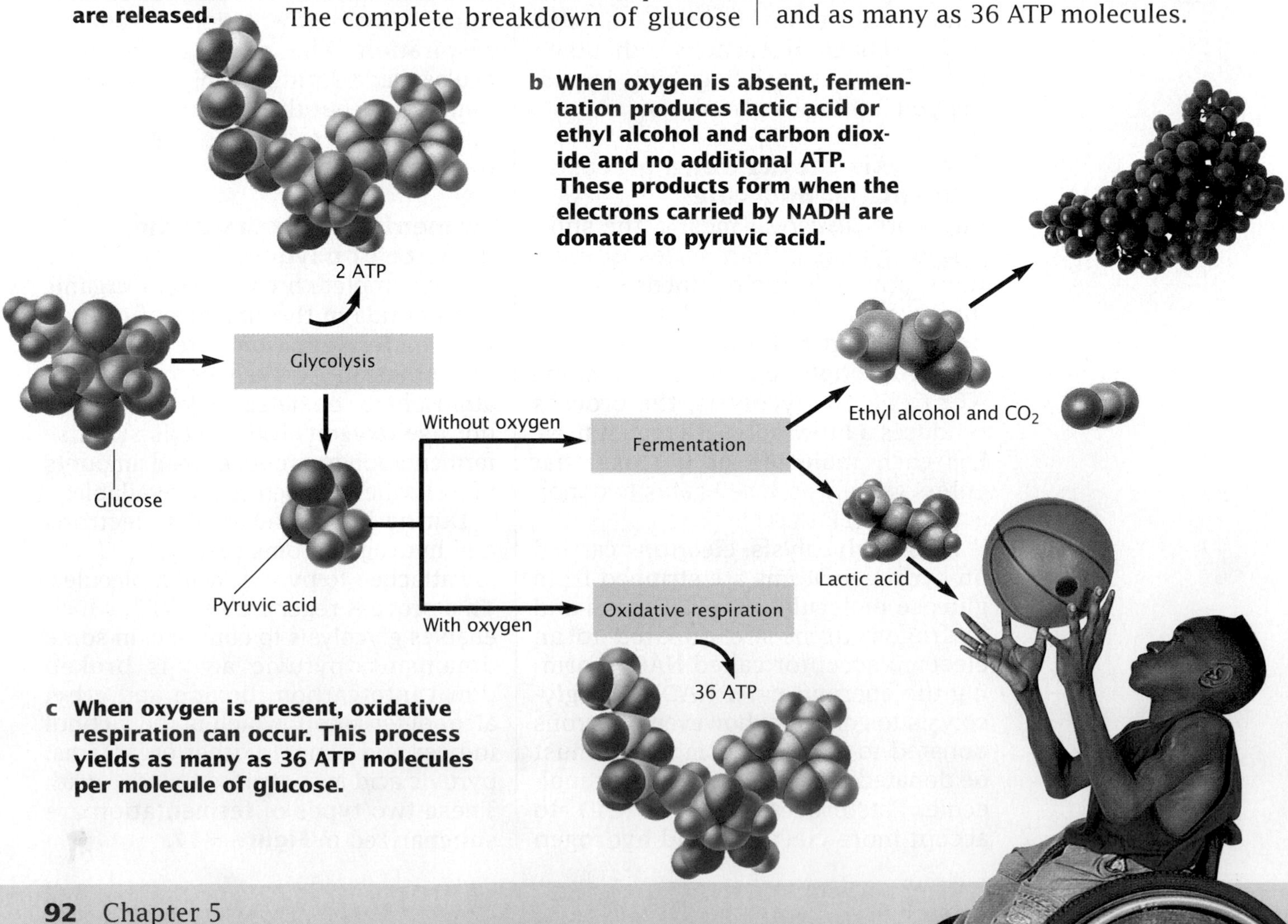

Figure 5-18

a Both cellular respiration and fermentation begin with glycolysis, in which glucose is broken down into pyruvic acid and two ATP molecules are released.

b When oxygen is absent, fermentation produces lactic acid or ethyl alcohol and carbon dioxide and no additional ATP. These products form when the electrons carried by NADH are donated to pyruvic acid.

c When oxygen is present, oxidative respiration can occur. This process yields as many as 36 ATP molecules per molecule of glucose.

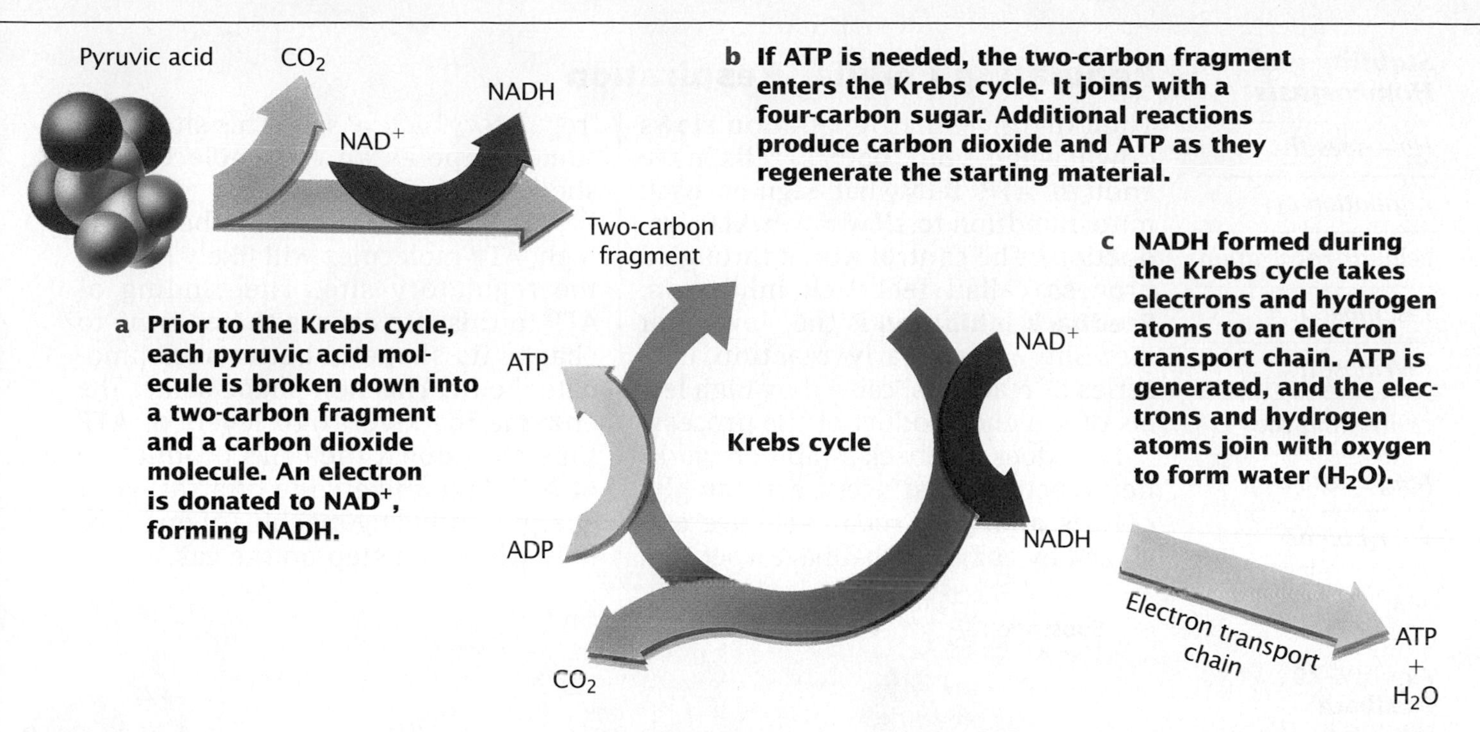

Pyruvic acid
CO_2
NADH
NAD^+
Two-carbon fragment
b If ATP is needed, the two-carbon fragment enters the Krebs cycle. It joins with a four-carbon sugar. Additional reactions produce carbon dioxide and ATP as they regenerate the starting material.
c NADH formed during the Krebs cycle takes electrons and hydrogen atoms to an electron transport chain. ATP is generated, and the electrons and hydrogen atoms join with oxygen to form water (H_2O).
a Prior to the Krebs cycle, each pyruvic acid molecule is broken down into a two-carbon fragment and a carbon dioxide molecule. An electron is donated to NAD^+, forming NADH.
ATP
NAD^+
Krebs cycle
ADP
NADH
Electron transport chain
ATP + H_2O
CO_2

Figure 5-19 **Oxidative respiration occurs in two major steps: the Krebs cycle and the electron transport chain.**

Oxidative respiration picks up where glycolysis leaves off

Figure 5-19 summarizes the major events of oxidative respiration. Inside the mitochondria, a carbon atom is removed from each pyruvic acid molecule produced by glycolysis. This reaction frees electrons that carry energy and releases carbon dioxide molecules. The electrons and a hydrogen atom combine with NAD^+, forming NADH (another carrier of electrons and hydrogen atoms). The NADH will be used at the end of oxidative respiration. After the two reactions, only a two-carbon fragment remains, as you can see in **Figure 5-19a**. If a cell has enough ATP, the two-carbon fragment is used in reactions that produce fat, which stores energy. If the cell needs ATP, the two-carbon fragment will continue on to the next step of oxidative respiration.

The Krebs cycle yields ATP, NADH, and carbon dioxide

Two-carbon fragments produced by the removal of carbon atoms from pyruvic acid enter a series of reactions called the Krebs cycle. This cycle was named for Sir Hans Krebs, the biochemist whose work in the 1930s revealed how the reactions work. During the Krebs cycle, summarized in **Figure 5-19b**, two carbon dioxide molecules, one ATP molecule, and more NADH molecules are made. The cycle also regenerates its starting molecule, enabling another two-carbon fragment to enter the cycle.

The electron transport chain makes more ATP

The electrons carried by molecules of NADH formed during the Krebs cycle are passed through a series of proteins in the membranes within mitochondria. This series of proteins, called the electron transport chain, uses the energy of these electrons to pump protons across the membrane. As in photosynthesis, protons become highly concentrated on one side of the membrane. The protons pass back through the membrane by diffusing through a protein channel. This movement of protons supplies the energy needed to attach a phosphate group to an ADP molecule, making ATP.

As **Figure 5-19c** shows, the protons and spent electrons from NADH are joined to oxygen at the end of the electron transport chain, forming water. Oxidative respiration requires oxygen because the electrons stripped from pyruvic acid must have a final home. Energy cannot be extracted from pyruvic acid without oxygen to accept spent electrons. Otherwise, the electron transport chain would become "clogged" with electrons.

Stability and Homeostasis

How does the regulation of cellular respiration resemble a thermostat controlling the temperature in your home?

Regulating Cellular Respiration

The rate of cellular respiration slows down when your body's cells have enough ATP. But what signals each mitochondrion to slow down ATP production? The control works through a process called feedback inhibition. **Feedback inhibition** is the slowing or stopping of an early reaction in a series of reactions, caused by high levels of the end product of the process.

How does feedback inhibition work? Key reactions that occur early in glycolysis and the Krebs cycle are catalyzed by enzymes that have a second, "regulatory" active site. This site is the same shape as an ATP molecule, as shown in **Figure 5-20**.

When the ATP level in the cell is high, ATP molecules will likely bind to the regulatory site. The binding of ATP to this site causes the enzyme to change its shape to better accommodate the fit. This new shape makes the enzyme inactive. High levels of ATP thus shut down the cell's production of ATP. Like the engine of a car, your energy-producing machinery operates only when you step on the gas.

Figure 5-20 Feedback inhibition slows or stops an early reaction in a series of chemical reactions when levels of the end product are high. Molecules of the end product bind to an enzyme's regulatory site.

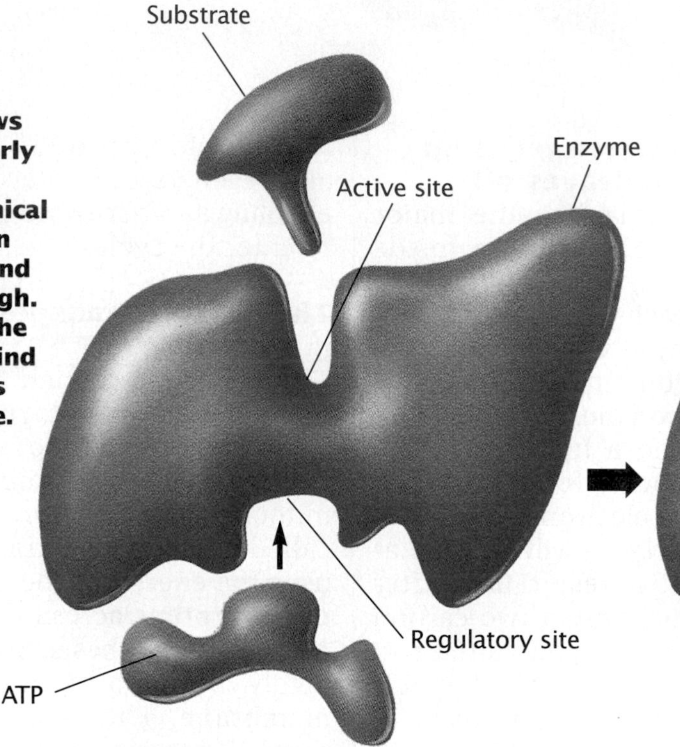

a When there is excess ATP in a cell, ATP binds to the regulatory site of an enzyme.

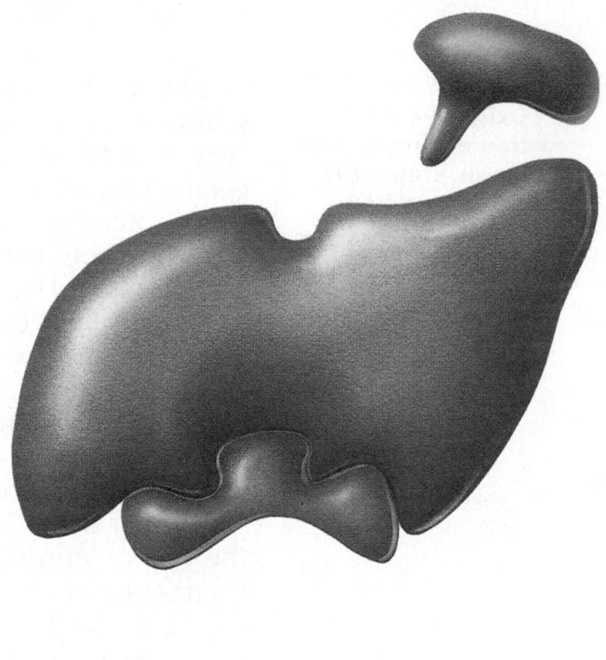

b The enzyme then changes shape. The active site is no longer able to catalyze its usual reaction, and no more ATP is made.

SECTION REVIEW

1. **How is cellular respiration important for living cells?**
2. **How do fermentation and oxidative respiration differ?**
3. **Why is oxidative respiration unable to occur without oxygen?**
4. **Explain how feedback inhibition enables cellular respiration to slow down when supplies of ATP are sufficient.**

CHAPTER

5 Highlights

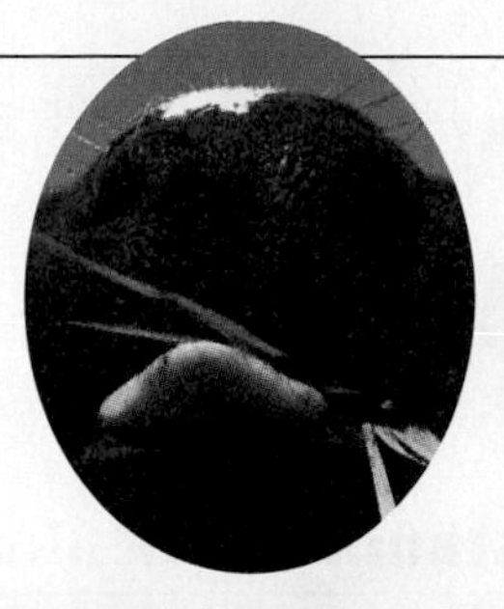

Grass is the mainstay of this cow's diet. Using what you've learned, can you explain the link between sunlight and dairy products?

	Key Terms	Summary
5-1 Cells and Chemistry	metabolism (p. 78) activation energy (p. 78) enzyme (p. 79) catalyst (p. 79) substrate (p. 80) active site (p. 80)	• The energy needed for life is absorbed and released through chemical reactions. • Enzymes make chemical reactions occur rapidly enough to sustain life. • Enzymes act on specific substrates. Enzyme activity is affected by heat, acidity, and substrate concentration.
5-2 Cells and Energy	ATP (p. 83) photosynthesis (p. 84) cellular respiration (p. 84)	• The energy for life is released gradually through a series of chemical reactions. • The energy released is stored temporarily in a molecule called adenosine triphosphate, or ATP. • Photosynthesis captures energy needed for life from light. • Cellular respiration releases energy stored in organic molecules.
5-3 Photosynthesis	pigment (p. 86) chlorophyll (p. 87)	• Photosynthesis involves a complicated series of chemical reactions. • Plants, algae, and some bacteria contain the pigment chlorophyll. • Light striking chlorophyll boosts electrons to higher energy levels. As electrons pass through a series of reactions, their energy is used to produce organic molecules.
5-4 Cellular Respiration	glycolysis (p. 90) oxidative respiration (p. 90) pyruvic acid (p. 91) fermentation (p. 91) feedback inhibition (p. 94)	• Living things obtain energy from organic molecules through cellular respiration. • In the absence of oxygen, fermentation converts the pyruvic acid from glycolysis to lactic acid or ethyl alcohol. With oxygen present, the Krebs cycle and the electron transport chain produce most of the ATP needed for life. • Feedback inhibition controls the reactions of cellular respiration.

Enzymes, such as carbonic anhydrase, hasten chemical reactions.

Many living things depend on energy stored in plants.

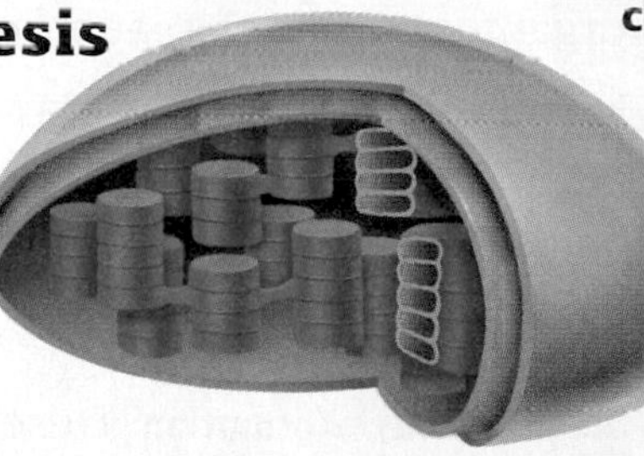

Photosynthesis begins when light strikes a chloroplast.

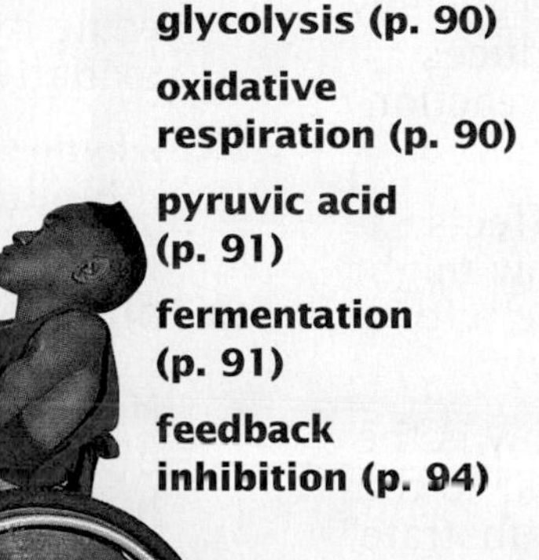

Cellular respiration releases the energy necessary to carry out life activities.

CHAPTER 5

review

Understanding Vocabulary

1. For each set of terms, complete the analogy.
   a. reactants : changed :: enzyme : ____
   b. ATP : three phosphate groups :: ADP : ____
   c. photosynthesis : stores energy :: cellular respiration : ____
2. For each pair of terms, explain the differences in their meanings.
   a. reactants, products
   b. substrate, enzyme
   c. glycolysis, fermentation

Understanding Concepts

3. **Relating Concepts** Copy the unfinished concept map below onto a sheet of paper. Then complete the concept map by writing the correct word or phrase in each box containing a question mark.

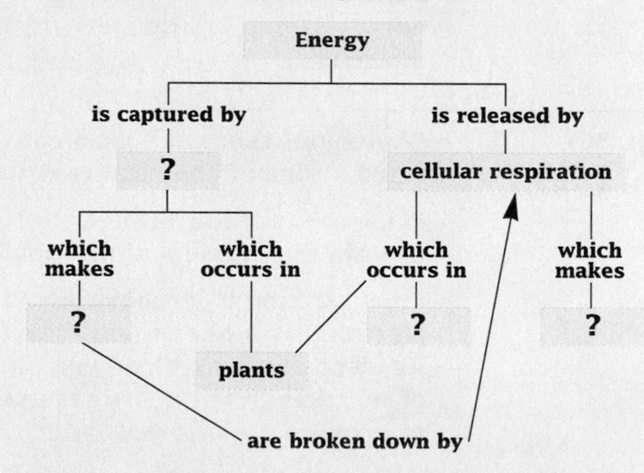

4. **Comparing Functions** In terms of energy, how does a chemical reaction that produces glucose compare with a chemical reaction that breaks down glucose?
5. **Inferring Conclusions** Temperature affects how quickly an enzyme reacts. How might freezing food affect the enzyme activity of organisms that cause food to decay?
6. **Comparing Structures and Functions** How is the interaction of a lock and key similar to the interaction of an enzyme and its substrate?
7. **Inferring Relationships** Explain why enzymes are needed to control chemical reactions within cells.
8. **Organizing Information** Describe the difference in the amount of activation energy required to start a chemical reaction with an enzyme and without an enzyme. Include a graph to illustrate your answer.
9. **Summarizing Information** Explain how cells use energy. Give two examples.
10. **Inferring Relationships** What is ATP? Why is it important to a cell?
11. **Interpreting Processes** Describe the role of chlorophyll in photosynthesis.
12. **Recognizing Relationships** How do plant cells use the ATP and NADPH that forms during the second phase of photosynthesis?
13. **Relating Concepts** What is the role of water in photosynthesis? What happens to the protons after the electrons are removed?
14. **Organizing Information** Indicate at which stage of photosynthesis each event occurs: (a) ATP is made when protons flow through a protein channel, (b) electrons within chlorophyll molecules are boosted to a higher energy level, (c) electrons are used to make NADPH, (d) carbon atoms are captured from carbon dioxide in the air and used to make organic molecules that store energy.
15. **Inferring Conclusions** Do plants perform cellular respiration? Support your answer.
16. **Evaluating Outcomes** Which process is more efficient—fermentation or oxidative respiration? Justify your answer.
17. **Summarizing Information** Describe the net gain of ATP molecules from one glucose molecule by glycolysis, fermentation, and oxidative respiration.
18. **Inferring Relationships** How does feedback inhibition control the rate of cellular respiration?
19. **BUILDING ON WHAT YOU HAVE LEARNED** In Chapter 3 you read about proton pumps. How does a proton pump function during photosynthesis? How does a proton pump function during oxidative respiration?

Interpreting Graphics

20. Scientists have discovered that the rate of photosynthesis is affected by both the intensity of light and temperature. Study the following two graphs, and then answer the questions below.

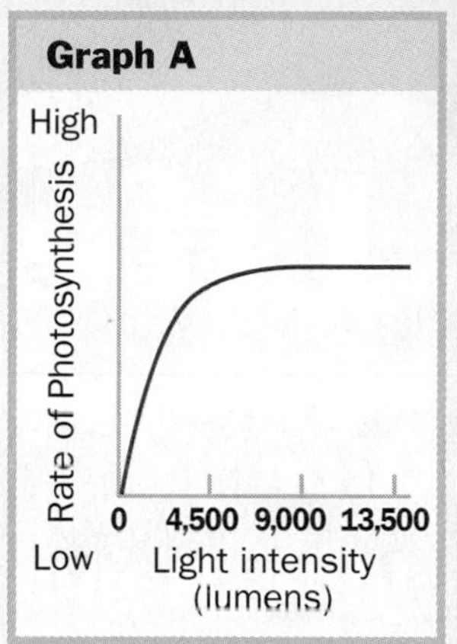

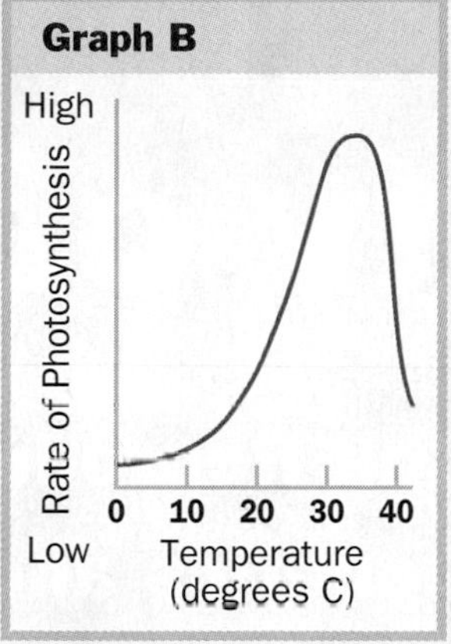

a. At what light intensity does the rate of photosynthesis level off?
b. In what range of temperatures does the rate of photosynthesis increase?

Reviewing Themes

21. ***Evolution***
How did the appearance of a new kind of photosynthesis about 3 billion years ago change Earth's atmosphere?

22. ***Evolution***
What evidence suggests that glycolysis is a more ancient process than oxidative respiration?

23. ***Stability and Homeostasis***
After strenuous exercise, an athlete often has sore muscles and breathes heavily. How does this muscle soreness and heavy breathing relate to the production of lactic acid by fermentation?

Thinking Critically

24. **Distinguishing Relevant Information** Scientists have recently discovered a cellular substance that has a shape and structure very similar to that of pyruvic acid. How might the presence of this substance in a cell affect the process of oxidative respiration?

25. **Justifying Conclusions** Would being able to photosynthesize eliminate humans' need to eat food? Why or why not?

Life/Work Skills

26. **Finding Information** Today, scientists know a great deal about the chemical processes of photosynthesis. Prior to the seventeenth century, however, scientists knew very little about how plants manufactured organic compounds. Use library references or search an on-line database to find information about the discoveries that led to the understanding of the process of photosynthesis. Look up the contributions made by English clergyman and chemist Joseph Priestley, Dutch physician and naturalist Jan Ingenhousz, Swiss chemist Nicolas de Saussure, German physicist Julius Robert von Mayer, and American biochemist Melvin Calvin. Summarize your findings in a poster or written report.

Activities and Projects

27. **Health Connection** Use library references or interview a medical specialist to find out about enzyme deficiency disorders in humans such as phenylketonuria (PKU) and alkaptonuria. Write a report that describes the symptoms and cause of the disorder, based on enzyme function. For PKU, also include information about how this disorder can be treated.

28. **Social Studies Connection** Yogurt is a fermented dairy product with an unusual history. Use library resources to find out how fermentation has been used in the production of yogurt and other commodities, such as cheese, soybean products, and beverages. Also find out which regions of the world are the largest producers of these commodities. Show these regions on a map of the world. Summarize your findings in a written report.

Discovering Through Reading

29. Read the article "When Science Imitates Nature: Using Artificial Photosynthesis to Harness Solar Energy" by Hunter Whitney in *Omni*, May 1994, page 25. What was Giacomo Ciamician's dream? Describe one obstacle that scientists have not yet been able to overcome in perfecting artificial photosynthesis.

UNIT Two

Continuity of Life

Human chromosomes

YOUR GENETIC COMPUTER

When you use a computer to type a paper, each letter is translated into a unique code that the computer reads. Using 0s and 1s, thousands of character combinations are made. But these numbers are small compared with the millions of instructions made from the four nucleotides that are repeated in your DNA.

Disorderly Conduct

Genes made of these nucleotides copy and transfer codes that determine how your body is built and how it works. Sometimes a gene gets damaged—part of the code is left out or changed. Called mutations, these changes can cause genetic disorders or uncontrolled cell division leading to cancer.

Knowing the Code

Scientists who study gene sequencing now know the code for many genes in the body. Gene technology is being used to pinpoint and correct errors in existing code and to identify factors that trigger mutations. Techniques such as gene transference are no longer science fiction but are scientific reality.

LOOKING AHEAD

- How are genes passed from parent to child? See **A Closer Look: Meiosis,** page 111.
- How can mutations enable a species to change over time? See page 127.
- How can knowledge of genetics help people make personal decisions? See **Career Opportunities: Genetic Counseling,** page 131.
- What benefits could you receive because of gene technology? See **Science, Technology, and Society: What Is the Future of Genetic Engineering?** pages 170–171.

CHAPTER 6

REVIEW
- nucleotide (Section 2-3)
- cell (Section 3-1)
- surface-area-to-volume ratio (Section 3-1)

Cell Reproduction

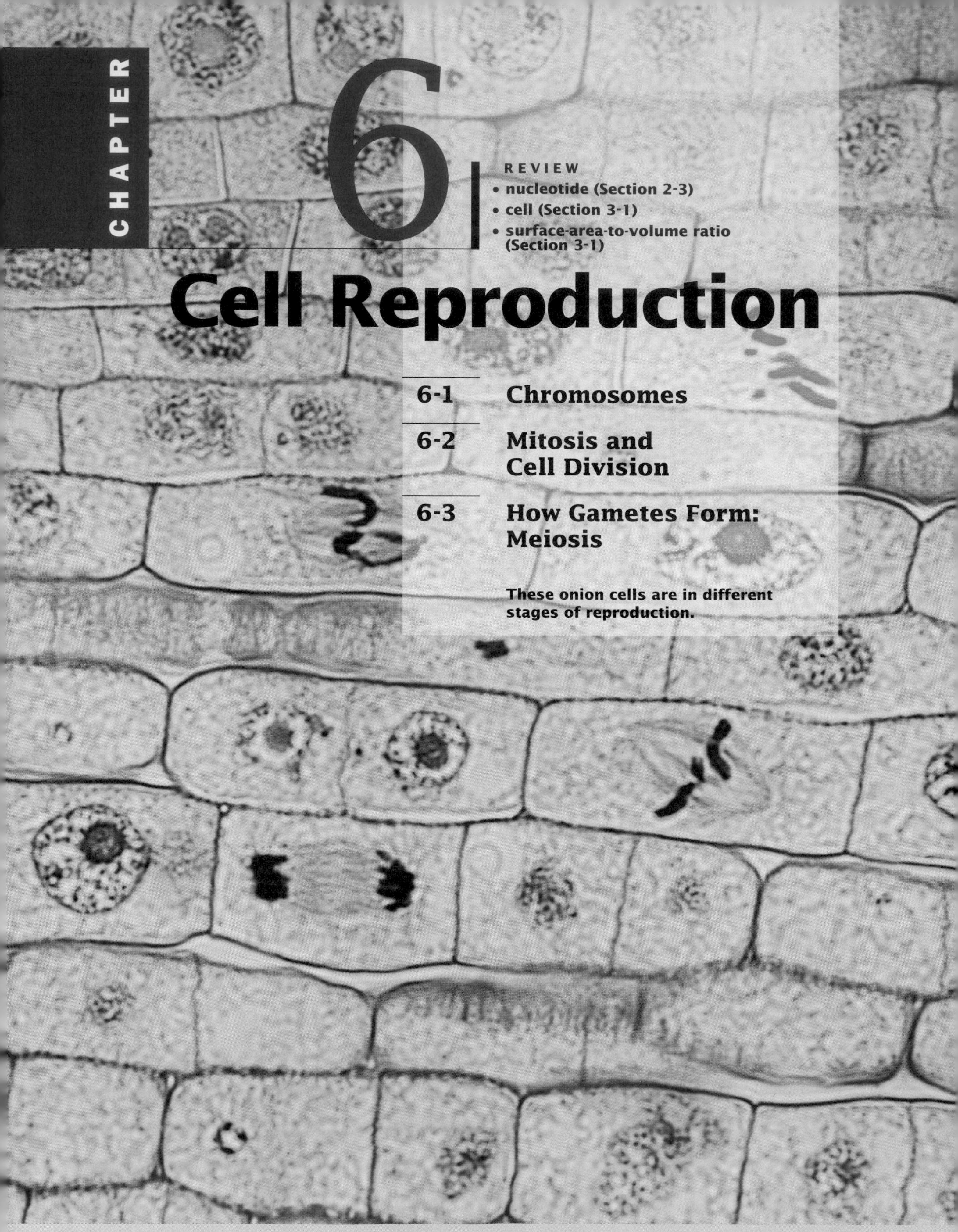

These onion cells are in different stages of reproduction.

A typical chromosome in your body contains thousands of genes. Each gene is a segment of DNA that codes for a particular trait. This information was used to assemble tissues and organs while you were an embryo. Now the information regulates the functions of the billions of cells in your body.

6-1 Chromosomes

OBJECTIVES

1. **Define *chromatin.***
2. **Describe the structure of a chromosome.**
3. **Distinguish between diploid and haploid cells.**
4. **Explain the significance of sex chromosomes.**

Chromosome Structure

The growth and development of your cells are carefully controlled by a set of instructions found inside the cell nucleus. This set of instructions, encoded in long molecules of DNA, is what determines that you are a human and not a different kind of organism, such as a housefly or a carrot. Molecules of DNA might look like strands of spaghetti piled on a plate, but on closer inspection you can see that they are more than a tangled mass. These long molecules are wound around chunks of protein. When cells divide, this material is tightly compacted, creating bodies of DNA and protein called **chromosomes**, as shown in **Figure 6-1**.

DNA coils to form a chromosome

In the century since their discovery, we have learned a great deal about the structure and function of chromosomes. Eukaryotic chromosomes are made of **chromatin** *(KROH muh tihn)*, a complex of DNA and protein. The DNA of a chromosome exists as one very long double-stranded thread that curls up tightly to form a compact rod-shaped body. A typical human chromosome contains about half a billion (5×10^8) nucleotides in its thread of DNA. If the strand of DNA from a single chromosome were laid out in a straight line, it would be about 5 cm (2 in.) long. This is much too long to fit into a cell. But when the strand is wound around proteins, it can be compacted into a smaller, more manageable structure, just as a piece of thread can be neatly stored by wrapping it around a wooden spool. The complex structure of a chromosome is shown in *A Closer Look: Chromosome*, on the next page.

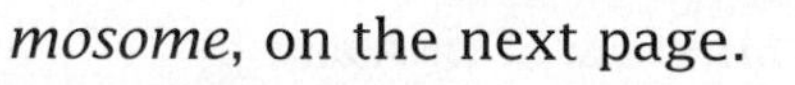

Figure 6-1
A chromosome is a library of genetic information. Each chromosome consists of one long strand of DNA wrapped around chunks of protein.

Chromosome

In eukaryotes, chromosomes are contained in the nucleus. Each chromosome contains one long DNA molecule in which two parallel strands are twisted like a spiral staircase.

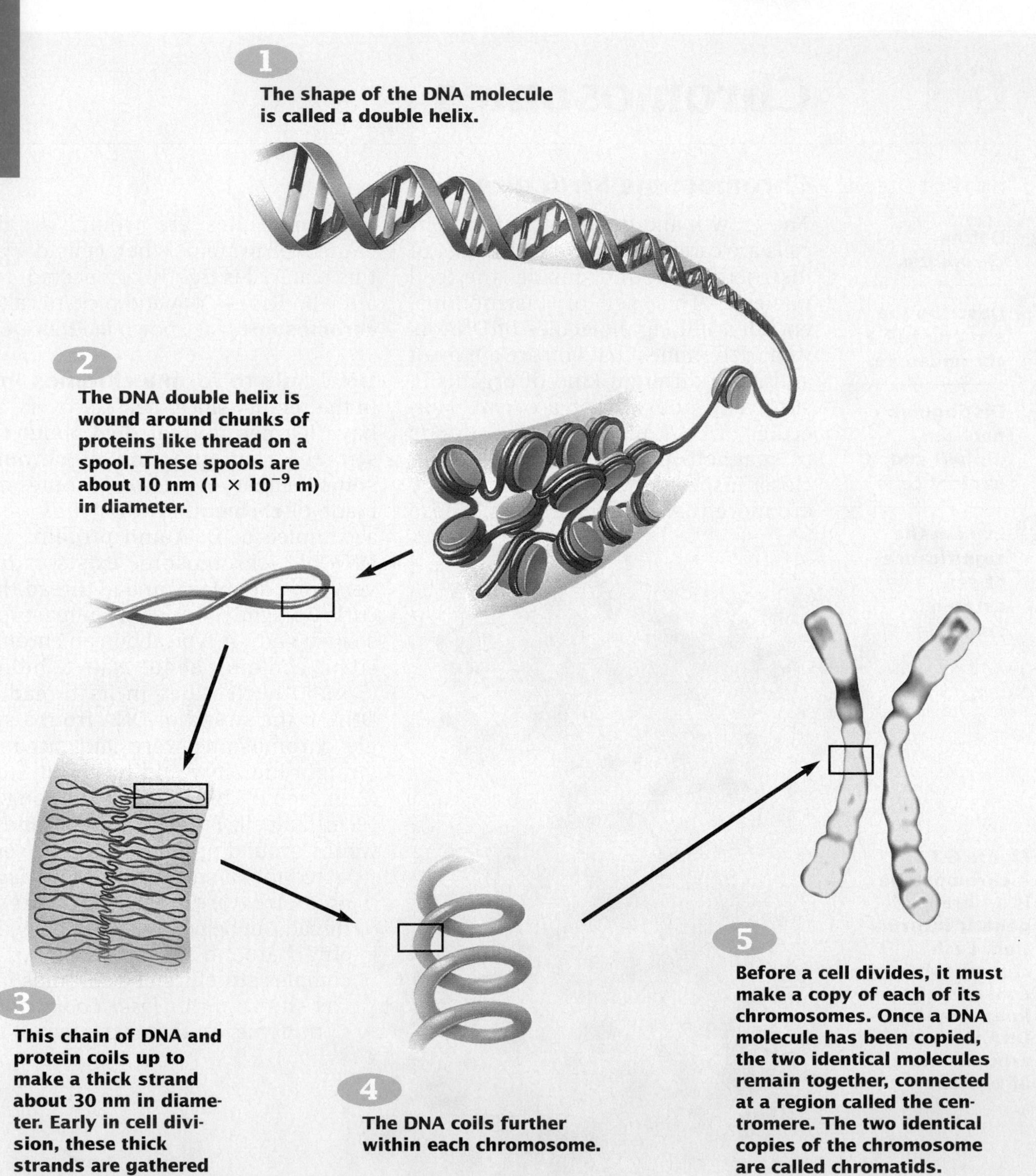

1
The shape of the DNA molecule is called a double helix.
2
The DNA double helix is wound around chunks of proteins like thread on a spool. These spools are about 10 nm (1 × 10⁻⁹ m) in diameter.
3
This chain of DNA and protein coils up to make a thick strand about 30 nm in diameter. Early in cell division, these thick strands are gathered into loops, which are wound like a corkscrew.
4
The DNA coils further within each chromosome.
5
Before a cell divides, it must make a copy of each of its chromosomes. Once a DNA molecule has been copied, the two identical molecules remain together, connected at a region called the centromere. The two identical copies of the chromosome are called chromatids.

Chromosome Number

The chromosomes contained within a cell can be extracted and stained to reveal the set of chromosomes. A set of chromosomes is called a **karyotype** *(KAH ree oh typ)*. If you could look inside almost any cell in your body, you would find 46 chromosomes that are similar to those in the karyotype shown in **Figure 6-2**. Sorted by size and shape, your chromosomes can be arranged into 23 pairs. The two members of each pair look very similar. In fact, they are almost identical copies of each other, containing the same genes in the same order. When a human cell contains 23 pairs of chromosomes, it is said to be **diploid** *(DIH ployd)*. Practically all the cells that make up your body are diploid. By contrast, the chromosomes in your gametes (eggs or sperm) occur singly. These cells, which contain only one copy of each chromosome, are said to be **haploid** *(HAHY ployd)*. The symbol commonly used for the diploid number is $2n$, and that used for the haploid number is n. For example, in humans, $2n = 46$ and $n = 23$.

Sex chromosomes determine whether you are male or female

One of your 23 pairs of chromosomes determines whether you are male or female. These chromosomes are called **sex chromosomes** and are designated as X and Y. If you are female, you have two X chromosomes. If you are male, you have one X chromosome and a smaller Y chromosome. Because females have two X chromosomes, they can only produce eggs that contain an X chromosome. Males, however, can produce sperm that contains either an X or a Y chromosome. Thus, if you're female, you've received an X chromosome from your father. If you're male, you've inherited your father's Y chromosome. Study the karyotype in **Figure 6-2**. Do these chromosomes belong to a male or to a female? The members of each pair of sex chromosomes are separated during a process called meiosis *(my OH sihs)*, which you will learn about later in this chapter.

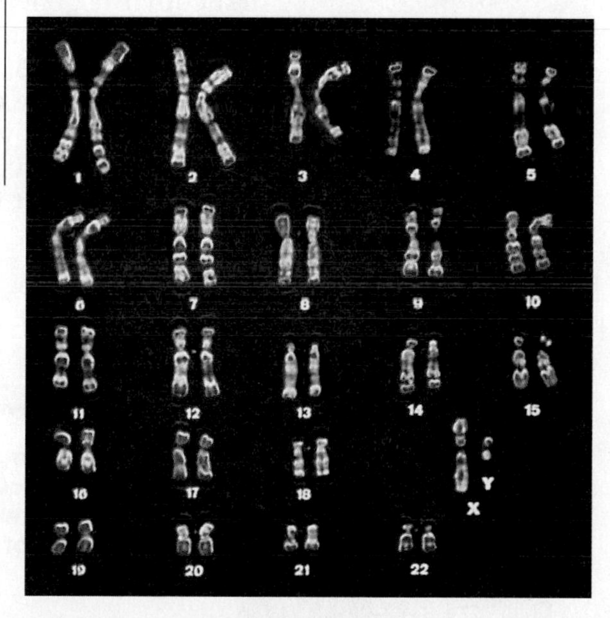

Figure 6-2

a Like this human white blood cell, almost every cell in the human body . . .

b . . . contains 46 chromosomes. They occur in 23 pairs whose members have similar shapes, sizes, and information.

SECTION REVIEW

1. **At what point does chromatin coil up to form a chromosome?**
2. **What do you think would happen to DNA if it were not wrapped around proteins?**
3. **Which cells in your body are diploid? How many chromosomes do they have?**
4. **What are sex chromosomes? Which sex chromosomes do your cells contain?**

As cells busily carry out the functions of life, they grow and develop. You learned in Chapter 3 that cells operate most efficiently when they have a high surface-area-to-volume ratio. When most cells reach a certain size, they either stop growing or divide into two cells. Cell reproduction is essential for the growth and development of an organism.

6-2 Mitosis and Cell Division

OBJECTIVES

1. **Describe cell reproduction in bacteria and eukaryotes.**
2. **Summarize the events of mitosis and cytokinesis.**
3. **Define *cell cycle.* Describe a cell's activities during interphase.**
4. **Explain how cancer arises.**

Cell Reproduction Differs in Bacteria and Eukaryotes

When a cell reaches a certain size, it will either stop growing or divide. Because chromosomes carry an organism's genetic information, they must be transmitted in an orderly fashion during cell reproduction. Just as a contractor needs a complete blueprint to build a house, many cells need a complete set of genetic information in order to function. If a cell is to divide and produce two new healthy cells, it must first copy the genetic information contained within its chromosomes. Then, as the cell divides, each new cell will receive a complete copy of the genetic information. As you will see, the process of cell division in bacteria is different from that in eukaryotic cells.

Bacteria simply split

In bacteria, the process of cell division appears fairly simple. First a bacterium becomes larger. Then its single circle of DNA attaches to the inner surface of the cell membrane and makes a copy of itself. The copy of the chromosome attaches to a different part of the inner wall of the bacterium. Next, new membrane and wall materials begin to form between the two chromosomes, near the middle of the cell. The long cell pinches inward at this point, and eventually the cell is split into two cells. Each cell contains a circle of DNA and is an independent, living bacterium. This process is illustrated in **Figure 6-3**.

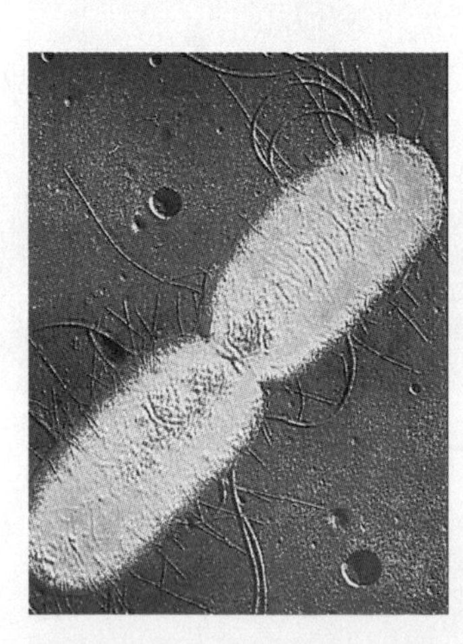

Figure 6-3
This bacterium is dividing. The process will be completed in about 20 minutes.

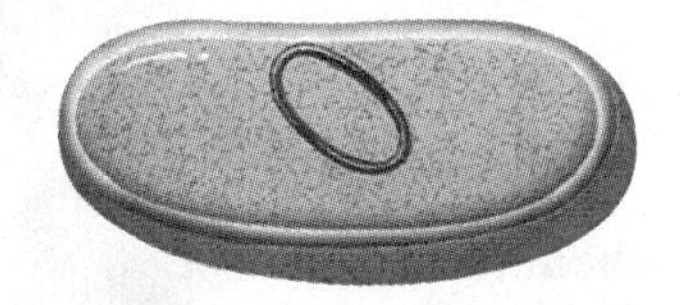

a Starting at one point of its circular chromosome, a bacterium makes a copy of its DNA.

b Side by side, the two copies of hereditary information attach to the inner wall of the bacterium.

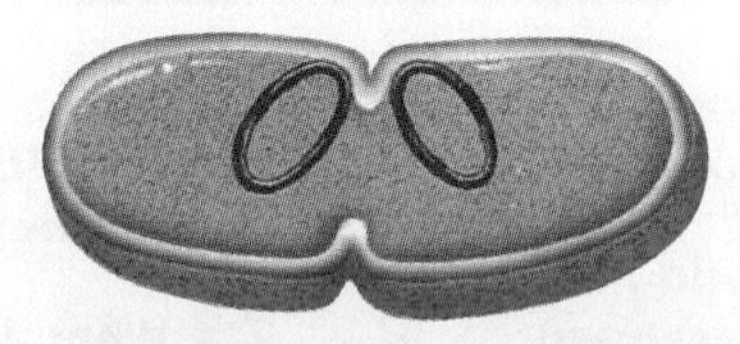

c A new membrane and cell wall form between the DNA copies, gradually pinching inward.

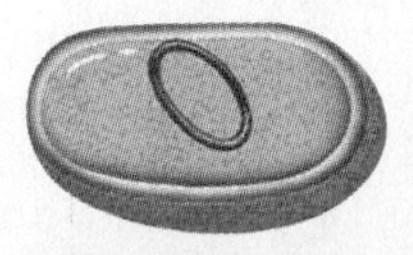

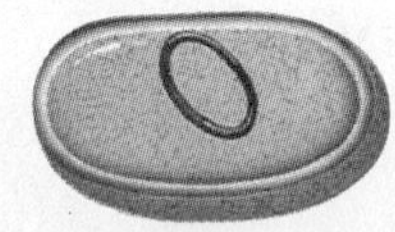

d Eventually, the cell is split into two cells. Each cell contains its own DNA and is a distinct living bacterium.

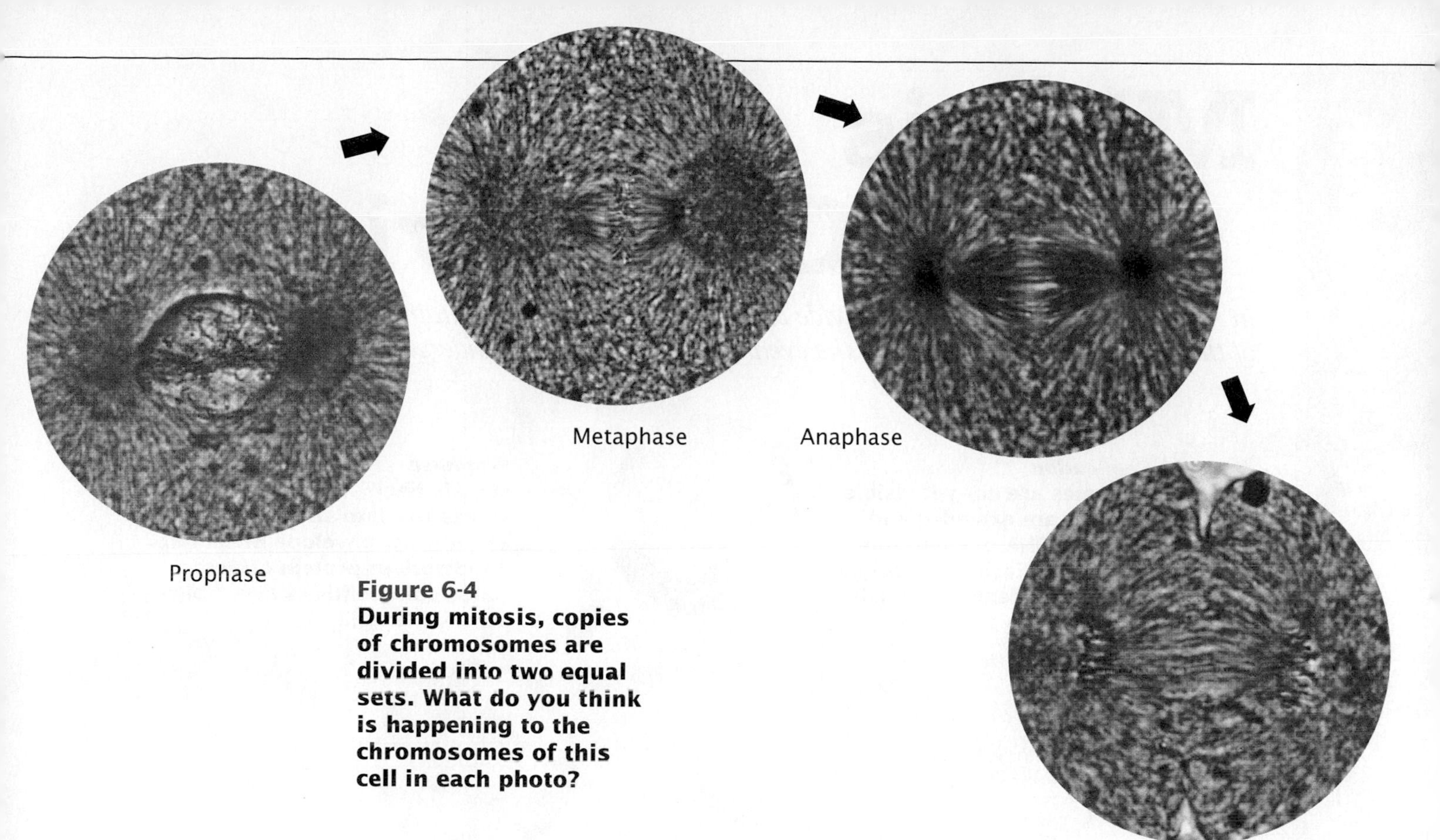

Figure 6-4
During mitosis, copies of chromosomes are divided into two equal sets. What do you think is happening to the chromosomes of this cell in each photo?

How Eukaryotic Nuclei Divide: Mitosis

Eukaryotic cells carry far more DNA than prokaryotic cells. During cell division, the DNA in eukaryotic cells is packaged into tightly wound chromosomes. Cell division plays a major role in the development and maintenance of the tissues of eukaryotic organisms. Look at your hand and try to imagine how many millions of cells divided to form your skin, bones, tendons, and the actively growing regions of your nails.

A typical human cell contains 46 chromosomes. Eukaryotic chromosomes are contained within the nucleus. Because eukaryotic cells have more DNA than prokaryotic cells have, and because the DNA is confined within a nucleus, eukaryotic cell division is quite different from bacterial cell division. First each chromosome must be copied exactly. Then the chromosomes must be sorted out precisely so that each new cell gets a complete set. Finally the cell itself can divide in half.

Two processes have evolved that enable eukaryotic cells to divide successfully: mitosis and cell division. **Mitosis** *(my TOH sihs)* is the process by which the nucleus of a eukaryotic cell divides to form two nuclei, each containing a complete set of the cell's chromosomes. Mitosis can be broken down into four distinct phases: prophase, metaphase, anaphase, and telophase. These steps are shown in **Figure 6-4**.

In many cells, mitosis is followed by cell division, also called **cytokinesis** *(syt oh kuh NEE sihs)*. During cell division, the cell divides into two cells, each with its own nucleus. Plant cells, which have strong cell walls that resist being pinched in like the membranes of animal cells, form a new cell wall in the center of the cell. This new cell wall divides the cell in half, like a partition dividing a room.

The end result of mitosis and cytokinesis is two cells with the same genetic information, where only one cell existed before. To follow the events of eukaryotic cell division, study the steps in *A Closer Look: Mitosis and Cell Division*, on the next page.

Mitosis and Cell Division

In mitosis, eukaryotic cell nuclei divide to form two nuclei, each containing a complete set of the cell's chromosomes. Follow the events as this animal cell undergoes mitosis.

DNA Replication

Chromosomes are not yet visible because they are extended and uncoiled. The DNA of each chromosome is copied. Each chromosome consists of two identical strands.

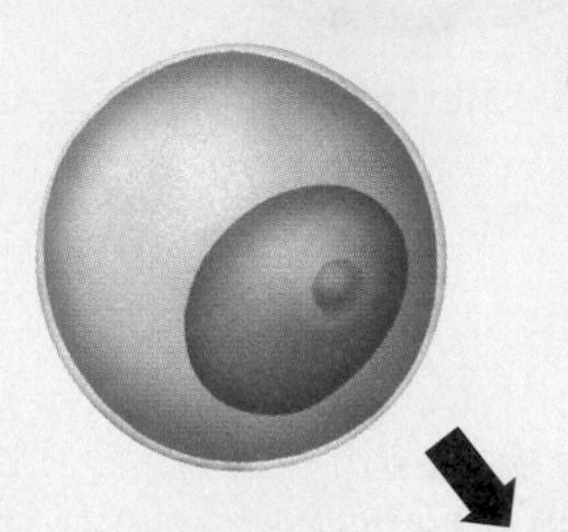

2

Prophase

Mitosis begins. The chromosomes coil into short, fat rods. The nuclear envelope breaks up. A network of protein cables called spindle fibers assembles across the cell.

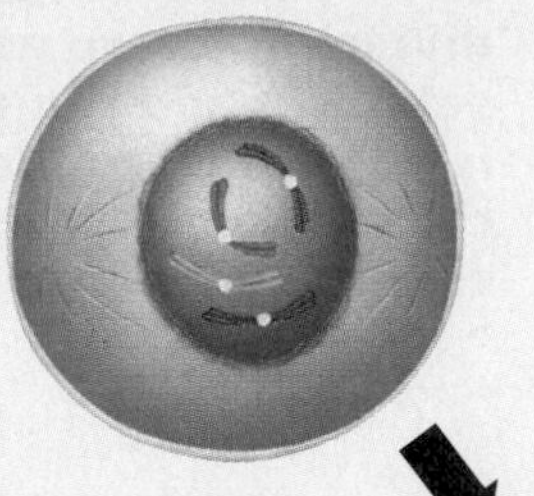

Metaphase

Chromosomes attach to the spindle fibers and line up in the center of the cell.

4

Anaphase

Each chromatid separates from its identical copy. Chromosomes are reeled to opposite sides of the cell. The spindle fibers begin to break down.

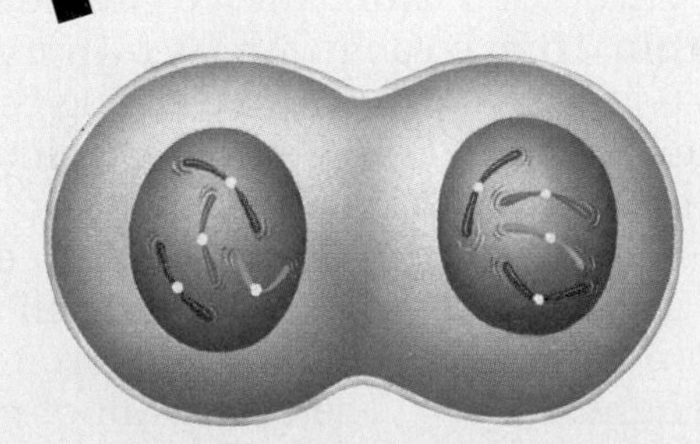

5

Telophase

Each side of the cell now has a complete set of chromosomes. A nuclear envelope forms around each new set of chromosomes. The chromosomes uncoil so that proteins can be built. The spindle fibers disappear.

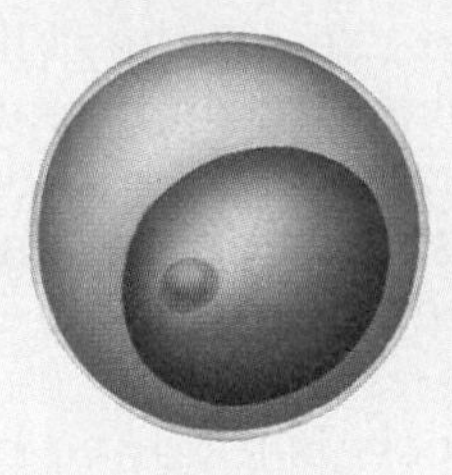

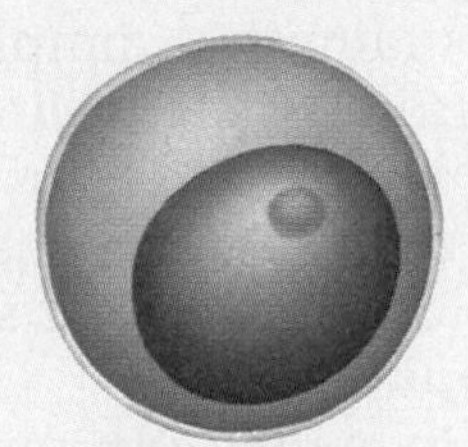

Cytokinesis

The cytoplasm is pinched in half, forming two new cells. Each new cell contains identical DNA. After growth and replication, these cells may divide again.

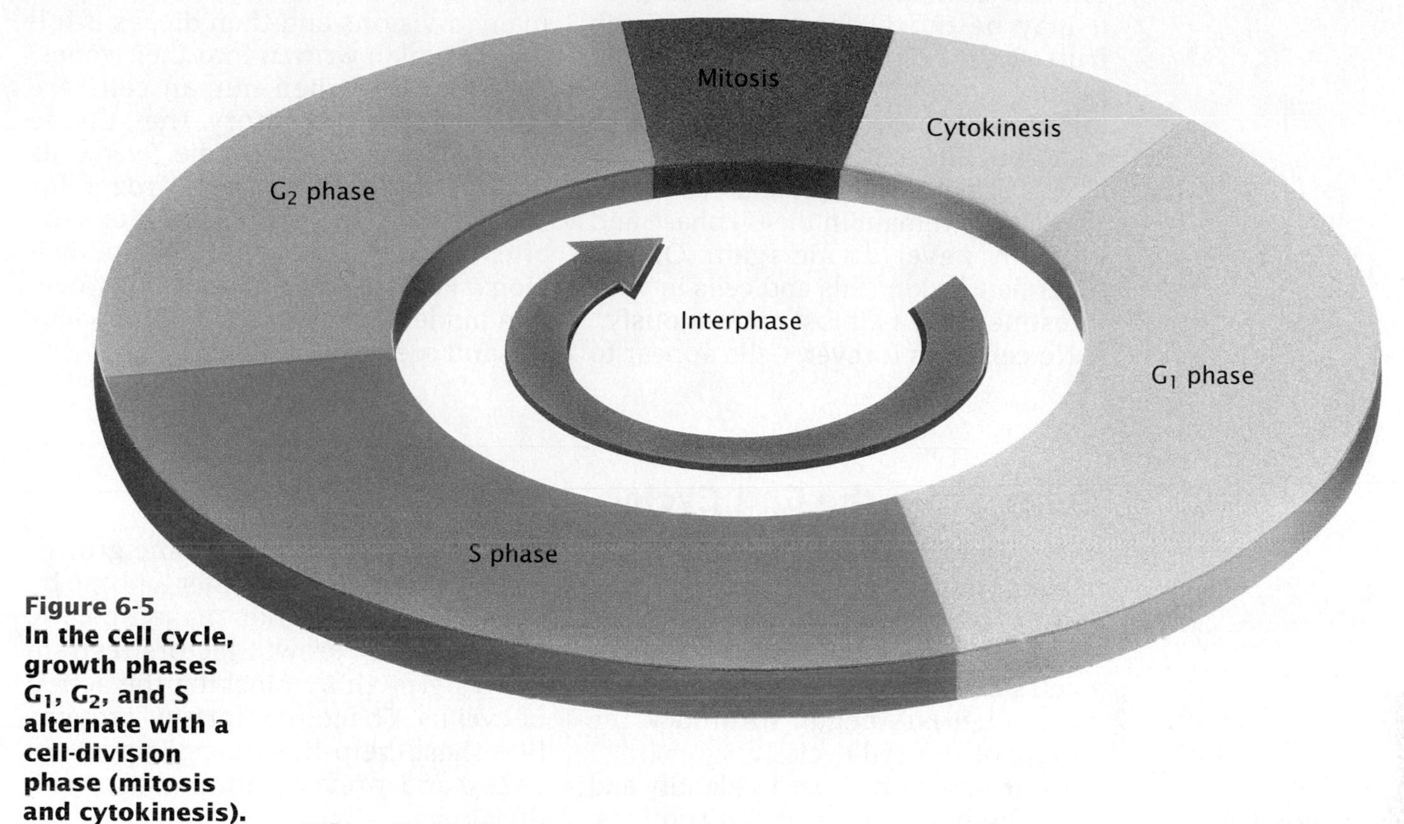

Figure 6-5
In the cell cycle, growth phases G_1, G_2, and S alternate with a cell-division phase (mitosis and cytokinesis).

The Cell Cycle

Mitosis and cytokinesis are just small portions of a cell's life span. Most of a cell's lifetime is spent growing. This cycle of growth and division is called the **cell cycle**. For convenience, the cell cycle is divided into the five phases shown in **Figure 6-5**. Identify the phases of the cell cycle as you follow a typical cell through its life.

Interphase is the most active part of the cell cycle

A newly formed cell begins its life by undergoing a period of growth and activity called interphase. During interphase, the new cell is hard at work carrying out the tasks of life: taking in food, converting energy, discarding wastes, growing, and differentiating. In addition, it is making copies of its genetic material and duplicating its organelles.

Interphase occurs in three distinct stages. First there is a gap in the cell cycle, designated the G_1 phase (G_1 stands for first gap). During this phase, the cell undergoes intense growth. In fact, the cells of many organisms spend most of their lives in this phase.

Next the cell may enter the S phase (*S* stands for synthesis) of the cycle. During this phase, the cell makes a precise copy of the DNA molecule in each of its chromosomes. Once copied, the two identical molecules remain together, joined by the centromere. The S phase is followed by another gap in the cycle, called the G_2 phase, the final stage of interphase. Here, the cell continues to grow and begins to make preparations for cell division. The cell is also busy duplicating its organelles.

Once the G_2 phase is finished, the cell has successfully completed interphase. It now enters the M phase (mitosis). During this phase the two copies of each chromosome are separated. When the cell reaches the final phase of the cell cycle, the C phase (cytokinesis), it divides in half and becomes two new, identical cells.

Cell cycle lengths differ among cell types

The time it takes to complete a cell cycle varies greatly among organisms. In plants, it usually lasts between 10 and 30 hours. In animals, the cell cycle can take between 18 and 24 hours, but it may be much shorter. Embryonic fruit flies hold the record for the fastest known animal cell cycle—8 minutes. Some cells in humans, such as nerve cells and some types of muscle cells, never complete a cell cycle. These cells remain in the G_1 phase and typically never divide again. On the other hand, skin cells and cells in your intestine divide almost continuously.

No cell lives forever. Cells appear to be programmed to undergo only so many divisions and then die, as if following a plan written into their genes. For example, when human cells are cultured in a laboratory, they divide about 50 times and then die. Even cells that are thawed after being frozen for years under laboratory conditions die after reaching a certain number of divisions. Perhaps cells in your body contain hidden "hourglasses" whose grains of sand are cell divisions.

Controlling the Cell Cycle

Some of the most intriguing questions in biology surround the cell cycle. One of them has challenged scientists since the discovery of cells—What controls a cell's cycle? Not until very recently have scientists begun to unlock the secrets of the cell cycle.

Using new techniques to identify and trace the movements of cell proteins, researchers have discovered that animal cells use a class of proteins called cyclins to help control different phases of the cell cycle. For example, S-cyclin helps stimulate DNA replication in the S phase. M-cyclin helps trigger mitosis. Cyclins are influenced by another class of proteins, called growth factors. Growth factors supervise and monitor a cell's progress through the phases of the cell cycle. For example, some growth factors direct a cell to enter a phase by triggering genes that make specific cyclins. Other growth factors restrain a cell's growth by blocking the action of cyclins. The joint efforts of proteins like these help keep a cell's cycle in check and prevent unnecessary cell division.

Cancer arises from a faulty cell cycle

The mechanisms controlling the cell cycle occasionally become damaged and fail. The effects can be disastrous. Without such regulation, a cell begins to grow and divide without restraint, like the cell in **Figure 6-6a**. The result is a clump of cells called a **tumor**, as shown in **Figure 6-6b**. A tumor may stay in the

Figure 6-6
a This epithelial cell (×110) has a faulty cell cycle and has grown large and misshapen.

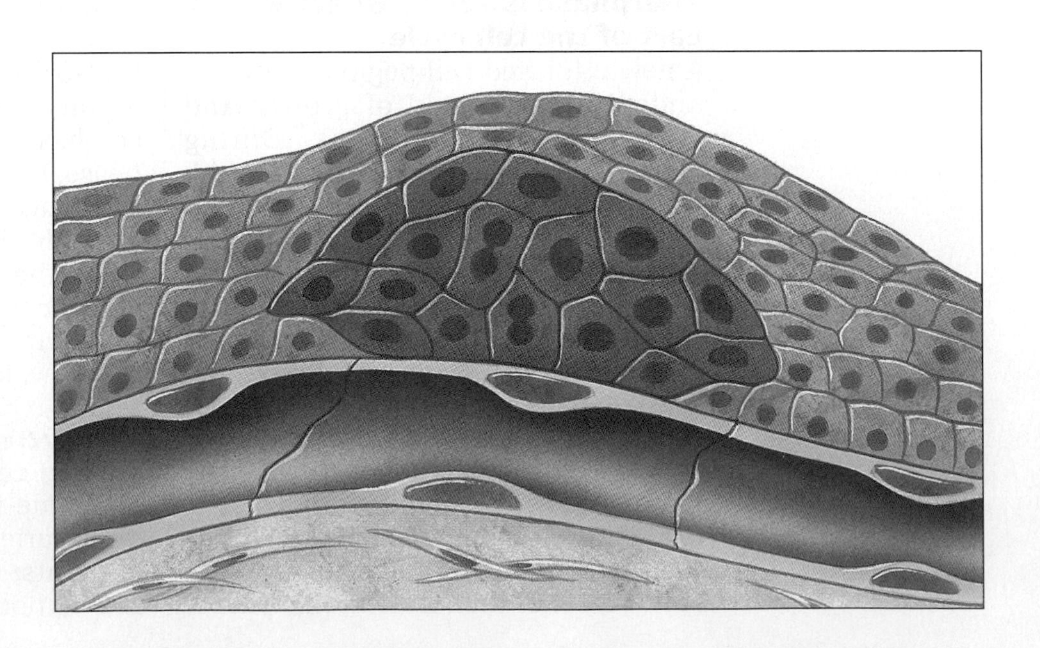

b A tumor forms when a single cell divides at an uncontrolled rate.

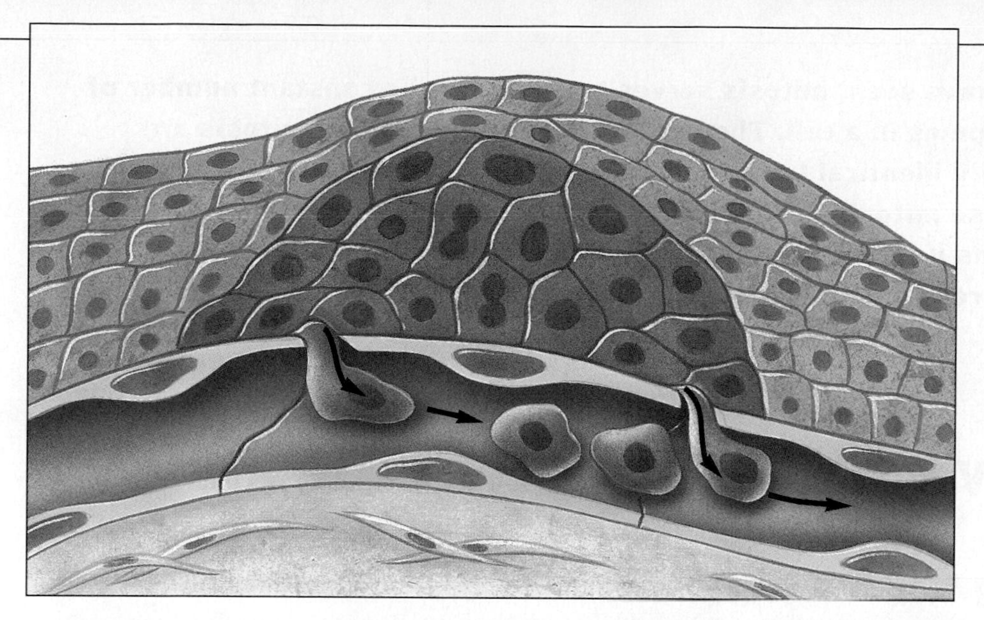

c Cells from the tumor may break away and spread to other parts of the body.

region in which it originally formed. In this case the tumor is usually harmless and is described as benign *(bee NEYN)*. Or, as shown in **Figure 6-6c**, it may continue to grow and invade other tissues, in which case the tumor is described as cancerous. **Cancer** is a term used to describe a disease characterized by uncontrolled cell division and growth. In essence, cancer results from a malfunction of the cell cycle.

A faulty cell cycle can produce many changes in a cell, making it significantly different from a normal one. For example, a cancer cell is typically very round, making it more mobile than a normal cell. And unlike a normal cell, which can usually only grow on a solid surface, a cancer cell can also grow in liquids. A cancer cell also has abnormal proteins and lipids in its outer membrane, which may prevent it from reacting to signals in its environment.

How does a normal, healthy cell become a life-threatening cancer cell? It isn't easy. Several independent changes must accumulate in the cell, such as loss of regulated cell division. In addition, the cell must be able to maneuver around and break through protective membranes in tissues. It also needs to establish a foothold in other tissues and to maintain a supply of oxygen and nutrients.

What causes these changes in a cell? By examining many cancers, scientists have discovered that cancer cells have defects in the genes regulating cell division. The cell cycle is regulated in several ways, and all must be damaged for cancer to occur. A change in a gene is called **mutation**. Because changes to genes usually occur in body tissues such as lungs or skin, cancer-causing mutations are usually not hereditary. Occasionally cell-cycle control genes become mutated in eggs and sperm, leading to an increased likelihood of developing cancer later in life.

Factors that are suspected of contributing to the development of cancer include environmental agents such as radiation and cancer-causing chemicals present in cigarette smoke. This is why some of the more significant risk factors are linked to lifestyle, including the use of tobacco products and exposure to ultraviolet (UV) radiation. Recently there has been a great deal of interest in the effects of diet and obesity on cancer rate. Although it is impossible to avoid exposure to all risk factors, it is prudent to minimize or eliminate exposure to those that you *can* control, particularly smoking.

SECTION REVIEW

1. **How is cell division in bacteria similar to that in eukaryotes? How does it differ?**
2. **How does mitosis ensure that a new cell resembles its parent cell?**
3. **What is the cell cycle? Are the cell cycles of all organisms the same? Explain.**
4. **What is cancer? What are some of the factors that may cause a cell to become cancerous?**

As you have seen, mitosis serves to maintain the constant number of chromosomes in a cell. The two new cells produced by mitosis are genetically identical to the parent cell. Although some organisms are able to use mitosis and cell division to reproduce, most multicellular organisms use another type of nuclear division to produce specialized cells for reproduction.

6-3 How Gametes Form: Meiosis

OBJECTIVES

1. **Define *gamete*, and state its function in sexual reproduction.**
2. **Explain how meiosis maintains chromosome number throughout generations.**
3. **Summarize the events of meiosis.**
4. **Define *crossing-over*, and explain its impact on evolution.**

Making Haploid Cells

At the beginning of this chapter, you learned that most of your cells are diploid and have 46 chromosomes. Gametes are the exception; they are haploid and have 23 chromosomes. A **gamete** is a reproductive cell—an egg or a sperm. Gametes are found only in organisms that reproduce sexually, as do most animals. In sexual reproduction, two gametes fuse to form the first cell of a new individual, as shown in **Figure 6-7**. If gametes were diploid instead of haploid, the new individual would have twice as many chromosomes as its parents. Imagine how this number would continue to increase with each new generation.

Gametes are formed by a type of nuclear division called meiosis *(my OH sihs)*. **Meiosis** is a two-stage form of nuclear division in which the chromosome number is halved. The first meiotic division (meiosis I) reduces the number of chromosomes by half. The second meiotic division (meiosis II) separates identical copies of chromosomes. Like mitosis, meiosis is usually followed by cell division. Study the details of meiosis in *A Closer Look: Meiosis*, on the next page.

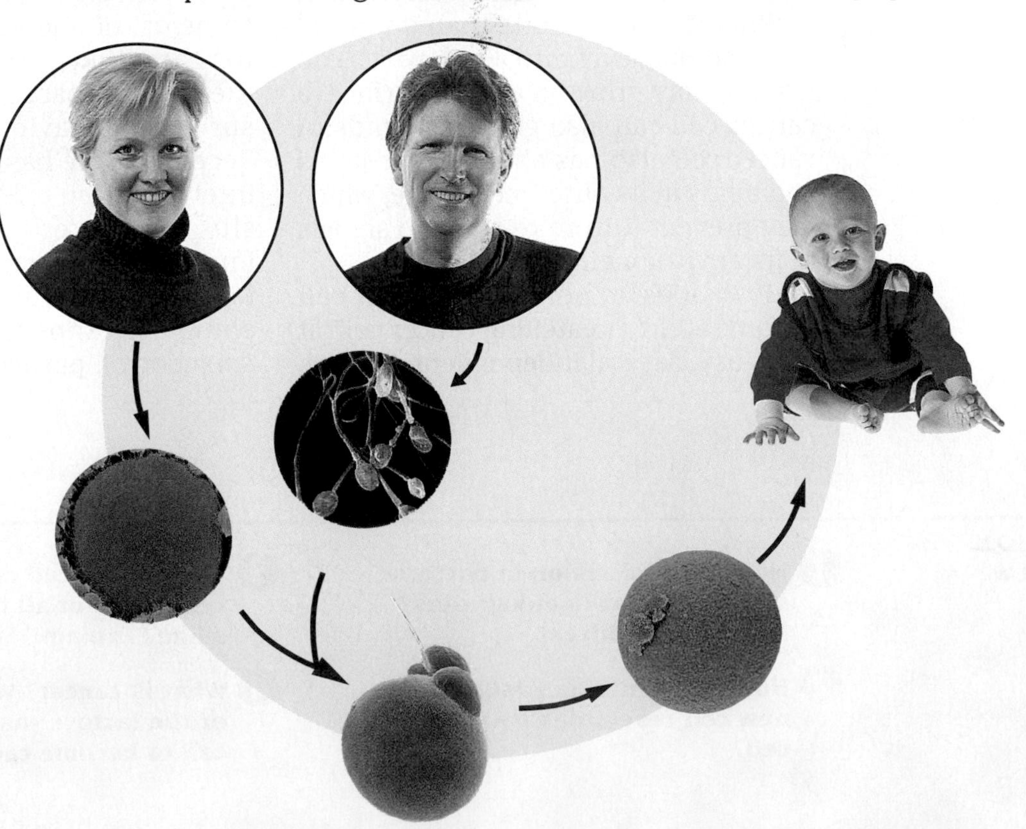

Figure 6-7
Most of your life cycle is spent as a diploid organism. Only certain cells—gametes—are haploid.

Meiosis

In meiosis, eukaryotes produce reproductive cells that contain one-half as many chromosomes as the parent cell.

A Closer Look

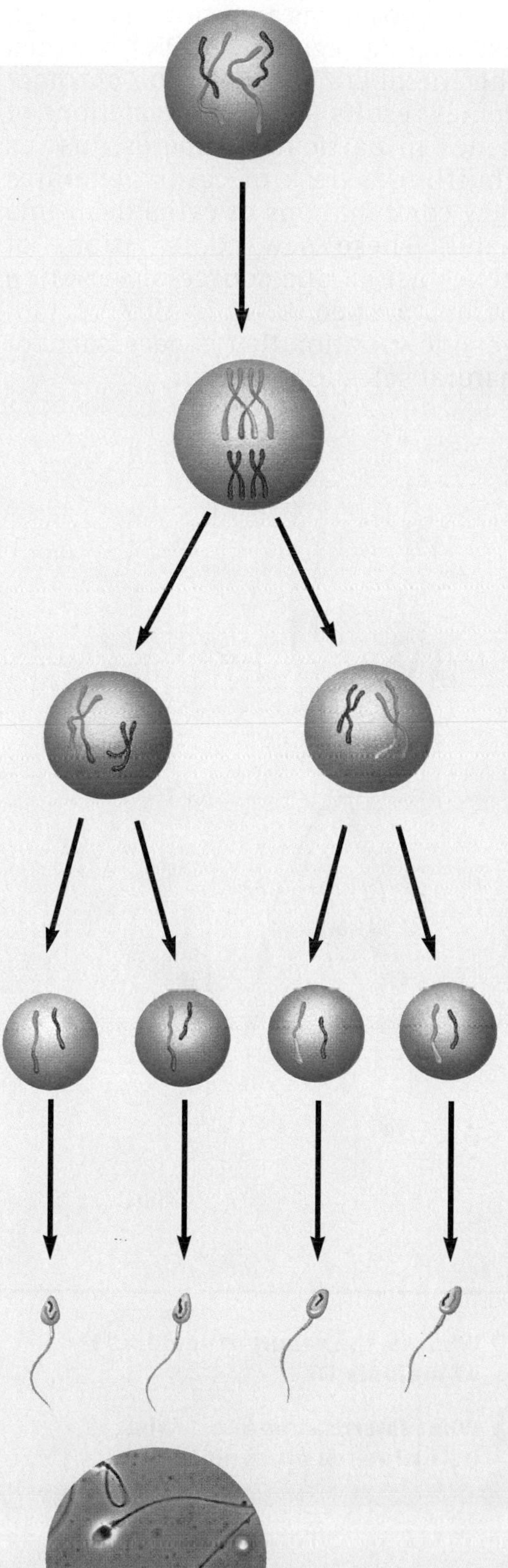

1 The diploid ($2n$) cell nuclei shown here each have four chromosomes. One member of each chromosome pair is from one parent; the other is from the other parent.

2 First the amount of DNA doubles. Then the similar chromosomes pair with one another. Occasionally, paired chromosomes will exchange segments in a process called crossing-over.

3 The first meiotic division (meiosis I) separates the members of each pair of chromosomes. Meiosis I reduces the number of chromosomes in the cell by one-half.

4 The second meiotic division (meiosis II) separates the two copies of each chromosome (the chromatids). Cell division now occurs, producing four haploid (n) cells.

5 During sperm formation, the four cells containing these nuclei develop heads and tails. During egg formation, only one of the cells containing these nuclei becomes a mature egg; the other three cells usually die.

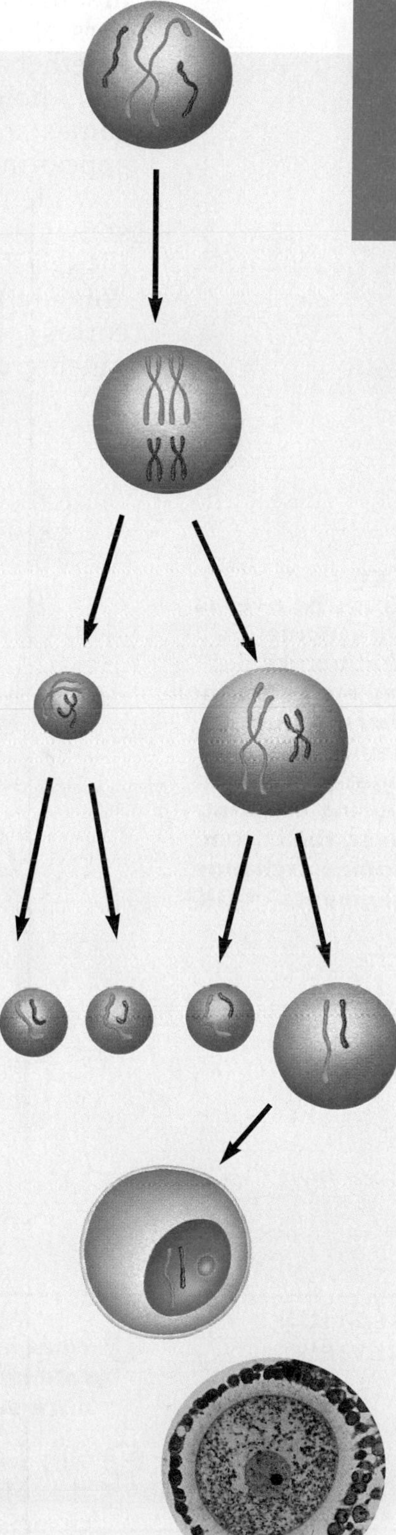

The Importance of Crossing-Over

When two haploid gametes unite in fertilization, the resulting cell is diploid. It received one of each type of chromosome from the egg of the female parent and one of each type from the sperm of the male parent. During the first division of meiosis, the two members of each chromosome pair move together and line up side by side. Proteins hold the two pairs of chromosomes so closely together that they appear indistinguishable.

While paired together, the chromosomes may exchange segments of DNA, as the two chromosomes are doing in **Figure 6-8**. This exchange of corresponding segments of DNA is called **crossing-over**. The importance of crossing-over lies in its consequences: the two sets of chromosomes are no longer identical. Each chromosome now contains genetic material from each parent.

From the perspective of evolution, crossing-over has an enormous impact. Exchanging segments of DNA between the members of a pair of chromosomes results in new combinations of genes in particular gametes, just as shuffling a deck of cards generates new combinations of cards dealt in a hand. These new combinations of genes act as one source of variation within a species. It is this sort of genetic variation that is necessary for natural selection to occur.

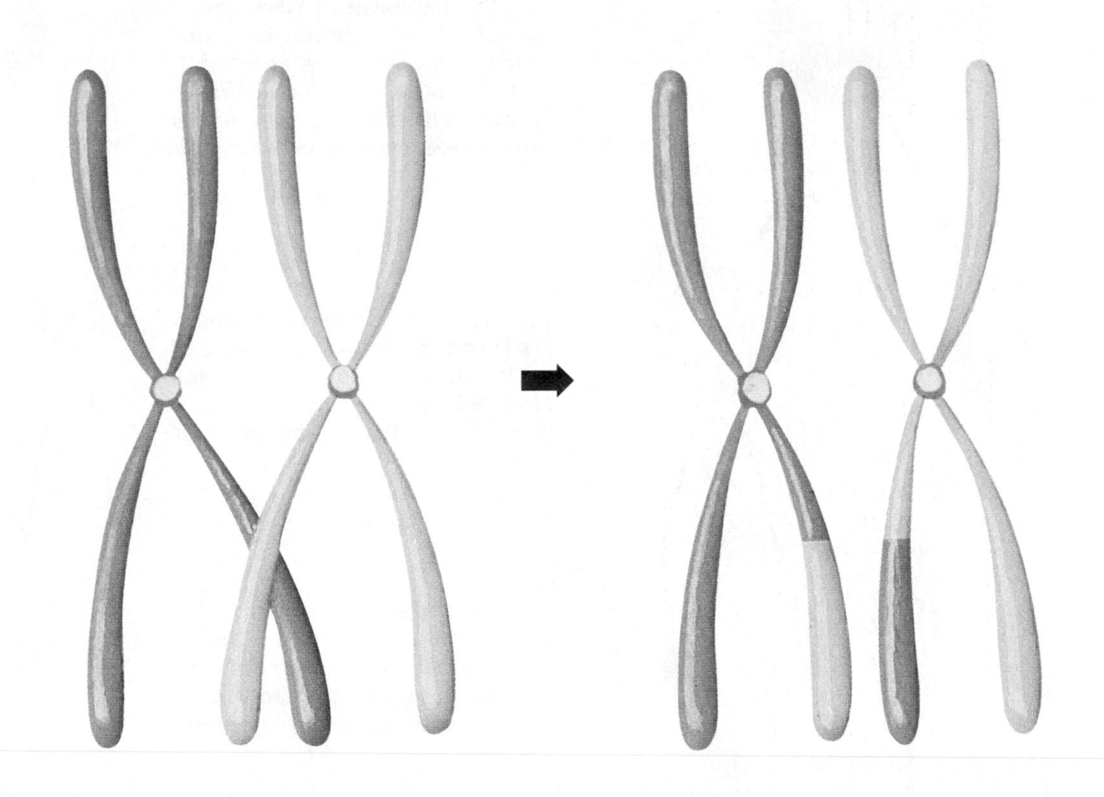

Figure 6-8
Crossing-over is a phenomenon that occurs during meiosis, when pairs of chromosomes are held tightly together. During crossing-over, the chromosomes exchange segments of DNA.

SECTION REVIEW

1. **What is a gamete? What is its role in organisms that reproduce sexually?**
2. **Why will meiosis ensure that offspring have the same chromosome number as parents?**
3. **What is the result of meiosis I? of meiosis II?**
4. **What is crossing-over? What is its influence on evolution?**

CHAPTER

6 Highlights

These identical twins grew from the same fertilized egg cell, which separated into two independent cells after its first division.

6-1 Chromosomes

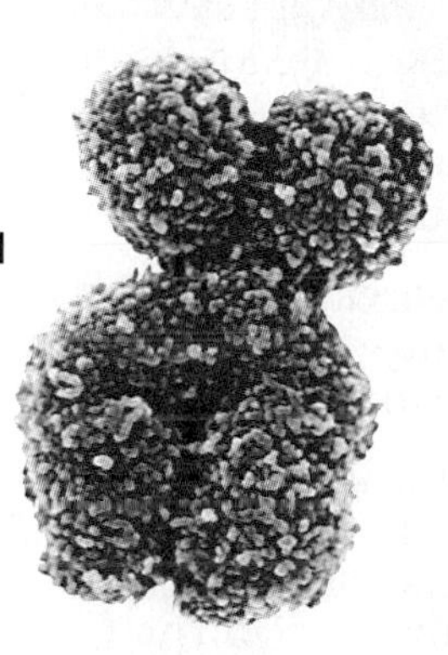

Chromosomes are composed of DNA and associated proteins.

Key Terms

chromosome (p. 101)
chromatin (p. 101)
karyotype (p. 103)
diploid (p. 103)
haploid (p. 103)
sex chromosome (p. 103)

Summary

- Chromosomes carry genetic information from parent to offspring.
- A eukaryotic chromosome is made of DNA wrapped around protein. A prokaryotic chromosome is a single circle of DNA.
- Before a eukaryotic cell divides, a chromosome condenses into a compact rod-shaped body that can be easily moved around the cell.

6-2 Mitosis and Cell Division

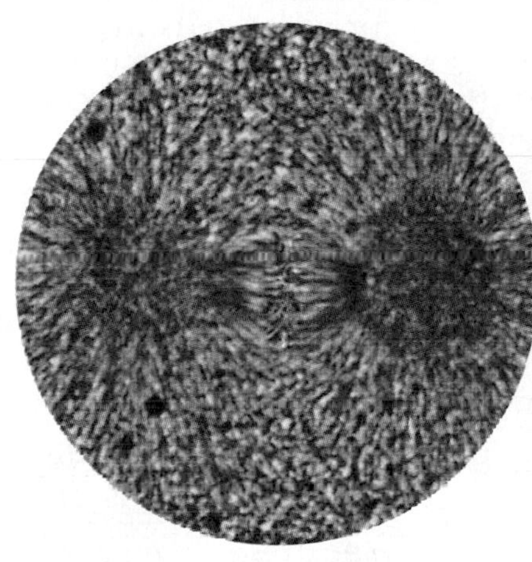

This cell is in the middle of mitosis; its chromosomes have been copied and are preparing to separate.

Key Terms

mitosis (p. 105)
cytokinesis (p. 105)
cell cycle (p. 107)
tumor (p. 108)
cancer (p. 109)
mutation (p. 109)

Summary

- Before it can divide, a cell must make a copy of its DNA.
- Bacteria divide by splitting in half. Eukaryotic cells divide through mitosis (division of the nucleus) followed by cytokinesis (division of the cytoplasm).
- In mitosis, the nucleus of a cell is divided into two nuclei, each with the same number of chromosomes as the parent cell. In cytokinesis, the cytoplasm forms two distinct cells.
- A cell undergoes an interval of growth and division called the cell cycle.
- Cancer is a life-threatening disease that results from unregulated cell growth.

6-3 How Gametes Form: Meiosis

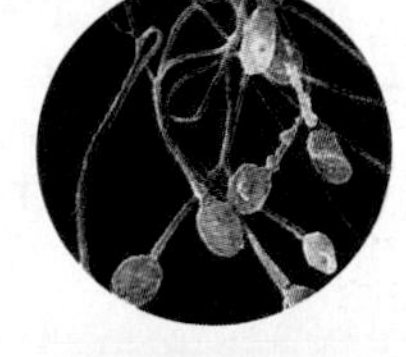

Meiosis results in the formation of haploid gametes: eggs or sperm.

Key Terms

gamete (p. 110)
meiosis (p. 110)
crossing-over (p. 112)

Summary

- A gamete is a haploid reproductive cell, such as an egg or a sperm cell.
- Meiosis is a type of nuclear division that reduces the number of chromosomes in a cell by half.
- Crossing-over is the exchange of segments of DNA between homologous chromosomes. It is a source of genetic variation.

CHAPTER 6

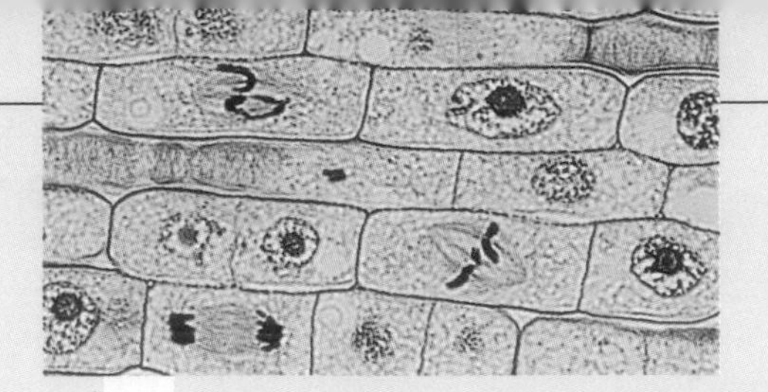

review

Understanding Vocabulary

1. For each pair of terms, complete the analogy.
   a. 23 chromosomes : haploid :: 46 chromosomes : ____
   b. XX chromosomes : female :: XY chromosomes : ____
   c. nuclear division : mitosis :: cell division : ____

2. Using each set of words below, write one or more sentences summarizing information learned in this chapter.
   a. chromatin, chromosomes, karyotype
   b. prophase, metaphase, anaphase, telophase
   c. mitosis, cytokinesis, cell cycle, interphase

Understanding Concepts

3. **Relating Concepts** Construct a concept map that describes cell division. Try to include the following terms in your map: cell cycle, interphase, nucleus, chromosomes, cytokinesis, mitosis, and metaphase.

4. **Identifying Structures** What is chromatin? How does its appearance change before a cell divides?

5. **Comparing Structures** How does the shape of a chromosome change prior to cell division?

6. **Comparing Processes** What is cytokinesis? How does it differ in plant and animal cells?

7. **Recognizing Relationships** What are sex chromosomes? Why are they significant in cell reproduction?

8. **Summarizing Data** Who determines the sex of a child—the mother or the father? Explain your answer.

9. **Relating Conclusions** Why must a cell make a copy of the genetic information contained within its chromosomes before it divides?

10. **Summarizing Information** Describe what happens during each of these stages of mitosis: prophase, metaphase, anaphase, telophase.

11. **Identifying Functions** Describe the events that result in the formation of two complete nuclei in a eukaryotic cell. What usually happens, following division of eukaryotic nuclei in most cells?

12. **Recognizing Relationships** What are homologous chromosomes? Are they present in prokaryotic or eukaryotic cells?

13. **Organizing Information** What happens to a cell during interphase? Summarize the events that take place during the G_1 phase, the S phase, and the G_2 phase. Explain why interphase is the most active part of the cell cycle.

14. **Comparing Functions** Why do you think it is important for some cells in your body, such as skin cells and those that line your intestine, to have shortened cell cycles and divide almost continuously?

15. **Comparing Structures** Describe three ways cancer cells are different from normal cells.

16. **Summarizing Information** What is a gamete? How does it function in sexual reproduction?

17. **Inferring Conclusions** Why is meiosis sometimes called reduction division?

18. **Analyzing Results** If you are observing a dividing living cell and you see homologous chromosomes paired along the equator of the cell, are you observing mitosis or meiosis? Which stage of nuclear division is occurring?

19. **BUILDING ON WHAT YOU HAVE LEARNED** In Chapter 3 you learned that the surface-area-to-volume ratio in a cell is very important to the cell's survival. Based on what you know about that relationship, explain this statement: When a cell reaches a certain size, it will either stop growing or it will divide.

Interpreting Graphics

20. Look at the figure below of the human life cycle. Indicate the number of chromosomes at each stage of the life cycle.

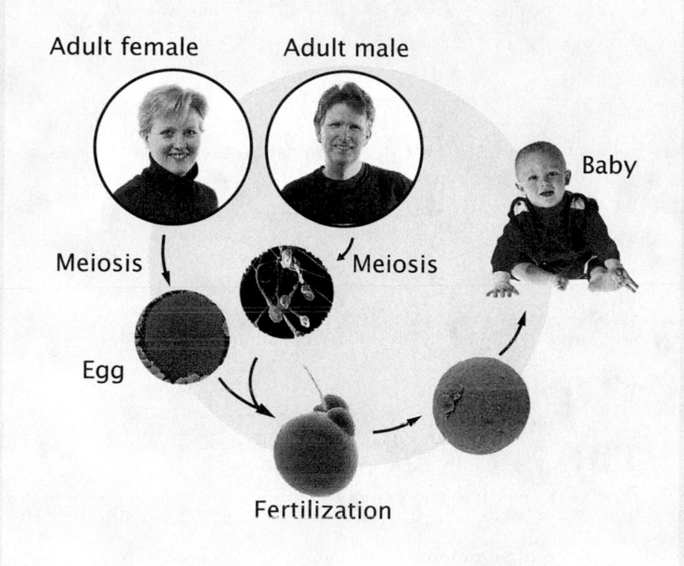

Reviewing Themes

21. ***Heredity***
How does mitosis promote genetic consistency?

22. ***Evolution***
Bacteria simply split, but eukaryotic cells reproduce by mitosis. How has the evolution of mitosis in eukaryotic cells aided their reproduction?

23. ***Evolution***
How does crossing-over during meiosis contribute to genetic diversity?

Thinking Critically

24. **Distinguishing Fact From Opinion** A classmate, after studying the cell cycle, concluded that the same cell goes through the cell cycle more than once. Do you agree or disagree? Support your answer.

25. **Evaluating Results** A genetic engineer was conducting studies on a plant with a haploid number of 12. After two gametes were joined together, the offspring had 36 chromosomes. What do you think happened to cause this result?

Life/Work Skills

26. **Applying Technology** Physicians may offer the use of amniocentesis or chorionic villus sampling to women who are pregnant. Investigate how these procedures are performed, and what technology is required. With a classmate, research the information that can be obtained by these two methods.

27. **Communicating** Research the link between cigarette smoking and different kinds of cancer. Communicate your results to your classmates by making an oral presentation that is supported by graphics, such as photos, charts, or graphs that show the number of cancer cases over a period of time.

Activities and Projects

28. **Career Connection: Cytotechnologist** A cytotechnologist is a person trained to observe abnormalities in cells that might indicate cancer or some other disease. Use library references to find out about becoming a cytotechnologist. What kind of equipment do cytotechnologists use? Where do they work? What kind of and how much education is required? Prepare a written report summarizing your findings.

Discovering Through Reading

29. Read the article "Strange Genes" in *Discover*, January 1996, p. 33. What is the "werewolf syndrome"? What common disorders are geneticists hoping to alleviate?

30. Read the article "By a Thousand Cuts" by Lori Oliwenstein in *Discover*, February 1993, pages 24–25. What is cellular senescence? What are telomeres? What happens to the telomeres each time a cell divides? What is the function of the enzyme telomerase, and how does its function contribute to cells becoming cancer cells?

31. Read the article "Stopping Cancer in Its Tracks" by J. Madeleine Nash in *Time*, April 25, 1994, pages 54–61. How does the gene p53 stop the formation of tumors?

CHAPTER 7

REVIEW
- karyotype (Section 4-1)
- cystic fibrosis (Section 4-3)
- meiosis (Section 6-3)
- crossing-over (Section 6-3)

Genetics and Inheritance

The enormous diversity in the human population is a result of genetic interactions.

Try to recall the last news story about genetics that you read or heard. Did it describe a new kind of genetically altered food or drink now available in grocery stores? Did it hail the discovery of a gene linked to cancer? Such stories are becoming commonplace these days. However, the events leading up to our current knowledge are not much older than your great-grandparents.

7-1 The Work of Gregor Mendel

OBJECTIVES

1. **Outline the garden-pea experiments performed by Gregor Mendel.**
2. **Explain how Mendel's approach enabled him to reach his conclusions.**
3. **Define the following terms: gene, allele, dominant, recessive, homozygous, heterozygous, genotype, and phenotype.**
4. **Relate Mendel's two laws of inheritance to the behavior of chromosomes during meiosis.**

Gregor Mendel and the Garden Pea

The scientific study of heredity is called **genetics** *(juh NEHT ihks)*. Its story begins in the 1860s in the gardens of a monastery in Brünn, Austria (which is now Brno in the Czech Republic). There, a monk named Gregor Mendel, shown in **Figure 7-1**, was performing a series of experiments on ordinary garden peas, *Pisum sativum*. Trained in the sciences and mathematics, Mendel was curious about patterns of heredity and was studying traits in generations of pea plants. Like others before him, he had noticed that when a pea plant with purple flowers was bred, or crossed, with a pea plant with white flowers, all the offspring had purple flowers. Then, if two of these purple-flowered offspring were crossed, the white flowers reappeared in some of the pea plants in the next generation.

Earlier researchers had tried to unravel this mystery of heredity in garden peas. Mendel repeated their efforts but added an important twist—he used a mathematical approach and counted the number of kinds of offspring in each generation of pea plant. Quantitative approaches to science—those that include measuring and counting—were just becoming popular in Europe, so Mendel was on the cutting edge of research.

You might be wondering why Mendel and others studied the garden pea. The garden pea is a good subject for genetic study. These small, delicate vines are easy to grow, produce large numbers of offspring, and mature quickly. Garden-pea plants also have their male and female reproductive parts enclosed in the same flower, as shown in **Figure 7-1b**. This meant that Mendel could allow a flower to fertilize itself through a process called self-fertilization. Or he could intervene and transfer the pollen from one flower to another flower on a different plant, a process called cross-fertilization.

Figure 7-1

a Modern genetics began with the garden-pea experiments performed by Gregor Mendel in the 1860s.

b The flower of the garden pea has both male and female reproductive parts, enabling the flower to fertilize itself.

Mendel's Experiments

Mendel carried out his experiments with garden peas in three steps, illustrated below in **Figure 7-2**.

Step 1

First Mendel produced pure-breeding strains of pea plants. He did this by allowing a population of pea plants to self-fertilize for many generations. He then collected and grew the seeds from these plants. These seeds gave rise to offspring that produced only one form of a particular trait. For example, the purple-flowering strain produced only purple flowers and the white-flowering strain produced only white flowers. Mendel called these pure-breeding offspring the parental generation, or, in abbreviated form, P generation.

Step 2

Next Mendel crossed the two different varieties in the P generation. For example, he took pollen from the anthers of purple flowers and placed it on the stigmas of white flowers. He then collected and grew the seeds from these plants. Mendel called the offspring that grew from the seeds the first filial generation, or F_1 generation for short. In the F_1 generation, Mendel found only purple flowers. White flowers did not appear.

Step 3

Finally Mendel allowed the F_1 generation plants to self-fertilize. As before, he grew the seeds. He called the offspring produced by these seeds the second filial generation, or F_2 generation for short. After noticing that white flowers reappeared in the F_2 generation, Mendel applied his mathematical approach. Out of the 929 F_2 individuals, he counted 705 plants with purple flowers and 224 plants with white flowers. He calculated that approximately three-fourths of the F_2 plants had purple flowers, a 3:1 ratio.

Mendel studied more than flower color in garden peas. He studied other traits that, like the purple and white flower color, appeared in contrasting forms. Several of these traits are listed in **Table 7-1**. For each pair of traits, Mendel counted the number of F_2 individuals exhibiting contrasting forms of each trait.

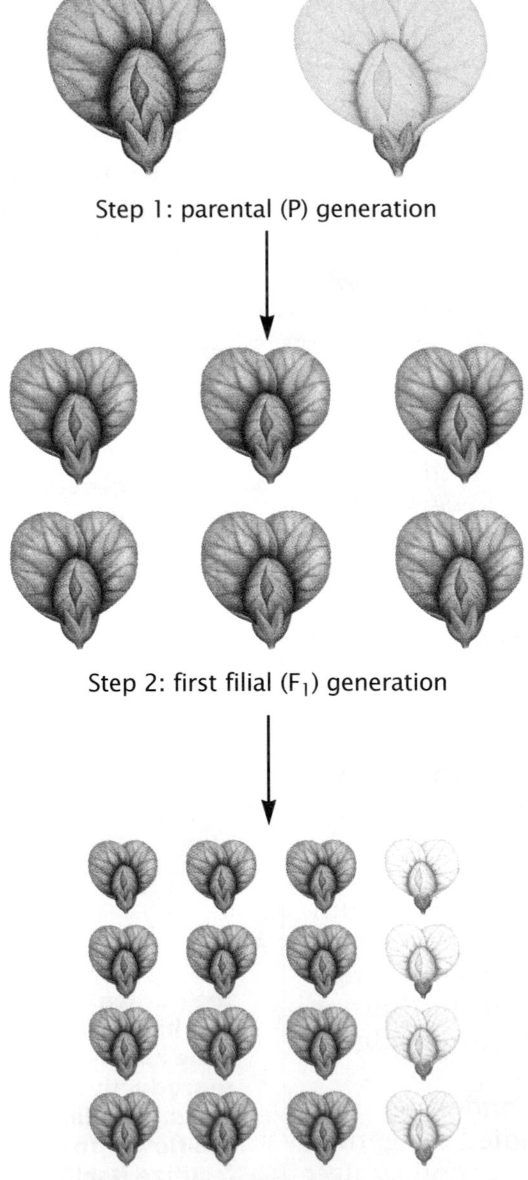

Figure 7-2
Mendel's garden-pea experiments consisted of three steps, each of which involved either self-fertilization or cross-fertilization.

Table 7-1 Contrasting Traits in Pea Plants

Trait	Contrasting forms of traits		F_2 generation ratio
Flower color	Purple	White	705:224
Seed color	Yellow	Green	6,022:2,001
Seed shape	Round	Wrinkled	5,474:1,850
Pod color	Green	Yellow	428:152

Mendel developed a model to explain his results

Each time Mendel counted the number of F_2 individuals exhibiting contrasting forms of a trait, he obtained a 3:1 ratio. Surely this ratio must mean something important. To explain his results, Mendel came up with a simple model, a set of rules that could be used to accurately predict patterns of heredity. These rules summarize Mendel's ideas about inheritance.

1. Parents transmit information about traits to their offspring. Mendel called this information "factors."

2. Each individual has two factors for each trait, one from each parent. The two factors may or may not have the same information. If the two factors do have the same information (for example, if both have information for purple flowers), the individual is said to be **homozygous** *(hoh muh ZY guhs)* for the trait. If the two factors have different information (for example, one factor codes for purple flowers and the other for white flowers), the individual is said to be **heterozygous** *(heht uh roh ZY guhs)*.

3. The alternative forms of a factor are called **alleles** *(uh LEELS)*. The many alleles that an organism possesses make up its **genotype** *(JEE nuh typ)*. An organism's physical appearance, which is determined by its alleles, is called its **phenotype** *(FEE nuh typ)*.

4. An individual possesses two alleles for each trait. One allele is contributed by the female parent, and the other is contributed by the male parent. The two alleles in each pair are not affected by each other. They are passed on when an individual matures and produces gametes (eggs and sperm). During the formation of gametes, the paired alleles segregate (separate) randomly so that a gamete receives a copy of one allele or the other.

5. The presence of an allele does not ensure that the trait will be expressed in the individual that carries it. In heterozygous individuals, only the **dominant** allele achieves expression. The **recessive** allele is present but remains unexpressed. In Mendel's F_1 generation, purple flower color was caused by a dominant allele. For every pair of contrasting forms of a trait—tall versus short, or green seeds versus yellow seeds—the allele for one form of the trait was always dominant and the allele for the other form of the trait was always recessive.

Visualizing Mendel's Model

Heredity

Explain how a form of a trait that appears in both parents could be absent from their children.

Mendel's model can be understood easily by diagramming Mendel's crosses. For example, consider Mendel's cross of parental (P) purple-flowering pea plants with parental white-flowering pea plants. For the sake of convenience, letters are used to represent alleles. The recessive allele is represented by the lower case letter "w" (for "white"), and the dominant allele is represented by the corresponding upper case letter "W." A plant that is true-breeding for the recessive white flower color would be designated "ww," and a plant that is true-breeding for the dominant purple flower color would be designated "WW." A heterozygote would be designated "Ww."

A simple diagram called a Punnett square can help you visualize crosses. Named for the British geneticist Reginald Punnett, a Punnett square is also a handy device for predicting the results of a cross.

In a Punnett square, the symbols for all the possible alleles carried by male gametes (sperm) are arranged along the top of the square, while all the possible alleles carried by the female gametes (eggs) are shown along the left side. By combining the symbol for an allele carried by a male gamete with the symbol for the allele carried by the female gamete, all the possible gamete combinations can be predicted, as shown in **Figure 7-3**.

Figure 7-3 This Punnett square illustrates a cross between two true-breeding varieties of garden pea plants. This was Step 2 of Mendel's experiment.

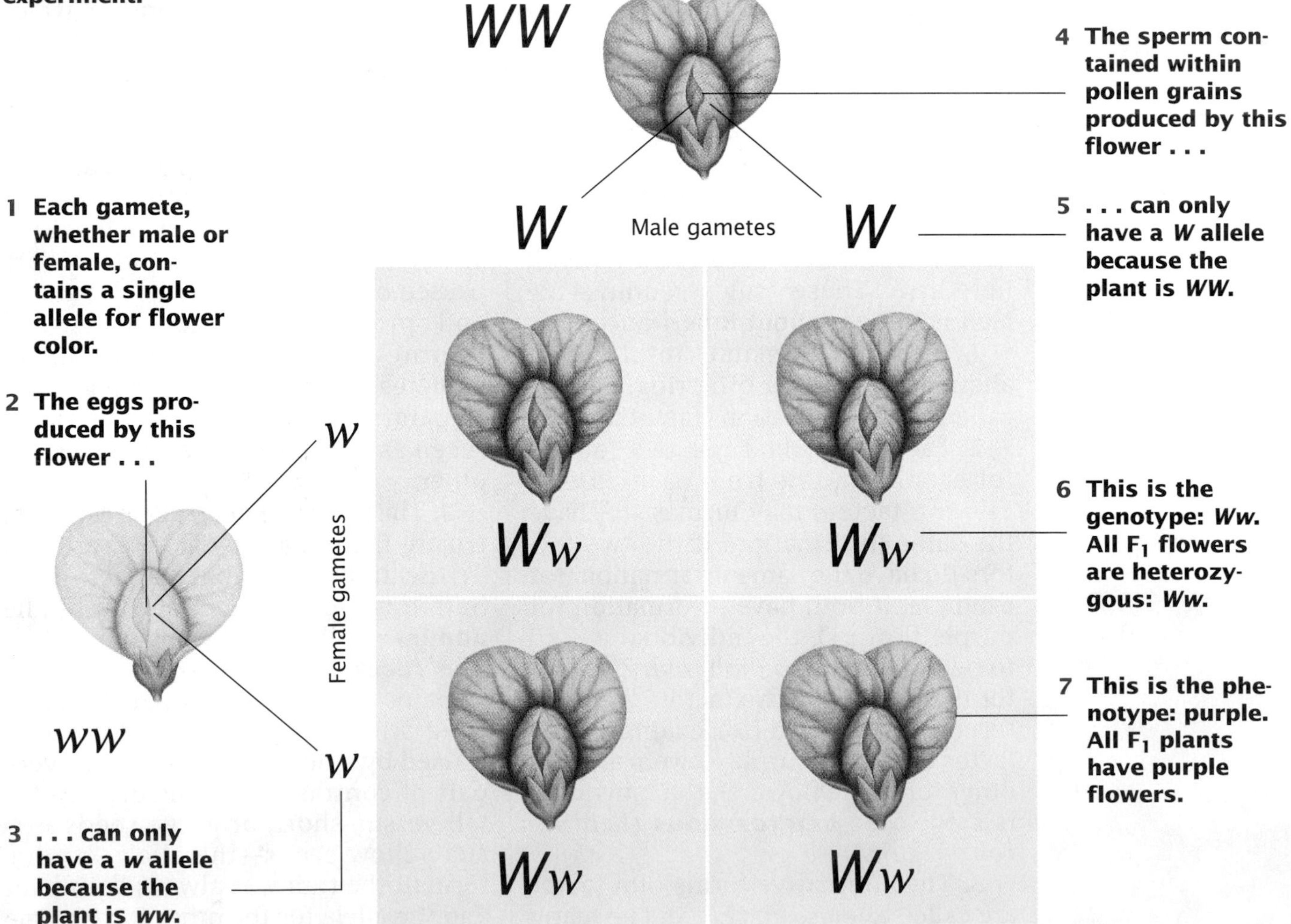
WW
W
Male gametes
W
Female gametes
w
w
ww
Ww
Ww
Ww
Ww
1 Each gamete, whether male or female, contains a single allele for flower color.
2 The eggs produced by this flower . . .
3 . . . can only have a w allele because the plant is ww.
4 The sperm contained within pollen grains produced by this flower . . .
5 . . . can only have a W allele because the plant is WW.
6 This is the genotype: Ww. All F1 flowers are heterozygous: Ww.
7 This is the phenotype: purple. All F1 plants have purple flowers.

Crossing two heterozygous plants

When heterozygous *(Ww)* F_1 individuals with purple flowers are allowed to self-fertilize, what will the resulting F_2 individuals be like? Let's predict what we would expect to find using Mendel's rules. To make a prediction, consider what scientists call **probability**. Probability is simply the likelihood that something will happen. For example, when you toss a coin, the probability that the coin will land with "heads" up is 50 percent, or one-half, because the coin, is just as likely to fall showing "tails." Similarly, because of the segregation of alleles, the probability that an egg or sperm cell in one of Mendel's heterozygous F_1 pea plants *(Ww)* will contain the allele *W* is 50 percent, or one-half, just as in a coin toss. Obviously, the reverse is also true: the probability that a gamete will contain the allele *w* is also 50 percent or one-half.

Study the Punnett square for a cross involving two purple-flowered heterozygous plants *(Ww)* shown in **Figure 7-4**. In this particular cross, one-half of the male and female gametes carry the allele for purple flowers *(W)*. The other one half of the gametes carry the allele for white flowers *(w)*. You can see that Mendel's model clearly predicts that 75 percent of the F_2 generation will have purple flowers and 25 percent will have white flowers, a 3:1 ratio. The Punnett square also shows the genotypes found in the F_2 generation. Twenty-five percent of the F_2 is homozygous *(ww)* with white flowers, 50 percent heterozygous *(Ww)* with purple flowers, and the remaining 25 percent is homozygous *(WW)* with purple flowers. The 3:1 ratio Mendel had repeatedly observed is the expression of an underlying 1:2:1 ratio of genotypes in which the heterozygotes look like one of the homozygotes.

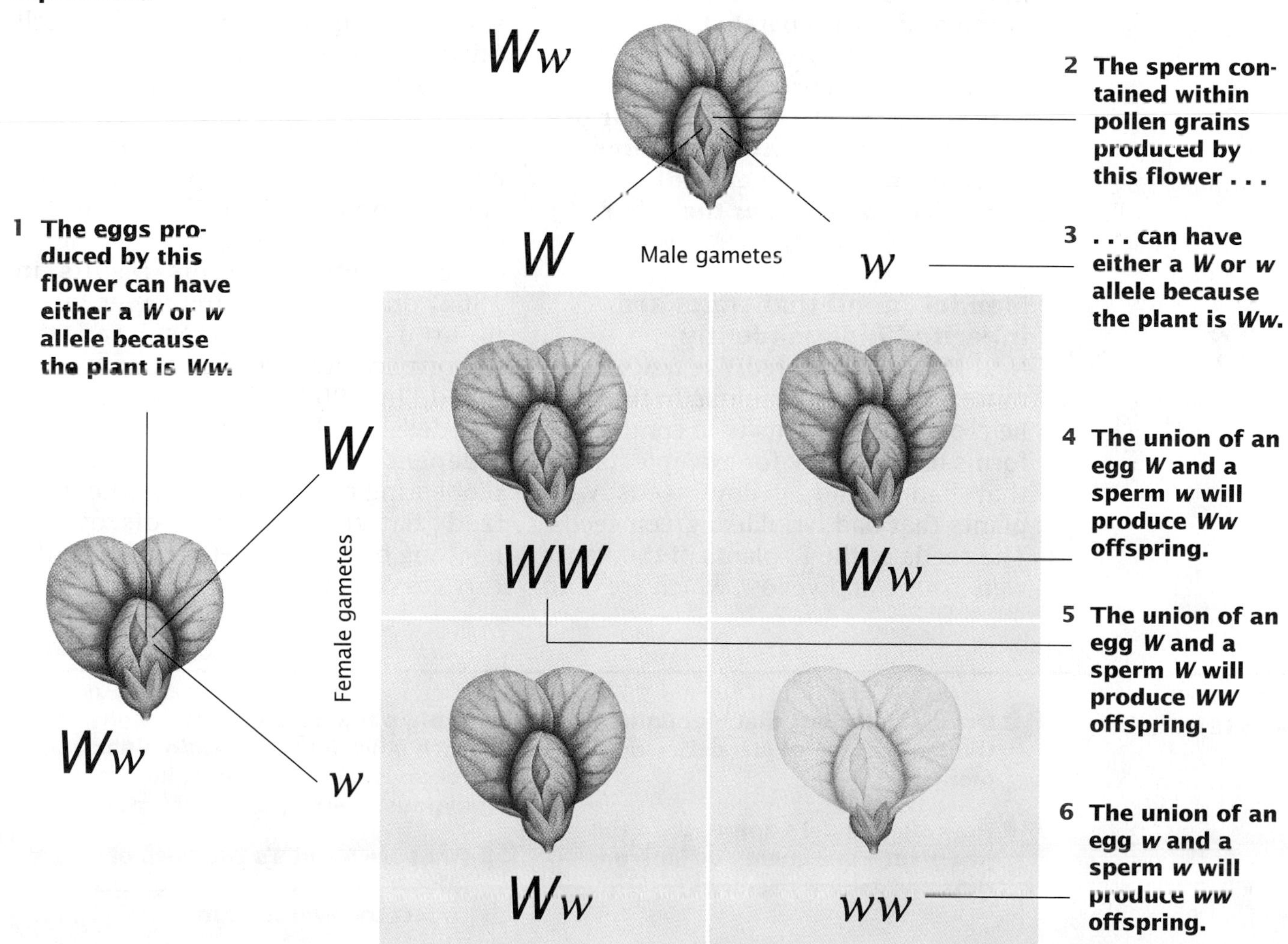

Ww
2 The sperm contained within pollen grains produced by this flower . . .
W
Male gametes
w
3 . . . can have either a W or w allele because the plant is Ww.
1 The eggs produced by this flower can have either a W or w allele because the plant is Ww.
W
Female gametes
WW
Ww
4 The union of an egg W and a sperm w will produce Ww offspring.
5 The union of an egg W and a sperm W will produce WW offspring.
Ww
w
Ww
ww
6 The union of an egg w and a sperm w will produce ww offspring.

Figure 7-4 This Punnett square illustrates a cross between two heterozygous F_1 individuals. This was Step 3 of Mendel's experiment.

Evolution

How does the law of independent assortment provide a source of variation from one generation to the next?

Mendel's Conclusions

Studying the results of his crosses, Mendel realized that in each case one form of each trait did not appear in the F_1 generation. The missing form of each trait then reappeared in approximately one-fourth of the plants in the F_2 generation. Every F_2 generation consisted of the contrasting trait forms in an approximate 3:1 ratio. Surely, this ratio was significant.

Mendel saw that these results could be explained if a pea plant had two hereditary factors for each trait. He concluded that a plant receives one hereditary factor from one parent and the other factor from the other parent. In turn, the plant passes on one of its two factors to each of its offspring. Today this conclusion is referred to as Mendel's first law, or the law of segregation. The law of segregation describes the behavior of chromosomes during meiosis, when homologous chromosomes and then chromatids are separated.

The law of segregation states that the members of each pair of alleles separate when gametes are formed. A gamete will receive one allele or the other.

Mendel found that traits are inherited independently

Next Mendel wondered, Are traits transmitted as individual units? To find out, he crossed pea plants with contrasting forms of two traits: for example, plants that had round, yellow seeds with plants that had wrinkled, green seeds. The seeds of the F_1 plants of this cross were round and yellow, which are both dominant traits. When these plants self-fertilized, the F_2 generation had four different phenotypes: 315 had round, yellow seeds, 101 had wrinkled, yellow seeds, 108 had round, green seeds, and 32 had wrinkled, green seeds. Mendel calculated that these numbers represent a 9:3:3:1 ratio. These results told him that the genes for two different traits—seed color and seed form—don't necessarily stay together. Half of the time they separate and join in new combinations. This idea is the basis for Mendel's second law, the law of independent assortment.

The law of independent assortment states that two or more pairs of alleles segregate independently of one another during gamete formation.

The law of independent assortment reflects the way homologous chromosomes pair up in the middle of the cell during meiosis I. This law applies to traits whose genes are found on different pairs of homologous chromosomes or far apart on the same chromosome pair. (Keep in mind that chromosomes had not yet been discovered in Mendel's time.)

Mendel published his results in 1866. Unfortunately, his paper failed to arouse much interest, and the importance of his work was not recognized. In 1900, sixteen years after Mendel's death, several investigators independently rediscovered Mendel's pioneering paper. Scientists soon realized that chromosomes, discovered not long before Mendel's death, are the carriers of heredity.

SECTION REVIEW

1. **Draw a flowchart that highlights the three steps of Mendel's experiments.**
2. **How did Mendel's approach to the pea-plant experiments differ from those of earlier scientists?**
3. **Explain the difference between each pair: gene, allele; dominant, recessive; homozygous, heterozygous; genotype, phenotype.**
4. **What are Mendel's two laws of inheritance? How do these laws reflect the events of meiosis?**

For more than half a century after Mendel's work was rediscovered, scientists sought a deeper understanding of genetics. Soon they discovered that most traits follow patterns of inheritance far more complex than those proposed by Mendel. Since then, genetics has moved to the forefront of scientific research.

7-2 Patterns of Inheritance

OBJECTIVES

1. **Recognize the relationship between the laws of probability and inheritance.**
2. **Use a Punnett square to predict the outcome of a cross.**
3. **Describe three other factors that influence inheritance.**

Predicting the Outcome of a Cross

Most of Mendel's conclusions are based on the fact that genes combine according to the rules of probability. When a pea plant reproduces, its pair of alleles for each gene segregate into separate cells during meiosis. As a result, the probability that an egg or a sperm cell will contain one or the other allele is 50 percent or one-half.

A simple Punnett square neatly illustrates a monohybrid *(mah noh HY brihd)* cross, which is a cross focusing on only one trait, such as flower color. **Figure 7-5** shows a monohybrid cross between a homozygous purple-flowering plant and a homozygous white-flowering plant and reveals the possible offspring.

Figure 7-5
This Punnett square illustrates a monohybrid cross between a homozygous purple-flowering plant and a homozygous white-flowering plant.

1 This purple flower is homozygous dominant: *WW*. Each gamete it produces has a *W* allele.

W *W*

2 This white flower is homozygous recessive: *ww*. Each gamete it produces has a *w* allele.

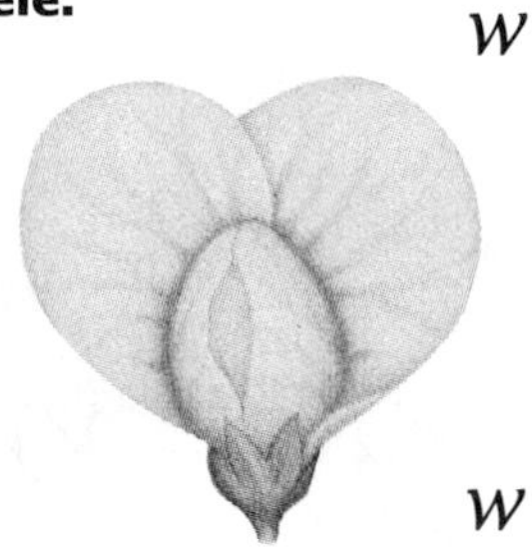

w

w

3 All the offspring of this cross have purple flowers. Their genotype is *Ww*.

Visualizing a Dihybrid Cross

The Punnett square in **Figure 7-6** illustrates a dihybrid *(dy HY brihd)* cross. A dihybrid cross shows two traits of interest. Notice that the Punnett square for a dihybrid cross contains 16 boxes. These additional boxes are necessary to account for all possible gene combinations.

To illustrate a dihybrid cross, reconstruct one of Mendel's experiments, in which he studied the inheritance of two traits. Mendel crossed plants that were homozygous for yellow, round seeds with plants that were homozygous for green, wrinkled seeds. The F_1 generation offspring produced yellow and round seeds, indicating that these were the dominant traits. Thus, let Y = yellow and y = green, and R = round and r = wrinkled. When these seeds were grown into plants that were allowed to self-fertilize, the offspring shown in **Figure 7-6** resulted. Recall Mendel's law of independent assortment, which states that members of pairs of alleles segregate independently into gametes. This means that the parental plants shown in **Figure 7-6** will produce four kinds of gametes, which are arranged along the top and the left side of the Punnett square. When the boxes of the Punnett square are filled, each box will hold four symbols. How many phenotypes are shown among the offspring of the dihybrid cross in **Figure 7-6**?

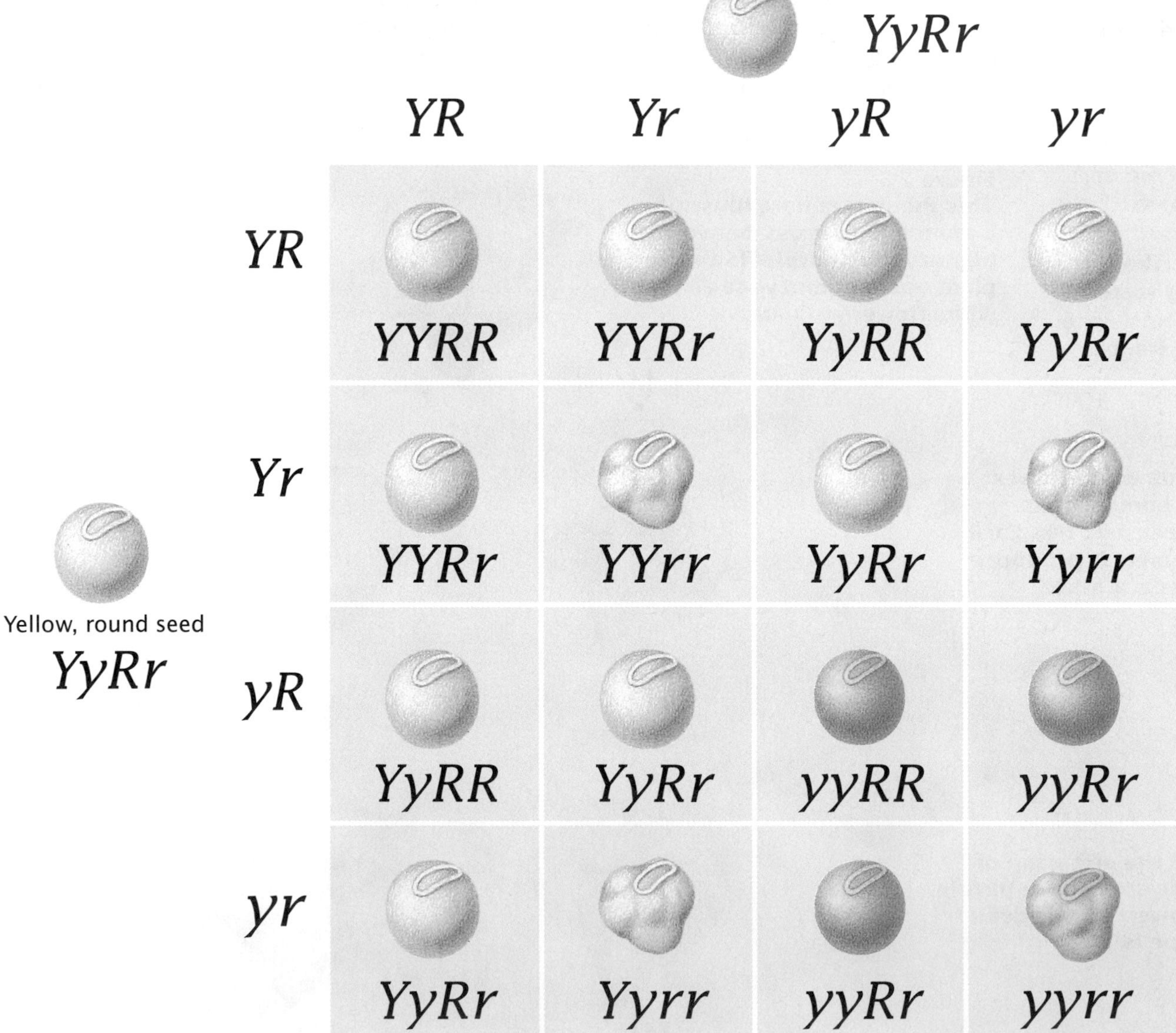

Figure 7-6 This Punnett square illustrates a dihybrid cross between two heterozygous plants that have yellow, round peas. Four phenotypes will be found among the offspring of this cross. Mendel calculated a 9:3:3:1 ratio of these phenotypes.

Other Factors That Influence Heredity

Since Mendel's time, scientists have discovered many genes that do not follow the simple pattern of inheritance described by Mendel. Most of the time, genes exhibit more complex patterns of inheritance, such as those described below.

Two traits appear to blend in incomplete dominance

In many cases, an individual has a trait that appears to be an intermediate form of the traits displayed by the two parents. This phenomenon is called incomplete dominance. An example of incomplete dominance is found in snapdragons. A cross between a red-flowered snapdragon and a white-flowered snapdragon will result in offspring with pink flowers, as shown in **Figure 7-7**. As dominant traits in snapdragons, red flowers and white flowers are homozygous. Pink flowers are heterozygous. Heterozygous flowers are pink because they are unable to produce enough red pigment to make their petals appear red.

Figure 7-7
When red-flowering snapdragons are crossed with white-flowering snapdragons, the offspring will produce pink flowers, an example of incomplete dominance.

Two traits are fully displayed in codominance

In some cases, both genes in a heterozygote are fully expressed, a phenomenon called codominance. Codominance can affect coat color in horses. When a horse that is homozygous for red coat color is crossed with a horse that is homozygous for white coat color, the offspring are heterozygous and have roan coats. A roan coat is one in which the color is a mixture of two colors, such as red hairs mixed with white hairs, as shown in **Figure 7-8**.

Varieties of a trait arise from multiple alleles

Some traits are controlled by genes that have more than two alleles. In such cases, a trait is said to be controlled by multiple alleles. Combinations of any two of these alleles can produce different phenotypes. An example of multiple alleles can be seen in the human ABO blood groups. The four blood groups are governed by three alleles—I^A, I^B, and i. Recall that a person can have only two alleles for any gene. **Table 7-2** shows the possible phenotypes and genotypes of human ABO blood groups.

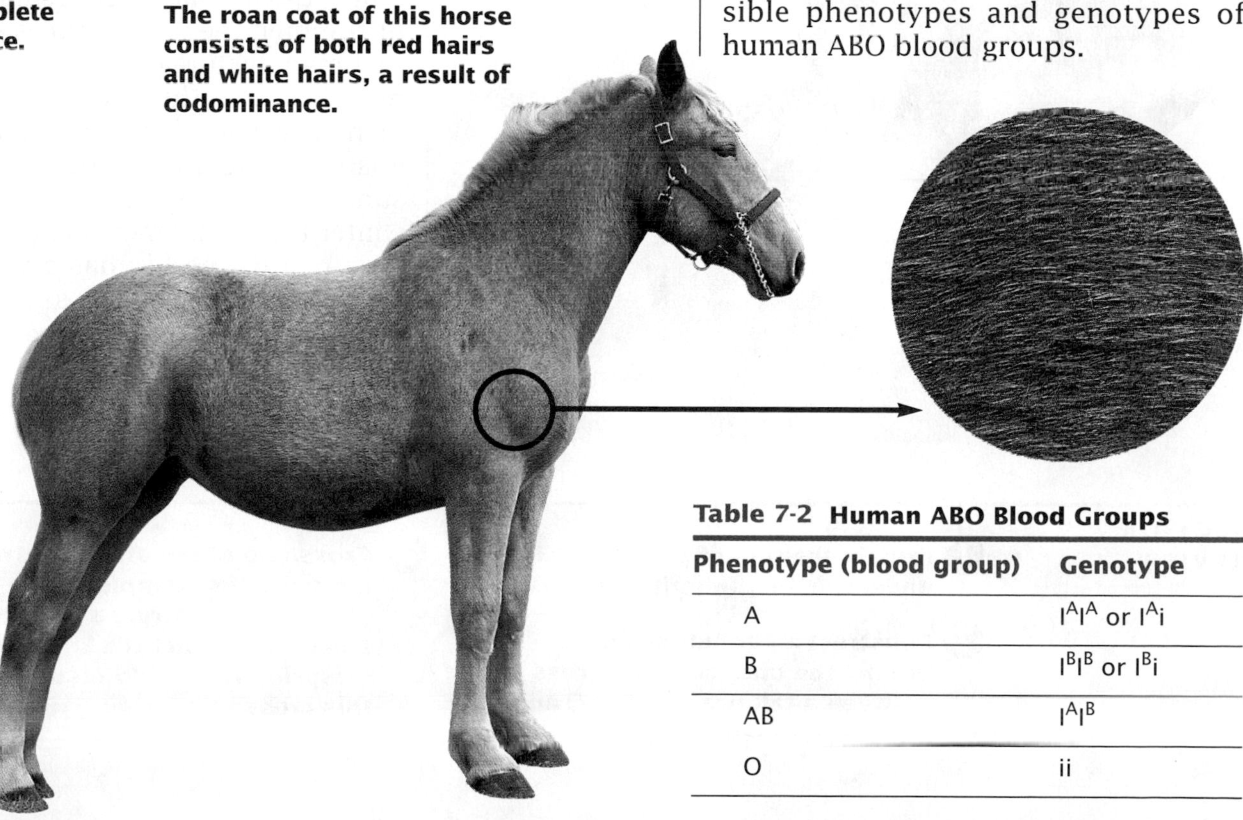

Figure 7-8
The roan coat of this horse consists of both red hairs and white hairs, a result of codominance.

Table 7-2 Human ABO Blood Groups

Phenotype (blood group)	Genotype
A	I^AI^A or I^Ai
B	I^BI^B or I^Bi
AB	I^AI^B
O	ii

Figure 7-9
Many of your distinguishing features—your height, weight, and skin color—are examples of polygenic traits, which are influenced by many pairs of genes.

Figure 7-10
a Conditions in the environment can affect an organism's genes and therefore its appearance. The arctic fox's red coat enables it to blend into the summer background, . . .

b . . . while in winter the fox blends into the snowy landscape.

Some traits are controlled by many genes

The term polygenic is used to describe a trait that is controlled by more than one pair of genes. The genes for a polygenic trait may be scattered along the same chromosome or located on different chromosomes. Due to independent assortment and crossing-over, many different combinations appear among the offspring. Each combination influences the trait to a different degree. Familiar examples of polygenic traits in humans include height, weight, body build, and hair and skin color, as shown in **Figure 7-9**. Characteristics such as these all have degrees of intermediate conditions between one extreme and the other.

Gene expression can be affected by environmental conditions

An individual's phenotype often depends on surrounding environmental conditions. For example, the color of an arctic fox is affected by temperature. During the warm temperatures of summer, the fox produces enzymes that manufacture pigments. These pigments darken the fox's coat to a reddish brown, as shown in **Figure 7-10a**, enabling the fox to blend into the summer landscape. But during the cold winter these enzymes do not function. As a result, the fox has a white coat that blends into the snowy background, as shown in **Figure 7-10b**.

SECTION REVIEW

1. **Explain the role of probability in understanding inheritance.**
2. **Construct a Punnett square to predict the offspring of a cross between a tall pea plant *(TT)* and a short pea plant *(tt)*.**
3. **Cross two of the offspring from the cross just completed in (2) above. Construct a Punnett square to predict the kinds of offspring that could result from this cross.**

Many genes in human populations code for metabolic processes that are far more important than characteristics such as hair color and eye color. Some of the most devastating human disorders are caused by rare genes that are the result of genetic mutations. Many of these genes are passed from parent to child in patterns of inheritance described by Mendel.

7-3 Human Genetic Disorders

OBJECTIVES

1. **Explain what mutations are and how they can affect an organism.**
2. **Describe the patterns of inheritance for three genetic disorders.**
3. **Explain a technique used to identify people at risk of passing genetic disorders to their children.**
4. **Discuss the possibilities that exist for curing genetic disorders.**

Mutations Are Changes in Genes

In order to develop and function properly, your body requires specific proteins that are built according to instructions in your genes. In some cases, genes are sometimes damaged or copied incorrectly. A change in a gene is called a **mutation** *(myoo TAY shuhn)*. Mutations are a source of the variation a species needs in order to adapt to changing conditions and thus to evolve over time. A mutation's effect can vary. Most mutations are harmful or neutral, such as the one shown in **Figure 7-11**. Since mutations change genes at random, the chance that a mutation will improve an individual is very slim. For instance, imagine that you randomly changed a part of a blueprint for the construction of an airplane. Some changes, such as the color of the passenger seats, would not matter. Other changes, such as the size of the jet engines, are more serious.

Most mutations are recessive

Most mutations are recessive and therefore are not expressed in the heterozygous condition. Although mutations are rare, everyone is probably heterozygous for a number of genetic mutations. It is unlikely, however, that a person who is heterozygous for a recessive mutation will mate with someone who is heterozygous for the same mutation. Therefore, there is little chance that his or her offspring would be homozygous for the recessive mutation, which could be a harmful and unfortunate condition.

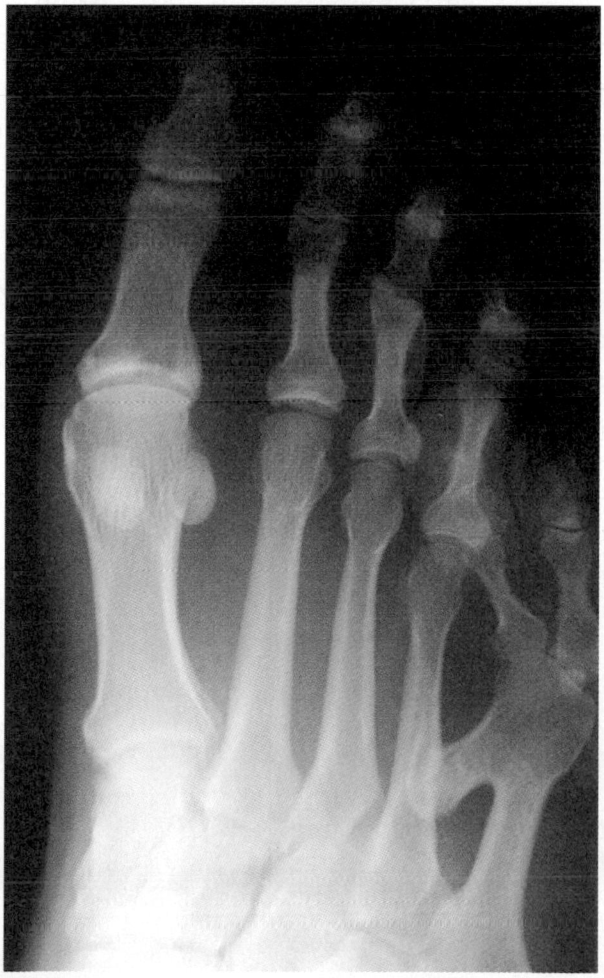

Figure 7-11
A genetic mutation causes some humans to have six toes, as shown in this X ray.

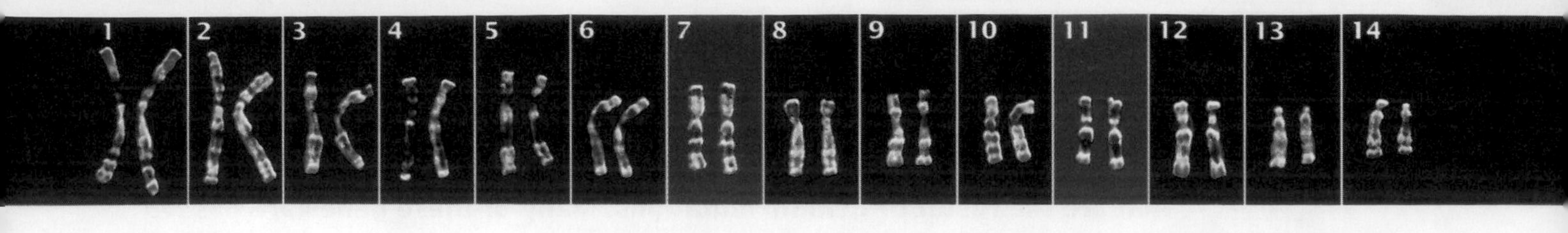

Figure 7-12
The chromosomes highlighted in blue play a role in the genetic disorders described on these pages. Staining techniques reveal the unique banding patterns, shown above, that help researchers identify each chromosome.

Genetic Disorders

Particular mutations have established a foothold in human populations. The harmful effects that some mutations produce are called **genetic disorders**. Many genetic disorders are the result of mutations in a single gene in one of the 46 human chromosomes, shown in **Figure 7-12**. In most cases, a genetic disorder is recessive. A person with a recessive genetic disorder has inherited a defective gene from both parents. Other genetic disorders are caused by dominant alleles. A single copy of the defective gene is enough to cause the disease.

Cystic fibrosis is caused by a faulty transport protein

Cystic fibrosis is a recessive genetic disorder caused by a mutation in a gene that codes for a protein responsible for transporting chloride ions. In the homozygous condition, this mutation causes cells in the respiratory system and in the digestive system to produce misshapen chloride transport proteins that are unable to function properly. In these individuals, such as the girl shown in **Figure 7-13**, mucus accumulates in the lungs and pancreas, clogging ducts needed for these organs to function properly. People with cystic fibrosis have difficulty breathing and cannot properly digest their food. Most of the current treatments for cystic fibrosis relieve the symptoms of the disease but do not cure it.

Researchers have isolated the mutated gene that causes cystic fibrosis. Isolation of the gene has led to the possibility of gene transfer therapy, in which a healthy copy of the mutated gene is transferred into the lungs of cystic fibrosis patients to cure the disease.

	F	*f*
F	*FF*	*Ff*
f	*Ff*	*ff*

Figure 7-13
Cystic fibrosis is caused by a recessive gene on chromosome 7. If both parents carry a copy of the cystic fibrosis gene (*f*), their child has a one-in-four chance of developing the disease.

Sickle cell anemia results from faulty hemoglobin

Sickle cell anemia is a recessive genetic disorder caused by a mutation in a gene that codes for hemoglobin, the protein in red blood cells that transports oxygen throughout the body. In the homozygous condition, this mutation causes the body to build an abnormal form of hemoglobin. When the faulty hemoglobin is inside a red blood cell, it gives the cell a deformed, sickle-shaped appearance, as shown in **Figure 7-14**. The sickled cells get caught in small blood vessels and restrict blood flow to tissues. The sickled cells are also very fragile and break down easily. This leaves the person with significantly fewer red blood cells than average, a condition called anemia. Symptoms of sickle cell anemia include fatigue, headaches, muscle cramps, and sometimes kidney or heart failure.

The sickle cell mutation in the hemoglobin gene apparently first arose in central Africa centuries ago. Up to 45 percent of the population in central Africa is heterozygous for the sickle cell gene. Evolution has favored the sickle cell gene in central Africa because heterozygous people are more resistant to malaria, a deadly tropical disease caused by parasites in red blood cells.

	S	*s*
S	*SS*	*Ss*
s	*Ss*	*ss*

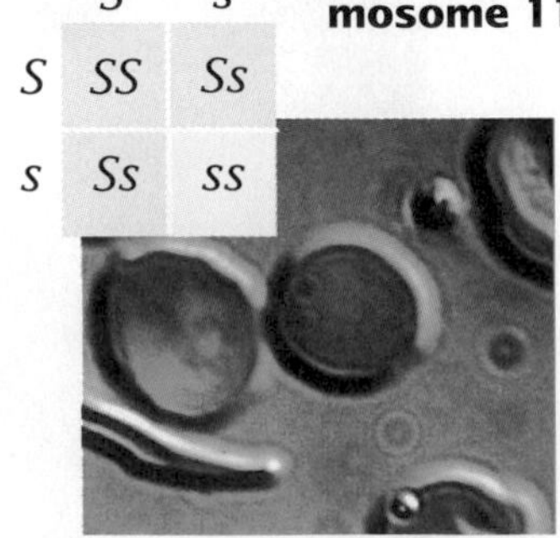

Figure 7-14
Sickle cell anemia is caused by a recessive gene on chromosome 11. If each parent carries a copy of the sickle cell gene, there is a one-in-four chance that their child will have sickle cell anemia.

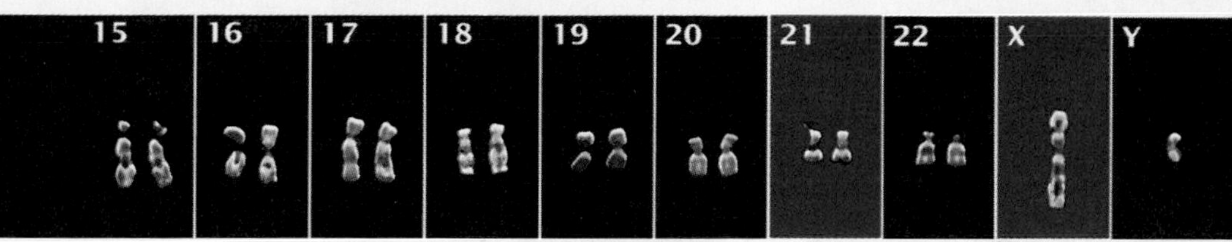

Figure 7-15
A mother who carries the gene for hemophilia may pass it on to her sons and daughters. However, only the sons who inherit the gene will develop the disorder. With proper treatment, hemophiliacs, like this boy, can lead active lives.

Hemophilia results from a faulty blood-clotting protein

Hemophilia is a recessive genetic disorder caused by a mutation on the X chromosome. This mutation causes the body to build a defective form of a protein needed to clot blood. Usually when body tissues are cut, the blood in the immediate area of the cut forms a clot that seals the wound. In an individual with hemophilia, such as the boy in **Figure 7-15**, small cuts are difficult to heal and internal bleeding can be fatal. Long-term sufferers are susceptible to serious arthritis, arising from bleeding in joint cavities, and other disorders. Fortunately, treatments are available, such as injections of the correct form of the faulty protein.

The proteins involved in blood clotting are encoded in dozens of genes. Two of these genes are found on the X chromosome. Any male who inherits a mutant copy of either gene will develop hemophilia because his other sex chromosome, a Y chromosome, lacks a gene for the blood-clotting protein. A pattern of heredity in which traits are transmitted by genes located on the X chromosome is called **sex-linked inheritance**. Traits that are determined by genes located on the X chromosome are said to be sex-linked traits.

Down syndrome arises from three copies of chromosome 21

Down syndrome is an example of a genetic disorder caused by an extra chromosome. Whereas most people have two copies of each chromosome, individuals with Down syndrome have three copies of chromosome 21, as shown in **Figure 7-16**. The features that characterize Down syndrome include extra folds in the upper eyelids, a broad and somewhat flattened nose, short stature, and, most important, varying degrees of mental retardation. These characteristics were first described in 1866 by a researcher named J. Langdon Down. Although people with Down syndrome are physically challenged, they are able to lead active lives and to make positive contributions to society.

Down syndrome occurs in about 1 of every 1,000 births. This syndrome is much more common among children born to older mothers. In mothers older than 45, the risk is as high as 1 in 16 births. Pregnant women over the age of 35 are usually informed of medical procedures in which fetal cells are extracted and used to prepare a fetal karyotype. In one technique, amniocentesis, a small amount of the fluid surrounding the fetus is removed and fetal cells are collected. In another technique, chorionic villi sampling, a tiny piece of embryonic membrane is removed. In addition to Down syndrome, amniocentesis can reveal hemophilia, sickle cell anemia, and cystic fibrosis.

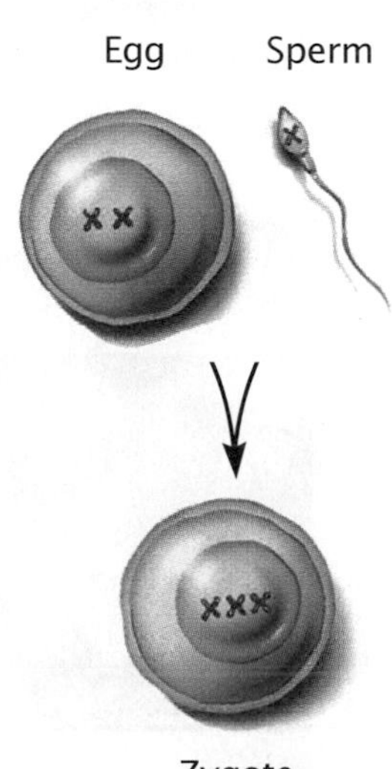

Figure 7-16
When chromosome 21 fails to separate properly during meiosis in either parent, the child can receive three copies of the chromosome, a condition that causes Down syndrome.

Figure 7-17
a Two boys in this family have hemophilia, but no one else does. How can this happen?

Genetic Counseling and Gene Therapy

Most genetic disorders are caused by recessive genes. Therefore, a gene that causes a genetic disorder could be present but not expressed in a person who is heterozygous for that gene. These heterozygous individuals are said to be carriers for the disorder. How can a person find out if he or she is a carrier for a genetic disorder? If one family member is affected by a genetic disorder, it may be possible that other family members are heterozygous for the disorder. In these cases, physicians may recommend genetic counseling to family members who wish to find out if they are at risk of having children with a genetic disorder. In some cases, a genetic counselor can prepare a family **pedigree** *(PEHD uh gree)*, a record that shows inheritance patterns over several generations, as shown in **Figure 7-17**. Noting relatives with a genetic disorder enables a genetic counselor to determine an individual's chances of being a carrier for that disorder. With such information, a person can decide whether or not to have children.

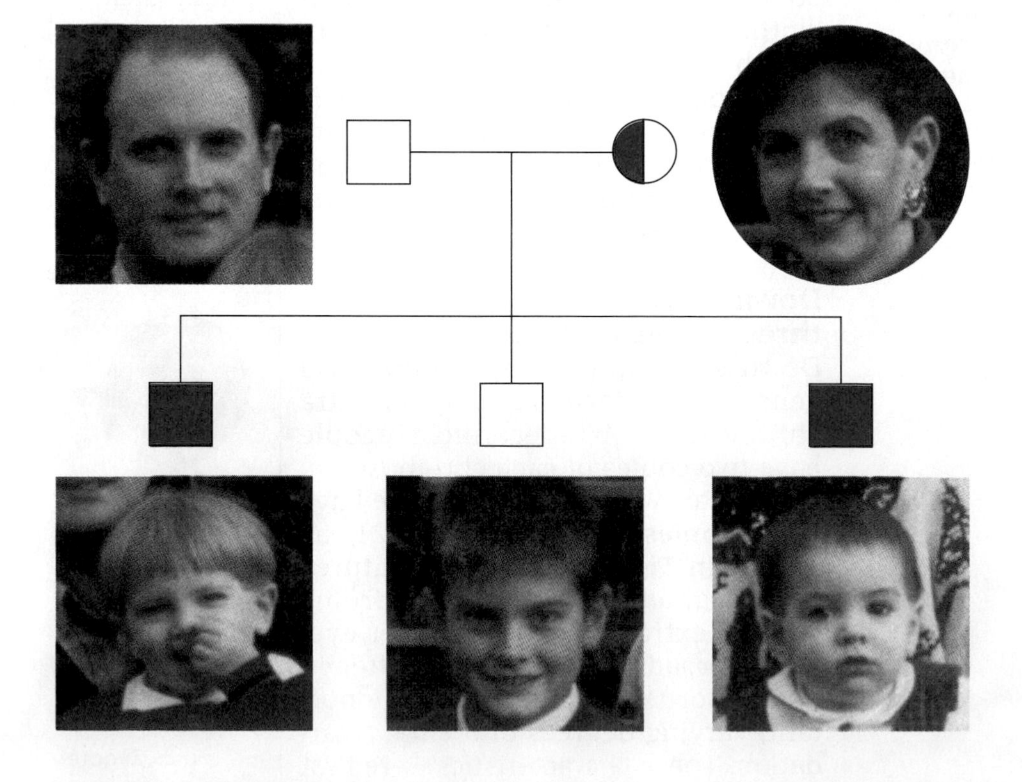

b This pedigree shows that the mother carries the gene for hemophilia (shaded), and that two of her sons have received this gene.

Career Opportunities

Genetic Counseling

What would it be like to . . .

- educate people and families about inherited diseases?
- play a significant role in bringing healthy children into the world?

These are just a few of the opportunities open to people who work in the field of genetic counseling.

Genetic Counseling: What Is It?

Genetic counseling is a service for people seeking information about a possible genetic disorder or disease that has affected one or more relatives. Might it affect them or their children? What treatment is available? In responding to such questions, a genetic counselor provides information that could help individuals make important decisions about having children.

Science/Math Career Preparation

High school
- Biology
- Chemistry
- Physics

College
- Biology
- Genetics
- Chemistry
- Psychology

Graduate school
- Master's degree is recommended

Employment Outlook

As our knowledge about human genetics and reproductive technologies grows more sophisticated, the demand for genetic counselors will quickly increase.

Career Focus: Diana Punales-Morejon

Genetic Counselor at a medical center in New York

"I was born in Cuba and didn't learn to speak English until the first grade. Learning another language was difficult but worth the effort. Being bilingual enables me to talk to many different people and has helped me advance my career. I went to college to become a biology teacher. However, after graduating, I worked as a researcher in a biomedical laboratory. Although it was challenging, I missed working directly with people. That's when I heard about genetic counseling, a field that requires an understanding of genetics and the desire to help people. It sounded perfect."

What I Do: "As a genetic counselor, I advise people who are at risk of having children with genetic disorders or birth defects. I answer questions, explain common misconceptions, and administer tests that detect genetic disorders in fetuses. One of my favorite things about being a genetic counselor is telling an expecting couple that their child will be healthy. However, it's never easy telling a couple that tests show their child will be affected with a genetic disorder. Dealing with these sensitive matters is difficult, but it can be one of the most rewarding things about my job."

To Find Out More

1. Write for more information:
   Is a Career in Genetic Counseling in Your Future?
   National Society of Genetic Counselors
   233 Canterbury Drive
   Wallingford, PA 19086

   Solving the Puzzle: Careers in Genetics
   The Genetics Society of America & The American Society of Human Genetics
   9650 Rockville Pike
   Bethesda, MD 20814-3998

2. Visit your school counselor and library to find out what colleges and universities offer programs in genetic counseling. Review the prerequisites and course requirements. Make a poster showing the range of job opportunities in this field.

Finding Cures for Genetic Disorders

Most genetic disorders cannot be cured. However, therapy is available for some genetic disorders if they are diagnosed early enough. For example, phenylketonuria (PKU) is a recessive disorder in which affected individuals lack an enzyme that converts the amino acid phenylalanine into the amino acid tyrosine. In these individuals, phenylalanine builds up in the body, leading to mental retardation. If PKU is diagnosed shortly after birth, the infant can be placed on a low-phenylalanine diet. Such a diet ensures that the infant gets enough phenylalanine to make proteins but not enough to cause damage. The patient must stay on the diet as directed in order to control the levels of phenylalanine. Because this genetic disorder is easily detected after birth by inexpensive laboratory tests, many states have extensive screening programs to check newborns for PKU.

Techniques in genetic engineering can make it possible to cure genetic disorders by replacing copies of defective genes with copies of healthy ones. One of the first gene therapy attempts involved two young girls, shown in **Figure 7-18**, who suffered from an immune-system disorder caused by a defective gene. The defective gene failed to produce an important immune-system enzyme. Doctors took bone marrow cells from the girls and replaced the defective gene with a normal copy. Later, after the genetically engineered cells were returned to the girls' bones, they began to produce the missing enzyme. Five years after their gene therapy, the girls are doing well, although they are still on medication. Doctors monitor them closely, looking for signs that the gene therapy was effective. Unfortunately, gene therapy as a simple and foolproof cure for a disorder is still in the future.

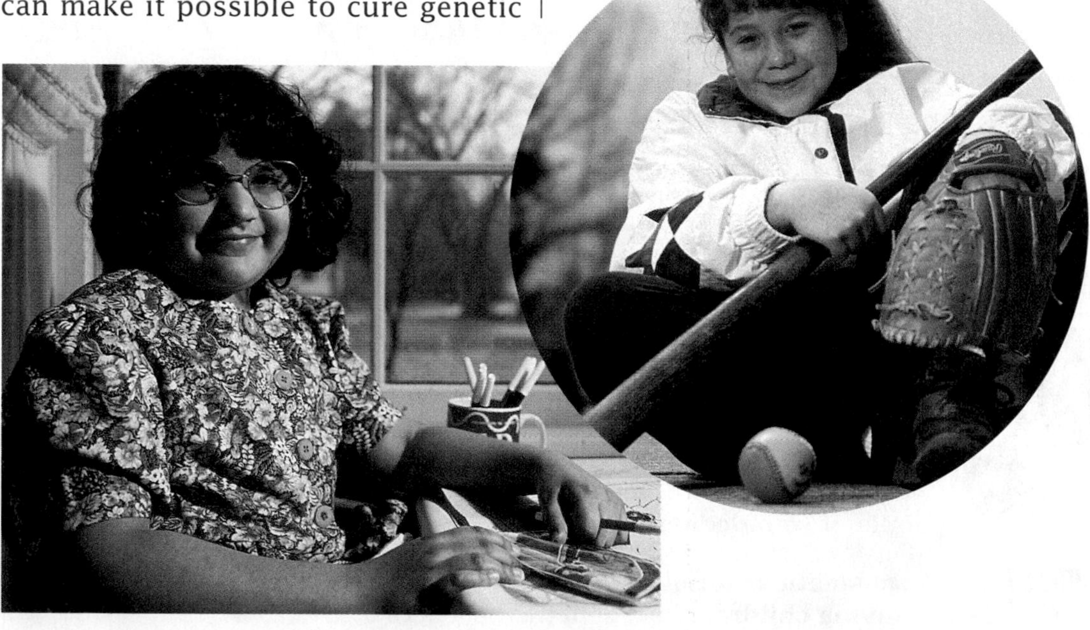

Figure 7-18 Ashanthi DeSilva (left) and Cynthia Cutshall (right) were among the first patients to receive gene therapy.

SECTION REVIEW

1. **What is a mutation? Why are most mutations recessive?**
2. **Create a table that contains information about three genetic disorders.**
3. **Describe a technique used to identify people at risk of passing genetic disorders to their children.**
4. **How are genetic engineering techniques providing hope for people with genetic disorders?**

CHAPTER 7 Highlights

These children are brothers. Even though they have the same parents, they look very different from each other.

7-1 The Work of Gregor Mendel

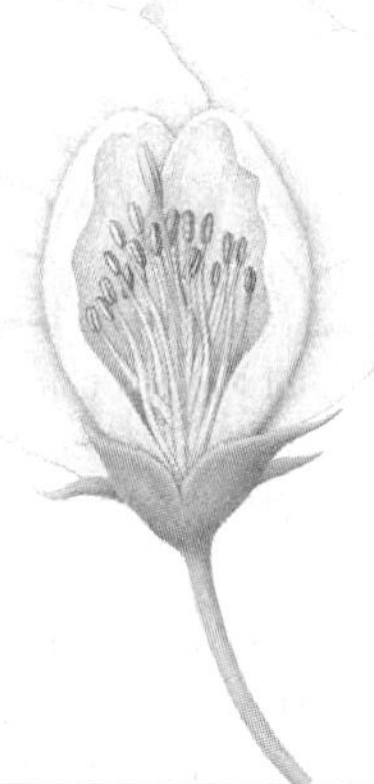
Mendel unlocked the secrets of heredity by studying pea plants.

Key Terms

genetics (p. 117)
homozygous (p. 119)
heterozygous (p. 119)
allele (p. 119)
genotype (p. 119)
phenotype (p. 119)
dominant (p. 119)
recessive (p. 119)
probability (p. 121)

Summary

- Gregor Mendel's experiments with pea plants marked the beginning of genetics.
- Using a mathematical approach, Mendel concluded that each trait is governed by two factors, one of which is received from each parent. An individual passes either one of its factors to each of its offspring.
- Mendel's ideas about heredity can now be interpreted using modern genetic terminology.
- Mendel's laws of inheritance reflect the events of meiosis.

7-2 Patterns of Inheritance

Patterns of heredity are passed from generation to generation.

Summary

- Probability and Punnett squares are two techniques used to predict the outcome of genetic crosses.
- A monohybrid cross studies the inheritance of one trait. A dihybrid cross studies the inheritance of two traits.
- The inheritance of most traits is complex and can be affected by many different influences.

7-3 Human Genetic Disorders

Occasionally a harmless mutation causes a person to have six toes.

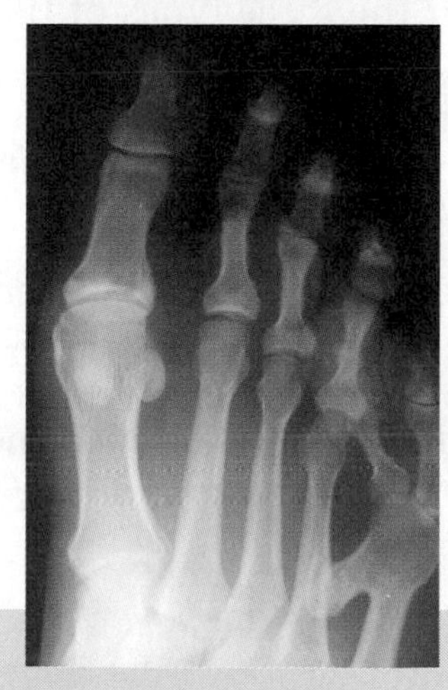

Key Terms

mutation (p. 127)
genetic disorder (p. 128)
sex-linked inheritance (p. 129)
pedigree (p. 130)

Summary

- A change in a gene, called a mutation, can cause inherited diseases called genetic disorders.
- Examples of genetic disorders include cystic fibrosis, sickle cell anemia, hemophilia, and Down syndrome.
- Genetic counseling can help identify people at risk of having children with genetic disorders.
- Therapies are available to treat some genetic disorders, if they are diagnosed early enough.

CHAPTER 7

review

Understanding Vocabulary

1. For each pair of terms, explain the differences in their meanings.
   a. parental generation, filial generation
   b. gene, allele
   c. dominant, recessive
2. For each set of terms, complete the analogy.
   a. Mendel : heredity :: Darwin : ____
   b. segregation : one pair of alleles :: independent assortment : ____
   c. one trait : monohybrid cross :: two traits : ____

Understanding Concepts

3. **Relating Concepts** Construct a concept map describing the study of genetics. Try to include the following terms in your concept map: chromosomes, factors, traits, dominant, recessive, genotype, alleles, and heredity.
4. **Organizing Information** Identify the three steps of Mendel's experiments with garden-pea plants. Summarize the events and results of each step. In your answer, include a diagram that shows the passage of traits from one generation to the next.
5. **Summarizing Information** Contrast Mendel's two laws: the law of segregation and the law of independent assortment. Explain how these two laws relate to the behaviors of chromosomes during meiosis.
6. **Inferring Conclusions** How did Mendel explain the results of his experiments that led to the law of segregation?
7. **Inferring Relationships** What is probability? Explain why the probability that an egg or a sperm cell will contain one or the other allele for each gene is 1/2.
8. **Summarizing Information** What is a mutation? Explain why mutations are seldom expressed in a heterozygous individual.
9. **Predicting Outcomes** In rabbits, the allele for black fur *(B)* is dominant over the allele for brown fur *(b)*. In order for all the offspring to have brown fur, what would the genotype and phenotype of the male and female parents have to be? Draw a Punnett square to support your answer.
10. **Inferring Relationships** How does the environment affect this organism's phenotype?

11. **Predicting Outcomes** A plant breeder wants to produce only pink snapdragons. Can he do so by crossing pink snapdragons and white snapdragons? Explain your answer.
12. **Comparing Functions** Compare incomplete dominance with codominance. Give an example of each pattern of heredity.
13. **Identifying Variables** Is it possible for a woman with type AB blood and a man with type B blood to have a child with type O blood? What possible blood types can their children have? Use a Punnett square to support your answer.
14. **Summarizing Information** What is a polygenetic trait? Give an example.
15. **Inferring Relationships** In Japanese four o'clocks, red flowers show incomplete dominance with white flowers. What color flowers will offspring in the F_1 generation have? Explain your answer. Give the genotypes for the three types of flowers.
16. **Predicting Outcomes** A rough-coated male guinea pig is crossed with a homozygous smooth-coated female *(rr)*. Their offspring are rough-coated and smooth-coated in a 1:1 ratio. Use a Punnett square to calculate the genotype of the male parent.
17. **BUILDING ON WHAT YOU HAVE LEARNED** Based on what you learned in Chapter 6 about the formation of male and female gametes, explain sex inheritance in humans.

Interpreting Graphics

18. The Punnett square below shows the cross between two parents heterozygous for two traits. In watermelons, the allele for solid green color *(G)* is dominant over the allele for striped pattern *(g)*. Short shape *(S)* is dominant over long shape *(s)*.

a. What are the genotype and phenotype of each parent?
b. What are the genotype and phenotype of the offspring in block *a*? in block *b*?
c. How many different genotypes are possible among the offspring of this cross?
d. How many different phenotypes are found among the offspring? What is the phenotypic ratio?

GgSs × *GgSs*	*GS*	*Gs*	*gS*	*gs*
GS	*GGSS*	*GGSs*	*GgSS*	*GgSs*
Gs	(a)	*GGss*	*GgSs*	*Ggss*
gS	*GgSS*	(b)	*ggSS*	*ggSs*
gs	*GgSs*	*Ggss*	*ggSs*	*ggss*

Reviewing Themes

19. ***Heredity***
Two tall pea plants are crossed. Both plants are heterozygous *(Tt)* for height. How does this information help you predict the genotypic and phenotypic ratios for the offspring?

20. ***Evolution***
Explain how mutations contribute to the evolution of a species over time.

Thinking Critically

21. **Evaluating Conclusions** The offspring of two short-tailed cats exhibit the following phenotypic ratio: 1 tailless:2 short tail:1 long tail. Based on this information, their owner has concluded that tail length is an example of incomplete dominance. Is the owner correct? Include a Punnett square to support your answer.

Life/Work Skills

22. **Finding Information** Colorblindness in humans is an example of sex-linked inheritance. Use references to find out about the different kinds of colorblindness and how this disorder is inherited. Draw Punnett squares to illustrate your findings.

Activities and Projects

23. **History Connection** The genetic disorder hemophilia was important to European history because it occurred in certain powerful royal families that were descendants of Queen Victoria. Queen Victoria was heterozygous for the trait and passed the disorder to her children and grandchildren. Research the presence of hemophilia in Queen Victoria's family. Draw a human pedigree that shows how this disorder was passed from generation to generation.

24. **Mathematics Connection** Probability and the law of independent assortment also apply to games of chance. Many states have lottery drawings in which a person selects five numbers from the numbers 1 to 20. To win, all five numbers must match the randomly selected winning numbers. Imagine that balls numbered from 1 to 20 are placed in a jar. Your chances of matching any number in the jar, say 7, are 1 chance in 20 chances. If you have two identical jars, the chance of drawing a 7 from each jar is $1/20 \times 1/20 = 1/400$, or 1 in 400 chances. If you had five identical jars, what is the chance that the numbers 3, 9, 12, 13, and 18 would be selected—in that order?

Discovering Through Reading

25. Read the article "Micro Gets Macro" by Josie Glausiusz in *Discover*, November 1995, page 40. What are microsatellites, and how are they used by gene mappers?

26. Read the article "Ascent of the Dog" by Rosie Mestel in *Discover*, October 1994, pages 90–98. How does Jasper Rine hope to study the inheritance of traits among different breeds of dogs?

CHAPTER 8

REVIEW
- protein structure (Section 2-3)
- nucleic acid (Section 2-3)
- ribosome (Section 3-3)
- chromosome (Section 6-1)
- cell division (Section 6-2)

How Genes Work

Helical strands of DNA, like the one shown in this model, contain the genes that determine physical traits and regulate body functions.

Mendel showed that traits are passed from parents to offspring in the form of discrete packages of genetic material that sort independently of each other. Then scientists set out to determine how a gene actually operates.

8-1 Understanding DNA

OBJECTIVES

1. **List the events that led scientists to identify the genetic material.**
2. **Summarize how scientists determined the structure of DNA.**
3. **Describe the structure of the DNA molecule.**
4. **Summarize the process of DNA replication.**

How Scientists Identified the Genetic Material

Scientists set out to identify the chemical substances in chromosomes that store and transmit genetic traits. Was it DNA or protein? Answers came from medical studies involving two forms, or strains, of the bacterium *Pneumococcus.* One strain is **virulent** *(VEER yoo luhnt)*, or disease-causing, and causes pneumonia. The other strain is nonvirulent, or harmless.

Griffith studies bacteria

In 1928 in London, a bacteriologist named Frederick Griffith was working out a way to identify two strains of *Pneumococcus.* He found that one strain formed rough colonies in a petri dish. The other strain formed smooth colonies. To determine which bacteria were virulent, he injected the two different strains into mice. The bacteria from the smooth colonies were virulent and killed mice **(Figure 8-1a)**. The bacteria from the rough colonies were nonvirulent and didn't harm the mice **(Figure 8-1b)**.

Griffith found that if he killed the virulent bacteria with heat, they lost their ability to cause disease in mice **(Figure 8-1c)**. However, heat-killed virulent bacteria (now made harmless) mixed with nonvirulent bacteria (also harmless) had deadly results. Mice injected with these two seemingly harmless bacterial strains became ill and died **(Figure 8-1d)**. Furthermore, when Griffith examined bacteria taken from the dead mice, he found living virulent bacteria even though the only living bacteria that had been injected were the nonvirulent ones. Nonvirulent bacteria had changed to virulent bacteria by taking in something from the virulent bacteria. Griffith had discovered **transformation**, a process in which a bacterium takes up foreign DNA.

Figure 8-1
Griffith's experiment showed that nonvirulent bacteria could acquire the ability to cause disease from virulent bacteria.

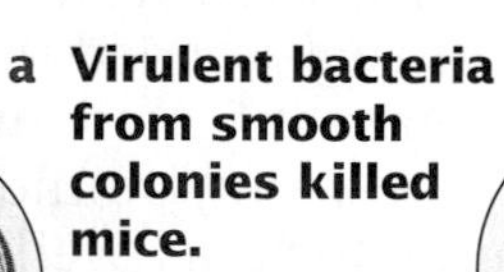

a Virulent bacteria from smooth colonies killed mice.

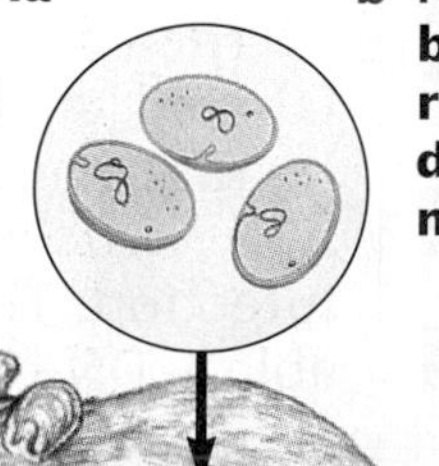

b Nonvirulent bacteria from rough colonies did not harm mice.

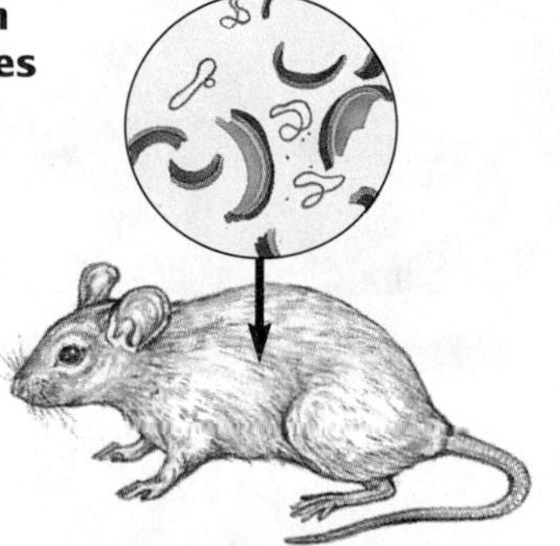

c Virulent bacteria killed with heat did not harm mice.

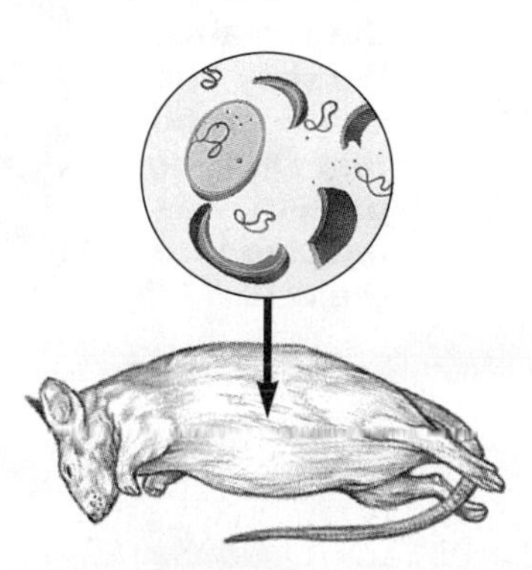

d A mixture of nonvirulent bacteria and heat-killed virulent bacteria killed mice.

Avery identifies the agent of transformation

In 1944 in New York, a biologist named Oswald Avery was very interested in Griffith's discovery. Avery wanted to identify the substance that made the nonvirulent bacteria become virulent. Was this substance DNA or protein? Avery and two colleagues modified Griffith's experiment. They extracted DNA from bacteria in smooth colonies and added it to bacteria in rough colonies. Some of the bacteria that grew from this mixture formed smooth colonies.

In a separate experiment, Avery found that when he added protein-destroying enzymes to bacteria, transformation still occurred. However, when DNA-destroying enzymes were added to bacteria, transformation did not occur. Avery's work provided clear evidence that DNA was the genetic material in these bacteria.

Figure 8-2
a Hershey and Chase used radioactive elements to label the DNA and the protein coats of viruses.

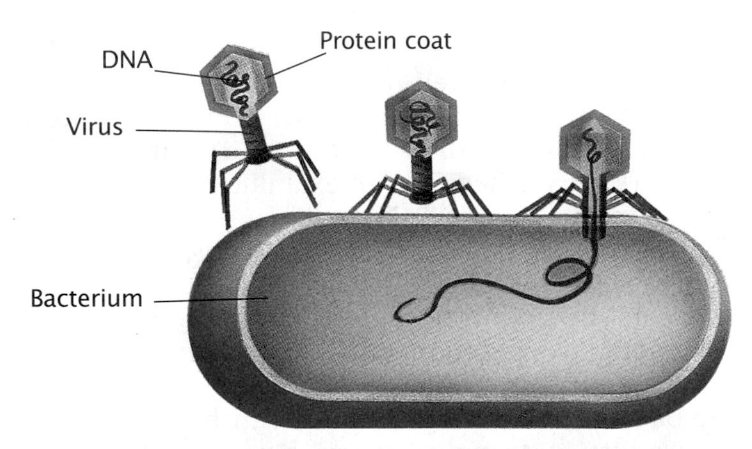

b They infected bacteria with the radioactive viruses. When sufficient time had passed, Hershey and Chase used a blender to detach the remains of the viruses.

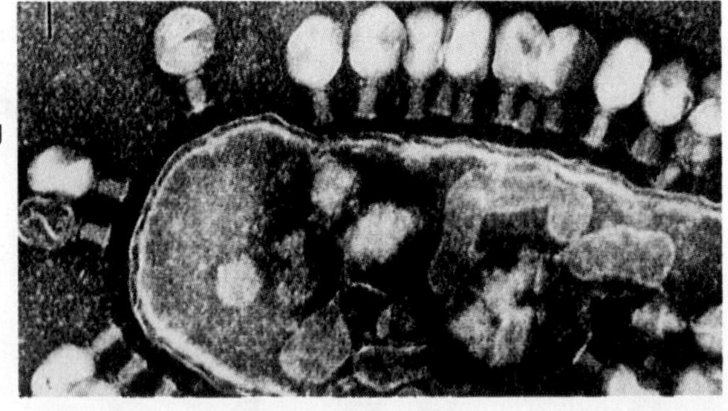

c After examining the remains, Hershey and Chase found that DNA, not protein, was injected into the cell.

Hershey and Chase confirm that DNA is the genetic material

At first, Avery's results were not widely appreciated. Many biologists were reluctant to give up the idea that protein was the genetic material. In 1952, however, Alfred Hershey and Martha Chase, two scientists at Cold Spring Harbor Laboratory on Long Island, New York, performed an experiment using viruses that infect bacteria. These viruses attach to the surfaces of bacteria and inject their hereditary information into the cells like tiny hypodermic needles. Once inside the bacteria, this hereditary information directs the production of hundreds of new viruses. When the new viruses are mature, they burst out of the infected bacteria and attack new cells. These bacteria-infecting viruses have a very simple structure: a core of DNA surrounded by a protein coat.

To identify the hereditary material, Hershey and Chase did two experiments using viruses with radioactive labels, as summarized in **Figure 8-2**. One experiment used viruses with radioactive sulfur (^{35}S), and one used viruses with radioactive phosphorus (^{32}P). Hershey and Chase used one kind of virus to infect one batch of bacteria and the other kind of virus to infect another batch of bacteria. After allowing enough time for the viruses to attach to the bacteria and inject their genetic material, Hershey and Chase put the two batches of bacteria separately into an ordinary food blender to detach the protein coats. When they examined the remains of the bacteria infected by the ^{35}S-labeled viruses, they found the protein outside the bacteria, but it was not found in any of the new viruses that resulted from this infection. When they examined the remains of the bacteria infected by the ^{32}P-labeled viruses, they found the DNA inside the bacteria as well as in the new viruses that resulted from this infection. The conclusion was undeniable—DNA is the hereditary material.

How Scientists Determined the Structure of DNA

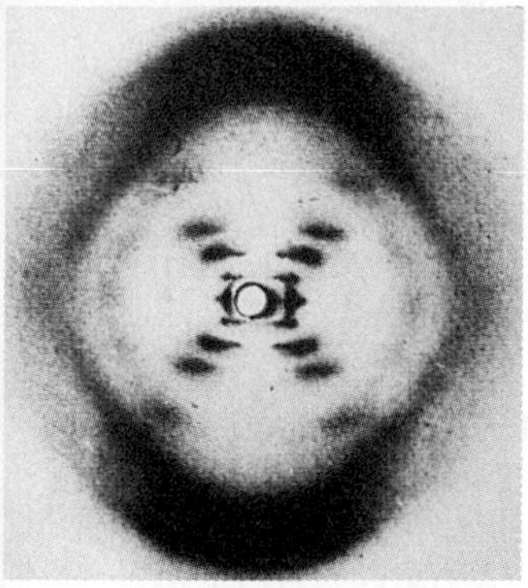

Scientists knew that DNA was composed of subunits called **nucleotides** *(NOO klee oh tydz).* A nucleotide has three parts: a sugar called deoxyribose, a phosphate group, and a base. The sugar and phosphate group are the same in every nucleotide, but the base can be one of four kinds. The two larger bases, adenine and guanine, are called **purines** *(PYUR eenz).* The two smaller bases, cytosine and thymine, are called **pyrimidines** *(py RIHM uh deenz).*

In 1949 Erwin Chargaff, a biochemist working at Columbia University in New York City, made a key discovery about the structure of DNA by analyzing DNA samples from different organisms. He found that the amount of adenine in a DNA molecule always equals the amount of thymine (A = T). Likewise, the amount of guanine always equals the amount of cytosine (G = C). These observations, now known as Chargaff's rules, suggested a regularity in the composition of DNA.

The value of Chargaff's rules became clear when Rosalind Franklin, a chemist working at King's College in London, began studying the structure of DNA using X-ray diffraction. Franklin's X-ray diffraction images, one of which is shown in **Figure 8-3**, suggested that the DNA molecule resembles a tightly coiled spring, a shape called a helix.

Figure 8-3
This X-ray diffraction photograph taken by Rosalind Franklin provided evidence that DNA was shaped like a helix.

Watson and Crick build a model showing DNA's structure

In the early 1950s a young American scientist, James Watson, went to Cambridge, England, on a research fellowship. At Cavendish Laboratories he worked with Francis Crick, a British physicist interested in DNA. Together, Watson and Crick attempted to construct a model of DNA. They applied the clues provided by Chargaff's rules and Franklin's X-ray diffraction studies. Using brass and wire models of the bases, sugars, and phosphate groups, Watson and Crick deduced that the structure of the DNA molecule is a **double helix**, a spiral staircase composed of two strands of nucleotides whose bases face each other. The double helix is held together by weak hydrogen bonds between the bases. Adenine can form hydrogen bonds only with thymine, and guanine can form hydrogen bonds only with cytosine. **Figure 8-4** shows the structure of DNA. Francis Crick and James Watson were awarded the Nobel Prize in 1962 for their development of this model.

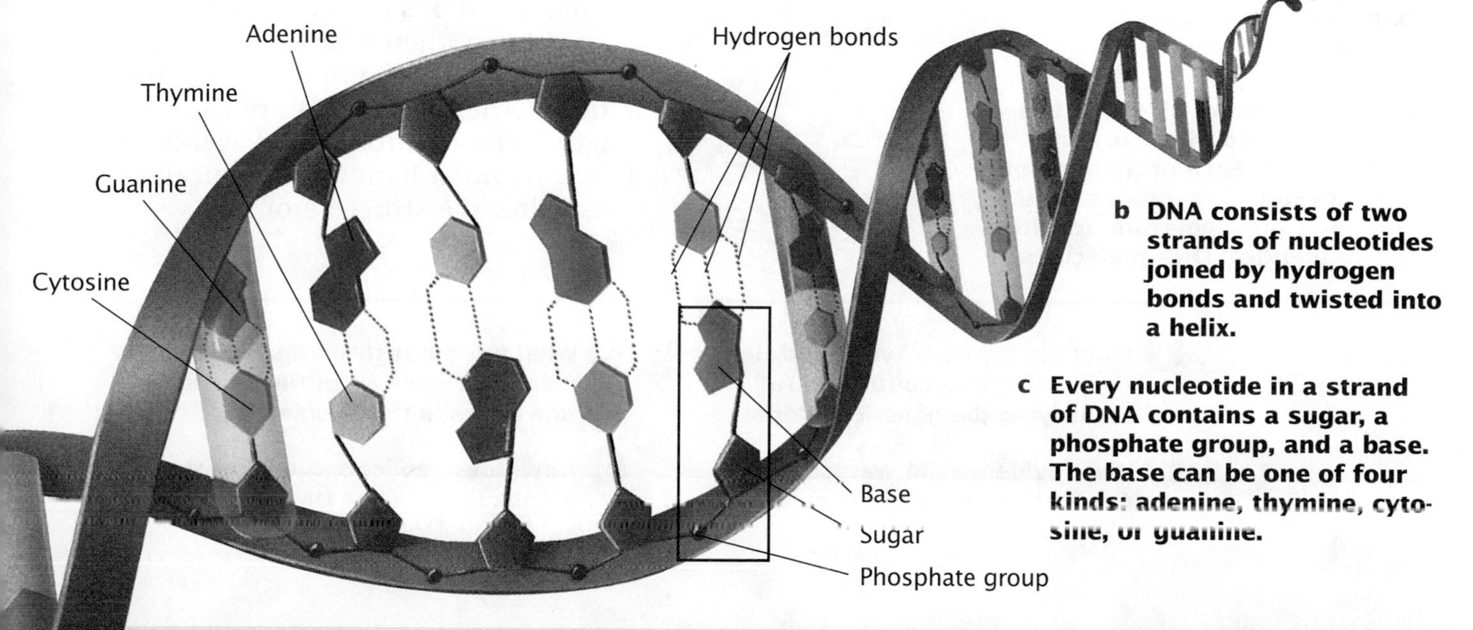

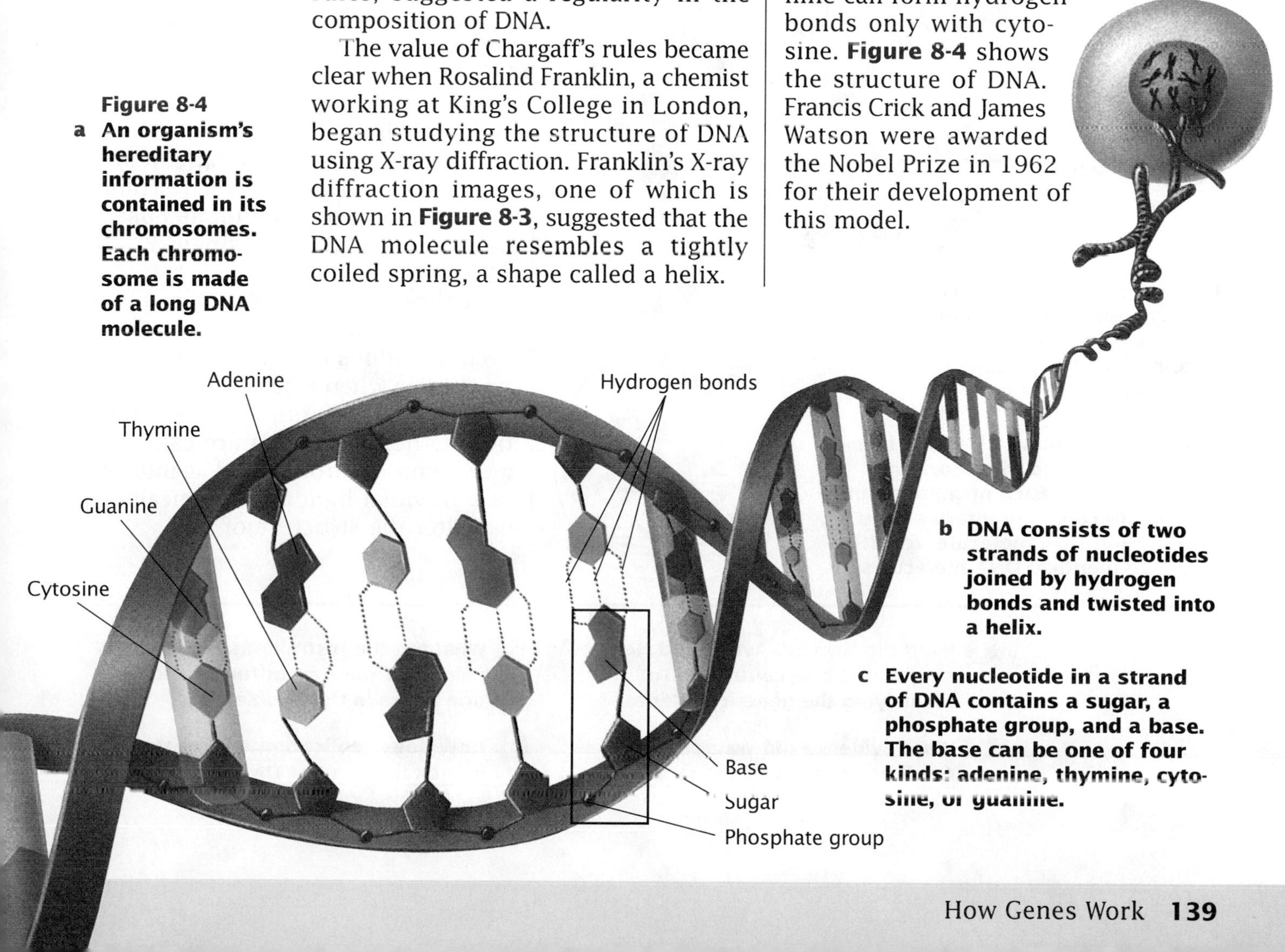

Figure 8-4
a An organism's hereditary information is contained in its chromosomes. Each chromosome is made of a long DNA molecule.

b DNA consists of two strands of nucleotides joined by hydrogen bonds and twisted into a helix.

c Every nucleotide in a strand of DNA contains a sugar, a phosphate group, and a base. The base can be one of four kinds: adenine, thymine, cytosine, or guanine.

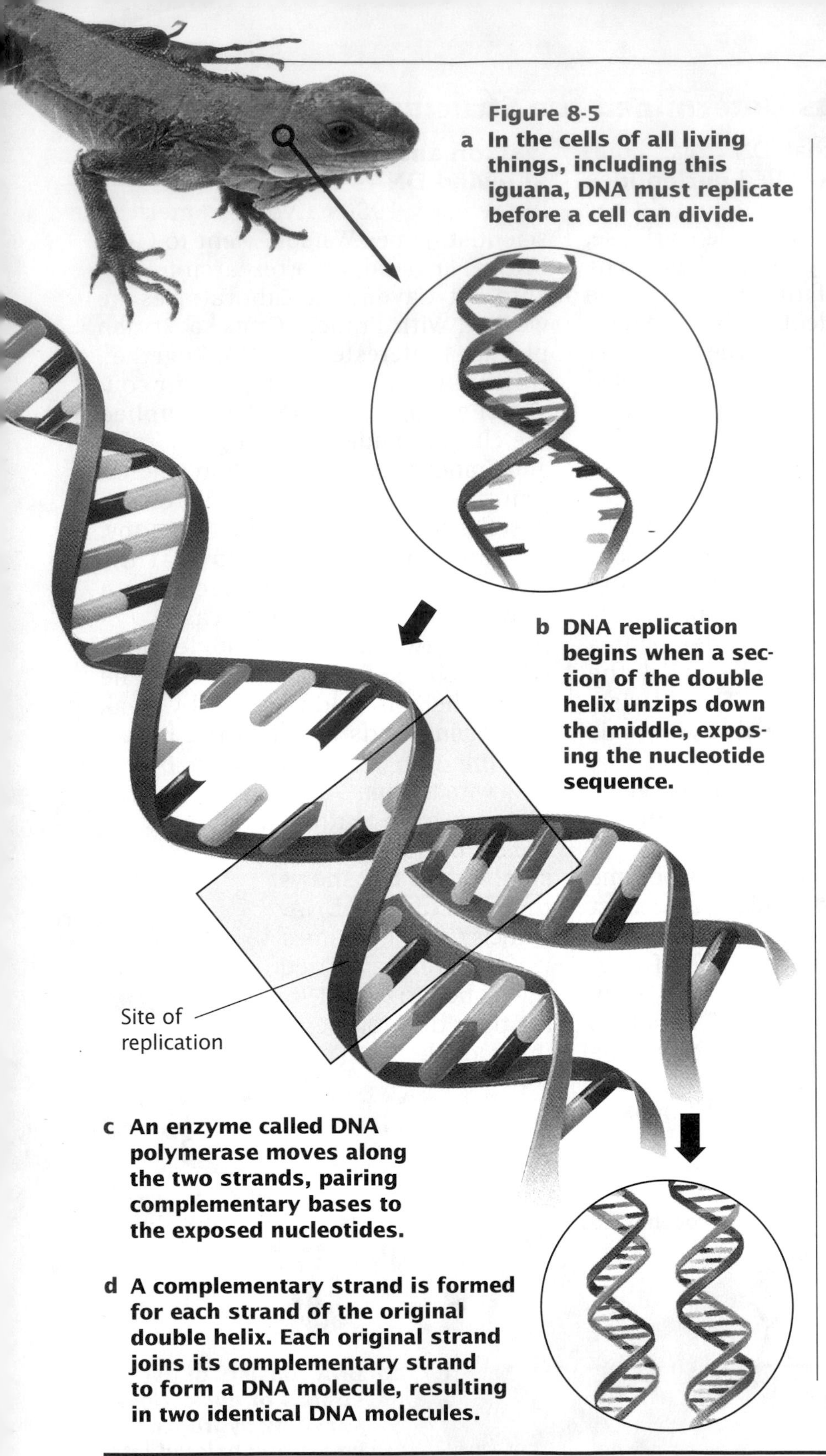

Figure 8-5
a In the cells of all living things, including this iguana, DNA must replicate before a cell can divide.

b DNA replication begins when a section of the double helix unzips down the middle, exposing the nucleotide sequence.

c An enzyme called DNA polymerase moves along the two strands, pairing complementary bases to the exposed nucleotides.

d A complementary strand is formed for each strand of the original double helix. Each original strand joins its complementary strand to form a DNA molecule, resulting in two identical DNA molecules.

How DNA Is Copied

Before a cell can successfully divide, its genetic information must be copied precisely. The process of copying DNA is called **replication**. To be successful, a model of DNA must be able to explain how replication could take place. Watson and Crick's double helix model readily suggested a mechanism. According to their model, the two strands of the double helix are complementary to each other; that is, the sequence of bases on one strand determines the sequence of bases on the other strand. For example, if the sequence of one strand of a DNA molecule is ATTGCAT, the sequence of the partner strand *must* be TAACGTA. Watson and Crick pointed out that if the two strands of the DNA molecule were separated, each strand would expose all of the necessary information to build two identical strands.

The actual process of replication, illustrated in **Figure 8-5**, is complex and requires the assistance of several enzymes. One enzyme is responsible for breaking the hydrogen bonds between base pairs and untwisting the two strands. Other enzymes work to shuttle the correct complementary nucleotides to the exposed strands. Replication occurs simultaneously at many points along a single human DNA molecule.

Replication preserves the sequence of bases in an organism's DNA. However, mutations can occur. Some mutations arise when a base is accidentally inserted or left out during replication. Other mutations are caused by **mutagens**, environmental agents such as ultraviolet light and chemicals that can alter the structure of DNA.

SECTION REVIEW

1. **What did Griffith, Avery, and Hershey and Chase contribute to identifying the genetic material?**
2. **What evidence did Watson and Crick use to deduce the structure of DNA?**
3. **What are the restrictions on the pairing of the four different nucleotide bases in the double helix?**
4. **How does replication ensure that identical copies of DNA are made?**

How is the information in DNA used? Scientists now know that DNA directs the construction of proteins. Proteins help determine the structure of cells and increase the rates of chemical reactions, such as those that occur during metabolism and photosynthesis.

8-2 How Proteins Are Made

OBJECTIVES

1. **Summarize the process of gene expression.**
2. **Describe how RNA is made.**
3. **Explain why the genetic code is said to be universal.**
4. **Explain the relationships among codons, anticodons, and amino acids.**

The Transfer of Genetic Information

Once scientists understood the structure of DNA, they were able to study how a specific protein is built from information found in the DNA of one gene. Scientists now know that DNA is used as a blueprint to make a similar molecule called **ribonucleic acid**, or RNA for short. This RNA molecule then directs the formation of proteins. **Gene expression** is the use of the genetic information in DNA to make proteins. Gene expression takes place in two stages. The first stage is called **transcription**. During transcription, an RNA copy of a gene is made. During **translation**, the second stage of gene expression, three different kinds of RNA work together to assemble amino acids into a protein molecule. The process of gene expression is summarized in **Figure 8-6**.

Through gene expression, the information encoded in DNA directs all cellular activities. For example, when you eat carbohydrates, such as those found in a bowl of cereal or a slice of bread, certain genes direct the production of a protein called insulin. Insulin helps your body maintain its blood-sugar level. Other genes direct the production of thousands of other structural proteins and enzymes.

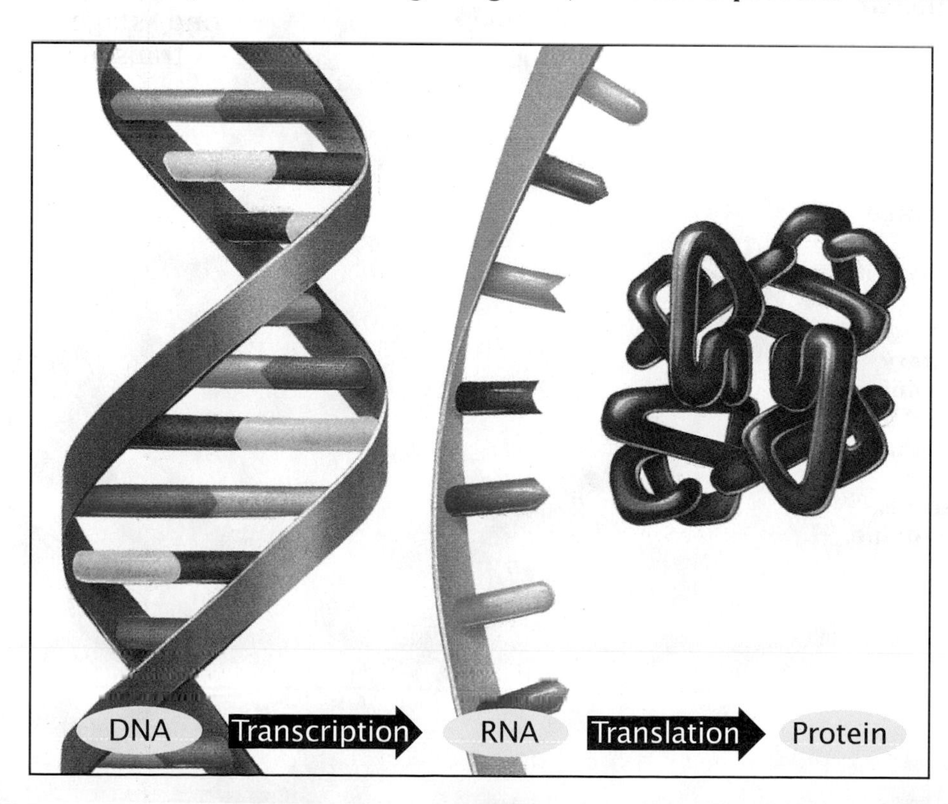

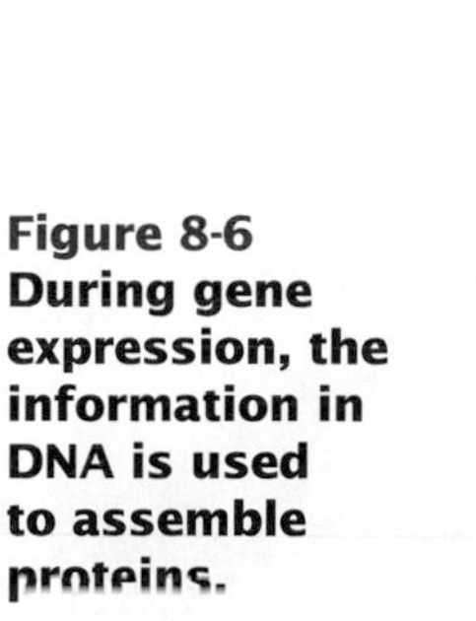

Figure 8-6 During gene expression, the information in DNA is used to assemble proteins.

How DNA Makes RNA

Transcription is the process by which genetic information encoded in DNA is transferred to an RNA molecule. Genetic information must be copied because DNA cannot leave the nucleus. Just as an architect protects building plans from loss or damage by keeping them in a central place, your cells protect genetic information by keeping the DNA safe within the nucleus. Instead of sending out the DNA, copies of genes are sent out into the cell to direct the assembly of proteins. These working copies of genes are single strands of RNA.

RNA is chemically similar to DNA except that its sugar, ribose, has an additional oxygen atom, and the base thymine (T) is replaced by a structurally similar base called uracil (U). RNA occurs in three different forms. **Figure 8-7** shows transcription of one form of RNA, messenger RNA (mRNA). Just as monks once copied manuscripts by faithfully transcribing each letter, enzymes in your cells' nuclei make mRNA copies of your genes by copying their nucleotides.

Like DNA replication, transcription is a complex process that requires the help of enzymes. Transcription begins when the enzyme RNA polymerase binds to a region of DNA called a promoter. Once bound to a promoter, the RNA polymerase unwinds the double helix and separates a section of the two DNA strands. As shown in **Figure 8-7**, RNA polymerase moves along one strand of the DNA like a train on a track. As it proceeds, complementary base-pairing takes place between the DNA strand and nucleotides. RNA polymerase joins these nucleotides one right after the other, forming a chain of single-stranded RNA. This process continues until a specific "stop" sequence in the DNA is reached. The polymerase then detaches from the DNA, and the RNA strand drifts free. In eukaryotes, after transcription of a gene is finished, the mRNA passes out of the nucleus and into the cytoplasm through pores in the nuclear membrane. There the second stage of gene expression, translation, takes place.

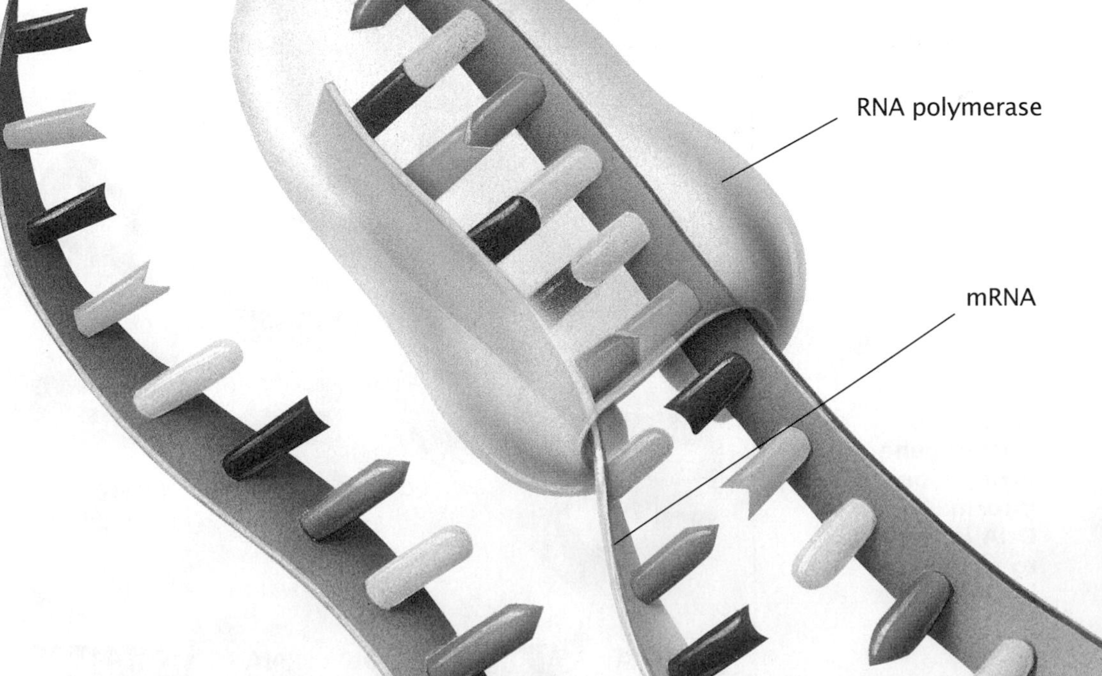

Figure 8-7

a During transcription, a portion of the double helix unwinds, exposing a sequence of genetic information.

b Then the enzyme RNA polymerase binds to the exposed bases. It moves along one strand, pairing complementary bases and joining them to build a strand of RNA. RNA contains four bases: adenine, guanine, cytosine, and uracil.

Figure 8-8
In the genetic code, each amino acid is coded for by three mRNA bases arranged in a specific sequence.

a **The first base in a codon is found along the left side of the chart.**

b **The second base is at the top of the chart.**

c **The third base in the codon is found along the right side of the chart.**

1st ↓ / 2nd →	U	C	A	G	
U	Phenylalanine Phenylalanine Leucine Leucine	Serine Serine Serine Serine	Tyrosine Tyrosine *Stop* *Stop*	Cysteine Cysteine *Stop* Tryptophan	U C A G
C	Leucine Leucine Leucine Leucine	Proline Proline Proline Proline	Histidine Histidine Glutamine Glutamine	Arginine Arginine Arginine Arginine	U C A G
A	Isoleucine Isoleucine Isoleucine Methionine	Threonine Threonine Threonine Threonine	Asparagine Asparagine Lysine Lysine	Serine Serine Arginine Arginine	U C A G
G	Valine Valine Valine Valine	Alanine Alanine Alanine Alanine	Aspartic acid Aspartic acid Glutamic acid Glutamic acid	Glycine Glycine Glycine Glycine	U C A G

The Genetic Code

Evolution

How is the DNA of all organisms similar?

How is mRNA translated into the sequence of amino acids that make up proteins? How can the four nucleotide bases found in mRNA carry instructions to build the thousands of proteins your body needs? Every three nucleotides in mRNA specify a particular amino acid. Each nucleotide triplet in mRNA is called a **codon** *(KOH dahn)*. The order of bases in a codon determines which amino acid will be added to a growing protein chain. In turn, the order of amino acids will determine the structure and function of a protein.

To learn more about how mRNA directs amino acids to join in a specific order, biologists performed laboratory experiments using artificial mRNA to direct protein production. An mRNA molecule that contained only the nucleotide uracil (U), for example, made a protein that consisted entirely of the amino acid phenylalanine. This information told scientists that the codon "UUU" codes for the amino acid phenylalanine. These experiments ultimately revealed the genetic code, which is shown in **Figure 8-8**. The **genetic code** is the correspondence between nucleotide triplets in DNA and the amino acids in proteins. Any of the four bases (U, C, A, G) found in mRNA can occur at any of the three positions of a codon. Thus, 64 different possible three-letter codons ($4 \times 4 \times 4 = 64$) are possible in the genetic code. Since there are 64 possible codons but only 20 different amino acids, more than one codon may specify a single amino acid. Codons known as "stop" codons specify the end of an amino-acid sequence.

The genetic code is the same in nearly all organisms, so it is said to be universal. For example, the code for phenylalanine is the same in bacteria and humans. The universality of the genetic code supports the view that the code had originated at least by the time bacteria evolved, over 3.5 billion years ago. The genetic code may be considered the oldest of languages.

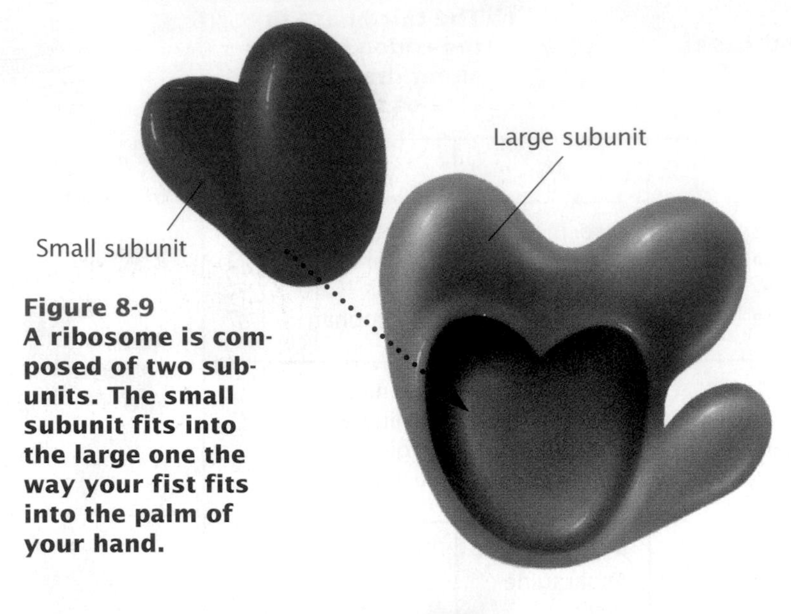

Figure 8-9
A ribosome is composed of two subunits. The small subunit fits into the large one the way your fist fits into the palm of your hand.

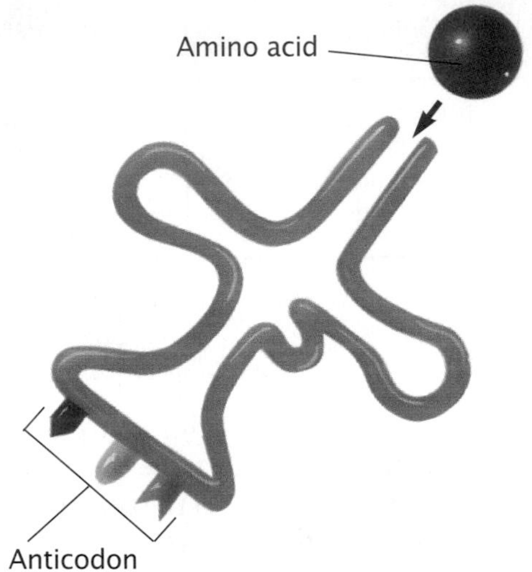

Figure 8-10
A transfer RNA molecule is a chain of RNA about 80 nucleotides long, folded into a compact shape. At one end of the molecule is a three-nucleotide sequence called the anticodon. At the opposite end is an amino acid.

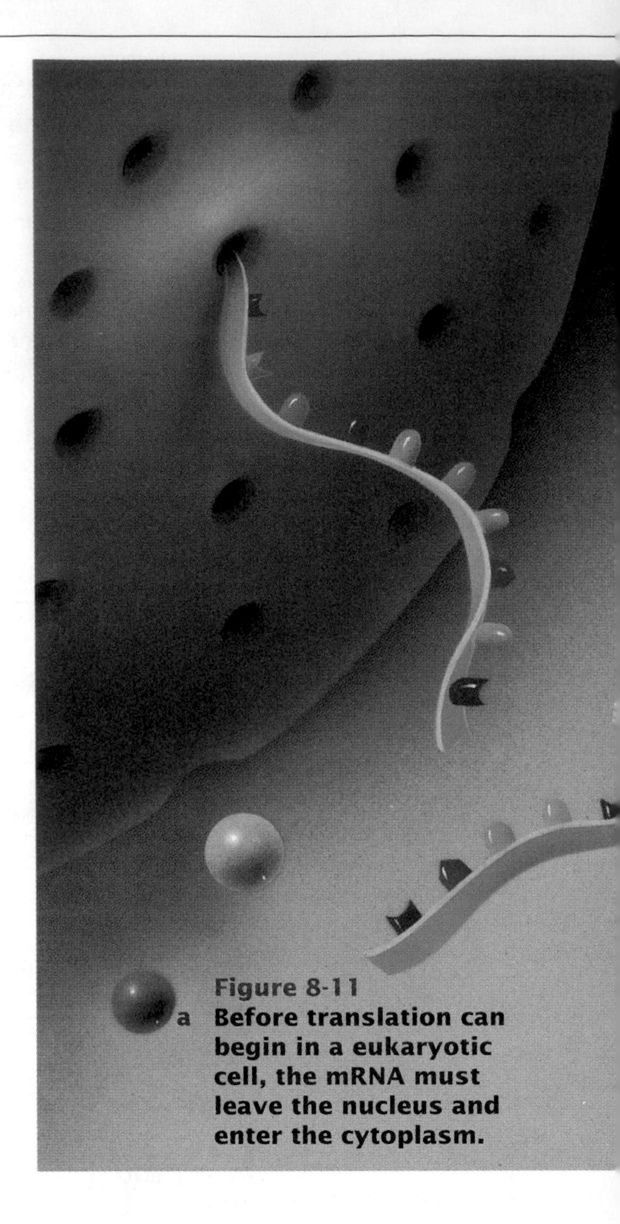

Figure 8-11
Before translation can begin in a eukaryotic cell, the mRNA must leave the nucleus and enter the cytoplasm.

How RNA Makes Proteins

Translation is the process by which the genetic message is deciphered. During translation, the mRNA works with two other types of RNA to build proteins by joining amino acids. Translation occurs on ribosomes, which are complex organelles that contain a special kind of RNA called ribosomal RNA (rRNA). **Figure 8-9** shows that each ribosome consists of two subunits. During translation, an mRNA molecule passes between the two subunits of a ribosome.

Translation begins when mRNA binds to a specific site on the small ribosomal subunit. The large ribosomal subunit attaches to form a complete ribosome with a strand of mRNA running through it. Just as factories use blueprints to direct the assembly of cars, ribosomes use mRNA to direct the assembly of proteins.

The pocket, or dent, in the small ribosomal subunit has just the right shape to bind a third kind of RNA molecule, transfer RNA (tRNA), shown in **Figure 8-10**. Transfer RNA carries amino acids to the ribosome. On one end of the transfer RNA molecule is a three-nucleotide sequence called an **anticodon**. Anticodons are complementary to mRNA codons.

Like the address on an envelope, the anticodon ensures that an amino acid is delivered to its proper "address" on the mRNA as a protein is being assem-

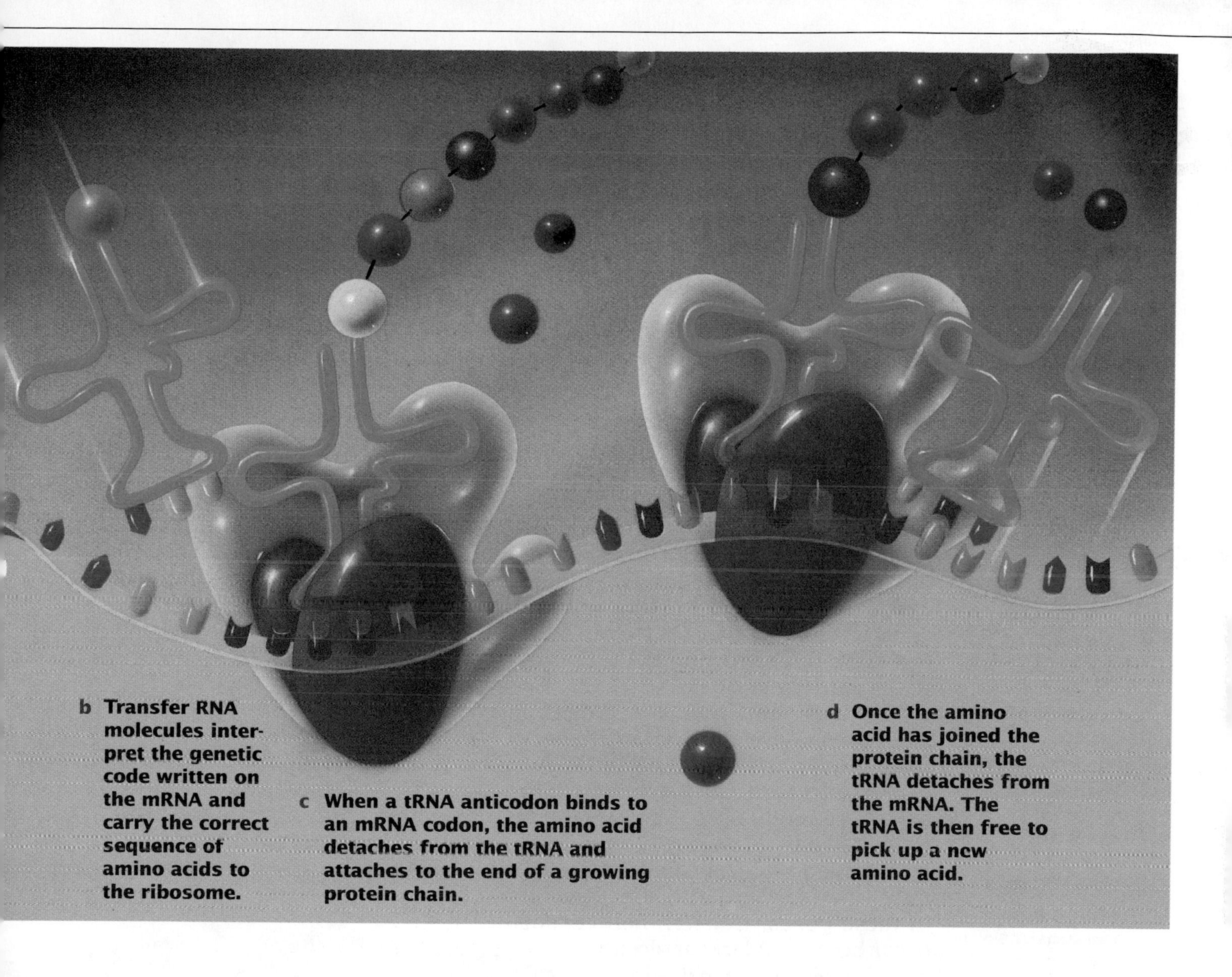

b Transfer RNA molecules interpret the genetic code written on the mRNA and carry the correct sequence of amino acids to the ribosome.

c When a tRNA anticodon binds to an mRNA codon, the amino acid detaches from the tRNA and attaches to the end of a growing protein chain.

d Once the amino acid has joined the protein chain, the tRNA detaches from the mRNA. The tRNA is then free to pick up a new amino acid.

bled. And just as a mail carrier checks the address before delivering a letter, so the small ribosomal subunit checks the tRNA anticodon to see that it is complementary to the mRNA codon.

As the mRNA passes along the ribosome, one tRNA after another is selected to match the sequence of mRNA codons. Amino acids are added to the growing protein chain until the end of the mRNA sequence is reached as seen in **Figure 8-11**. At this point, a "stop" codon is encountered for which there is no corresponding anticodon on any tRNA molecule. With nothing to fit into the tRNA site, the ribosome complex falls apart and the newly assembled protein is released into the cell. A cell's cytoplasm contains thousands of protein-making ribosomes.

SECTION REVIEW

1. **Outline the path taken by the genetic information as proteins are made.**
2. **How does a cell make RNA?**
3. **What is the genetic code? Why is it said to be universal?**
4. **How do codons and anticodons interact to build proteins? How will a mutation that causes the codon UGU to be changed to UGG affect the amino-acid sequence specified by a strand of mRNA?**

The translation of a gene into a protein is only part of gene expression. Every cell must also be able to regulate the use of particular genes. Just as a conductor controls when the different instruments in an orchestra play, a cell determines when particular genes are transcribed by controlling when genes are switched "on" and "off."

8-3 Regulating Gene Expression

OBJECTIVES

1. **Explain why cells must regulate gene expression.**
2. **Summarize how a gene can be switched off and on.**
3. **Distinguish between exons and introns.**
4. **Explain how transposons affect gene expression.**

Switching Genes On and Off

Cells control the expression of their genes by determining when individual genes are to be transcribed. Each gene possesses special regulatory sites, which act as points of control. Specific regulatory proteins within the cell bind to these sites, switching transcription of the gene on or off.

The best understood regulatory mechanisms are those used by bacteria. Some genes in bacteria are expressed nearly all the time, while others are rarely used. Genes that are expressed only occasionally are said to be switched off. They are transcribed only when the proteins are needed. In these genes, transcription cannot occur because a large molecule called a **repressor protein** is bound to the DNA in front of the genes, as shown in **Figure 8-12a**. The repressor protein blocks transcription by preventing the RNA polymerase from moving along the gene. If someone placed a brick wall between your chair and desk, you could not begin your work until the wall was removed. In the same way, transcription cannot begin until the repressor protein is removed.

Transcription begins when an inducer is present

For transcription to begin, molecules called **inducers** must bind to the repressor protein. The binding causes the repressor protein to change its shape so that it no longer fits the DNA. As a result, the repressor protein falls off, removing the barrier to transcription. When this happens, the gene is switched on.

In bacteria, gene expression controls the digestion of lactose, the sugar

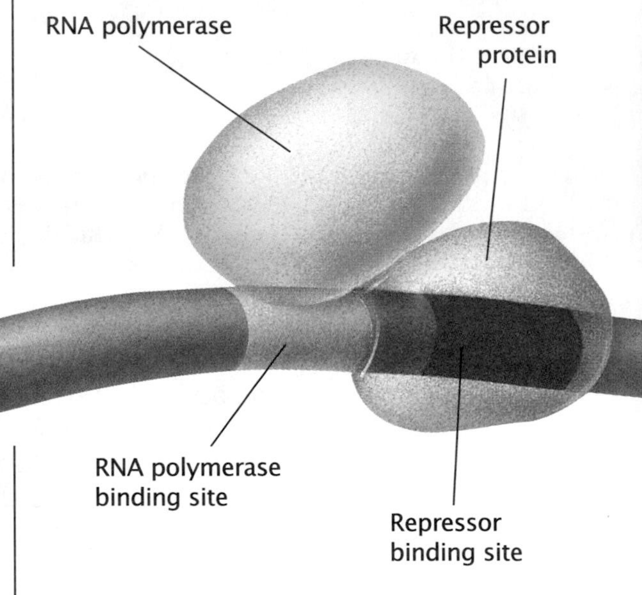

Figure 8-12
a **A repressor protein is bound to DNA, blocking RNA polymerase from its binding site and preventing transcription.**

found in milk. When a bacterium encounters lactose, the lactose acts as an inducer by binding to the repressor protein and altering its shape. The altered repressor protein then detaches from the DNA molecule, which allows the RNA polymerase to transcribe the genes needed to produce the enzyme responsible for digesting lactose.

Cell Structure and Function

What processes provide a cell with the proteins it needs to function properly?

Activators help DNA unwind

Because RNA polymerase binds to only one strand of the DNA double helix, it is necessary that the DNA molecule partially unwind to expose the bases at the promoter site. In many genes, this unwinding cannot take place without the help of a regulatory protein called an activator. An activator binds to the DNA in this region and helps it unwind. Cells are able to switch genes "off" by binding signal molecules to the activator protein. This prevents the activator from binding to the DNA molecules. As a result, the double helix cannot unwind and the genes cannot be transcribed.

Activators enable a cell to carry out a second level of control. When a bacterium already has plenty of energy, the level of another signal molecule—a special "I-need-sugar" signal molecule—decreases. Without being prodded by this signal molecule, the activator protein cannot bind to the DNA molecule's unwinding site. As a result, the genes are not transcribed, even though the repressor protein is not blocking the RNA polymerase.

Eukaryotic cells have many ways to regulate genes

Gene regulation is more complex in humans and other eukaryotic organisms than in bacteria, even though the same basic mechanisms are used. Because a nuclear envelope physically separates transcription from translation in a eukaryotic cell, more opportunities for regulating gene expression exist. Scientists have found mechanisms that occur after mRNA leaves the nucleus and after translation, when the protein is functional.

As in bacteria, most gene regulation in eukaryotes works by controlling the onset of transcription. However, eukaryotic cells employ a more flexible system. The decoding of each gene is controlled by proteins called transcription factors. Transcription factors help arrange polymerases in the correct position on the promoter. A gene can be influenced by many different factors. Sometimes one kind of transcription factor has several forms that bind to regions of DNA called enhancers. An enhancer is a sequence of DNA located thousands of nucleotide bases away from the promoter. When the strand of DNA is looped in the correct manner, factors bound to enhancers are able to activate factors bound to the promoter.

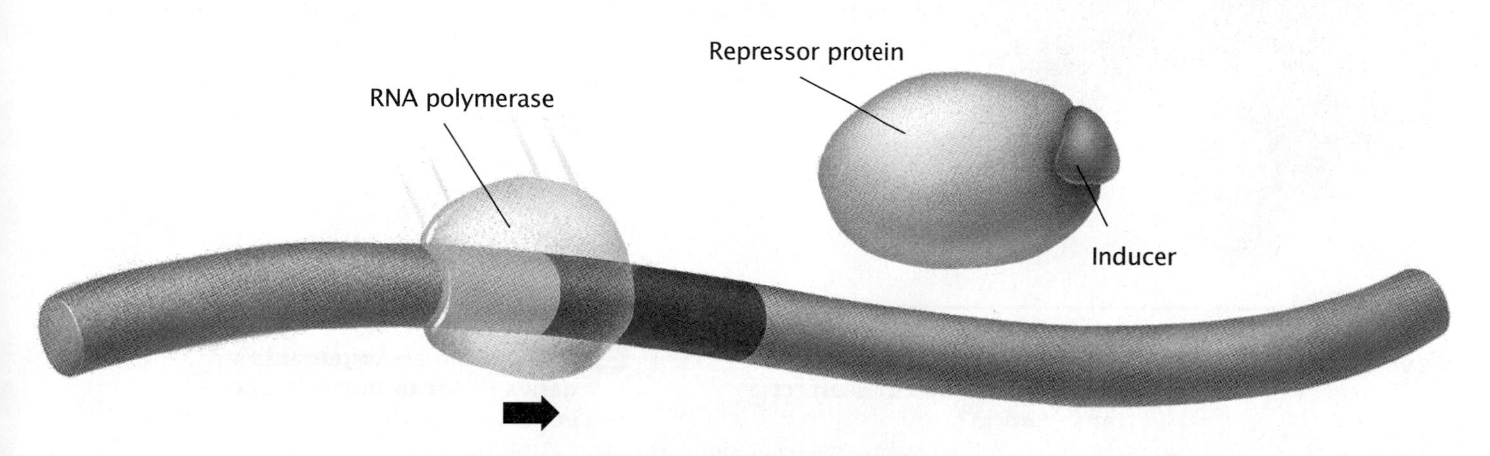

b When an inducer is present, it binds to the repressor protein and changes its shape. As a result, the repressor protein falls off the DNA, enabling RNA polymerase to bind to the DNA strand and begin transcribing genes.

Architecture of the Gene

While it is tempting to think of a gene as a single, uninterrupted stretch of DNA that codes for proteins, this simple arrangement occurs only in bacteria. In eukaryotes, a gene contains a series of sequences called exons and introns. **Exons** are the portions of a gene that actually get translated into proteins. They are interrupted by noncoding portions of the DNA called **introns**. In most genes, the introns far outweigh the protein-encoding exons. In fact, in most cases, less than 10 percent of a human gene consists of exons. Like cars on a rural highway, exons are scattered here and there within genes. Introns are separated from exons during transcription.

Figure 8-13 Barbara McClintock received a Nobel Prize in 1983 for her discovery of transposons in corn.

Exons play a role in evolution

How could such a complicated system have survived the process of natural selection? The answer is that it adds evolutionary flexibility. Each exon encodes a different part of a protein. One exon may influence which molecules an enzyme is able to recognize, while another may determine whether a protein will respond to particular signal molecules. By possessing introns and exons, cells can shuffle exons between genes and create new combinations. Natural selection probably favored the intron-exon system of organization because it enables cells to manufacture many different proteins by juggling exons between genes. The many thousands of proteins that occur in human cells appear to have arisen from only a few thousand exons!

Genes can jump to new locations

A few genes in chromosomes have the ability to move from one location in the chromosome to another. These genes are known as **transposons** *(tranz POH zahnz)*. Once every few thousand cell divisions, a transposon jumps to a new location in the same chromosome or in a different chromosome. Transposons often inactivate the genes they jump into, creating mutations. Barbara McClintock, a geneticist working in the Cold Spring Harbor Laboratory on Long Island, New York, discovered transposons during her studies of corn. The spotted and streaked patterns seen in the corn in **Figure 8-13** result from the interactions of transposons that control its kernel pigments.

SECTION REVIEW

1. **How do repressor proteins and inducer molecules affect transcription?**
2. **How do activator proteins help RNA polymerase bind to a strand of DNA?**
3. **How do the arrangements of genes differ in bacteria and eukaryotes?**
4. **What effect do transposons have on other genes?**

CHAPTER

8 Highlights

Watson and Crick built this model of DNA in 1953.

	Key Terms	Summary
8-1 Understanding DNA 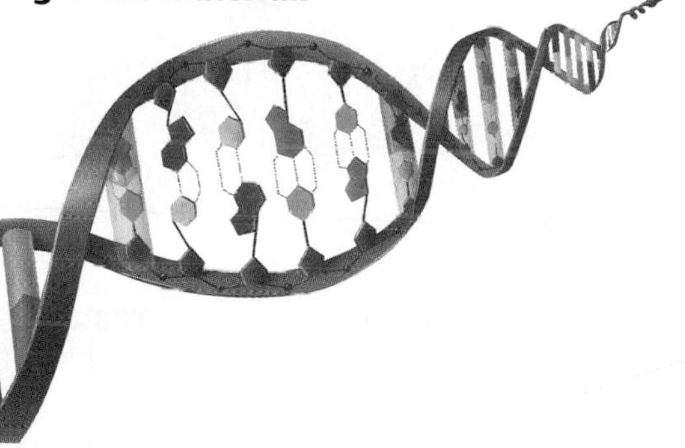 It took scientists many years to prove that DNA is the genetic material.	virulent (p. 137) transformation (p. 137) nucleotide (p. 139) purine (p. 139) pyrimidine (p. 139) double helix (p. 139) replication (p. 140) mutagen (p. 140)	• Griffith, Avery, and Hershey and Chase performed experiments that helped show that DNA is the hereditary material. • DNA is composed of subunits called nucleotides. Each nucleotide contains a sugar, a phosphate group, and one of four bases. • Watson and Crick showed that the DNA molecule is a double helix. • Before cell division, DNA copies itself in a process called replication. The DNA separates into two strands, and new complementary bases attach to the exposed base.
8-2 How Proteins Are Made 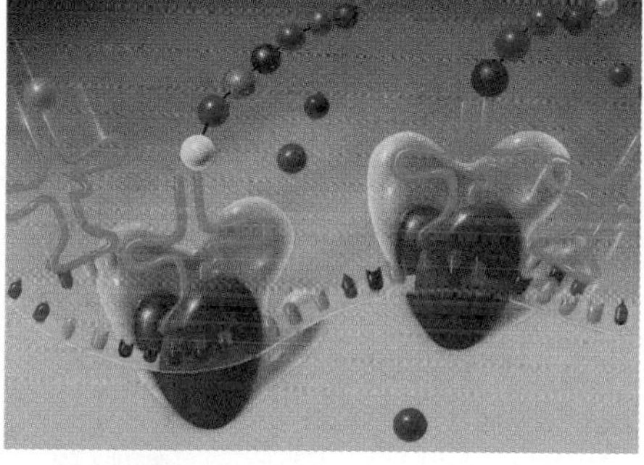 Proteins are made when amino acids are assembled by mRNA and tRNA molecules at a ribosome.	ribonucleic acid (p. 141) gene expression (p. 141) transcription (p. 141) translation (p. 141) codon (p. 143) genetic code (p. 143) anticodon (p. 144)	• During transcription, the genetic message from DNA is transferred to RNA. • During translation, RNA directs the production of specific proteins encoded by genes. • Each group of three nucleotides in mRNA is called a codon. Codons specify amino acids. • A three-nucleotide sequence on a tRNA molecule is called an anticodon. Anticodons complement the nucleotide sequence in mRNA codons.
8-3 Regulating Gene Expression Cells can control gene activity. When an inducer binds to the repressor molecule, it falls from the DNA, allowing transcription to begin.	repressor protein (p. 146) inducer (p. 146) exon (p. 148) intron (p. 148) transposon (p. 148)	• In prokaryotes, transcription is regulated by repressor proteins. A repressor protein blocks RNA polymerase from transcribing a gene. • In eukaryotes, genes are fragmented. Exons are the portions of a gene that are translated into proteins. Introns are noncoding regions of DNA. • Transposons are genes that can jump to new locations in chromosomes.

CHAPTER 8

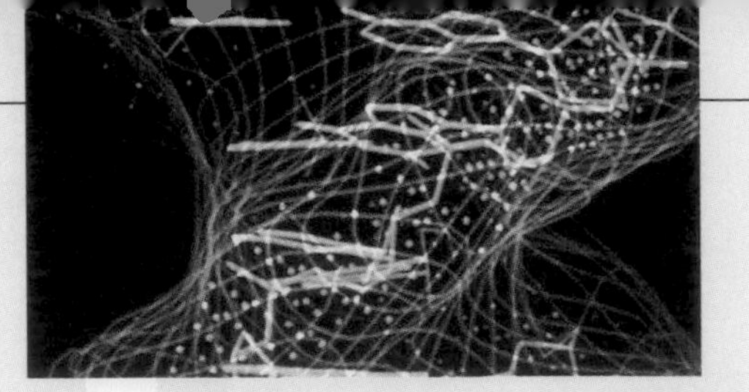

review

Understanding Vocabulary

1. For each pair of words, explain the differences in their meanings.
   a. gene, nucleotide
   b. transcription, translation
   c. repressor protein, inducer

2. For each set of terms, complete the analogy.
   a. Griffith : transformation :: Avery : ____
   b. process in which a bacterium takes up foreign DNA : transformation :: process of copying DNA : ____
   c. thymine : DNA :: uracil : ____

Understanding Concepts

3. **Relating Concepts** Construct a concept map describing the process of translation. Try to use the following terms in your map: codons, rRNA, tRNA, amino acids, RNA, ribosome, anticodons, mRNA, and protein.

4. **Inferring Conclusions** In Frederick Griffith's experiments, how did nonvirulent strains of *Pneumococcus* bacteria become virulent?

5. **Relating Conclusions** Describe how Oswald Avery and his associates determined that DNA is the genetic material in cells.

6. **Summarizing Information** Summarize the experiments of Hershey and Chase that confirmed DNA as the genetic material.

7. **Identifying Structures** What are the three parts of a nucleotide?

8. **Comparing Structures** What are purines? pyrimidines? Give an example of each.

9. **Summarizing Information** Summarize Chargaff's rules.

10. **Inferring Relationships** What contribution did Rosalind Franklin make toward discovering the structure of DNA?

11. **Relating Conclusions** Why is the genetic code considered to be the oldest of languages?

12. **Predicting Patterns** What would be the complementary base sequence for the following sequence of bases on an unzipped strand of DNA: T T A C G C T A? Explain your answer.

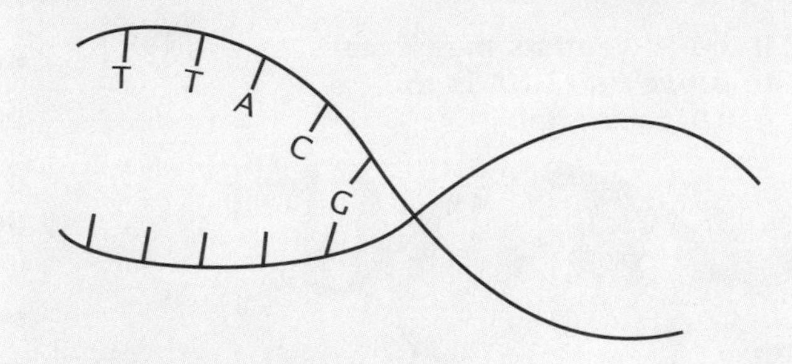

13. **Organizing Information** What is gene expression? Identify and describe the two stages of gene expression.

14. **Inferring Relationships** Where and how do the processes of transcription and translation each occur?

15. **Summarizing Information** Explain how codons specify a particular amino acid.

16. **Comparing Functions** What are the names and the roles of the three types of RNA? Where in the cell does each type of RNA function?

17. **Identifying Functions** What is a repressor protein? How does it prevent transcription?

18. **Identifying Structures** What are promoters, activators, and enhancers? Where is each located on a gene?

19. **Inferring Conclusions** How would the operation of RNA polymerase be affected if the repressor protein were not bound to the proper site on a gene?

20. **Summarizing Information** Describe what happens when a "stop" codon is encountered during protein assembly.

21. **BUILDING ON WHAT YOU LEARNED** In Chapter 6 you learned about mitosis and cell division. Why is it important that DNA replication occur before mitosis and cell division?

Interpreting Graphics

22. The figure below shows the events of translation. Recall that translation follows transcription and is the process by which a protein molecule is assembled according to the mRNA code.

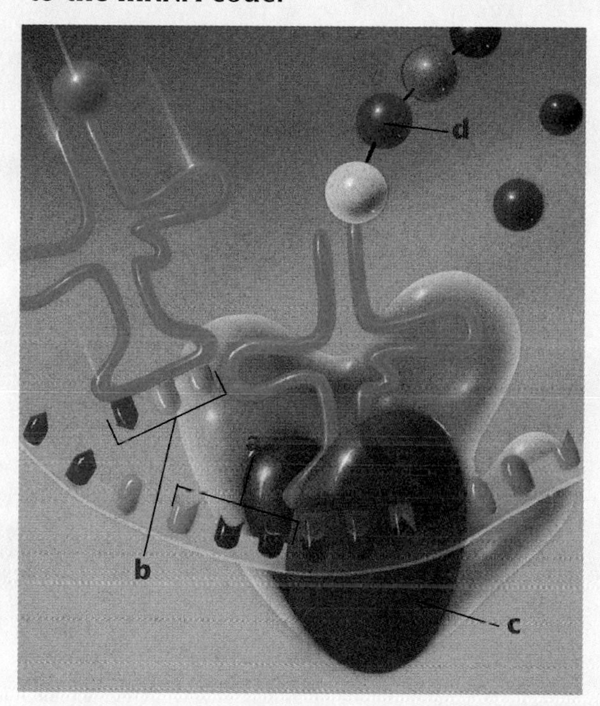

a. What does *a* represent in the model?
b. Part *c* represents the organelle inside the cell where translation occurs. What is the name of this organelle?
c. What is the relationship between *a* and *b* during translation?
d. What structure is represented by *d*?

Reviewing Themes

23. ***Evolution***
How does the genetic code support the theory that all organisms have a common ancestry?

24. ***Evolution***
How does natural selection favor the intron-exon system of organization within chromosomes?

Thinking Critically

25. **Distinguishing Relevant Information** After messenger RNA has assisted in the formation of necessary proteins within the cell, it breaks down into free-floating nucleotides. What problems would occur inside the cell if mRNA did not break down after protein synthesis?

Life/Work Skills

26. **Finding Information** Find out more about Barbara McClintock, who received a Nobel Prize in 1983 for her discovery of transposons in corn. Write a report about her life and include when she began to study the "jumping" gene that causes irregular coloration in corn.

27. **Communicating Effectively** Find out the genetic link between "mad cow disease" and similar diseases that afflict humans. What are prions? Why are prions considered to be very different from bacteria and viruses that cause disease? How do prions cause disease? Report to your class on your findings.

Activities and Projects

28. **Language Arts Connection** As you learned in this chapter, proteins are compounds that consist of one or more chains of amino acids. Do library research to find the origin of the word *protein* and to determine why it is used to describe amino-acid chains.

29. **Unit Focus** Read J. D. Watson's book *The Double Helix*, which provides a day-by-day account of how he and Crick developed the double-helix model of DNA. Write a book report, and share your favorite passages from Watson's book with your classmates.

Discovering Through Reading

30. Read the article "Gene Defects and Therapies: An Update on the Expanding Nature of Science," in *American Biology Teacher*, February 1995, pages 70–75. What will dissection of the human genome lead to? What approaches do geneticists want to use in treating hereditary and medical diseases?

31. Read "Hidden Costs of a Clean Inheritance," in *New Scientist*, May 1994, pages 14–15. What is germ-line therapy? What do scientists think its therapeutic value will be? What outcomes do its critics fear?

32. Read "Artificial RNA Enzymes: Big and Fast," in *Science News*, July 22, 1995, page 53. What are ribozymes? What is a catalytic domain? How did Bartel and Szostak create molecular test-tube evolution? Why did researchers try to create ribozymes?

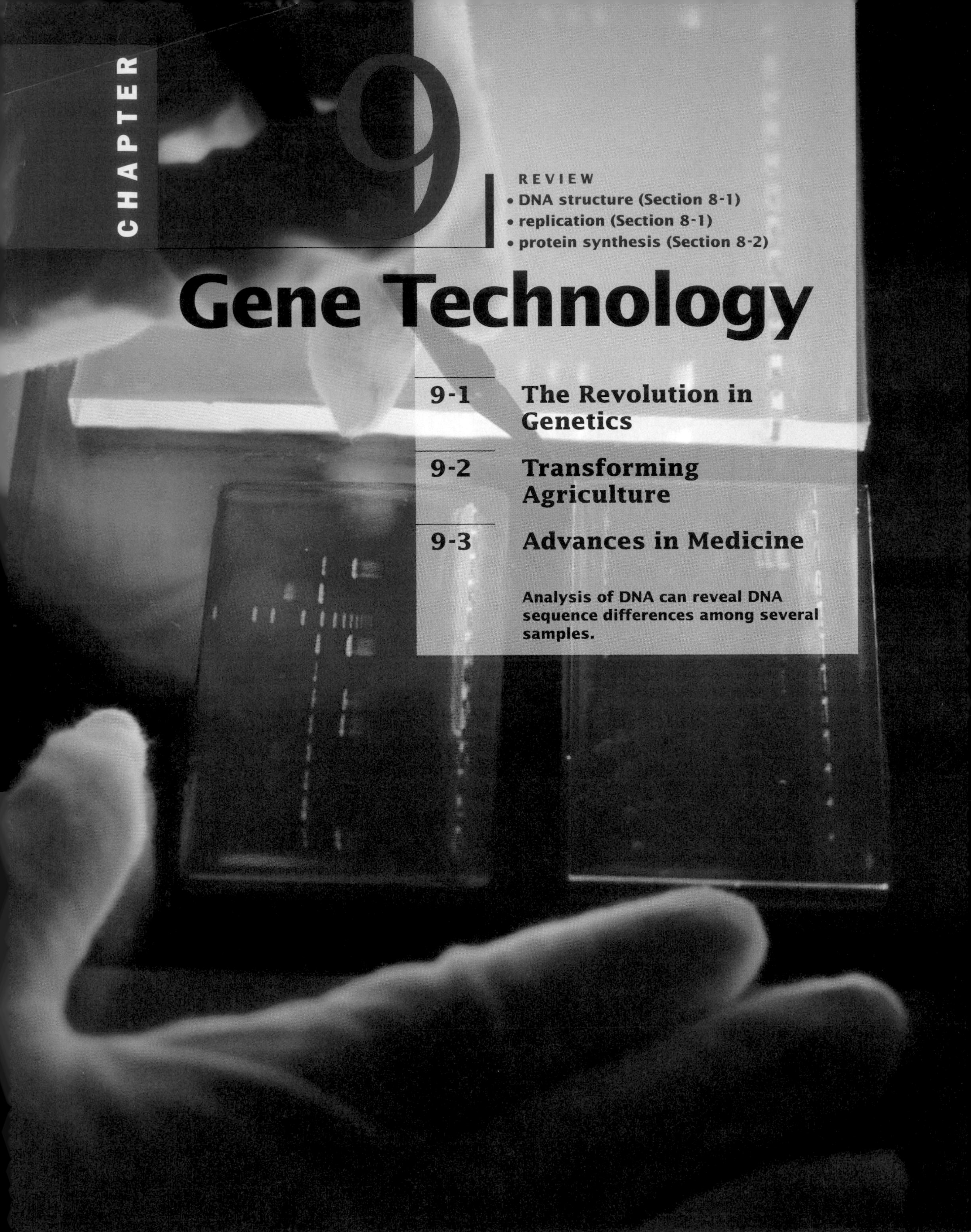

CHAPTER 9

REVIEW
- DNA structure (Section 8-1)
- replication (Section 8-1)
- protein synthesis (Section 8-2)

Gene Technology

Analysis of DNA can reveal DNA sequence differences among several samples.

The basic mechanisms of gene expression have been understood since the 1950s. Since then, scientists, like this one, have worked to transfer genes under laboratory conditions. In order to perform effective gene transfers, they needed to be able to transfer genes in a controlled fashion. Within the last 20 years, scientists have successfully developed the technology that has made gene transfers common.

9-1 The Revolution in Genetics

OBJECTIVES

1. Outline the genetic engineering experiment performed by Cohen and Boyer.
2. Explain the four stages involved in a gene transfer experiment.
3. Describe how scientists use restriction enzymes.
4. Summarize some of the uses of genetic engineering.

What Is Genetic Engineering?

In 1973 Stanley Cohen and Herbert Boyer, two geneticists at the University of California at San Francisco, began an experiment that would revolutionize genetics. Cohen and Boyer set out to insert a gene from an African clawed frog into a bacterium. They were hoping that the bacterium would then make the molecule encoded by the frog gene.

To begin their experiment, Cohen and Boyer isolated the gene that coded for frog ribosomal RNA. Then they added that gene to bacterial DNA. This procedure produced the first living cells that had DNA from a foreign organism added to their own DNA. **Recombinant DNA** is a molecule formed when fragments of DNA from two or more different organisms are spliced together in a laboratory. Did the bacteria with the recombinant DNA use the frog gene? Yes! The genetically engineered bacteria produced frog rRNA, just as Cohen and Boyer had hoped. **Figure 9-1** outlines the stages of their experiment.

This experiment ushered in a new age in biology, one that explores the possibilities of moving genes from one organism to another. Over a period of 20 years, the increasing ability of researchers to transfer DNA from one organism into another has revolutionized genetics. **Genetic engineering** is the process of moving genes from the chromosomes of one organism to those of another. Today, for example, specific genes from human chromosomes can be routinely transferred into bacteria.

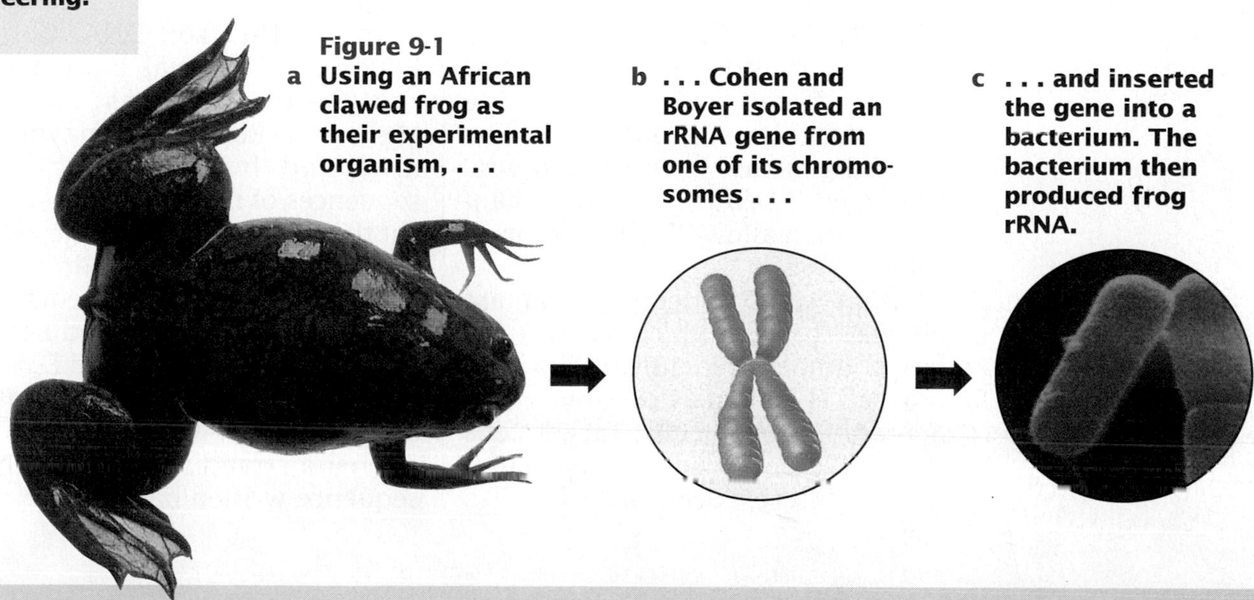

Figure 9-1
a Using an African clawed frog as their experimental organism, . . .
b . . . Cohen and Boyer isolated an rRNA gene from one of its chromosomes . . .
c . . . and inserted the gene into a bacterium. The bacterium then produced frog rRNA.

Figure 9-2
a In the first stage of their experiment, Cohen and Boyer obtained chromosomes from a frog . . .

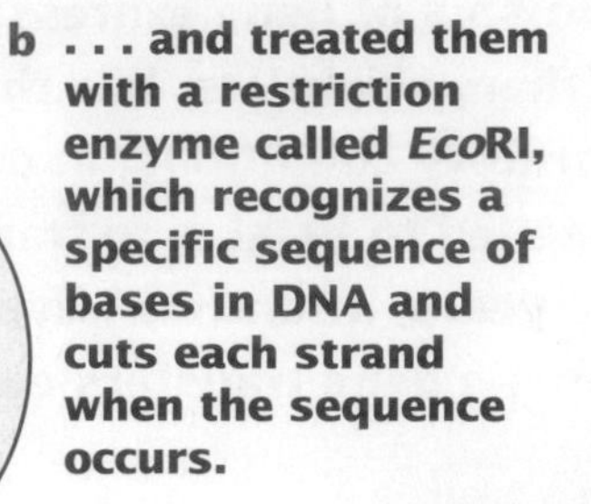

b . . . and treated them with a restriction enzyme called *Eco*RI, which recognizes a specific sequence of bases in DNA and cuts each strand when the sequence occurs.

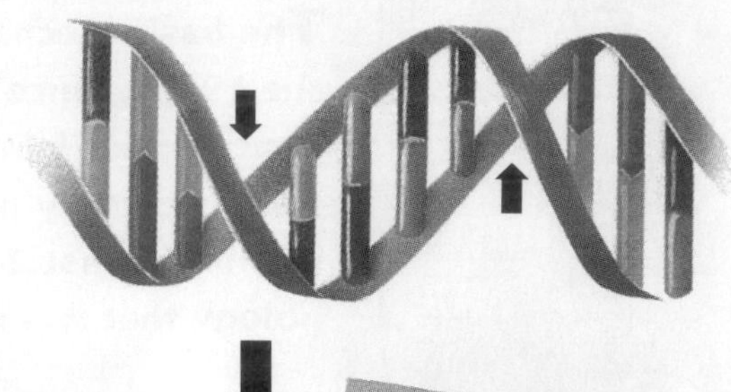

c These cuts resulted in fragments of DNA with single-stranded tails, or "sticky ends."

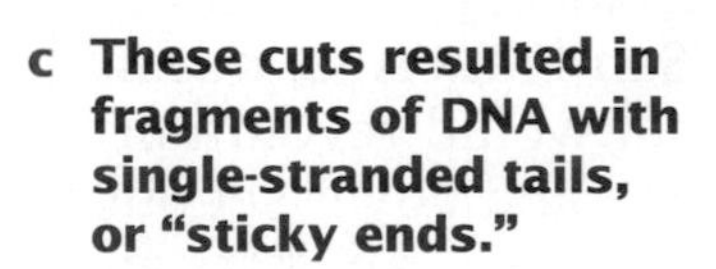

How to Move a Gene From One Organism to Another

If you were a scientist working in a genetic engineering lab, how would you go about moving a gene? First you would need to find a source chromosome that contains the gene you wish to isolate. Next you would need to select a target cell into which the gene could be moved. And finally, you would need to figure out a way to transfer the isolated gene into the target cell. Every gene transfer experiment is performed in four distinct stages:

1. *Cleaving DNA* The source chromosome is cut into fragments of DNA.
2. *Producing recombinant DNA* The DNA fragments containing the desired gene are inserted into viral or bacterial DNA. The recombinant DNA is then allowed to infect target cells.
3. *Cloning cells* Infected cells are allowed to reproduce. Growing a large number of identical cells from one cell is known as **cloning**.
4. *Screening target cells* Target cells that have received the particular gene of interest are isolated.

Each of these stages will be explained in greater detail using the Cohen-Boyer experiment as an example.

Enzymes cut DNA into pieces with sticky ends

The first stage in any genetic engineering experiment involves cutting the source chromosome into fragments to isolate the gene you wish to transfer. In their experiment, Cohen and Boyer used chromosomes from *Xenopus laevis*, the African clawed frog.

To cut the frog DNA, Cohen and Boyer used special enzymes called **restriction enzymes**, as shown in **Figure 9-2**. Restriction enzymes recognize and bind to specific short sequences of nucleotide bases and then cut the DNA at a specific site within that sequence. There are hundreds of these enzymes, each of which cuts at a specific sequence. Cohen and Boyer used a restriction enzyme called *Eco*RI, which cuts DNA whenever it encounters the sequence CTTAAG. The sequence of the opposite strand is GAATTC, the same sequence written backward.

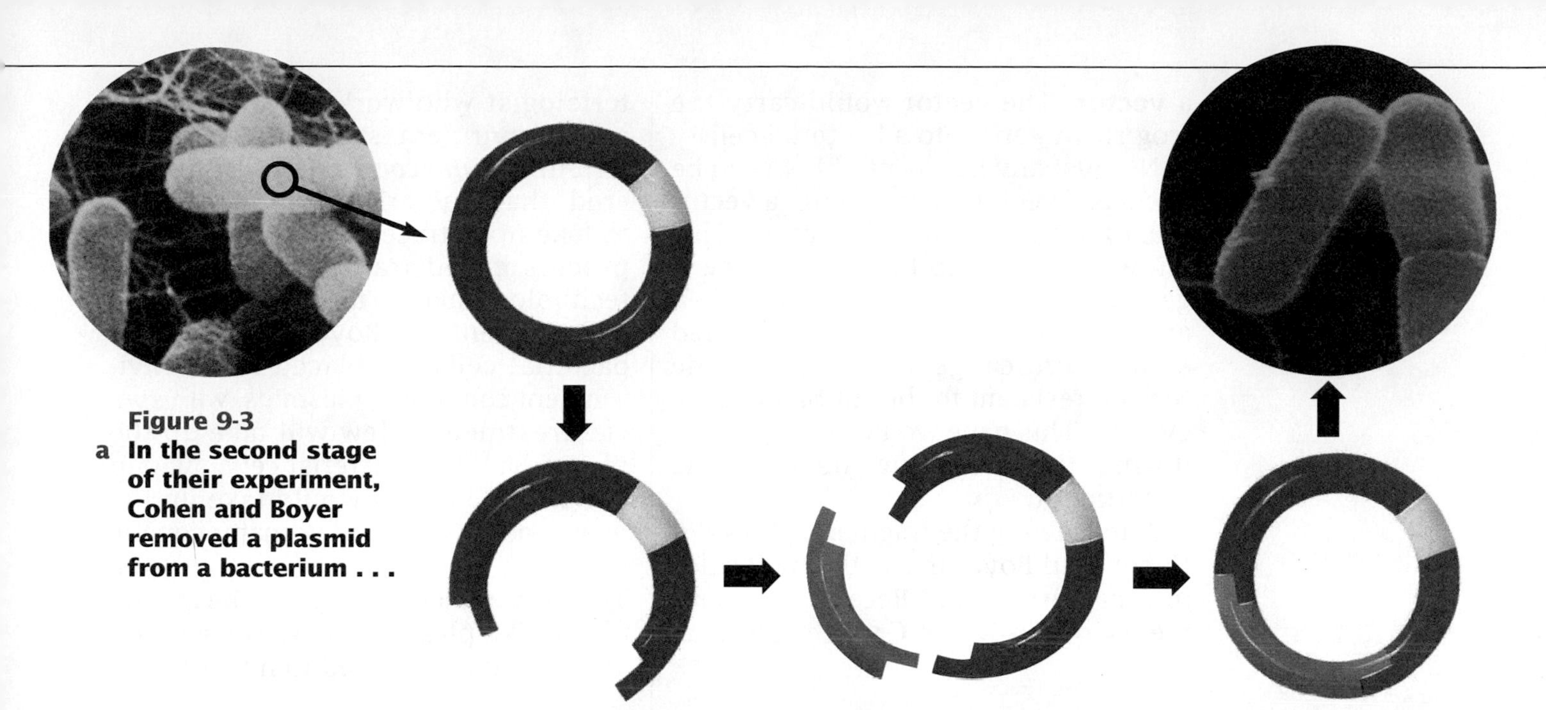

Figure 9-3

a In the second stage of their experiment, Cohen and Boyer removed a plasmid from a bacterium . . .

b . . . and treated it with the restriction enzyme *Eco*RI. The enzyme cut the plasmid, leaving sticky ends.

c Then they mixed the frog DNA with the plasmid. The fragments and the plasmid had complementary sticky ends that paired together like pieces of a puzzle.

d The resulting recombinant DNA molecules were used as vectors to infect bacterial cells.

Evolution

What evidence revealed by gene technology suggests that all organisms are related?

Note that restriction enzymes do not make a straight cut through both strands of DNA. Instead, the cut is offset a few bases. For example, in the sequence CTTAAG, *Eco*RI cuts each strand between the A and G, a site four bases apart on both strands, creating DNA fragments with single-stranded "sticky ends." One of the fragments that Cohen and Boyer cleaved from the frog DNA contained the rRNA gene that they wanted to transfer.

Because the two single-stranded sticky ends have complementary sequences, they can pair back up and seal the break (which is why they are called sticky), *or they can pair with any other DNA fragment cut by the same enzyme.* DNA cut by the restriction enzyme *Eco*RI can be joined to DNA from any other organism that has also been cut by *Eco*RI because they have complementary sticky ends. At this point, the DNA segments are held together by hydrogen bonds between complementary nucleotide bases. The bound fragments are then treated with an enzyme called ligase, which seals the breaks in the DNA molecule.

DNA fragments are inserted into a vector

Cohen and Boyer had to figure out a way to get a fragment of frog DNA into a bacterial cell. They used **plasmids**, which are small, circular DNA molecules found in the cytoplasm of bacteria. Plasmids are independent from the bacterial chromosome but can carry genes necessary for the cell's survival, as you will see in a moment. After isolating plasmids from bacterial cells, Cohen and Boyer exposed them to the restriction enzyme *Eco*RI. Because each plasmid had only one sequence that *Eco*RI could recognize, it was cut in only one place. As a result, the plasmids opened up with the same sticky ends as the frog DNA fragments.

Now Cohen and Boyer mixed the frog DNA fragments and the cut plasmids. Because all the sticky ends were complementary, they joined in a variety of combinations. Cohen and Boyer hoped that a fragment of frog DNA would join with a cut plasmid and then reform as a circular molecule, as shown in **Figure 9-3**. This combination would serve as a delivery agent called a **vector**. The vector would carry the frog rRNA gene into a bacterial cell.

Not just any fragment of DNA can be used as a vector. To be useful, a vector must have a gene for replication. This allows the DNA to be copied once it gets into a new cell. The vector Cohen and Boyer sought also contained another critical gene—one that made the cell resistant to the antibiotic tetracycline. This gene would be important during the screening stage of the experiment.

After mixing the fragments, how did Cohen and Boyer insert the molecules into bacterial cells? Recall the experiments of Frederick Griffith, the bacteriologist who worked with virulent and nonvirulent strains of the bacterium *Pneumococcus*. Griffith discovered that bacteria have the ability to take up extracellular bits of DNA, a process called transformation. Gene technology makes use of transformation. Cohen and Boyer knew that if bacterial cells are placed in an environment containing plasmids, with special treatment, a few will take up the plasmids. When bacterial cells take up the plasmids, they would take up the new gene as well. Today many genetic engineers also use viruses as vectors. The DNA in these viruses can be manipulated like plasmids, placed back into viruses, and then used to infect cells.

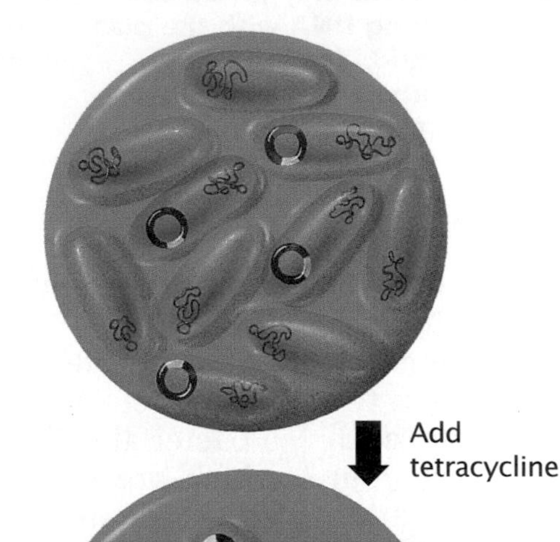

Figure 9-4

a In the final stages of their genetic engineering experiment, Cohen and Boyer grew bacteria in dishes that contained the antibiotic tetracycline.

b Only bacteria that contained the plasmids with the genes for resistance to tetracycline survived.

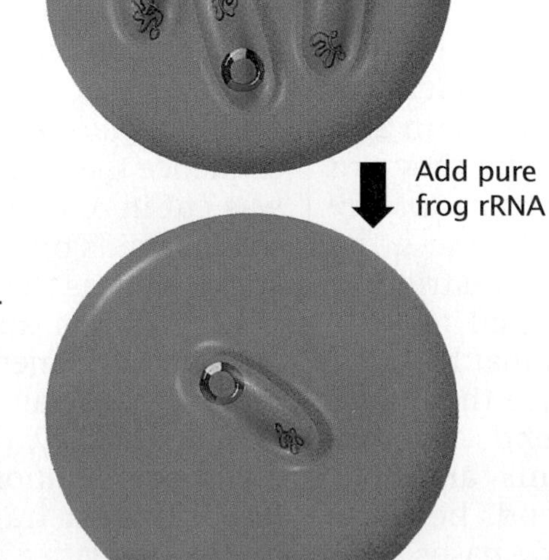

c Some of these plasmids also possessed the gene for frog rRNA. Cohen and Boyer then used radioactive probes to locate the cells that had taken up the gene.

Bacterial cells with plasmids are cloned

In the third stage of their experiment, Cohen and Boyer let the bacterial cells reproduce. As the bacteria reproduced, the plasmids inside the cells made many identical copies of themselves, called clones. Because each plasmid carried different fragments of DNA, thousands of different clones were produced. Such clones are sometimes referred to as a gene library. The entire gene library had to be searched to find the one clone that contained the desired gene.

Cells are screened to locate the desired gene

In the fourth stage of their experiment, Cohen and Boyer needed a way to eliminate bacterial cells that hadn't taken up plasmids. Recall that they used plasmids that contained a gene for resistance to the antibiotic tetracycline. By growing the bacterial cells in dishes that contained the antibiotic tetracycline, Cohen and Boyer ensured that a cell lacking a plasmid would die. This screening procedure, illustrated in **Figure 9-4a** and **b**, left cells that at least contained plasmids.

Now Cohen and Boyer were faced with a more difficult problem—locating the cells that contained plasmids with the frog rRNA. Remember, a plasmid could consist of any number of combinations of DNA fragments.

They began with what they knew—somewhere in the colonies of bacteria was a gene for frog rRNA. If they made a short single strand of radioactive frog rRNA, they could use it as an indicator or probe to find the gene. Patient searching revealed a colony with cells containing DNA that bound to the radioactive frog rRNA. These are the cells they sought—bacterial cells that carried the frog gene. **Figure 9-4c** illustrates the final step of the screening procedure.

Safety Considerations

Genetic engineering techniques are now being used for a wide variety of commercial applications. Like any new technology, genetic engineering and its products have been sources of controversy. Critics have raised questions about potential risks and hazards.

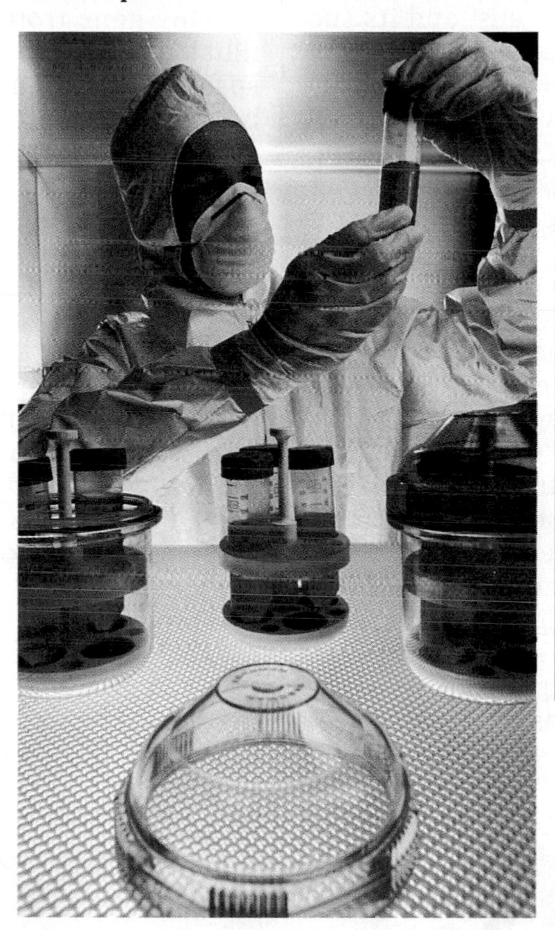

Some question the safety of certain procedures. Others worry about the effects of genetically engineered organisms on the environment. For example, certain bacteria have been engineered to break down contaminants in water and soil. Could these bacteria have an adverse effect on the environment at some time in the future? Or could bacteria be genetically engineered to create new microorganisms for biological warfare?

Recognition of the possible undesirable effects has led to strict regulation of how gene engineering technology is used, as shown in **Figure 9-5**. Laboratories must be equipped with facilities to ensure physical containment of genetically engineered organisms. In addition, scientists follow strict rules stating that microorganisms used in genetic engineering should be incapable of surviving outside a laboratory. Use of these specialized strains of microorganisms, and the rigorous testing of their products, helps ensure that genetic engineering poses no threat to the natural environment.

Figure 9-5
The wearing of safety garb protects both the lab worker and the genetic samples, which could easily become contaminated by the bacteria, viruses, and enzymes on the worker's skin.

SECTION REVIEW

1. **Describe the first experiment that successfully transferred a gene from a frog to a bacterium.**
2. **Define the term *genetic engineering*. List the four distinct stages involved in a gene transfer.**
3. **How are restriction enzymes used in gene transfer experiments?**
4. **How do scientists ensure that microorganisms used in genetic engineering experiments will not cause damage?**

One of the greatest successes of genetic engineering has been the manipulation of genes in crop plants and livestock. Gene transfers have made crop plants more resistant to disease, herbicides, and insects. Genetic engineering has also been used to produce hormones that increase the efficiency of milk production in dairy cows and increase the growth rate of livestock.

9-2 Transforming Agriculture

OBJECTIVES

1. **Explain the role of the Ti plasmid in agricultural research.**
2. **Describe how herbicide-resistant genes in crop plants can benefit the environment.**
3. **Explain how genetic engineering techniques are used to improve crop yields.**
4. **Describe how gene transfers are used to make livestock more productive.**

Getting Genes Into Eukaryotic Cells

The gene transfer method used by Cohen and Boyer is useful for getting genes into bacterial cells, and its success has been far-reaching. For example, many kinds of human proteins used to make drugs are inexpensively manufactured in large amounts using bacteria. However, this method is ineffective for getting genes into eukaryotic cells because the vectors used cannot enter eukaryotic cells. How do scientists get desired genes into eukaryotic cells? Clearing this hurdle was essential before gene technology could be useful in agriculture.

One method that enabled scientists to transfer genes into plants uses a bacterium that causes a type of tumor in plants. This bacterium contains the Ti plasmid, a circular molecule of DNA that possesses the genes responsible for the tumor. When the bacterium containing the Ti plasmid infects a plant cell, the Ti plasmid inserts itself into the plant cell's chromosomes. To transform the Ti plasmid into an effective genetic engineering vehicle, scientists first remove the tumor-causing gene from the Ti plasmid, as illustrated in **Figure 9-6**. The vacant space in the now harmless plasmid can be filled by the desired gene. Unfortunately, bacteria carrying the Ti plasmid cannot be used to insert DNA into plants that produce cereal grains, such as corn, rice, and wheat.

Researchers are developing powerful new techniques for introducing useful genes into these plants, such as shooting the cells with a DNA particle gun. In this procedure, tiny metal pellets are coated with DNA that contains the desired gene. Using a shotgun mechanism, these pellets are "shot" into plant cells. Inside some of the plant cells, the DNA inserts itself into one of the chromosomes. When these plant cells divide, they produce a clone of cells that can give rise to a new plant with the desired gene. Recently, this technique has been used to insert genes into the DNA in chloroplasts.

Figure 9-6

a **One method of transferring genes into eukaryotic cells uses the Ti plasmid.**

b **The gene that causes tumors can be removed and replaced with a desired gene.**

c **When the plasmid is put back into a bacterium, the new gene is carried along when it infects a plant cell.**

Resistance to Herbicides

A recent improvement in agriculture has been the development of crop plants that are resistant to the chemical glyphosate, a powerful weed killer, or herbicide, that kills most actively growing plants. Glyphosate kills plants by destroying an enzyme needed to make certain amino acids. Scientists found a kind of bacterium in which this enzyme is resistant to glyphosate. They then isolated the gene that codes for the enzyme. Shooting the gene in like a bullet, they transferred the glyphosate-resistant gene into crop plants such as the wheat plant shown in **Figure 9-7**.

This advance is of great interest to farmers. The farmer simply treats a field with glyphosate, and all growing plants die except the crop, which is resistant to glyphosate. After it is applied, glyphosate is quickly broken down in the environment. Glyphosate is not harmful to humans because our bodies do not make the amino acids that glyphosate affects. These qualities make it a great improvement over most commercial herbicides, which can be highly toxic. In addition, much of the erosion of fertile topsoil could be prevented if cropland did not have to be intensively cultivated to remove weeds.

Figure 9-7
Scientists have made crop plants such as this wheat resistant to herbicides.

Nitrogen Fixation

Nitrogen is an element that plants must have to make proteins and DNA. The most abundant source of nitrogen in the environment is the atmospheric gas N_2. However, plants cannot obtain nitrogen from the air. Bacteria living in the roots of plants such as soybeans, peanuts, and clover convert N_2 gas from the atmosphere into a form that plants can use. **Figure 9-8** shows these bacteria. The process of converting nitrogen into a form that plants can use is called **nitrogen fixation**.

Because crops use nitrogen rapidly, most farmers add high-nitrogen fertilizers to the soil. Worldwide, farmers applied more than 65 million metric tons of nitrogen fertilizers in 1990. Since high-nitrogen fertilizers are costly, using them adds a considerable expense to a farmer's budget. Farming would be much cheaper if major crops such as wheat, rice, and corn could be genetically engineered to carry out nitrogen fixation. The bacterial genes for nitrogen fixation have been successfully inserted into plants. However, these genes do not seem to function properly in their new hosts. Many experiments are being performed to find a way to overcome this difficulty.

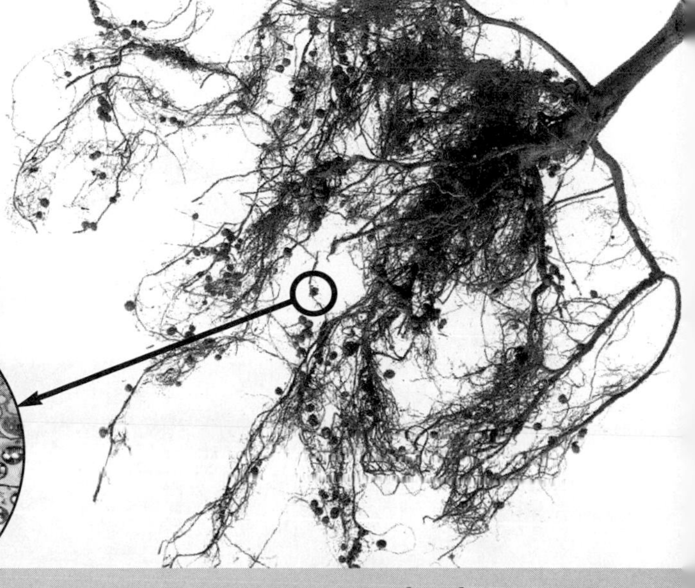

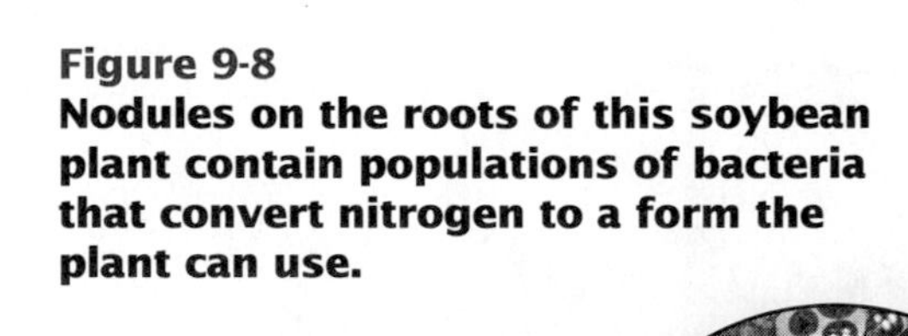

Figure 9-8
Nodules on the roots of this soybean plant contain populations of bacteria that convert nitrogen to a form the plant can use.

Resistance to Insects

Genetic engineering techniques can also make crops resistant to destructive insects, such as the locust and other pests. Resistant crops would not need to be sprayed with insecticides, which are expensive and can be harmful to the environment. Today more than 40 percent of all chemical insecticides are used to kill insects that eat cotton plants. Biologists have produced cotton plants that are resistant to these pests so that insecticides are not needed.

One successful approach to making plants resistant to insects uses bacteria that produce enzymes which kill destructive caterpillars. When the genes coding for these enzymes are inserted into tomato plants, for example, the new enzymes make the plants highly toxic to tomato hornworms. Hornworms that eat any of the tomato plants quickly die.

Many pests attack the roots of important plants. To combat these pests, genetic engineers are introducing the same insect-killing enzyme into a bacterium that colonizes the roots of crop plants. If an insect eats these roots, it will consume the bacteria and be killed. **Figure 9-9** shows how genetic engineering could improve a cotton plant.

Figure 9-9
Plants, such as this cotton plant, could be genetically engineered . . .

a . . . to produce an enzyme that kills pests so that highly toxic pesticides would not be needed, . . .

b . . . to resist glyphosate so that highly toxic herbicides would not be needed, . . .

c . . . and to carry out nitrogen fixation so that nitrogen fertilizers would not need to be added to the soil.

Genetic Engineering in Livestock

Genetic engineering can also produce bigger, more productive livestock. For example, injecting growth hormone, genetically engineered in bacteria, into dairy cows, such as the ones shown in **Figure 9-10**, increases their milk production. The milk is no different from milk produced by other cows. Bovine growth hormone is the natural signal that increases milk production in cows. In this case, the hormone is used in higher amounts than occur naturally.

Prior to the development of genetic engineering, growth hormone was extracted from the pituitary glands of dead cows. Today the hormone is obtained in the laboratory using genetic engineering. The gene containing the instructions for producing growth hormone has been introduced into bacteria. These bacteria produce growth hormone, which is then injected into cows.

Biologists have introduced extra copies of the genes that code for growth hormones into the chromosomes of cows and hogs. These attempts are likely to create new, leaner, fast-growing cattle and hogs.

Figure 9-10
These cows produce more milk after being injected with genetically engineered growth hormone.

SECTION REVIEW

1. **How is the Ti plasmid used to insert genes into plant cells? What are its shortcomings as a genetic engineering tool?**
2. **How can genetic engineering make crop plants resistant to glyphosate?**
3. **How can genetic engineering reduce the use of insecticides?**
4. **What effects can genetic engineering have on livestock?**

Much of the excitement surrounding genetic engineering has focused on its potential to aid in preventing and curing illnesses. Major advances have been made in the production of proteins that can treat illnesses, in the development of new vaccines to combat diseases, and in the replacement of defective genes with healthy ones.

9-3 Advances in Medicine

OBJECTIVES

1. **Describe how transferring human genes into bacteria can benefit human health.**
2. **Explain how gene transfers could be used to combat genetic disorders.**
3. **Describe three uses of DNA profiling.**
4. **Summarize the goals of the Human Genome Project and the ethical issues it raises.**

Making Miracle Drugs

Many human illnesses occur because our bodies fail to make critical proteins. One example of such an illness is diabetes. In one form of diabetes, a person cannot make a protein called insulin. As a result, they cannot regulate the levels of sugar in their blood. Diabetes can be treated if the body is supplied with insulin.

Until recently, it was not practical to use proteins as drugs. The body contains only small amounts of proteins such as insulin, making them difficult and expensive to obtain. In most cases, proteins had to be obtained from tissues of cows or pigs. Now genetic engineering can be used to provide large quantities of proteins in a short period of time. Genes that produce medically important proteins can be inserted into bacteria. Because bacteria can be easily grown in bulk, large amounts of the desired protein can be inexpensively prepared. **Figure 9-11** shows how genetic engineering is used to produce insulin.

Some proteins now being produced by genetically engineered bacteria dissolve blood clots, helping to prevent heart attacks and strokes. Genetically engineered bacteria are also used to produce proteins that prevent high blood pressure and that help regulate kidney function. Genetically engineered vaccines that are safer and less expensive than traditional vaccines are also being produced.

Figure 9-11

a **To manufacture large amounts of insulin, scientists insert the gene for insulin into a bacterial plasmid.**

b **Bacteria containing the plasmids are cloned and produce insulin.**

c **The insulin can be collected and used to treat diabetics like this young man.**

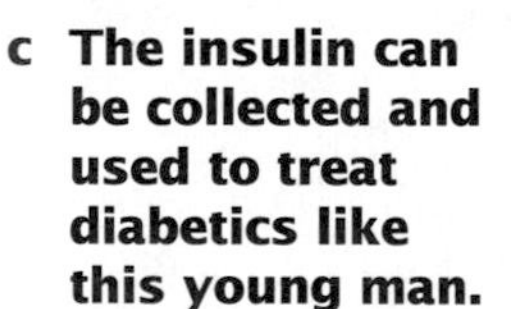

Figure 9-12

a To produce a vaccine against genital herpes, the virus that causes genital herpes is obtained from the cells of an infected person.

b A DNA fragment containing the gene that codes for the herpes surface protein . . .

c . . . is inserted into a harmless cowpox virus. The gene causes the virus to make the herpes surface proteins.

d A person injected with the engineered virus containing the desired gene can make antibodies against the genital herpes virus.

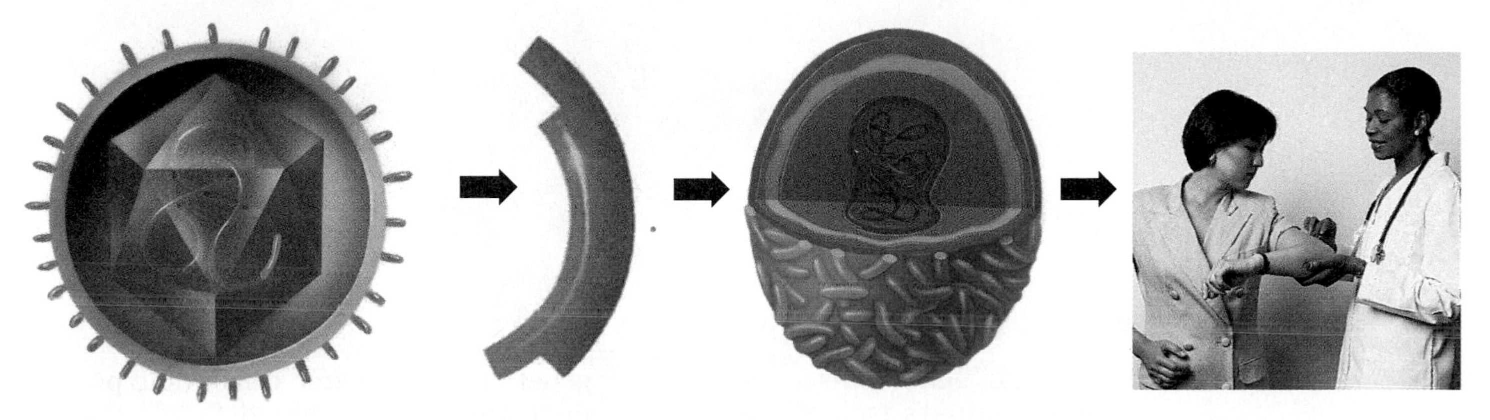

Making Vaccines

A **vaccine** is a harmless version of a disease-causing microbe (bacterium or virus) injected into animals or people so that their immune systems will develop defenses against the disease. The injected version serves as a model for the body, which responds by making defensive proteins called **antibodies**. If a vaccinated animal is exposed to the disease-causing microbe, it will immediately begin large-scale production of microbe-attacking antibodies. The antibodies will stop the growth of the microbe before the disease can develop.

Traditionally, vaccines have been prepared either by killing the microbe or by preventing its growth. This ensures that injecting a vaccine into your body will not make you sick. The problem with this approach is that any failure in the killing or weakening of the disease-causing microbe can result in the disease being introduced into the very patients in need of protection. While the majority of such vaccines are extremely safe, a tiny percentage of vaccinated individuals contract the disease from the vaccine. This small, but real, danger is one reason why rabies vaccines are administered only when a person has actually been bitten by an animal suspected of carrying rabies.

Harmless microbes can be made into vaccines

Using genetic engineering techniques, genes from disease-causing microbes can be inserted into harmless bacteria or viruses. These harmless microbes can be used to stimulate your body to make disease-attacking antibodies. For example, a vaccine against the genital herpes virus, which produces small blisters on the genitals, has been made using genetic engineering techniques, as shown in **Figure 9-12**. Genes that encode the surface proteins of the herpes virus are inserted into a harmless virus. This viral vector carries the genes "piggyback" into the human body, stimulating the person's immune system to make protective antibodies against the virus. Individuals who are injected with the vaccine are able to fight infections from the herpes simplex virus type 2.

Genetically engineered vaccines also offer hope for other diseases. A vaccine against the hepatitis B virus, which causes a fatal inflammation of the liver, is also available. Scientists are also working to produce a vaccine that will protect people against malaria. Transmitted by mosquitoes, malaria affected more than 500 million people worldwide in 1994.

Figure 9-13
White blood cells containing tumor necrosis factor are attacking this cancer cell.

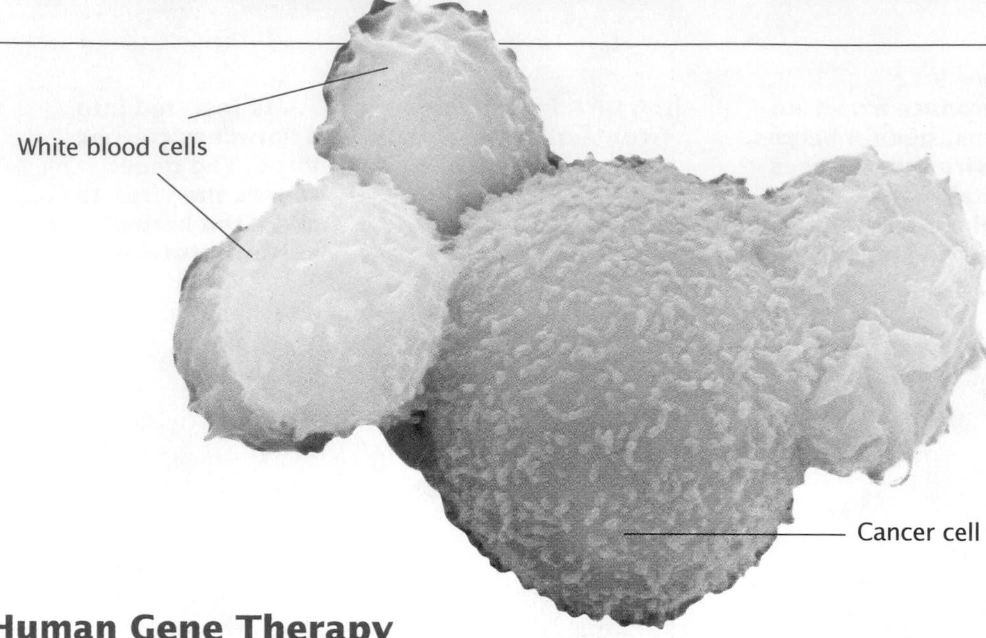

Human Gene Therapy

Most human genetic diseases are due to an individual's lack of a normally functioning copy of a particular gene. The problem usually arises when both of the individual's parents contribute defective copies of a gene. One obvious way to cure such disorders is to give the person a working copy of the defective gene. Until recently, this approach was not practical for three reasons. First, the defective gene was difficult to identify and isolate. Second, it was hard to transfer a "healthy" copy into the cells of body tissues that use it. Finally, it was necessary to find a way to keep the altered cells or their offspring alive in the body for a long time. With genetic engineering, it is now possible to overcome these difficulties. Gene transfers are being attempted as a way of combating a variety of genetic disorders, including cystic fibrosis and muscular dystrophy.

Heredity

If a woman with genetically altered bone-marrow cells has children, will the genetic change be passed on to them?

Human gene therapy may prove successful

In 1995, doctors reported that the first three infants to undergo gene therapy appeared to be responding to the treatment. In 1993, the newborns were given a gene that they were missing at birth. People born without this gene develop a fatal blood disease. In the therapy, doctors inserted the gene into certain blood cells taken from the newborns' umbilical cord. These cells were then injected into the infants' bloodstream. The hope was that the genetically engineered cells would produce the missing blood factor that would protect them from the disease. Two years later, the children are found to be producing a blood factor necessary for survival.

However, it is too soon to declare the children entirely cured of the disease. They are also being treated with medication, so doctors are unable to tell whether the drug or the gene therapy is keeping the children alive. While it is too early to say the technique has been completely successful, the signs are positive. If the technique works over the long run, it might be used to treat common blood disorders like hemophilia and sickle cell anemia, as well as diseases like AIDS.

Human gene transfers may also help in the battle against cancer

All humans have white blood cells that secrete a protein called TNF (tumor necrosis factor) into the blood. TNF attacks and kills cancer cells like the one shown in **Figure 9-13**. Unfortunately, TNF cannot work unless it encounters a cancer cell. This does not often happen. Recently, genetic engineers developed a way to add the gene encoding TNF to a kind of white blood cell that is effective at locating cancer cells but not effective at harming them. Armed with this new TNF weapon, however, these white blood cells will become more like cruise missiles homing in on cancer cells.

Biotechnology

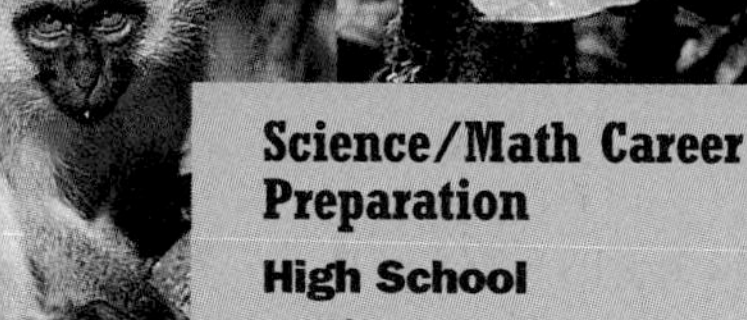

What would it be like to . . .

- search the rain forest for new medicines?
- study the effects of chemicals on living cells?
- design new foods and drugs?

These are just a few of the daily adventures experienced by people working in the field of biotechnology.

Biotechnology: What Is It?

Biotechnologists work in areas such as developing drugs, using enzymes and catalysts to increase the yields of reactions, and developing chemical and genetic engineering processes that are safe for the environment. Advances in the field of biotechnology have resulted in the production of synthetic hormones and in crops that are tastier, that are more disease resistant, and that have a longer shelf life.

Science/Math Career Preparation

High School
- Chemistry
- Physics
- Calculus

College
- Chemistry/biochemistry
- Molecular biology
- Statistics
- Genetics
- Organic chemistry
- Physics
- Anatomy and physiology

Graduate
- Master's degree is suggested

Employment Outlook:
Accelerated growth in this field is predicted as the number of research areas in biotechnology increases.

Career Opportunities

Career Focus: Dr. Eloy Rodriguez

Biochemist at Cornell University, Ithaca, New York

"It was a good thing that I ignored a school counselor's advice to go to technical school, which was standard advice for many minority students at the time. I grew up in Hidalgo County, Texas, the county with the lowest average income per person in the country. When I was a young boy, my grandfather would take me on long walks through nearby fields, where he taught me everything he knew about the different plants and insects we found there. Ever since then, I have been fascinated by the life around me. I have made this passion my career."

What I Do: "I have helped develop a new scientific discipline called zoopharmacognosy—the chemistry of plants that are used medicinally by animals. One plant that I studied in the rain forest of Africa is used by wild chimpanzees to remove parasites. By studying the plant's chemicals, I discovered a drug that is effective against human intestinal parasites, fungi that cause skin infections, and possibly cancer."

To Find Out More

1. For further information, write to:
   American Society for Biochemistry and Molecular Biology
   9650 Rockville Pike
   Bethesda, MD 20014
2. See your school counselor to find out what colleges and universities offer programs in biotechnology. Review the prerequisites and course requirements. Make a poster showing the range of job opportunities in this field.
3. Find articles that describe how new drugs and enzymes are created using computers. Prepare a report for your class.

The rosy periwinkle is the source of two anticancer drugs.

Identifying Sequences in Genes

The knowledge that restriction enzymes cut DNA at specific sites has produced powerful tools that have great effects beyond the biological sciences. Because every person's DNA is unique, analysis of DNA from tissues can identify an individual. In this procedure, DNA samples are taken from the blood of a person suspected of committing a crime and from a tissue sample (such as blood, semen, or bone) found at the scene of the crime. Then the DNA in each sample is treated and developed to produce a piece of film characterized by columns of short black bands. This pattern of bands is a **DNA fingerprint**. If the patterns of bands match, an individual can be positively identified.

These techniques for DNA analysis can link a suspect to a crime, confirm family relationships, and identify the remains of deceased individuals. DNA analysis is also valuable for identifying the genes that cause genetic disorders. Alleles for Huntington's disease, sickle cell anemia, and a number of other genetic disorders can now be detected in this way.

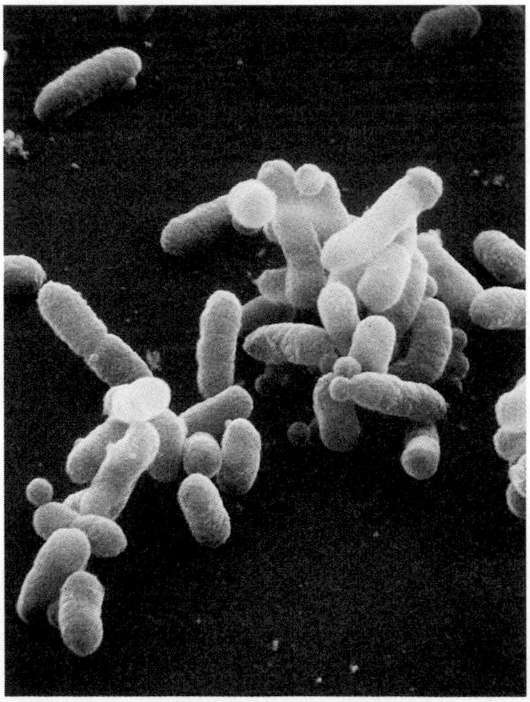

Figure 9-14
In 1995 biologists successfully sequenced all the genes in the bacterium *Haemophilus influenzae*, which causes bacterial meningitis. This is one of the first bacteria to be completely sequenced and biologists hope to use this "blueprint" to develop a vaccine.

DNA can be copied in a test tube

In 1985, Kary B. Mullis, a biochemist working for a biotechnology corporation in California, discovered a simple way to make unlimited copies of a gene. This process is known as polymerase chain reaction (PCR). The starting materials for PCR include a fragment of DNA, DNA polymerase, and a supply of the four nucleotides—adenine, thymine, guanine, and cytosine. When these ingredients are heated in a test tube, millions of copies of the molecule can be made in just a few hours. PCR has revolutionized biological research.

Scientists are mapping the human genome

Gene sequencing is the process of determining the order of nucleotide bases within a gene. Gene sequencing has been successfully completed on several species of bacteria, including *Haemophilus influenzae*, shown in **Figure 9-14**. Genetic engineers are attempting to locate, catalog, and sequence every gene on the 23 human chromosomes. Biologists call the entire collection of genes the **human genome**. The effort to sequence every human gene is called the Human Genome Project. Because the human genome contains approximately 3 billion nucleotides, the project is not a small one. This international project was launched in 1990 and is expected to cost several billion dollars.

SECTION REVIEW

1. **Explain how genetic engineering can be useful in the treatment of human illnesses such as diabetes.**
2. **How could human genetic therapy cure genetic disorders such as cystic fibrosis?**
3. **Describe how genetic engineering techniques could be used to help determine a suspected criminal's innocence or guilt.**
4. **What are the goals of the Human Genome Project?**

CHAPTER 9

Highlights

Gene therapy offers hope for patients with genetic disorders, such as this child with cystic fibrosis.

	Key Terms	Summary
9-1 The Revolution in Genetics	recombinant DNA (p. 153) genetic engineering (p. 153) cloning (p. 154) restriction enzyme (p. 154) plasmid (p. 155) vector (p. 156)	• Over the last 20 years, genetic engineers have learned how to transfer genes from one organism to another. • Every gene transfer starts by cleaving DNA into small fragments using restriction enzymes. • Fragments containing the desired gene are then transferred to a target cell using a vector. • After removing uninfected cells, scientists search for cells that have taken up the desired gene.
9-2 Transforming Agriculture	nitrogen fixation (p. 159)	• Genetic engineers have manipulated the genes of certain kinds of crop plants to make these plants resistant to herbicides and destructive pests. • Genetic engineers are looking for a way to transfer genes for nitrogen fixation into crop plants. • Adding genetically engineered bovine growth hormone to the diet of livestock increases milk production in dairy cows and increases the weight of cattle and hogs.
9-3 Advances in Medicine	vaccine (p. 163) antibody (p. 163) DNA fingerprint (p. 166) human genome (p. 166)	• Genetic engineering techniques can be used to manufacture proteins, such as insulin, and vaccines. • Some human genetic disorders are being treated by inserting copies of healthy genes into individuals lacking them. • Scientists use DNA fingerprinting to reveal DNA patterns that vary from one individual to the next. • The Human Genome Project seeks to locate, catalog, and record the nucleotide base sequence of every human gene.

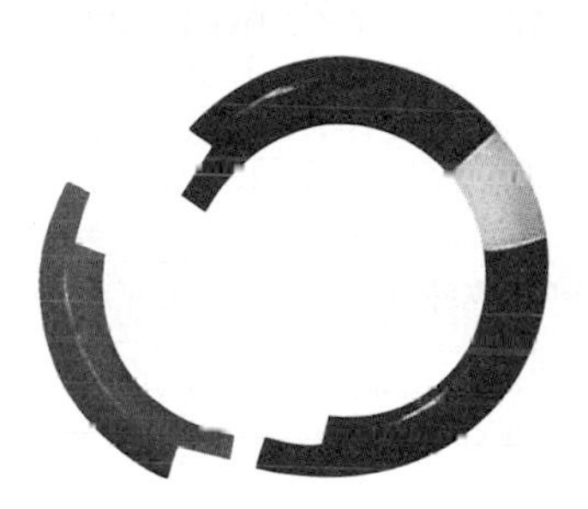

Bacterial plasmids make genetic engineering possible.

Cows injected with genetically engineered growth hormone produce more milk.

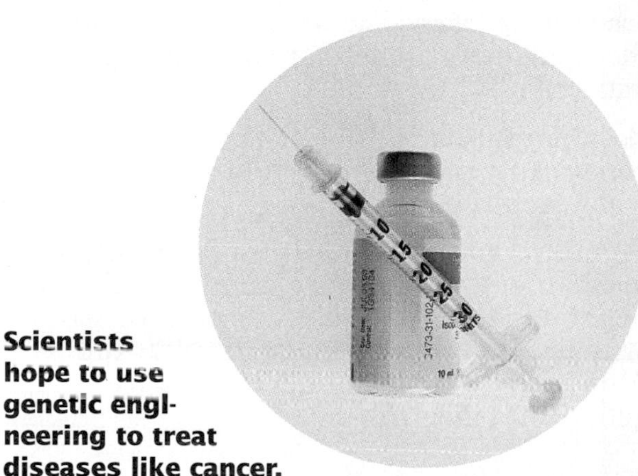

Scientists hope to use genetic engineering to treat diseases like cancer.

CHAPTER 9

review

Understanding Vocabulary

1. For each pair of terms, explain the difference in their meanings.
   a. recombinant DNA, restriction enzyme
   b. vector, plasmid
   c. glyphosate, nitrogen fixation

2. Using each set of words below, write one or more sentences summarizing information learned in this chapter.
   a. genetic engineering, cloning, restriction enzyme
   b. plasmid, vector, recombinant DNA, *Eco*RI
   c. gene sequencing, human genome, DNA fingerprint

Understanding Concepts

3. **Relating Concepts** Construct a concept map describing genetic engineering. Try to include the following terms in your map: plasmids, livestock, DNA, medical research, drugs, agriculture, restriction enzymes, and Ti.

4. **Inferring Relationships** What is recombinant DNA? How did Cohen and Boyer recombine DNA in bacteria?

5. **Inferring Conclusions** Explain why Cohen and Boyer's experiments ushered in a "new age" of biology.

6. **Predicting Outcomes** How did Cohen and Boyer use their knowledge of transformation to predict that bacteria would take up plasmids containing recombinant DNA?

7. **Identifying Functions** What is the function of the enzyme ligase in forming recombinant DNA?

8. **Identifying Structure and Function** What are plasmids? How do they function in the formation of recombinant DNA?

9. **Applying Information** You are a paleontologist with experience in using DNA fingerprinting techniques. A colleague has discovered a fossilized bone that contains small amounts of DNA. What method will yield sufficient genetic material for DNA analysis?

10. **Interpreting Graphs** The following graph shows the results of an experiment with populations of natural and engineered bacteria, both of which were able to convert a harmful chemical to harmless byproducts. Each population had 40 mL of the chemical added on the first day. The graph shows how much of the chemical remained after almost three days. Approximately how much of the chemical did each population of bacteria convert by the end of the test?

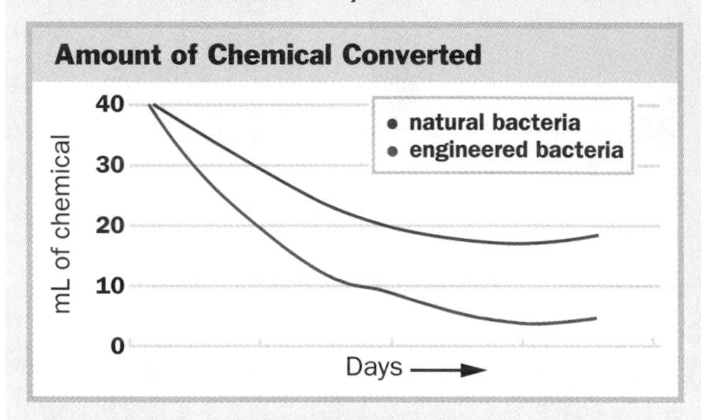

11. **Identifying Functions** How does a restriction enzyme, such as *Eco*RI, create DNA fragments with "sticky" ends?

12. **Recognizing Relationships** How did the bacterial cell's resistance to the antibiotic tetracycline contribute to the results of Cohen and Boyer's experiments?

13. **Identifying Functions** Explain how scientists introduce genes into plants in which the Ti plasmid cannot be used.

14. **Inferring Conclusions** How can herbicide-resistant genes in crop plants benefit the environment? Give an example.

15. **Summarizing Information** Describe two ways that genetic engineers are attempting to make plants resistant to insects.

16. **Identifying Functions** Identify two types of genetically engineered medicine, and tell what each type does.

17. **BUILDING ON WHAT YOU HAVE LEARNED** In Chapter 8 you learned about the double-helix model of DNA that was first constructed by Watson and Crick. How did Watson and Crick's model of DNA lead to genetic engineering?

Interpreting Graphics

18. The figures below show four distinct stages of a gene transfer experiment.

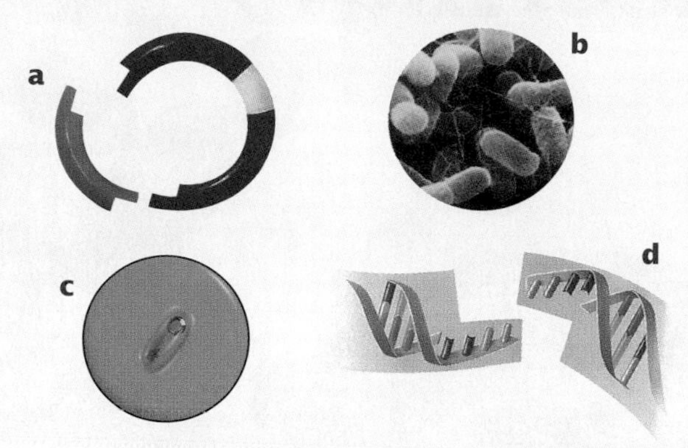

a. How should the figures be ordered to show the proper sequence of events?
b. What is occurring in each step of the illustration?

Reviewing Themes

19. ***Heredity***
How are organisms produced by genetic engineering different from organisms produced by sexual reproduction?

20. ***Evolution***
Natural selection is a process of evolution in which the members of a population who are best able to adapt to their environment survive and produce offspring. How is natural selection affected by genetic engineering?

21. ***Interdependence***
Why would farmers be interested in genetically engineered crop plants that are resistant to the herbicide glyphosate?

Thinking Critically

22. **Evaluating Opinions** The federal government of the United States has strict regulations that require researchers to contain genetically engineered organisms inside a laboratory and to ensure that the organisms could not survive outside the laboratory if they did escape. Why do you think the government has these strict regulations?

23. **Distinguishing Fact From Opinion** A judge presiding over a highly publicized murder trial dismissed the prosecution's request to admit DNA fingerprints as evidence, calling it "unproven scientific mumbo-jumbo." Do you agree with the judge? Explain your answer.

Life/Work Skills

24. **Finding Information** Use library references or search an on-line database to find out about American plant breeder Luther Burbank and the work he did on mass selection. Prepare a visual summary of your findings to share with your class.

25. **Working Cooperatively** Hold a debate with your classmates regarding the safety questions raised by the potential release of genetically engineered plants, bacteria, and animals into the environment.

Activities and Projects

26. **Agriculture Connection** Search an on-line data base to find out about vegetable vaccines. Researchers at Texas A & M University are testing genetically-engineered potatoes containing a vaccine that they hope will protect people from common gut infections. What are some of the advantages of these vegetable vaccines?

27. **Economics Connection** Use library references or contact seed companies to find out how customer demand for new plants and seeds is met.

Discovering Through Reading

28. Read the article "The New Science of Identity" by Cassandra Franklin-Barbajosa in *National Geographic*, May 1992, pages 112–124. Summarize five ways DNA fingerprinting is being used today to answer questions about identity and heredity.

29. Read the article "Invasion of the Shapechangers" by Stephen Day in *New Scientist*, October 1995, pages 30–35. What are Hox genes? How are scientists manipulating these genes? What do they hope to discover from them?

30. Read the article "Swallowing *Shigella*" by John Travis in *Science News*, May 11, 1996, pages 302–303. What is the third vaccine revolution? What are the alternatives to using *Shigella* as a vaccine carrier? How do DNA vaccines work?

31. Read the article "Computing with DNA" by Ivars Peterson in *Science News*, July 13, 1996, pages 26–27. What advantages and disadvantages do DNA-based computers have? What proposals have been made to cope with errors?

What Is the Future of Genetic Engineering?

From food to pharmaceuticals, genetic engineering is affecting our lives. Science and industry are eager to explore the possibilities of this revolutionary technology. Will society be responsible enough to handle the powerful tools that science has developed?

Genetic engineering has produced many useful pharmaceutical and agricultural products.

Joining Genes

Every living organism has genes made of DNA. Genes contain the information to build proteins, the structurally and metabolically important molecules required for all life processes. In genetic engineering, a gene from one organism is spliced into the DNA of another organism, where it builds a protein that wasn't there before.

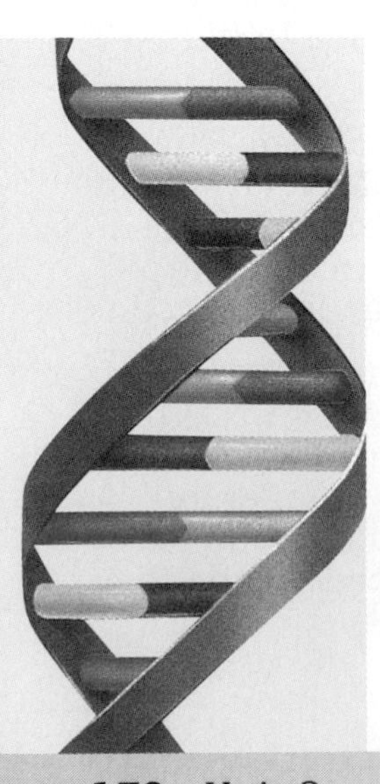

Genetic Engineering at Work

Since its development over 20 years ago, genetic engineering has yielded many useful pharmaceutical and agricultural products. Today, bacteria are routinely turned into microscopic factories that produce large supplies of medically important human proteins, such as insulin and growth hormone. Genetically engineered drugs now offer hope for treating diseases such as cystic fibrosis and cancer. Every year, agriculturally important plants, animals, and bacteria are given new genes that increase crop yield, enhance the nutritional value of food, increase productivity, and reduce the need for toxic pesticides and herbicides.

What Will the Future Bring?

What can we expect from genetic engineering in the future? Agriculture can continue to look forward to the development of new varieties of crops that are resistant to herbicides, insects, viruses, and changes in temperature. For example, in 1995, the United States government approved commercial use of a variety of corn that contains genes from the bacterium *Bacillus thuringiensis.* When these genes are expressed, they produce high levels of a protein that is toxic to major pests such as corn borers and corn earworms.

Scientists are also now working on new vaccines called DNA vaccines. These vaccines contain a portion of the DNA from an organism that causes disease. Once injected into a host, the host's cells take up the DNA and express the foreign proteins on their surface—just as occurs in a natural infection, but without causing disease. This stimulates the host's immune system to provide protection against the disease. It is hoped that these vaccines will someday be able to protect us against allergies and against diseases such as AIDS and malaria.

Technique

Polymerase Chain Reaction (PCR)

PCR is used to make exact copies of a sample of DNA. The DNA sample to be copied is placed in a test tube with short pieces of DNA called primers, a surplus of nucleotides, and the enzyme DNA polymerase. First the two DNA strands in the sample are separated by heating the sample to a high temperature (typically 94°C). Then the temperature is lowered, and the primers form complementary base pairs with the DNA sample. DNA polymerase then uses the surplus nucleotides to form the new strand of DNA. The temperature is raised again so that the strands will separate, and the cycle begins again. By repeating this many times, huge amounts of DNA can be made from a very small sample.

Applications: locating and cloning genes from an organism, diagnosing genetic diseases, and producing large amounts of DNA for research or for use in crime investigations

User: research biologists, forensic analysts, medical geneticists

In a test tube, strands of a DNA sample are separated by heat. Using nucleotides as building blocks, DNA polymerase makes a complementary copy of the DNA. Once a copy is made, the DNA strands are separated by heat, and each strand is copied again.

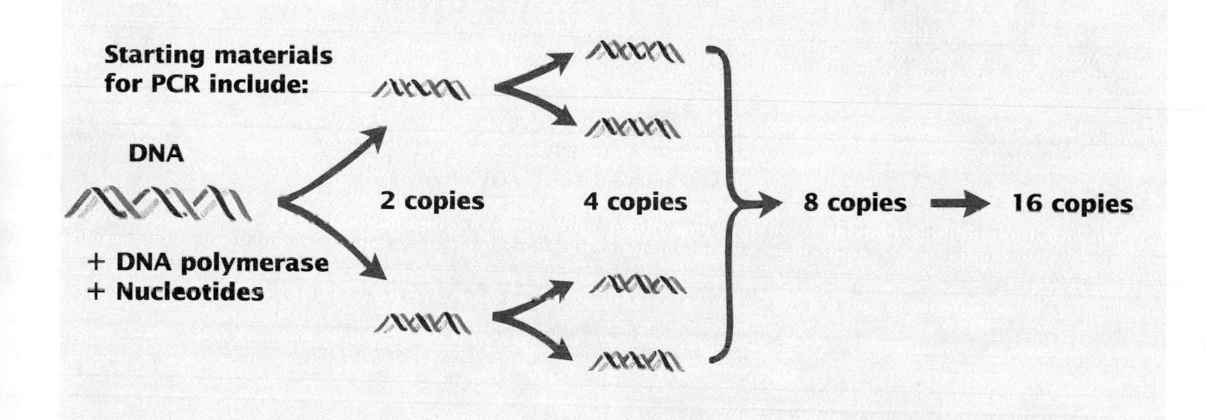

Analyzing the Issues

Should genetic engineering research include humans?

1. Do ethical guidelines exist to help scientists determine what kinds of human genetic research should be permitted? Do library research or use an on-line database to find out. Describe additional guidelines that you would propose.

Is genetic engineering safe?

2. Use library references and on-line databases to learn about the regulations that govern genetic engineering. Design and distribute a questionnaire that examines your classmates' knowledge of the safety of genetic engineering.

Can gene therapy become a reality?

3. Many of the early results of human gene therapy have been disappointing. Read "Promises, Promises," in *Newsweek*, October 9, 1995, pages 60–62. What kind of obstacles must be overcome if gene therapy is to succeed?

Do genetically engineered foods pose health risks?

4. Find out what genetically engineered foods are currently available in the marketplace. How are new genetically engineered food products tested to ensure their safety?

CHAPTER 10

REVIEW
- species (Section 1-3)
- evolution (Section 1-3)
- sickle cell anemia (Section 7-3)
- DNA structure (Section 8-1)
- the term *allele* (Glossary)

Evolution and Natural Selection

The giant tortoises and other unique animals and plants of the Galápagos Islands helped inspire Charles Darwin's theory of evolution.

The Earth is rich in life. About 1.5 million species have been named, and millions more are thought to exist. Where did all of these species come from? Was each species separately created, or can species change and give rise to other species? It was not until 1859, through the work of Charles Darwin, that convincing answers were provided to these questions.

10-1 Charles Darwin

OBJECTIVES

1. **Summarize Darwin's beliefs about the origin of species before he sailed around the world.**
2. **Identify two observations from Darwin's voyage that led him to question his beliefs.**
3. **Describe the two major ideas Darwin put forth in *The Origin of Species.***

Voyage of the *Beagle*

On December 27, 1831, H.M.S. *Beagle* sailed from England to survey the coast of South America. On board as the ship's unpaid naturalist was Charles Darwin, a 22-year-old who had just graduated from Cambridge University. Darwin's observations during his five years at sea have changed the way we think of ourselves and our world.

The son of a wealthy doctor, Darwin was not an attentive student, spending more time outdoors than in school. As a medical student in Edinburgh, he was horrified by operations, which were performed without anesthetic. For two years Darwin skipped lectures to spend time collecting biological specimens. In desperation, his father sent him to Cambridge University to train to be a minister. After graduating from Cambridge in 1831, Darwin was recommended by one of his professors for the position on the *Beagle*. Follow the voyage of the *Beagle* in **Figure 10-1**.

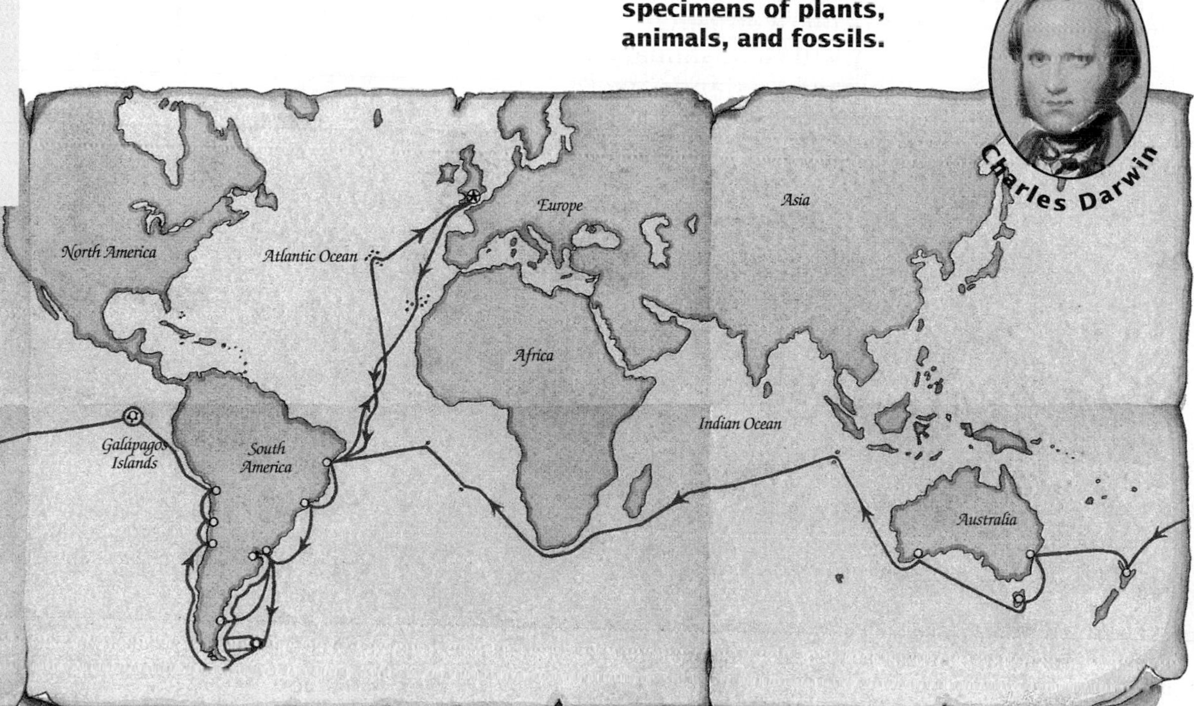

Figure 10-1
Below you can see the course of the *Beagle*, the ship in which Darwin sailed around the world. On this voyage, Darwin collected thousands of specimens of plants, animals, and fossils.

When the *Beagle* sailed from England, Darwin, like nearly all scientists of his time, accepted the view that God was responsible for the creation of all species of organisms. According to this view, God designed each species of animal and plant to match its habitat. This explained why, for example, polar bears, which live in a very cold climate, have white fur for camouflage and a thick layer of insulating fat beneath the skin. In addition, most scientists thought that species were unchanging.

During his journey, Darwin often left the ship to collect specimens of animals, plants, and fossils. He made careful observations and recorded them in a journal. What Darwin saw on his voyage led him to doubt the idea that species were constant.

Darwin's Finches

Darwin repeatedly saw patterns in how kinds of animals and plants differed, patterns suggesting that species changed over time and gave rise to new species. On the Galápagos Islands, 1,000 km (600 mi) from the coast of Ecuador, Darwin collected several species of finches. All of these species were similar, but each was specialized to catch food in a different way, as shown by the different shapes of the birds' bills in **Figure 10-2**. Some species had thick, sturdy bills for cracking open tough seeds. Others had slender bills for catching insects.

All of the species of finches closely resembled one species of South American finch. In fact, all of the plants and animals of the Galápagos Islands were very similar to those of the nearby coast of South America. If each one of these plants and animals had been created to match the habitat on the Galápagos Islands, why did they not resemble the plants and animals of islands with similar environments that lie off the coast of Africa? Why did they instead resemble those of the adjacent South American continent? Darwin felt that the simplest explanation was that a few organisms from South America must have migrated to the Galápagos Islands in the past. These few kinds of animals and plants then changed over the years that they lived in their new home, giving rise to many new species. Change in species over time is known as evolution.

Figure 10-2
The blue-black grassquit (inset), native to the Pacific coast from Mexico to Chile, is thought to be very similar or identical to the ancestor of the Galápagos finches (below). Darwin attributed the differences in bill size and feeding habits among these finches to evolution that occurred after their ancestor migrated to the Galápagos Islands.

a The woodpecker finch captures insects with its grasping bill.

b The crushing bill of the large ground finch enables it to feed on seeds.

c The cactus finch uses its probing bill to feed on cactuses.

Figure 10-3
In 1859 Darwin published his famous book, *The Origin of Species.* He accomplished much of his work in his study at Down House in Kent, England (above right). Darwin is shown at age 73 (above).

Darwin's Mechanism for Evolution

Darwin returned to England in 1836. He got married and wrote several books. He accomplished much of his work in Down House in Kent, England, shown in **Figure 10-3**. For 20 years, Darwin gathered evidence supporting his ideas about evolution, but he did not publish them. Then in 1858 another biologist, Alfred Russel Wallace, sent Darwin an essay that presented these same ideas. This prompted Darwin to finally publish his work.

When Darwin's book *On the Origin of Species by Means of Natural Selection* appeared in November of 1859, it stirred up great controversy. Darwin's conclusion that species changed over time and gave rise to new species contradicted the prevailing beliefs that God created all species and that species did not change. Furthermore, the implication that apes were close relatives of humans was unacceptable to many people.

In *The Origin of Species,* as the book is commonly known, Darwin not only presented much evidence that evolution occurred but also proposed that natural selection was its mechanism. Organisms with traits well suited to an environment are more likely to survive and produce more offspring than organisms without these favorable traits. This process is called **natural selection**.

Because Darwin presented a mechanism as well as evidence for evolution, his arguments were compelling. His views were soon accepted by biologists around the world. Since the discovery of Mendel's ideas about genetics in the early 1900s, genetic principles have been added to Darwin's ideas, forming the modern theory of evolution.

SECTION REVIEW

1. **What were Darwin's views about the origin of species before he sailed on the *Beagle*?**
2. **Describe two observations Darwin made on his voyage that led him to doubt these views.**
3. **Explain the two major ideas in *The Origin of Species.***
4. **How did Darwin explain the similarities and differences between the species of Galápagos finches?**

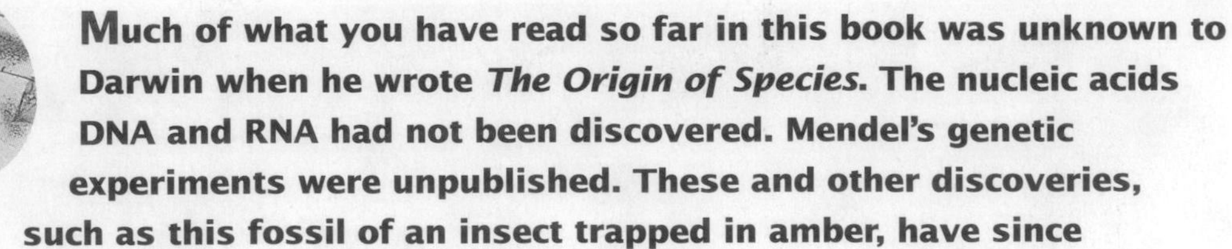

Much of what you have read so far in this book was unknown to Darwin when he wrote *The Origin of Species.* The nucleic acids DNA and RNA had not been discovered. Mendel's genetic experiments were unpublished. These and other discoveries, such as this fossil of an insect trapped in amber, have since contributed additional evidence for evolution.

10-2 The Evidence for Evolution

OBJECTIVES

1. **Describe the conditions necessary for fossils to form.**
2. **List one example from the fossil record indicating that evolution has occurred.**
3. **Explain how comparisons of organisms can reveal evidence of evolution.**
4. **Describe the important evidence for evolution found in proteins and DNA.**

Understanding the Fossil Record

More than a century has passed since Darwin's death in 1882. During this period, a great deal of new evidence has accumulated supporting the theory of evolution, much of it far stronger than that available to Darwin and his contemporaries. This evidence has come from a variety of sources, including studies of fossils, comparisons of the structures of organisms, and the rapidly expanding knowledge about DNA and proteins.

Fossils are any traces of dead organisms

What are fossils? Most people think of fossils as shells or old bones. Actually, fossils are any traces of dead organisms. Tracks of dinosaurs, footprints of human ancestors, insects trapped in sticky tree sap, impressions of leaves or skin, and animals buried in tar are all fossils. For fossils to form, very special conditions are necessary. If a skeleton or shell is to fossilize, for example, the dead animal must be buried by sediment. Burial usually occurs on the ocean floor, in swamps, in mud, or in tar pits. Calcium in the bone or in the shell is slowly replaced by other, harder minerals. Unless the sediment is very fine and no oxygen is present to promote decay, soft tissues, such as those found in skin or muscle, do not fossilize. Thus, the fossil record provides only a glimpse into life's history.

How Fossils Are Dated

Since the late 1940s scientists have been able to determine the ages of rocks and fossils by measuring the amount of radioactive decay, or breakdown, of radioactive atoms in the rock. A radioactive atom contains an unstable combination of protons and neutrons. Since it is unstable, a radioactive atom will eventually change into a more stable atom of another element. For example, carbon-14, a rare form of carbon found in tiny amounts in living things, decays into nitrogen. The term **half-life** describes how long it takes for one-half of the radioactive atoms in a sample to decay. For example, the half-life of carbon-14 is 5,730 years. Thus, a sample that initially contained 12 g of carbon-14 will have 6 g of carbon-14 left after 5,730 years and 3 g of carbon-14 left after 11,460 years. Since carbon-14 decays relatively rapidly, other isotopes with longer half-lives are more often used to date fossils.

Because the rate of decay of a radioactive element is constant, scientists can use the amount of radioactive element remaining in a rock or fossil to determine its age. This technique is called **radioactive dating**.

Evolution can be a very slow process; the transformation of one species into another by natural selection requires thousands of years. Using radioactive dating, scientists have determined that the Earth is about 4.5 billion years old, ancient enough for all species to have been formed through evolution.

Stability and Homeostasis

Explain why the evolution of one species from another can be considered an instance of instability.

Transitional forms link new species to old

Because new species form from existing species, Darwin predicted that transitional forms, intermediate stages between older and newer species, would be found in the fossil record. When *The Origin of Species* was published, no intermediates had been found. Darwin recognized that this was a weakness in his argument. But there are now many good examples of evolutionary transitions. For instance, modern whales are the descendants of four-legged land animals that are also the ancestors of horses and cows. As you can see in **Figure 10-4**, fossil intermediates between modern whales and their 60-million-year-old ancestor reveal a history of slow transformation. Over time, the hind limbs became smaller and smaller, until eventually they were lost entirely. Modern whales have a pelvis, but no rear legs are attached to it. A detailed picture of whale evolution is emerging from fossil finds, making it clear that a progressive reduction in the hind limbs was a central feature of whale evolution.

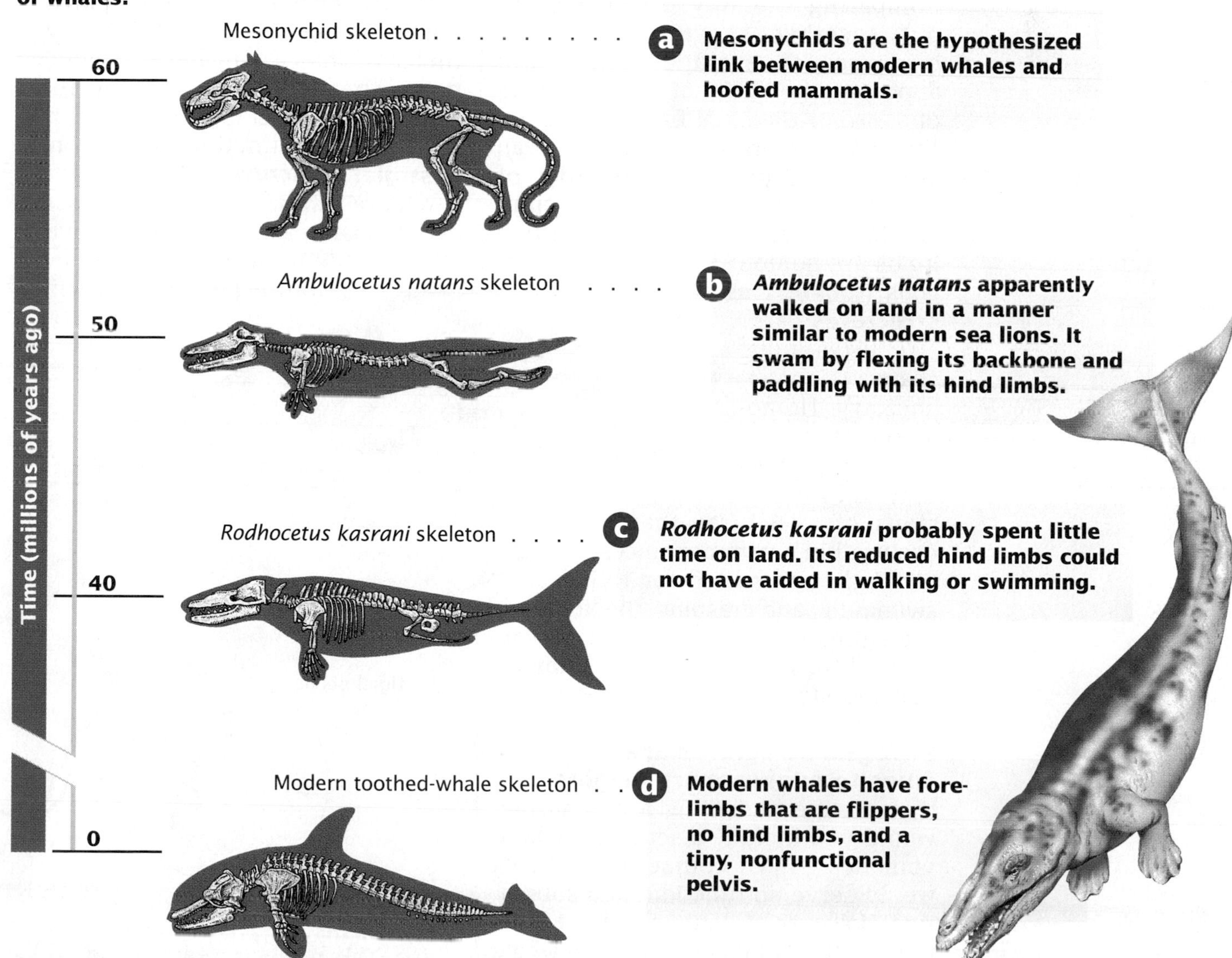

Time (millions of years ago)
60
50
40
0
Mesonychid skeleton
a Mesonychids are the hypothesized link between modern whales and hoofed mammals.
Ambulocetus natans skeleton
b Ambulocetus natans apparently walked on land in a manner similar to modern sea lions. It swam by flexing its backbone and paddling with its hind limbs.
Rodhocetus kasrani skeleton
c Rodhocetus kasrani probably spent little time on land. Its reduced hind limbs could not have aided in walking or swimming.
Modern toothed-whale skeleton
d Modern whales have forelimbs that are flippers, no hind limbs, and a tiny, nonfunctional pelvis.
Rodhocetus kasrani

Figure 10-4 Recent fossil discoveries have revealed some of the steps in the evolution of whales.

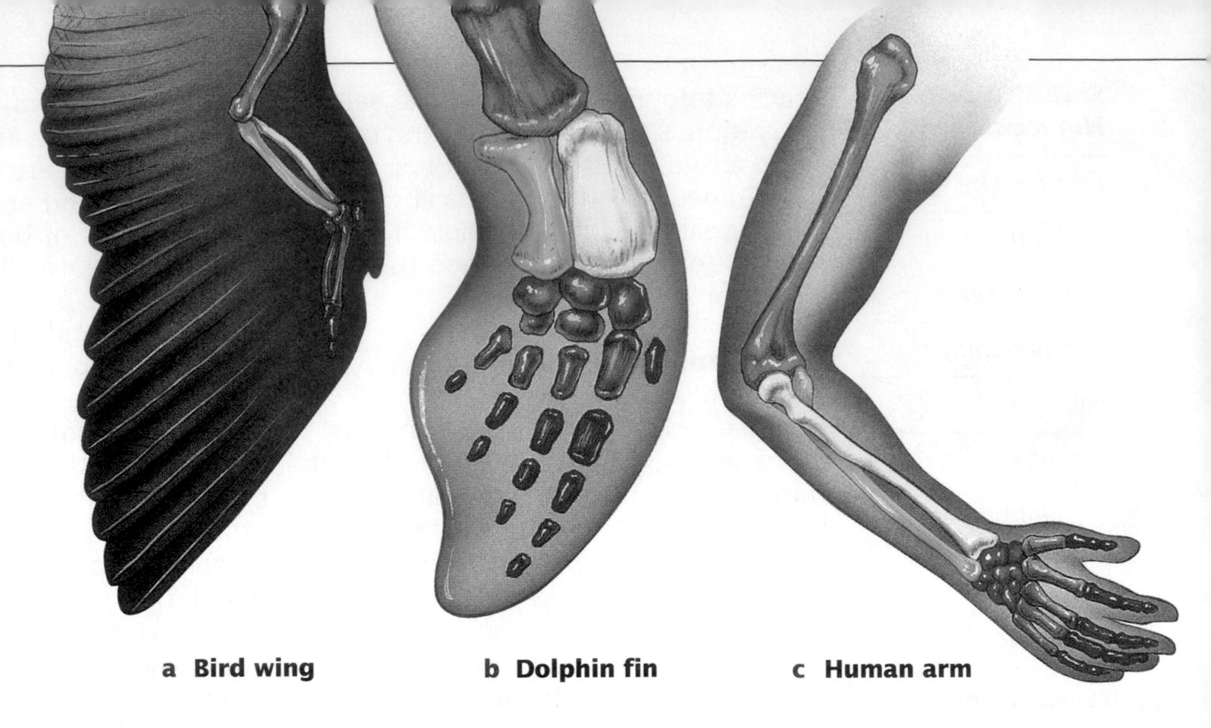

Figure 10-5
The bones in the front limbs of a bird, a dolphin, and a human are homologous structures. Homologous bones are shown in the same color on each diagram.

Comparing Organisms

Comparing the way organisms are put together provides important evidence for evolution. Your arm appears quite different from the wing of a bird or the front fin of a dolphin. Yet if you examine **Figure 10-5**, you can see that the position and order of bones in these limbs are very similar. Biologists say that these three limbs are homologous. **Homologous structures** are structures that share a common ancestry. Homologous structures are similar because they are modified versions of structures that occurred in a common ancestor. Although suited for flying, swimming, and grasping, the limbs of the animals above are modified versions of the front fins of their common fish ancestor.

Vestigial structures are clues to evolutionary origins

If you were designing a submarine, would you include a set of wheels in your design? Of course not. Wheels would serve no function on a submarine. However, structures without function are found in living things. A whale propels itself with its powerful tail and has no need for hind limbs or the pelvis to which they attach. Nevertheless, whales still have a reduced pelvis that serves no apparent function, as shown in **Figure 10-6**. Structures with no function are known as **vestigial structures**. Vestigial structures are remnants of an organism's evolutionary past. The whale's pelvis is evidence of its evolution from four-legged, land-dwelling mammals.

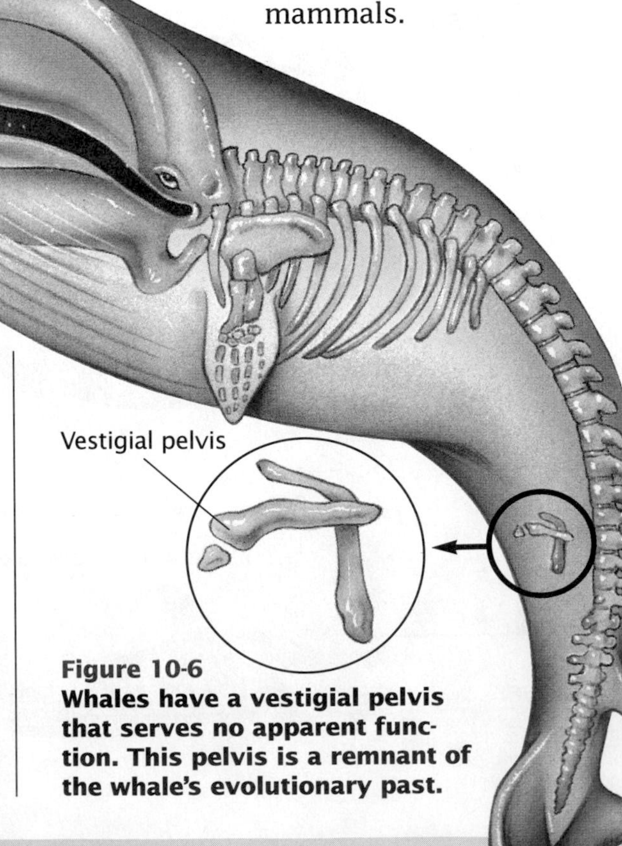

Figure 10-6
Whales have a vestigial pelvis that serves no apparent function. This pelvis is a remnant of the whale's evolutionary past.

Developmental patterns show evolutionary relationships

Much of our evolutionary history can be seen in the way human embryos develop. Early in development, human embryos and embryos of all other vertebrates are strikingly similar, as shown in **Figure 10-7**. In later stages of development, a human embryo develops a coat of fine fur. The similarity of these early developmental forms strongly suggests that the process of development has evolved. New instructions on how to grow have been added to old instructions inherited from ancestors.

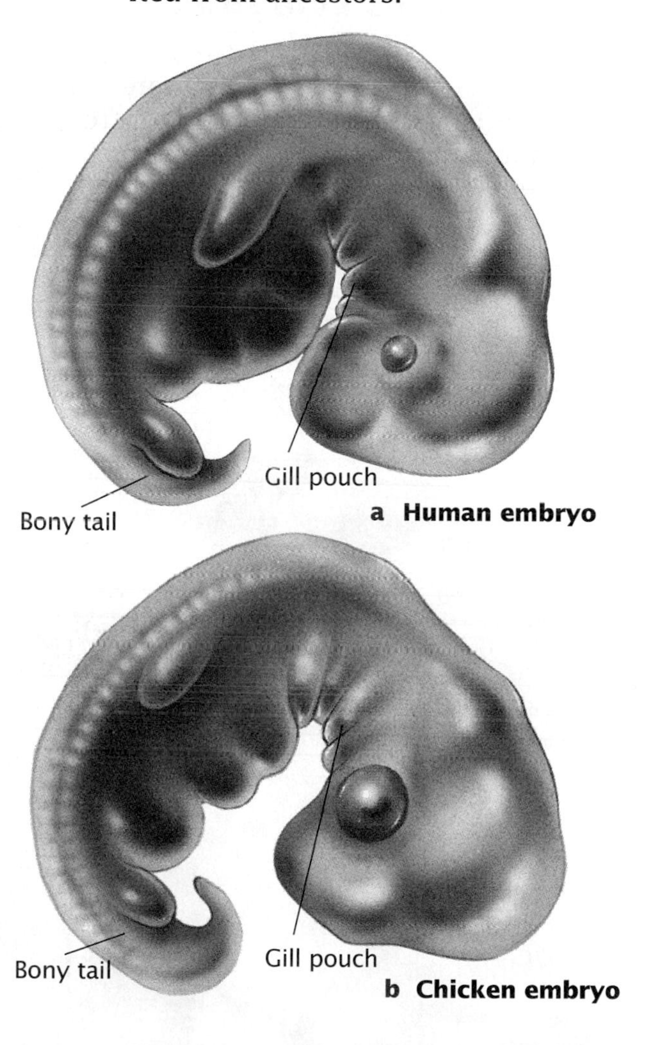

Figure 10-7
The five-week-old human embryo (a) and the four-day-old chicken embryo (b) each have a bony tail and gill pouches similar to those of fishes.

DNA and proteins contain evidence of evolution

Although complete fossil histories for living organisms are rare, an organism's history is written in the sequence of nucleotides making up its DNA. If species have changed over time, their genes also should have changed. The theory of evolution predicts that genes will accumulate more alterations in their nucleotide sequences over time. Thus, if we compare the genes of several species, closely related species will show more similarities in nucleotide sequences than will distantly related species. Closely related species also will show more similarities in the amino acid sequences in their proteins. This is because the amino acid sequence in a protein reflects the nucleotide sequence of the gene coding for that protein.

For example, to see how closely related chimpanzees, dogs, and rattlesnakes are to humans, scientists examined the sequence of amino acids in the protein cytochrome *c*, an essential participant in cellular respiration. They found that human cytochrome *c* and chimp cytochrome *c* are identical in all 104 amino acids. This high degree of similarity indicates our very close kinship to chimpanzees. A dog's cytochrome *c* differs from human cytochrome *c* in 13 amino acids, indicating that dogs are fairly distant relatives. But dogs are more closely related to us than are rattlesnakes, whose cytochrome *c* differs from ours in 20 amino acids. In most cases, the evolutionary relationships indicated by DNA or protein sequences confirm those suggested by comparative anatomy and by developmental patterns.

SECTION REVIEW

1. **Why is it unlikely that you will be fossilized?**
2. **Explain why transitional species, such as the ancestors of modern whales, are crucial evidence for evolution.**
3. **How does the whale's vestigial pelvis provide evidence in support of evolution?**
4. **Explain how sequences of amino acids in proteins can be used to reveal relationships among organisms.**

Has natural selection affected your life directly? Yes, because your body has been shaped by natural selection. For example, the ability of your eyes to focus, the way your hands grip objects, your upright posture, your large brain, the color of your hair, and numerous other characteristics are all results of evolution by natural selection.

10-3 Natural Selection

OBJECTIVES

1. **Describe how natural selection occurs.**
2. **Summarize the effects of natural selection on the peppered moth and on the sickle cell allele.**
3. **Describe how natural selection can lead to the formation of new species.**
4. **Contrast the hypotheses of punctuated equilibrium and gradualism.**

How Natural Selection Causes Evolution

Darwin not only demonstrated that evolution has occurred but also proposed its principal mechanism—natural selection. The key factor in natural selection is the environment. The environment presents challenges that individuals with particular traits can better overcome. Thus, the environment "selects" which organisms will survive and reproduce more often. Traits possessed by organisms successful at survival and reproduction are more likely to be transmitted to the next generation. These traits, therefore, will become more common. Compare the modern giraffe in **Figure 10-8** with its short-necked ancestors in **Table 10-1**. This table explains in detail how natural selection could drive the evolution of long-necked giraffes from short-necked ancestors.

Figure 10-8
How did giraffes evolve long necks? Natural selection is the mechanism by which their long necks probably evolved.

Table 10-1 The Process of Natural Selection

Step	Explanation	Example
Variation is the raw material for natural selection.	Every species contains genetic variation: individuals differ because they carry different alleles for certain traits. As you learned in Chapter 7, mutation is the source for new variation. In addition, sexual reproduction and crossing-over produce individuals with unique combinations of alleles.	Giraffes were born with alleles for varying neck lengths. Some had longer necks than others.
Living things face a constant struggle for existence.	Organisms produce more offspring than can survive. These offspring emerge into a hostile world where they must evade predators and compete with other individuals for limited supplies of food and living space.	Giraffes with longer necks could reach the leaves in tall trees. Those with shorter necks could not.
Only some individuals survive and reproduce.	Some individuals are better able to survive the challenges of life than others. Perhaps a particular allele makes them more drought tolerant, or more efficient, or more resistant to disease. These individuals are more likely to survive and produce offspring.	The giraffes with longer necks were better at getting food than were giraffes with shorter necks. Consequently, long-necked giraffes produced more offspring than did giraffes with short necks.
Natural selection results in genetic change.	Each generation consists of the offspring of individuals that successfully reproduced. Thus, it contains an increased proportion of individuals with traits that promote survival and reproduction than did the previous generation. The same is true for subsequent generations. Over time, the alleles for successful traits will increase in frequency, while alleles for traits that reduce the chances of survival and reproduction will decline in frequency.	Since more long-necked than short-necked giraffes were being born, long-necked giraffes became common, and short-necked giraffes became rare. The average neck length increased. Eventually, long-necked giraffes replaced short-necked giraffes.
Species adapt to their environment.	Selection tends to make a population better suited to its environment. The environment determines the direction of genetic change. An allele favored in one environment may not be favored in another.	Long-necked giraffes are well suited for browsing on the foliage of tall trees, which is out of the reach of most other animals.

The Peppered Moth: Natural Selection in Action

Over many generations natural selection gradually changes a species in response to the demands of its environment. **Adaptation** is the process by which a species becomes better suited to its environment. The word *adaptation* can also refer to any change in a trait that increases the likelihood that an organism will survive or reproduce. For an example of natural selection's effects, look closely at the light and dark peppered moths in **Figure 10-9**. Until the 1850s, dark gray peppered moths were rare and were treasured by British butterfly and moth collectors. Almost all peppered moths were pale. Around 1850, however, dark peppered moths started to become more common, usually in heavily industrialized areas. By 1950, peppered moth populations living near industrial centers consisted almost entirely of dark individuals.

Figure 10-9 Studies of light and dark peppered moths provide an example of adaptation.

Why did the dark peppered moths become more common? Darwin's theory of evolution by natural selection suggests a hypothesis. The color change coincided with a great increase in the number of factories in England. Pale tree trunks were blackened by heavy pollution from these factories. Perhaps dark moths sitting on soot-darkened bark escaped being eaten by birds because it was hard for the birds to see the dark moths against the dark background. Light-colored moths, on the other hand, would have stood out against a dark background and would have been easily spotted by hungry birds. H.B.D. Kettlewell, a British biologist, tested this hypothesis in the late 1950s. **Table 10-2** describes Kettlewell's experiments.

Table 10-2 How Kettlewell Demonstrated Natural Selection in Peppered Moths

	What Kettlewell did	What Kettlewell found
Step 1	Kettlewell knew that coloration was an inherited trait. He raised large numbers of both light and dark moths in the laboratory. He released equal numbers of light and dark moths into a forest near Birmingham, England, where trees were blackened by soot. For identification, each of these moths was marked with a dot on the underside of its wings.	Kettlewell set out rings of traps to recapture moths that survived. He found that two-thirds of the recaptured moths were dark. More moths that matched the dark tree trunks had survived.
Step 2	Again using marked moths, Kettlewell released equal numbers of light and dark moths into an unpolluted forest in Dorset, England. Trees here were light gray, not black.	When Kettlewell set out traps, two-thirds of the moths that were recaptured were light. Again, more moths that matched the color of the tree trunks survived. 
Step 3	Kettlewell set up hidden cameras in both forests to record the capture of moths by birds.	Films showed that birds were more likely to capture light moths on the dark trunks near Birmingham. In Dorset, birds were more likely to eat the dark moths, which were easily seen against the light trunks.

The Puzzle of Sickle Cell Anemia

Sickle cell anemia, as you learned in Chapter 7, is a hereditary disease that affects hemoglobin molecules, the proteins in human blood that carry oxygen. Defective hemoglobin molecules deform the red blood cells, as shown in **Figure 10-10**. People who are homozygous for the defective sickle cell allele have sickle cell anemia and often die at an early age. Heterozygous individuals, who have both a defective and a normal allele, are healthy. The disease is now known to have originated in central Africa, where 1 person in 100 is homozygous for the defective allele and thus has the disease. In the United States, however, only 1 African American out of every 500 has sickle cell anemia, and the disease is almost unknown in other races.

Why has natural selection not acted against the sickle cell allele in Africa and reduced its frequency? Why is this potentially fatal allele common there?

The sickle cell allele confers an unexpected advantage in Africa

The defective allele is common in central Africa because people who are heterozygous for the sickle cell allele are much less susceptible to malaria. Over 1 million people die of malaria each year, most of them in central Africa and tropical Asia.

In central Africa, the number of deaths from sickle cell anemia is far lower than would occur from malaria if heterozygous individuals were not malaria resistant. One in five individuals is heterozygous for the sickle cell allele and survives malaria. But only 1 person in 100 is homozygous for the allele and dies of anemia. Stated simply, natural selection has favored the sickle cell allele in central Africa because the payoff in survival of heterozygotes is greater than the price in death of homozygotes.

You can see that natural selection is acting on the sickle cell allele in opposite directions. On one hand, selection tends to eliminate the sickle cell allele because of its lethal effects on homozygotes. On the other hand, selection tends to favor the sickle cell allele because it protects heterozygotes from malaria. Biologists use the term **balancing selection** to refer to the situation in which two opposing selective forces affect the frequency of an allele in a population.

Selection does not favor the sickle cell allele in the United States

Why is the frequency of individuals who are homozygous for sickle cell anemia so much lower among African Americans? After all, most African Americans are descended from Africans who arrived in this country hundreds of years ago. While interbreeding has occurred, much of the answer lies in the distribution of malaria. Malaria is extremely rare in North America today. The protection against malaria that is associated with being heterozygous for the sickle cell allele offers no real advantage in the United States. As a result, the allele for sickle cell anemia has become far less common in the United States than it is in central Africa. Biologists call unopposed selection **directional selection**. Directional selection moves the frequency of a particular allele (and that of the trait the allele produces) in one direction.

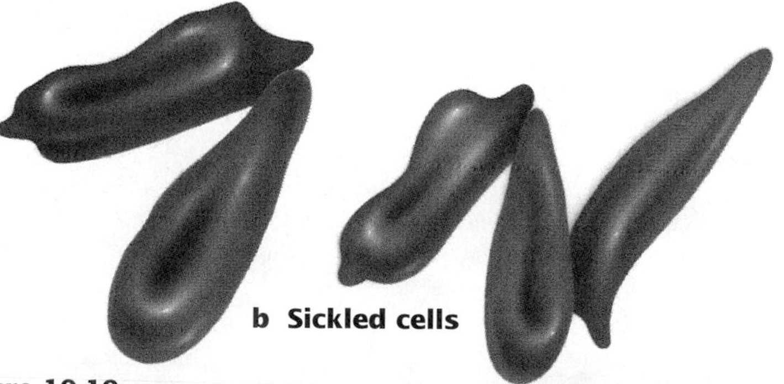

Figure 10-10
When oxygen levels in the blood are high, the red blood cells of people homozygous for the sickle cell allele appear normal (a). But when oxygen levels are low, some of their cells collapse into a sickle shape (b) and cannot function normally. In people heterozygous for the sickle cell allele, only a few red blood cells sickle at low oxygen levels. These individuals are malaria resistant.

How Species Form

Because natural selection favors changes that increase an organism's chances of surviving and reproducing, it will continuously shape a species to improve the fit between the species and its environment. Recall from Chapter 1 that a species is a group of individuals that can interbreed and produce fertile offspring but that cannot breed with any other such group. When populations of a species are found in several different kinds of environments, natural selection will act to make each population suited to its particular environment. Populations in different places thus become increasingly different, as each population becomes better suited to its particular environment.

Over time, separate populations of a species can eventually become quite distinct if their environments differ enough. These populations form what biologists call ecological races, as shown by the example of the seaside sparrows in **Figure 10-11**. **Ecological races** are populations of the same species that differ genetically because they have adapted to different living conditions.

Figure 10-11

a Dusky seaside sparrow
Ammodramus maritimus nigrescens
This ecological race is now extinct. The last individual died in captivity in 1987. This race occurred only near Merritt Island and the marshes of the St. John's River near Titusville, Florida. It differed from the other seaside sparrows in having a very dark gray back and a light, streaked breast.

b Gulf seaside sparrow
Ammodramus maritimus fisheri
This sparrow is found along the Gulf Coast, but not along the Atlantic Coast. It differs from other ecological races of seaside sparrows in its more orange color. Its range (shown in yellow) overlaps that of the seaside sparrow in Figure 10-11c.

Members of different ecological races are not yet different enough to belong to different species, but they have taken the first step. The differences among human races, while they may seem large to some of us, are actually very small in an evolutionary sense. Notice that each sparrow in **Figure 10-11** has a three-word scientific name; the third word of the name indicates a subspecies, a distinct group within a species.

Ecological races form new species

Ecological races often become increasingly different. The accumulation of differences between species or populations is called **divergence**. Divergence occurs because natural selection favors different survival strategies in different environments. Eventually, races can accumulate so many differences that biologists consider them separate species.

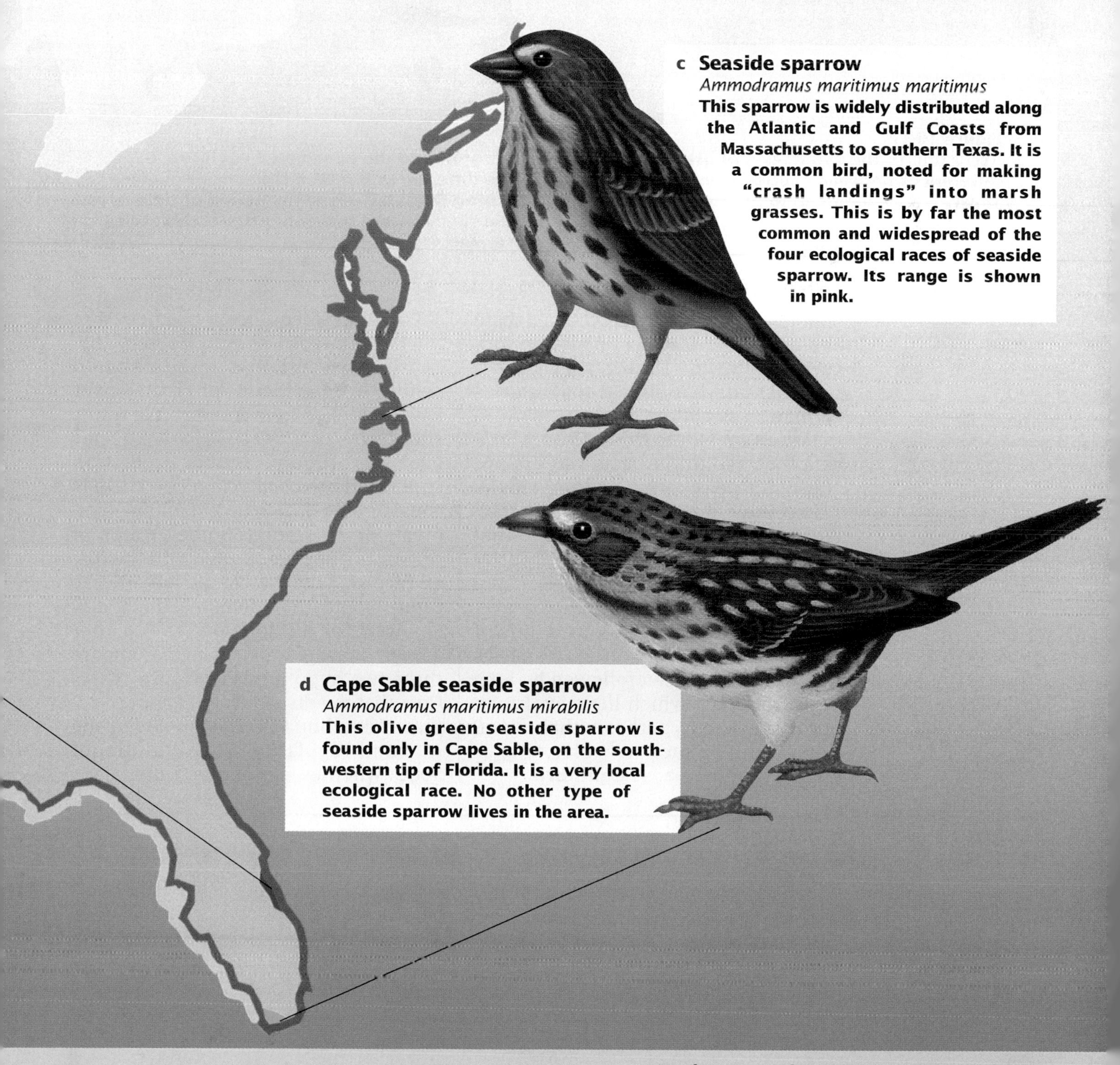

c Seaside sparrow
Ammodramus maritimus maritimus
This sparrow is widely distributed along the Atlantic and Gulf Coasts from Massachusetts to southern Texas. It is a common bird, noted for making "crash landings" into marsh grasses. This is by far the most common and widespread of the four ecological races of seaside sparrow. Its range is shown in pink.
d Cape Sable seaside sparrow
Ammodramus maritimus mirabilis
This olive green seaside sparrow is found only in Cape Sable, on the southwestern tip of Florida. It is a very local ecological race. No other type of seaside sparrow lives in the area.

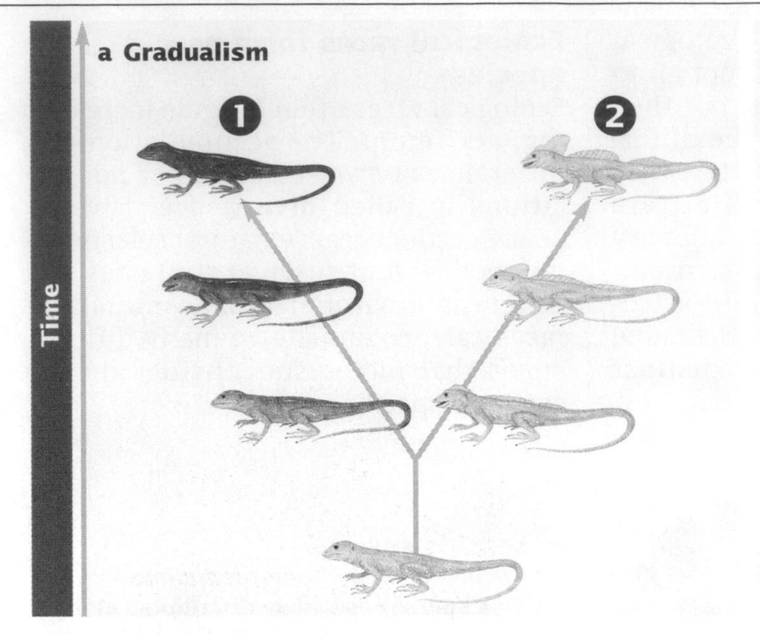

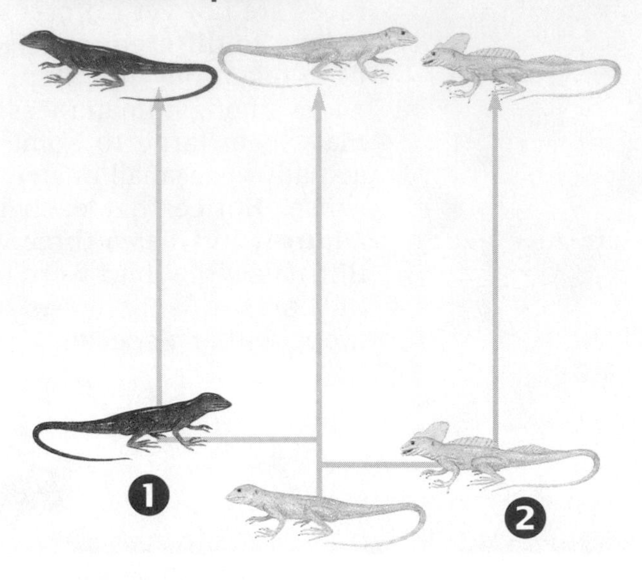

Figure 10-12

a According to the hypothesis of gradualism, the differences between species 1 and 2 accumulated slowly over a long period of time.

b According to the punctuated equilibria hypothesis, the differences between the species evolved rapidly during the relatively short period of time in which the species were forming. After each species formed, it did not change for a long period of time.

Does Evolution Occur in Spurts?

Biologists are now engaged in a debate about the rate at which evolution proceeds. Following Darwin's lead, most biologists have assumed that species formation is a slow, gradual process that goes on all the time. The hypothesis that evolution occurs at a slow, constant rate is known as **gradualism**.

Recently, some biologists have challenged gradualism, arguing that species formation occurs rapidly after major environmental upheavals. (Keep in mind that a rapid occurrence in geological time lasts many thousands of years.) Short periods of rapid species formation have been followed by long periods during which little evolution occurred. The hypothesis that evolution occurs at such irregular rates is known as **punctuated equilibria**.

Punctuated equilibria predicts that the fossil record should be very discontinuous and that fossils should exhibit little evidence of change over long periods of time. Transitional forms should be rare because new species evolve so rapidly. Is this prediction supported by fossil evidence? There is considerable disagreement among biologists on this point. Some groups of organisms appear suddenly in the fossil record, as if they had evolved very rapidly, and then remain almost unchanged for millions of years. Other groups show gradual change. Compare these two hypotheses in **Figure 10-12**. Note that the disagreement about punctuated equilibria is a debate about the rate and regularity of evolution, not about whether evolution occurs.

SECTION REVIEW

1. **Describe the steps of natural selection.**
2. **The British have instituted pollution controls on factories. How do you think this will affect the evolution of the peppered moth?**
3. **Describe how natural selection can lead to the formation of new species.**
4. **Contrast the hypothesis of punctuated equilibria with that of gradualism.**

CHAPTER

10 Highlights

Using radioactive dating, scientists found that this fossil of a duck-billed dinosaur skull is about 125 million years old.

10-1 Charles Darwin

The observations of a young Charles Darwin changed our ideas about the world.

Key Terms

natural selection (p. 175)

Summary

- Darwin's observations during his voyage led him to doubt his beliefs about the origin of species.
- Darwin proposed that new species formed by the slow transformation of existing species.
- Darwin proposed that natural selection caused evolution.
- Natural selection occurs because some organisms have traits that increase their ability to survive or produce offspring.

10-2 The Evidence for Evolution

Fossils show that modern whales evolved from land-dwelling animals with four limbs.

Key Terms

half-life (p. 176)

radioactive dating (p. 176)

homologous structure (p. 178)

vestigial structure (p. 178)

Summary

- Fossils are preserved traces of dead organisms. Fossils form only under specific conditions.
- Radioactive dating enables us to determine the ages of fossils.
- The fossil record provides evidence that older species gave rise to more recent species.
- Organisms with homologous structures share common ancestry.
- Closely related organisms have more similarities in their DNA than do distantly related organisms.

10-3 Natural Selection

Dark peppered moths became increasingly common when pollution darkened forest trees in England.

Key Terms

adaptation (p. 182)

balancing selection (p. 183)

directional selection (p. 183)

ecological race (p. 184)

divergence (p. 185)

gradualism (p. 186)

punctuated equilibria (p. 186)

Summary

- Adaptation is the process by which organisms become better suited to their environments.
- In response to the darkening of tree trunks by pollution, some peppered moth populations evolved from cream colored to dark gray.
- In Africa, the sickle cell allele is favored because it produces resistance to malaria in heterozygotes.
- Gradualism is the hypothesis that evolution occurs at a constant rate. Punctuated equilibria is the hypothesis that evolution occurs at an irregular rate.

CHAPTER 10

review

Understanding Vocabulary

1. For each pair of terms, explain the differences in their meanings.
   a. evolution, natural selection
   b. homologous structures, vestigial structures
   c. balancing selection, directional selection

2. Using each set of words, write one or more sentences summarizing information learned in this chapter.
   a. fossil, half-life, radioactive dating
   b. natural selection, environment, trait, adaptation, survival
   c. natural selection, ecological race, divergence, new species

Understanding Concepts

3. **Relating Concepts** Construct a concept map summarizing the study of evolution. Try to include the following terms in your concept map: Darwin, adaptation, fossils, natural selection, vestigial structures, and nucleotide sequences.

4. **Summarizing Information** Each species of Galápagos finch has a beak suited for eating a particular type of food. How would Darwin have explained this pattern before his voyage on the *Beagle*? How would he have explained it after writing the *Origin of Species*?

5. **Recognizing Patterns** The plants and animals of the Galápagos Islands closely resemble those of the South American mainland. What did Darwin conclude from this pattern of resemblance?

6. **Relating Concepts** Explain the difference between evolution and natural selection.

7. **Comparing Structures** What is a fossil? Give examples of three different kinds of fossils. Why are hard parts of organisms, such as bones and shells, more likely to be fossilized than soft parts, such as muscles and skin?

8. **Interpreting Processes** What does the term *half-life* mean? Explain the process of radioactive dating.

9. **Recognizing Relationships** The Earth is thought to be about 4.5 billion years old. Explain the importance of this information for the theory of evolution.

10. **Interpreting Data** Use the data in the table below to answer the following questions. Which species is most closely related to humans? Which species is most distantly related to humans? Explain your answers.

Species	Number of amino acid differences compared with human hemoglobin
Humans	0
Species A	17
Species B	39
Species C	8

11. **Recognizing Relationships** Why is variation among individuals so important for the process of natural selection? What are the main sources of this variation?

12. **Analyzing Methods** What was Kettlewell's hypothesis to explain the evolution of dark coloration in peppered moths? Describe the methods he used to test this hypothesis.

13. **Predicting Outcomes** Suppose that scientists discover an effective vaccine for the disease malaria. How would you expect the vaccine to affect the frequency of the sickle cell allele in central Africa over several generations? Explain your answer.

14. **Summarizing Information** What is an ecological race? How does an ecological race form? What is the relationship between an ecological race and a species?

15. **BUILDING ON WHAT YOU HAVE LEARNED** In Chapter 8 you learned that DNA is a molecule made up of the nucleotides that code for amino acids. What can biologists learn about two species by comparing their DNA sequences?

Interpreting Graphics

16. The graph below shows average shell length for three species of clams. The data begin with the first appearance of each species in the fossil record and end with the species' last appearance. Which hypothesis about evolutionary rates—gradualism or punctuated equilibria—do the data support? Explain your answer.

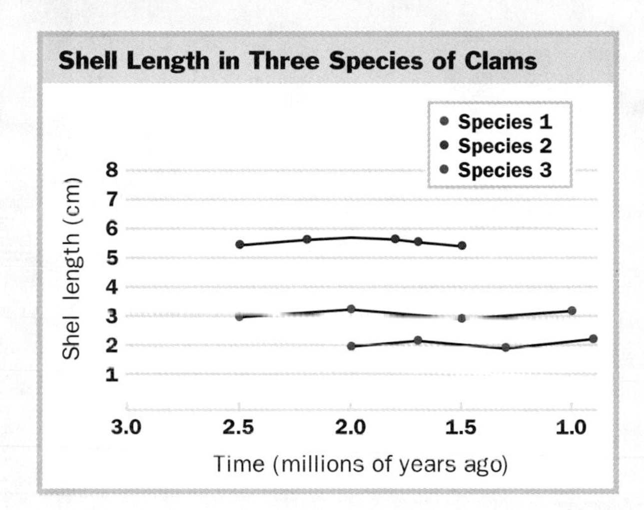

Reviewing Themes

17. ***Evolution***
The bones in the human arm are very similar to the bones in a bird's wing and a dolphin's flipper. What does this similarity in structure reveal about the ancestry of these animals?

18. ***Evolution***
What is sickle cell anemia? What unexpected advantage do heterozygous individuals have in central Africa? Why don't heterozygotes have the same advantage in the United States?

Thinking Critically

19. **Evaluating an Argument** Here is a student's explanation of the evolution of the peppered moth. "At first, all peppered moths were light colored. When pollution began to darken the tree trunks, some light colored moths turned a little darker. These darker moths survived and passed on their darker coloration to their offspring. The offspring became even darker, and their offspring were darker still. This continued until all peppered moths matched the color of the tree trunks." What is wrong with this explanation?

Life/Work Skills

20. **Finding Information** Many species of destructive insects have evolved resistance to the pesticides once used to control them. Use references available in a library or search an on-line database to gather information about pesticide resistance in insects. Write a report that explains the causes of pesticide resistance, describes its costs to humans, and explains some of the alternative pest-control methods being researched.

Activities and Projects

21. **Career Connection: Paleontologist** A paleontologist is a scientist who excavates and studies fossils and other remains. Use library references or search an on-line database to find out what a paleontologist does in his or her day-to-day job. Where do most paleontologists work? What kind of educational background is required?

22. **Math Connection** Study the following table, which shows the half-lives of radioactive elements. Then answer the questions.

a. How old is a rock that has half the amount of uranium-238 that it originally had?
b. How old is a rock that has approximately one-quarter of the amount of potassium-40 that it originally had?

Radioactive isotope	Half-life	Stable product
Carbon-14	5,730 years	Carbon-12
Uranium-235	704 million years	Lead-207
Potassium-40	1.25 billion years	Argon-40
Uranium-238	4.5 billion years	Lead-206

Discovering Through Reading

23. Read the article "No Longer Human" by Lori Oliwenstein in *Discover*, December 1992, pages 34–35. What was the original source of HeLa cells, and how did they travel from lab to lab? Why do some researchers claim that HeLa cells have become a new species? What are the arguments against the cells being a new species?

CHAPTER 11

REVIEW
- biological macromolecules (Section 2-3)
- theory of endosymbiosis (Section 3-3)
- structure of RNA (Section 8-2)
- natural selection (Section 10-3)

History of Life on Earth

Scientists have reconstructed the history of life from fossilized remains like these dinosaur bones.

With life all around us, it is difficult to imagine a time when there was no life on Earth, when no plants grew and no fishes swam in the sea. The Earth is much older than life, however. The study of radioactive decay in rocks reveals that the Earth is some 4.5 billion years old, about 1 billion years older than the oldest fossils. Where did life come from?

11-1 Origin of Life

OBJECTIVES

1. **Contrast the three explanations for the origin of life.**
2. **Describe the importance of the Miller-Urey experiment.**
3. **Summarize the reasons scientists think RNA, not DNA, was the first genetic material.**
4. **Describe how the first cells might have evolved.**

How Did Life Begin?

There were no witnesses to the origin of life, and you cannot go back in time to see for yourself. In principle, there are at least three ways life could have begun:

1. *Extraterrestrial origin* Some scientists hypothesize that life originated on another planet outside our solar system. Life was then carried here on a meteorite or an asteroid and colonized the Earth. How life arose on other planets, if it did, is a question we cannot hope to answer soon.

2. *Creation* Many people believe that life was put on Earth by divine forces. In this view, common to many of the world's religions, the forces leading to life cannot be explained by science.

3. *Origin from nonliving matter* Most scientists think that life arose on Earth from inanimate matter. First, random events produced stable molecules that could reproduce themselves. Then natural selection favored changes in these molecules that increased their rate of reproduction, leading eventually to the first cell.

This chapter will examine the third option. The first two possibilities are not considered because they are not testable and thus fall outside the realm of science. The first step in the origin of life from nonliving matter is the origin of the Earth. The Earth formed about 4.5 billion years ago. Three steps in its formation are shown in **Figure 11-1**.

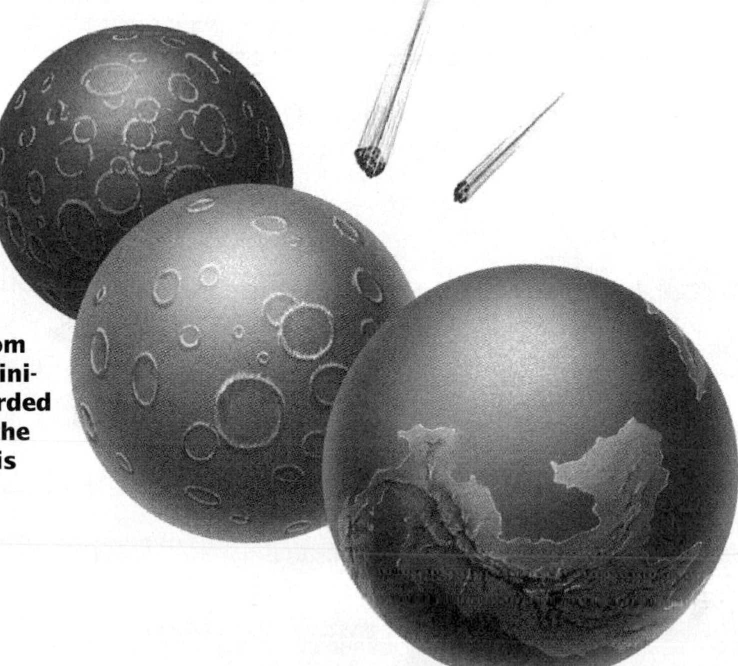

Figure 11-1
About 4.5 billion years ago, the Earth and the rest of the solar system condensed from a cloud of dust and debris. The Earth was initially uninhabitable because it was bombarded by debris left over from the formation of the solar system. About 4 billion years ago this bombardment tapered off, and the oceans formed shortly afterward.

Origin of Life's Chemicals

For life to have arisen from nonliving matter on the early Earth, the chemicals that make up living things must have been present. Recall from Chapter 2 that all organisms are composed of the same four kinds of macromolecules: proteins, lipids, carbohydrates, and nucleic acids. But where did these macromolecules, the building blocks of life, come from?

In a letter to a friend, Charles Darwin speculated that life began in "a warm little pond." In the 1920s, the Russian scientist A. I. Oparin and the British geneticist J.B.S. Haldane independently proposed a more elaborate version of Darwin's idea. They suggested that the Earth's oceans once contained large amounts of simple molecules and that these molecules spontaneously came together to form the macromolecules that compose living things.

Oparin and Haldane proposed that the early Earth's atmosphere lacked oxygen. They believed that it was instead rich in nitrogen gas (N_2) and hydrogen-containing gases such as water vapor (H_2O), methane (CH_4), hydrogen gas (H_2), and ammonia (NH_3). Electrons in these gases would have been frequently pushed to higher energy levels by photons from the sun and by electrical energy from lightning. Today, such high-energy electrons are quickly soaked up by oxygen in the atmosphere (which is 21 percent oxygen gas) because oxygen atoms have a great attraction for such electrons. But early in Earth's history, there was no oxygen in the air, and a wealth of high-energy electrons would have been available to help form complex organic molecules.

Life's building blocks can form spontaneously

Oparin's and Haldane's hypothesis was first tested in 1953 by Stanley Miller and Harold Urey of the University of Chicago. **Figure 11-2** explains the reasoning behind their experiment, and **Figure 11-3** shows the laboratory setup they used to simulate conditions on the young Earth. The brownish mixture that resulted from this experiment contained a variety of organic molecules and, most important, six of the twenty amino acids that organisms use to make proteins. Similar experiments by Miller and other scientists have produced a variety of biologically important molecules, including many of the amino acids found in living things, nucleotides from DNA and RNA, lipids, carbohydrates, and ATP.

Figure 11-2
The results of Miller and Urey's experiment were startling. Other scientists accepted these results, however, because of the careful planning and execution by the experimenters. Follow some of the steps in the experiment below.

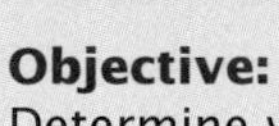

Objective:
Determine whether organic molecules that make up living things could form under conditions found on early Earth.

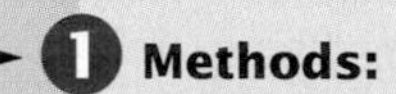

1 Methods:
- Mix molecules found in early atmosphere.
- Stimulate reactions with heat and electricity.
- Collect and analyze products of reactions.

2 Key Assumptions:
- Experiment matches conditions on early Earth—in temperature and chemical composition.
- Experiment has not been contaminated by bacteria, which could also produce organic molecules.

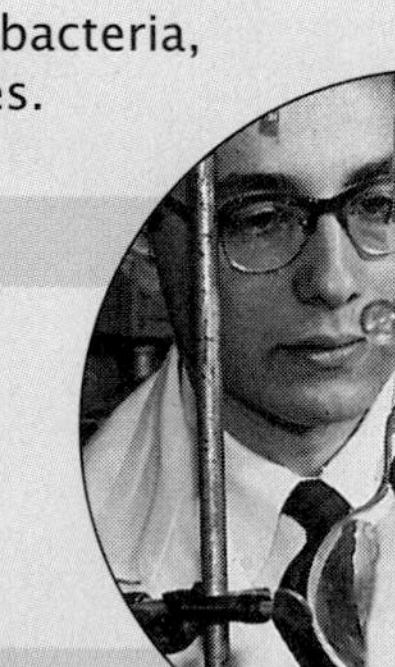

3 Results:
Complex mixture of organic molecules forms; contains amino acids.

4 Conclusions:
Organic molecules could have formed on early Earth before life existed. Thus, a key step in the origin of life is possible.

Figure 11-3
Stanley Miller and Harold Urey used an apparatus like this to simulate the conditions on the early Earth.

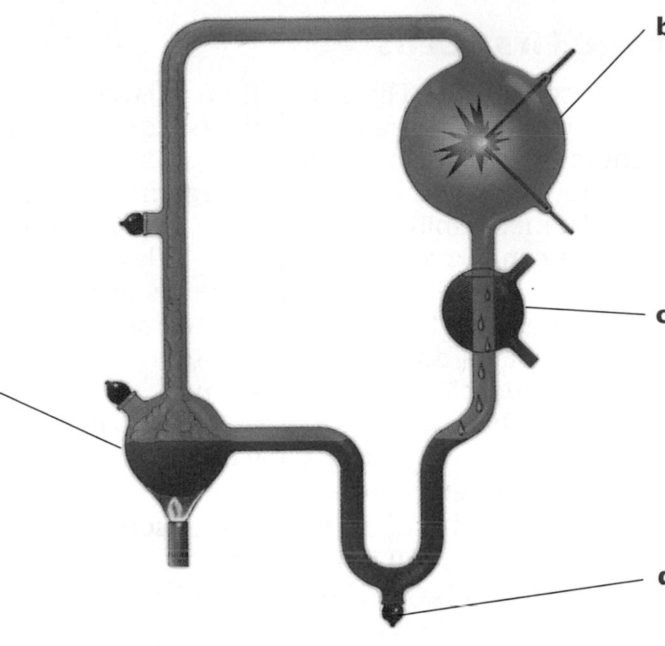

a **This flask contains water and represents the ocean. Heating causes water to evaporate and move up a glass tube to . . .**

b **. . . the vessel representing the atmosphere. It contains a mixture of hydrogen gas, methane, ammonia, and water vapor. Sparks discharged into the mixture mimic the effects of lightning, stimulating reactions among the gases.**

c **Cold water circulates over the tubing, causing water vapor to condense, simulating rain. Thus, the products of any reactions in the "atmosphere" are carried back into the "ocean," where they can continue to react.**

d **Miller could remove samples for analysis through this valve.**

These experiments show that the components of the macromolecules found in living things could have formed spontaneously on the early Earth.

Recent discoveries suggest that life was present within a few hundred million years of Earth's origin. Also, the intense solar radiation reaching the Earth in the absence of oxygen would have depleted the ammonia and methane in the atmosphere. So there would have been less of these gases dissolved in the ocean than in Miller and Urey's experimental mixture. Some scientists now speculate that spontaneous processes like those in the Miller-Urey experiment may have occurred primarily within the tiny bubbles that are abundant at the ocean's surface. The bubbles would have concentrated the chemicals and thereby speeded the reactions. They also would have protected these reactions from solar radiation.

RNA was probably the first genetic molecule

Miller and Urey's experiment and other similar experiments show that the basic building blocks of life can assemble spontaneously. But there was still a long way to go before life could arise. How did amino acids join together to form the chains we call proteins? And how did nucleotides join to form long chains of DNA and RNA?

As you learned in Chapter 8, cells link amino acids together into proteins, but only according to instructions encoded in DNA and carried in RNA. Similarly, cells synthesize DNA and RNA, but only with the aid of enzymes, which are proteins. So how could proteins form without DNA and RNA, and vice versa?

Although nucleic acids are only produced by living things today, small chains of RNA will form spontaneously under conditions that were probably found on the early Earth. Also, Thomas Cech of the University of Colorado and Sidney Altman of Yale University have recently discovered that RNA molecules can catalyze chemical reactions. In your cells, for example, RNA molecules participate in the synthesis of other RNA molecules.

The discovery that RNA can act as a catalyst suggests the following hypothesis. First, RNA nucleotides formed from simple gas molecules. Nucleotides then assembled spontaneously into small chains. These small chains were able to catalyze chemical reactions, such as synthesis of protein, and were able to make copies of themselves. Once self-replicating molecules like these appear, natural selection and evolution are possible. Molecules that can replicate faster or more efficiently will become more common than slower-replicating or less-efficient molecules.

Origin of the First Cells

How did the first cells form? The crucial feature that separates the cell from its environment is the cell membrane, which contains lipids, as you learned in Chapter 3. If you mix a lipid such as oil with water, you can see an important lipid property: small spherical bubbles of oil appear in the water because the oil molecules do not mix with the water molecules. You can see a similar effect if you shake a bottle of oil and vinegar salad dressing, as shown in **Figure 11-4a**. Scientists think that tiny spheres of lipid may have been the first stage in the origin of the cell. When mixed with water, certain lipids will form a bubble or droplet that has a double-layered membrane much like the lipid bilayer of the cell membrane. **Figure 11-4b** shows some of these droplets.

The early oceans probably contained numerous small lipid droplets, each one forming and then dispersing. Over millions of years, droplets that could survive longer by taking in molecules and energy from their surroundings would have become more common than the here-today-gone-tomorrow kind. When a means arose to transfer this ability to "offspring" droplets, probably through self-replicating RNA, life had begun.

If RNA was the first genetic material, when did DNA evolve? Most scientists think DNA evolved after simple cells had arisen. The advantage of DNA over RNA as a genetic material may have been that DNA ensured the safety of the hereditary information by storing it in a central, protected location.

Our vision of the origin of life is incomplete. No scientist has been able to create life from nonlife in the laboratory. Yet many of the important steps that might have led to living cells have been worked out. Also remember that scientists have been performing experiments regarding the origin of life for fewer than 50 years. Simple molecules were combining in the ancient seas for millions of years before life resulted.

Figure 11-4
Recall from Chapter 3 that lipids, which are nonpolar molecules, will not mix with polar molecules such as water.

a In a bottle of oil-and-vinegar salad dressing, lipids cluster into droplets to minimize contact with water.

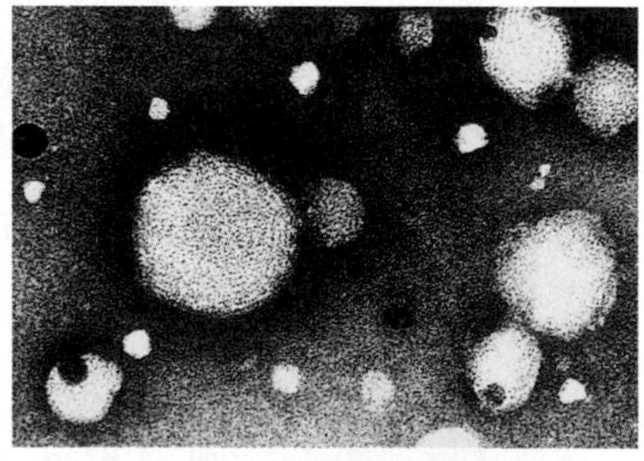

b Similar, but smaller, droplets form when lipids are combined with water in the laboratory.

SECTION REVIEW

1 Explain why the hypothesis that life arose from nonliving matter is a scientific hypothesis.

2 What were the results of the Miller-Urey experiment? Why was this experiment so important?

3 Describe the evidence that suggests RNA was the first genetic molecule.

4 Describe one hypothesis that explains how the cell membrane might have evolved.

Soon after the Earth became hospitable to living things, life arose in the ancient seas. The first organisms to appear on the planet were bacteria. These early bacteria are the ancestors of modern bacteria and of all the many different kinds of organisms living today, including you.

11-2 Early Life in the Sea

OBJECTIVES

1. **Recognize the great age of the Earth.**
2. **Compare and contrast the two major groups of bacteria.**
3. **Identify the major change in the early atmosphere caused by bacteria.**
4. **Describe the evolutionary relationships between prokaryotes and eukaryotes.**

Earliest Life: Bacteria

Several independent studies estimate the Earth's age as 4.5 billion years. So far, no fossils have been found in the oldest known rocks, which are about 3.8 billion years old. The oldest fossils that have been discovered occur in 3.5-billion-year-old rocks that were once sediments on the ocean floor. The tiny fossils that have been found in these ancient rocks were bacteria.

Biologists separate the bacteria into two groups that differ in the composition of their cell walls and cell membranes and in the structure of some of their proteins. Because the differences between these two groups are so great, it is likely that they diverged early in the history of life. One group is the **eubacteria** *(YOO bak TIHR ee ah)*, or "true bacteria," the most common bacteria today. Most living bacteria, including those that cause disease and decay, are eubacteria. Eukaryotic cells contain mitochondria and chloroplasts, which are probably descendants of ancient eubacteria.

Representatives of the second group of bacteria, the **archaebacteria** *(AHR kee bak TIHR ee ah)*, or "ancient bacteria," are shown in **Figure 11-5**. Today most kinds of archaebacteria live in extreme environments such as very salty lakes, hot springs, and swamps. However, there are archaebacteria in your intestine and in the stomachs of sheep, cattle, and other hoofed mammals. The RNA polymerase (the enzyme that copies DNA into messenger RNA) and ribosomes of archaebacteria are much more like those of eukaryotes than are those of eubacteria. Most biologists now think that archaebacteria are direct ancestors of eukaryotes.

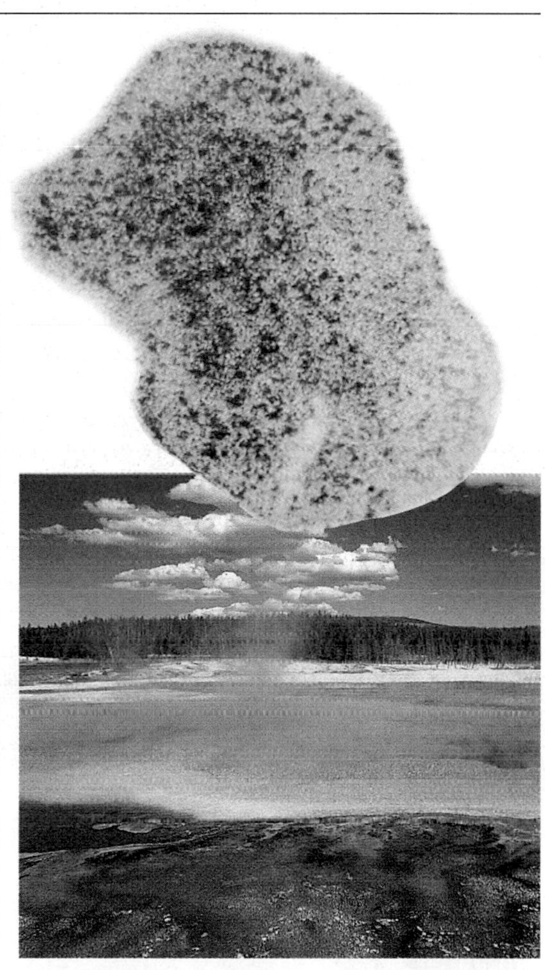

Figure 11-5
The archaebacterium *Sulfolobus* thrives at temperatures of up to 90°C (194°F). That is how it can live in this hot spring in Yellowstone National Park.

Though small, bacteria changed the atmosphere

The modern atmosphere is about 21 percent oxygen. Where did this oxygen come from, since the early Earth's atmosphere lacked oxygen? It was produced by bacteria. About 3 billion years ago, a group of photosynthetic eubacteria known as **cyanobacteria** *(sy uh noh bak TIHR ee ah)* evolved. By carrying out photosynthesis, cyanobacteria released oxygen gas into the oceans. After hundreds of millions of years, when the waters of the ancient oceans had soaked up all the oxygen they could hold, the oxygen produced by photosynthesis began to bubble out of the oceans and into the air. Over the billions of years that followed, more and more oxygen was added to the air until the modern composition of the atmosphere was achieved. As you will see in the following section, this change in the atmosphere had an important influence on the course of evolution.

Dawn of the Eukaryotes

For about 2 billion years, bacteria were the only living things on Earth. About 1.5 billion years ago, the first eukaryotic cells evolved from bacteria. Fossils of early eukaryotes show that they were much larger than bacteria and that they had internal membranes.

Based on evolutionary relationships, biologists classify living things into six great kingdoms: eubacteria, archaebacteria, protists, plants, animals, and fungi. Study the representatives of these six groups in **Figure 11-6**. Archaebacteria and eubacteria are prokaryotic. Of the four eukaryotic kingdoms, the most diverse by far is the kingdom that first arose from bacteria, the protists. Protists are important to the history of life because they are the ancestors of the three other eukaryotic kingdoms. Some protists and most members of the other three eukaryotic kingdoms are made of many cells. Having more than one cell is known as **multicellularity**. Multicellularity is a relatively recent evolutionary event. The first known fossils of multicellular organisms are found in 630-million-year-old rocks, which are nearly 1 billion years younger than the first eukaryotes. These early multicellular organisms were animals that did not have hard shells or bones and, as a result, were not well preserved as fossils. Many appeared to have been very flat and thin, like pancakes, and they probably floated on the surface of the ancient seas.

Figure 11-6 This diagram shows the evolutionary relationships among the six kingdoms of organisms.

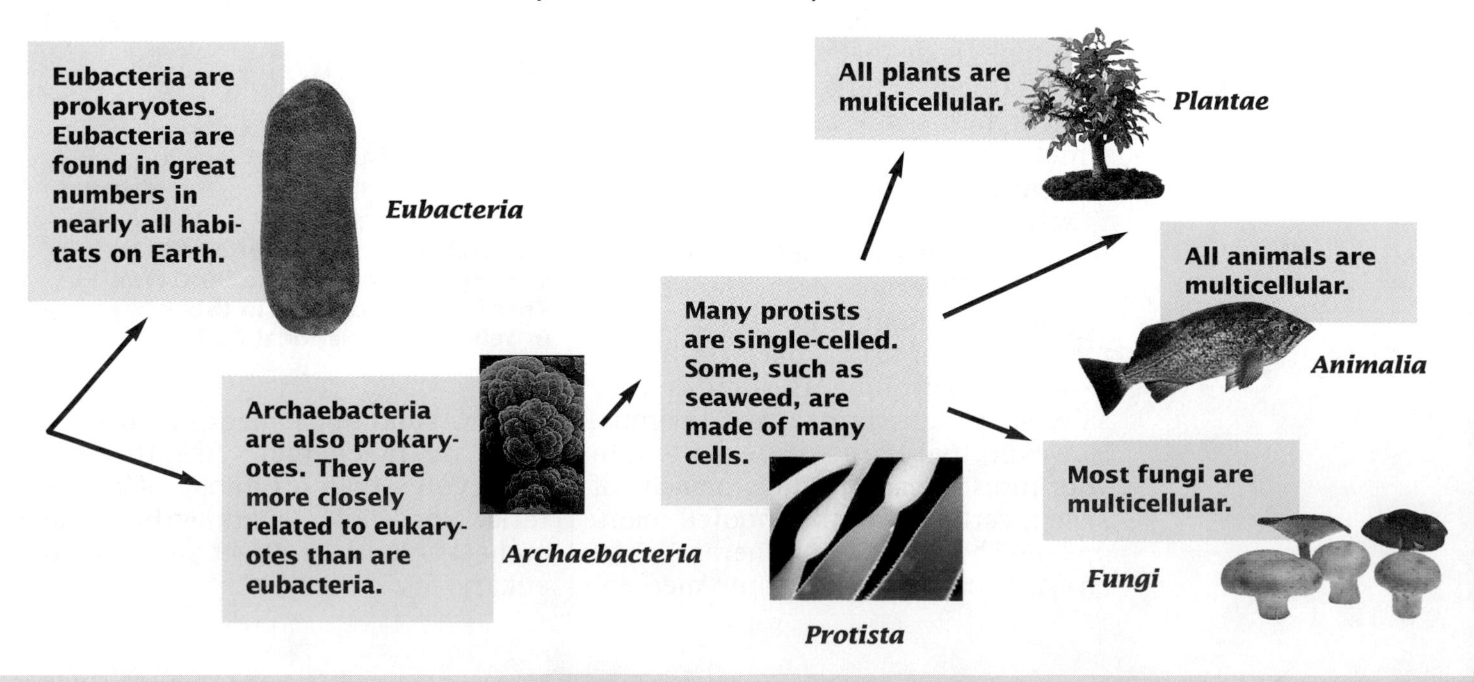

Eubacteria are prokaryotes. Eubacteria are found in great numbers in nearly all habitats on Earth.
Eubacteria
Archaebacteria are also prokaryotes. They are more closely related to eukaryotes than are eubacteria.
Archaebacteria
Many protists are single-celled. Some, such as seaweed, are made of many cells.
Protista
All plants are multicellular.
Plantae
All animals are multicellular.
Animalia
Most fungi are multicellular.
Fungi

Figure 11-7 This artist's rendition shows some of the unusual Burgess Shale animals. Since these animals are known only from fossils, their true coloration is unknown.

Life Blooms in the Ancient Seas

The appearance of multicellular animals led to a great blossoming of life in the Earth's oceans. Almost all of the major groups of multicellular organisms that survive today, except plants, originated during a very short period of time, from about 550 million years ago to about 500 million years ago, a time period known as the Cambrian period.

Biologists further divide each kingdom into major groups called **phyla** (singular, phylum). Among the important animal phyla are mollusks (including snails, slugs, clams, and octopuses), sponges, and arthropods (including insects, crabs, lobsters, and spiders). Almost all of the major animal phyla that exist today evolved during the Cambrian period.

Animals unlike any living today also originated during the Cambrian. A rich collection of Cambrian fossils has been discovered on a rocky mountain slope in eastern British Columbia, Canada, in a rock formation called the Burgess Shale. There you can see the remains of many bizarre "oddball" animals that are members of extinct phyla, such as those illustrated in **Figure 11-7**. For unknown reasons, many of the phyla represented in the Burgess Shale became extinct before the end of the Cambrian period.

Large numbers of species die out during mass extinctions

At least five times during the history of life, a large percentage of existing species have become extinct within a short period of time. These large extinction events are called **mass extinctions**. The first well-documented mass extinction occurred about 440 million years ago, some 60 million years after the end of the Cambrian period. Scientists are not sure what caused most of these mass extinctions.

SECTION REVIEW

1. **A human lifetime is about 75 years. Calculate the number of human lifetimes in 4.5 billion years.**
2. **Identify two differences between eubacteria and archaebacteria.**
3. **Describe how cyanobacteria changed the composition of the atmosphere.**
4. **Explain why the following statement is true: A human cell is descended from both eubacteria and archaebacteria.**

What was the Earth like 500 million years ago? It was very different from today's Earth. Even though the seas teemed with life, the dry land was uninhabited, as it had been for the Earth's entire history. The great variety of terrestrial life that exists today evolved only in the last 400 million years.

11-3 Invasions of the Land

OBJECTIVES

1. **Explain how ozone was critical to the development of life on Earth's surface.**
2. **Recognize how the relationship between plants and fungi enabled both to invade the land.**
3. **Identify the importance of flight in the evolution of insects.**
4. **Describe the role that insects and plants played in each other's evolutionary success.**

The Importance of Ozone

Until just over 400 million years ago, there was no life on the dry, rocky surface of the land because high levels of ultraviolet light from the sun bombarded the Earth. Ultraviolet light damages DNA. So much ultraviolet light was reaching Earth that life could not survive out of water, which absorbs ultraviolet rays.

Life was able to move onto land because of a change in the atmosphere. Recall that photosynthesis carried out by cyanobacteria added oxygen gas to the atmosphere. As large amounts of oxygen diffused into the upper atmosphere, ultraviolet rays broke apart some of the oxygen molecules, which then recombined to form **ozone**. Ozone (O_3), shown in **Figure 11-8**, is a gas that has the remarkable and very fortunate property of absorbing ultraviolet radiation. In the upper atmosphere, ozone acted like a great shield, blocking out ultraviolet radiation. By about 400 million years ago, enough ozone had formed in the atmosphere to make the Earth's surface a safe place to live.

In this context, you can understand why so many people are worried about the ongoing destruction of Earth's ozone shield by industrial chemicals. You will learn more about the damage to the ozone layer in Chapter 16.

Figure 11-8

a Ozone, composed of three oxygen atoms, protects Earth's surface by absorbing ultraviolet radiation in the upper atmosphere.

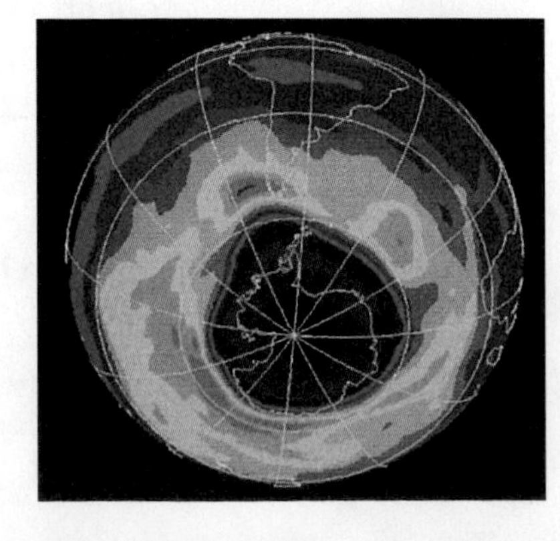

b The protective layer of ozone is being destroyed by industrial chemicals. The dark center of this satellite image indicates an area of very low ozone concentration over Antarctica.

Plants and Fungi Colonize Land

The first living things to populate the surface of the land were plants and fungi. Together, plants and fungi solved a particularly difficult challenge—surviving on bare rock. Each brought to this task a unique ability. Plants, which evolved from green algae, were able to carry out photosynthesis. However, they could not extract minerals from the soil-less, rocky surface of ancient Earth. Fungi, which evolved from protists, could not make sugars from sunlight but were adept at absorbing minerals.

The solution to the challenge of living on dry land was a unique biological partnership between plants and fungi called **mycorrhizae** *(MY koh REYE zee)*. Mycorrhizae are close associations between the roots of plants and fungi. Fungi actually grow on or into the plant root and then branch out into rock or soil, as shown in **Figure 11-9**. The fungi provide the plant with minerals absorbed from rock or soil. The plant provides food to the fungi. This kind of "you-help-me-and-I-help-you" partnership is called **mutualism**. Fossils show that the earliest plants, which lived about 410 million years ago, had mycorrhizae. The partnership between plants and fungi continues today. Indeed, 80 percent of plant species have mycorrhizae associated with their roots.

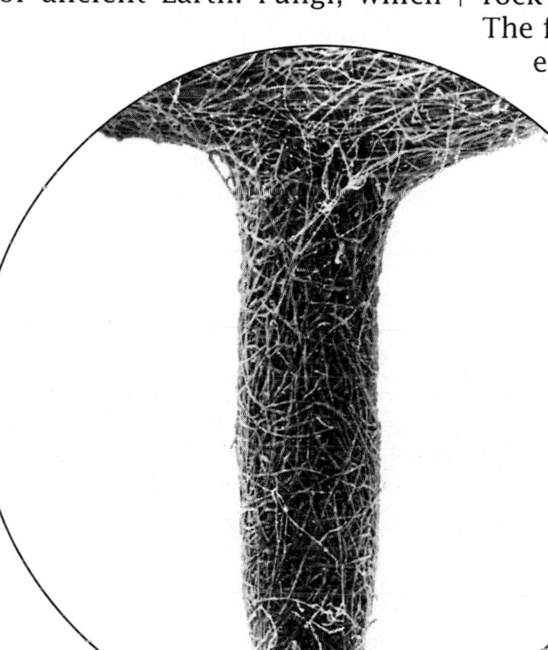

Figure 11-9
a Eighty percent of living plant species have mycorrhizae associated with their roots.

b This is a single branch root. The hairlike strands on the branch root are fungi, which help the plant absorb minerals in return for food from the plant.

Invasion of the Arthropods

The fossil record reveals that plants covered the surface of Earth within 80 million years of their initial invasion. Animals soon followed plants onto land. The first animals to leave the water were the arthropods, a kind of animal with a hard body covering and jointed legs. Crabs and lobsters are examples of existing aquatic arthropods. Some of the first arthropods to live on land resembled scorpions, carnivorous relatives of spiders with two large pincers on their front legs and a venomous stinger at the end of their tail. The arthropod invasion of the land and other major events in Earth's history are illustrated in **Figure 11-10** on the following pages.

Figure 11-10
Scientists have pieced together a detailed picture of the history of life from the fossil record. The upper timeline illustrates some of the major events in the history of life on Earth. The timeline below shows a more detailed view of the events of the last 550 million years.

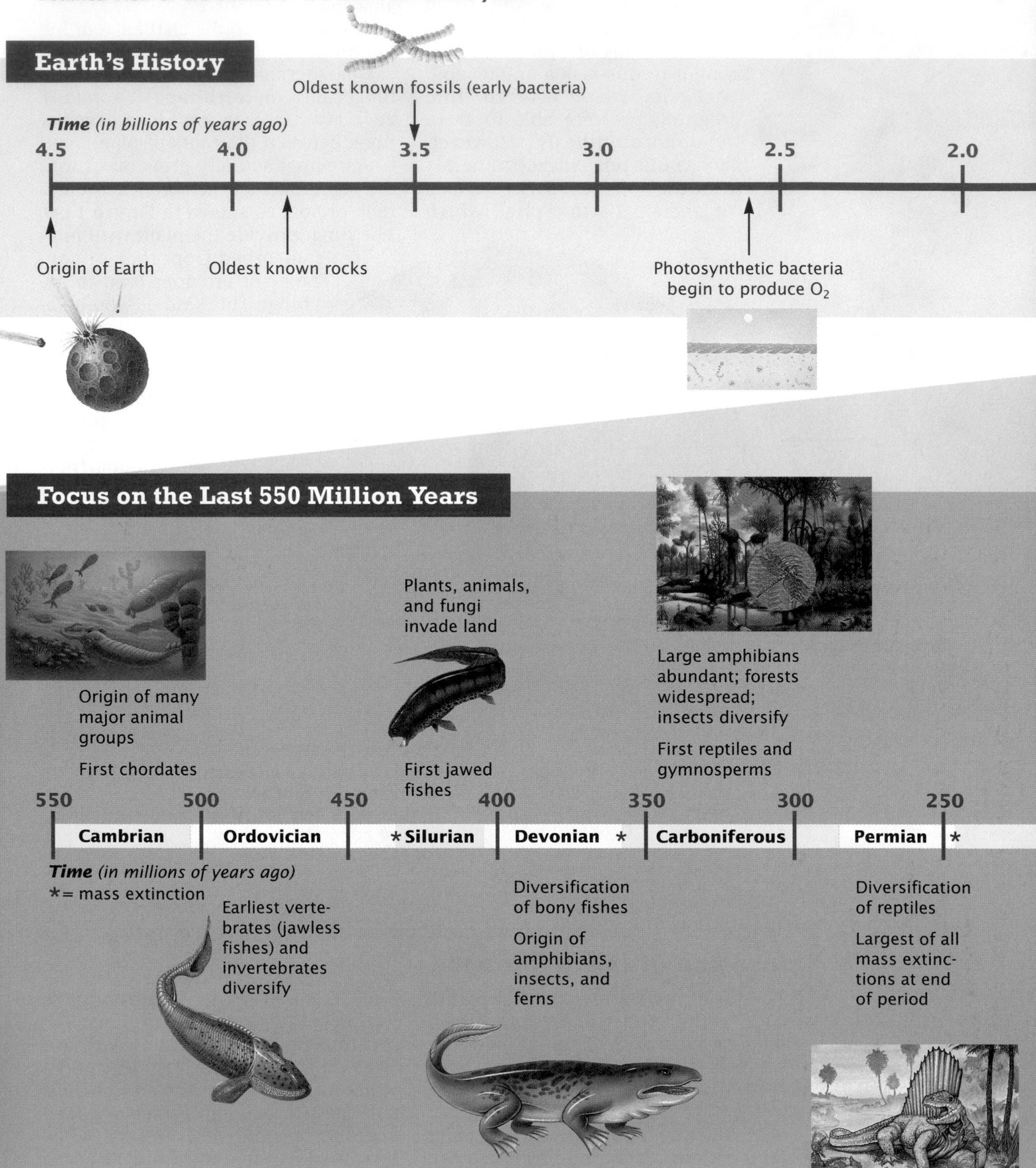

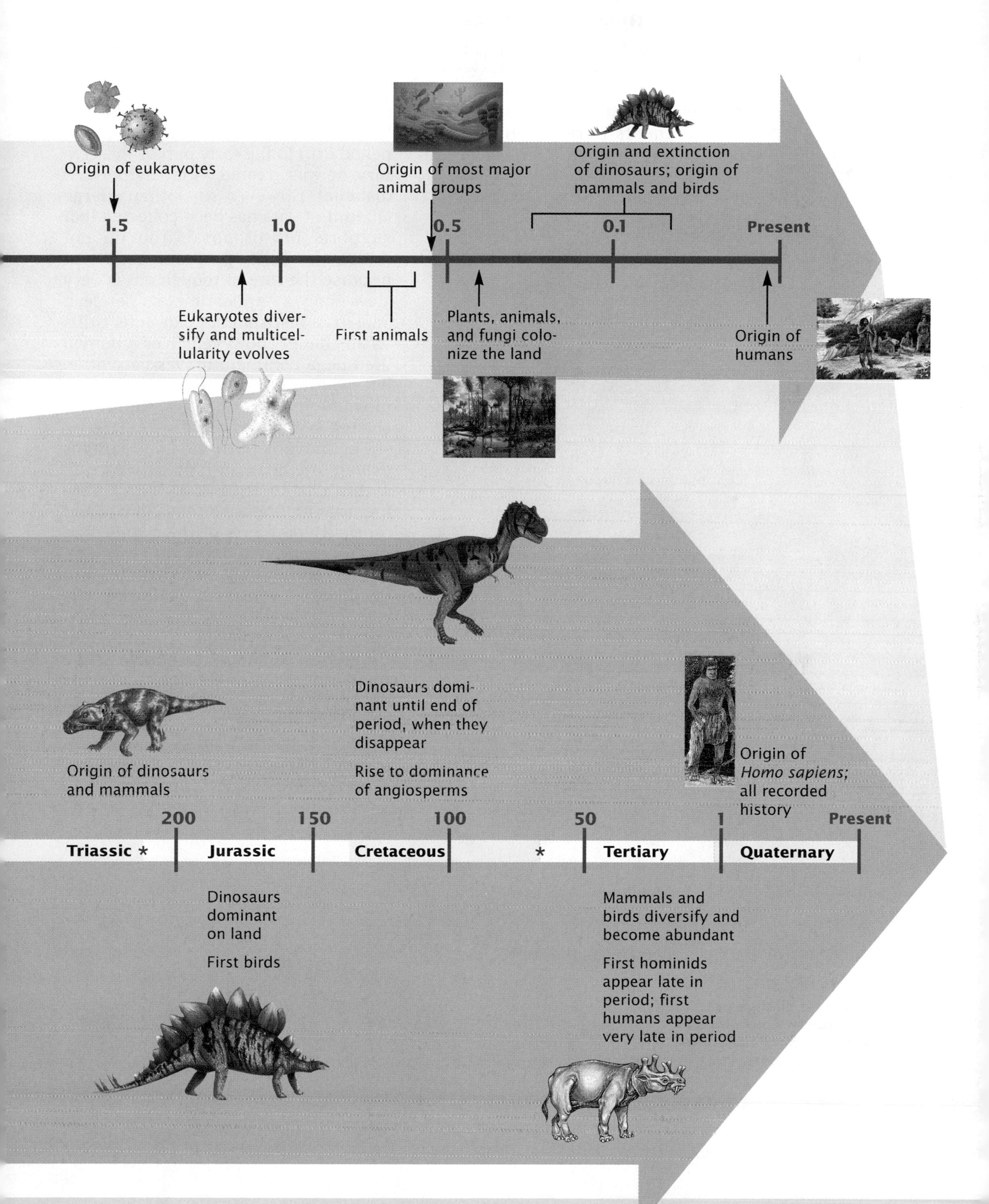

Origin of eukaryotes
Origin of most major animal groups
Origin and extinction of dinosaurs; origin of mammals and birds
1.5
1.0
0.5
0.1
Present
Eukaryotes diversify and multicellularity evolves
First animals
Plants, animals, and fungi colonize the land
Origin of humans
Dinosaurs dominant until end of period, when they disappear
Rise to dominance of angiosperms
Origin of dinosaurs and mammals
Origin of *Homo sapiens*; all recorded history
200
150
100
50
1
Present
Triassic *
Jurassic
Cretaceous
*
Tertiary
Quaternary
Dinosaurs dominant on land
First birds
Mammals and birds diversify and become abundant
First hominids appear late in period; first humans appear very late in period

The Drifting Continents

When the first plants and animals colonized the land, the world was very different. One of the most significant differences is shown in **Figure 11-11**—the continents were not in the same positions as they are today. The continents have been slowly moving for at least the last 2 billion years, and they continue to move. This movement is called **continental drift**. When Alfred Wegener proposed the idea of continental drift in the early part of this century, he was greeted with laughter and disbelief. However, since then a large amount of data has been collected that supports his proposal. You do not notice the drifting of our continent because the rate of movement is very slow—only a few centimeters per year. Sensitive instruments can detect this small change, however. Over a human lifetime, a continent moves very little. But over millions of years, these small movements add up to a significant change in position.

The movement of the continents is driven by heat and pressure deep within the Earth, but it had a very important influence on the climate and on the inhabitants of the planet. For instance, about 300 million years ago, all the continents joined together into one massive continent. Following this event, the swamps that had been widespread were replaced by drier environments. Relatives of today's pine trees prospered in the dry conditions because they were able to deal with shortages of water. Water-loving plants that had thrived in swamps were unable to tolerate the dry conditions, however, and they disappeared.

385 million years ago

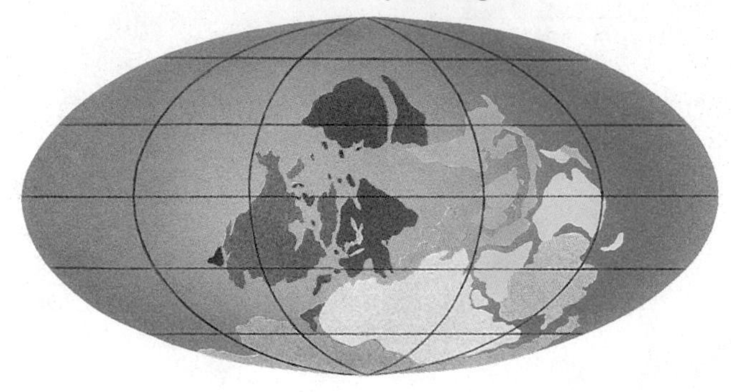

265 million years ago

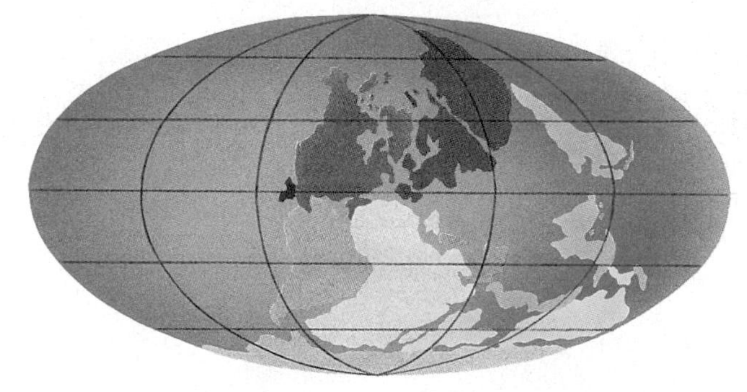

175 million years ago

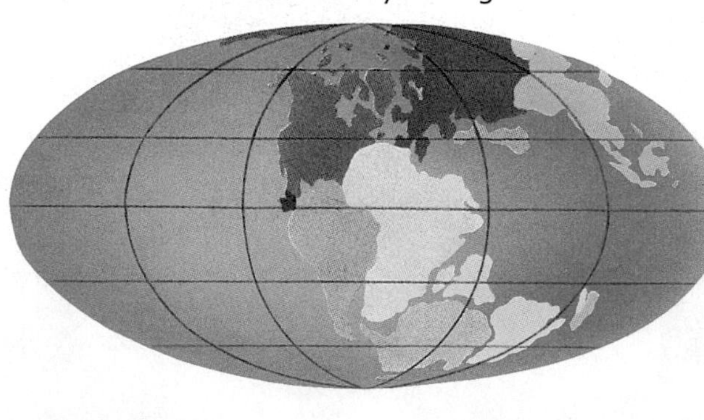

Present

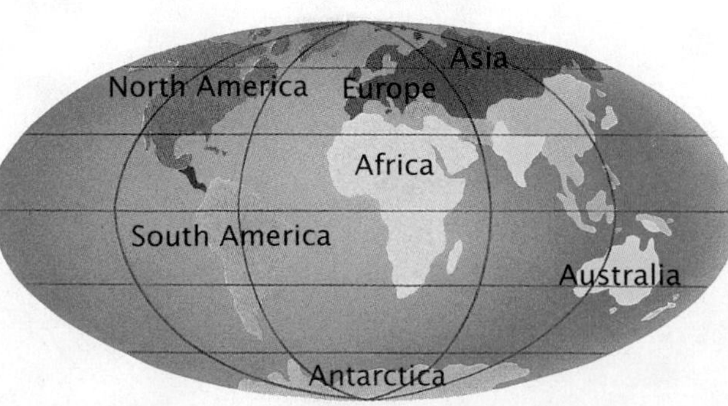

Figure 11-11
The continents change position very slowly. Four stages in the movement of the continents are shown here.

Insects Were the First Flying Animals

Interdependence

Is the relationship between insects and flowering plants a case of mutualism? Why or why not?

From the initial arthropod invaders of the land, a unique kind of terrestrial arthropod soon evolved—insects. Insects, such as the dragonfly in **Figure 11-12**, would eventually become the largest and most diverse group of animals ever. Today there are more than 200 million individual insects alive at any one time for each person on Earth. In addition, more than half of the animal species discovered so far are insects. What is special about the insects? From fossils, we know that insects were the first animals to evolve wings.

Flying opened up the world's entire surface to insects, enabling individual insects to patrol the countryside in search of food, mates, or nesting sites. A flying insect can also transport objects long distances. Plants with flowers have benefited from this ability. With sweet nectars or other food, the flower of a plant attracts insects. When the insect lands on the flower, pollen containing the plant's male gametes is brushed onto the insect. After the insect has finished feeding, it flies off, carrying pollen that will pollinate other flowers of the same species that the insect visits. The relationship between insects and flowering plants has been an important factor in the evolution of both groups for well over 100 million years.

Figure 11-12 About 370 million years ago, forests might have looked like the one illustrated here. Some dragonflies that lived at this time, although similar in structure to modern dragonflies, had wingspans of more than 1 m (3 ft).

SECTION REVIEW

1. **Why is ozone important for life on dry land?**
2. **Explain how plants and fungi were able to move onto land together.**
3. **What advantages does a flying insect have over a non-flying insect?**
4. **Describe the relationship between insects and flowering plants.**

Think of a common animal. Was the animal you thought of a horse? an elephant? a guppy? a cat? The animal you thought of was probably a vertebrate. Vertebrates are the animals most familiar to us, not only because we are vertebrates but also because all land animals bigger than your fist are vertebrates.

11-4 Parade of Vertebrates

OBJECTIVES

1. **Identify the defining characteristic of vertebrates.**
2. **List three adaptations that enabled amphibians to colonize land.**
3. **Describe some of the reptilian features for life on land.**
4. **Identify three differences between mammals and reptiles.**

Animals With Backbones

Along with the organisms you saw in **Figure 11-7**, the Burgess Shale also contains fossils of a 5 cm (2 in.) wormlike animal that is the earliest known chordate. Members of this phylum have a flexible rod of cartilage known as a **notochord** that extends along the back. In most living chordates, the notochord exists for only a short time during early embryonic development and is then replaced by the vertebral column, or backbone. Chordates with a vertebral column are called **vertebrates**.

The first vertebrates lacked jaws

The earliest vertebrates, which appeared about 500 million years ago, were jawless fishes with bony skeletons. These small fishes appear to have fed in a head-down position, with their fins helping to keep them upright while they sucked up organic particles from the bottom. For about 100 million years, jawless fishes were the only vertebrates. Today, only two groups of jawless fishes remain. These are the eel-like, parasitic lampreys and the scavenging hagfishes. Two examples of jawless fishes are shown in **Figure 11-13**.

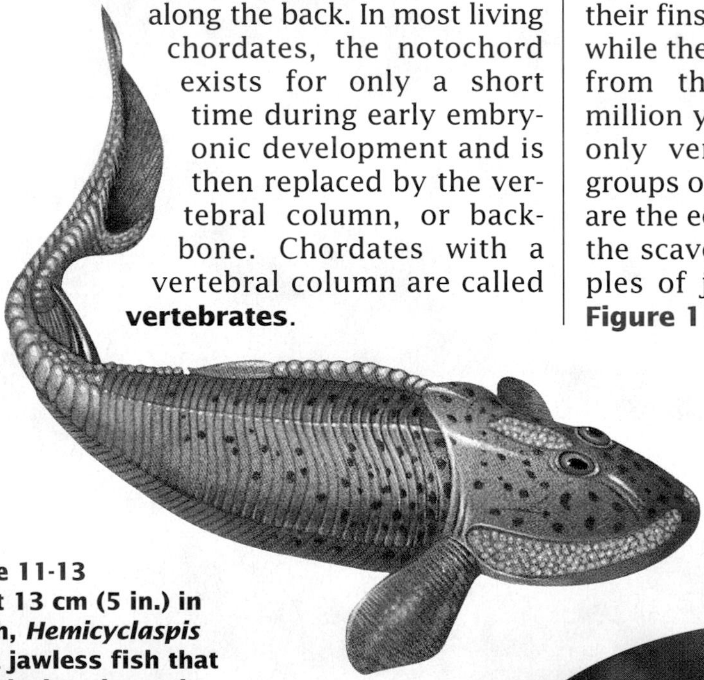

Figure 11-13

a About 13 cm (5 in.) in length, *Hemicyclaspis* was a jawless fish that lived during the early Devonian period, about 400 million years ago.

b This lamprey is a modern-day jawless fish. It attaches to other fishes with its suckerlike mouth and feeds on their flesh and blood.

Figure 11-14
Like modern sharks, the early shark *Cladoselache* was a predator. It grew to 1.8 m (6 ft) in length and lived about 360 million years ago.

Jaws evolved about 400 million years ago

Scientists know that the first fishes with jaws evolved approximately 400 million years ago. These jawed fishes rapidly replaced the jawless fishes in the oceans. Jaws enabled fishes to bite instead of suck and thus become efficient predators. Early jawed fishes dominated the seas for 50 million years before being replaced by swifter swimmers, the sharks and bony fishes.

Shark skeletons are not made of bone

From the early jawed fishes evolved a group of very efficient predators, the sharks. Shark skeletons are made of cartilage rather than bone, making sharks lighter and more buoyant than the early jawed fishes. Sharks also have large, strong, mobile fins, which allow them to swim fast and to change their direction quickly in the water. The first sharks evolved about 400 million years ago. **Figure 11-14** illustrates one of the early sharks.

Bony fishes are versatile and abundant

Sharks have been largely replaced by more versatile fishes, the bony fishes. Today's bony fishes include trout, catfish, perch, bass, guppies, salmon, and most other familiar fishes. Almost one-half of all living vertebrate species are bony fishes. As their name indicates, bony fishes have a skeleton of bone instead of cartilage. This very successful and diverse group first appeared more than 400 million years ago. **Figure 11-15** shows one modern bony fish.

Although bony fishes are now abundant in both fresh water and salt water, they probably first appeared in fresh water. The early bony fishes were small and had gills to absorb oxygen from water and lungs to absorb oxygen from the air. In the ancestors of most modern bony fishes, lungs evolved into the swim bladder, a gas-filled sac used to regulate buoyancy. Amphibians and all other land-dwelling vertebrates are the descendants of some of the early air-breathing bony fishes.

Figure 11-15
Pink salmon are among the few bony fishes that can live in both salt water and fresh water.

Amphibians, though adapted to life on land, reproduce in moist environments

The first vertebrates on land were amphibians, which evolved from bony fishes about 350 million years ago.

Three crucial adaptations enabled amphibians to become successful on land. First, amphibians absorb oxygen from the air with lungs. Second, in amphibians, blood flows from the heart to the lungs, where it picks up oxygen. The blood then returns to the heart to be repumped throughout the body. In a fish, by contrast, the heart pumps blood to the gills to absorb oxygen, but the blood does not return to the heart for repumping. Thus, in an amphibian, oxygen-rich blood flows more rapidly to the muscles and organs than it does in a fish. Third, amphibians walk on four sturdy limbs, which evolved from the fins of their fish ancestors.

Although they live on land, amphibians constantly lose water through their moist skins, through which they also absorb oxygen. Hence they generally must remain in moist places. Most amphibians also must lay their eggs in water or in moist environments. Modern amphibians include toads, salamanders, and frogs, such as the tree frog in **Figure 11-16**.

Figure 11-16
The tropical red-eyed tree frog has suction pads on its toes, enabling it to cling to vertical surfaces.

Figure 11-17
***Dimetrodon*'s sharp "steak-knife" teeth reveal that it was a predator. The large "sail" along the back may have helped regulate body temperature. Relatives of dimetrodon were the ancestors of mammals.**

Reptiles reproduce out of water

Unlike amphibians, reptiles thrive in dry climates because their skin and their eggs are largely watertight. Examples of living reptiles include snakes, lizards, turtles, and crocodiles.

Reptiles evolved from amphibians about 300 million years ago. During the following 50 million years, these earliest reptiles gave rise to a variety of forms, such as *Dimetrodon*, shown in **Figure 11-17**. Reptiles gradually replaced the amphibians as the dominant land vertebrates. Reptiles enjoyed a special advantage over amphibians during this period, because dry conditions were widespread. About 240 million years ago, at the end of the Permian period, another mass extinction occurred. Scientists estimate that about 96 percent of all species became extinct.

Those reptiles lucky enough to survive found themselves in a world full of food that no one else was eating and space that was no longer occupied by others. A great burst of evolution occurred among the reptiles. The most famous and most impressive reptiles, the dinosaurs, arose about 220 million years ago, soon after the Permian mass extinction. You can read more about the dinosaurs in the *Tour of a Dinosaur* on page 207.

Dinosaur

Brachiosaurus is the largest dinosaur for which a complete skeleton has been assembled. Its fossils were first discovered in East Africa in 1907 and have also been found in western North America. Brachiosaurus lived during the Jurassic period, which lasted from 213 million to 144 million years ago.

The nostrils of *Brachiosaurus* were on top of its head.

Its simple teeth suggest that *Brachiosaurus* fed on plants.

A long, flexible neck was characteristic of sauropods. Each of the 15 neck vertebrae of *Brachiosaurus* was as long as 1 m (3 ft).

Habitat
Brachiosaurus lived on land and fed on the foliage of tall trees, like modern giraffes do.

Evolutionary Relationships
Dinosaurs are classified into two major groups based on the structure of the pelvis: the lizard-hipped dinosaurs (including *Brachiosaurus*) and the bird-hipped dinosaurs. Within the lizard-hipped dinosaurs, *Brachiosaurus* belongs to a group known as sauropods, which includes the largest land animals that have ever lived. Other sauropods are *Diplodocus* and *Apatosaurus*.

Brachiosaurus's tracks show a narrow stance, telling us that its legs were positioned beneath its body (unlike the legs of modern reptiles) and directly supported its immense weight. Since these tracks show no marks of a dragging tail, *Brachiosaurus* probably held its tail off the ground as it walked.

Size
From nose to tip of tail, *Brachiosaurus* was about 23 m (75 ft) long. It was about 12.5 m (41 ft) tall, as high as a four-story building. If you stood next to its skeleton in a museum, you would not even reach its knee. *Brachiosaurus* weighed about 73,000 kg (80 tons), about as much as 1,000 people.

Evolution

What effect did the extinction of the dinosaurs have on the evolution of mammals?

Mammals have hair and produce milk

Mammals appeared at about the same time as the dinosaurs. Mammals, such as the zebras shown in **Figure 11-18**, are animals with hair that produce milk to feed their young.

Another feature of mammals is the four-chambered heart. This type of heart is more efficient than the three-chambered heart found in most living reptiles. Mammals are also **endotherms**. Endotherms are able to regulate their body temperature through internal mechanisms. **Ectotherms**, on the other hand, cannot regulate their temperature internally and must absorb heat from their surroundings. Living reptiles and amphibians are ectotherms.

For more than 150 million years, mammals and dinosaurs coexisted. During this time, mammals were small and did not evolve into many species.

Figure 11-19
a Hummingbirds, such as this bee hummingbird, are among the smallest birds. Some are only 5 cm (2 in.) long.

b The ostrich is the largest bird, standing almost 1.8 m (6 ft) tall. Although the ostrich cannot fly, it has feathers like all other birds.

Birds are the descendants of dinosaurs

Although dinosaurs are extinct, descendants of small, insect-eating dinosaurs are still with us today. These descendants are birds, which evolved about 150 million years ago. Bird feathers evolved from the same scales that protected the dinosaurs so well. Feathers are one of the features that enable birds to fly. Like mammals, birds are endotherms with four-chambered hearts. Two modern-day birds are shown in **Figure 11-19**.

Birds, mammals, and dinosaurs coexisted until the mass extinction 65 million years ago, in which about two-thirds of land animals, including the dinosaurs, became extinct. The post-dinosaur world offered many opportunities for evolution. In response, mammals and birds diverged rapidly and filled the nearly empty world.

Figure 11-18
All mammals, including these zebras, produce milk to feed their young.

SECTION REVIEW

1. **Explain why you are considered a vertebrate.**
2. **Describe three adaptations that enable amphibians to live on land.**
3. **Compare the reptilian adaptations for living on land with the adaptations of amphibians.**
4. **Describe two differences between mammals and reptiles.**

CHAPTER

11 Highlights

There are many stars similar to the sun in our galaxy. Could life exist on another planet?

	Key Terms	Summary
11-1 Origin of Life		• Life arose from nonliving matter present on the early Earth. • Miller and Urey showed that simple molecules could react to form some of life's building blocks. • RNA can catalyze chemical reactions. • RNA was the first self-replicating molecule. DNA evolved later.
11-2 Early Life in the Sea	eubacteria (p. 195) archaebacteria (p. 195) cyanobacteria (p. 196) multicellularity (p. 196) phylum (p. 197) mass extinction (p. 197)	• The Earth is 4.5 billion years old. • The oldest fossils are 3.5-billion-year-old bacteria. Eukaryotes evolved from prokaryotes about 1.5 billion years ago. • Multicellular organisms arose about 630 million years ago. • There have been at least five mass extinctions that wiped out many of the Earth's inhabitants.
11-3 Invasions of the Land	ozone (p. 198) mycorrhizae (p. 199) mutualism (p. 199) continental drift (p. 202)	• By 400 million years ago, enough ozone had formed to make life on land possible. • Plants and fungi invaded the land about 400 million years ago. • Arthropods, including insects, followed plants and fungi onto land.
11-4 Parade of Vertebrates	notochord (p. 204) vertebrate (p. 204) endotherm (p. 208) ectotherm (p. 208)	• Chordates have a rod of cartilage called the notochord that runs along the back. • In most vertebrates, the vertebral column replaces the notochord. • Vertebrates include sharks, bony fishes, amphibians, reptiles, birds, and mammals. • The first vertebrates were fishes without jaws. • The dinosaurs died out about 65 million years ago.

Lipids may have formed protective coatings around the earliest molecules.

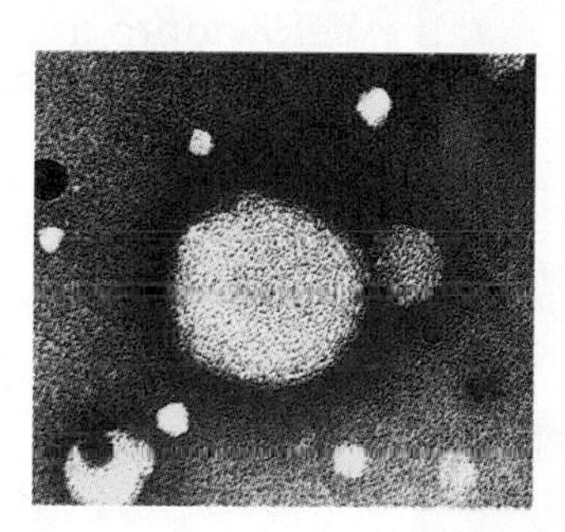

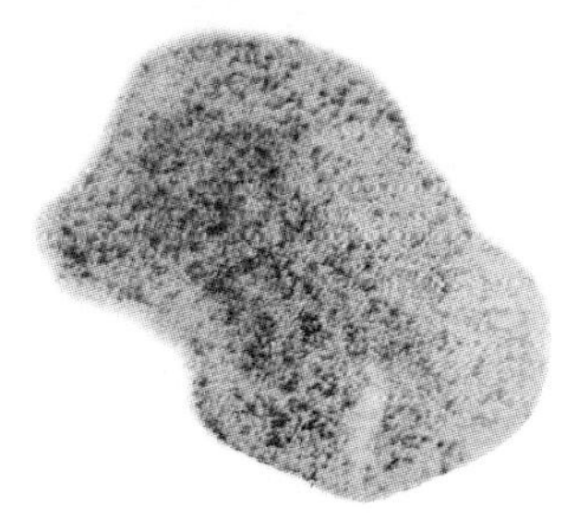

Archaebacteria are thought to be the direct ancestors of eukaryotes.

Ozone made it possible for life to move onto land. The thinning of the ozone layer is threatening all life.

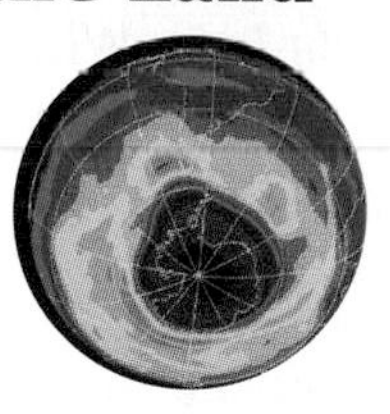

Birds are the descendants of the dinosaurs.

CHAPTER 11

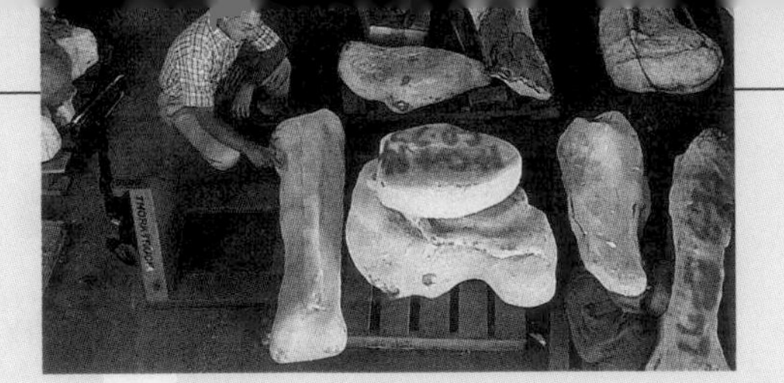

review

Understanding Vocabulary

1. For each set of terms, complete the analogy.
   a. ancestors of eukaryotes : archaebacteria :: ancestors of mitochondria and chloroplasts : ____
   b. first life on land : plants and fungi :: first animals on land : ____
   c. evolved from fishes : amphibians :: evolved from dinosaurs : ____

2. For each pair of terms, explain the differences in their meanings.
   a. eubacteria, archaebacteria
   b. mycorrhizae, mutualism
   c. notochord, vertebrate

Understanding Concepts

3. **Relating Concepts** Construct a concept map describing life's origin and earliest forms. Try to include the following terms in your map: non-life, Protista, Archaebacteria, plants, chloroplasts, Fungi, ancestors, and kingdoms.

4. **Summarizing Information** Describe the three explanations for the origin of life. Why is only one explanation considered within the realm of science?

5. **Recognizing Relationships** According to Oparin and Haldane, what was the atmosphere of the early Earth composed of?

6. **Interpreting Processes** What is the importance of Thomas Cech's and Sidney Altman's work to understanding the origins of life on Earth?

7. **Summarizing Information** What characteristic of lipids may have allowed the cell membrane to form?

8. **Summarizing Information** How old is the Earth? How old are the oldest known fossils? What type of organisms are represented by these fossils?

9. **Comparing Structures** Identify two differences between archaebacteria and eubacteria. Where could you find examples of eubacteria? of archaebacteria?

10. **Recognizing Relationships** When did cyanobacteria first appear? What important effect did they have on the environment of the early Earth?

11. **Organizing Information** Copy the unfinished table below onto a sheet of paper. Then fill in the blanks in each column.

	Prokaryotic or eukaryotic	Unicellular or multicellular
Archaebacteria		unicellular
Eubacteria	prokaryotic	
Protista		
Fungi		multicellular
Plantae		multicellular
Animalia	eukaryotic	

12. **Identifying Functions** Describe some of the benefits of flight for an insect.

13. **Recognizing Relationships** When a bee visits the flower of a plant, how does the bee benefit? What does the plant gain?

14. **Recognizing Relationships** Explain why reptiles are absent from extremely cold environments, such as northern Alaska, that are inhabited by mammals and birds.

15. **Analyzing Conclusions** What makes a vertebrate different from other animals?

16. **Organizing Information** Arrange the following animals in the correct order in which they first appeared on Earth: Include a key characteristic about each animal in your list.
    a. reptiles
    b. arthropods
    c. bony fishes
    d. jawless fishes
    e. insects
    f. mammals
    g. amphibians
    h. birds

17. **BUILDING ON WHAT YOU HAVE LEARNED** As you learned in Chapter 2, all living things make proteins from the same 20 kinds of amino acids. Explain how this fact supports the idea that all life shares a common ancestor.

Interpreting Graphics

18. The diagram below shows the apparatus used by Miller and Urey in their 1953 experiment.

a. What is the purpose of discharging electricity into the flask labeled *c*?
b. Why was it important for Miller and Urey to sterilize the apparatus to kill any bacteria in it?

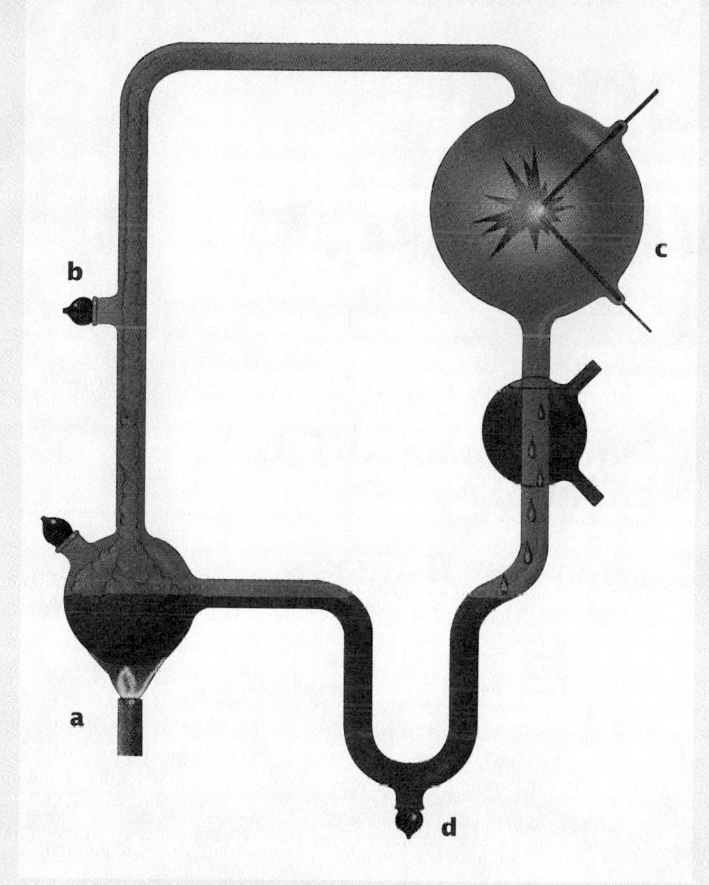

Reviewing Themes

19. ***Evolution***
Why are amphibians sometimes called transitional land-dwellers?

20. ***Interdependence***
What are mycorrhizae? What does each partner gain from this relationship? What role did mycorrhizae play in the colonization of land?

Thinking Critically

21. **Distinguishing Fact From Opinion** In an ongoing debate with environmentalists, a politician stated, "The thinning ozone layer does not affect life on Earth." Explain the error in the politician's statement.

Life/Work Skills

22. **Finding Information** Scientists have determined the past positions of the continents, but how will the Earth look millions of years from now if the continents continue to move in the same direction and at the same rate? Use library references or search an on-line database to find information about the projected positions of the continents. Draw maps that show how the Earth is expected to look 5 million, 10 million, and 50 million years from now. Share what you have learned with your class.

23. **Using Technology** Radioactive dating is one technique for determining the age of a fossil or artifact. Some others are electron spin resonance and thermoluminescence. Use library resources or search an on-line database to find out more about these methods of dating fossils. Find out how each method is carried out. What are the advantages of each method? What are the limitations of each method? Prepare a report that summarizes what you have learned.

Activities and Projects

24. **Language Arts Connection** *Saurus* is the Greek word for lizard. Find out what the names of these dinosaurs mean: *Brontosaurus, Tyrannosaurus, Stegosaurus, Gorgosaurus.* How does each dinosaur's name reflect its habits and structure? Why do scientists no longer use the name *Brontosaurus*?

Discovering Through Reading

25. Read the article "The Origin of Life" by Anthony Mellersh in *Natural History*, June 1994 (vol. 103, no. 6), pages 10–13. What two characteristics of RNA suggest that it might have been the first living molecule?

26. Read the article "Nesting Dinosaur Discovered in Mongolia," in *Science News*, January 6, 1996, page 7. Why is the fossil discovered by Mark Norell and his colleagues significant? What does *Oviraptor*'s name mean? Why did its discoverers give it this name? How have recent discoveries changed scientists' view of *Oviraptor*?

CHAPTER 12

REVIEW
- DNA structure (Section 8-1)
- use of DNA and proteins to study evolution (Section 10-2)
- radioactive dating (Section 10-2)
- natural selection (Section 10-3)

Human Evolution

Early humans drew these images deep in a French cave about 20,000 years ago.

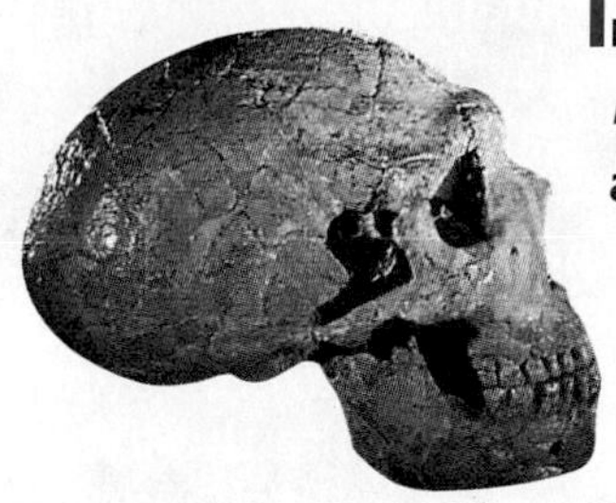

In 1871 Darwin published a second controversial book, ***The Descent of Man***. **In this book, he argued that humans were the product of evolution and were most closely related to the African apes—the gorilla and the chimpanzee. Although little fossil evidence existed in 1871 to support Darwin's case, numerous fossil discoveries made since Darwin's death, such as the skull shown at left, strongly support his hypothesis.**

12-1 Primates

OBJECTIVES

1. **List two distinctive features of primates.**
2. **Describe one adaptation of modern prosimians to nighttime activity.**
3. **Identify two differences between monkeys and prosimians.**
4. **Recognize the close evolutionary relationship between humans and apes.**

Evolution of Primates

Look closely at your hand. You have five flexible fingers. Animals with five flexible fingers are called **primates**. Monkeys, apes, and humans are examples of primates. All primates are mammals, animals that have hair and nurse their offspring with milk. Primates most likely evolved from small, insect-eating, rodentlike mammals that lived about 60 million years ago.

Like nearly all of today's primates, the earliest primates dwelt in trees. Natural selection favored traits that enabled these early primates to capture insects while scampering through the trees. Unlike rodents, which have clawed feet, primates have grasping hands and feet. Primates can grip objects, hang from branches, and seize food. Some primates can even use tools. Compare your hand with the primate hands shown in **Figure 12-1**. The hand of a primate can grasp an object because it has an **opposable thumb**. An opposable thumb stands out at an angle from the other fingers and can be bent toward them to grip an object. All primates except for humans also have opposable big toes.

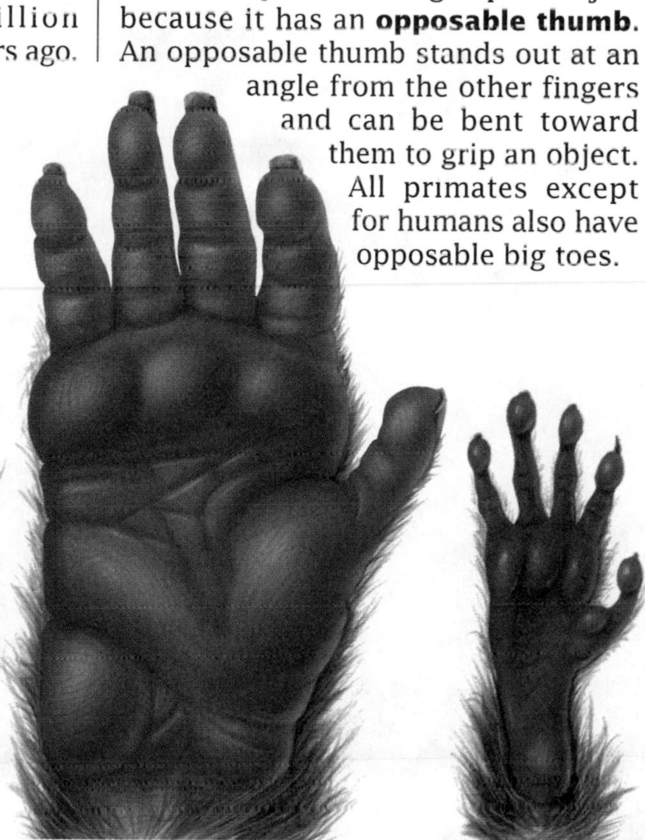

Figure 12-1
Primates have five flexible fingers on each hand.

a Chimpanzee hand **b Baboon hand** **c Lemur hand**

Figure 12-2
The large eyes of this loris reveal that it is a nighttime hunter. Native to India, Sri Lanka, and Southeast Asia, lorises creep slowly through the trees, hunting insects and small animals. They grow to about 25 cm (10 in.) in length.

Another adaptation for living in trees is the position of the eyes in the skull. The eyes of rats and squirrels are in the sides of their heads.The fields of vision of their eyes do not overlap. The eyes of primates, in contrast, are located in the front of the head, like the eyes of the loris in Figure 12-2. Each eye of a primate sees a slightly different view of the same scene. The brain merges the two views in perceiving the distances to objects. This type of vision is called **binocular vision**. The ability to judge distance is advantageous when jumping from one branch to another or when stalking prey.

Other mammals, such as cats, have binocular vision, but only primates have both binocular vision and grasping hands. These features require an enlarged brain to process information from the eyes and to coordinate hand movements. Large brains and increased intelligence have become hallmarks of the primates.

Prosimians were the first primates

The first primates were **prosimians** (meaning "before monkeys"). Fossils 30 million to 40 million years old show that prosimians were common in North America, Europe, Asia, and Africa. Only a few species of prosimians, such as the loris shown in Figure 12-2, survive today. Many of these surviving species are nighttime hunters. You can tell by the disproportionately large size of their eyes. Large eyes are necessary to capture what little light is available at night or in dark forests. All 24 surviving species of lemurs, cat-sized prosimians with long tails for balancing, live on Madagascar, an island about 400 km (250 mi) off the east coast of Africa. Today the native vegetation of Madagascar is being rapidly destroyed by an expanding human population. As the lemurs' forest home disappears, many of these unique prosimians may become extinct in the wild.

Anthropoids Are Day-Active Primates

About 35 million to 40 million years ago, a revolutionary change occurred in how primates lived—they became active during daytime. How do we know this change occurred? Fossil skulls of primates that lived at this time have much smaller eye sockets than do earlier prosimian skulls, suggesting that they were active during the day. The new day-active primates were **anthropoids**. Monkeys, apes, and humans are the existing anthropoids.

Since daytime activity places different demands on the eye, many changes in eye design probably evolved at this time. One of these changes was the development of color vision. Of course, we can't actually examine 35-million-year-old anthropoid eyes to learn if they could see color. Soft tissues such as those of the eye are rarely preserved as fossils. Instead, we infer that early anthropoids had color vision because all living anthropoids see color. Therefore, color vision probably arose early in anthropoid history.

Figure 12-4 This spider monkey, like many other New World monkeys, has a long tail that functions as a limb for grasping branches.

The brains of anthropoids are larger than prosimian brains. Larger brain size seems to be associated with the more complicated behavior patterns of anthropoids. Anthropoids replaced prosimians rather rapidly. In the fossil record, prosimian fossils become rare as anthropoid fossils become common. Scientists don't know exactly why prosimians declined while anthropoids became more abundant, but they suspect that the larger brains and color vision of anthropoids contributed to their success.

Figure 12-3 Monkeys, such as the adult baboon (below left), take care of their young for a longer time than do most other mammals.

Monkeys have complex social interactions

Monkeys are anthropoids with tails. Two groups of monkeys occur today. Old World monkeys, such as the baboons in **Figure 12-3**, live in Asia and Africa. New World monkeys, such as the spider monkey in **Figure 12-4**, inhabit Central America and South America. Monkeys feed mainly on fruits and leaves rather than on insects. They live in groups in which complex social interactions occur. Monkeys tend to care for their young for a longer time than most other mammals, except for humans and apes. This long period of dependency seems to be necessary for the development of the large brains of monkeys, apes, and humans.

Apes

Unlike monkeys, apes lack tails. Apes also have larger brains than monkeys. The existing apes are the gibbons, the orangutans, the gorillas, the bonobos *(BAHN uh bohs)*, and the chimpanzees. With their large brains, apes are capable of learning a greater variety of behaviors than any mammal except human beings. Once common, apes are rare today. Their natural forest habitat has largely disappeared as the growing human population has cleared woodland to make way for farms. Modern apes are confined to relatively small areas in Africa and Asia. No apes have ever been native to North America or South America.

DNA sequences reveal our close kinship to chimpanzees

Modern apes are not our direct ancestors. Humans share a common ancestor with the living apes, but the apes from which humans descended are extinct. Nevertheless, much can be learned about our own evolution by studying the modern apes, such as the chimpanzees in **Figure 12-5** below. For example, studies of DNA and protein sequences have revealed a great deal about the relationships among the living apes and about their relationship to humans. You learned in Chapter 10 how differences in nucleotide sequences in DNA and differences in amino acid sequences in proteins can be used to determine evolutionary relationships. If the average rates of change in these sequences are known, it is possible to calculate approximately when evolutionary divergences occurred. When DNA from apes and humans is compared, gibbon DNA shows the greatest number of differences in nucleotide sequence. Scientists think the species that gave rise to gibbons diverged from the common ancestor of humans and the other apes about 10 million years ago, as shown in **Figure 12-6**. The next ape to evolve, the ancestor of the orangutan, split off approximately 8 million years ago.

Humans shared a common ancestor with gorillas, chimpanzees, and bonobos until about 7 million years ago. At this time, the gorilla's ancestor diverged from the common ancestor of chimpanzees, bonobos, and humans. The ancestor of humans did not begin to diverge from the ancestor of chimpanzees and bonobos until about 5 million years ago. Because this divergence began so recently, the genes of humans and chimpanzees have not had time to accumulate many differences. Overall, the nucleotide sequences of human and chimpanzee genes differ by only about 1.6 percent. As a result, most of the proteins encoded by your genes are very similar or even identical to the corresponding proteins in a chimpanzee.

Figure 12-5
The physical structure, DNA sequences, and protein sequences of humans are more similar to those of bonobos and chimpanzees, such as this female and her offspring, than to those of any other living species.

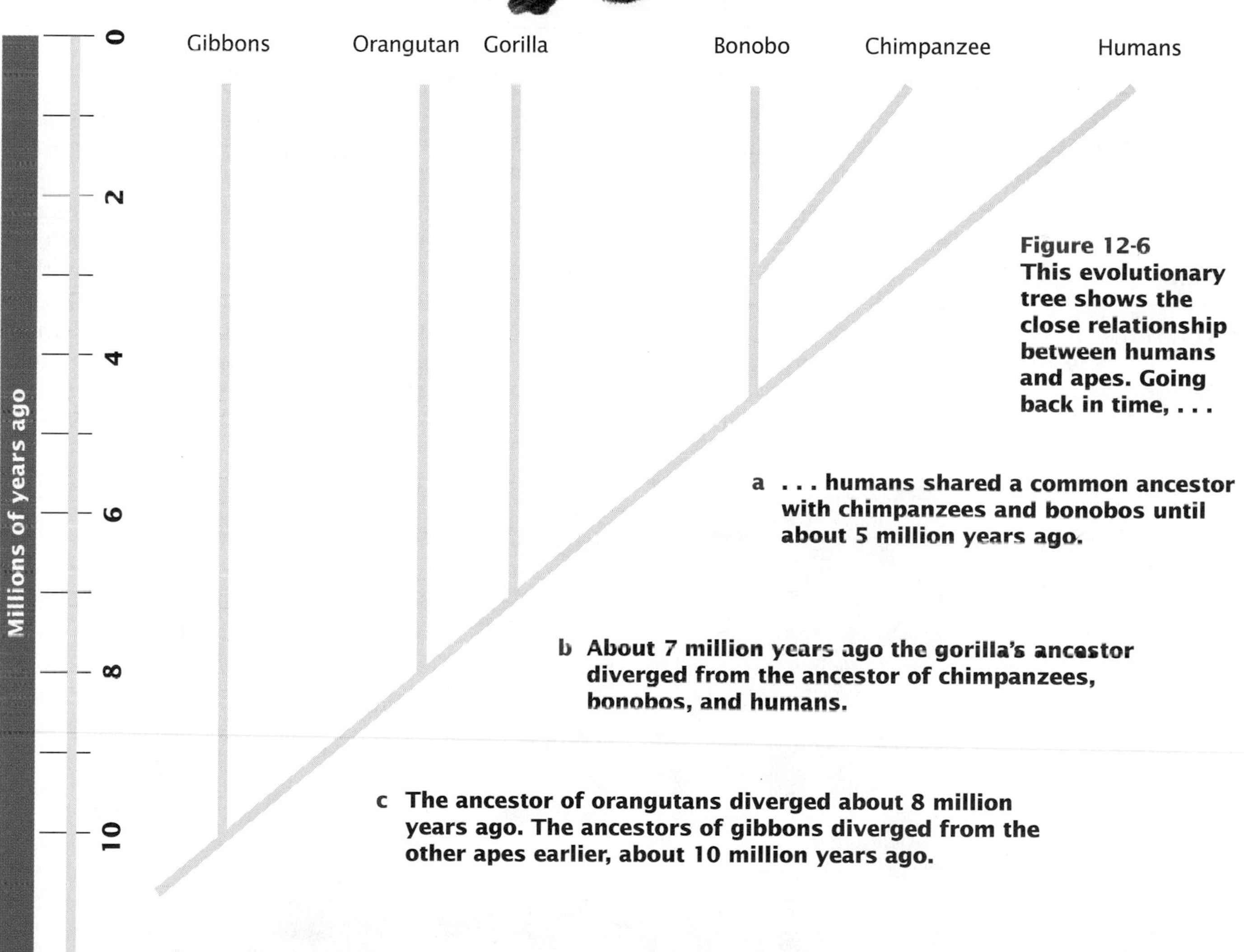

Figure 12-6
This evolutionary tree shows the close relationship between humans and apes. Going back in time, . . .

a . . . humans shared a common ancestor with chimpanzees and bonobos until about 5 million years ago.

b About 7 million years ago the gorilla's ancestor diverged from the ancestor of chimpanzees, bonobos, and humans.

c The ancestor of orangutans diverged about 8 million years ago. The ancestors of gibbons diverged from the other apes earlier, about 10 million years ago.

SECTION REVIEW

1. **Describe two distinctive primate features.**
2. **How could you distinguish the skulls of day-active primates from those of night-active primates?**
3. **Contrast the characteristics of monkeys with those of prosimians.**
4. **Which apes are most distantly related to humans? Explain your answer.**

Writing, sculpting, and dialing a telephone—these are just three of the many tasks you can do with your hands. Your hands are free to do these tasks because you walk on two legs instead of on four. Upright walking was one of the first adaptations to evolve among our ancestors after the human and chimpanzee lines diverged about 5 million years ago.

12-2 Evolutionary Origins of Humans

OBJECTIVES

1. **Contrast the characteristics of chimpanzees with those of humans.**
2. **List three reasons why hominid fossils are rare.**
3. **Describe evidence that human ancestors walked upright before their brains enlarged.**
4. **Compare and contrast australopithecines with modern humans.**

Searching for the Fossil Relatives of Humans

No one could mistake a human for a chimpanzee, despite their 98.4 percent genetic similarity. That 1.6 percent genetic difference—accumulated in the 5 million years since the lines that gave rise to humans and chimps began to separate—results in substantial differences in physical appearance. The chimp is covered with long hair, and its skull, teeth, limbs, and feet also differ from those of a human. Moreover, chimps walk on four legs, while humans walk on two. The earliest primates to walk upright diverged from the chimpanzee line about 5 million years ago. Very little is known about these ancient human ancestors. Scientists like the Leakey family, shown in **Figure 12-7**, dig into the fossil record to find out what these early relatives were like. Humans and their closest fossil relatives are known as **hominids** *(HAHM uh nihdz)*.

Figure 12-7
For more than 60 years, members of the Leakey family have been searching for fossils in East Africa.

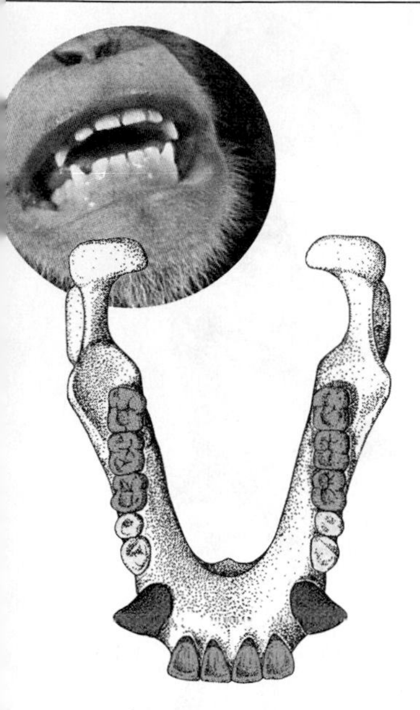
Chimpanzee jaw

Table 12-1 Comparison of Chimpanzee and Human Jaws

Chimpanzee	Human
U-shaped tooth arrangement; molars in two parallel rows	Arc-shaped tooth arrangement; molars not parallel
Space between incisors and canines	No space between incisors and canines
Relatively long canines	Short canines

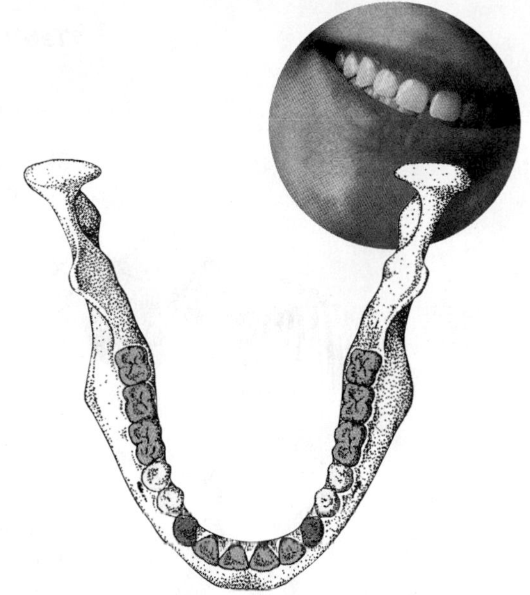
Human jaw

Molars Premolars Canines Incisors

Hominid fossils are rare and hard to find

As you learned in Chapter 10, fossils do not form very often. Conditions have to be right, or a dead body will simply decay, leaving no evidence of what it looked like. Not surprisingly, scientists searching for hominid fossils face many difficulties. Until the development of agriculture about 10,000 years ago, hominids seem to have been rare. For example, early hominids left few footprints relative to other animals. Add to this difficulty the fact that only individuals that died on the shores of a lake or swamp or that fell into a mud pit would be covered by sediment, a necessary step in fossilization. Furthermore, fossils that formed 3 million to 5 million years ago are embedded within rock formations deep in the ground. These fossils can be found only if the rock formation is exposed by wind, water, or earth movements. Once exposed, however, fossils are quickly destroyed by erosion. Since most hominid fossils are only fragments, scientists must reassemble them like pieces of a jigsaw puzzle.

How can the age of a fossil be determined? Only fossils less than about 40,000 years old can be dated directly with radioactive dating. And we cannot usually date the rock in which the fossil is found, because fossil-bearing rock is composed of little bits of rock broken off from other rocks. Therefore, radioactive dating would yield the age of the original rock that broke into bits, not the rock that formed much later around the fossil. So how do scientists find the age of a fossil? First, they know that volcanic rock can be dated with great accuracy. If a nearby volcano erupted within a few thousand years of the time when the fossil was deposited, scientists can date the rock that formed as a result of the eruption. A fossil embedded nearby can be said to have a similar age.

How are hominid fossils recognized?

Despite the difficulties, scientists have discovered many hominid fossils. Once a fossil is discovered, how can a scientist tell if it is from a hominid or some other kind of animal? After all, fossils don't come with name tags. Identification is possible because different kinds of animals exhibit different details of structure. For instance, hominid jaws and teeth are quite different from those of chimpanzees, as shown in **Table 12-1**. Other structural differences are apparent in the skull, spine, pelvis, and other bones, as you will soon see. Only by painstaking comparisons can scientists distinguish hominid fossils from those of apes or other animals.

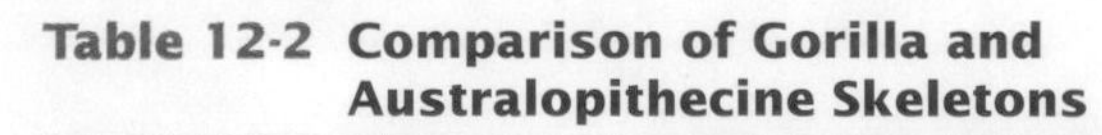

Table 12-2 Comparison of Gorilla and Australopithecine Skeletons

Gorilla	Australopithecine
Skull atop C-shaped spine	Skull atop S-shaped spine
Spinal cord exits near rear of skull	Spinal cord exits at bottom of skull
Arms longer than legs; arms and legs used for walking	Arms shorter than legs; only legs used for walking
Tall and narrow pelvis	Bowl-shaped pelvis, centering the body weight over the legs
Femurs (thigh bones) angled away from pelvis when walking upright	Femurs angled inward so legs are directly below body to carry its weight

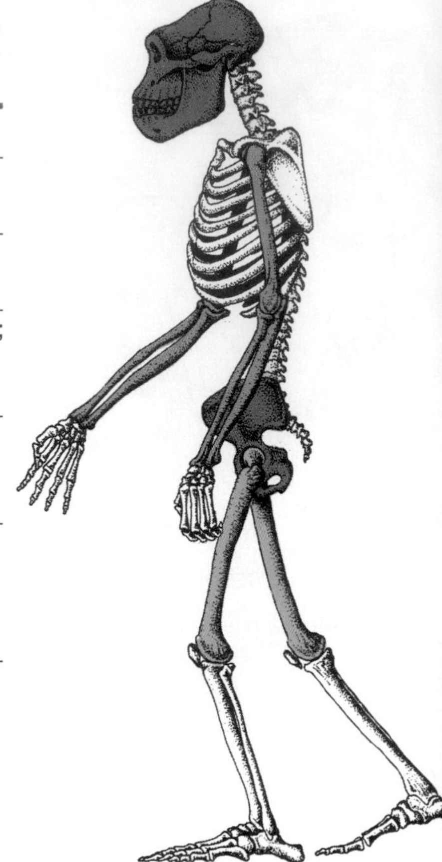

Characteristics of the Earliest Hominids

If you were looking for early hominid fossils, where would you search? In 1871 Charles Darwin considered this question and decided that Africa was the most likely place to find these fossils. How did he reach this conclusion? Darwin reasoned that because the closest relatives of humans, the chimpanzees and gorillas, lived in Africa, humans probably evolved there too.

As Darwin predicted, fossils of the earliest hominids were uncovered in eastern and southern Africa. Ranging in age from more than 4 million years old to 1 million years old, these fossils represent species that evolved after the human line split off from the line leading to chimpanzees and bonobos.

At least six species of these early hominids have been discovered, all belonging to the genus *Australopithecus.* Members of this genus are called australopithecines *(aw stray loh PIHTH uh seenz).* Even the earliest australopithecines show clear differences in structure from apes.

Australopithecines walked upright

Australopithecines were **bipedal**—that is, they walked upright on two legs. In fact, upright walking is a characteristic of all hominids. Although apes can walk upright for short distances, they do so with an awkward, waddling gait. How do we know that australopithecines that died millions of years ago walked upright? Comparisons of australopithecine skeletons with those of apes such as the gorilla have revealed how they walked, as shown in **Table 12-2**.

Most australopithecines also had larger brains than apes have. A chimp's brain occupies a volume of about 400 cm^3 (24 $in.^3$), about the size of an orange. Australopithecine brains were slightly larger, ranging from 400 to 550 cm^3 (24–34 $in.^3$). Though larger than an ape's brain, australopithecine brains were still much smaller than the brains of modern humans, which average 1,350 cm^3 (83 $in.^3$), about the size of a small melon.

Branches of the Hominid Evolutionary Tree

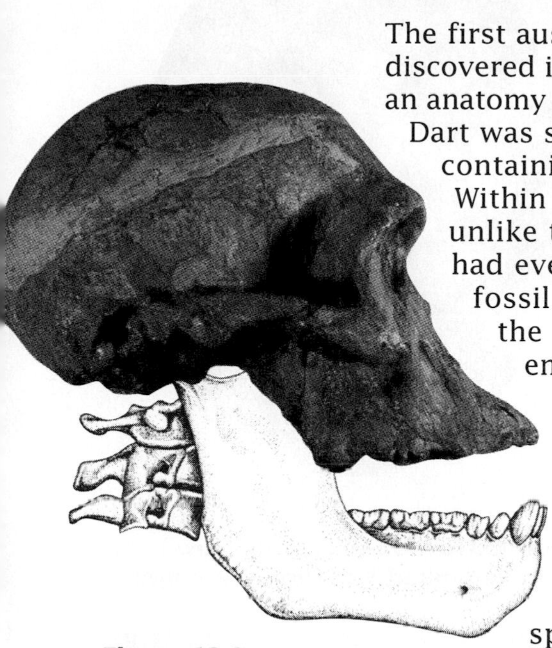

The first australopithecine fossil was discovered in 1924 by Raymond Dart, an anatomy professor in South Africa. Dart was sent a shipment of fossil-containing rock from a quarry. Within this rock was a skull unlike the skull of any ape Dart had ever seen. One detail of the fossil riveted Dart's attention: the rock in which it was embedded was from a geological formation thought to be at least a million years old. At that time, the oldest reported hominid fossils were only 500,000 years old.

Dart named the new species *Australopithecus africanus*, which means "southern ape from Africa." (This name is unfortunate, since we no longer think *Australopithecus* was an ape.) Dart argued that *A. africanus* was the direct ancestor of humans, the long-sought "missing link" between apes and humans. Most scientists at the time dismissed Dart's discovery as the skull of a young ape. Yet many *A. africanus* fossils have been discovered since 1924, and we now know that this species was a hominid, not an ape. **Figure 12-8** shows a skull of *A. africanus*. This species had jaws that are more rounded than those of an ape and teeth that resemble those of a human. The brain of *A. africanus* was larger than the brain of an ape, about 440 cm^3 (27 $in.^3$). *Australopithecus africanus* was bipedal and existed from 2.5 million to 3 million years ago. Whether *A. africanus* was a direct ancestor of humans is still a matter for debate.

Figure 12-8
The spinal cord of *A. africanus* exited from the bottom of its skull, indicating that this hominid walked upright. This species lived about 2.5 million to 3 million years ago. (The lower jaw and spine have been added by an illustrator.)

Another important find was the footprints of a group of bipedal animals that walked across wet volcanic ash about 3.5 million years ago in what is now northern Tanzania. The footprints, shown in **Figure 12-9**, were preserved when the ash hardened. They reveal small but very humanlike feet, lacking the ape's opposable toe. Our ancestors or very close relatives were walking upright only 1.5 million years after diverging from the chimpanzee line.

Figure 12-9
The footprints below show that bipedal hominids had evolved by about 3.5 million years ago.

Other australopithecines have been discovered

A stockier kind of australopithecine was unearthed in South Africa in 1938. Called *A. robustus*, this species had massive teeth and jaws. In 1959 Mary Leakey discovered the even more solidly built *A. boisei*, shown in **Figure 12-10**, in Tanzania. Nicknamed "nutcracker man" because of its massive jaws, *A. boisei* had a large, bony ridge running along the crest of its skull to anchor powerful jaw muscles. Excavations in 1985 turned up yet another australopithecine, a massively boned species resembling *A. boisei.*

Figure 12-10
***Australopithecus boisei*'s skull had a massive jaw (illustrated below) and a bony ridge along its crest.**

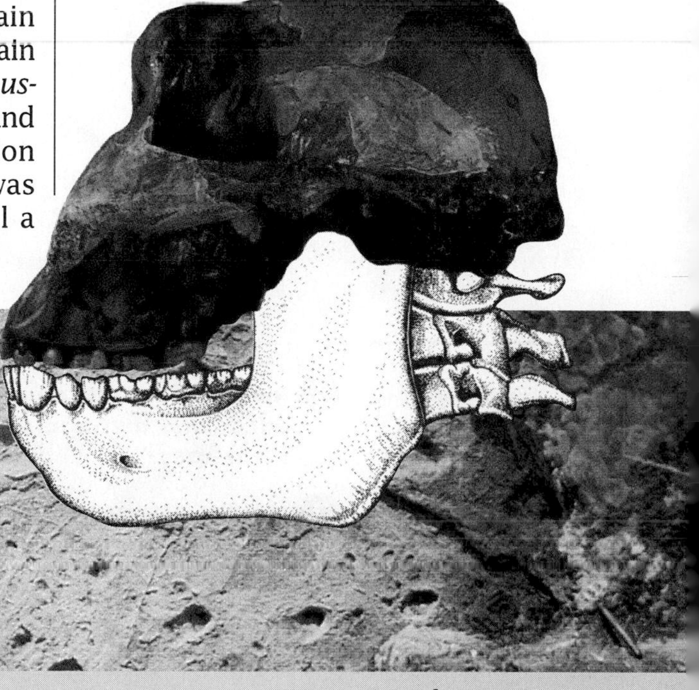

Discovery of the oldest hominids

In 1974 Donald Johanson went to the Afar Desert of Ethiopia in search of early hominid fossils. There he found the most complete, best preserved skeleton of a prehuman hominid ever discovered. Nicknamed "Lucy," the skeleton was over 40 percent complete, as shown in **Figure 12-11**. Lucy was found to be nearly 3 million years old. Johanson assigned the skeleton the scientific name *Australopithecus afarensis.* Since the discovery of Lucy, many other fossils of *A. afarensis*, some over 3.5 million years old, have been unearthed in Ethiopia and Tanzania.

Evolution

Reconstructions of Lucy usually show her as covered with hair, like an ape. What assumptions about evolution are made by showing Lucy in this way?

What do we know about Lucy's species? A reconstruction of how a living *A. afarensis* individual might have looked is shown in **Figure 12-12**. Lucy herself was small by modern standards, just over 1 m (3 ft 6 in.) tall. Other *A. afarensis* specimens range up to 1.5 m (5 ft) tall, with males much larger than females. The shape of Lucy's pelvis indicates that she was female, and the inward angle of her femur shows that she walked upright.

Despite being bipedal, *A. afarensis* was quite apelike in many respects.

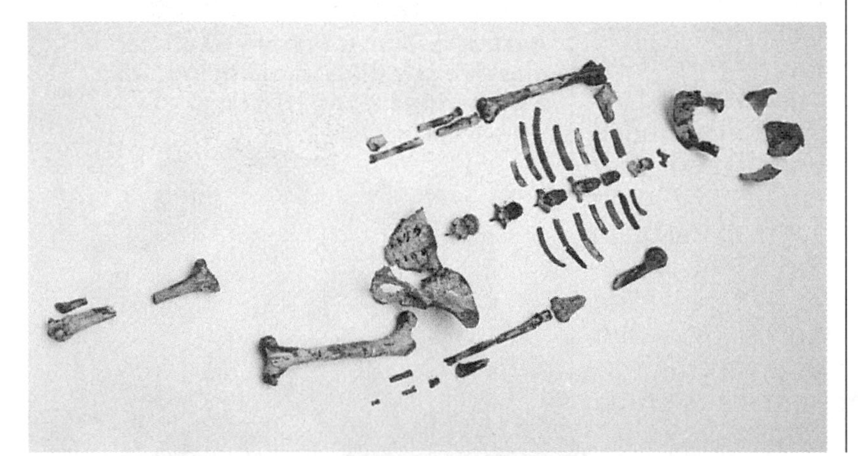

Figure 12-11
Because major portions of Lucy's limb bones and pelvis were preserved, scientists were able to determine that she walked upright.

Figure 12-12
This reconstruction illustrates how Lucy might have looked 3 million years ago.

With a protruding face and fairly large jaws, an *A. afarensis* skull resembles a chimpanzee skull more than a human skull. In size, *A. afarensis* teeth are intermediate between most ape and human teeth. For instance, the canines of *A. afarensis* were shorter than those of apes, but still longer than human canines. Most important, the brain of *A. afarensis* was only about 400 cm^3 (24 $in.^3$), the size of a chimpanzee brain. Johanson's discovery demonstrated that hominids walked upright before they evolved brains larger than those of apes.

Australopithecus afarensis is no longer the oldest known hominid. In 1995, a team led by Meave Leakey and Alan Walker discovered fossils of a hominid species that lived about 4 million years ago. Though only fragmentary, these fossils indicate a species that walked upright but had an apelike jaw. Leakey and Walker have named their find *Australopithecus anamensis.*

SECTION REVIEW

1. **Identify four structural differences that distinguish humans from apes.**
2. **Describe three difficulties you might face in trying to find hominid fossils.**
3. **Evaluate: "During human evolution, upright posture and large brain size evolved at the same time."**
4. **Describe three differences between *A. afarensis* and modern humans.**

Because of your large, complex brain, you are able to read this book, whereas a chimpanzee is not. Your brain is about three times larger than a chimp's brain and more than twice the size of the largest australopithecine brain. Large brains are a characteristic of our genus, *Homo*, and are responsible for our complex culture.

12-3 The First Humans

OBJECTIVES

1. **List the two key characteristics of *Homo habilis*.**
2. **Contrast the skeletal features of *Homo habilis* with those of *Homo erectus*.**
3. **Summarize the evidence that *Homo sapiens* evolved in Africa.**
4. **Describe evidence that Neanderthals had a complex culture.**

Evolution of the Genus *Homo*

Modern humans belong to the species *Homo sapiens*, the third and only surviving species of the genus *Homo*. All members of this genus are called humans. About 2 million years ago, the first member of our genus, *Homo habilis*, evolved from australopithecine ancestors. Between 1.8 million and 1.5 million years ago, *Homo habilis* was replaced by its larger-brained descendant, *Homo erectus*. In turn, *H. erectus* gave way to its even larger-brained descendant, *H. sapiens*.

There are many hypotheses describing which australopithecines gave rise to humans. **Figure 12-13** shows two hypotheses. Most researchers think that *Australopithecus afarensis* is the ancestor of all later hominids, including ourselves.

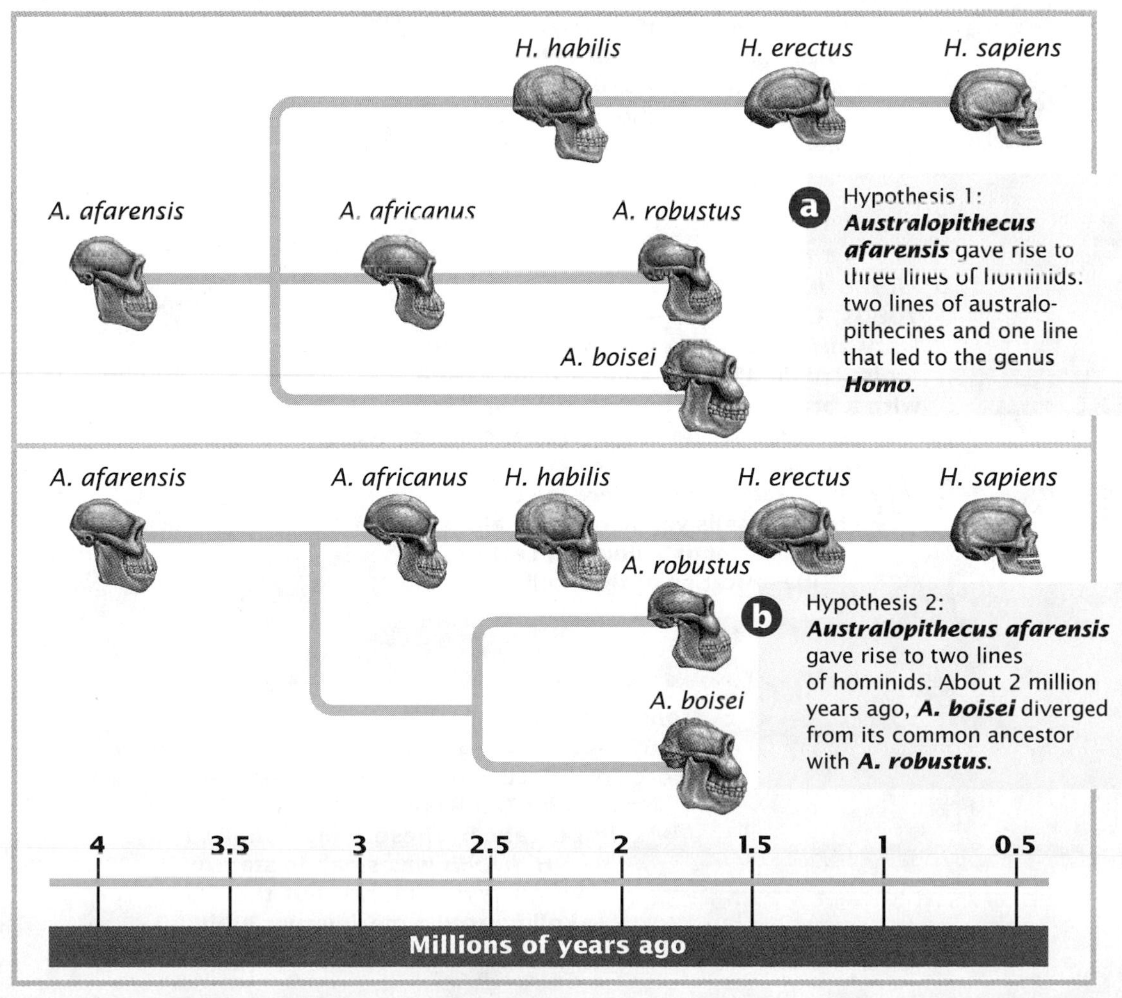

Figure 12-13 These evolutionary trees show two hypotheses of how *Homo sapiens* evolved from australopithecines.

Our African Origins

Since the publication of Darwin's *Descent of Man* in 1871, scientists have spread over the globe in search of fossil remains that tell the story of the origin of humans. Scientists have unearthed fossils, tools, and other remains belonging to *Homo habilis* and *H. erectus* the ancestors of our species. Look at the map in **Figure 12-14** to see where many of these finds were discovered.

Homo habilis evolved in Africa

The earliest fossils of *H. habilis* have been discovered in East Africa. These fossils are about 2 million years old. *Homo habilis* is known for its large brain; the brain of one specimen is about 775 cm^3 (47 $in.^3$), nearly twice the size of Lucy's brain. *Homo habilis* fossils have been found alongside crude, stone tools. Most scientists think that *H. habilis* fashioned and used these tools for hunting and scraping. *Homo habilis* existed from about 2 million to 1.5 million years ago, when it was replaced by *H. erectus.*

Homo erectus migrated out of Africa

Homo erectus was a larger species than *H. habilis*—about 1.5 m (5 ft) tall. It also had a large brain (about 1,000 cm^3, or 61 $in.^3$) and walked erect.

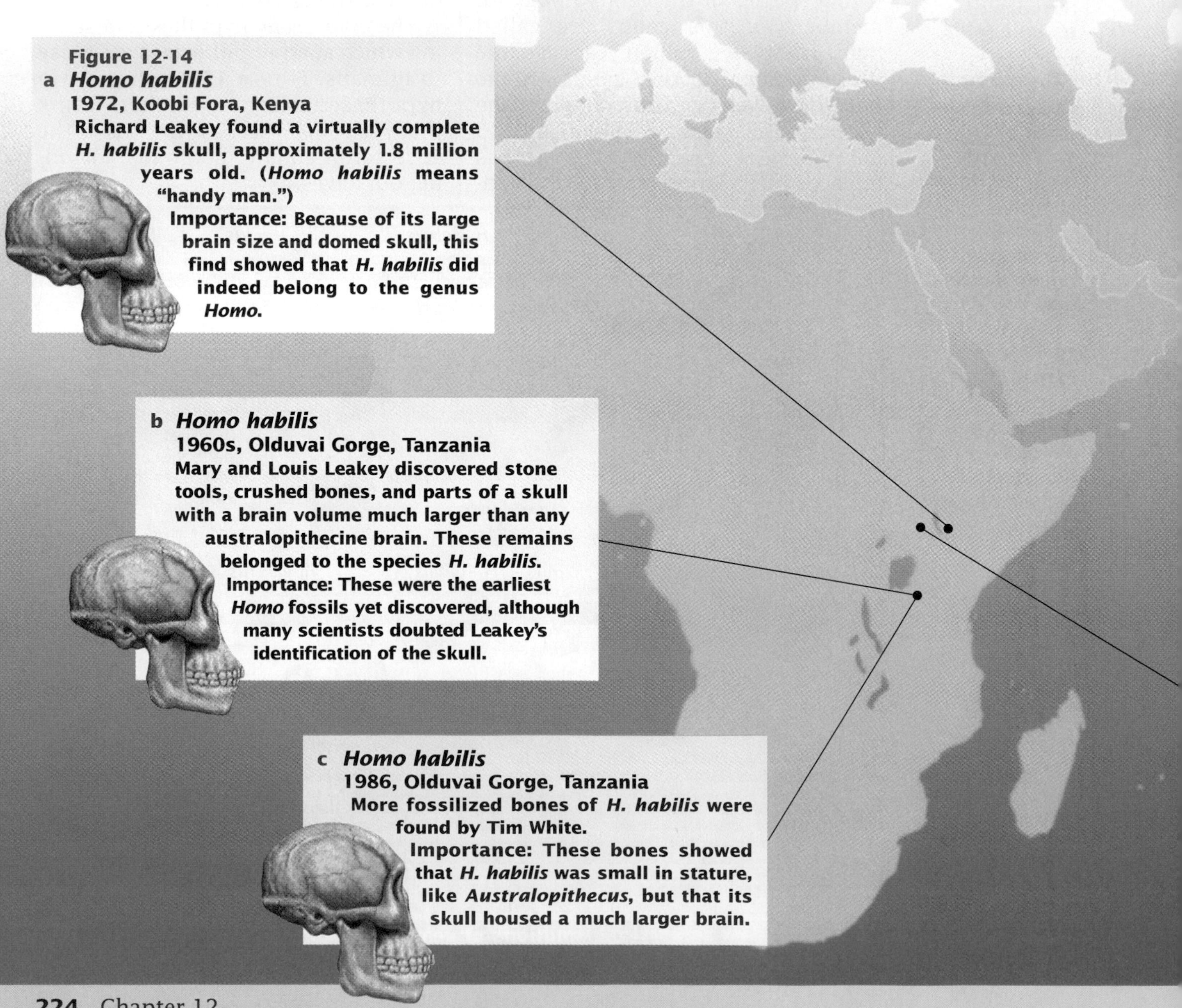

Figure 12-14

a *Homo habilis*
1972, Koobi Fora, Kenya
Richard Leakey found a virtually complete *H. habilis* skull, approximately 1.8 million years old. (*Homo habilis* means "handy man.")
Importance: Because of its large brain size and domed skull, this find showed that *H. habilis* did indeed belong to the genus *Homo*.

b *Homo habilis*
1960s, Olduvai Gorge, Tanzania
Mary and Louis Leakey discovered stone tools, crushed bones, and parts of a skull with a brain volume much larger than any australopithecine brain. These remains belonged to the species *H. habilis*.
Importance: These were the earliest *Homo* fossils yet discovered, although many scientists doubted Leakey's identification of the skull.

c *Homo habilis*
1986, Olduvai Gorge, Tanzania
More fossilized bones of *H. habilis* were found by Tim White.
Importance: These bones showed that *H. habilis* was small in stature, like *Australopithecus*, but that its skull housed a much larger brain.

Where did *H. erectus* originate? Since its immediate ancestor, *H. habilis*, lived in Africa, it should come as no surprise that the earliest *H. erectus* skulls have been found in Africa. *Homo erectus* first evolved in Africa about 1.8 million years ago. Its appearance marked the beginning of an expansion of human populations across the globe. Far more successful than *H. habilis*, *H. erectus* quickly became widespread and abundant in Africa and had migrated into Asia and Europe by at least 1 million years ago.

Homo erectus probably lived in small groups of 20 to 50 individuals. We know how these early humans lived because scientists have found the remains of their living places. Some *Homo erectus* groups lived in caves, and there is some evidence that they also built crude wooden shelters. They successfully hunted large animals, butchered them using flint and bone tools, and cooked them over fires. *Homo erectus* living sites in China contain the remains of horses, bears, elephants, deer, and rhinoceroses. *Homo erectus* survived for more than 1 million years, longer than any other species of human. These very adaptable humans disappeared in Africa and Europe only about 300,000 years ago, while modern humans were evolving. Interestingly, *H. erectus* may have survived much longer in Asia, until about 250,000 years ago. *Homo erectus* was clearly our immediate ancestor. From the neck down, we are almost identical to this early human.

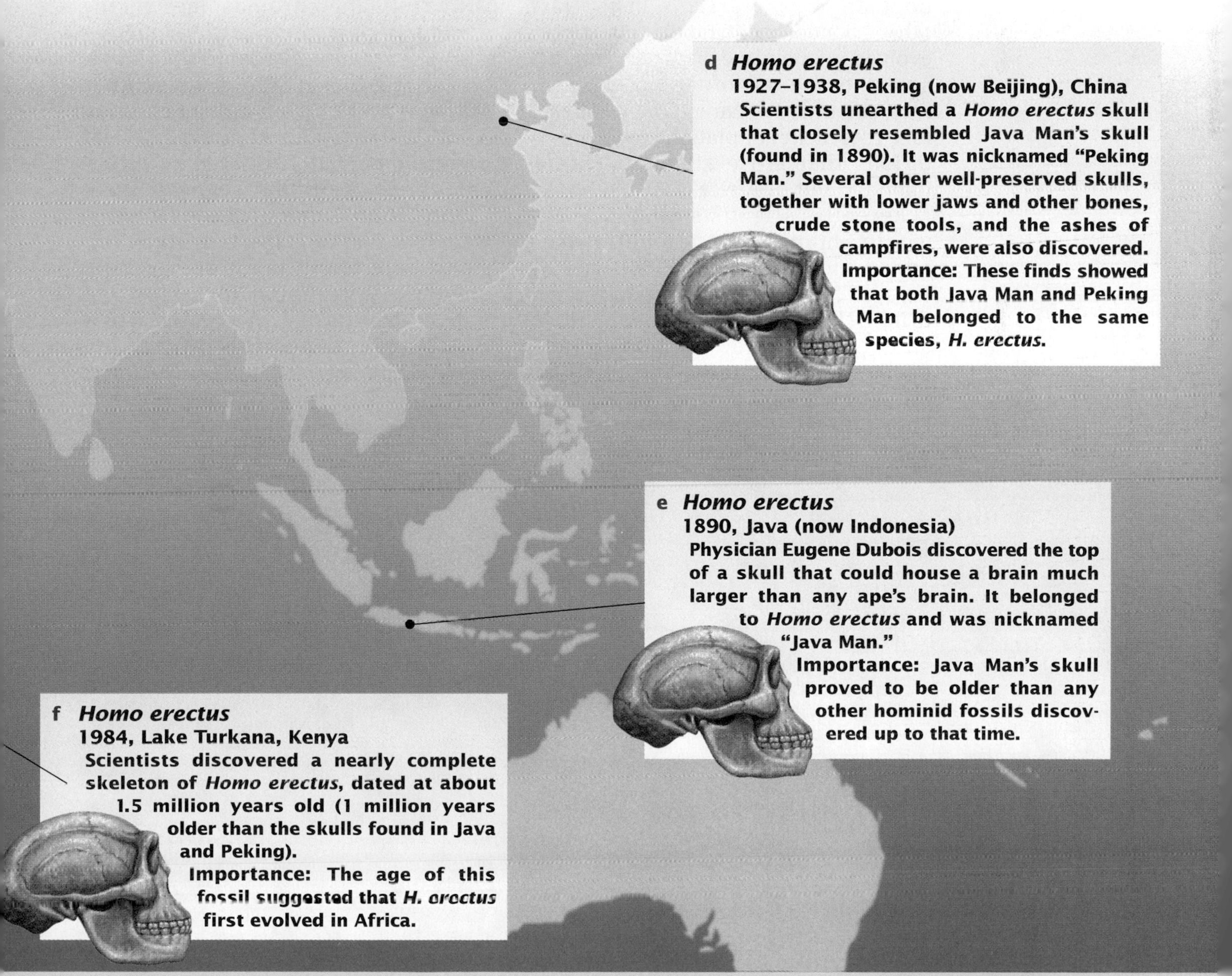

d *Homo erectus*
1927–1938, Peking (now Beijing), China
Scientists unearthed a *Homo erectus* skull that closely resembled Java Man's skull (found in 1890). It was nicknamed "Peking Man." Several other well-preserved skulls, together with lower jaws and other bones, crude stone tools, and the ashes of campfires, were also discovered.
Importance: These finds showed that both Java Man and Peking Man belonged to the same species, *H. erectus*.

e *Homo erectus*
1890, Java (now Indonesia)
Physician Eugene Dubois discovered the top of a skull that could house a brain much larger than any ape's brain. It belonged to *Homo erectus* and was nicknamed "Java Man."
Importance: Java Man's skull proved to be older than any other hominid fossils discovered up to that time.

f *Homo erectus*
1984, Lake Turkana, Kenya
Scientists discovered a nearly complete skeleton of *Homo erectus*, dated at about 1.5 million years old (1 million years older than the skulls found in Java and Peking).
Importance: The age of this fossil suggested that *H. erectus* first evolved in Africa.

Figure 12-15
Scientists disagree about how and when different racial groups, some of which are represented by the people above, evolved.

The Origin of *Homo sapiens*

Scientists have long debated the issue of where *Homo sapiens* evolved. Some scientists argue that our species evolved simultaneously from different groups of *H. erectus* that lived in different areas of the Old World. According to these scientists, the differences among racial groups, such as among the young people in **Figure 12-15**, reflect the descent of each racial group from a different population of *H. erectus*. In contrast, other scientists contend that our species evolved in one place, in Africa, and then spread to the rest of the world. Therefore, the unique characteristics of each race represent adaptations to local conditions that arose after our species migrated out of Africa.

Recently, scientists studying DNA within human mitochondria have added fuel to the fire of this controversy. They examined the mitochondrial DNA sequences in people from all over the world. The human races evolved only recently, and there has not been enough time for many differences in their mitochondrial DNA sequences to accumulate. Since DNA accumulates mutations over time, the oldest DNA should show the largest number of mutations. It turns out that the greatest number of different mitochondrial DNA sequences occur among modern Africans. This result suggests that humans have been living in Africa longer than on any other continent and that *Homo sapiens* originated in Africa.

A clearer analysis is possible using DNA from the chromosomes in the nucleus. Some segments of nuclear DNA are far more variable than mitochondrial DNA. When a variable segment of DNA from human chromosome 12 was analyzed in 1996, a clear pattern emerged. A total of 24 different versions of this segment were found. Twenty-one of these versions were found in Africans; only three versions were found in Europeans, and only two were found in Asians and Native Americans. Although these new data strongly support the African origin of *Homo sapiens*, this conclusion is still quite controversial, and there is no firm consensus among researchers as to where our species originated.

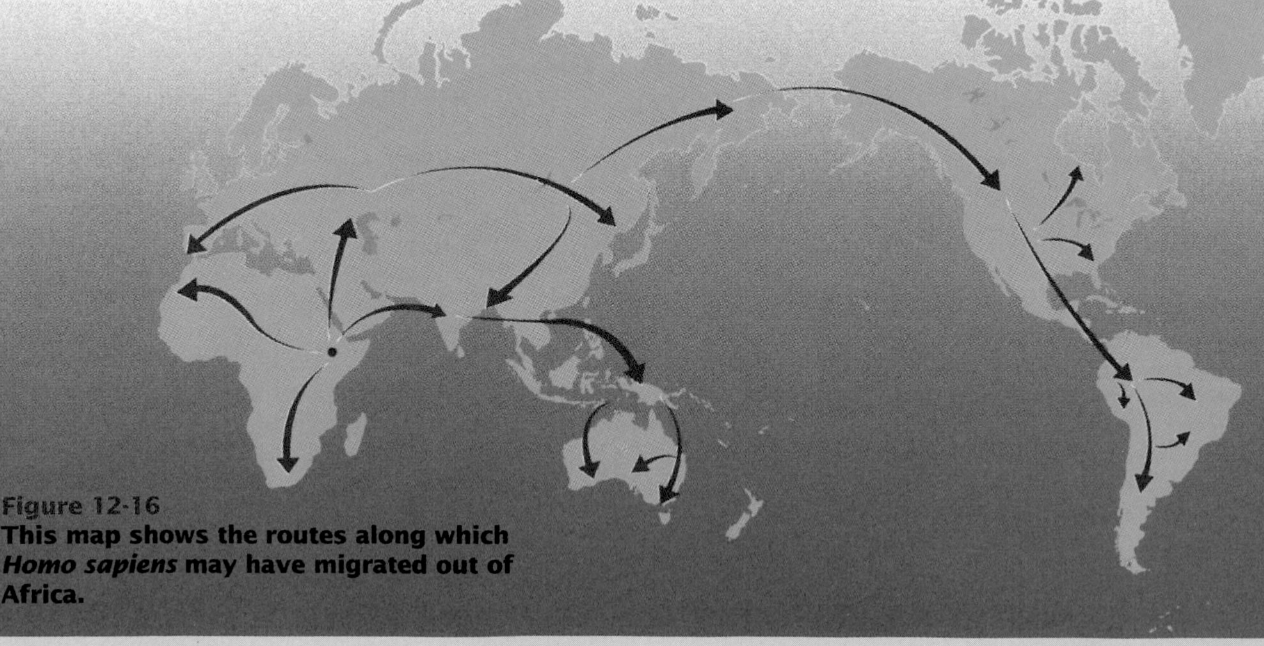

Figure 12-16
This map shows the routes along which *Homo sapiens* may have migrated out of Africa.

Figure 12-17 Fossil evidence shows that Neanderthals had a complex culture. They hunted, built shelters for their families, buried their dead, and made a variety of tools.

Neanderthals were the first *Homo sapiens* in Europe

Homo sapiens first appeared in Europe about 130,000 years ago. The first fossil of one of these individuals was found in 1856 in the Neander Valley of Germany. Hence, these early humans were called Neanderthals (*nee AN dur THALZ*; the word *thal* means "valley" in Old German). By 70,000 years ago, Neanderthals were common in Europe and in parts of the Middle East. They made diverse stone tools, including scrapers, spearheads, and hand axes. Neanderthals were more powerfully built, shorter, and stockier than humans today. Their skulls were massive, with protruding faces and bony brow ridges. Their brains were even larger than those of modern humans. They lived in huts, caves, or open-air sites, as shown in **Figure 12-17**. Neanderthals took care of their injured and sick and buried their dead, often placing food and weapons with the dead bodies. Such attention to the dead suggests that Neanderthals may have believed in a life after death. This is the first evidence of the kind of symbolic thought processes that are characteristic of modern humans.

Modern *Homo sapiens* evolved in Africa and replaced the Neanderthals

A new version of *Homo sapiens* essentially identical to modern humans appeared in Africa at least 100,000 years ago. By about 40,000 years ago, these modern humans had migrated out of Africa and had replaced the Neanderthals of southwest Asia. The modern humans then spread across Europe, coexisting and possibly interbreeding with Neanderthals.

By about 34,000 years ago, the Neanderthals had disappeared. The modern humans that replaced them had skulls identical to those of present-day humans. In fact, one of these early modern humans would look just like you or me (groomed and wearing modern clothes, of course).

These modern humans used tools made of bone, horn, and stone, some of which are shown in **Figure 12-18**. Their preserved camps and burial grounds show that they had complex

Figure 12-18 Modern *Homo sapiens* used these stone tools about 10,000 years ago.

social organization. And the shape of certain regions of the skull suggests that modern *H. sapiens* had full language capabilities.

At the time anatomically modern *H. sapiens* moved into Europe, the world was cooler than it is now. A huge sheet of ice extended from the Arctic into northern Europe; southern Europe was covered with grasslands inhabited by large herds of grazing animals. Modern *H. sapiens* painted pictures of these animals deep within caves throughout Europe, as shown in **Figure 12-19**.

Modern *H. sapiens* eventually spread throughout Asia and Australia and entered North America, which they reached at least 13,000 years ago. At that time, a land bridge connected Siberia and Alaska, and it was possible to walk from Asia to North America. By 10,000 years ago, about 3 million people lived throughout the entire world.

Like all other living things, humans are the product of evolution. Our evolution has been marked by a progressive increase in brain size. As a result, humans are the only animal able to make complex tools, a capability that has enabled us to change the world around us. Although other animals are capable of conceptual thought, we humans have extended this ability until it has become the hallmark of our species. We use symbolic language and can use words to shape concepts out of experience. This ability has led to the accumulation of knowledge that can be transmitted from one generation to the next. Thus, we have what no other animal has ever had, cultural evolution. Through culture, we have found ways to mold our environment to our needs. We control our biological future in a way never before possible—an exciting potential, but one that carries weighty responsibility.

Figure 12-19 The painting below was discovered in a cave in Lascaux, France, about 50 years ago. It was painted by *Homo sapiens* individuals 17,000 to 20,000 years ago.

SECTION REVIEW

1. **Identify two characteristics of *Homo habilis*.**
2. **Contrast two features of the skeletons of *Homo habilis* and *Homo erectus*.**
3. **How does mitochondrial DNA analysis support the hypothesis that *Homo sapiens* evolved in Africa?**
4. **Describe some of the cultural traits of Neanderthals.**

CHAPTER

12 Highlights

"I don't know what it means, but I like the look of it."

	Key Terms	Summary
12-1 Primates	**primate (p. 213)** **opposable thumb (p. 213)** **binocular vision (p. 214)** **prosimian (p. 214)** **anthropoid (p. 215)**	• **Prosimians, monkeys, apes, and humans are primates. They have opposable thumbs and binocular vision.** • **The first primates were prosimians.** • **Anthropoids evolved about 35 million to 40 million years ago. Apes, monkeys, and humans are anthropoids.** • **Humans are most closely related to chimpanzees and bonobos. The DNA nucleotide sequences of humans and chimpanzees differ by 1.6 percent.**
12-2 Evolutionary Origins of Humans	**hominid (p. 218)** **bipedal (p. 220)**	• **Humans and their closest fossil relatives are known as hominids.** • **Fossils show that upright walking evolved soon after the evolutionary line leading to modern humans diverged from the line leading to chimpanzees.** • ***Australopithecus afarensis* walked upright but had an ape-sized brain.**
12-3 The First Humans		• **The genus *Homo* evolved about 2 million years ago in Africa. The first member of our genus was *Homo habilis*.** • ***Homo habilis* was later replaced by *Homo erectus*. *Homo erectus* migrated out of Africa to Europe and Asia.** • **Analysis of mitochondrial and nuclear DNA suggests that *Homo sapiens* evolved in Africa.** • **Modern *Homo sapiens* evolved in Africa at least 100,000 years ago and migrated to Europe about 40,000 years ago.**

Many New World monkeys, such as this spider monkey, have long tails that can be used as an extra limb.

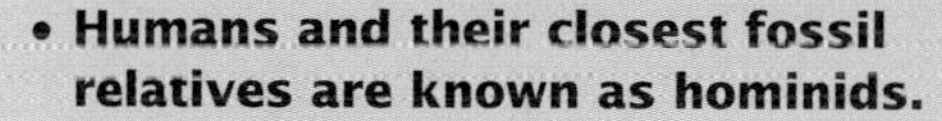

Because of its massive jaws, *Australopithecus boisei* was nicknamed "nutcracker man."

Neanderthals hunted, built shelters, cared for their sick, and buried their dead.

CHAPTER 12

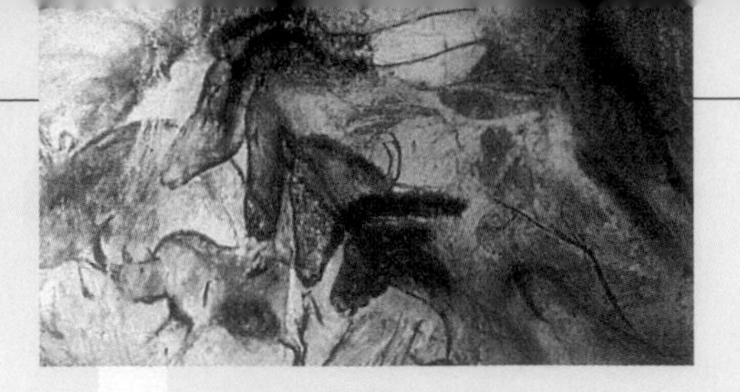

review

Understanding Vocabulary

1. Identify the word or phrase that does not fit the pattern, and explain why.
   a. baboon, chimpanzee, gorilla, gibbon
   b. anthropoid, *Homo sapiens*, gorilla, lemur
   c. migrated to Europe and Asia, walked erect, controlled fire, brain about 400 cm^3 (29 $in.^3$)

2. For each pair of terms, explain the differences in their meanings.
   a. prosimian, anthropoid
   b. Old World monkeys, New World monkeys
   c. *Homo habilis, Homo erectus*

Understanding Concepts

3. **Relating Concepts** Construct a concept map describing primates. Try to include the following terms in your concept map: monkeys, five fingers, prosimians, apes, grasping, *Homo erectus, Homo sapiens*, and Africa.

4. **Summarizing Information** Describe two characteristics that make primates unique. How are these characteristics advantageous for living in trees?

5. **Recognizing Relationships** The eyes of an ocelot are relatively large in relation to the size of its head. What might the large eyes of an ocelot suggest about this animal?

6. **Summarizing Information** Copy the following chart onto a sheet of paper. Then fill in the blanks in the chart.

	Prosimians	Monkeys
Opposable thumb		Yes
Binocular vision		
Tail	Yes	
Color vision	No	

7. **Comparing Structures** What features of a chimpanzee's skeleton indicate that chimps cannot walk upright like humans?

8. **Analyzing Methods** Lucy's skeleton was found on the shore of an ancient lake. Explain why this is a likely place to find fossils.

9. **Analyzing Conclusions** Why did Charles Darwin conclude that fossils of human ancestors would likely be found in Africa?

10. **Summarizing Information** Suppose that Donald Johanson had found only Lucy's skull. From just this evidence, could he conclude that she walked upright? Explain your answer.

11. **Comparing Structures** Match the hominid species in the first column with its description in the second column.

a. *Australopithecus afarensis*	1. migrated to North America
b. *Australopithecus anamensis*	2. had a chimpanzee-sized brain
c. *Homo habilis*	3. first hominid to use fire
d. *Homo erectus*	4. oldest known hominid
e. *Homo sapiens*	5. had a brain of about 775 cm^3

12. **Comparing Hypotheses** What are the two hypotheses for the origin of *Homo sapiens*? How does evidence from DNA sequences support the "African origins" hypothesis?

13. **Sequencing Information** Put the following hominid species in correct evolutionary order: *Homo erectus, Australopithecus anamensis, Homo habilis, Australopithecus afarensis, Homo sapiens.*

14. **Analyzing Conclusions** Chimpanzees sometimes strip the leaves from a twig and use it as a tool to capture termites. How does the way a chimpanzee uses a tool compare to the way humans use tools?

15. **Summarizing Information** Where did the Neanderthals live? How do they differ from modern *Homo sapiens*? Describe what is known about the culture of the Neanderthals.

16. **BUILDING ON WHAT YOU HAVE LEARNED** Based on what you learned about fossils in Chapter 10, explain how scientists could have made mistakes in some of their conclusions about human evolution.

Interpreting Graphics

17. Look at the evolutionary tree below.

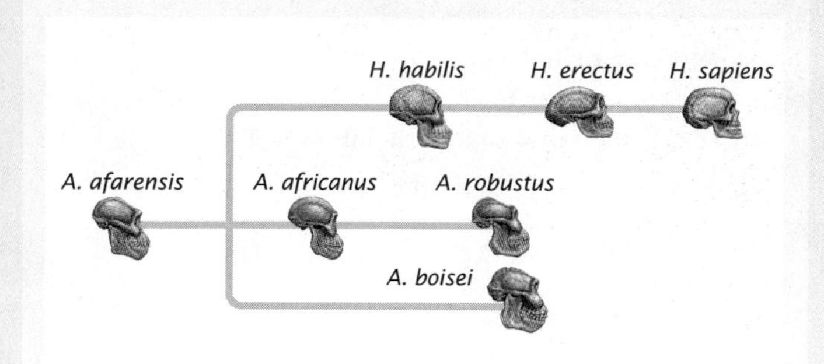

a. Which species is the common ancestor of all the other species?
b. Which species is the direct ancestor of *A. robustus*?
c. Which species exists today?

Reviewing Themes

18. ***Interdependence***
Describe two instances in which human actions are negatively affecting wild primates.

19. ***Evolution***
The DNA nucleotide sequences of chimpanzees and humans differ by only 1.6 percent. The DNA sequences of gorillas differ from those of humans by about 2.3 percent. What do these differences reveal about the evolutionary relationship among these primates?

Thinking Critically

20. **Distinguishing Relevant Information** Here is a scientist's hypothesis about the evolution of upright walking in hominids. "About 5 million years ago, the climate was warming, causing forests to transform into grasslands. Some primates began to spend less time in trees and more time on the ground. Ground-dwelling primates that could stand upright could gather more food because their hands were free. The primates that remained in the trees had grasping hands. Over time, upright primates left more offspring than primates that walked on all fours." Which information is irrelevant to the evolution of upright walking? Justify your answer.

Life/Work Skills

21. **Using Technology** You and your classmates are in charge of specimens at a natural history museum. A university professor, who has returned from a successful dig revealing the bones of three new species of dinosaur, wishes to present your museum with two entire skeletons. This would be a valuable asset and would bring many people and more money to the museum. The professor, however, wants you to transport and clean the fossils. Investigate the techniques paleontologists use to transport and clean valuable specimens so that they can be studied, displayed, and preserved. Check the library for the information you need, and send a report to the museum director about your plans.

Activities and Projects

22. **Health Connection** The cause of 80 percent of human backaches cannot be diagnosed. Some scientists and doctors have suggested that back problems are the result of strain caused by upright posture. Use references available in a library or search an on-line database to gather information about health problems caused or aggravated by our upright posture. In addition to back pain, consider other common problems, such as bunions, fallen arches, and knee injuries. Prepare a poster summarizing what you have learned.

23. **Career Connection: Museum Technician** Museum technicians are professionals who work in museums and help to prepare, maintain, and install exhibits, including fossil exhibits of dinosaurs and other organisms. Find out what a museum technician does on a day-to-day basis and what kind of educational background is required. For information, write to the American Association of Museums, 1005 Thomas Jefferson Street NW, Washington, DC 20007.

Discovering Through Reading

24. Read the article "The Farthest Horizon: The Dawn of Humans" by Meave Leakey in *National Geographic*, September 1995, pages 38–51. What evidence from fragments of leg bones discovered at the Kanapoi site suggest that this hominid, *Australopithecus anamensis*, walked upright?

CHAPTER 13

REVIEW
- testing hypotheses (Section 1-2)
- meiosis (Section 6-2)
- natural selection (Section 10-3)

Animal Behavior

Male bighorn sheep fight for control of groups of females by charging each other and butting heads.

A squirrel buries a nut. A lizard spreads its collar of skin when threatened. A hungry baby cries. These are examples of animal behavior. Over the last few decades, scientists have greatly increased their understanding of animal behavior by considering it from an evolutionary perspective.

13-1 Evolution of Behavior

OBJECTIVES

1. **Distinguish between "how" and "why" questions about behavior.**
2. **Explain how natural selection affects behavior.**
3. **Describe a case in which learning helps shape a behavior.**
4. **Explain the importance of Tryon's experiment on learning in rats.**

What Is Behavior, and How Is It Studied?

A **behavior** is an action or series of actions performed by an animal in response to a stimulus. The stimulus might be something in the environment, such as a sound, a smell, or a subtle change in the facial expression of another individual. The stimulus can also be the internal state of the animal, such as whether it has recently eaten. Behavior is crucial to an animal's survival because it is often the first response to a change in the environment.

Scientists studying behavior investigate two kinds of questions—"how" questions and "why" questions. "How" questions are about the mechanisms of the behavior within the animal—how the behavior is triggered, controlled, and performed. For instance, consider a squirrel burying a nut, as shown in **Figure 13-1**. "How" questions about this behavior include "How does a squirrel select which nuts to bury?" and "How does it choose where to bury nuts?" A scientist seeking to answer the first question might offer nuts of different kinds and sizes to a squirrel to see which it prefers.

Answering the "how" questions provides only a partial understanding of a behavior. Scientists also must answer the "why" questions. "Why" questions concern the reasons the behavior exists. "Why do squirrels bury nuts?" is an example of a "why" question about behavior. **Figure 13-1** lists some "how" questions and some "why" questions about the squirrel's behavior. The Dutch biologist Niko Tinbergen, who received the Nobel Prize in 1973 for his work on behavior, emphasized that "why" questions are really questions about the evolution of behavior. To explain why a behavior exists, it is necessary to understand why the behavior evolved and why it is maintained in the species. Therefore, it is necessary to learn how natural selection acts on the behavior.

Figure 13-1 Gaining a full understanding of a behavior requires answering "how" questions, which concern the behavior's mechanisms, and "why" questions, which concern its evolution.

How?
- *How does a squirrel select which nuts to bury?*
- *How does a squirrel choose where to bury nuts?*
- *How does this behavior develop?*
- *What environmental conditions affect the performance of the behavior?*

Why?
- *What advantage does a squirrel gain from burying nuts?*
- *What is the evolutionary history of this behavior?*

How Natural Selection Shapes Behavior

In Chapter 10 you learned how natural selection causes evolution. Recall that natural selection favors features that improve the likelihood that an individual will survive and reproduce. Over time, features that are advantageous become more common, while disadvantageous features become less common and may disappear.

To see how an understanding of natural selection can help answer a "why" question about behavior, it helps to consider a real example. In East Africa, lions live in small groups called prides. Each pride contains several adult females, several youngsters (called cubs), and one or more adult males. The adult males, which are usually brothers or other relatives, dominate the pride and father all the cubs. They also defend the pride against other males. A group of males usually can control a pride for only a couple of years before being forced out by younger males. When this happens, the new males often kill all the young cubs in the pride. Why do the new males behave in this way?

To explain why this behavior exists, you need to know how a male lion benefits from performing the behavior. Consider a male lion who has just taken over a pride. Because he likely will be in the pride for only a few years before he is forced out by a younger male, he has a very short time in which to reproduce. But female lions with cubs will not breed until their cubs are grown, which may take more than two years. If a female's cubs die, however, she will mate again almost immediately. These observations suggest the evolutionary hypothesis for cub-killing behavior explained in **Figure 13-2**.

Note that this explanation does not imply that male lions are aware that they are killing the offspring of other males or that they somehow calculate how they will benefit from performing this behavior.

Figure 13-2 After taking over a pride, a male lion usually kills the young cubs in the pride. This diagram shows one hypothesis of how this behavior benefits the male lion.

Hypothesis
A new male lion kills the young cubs in the pride so that . . .

. . . female lions will mate with him and have his offspring.

Predictions

- Male lions that kill young cubs after taking over a pride have more offspring than males that let cubs live.
- Males are unlikely to kill cubs after they have been in residence in a pride for several months—they would be killing their own offspring.

Natural selection favors traits that benefit individuals

Most scientists now agree that natural selection favors traits that contribute to the survival and reproduction of individuals. The idea that natural selection favors traits that benefit groups or species is most often expressed as the claim that a trait has evolved because it "ensures the survival of the species." The cub-killing behavior of lions and many other behaviors show that this belief is false. The actions of male lions increase the already high death rate among cubs and actually reduce the likelihood that the species will survive. Because natural selection favors traits that benefit individuals, an animal usually will behave in ways that are most likely to ensure its own survival and increase its likelihood of reproducing.

The Genetic Basis of Behavior

Do genes determine behavior, or do animals learn how to behave from experience? Over the last century, this question has been very controversial. Some scientists argued that most behaviors were genetically programmed and appeared in the same form in different individuals of a species. Other scientists claimed that behaviors were shaped by an animal's experience, not controlled by its genes. The conflict between these two views is largely settled now. Many studies have shown that both heredity and learning play significant roles, often interacting to shape the final behavior.

Behaviors can be inherited

From years of observation and experimentation, biologists have learned that many kinds of animal behaviors are influenced by genes. One behavior demonstrated to have a genetic basis is nest building in lovebirds. These small parrots construct their nests from vegetation that they collect and carry back to the nesting site. One species, Fischer's lovebird, carries its nesting material in its beak. A second, closely related species, the peach-faced lovebird, carries its nesting material in the feathers near its tail.

In the 1960s, biologist William Dilger created hybrids by crossing peach-faced lovebirds with Fischer's lovebirds. As you can see in **Figure 13-3**, the hybrids' nesting behavior resembled the behaviors of both their parents. Though the hybrids usually tried to place nesting material in their feathers, they were rarely successful and usually ended up carrying the nesting material in their beak.

Figure 13-3 William Dilger crossed individuals of two lovebird species and showed that nest-building behavior in these birds has a genetic basis. The hybrid offspring exhibited behaviors of both their parents.

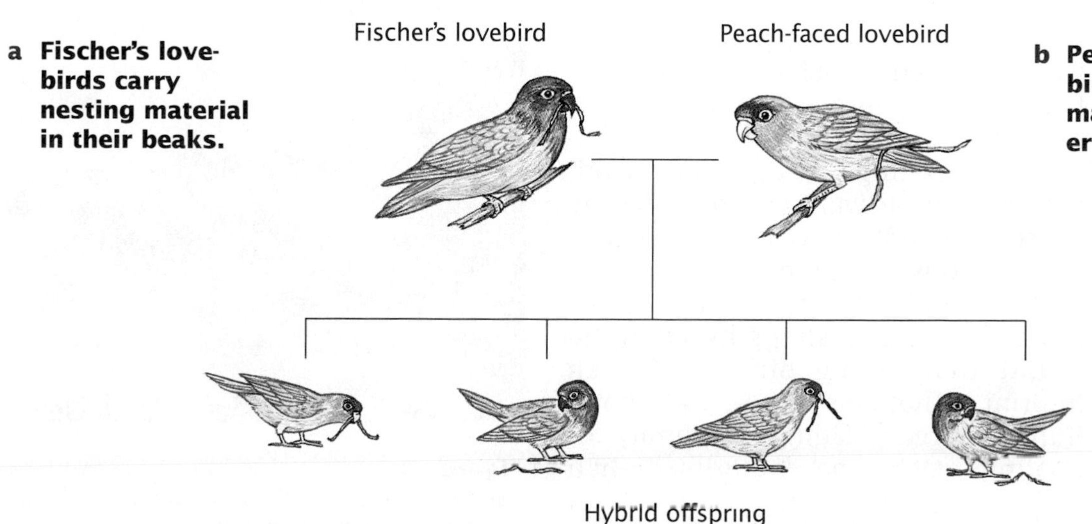

a Fischer's lovebirds carry nesting material in their beaks.

b Peach-faced lovebirds carry nesting material in the feathers near their tail.

c Hybrids of these two species try to insert nesting material into the feathers near their tail, but they usually don't succeed and must carry the material in their beaks.

Figure 13-4
How long a male cricket sings is controlled genetically. In William Cade's experiment, crickets in the initial group varied in how long they sang. By allowing only the males that sang for the greatest or least amount of time to breed, Cade produced two distinct groups.

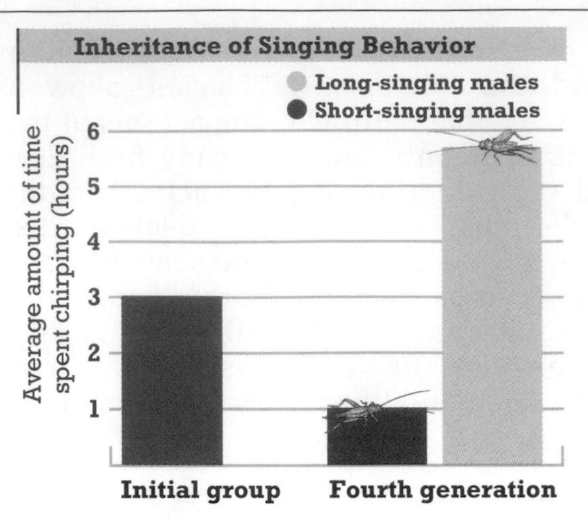

Another behavior that has a genetic basis is chirping in crickets. Males chirp to attract females. Biologist William Cade noticed that males differed in how much time they spent chirping per night. Some males did not chirp at all, some chirped for just a few hours, and a few chirped for almost the entire night. To determine whether this difference in behavior was inherited, Cade selected two groups of crickets. One group consisted of the males that spent the most time chirping. The second group contained the males that spent the least amount of time chirping. He allowed males in both groups to mate and then examined their male offspring. Of these offspring, only the extreme individuals (the males that sang for the longest and shortest times) were permitted to mate. Cade continued this procedure through four generations of crickets. As you can see in **Figure 13-4**, by the fourth generation the two groups differed significantly in the amount of time they chirped. Because this behavioral difference was transmitted from one generation to the next, it clearly had a genetic basis.

Learned Behaviors Are Modified by Experience

Behaviors can be influenced by genes. But to what degree can behavior be modified by experience? The development of behaviors through experience is called **learning**. In many animals, learning is very important in determining the final shape of a behavior. For example, young geese and ducks follow their mother as she leads them to water, helps them find food, and keeps them out of danger. However, a duckling or gosling must learn to recognize its mother. This type of learning is called **imprinting**. Nobel Prize–winning animal behaviorist Konrad Lorenz discovered that imprinting takes place during a short period following hatching. During this period, the young birds will learn to follow any object—including toy wagons, boxes, and balloons—that they are associated with. When Lorenz raised a group of newly hatched goslings by hand, he found that young birds will even imprint on humans. **Figure 13-5** shows Konrad Lorenz leading his "family" of goslings. Once the young birds had imprinted onto Lorenz, they preferred to follow him, even when given the opportunity to follow a member of their own species.

Figure 13-5
These goslings followed Konrad Lorenz as if he were their mother.

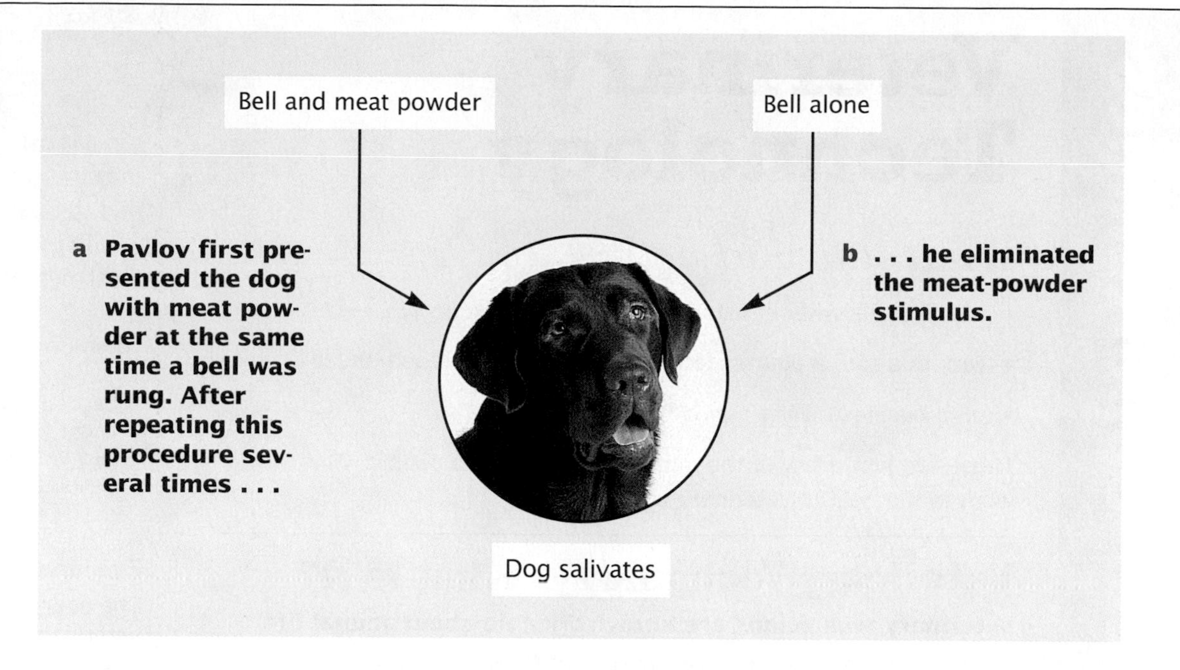
Bell and meat powder
Bell alone
a Pavlov first presented the dog with meat powder at the same time a bell was rung. After repeating this procedure several times . . .
b . . . he eliminated the meat-powder stimulus.
Dog salivates

Figure 13-6
In Ivan Pavlov's experiments, dogs learned to associate the ringing of a bell with food. They would salivate each time the bell rang, whether food was present or not.

One of the most famous studies of learning is Russian psychologist Ivan Pavlov's work on salivation behavior in dogs, which he carried out in the late 1890s and early 1900s. **Figure 13-6** is an overview of Pavlov's methods. If meat powder (stimulus) is presented to a hungry dog, the dog will salivate (response). Pavlov also presented the dog with a second, unrelated stimulus—a ringing bell—at the same time that meat powder was blown into the dog's mouth. As expected, the dog salivated. After repeated trials, the dog would salivate in response to the bell alone, having learned to associate the unrelated stimulus (the ringing of the bell) with the meat stimulus.

In another famous set of experiments, the American psychologist B. F. Skinner studied learning in rats by placing them in a "Skinner box," illustrated in **Figure 13-7**. Once inside, the rat would explore the box. Occasionally, it would accidentally press a lever, and a pellet of food would appear. At first, the rat would ignore the lever and continue to move about, but it soon learned to press the lever to obtain food. When it was hungry, it would spend all of its time pushing the lever. This sort of trial-and-error learning is of major importance to most vertebrates.

Figure 13-7
When the rat presses the lever it has its paw on, it receives a reward of food. A rat placed in this Skinner box quickly learns to press the lever when it is hungry.

Career Opportunities

Veterinary Technology

What it would be like to . . .

- help animals without attending veterinarian school?
- work in a zoo, a pharmaceutical company, or a research lab?
- run a business while caring for animals?

These are just a few of the opportunities open to people who work in the field of veterinary technology.

Science/Math Career Preparation

High School
- Biology
- Chemistry
- Math

Vocational School/College
- Two-year associate's degree at a school accredited by the American Veterinary Medical Association

Veterinary Technology: What Is It?

Veterinary technicians are knowledgeable about animal life processes and assist in the care and handling of animal patients. They play important supportive roles in diagnostic, medical, and surgical procedures. Veterinary technicians are also skilled in routine laboratory and clinical procedures. Many work in private veterinary practices where they may also help out in the office.

Employment Outlook:
The ever-broadening scope of veterinary medicine has created a shortage of trained veterinary technicians. Also, as the number of family pets rises, the demand for veterinary technicians is increasing quickly.

Career Focus: Jennifer Dusek

Veterinary technician at a veterinary clinic

"I've loved animals for as long as I can remember. I was always bringing home hurt animals—birds, chickens, squirrels, dogs, and even a pig. When I was in high school, I helped out at a veterinary clinic feeding animals and cleaning kennels. It's a good way to get experience, and vets are always happy to have the help. I looked into becoming a veterinarian but didn't feel comfortable with the amount of school required. I wanted practical training. I heard about a vocational school that offered courses in veterinary technology, so I enrolled."

What I Do: "Today I'm a manager of a veterinary clinic, where I've been working for 10 years. As a veterinary technician I'm responsible for a variety of duties, from monitoring a patient during surgery to taking X rays. For me, every day is busy, exciting, and truly rewarding. I also work in a wildlife rehabilitation center nursing injured animals back to health. I visit elementary schools and bring one or two animals along. The kids love learning about animals, especially when they get to use the stethoscope to listen to the animals' hearts."

To Find Out More

1. For further information, write to:
American Veterinary Medical Association
930 N. Meachum Road
Schaumburg, IL 60172

2. Many states require veterinary technicians to be registered or certified. Find out the specific registration provisions in your state.

3. See your school counselor to find out what colleges and institutions offer programs for veterinary technicians. Review the prerequisites and course requirements. Make a poster showing the range of job opportunities in this field.

Genes and Learning Interact to Shape Behaviors

Most behavioral biologists have come to think that animal behavior has both learned and genetic components. An example is provided by a famous 1940s experiment by the biologist Robert Tryon. Tryon studied rats' ability to find their way through a maze with many blind alleys and only one exit, where a reward of food awaited. It took awhile, as false avenues were tried and rejected, but eventually some individuals learned to finish the maze while making few incorrect turns. Other rats never seemed to learn the correct path. Tryon bred the "maze-bright" rats with one another, establishing a colony from the fast learners. He also established a "maze-dull" colony by breeding the slowest-learning rats with each other. He then tested the offspring in each colony to see how quickly they learned the maze. The offspring of maze-bright rats learned even more quickly than their parents had, while the offspring of maze-dull parents were even poorer at maze learning. After Tryon repeated this procedure over several generations, two behaviorally distinct types of rats with very different maze-learning ability resulted, as shown in **Figure 13-8**. Clearly, the ability to learn the maze was to some degree hereditary, controlled by genes passed from parent to offspring. It is important to note that the genes were specific for this particular behavior. When Tryon tested the ability of the two groups of rats to complete a different kind of maze, he found no difference between maze-dull and maze-bright rats.

Figure 13-8 In rats, the ability to learn a maze is at least partly inherited.

Inheritance of maze-running ability

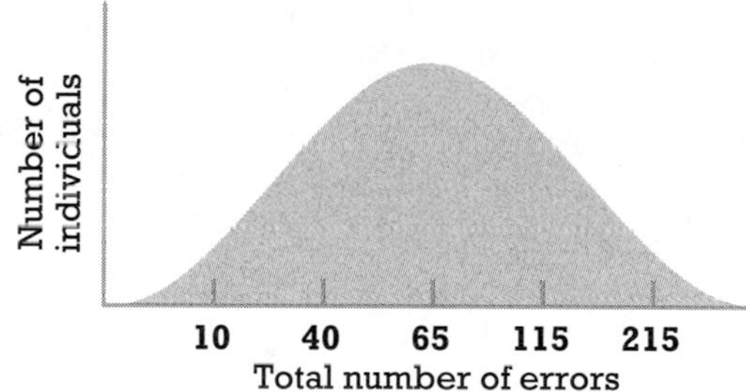

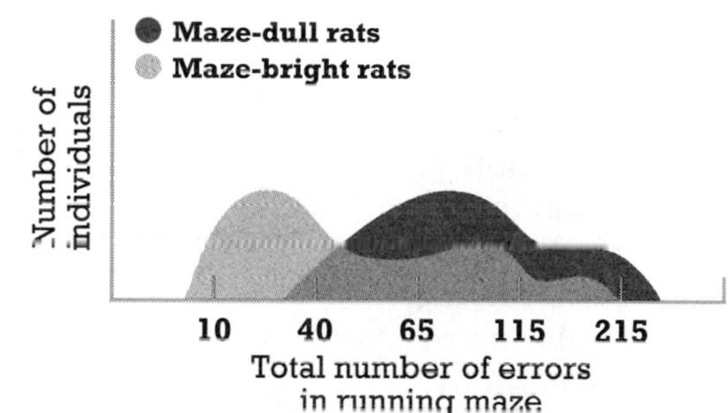

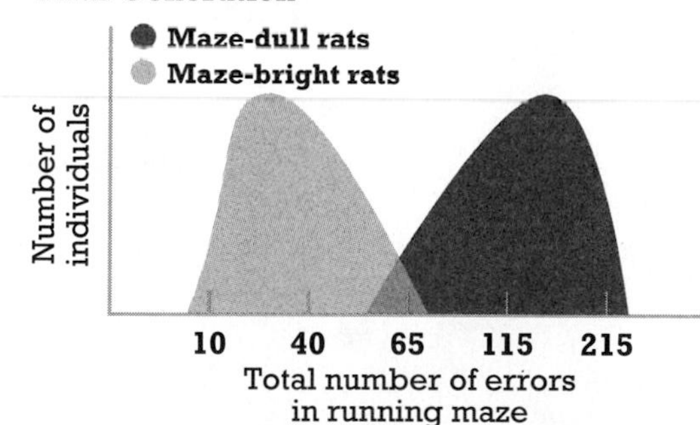

SECTION REVIEW

1. **Propose a hypothesis to explain why squirrels bury nuts. How would you test your hypothesis?**
2. **Construct an argument to explain why female lions do not kill their cubs when new males take over a pride.**
3. **What is imprinting?**
4. **Describe Tryon's experiment, and explain how it shows the interaction between genes and learning.**

Most biologists have concluded that many aspects of animal behavior are influenced by genes. Because natural selection acts on genes, animal behavior should be, at least to some extent, the result of evolution. When animal behavior is studied in nature, behaviors characteristic of particular animals are, by and large, suited to their way of life.

13-2 Kinds of Behavior

OBJECTIVES

1. **List three types of behavior.**
2. **Explain the function of animal signals.**
3. **Summarize how sexual selection affects the evolution of behavior.**
4. **Describe the evolution of altruism by kin selection.**

Categories of Behavior

Behavior is an animal's most immediate way of dealing with its environment. Because the environment is complex and can change rapidly, most animals have several different kinds of behavior, each suited to a particular situation. For instance, a squirrel may perform one kind of behavior when it finds a nut on the ground—it digs a hole. It performs a completely different behavior when a snake approaches—it runs for shelter—because digging a hole would not help it escape from the snake. Biologists have classified the behaviors animals perform into several broad categories. **Table 13-1** lists some of these categories and gives an example of each. The rest of this section examines some kinds of animal behavior in more detail.

Table 13-1 Some Types of Animal Behaviors

Type of behavior	Function	Example
Parental care	Ensure survival of young	This robin is feeding an insect to its offspring.
Courtship behavior	Attract a mate	The peacock spreads his large, brightly colored tail to attract females.
Defensive behavior	Protect individual from predators	A hognose snake flips onto its back and plays dead when threatened.

Foraging behavior	Locate, obtain, and consume food 	A sea otter opens mussels and clams by smashing them against a rock it carries on its chest.
Migratory behavior	Move to a more suitable environment for living or reproduction	Monarch butterflies migrate thousands of kilometers, from the United States to central Mexico. 
Territorial behavior	Protect a resource for exclusive use	These wolves belong to a pack that defends its territory from other packs. 

Evolution

Many songbirds migrate north in the spring and south in the fall. Identify some benefits and costs of this behavior.

How Animals Communicate

If you approach an unfamiliar dog and it growls, you know that it may bite if you come any closer. You have responded to a signal given by the dog. Animals use signals to influence the behavior of other animals. A signal can be a posture, call, movement, scent, color, or facial expression. Animals send and receive signals through all of the senses familiar to us—sight, hearing, smell, touch, and taste. Some fishes communicate through a sense we do not have—sensitivity to weak electrical fields.

Natural selection has shaped animal signals so that they reach the intended receiver efficiently and stimulate a response. To be transmitted efficiently, a signal must be able to travel through the environment from sender to receiver. A signal must also be recognizable to the receiver, or it won't have any effect on behavior. When a male frog like the one shown in **Figure 13-9** seeks to attract a mate, he emits a loud call. This signal is a very effective way to reach female frogs. The call carries a long distance, reaching even far-off females. At night, when many frogs are active, visual signals such as colors and movements are visible from only a short distance away and would therefore be much less effective at attracting a mate.

Figure 13-9
This male frog (left) is calling for a mate. A female (right) has been attracted by his call.

Figure 13-10 Animals communicate in many ways. Here are three examples.

Because an animal faces a variety of social situations in which signals are needed, it usually has several different signals. Each signal is suited to a different social situation. To appreciate this conclusion, you need only to think about your own behavior. Would you try to attract the attention of someone far down the street with subtle facial expressions and whispers? Of course not. You would probably shout and perhaps wave your arms over your head. However, these signals wouldn't be suitable for a face-to-face conversation with a friend. **Figure 13-10** shows several examples of animal signals and describes their functions.

a A male sparrow sings to attract females and to warn other males against entering his territory.

Humans communicate with complex symbolic language

Among animals, vocal communication is most developed in the primates. Many primates have a "vocabulary" that allows individuals to communicate the identity of specific predators, such as eagles, leopards, and snakes. Chimpanzees and gorillas can learn to recognize and use a large number of symbols to communicate abstract concepts. They cannot talk, however, because they cannot produce the sounds of speech. Furthermore, the ability to assemble symbols into new sentences requires very complex brain structure, a complexity that has evolved only in humans.

Language develops at an early age in humans. Human infants learn language by trial and error during the "babbling baby" phase of childhood. Humans learn a particular language while young. Children who have not heard certain consonant sounds as infants usually cannot distinguish or produce them as adults. That is why most Americans never master the throaty French \r\, whereas speakers of French typically replace the English \th\ with \z\ or \s\. Children quickly and effortlessly learn a vocabulary of thousands of words. This rapid-learning ability seems to be genetically programmed.

Although language is the primary form of human communication, much evidence suggests that odor and other nonverbal signals (body language) may also be important. Think of how much you learn by a quick look at the expression on the face of another person. Among humans, many forms of communication may occur simultaneously.

b When scout ants find food, they lay down a chemical trail that leads other members of the nest to the food. These ants are following such a trail.

c Using rhythmic head bobs and pulsations of their throat fan, male anoles signal to other members of their species.

Choosing a Mate

When ready to mate, animals produce signals to communicate with potential partners. Each species usually produces a unique courtship signal. This helps ensure that individuals don't mate with individuals of another species. For example, the flash patterns of fireflies are coded for species identity. A female recognizes males of her own species by the number of flashes in his pattern, and she will ignore any male with a different flash pattern. The chemical produced by a female silk moth attracts only males of her own species. Many species of insects, amphibians, and birds produce unique sounds or songs to attract mates. A white-crowned sparrow will respond to the song of another white-crowned sparrow, but totally ignore the song of a song sparrow.

Selection of a mate involves much more than just finding an individual of the same species. Females, and to a lesser extent males, are often very selective about which potential partner they mate with. Female frogs, for instance, have been observed "shopping around" among calling males. A female will sit near a male, listen to his call for several minutes, and then move on to another male and listen to his call. She may evaluate several males before choosing one (usually the largest) with which to mate.

Mate choice can drive evolution

What characteristics do animals use in choosing a mate? When Charles Darwin considered this question more than a century ago, he made an important discovery about evolution. Darwin noticed that males often had extreme characteristics that they used in their courtship displays. Take, for example, a widowbird, shown in **Figure 13-11**. During the breeding season, the male widowbird grows an extremely long tail, up to 0.5 m (19 in.) in length. The female bird, also shown in **Figure 13-11**, has a short tail, only about 7 cm (3 in.) long. How has this difference between the sexes evolved? The long tail of the male widowbird cannot be essential for survival, since the female bird survives quite well without it. In fact, the male's long tail probably reduces his chances of survival by making him more visible to predators.

Figure 13-11 The tail of the male widowbird (top) can be more than three times his body length. The female (left) has a short tail.

Darwin recognized that such traits could have evolved if they helped males attract mates. He proposed the mechanism of **sexual selection** to account for the evolution of traits like the widowbird's tail. Sexual selection is an evolutionary mechanism in which traits that enhance the ability of individuals to acquire mates increase in frequency. Even traits that have a negative effect on survival can evolve in this way, provided that their benefits to reproduction are high enough. For example, the long tail of the male widowbird probably evolved because males with this trait were able to attract more mates than males with short tails. Even though long-tailed males were more likely to be captured by predators, they still fathered more offspring than short-tailed males because of their greater attractiveness to females.

What kinds of traits are advantageous for acquiring mates? Because males usually compete among themselves for the chance to mate with a female, sexual selection has often favored traits that make males more intimidating or better at combat. Examples of such traits are large body size, antlers of deer, manes of lions, and horns of bighorn sheep. **Figure 13-12** shows several examples of male traits favored by sexual selection. Competition among males can also take subtle forms. A male can gain a reproductive advantage over other males by interfering with their reproduction. In some species of worms, butterflies, and snakes, for instance, the male seals the female's reproductive tract after mating so that no other males can mate with her.

Figure 13-12 What characteristics are favored by sexual selection? Three examples are shown below.

a A large rack of antlers helps a male deer intimidate rivals and attract mates.

b Swordtails are common aquarium fish. Females prefer males with longer swords attached to the tail.

c Gorillas live in groups of many females dominated by one large adult male. The dominant male, known as a silverback, must be large to fight off attacks by other males. Males, on average, are 50 percent larger than females.

Figure 13-13
Young scrub jays often will help their parents raise offspring. This is an example of altruism.

Figure 13-14
This honeybee is performing an extremely altruistic behavior for the benefit of her hive. She is stinging an attacker, even though this will result in her death.

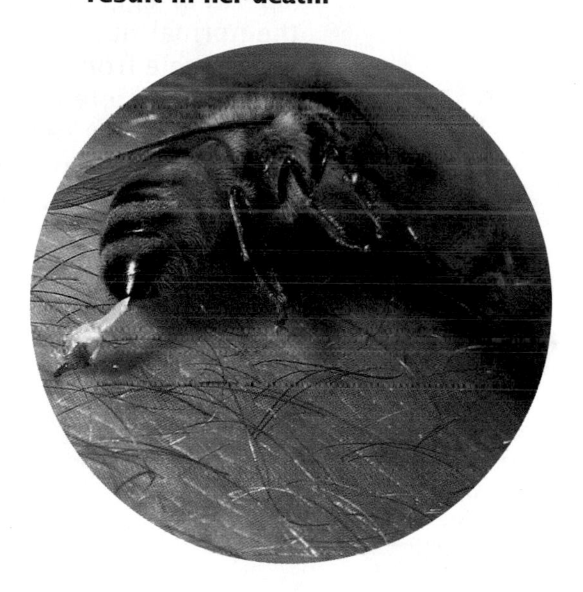

The Evolution of Self-Sacrificing Behavior

When a young scrub jay reaches maturity, it may not leave its parents' nest and begin to have its own offspring, as do young birds in most other species. Instead, the young scrub jay, shown in **Figure 13-13**, often remains with its parents and helps them raise additional offspring. This behavior is an example of altruism. **Altruism** is self-sacrificing behavior. Altruism is often thought of as heroic human behavior, such as jumping into a river to save a drowning person. But altruism is not limited to humans; it occurs in many other kinds of animals. In many species it is an important aspect of cooperation on which the survival of the whole group depends. The social insects (ants, bees, wasps, and termites) exhibit extreme forms of altruism. Most of the individuals in a beehive are female workers that never reproduce. They care for and defend the only female that does reproduce, the queen, and help raise her offspring. If the hive is attacked, the workers sacrifice themselves to protect the hive, as shown in **Figure 13-14**.

Altruism usually occurs among relatives

How can altruism evolve? How can a trait producing an altruistic behavior be passed from generation to generation if individuals that have the trait never leave any offspring? The most widely accepted answer was proposed by the British biologist W. D. Hamilton in 1964. This mechanism is called **kin selection**. Kin selection occurs when an individual helps relatives (kin) reproduce rather than producing its own offspring. **Figure 13-15** explains the logic underlying kin selection. To understand kin selection, you need to think about meiosis and sexual reproduction. Hamilton realized that relatives share genes in common because of the way gametes form during meiosis and join together during sexual reproduction. Recall that each gamete (sperm or egg) contains only one-half the normal number of chromosomes. One gamete from the male parent joins with one gamete from the female parent to form the fertilized egg, which develops into a new individual. As a result, one-half of an individual's genes come from its mother and one-half come from its father. In other words, each individual shares one-half of its genes with each parent. Other relatives also share genes. On average, each individual shares one-half of its genes with a full brother or sister, and one-fourth of its genes with an uncle, aunt, half brother, or half sister. The more distantly related two individuals are, the fewer genes they share.

Because relatives have genes in common, an individual can sometimes transmit more copies of its genes to the next generation by helping relatives reproduce than by producing its own offspring. Kin selection explains why worker bees do not reproduce and even give up their lives for the benefit of the hive. They are helping the queen, who is their mother, reproduce. It also explains why a scrub jay would remain with its parents and help raise its brothers and sisters.

Figure 13-15
Each *X* in this pedigree of a scrub jay family represents a chromosome. For simplicity, only one pair of chromosomes for each individual is shown here.

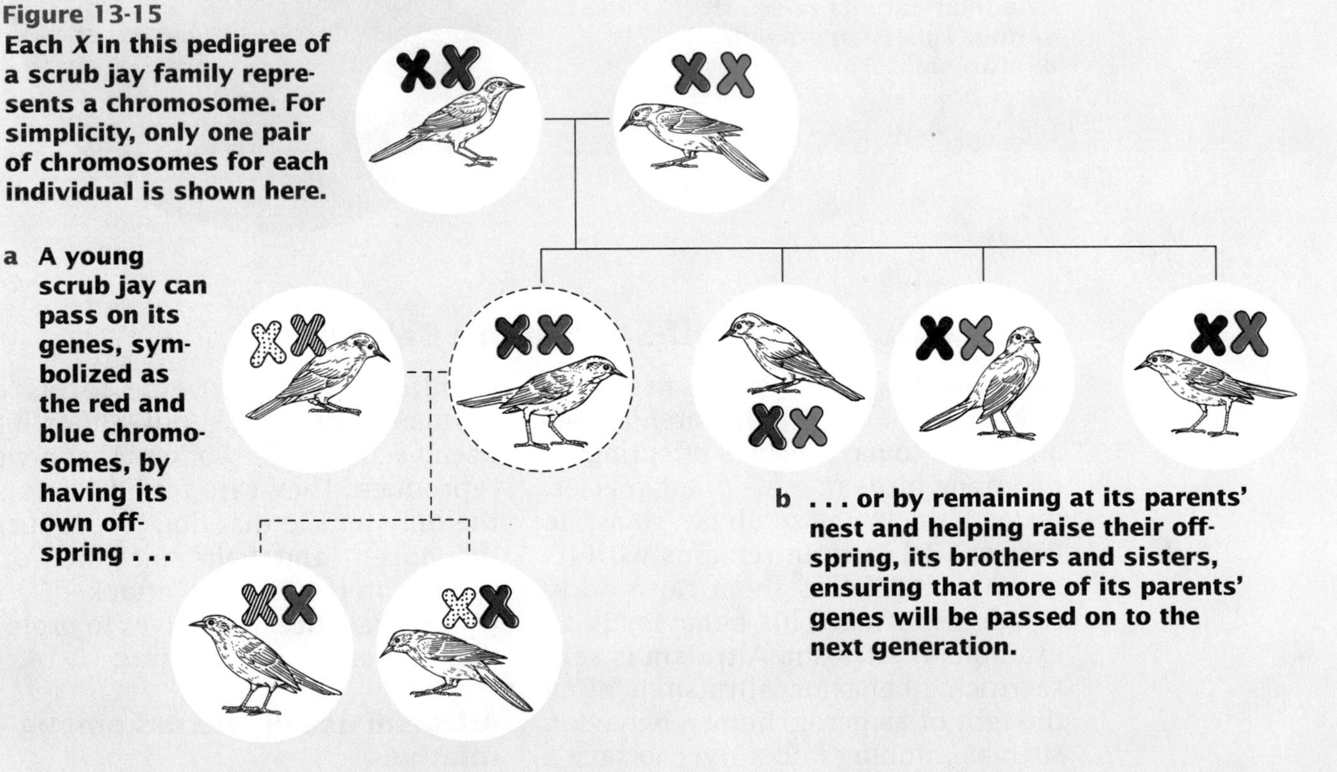

a A young scrub jay can pass on its genes, symbolized as the red and blue chromosomes, by having its own offspring . . .

b . . . or by remaining at its parents' nest and helping raise their offspring, its brothers and sisters, ensuring that more of its parents' genes will be passed on to the next generation.

SECTION REVIEW

1. **What is the function of territoriality?**
2. **List three human nonverbal signals, and describe their functions.**
3. **Explain how sexual selection accounts for the long tail of the widowbird.**
4. **Why is the evolution of altruism a puzzle? How does kin selection help solve this puzzle?**

CHAPTER

13 Highlights

"Ernie! Look what you're doing— take those shoes off!"

Animal Misbehavior

	Key Terms	Summary
13-1 Evolution of Behavior	behavior (p. 233) learning (p. 236) imprinting (p. 236)	• Biologists ask two kinds of questions about behavior. "How" questions concern what stimulates a behavior, how it is controlled, and how it is performed. "Why" questions are about why a behavior exists. • Behaviors contribute to an animal's survival and reproduction. Behaviors evolve through natural selection. • Natural selection favors behaviors that are beneficial to individuals. • Learning is the alteration of behavior by experience. Genes and learning interact to shape behavior.
13-2 Kinds of Behavior	sexual selection (p. 244) altruism (p. 245) kin selection (p. 246)	• Animals use signals to communicate. Signals can be movements, colors, scents, calls, electrical impulses, or physical contact. • Darwin proposed that traits enabling an individual to attract more mates will evolve, even if they have a negative impact on survival. He called this evolutionary mechanism sexual selection. • Altruism is self-sacrificing behavior. W. D. Hamilton showed that altruism directed toward relatives can be favored by natural selection.

Lions live in cooperative groups called prides.

Female swordtails prefer males with long swords on their tail.

CHAPTER 13

review

Understanding Vocabulary

1. For each pair of terms, explain the differences in their meanings.
   a. kin selection, sexual selection
   b. behavior, stimulus
   c. territoriality, altruism

2. For each set of terms, complete the analogy.
   a. Pavlov : dogs :: Tryon : ____
   b. "how" question : mechanisms :: "why" question : ____
   c. stimulus : meat powder :: response : ____

Understanding Concepts

3. **Relating Concepts** Construct a concept map describing animal behavior. Try to use the following terms in your map: behavior, stimulus, natural selection, environment, individuals, kin selection, sexual selection.

4. **Inferring Conclusions** In his studies of black-headed gulls, Niko Tinbergen observed that the gulls removed the empty shells from the nest after the chicks had hatched. Gulls that did not remove the empty shells from the nest lost more chicks to predators. The white inside part of the eggshell attracted predators to the camouflaged nest. Explain how natural selection might favor the removal of empty shells from the nest.

5. **Summarizing Information** List two "how" questions and two "why" questions about the behavior described in question 4.

6. **Recognizing Relationships** Explain William Cade's experiment on crickets. What evidence led him to conclude that the duration of singing in crickets had a genetic basis?

7. **Analyzing Methods** Could Tryon have concluded that the maze-bright rats were better at all kinds of learning than the maze-dull rats? Explain your conclusion.

8. **Inferring Relationships** What is trial-and-error learning? Give one example. Explain how trial-and-error learning can be beneficial to an animal. What are some of the possible drawbacks of trial-and-error learning?

9. **Analyzing Relationships** Assume that natural selection acts on the behavior of territoriality. Identify some of the possible advantages to defending a territory in spite of the fact that doing so consumes energy and increases the risk of injury.

10. **Comparing Functions** Give three examples of different types of animal signals. Explain the function of animal signals in terms of natural selection.

11. **Inferring Relationships** Based on your personal observations of dogs, describe some of the signals dogs use when they meet another dog and when they meet their owner. How do you think these signals influence the behavior of the other dog and the owner?

12. **Summarizing Information** What is sexual selection? Describe how it affects the evolution of behavior.

13. **Comparing Functions** Describe the kinds of male traits that are advantageous for acquiring mates. Give three examples.

14. **Analyzing Methods** A scientist is studying sexual selection in widowbirds. He captures several male widowbirds, cuts off part of their long tail, and releases the birds. He finds that these males mate with half as many females as males that weren't captured. He concludes that female widowbirds prefer males with longer tails. What is an alternative explanation for this observation? How can the scientist alter his experiment to determine which explanation is correct?

15. **Recognizing Relationships** What is altruistic behavior? Give an example of altruistic behavior. Describe W. D. Hamilton's explanation for the evolution of altruistic behavior.

16. **BUILDING ON WHAT YOU HAVE LEARNED** In Chapter 10 you learned how natural selection causes evolution. Summarize the mechanism of natural selection and explain why behaviors are just as important to survival and reproduction as physical features are.

Interpreting Graphics

17. The illustrations below show a chimpanzee that was placed in a room for the first time. Notice the boxes in the room and the bananas hanging over the chimp's head. The type of behavior exhibited by the chimpanzee is called insight.

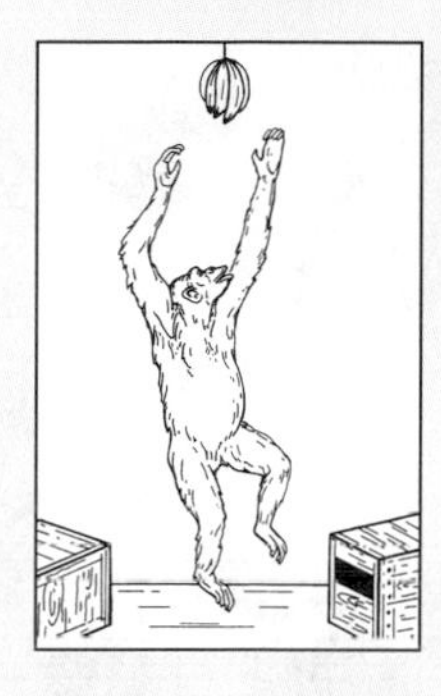

Which statement best describes the chimpanzee's behavior?

a. The chimpanzee stacked boxes to reach the bananas based on past experience.
b. The chimpanzee learned to stack the boxes to reach the bananas.
c. The chimpanzee figured out how to reach the bananas by trial and error.

Explain why you chose the statement you did.

Reviewing Themes

18. ***Homeostasis and Stability***
How does a dog's behavior on a hot day help it control its body temperature?

Thinking Critically

19. **Recognizing Logical Inconsistencies** "A child's behavior closely resembles the behavior of its parents. Therefore, most human behaviors are genetically controlled." Explain the logical flaw in this argument.

Life/Work Skills

20. **Working Cooperatively** Prepare a chart with your classmates. On one side list some of the behavior patterns you have read about in this chapter. Now think about human behaviors that are similar to the animal behaviors you just listed. Can you think of an example of human territoriality? human aggression? Think about whether your behavior changes when you are in a group, or with just a few friends. How do you behave with a large group of people you don't know? How are some of our reactions and behaviors similar to the ones you have read about in this chapter?

Activities and Projects

21. **Career Connection: Psychologist** A psychologist studies human behavior and helps people improve their mental health. Find out what a psychologist does on a daily basis. What kind of educational background is required? Where do psychologists usually work?

22. **Art Connection** Honeybees have a sophisticated form of communication many people call the "honeybee dance." Use library references or an on-line database to research how honeybees dance to communicate. Make posters that summarize the honeybee dance. Share the posters with your classmates.

Discovering Through Reading

23. Read the article "Leafcutters: Gardeners of the Ant World" by Mark W. Moffett in *National Geographic*, July 1995, pages 98–111. Explain the ant's relationship with fungi. What positive effect do leafcutter ants have on their environment?

24. Read the article "Scent of a Man" by Sarah Richardson in *Discover*, February 1996, pages 26–27. What are the functions of MHC genes? How did the women in Claus Wedekind's experiment respond to the different T-shirts? How might this response have been favored by natural selection?

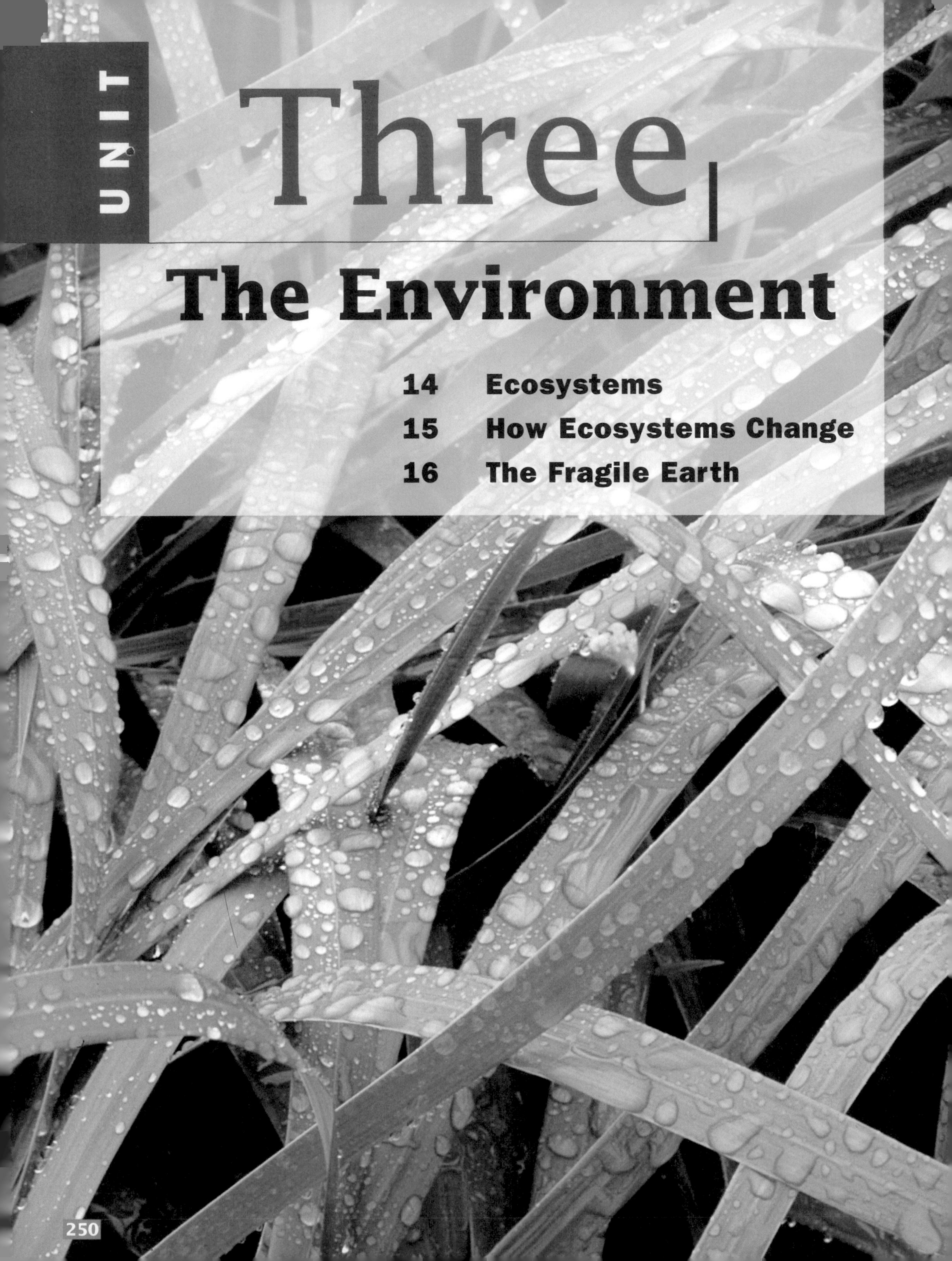

UNIT

Three

The Environment

A GLOBAL GREENHOUSE

Whatever your home address, your universal address is Earth. Each day, CO_2 and other gases in the air above Earth trap most of the sunlight that pours down, allowing the Earth to remain warm enough to sustain life. As the CO_2 admits sunlight and retains heat, it works like a greenhouse!

A World of Convenience

Inside the global greenhouse, you and your family enjoy the conveniences of electronic communication and entertainment, public and private transportation, and a broad array of appliances and labor-saving devices. Ever greater amounts of coal, oil, wood, and other hydrocarbons are burned each time someone flips a switch.

An Uncertain Future

The burning of these common hydrocarbon fuels releases more CO_2 into Earth's atmosphere. As CO_2 levels rise, it's like raising Earth's thermostat—more heat is retained. Scientists predict that Earth's temperature might increase as much in 100 years as it has since the last ice age!

LOOKING AHEAD

- What might someone like Carmen Cid do for you or your future children? See **Career Opportunities: Environmental Science,** page 267.
- How might you and other life-forms be affected if the Earth gets much warmer? See page 293.
- How can you help solve environmental problems? See pages 301–304.
- How can the loss of atmospheric ozone be slowed? See **Science, Technology, and Society: Can the Ozone Layer Be Saved?** pages 308–309.

CHAPTER

14

REVIEW
- metabolism (Section 5-1)
- photosynthesis (Section 5-3)
- cellular respiration (Section 5-4)

Ecosystems

This desert vegetation is growing in Organ Pipe Cactus National Monument, Arizona.

You cannot pick up a newspaper today without seeing news about the environment. Environmental issues are important because we all have to live in a world we seem to be destroying. We need detailed knowledge of how the world works so that we can prevent further abuse to our planet and perhaps begin to repair the damage we have already done.

14-1 What Is an Ecosystem?

OBJECTIVES

1. **Identify the components of an ecosystem.**
2. **Discuss the flow of energy through ecosystems.**
3. **Identify the different trophic levels in an ecosystem.**
4. **Explain why ecosystems can contain only a few trophic levels.**

State of Our World

Nearly 6 billion humans live on Earth. Scientists estimate that at least 10 million other species share the world with us. Yet we seem to be rapidly destroying our planet's ability to support us and its other inhabitants. Here are a few changes humans will make to the Earth *in the next year.* About 17 million hectares (40 million acres) of forest will be burned down or cut, like the forest shown in **Figure 14-1**. That is an area almost as large as the state of Washington. The world's human population will increase by about 88 million people, an amount nearly equal to the population of Mexico. Most of these 88 million people will be born in the world's poorest countries. Worldwide, there are already nearly 500 million people who are not adequately nourished.

For centuries people have believed that the environment was there for them to use as they saw fit. But today's environmental problems teach us that the environment is not a passive stage on which we can act as we please. Rather, we share the environment with other organisms, resulting in a complex network of interactions on which we all depend. Changes made to the environment can have serious consequences, not all of which are predictable.

Figure 14-1
This section of forest on the Quinleute Reservation in Washington has been logged. Although the trees will provide many useful products, cutting the forest has a devastating effect on the organisms that live there.

Figure 14-2
The inhabitants of this soil ecosystem include earthworms, insects, snakes, moles, bacteria, and a variety of plants and fungi.

Ecology and Ecosystems

In 1866, the German biologist Ernst Haeckel gave a name to the study of how organisms fit into their environment. He called it **ecology**, from the Greek words *oikos*, meaning "house, place where one lives," and *logos*, meaning "study of." Ecology, then, is the study of the "house" in which we live. Most of our environmental problems could be avoided if we treated the world in which we live the same way we treat our own homes.

Ecology is the study of the interactions of organisms with one another and with their physical environment. The organisms that live in a particular place, such as a forest, are known as a **community**. Ecologists, the scientists who specialize in ecology, call the physical location of a community its **habitat**. You can think of a habitat as a neighborhood and of a community as the residents of the neighborhood. The sum of the community and habitat is called an ecological system, or ecosystem. An **ecosystem** is a self-sustaining collection of organisms and their physical environment.

Imagine that you could collect every organism living in an ecosystem. **Figure 14-2** shows just some of the inhabitants of a soil ecosystem. If you visited a tropical rain forest, you could collect many more species than occur in this soil ecosystem. As many as 100 species of trees can be found in a patch of South American rain forest about the size of two football fields. The **diversity** of an ecosystem is a measure of the number of species living there. Tropical rain forests are the most diverse terrestrial ecosystems.

Why Study Ecology?

You study ecology because you need to know about the place in which you live. If you are going to prevent pollution, conserve resources, and help preserve the world for your children to live in, then you need to know how your world works—just as you need to study how any complex machine works in order to keep it running properly. Do you think a car would run for long if its owner had no idea of the need for water, oil, and gasoline? Remember that there is a fundamental difference between a car and an ecosystem, however. A car that receives no care from its owner will eventually break down. Ecosystems, on the other hand, normally function without human tending. But some ecosystems now require our help because we have caused them great damage.

Figure 14-3 An ecological model can predict how an ecosystem, such as this pond, responds to disturbance.

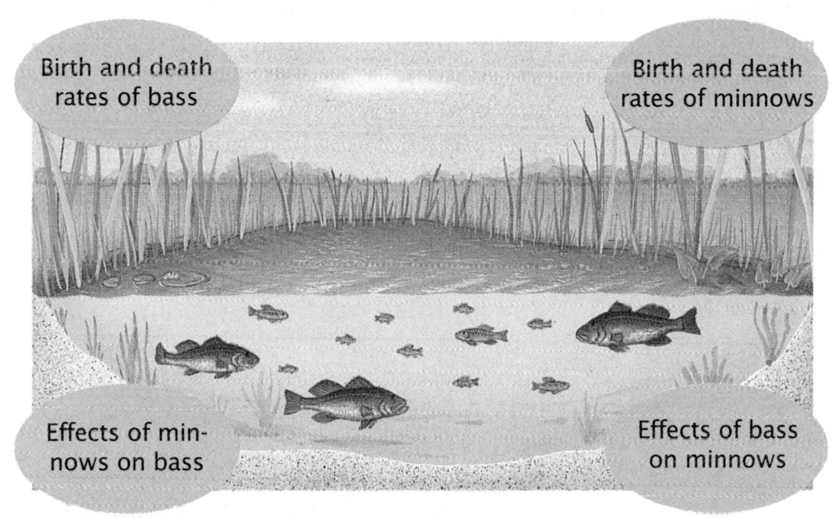

a A scientist making a model of this ecosystem might consider the factors shown in the yellow ovals.

b The model predicts that a disturbance, such as a disease that affects minnows, would cause the number of bass to decrease.

Ecosystems are very complex

It is very difficult to understand how an ecosystem works because it can contain hundreds or even thousands of interacting species. Nevertheless, you can gain a basic understanding of how an ecosystem works by asking two questions: Where does the energy come from that is needed by particular animals and plants? and How do organisms in ecosystems acquire adequate amounts of the minerals and other inorganic substances they need?

Answering these questions will give you a pretty good idea of how an ecosystem normally works. You will then be in a position to ask how an ecosystem might be expected to respond to a disturbance. To make such predictions, ecologists build a model, a simplified version of the ecosystem. An ecosystem model consists of a series of hypotheses that describe how the ecosystem functions: how energy moves through the ecosystem, how species interact, and so on. In some cases, ecologists express their models as mathematical equations and then solve these equations for various situations. **Figure 14-3** is a schematic representation of an ecological model.

Because ecosystems are so complex, no model can consider all of the factors that affect an ecosystem. Nevertheless, seeing what happens to the ecosystem model when a variable is changed helps ecologists predict what might happen if some component of the real ecosystem were altered. If, for example, one species in an ecosystem becomes extinct or the amount of available energy declines, an ecological model can help scientists predict the possible consequences. Ecological models may allow you to look into the future, but your vision can be only as accurate as the information and assumptions used to build the model.

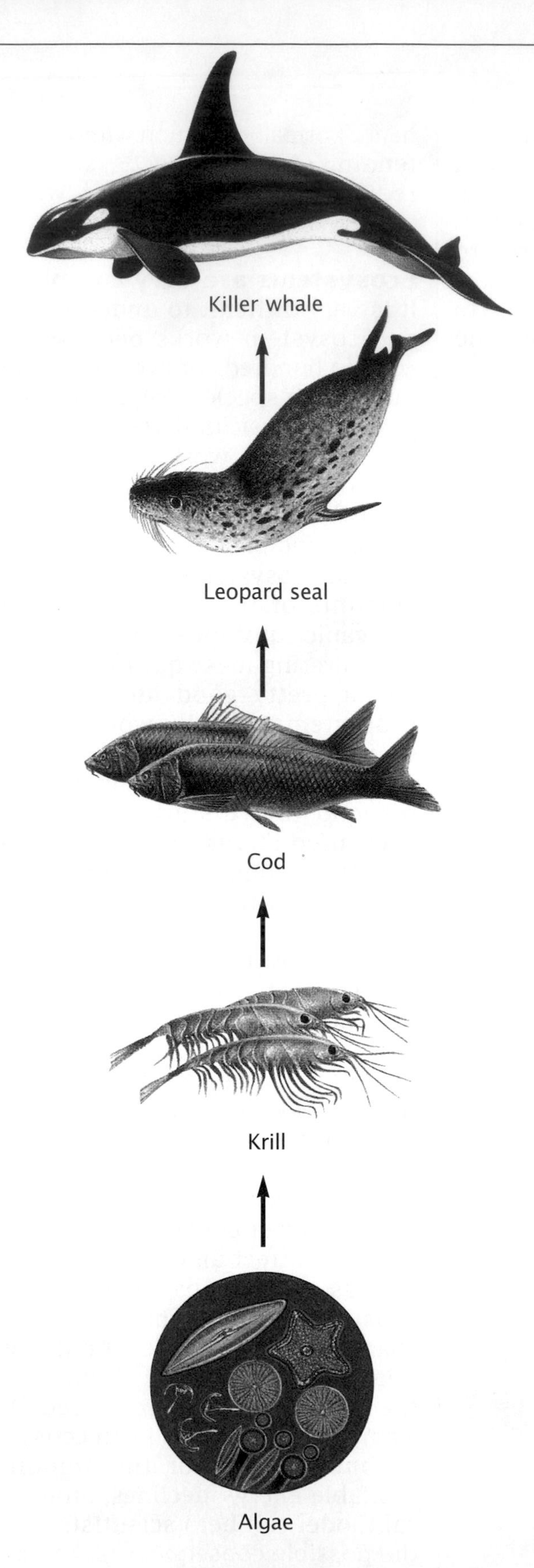

Figure 14-4
This food chain shows a path of energy transfer in an ocean ecosystem. Algae, which are photosynthetic, are the producers in this ecosystem. Krill, cod, leopard seals, and killer whales are consumers.

Energy in Ecosystems

An ecosystem needs energy because its living members use energy. A flower blooming, a squirrel running along a tree branch, a worm burrowing through the soil—every action of even the tiniest creature requires energy. The most important factor determining how many and what kinds of organisms live in an ecosystem is the amount of energy available.

How do the organisms in an ecosystem obtain energy? Life exists because organisms called **producers** take in energy from their surroundings and store it in complex molecules. Nearly all producers are photosynthetic; they capture the sun's energy to synthesize carbohydrates. Plants, some bacteria, and algae are producers. All other organisms are called **consumers**. They obtain their energy by consuming other organisms. Animals, most protists and bacteria, and all fungi are consumers. The food chain in **Figure 14-4** shows how energy is transferred from producers to consumers in an ocean ecosystem.

Each ecosystem contains consumers called **decomposers**. Decomposers obtain energy by consuming organic wastes (feces, urine, fallen leaves) and dead bodies. Fungi, such as the mushrooms shown in **Figure 14-5**, and some species of bacteria, are decomposers.

Figure 14-5
These scarlet waxy cap mushrooms are decomposers.

Interdependence

If pollution drastically reduced the number of algae in the food web shown below, would killer whales be affected? Explain.

Energy flows from producers to consumers

To follow the movement of energy through an ecosystem, ecologists assign each organism to a **trophic** (feeding) **level**. All members of a trophic level are the same number of energy-transferring steps away from the sun. Producers, such as plants, are in the first trophic level. Plants and other organisms that make their own food are called **autotrophs**. Animals that eat plants are in the second trophic level. Animals that feed on plant eaters are in the third trophic level.

Creatures in the second trophic level are **herbivores** (plant eaters). Cows, caterpillars, elephants, and ducks are herbivores. All organisms at the third trophic level or above are **carnivores** (flesh eaters). Carnivores at the third level feed on herbivores, while carnivores above the third trophic level feed on other carnivores. Tigers, hawks, weasels, pelicans, and killer whales are carnivores. Some animals cannot be classified as either carnivores or herbivores. **Omnivores**, such as bears and humans, eat both plants and animals. Because they cannot make their own food, organisms in trophic levels above the first trophic level are **heterotrophs**.

Most animals feed at more than one trophic level and feed on several different species at each trophic level. As shown in **Figure 14-6**, energy moves through an ecosystem in a complex network of feeding relationships called a food web. Notice that the food chain in **Figure 14-4** is just one part of this food web.

Figure 14-6
This food web shows how energy flows through an ocean ecosystem as one organism is eaten by another.

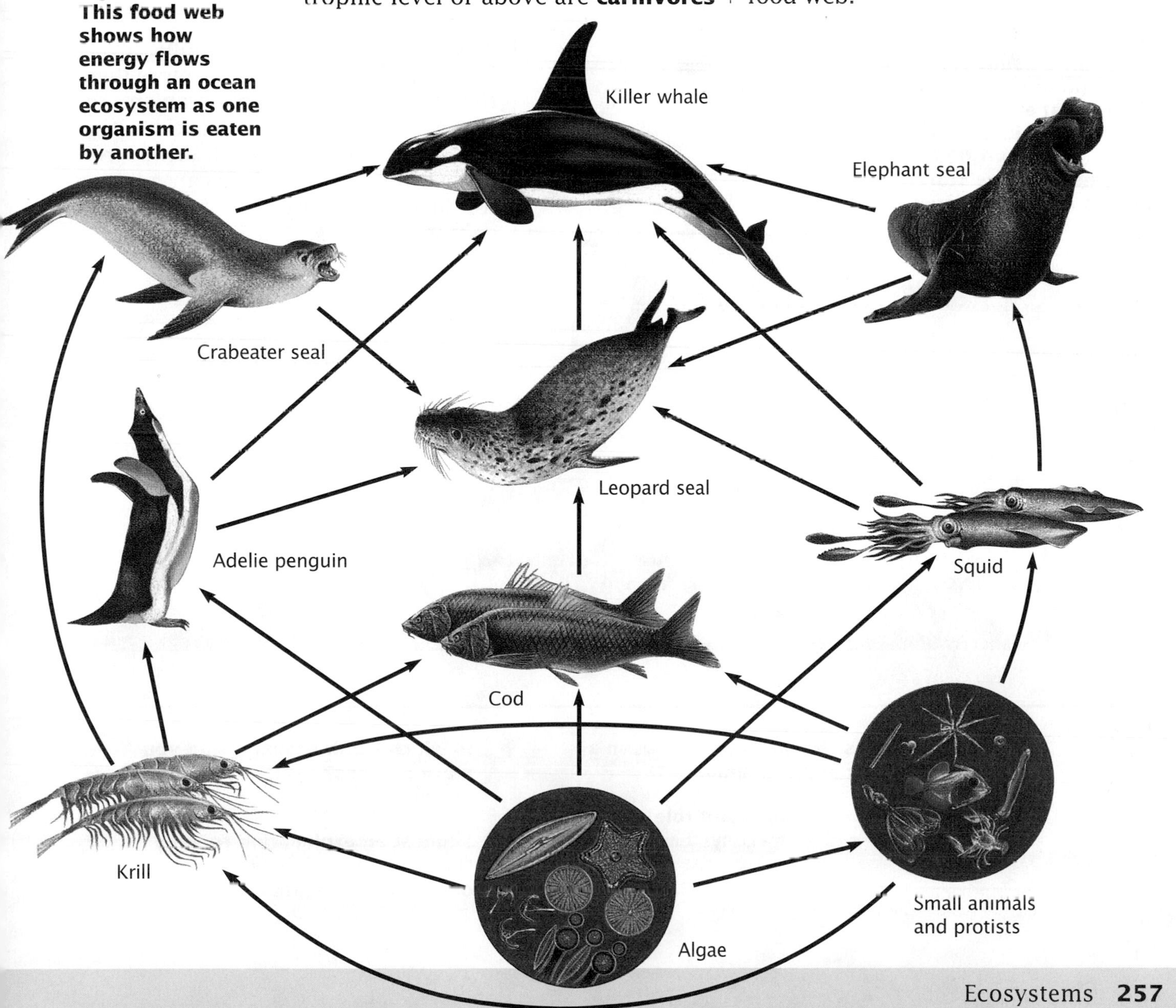

Matter, Energy, and Organization

More food would be available for the growing human population if humans did not eat meat. Explain why.

How Many Trophic Levels Can an Ecosystem Contain?

A plant absorbs energy from the sun and uses it to make carbohydrates such as cellulose, the major component of cell walls. Only about one-half of the energy captured by a plant becomes part of the plant body, however. Part of the remaining energy is stored in ATP made during cellular respiration. Most of the remaining energy escapes as heat. Similar losses of energy occur at each trophic level of an ecosystem.

In the 1950s, the ecologist Howard Odum determined how much energy was present at each trophic level in a Florida stream ecosystem. He captured animals and plants, measured their energy content, and built a model of how energy passed through the ecosystem. Odum found that when a herbivore eats a plant, only about 10 percent of the energy present in the plant's molecules ends up in the herbivore's molecules. The other 90 percent of the energy is "lost," some as the cost of doing work (breathing, moving, chewing) and much more as heat. Likewise, when a carnivore eats the herbivore, only 10 percent of the energy in the herbivore goes toward making carnivore molecules. At each trophic level, the energy stored in the organisms is about one-tenth that of the level below it.

One way ecologists show the declining amount of energy at higher trophic levels is by drawing a diagram called an ecological pyramid, shown in **Figure 14-7**. Because energy diminishes at each successive trophic level, few ecosystems can contain more than five trophic levels. Also, organisms at higher trophic levels tend to be less numerous than those at lower trophic levels. On the African plains, for instance, there are about 1,000 zebras, gazelles, and wildebeest for each lion.

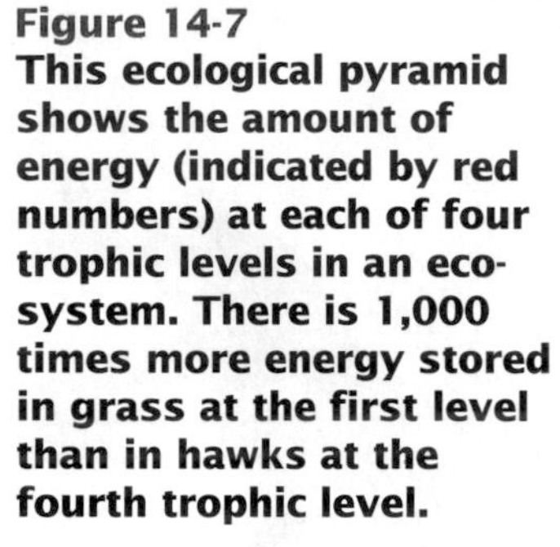

Figure 14-7
This ecological pyramid shows the amount of energy (indicated by red numbers) at each of four trophic levels in an ecosystem. There is 1,000 times more energy stored in grass at the first level than in hawks at the fourth trophic level.

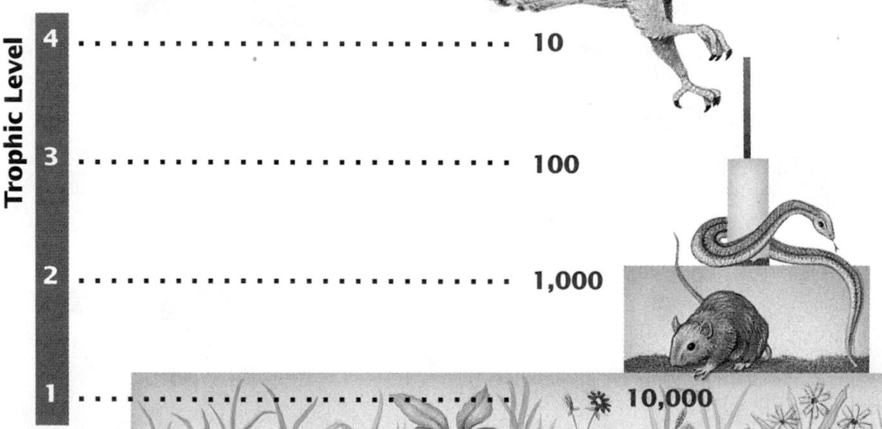

SECTION REVIEW

1. **What is the difference between a habitat and a community?**
2. **Explain the important role producers play in ecosystems.**
3. **In which trophic level would you place humans? Explain your answer.**
4. **Suggest an explanation for why there are fewer lions than zebras on the African plains.**

Newspapers, aluminum cans, plastic bottles, and many other products can be recycled and used again. Likewise, the physical components of an ecosystem are used again and again. Ecologists call this continual reuse "cycling." Materials that cycle within ecosystems include nitrogen, water, and carbon.

14-2 Cycles Within Ecosystems

OBJECTIVES

1. **Describe the results of Bormann and Likens's experiments.**
2. **Describe how nitrogen and water are recycled within ecosystems.**
3. **Explain the role of plants in the cycling of materials within ecosystems.**
4. **Trace the pathway of carbon between living organisms and the environment.**

Nutrient Cycles

Unlike energy, which flows through an ecosystem, nutrients such as calcium and nitrogen circulate within an ecosystem. To study how nutrients cycle, a team of scientists led by ecologists Herbert Bormann and Gene Likens carried out studies at Hubbard Brook in New Hampshire beginning in the 1960s. Bormann and Likens wanted to determine if rainwater removed nutrients from ecosystems. They built small dams so that they could measure how much water left the ecosystem in the stream at the base of the valley. They found that water leaving the ecosystem contained few nutrients. They concluded that the trees very efficiently prevented nutrients from leaving the ecosystem.

Having built an ecological model, Bormann and Likens were able to make predictions about what would happen if the ecosystem were disturbed. For example, they knew that nutrients such as calcium were held by the trees of the forest. Their ecological model predicted that much more calcium would be lost if the trees were cut down.

Bormann and Likens test their model of nutrient cycling

To test their ecological model, Bormann and Likens cut the trees and vegetation from one portion of the forest. For several years they monitored the levels of calcium and other nutrients in the water that drained from the forest into Hubbard Brook. The ecosystem lost the ability to retain nutrients. The runoff of calcium, for example, was six times greater than it had been before the trees were cut, as shown in **Figure 14-8**. Other studies have confirmed that minerals and other nutrients pass from organisms to their habitat and back again in delicate cycles that are easily disturbed.

Figure 14-8 Herbert Bormann and Gene Likens found that removing the vegetation from an ecosystem greatly reduced its ability to recycle nutrients such as calcium.

Bacteria play a key role in the nitrogen cycle

Organisms must have nitrogen to produce proteins and nucleic acids. **Figure 14-9** shows how nitrogen cycles through an ecosystem in the northeastern United States. As you learned in Chapter 9, most living things cannot use the nitrogen gas in the air. The two nitrogen atoms of a molecule of nitrogen gas are held together by a strong chemical bond that is difficult to break. The variety of life found on Earth is possible only because a few kinds of bacteria have enzymes that can break this strong bond. Nitrogen atoms are then free to bond with hydrogen atoms to form ammonia molecules. Conversion of nitrogen gas to ammonia is called **nitrogen fixation**. Ammonia is a form of nitrogen that plants can absorb and use to make proteins. Since animals cannot absorb nitrogen from the soil, they must obtain nitrogen by eating plants or other animals.

Nitrogen-fixing bacteria live in the soil or within the roots of plants such as peas, clover, alfalfa, beans, and alder trees. Because nitrogen is such an important nutrient, the growth of plants in ecosystems is often severely limited by shortages of nitrogen in the soil.

When an organism dies, the nitrogen in its body is released by decomposers. Animal wastes, such as dung and urine, and plant materials, such as leaves and bark, also contain nitrogen. These materials are also broken down by decomposers. Thus, decomposers play a vital role in ecosystems by returning nitrogen to the soil. Some kinds of bacteria absorb nitrogen compounds from the soil and convert them to nitrogen gas, thereby completing the nitrogen cycle.

Figure 14-9 White-tailed deer and alder play important roles in the nitrogen cycle of an ecosystem found in the northeastern United States.

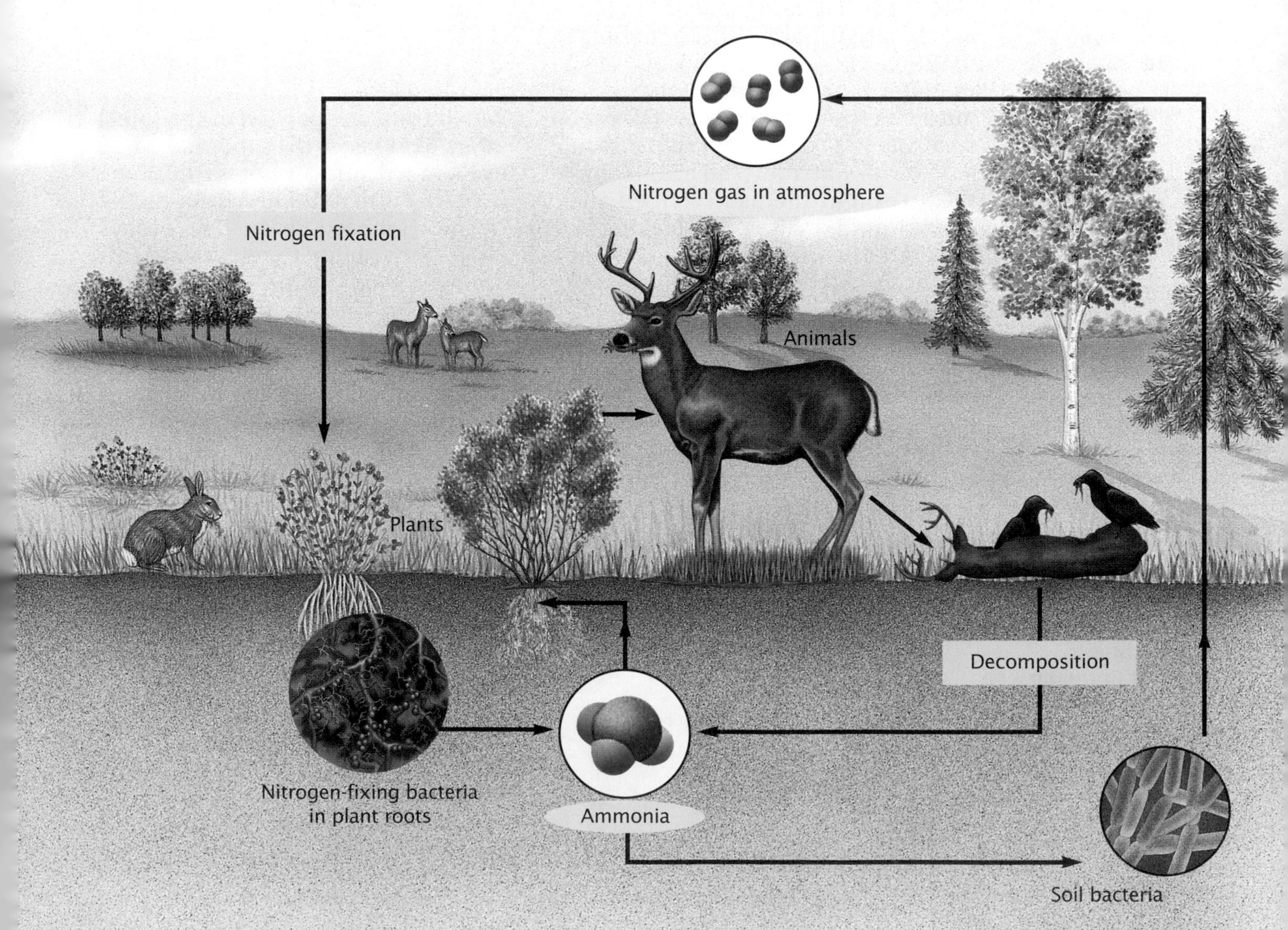

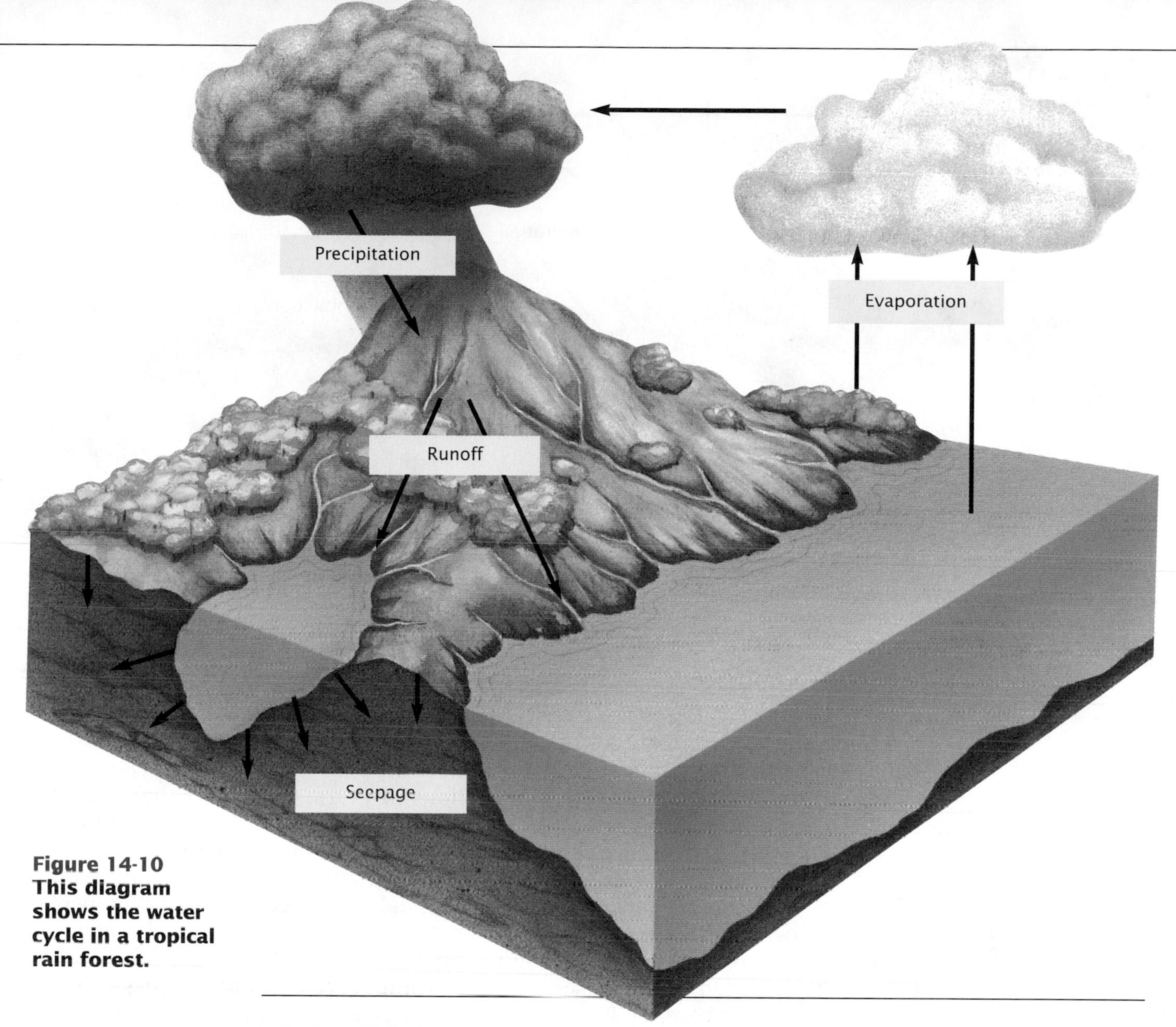

Figure 14-10
This diagram shows the water cycle in a tropical rain forest.

Water Cycle

Water is perhaps the most important nonliving component of an ecosystem. To a large degree, availability of water determines the diversity of organisms in an ecosystem. For example, fewer species live in the desert, where there is little water, than in the tropical rain forest, where water is plentiful. Water is constantly cycling within ecosystems. The water cycle is illustrated in **Figure 14-10**.

Plants play an important role in the water cycle

In tropical rain forests, where there are dense concentrations of trees and other plants, more than 90 percent of the moisture that enters the ecosystem passes through plants and evaporates from their leaves. In a very real sense, these plants create their own rain.

When forests are cut down, the water cycle is broken. Moisture cannot be returned to the atmosphere by plants. Instead, water drains into streams and rivers and eventually flows into the ocean. Moreover, without protection from the roots of trees and other plants, the soil is easily carried away by runoff. As a result, nutrient cycles also are broken. Because neither water nor nutrients can cycle in a forest ecosystem after the trees are cut down, extensive cutting can convert lush forests into deserts. Tragically, such a transformation is presently occurring in many tropical rain forests.

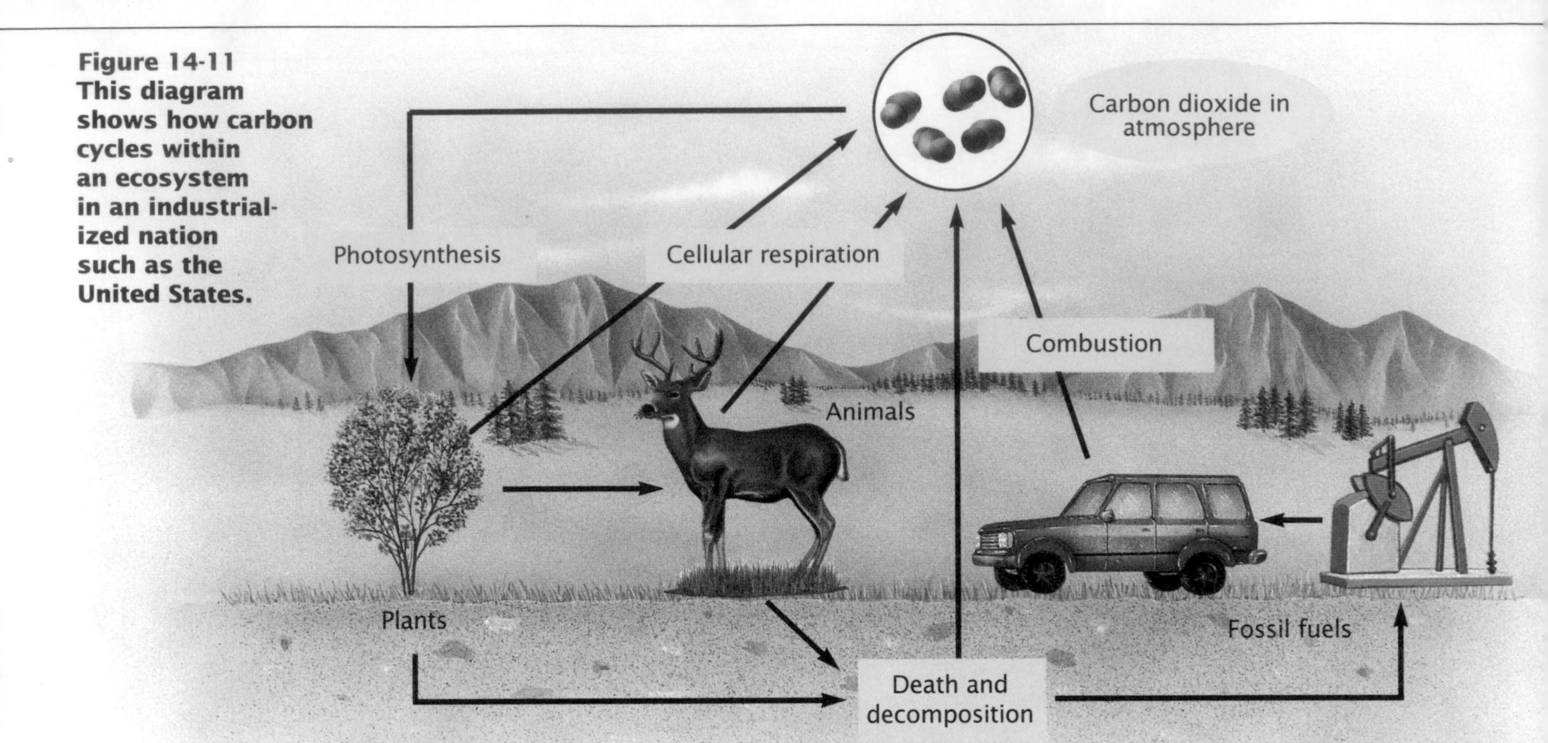

Figure 14-11 This diagram shows how carbon cycles within an ecosystem in an industrialized nation such as the United States.

Carbon Cycle

Like water, carbon also cycles between the nonliving environment and organisms. The Earth's atmosphere contains carbon in the form of carbon dioxide. Plants use carbon dioxide to build organic molecules during photosynthesis. Consumers obtain energy-rich molecules that contain carbon by eating plants or other animals. As these molecules are broken down, carbon dioxide is produced and released into the Earth's atmosphere. Cellular respiration by decomposers and photosynthetic organisms also returns carbon dioxide to the atmosphere. **Figure 14-11** shows how carbon cycles within an ecosystem.

Humans are overloading the carbon cycle

Large amounts of carbon are tied up in wood and may stay trapped there for hundreds of years, returning to the atmosphere only when the wood decomposes or is burned. Over millions of years, organisms that become buried in sediment may be gradually transformed into fossil fuels such as coal, oil, and natural gas. The carbon originally trapped by these organisms is not released back into the atmosphere until fossil fuels are burned. By burning large amounts of fossil fuels, humans are increasing the concentration of carbon dioxide in the atmosphere. Carbon dioxide traps heat from the sun within the atmosphere, much like glass panes trap the sun's heat in a greenhouse. The ability of gases such as carbon dioxide to retain the sun's heat, and in so doing to warm the atmosphere, is called the **greenhouse effect**. You will learn more about the greenhouse effect in Chapter 16.

SECTION REVIEW

1. **Explain the significance of Bormann and Likens's experiments at Hubbard Brook.**
2. **Explain the role of nitrogen-fixing bacteria in the nitrogen cycle.**
3. **How does deforestation affect the water cycle?**
4. **Describe how human interference in the carbon cycle may be causing an increase in global temperatures.**

All ecosystems are connected. The destruction of the rain forests in Brazil, for example, will affect you and everyone else in the world. If humans are going to take action to preserve the world for future generations, it is important that we understand what the major ecosystems of the world are like and how they function.

14-3 Kinds of Ecosystems

OBJECTIVES

1. **Identify the importance of plankton in freshwater ecosystems.**
2. **Identify the factors that determine the type of ecosystem found in a particular area.**
3. **Contrast the seven major biomes.**
4. **Identify the major ocean ecosystems.**

Freshwater Ecosystems

Freshwater ecosystems include lakes, ponds, and rivers. These ecosystems are very limited in area. Inland lakes cover 1.8 percent of the Earth's surface, and rivers and streams cover about 0.3 percent. Although small in total area, freshwater ecosystems support a rich array of life, including fishes, amphibians, insects, turtles, crocodiles, and many plants. A diverse biological community of microscopic organisms called **plankton** lives near the surface of lakes and ponds. Plankton contain photosynthetic organisms that are the base of aquatic food webs. All freshwater habitats are strongly connected to land ecosystems. Nutrients flow from terrestrial ecosystems into freshwater ecosystems. In addition, many land animals come to the water to feed, drink, or reproduce. **Figure 14-12** shows a freshwater ecosystem.

Ponds and lakes usually have three zones in which organisms occur: a shallow "edge" zone, an open-water surface zone, and, in deep lakes and ponds, a deep-water zone to which little light penetrates.

Figure 14-12
Freshwater ecosystems, such as this pond, support a variety of plants and animals.

Terrestrial Ecosystems

Major biological communities that occur over wide areas on land are called **biomes**. The seven major biomes are (1) tropical rain forests, (2) savannas, (3) deserts, (4) temperate grasslands, (5) deciduous forests, (6) coniferous forests, and (7) tundra. These biomes differ remarkably from one another because they evolved in areas that have very different physical characteristics. The kinds of animals and plants that live in a biome depend on the physical nature of the habitat: the soils, the terrain, and the climate. The following boxes contain information about the seven major biomes.

Tropical Rain Forests

- *Climate:*
  Warm and moist, with little variation in either rainfall or temperature
- *Yearly precipitation:*
  250 cm (100 in.)

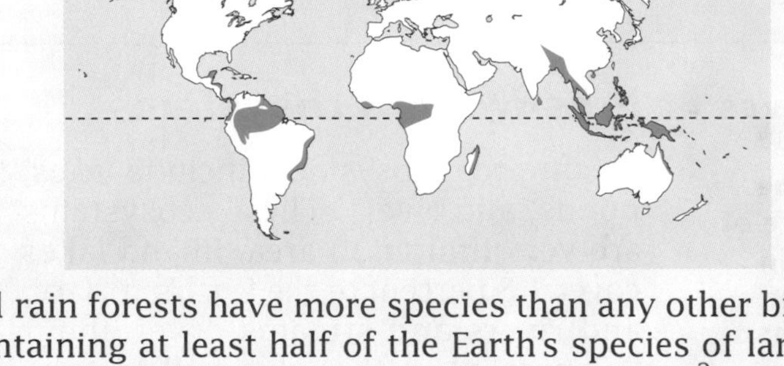

Tropical rain forests have more species than any other biome, probably containing at least half of the Earth's species of land-dwelling organisms. In a single square mile (just over 2.5 km^2) of tropical rain forest in Peru or Brazil, there may be 1,500 or more species of butterflies, twice the total number of butterfly species found in the United States and Canada combined. Although tropical rain forests are noted for lush growth, they exist mainly on quite infertile soils. Most of the nutrients are held within the plants themselves. Today tropical rain forests are being destroyed so rapidly that most will be gone within 40 years. Their destruction will lead to the extinction of large numbers of species.

Savannas

- *Climate:*
  Hot, with alternating wet and dry seasons
- *Yearly precipitation:*
  90–150 cm (36–60 in.)

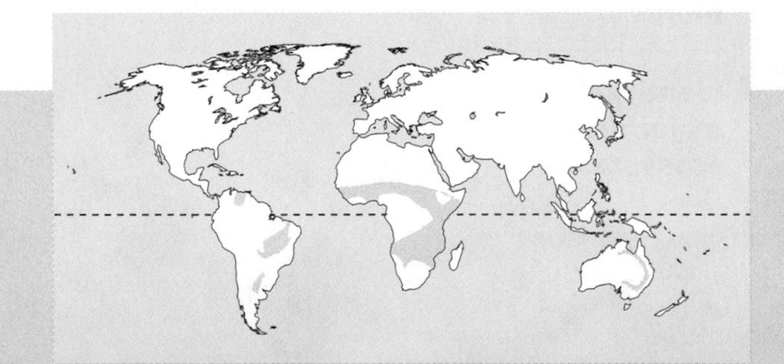

The world's great dry grasslands, called savannas, are found in tropical areas that have relatively low annual precipitation or prolonged annual dry seasons. There are wider extremes in temperature during the year in savannas than in tropical rain forests, and there is seasonal drought. These factors have led to the evolution of an open landscape with widely spaced trees. The huge herds of grazing mammals and their predators that inhabit the savannas of Africa are well known and spectacular. Typical inhabitants of the African savanna are lions, cheetahs, zebras, rhinoceroses, and gazelles.

Deserts

- *Climate:*
  Very dry, often hot
- *Yearly precipitation:*
  20 cm (8 in.)

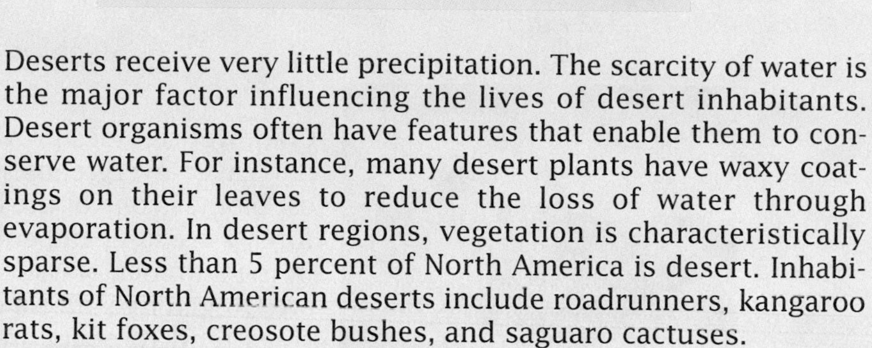

Deserts receive very little precipitation. The scarcity of water is the major factor influencing the lives of desert inhabitants. Desert organisms often have features that enable them to conserve water. For instance, many desert plants have waxy coatings on their leaves to reduce the loss of water through evaporation. In desert regions, vegetation is characteristically sparse. Less than 5 percent of North America is desert. Inhabitants of North American deserts include roadrunners, kangaroo rats, kit foxes, creosote bushes, and saguaro cactuses.

Temperate Grasslands

- *Climate:*
  Dry, hot summers and cold winters
- *Yearly precipitation:*
  10–60 cm (4–24 in.)

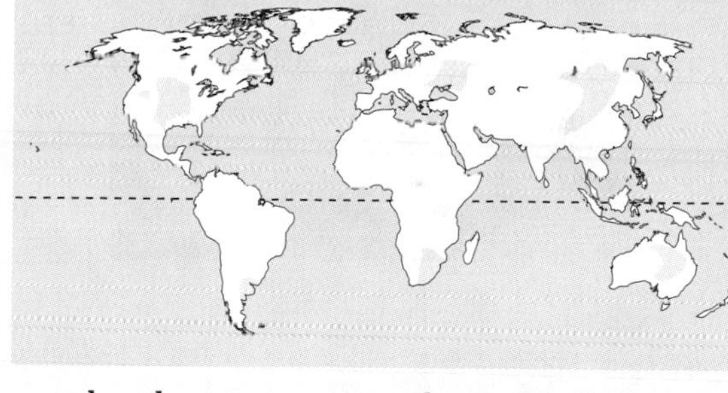

Temperate grasslands once covered much of the interior of North America and were widespread in Europe, Asia, and South America. Much of the rich agricultural land in the United States and Canada was once covered by prairie, another name for temperate grassland. Temperate grasslands are often populated by herds of grazing mammals. In North America, the prairies were once inhabited by huge herds of bison.

Temperate Deciduous Forests

- *Climate:*
  Warm summers and cool to cold winters
- *Yearly precipitation:*
  75–250 cm (30–100 in.)

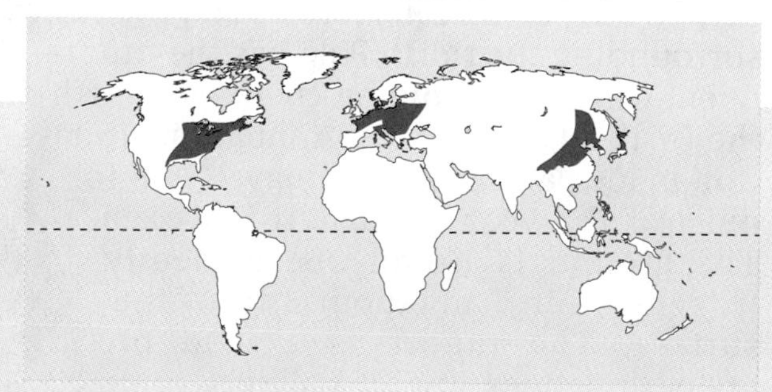

Relatively mild climates and plentiful rainfall promote the growth of deciduous forests (deciduous trees shed their leaves all at once in the fall). Precipitation in temperate deciduous forests is usually distributed fairly evenly throughout the year. In North America, deciduous forests are home to deer, bears, beavers, and raccoons. The trees are hardwoods such as oak, hickory, and beech, and shrubs and herbs grow on the forest floor.

Coniferous Forests

- *Climate:*
  Cool, short summers and cold, long winters
- *Yearly precipitation:*
  20–60 cm (8–24 in.)

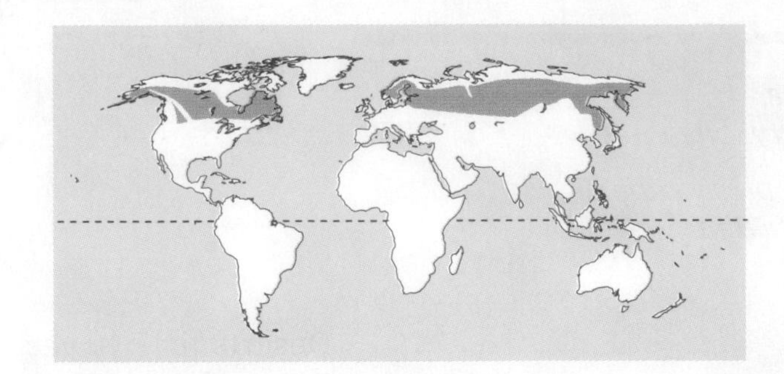

Cold, wet climates provide ideal conditions for conifers, needle-leaved evergreens such as pine trees. The coniferous forest biome is one of the largest on Earth. Winters in coniferous forests are long and cold, and most of the precipitation falls in the summer. Marshes, lakes, and ponds are common and are fringed by willows or birches. Most of the few species of trees tend to occur in dense stands of one or only a few species. Many large mammals live in this biome, including herbivores such as elk, moose, and deer. Carnivores of coniferous forests include wolves, bears, lynxes, and wolverines.

Tundra

- *Climate:*
  Cold, long winters and cool, short summers, with little precipitation
- *Yearly precipitation:*
  25 cm (10 in.)

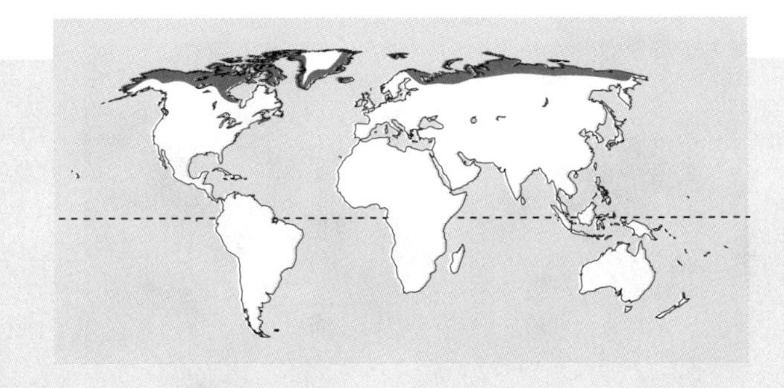

Between the coniferous forests and the permanent ice surrounding the North Pole lies the tundra. This open, wind-swept biome covers one-fifth of the Earth's land surface. Annual precipitation is low, and water is unavailable for most of the year because it is frozen. The ground is usually permanently frozen within 1 m (about 3 ft) of the surface. The tundra is a land of grasses, sedges, dwarf willows, and mosses. Animals living here include foxes, lemmings, owls, and caribou.

Environmental Science

What would it be like to . . .

- preserve places of great natural beauty for future generations?
- diagnose "sick" ecosystems and prescribe treatments to help restore their health?

These are just a few of the opportunities open to people who work in the field of environmental science.

Science/Math Career Preparation

High school
- Biology
- Chemistry

College
- Courses vary according to field of specialization

Graduate school
- Master's degree is suggested

Career Opportunities

Environmental Science: What Is It?

Environmental science focuses on one aspect of the natural world. For example, ecology and forestry are specialized environmental subjects, as are oceanography, geology, and meteorology. Traditionally, environmental scientists have worked as researchers or teachers. But today, exciting opportunities are opening for environmental scientists in many industries and businesses.

Employment Outlook:
As concerns about pollution and threatened plant and animal species heighten, opportunities in environmental science will increase significantly.

Career Focus: Dr. Carmen Cid

Ecologist at Eastern Connecticut State University

"I've been interested in biology ever since I was a little girl living in Cuba. My interest intensified in high school in New York, thanks to my high school biology teacher. We were a tough audience of inner-city teenagers, but she demanded our respect and cooperation. In college, I fell in love with the outdoors. Most of the time, class was held in nearby woods and wetlands. Having always lived in large, crowded cities, this class gave me my first opportunity to visit the unspoiled environment of a forest. My attitude toward nature changed when I began studying plants as living, growing, competing organisms."

What I Do: "Being an ecologist is like being a detective—you need to study a scene, search for clues, and test your ideas with investigations. I design experiments to discover how different species coexist in a forest. One project examined a species of short-lived annual plant that grew in patches throughout forests. Why did they grow in isolated, scattered clusters? Once I discovered the answers to such questions, I could solve the puzzle of species interaction in the forest."

To Find Out More

1. For further information, write to:
   Ecological Society of America
   Center for Environmental Studies
   Arizona State University,
   Tempe, AZ 85287-3211
2. Visit your school counselor to find out what universities offer programs in environmental science. Review the prerequisites. Make a poster showing the range of job opportunities in this field.
3. Find articles that describe an environmental issue being debated in your community. Prepare a report that describes the area in conflict and causes for concern. With your class, make a list of suggestions to help solve the problem.

Ocean Ecosystems

Nearly three-quarters of the Earth's surface is covered by ocean. Three types of ocean ecosystems are shown in **Figure 14-13**. Shallow ocean waters are small in area but contain most of the ocean's diversity. Many fishes swim in the open ocean surface, feeding on plankton. Photosynthetic plankton account for about 40 percent of all photosynthesis on Earth. There is increasing evidence that pollution is harming photosynthetic plankton. If significant numbers of these organisms are destroyed, the oxygen you breathe will be slowly depleted from the Earth's atmosphere. The deep ocean waters are cold and dark. Among the few residents of the deep ocean are some of the most bizarre organisms found on Earth. Many organisms in the deep ocean have light-producing body parts that they use to attract mates or lure prey. Because photosynthesis cannot occur in the deep ocean, most of these organisms prey on other deep-sea residents or scavenge the dead bodies of organisms that have drifted down from above.

Figure 14-13 Each type of ocean ecosystem supports different kinds of organisms. This representation is not drawn to scale.

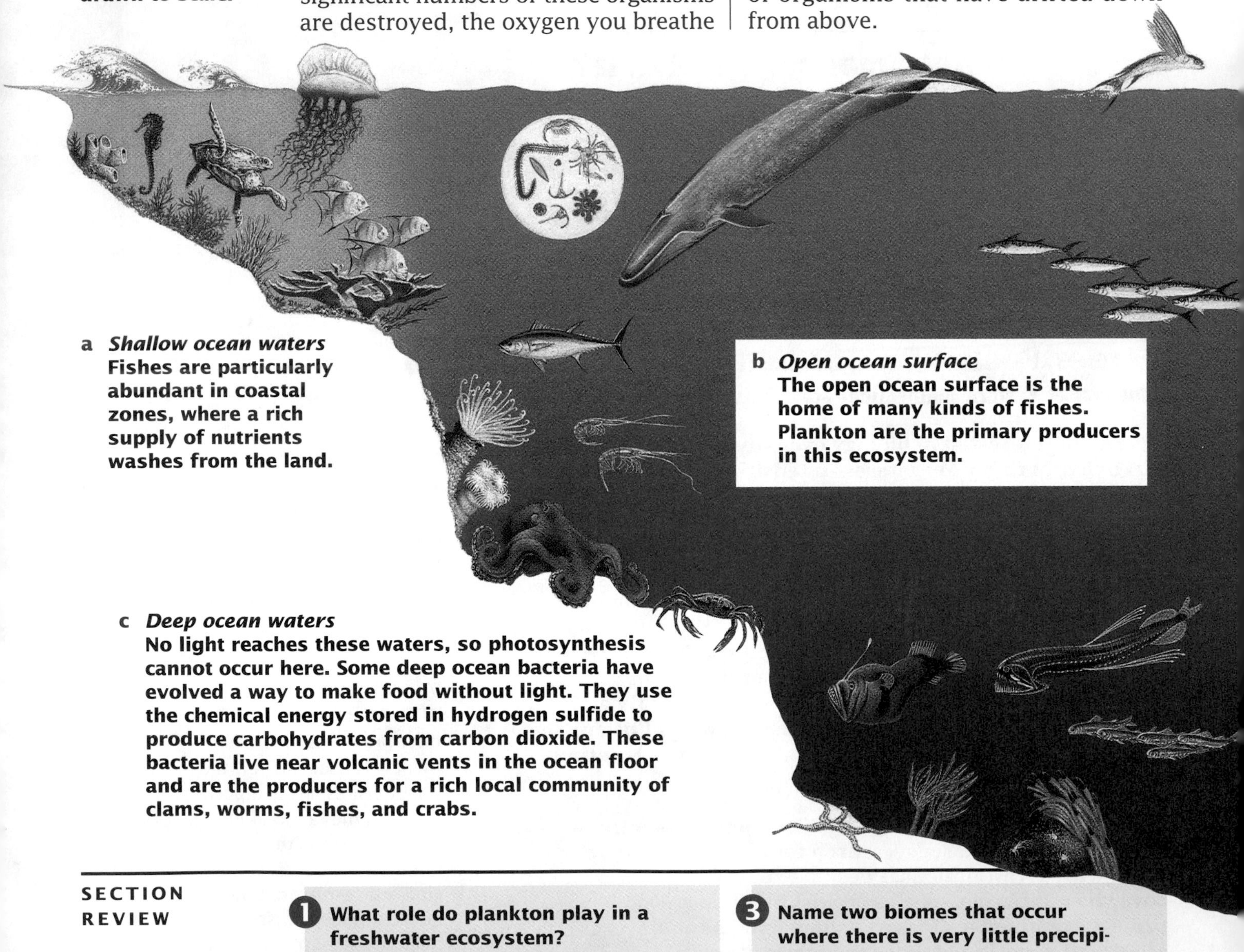

a *Shallow ocean waters*
Fishes are particularly abundant in coastal zones, where a rich supply of nutrients washes from the land.

b *Open ocean surface*
The open ocean surface is the home of many kinds of fishes. Plankton are the primary producers in this ecosystem.

c *Deep ocean waters*
No light reaches these waters, so photosynthesis cannot occur here. Some deep ocean bacteria have evolved a way to make food without light. They use the chemical energy stored in hydrogen sulfide to produce carbohydrates from carbon dioxide. These bacteria live near volcanic vents in the ocean floor and are the producers for a rich local community of clams, worms, fishes, and crabs.

SECTION REVIEW

1 What role do plankton play in a freshwater ecosystem?

2 List two reasons why tropical rain forests do not occur in the United States.

3 Name two biomes that occur where there is very little precipitation.

4 Describe how organisms living deep in the ocean obtain their food.

CHAPTER

14 Highlights

If the tropical rain forests disappear, so will this butterfly from Trinidad.

14-1 What Is an Ecosystem?

Decomposers, such as these mushrooms, are an important part of every ecosystem.

Key Terms

- ecology (p. 254)
- community (p. 254)
- habitat (p. 254)
- ecosystem (p. 254)
- diversity (p. 254)
- producer (p. 256)
- consumer (p. 256)
- decomposer (p. 256)
- trophic level (p. 257)
- autotroph (p. 257)
- herbivore (p. 257)
- carnivore (p. 257)
- omnivore (p. 257)
- heterotroph (p. 257)

Summary

- Ecology is the study of how living things fit into their environment.
- An ecosystem is a group of interacting organisms and their physical environment.
- Producers capture energy and store it in complex molecules. Consumers obtain energy by feeding on producers or other consumers. Decomposers eat dead organisms, animal wastes, fallen leaves, twigs, and other debris.
- Autotrophs, such as plants, are organisms that make their own food. Heterotrophs cannot make their own food and must obtain it by eating other organisms.

14-2 Cycles Within Ecosystems

Plants play an important role in the water cycle.

Key Terms

- nitrogen fixation (p. 260)
- greenhouse effect (p. 262)

Summary

- Materials such as water, nitrogen, and carbon move through ecosystems in cycles.
- Some bacteria absorb nitrogen gas and convert it to ammonia. The process of transforming nitrogen gas into ammonia is known as nitrogen fixation.
- The burning of fossil fuels releases large amounts of carbon dioxide into the atmosphere. Carbon dioxide in the atmosphere retains heat from the sun, a phenomenon known as the greenhouse effect.

14-3 Kinds of Ecosystems

The tundra is one of the seven major biomes.

Key Terms

- plankton (p. 263)
- biome (p. 264)

Summary

- Photosynthetic plankton are the basis of the food web in aquatic ecosystems.
- On land, there are seven major types of communities, which are called biomes.
- There are three major types of ecosystems found in the ocean.

CHAPTER 14

review

Understanding Vocabulary

1. For each set of terms, identify the term that does not fit and explain why.
   a. producers, consumers, herbivores, omnivores
   b. water, energy, carbon, nitrogen
   c. photosynthesis, cellular respiration, decomposition, combustion

2. For each set of terms, complete the analogy.
   a. organisms that live in a particular place : community :: physical location of a community : ____
   b. Plants and other organisms that make their own food : autotrophs :: organisms that cannot make their own food : ____
   c. movement of energy through an ecosystem in a complex network of feeding relationships : food web :: a group of organisms whose energy sources are the same number of steps away from the sun : ____

Understanding Concepts

3. **Relating Concepts** Construct a concept map that describes energy flow in ecosystems. Try to include the following terms in your concept map: ecological pyramid, producers, community, ecosystems, habitat, consumers.

4. **Identifying Components** What is an ecosystem? Describe its two basic components.

5. **Summarizing Information** What is an ecological model? How are ecological models used by ecologists?

6. **Comparing Functions** Differentiate between the roles of producers and consumers in the transfer of energy in an ecosystem. Give an example of each type of organism.

7. **Inferring Conclusions** Figure 14-7 shows an ecological pyramid for a very simple ecosystem. Suppose that sheep were introduced into this ecosystem. Explain some possible effects of this introduction on each species in the ecosystem.

8. **Recognizing Relationships** Explain why most ecosystems usually can contain no more than five trophic levels.

9. **Analyzing Methods** Bormann and Likens first determined how efficiently nutrients were recycled in an undisturbed ecosystem. Explain how this ecosystem served as a control group for their experiment.

10. **Organizing Information** Imagine an atom of nitrogen that is in a protein in the leaf of a plant. Trace the steps of the nitrogen cycle this atom must pass through in order to reach the atmosphere.

11. **Recognizing Relationships** What role do plants play in the water cycle in a tropical rain forest? How do the plants affect the climate of the forest?

12. **Summarizing Information** What is the greenhouse effect? How has the burning of fossil fuels changed the atmosphere? What is the consequence of this change?

13. **Recognizing Relationships** Evaluate this statement: Photosynthetic plankton play the same role in aquatic ecosystems as plants do in terrestrial ecosystems.

14. **Summarizing Information** Copy the following table onto a sheet of paper. Complete the table by filling in the missing entries.

Biome	Annual precipitation	Found in United States?
Tropical rain forest		
	10–60 cm	Yes
Desert		
	90–150 cm	No

15. **Recognizing Relationships** What effect do photosynthetic marine plankton have on the composition of the atmosphere?

16. **BUILDING ON WHAT YOU HAVE LEARNED** In Chapter 5 you learned about photosynthesis and cellular respiration. Using what you learned in this chapter, explain this observation about the carbon cycle. Plants produce and use carbon dioxide, while animals only produce carbon dioxide.

Interpreting Graphics

17. Look at the food chain of a marine ecosystem shown below.

a. What is the producer in this ecosystem?
b. Which trophic level is the killer whale in?
c. Which organism should be least common? Explain your answer.

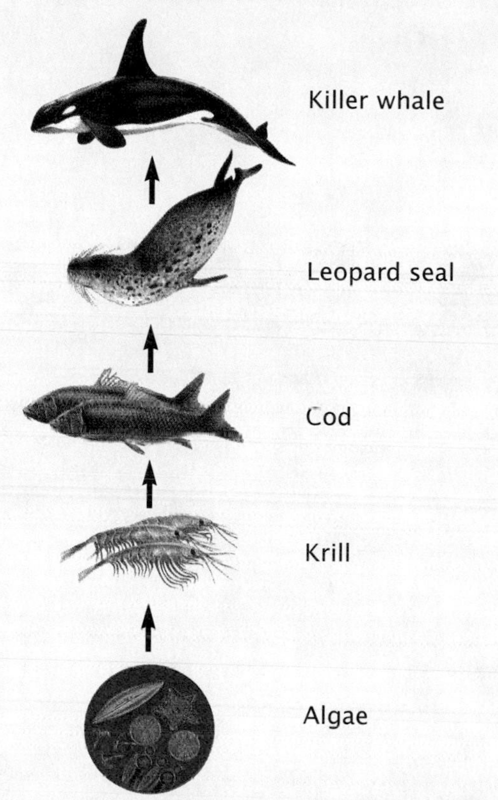

Reviewing Themes

18. ***Interdependence***
If destruction of tropical rain forests continues at its present rate, most of the world's rain forests will be gone within 40 years. What are two likely effects of the destruction of tropical rain forests?

Thinking Critically

19. **Evaluating Methods** Ecologist often use models to help them understand ecosystems. What is a model? Why are models useful to ecologists? What are some possible drawbacks of models?

Life/Work Skills

20. **Finding Information** In the last few decades, scientists have noticed dramatic declines in the populations of some kinds of amphibians. Use library resources or search an on-line database to find out more about disappearing amphibians. What amphibians are affected, and where do they live? What are some of the hypothesized causes of the decline in amphibian populations? Write a report that summarizes what you have learned.

21. **Using Technology** Computer mapping, remote sensing, and satellite imaging are important tools for evaluating the health of the environment. Use library references or search an on-line database to find out how these technologies are helping scientists monitor the environment. What agencies or businesses would require the kind of information that these technologies provide? Write a report that summarizes your findings. Try to find a map created by one of these methods that shows an area around your city or town.

Activities and Projects

22. **Math Connection** Scientists introduced a breeding pair of one species into an ecosystem. This species doubles its population size every year. Another species, native to the ecosystem, maintains its numbers at 500 individuals from one year to the next. How many years will it take for the introduced species to outnumber the native species?

Discovering Through Reading

23. Read the article "Why Trees Need Birds" by Sharon Begley in *National Wildlife*, August/September 1995, pages 42–45. What idea did Robert Marquis and Christopher Whelan set out to test? Describe their experiments. What is the importance of their discovery? What are the two main explanations for the decreasing numbers of some songbirds?

24. Read the article "Collapse of a Food Chain" by Kathy A. Svitil in *Discover*, July 1995, pages 36–37. What is the cause of fewer fish and birds in California's coastal waters? What is the cause of the decline in the zooplankton population?

CHAPTER 15

REVIEW
- natural selection (Section 10-3)
- structure of ecosystems (Section 14-1)
- diversity (Section 14-1)
- biomes (Section 14-3)

How Ecosystems Change

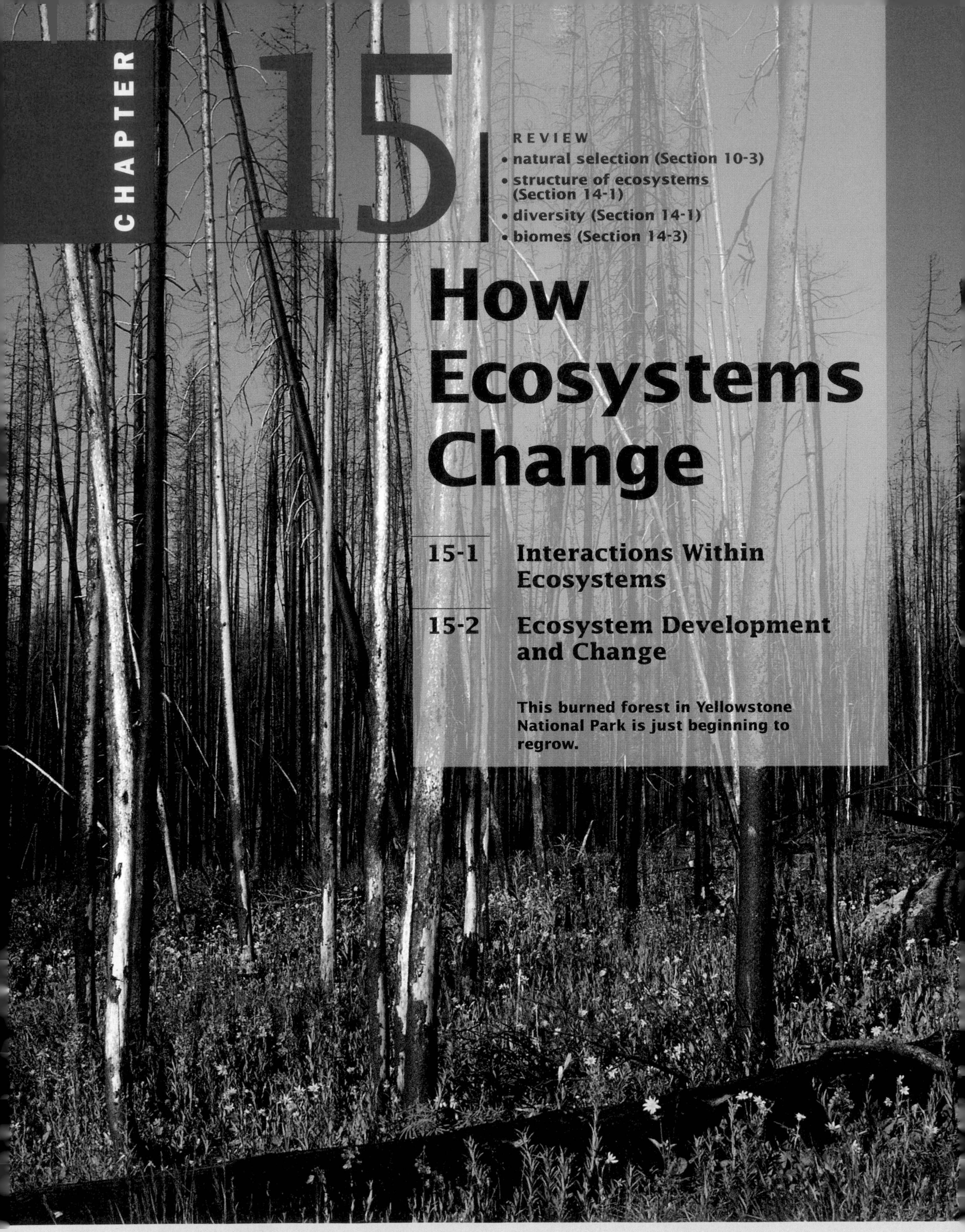

This burned forest in Yellowstone National Park is just beginning to regrow.

Every ecosystem on Earth, whether it is frozen tundra or tropical rain forest, is a complex network of interacting species. To preserve the Earth's fast-disappearing ecosystems, it is essential to understand the nature of these interactions and how they are shaped by natural selection and the physical environment.

15-1 Interactions Within Ecosystems

OBJECTIVES

1. **Recognize the role of coevolution in shaping the structure of ecosystems.**
2. **Relate the characteristics of flowers to the coevolution of flowering plants with insects.**
3. **Describe how plants and herbivores have coevolved.**
4. **Contrast parasitism, mutualism, and commensalism.**

Evolution and Ecosystems

Species evolve in response to the challenges posed by their environments. As a result, animals, plants, and other organisms in an ecosystem have characteristics that fine-tune them for living where they do. For example, many plants in desert ecosystems have thick, waxy coatings on their leaves that help them retain water. For the same reason, desert animals often hide underground during the hottest part of the day; an animal's behavior is just as much an adaptation to its environment as are its physical characteristics.

An organism's survival and reproduction also depend on interactions with other living members of its ecosystem, including members of other species. **Figure 15-1** shows an interaction between members of two species of fish. Species evolve not only in response to the physical environment but also in response to other species. For instance, many plant species have evolved tough leaves that protect them against being eaten by herbivores. Of course herbivores evolve too, and many have evolved flatter, larger teeth that are better suited to grinding the very tough leaves they eat. Cows and horses have teeth such as these. **Coevolution** occurs when two or more species evolve in response to each other.

Figure 15-1
The cleaner wrasse removes parasites from the gills and mouths of larger fish. The wrasse benefits by gaining a food source, and the fish it cleans benefit by having harmful parasites removed.

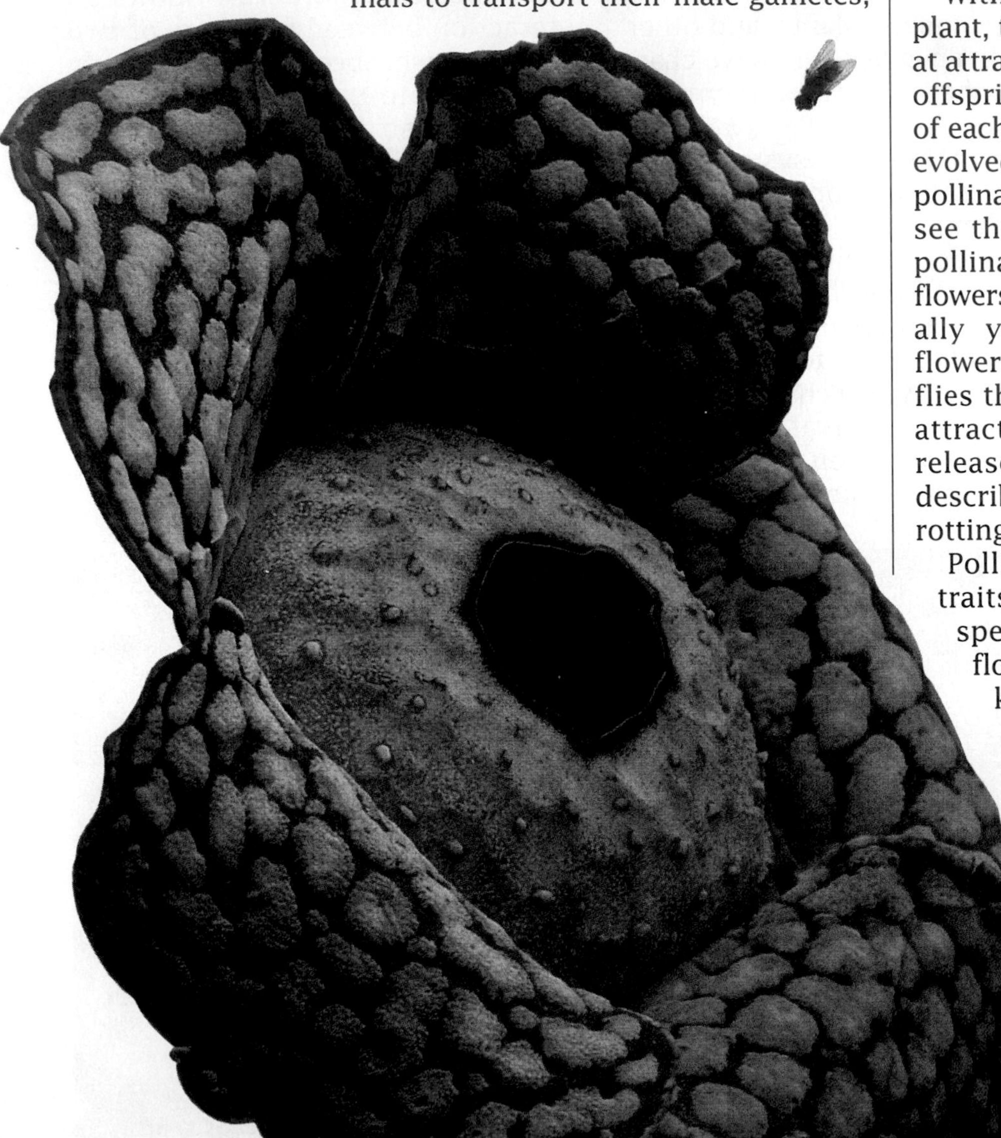

Figure 15-2
The odor of this giant *Rafflesia* flower attracts flies. The flies will pollinate it and then transfer pollen from this flower to another *Rafflesia* flower.

Coevolution Shapes Species Interactions

Coevolution shapes the many ways that animals and plants interact within ecosystems, creating a complex web of interactions among the organisms in each ecosystem. For example, producers, herbivores, and carnivores in a mature forest ecosystem have adjusted to one another over long periods of evolutionary time.

One of the most dramatic examples of coevolution is the side-by-side evolution of flowering plants with insects. Plants are immobile and cannot actively search for mates. Instead of moving to seek a mate, many species of flowering plants rely on animals to transport their male gametes, which are located within grains of pollen. Insects are attracted to a plant's flowers by their colors, odors, and the food that they provide, which is usually sweet nectar. As the insect feeds, pollen attaches to its body. After feeding at the flowers of one individual, the pollen-covered insect flies to flowers of another individual. At this next feeding stop, some of the insect's load of pollen may be rubbed onto the female reproductive structures of the flowers. Fertilization is now able to take place. Animals that carry pollen from flower to flower are known as pollinators.

Within each species of flowering plant, those individuals that are better at attracting pollinators will leave more offspring. Thus, the attractive features of each species of flowering plant have evolved to match the preferences of its pollinators. For instance, bees cannot see the color red, and plant species pollinated by bees rarely have red flowers. Bee-pollinated flowers are usually yellow or blue. The *Rafflesia* flower in **Figure 15-2** is pollinated by flies that feed on dead organisms. To attract its pollinators, this flower releases a scent that has been described as resembling the stench of rotting flesh.

Pollinators, in turn, have evolved traits that enable certain species to specialize on particular species of flowers when seeking nectar. Some kinds of honeysuckles are pollinated by hawk moths and produce nectar at the bottom of very long, tube-shaped flowers. Hawk moths can reach the bottom of the flower with their long tongue.

Avoiding Being Eaten: Plants and Herbivores

Being eaten is not beneficial for plants. Herbivores can kill plants by feeding on them, just as carnivores kill their prey. Therefore, characteristics that enable plants to protect themselves from herbivores are favored by natural selection. The biological structures of many of the Earth's ecosystems have been determined largely by the ways plants avoid being eaten and herbivores succeed in eating them.

Figure 15-3
The spines of this golden barrel cactus protect it from herbivores.

Plants defend themselves from herbivores

Touch the stem of a rosebush and you might experience a familiar plant defense against herbivores. Many species of plants, such as roses and the cactus in **Figure 15-3**, employ physical defenses such as thorns, prickles, sticky hairs, and tough leaves. The most crucial plant defenses, however, are chemical. Virtually all plant species produce chemicals that protect them against herbivores. Some of these defensive chemicals are poisons that kill the animal that eats the plant, while other chemicals simply make the plant taste bad. Most herbivores learn to avoid plants that have defensive chemicals. Anyone who has broken out in a rash from contact with poison ivy, shown in **Figure 15-4**, has been the victim of plant chemical defenses. Poison ivy produces a substance that causes severe blistering in many people.

As a rule, each group of closely related plant species has a unique battery of chemical defenses. The plants of the mustard family, for instance, produce a group of defensive chemicals called mustard oils. Mustard oils are the source of the pungent aromas and tastes characteristic of such plants as mustard, cabbage, radish, capers, and horseradish. The same tastes that we enjoy signal the presence of chemicals that are toxic to many groups of insects.

Many herbivores overcome plant defenses

Over time, some herbivores have evolved ways to overcome the chemical defenses of plants. Because different plant species produce different chemicals, coevolution has resulted in a very specialized pattern of feeding; certain kinds of herbivores feed exclusively on particular kinds of plants.

Cabbage butterflies provide a good example of how counteractive measures have evolved in response to the chemical defenses of plants. Although most insects avoid plants of the mustard family, the caterpillars of cabbage butterflies eat these plants voraciously. These caterpillars are able to eat plants of the mustard family because they have evolved the ability to break down mustard oils into harmless chemicals. As a result of this evolutionary breakthrough, cabbage butterflies have been able to use a new food resource—plants of the mustard family—without competition from other insect herbivores.

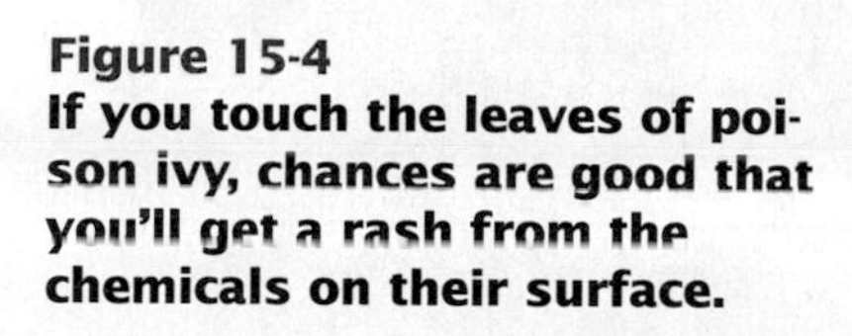

Figure 15-4
If you touch the leaves of poison ivy, chances are good that you'll get a rash from the chemicals on their surface.

Three Types of Close Species Interactions

Three types of species interactions involve particularly close relationships among the participants. **Symbiosis** is a close, long-term association between two or more species. The three types of symbiotic relationships are parasitism, mutualism, and commensalism.

Evolution

Variation is essential for natural selection. What role did variation play in the long-term survival of rabbits in Australia?

Parasites and their hosts coevolve

Worldwide, between 200 million and 300 million people suffer from malaria. People who have malaria play host to a single-celled parasite that was injected into their blood by the bite of a mosquito. Parasites obtain nutrition by feeding on their host. How is a parasite different from a predator? A parasite usually does not kill its host, and it is usually smaller than the organism on which it feeds. The relationship between a parasite and its host is called **parasitism**.

Both host and parasite can coevolve in response to each other. In 1859, 12 rabbits were introduced to Australia. No rabbits occur naturally on the Australian continent. By the 1940s there were millions of rabbits swarming the countryside. To control the rabbit population, the Australian government introduced the viral disease myxomatosis *(mihk suh muh TOH suhs)* from South America. At first, myxomatosis was very deadly to the rabbits. In some areas, more than 90 percent of the rabbits were killed by the virus, and nearly all infected individuals died. Soon, however, scientists monitoring the introduction noticed that the disease was becoming less deadly.

Tests on the rabbits and the virus showed that both were evolving. Because rabbits resistant to myxomatosis were more likely to survive and reproduce, the rabbit population contained a greater proportion of resistant individuals after each generation. Also, instead of becoming more deadly to overcome the rabbit's resistance, the virus had actually become less virulent, or deadly. If a virus kills its host rabbit too quickly, that rabbit cannot spread the virus to other rabbits. A virus that "allows" its hosts to live longer is able to infect more hosts. However, some recent evidence suggests that the virus's virulence may be increasing.

All parties benefit in mutualism

Not all coevolution involves antagonistic relationships, such as those between plants and herbivores or between parasites and their hosts. **Mutualism** is a symbiotic relationship in which all participating species benefit. For example, a lichen, such as the British soldier lichen in **Figure 15-5**, is a mutualistic partnership between a fungus and a green alga or a cyanobacterium. The fungus absorbs nutrients for both partners from the surface on which the lichen is growing. The alga or cyanobacterium carries out photosynthesis to provide food for itself and its fungal partner. Mycorrhizae, which you read about in Chapter 11, are mutualistic associations between plants and fungi. In Chapter 14, you learned about

Figure 15-5
The British soldier lichen is composed of a fungus and an alga living in a mutualistic partnership.

Figure 15-6
Stinging anemones (the yellow, flowerlike objects) live on the claws of this female boxing crab. She guards herself and the red eggs she is carrying by using the anemones like boxing gloves, jabbing them at attacking predators. This is an example of a commensal relationship.

the relationship between nitrogen-fixing bacteria and the roots of plants such as peas and beans. These bacteria provide the plants with a source of nitrogen. In exchange, the nitrogen-fixing bacteria receive a place to live (swellings on the plant roots) and sugars produced by the plant.

Mutualism also has played an important role in the coevolution of humans and the microbes that live in our intestines. Within the human large intestine live immense colonies of the bacterium *Escherichia coli*. These bacteria have ready access to food while providing us with vitamin K, which is necessary for blood to clot. Animals that lack these bacteria, such as birds, must consume food that contains the necessary amounts of vitamin K.

Commensalism is taking without harming

Commensalism is an ecological relationship in which one species benefits and the other is not obviously affected. A very intriguing example of commensalism is the boxing crab, which is described in **Figure 15-6**. The crab benefits from the protection anemones provide, and the anemones apparently are not harmed or helped.

The crusty growths seen on the back of the gray whale in **Figure 15-7** are actually small animals called barnacles. Barnacles hitch a ride on the whale. In doing so, they gain protection from predators and transportation to new sources of food (tiny animals they filter from the water). Apparently, the whale neither benefits nor is harmed by the presence of barnacles.

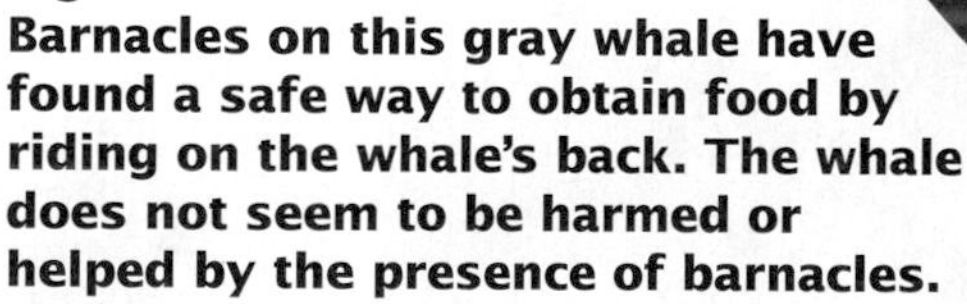

Figure 15-7
Barnacles on this gray whale have found a safe way to obtain food by riding on the whale's back. The whale does not seem to be harmed or helped by the presence of barnacles.

SECTION REVIEW

1. **Describe one example of coevolution.**
2. **Do you think that the drab flowers of some grasses and trees are pollinated by insects? Explain.**
3. **How have caterpillars and their food plants coevolved?**
4. **How could you determine whether the relationship between two species is mutualism or commensalism?**

Our effect on the environment now extends to all parts of the globe, to every living thing in the world. Whether the world's ecosystems can survive depends on how they respond to human interference. The relationships among organisms in an ecosystem determine how the ecosystem responds to change.

15-2 Ecosystem Development and Change

OBJECTIVES

1. **Describe the components of an organism's niche.**
2. **Describe the process of succession.**
3. **Relate the stability of an ecosystem to its diversity, and explain two factors that affect diversity.**
4. **Explain the three main ways that humans disrupt ecosystems.**

Ecosystem Lifestyles

Changes are a natural part of the history of any ecosystem. However, to save ecosystems threatened by human interference, it is essential to first understand that every organism in an ecosystem plays a role in the ecosystem. An organism might be prey for one species and predator of another. The sum of an organism's interactions with its physical environment and with other organisms is its **niche**.

A niche describes how an organism lives, its role in the ecosystem. A niche can be described as the position a species occupies in the movement of energy through the ecosystem. The niche of grass growing in a meadow, for example, is that of producer, while the niche of deer that eat the grass is that of herbivore. The niche of an organism also includes the climate it prefers, the time of day it feeds, the time of year it reproduces, what it likes to eat, and where it finds its food. Each species in an ecosystem, like the jaguar shown in **Figure 15-8**, has a unique niche. The total niche that an organism could potentially use within an ecosystem is that organism's **fundamental niche**.

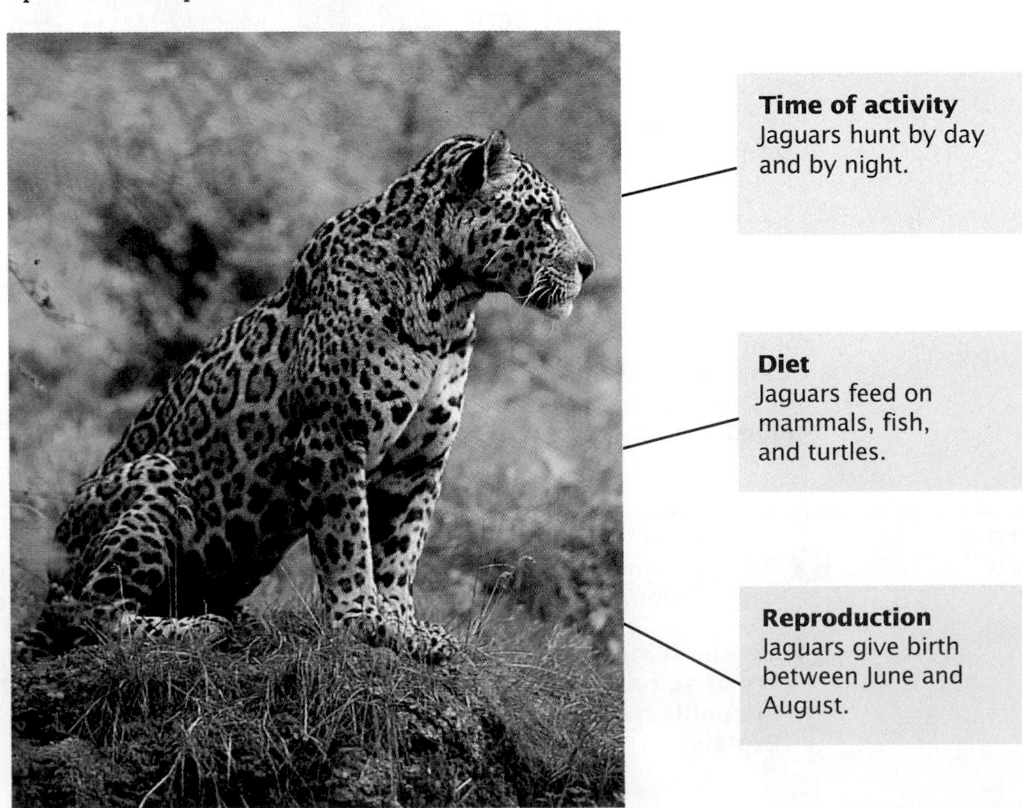

Time of activity
Jaguars hunt by day and by night.
Diet
Jaguars feed on mammals, fish, and turtles.
Reproduction
Jaguars give birth between June and August.

Figure 15-8
Every organism has a niche, or role, in its ecosystem. The jaguar's niche includes all the ways it interacts with its environment. This diagram shows just three aspects of the jaguar's niche.

Figure 15-9

a The barnacle *Chthamalus stellatus* can live in both shallow and deep water on a rocky coast. These areas are its fundamental niche.

b The barnacle *Balanus balanoides* prefers to live in deep water, which is its fundamental niche.

c When the two barnacles live together, *Chthamalus* is restricted to shallow water, its realized niche. What is the realized niche of *Balanus*?

Competing Organisms Coevolve

Sometimes organisms are unable to occupy their entire fundamental niche because another species already occupies part of it. Along the coast of Scotland, for example, the barnacle *Chthamalus* is able to live on rocks in both shallow and deep water. It is not usually found in deep water, but another species of barnacle, *Balanus*, does live at these depths, as shown in **Figure 15-9**. In the 1950s, ecologist Joseph Connell demonstrated that this pattern was the result of competition between the two species. *Balanus* individuals grow faster and are larger as adults. Connell showed that *Chthamalus* individuals could live on rocks in deep water if he removed nearby *Balanus* individuals. If *Balanus* individuals were not removed, they grew over *Chthamalus* individuals or crowded them off the surface of the rocks.

Situations in which two or more organisms attempt to use the same scarce resource are called **competition**. The two kinds of barnacles in Connell's experiment competed for living space. Competition often prevents an organism from occupying all of its fundamental niche. That part of a fundamental niche that an organism actually occupies as a direct result of competition is called its **realized niche**. The fundamental niche of the barnacle *Chthamalus* extended down to deep water. Its realized niche was restricted by competition to shallow water.

Competition can cause changes in an ecosystem

If two species compete intensely for the same resource, one species usually wins. The losing species may be eliminated from the ecosystem. The process in which one species is outcompeted and dies out within an ecosystem is called **competitive exclusion**. Competitive exclusion is rare in most ecosystems, however, because natural selection tends to favor evolutionary changes that decrease competition. As a result, competing species reduce their use of common resources, and their niches become less similar.

In the 1950s the Princeton ecologist Robert MacArthur showed how potential competitors often compete very little because of subtle differences in their niches. MacArthur studied five species of warblers (small insect-eating songbirds) that fed in the same spruce trees. Each species spent most of its time feeding in a different part of the tree and concentrated on a different part of the available resources. As a result, competition among the species was reduced. The result of this pattern of resource use is that the ecosystem supports many more ways of life, which makes the ecosystem more complex.

Competition and Ecosystem Development

Competition plays an important role in how ecosystems develop. The role of competition in the development of ecosystems is most easily seen when a serious disruption creates a new habitat. New habitats are formed when a volcano forms a new island, when a glacier recedes and exposes bare soil, or when a fire burns all the vegetation in an area. In every case scientists have been able to study, the empty habitat is quickly occupied. The first organisms to move into a burned patch of forest are small, fast-growing plants—you would probably call them "weeds." On bare rock, the early "settlers" are mosses and lichens, which are able to eke out a living under harsh conditions.

Figure 15-10
a This meadow is in an early stage of succession.

b Later in succession, trees and shrubs move in.

c After many years, a forest of deciduous trees becomes established.

Competition drives change in a developing ecosystem

The initial colonists do not remain in the ecosystem very long, however, because their pioneering efforts soon make the habitat more hospitable. As a result of this improvement, later arrivals soon outcompete and replace the original inhabitants. This second wave of immigrants is replaced in turn by still other species that are better able to survive in the new environment. As the ecosystem matures, niches become more and more finely subdivided, and species become more and more interdependent. Hence, the diversity of the ecosystem usually increases as the ecosystem matures.

The regular progression of species replacement in a developing ecosystem is called **succession**. When succession takes place on land where there was no previous growth, it is called **primary succession**. When it occurs in areas where there has been previous growth, such as in abandoned fields or forest clearings, it is called **secondary succession**. **Figure 15-10** shows an example of secondary succession. Succession does not continue indefinitely. Eventually, if the ecosystem is undisturbed for a long time, a community that is resistant to change results. Although no two episodes of succession are exactly alike, the progression of species replacement tends to result in similar communities developing in similar physical conditions. That is why biomes such as tropical rain forests are so similar wherever they occur.

Ecosystem Stability

Why does succession stop? The last stage in a series of successional stages is able to absorb disruption without major change better than its predecessors could. Early successional stages, for instance, are easily changed by the invasion of new species. In contrast, the final stage of succession is able to resist invasion by potential competitors. The ability of an ecosystem to resist change in the face of disturbance is known as **stability**. During succession, early stages show low stability, later stages are more stable, and the final community is the most stable.

Figure 15-11
a The sea star in the photograph above is prying open a mussel. The sea star is a keystone species in a 15-species ecosystem that was studied by Robert Paine.

What factors promote stability?

What makes some ecosystems more stable than others? Most ecologists now agree that more diverse ecosystems tend to be more stable than less diverse ecosystems. A more diverse ecosystem contains a more complex web of interactions among species than a less diverse ecosystem does. Alternate links in the web of species interactions are more likely to be available to compensate for disruptions such as the loss of a species.

Even ecosystems that are very diverse contain points of vulnerability, however. A species whose niche affects many others in the ecosystem and that cannot be readily replaced if lost is called a **keystone species**. Because keystone species are the focus of many biological interactions, these species represent points where the web of species interactions can come unraveled.

In the 1960s, ecologist Robert Paine of the University of Washington discovered an excellent example of a keystone species. Paine worked on a 15-species ecosystem along the Washington coast. As described in **Figure 15-11**, when he removed all the sea stars from this ecosystem, one species of mussel that the sea star ate began to thrive and outcompeted many other species in the ecosystem. The number of species in this ecosystem fell from 15 to 8. The sea star in Paine's ecosystem was a keystone species.

To preserve natural ecosystems, it is essential to promote their biological diversity. It is very important to realize that diverse ecosystems can be damaged if key species are lost.

b When the sea stars were removed, the mussels thrived. They out-competed other inhabitants of the ecosystem, thereby reducing the total number of species in the ecosystem from 15 to 8.

Why Are Some Ecosystems More Diverse Than Others?

What determines the diversity of an ecosystem? Two key factors are important: the size of the ecosystem and its latitude (distance from the equator).

Larger ecosystems support more species

Ecosystems that occupy large areas and are not subdivided into isolated patches will usually contain a wider variety of physical habitats than small ecosystems. Large ecosystems therefore usually support more species than small ecosystems, as shown in **Figure 15-12**. Geography often acts to restrict the size of an ecosystem. An island forest can be only as big as the island. Even on the continents, terrain or human activity can limit the size of an ecosystem. A river can isolate the inhabitants on one side from those on the other side, and so can a road or a fence. These divisions have little effect on organisms that can readily cross them (for example, eagles that fly over fences), but they serve to isolate organisms that cannot (deer that are unable to jump high fences).

Reducing the area of an ecosystem reduces the variety of physical habitats it contains. Thus, the number of species the ecosystem can support declines. Today human activity is causing ecosystems to shrink or disappear, resulting in the extinction of many species.

Figure 15-12 Larger ecosystems generally support more species.

a In the Caribbean, more species of reptiles and amphibians live on large islands like Cuba than on smaller islands like Jamaica.

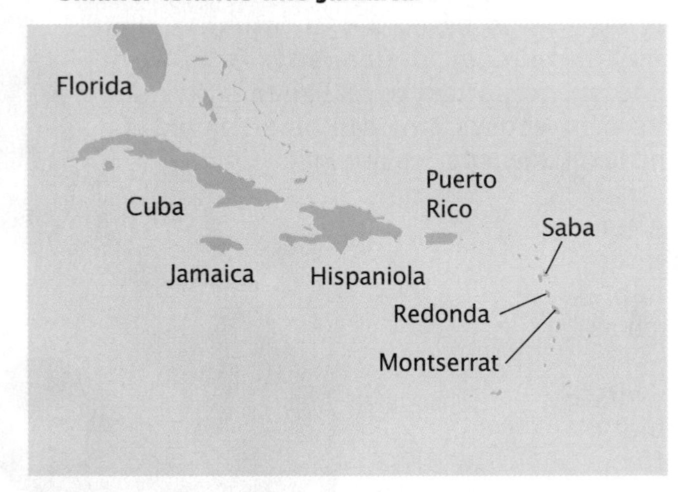

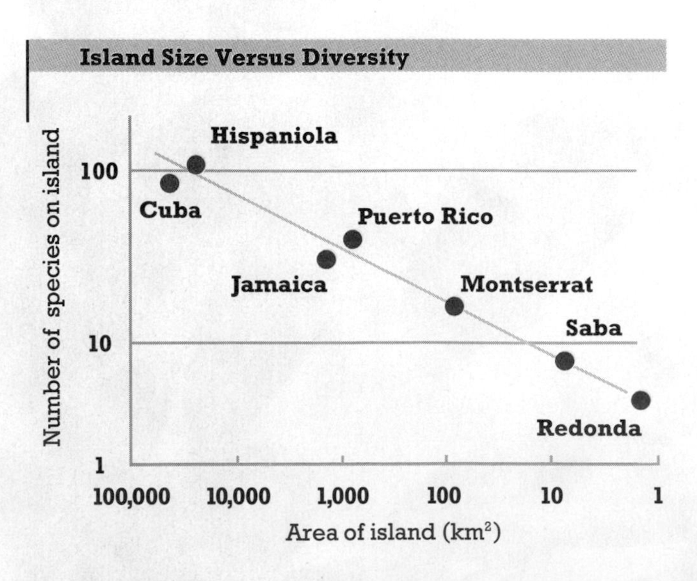

b The knight anole is one of the Cuban species.

Table 15-1 The Effects of Latitude on Diversity

Latitude	Number of bird species	Location
70° N	26	Northern Alaska
30° N	153	Southwest Texas
10° N	600	Central America

Latitude affects diversity

Latitude has a great influence on the diversity of ecosystems because both moisture and temperature vary with distance from the equator. For example, as you can see in **Table 15-1**, there are almost 30 times more bird species in Central America than there are in Alaska.

The tropics, the regions closest to the equator, have the highest species diversity for two reasons. First, latitude helps determine the length of the growing season. The greater the amount of food produced by plants and other producers, the more consumers an ecosystem can support. In the tropics, where ample sunlight, warm temperatures, and generous rainfall occur throughout the year, the growing season never ends.

Second, latitude plays a major role in determining climatic stability. The tropical climate does not vary much from season to season or from year to year. The unchanging physical conditions in the tropics have provided a long evolutionary window for specialized relationships to coevolve. In temperate or arctic regions, by contrast, weather can vary a great deal from one year to the next and does vary from season to season. Conditions do not usually persist long enough for coevolution to foster as many relationships as in the tropics.

How Humans Disrupt Ecosystems

The most important single influence on natural ecosystems today is human activity. Humans are disturbing ecosystems on a greater scale and to a greater degree than ever before. Humans alter ecosystems in three principal ways: by disrupting the physical habitat, by decreasing species diversity, and by destroying interactions among species.

Ever since humans learned to grow crops, about 10,000 years ago, natural habitats have been altered and replaced with habitats of human construction. Forests are cut down, swamps are drained and filled, and rivers are diverted, all to make room for buildings, parking lots, roads, and farms. Now humans are changing not only local habitats but the whole globe. Burning of fossil fuels releases carbon dioxide, leading to a warmer climate worldwide. At the same time, industrial chemicals are destroying the Earth's ozone shield. These human-caused changes reduce the ability of natural habitats to support living things.

Altering habitats in drastic ways often eliminates native organisms that have evolved to fit the natural habitats. For instance, conversion of forest to farmland or pasture reduces the number of species from hundreds or thousands to only a few. Similarly, massive logging of virgin forests followed by the planting of a single species of tree as a future lumber "crop" diminishes the diversity of forest ecosystems.

Intentionally or unintentionally, human actions often disrupt the kinds of species interactions that promote diversity. Removal of predators often reduces diversity, as you saw in the earlier keystone species example. The intentional or accidental introduction of exotic species from other parts of the world, such as the fire ants shown in **Figure 15-13**, can disrupt competitive balances that have coevolved among native species. Freed from the controls imposed by their natural predators and parasites, exotic species often easily outcompete native species and may drive them to extinction.

Figure 15-13
This is a fire-ant mound. Brazilian fire ants were accidentally introduced into Mobile, Alabama, in the 1920s. They are now found throughout the South. Fire ants are very aggressive, and they have had a serious negative impact on many native species.

SECTION REVIEW

1. **Describe the niche of humans.**
2. **What features of weeds make them well suited for their role in early successional stages?**
3. **Explain what makes high-diversity ecosystems more stable than low-diversity ecosystems.**
4. **Why do introduced species often have an advantage over their native competitors?**

CHAPTER

15 Highlights

Naturally occurring fires are part of the process of succession in many ecosystems.

15-1 Interactions Within Ecosystems

This boxing crab has a commensal relationship with the anemones it is carrying.

Key Terms

coevolution (p. 273)
symbiosis (p. 276)
parasitism (p. 276)
mutualism (p. 276)
commensalism (p. 277)

Summary

- Species evolve in response to other living members of their ecosystems. This process, called coevolution, shapes the species interactions in an ecosystem.
- Insects and flowers coevolve, as do plants and the herbivores that attempt to eat them.
- A symbiosis is a close, long-term relationship between species.
- In parasitism, one species (the parasite) lives on or in another species (the host). Mutualism is a symbiotic relationship in which all parties benefit. Commensalism is a relationship in which one species benefits and the other is neither helped nor harmed.

15-2 Ecosystem Development and Change

The sea star in the 15-species ecosystem studied by Robert Paine is an example of a keystone species.

Key Terms

niche (p. 278)
fundamental niche (p. 278)
competition (p. 279)
realized niche (p. 279)
competitive exclusion (p. 279)
succession (p. 280)
primary succession (p. 280)
secondary succession (p. 280)
stability (p. 281)
keystone species (p. 281)

Summary

- An organism's niche is the sum of all its interactions in its environment, including interactions with other organisms.
- Competition occurs when organisms attempt to use the same scarce resource.
- Succession, the regular progression of species replacement in a developing ecosystem, is driven by competition.
- Stability is the ability of an ecosystem to resist change. More diverse ecosystems tend to be more stable than less diverse ecosystems.
- An ecosystem's diversity is partly determined by its latitude and its size.
- Human activities disrupt ecosystems in three main ways: by altering natural habitats, by reducing species diversity, and by disrupting interactions among species.

CHAPTER 15

review

Understanding Vocabulary

1. For each pair of terms, explain the differences in their meanings.
   a. parasitism, mutualism
   b. fundamental niche, realized niche
   c. competition, competitive exclusion

2. Using each set of terms below, write one or more sentences summarizing information learned in this chapter.
   a. symbiosis, benefit
   b. habitat, succession
   c. predation, competition

Understanding Concepts

3. **Relating Concepts** Construct a concept map that illustrates the relationship between stability and diversity in an ecosystem.Try to include the following terms in your concept map: ecosystem, diversity, stability, human impact, ecosystem size, and latitude.

4. **Summarizing Information** What is coevolution? Give an example of coevolution. How is coevolution different from evolution in response to the physical environment?

5. **Inferring Conclusions** Bee-pollinated flowers are rarely red, while bird-pollinated flowers often are. What does this fact indicate about the sensitivity of bees and birds to the color red? Explain your answer.

6. **Comparing Functions** Identify and describe two ways plants defend themselves from herbivores.

7. **Recognizing Relationships** Crocodile birds eat leeches and food particles from the mouths of crocodiles. What type of symbiosis is exhibited by this relationship? Explain your answer.

8. **Inferring Conclusions** Can an organism's realized niche be larger that its fundamental niche? Explain your answer.

9. **Organizing Information** Copy the chart below onto a separate sheet of paper. Complete the chart so that it summarizes the effects of the four kinds of species interactions on each participating species. Place a "+" (if the interaction is beneficial), a "–" (if the interaction is harmful), or a "0" (if the interaction is neutral) in each blank. The first row of the table has been completed for you.

Interaction	Effect on 1st species	Effect on 2nd species
Parasitism	+	–
Mutualism		
Competition		
Commensalism		

10. **Summarizing Information** What effect does *Balanus* have on *Chthamalus*? Does *Chthamalus* have any effect on *Balanus*? Explain your answer.

11. **Predicting Outcomes** Why would competition between two animals of the same species likely be more intense than competition between two animals of different species?

12. **Analyzing Information** A lawn is not a stable community, yet it usually does not undergo succession. Explain why.

13. **Recognizing Relationships** Explain two reasons why ecosystems in the tropics usually have more species than ecosystems in the temperate regions.

14. **Summarizing Information** What is a keystone species? Why is it particularly important to its ecosystem? Give an example of a keystone species.

15. **Inferring Conclusions** How are exotic species, such as fire ants, able to spread rapidly when introduced to a new environment?

16. **BUILDING ON WHAT YOU HAVE LEARNED** In Chapter 14 you learned that habitat is the physical location where an organism lives in an ecosystem. How does an organism's habitat differ from its niche?

Interpreting Graphics

17. The two graphs below summarize an experiment in which two types of single-celled organisms, *Paramecium aurelia* and *Paramecium caudatum*, were grown alone and together. Study the graphs and answer the question.

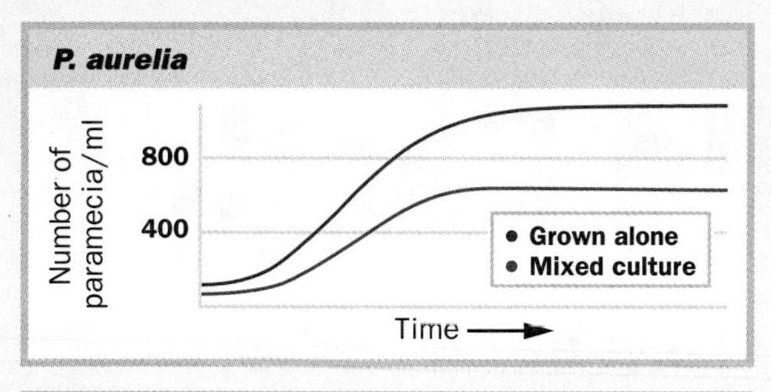

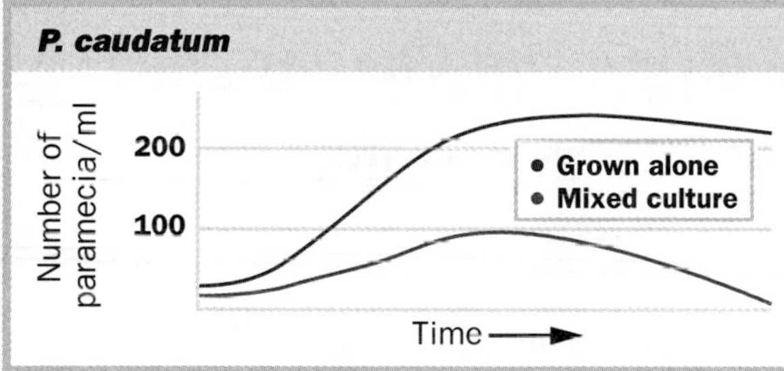

Which statement is most consistent with the data summarized in the graphs?

a. The two types of paramecium grew equally well alone and together.
b. The two types of paramecium competed for the same food source, causing the decline in the *P. caudatum* population.
c. *Paramecium aurelia* was unaffected when grown together with *P. caudatum*.

Reviewing Themes

18. ***Evolution***
Why was the disease myxomatosis introduced into the Australian rabbit population? What evidence suggests that the virus that causes myxomatosis coevolved with the rabbits?

Thinking Critically

19. **Evaluating Results** A scientist studying termites discovered a single-celled organism living inside the termite's digestive system. When the scientist removed the organism from a group of termites, the termites and the single-celled organism died. What kind of relationship existed between termites and the single-celled organism? Explain your answer.

Life/Work Skills

20. **Communicating** Design a "wanted" poster that alerts the public to the problems caused by introduced, or exotic, species, such as the Brazilian fire ant. Include information about exotic species that are problems in your area. These species might include zebra mussels, purple loosestrife, or kudzu. Include a photo or drawing of the introduced species, information on how it damages the environment, where the organism comes from, and a summary of efforts to control it. Add any additional information that you think is interesting and important.

Activities and Projects

21. **Unit Focus** Study historical records and documents about the area where your town or city is located. What kind of environment existed there 300, 200, 100, and 50 years ago? What will the environment be like 100 years from now? Prepare a report to share your findings with your class.

22. **Career Connection: Ecologist** Use library references or talk to a career counselor to find out about becoming an ecologist. Ecologists study the relationships of living organisms to each other and to their physical environment. What kind of educational background is required? Where and for whom do ecologists work? For additional information, write to the Ecological Society of America, Department of Botany and Microbiology, Arizona State University, Tempe, AZ 85281.

Discovering Through Reading

23. Read the article "The Secret Life of Backyard Trees" by Kevin Krajick in *Discover*, November 1995, pages 92–101. What surprise did Neville Winchester discover the first time he climbed a tree in an old-growth forest?

24. Read the article "A New Ant on the Block" by Karen Schmidt in *New Scientist*, November 4, 1995, pages 28–31. How did Brazilian fire ants spread to the United States? What effects have they had on native organisms? What new control methods for fire ants are being researched?

CHAPTER 16

REVIEW
- ozone (Section 11-2)
- ecological models (Section 14-1)
- carbon cycle (Section 14-2)
- how humans disrupt ecosystems (Section 15-2)

The Fragile Earth

Every day, Americans throw away nearly 500 million kg (about 500,000 tons) of garbage.

Imagine every plastic wrapper, cup, and container you have ever thrown away piled up beside you. All of this plastic is still in the environment. Plastics are just one of the many kinds of chemicals and other substances that humans are adding to ecosystems, sometimes with damaging effects to our health and the health of ecosystems.

16-1 Planet Under Stress

OBJECTIVES

1. **List some ways that human health is affected by pollution.**
2. **Explain the cause and the effects of acid rain.**
3. **Identify the causes and effects of ozone depletion.**
4. **Describe why global temperatures are rising.**

Humans Have Damaged the Environment

Our world is a patchwork of interconnected ecosystems. Because of this interdependence, damage done to any one ecosystem can have ill effects on others. Burning sulfur-rich coal in Missouri kills trees in Canada; dumping refrigerator coolants in California destroys atmospheric ozone over Antarctica and leads to higher rates of skin cancer in Paris. Biologists call such widespread effects on our world *global change.*

How much trash do you think you throw away? On average, each American tosses out almost 2 kg (4 lb) of unwanted paper, metals, glass, plastics, food, and other items *every day.* We cannot banish our wastes from existence—garbage does not just disappear when you haul it to the curb or drop it into a dumpster. We release our wastes into the environment—the water, the air, and the soil, as shown in **Figure 16-1**. **Pollution** is anything potentially harmful that humans add to the environment. Automobile exhaust is an example of pollution, as are poisonous chemicals dumped into rivers and heat from the cooling towers of nuclear power plants. The substance added to the environment is called a pollutant. Automobile exhaust, for example, contains a variety of pollutants. Until recently, people felt that the environment could absorb and dilute pollution without suffering ill effects. It is now apparent that this was a mistaken belief.

Pollution can drastically damage the health of ecosystems as well as the health of human beings. Some pollutants are poisonous. Others are carcinogens. A **carcinogen** is a substance that causes cancer. Examples of carcinogens include industrial chemicals such as dioxin, benzene, and carbon tetrachloride (a solvent formerly used in dry cleaning).

Figure 16-1
The chemicals that these smokestacks are pouring into the air are pollutants.

Pollution's Toll

Ecosystems cannot absorb chemical punishment indefinitely. Too much pollution disrupts the delicate and complex web of relationships that binds the world's ecosystems together. In eastern Europe, a century of unrestrained pollution has destroyed forests and rendered lakes nearly lifeless. Eighty percent of Poland's deep wells are polluted, and one-fourth of its soil is far too contaminated for safe farming. One-third of Bulgaria's forests are damaged or dying. Intensive efforts are underway to reverse this damage, but no one knows if they will succeed.

Pollution endangers our water supply

Humans need water for drinking, irrigation, and industry, yet we have a very casual attitude toward water pollution. Every day, wastes are poured down the sink, flushed down the toilet, or dumped into rivers and lakes without considering where they will end up. For instance, while putting out a fire in a chemical warehouse in Basel, Switzerland, in November 1986, firefighters washed 27 metric tons (30 tons) of mercury and pesticides into the Rhine River. A deadly wall of poisonous pollutants flowed down the Rhine, killing everything as it moved through Germany and Holland to the sea.

The poisoning of the Rhine is just one example of a worldwide problem. In 1989 an oil tanker, the *Exxon Valdez*, ran aground off the coast of Alaska. If the *Exxon Valdez* had been loaded no higher than the waterline, little oil would have spilled—but it was loaded much higher than that. The weight of the above-waterline oil forced thousands of tons of oil out of a rip in the ship's hull. The spilled oil fouled about 1,600 km (1,000 mi) of coastland and killed fishes, marine mammals, and birds. **Figure 16-2** shows the *Exxon Valdez*.

The high productivity of modern agriculture is based on the widespread use of insecticides to kill insect pests, herbicides to control weeds, and fertilizers to enrich the soil. Rainfall washes these chemicals, some of which are carcinogenic or toxic, into rivers, lakes, and the ocean. It is no coincidence that one of the highest rates of cancer in the United States is seen in the Mississippi Delta region. Here the river draining our country's agricultural heartland empties into the sea, carrying with it herbicides, insecticides, and industrial chemicals.

Figure 16-2
a Oil that spilled from the *Exxon Valdez* (the larger ship in this photo) is visible as a dull film on the surface of the water.

b This seabird was killed by oil that spilled from the *Exxon Valdez*. Thousands of other seabirds, sea otters, and other marine mammals were also killed.

Acid rain threatens forests and lakes

Many coal-burning power plants use high-sulfur coal because it is cheap and plentiful. The smoke produced when high-sulfur coal is burned smells bad (like rotten eggs), blackens buildings, and kills local trees. Smokestacks more than 65 m (210 ft) tall were introduced as a way to burn high-sulfur coal without public outcry. The intent of those who designed plants with tall stacks was to release the sulfur-rich smoke high in the atmosphere, where winds would disperse and dilute it.

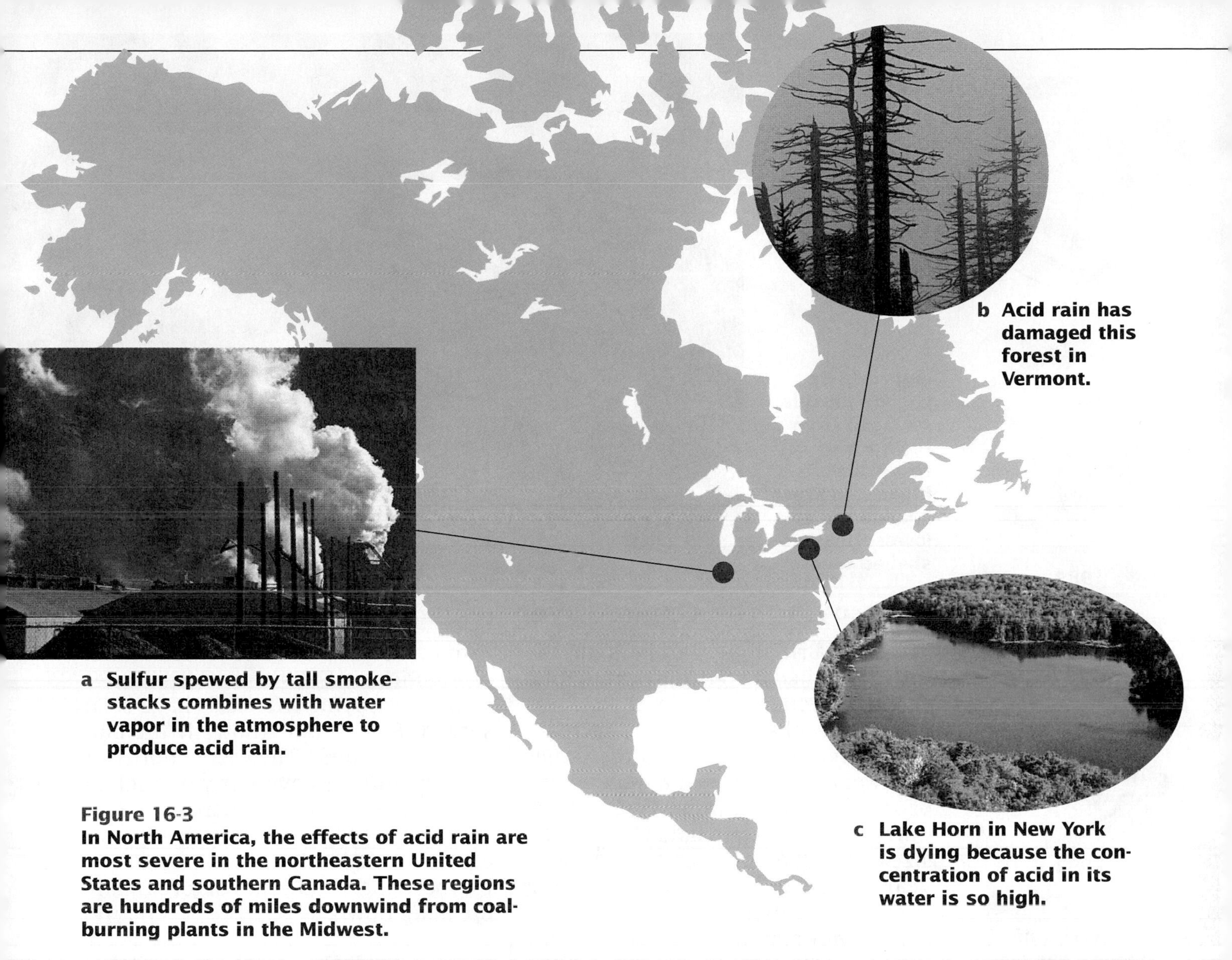

a **Sulfur spewed by tall smokestacks combines with water vapor in the atmosphere to produce acid rain.**

b **Acid rain has damaged this forest in Vermont.**

c **Lake Horn in New York is dying because the concentration of acid in its water is so high.**

Figure 16-3
In North America, the effects of acid rain are most severe in the northeastern United States and southern Canada. These regions are hundreds of miles downwind from coal-burning plants in the Midwest.

In the 1970s, it became clear to ecologists that tall stacks were not eliminating the problems of sulfur-rich coal, just exporting the ill effects elsewhere. In the upper atmosphere, sulfur released by smokestacks combines with water vapor to produce sulfuric acid. When the water vapor later condenses and falls back to the surface as rain or snow, it carries the sulfuric acid with it. Since moisture can travel great distances high in the atmosphere, the acid is far from its source when it falls. Beginning in the 1970s, ecologists reported that the lakes of Sweden were beginning to die. These lakes could no longer support life. Also dying were the trees of the great Black Forest of Germany. **Figure 16-3** shows some areas of North America damaged by acid rain.

What was causing lakes and forests to die? Scientists discovered that the rain in these areas was unusually acidic, a phenomenon called **acid rain** (although it is more correctly called acid precipitation, since snow can be acidic as well). When an **acid** is dissolved in water, the resulting solution has a higher concentration of hydrogen ions (H^+) than does pure water. Scientists describe the acidity of a solution using a logarithmic value called **pH**. A solution with a low pH has a high concentration of hydrogen ions. Rainwater normally has a pH value of 5.6. Pure water has a pH of 7.0. Rainfall in some areas of the northeastern United States, however, has a pH value of 3.8, almost 100 times more acidic than the typical value for the rest of the country.

1980

1989

1992

Figure 16-4 **This series of images was taken by a satellite over the South Pole. The colors were added by computer and indicate the levels of ozone in the atmosphere. Black, purple, and blue indicate the lowest ozone levels. The ozone hole is visible as a blue or purple blotch in the center of each photo.**

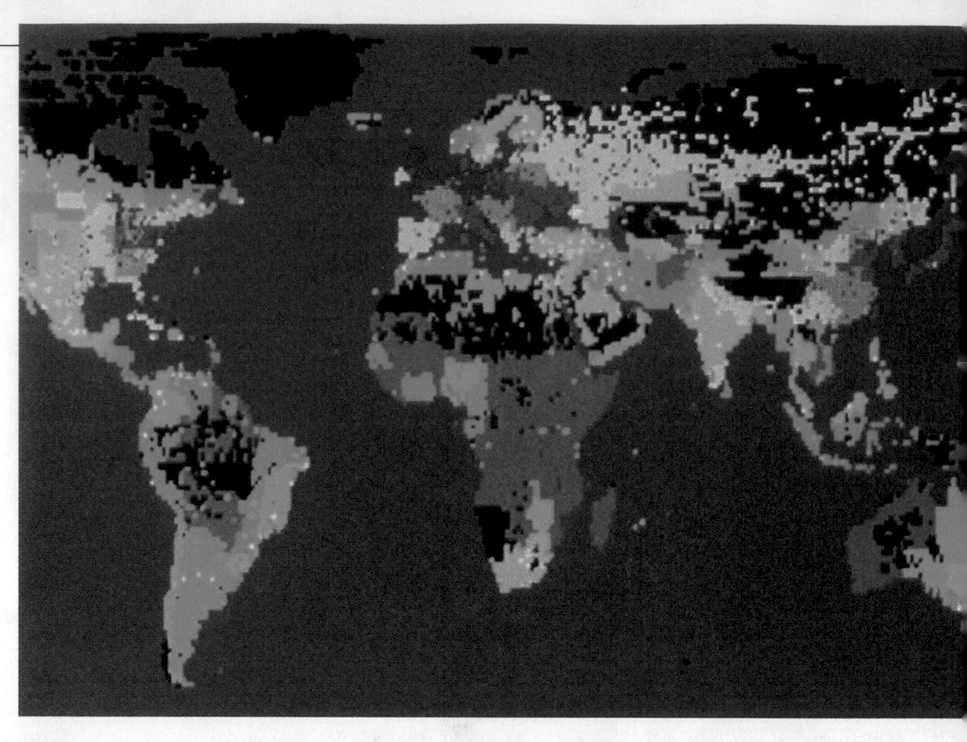

Figure 16-5a **Carbon dioxide released from the burning of fossil fuels is highest in areas shaded red and lowest in areas shaded blue.**

Destroying the Ozone Layer

In 1985, British researchers in the Antarctic discovered that the ozone concentration over the South Pole was much lower than expected. A decrease in the amount of ozone in the upper atmosphere has caused a "hole" in the ozone layer, as shown in **Figure 16-4**. Currently, the ozone hole is about the size of North America and grows every year. Ozone destruction is occurring over areas other than Antarctica. A similar but smaller zone of very low ozone levels has appeared over the Arctic. And the ozone layer over most of the globe is thinning.

What is destroying the ozone? The major culprit is a class of chemicals that scientists once thought were harmless—**chlorofluorocarbons** (CFCs). Since their invention in the 1920s, CFCs have been manufactured for use as the coolant in refrigerators and air conditioners, as the propellant in aerosol cans, and as the foaming agent in the production of plastic-foam containers.

Eventually CFCs escaped into the atmosphere and accumulated there. High in the atmosphere, CFCs began to attack ozone molecules. Just as an enzyme carries out a reaction in your cells without being changed itself, so CFCs catalyze the conversion of ozone (O_3) into oxygen (O_2).

International agreements to ban the production of CFCs and several other ozone-destroying chemicals have been signed. But no one knows if the ban has come in time. The vast majority of CFCs produced have not yet reached the Earth's upper atmosphere. Furthermore, CFCs are long-lived molecules; some kinds can destroy ozone for more than 100 years before finally breaking down.

Ozone depletion leads to health problems

The ozone layer blocks ultraviolet (UV) radiation. Ozone depletion is frightening because exposure to high levels of UV radiation can cause severe health problems. In humans, exposure to UV radiation causes skin cancer and cataracts, an eye disorder that can lead to blindness if not treated. Experts estimate that each 1 percent drop in atmospheric ozone leads to a 3 percent increase in the incidence of skin cancer. Thus, the drop of just over 6 percent in ozone concentration that has already occurred is estimated to have caused an increase of as much as 18–20 percent in skin cancer. In addition, productivity of the ocean's plankton, the basis of all oceanic food webs, has declined as a result of increased levels of UV radiation.

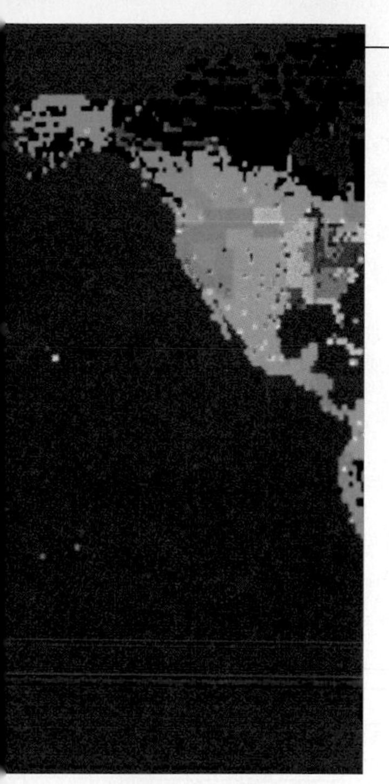

Global Warming

For more than 200 years, our industrial society has grown on a diet of cheap energy. Much of this energy has been obtained by the burning of fossil fuels. **Figure 16-5a** shows that most of the carbon dioxide released by burning fossil fuels comes from industrialized nations. When something is burned, its molecules are broken apart. They combine with oxygen, freeing energy as a result. Coal, oil, and natural gas are the remains of ancient organisms and are rich in carbon. When fossil fuels are burned, this carbon combines with oxygen to produce carbon dioxide. Centuries of fossil-fuel burning has released large quantities of carbon dioxide into the atmosphere.

Plants use carbon dioxide to make carbohydrates. Unfortunately, too much carbon dioxide is harmful. As you learned in Chapter 14, carbon dioxide absorbs solar energy, trapping heat in the atmosphere. That is one reason why the Earth is warm and the moon (which has no atmosphere) is very cold. Most scientists think that the increased levels of carbon dioxide in the atmosphere are causing increases in global temperatures, or **global warming**. **Figure 16-5b** compares the release of carbon dioxide from all sources with the change in global temperatures. Today the average global temperature is about 0.6°C (1°F) higher than it was a century ago. Nearly all scientists think this temperature increase is the result of human modification of the atmosphere.

Figure 16-5b The blue line indicates change in average global temperature compared with the average from 1951–1980. The red line indicates carbon dioxide levels.

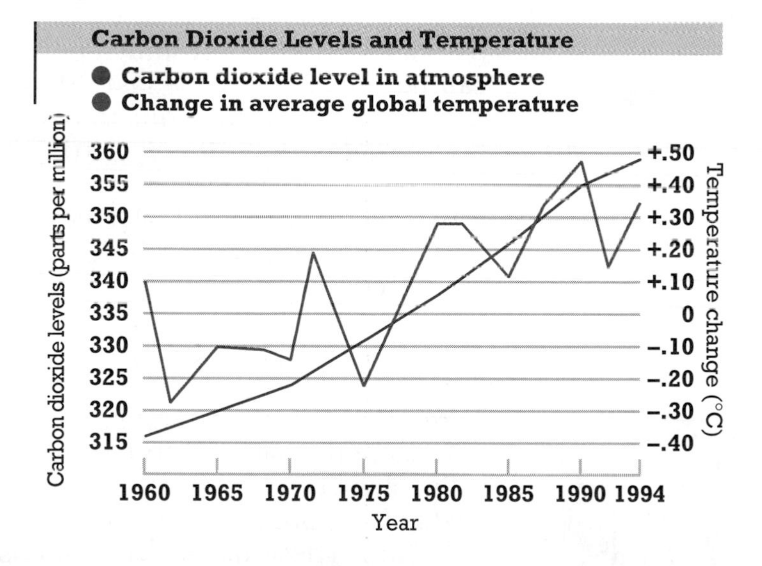

Global warming may have many serious consequences

Most scientists studying the problem expect the world's average annual temperature to rise 1.5–4.5°C (3–8°F) by the year 2100. Based on this, the city of Dallas, Texas, would have 78 days each year in which the temperature would be greater than 38°C (100°F), compared with only 19 such days currently.

That 1.5°C–4.5°C change is approximately equal to the temperature change from the last ice age, 15,000 years ago, until now. However, the change in temperature is now occurring over a period of less than 100 years. Obviously, this kind of climatic change will vastly change patterns of rainfall and temperature, which will have a major and unpredictable impact worldwide. Rising temperatures may cause partial melting of the polar icecaps, which would raise sea levels and threaten many of the world's major cities with flooding.

SECTION REVIEW

1. **Define *carcinogen*, and give one example.**
2. **Describe the cause of acid rain.**
3. **Explain two health effects of ozone depletion.**
4. **How does the destruction of tropical rain forests contribute to global warming?**

In the United States today, practically everyone, from individual citizens to large corporations, participates in recycling in some way. Seventy percent of the glass in every glass bottle was present in some previous glass container. As you will see in this section, recycling is just one way we are meeting the challenges posed by our environmental problems.

16-2 Meeting the Challenge

OBJECTIVES

1. **Explain two ways pollution has been reduced in the United States.**
2. **Identify alternatives to fossil fuels, and describe some benefits of each.**
3. **Contrast nonrenewable resources with renewable resources.**
4. **Explain the causes and consequences of the human population explosion.**

Reducing Pollution

The pattern of global change overwhelming our world is very disturbing. Human activities are placing severe stress on ecosystems worldwide, and we must quickly find ways to reduce the harmful impact. There are five areas in which it is particularly important to find solutions: reducing pollution, finding enough energy, preserving irreplaceable resources, curbing human population growth, and reducing consumption of natural resources.

Governments all around the world are now making serious efforts to reduce pollution. In some cases, such as CFC production, the efforts involve international agreements to reduce or stop production of the pollutant. In most cases, however, pollution problems are national or local and require action by individual governments. Individuals can also have an impact. The man shown in **Figure 16-6** hopes to improve the quality of the water he and his family drink.

Two approaches, both effective, have been taken to curb pollution in this country. The first approach is to pass laws limiting how much pollution can be released. In the last 20 years, laws have begun to significantly curb the spread of pollution by setting stiff standards for what can be released into the environment. All new cars must have effective catalytic converters to reduce the amount of pollution they release. Similarly, the Clean Air Act of 1990 requires that power plants install scrubbers on their smokestacks to restrict sulfur emissions.

A second approach to curbing pollution has been to directly increase the costs of polluting by placing a "tax" on pollution. These taxes, often imposed as "pollution permits," are becoming an increasingly important part of laws that regulate pollution. They are a key element of the 1990 Clean Air Act.

Figure 16-6
To solve our environmental problems, action is needed from both governments and individuals.

Finding Enough Energy

Our dependence on fossil fuels has led to serious pollution problems and has resulted in rapid draining of fuel supplies. The known reserves of oil and natural gas will be nearly depleted by the middle of the next century. Fossil fuels are nonrenewable energy sources. **Nonrenewable resources** do not replenish themselves naturally, whereas **renewable resources** do. Trees are a renewable resource. New trees can be grown to replace those cut down.

Interdependence

Trees are a renewable resource, but a forest ecosystem is nonrenewable. Explain why.

Alternatives to fossil fuels do exist

Nuclear power—capturing the energy released when radioactive atoms break apart—is one alternative to fossil fuels. More than 70 percent of France's electricity is produced by nuclear power plants. For all its promise of plentiful energy, nuclear power presents three areas of concern that must be addressed if it is to provide energy for our future. These areas are safe operation, waste disposal, and security.

New nuclear power plant designs are much safer than those of the past. The best of these designs virtually eliminates the possibility of loss-of-coolant explosions, such as the one that occurred in 1986 at Chernobyl in Ukraine. Spent nuclear fuel is very radioactive, and the power plants themselves eventually wear out. Spent fuel and components of power plants will remain dangerously radioactive for thousands of years. For this reason, wastes must be disposed of in a safe manner, often by burying them in a remote location. Finally, security is an issue because spent nuclear fuel can be used to recover plutonium, which can be used to make atomic weapons.

It is important to develop alternatives to nonrenewable energy sources like oil, coal, and gas. Solar energy and wind power are two promising sources of energy that are not in limited supply.

Energy conservation can reduce reliance on fossil fuels

The most cost-effective way to meet our future energy needs is conservation, using energy more efficiently. As much as 75 percent of the electricity used in the United States and Canada is wasted through the use of inefficient appliances, according to scientists at the Lawrence Berkeley Laboratory in California. The use of highly efficient motors, lights, heaters, air conditioners, refrigerators, and light bulbs, like the one shown in **Figure 16-7**, could lead to large energy savings.

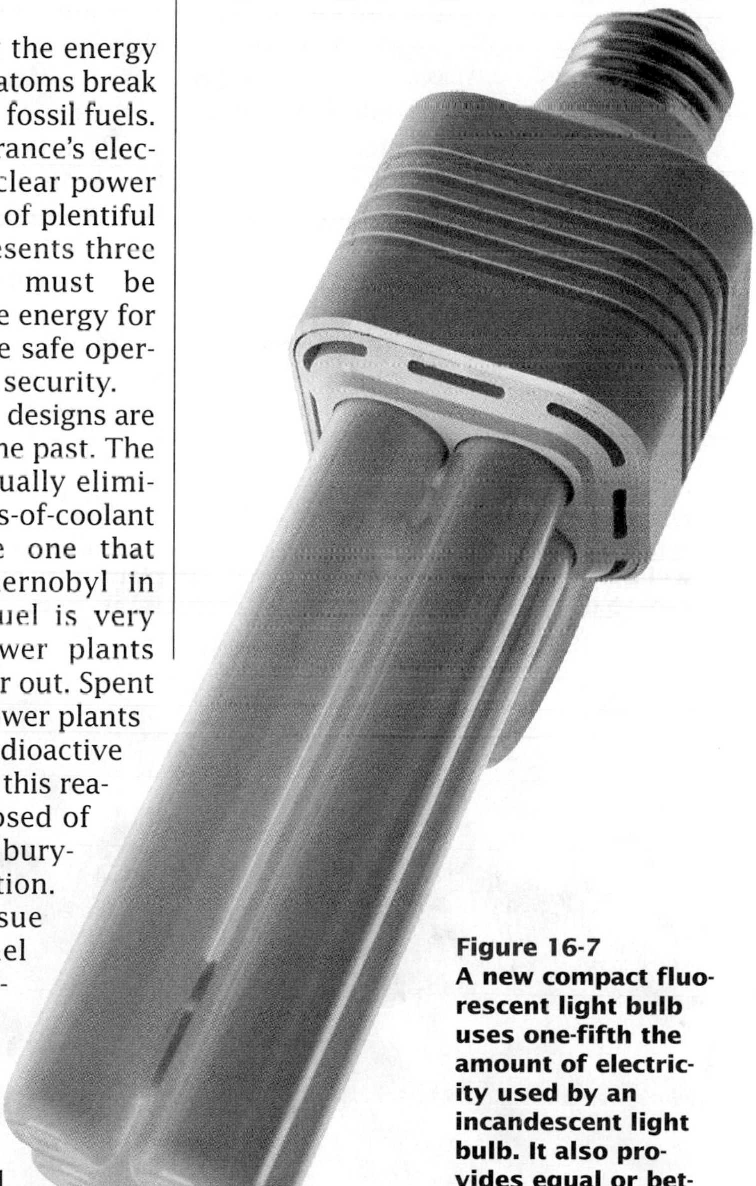

Figure 16-7
A new compact fluorescent light bulb uses one-fifth the amount of electricity used by an incandescent light bulb. It also provides equal or better light and lasts up to 13 times longer.

Conserving Nonrenewable Resources

While a polluted stream can be cleaned up, no one can restore an extinct species. Worldwide, three sorts of nonrenewable resources merit particular attention.

Topsoil is the basis for agriculture

One of the great strengths of the United States over the last two centuries has been its strong agriculture. The United States has been able to grow extraordinary amounts of crops because of its particularly fertile soils. These soils have accumulated slowly during hundreds of thousands of years. Although it takes hundreds of years for an inch of topsoil to form, we are allowing topsoil to be lost at a rate of inches each decade. By intensively cultivating crops—repeatedly turning the soil over to eliminate weeds—we permit rain and wind to carry topsoil away. The world has lost one-fourth of its topsoil since 1950. Each year, nearly 22 billion metric tons (24 billion tons) of topsoil is blown or washed off the world's farmland. **Figure 16-8** shows how some farmers have implemented contour farming in an effort to prevent the erosion of topsoil. Another method of soil conservation, minimum till cultivation, is now practiced on more than half of the farmland in the United States. It greatly reduces the amount of soil lost to erosion.

Ground-water supplies are being depleted

A second resource that we are depleting is ground water. Ground water is water trapped beneath the soil, largely in porous rock. This water seeped into its underground reservoirs very slowly during the last 12,000 years. We use ground water in thousands of ways, especially for irrigating crops and in our homes. But we should not waste it, for it accumulates very slowly.

Today there is very little control over the use of ground water, and much of what is used is wasted to water lawns, wash cars, and run fountains. A great deal more ground water is inadvertently polluted by poor disposal of chemical wastes. Once pollution enters the ground water, there is no effective means of removing it.

Figure 16-8

a When trees and vegetation are cleared from land that will become a field, erosion can occur.

b In an effort to save land that has begun to erode, farmers can plant rows on different levels, or "contours."

c In the long run, contour farming limits erosion.

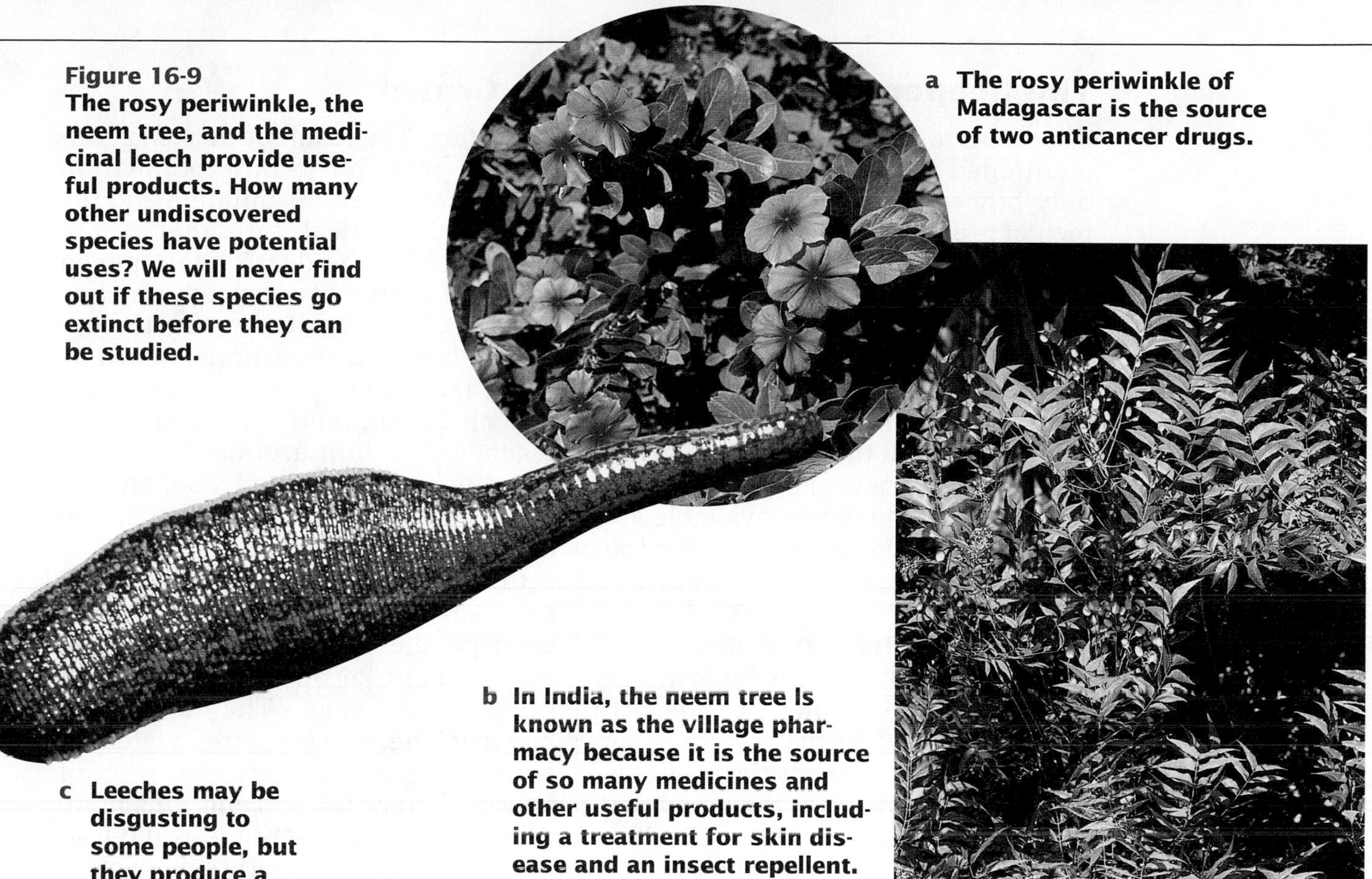

Figure 16-9
The rosy periwinkle, the neem tree, and the medicinal leech provide useful products. How many other undiscovered species have potential uses? We will never find out if these species go extinct before they can be studied.

a The rosy periwinkle of Madagascar is the source of two anticancer drugs.

b In India, the neem tree is known as the village pharmacy because it is the source of so many medicines and other useful products, including a treatment for skin disease and an insect repellent.

c Leeches may be disgusting to some people, but they produce a chemical that prevents blood from clotting. This chemical is now used in treating injuries and some diseases.

Species are disappearing

During the last 20 years, about half of the world's tropical rain forests have been destroyed—burned to make pasture and farmland or cut for timber. Each year the rate of loss increases; nearly 2,000 hectares (4,800 acres) of forest are cut each minute, an area larger than Indiana each year. At this rate, most of the world's rain forests will be gone in about 40 years. It is estimated that one-fifth or more of the world's species of animals and plants will become extinct, more than a million species lost, if the rain forests are destroyed.

Why saving species is important

As you learned in Chapter 15, removing even one species from an ecosystem can seriously disrupt the workings of that ecosystem. Moreover, as species disappear, so do our chances to learn about them and their potential benefits. Like burning a library without reading the books, we do not know what we are wasting. All we can be sure of is that we cannot retrieve it.

Many important and useful species are discovered each year. For instance, the rosy periwinkle, a plant native to Madagascar, has been used to develop two drugs used to treat leukemia, a type of cancer that affects white blood cells. A child with Hodgkin's disease (a form of leukemia) has a 90 percent chance of survival if treated with these drugs. Without the drugs, the child would have only a 5 percent chance of living.

With the advent of genetic engineering, scientists now have the ability to transfer desirable genes—ones that confer resistance to pests or spur more rapid growth—from one species to another. Genetic engineering could lead to major improvements in agriculture—but only if the vast library of genes contained in the world's species is there to be searched.

Figure 16-9 shows three useful species. No one knows how many other useful species have already disappeared or how many more will disappear if extinction continues at its present rate.

The Deeper Problem: Population Growth

If we were to solve the many problems mentioned in this chapter, we would only buy time to address the fundamental problem: there are too many of us. **Figure 16-10** shows the growth of the world's human population over the last several thousand years.

Ten thousand years ago, when agriculture was first developed, there were about 3 million people on Earth, distributed over all the continents except Antarctica. As new, more dependable sources of food became available as a result of agriculture, the human population began to grow more rapidly.

What factors triggered the human population explosion?

A population will grow when the birth rate (number of births per thousand people per year) exceeds the death rate (number of deaths per thousand people per year). From the time agriculture was introduced, approximately 10,000 years ago, until about 1650, the birth rate was only slightly higher than the death rate. The result of this small difference was fairly slow population growth. With better sanitation and the improved medical care that began around 1650, the death rate plunged while the birth rate remained relatively constant. As a result, the human population began to grow more rapidly. By 1800, there were about 1 billion people on Earth, and the population reached 2 billion around 1930. The world population passed 5 billion people in early 1987 and will hit 6 billion before the year 2000.

Currently, the death rate (as a world average) is about 9 deaths per thousand people per year, and the birth rate is about 24 births per thousand people per year. The difference between these two figures yields an annual population growth rate of approximately 1.5 percent. This number may seem small, but it would lead to a doubling of the world's population in only 45 years. About 88 million more people are born each year than die. About 240,000 people are added to the world population each day, about 170 each minute. Our world cannot continue to support such growth over such a short period of time without sharply reducing the quality of life.

Figure 16-10
In the past, the human population grew very slowly. Technology has increased the average life span and decreased infant mortality to the point where the human population can now double in just 45 years.

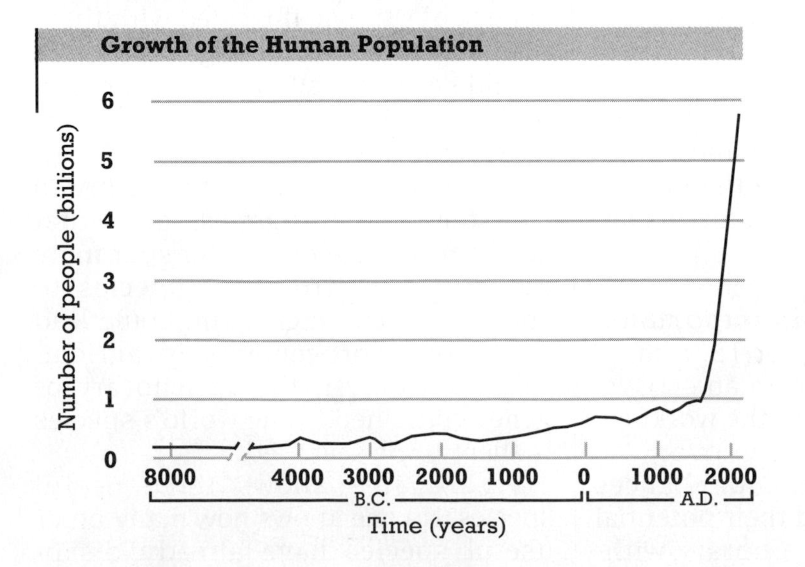

The human population continues to grow

Population growth is occurring most rapidly in the developing countries, a category that includes most of the countries in Asia, Africa, and Latin America. Population growth is much slower or has stopped in the industrialized countries, which include the United States, Japan, all the countries of Europe, Russia, Canada, New Zealand, and Australia. For example, Kenya's population is growing at about 3.3 percent per year. The United States population is growing by about 0.7 percent, and the populations of Germany and Russia are actually shrinking. **Figure 16-11** shows the current and projected populations of several countries.

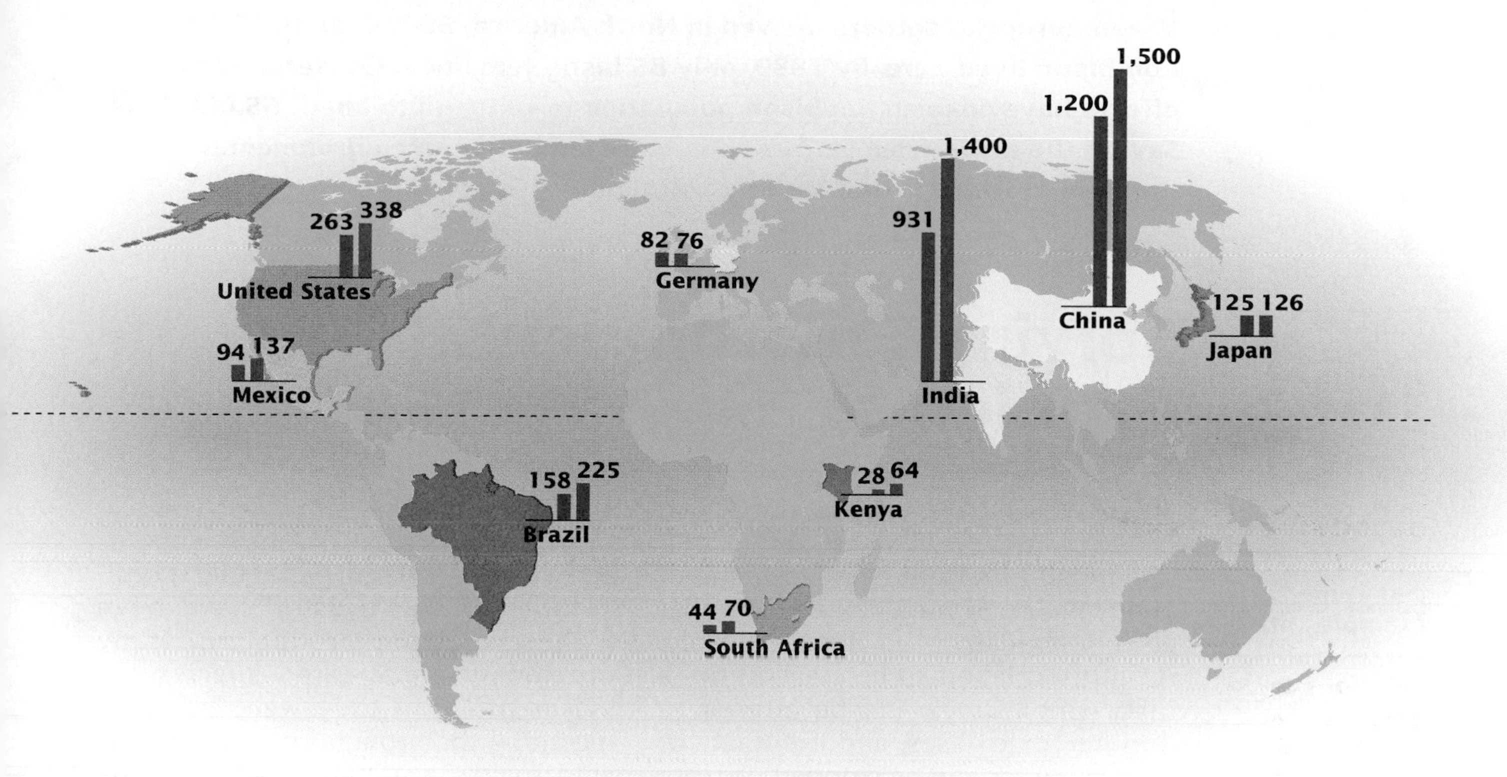

Figure 16-11 This chart shows the current population (blue bar) and projected population in 2025 (red bar) of selected countries. All values are in millions of people.

Many countries are devoting considerable attention to slowing the growth rates of their populations, and there are genuine signs of progress. For example, by encouraging families to have only two children, Thailand reduced its population growth rate from 3.2 percent to 2.4 percent between 1960 and 1994. In Mexico over the last 30 years, the average number of children per family has decreased from five to fewer than three.

The United Nations estimates that if these efforts continue, the world's population will peak by the middle of the twenty-first century at somewhere between 8 billion and 13 billion people. Can the Earth support this many people? Scientists agree that the world cannot support 5.8 billion people, its current population, if current resource consumption patterns continue. Finding ways to reduce consumption, improve technology, and restrain human population growth are the greatest tasks we will face in the coming years. The quality of life available for your children in the next century will depend to a large extent on our success.

SECTION REVIEW

1. **Describe two approaches that the United States government has taken to reduce pollution.**
2. **Explain two reasons why we should reduce our dependence on fossil fuels.**
3. **Give an example of a nonrenewable resource and a renewable resource.**
4. **What caused the rapid population growth that began around 1650?**

When European settlers arrived in North America, 60 million to 125 million bison lived here. By 1889, only 85 bison remained. Conservation efforts have enabled the bison population to rebound to about 65,000. Saving the bison from extinction is one example of an environmental problem in which action led to a solution.

16-3 Solving Environmental Problems

OBJECTIVES

1. **List some examples of successful solutions to environmental problems.**
2. **List the five basic elements necessary to solve any environmental problem.**
3. **Recognize your role in solving environmental problems.**

Environmental Problems Can Be Solved

The most important fact to remember about the environmental crisis is that each of its many problems is solvable. A polluted lake can be cleaned up, a dirty smokestack can be altered to remove toxic gases, and the waste of key resources can be stopped. Solving each problem requires a clear understanding of the problem and a commitment to doing something about it. The extent to which American families recycle aluminum cans and newspapers shows that people want to become part of the solution rather than remaining part of the problem. Progress is being made. Newspapers carry success stories daily. In the United States since 1970, 18 national parks have been established. Forty-two million hectares (103 million acres) of land have been protected as wilderness, and 14 million hectares (34 million acres) of farmland particularly vulnerable to soil erosion have been withdrawn from production. Many previously endangered species are better off than they were in 1970, including the pronghorn antelope, the wild turkey, the bald eagle, and the peregrine falcon, which is shown in **Figure 16-12**.

Pollution control efforts have been particularly successful. Emissions of sulfur dioxide, carbon monoxide, and soot were 200 million tons in 1970, and they have been reduced by more than 30 percent. The release of some toxic chemicals into the environment (notably the insecticide DDT and carcinogens such as PCBs and dioxin) has been banned outright. The Environmental Protection Agency estimates that private firms and public agencies are spending more than $100 billion per year on pollution control, more than double the figure of 10 years ago and five times the figure of 1970. In the same period, the population of the United States increased by more than 60 million people and the number of cars grew by about 60 million vehicles. Had this progress not been made, environmental quality in the United States would almost certainly have declined dramatically.

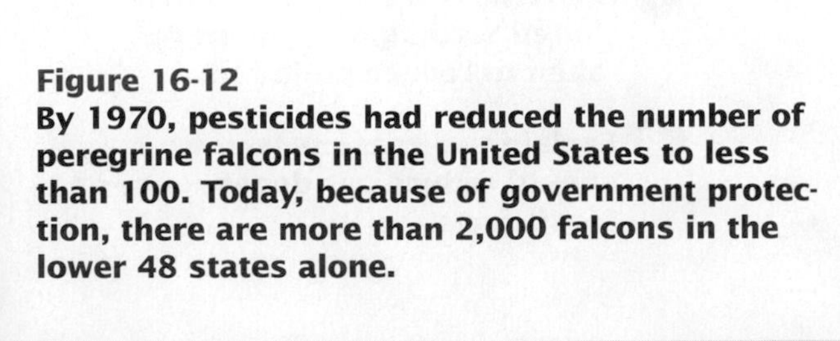

Figure 16-12
By 1970, pesticides had reduced the number of peregrine falcons in the United States to less than 100. Today, because of government protection, there are more than 2,000 falcons in the lower 48 states alone.

Steps Toward Saving the Environment

There are five steps to solving an environmental problem: assessment, risk analysis, public education, political action, and follow-through. **Table 16-1** outlines these steps, using a lake damaged by pollution as an example.

Table 16-1 Steps Toward Solving an Environmental Problem

Step	Plan of action	What you can do
Assessment	Data must be collected and experiments performed in order for scientists to construct an ecological model of an ecosystem. The model predicts how the environment will respond to changes.	You can volunteer to test the water of a nearby lake to determine the degree of damage done by chemical pollution. You also can record changes in the numbers of animals and plants found in and around the lake.
Risk analysis	Using the ecological model, scientists can predict the effects of environmental intervention. Scientists will evaluate the potential for solving the environmental problem as well as the potential for any adverse effects of the proposed solution.	A local college or university can test the water samples that you collect. Scientists there can suggest ways to decrease the amounts of chemicals entering the lake.
Public education	When a clear choice can be made, the public must be informed. This involves explaining the problem in terms people can understand. Costs and expected results of each alternative should be presented and explained.	The people of your town or city will have to decide whether they are willing to pay for the ditches necessary to divert agricultural runoff from the lake. You could go to a city-council or town meeting and emphasize the importance of saving the lake.
Political action	The public, through its elected officials, selects a course of action and implements the plan. Individuals can have a major impact by exercising their right to vote. Many voters do not understand the magnitude of what they can achieve by writing letters and supporting special-interest groups.	It is important to start learning about environmental issues now so that you can make good decisions when you are old enough to vote. Voters in your town or city will need to ask themselves if the potential benefits of preventing water pollution are worth the costs.
Follow-through	The results of any action taken should be monitored carefully to see if the environmental problem is being solved and to evaluate and improve the initial assessment and modeling of the problem. We learn by doing.	If the town does agree to channel runoff water away from the lake, you can volunteer to continue collecting water samples. From these samples, scientists can determine how well the lake is recovering.

Figure 16-13
These environmental organizations and magazines can provide more information on how to help the environment.

National Audubon Society
950 Third Avenue
New York, New York 10022
http://www.audubon.org/audubon/

World Wildlife Fund
1250 24th Street NW, Suite 500
Washington, D.C. 20037
http://www.envirolink.org:80/orgs/wqed/wwf_home.html

Natural Resources Defense Council
1350 New York Avenue, NW
Suite 300
Washington, D.C. 20005
http://www.nrdc.org/nrdc/

What You Can Contribute

Can you help solve environmental problems? It may seem that the solutions to these problems are the responsibility of far-off scientists or government agencies. Not so. Your actions directly affect the environment. There are many simple and inexpensive ways you can help solve the environmental problems described in this chapter. Some ways are listed on the following pages. You can find out more from your teacher; from books and magazines; from local, state, and federal government agencies; and from local, national, and world environmental organizations. **Figure 16-13** shows some magazines that contain useful information and some environmental organizations you can contact.

Helping conserve energy

The United States consumes more energy than any other country. The United States consumes 25 percent of the world's energy but has less than 5 percent of the world's population. Nearly 90 percent of our energy is derived from nonrenewable fossil fuels—coal, oil, and natural gas. Extracting fossil fuels from the Earth causes significant environmental damage. Transporting and refining these fuels is also damaging, as the *Exxon Valdez* disaster shows. Burning fossil fuels for energy creates pollution; an automobile emits its own weight in pollutants each year. Global warming is a direct result of our massive appetite for fossil fuels.

Here are some ways you can reduce your consumption of energy.

- Turn off all lights and appliances when you leave a room.
- Set the water-heater temperature at 54°C (130°F).
- Use compact fluorescent light bulbs. Although more expensive to buy than incandescent bulbs, compact fluorescent bulbs use one-fourth as much energy and last 10 to 13 times longer. Thus, they save money in the long run.
- Reduce your use of automobiles. Ride a bike, walk, or take public transportation (buses or trains).
- If you drive a car, make sure the tires are properly inflated; this can increase fuel efficiency by 5 percent. A tuneup can increase fuel efficiency by 9 percent.

Helping reduce pollution and waste

Each year, Americans generate 138 billion kilograms (153 million tons) of garbage. This mountain of trash is dumped into landfills or is burned, creating air pollution. Moreover, much of this waste could be reused or recycled. Here are some ways to help reduce waste and pollution.

- Recycle as much as possible. Aluminum products (cans, foil, and pie plates), glass containers, office and school paper, newspaper, and many kinds of plastic can be recycled. Many communities, school groups, and private organizations collect recyclable materials. Also, whenever possible, buy products that contain recycled material.
- Properly dispose of toxic and hazardous materials. These materials include many household products (insecticides, oven cleaner, furniture cleaners, oil-based paint), used automobile batteries, and used motor oil. To avoid contaminating ground water, no toxic or hazardous substance should be poured down the drain or thrown into the trash. Many communities will collect and properly dispose of your hazardous and toxic materials.

Helping save water

The Unites States is the world's biggest consumer of water. In many of the Western states, water is already in short supply. By the year 2000, most of the United States west of the Mississippi River is projected to face water shortages. Here are some ways you can reduce your water use.

- Turn off faucets and taps when not in use. Letting a faucet run while brushing your teeth wastes 11–19 L (3–5 gal.) per minute.
- One flush of the toilet uses 19–26 L (5–7 gal.). Fill a plastic bottle with water, and add a few stones for ballast. Placing the bottle in the toilet tank saves 4–8 L (1–2 gal.) per flush.
- Showers account for as much as 32 percent of home water use. Inexpensive, low-flow shower heads are easy to install, increase water pressure, and reduce shower-water use by 50 percent. Taking shorter showers also helps reduce water use.

Figure 16-14
Riding your bicycle for transportation helps save energy and reduce pollution.

Figure 16-15
By recycling newspapers, you can help reduce the amount of material added to your local landfill.

Figure 16-16
a We can all do our part to solve the Earth's environmental problems. Members of a community transformed a vacant lot in New York City into a community garden.

b The garden was once similar to this vacant lot across the street.

Why Learning About Ecology Is Important

Interdependence

How can your decision to walk or ride a bike to school affect global temperatures?

Humans rely on the Earth's ecosystems for food and for all of the other materials on which our civilization depends. It has been said that we do not inherit the Earth from our parents but borrow it from our children. We must preserve for them a world in which they can live. Each of us can contribute to solving today's environmental problems. Although it may seem that your actions will have a very small effect, the cumulative efforts of large numbers of people can result in significant changes. For instance, the difference between the two photographs in **Figure 16-16** shows the impact that students like you can make by joining a community effort to improve your neighborhood.

The most important way you can contribute to solving environmental problems is to make a very serious effort to *understand* the environment. You cannot preserve what you do not understand. Although solving the world's environmental problems will take the efforts of many different kinds of people, including politicians, economists, and engineers, the issues are largely biological. Your knowledge of ecology is an essential tool that you will need in order to contribute to the effort.

SECTION REVIEW

1. **Describe two examples of reductions in pollution that have occurred in the United States since 1970.**
2. **List the five essential steps to solving environmental problems.**
3. **Describe two things you can do to help save the environment.**
4. **Describe the advantages and disadvantages of using an ecological model to help plan a solution to an environmental problem.**

CHAPTER 16

Highlights

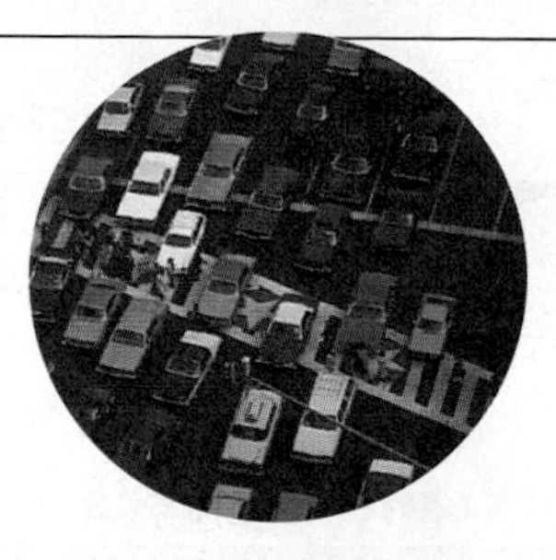

Automobiles are a major source of the carbon dioxide that causes global warming.

16-1 Planet Under Stress

Water pollution and acid rain threaten our water supply and the inhabitants of rivers, lakes, and streams.

Key Terms

pollution (p. 289)
carcinogen (p. 289)
acid rain (p. 291)
acid (p. 291)
pH (p. 291)
chlorofluorocarbon (p. 292)
global warming (p. 293)

Summary

- Human activities are causing changes to the entire planet.
- Pollution is anything potentially harmful that humans add to ecosystems.
- Sulfur-containing pollution from coal-burning plants mixes with water in the atmosphere, causing acid rain. Acid rain is killing forests and lakes.
- The protective ozone shield that surrounds our planet is being destroyed by chemicals called chlorofluorocarbons.
- The world's climate is warming as large amounts of carbon dioxide are released into the Earth's atmosphere.

16-2 Meeting the Challenge

The neem tree is a source of medicines, insect repellents, and many other useful products.

Key Terms

nonrenewable resource (p. 295)
renewable resource (p. 295)

Summary

- Humans need to reduce pollution, ensure adequate energy supplies, and conserve nonrenewable resources.
- Nuclear power is one alternative to fossil fuels, although there are significant safety problems associated with its use.
- We must act now to save nonrenewable resources such as topsoil, ground water, and species diversity.
- The exploding human population is the single greatest environmental problem.

16-3 Solving Environmental Problems

You can take action to improve the environment.

Summary

- Almost all environmental problems can be solved.
- Every citizen can help by reducing consumption of resources and by learning about ecology.

CHAPTER 16

review

Understanding Vocabulary

1. For each set of terms, complete the analogy.
   a. acid rain : sulfuric acid :: global warming : ____
   b. less atmospheric ozone : skin cancer :: more carbon dioxide : ____
   c. tree : renewable :: gasoline : ____

2. For each pair of terms, explain the differences in their meanings.
   a. pollutant, carcinogen
   b. ozone depletion, global warming
   c. renewable resources, nonrenewable resources

Understanding Concepts

3. **Relating Concepts** Construct a concept map that describes how humans can conserve the environment. Try to include the following terms in your concept map: environment, pollution, renewable, nonrenewable, acid rain, industrial chemicals, trees, topsoil, ground water, species, and fossil fuels.

4. **Recognizing Relationships** What problems were tall smokestacks designed to eliminate? Did they eliminate these problems? Explain why or why not.

5. **Recognizing Patterns** What is acid rain? Where in the United States is acid rain most severe?

6. **Summarizing Information** List three uses of CFCs.

7. **Analyzing Conclusions** Although production of CFCs is scheduled to stop by 2006, scientists think that CFCs will continue to destroy ozone for several decades afterward. Explain this prediction.

8. **Comparing Processes** What is the difference between the greenhouse effect and global warming? How are the two processes related?

9. **Organizing Information** Copy the following chart onto a separate piece of paper. Complete the chart by filling in the blanks.

Problem	Main cause	Consequence
Ozone depletion		
	Sulfuric acid	
Global warming		

10. **Summarizing Information** What two general approaches has the United States government taken to reduce pollution? How do these two approaches differ?

11. **Recognizing Relationships** How might widespread use of efficient appliances help reduce global warming? Give an example of one of these appliances.

12. **Contrasting Concepts** What is the difference between a renewable and a nonrenewable resource? Why is topsoil considered a nonrenewable resource?

13. **Interpreting Data** This chart shows the growth of the human population since 1800. How long did it take for the population to double from 1 billion to 2 billion? from 2 billion to 4 billion?

Population size	Year reached
1 billion	1800
2 billion	1930
3 billion	1960
4 billion	1975
5 billion	1987

14. **Summarizing Information** List three steps you can take to help reduce energy use. List three things you can do to help reduce pollution.

15. **BUILDING ON WHAT YOU HAVE LEARNED** In Chapter 14 you learned about ecological models. Explain how ecological models help scientists understand and plan solutions for environmental problems.

Interpreting Graphics

16. A biologist interested in population growth conducted an experiment in which she grew yeast in a closed ecosystem. A closed ecosystem is one with only a limited amount of space and resources, including food. The illustrations below show how the population changed over a period of five days. Each dot represents a single yeast cell. Answer the following question about the data.

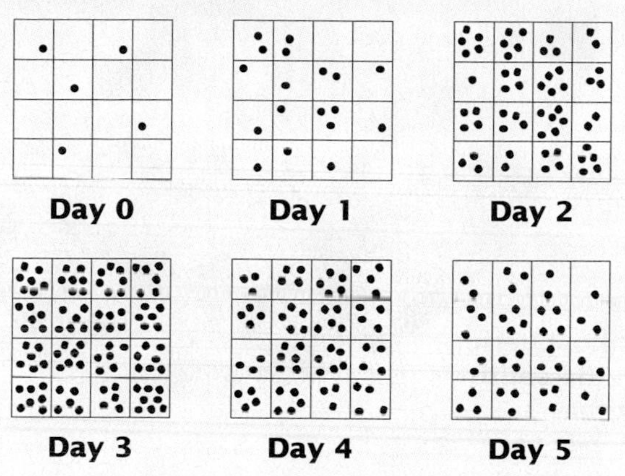

How do you think the population size will change between day 5 and day 10?

a. The population will drop slightly and then level off to the appropriate number of yeast cells for the ecosystem.
b. The population will drop slightly and then increase again as part of the natural cycle of yeast population.
c. The population size will decrease, possibly to 0, due to consumption of all available food and accumulation of wastes.

Explain your choice.

Reviewing Themes

17. ***Interdependence***
Describe two ways people in the United States would benefit from helping to protect tropical rain forests.

Thinking Critically

18. **Recognizing Logical Inconsistencies** A candidate for mayor of a large city proposed that the city stop spending money to encourage people to ride public transportation. He argued that the money was no longer needed because air-pollution levels in the city had dropped significantly. His opponent disagreed. Identify a logical flaw in the candidate's argument that the opponent can use to support his case. What additional data might help voters decide who was right?

Life/Work Skills

19. **Finding Information** The oil spill from the *Exxon Valdez* is one of the largest environmental disasters in human history. Although immediate efforts to contain the oil and minimize the damage were somewhat successful, no one knows what the long-term effects of this disaster will be. Use library references or search an on-line database for information about the *Exxon Valdez* oil spill and its consequences. What continuing efforts are underway to monitor and correct long-term effects on wildlife and the environment?

20. **Communicating** Contact your state division of wildlife or natural resources. Find out which animals and plants in your state are on the endangered species list, and select one of the animals or plants to investigate. Find out all you can about the organism, such as where it lives, what it eats, its predators, its reproduction rate, and what efforts are being made to save it. Summarize your findings, including a photo or a drawing of the organism, in a poster to share with your class.

Activities and Projects

21. **Mathematics Connection** Obtain census records from your town or community for the last 100 years, or longer if possible. Summarize the figures on a bar graph to share with your class. Predict what the population will be in your community 50 years from now and 100 years from now. What factors helped you make your predictions?

Discovering Through Reading

22. Read the article "The Alarming Language of Pollution" by Daniel Glick in *National Wildlife*, April/May 1995, pages 38–45. What developmental effects do pollutants such as dioxins have on animals such as gulls and turtles?

23. Read the article "The Return of the Peregrine" by Frank Graham in *Audubon*, October 1995, pages 20 and 22. What effect did DDT have on peregrine falcons? Describe two steps taken to help protect the peregrine falcon.

Can the Ozone Layer Be Saved?

Once the international community accepted the fact that the ozone hole was a direct result of human activity, solutions were sought to reverse the damage. But questions still remain: will these solutions work and can they be implemented in time?

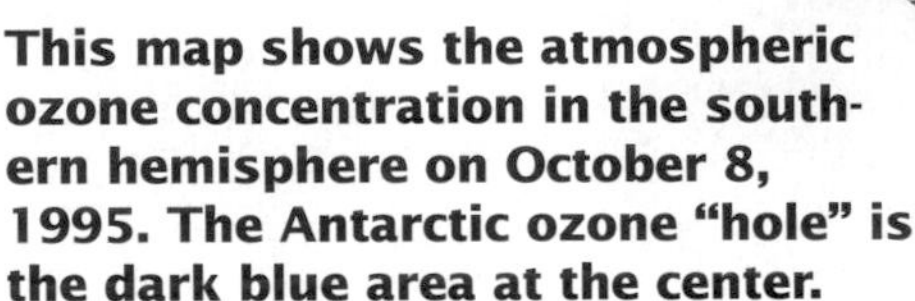

This map shows the atmospheric ozone concentration in the southern hemisphere on October 8, 1995. The Antarctic ozone "hole" is the dark blue area at the center.

The Problem

The ozone layer absorbs harmful ultraviolet radiation, preventing it from reaching the Earth's surface. The ozone "hole" is a thinning of the ozone layer over the Antarctic caused primarily by the release of synthetic chemicals called CFCs (chlorofluorocarbons). Further destruction of atmospheric ozone will cause extensive health and environmental problems.

A Study of Ozone Destruction

After the first reports of the ozone hole were published in 1985, an international meeting, called the Vienna Convention, initiated an effort to save the ozone layer. Ultimately, 130 nations pledged to reduce the use of CFCs, share information, and cooperate in monitoring the ozone hole through the United Nations–sponsored Intergovernmental Panel on Climate Change (IPCC). Because a great deal of information about ozone levels comes from satellites, NASA established the Ozone Trends Panel in 1986 to evaluate data collected during satellite studies and to recommend solutions to the problem of ozone destruction.

In 1986 and 1987, extensive research produced a clear picture not only of the size of the ozone hole, but the cause. Measurements of ozone and other gases were taken from balloons floating above the Antarctic, from airplanes flying through the hole, and from satellites orbiting the Earth. These measurements showed increasing amounts of chlorofluorocarbons, which contain chlorine. One chlorine atom can destroy 100,000 ozone molecules. The results were clear: CFCs and other synthetic chemicals were responsible for significant ozone destruction.

A Time for Action

In 1987, 149 countries approved the Montreal Protocol, an agreement calling for drastic reductions in CFC production. Additional meetings among climatologists, other scientists and policy makers in 1990 and 1992 amended the Montreal Protocol to require complete elimination of CFCs by the year 2000. Limits on other ozone-destroying chemicals were also imposed. Production of new CFCs in the United States ceased on January 1, 1996.

The Good News...

Due to public outcry and boycotts, some reductions in CFC use occurred even before the Montreal Protocol took effect. Manufacturers stopped using CFCs in aerosol products, and many fast-food restaurants switched from foam containers, produced with CFCs, to paper wrappers.

Industrialized nations have developed alternatives to CFCs. Scientists predict that if the Montreal Protocol is strictly followed, the ozone hole will begin to repair itself after the turn of the century.

...And the Bad News

Some industries have stockpiled CFCs. Although production will cease, the use of CFCs will continue for some time. In addition, many countries, such as those of the former Soviet Union, might not comply with the Montreal Protocol and may continue to make CFCs indefinitely. This noncompliance has already created a black market for these chemicals. Large amounts of CFCs are smuggled from these countries into other countries, including the United States, and are sold illegally.

Technique

Climate Modeling

Atmospheric measurements of ozone-destroying chemicals and images of the ozone hole are gathered on a daily basis. These data are interpreted with computers in order to design models. A model seeks to provide answers to a "what if" scenario.

In climate modeling, computers analyze the data from satellites to predict our climate. The accuracy of climate modeling depends on the accuracy of the data gathered and our understanding of the complex environmental relationships among climate, land, and ocean. Scientists used such models to correctly predict the temperature changes caused by the eruption of Mount Pinatubo in the Philippines in 1991.

Applications: weather forecasting; monitoring, and predicting climate change; predicting future trends in environmental and human health problems

Users: meteorologists; epidemiologists; atmospheric, environmental and marine scientists

Analyzing the Issues

1 Why did CFC use stop only *after* the ozone hole formed?

Researchers predicted the destructive nature of CFCs long before the ozone layer was damaged. In 1995, the Nobel Prize in chemistry was awarded to Paul Crutzen, Mario Molina, and F. Sherwood Rowland for their work on ozone. Research what these scientists contributed to our understanding of the ozone hole.

2 Are the alternatives to CFCs safe?

The Montreal Protocol calls for the elimination of production of CFCs by the year 2000. Although CFC alternatives such as HFC-134a and HFC-23 do not destroy ozone, they may present different environmental problems. Research these alternatives and the risks involved with their use.

3 How can the Montreal Protocol be enforced?

To effectively reduce the hole in the ozone, we must stop the production of ozone-destroying chemicals on a global level. Several nations allow the continued production and use of CFCs. What countries are not complying with the Montreal Protocol? What could be done to encourage compliance?

4 How can existing supplies of CFCs be destroyed safely?

Because CFCs are chemically stable and unreactive, it is difficult to dispose of them in an environmentally friendly way. In January of 1996, a new method of eliminating CFCs was introduced. Read "CFCs Transformed Into Pillar of Salt," in the January, 1996 *Science News*, page 36. How does this method of CFC disposal work?

UNIT

Four

Diversity of Life

Sheep grazing on South Island, New Zealand

FOOD FOR THOUGHT

The next time you eat, think about your food. Is it from a plant that received artificial fertilizer or one that was nourished with compost and mulch? Is it from an animal that was grazed on grass or from one that was fed a fortified formula?

Getting the Job Done

Farmers need high yields per acre and healthy crops and livestock in order to feed millions of consumers like you. For decades, they turned exclusively to commercial fertilizers, herbicides, and pesticides to help them. Now some chemicals are suspected of fouling streams, polluting ground water, and endangering other forms of life.

Back to Basics

"Sustainable agriculture" is a form of agriculture that aims to improve the soil by cooperating with nature. It is helping some farmers achieve impressive yields with fewer chemicals and less machinery, and it is producing healthier conditions for workers, plants, and beneficial bacteria and insects.

LOOKING AHEAD

- How do beneficial bacteria help cycle nutrients among plants, animals, and the soil? See page 336.
- What role can composting play to sustain trees, flowers, or crops? See page 423.
- How are new drugs to fight harmful bacteria and viruses being tested for human use? See **Science, Technology, and Society: New Drug Development—How Are New Drugs Tested?** pages 354–355.
- How might art and science work together to beautify an urban landscape? See **Career Opportunities: Landscape Architecture,** page 425.

CHAPTER 17

REVIEW
- characteristics of living things (Section 2-1)
- natural selection (Section 10-1)
- six kingdoms (Section 11-2)

Enoplotrupes sharpi—Thailand

Sternocera aequisignata—Thailand

Sphingnotus miriabilis—Papua New Guinea

Classifying Living Things

Scientists give each kind of organism on Earth a unique name.

Chrysocarabus auronitens—France

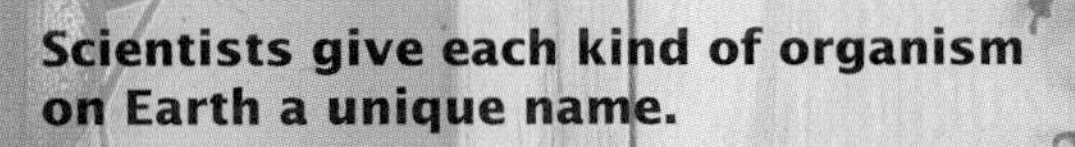

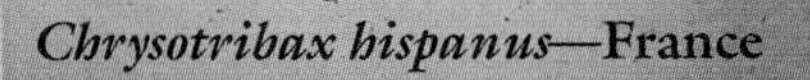

Chrysotribax hispanus—France

Eupholus bennetti—Papua New Guinea

Euchroma gigantea—Brazil

Chrysochroa buqueti—Malaysia

Plusiotis gloriosa—USA

Do the words *inu*, *chien*, and *hund* mean anything to you? These are Japanese, French, and German words for "dog." Could you conduct a conversation about dogs with someone who spoke only one of these languages? Scientists—regardless of their native language—can communicate about dogs because they have the same name for the dog: *Canis familiaris*.

17-1 The Need for Naming

OBJECTIVES

1. **Explain why scientists use scientific names instead of common names.**
2. **Describe the scientific system of naming organisms.**
3. **Explain why scientific names are in Latin.**
4. **Evaluate the role of Linnaeus in creating the modern system of naming organisms.**

The Importance of Scientific Names

What do you and your classmates call the "bug" shown in **Figure 17-1**? You will probably have many common names for this animal, including sow bug, pill bug, wood louse, roly-poly, and potato bug. If some of your classmates are from other countries, you could collect an even longer list of names.

But if you ask a biologist to name this creature, you will receive only one answer: *Porcellio scaber*. Each kind of organism on Earth is assigned a unique two-word **scientific name**. *Porcellio scaber* is the scientific name of the animal shown below. *Homo sapiens*, which means "wise man," is our species' scientific name. All biologists, regardless of their native language, use scientific names when speaking or writing about organisms.

Most organisms also have common names. Why don't scientists use them? Although adequate for everyday use, common names may be confusing in scientific communication. As you have seen, an organism can have more than one common name. In addition, science requires international cooperation and communication, and an organism rarely has the same name in different languages. Finally, one common name often refers to more than one kind of living thing. The plant we know as corn in North America is called maize in Great Britain. To a resident of Britain, corn is the plant we call wheat. When a biologist writes a scientific paper on *Zea mays*, however, other scientists know the subject of the paper is the American "corn" plant. Using scientific names enables scientists, no matter what language they speak, to exchange information about an organism and to be certain that they are referring to the same living thing.

Figure 17-1
These are some of the common names for this animal (240×). Scientists have assigned it a single scientific name: *Porcellio scaber*.

What's in a Scientific Name?

The first word in a scientific name describes the organism in a general way. The second word identifies the exact kind of living thing. We use a similar naming system in our everyday speech. For instance, your last name identifies your family, while your first name specifies exactly who you are.

The first word of a scientific name is the name of the **genus** *(JEE nuhs)* to which the organism belongs. (The plural of genus is *genera.*) A genus is a group of organisms that share major characteristics. For example, all oak trees produce acorns. Therefore, all oak trees are assigned to the genus *Quercus*, which means "oak" in Latin.

There are dozens of different kinds of oak trees. One kind is tall, while another is low and spreading. Leaves and acorns of different oak trees can vary in size and shape, as shown in **Table 17-1**. The second word in a scientific name identifies one particular kind of organism within the genus. For example, *Quercus rubra* is the red oak, and *Quercus phellos* is the willow oak. Scientists call each different kind of organism a **species** *(SPEE sheez).* (The plural of species is *species.*) The correct name for an organism must include *both* parts of its scientific name. The red oak is properly called *Quercus rubra*, not just *rubra*.

Table 17-1 Comparison of Red Oak and Willow Oak

	Red oak	Willow oak
Genus name	*Quercus*	*Quercus*
Scientific name	*Quercus rubra*	*Quercus phellos*
Traits	Acorns about 25 mm (1 in.) long	Acorns about 15 mm (0.5 in.) long
	Common in open Northeastern forests; tolerant of city soot and cold temperatures	Popular shade tree found in the South; grows well in rich, moist soil
	Lobed leaves	Unlobed, narrow leaves

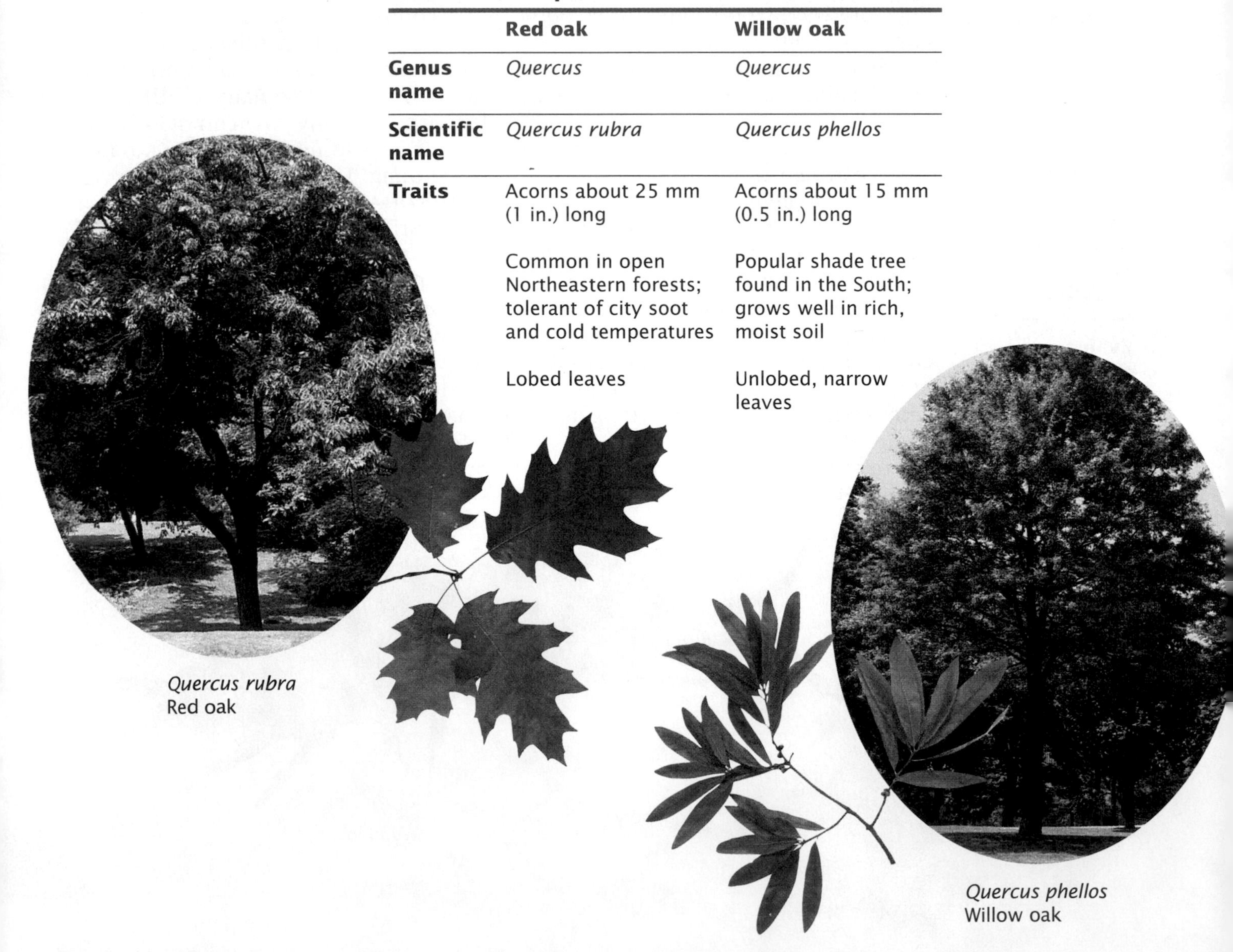

Quercus rubra
Red oak

Quercus phellos
Willow oak

Figure 17-2

a **The second word of a scientific name is often descriptive of an organism or its distribution. The green anole lizard *Anolis carolinensis* and the chickadee *Parus carolinensis* are both found in North Carolina and South Carolina.**

Scientific names must conform to a set of rules

The name given to a newly discovered species must conform to rigorous rules formulated by authorized groups of scientists. All scientific names must consist of Latin words or of terms constructed according to the rules of Latin grammar. Two different organisms cannot be assigned the same name. Because all members of a genus share their genus name, the second word in the name must be different. Only one member of the genus *Homo* can be given the name *sapiens*. Organisms in different genera cannot have the same genus name but can share the second word of their scientific names. For example, the green anole lizard *Anolis carolinensis* and the chickadee *Parus carolinensis*, shown in **Figure 17-2a**, share the name *carolinensis* because they both occur in North Carolina and South Carolina.

When choosing a name for a species, biologists often pick a name that describes the appearance or distribution of an organism. One of the most vivid names belongs to the ferocious dinosaur *Tyrannosaurus rex*, shown in **Figure 17-2b**. *Tyrannosaurus rex* means "tyrant-lizard king," a fitting name for a carnivore with teeth 15 cm (6 in.) long. Sometimes scientific names are a tribute to the discoverer of a species or to an admired colleague or teacher of the discoverer. The frog *Rhinoderma darwinii*, shown in **Figure 17-2c**, the lizard *Liolaemus darwinii*, and the bird *Rhea darwinii* are three of the many species named to honor Charles Darwin.

b ***Tyrannosaurus rex*, which means "tyrant-lizard king," was named for its enormous teeth and tremendous size. This dinosaur measured about 15 m (49 ft) in length.**

c **The frog *Rhinoderma darwinii* was named to honor Charles Darwin.**

Figure 17-3
If you asked scientists from around the world what they might call the animal in Figure 17-1, they would all agree on one name: *Porcellio scaber*.

"טחבית *(tah hah VEET)* or *Porcellio scaber*"
—Jacob Cohen, Molecular Evolution Biologist
Israel

Why are scientific names in Latin?

Why are scientific names written in a language that is no longer spoken? Why aren't they in English or Russian or Chinese? In the Middle Ages, when scientists began to name organisms, Latin was used in academic circles. Hence, scientists and other scholars found it easier to communicate with each other in Latin. Latin was the language of the scholar and was used for all spoken and written communication.

Although scientists no longer communicate with each other in Latin, it is easier and more logical to retain the Latin names for living things than to rename all 1.4 million known organisms in a more modern language. **Figure 17-3** shows how useful a Latin name can be when scientists from around the world communicate with each other.

Linnaeus devised the two-name system

The modern system of naming organisms was the brainchild of Swedish botanist Carl Linnaeus. In Linnaeus's day, organisms were given very long Latin names (sometimes more than 15 words), which were often changed according to the whims of particular scientists. Linnaeus assigned a standard, two-word Latin name to each organism known in his time.

Writing a scientific name is simple

When you write a scientific name, always capitalize the genus name. Begin the second word with a lowercase letter. Both parts of a scientific name are underlined or written in italics. The scientific name for humans can be written as either Homo sapiens or *Homo sapiens*. After the first use of the full scientific name, the genus name can be abbreviated as a single letter if the meaning is clear. For example, since *Homo sapiens* has just been mentioned, *H. sapiens* is acceptable.

"Cloporte *(klo PORT)* or *Porcellio scaber*"
—Paul Melançon, Chemist and biochemist
Canada

"Mangy sow bug or *Porcellio scaber*"
—Maria Alma Solis, Entomologist
United States

"КОЛОРАДСКИЙ ЖУК *(koh loh RAD skee JOOK)* or *Porcellio scaber*"
—Andrei Lapenis, Earth systems scientist
Russia

"Cucaracha *(koo kah RAH chah)* or *Porcellio scaber*"
—Ernest H. Williams, Jr., Marine biologist
—Lucy Bunkley-Williams, Microbiologist
Puerto Rico

SECTION REVIEW

1. **List and explain the problems associated with using common names to identify living things.**
2. **Why is it necessary to use both words of a scientific name to correctly identify an organism?**
3. **Explain the advantages of using Latin for scientific names.**
4. **Evaluate the accuracy of this statement: "Linnaeus was the first scientist to give species Latin names."**

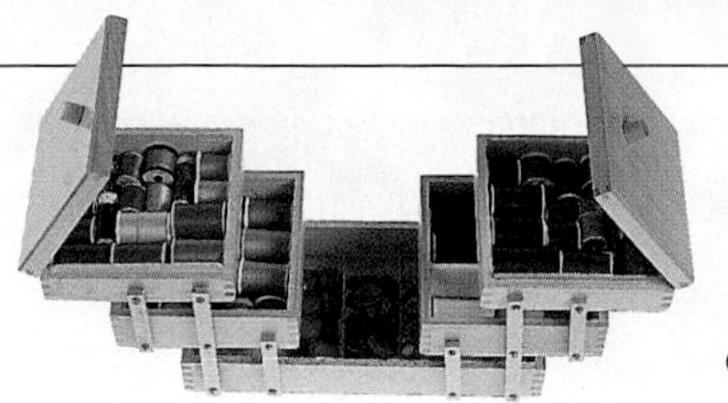

Because sewing supplies in a sewing box are organized, you can go straight to the compartment where cotton thread is kept to find red cotton thread, instead of searching the entire box. Living things have been similarly organized by scientists, but not for convenience only. Living things are organized into groups to reflect evolutionary relationships.

17-2 Classification: Organizing Life

OBJECTIVES

1. **Describe the system scientists use to classify organisms.**
2. **How does the classification of living things reflect their evolutionary history?**
3. **Distinguish between the methods of classification used by taxonomists.**
4. **Define the term *species*.**

Classification of Living Things

Humans have classified organisms for thousands of years. The Greek philosopher Aristotle (384–322 B.C.) grouped animals by their physical similarities. Some of those groupings are still recognized today, such as the mammals. In the Middle Ages, herbalists used plants to treat disease, and they needed to identify which plants were poisonous and which could heal. Herbalists produced manuals of plant types, called "herbals," like the one shown in **Figure 17-4**. In herbals, plants were organized by their medicinal uses.

Today biologists classify organisms based on their physical, genetic, biochemical, and behavioral similarities. These similarities reveal evolutionary relationships. The classification of organisms is based on decisions made by many scientists using available information. Not all scientists agree on these classifications, and additional information or new technology can cause them to reevaluate their conclusions. The science of classifying living things is called **taxonomy**. Taxonomists are scientists who examine, classify, and argue about where organisms fit in a group.

The system of classification that biologists have developed over the 250 years since Linnaeus wrote his fundamental works is a hierarchical system. In a hierarchical system of classification, species are assigned to genera, genera are assigned to families, and families are assigned to groups of increasing size. A similar system is used in the U.S. Army. Each soldier belongs to a squad containing about nine soldiers. Four squads are organized into a platoon. Each platoon belongs to a company of about 150 people, and so on.

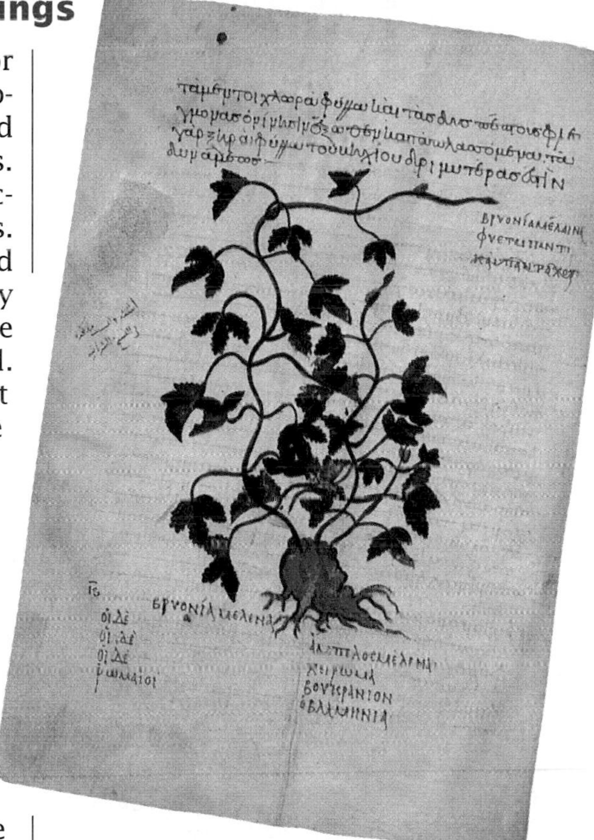

Figure 17-4
This page from a manual called a herbal was published in Turkey in the tenth century. The Greek text classifies the plant illustrated according to its medicinal uses.

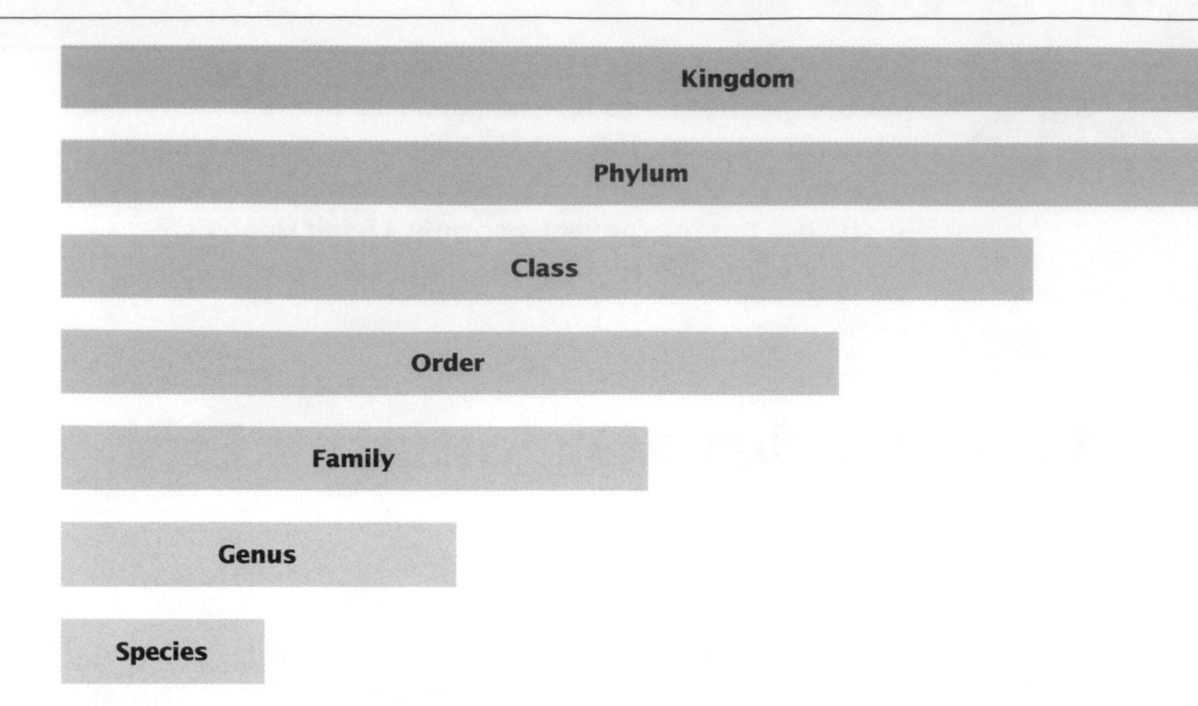

Figure 17-5
The biological hierarchy of classification is made of seven different levels.

Evolution

How is the hierarchy of classification based on evolutionary relationships among organisms?

Organisms are classified by similarity

In the army, soldiers are assigned to a group without regard to their appearance; there are no companies of only tall soldiers or of only blond soldiers, for instance. In a biological classification, however, organisms are assigned to a group because they share distinctive characteristics with the other members of that group. The biological hierarchy of classification has seven different levels: kingdom, phylum, class, order, family, genus, and species, as shown in **Figure 17-5**.

The smallest group in biological classification is the species. Similar species are collected into a genus. Similar genera are united into a **family**. Families that are alike are combined into an **order**. Similar orders are collected into a **class**. Classes are united into a **phylum** *(FEYE luhm)*. Finally, similar phyla are collected into a **kingdom**. The term *division* was formerly substituted for *phylum* in classifications of plants, bacteria, and fungi. **Table 17-2**, at right, illustrates the classification of our species, *Homo sapiens*.

The more classification categories two species share, the more traits they have in common. For instance, a house cat and a guppy belong to only two of the same categories: the kingdom Animalia and the phylum Chordata. The guppy and the cat both have bony skeletons, notochords, and nerve cords. But only the cat breathes with lungs, is covered with fur, walks on four legs, and nurses its young with milk. The guppy breathes with gills, has fins and scales, and does not nurse its young. The house cat, *Felis cattus*, and the mountain lion, *Felis concolor*, belong to six of the same classification categories: kingdom Animalia, phylum Chordata, class Mammalia, order Carnivora, family Felidae, and genus *Felis*.

If more specific groups are needed, the seven classification categories can be subdivided into smaller units. For example, the phylum Chordata includes all animals that develop notochords. The great majority of the members of Chordata also have backbones. To reflect this difference, animals with backbones are placed into their own subphylum—Vertebrata.

Table 17-2 The Classification of Modern Humans

Homo sapiens | *Homo erectus* | *Australopithecus* | Gorilla | Elephant | Fish | Snake | Sea star | Earthworm | Snail

Kingdom Animalia		Chordates, sea stars, earthworms, snails, jellyfish, sponges, clams, and insects are members of the kingdom Animalia.
Phylum Chordata		Mammals, fishes, reptiles, birds, and amphibians are chordates, members of the phylum Chordata.
Class Mammalia		Primates and elephants, along with cats, dogs, horses, kangaroos, whales, bats, seals, dolphins, and many others, belong to the class Mammalia.
Order Primates		Members of the family Hominidae, along with prosimians, monkeys, and apes such as the gorilla, make up the order Primates.
Family Hominidae	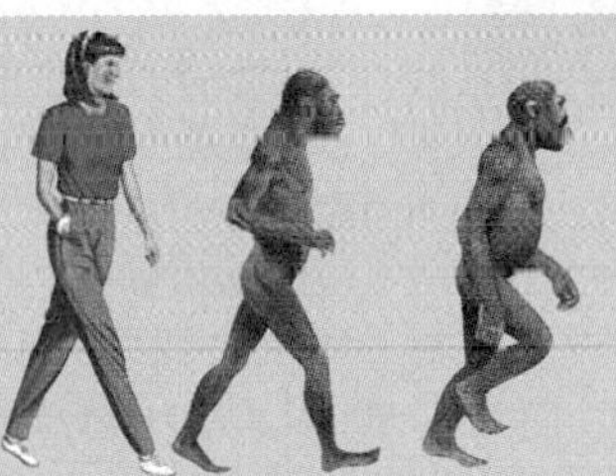	The family Hominidae includes the genus *Homo* and the extinct genus *Australopithecus*.
Genus *Homo*	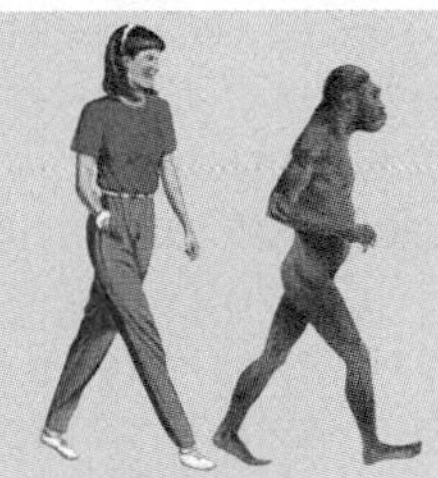	*Homo sapiens* belongs to the genus *Homo*, along with the extinct species *Homo habilis* and *Homo erectus*, which is shown here.
Species *Homo sapiens*		Modern humans belong to the species *Homo sapiens*.

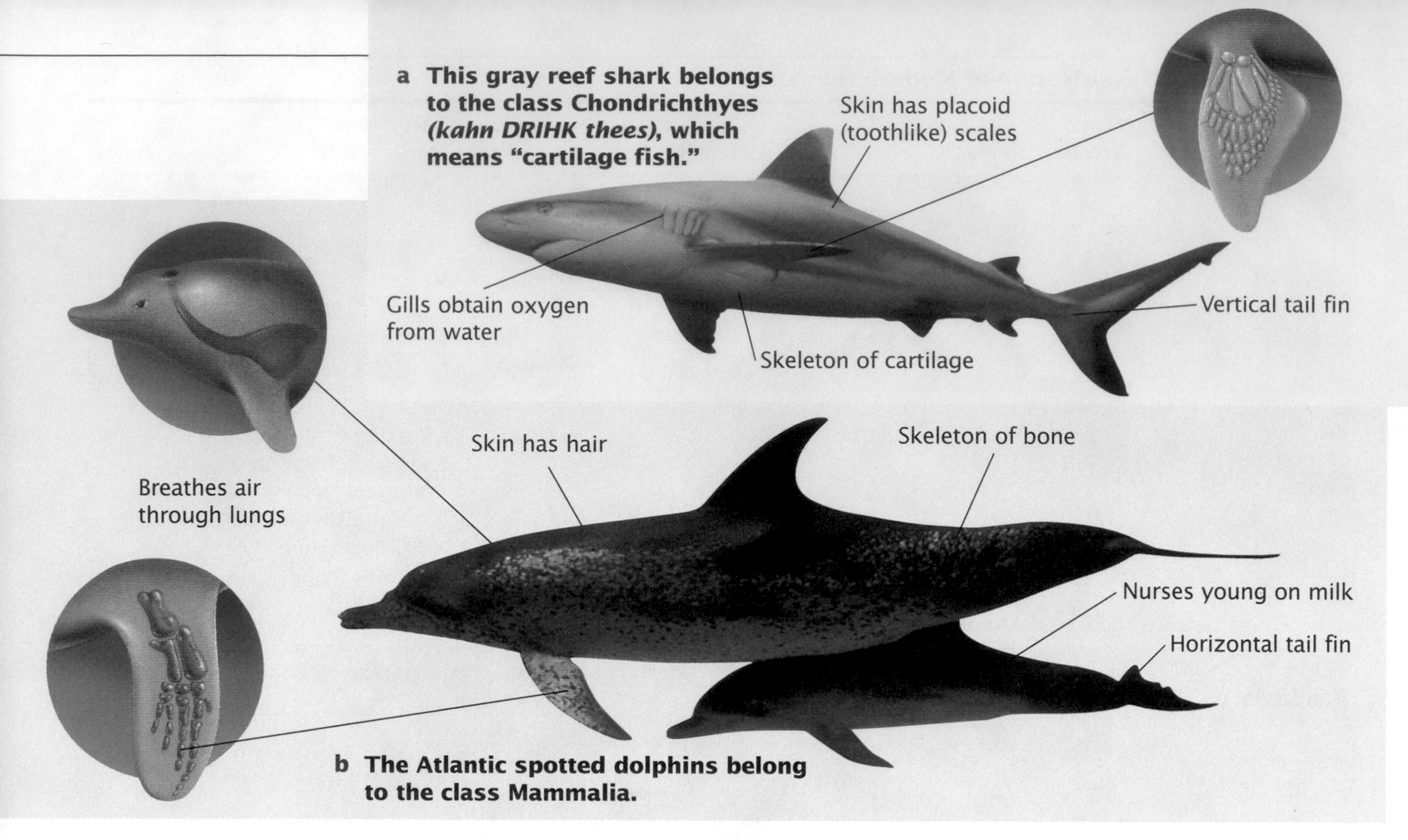

Figure 17-6 Although the shark and the dolphin look somewhat alike, they belong to different classes. This is an example of convergent evolution.

Classification and Evolution

The biological hierarchy of classification is based on the fact that different degrees of similarity exist among organisms. For instance, mountain lions more closely resemble house cats than do guppies. For Darwin, classification provided strong evidence supporting evolution. Organisms are similar because they descended from a common ancestor. The more similarities two organisms share, the more recently they shared a common ancestor. For instance, the ancestor of the house cat diverged from that of the mountain lion just a few million years ago. In contrast, the ancestor of guppies and the far-distant ancestor of cats diverged about 390 million years ago. Thus, the more classification categories two organisms share, the more closely related they are.

Similarity does not guarantee close relationship

Compare the two ocean-dwelling animals in **Figure 17-6**. Both have streamlined bodies, paddlelike fins, and flattened tails. Would you say these organisms are closely related? Many people classified both as "fish." Although both belong to the subphylum Vertebrata, they are now placed in different classes.

The shark and dolphin are an example of **convergent evolution**. In convergent evolution, organisms evolve similar structures independently, often because they live in similar habitats. Similar features that evolved through convergent evolution are known as **analogous structures**. Homologous structures, which you read about in Chapter 10, are similar because of common ancestry. Analogous structures are similar because of similar selection of two or more species.

Convergent evolution creates problems for scientists trying to classify species, because it means that similar appearance does not guarantee common ancestry. Taxonomists, therefore, view a classification as a hypothesis of relationships between organisms. Like any hypothesis, a classification can be tested against the available evidence and disproved or supported. Because the number of differences between sharks and dolphins far exceeds the number of similarities, it is easy to reject the hypothesis that these animals are close relatives.

Methods of Taxonomy

Matter, Energy, and Organization

How could methods of obtaining energy be used as derived characters in a cladogram of the six kingdoms?

The example of the shark and dolphin illustrates the difficulty in determining which similarities will be useful when classifying an organism. There are two alternative methods of choosing which similarities are important. The first method is cladistics *(kluh DIHS tihks)*, from the Greek word *klados*, meaning "branch." The scientists who use cladistics seek to determine the order in which evolutionary lines diverged, or branched. To do so, they consider only a restricted set of characters of the organisms they want to classify. Organisms are assigned to a group because they share unique characters not found in any other organisms. These unique characters are called **derived characters**. For example, all species of mammals share the derived characters of hair and the ability to produce milk. Using patterns of shared derived characters, scientists construct branching diagrams called **cladograms**, which show the evolutionary relationships among groups of organisms. A cladogram of the major groups of plants is shown in **Figure 17-7**.

The second method for classifying organisms is phenetics *(fuh NEHT ihks)*. Scientists using phenetics consider as many characters of organisms as possible and classify organisms into groups based on overall degree of similarity. Phenetics does not attempt to reconstruct the relationships among organisms, just to produce groups that can be named.

Both cladistics and phenetics have drawbacks. Phenetic classifications do not always reveal the relationships among organisms. And while cladistics is able to reconstruct the sequence of evolutionary divergence, it does not reveal the amount of difference between groups.

The taxonomy used in this book and by many biologists reflects both overall similarity and the sequence of evolutionary divergence—a compromise between phenetics and cladistics.

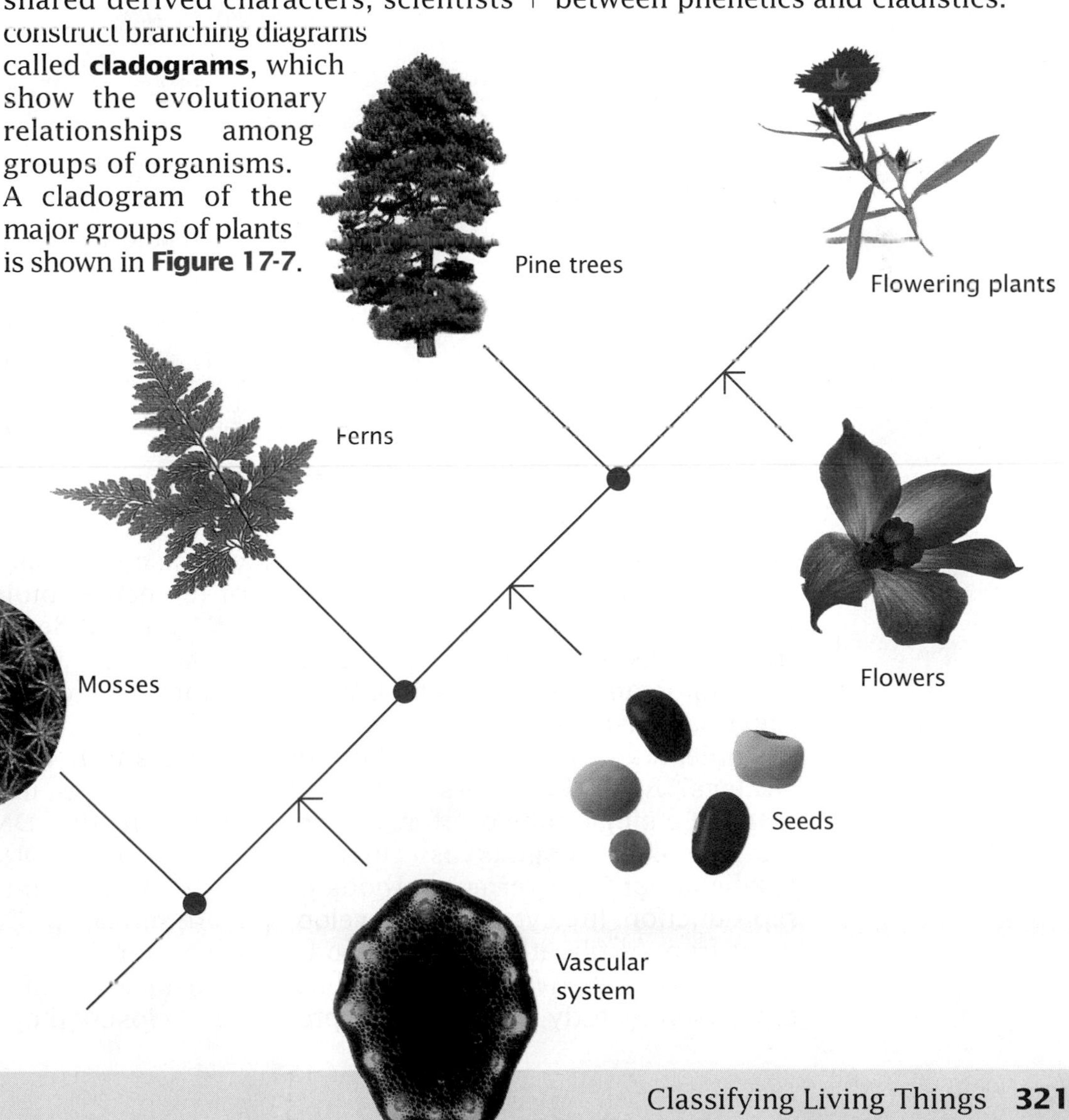

Figure 17-7
This cladogram shows the evolutionary relationships among the selected phyla of plants. A derived character distinctive to each group is marked on the cladogram. These derived characters indicate the divergence of one group from the common ancestor shared with the others.

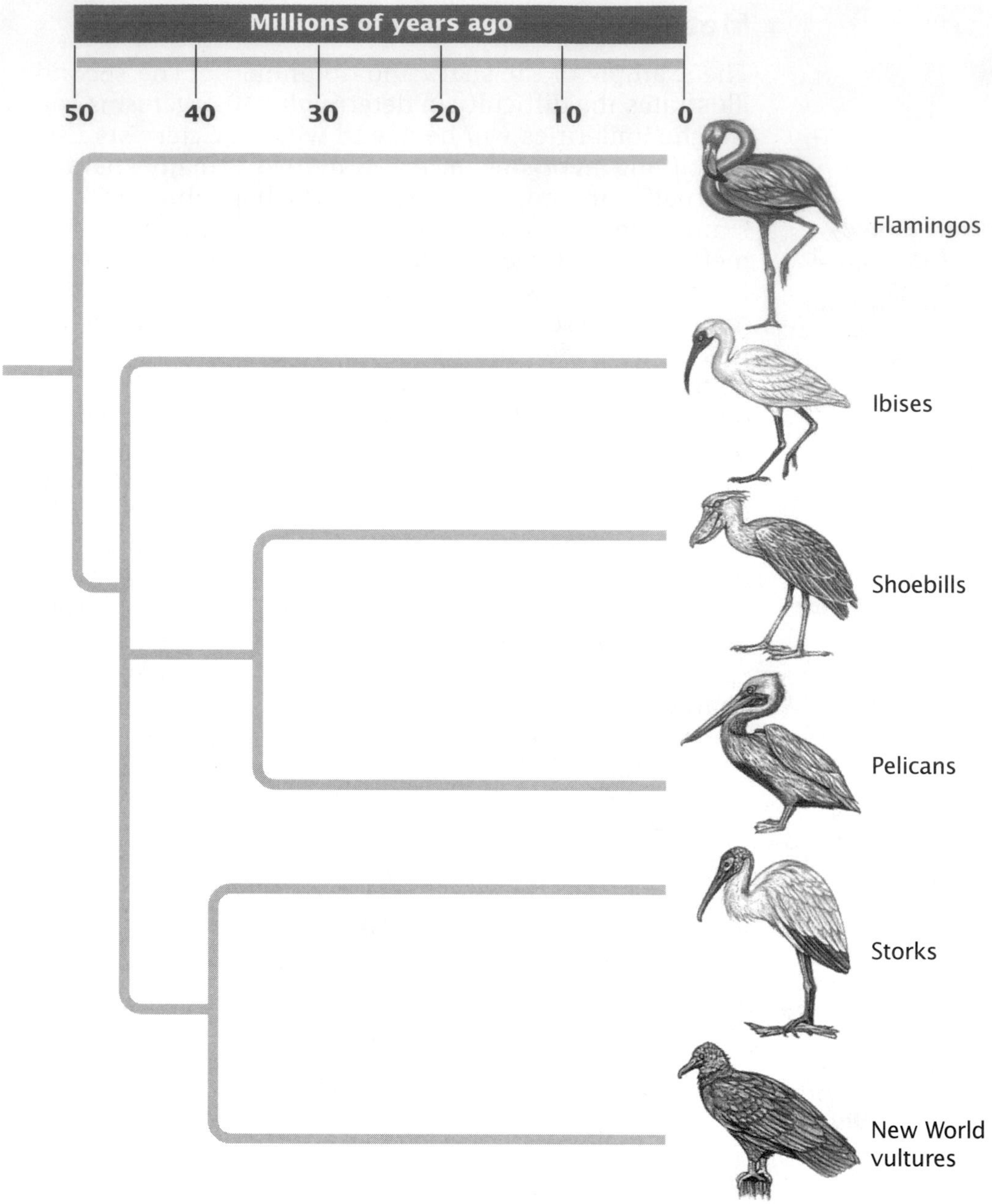

Figure 17-8 Comparison of the DNA sequences of these birds reveals their evolutionary relationships. For example, the DNA sequences of flamingos differ by the greatest amount. Their ancestor diverged from the other birds about 50 million years ago.

Millions of years ago

50 40 30 20 10 0

Flamingos

Ibises

Shoebills

Pelicans

Storks

New World vultures

Taxonomy and Technology

What characteristics of organisms do you think might be useful for determining evolutionary relationships? As you might expect, biologists have traditionally compared the appearances of organisms in order to discover the relationships among them. For example, does the animal have four or six legs? Are the flowers of a plant made of a single tube or of many separate petals? Biologists also consider the behavioral patterns, methods of reproduction, life cycles, and development from fertilization to adulthood.

Technological advances have enabled biologists to study the genes that produce the traits used to classify organisms. Taxonomists use techniques of molecular biology to compare the DNA nucleotide sequences of different organisms. Comparisons of DNA sequences are especially important for the taxonomist because mutations—changes in DNA—are random events. As time passes, more mutations tend to occur in the DNA of a particular species. Thus, DNA acts as a "molecular clock." As shown in **Figure 17-8**, the more similar the DNA sequences of two species, the more recently their common ancestor must have lived, and the more closely they are related.

What Is a Species?

Why do we say that house cats belong to one species while mountain lions belong to another? What is a species? In one sense, a species is just a level in the classification system to which scientists assign very similar organisms. But in a more profound sense, a species is the basic unit of evolution. Over time, species change and give rise to new species in a process known as speciation. Indirectly, speciation gives rise to new genera, new families—all of the so-called higher classification categories. A new genus, for example, is not produced by the transformation of all members of an existing genus. A new genus is formed when one species in the "old" genus accumulates enough changes to be considered not only a new species but also a member of a new genus.

Figure 17-9
The zebra and horse belong to the same genus *(Equus)* but are members of different species. When they mate, they produce sterile offspring, known as a "zebroid," like the one below.

Biologists have traditionally defined a species as organisms that are able to interbreed with each other to produce fertile offspring and that usually do not reproduce with members of other groups. This definition works well for most animals. For example, the horse and the zebra belong to different species. Although they can mate, the resulting offspring, the "zebroid" shown in **Figure 17-9**, is sterile.

Reproductive barriers between species are not always perfect. Indeed, the traditional definition of species does not work nearly as well for plants because hybrids of different species are often fertile. **Hybrids** are offspring that result from interbreeding by individuals of different species. Coyotes, dogs, and wolves are all separate species in the genus *Canis*. Interbreeding between dogs, coyotes, and wolves produces fertile hybrids.

A species is a unique kind of organism

Despite these complications, the classification of organisms into species has worked fairly well. The word *species* means "kind" in Latin. Therefore, a species is basically a unique kind of organism. Members of a species share at least one inherited characteristic not found in other similar organisms. The characteristic that sets a species apart might be the shape of a leaf, beak, flower, or tooth. In sexually reproducing species, this distinctive characteristic is maintained from generation to generation because members of different species do not usually interbreed.

SECTION REVIEW

1. **Explain why two species in the same genus must also belong to the same family.**
2. **Explain how convergent evolution can make classification difficult.**
3. **Describe how cladistics classifies organisms.**
4. **Why aren't the breeds of domestic dogs considered separate species?**

The ability to move is probably one trait that you associate with animals, while the ability to photosynthesize is characteristic of plants. *Euglena* is a single-celled organism that is able to move and to perform photosynthesis. Is *Euglena* a plant or an animal? There are a number of organisms that cannot be categorized as animals or plants.

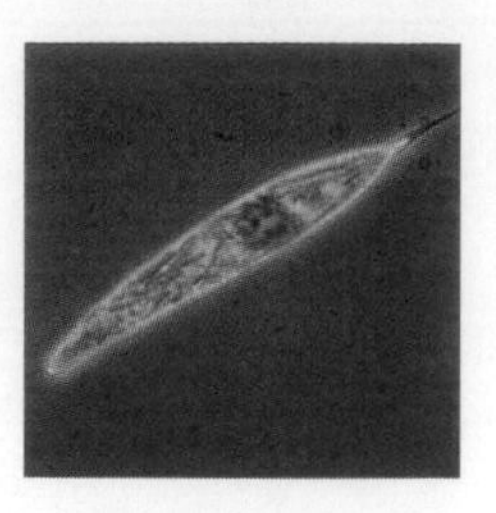

17-3 Six Kingdoms

OBJECTIVES

1. **Discuss the weaknesses of Linnaeus's two-kingdom classification system.**
2. **Differentiate between the six kingdoms of organisms.**
3. **Identify two characteristics of members in each of the six kingdoms.**

Six-Kingdom System

Following Linnaeus's lead, biologists of his time classified every living thing into either kingdom Plantae or kingdom Animalia. This seemed logical because most familiar organisms are either plants or animals. If it is green, has leaves, and is rooted in soil, it is a plant. If it is furry, slimy, or scaly and it runs, swims, crawls, or slithers, it is an animal. However, numerous living things do not quite fit either description. For example, where would a mushroom fit? It doesn't seem like an animal. On the other hand, is it a plant if it doesn't have leaves and isn't green?

If you had to fit a mushroom into one of the two kingdoms—Plantae or Animalia—you would most likely place it in kingdom Plantae. That is what Linnaeus did. After all, a mushroom does look a little like a plant. It grows in the soil and doesn't move from one place to another.

Does this mean that any organism that doesn't move from place to place is a plant? Sea anemones, such as the one in **Figure 17-10**, are organisms that are firmly attached to one spot and look like flowers. They eat other organisms, however, and are animals. Since Linnaeus's time, biologists have learned a great deal about the structure and function of living things. This information has enabled them to make increasingly precise distinctions among the major groups of organisms. Most biologists now use a six-kingdom system of classification. The six kingdoms are Archaebacteria, Eubacteria, Protista, Fungi, Plantae, and Animalia.

Figure 17-10
Like plants, sea anemones cannot move from place to place. However, they are classified as animals. They are heterotrophs, using tentacles to paralyze and capture small fishes.

Matter, Energy, and Organization

Which traits of bacteria are shared by all living things?

The Six Kingdoms of Living Things

All prokaryotes, also called bacteria, are in the kingdoms Archaebacteria *(ahr kee bak TIHR ee uh)* or Eubacteria *(yoo bak TIHR ee uh).*

The bacteria represent the most ancient groups on Earth. They have adapted to almost every environment, including those with extremes of acidity, temperature, and salt content. In addition, bacteria have evolved more ways to obtain nutrients than have all eukaryotic organisms combined.

Bacteria
- are unicellular
- lack cell nuclei
- lack membrane-bound organelles
- have more methods of obtaining nutrients than eukaryotes
- are the oldest forms of life

Archaebacteria
- found in extreme environments
- gave rise to eukaryotes

Eubacteria
- common environments
- gave rise to eukaryotic cell organelles

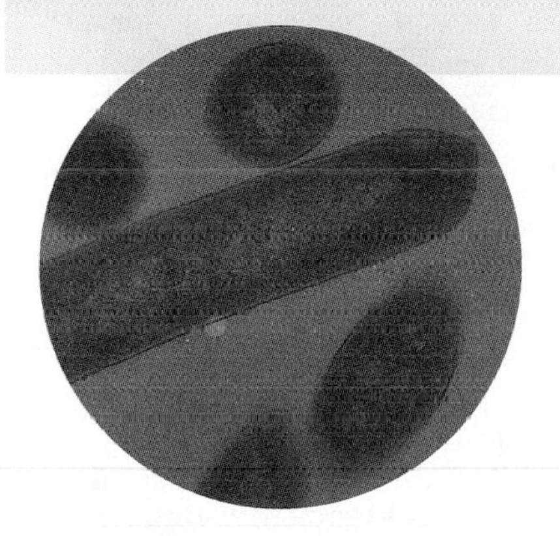

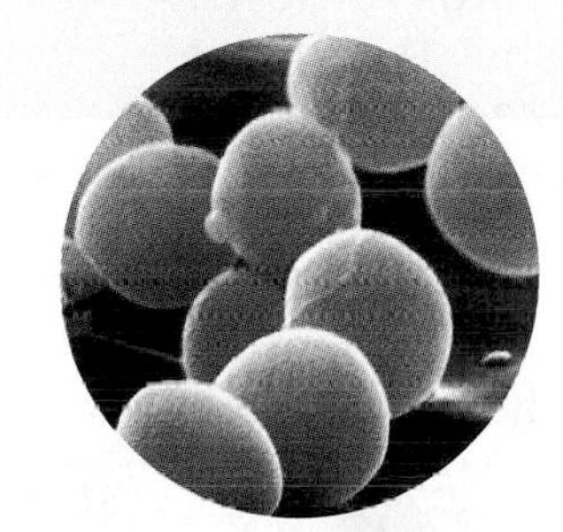

Figure 17-11
a Archaebacteria, like this one, have been found in extreme environments.
b These eubacteria can live on human skin and inside the human body.

All bacteria lack cell nuclei. Bacteria are 0.20 to 500 μm in size and exist in almost every place on Earth.

But during the last decade, studies of DNA have revealed such fundamental differences between the two groups of bacteria that many scientists now assign them to different kingdoms instead of simply the kingdom Monera. Monera used to comprise all bacteria.

It is now believed that these two very different groups evolved from a common ancestor about 4 billion years ago!

The archaebacteria, one of which can be seen in **Figure 17-11**, evolved before oxygen filled our atmosphere and now are found principally in extreme environments, obtaining energy from sulfur, salt, hydrogen, and carbon dioxide. Fewer than 100 species have been recognized so far, but recent research suggests that these bacteria exist in far larger numbers than was previously believed. Archaebacteria are believed to be the ancestors of the protists, the first eukaryotes.

The eubacteria, one of which is seen in **Figure 17-11**, contain most of the common bacteria that share our world. They are an extremely diverse group, containing both autotrophic and heterotrophic forms. Approximately 5,000 species have been characterized so far, but undoubtedly many more exist. Eubacteria are believed to be the ancestors of mitochondria and chloroplasts, organelles within eukaryotic cells.

Kingdom Protista is a catchall kingdom

All the multicellular eukaryotes not classified as plants, animals, or fungi are assigned to the kingdom Protista. Plants, animals, and fungi probably descended from ancient protists. One of the protists can be seen in **Figure 17-12**.

Protists include protozoa, such as *Amoeba* and *Paramecium*, and algae, such as seaweeds and kelps. Slime molds and water molds also belong to kingdom Protista.

Figure 17-12
This colony of Volvox (50×) belongs to the kingdom Protista.

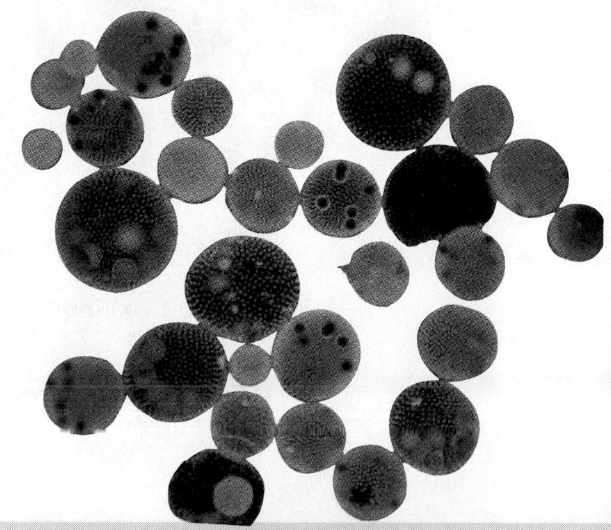

Kingdom Fungi contains the mushrooms

Mushrooms, yeast, and molds are members of the kingdom Fungi. Difficult to classify in a two-kingdom system, a mushroom, like the one in **Figure 17-13**, fits neatly into kingdom Fungi in the six-kingdom system. Instead of roots, stems, and leaves, fungi are made of thin filaments that penetrate the soil or decaying organisms, absorbing nutrients from them (absorption). Fungi do not contain chloroplasts and cannot make their own food by photosynthesis. Fungi are so different from all other living organisms that they are clearly best classified as a separate kingdom.

Figure 17-13
This mushroom belongs to the kingdom Fungi. It secretes enzymes that break down organic matter, which it then absorbs.

Kingdom Plantae contains multicellular photosynthetic organisms

This kingdom includes only terrestrial multicellular organisms that obtain their nutrients by photosynthesis. These are the organisms that most people would recognize as "plants," such as mosses, ferns (shown in **Figure 17-14**), flowers, and trees. Nearly all plants occur on dry land, but a few grow submerged in fresh water, and a very few grow at the edges of the sea. Plant cells have cell walls made of the polysaccharide cellulose. Because some green algae are so similar to plants in this and other features, they have been identified as the ancestral group for this kingdom.

Figure 17-14
The leaves of this Dallas fern contain chloroplasts, which carry out photosynthesis.

Kingdom Animalia has about 1 million known species

The first members of this kingdom, like the demon stinger in **Figure 17-15**, evolved in the ocean. The largest number of animal phyla are still found only in the sea. Like plants and most fungi, organisms in kingdom Animalia are multicellular. Animals, however, do not photosynthesize. Their cells do not have cell walls. Nearly all animals have some kind of nervous system, although it might be a very simple one.

Figure 17-15
This demon stinger, which measures about 8 cm (3.2 in.) across, looks like a plant but belongs to kingdom Animalia. It hides in the sand in the South Pacific Ocean with only its mouth and eyes exposed. There it lies waiting for crabs and small fish to cross its path.

SECTION REVIEW

1. **Explain why it would be difficult to classify a mushroom using the two-kingdom classification system.**
2. **How can you recognize an animal?**
3. **List two characteristics of protists.**
4. **What are two differences between plants and fungi? What are two characteristics they share?**

CHAPTER

17 Highlights

Interbreeding among members of the genus *Canis* produces fertile hybrids, such as this hybrid of a dog and a wolf.

	Key Terms	Summary
17-1 The Need for Naming	scientific name (p. 313) genus (p. 314) species (p. 314)	• To ensure accurate communication of information, biologists assign a unique two-word scientific name in Latin to each organism. • The first word of a scientific name is the name of the genus to which the organism belongs. The first and second words identify the kind of organism within the genus. • The system of scientific names used today was developed by Linnaeus in the eighteenth century.
17-2 Classification: Organizing Life	taxonomy (p. 317) family (p. 318) order (p. 318) class (p. 318) phylum (p. 318) kingdom (p. 318) convergent evolution (p. 320) analogous structure (p. 320) derived character (p. 321) cladogram (p. 321) hybrid (p. 323)	• Scientists classify organisms into a hierarchical system of groups within groups. • Linnaeus classified organisms into groups according to their similarities. Since Darwin, some classifications have reflected the evolutionary relationships among organisms. • Analogous structures are similar but do not share a common ancestry. Convergent evolution is the evolution of similar structures in unrelated organisms. • A species is a group of organisms with unique characteristics.
17-3 Six Kingdoms		• Linnaeus classified all living things into either the plant kingdom or the animal kingdom. Today many biologists use a six-kingdom classification system. • The six kingdoms are Archaebacteria, Eubacteria, Protista, Fungi, Plantae, and Animalia.

The scientific name for the red oak is *Quercus rubra.*

In the Middle Ages, herbalists classified plants according to their medicinal uses.

The sea anemone is one example of an organism that looks like a plant but is classified as an animal.

CHAPTER 17

review

Understanding Vocabulary

1. For each pair of terms, explain the differences in their meanings.
   a. genus, species
   b. phylum, division
   c. analogous structure, homologous structure

2. For each set of terms, complete the analogy.
   a. animal : phylum :: plant : ____
   b. common ancestry : homologous structure :: similar features evolved independently : ____
   c. ancestors of the first eukaryotes : archaebacteria :: ancestors of mitochondria and chloroplasts : ____

Understanding Concepts

3. **Relating Concepts** Construct a concept map that describes how a scientific name is written. Try to include the following terms in your map: scientific name, first word, second word, Latin, capitalized, genus, species, scientists, lowercase, communication, unrelated organisms.

4. **Organizing Information** Give three reasons why using an organism's common name is not clear enough for scientific communication.

5. **Comparing Relationships** What relationship can you guess when you see these names: *Oecanthus californicus* and *Oecanthus fultoni*?

6. **Inferring Conclusions** Determine whether the following statement is true or false: "The more classification categories two organisms share, the more traits they have in common." Explain your answer and give an example.

7. **Summarizing Information** What are the two parts of a scientific name? Give an example of each.

8. **Inferring Relationships** What kinds of information can you find out about an organism from its scientific name?

9. **Organizing Information** What are two rules that must always be followed when giving an organism a scientific name?

10. **Inferring Relationships** What is a cladogram? How does it help scientists classify organisms?

11. **Summarizing Information** Who was responsible for devising the two-name system of scientific names? Describe the contribution made by this individual.

12. **Comparing Relationships** How is the classification of living things similar to the way an army is organized? How is it different?

13. **Organizing Information** Arrange the following classification terms in order from least specific to most specific: family, genus, order, kingdom, species, class, phylum.

14. **Inferring Relationships** Look at the organisms below. Use the information in your chapter to assign as many appropriate biological terms as you can to each organism.

15. **Predicting Outcomes** What is convergent evolution? Explain why it can create problems for taxonomists.

16. **Summarizing Information** Name the six kingdoms of living organisms, and give at least two characteristics of each.

17. **BUILDING ON WHAT YOU HAVE LEARNED** Based on what you learned in Chapter 12, which pair of organisms—humans and chimpanzees or chimpanzees and prosimians—are more closely related?

Interpreting Graphics

18. Examine the cladogram shown below.

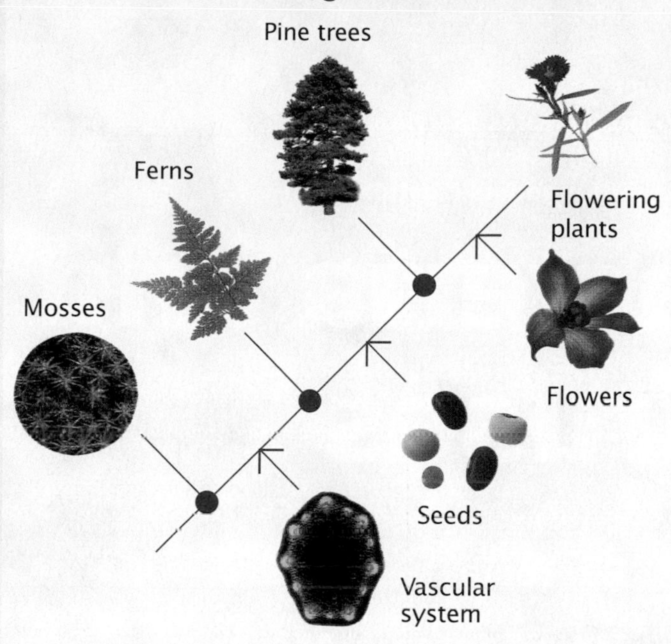

a. What is the derived character possessed by ferns?
b. What derived characters are shared by pine trees and flowering plants?
c. Which type of plant is most ancient?
d. Which type of plant evolved most recently?
e. Which type of plant lacks a vascular system?

Reviewing Themes

19. ***Heredity***
How does DNA act like a "molecular clock"?

20. ***Evolution***
What features common to both the shark and the dolphin suggest that convergent evolution has occurred?

Thinking Critically

21. **Evaluating Viewpoints** Explain this statement: Diversity is the result of evolution; classification systems are the inventions of humans.

22. **Recognizing Verifiable Facts** In the laboratory, a scientist studied two identical-looking daisies that belonged to the genus *Aster*. The two plants produced fertile hybrids in the laboratory, but they never interbreed in nature because one plant flowers only in the spring and the other only in autumn. Are the plants the same species? Explain your answer. What could the scientist do to be certain about whether the plants are the same or different species?

Life/Work Skills

23. **Allocating Resources** You and your classmates have been allotted $500 to buy enough laboratory specimens to demonstrate the six kingdoms of life in a high school biology laboratory. Check scientific supply catalogs to get ideas of what is available and the prices of those items. Think about whether you would select prepared microscope slides, preserved specimens, or living organisms and about the advantages and drawbacks of each. Make a poster of your choices to "sell" your biology teacher on your ideas.

24. **Using Technology** Work with two or three of your classmates to develop a system of classification using computer software (spreadsheet or word processor) to organize your favorite books, movies, TV programs, sports teams, music, or performers. For each system of classification, include at least three levels of distinguishing characteristics. Make a poster of the classification system to share with your class.

Activities and Projects

25. **Language Arts Connections** A mnemonic is a code used to aid memory. In the mnemonic "King Philip came over from Geneva, Switzerland," the first letter of each word is the same as the first letter of each of the seven levels of classification. Design your own mnemonics to help you remember the six kingdoms and two characteristics of each.

26. **Reading Connection** Read the book *Life on Earth: A Natural History* by David Attenborough, which describes the diversity of life on Earth and how organisms relate to and interact with one another. Write a book report, and share with your class the ways in which some life-forms relate to their unique geographical habitats.

Discovering Through Reading

27. Read the article "Why Cladistics?" by Gaffney, Dingus, and Smith in *Natural History*, June 1995, pages 33–35. What did cladistics reveal about the evolution of birds? What does cladistic analysis reveal about the fossil record?

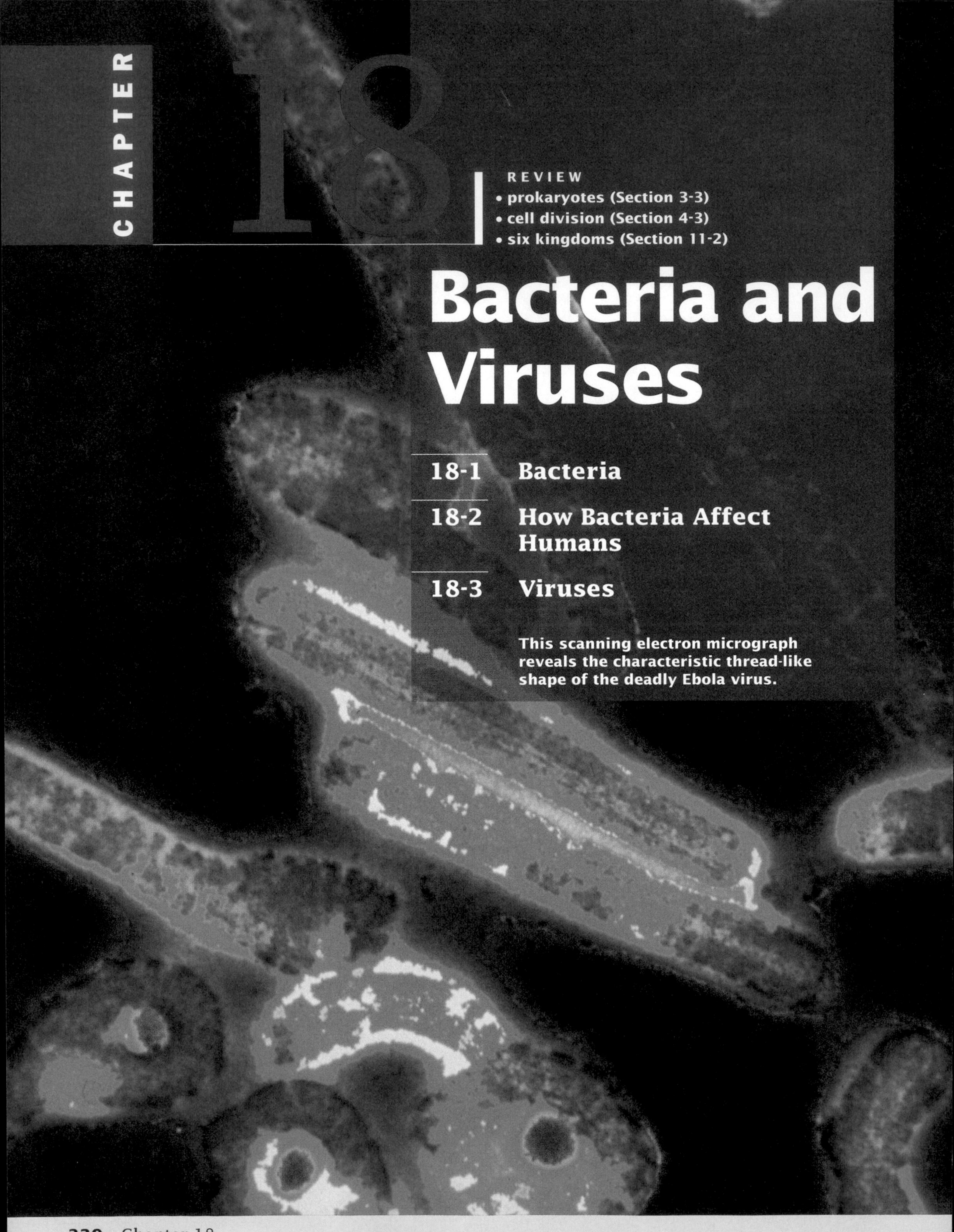

CHAPTER 18

REVIEW
- prokaryotes (Section 3-3)
- cell division (Section 4-3)
- six kingdoms (Section 11-2)

Bacteria and Viruses

This scanning electron micrograph reveals the characteristic thread-like shape of the deadly Ebola virus.

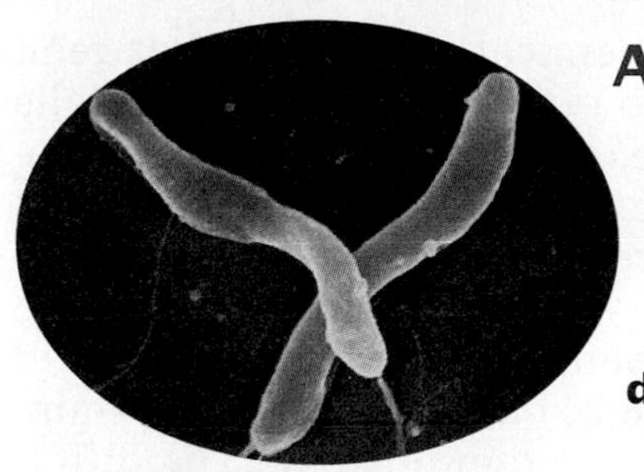

Although you do not see them, bacteria affect your life in many ways. In your large intestine, bacteria synthesize vitamin K, which is essential to good nutrition. Bacteria thrive in your mouth and may cause cavities in your teeth. Bacteria also cause many diseases, including cholera, typhus, pneumonia, sexually transmitted diseases, and tuberculosis.

18-1 Bacteria

OBJECTIVES

1. **Draw the structure of a bacterial cell.**
2. **Differentiate between a bacterial cell and a eukaryotic cell.**
3. **Explain how bacteria reproduce.**
4. **Distinguish between the diverse ways bacteria obtain nutrition.**

Bacteria Are Small and Successful

In many ways, bacteria are the most successful organisms on Earth. The earliest known fossils are of 3.5-billion-year-old bacteria, making them the oldest group of organisms. Today, bacteria can be found living almost everywhere on the globe, even in some very hostile habitats. Certain bacteria live beneath more than 400 m (1,200 ft) of ice in Antarctica, and others live near deep-sea volcanic vents where temperatures reach 360°C. Bacteria have even been found thriving miles below the Earth's surface. Bacteria also occur in great abundance. One gram (0.035 oz) of rich soil can contain about 2.5 billion bacteria.

It is obvious that bacteria are very small, otherwise you could see them without a microscope. To get an idea of how small bacteria are, look at **Figure 18-1**. A human red blood cell dwarfs the *Escherichia coli* bacterium that is normally found in the human intestine. If you made a chain of *E. coli* bacteria laid end to end, the chain would have to be more than 250 bacteria long just to be visible to the unaided eye.

Most species of bacteria are one of three different shapes: spherical, spiral, or rod-shaped. **Figure 18-1** clearly illustrates these shapes. *Streptococcus* bacteria that cause strep throat are spherical, for example. Spherical bacteria often link to form long chains of cells.

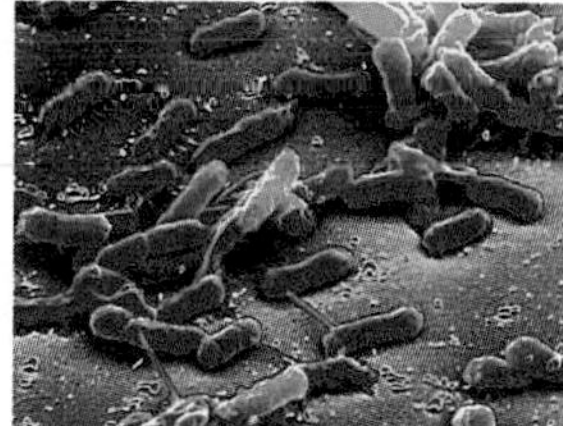

a These rod-shaped bacteria, *Escherichia coli*, synthesize vitamin K.

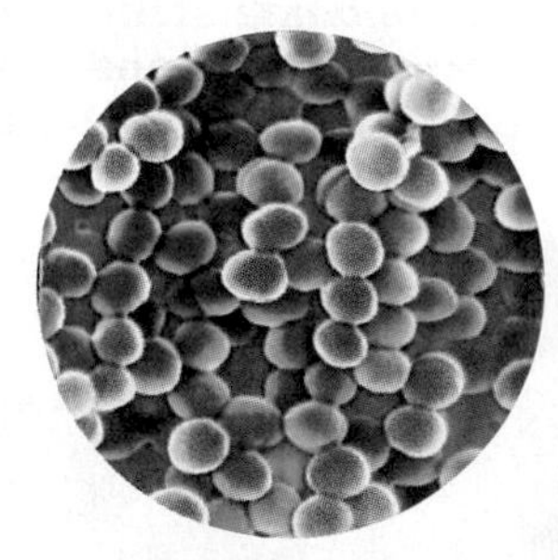

b These spherical bacteria, *Staphylococcus aureus*, cause skin infections.

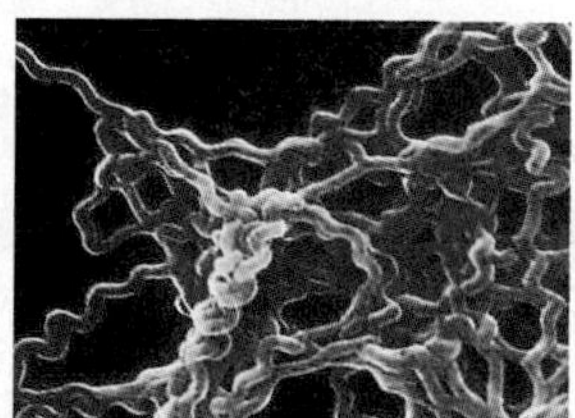

c This spiral bacterium, *Leptospira* sp., can infect the liver or the brain.

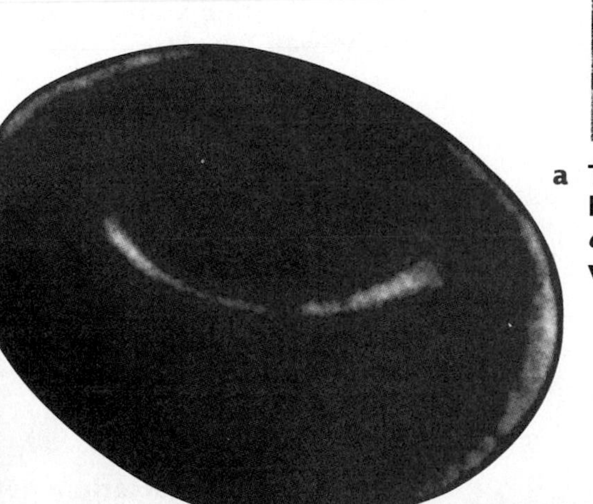

Figure 18-1
A human red blood cell (250×) is larger than the bacteria shown above and to the right.

Cell Structure and Function

How does the structure of the bacterial cell wall affect its susceptibility to antibiotics?

The structure of a bacterial cell is simple

All bacteria are prokaryotes, members of either the kingdom Archaebacteria or the kingdom Eubacteria. Recall from Chapter 3 that prokaryotes lack cell nuclei, chromosomes, and membrane-bound organelles, such as mitochondria, Golgi bodies, and chloroplasts. You can see the internal and external structure of a bacterial cell in the *Tour of a Bacterium* on the next page.

Bacteria have one of two kinds of cell walls

Like a cell from your body, a bacterial cell is enveloped by a cell membrane made of a double layer of lipids. Unlike your cells, bacterial cells have an outer cell wall composed of polysaccharides. In some bacteria, this cell wall is surrounded by yet another layer, which is made up of polysaccharides bound to molecules of lipid.

The chemical difference between these two types of bacterial cell walls is revealed using a special staining procedure called **Gram staining**. In this procedure, a series of dyes is added to a sample of bacteria as a microscope slide is being prepared. As shown in **Figure 18-2**, bacteria are either Gram-negative or Gram-positive. The difference between Gram-negative and Gram-positive bacteria is important in diagnosing and treating diseases caused by bacteria. Gram-negative bacteria are unaffected by many antibiotics, the drugs used to treat bacterial infections. These antibiotics cannot penetrate the additional layer outside the cell wall.

As you learned in Chapter 4, bacteria reproduce by splitting in two. Except for occasional mutations, each new cell is exactly like the parent cell. Genetic material in some cases is transferred from one bacterium to another, as in the case of **conjugation**. Conjugation does not increase the number of bacteria, but it does give them more genetic possibilities.

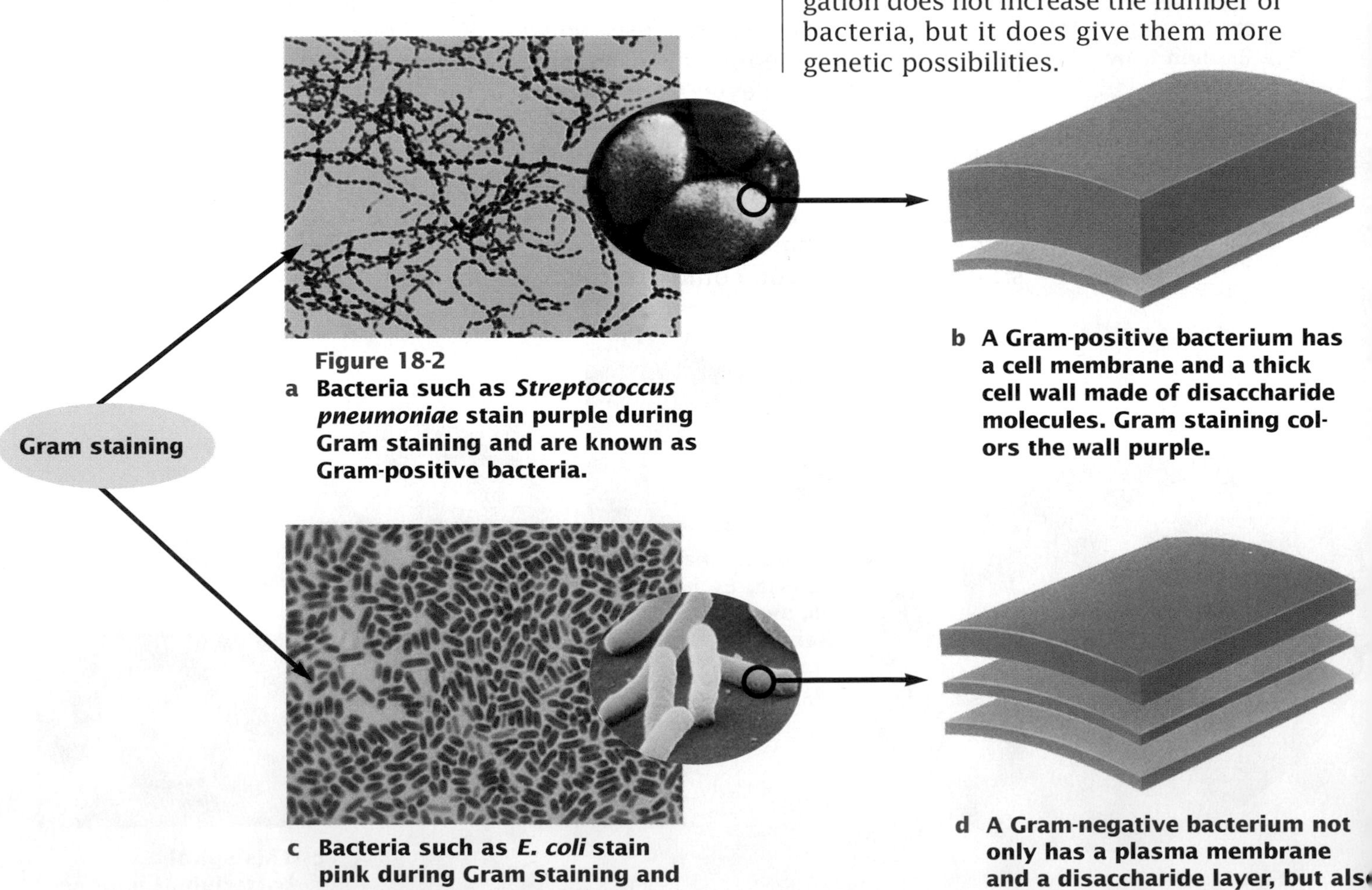

Figure 18-2

a Bacteria such as *Streptococcus pneumoniae* stain purple during Gram staining and are known as Gram-positive bacteria.

b A Gram-positive bacterium has a cell membrane and a thick cell wall made of disaccharide molecules. Gram staining colors the wall purple.

c Bacteria such as *E. coli* stain pink during Gram staining and are known as Gram-negative bacteria.

d A Gram-negative bacterium not only has a plasma membrane and a disaccharide layer, but als has an outer layer that does not retain the purple stain.

Bacterium

This bacterium, Escherichia coli, *is perhaps the most thoroughly studied organism in the world. It is used in genetic engineering and in studies of protein synthesis and the control of gene expression.*

Tour

Evolutionary Relationships

Bacteria are the oldest and most successful living organisms on Earth. Bacteria, which are prokaryotic cells, existed for 2 billion years before eukaryotic cells appeared. *E. coli* belongs to the kingdom Eubacteria.

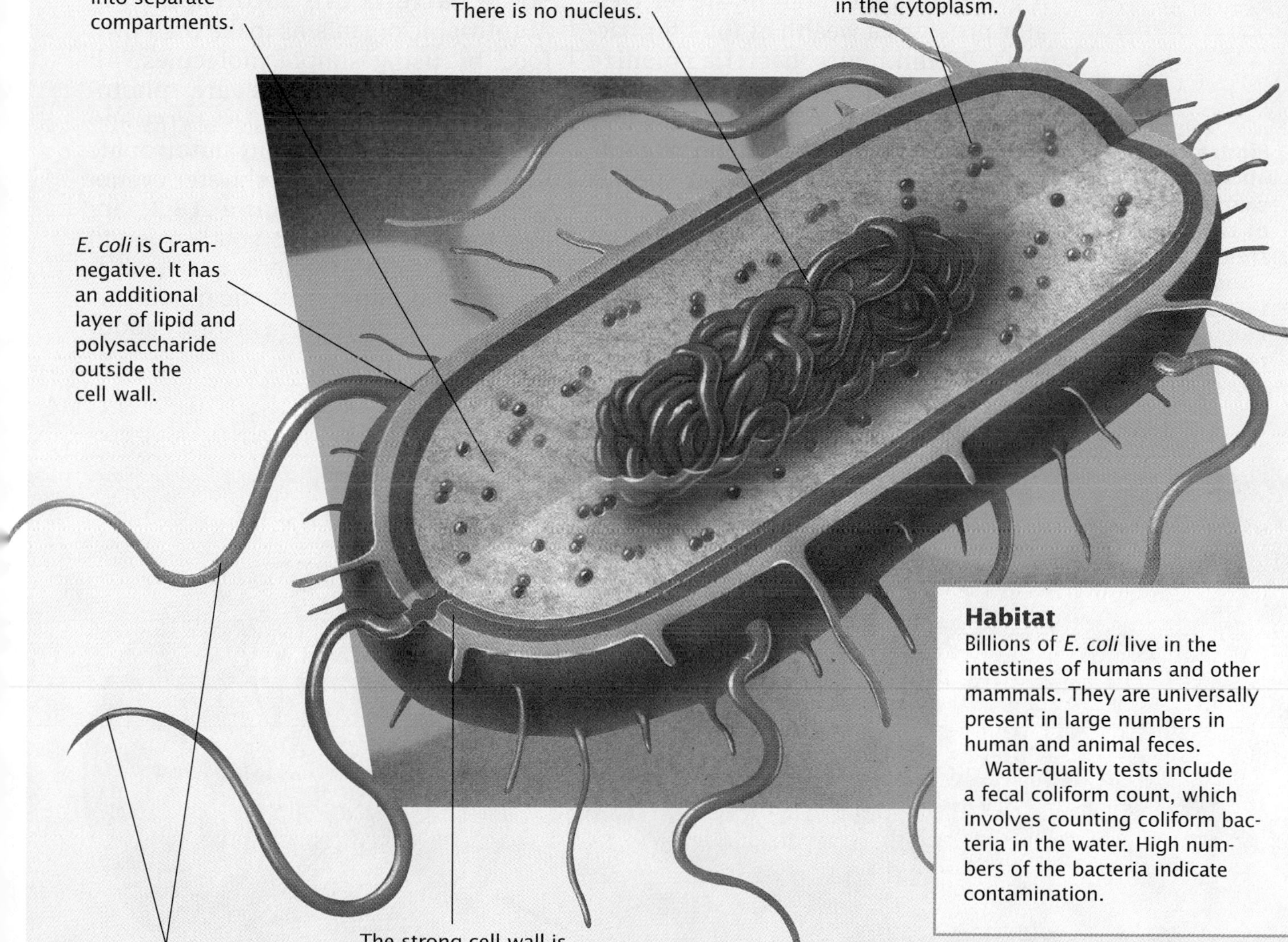

The cytoplasm is not divided into separate compartments.

All of the genes are located on a single molecule of DNA. There is no nucleus.

Proteins are made on ribosomes that float free in the cytoplasm.

E. coli is Gram-negative. It has an additional layer of lipid and polysaccharide outside the cell wall.

Flagella rotate like propellers to drive the cell through its environment.

The strong cell wall is a network of polysaccharides cross-linked by short chains of amino acids.

Habitat

Billions of *E. coli* live in the intestines of humans and other mammals. They are universally present in large numbers in human and animal feces.

Water-quality tests include a fecal coliform count, which involves counting coliform bacteria in the water. High numbers of the bacteria indicate contamination.

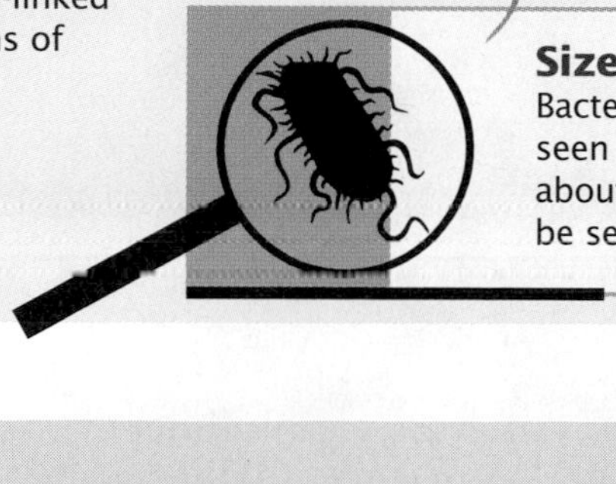

Size

Bacteria are single cells that are too small to be seen without a microscope. This one, *E. coli*, is about 1 μm in size. A line of 250 *E. coli* could just be seen by the naked eye.

Bacteria reproduce rapidly

Bacteria, unlike most other organisms, do not undergo mitosis or meiosis. Instead, a bacterium first duplicates its DNA so that there is enough DNA for two cells. Then the bacterium splits into two identical cells. Each cell receives one molecule of DNA and some cytoplasm, a process you saw in Chapter 4. Some kinds of bacteria are able to divide as many as five times in an hour. The time required for a cell to divide is called the generation time.

If you were to place a single bacterium into a culture dish containing an abundant supply of food, you could find more than 600,000 bacteria in the dish after only 4 hours. After 6 hours, the bacterial population of the dish could reach 476 million.

Matter, Energy, and Organization

Bacteria use many different energy sources. How does this enable them to live in a wide variety of places?

How Bacteria Obtain Nutrition

A glass of milk left out of the refrigerator provides a wealth of food for bacteria. Within hours, bacteria colonize the milk and break down its supply of sugar, causing the milk to curdle. Other species of bacteria feed on organic material in sewage. Still other species consume industrial products such as nylon and pesticides. Some bacteria are able to metabolize petroleum and may be used to clean oil spills. A major reason for the success of bacteria is the wide variety of foods they can use.

Some bacteria are autotrophs

Autotrophic organisms make their own food by using simple molecules. All autotrophic eukaryotes are photosynthetic: they capture solar energy and use it to make food. Many autotrophic bacteria, such as the freshwater cyanobacteria shown in **Figure 18-3**, are also photosynthetic. As you learned in Chapter 11, cyanobacteria were probably the first photosynthetic organisms to produce oxygen, starting about 3 billion years ago.

Figure 18-3
a Microscopic examination of the water in freshwater ponds, like this one in Pennsylvania, reveals . . .

b . . . that photosynthetic cyanobacteria such as these are a major component of the pond's ecosystem.

Not all autotrophic bacteria are photosynthetic, however. In Chapter 14 you studied the communities that surround volcanic vents on the ocean floor, as shown in **Figure 18-4**. The organisms in these very deep ocean environments cannot perform photosynthesis because no light reaches them. Instead, these bacteria are able to use the energy stored in the inorganic compound hydrogen sulfide. Similarly, some kinds of soil-dwelling bacteria obtain energy from ammonia. And bacteria that live in swamps use methane. These bacteria use a process called **chemosynthesis** to make complex organic molecules from the energy in inorganic molecules. All chemosynthetic organisms are prokaryotes.

Heterotrophic bacteria are consumers

Most bacteria cannot make their own food and are therefore heterotrophs. Many feed on dead animals and animal wastes; dead plants; and fallen leaves, branches, and fruit. Other types of heterotrophic bacteria are parasites. Parasitic bacteria cause many diseases, as you will see later in the chapter.

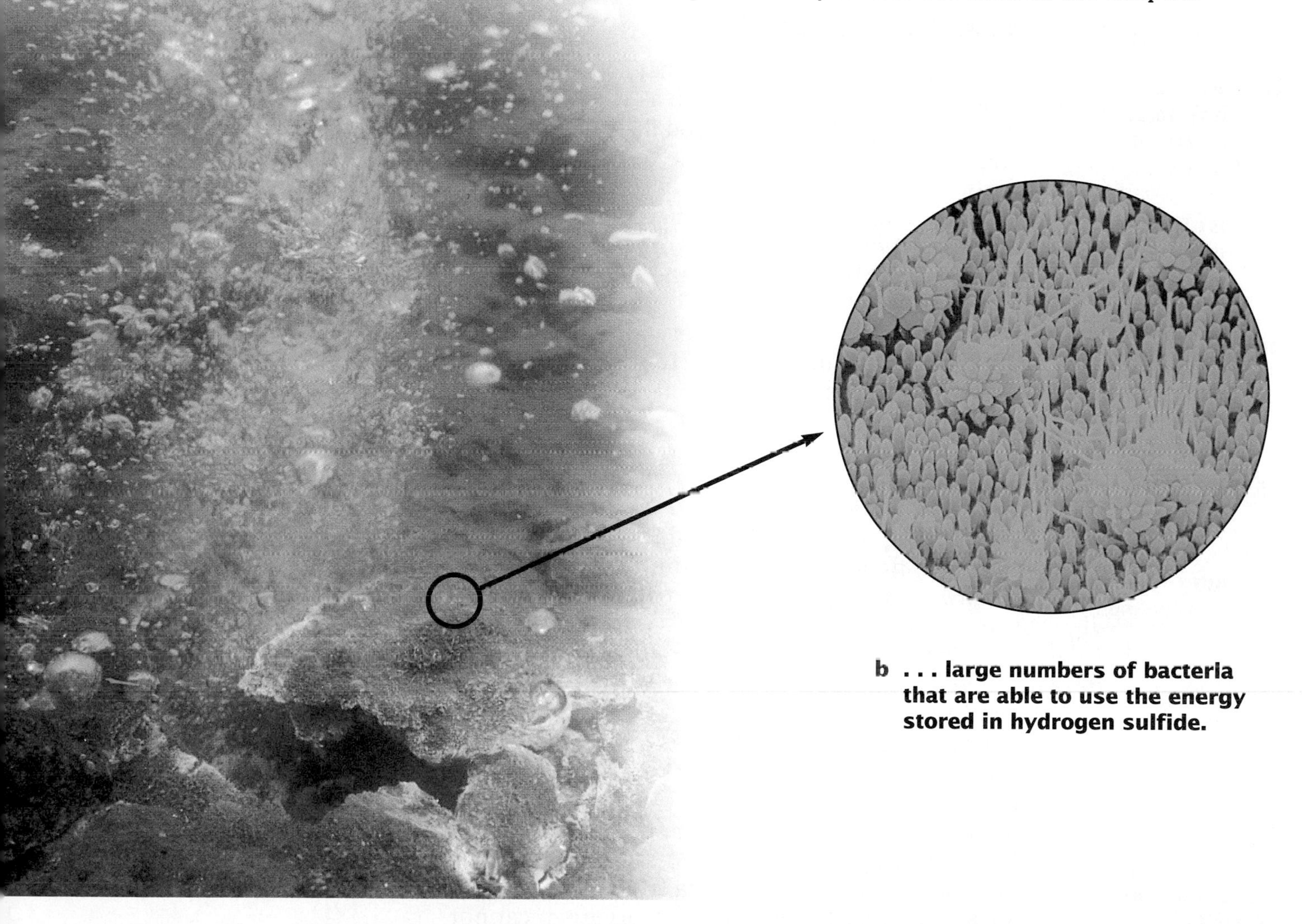

Figure 18-4
a On the ocean floor, volcanic vents are surrounded by organisms that receive nourishment from . . .

b . . . large numbers of bacteria that are able to use the energy stored in hydrogen sulfide.

SECTION REVIEW

1. **How does Gram staining help a doctor prescribe treatment for a bacterial infection?**
2. **List three structures found in a eukaryotic cell that are not present in a bacterial cell.**
3. **How do bacteria reproduce?**
4. **Name and explain the two ways in which autotrophic bacteria obtain energy.**

The disease tuberculosis is caused by the bacterium *Mycobacterium tuberculosis*. Tuberculosis patients were once treated with streptomycin, an antibacterial drug that is produced by bacteria of the genus *Streptomyces*. This example shows how bacteria are both harmful and beneficial to humans.

18-2 How Bacteria Affect Humans

OBJECTIVES

1. **Describe three beneficial effects of bacteria.**
2. **List five human diseases caused by bacteria.**
3. **Summarize three ways to prevent bacterial diseases.**
4. **Evaluate the importance of antibiotics in fighting bacterial diseases.**

Beneficial Bacteria

Although you probably know that some bacteria can cause harm—such as when they destroy food or make you ill—you might not be fully aware of the tremendous benefits bacteria provide. For instance, bacteria maintain crucial links in nutrient cycles that make essential elements such as nitrogen and sulfur available to plants and, indirectly, to humans. Bacteria are also very important in the manufacture of food and life-saving drugs. Without bacteria, life as we know it could not exist.

Decomposers are nutrient recyclers

Try to imagine what life would be like if all the dead plants and animals in the world never decayed! Garbage and waste would pile up and every bit of land on Earth would be filled with material that could not be used and would never go away.

Recall from Chapter 14 that decomposers are organisms that return nutrients to the environment by breaking down organic matter, like the leaf litter shown in **Figure 18-5**. When a plant or animal dies, bacteria from the air and soil settle on the dead organism. The bacteria begin to grow, releasing carbon dioxide, water, nitrogen, phosphorus, and sulfur—nutrients that plants need in order to grow. Without decomposers, most of these nutrients would be locked away in the bodies of dead organisms.

Nitrogen-fixing bacteria enrich the soil

Because nitrogen is a component of proteins, plants cannot photosynthesize, grow, or reproduce without it. Much of the nitrogen in a form usable by plants is made available by nitrogen-fixing bacteria. Nitrogen-fixing bacteria transform atmospheric nitrogen, which cannot be absorbed by plants, into ammonia, a nitrogen compound that plants can absorb. No other organisms have this ability.

Nitrogen-fixing bacteria are found in the soil, in aquatic ecosystems, and within the roots of some plant species. The legumes, plants including the peas and beans, contain nitrogen-fixing bacteria in swellings on their roots. These bacteria enable legumes to grow in nitrogen-poor soils where other plants cannot.

Figure 18-5 Without decomposers, nutrients would not be recycled.

Helpful Bacteria

- *Fix nitrogen*
- *Recycle nutrients*

Figure 18-6
a Without the billions of bacteria that share our world, life as we know it could not exist.

- *Used to make certain foods and aid in digestion*
- *Clean up oil and toxic spills*
- *Used to make antibiotics, vitamins and drugs*
- *Used as natural pesticides*

Harmful Bacteria

- *Cause disease*
- *Contaminate food*

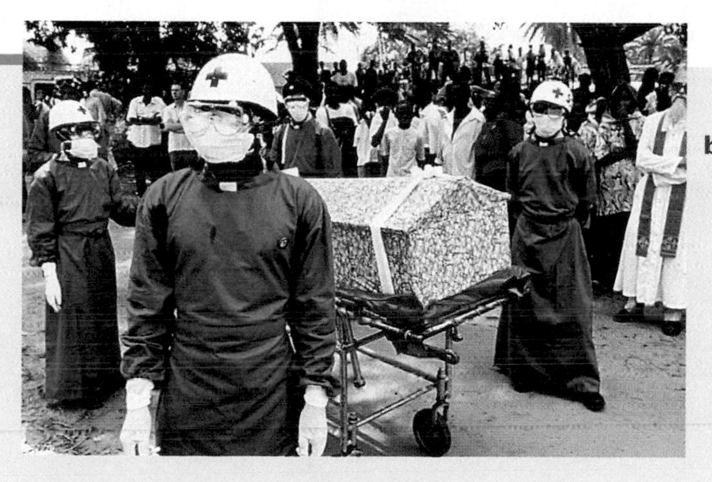

b Unfortunately, bacteria do not always interact with humans in a helpful way.

Interdependence

How would an ecosystem be affected if its nitrogen-fixing bacteria were destroyed?

Bacteria are used to manufacture food and drugs

When you eat yogurt or olives, you are eating food that is the product of bacterial decomposition. Humans have learned that decomposition is occasionally beneficial because it adds flavor to food. For example, bacteria convert cabbage and cucumbers into tangy sauerkraut and pickles.

Modern technology is taking advantage of the enormous genetic diversity among the bacteria. Using genetic engineering technology, biologists can now "reprogram" bacteria to manufacture any protein for which a gene has been isolated. For example, bacteria now produce most of the insulin needed by diabetics in the United States. Before genetic engineering, insulin had to be isolated from the pancreases of animals killed in slaughterhouses. Some of the other ways we are affected by bacteria are listed in **Figure 18-6**.

Bacteria and Disease

Your body occasionally becomes a temporary home for parasitic bacteria, with serious effects—they cause disease or infection. A disease-causing agent is called a **pathogen**. Pathogenic bacteria are harmful because they damage their host's tissues. This damage results either from direct attacks on the host's cells or from poisonous substances called toxins that many bacteria release.

How are bacterial diseases actually transmitted? Each species of pathogenic bacteria has a characteristic way of being carried to new hosts: in water, in the air, in food, by insects, or by direct human contact. **Table 18-1** lists some bacterial diseases and their modes of transmission.

Bacteria cause tooth decay and ulcers

One of the most widespread of human diseases caused by bacteria is dental caries (decay leading to cavities), which arises in the film on our teeth. Diets high in sugar promote dental caries because the bacteria ferment the sugars into lactic acid. The acid causes a loss of calcium, followed by breakdown of the proteins in tooth enamel. Scientists have recently found that most common stomach ulcers are also the result of bacterial infection. It was once thought that excessive stress caused ulcers, but it is possible that stress creates an environment in the stomach that is beneficial to the bacteria's growth.

Table 18-1 Examples of Bacterial Diseases

Disease	Mode of transmission	Symptoms
Tuberculosis	Airborne water droplets	Fatigue, persistent cough, bleeding in lungs; can be fatal
Diphtheria	Airborne water droplets	Fever, sore throat, fatigue
Syphilis	Sexual contact	Lesions, CNS damage if untreated
Bubonic plague	Fleas	Swollen glands, bleeding under skin; often fatal
Typhus	Lice	Rash, chills, fever; often fatal
Tetanus	Wounds, skin breaks	Severe, prolonged muscle spasms
Cholera	Contaminated water	Severe diarrhea, vomiting; often fatal
Typhoid	Contaminated water and food	Headaches, fever, diarrhea, rash; often fatal
Leprosy	Personal contact	Nerve damage, skin lesions, tissue degeneration
Lyme disease	Ticks	Rash, pain, swelling in joints

Figure 18-7 **Just in case the chicken you are preparing is contaminated with *Salmonella*, you should use hot, soapy water to wash the surfaces that have been touched by the raw chicken.**

Food can be contaminated by bacteria

Although certain bacteria can produce yogurt and cheese, most bacteria in food are not helpful. Pathogenic bacteria can contaminate foods and cause food poisoning.

One of the most dangerous kinds of food poisoning is botulism, which is caused by a toxin released by the bacterium *Clostridium botulinum*. Consumption of less than one-millionth of a gram of this toxin causes paralysis and death. *C. botulinum* normally lives in the soil, but it can grow in canned foods that have not been properly sterilized. Because oxygen kills botulism bacteria, they cannot grow in fresh and frozen foods, which contain oxygen.

Another type of food poisoning is caused by *Salmonella* bacteria found in pork, eggs, poultry, and other foods. The symptoms of *Salmonella* food poisoning are diarrhea, vomiting, and abdominal cramps. A serious infection can cause dehydration, a drastic loss of water from the body. In the very young or in the elderly, severe dehydration can be fatal.

Food poisoning has also been linked to a newly discovered strain of the bacterium *Escherichia coli*. This bacterium normally lives in the intestines of mammals, including humans, and synthesizes vitamin K for us. The new strain of *E. coli*, however, has caused serious outbreaks of severe food poisoning and even death in people who unknowingly ate undercooked hamburgers from a fast-food restaurant. New regulations concerning cooking temperatures in restaurants make it safer to eat out.

Taking the following precautions can mean the difference between a good meal and a trip to the hospital: thaw frozen meats and poultry in the refrigerator or microwave; wash working surfaces, utensils, and hands with hot, soapy water after contact with raw meats, as shown in **Figure 18-7**; and, most important, cook meat thoroughly until there are no pink juices left.

Heat and cold help protect food from bacterial contamination

Bacterial contamination of food can be reduced by either heat or cold. **Pasteurization** involves heating food to a temperature that kills most bacteria. Pasteurization eliminates the possibility of contracting diseases such as botulism from canned food or tuberculosis from milk.

Cooling food to a few degrees above freezing reduces bacterial contamination. Although bacteria are not killed, their rate of growth is greatly slowed. Most *Salmonella* infections are caused by food that has been left out of the refrigerator too long, allowing bacteria to grow to significant numbers.

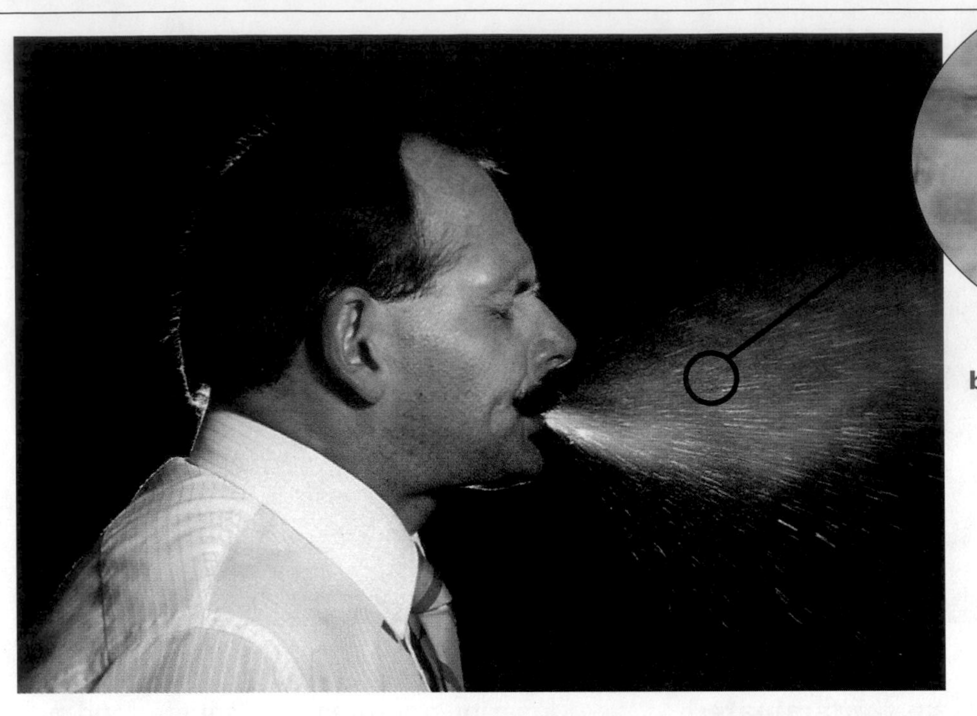

b **This view shows the bacteria that can be found in a single water droplet of a sneeze. *Mycobacterium tuberculosis,* the bacterium that causes the disease tuberculosis, can be transmitted in airborne water droplets such as these.**

Figure 18-8
a **A sneeze sends thousands of bacteria into the air. Some of these bacteria may be pathogens.**

Bacteria travel through the air

A sneeze or cough sprays out a shower of tiny droplets, as you can see in **Figure 18-8**. Visible in the circular photograph are some of the 10,000 to 100,000 bacteria carried in the droplets produced by a single sneeze. The bacteria that cause diphtheria, scarlet fever, whooping cough, and tuberculosis drift through the air in droplets such as these. Most airborne diseases affect the respiratory system of their victims.

By far the most serious airborne disease is tuberculosis (also called TB), caused by the bacterium *Mycobacterium tuberculosis.* Three million people die of this disease annually. In fact, for most of this century, TB has been the leading worldwide cause of death from a single infectious agent. In the 1970s and 1980s, TB was not a serious health problem in the United States. However, in the last decade there has been a marked increase in cases, many of which are resistant to the best available antituberculosis medications. Drug-resistant pneumonia is also on the rise.

Bacteria are carried by animals

Lyme disease is an example of a disease caused by a bacterium that lives inside several species of ticks. When an infected tick, like the one in **Figure 18-9**, bites a human to feed on blood, the bacteria can be transferred to the human, who may then develop Lyme disease.

Ticks carrying the bacteria are spread by deer, rodents, and other mammals as well as by birds. The first symptom of Lyme disease is usually a red rash that begins at the site of the tick bite and spreads outward. If left untreated, muscle pains, joint inflammation, and, in advanced stages, brain damage can occur.

Lyme disease is treated with various antibiotics. With early treatment, a patient can be cured before any severe damage occurs. The best "cure" is prevention. Cover up when you journey into wooded, grassy, or brushy areas. Wear long sleeves and long pants tucked into socks. Remove any attached ticks according to a physician's instructions.

Figure 18-9
Deer ticks have eight legs and a flat body. Like lobsters, crabs, and insects, they have a hard outer covering that must be periodically shed as the animal grows.

Since ticks are unable to fly, they must wait for a host to wander close to them. They have poor vision and rely on their keen sense of smell. They are especially sensitive to carbon dioxide and butyric acid, a rancid-smelling chemical occurring on the skin of many animals, including humans. These substances immediately cause the hungry tick to move toward the odor. A biting tick injects its victim with not only the bacteria that cause Lyme disease, but with chemicals that suppress the victim's immune system.

Controlling Bacterial Diseases

Since the 1940s, when antibiotics like penicillin began to be widely used, most Americans have had little experience with serious bacterial diseases. Advances in medicine and sanitation have brought freedom from illnesses that had been commonplace for centuries.

Sanitation and hygiene help prevent bacterial diseases

Cholera is a serious disease transmitted through polluted water. Cholera bacteria produce a strong toxin that causes acute diarrhea and vomiting, which can lead to rapid dehydration. If untreated, this severe loss of water can be fatal within 24 hours. Cholera bacteria are spread in drinking water that has been contaminated by the feces of infected individuals. Any fish that are caught in contaminated water can also carry cholera bacteria. In 1991 a large cholera epidemic struck Peru and rapidly spread to the rest of South America and Central America, as shown in **Figure 18-10**. At least 1,700 people died in the first year of this devastating outbreak.

In the industrialized countries, drinking water is filtered and then purified with chlorine, a chemical that kills bacteria. Similarly, sewage is collected and treated to remove pathogens before it is discharged into rivers or the ocean. Thus, cholera, typhoid, and other diseases of contaminated water are almost unknown in the industrialized countries.

Unfortunately, these clean conditions are not characteristic of some of the developing countries. Because of

Figure 18-10
In 1991 a large cholera epidemic struck Peru. The bacteria that cause cholera can exist for long periods of time in aquatic environments, making contaminated water very dangerous. These Peruvian children are waiting for uncontaminated water to be brought to them.

Figure 18-11
The preparation of vaccines involves inactivation of the disease-causing organism and its subsequent introduction into the body.

a **Whooping cough vaccine is prepared by first culturing the bacteria responsible for whooping cough.**

b **The bacteria, *Bordetella pertussis*, are then killed using heat or chemicals.**

c **The dead bacteria are purified and cleansed, and the vaccine is packaged.**

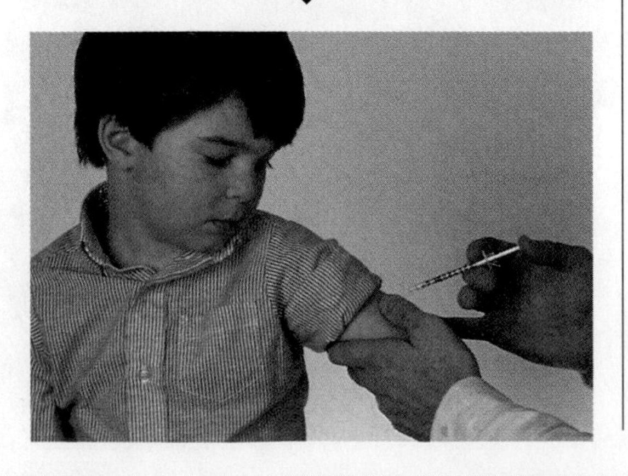

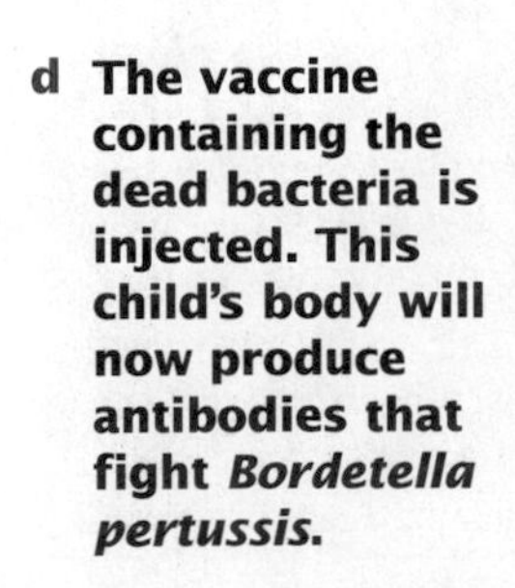

d **The vaccine containing the dead bacteria is injected. This child's body will now produce antibodies that fight *Bordetella pertussis*.**

poverty, many of these countries cannot provide clean drinking water for all of their citizens. In any nation, natural or human-caused disasters, such as floods, earthquakes, or wars, can destroy or interrupt water treatment. These situations create conditions that are ideal for water contamination. Each year about 25 million people around the world die from typhoid, cholera, and other diseases caused by contaminated water.

Vaccination stimulates the body's defenses

Preventing the contamination of food and drinking water are only two ways to protect yourself from infection. Vaccinating against disease is another way. Before you started school, you were probably vaccinated against bacterial diseases such as diphtheria, whooping cough, and tetanus. A **vaccine** is a solution that contains pathogens or their toxins that have been made harmless, usually by treatment with heat or chemicals or by genetic engineering. How does a vaccine protect you against pathogenic bacteria? As shown in **Figure 18-11**, a vaccine against whooping cough, for example, is a suspension of whooping cough bacteria that have been killed with heat or chemicals. These dead bacteria cannot cause disease, but once inside your body they stimulate your immune system to make defenses against them. Once you have been vaccinated against whooping cough, your immune system has defenses ready to destroy any live whooping cough bacteria before they have a chance to make you ill. Scientists have learned that many vaccinations, such as those for diphtheria, tetanus, and whooping cough, require more than one dose. This is because our bodies require more stimulation from some vaccines in order to provide us with a strong defense against those diseases. So it is very important that all people receive their booster vaccinations on schedule to ensure protection against those disease-causing organisms. Scientists are trying to develop vaccines to protect us against diseases for which we currently do not have vaccines.

Figure 18-12 Alexander Fleming's discovery of penicillin is one of the most important medical milestones of the century. Millions of lives have been saved because of his observations. The fungus that produces penicillin is shown below.

Antibiotics are used to treat bacterial diseases

Before the 1940s, doctors had few treatments for bacterial diseases. Whether a patient recovered or died often depended more on the type of disease and the strength of the patient than it did on the efforts of the doctor. This grim situation changed because of a chance event in 1928. In that year, Alexander Fleming, the British physician shown in **Figure 18-12**, found a blue-green mold growing on one of his bacterial samples. At first Fleming was angry because his experiment had been contaminated. Before throwing out the sample, however, he noticed that bacteria did not grow near the mold. The mold was apparently releasing a chemical that was poisonous to bacteria. Fleming isolated this substance and named it **penicillin**, after the *Penicillium* mold that produced it.

In the early 1940s, scientists following up on Fleming's discovery showed that penicillin was a very effective treatment for many bacterial diseases. Antibacterial drugs such as penicillin are called **antibiotics**. There have been many other antibiotics discovered since penicillin; tetracycline and streptomycin are two common examples. Some antibiotics prevent bacteria from making new cell walls. Unfortunately, antibiotics have been so successful in alleviating our illnesses that some doctors prescribe them when they are unnecessary. Most sore throats, upset stomachs, and other minor problems go away by themselves. Our own immune systems can usually deal with attacks very effectively.

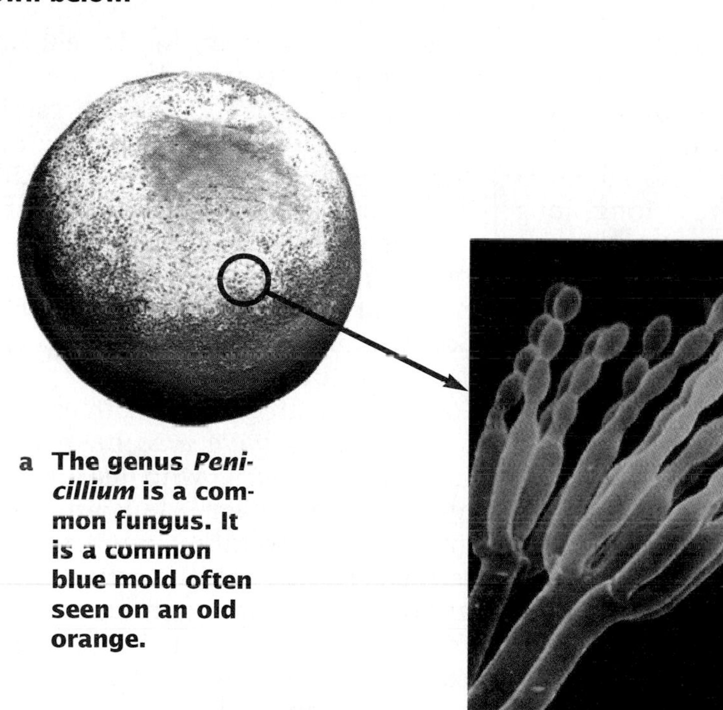

a **The genus *Penicillium* is a common fungus. It is a common blue mold often seen on an old orange.**

b **The organism is microscopic and reproduces by forming spores on the ends of fungal branches. It is easy to culture, and modern manufacturing methods can produce large quantities of penicillin.**

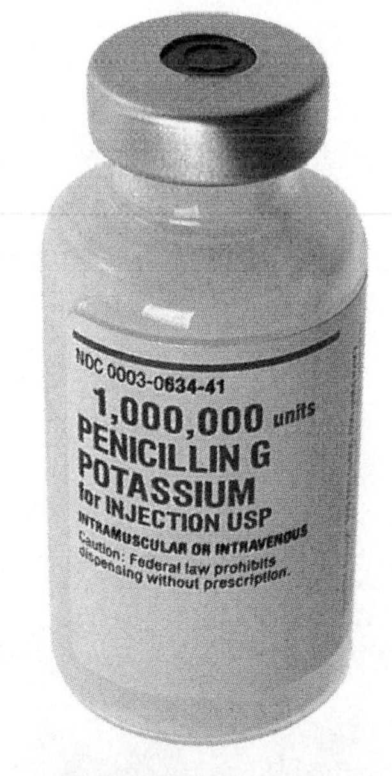

c **Penicillin is still one of the most common medicines used today to combat bacterial infections.**

Figure 18-13 Bacteria acquire resistance through natural selection or by receiving the resistance gene through conjugation.

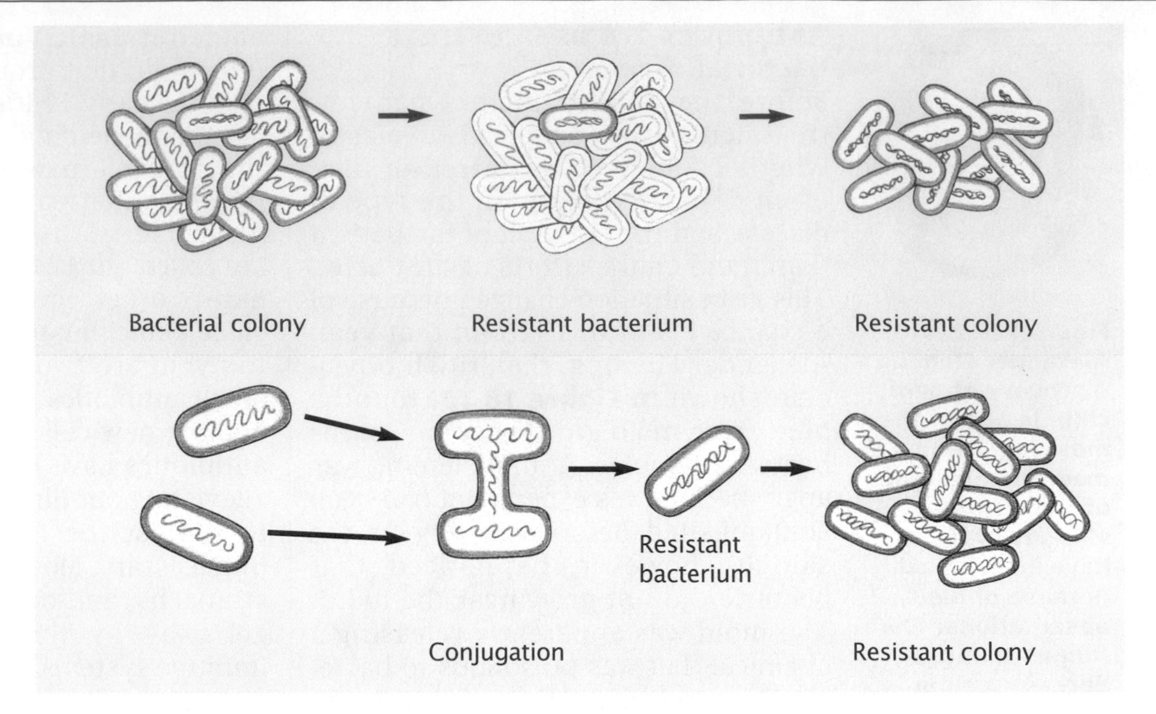

Bacteria are fighting back

Until the late 1980s, many scientists thought that we had conquered most infectious bacterial diseases and that the rest would soon be controlled. But bacteria have been here a long, long time. They reproduce rapidly, as we have seen, and they mutate. And they become resistant to the drugs we use against them by a variety of methods. **Antibiotic resistance** evolves in bacteria in a number of ways. For example, every time an antibiotic is used, it kills the majority of bacteria. But the few bacteria that remain alive after treatment with an antibiotic will reproduce, as you can see in **Figure 18-13**. These bacteria are resistant to the antibiotic, and they will pass along that resistance to the next generation. Just as we have seen insects develop resistance to pesticides, so bacteria have developed increasing numbers of resistant strains that are not affected by antibiotics.

Some cases of antibiotic resistance arise in patients who do not complete the full course of antibiotics prescribed by a physician. Without a full treatment, only the weaker bacteria are killed. Their stronger counterparts come back full force, and repeated antibiotic treatment encourages the rapid development of antibiotic-resistant strains. Some serious outbreaks of tuberculosis have been due to the failure of patients to complete antibiotic treatment.

Antibiotics have also been widely used by farmers, who learned that animals that were fed antibiotics gained more weight. Constantly challenged by antibiotics, the rapidly evolving bacteria give rise to new, resistant strains. Bacteria are the oldest and most successful group of organisms on Earth. Human beings are beginning to appreciate that the struggle to defeat disease-causing bacteria is going to be an ongoing battle.

SECTION REVIEW

1. **Describe one way in which bacteria are used to treat a disease.**
2. **Name four diseases caused by bacteria. Identify the mode of transmission for each disease.**
3. **Explain why pasteurization is effective in preventing bacterial contamination.**
4. **Why is it important to finish the entire course of antibiotics prescribed by a physician?**

The World Health Organization estimates that more than 10 million people throughout the world are infected by HIV, the virus that causes AIDS. The number of infected people grows rapidly each year. Finding a cure for AIDS or a vaccine against HIV will require a clear understanding of what viruses are, how they reproduce, and how they affect their hosts.

18-3 Viruses

OBJECTIVES

1. **Draw the structure of a virus.**
2. **Justify this statement: Viruses are not living organisms.**
3. **Describe viral reproduction.**
4. **List four diseases that are caused by viruses.**

What Is a Virus?

Think back to the last time you had a cold or the flu. Did you ask the doctor for antibiotics to kill the flu or cold "bacteria" that were making you so miserable? Did the doctor explain that colds and flu are not caused by bacteria, so antibiotics would have no effect? Colds and flu are caused by **viruses**. Viruses are microscopic particles that invade the cells of plants, animals, fungi, and bacteria. Viruses often destroy the cells they invade.

Viruses are small, simple particles

If you could open a virus, what would you find inside? Would you find cytoplasm, ribosomes, and mitochondria like you would find in one of your own cells? No, because a virus is not a cell. A typical virus, such as the polio virus shown in **Figure 18-14**, is composed of a core of genetic material surrounded by a protein "coat." The protein coat protects the genetic material and enables the virus to invade its host cell. The *Tour of a Virus* on the next page shows the structure of the human immunodeficiency virus (**HIV**). This virus causes acquired immune deficiency syndrome (**AIDS**).

In many viruses, DNA is the genetic material. Other viruses have RNA instead. The viruses that cause AIDS, polio, and the flu are RNA viruses. Viruses are parasitic and can reproduce only inside the cells of their hosts. Because viruses are so small, viral genetic material has room for only a few genes, usually only genes coding for the protein coat and for enzymes that enable the virus to take over its host cell.

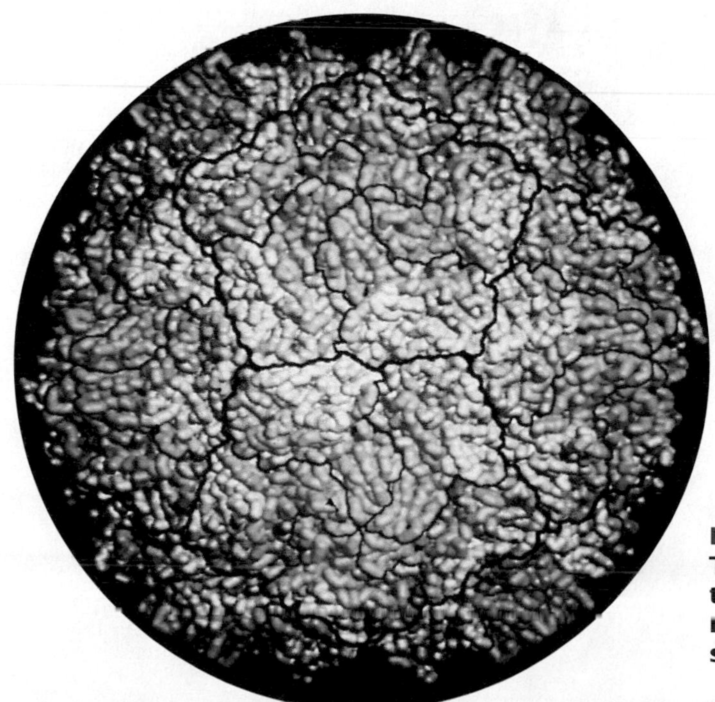

Figure 18-14
The computer-generated colors on this image of the polio virus represent the different proteins present on the surface of the virus.

Virus

The disease AIDS is caused by a virus called HIV, human immunodeficiency virus.

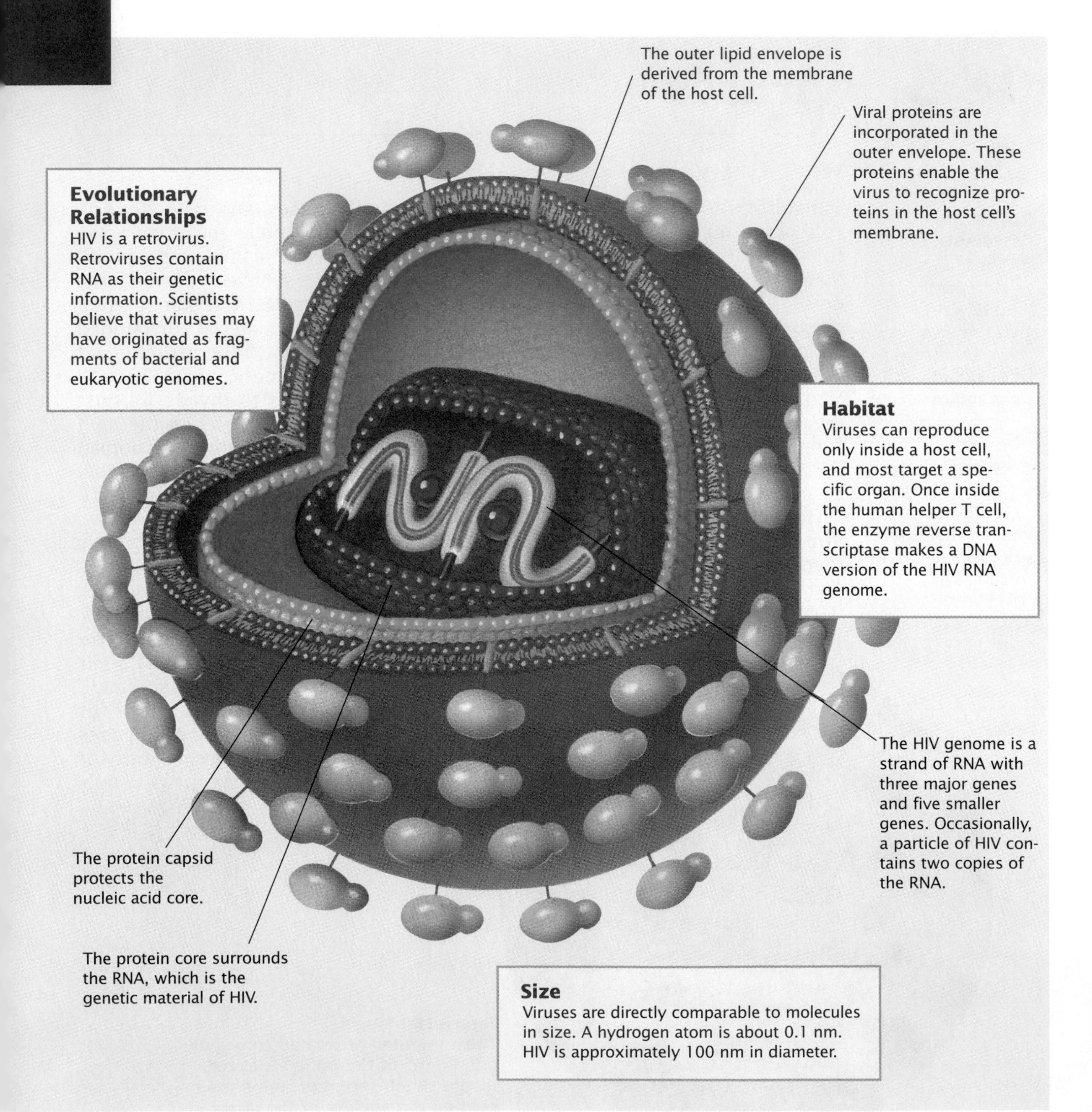

The outer lipid envelope is derived from the membrane of the host cell.
Viral proteins are incorporated in the outer envelope. These proteins enable the virus to recognize proteins in the host cell's membrane.
The HIV genome is a strand of RNA with three major genes and five smaller genes. Occasionally, a particle of HIV contains two copies of the RNA.
The protein capsid protects the nucleic acid core.
The protein core surrounds the RNA, which is the genetic material of HIV.

Evolutionary Relationships

HIV is a retrovirus. Retroviruses contain RNA as their genetic information. Scientists believe that viruses may have originated as fragments of bacterial and eukaryotic genomes.

Habitat

Viruses can reproduce only inside a host cell, and most target a specific organ. Once inside the human helper T cell, the enzyme reverse transcriptase makes a DNA version of the HIV RNA genome.

Size

Viruses are directly comparable to molecules in size. A hydrogen atom is about 0.1 nm. HIV is approximately 100 nm in diameter.

Viruses: living or nonliving?

Viruses do have some characteristics of living things. They have genetic material that is transmitted to future generations and that can change over time. Therefore, viruses are able to evolve. On the other hand, viruses lack three essential features of living things: they are not made of cells, they cannot make proteins by themselves, and they cannot use energy. Even though viruses can reproduce, they are able to do so only when inside living cells. Viruses are considered nonliving because they lack characteristics that biologists attribute to living things. Outside a host cell, a virus is an inert collection of chemicals. Nevertheless, because viruses are active inside living cells, the study of viruses is part of biology.

How Viruses Reproduce

To reproduce, a virus must insert its genetic material, which contains the instructions for making new viruses, into a host cell. This viral genetic material seizes control of its host cell and transforms it into a virus-making factory. All viruses reproduce by taking over the reproductive machinery of a cell. Follow how HIV reproduces in the example below.

HIV seizes control of host cells

HIV cannot infect a cell that it cannot enter. HIV is able to enter only those cells that have particular receptor proteins in their cell membranes. HIV recognizes these cells because the virus contains a protein that will bind to the receptor proteins on the cell membrane. You can think of the relationship between the viral protein and the cell membrane proteins as being like the relationship between a key and a lock. If the key (the viral protein) fits the lock (the cell proteins), then the cell is opened to invasion. Otherwise, the virus is locked out.

One host cell for HIV is a type of white blood cell known as the helper T cell. Helper T cells occur in the blood and in the lymphatic system. They play a very crucial role in the body's ability to fight infection. Follow the events that occur when HIV infects a helper T cell in **Figure 18-15**.

Figure 18-15 HIV enters a human helper T cell and then takes over the cell's machinery. The cell is directed to make HIV proteins and RNA. Newly made viruses then leave the cell and continue the cycle of infection.

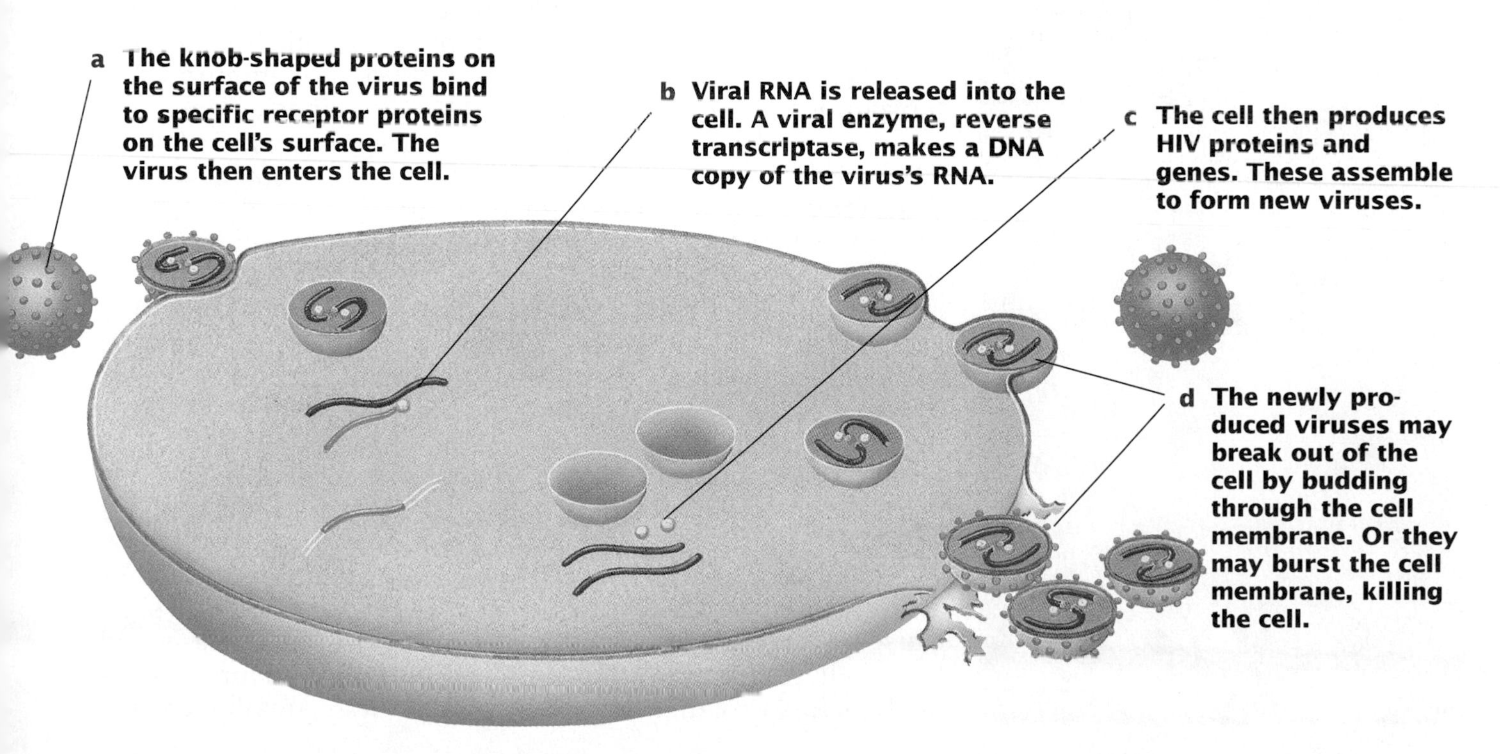

a The knob-shaped proteins on the surface of the virus bind to specific receptor proteins on the cell's surface. The virus then enters the cell.
b Viral RNA is released into the cell. A viral enzyme, reverse transcriptase, makes a DNA copy of the virus's RNA.
c The cell then produces HIV proteins and genes. These assemble to form new viruses.
d The newly produced viruses may break out of the cell by budding through the cell membrane. Or they may burst the cell membrane, killing the cell.

AIDS is fatal

HIV causes AIDS by killing helper T cells, which coordinate the body's defenses against pathogens. However, HIV may not begin to destroy large numbers of cells for a few months to over 10 years after infection. During this period, an infected person may experience only mild flulike symptoms or no symptoms at all. An infected person can transmit HIV to others during this period. Scientists now think that practically everyone infected with HIV will develop AIDS. Eventually, the virus begins to destroy a large percentage of an infected person's helper T cells. When the number of helper T cells has fallen to very low levels, the person is said to have AIDS. AIDS patients usually die of cancer or diseases a healthy immune system would defeat. You will read more about HIV transmission and AIDS prevention in Chapters 33 and 35.

Diseases Caused by Viruses

Viruses cause many serious human diseases in addition to AIDS, such as measles, shown in **Figure 18-16**, and influenza. Viruses also cause diseases in plants, insects, and even bacteria. Like pathogenic bacteria, pathogenic viruses are transmitted from host to host in characteristic ways. Some of the viral diseases listed in **Table 18-2** are airborne. A few, such as infectious hepatitis and polio, can spread through contaminated water. Insects also transport viruses. The yellow fever virus, common in tropical regions, is carried by mosquitoes.

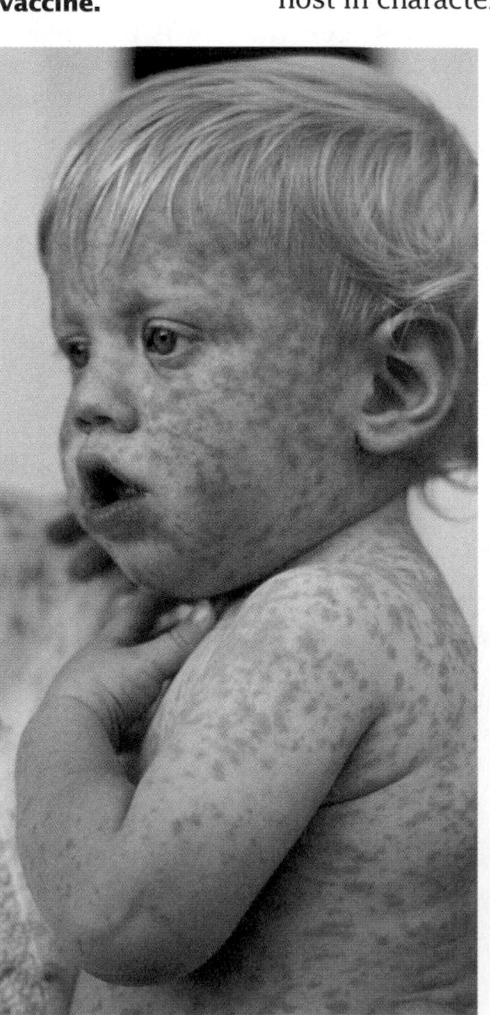

Figure 18-16 The telltale blotchy rash of measles has become relatively uncommon since the development of the measles vaccine.

There are defenses against viruses

Why don't physicians treat viral diseases such as colds, flu, and AIDS with antibiotics? Antibiotics work by interfering with cellular processes such as protein production or cell-wall synthesis, which do not occur in viruses. Moreover, because a virus uses its host cell for reproduction, it is very difficult to find any drugs that will destroy the virus without damaging the host.

A drug called azidothymidine (AZT) blocks an enzyme essential for DNA replication. Many AIDS patients are now being treated with AZT, which can prolong the lives of these patients. Unfortunately, AZT cannot cure AIDS; it only slows the course of the disease. Furthermore, AZT is very toxic and has numerous side effects.

Vaccination also protects against viral diseases

The only effective defense we now have against most viral diseases is a healthy immune system. Vaccination works to build up your immune system. Recall that vaccines against bacterial disease are composed of dead bacteria. But viruses are not "alive" and cannot be "killed." Instead, vaccines against viral diseases contain viruses made harmless by treatment with chemicals or by genetic engineering. These harmless viruses stimulate the immune system to create defenses against the harmful form of the virus.

Smallpox is no longer a killer

Eliminating smallpox from the world is vaccination's greatest triumph. Smallpox virus produced tiny pustules or sores (small "pox") on its victim's skin. These pustules developed scabs and often turned into permanent, disfiguring scars. Far worse, smallpox killed

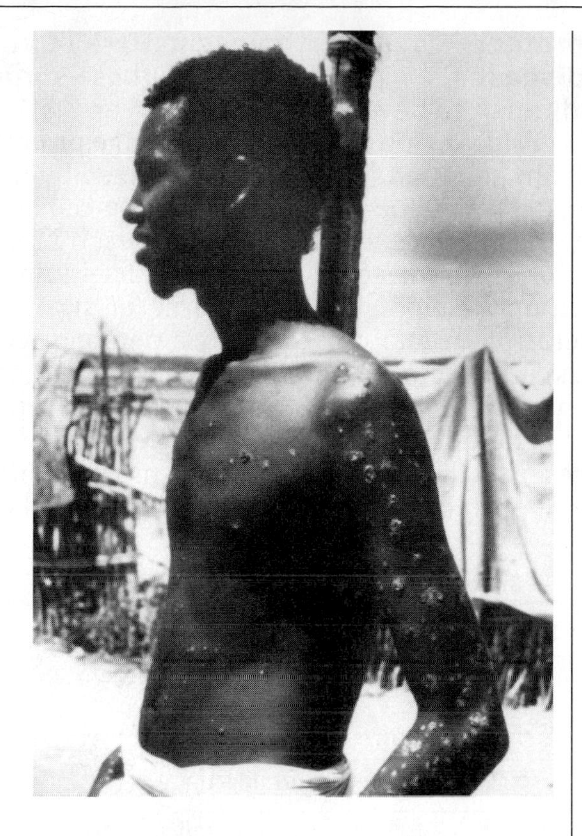

Figure 18-17
Ali Maow Maalin, of Merka, Somalia, contracted smallpox in 1977 at the age of 23. His was the last known case of smallpox contracted outside a laboratory.

half of those who contracted it. Because smallpox was so deadly, cures and means of prevention were sought for centuries. A vaccine was discovered in 1796, and efforts to extend its use increased with time.

Finally, vaccination against smallpox became commonplace in industrialized countries. As a result, smallpox rapidly disappeared from these countries. The last case in the United States occurred in 1949. Because the disease persisted in some of the poorest nations, the World Health Organization launched a worldwide vaccination campaign against smallpox. In 1967, the year this campaign was launched, 10 million to 15 million cases of the disease occurred. Just 11 years later, smallpox had been eliminated. The last known person to contract smallpox outside a laboratory is shown in **Figure 18-17**.

Emerging viral diseases

Many viruses live in animals without causing apparent harm, but they cause diseases when they infect other animals or humans. Among the most dangerous of these are a group of filoviruses (*filo-* from the Greek word for "thread," and *viron* from the Latin word for "poison") that have produced outbreaks of hemorrhagic fever in regions of central Africa and other parts of the world that received animals imported from Africa. These viruses attack human connective tissue, leading to massive bleeding (hemorrhaging). With mortality rates ranging from 30 to 90 percent, these filoviruses are among the most lethal infectious agents known. The outbreak of Ebola virus in Zaire, shown in **Figure 18-18**, in the spring of 1995 was confined to a few villages and fortunately did not reach the nearby city of Kinshasa, which has a population of 4 million.

No one knows the identity of the central African animal that harbors the Ebola virus. It does not appear to be chimpanzees, because the virus kills them as well. It might be another primate, a mosquito, or even an elephant. Intensive efforts are being made to find the carrier so that an attempt can be made to control the disease. Filoviruses could be a very real threat to worldwide human health if they ever achieve widespread dissemination, as the AIDS virus has done.

Figure 18-18
Two boys waiting for the body of a relative at a cemetary in Kikwit, Zaire, in May 1995 try to protect themselves by covering their faces. Ebola virus killed over 200 people in central Africa before the epidemic ended.

Evolution

Why is it difficult for researchers to develop effective vaccines against those viruses that change the shapes of their surface proteins?

Vaccines do not protect against all viral diseases

Why has it been so difficult to develop a vaccine against HIV? And why do millions of us suffer from colds and flu each year? Unfortunately, vaccination is effective only against viruses with nonvarying proteins on their surfaces, proteins that remain unchanged generation after generation. Smallpox virus, measles virus, and polio virus have surface proteins that remain the same from one viral generation to the next. HIV, cold viruses, and flu viruses, however, have surface proteins whose genes mutate often. As a result, the shape of these surface proteins often changes over just a few generations. Such a change produces a form of virus not recognized by the immune system. Although you can be vaccinated against influenza, each kind of influenza vaccine produces immunity against just one form, or strain, of the virus. That is why a person vaccinated against the flu can get the flu many times in a lifetime, each time from a different strain. Of all the changeable viruses, HIV changes most rapidly. **Table 18-2** lists some common viral diseases.

Table 18-2 Examples of Viral Diseases

Disease	Transmitted by	Symptoms
Chickenpox	Airborne	Rash, fever
Measles	Airborne	Blotchy rash, high fever, congestion in nose and throat
Rabies	Bite of infected animal	Fever, paralysis; fatal unless treated
Rubella (German measles)	Airborne	Rash, swollen glands
Mumps	Airborne	Swollen salivary glands
Influenza (flu)	Airborne	Headache, muscle aches, sore throat, cough; historically, one of the great "killer" diseases
Smallpox	Airborne	High fever, pustules on skin; often fatal; now extinct
Infectious hepatitis	Contaminated food or water	Fever, chills, nausea, swollen liver, jaundice, pain in the joints
Polio	Contaminated food or water	Headache, stiff neck, possible paralysis
Yellow fever	Mosquitoes	Nausea, fever, aches, liver cell destruction; can be fatal
AIDS	Sexual contact, contaminated blood products, contaminated hypodermic needles and syringes	Immune system failure; fatal
Hemorrhagic fever: Ebola virus	Sexual contact, contact with blood or other body fluids	Fever, epithelial cell destruction, massive hemorrhaging; often fatal

SECTION REVIEW

1. **Draw a virus and label its parts.**
2. **Describe three ways in which viruses differ from living cells.**
3. **How does a virus reproduce?**
4. **Explain why a vaccination against influenza will not give you lifelong protection against the flu.**

CHAPTER

18 Highlights

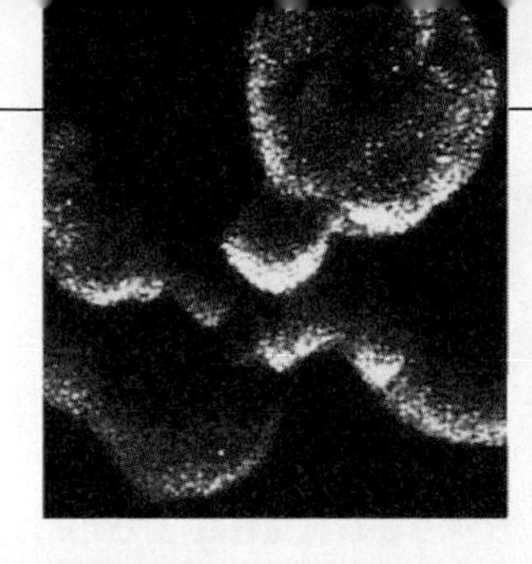
These are human immunodeficiency viruses seen through a scanning electron microscope.

	Key Terms	Summary
18-1 Bacteria Unlike eukaryotic cells, bacterial cells do not have a nucleus. 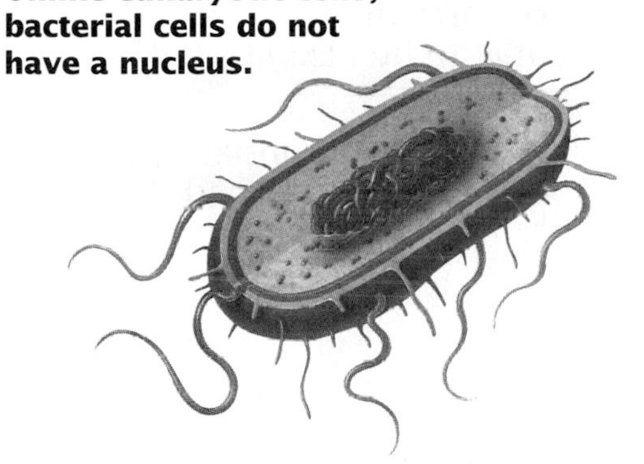	Gram staining (p. 332) conjugation (p. 332) chemosynthesis (p. 335)	• Bacteria, the oldest group of organisms on Earth, have cells much smaller than those of eukaryotes. • Most bacteria are spherical, spiral, or rod-shaped, and they have one of two types of cell walls. • Bacterial cells lack membrane-bound organelles and chromosomes. They have ribosomes, cytoplasm, and one molecule of DNA. • The kingdoms Archaebacteria and Eubacteria contain autotrophs and heterotrophs.
18-2 How Bacteria Affect Humans The discovery of penicillin by Alexander Fleming is one of the great scientific and medical achievements of this century. 	pathogen (p. 338) pasteurization (p. 339) vaccine (p. 342) penicillin (p. 343) antibiotic (p. 343) antibiotic resistance (p. 344)	• Beneficial bacteria include decomposers and nitrogen-fixing bacteria. Bacteria also make certain foods and drugs. • Vaccines help the body's immune system resist infection, thereby reducing the incidence of many bacterial diseases. • Antibiotics are antibacterial drugs. Alexander Fleming discovered the first antibiotic, penicillin, in 1928. • Overuse of antibiotics has encouraged the development of resistant bacteria.
18-3 Viruses Viruses are responsible for many diseases, including polio, herpes, AIDS, and the common cold. 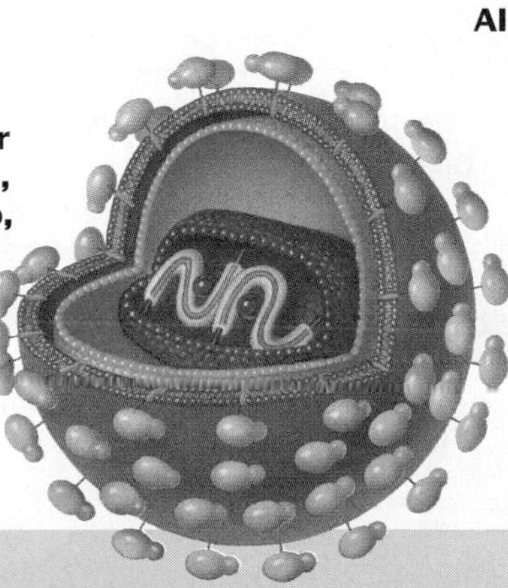	virus (p. 345) HIV (p. 345) AIDS (p. 345)	• Viruses are not considered living organisms. They are small particles that invade cells. • Viruses consist of RNA or DNA surrounded by a coat of protein. • A virus can only reproduce by controlling a cell. • Emerging viruses such as HIV and Ebola virus can appear and disappear without warning or can become a long-term world health problem.

CHAPTER 18

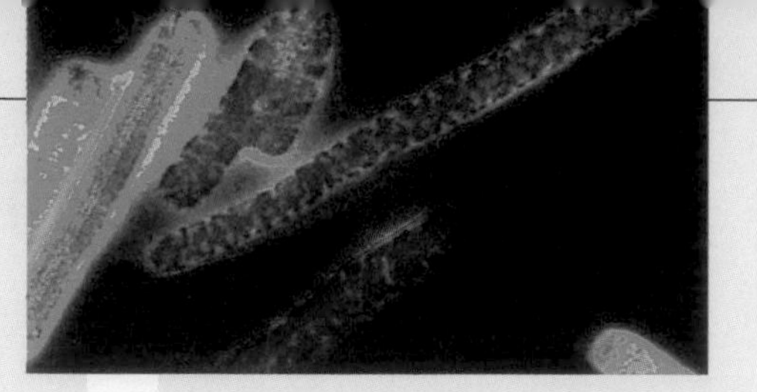

review

Understanding Vocabulary

1. For each set of terms, complete the analogy.
   a. bacteria : living :: viruses : ______
   b. refrigeration : cold :: pasteurization : ______
   c. pathogen : harmful :: vaccine : ______

2. Using each set of words below, write one or more sentences summarizing information learned in this chapter.
   a. cell walls, Gram staining, diseases, bacteria
   b. heterotrophic, nutrition, photosynthetic, bacteria, autotrophic, chemosynthesis
   c. bacterial diseases, pasteurization, vaccine, sanitation, hygiene

Understanding Concepts

3. **Relating Concepts** Construct a concept map describing the relationships of bacteria and viruses to diseases. Use as many terms as necessary from the vocabulary list. Try to include the following terms in your map: bacteria, viruses, pathogen, antibiotics, benefits, and vaccination.

4. **Comparing Structures** Look at the two representations below. What are they? How do their structures differ? How are they the same?

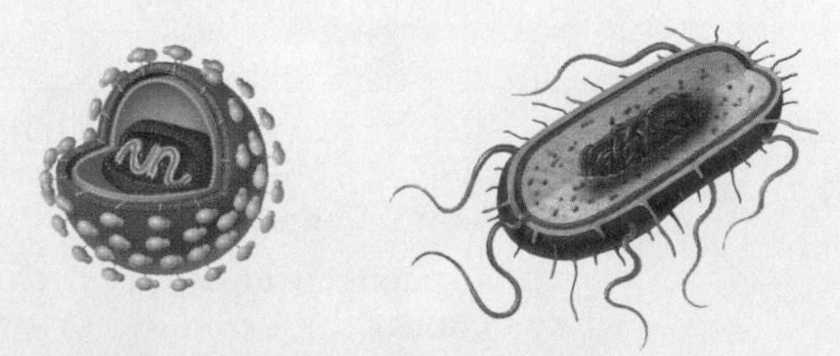

5. **Interpreting Conclusions** How can a vaccine protect you against a disease?

6. **Applying Information** The synthetic drug AZT is used by many people who have AIDS. Does AZT cure AIDS? Explain your answer.

7. **Analyzing Conclusions** Explain why antibiotics work against bacterial diseases but not against viral diseases.

8. **Summarizing Information** Explain why tuberculosis is once again becoming a serious concern to public-health officials.

9. **Analyzing Functions** Describe the process by which bacteria develop resistance to antibiotics.

10. **Recognizing Relationships** Even though bacteria can be very harmful to humans, they are also essential to life. Explain why

11. **Inferring Relationships** Match each term on the left with the most appropriate term on the right. After you have matched them, explain the relationships between them.

a. *Escherichia coli*	1. Lyme disease
b. HIV	2. vitamin K
c. ticks	3. photosynthetic
d. conjugation	4. AIDS
e. autotrophic eukaryotes	5. immunization
f. bacterial decomposition	6. genetic exchange
g. vaccination	7. pickles

12. **BUILDING ON WHAT YOU HAVE LEARNED** Which of the five properties of life described in Chapter 2 would have to be changed before the viruses could be considered alive?

13. **Interpreting Data** Lab technicians routinely test for the presence of *E. coli* to ensure that municipal water supplies are not polluted by sewage. The number of *E. coli* bacteria found in water samples may indicate the amount of sewage present in the water.

Three-Year Fecal Coliform Averages (bacterial colonies per 100 mL of water)

	1992	1993	1994
Site 1	15	13	12
Site 2	5	16	35
Site 3	88	92	6

Which statement is most consistent with the data?
a. The water at Site 1 is safe to drink.
b. The amount of sewage decreased each year at Sites 1, 2, and 3.
c. In 1993, a factory at Site 3 was fined for releasing untreated sewage.
d. Water samples collected from Site 1 contain the largest amounts of sewage.

Interpreting Graphics

14. Examine the figure shown below. Match each numbered event in the figure with the sentence that best describes it.

a. Viral RNA is released into the host cell.
b. Viruses are released.
c. New viruses are made.
d. A virus attaches to the host cell surface.

Arrange the events in the correct sequence. Using your imagination, describe some ways in which a drug or vaccine could affect events shown in the figure below.

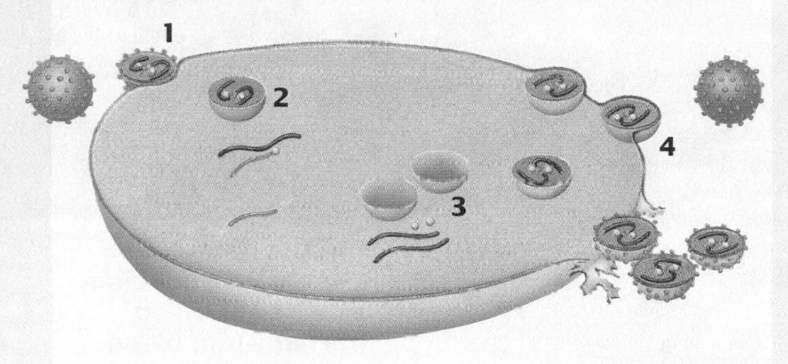

Reviewing Themes

15. ***Interdependence***
Even though bacteria can be very harmful to humans, they are also essential to life as we know it. Explain why.

16. ***Interdependence***
HIV attacks and destroys helper T cells. How does having fewer helper T cells affect the human immune system?

17. ***Evolution***
Hundreds of thousands of people get flu shots each year. Yet many of them still get the flu. Why are flu shots often ineffective?

18. ***Cell Structure and Function***
How do the cell walls of Gram-positive bacteria differ from those of Gram-negative bacteria? How does this difference affect the effectiveness of antibiotics used to treat bacterial infections?

Thinking Critically

19. **Distinguishing Fact From Opinion** A reporter wrote, "The ongoing development of new antibiotics means that humans will soon be free of bacterial diseases." Explain the error that the reporter has made.

Life/Work Skills

20. **Applying Technology** Farm managers have discovered that mixing low doses of antibiotics into cattle feed makes animals grow larger. Biologists claim that this practice encourages the development of resistant strains of bacteria. In your opinion, is this use of antibiotics justified? Use references available in a library or search an on-line database to write a report summarizing your opinion. Include facts and examples to support your conclusions.

Activities and Projects

21. **Social Studies Connection** The global flu epidemic of 1918–1919 claimed 20 million lives. Use references available in a library or search an on-line database to find resources for writing a short description of the steps taken to prevent further epidemics. How long was it before another epidemic occurred? How did it differ from that in 1919?

22. **Career Connection: Microbiological Technician** Technicians in this field study bacteria, fungi, and viruses and their relationships to disease, agriculture, genetic engineering, and pollution. Use references in the library or search an on-line database to discover the many job possibilities in this area.

23. **Health Connection** Make a poster showing a list of behaviors that could reduce the spread of colds and flu in your school. Write to the Association for Practitioners in Infection Control, 505 East Hawley, Mundelein, IL 60060, for more information about the importance of hand washing and other sanitation issues mandated for those who work in the food industry.

Discovering Through Reading

24. Read the article "Revenge of the Killer Microbes" in *Time*, September 12, 1994, pages 63–69. Describe four human behaviors that have encouraged the development of antibiotic-resistant bacteria.

Science, Technology, and Society

New Drug Development—How Are New Drugs Tested?

What happens when you or a family member is seriously ill? If you're lucky, a drug is available to treat the illness or disease. A streamlined Food and Drug Administration approval process now allows new drugs that offer the promise of saving lives to reach the market faster.

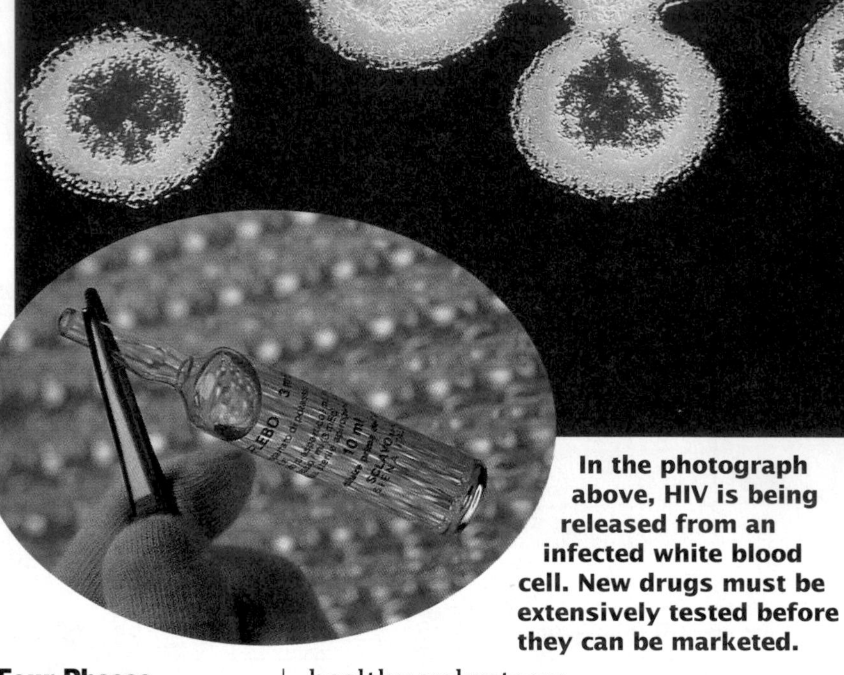

In the photograph above, HIV is being released from an infected white blood cell. New drugs must be extensively tested before they can be marketed.

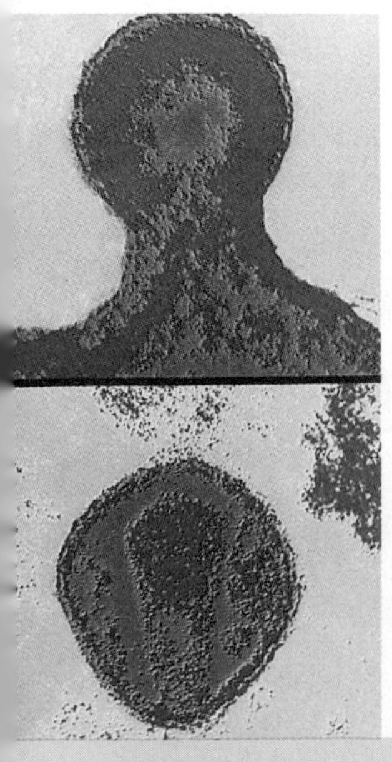

Drug Studies Include Four Phases
Before a new drug can be sold, its sponsor—a pharmaceutical company, a research organization, or a public or private agency—must first conduct studies to show that the drug is safe and effective. The results of these studies are submitted to the Food and Drug Administration (FDA) for approval. If the FDA determines that the drug is both safe and effective, the drug may then be approved for sale. The FDA currently requires drugs to be tested in four phases, each designed to answer specific questions about the drug.

Is the drug safe to use on humans? During Phase I, investigators seek to answer questions about how the drug is absorbed and excreted by the body and seek to discover what effects (including short-term side effects) various doses of the drug have on the body. Investigators also determine the minimum effective dosage, the amount of the drug that is effective yet does not produce undesirable side effects. Phase I testing involves only a few individuals, usually healthy volunteers.

Does the drug work? If Phase I testing does not reveal unacceptable safety problems, Phase II testing can begin. Investigators look to see if the drug is effective in treating the disease or condition for which it is intended. Researchers also seek to discover if the drug has short-term side effects. These side effects may be very harmful, so it is important to know exactly what they are before many people use the drug.

How effective is the drug? By the time a drug reaches Phase III, researchers have shown that the drug does have a therapeutic effect. Phase III tests involve large numbers of people (usually several hundred individuals) who have the condition against which the drug is effective. These tests measure the extent of the drug's effectiveness and help determine what fraction of the population is helped by the drug. In some cases, Phase III testing may also examine the consequences of long-term use of the drug. After the drug has gained FDA approval, Phase IV studies evaluate the drug's long-term effectiveness. Phase IV studies are sometimes referred to as post-marketing studies. Phase IV studies obtain information from large numbers of individuals who are using the drug as directed by their physicians.

Accelerated Studies Speed Drug Availability

Testing a new drug from Phase I to Phase IV, can take more than ten years. When a new drug offers hope to patients with otherwise untreatable, life-threatening diseases, new FDA guidelines can provide access to the drug before it has received full FDA approval. In each case, the benefits of early use must be carefully weighed against the threat of potential dangers, and patients using the drug must be carefully monitored. If the drug is found to be ineffective or dangerous, it can then be removed from the market.

Technique

Experimental Design

Before a new drug study can begin, the Food and Drug Administration must approve its experimental design. The study must include two groups of individuals: a group that receives the drug and a control group that does not. Instead of taking the medication containing the drug, the control group is instead given a placebo *(pluh SEE boh)*. The placebo is identical to the medication in every way except one—it does not contain the drug.

To ensure valid results, drug studies are often conducted as double-blind studies. In a double-blind study, neither the doctors nor the patients know who receives the drug and who receives the placebo. Double-blind studies eliminate bias, the tendency of individuals to lean towards a particular conclusion based upon what they know. The removal of bias reduces the likelihood that a positive drug response could be due solely to a patient's wishful thinking.

Applications: Knowledge of experimental design is required by research scientists in many areas.

Users: Medical researchers, biostatisticians, marketing researchers

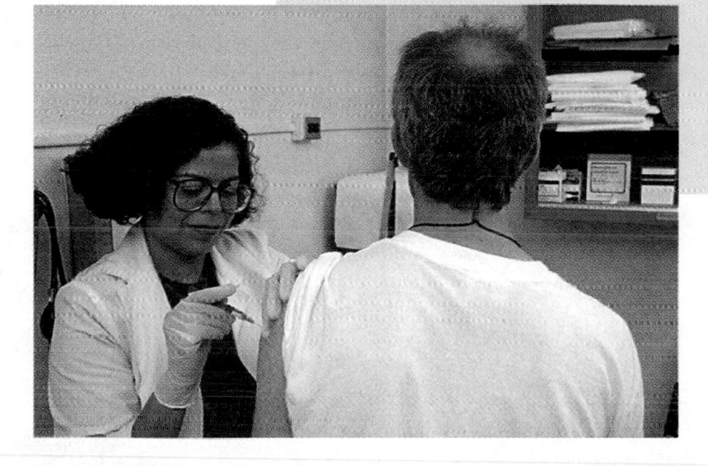

In most drug studies, some participants are given a placebo instead of the drug. The use of a placebo enables researchers to ensure that effects are due solely to the drug.

Analyzing the Issues

Why is it important that a new drug be thoroughly tested?

1. Research the use of the drug thalidomide in Europe in the early 1960s. For what purpose was the drug prescribed? What happened when the drug was taken by pregnant women? How do the consequences of thalidomide use support the need for drug testing?

When a new drug is tested, should *all* test participants be given the drug?

2. If a potential new drug is known to improve survival, should Phase II tests omit the use of placebos? Use references available in the library or research an on-line database to find information supporting your viewpoint. Write a report summarizing the benefits and drawbacks of using placebos.

Are study participants informed of potential risks and benefits?

3. By law, before individuals can participate in a drug study, they must give their informed consent, indicating that they are willing participants and that they understand the purpose of the study. What facts do you think participants in a new drug study should know before they agree to participate in the study?

How does the cost of drug testing influence which drugs are developed?

4. The Orphan Drug Act passed by the United States Congress in 1983 has helped to ensure that certain drugs will be developed and brought to market. What is an "orphan drug"? How did the Orphan Drug Act encourage pharmaceutical companies to develop orphan drugs?

CHAPTER

19

REVIEW
- evolution of protists (Section 11-2)
- autotrophs and heterotrophs (Section 14-1)
- archaebacteria and eubacteria (Section 17-3)
- the terms *endocytosis*, *exocytosis*, and *conjugation* (Glossary)

Protists

Diatoms are single-celled algae, which form a large part of both marine and freshwater plankton.

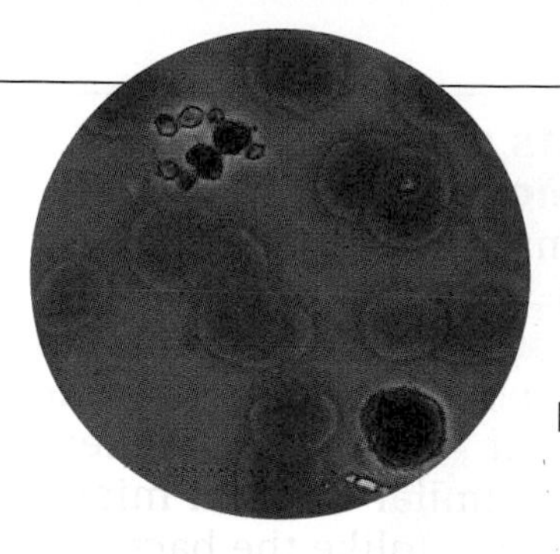

Did you know that malaria, a disease caused by a parasitic protist, currently infects up to 500 million human beings? Malaria kills more than 1 million people each year. Other protists cause sleeping sickness, dysentery, and giardiasis. Like bacteria and viruses, protists exert major influences on humans.

19-1 What Is a Protist?

OBJECTIVES

1. **Identify the characteristics shared by all members of the kingdom Protista.**
2. **Describe the evolutionary relationship between bacteria and protists.**
3. **Summarize the relationships of protists to the other three eukaryotic kingdoms.**

Characteristics of Protists

If you examine a drop of pond water under a microscope, you might see, swimming on the slide, a tiny one celled organism called *Euglena*, shown in **Figure 19-1**. To which kingdom does *Euglena* belong? It cannot be a member of the kingdoms Archaebacteria or Eubacteria because it has organelles and a nucleus. Its lack of a cell wall and its ability to move suggest that *Euglena* might be an animal—but it is single-celled, and all animals are multicellular. Also, if you look carefully, you can see that this *Euglena* has chloroplasts; it is photosynthetic. Scientists place organisms such as *Euglena* into the kingdom Protista. The kingdom Protista contains nearly all of the single-celled eukaryotes as well as multicellular algae.

Protists have complex cells

Like all protists, *Euglena* is eukaryotic. Its nucleus, which contains chromosomes, is separated from the cytoplasm by a membrane. Also, membrane-bound organelles, such as chloroplasts and mitochondria, are found in the cell's cytoplasm. Chloroplasts occur in autotrophic protists, such as *Euglena*, but they are lacking in heterotrophic protists. *Euglena* has the unique ability to shift between being autotrophic and heterotrophic. In abundant light, *Euglena* produces its own nutrients. When light is scarce, *Euglena* feeds on bacteria and smaller protists. Most protists, however, are either autotrophs or heterotrophs.

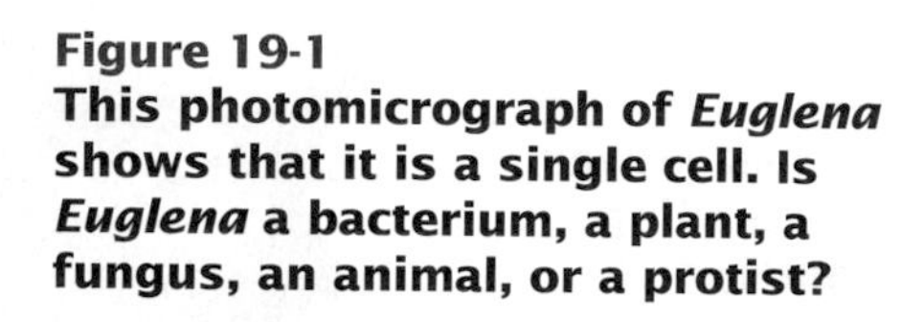

Figure 19-1
This photomicrograph of *Euglena* shows that it is a single cell. Is *Euglena* a bacterium, a plant, a fungus, an animal, or a protist?

Heterotrophic Protists

The word *carnivore* brings to mind fierce, sharp-toothed animals like lions, tigers, wolves, and sharks, and the word *herbivore* makes you think of plodding creatures like cows and sheep. You don't normally think of tiny one-celled protists as carnivores and herbivores, but most heterotrophic protists play one of these two ecological roles. In the plankton, for example, some heterotrophic protists "graze" on phytoplankton, and others prey on these grazers. Other heterotrophic protists are parasites or decomposers.

Slime molds and water molds superficially resemble fungi

If you look carefully through moist, decaying leaves on a forest floor, you might find a brightly-colored plasmodial slime mold, such as the one shown in **Figure 19-9a**. Such an organism spends much of its life as a mass of streaming material that moves through damp environments in the forest feeding on small particles of organic matter. If food or moisture becomes scarce, the streaming mass produces black spore-bearing bodies like those shown in **Figure 19-9a**. These spores can survive many years.

If your ancestors arrived in the United States from Ireland in the 1840s, they were probably fleeing the effects of a heterotrophic protist. From 1845 to 1847, a famine in Ireland killed more than 1 million people and forced another 3 million to emigrate; most came to the United States. This famine occurred because the potato crop (the staple of the Irish diet) was almost wiped out by an outbreak of late blight. Late blight is a disease caused by *Phytophthora infestans*, a protist that is a water mold, shown in **Figure 19-9b**.

Years ago, slime and water molds were placed in the kingdom Fungi. Now we know they are unrelated to the fungi and have different cell walls, composition, and structure.

Figure 19-9 Heterotrophic protists are extremely diverse. Two very different species are shown below.

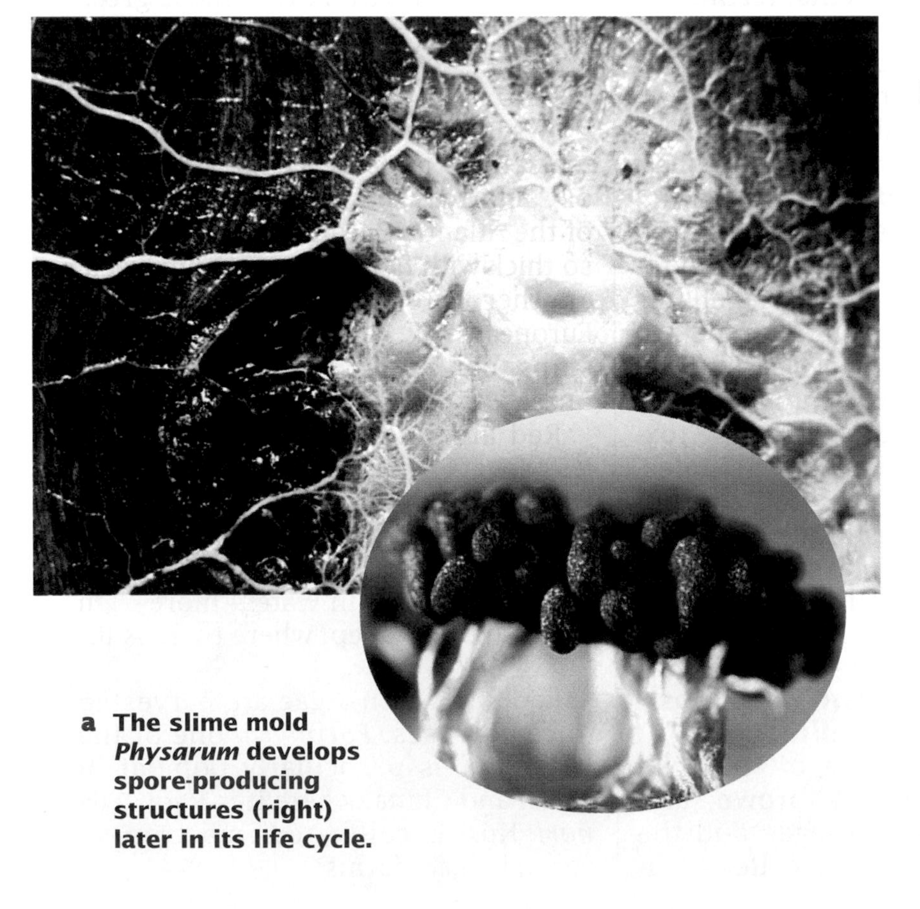

a The slime mold *Physarum* develops spore-producing structures (right) later in its life cycle.

b Water molds are microscopic organisms. The water mold *Phytophthora infestans*, which causes late blight of potatoes, has profoundly influenced history.

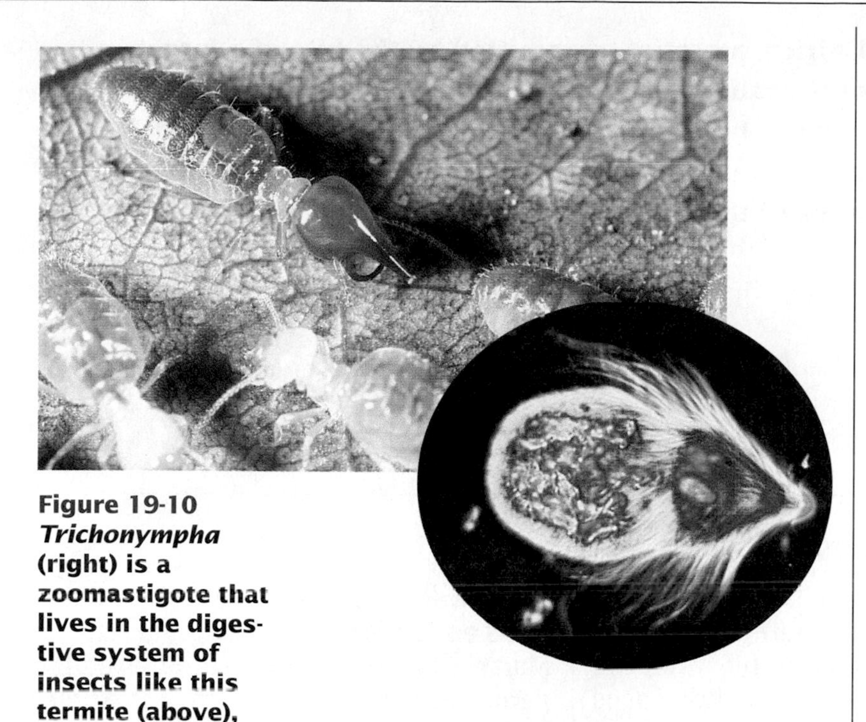

Figure 19-10
***Trichonympha* (right) is a zoomastigote that lives in the digestive system of insects like this termite (above), making it possible for the termite to eat wood.**

Protists include the ancestors of animals

About 300 years ago, scientists using the first microscopes saw what appeared to be tiny animals darting about in drops of water. They called these organisms protozoa, which means "first animals." Biologists now classify these organisms as protists and believe they are related to the ancestors of modern animals.

The ancestor of animals probably belonged to a group of protists known as the zoomastigotes *(ZOH mast ih gohtz)*. Zoomastigotes are one-celled, heterotrophic protists with at least one flagellum. Some zoomastigotes, such as the *Trichonympha* illustrated in **Figure 19-10**, are covered with flagella. If your house has ever been attacked by termites, you can blame *Trichonympha*. Without these protists living in their intestines, termites could not digest wood.

Paramecium and *Didinium*, shown in **Figure 19-11**, are members of a group of protists called ciliates. Ciliates are covered with numerous cilia, which function in locomotion and feeding. In some ciliates, the cilia have fused to form sheets that function as mouths or teeth.

Not all groups of protists are closely related. Indeed, based on recent molecular analysis of DNA, some biologists argue that ciliates are so different from all other protists that they should be placed in a separate kingdom of their own.

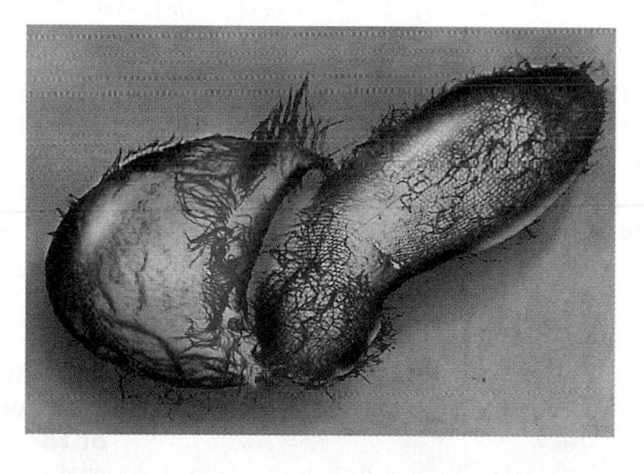

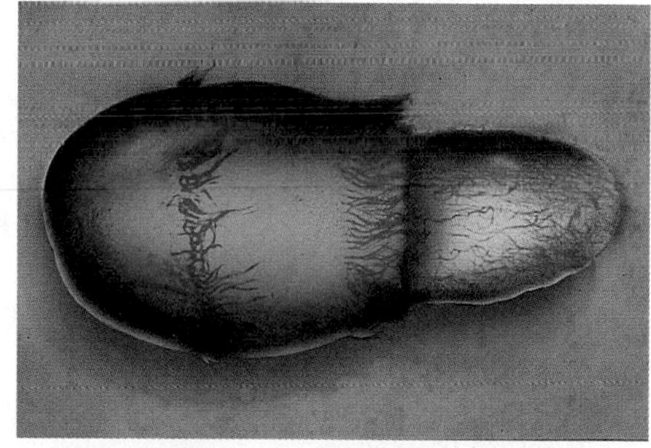

Figure 19-11
One protist, *Didinium*, attacks and swallows another protist, *Paramecium*.

SECTION REVIEW

1. **Explain why the statement "All protists are microscopic" is untrue.**
2. **Summarize the importance of algae in aquatic ecosystems.**
3. **Describe two pieces of evidence indicating that plants evolved from green algae.**
4. **Name two kinds of heterotrophic protists.**

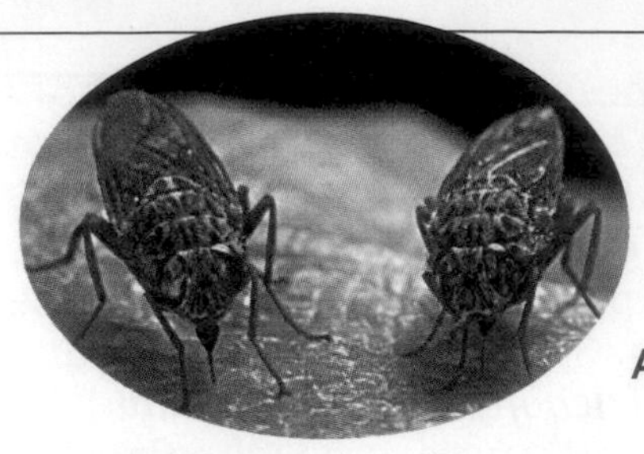

In central Africa, a protist transmitted by the bite of a large fly, the tsetse fly, causes the disease sleeping sickness. This disease is often fatal to humans and domestic animals such as cattle, sheep, and pigs. Livestock cannot be raised in much of central Africa because of the presence of sleeping sickness.

19-3 Diseases Caused by Protists

OBJECTIVES

1. **Explain how malaria is transmitted.**
2. **Evaluate the methods used to control malaria.**
3. **Explain why malaria control methods have not been entirely successful.**
4. **Name three human diseases caused by protists.**

Protists and Disease

Although many protists do not affect the health of human beings, millions of people are infected by, and die from, various protist-caused diseases each year. Disease-causing protists are transmitted mainly by insects and contaminated water.

Malaria is the great killer

Currently, from 300 million to 500 million people are infected with malaria. This disease kills more than 1 million people each year, making it one of the most serious human diseases. It is a leading cause of death among young children in tropical countries. This disease has affected the evolution of our species. As you learned in Chapter 10, the allele for sickle cell hemoglobin, a mutant form of hemoglobin, is prevalent in malaria-prone areas of Africa. This mutant hemoglobin produces resistance to the parasite that causes malaria.

A parasitic protozoan causes malaria

Malaria is caused by protozoa of the genus *Plasmodium*, which has a very complex life cycle. *Plasmodium* is carried between human hosts by mosquitoes, particularly mosquitoes of the genus *Anopheles*, shown in **Figure 19-12**. When a female mosquito feeds, she injects saliva into the wound to prevent her victim's blood from clotting. The mosquito's saliva transfers *Plasmodium* into the human bloodstream. (Male mosquitoes do not spread the disease because they feed on nectar instead of blood.) Once in the blood, the *Plasmodium* parasites are carried to the liver, where they reproduce. They then reenter the bloodstream, penetrate the red blood cells, and reproduce again. Every 48 or 72 hours, depending on the species of *Plasmodium*, a new generation of parasites bursts out of infected red blood cells, as shown in

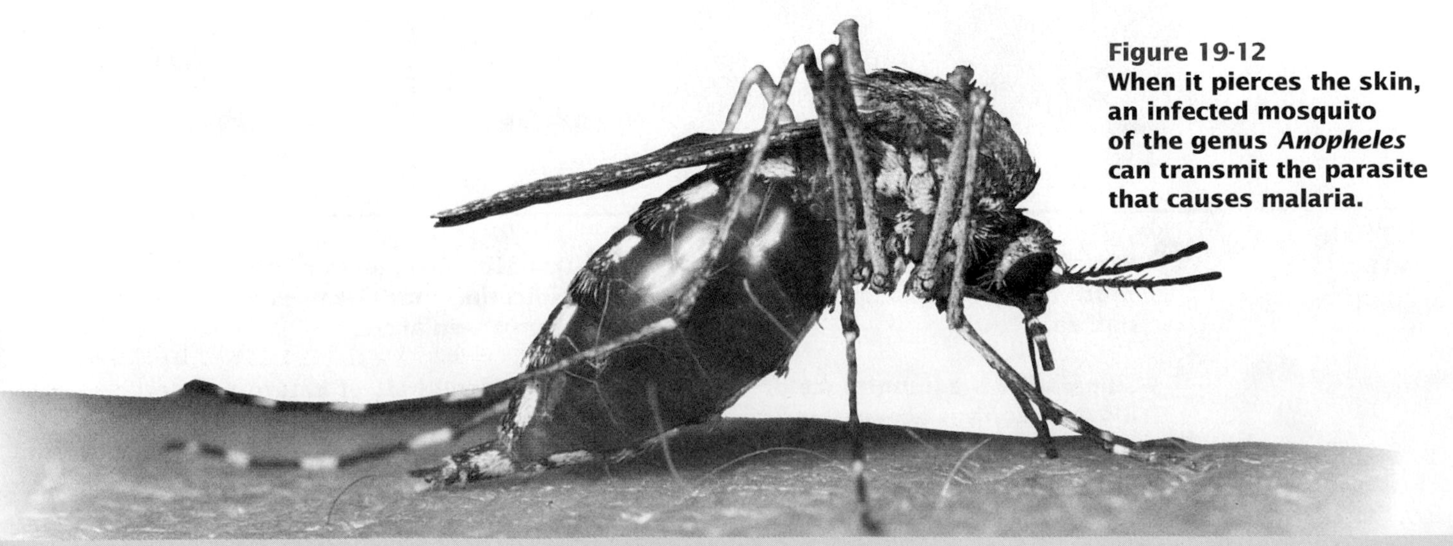

Figure 19-12
When it pierces the skin, an infected mosquito of the genus *Anopheles* can transmit the parasite that causes malaria.

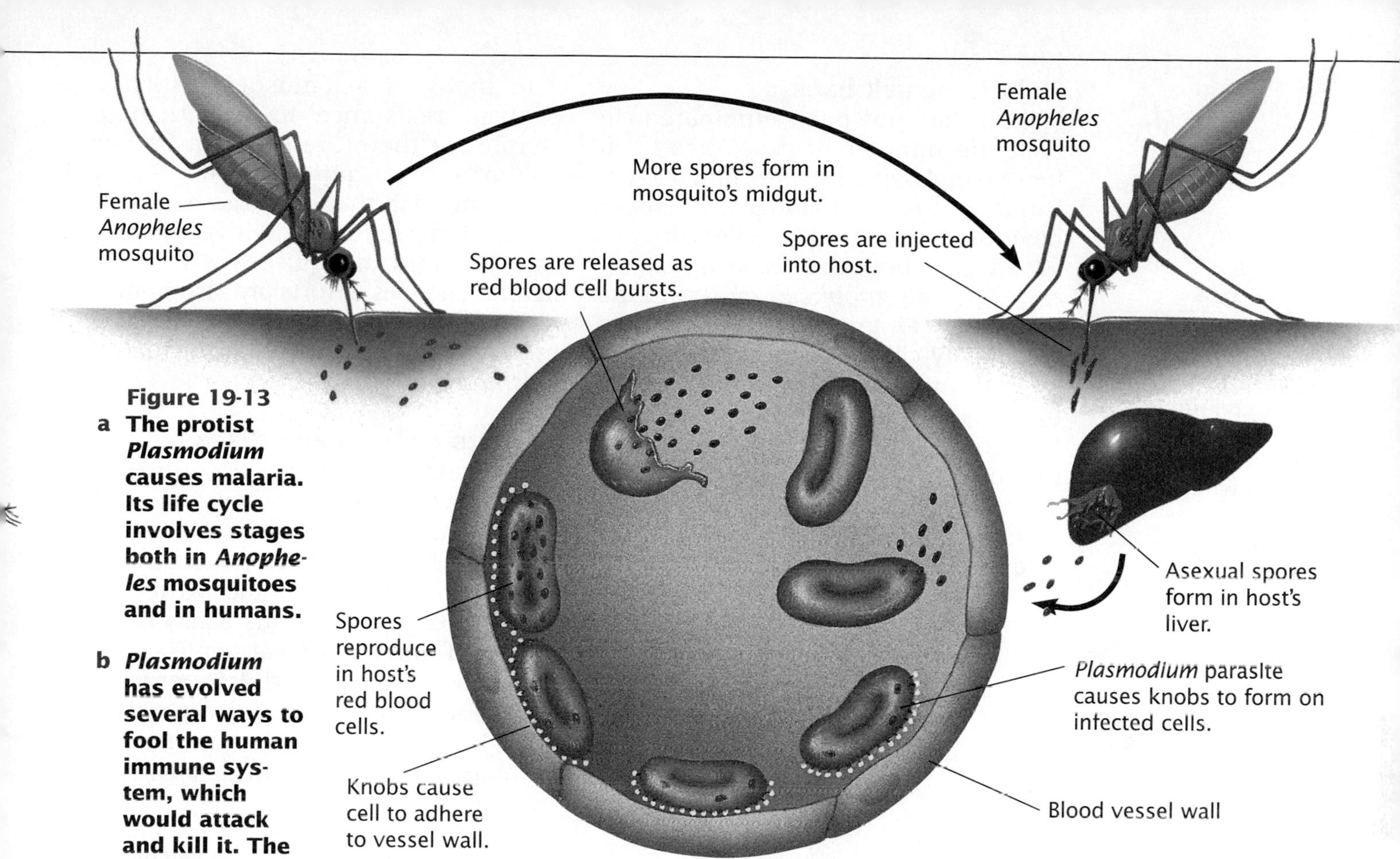
Female Anopheles mosquito
More spores form in mosquito's midgut.
Female Anopheles mosquito
Spores are injected into host.
Spores are released as red blood cell bursts.
Asexual spores form in host's liver.
Spores reproduce in host's red blood cells.
Plasmodium parasite causes knobs to form on infected cells.
Knobs cause cell to adhere to vessel wall.
Blood vessel wall

Figure 19-13
a The protist *Plasmodium* causes malaria. Its life cycle involves stages both in *Anopheles* mosquitoes and in humans.
b *Plasmodium* has evolved several ways to fool the human immune system, which would attack and kill it. The parasites change the proteins on an infected cell's surface every few generations to keep them invisible to the body's defenses.

Figure 19-13, and invades other red blood cells. Each of these outbreaks can destroy up to 40 percent of the host's red blood cells.

Female mosquitoes acquire *Plasmodium* when they bite infected humans. *Plasmodium* does not make the mosquito ill, so she is able to fly off and transmit the parasite to other humans.

The symptoms of malaria follow a cycle that corresponds to the reproduction of *Plasmodium* in the blood. When the parasites emerge from the blood cells, the host experiences high fever, delirium, and sweating. Severe chills follow the fever and can last until the next outbreak. Malaria sometimes weakens its victims so much that they die from other infections. Also, the drastic loss of blood cells causes anemia and can result in fatal brain or kidney damage.

Malaria was hit hard by treatment and prevention programs

In the 1600s, quinine, a bitter chemical derived from the bark of a tropical tree, was found to reduce the symptoms of malaria, although it did not always cure the disease. Derivatives of quinine, such as chloroquine and primaquine, are used today to treat and prevent malaria. If you travel to the tropics, you will probably have to take regular doses of one of these drugs to prevent malaria.

One way to reduce the number of cases of malaria is to reduce the size of mosquito populations. To control *Anopheles* mosquitoes, swamps and ditches where mosquitoes breed were drained. Powerful insecticides such as DDT were applied to kill mosquitoes. Through these two measures, malaria has been eradicated from the southern United States and southern Europe, where the disease was common until the 1940s. The World Health Organization introduced similar mosquito population control measures into parts of Africa and South America. Initially, these programs greatly reduced the worldwide incidence of malaria. Some optimists predicted the elimination of the deadly disease.

Malaria struck back

Malaria has not been eliminated. In fact, the number of cases has been increasing since the mid-1970s. Human efforts to combat the disease have been thwarted by evolutionary advances in both the *Plasmodium* parasite and the *Anopheles* mosquito. As you saw in **Figure 19-13**, the parasite frequently changes the protein knob configurations on the surfaces of infected blood cells, effectively making them invisible to the body's immune system. Additionally, *Plasmodium* parasites have developed resistance to some of the drugs that traditionally controlled them.

In many areas, mosquitoes have evolved resistance to insecticides. People are therefore exploring other methods of mosquito control, such as importing fish that eat large numbers of larval mosquitoes and investigating ways to fight mosquitoes with their natural enemies. Efforts are also underway to develop a malaria vaccine that protects individuals from infection by *Plasmodium.*

Figure 19-14 This computer model shows the regions of the world where malaria does occur.

Protists cause many diseases

In the less industrialized countries, one in five people have been infected with malaria, dysentery, or sleeping sickness. In the industrialized countries, we have little contact with protists that can kill us. Protists may make us sick if our water treatment facility breaks down or if we drink water from a stream contaminated by protists from the feces of infected humans or animals. But for us, a disease caused by protists is an oddity, not an everyday possibility. **Figure 19-14** shows how widespread malaria is in the world now. However, because of predicted climate changes, the range of disease-causing insects could increase. Examples of diseases caused by protists are listed in **Table 19-1**.

Table 19-1 Human Diseases Caused by Protists

Disease	Host	Organism
Amoebic dysentery	Humans	*Entamoeba*
Malaria	Humans	*Plasmodium*
Toxoplasmosis	Humans, cats	*Toxoplasma*
Giardiasis	Humans	*Giardia*
Sleeping sickness	Humans, tsetse flies	*Trypanosoma*
Chagas' disease	Humans, kissing bugs	*Trypanosoma*
Leishmaniasis	Humans, sand flies	*Leishmania*

SECTION REVIEW

1. **Why might the use of mosquito netting in malaria-prone areas decrease the number of malaria cases?**
2. **Describe two methods by which the incidence of malaria was reduced.**
3. **Explain why malaria has not been eliminated despite intense efforts to do so.**
4. **What are three human diseases caused by protists?**

CHAPTER

19 Highlights

Shirt's on fire...
Now IT'S OUT!
Humor at its lowest form.

Amoeba **and other protists are remarkably sensitive to their environments.**

19-1 What Is a Protist?

Key Terms

cilia (p. 358)

pseudopod (p. 358)

Summary

- The kingdom Protista contains the eukaryotes that are not plants, animals, or fungi.
- Both mitosis and meiosis occur among protists.
- Protists descended from bacteria. Plants, animals, and fungi are the descendants of different protist ancestors.

Euglena **is a unicellular protist that is photosynthetic.**

19-2 Protist Diversity

Key Terms

phytoplankton (p. 362)

Summary

- The kingdom Protista is the most diverse eukaryotic kingdom.
- Both autotrophs and heterotrophs are included among the protists.
- Most protists are unicellular, but some are multicellular.
- Green algae are the ancestors of plants.
- Zoomastigotes are probably the ancestors of animals.

Trichonympha **is a single-celled protist that lives in the guts of termites and digests wood.**

19-3 Diseases Caused by Protists

Summary

- Malaria is a disease caused by a parasitic protozoan in the genus *Plasmodium*. About 300 million to 500 million people are currently infected with the malaria parasite.
- *Plasmodium* is transmitted by mosquitoes, especially species in the genus *Anopheles.*
- Intense efforts to eliminate malaria initially showed promise. These efforts have failed to eliminate malaria because of the evolution of drug resistance in the *Plasmodium* parasite and pesticide resistance in mosquitoes.
- Other protists cause amoebic dysentery, giardiasis, and sleeping sickness.

The female *Anopheles* mosquito carries a protist called *Plasmodium*, which causes malaria.

CHAPTER 

19 review

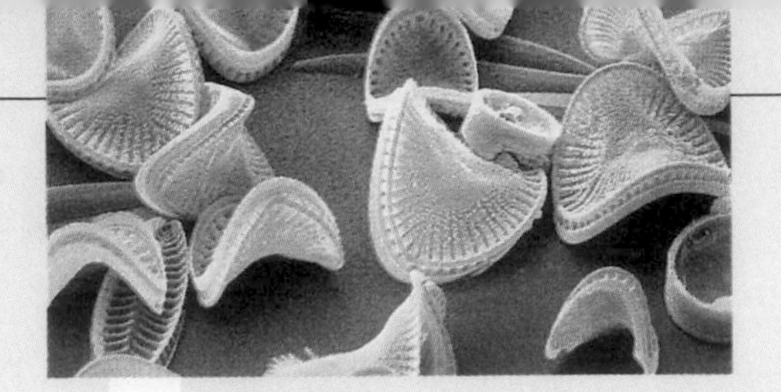

Understanding Vocabulary

1. Identify the word or phrase that does not fit the pattern and explain why.
   a. eukaryotic, chloroplasts, single-celled, multicelled, *Escherichia coli*
   b. flagella, pseudopods, cilia, algae
   c. kelp, *Amoeba*, *Salmonella*, slime mold

2. Using the words below, write one or more sentences summarizing information learned in this chapter.
   a. red tides, dinoflagellate, population explosion, toxins
   b. *Euglena*, nutrients, chloroplasts, heterotrophic
   c. *Paramecium*, protists, microscopic, algae 100 m long

Understanding Concepts

3. **Relating Concepts** Construct a concept map describing the protists. Try to include the following terms in your map: algae, red, malaria, protozoa, slime mold, autotrophs, plants, diseases, heterotrophs, animals.

4. **Inferring Relationships** The kingdom Protista is sometimes called a catchall kingdom. What does this term mean, and why is it used?

5. **Evaluating Information** What characteristic of the protist *Euglena* makes it unusual among both the protists and other organisms?

6. **Predicting Outcomes** Describe how one protist caused the major immigration of many people to the United States.

7. **Applying Information** How is a brown alga similar to a plant? How is it different?

8. **Inferring Conclusions** Explain why the *Euglena*'s flagellum and the *Paramecium*'s cilia are homologous structures.

9. **Predicting Outcomes** Describe two ways in which protists react to their environment, and explain what each reaction reveals about their way of obtaining nutrients.

10. **Inferring Relationships** Match each term on the left with the most appropriate term on the right. Explain the relationship between each pair.

a. flagellum	1. *Paramecium*
b. light	2. termites
c. *Trichonympha*	3. *Euglena*
d. cilia	4. phytoplankton
e. *Plasmodium*	5. *Anopheles*

11. **Recognizing Relationships** Describe the symptoms of malaria and relate them to what is happening inside the individual who is infected with the disease.

12. **Summarizing Information** Explain why blood cells damaged by *Plasmodium* are not swept away to the spleen for destruction, as are other damaged red blood cells.

13. **Comparing Structures** Draw and label a bacterial cell and a protistan cell.

14. **Summarizing Information** Name five diseases caused by protists.

15. **Identifying Functions** Explain why red algae, which obtain nutrients by photosynthesizing, can exist at depths in the ocean where little light is able to penetrate.

16. **Summarizing Functions** By what methods do protists reproduce? How do they exchange genetic information?

17. **Inferring Relationships** Why have the major diseases AIDS and malaria been so hard to combat?

18. **Inferring Relationships** Describe the characteristics of slime molds and water molds that have sometimes caused them to be classified with the fungi. Describe how they are different from fungi.

19. **Comparing Structures** Describe the methods of locomotion used by the structures flagella, pseudopodia, and cilia.

20. **BUILDING ON WHAT YOU HAVE LEARNED** Based on what you learned about bacteria and viruses in Chapter 18, explain five ways that bacteria, viruses, and protists help humanity, and explain five ways that they harm us.

Interpreting Graphics

21. Examine the figure shown below.

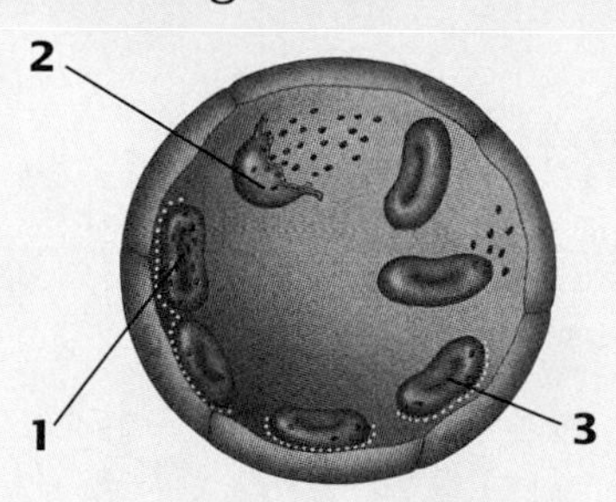

Match each numbered event in the figure with the sentence that best describes it.

a. *Plasmodium* causes protein knobs to form on the surface of infected red blood cells.
b. An infected *Anopheles* mosquito bites a human being.
c. Spores reproduce in a host's red blood cell.
d. Spores are released as the red blood cell bursts.

Arrange the events described above, in the correct sequence. Using your imagination, describe some ways in which a drug or vaccine could affect events shown in the figure above.

Reviewing Themes

22. ***Evolution***
In what ways are protists related to bacteria?

23. ***Interdependence***
Describe the role of autotrophic protists in the environment. What other living organisms depend on them?

24. ***Interdependence***
Trichonympha is a protist that lives in the guts of termites. Explain how this has been a favorable situation for the termite and an unfavorable one for humans.

Thinking Critically

25. **Evaluating Arguments** A classmate says, "Holes in the ozone layer that allow more radiation to reach the Earth's surface can't have any effect on phytoplankton." Do you agree or disagree with this statement? Explain your reasoning.

Life/Work Skills

26. **Communicating** You are in charge of a public health service in a seaside community. You are informed by the authorities that a red tide has occurred along your coast. What information would you want to communicate to tourists and others in your community? Research red tides, and learn about the possible health complications they can cause. Make a presentation before your class as if you were addressing the city council, and answer any questions they might have.

27. **Applying Technology** If you were in charge of trying to control the mosquito population in a small town and knew that repeated applications of the same pesticide kill fewer and fewer mosquitoes, what other alternatives would you recommend? What other information would you need to have before proposing alternatives?

Activities and Projects

28. **Health Connection** Create a poster outlining the many diseases caused by protists in the world. What medical efforts are being made to treat and control these diseases?

29. **History Connection** Do research in the library or search an on-line database to investigate the early attempts to describe and control malaria. What did people think caused malaria before Ross made his discoveries?

Discovering Through Reading

30. Read the article "Malaria Vaccine Raises a Dilemma" in *Science*, January 1995, pages 320–323. What have the clinical trials shown that has caused the research community to be divided in their endorsement of the vaccine?

31. Read the article "Super Vaccine Crawls Out of the Slime" in *New Scientist*, June 1994. How are researchers growing genetic material in a slime mold? Why are they using this method instead of cloning genes by conventional methods? Why are the cells of slime molds better to work with?

CHAPTER 20

REVIEW
- ratio of surface area to volume (Section 3-1)
- invasions of the land (Section 11-3)
- six kingdoms (Section 17-3)

Fungi and Plants

Fungi and plants grow side by side on the forest floor.

Fungi and the algal ancestors of plants were the first eukaryotes to invade land from the sea. Like the bracket fungi on this tree trunk, most fungi are terrestrial, as are the majority of plants. The primary role of fungi in the environment is to decompose organic matter; this is also the role of many bacteria. Fungi also affect human life by causing disease as well as by providing us with food and medicine.

20-1 Fungi

OBJECTIVES

1. **Identify the characteristics shared by most fungi.**
2. **Describe how fungi obtain nutrients.**
3. **Differentiate among the four groups of fungi.**
4. **Relate characteristics of fungi to their ecologic and economic importance.**

Characteristics of Fungi

The kingdom Fungi *(FUHN jy)* is a group of eukaryotic organisms, most of which are multicellular. Little is known about the origins of fungi. The oldest fossils that have been identified as fungi are 450–500 million years old. Fungi have many unique features, which strongly suggests that they are not closely related to any other kingdom of organisms.

Biologists once placed fungi in the plant kingdom, probably because many fungi seem to be more plant-like than animal-like. Most fungi are immobile, grow at both ends, and appear to be "rooted" like plants. However, fungi are as different from plants as they are from other living things. Compare familiar fungi, such as those in **Figure 20-1**, with a familiar plant. Careful study reveals several characteristics that distinguish fungi from plants.

Figure 20-1 Fungi exist in many forms, . . .

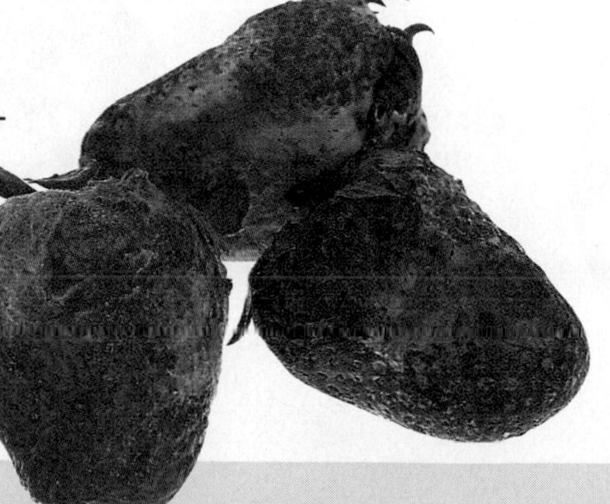

a . . . from the mold growing on these strawberries . . .

b . . . to this poisonous fly agaric mushroom . . .

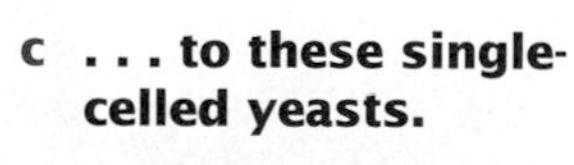

c . . . to these single-celled yeasts.

Fungi are heterotrophs

One obvious difference between fungi and plants is their color; fungi are not green like the leaves of plants. Certain parts of plants appear green because they contain chlorophyll, which enables them to capture energy from sunlight. Plants use that energy to make organic molecules from inorganic materials. As a result, plants are autotrophs (organisms that capture energy and make organic molecules from inorganic materials). Fungi do not contain chlorophyll and cannot capture energy from sunlight. As a result, fungi are heterotrophs (organisms that get energy and organic molecules by obtaining food from their surroundings).

The bodies of multicellular fungi consist of filaments

Unlike the bodies of most plants, which consist of many types of cells arranged in tissues, the body of a fungus consists of similar-looking cells arranged in long slender filaments. The filaments that make up the body of a fungus are called **hyphae** *(HY fee)*. Each hypha is a long string of individual cells that are separated by walls called septa. The septa of many kinds of fungi have holes in them. These holes enable the cytoplasm and nuclei of the cells in a hypha to flow from cell to cell. As hyphae grow, they branch repeatedly and form a tangled mass called a **mycelium** *(my SEE lee uhm)*. Yeasts, which are unicellular fungi, do not form either hyphae or a mycelium.

Often, the only parts of a fungus that you can see are its reproductive structures. This is the case with the *Aspergillus* mold seen on the fruit in **Figure 20-2**. The black fuzz that you recognize as the mold is a mass of reproductive structures. Most of the mycelium of the fungus lies within the fruit. In another type of fungi, some of the hyphae in the mycelium weave tightly together to form a mushroom, which is also a reproductive structure.

Cell walls of fungi contain chitin

The cells of fungi have cell walls made of a polysaccharide called **chitin** *(KY tihn)*. This tough material is also found in the exoskeletons of insects and other arthropods. Cells of plants, on the other hand, have cell walls made of cellulose, a different polysaccharide. Chitin is more resistant to decomposition by bacteria than is cellulose.

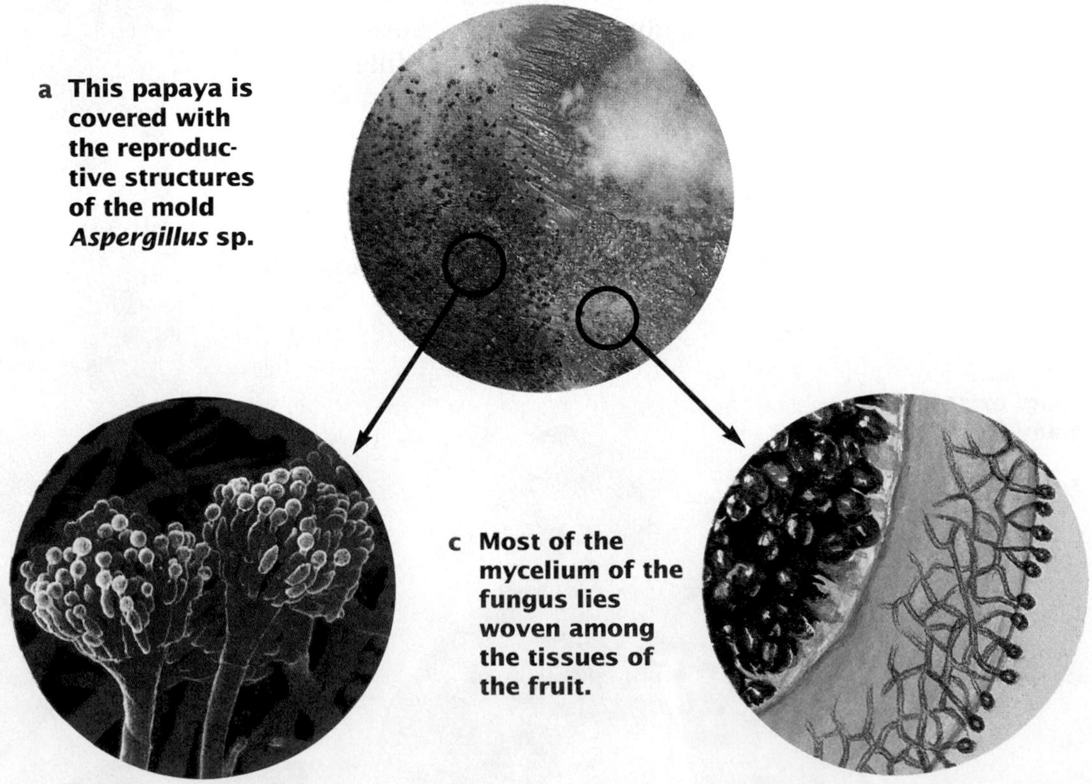

Figure 20-2
The parts of a fungus that you normally see are reproductive structures.

a This papaya is covered with the reproductive structures of the mold *Aspergillus* sp.

b The black fuzz consists of fungal spores produced by these structures.

c Most of the mycelium of the fungus lies woven among the tissues of the fruit.

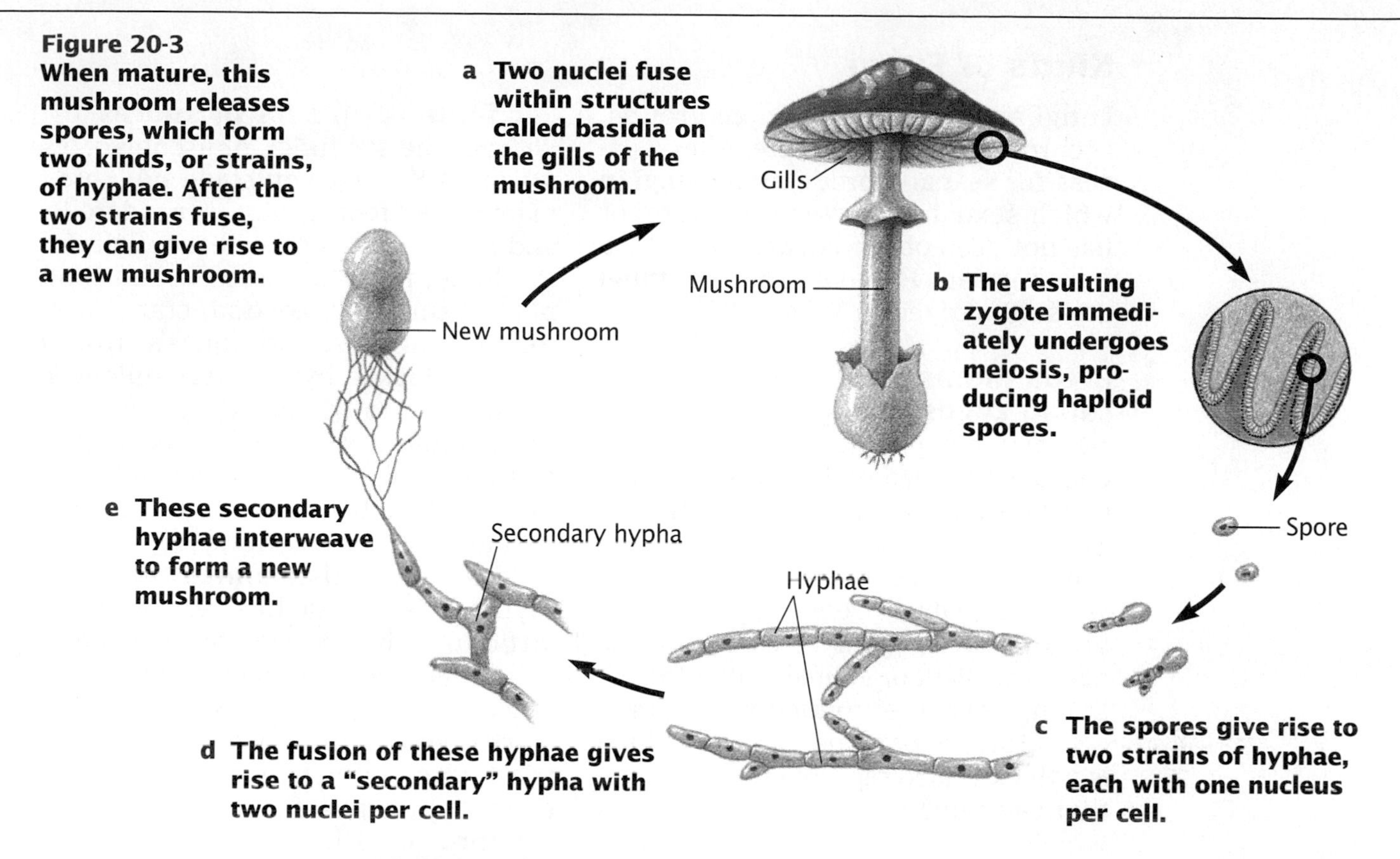

a Two nuclei fuse within structures called basidia on the gills of the mushroom.
Gills
Mushroom
b The resulting zygote immediately undergoes meiosis, producing haploid spores.
Spore
c The spores give rise to two strains of hyphae, each with one nucleus per cell.
Hyphae
d The fusion of these hyphae gives rise to a "secondary" hypha with two nuclei per cell.
Secondary hypha
e These secondary hyphae interweave to form a new mushroom.
New mushroom

Figure 20-3
When mature, this mushroom releases spores, which form two kinds, or strains, of hyphae. After the two strains fuse, they can give rise to a new mushroom.

Fungi digest food outside their bodies

You have seen mold growing on bread and fruit. Fungi use these items for food just as you do. Some fungi attack living plants and animals to obtain nutrients, while others attack dead organisms. Both methods provide fungi with a rich supply of nutrients (organic molecules, water, and minerals).

A fungus growing on a food source is digesting the food and absorbing nutrients. Like most animals, you digest food within your digestive system. However, fungi digest food outside their bodies. As the hyphae of a fungus spread through a food source, they secrete enzymes that break down organic matter. The hyphae then absorb the nutrients. The mycelium of a fungus may consist of many meters of hyphae. As a result, fungi have a very high ratio of surface area to volume and are efficient feeders.

Interdependence

How do the fungi and plants in a forest depend on each other?

Most fungi reproduce by spores

If you examine part of a patch of mold or mildew with a microscope, you will probably see many hyphae that are tipped with clusters or short strings of tiny, beadlike spheres. These tiny structures are reproductive cells called **spores**. A fungal spore can develop into a new individual.

Most fungal spores are well adapted to life on land. They can withstand dry conditions and a wide range of temperatures. Some fungal spores are light and easily carried to new sites by air currents, while others are sticky and cling to animals that carry them to new places. When a fungal spore lands on a food source with enough moisture, the spore rapidly grows into a hypha. If conditions remain favorable, the hypha continues to grow and branches repeatedly, forming a mycelium.

Fungi produce spores by both asexual and sexual reproduction. During asexual reproduction, spores form by cell division at the tips of certain hyphae. Sexual reproduction involves the combining of genetic materials from two strains of a fungus. In mushrooms, for example, this combining occurs just before a mushroom's stalk and cap develops. Spores are produced in gills, which are found under the cap of the mushroom. The spores are shed when the cap opens like an umbrella. The life cycle of a mushroom is shown in **Figure 20-3**.

Kinds of Fungi

Fungi are divided into three groups, according to the structures they possess for sexual reproduction. Fungi in which sexual reproduction is rare, or has not been observed, are placed in a fourth group. The four groups of fungi are summarized in **Table 20-1**.

Bread molds form structures called zygosporangia

You may have had unpleasant experiences with bread molds, which belong to the smallest group of fungi—the zygomycetes *(zy goh MY seets)*. This group includes the mold *Rhizopus stolonifer*, which frequently grows on bread. Other zygomycetes live on the decaying plant or animal matter in soil. During sexual reproduction, zygomycetes form a zygosporangium, which is a thick-walled reproductive structure that can remain dormant for months. When conditions are favorable, spore-producing stalks grow from the zygosporangium. When the spores reach maturity, they are often picked up by the wind and blown about.

Sac fungi form sacs of spores

Without the sac fungi, or ascomycetes *(as coh MY seets)*, you could not enjoy many baked foods, such as bread, rolls, and pizza. On the other hand, without sac fungi, many ornamental and food plants, such as boxwood, crape myrtles, cucumbers, and squash, would not be attacked by powdery mildews. In addition to the powdery mildews, the ascomycetes include yeasts, morels, truffles, and the *Penicillium* molds that produce the strong flavors of Camembert and Roquefort cheese.

The feature that unites the ascomycetes is the saclike reproductive structure that forms after genetic materials fuse together during sexual reproduction (the prefix *asco* means "sac"). Spores are produced inside these sacs, which eventually burst open. The spores of some ascomycetes are thrown as far as 30 cm (12 in.). Many species of fungi resemble ascomycetes but have not been observed to reproduce sexually. These species are placed in the fourth group of fungi.

Table 20-1 Groups of Fungi

Group	Sexual Reproduction	Examples	
Zygomycetes	Spores are produced on stalks that grow from a structure called a zygosporangium	Some of the black molds on bread and other foods	*Rhizopus stolonifer* on bread
Ascomycetes	Spores are produced in saclike reproductive structures	Yeast, mildews, morels, and truffles; fungi that cause Dutch elm disease and chestnut blight	Powdery mildew
Basidiomycetes	Spores form at tips of clublike reproductive structures	Toadstools, puffballs, mushrooms, bracket fungi, rusts, and smuts	*Amanita muscaria* mushroom
Deuteromycetes	Unusual, rare, or not observed	Many common molds; molds that produce antibiotics, and many disease-causing fungi	*Penicillium* sp.

Club fungi produce spores on clublike structures

Most organisms you call mushrooms belong to a third group of fungi, the basidiomycetes *(buh sid ee oh MY seets)*. This is the largest group of fungi and also includes bracket fungi, toadstools, puffballs, rusts, and smuts. These fungi form reproductive structures shaped like wooden bats, or clubs (the prefix *basidio* means "little club"). The clublike structures develop under the caps of mushrooms or under the "shelves" of bracket fungi. Spores form on the tips of the clubs. Many club fungi live on dead organisms.

Some fungi have no known form of sexual reproduction

Fungi with no known method of sexual reproduction are placed in a group called the deuteromycetes *(doot uh roh MY seets)*. This group is also known as the Fungi Imperfecti. Some of these fungi have great economic importance. *Penicillium notatum* produces the antibiotic penicillin. Two species of *Aspergillus* are used to ferment soy beans in the production of soy sauce. In addition, deuteromycetes cause many fruits and vegetables to rot, and they cause human skin diseases such as athlete's foot.

Fungi in Nature

There are very few places on land that at least one species of fungi does not call home. Carried by air currents, fungal spores may land almost anywhere. Thus, many molds and mushrooms seem to spring up wherever there is the right amount of food and moisture.

Although some fungi cause diseases of plants and animals, most fungi are beneficial to other organisms. Fungi decompose organic wastes and the remains of dead plants and animals, returning nutrients such as nitrogen and phosphorus to the soil. Imagine what the world would look like if fungi did not exist! Fungi are also key components in two important types of mutualistic associations—mycorrhizae and lichens. Similar symbiotic associations played a crucial role in the invasion of land by life from the sea.

Fungi form mycorrhizae with plants

As you read in Chapter 11, many fungi grow on or within the roots of certain plants, in symbiotic associations called mycorrhizae *(my koh RY zee)*. Mycorrhizal fungi played a critical role in the successful colonization of land by plants. Many of the earliest known plants formed mycorrhizae with fungi. The fungi provided the plants with water and minerals, and the plants provided the fungi with photosynthetically produced food.

One type of mycorrhizal fungus is shown in **Figure 20-4**. Recently, researchers have discovered that the destruction of mycorrhizal fungi by acid rain is playing a key role in the deaths of many forests. Without mycorrhizae, most forest trees are unable to absorb minerals from soil.

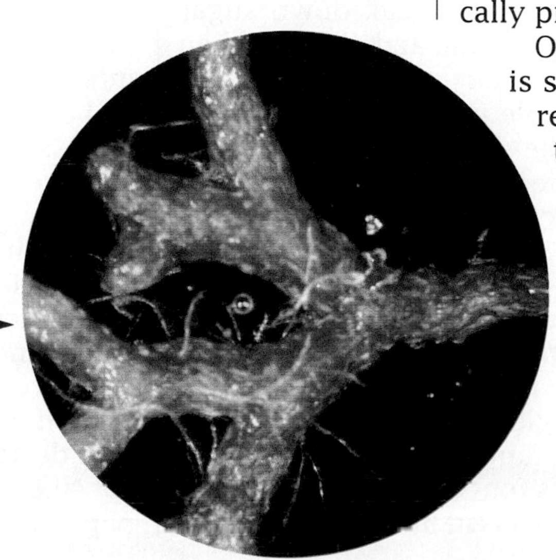

Figure 20-4
a These tiny mushrooms, which grow under ponderosa pine trees, are reproductive structures of a mycorrhizal fungus.

b Underground, the hyphae of this fungus surround and penetrate the roots of the trees.

Fungi form lichens with algae and cyanobacteria

If you have ever visited a rocky seacoast or a forest, you may have noticed rocks or logs with crusty orange, yellow, or gray-green patches, like those shown in **Figure 20-5**. These patches are lichens. A **lichen** *(LY kuhn)* is a mutualistic association between a fungus and either a green alga or a cyanobacterium. The fungus absorbs water and other nutrients from the environment. The alga uses these nutrients for photosynthesis. In turn, the alga produces organic molecules that the fungus uses as food.

Figure 20-5 Lichens are able to survive in extremely harsh environments. Chemicals released by these lichens help begin the breakdown of rock into soil.

Lichens can live in very harsh environments. For instance, they live in deserts and on the highest mountains. The secret to a lichen's success in a harsh environment seems to be its ability to dry out very quickly. When it is dry, a lichen is less likely to be damaged by very hot or very cold temperatures. Today, many lichens play an important role in ecosystems as pioneers, the first organisms to inhabit bare rock.

Because they absorb water and nutrients from the air, lichens are extremely susceptible to air pollution. For example, lichens are usually absent in and around cities, where automobile traffic and industrial activity produce pollutants that destroy chlorophyll. As a result, the alga or cyanobacterium of a lichen cannot photosynthesize for itself or for the fungus. Because of this sensitivity, scientists use lichens as indicators of air quality.

Fungi and Humans

Fungi benefit humans in many ways. Take the unicellular yeasts, for example. You could not enjoy freshly baked bread or pizza without *Saccharomyces cerevisiae*, better known as baker's yeast. Yeast cells break down sugar into carbon dioxide and alcohol when oxygen is in short supply. These substances enable us to make bread and alcoholic beverages. However, the greatest effect of fungi on humans is as disease-causing and decay-causing organisms.

Fungi cause diseases in humans and crop plants

Fungi cause many diseases in humans. Some examples of fungal diseases that you may be familiar with are ringworm and athlete's foot. These common skin diseases are caused by deuteromycetes. Histoplasmosis is a serious and sometimes fatal disease that is caused by the ascomycete *Histoplasma capsulatum*. This fungus grows in bird and bat droppings. When its spores are inhaled, they infect the lungs. New spores produced in the lungs may spread and infect the liver, central nervous system, and other internal organs. Many people also suffer from allergies caused by fungi.

Every year, farmers lose billions of dollars in crops due to the effects of fungal diseases. Powdery mildews, leaf spots, stem and root rots, sooty molds, rusts, and smuts are common names for several types of fungal diseases that affect crop plants. Fungal diseases also destroy millions of dollars of forest trees that provide us with lumber, paper, and many other products.

After harvest, millions of dollars of fruits, vegetables, and grains are destroyed by fungi that cause them to rot. As a result, growers and shipping companies must spend a great deal of money to treat or ship these crops in airtight or refrigerated containers to protect them. Molds such as *Rhizopus*, *Penicillium*, and *Aspergillus* commonly cause unpreserved foods to rot.

Fungi provide medically valuable compounds

Some of the same fungi that are harmful to organisms also have beneficial properties. As you read in Chapter 18, Alexander Fleming discovered the antibiotic effects of the fungus *Penicillium notatum* in 1928. Fleming's accidental discovery, shown in **Figure 20-6**, led to the development of the world's most widely used antibiotic—penicillin.

Other fungi have helped revolutionize the field of medicine. In 1972, the Swiss immunologist Jean Borel found a type of fungus that produces a substance capable of suppressing the immune system's response to transplanted organs. This substance, called cyclosporine, opened up a new frontier in surgery. Before it became available in 1979, fewer than half of all kidney transplants were successful. Now, with the widespread use of cyclosporine, the success rate has risen to 90 percent.

Yeasts are now popular subjects for genetic engineering. Although bacteria are excellent subjects for genetic engineering, they are prokaryotes. Yeast cells, which are eukaryotes, are more suitable models for developing genetic engineering techniques for plants and animals, which are also eukaryotes. Research with yeast may lead to new and innovative treatments for diseases such as cancer and AIDS.

Figure 20-6
In 1928, Alexander Fleming discovered that *Penicillium notatum* killed bacteria. Ten years later, the drug penicillin was purified. Penicillin is effective in curing several bacterial diseases, including pneumonia, scarlet fever, diphtheria, and many others.

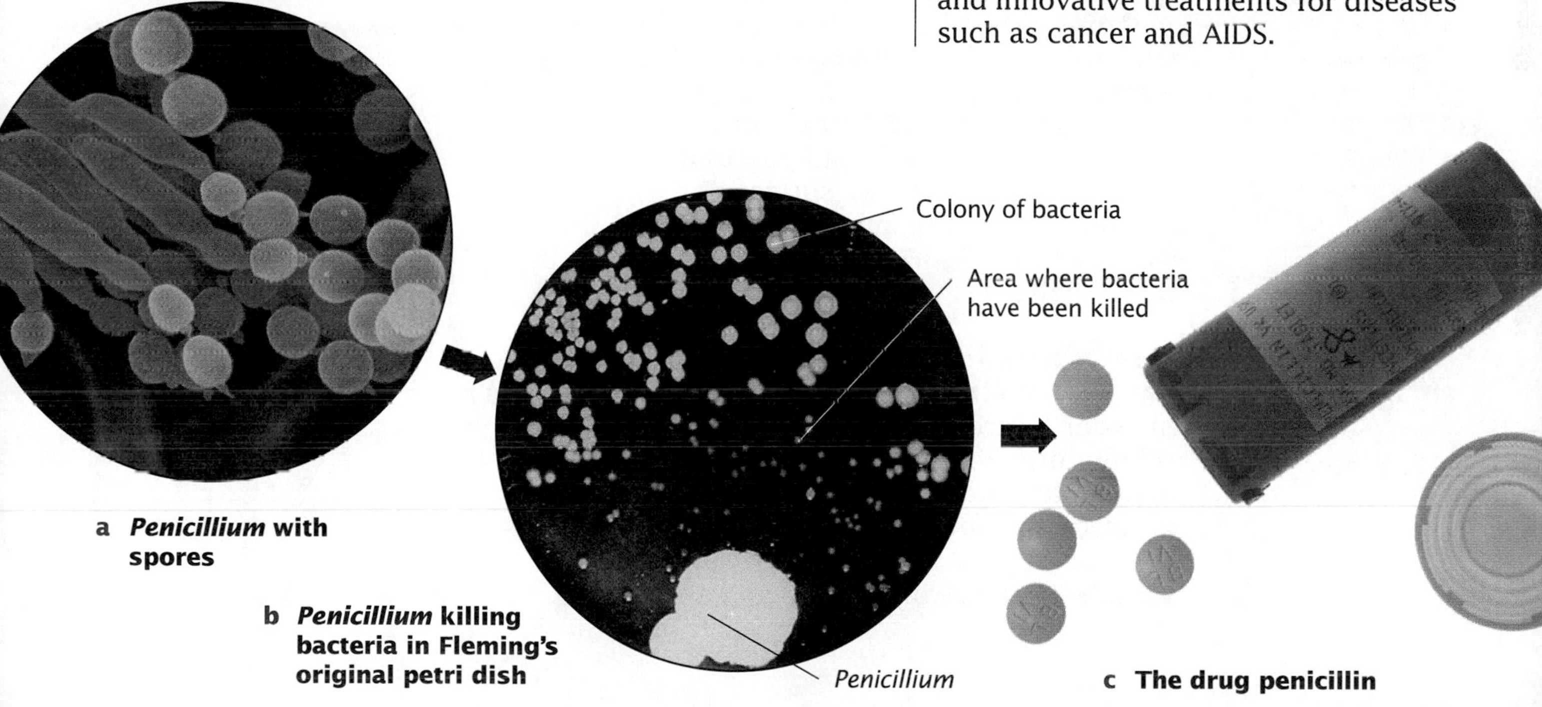

a *Penicillium* with spores

b *Penicillium* killing bacteria in Fleming's original petri dish

c The drug penicillin

SECTION REVIEW

1. **Describe four characteristics shared by all fungi.**
2. **How do fungi obtain nutrients?**
3. **Compare and contrast the methods of reproduction found in the four groups of fungi.**
4. **How might life on Earth be different if all fungi were to suddenly disappear from this planet?**

Plants are another group of mostly terrestrial, multicellular, eukaryotic organisms. We take for granted their presence almost everywhere on Earth. Plants probably evolved from an ancient form of green algae that grew at the edges of bodies of water. These algae invaded the land about 400 million years ago. The successful occupation of the land by plants involved adaptation to dry conditions and cooperation with fungi.

20-2 Early Land Plants

OBJECTIVES

1. **Describe the adaptations that enable plants to survive on land.**
2. **Define alternation of generations.**
3. **Distinguish nonvascular plants from vascular plants.**
4. **Relate characteristics of nonvascular and vascular plants to their habitats.**

Characteristics of Plants

Plants are photosynthetic, like their ancestors, green algae. Unlike fungi, plants have cells with cell walls made of the polysaccharide cellulose. And unlike both fungi and algae, most plants have bodies made up of many types of cells arranged in tissues.

Before early plants could survive on land, they had to overcome three challenges. First, they had to avoid drying out. Second, they had to reproduce on land. And third, they had to obtain minerals from the rocky surface. Fossil remains of early land plants, such as the one in **Figure 20-7**, show how these challenges were met.

Most plants have vascular tissues

In order to grow very large, a multicellular organism must have a way to distribute water and nutrients to all parts of its body. Like you, plants have a system of internal tubes called **vascular tissue**, which conducts water and other nutrients. Plants that have vascular tissue are called vascular plants. Most of the plants you are familiar with—including roses, houseplants, grasses, oak trees, and ferns—are vascular plants.

a A waterproof, waxy coating *reduced water loss.*
b Spores were produced at the tips of stems. The spores grew if they landed in a spot covered by dew, enabling the plant to *reproduce with very little water.*
c The formation of mycorrhizae with fungi enabled the plant to *absorb minerals from soil.*

Figure 20-7
The early land plant *Rhynia* evolved with features that enabled it to overcome the environmental challenges of life on land. This illustrated reconstruction of *Rhynia* is based on fossil evidence.

Figure 20-8
In a plant's life cycle, a haploid gametophyte alternates with a diploid sporophyte.

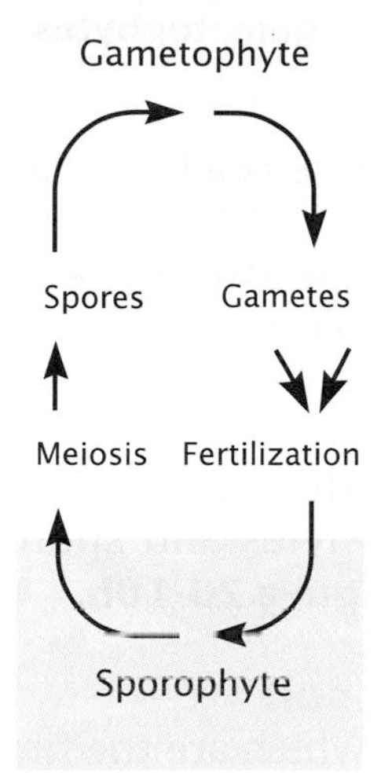

However, some small plants, such as mosses, lack vascular tissue. Plants that lack vascular tissue are called nonvascular plants. Nonvascular plants, which live in moist environments, transport water and nutrients by osmosis and diffusion, much as algae do.

Plants have adaptations for survival in dry conditions

The waxy, waterproof layer that covers the surface of most plants is called the **cuticle**. The cuticle helps prevent a plant from drying out by reducing the evaporation of water from leaves and stems. Today, only one group of nonvascular plants lacks a cuticle.

Plants also produce reproductive structures that resist drying. All nonvascular plants and some vascular plants produce spores, tiny drought-resistant structures that can be carried to new locations by wind. Most vascular plants produce **seeds**, plant embryos that are surrounded by protective coats.

Haploid and diploid generations alternate in the plant life cycle

In plants, gametes are produced by a haploid form of the plant called the **gametophyte** *(guh MEET uh fyt)*. When two gametes fuse they produce a diploid form, the **sporophyte** *(SPOHR uh fyt)*. In most plants, this is the form you see. Some of the sporophyte's cells undergo meiosis, producing haploid spores. Each spore then develops into a haploid gametophyte. Thus, in a plant's life cycle, a haploid stage alternates with a diploid stage, as shown in **Figure 20-8**. This cycle is referred to as **alternation of generations**.

Nonvascular Plants

The first successful land plants had no vascular tissue or seeds. Nonvascular plants transport materials by osmosis and diffusion, so they need a large supply of water to survive. Because many of these plants produce sperm that must swim to an egg to fertilize it, they also need a film of water for sexual reproduction. For these reasons, most nonvascular plants—mosses, hornworts, and the liverworts shown in **Figure 20-9**—grow close to the ground in moist, shady environments such as forests and swampy shorelines.

Nonvascular plants are relatively small—nearly all are less than 20 cm (8 in.) tall, and most are less than 2.5 cm (1 in.) tall. Most are anchored by rootlike structures called rhizoids *(REYE zoydz)*. The green, leaflike structures where photosynthesis occurs are only one or two cells thick. In most nonvascular plants, a waxy cuticle protects the upper surfaces of these structures. Water enters through pores on the lower surfaces, which are not covered by a cuticle.

Mosses are the most widespread nonvascular plants. Mosses often grow in dense carpets on forest floors and decaying logs or in cushionlike patches in sidewalk cracks. Most mosses grow in areas of high humidity, although some mosses are extremely drought-resistant.

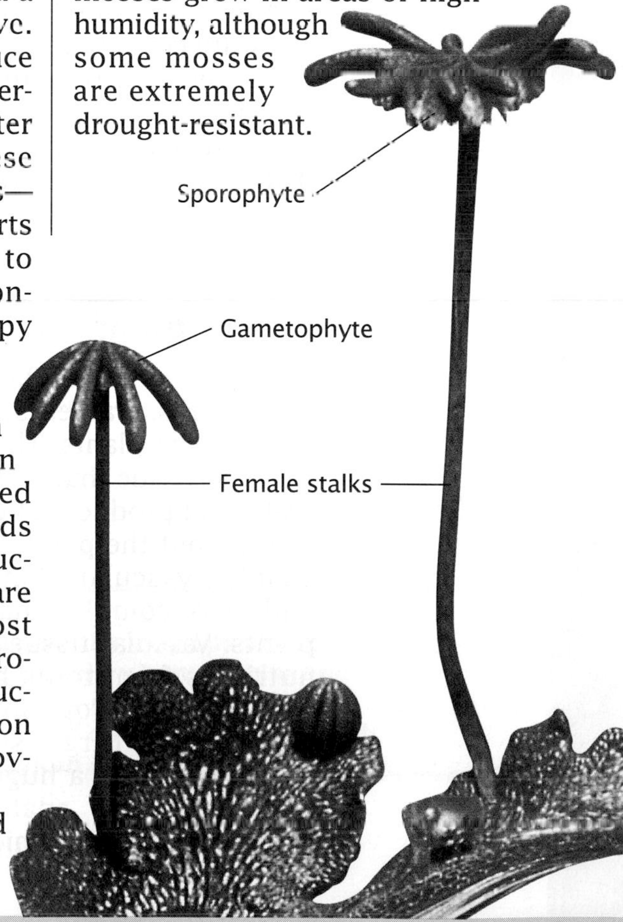

Sporophyte
Gametophyte
Female stalks

Figure 20-9
The gametophyte of the liverwort *Marchantia* produces gametes under separate, 2 cm high stalks. Very tiny sporophytes develop under the female stalks.

The evolution of the seed was a critical step in the spread of plants across the land. A seed protects an embryonic plant from drying out. It also stores carbohydrates, fats, and proteins that a rapidly growing young plant uses as a source of energy and as building blocks for growth. Some seeds develop within fruits that help distribute them in a variety of ways.

20-3 Seed Plants

OBJECTIVES

1. **Explain how the development of seeds enabled plants to be successful in more habitats.**
2. **Summarize the process of reproduction in gymnosperms.**
3. **Describe the role of flowers in the life cycle of angiosperms.**
4. **Identify several important families of angiosperms.**

Vascular Plants With Seeds

Although some early land plants such as ferns had vascular tissue, they were still confined to moist environments. In contrast, the plants with seeds were fully adapted to life on land and were able to live in dry places.

One of the evolutionary achievements of the seed plants is woody tissue, which lends strength to plants so that they can grow tall and compete for sunlight. Another development—more complex vascular tissue—enables water and nutrients to be carried to these new heights. Seed plants also developed features that freed them from needing water to reproduce—pollen, cones or flowers, and seeds.

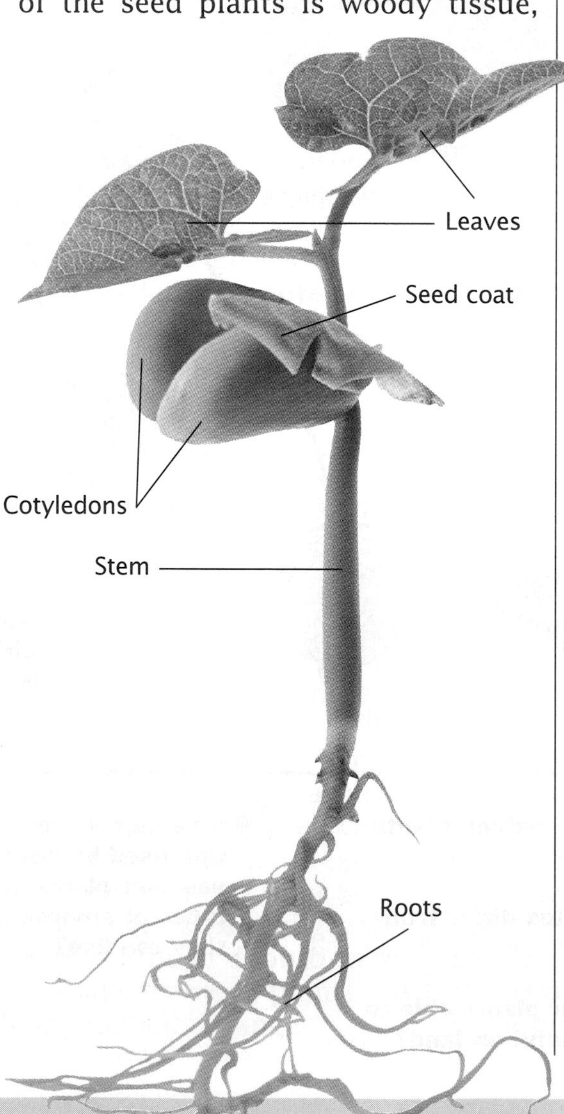

Figure 20-12
This bean seedling has just emerged from a seed. A seed is a storage container that houses a plant embryo. A tough outer seed coat protects the embryo from drying out. Food is stored in the cotyledons.

Seeds enable seed plants to survive in harsh environments

The seed is a remarkable evolutionary development that enables seed plants to survive unfavorable conditions. Each seed contains an embryo that can grow into a mature plant. Most seeds contain stored nutrients. The embryos of flowering plants have structures called **cotyledons** *(kaht uh LEED uhnz).* Cotyledons store food or help absorb food stored elsewhere in the seed.

A hard outer seed coat surrounds the embryo in a seed, protecting it from physical injury and drought. Because the seed coat provides protection from harsh conditions, a seed can lie dormant for years and still grow into a healthy plant when conditions improve. When a seed germinates (cracks open and starts to grow), the embryo first develops into a **seedling**, as shown in **Figure 20-12**.

Seeds may also help carry young plants away from their parents. For example, some seeds have hooks, feathery projections, or winglike projections that enable them to travel long distances on animals or with the wind.

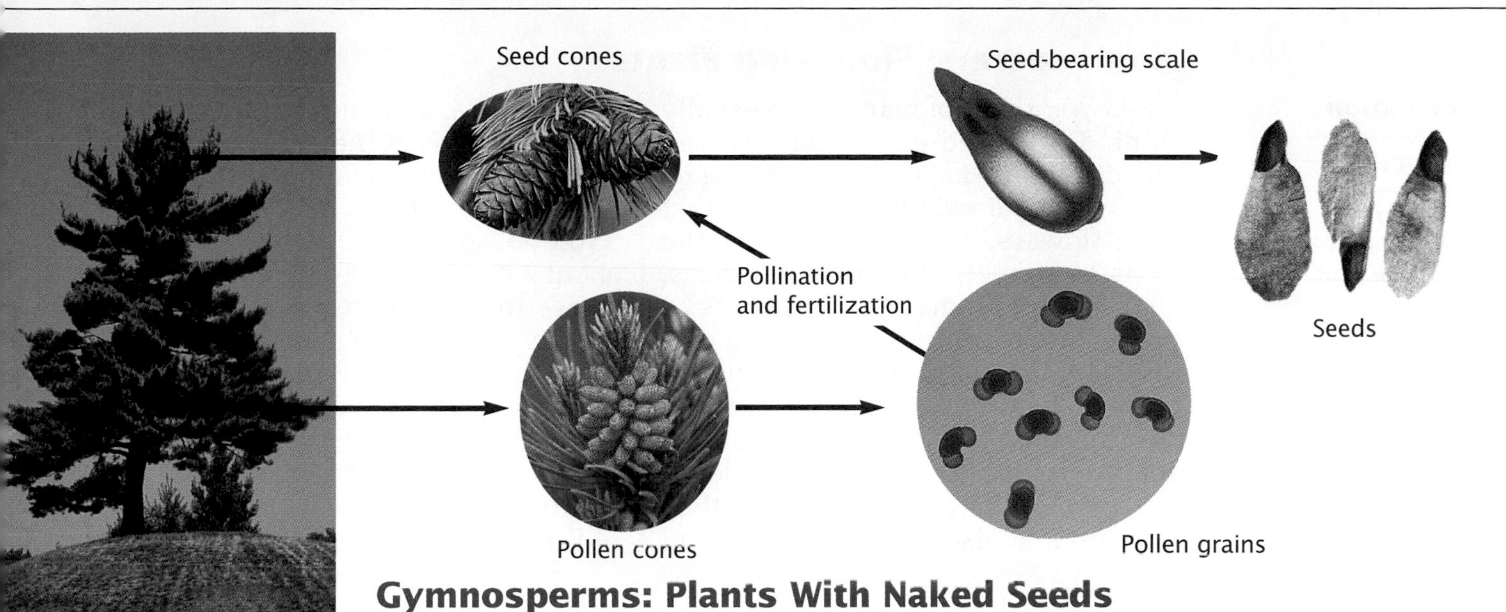

Figure 20-13
A pine tree is a gymnosperm that produces both pollen cones and seed cones. After pollination, new seeds are formed and fall to the ground. There, they germinate to form a new tree.

Gymnosperms: Plants With Naked Seeds

Gymnosperms were among the first seed plants. The word *gymnosperm* combines the Greek words for "naked" and "seed." **Gymnosperms** *(JIHM noh spurmz)* are plants with seeds that do not develop within a fruit. The seeds of most gymnosperms form in **cones**, which are clusters of scales or modified leaves. Egg-bearing structures form on the surfaces of scales.

Most familiar gymnosperms—firs, junipers, pines, and spruces—are conifers (trees that produce seeds in cones). The redwoods of California and the bald cypresses in Southern swamps are also conifers. A conifer that grows in Nevada—a 5,000-year-old bristlecone pine—may be the world's oldest tree. Another conifer—a giant sequoia that grows in California and is named for General Sherman of the Civil War—is the world's biggest tree. This tree is more than 84 m (276 ft) tall and 31 m (102 ft) around its base. The palmlike cycads of the tropical climates and the ginkgo trees growing along city streets are also gymnosperms, but they are not conifers and do not produce cones.

Pollen eliminates the need for water in sexual reproduction

Unlike earlier plants, gymnosperms have an innovation that enables sperm to reach the eggs without swimming—pollen. **Pollen grains** are tiny male gametophytes that are portable, unlike the gametophytes of earlier plants.

Conifers produce two types of cones—female (seed) cones and male (pollen) cones. Each type of cone produces a different kind of spore by meiosis. The spores produced by female cones develop into female gametophytes, each of which contains an egg. The spores produced by male cones develop into pollen grains, each of which contains a sperm.

Gymnosperms are wind pollinated

Gymnosperms produce huge quantities of pollen. A pair of air sacs helps each pollen grain drift on the wind. Pollen grains often settle on the surface of ponds and lakes, where they form a sticky yellow layer. A few pollen grains settle onto the scales of female cones. The transfer of pollen from where it is produced to where the eggs are located is called **pollination**. Wind pollination is the transfer of pollen by the wind.

Within a female cone, a pollen grain completes its development and grows a slender tube that will deliver a sperm to a female gametophyte. Fertilization occurs when the sperm fuses with an egg, forming a zygote. The zygote becomes the embryo of a seed. Most pine seeds have thin, flat wings, which help catch air currents that carry the seeds to new areas. When conditions are favorable, the seeds will sprout. Reproduction in gymnosperms is summarized in **Figure 20-13**.

Angiosperms: Flowering Plants

Evolution

What traits shared by angiosperms and gymnosperms suggest a close evolutionary relationship?

When you think of plants, you usually think of the seed plants that evolved most recently—angiosperms *(AN jee oh spurmz)*. **Angiosperms** are plants that have flowers, which help ensure the transfer of sperm to eggs. The colors, shapes, and aromas of the flowers of many angiosperms attract particular insects, birds, or other animals, which carry pollen from one plant to another.

In contrast to gymnosperms, angiosperms produce seeds that are enclosed in fruits. Fruits protect seeds as they develop and also help disperse them. For example, an animal that eats fruit may later deposit the seeds elsewhere.

Angiosperms have been extremely successful on land. Of the more than 250,000 species of vascular plants, about 235,000 species are angiosperms. These include herbs, grasses, vines, shrubs, and many trees—nearly all the plants you encounter daily. Some examples are shown in **Table 20-2**. Virtually all of our food is derived, directly or indirectly, from flowering plants. They are also valuable sources of timber, textiles, and medicines.

Table 20-2 Economically Important Families of Angiosperms

Families	Familiar examples
Liliaceae (lily)	daylily, tulip, asparagus, aloe vera
Poaceae (grass)	wheat, corn, rice, lawn grass
Lamiaceae (mint)	mint, rosemary, sage, coleus
Rosaceae (rose)	roses, apple, peach, plum
Fabaceae (legume)	beans, peas, peanut, acacias
Asteraceae (composite)	daisies, sunflowers, lettuce
Brassicaceae (mustard)	broccoli, turnip, cabbage

SECTION REVIEW

1. **How do seeds enable plants to survive in the harsh environments of deserts and tundras?**
2. **How do the male gametes of gymnosperms reach the female gametes?**
3. **How do you think a flower that is wind pollinated might differ from a flower that is pollinated by insects?**
4. **What do the grass family and the legume family have in common?**

CHAPTER

20 Highlights

This pitcher plant is a vascular plant that traps insects within its leaves. Enzymes break down an insect's tissues, and the plant absorbs its nutrients.

	Key Terms	Summary
20-1 Fungi  Fungi include molds, morels, and mushrooms, such as this poisonous fly agaric mushroom.	hypha (p. 374) mycelium (p. 374) chitin (p. 374) spore (p. 375) lichen (p. 378)	• Fungi are eukaryotic organisms made up of slender filaments called hyphae. Their cells have cell walls that contain chitin. • Fungi secrete enzymes that digest food outside their bodies, and then they absorb the nutrients. • Fungi are classified into three groups based on their sexual reproductive structures. Some fungi display no evidence of sexual reproduction and are placed into a fourth group. • While many fungi spoil food and cause disease, some are helpful to humans.
20-2 Early Land Plants  Mosses are the most widespread nonvascular plants. The mosses shown here are pioneers, which help break down bare rock into soil.	vascular tissue (p. 380) cuticle (p. 381) seed (p. 381) gametophyte (p. 381) sporophyte (p. 381) alternation of generations (p. 381) zygote (p. 382)	• Land plants have certain adaptations that enable them to live in a dry environment. These adaptations reduce water loss, move water and nutrients throughout the plant, and enable plants to reproduce in dry environments. • Land plants have a life cycle that involves alternation of generations between a haploid gametophyte and a diploid sporophyte. • Mosses and ferns, like their ancestors, the early land plants, are able to reproduce only when water is available for their sperm to swim through.
20-3 Seed Plants  Angiosperms are the most recently evolved of the land plants. They are unique in that they produce flowers and fruits.	cotyledon (p. 384) seedling (p. 384) gymnosperm (p. 385) cone (p. 385) pollen grain (p. 385) pollination (p. 385) angiosperm (p. 386)	• Seed plants are able to reproduce without water for sperm to swim through. Seeds can survive for long periods of time in unfavorable environments. • Gymnosperms are seed plants that do not produce seeds within fruits. Most form seeds in cones. • Angiosperms are seed plants that have flowers and produce seeds within fruits.

CHAPTER 20 review

Understanding Vocabulary

1. For each set of terms, complete the analogy.
   a. fungus : heterotroph :: plant : ____
   b. gamete-forming plant : gametophyte :: spore-forming plant : ____
   c. gymnosperms : cone :: angiosperms : ____
2. For each pair of terms, explain the difference in their meanings.
   a. hypha, mycelium
   b. mycorrhiza, lichen
   c. nonvascular, vascular

Understanding Concepts

3. **Relating Concepts** Construct a concept map that differentiates between nonvascular and vascular plants. Try to include the following terms in your concept map: vascular tissue, gymnosperms, angiosperms, mosses, ferns, sporophyte, gametophyte, alternation of generations.
4. **Summarizing Information** What characteristics are shared by most fungi?
5. **Summarizing Information** Explain why fungi were once considered plants. What features distinguish fungi from plants?
6. **Identifying Structures** Identify the major group of fungi represented by each fungus below, and describe how each reproduces.

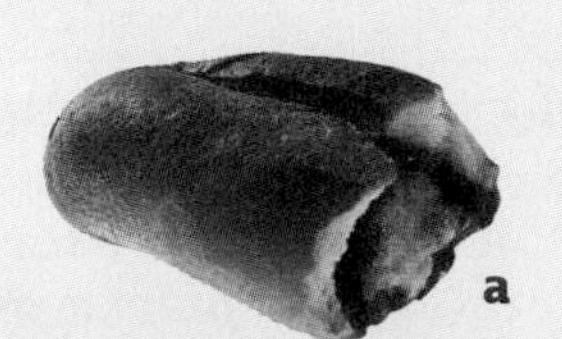

a

Rhizopus stolonifer on bread

b

Powdery mildew

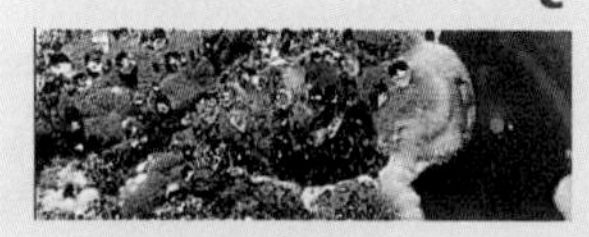

c

Penicillium sp.

d

Amanita muscaria mushroom

7. **Describing Processes** Describe how fungi obtain nutrients.
8. **Inferring Relationships** Explain why fungal spores are well adapted to life on land.
9. **Recognizing Relationships** Describe the relationship that occurs in a lichen.
10. **Inferring Relationships** Describe how fungi can be both helpful and harmful to you.
11. **Organizing Information** Make a table in which you identify three challenges that plants had to overcome to survive on land, and describe how plants adapted to meet each challenge.
12. **Summarizing Information** Describe alternation of generations in a plant's life cycle.
13. **Identifying Structures** Which structures are characteristics of most nonvascular plants, such as mosses?
14. **Inferring Relationships** Differentiate between vascular and nonvascular plants in terms of size and transportation of water and nutrients. Relate the characteristics of each group to how successful the group is on land.
15. **Identifying Structures and Functions** List the three main parts of a seed, and describe the function of each part.
16. **Recognizing Relationships** How did seeds enable the seed plants to become so successful on land?
17. **Summarizing Information** Describe the process of reproduction in gymnosperms.
18. **Comparing Processes** Describe the ways in which gymnosperms and angiosperms distribute pollen. Which method of pollen distribution is more efficient?
19. **Summarizing Information** Identify at least three economically important families of angiosperms. Name a member of each family, and explain why it is economically important.
20. **BUILDING ON WHAT YOU HAVE LEARNED** In Chapter 14 you learned the importance of recycling minerals and nutrients within an ecosystem. Describe the role that fungi play in the recycling of materials in the environment.

Interpreting Graphics

21. Study the diagram below, and answer the following questions.

a. What process is shown at A?
b. What structures are produced at B?
c. What structures are produced at C?
d. What process takes place at D?
e. Which plant in the life cycle is haploid?
f. Which plant in the life cycle is diploid?

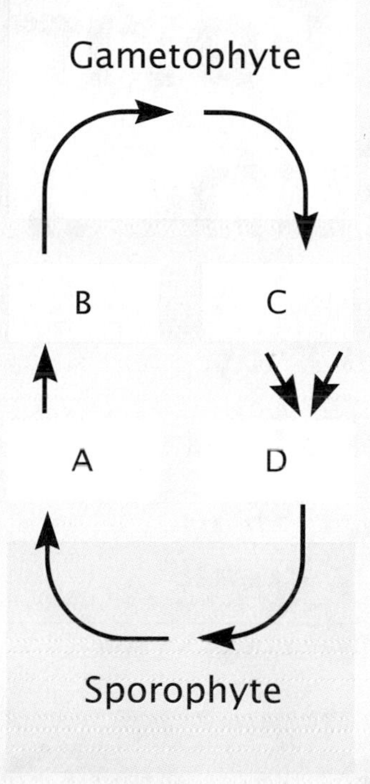

Reviewing Themes

22. ***Heredity***
Both gymnosperms and angiosperms require pollination. What, specifically, is accomplished by pollination? How is this event important to the reproduction of gymnosperms and angiosperms?

23. ***Matter, Energy, and Organization***
What energy source is used by the embryo in a seed before photosynthesis begins in the new plant?

Thinking Critically

24. Predicting Outcomes You want to grow some wildflower seeds that you have collected from the roadside. Why do you think including some of the soil from the original habitat is important to the survival of the plants that grow from the seeds?

Life/Work Skills

25. Finding Information Use library references or search an on-line database to find out about the importance of fungi as food or in the preparation or manufacture of food. You may want to investigate mushrooms eaten as food, fungi used to produce certain types of cheeses and soy sauce, fungi used in baking bread, and fungi used to produce wine. Summarize your findings in an oral report to the class.

26. Communicating Investigate the history and use of coal. How and when was coal formed? When did people first begin to use coal? Is coal still being formed today? Create a series of posters to summarize your findings for your classmates.

Activities and Projects

27. Career Connection: Agriculture Mushroom growers grow mushrooms to be sold as food in stores and restaurants. Some mushroom growers have large farms where they produce several kinds of common mushrooms. Other growers have small farms where they grow exotic or gourmet mushrooms on a small scale. Use library references or write to the Mushroom Growers Association, at 18 South Water Market, Chicago, IL 60608, to find out the kind of education and equipment necessary to become a mushroom grower.

28. History/Economics Connection Coniferous trees (gymnosperms) are among the oldest living organisms on Earth. The western parts of North America are home to some of the oldest and largest of these trees. Use library references or search an on-line database to find out about the historical distribution, age, size, and economic importance of trees such as pinyon pine, redwood, spruce, juniper, Douglas fir, and lodgepole pine. Prepare an oral and visual report of your findings to share with your class.

Discovering Through Reading

29. Read the article "A New Gold Rush Packs the Woods in Central Oregon" by Mike Pipske in *Smithsonian*, January 1994, pages 35–45. What are the causes of the "mushroom war"?

CHAPTER 21

REVIEW

- ratio of surface area to volume (Section 3-1)
- plant cells (Section 3-3)
- photosynthesis (Section 5-3)
- cotyledon (Section 20-3)

Plant Form and Function

21-1 The Plant Body

21-2 How Plants Function

21-3 Reproduction in Flowering Plants

Sunflowers always face the sun.

A plant cannot move about and capture food as an animal can. Instead, a plant's structure enables it to obtain nutrients from its environment. Most vascular plants consist of roots, stems, and leaves. Roots anchor plants and absorb water and minerals. Stems hold leaves high, enabling them to capture light energy and to take in carbon dioxide that is used during photosynthesis.

21-1 The Plant Body

OBJECTIVES

1. **Describe the structures and functions of roots and shoots.**
2. **Compare and contrast vascular tissue, ground tissue, and dermal tissue.**
3. **Describe the role of meristems in plant growth.**
4. **Distinguish between monocots and dicots.**

Roots

When a seed germinates, the first part of the seedling (young plant) to start growing is the root. **Figure 21-1** shows the root of a radish seedling. Roots are usually an underground part of a plant. They anchor the plant and absorb water and minerals from the soil. The roots of most plants grow no deeper than about 5 m (16 ft). However, the roots of some plants are very long. For example, mesquite trees have roots that can grow to more than 50 m (164 ft) long. Roots make up about one-third of the dry mass of a plant.

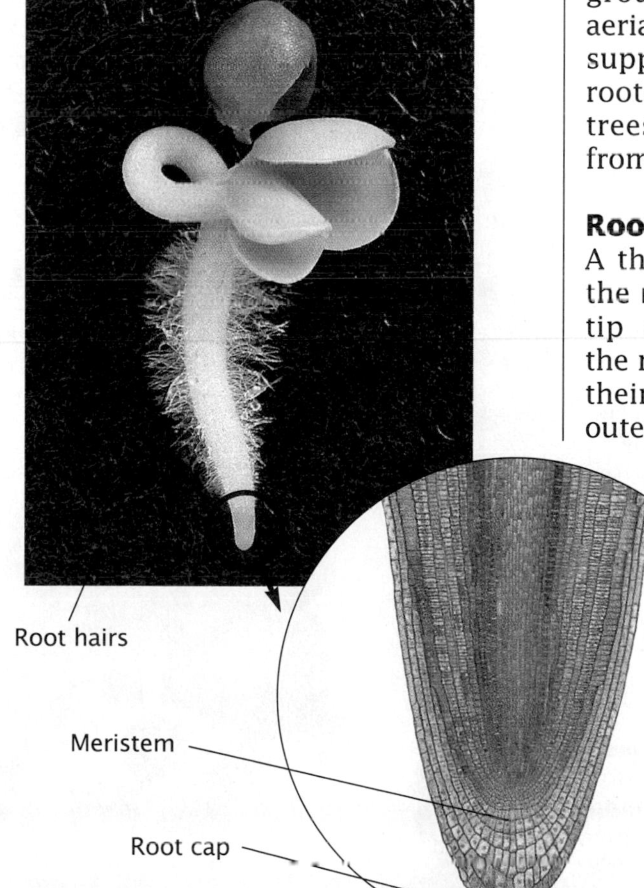

Figure 21-1 **This radish seedling's root has many root hairs, which greatly increase its surface area for absorption. A root cap protects the root tip, which contains a region of actively dividing cells called a meristem.**

The roots of some plants have specialized functions

Roots often store nutrients. For example, carrots and sweet potatoes have roots that store large amounts of carbohydrates. Some plants have aboveground roots such as prop roots or aerial roots. Prop roots give additional support to plants such as corn. Aerial roots fasten plants such as orchids to trees and absorb water and minerals from the air and surrounding surfaces.

Roots grow from their tips

A thimble-like cluster of cells called the **root cap** covers and protects the tip of a root. Cell division within the root tip causes roots to grow from their tips. As a root grows longer, the outer cells of the root cap are sloughed off and replaced by new cells from underneath.

In most roots, the region above the root cap has many slender projections called **root hairs**. Root hairs are so tiny that they can penetrate the spaces between soil particles to obtain water and minerals. Each root hair is a projection from a single cell. Root hairs greatly increase the surface area of a root.

Shoots

Figure 21-2
To survive in different environments, plants have evolved with different types of shoots.

The aboveground part of a plant is called the **shoot**. The shoots of most plants consist of leaves and stems and, in many cases, flowers and fruits. Leaves conduct photosynthesis. Stems support the leaves and enable them to receive sunlight. Compare the shoots of the plants shown in **Figure 21-2**.

Stems connect roots to leaves

The stems of a flowering plant support its leaves, flowers, and fruits, usually holding them off the ground. Stems also contain vascular tissue that transports substances between the roots and the leaves.

Stems vary greatly in shape and size from one plant species to another, as is evident when comparing the trunk of a redwood tree with a grass stalk. Some stems are adapted for storage. For example, cactuses can store a considerable amount of water inside their fleshy stems. Potatoes are underground stems that are swollen with nutrients stored as starch.

Leaves are the primary sites of photosynthesis

In Chapter 5, you learned that plants capture energy from sunlight and use it to make organic molecules through photosynthesis. Leaves are specialized to carry out photosynthesis. Most of the cells in a leaf are packed with chloroplasts, the chlorophyll-containing organelles where photosynthesis occurs.

Regardless of their size, most leaves are relatively thin and flat. This shape produces the highest ratio of surface area to volume. Thin, flat leaves help a plant efficiently capture the light energy and carbon dioxide needed for photosynthesis. A waxy outer layer called the cuticle prevents leaves from losing too much water and drying out. Carbon dioxide enters a leaf, and water and oxygen exit, through tiny pores called **stomata** *(STOH muh tuh).* The singular of *stomata* is *stoma.*

Inside a leaf, there are layers of photosynthetic cells. Bundles of vascular tissue also run through a leaf. These bundles are the veins you see when you look at a leaf. Vascular tissue moves water and dissolved minerals to photosynthetic cells and carries away the products of photosynthesis. You can read more about leaves in *A Closer Look at a Leaf* on page 393.

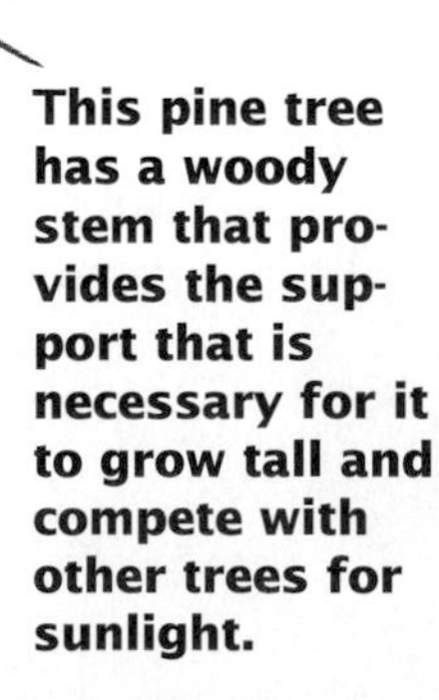

a This pine tree has a woody stem that provides the support that is necessary for it to grow tall and compete with other trees for sunlight.

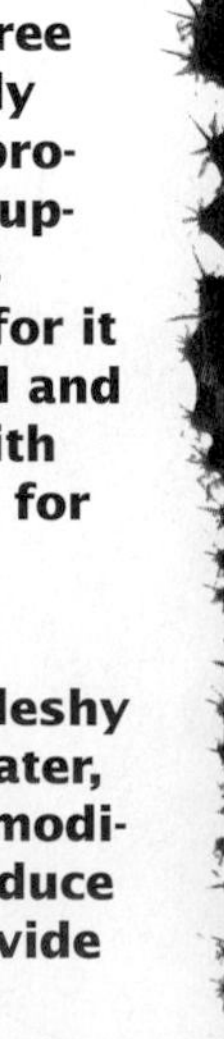

b This cactus has a fleshy stem that stores water, and it has spines (modified leaves) that reduce water loss and provide protection.

c The horizontal, green stems of this strawberry plant spread over the ground and ultimately may take root and establish a new plant.

Leaf

A leaf is not as simple as it might first appear. It has an intricate architecture well suited to its function—photosynthesis.

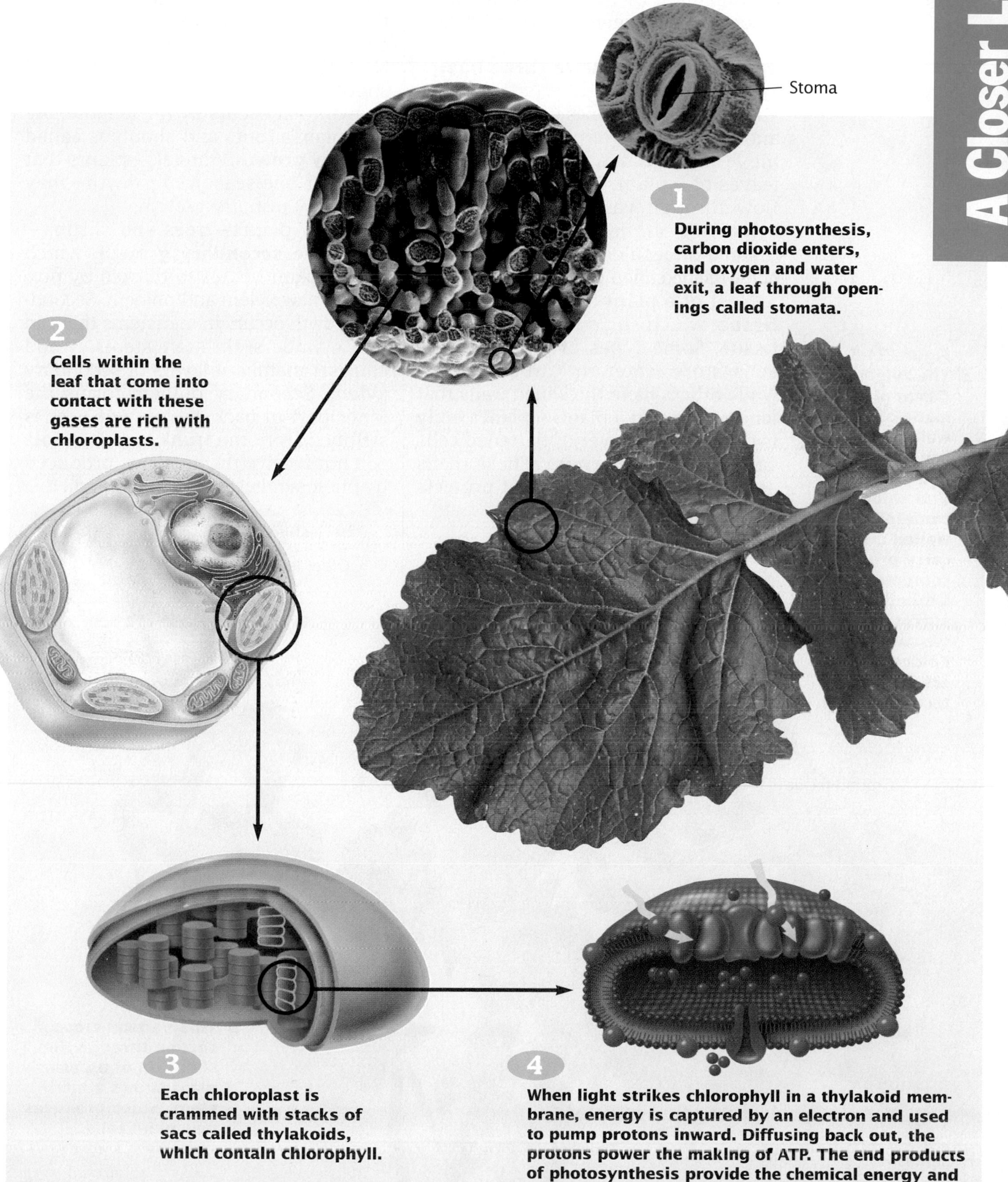

Stoma
1
During photosynthesis, carbon dioxide enters, and oxygen and water exit, a leaf through openings called stomata.
2
Cells within the leaf that come into contact with these gases are rich with chloroplasts.
3
Each chloroplast is crammed with stacks of sacs called thylakoids, which contain chlorophyll.
4
When light strikes chlorophyll in a thylakoid membrane, energy is captured by an electron and used to pump protons inward. Diffusing back out, the protons power the making of ATP. The end products of photosynthesis provide the chemical energy and building blocks needed for the plant's life.

Plant Tissues

The leaves, stems, and roots of plants are made of specialized tissues. The arrangement of the tissues in a nonwoody stem is seen in **Figure 21-3a**.

Vascular plants have three basic types of tissue

Vascular tissue enables water, minerals, and the products of photosynthesis to move through the roots, stems, and leaves of a plant. Water and minerals flow through vascular tissue called **xylem** *(ZY luhm)*. Sugars and other organic molecules move through vascular tissue called **phloem** *(FLOH ehm)*.

Most of a plant is made of **ground tissue**, which surrounds the vascular tissue. Some types of ground-tissue cells store water or carbohydrates, while others have thickened walls that lend support to a plant. A plant's body is covered by a layer of flattened cells called the **epidermis**. These cells secrete the waxy cuticle that protects a plant from water loss.

New tissue forms in regions of actively dividing cells

Plants grow at regions of cell division called **meristems** *(MEHR uh stemz)*. One type of meristem is shown in **Figure 21-3b**. Every time a cell in the meristem divides, one cell remains in the meristem and the other cell becomes a specialized cell as it matures. Meristems at the tips of roots and shoots enable plants to grow in length. Growth that occurs by the lengthening of a plant's roots and shoots is called **primary growth**. Annuals—plants that die after one season of growth—may have only primary growth.

Woody plants—trees and shrubs—also have **secondary growth**, which causes plant bodies to thicken by producing new xylem and phloem. Secondary growth occurs in meristems that run like cylinders through stems. Wood consists mainly of layers of secondary xylem. Secondary phloem forms the inner part of bark. If a ring of bark is stripped from the trunk, a tree's roots will not receive the nutrients produced by photosynthesis, and the tree will die.

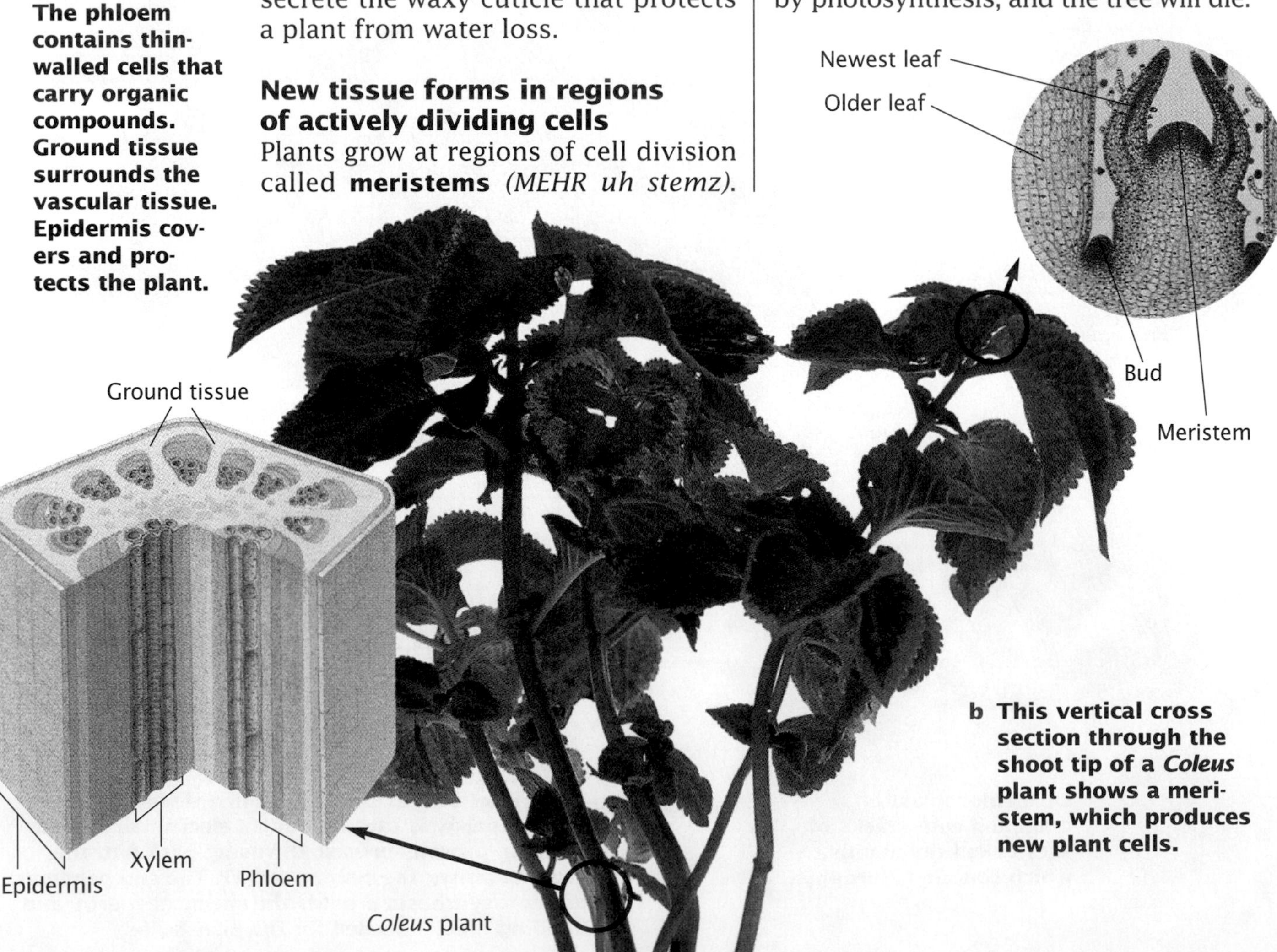

Figure 21-3

a The xylem of a *Coleus* plant is made of thick-walled cells that conduct water and minerals. The phloem contains thin-walled cells that carry organic compounds. Ground tissue surrounds the vascular tissue. Epidermis covers and protects the plant.

b This vertical cross section through the shoot tip of a *Coleus* plant shows a meristem, which produces new plant cells.

Table 21-1 Summary of Differences Between Monocots and Dicots

Monocots (lily)	Dicots (hibiscus)
Embryos have one cotyledon.	Embryos have two cotyledons.
Leaf veins are parallel.	Leaf veins are netlike.
Flower parts occur in threes or multiples of three.	Flower parts occur in fours or fives, or multiples of four or five.
Bundles of vascular tissue are scattered throughout ground tissue in the stem.	Bundles of vascular tissue often form a ring within the ground tissue in the stem.

Monocots and Dicots

As you read in Chapter 20, the embryos of flowering plants have cotyledons. The number of cotyledons differs in the two great groups of flowering plants: monocots (about 65,000 species) and dicots (about 170,000 species). Monocots include grasses, palms, lilies, and orchids. The crop plants wheat, corn, rice, oats, rye, and barley are also monocots. Dicots include many familiar nonwoody plants, shrubs, trees, and cactuses. Monocots and dicots are similar in structure and function, but they differ in distinct ways. Evidence suggests that monocots evolved from primitive dicots. **Table 21-1** shows some of the characteristic differences between most monocots and dicots.

SECTION REVIEW

1. **How do the functions of roots and shoots relate to their structures?**
2. **What are the functions of vascular tissue, ground tissue, and epidermis?**
3. **How do meristems enable a plant to grow?**
4. **If you had neither flowers nor seeds to look at, how would you determine whether a plant is a monocot or a dicot?**

Although a plant may not seem as complex as an animal, its internal structure is more complex than you might think. Its vascular system sends water, minerals, and organic compounds from one plant part to another. Like your body, a plant regulates its growth with hormones, chemicals that act as messengers to coordinate activities.

21-2 How Plants Function

OBJECTIVES

1. **Describe how water and other nutrients move within xylem and phloem.**
2. **Relate plant growth patterns to the action of plant hormones.**
3. **Describe the ways in which a plant can respond to seasonal changes.**

Cell Structure and Function

Explain how the structure of xylem enables the efficient transport of water and minerals.

How Nutrients Move Through Plants

In northern California, giant sequoia (redwood) trees can be up to 111 m (364 ft) tall. How do trees move water from roots deep in the soil up to branches and leaves at such heights? Water enters the roots of a plant by osmosis. But it takes more than just osmosis to move water up to the top of a tall tree!

Water is pushed and pulled through a plant

Several factors work together to move water through a plant. First, water continually enters a plant's roots by osmosis when enough water is present in the soil. The movement of more water into the roots pushes water into the xylem tissue that extends into the roots. Second, xylem cells form long narrow tubes. **Figure 21-4a** shows the xylem tissue of a plant's stem. The water in these tubes is attracted to the xylem cell walls. This attraction tends to pull water up the tubes. The movement of water up narrow tubes such as those in the xylem is called **capillary action**.

The major forces behind the movement of water in plants, however, are the attraction of water molecules for each other and the evaporation of water from leaves. As water evaporates from leaves, the water in the xylem moves upward because each water molecule pulls on those below it. Thus, the loss of water from leaves is responsible for the flow of water through a plant.

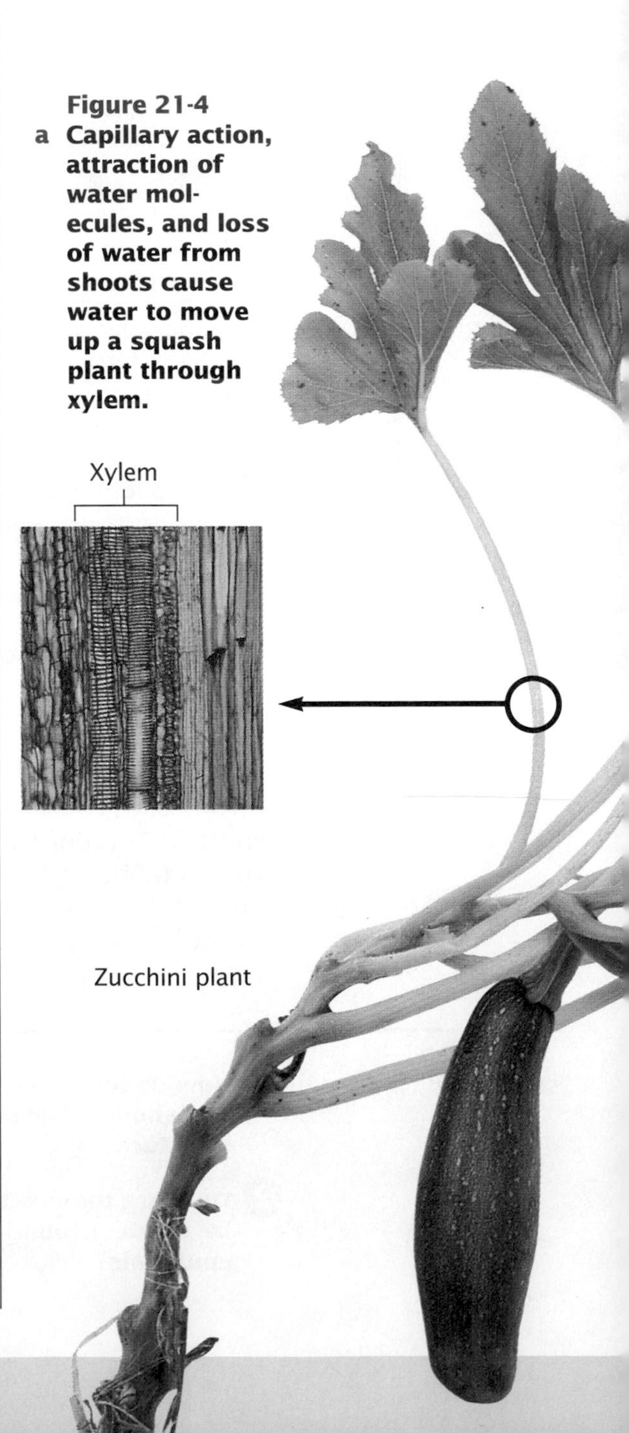

Figure 21-4
a Capillary action, attraction of water molecules, and loss of water from shoots cause water to move up a squash plant through xylem.

b **Water is lost from the aboveground portion of the plant through the process of transpiration.**

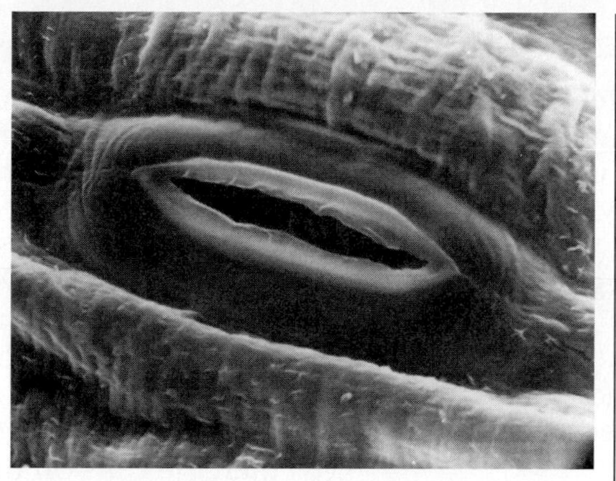

c **Stomata on the surfaces of leaves open and close to regulate water loss.**

Phloem

d **Within phloem, organic molecules made during photosynthesis are transported throughout a plant by the process of translocation.**

More than 90 percent of the water taken up by the roots of a plant is lost to the atmosphere as water vapor. The loss of water by a plant is called **transpiration**. Water is lost by a plant primarily through the stomata in its leaves, shown in **Figure 21-4b**. Water entering the plant's roots replaces the water lost through transpiration.

Stomata regulate water loss

The main way that plants control water loss is to close their stomata. Each stoma is formed by two pickle-shaped cells called guard cells. The structure of guard cells causes them to curve outward when they are swollen with water. As a pair of guard cells take in water, the stoma opens, as shown in **Figure 21-4c**. When the guard cells lose water, they collapse and the stoma closes. The stomata of a zucchini plant are open during the day and closed at night. Therefore, very little water is lost from its leaves at night.

Organic compounds are transported in another way

Organic compounds and some minerals move throughout a plant within the phloem. **Figure 21-4d** shows the phloem tissue from a plant. The contents of phloem cells are under great pressure. Yet individual molecules move within phloem tissue from areas of high concentration to areas of low concentration.

The transport of nutrients within phloem tissue is called **translocation**. Through translocation, organic molecules made in a plant's photosynthetic (green) parts move to other parts of the plant. Many organic molecules move to the actively growing regions of a plant. In some plants, such as carrots and potatoes, excess organic molecules accumulate in storage structures, where the molecules are converted to starch. When other parts of the plant need organic molecules, the starch molecules are broken down into smaller molecules and moved through the phloem. **Figure 21-4** summarizes the structures and processes that transport nutrients within a plant.

Factors Regulating Plant Growth

As a plant grows, leaves, stems, and reproductive structures form, and then seeds develop and mature. After they form, seeds usually remain dormant, or inactive, for a while. As a seed sprouts, the seedling's shoot grows up and its root grows down. Hormones control most of these events. Three examples of plant hormones are auxin *(AWK sihn)*, gibberellin *(jihb uhr EHL ihn)*, and ethylene.

Auxin stimulates the elongation of plant cells

Have you ever seen a plant bending toward light? Research has shown that this growth pattern occurs in many plants when auxin becomes more concentrated on the side of a stem away from the light's source. Auxin, which is produced in the tips of stems, causes cell walls to become more flexible. As a result, cells exposed to auxin grow longer. Auxin also stimulates root development; prevents the dropping of leaves, flowers, and fruits; and blocks the growth of buds along a stem.

If a stimulus causes auxin to concentrate on only one side of a stem, then the cells on that side will elongate more than the cells on the opposite side. As a result, the stem will bend away from the side where auxin is concentrated. Auxin causes a variety of growth responses that result in the bending of plants either toward or away from a stimulus such as light, gravitational pull, or touch. These growth responses are called **tropisms** *(TROH pihz uhmz).*

Figure 21-5
The plants on the right have greatly elongated stems because each was sprayed with water containing gibberellin. The plants on the left, which were sprayed with plain water, show normal development.

Gibberellin stimulates cell division and cell elongation

Gibberellin, which occurs naturally in plants, was named after the fungus *Gibberella*. In 1926, Japanese scientists reported that the stems of rice plants infected with the fungus grew abnormally long. A substance (gibberellin) extracted from the fungus caused the same effect in plants that were not infected by the fungus. Gibberellin not only greatly increases stem growth, as you can see in **Figure 21-5**, but also stimulates flower formation and the sprouting of seeds.

Ethylene stimulates fruit ripening

In ancient China, farmers ripened fruits in rooms that contained burning incense. Ethylene gas released by the burning of incense caused the ripening. Today ethylene is used commercially to ripen bananas, honeydew melons, and mangoes. These fruits are harvested when they are still green. In 1934, scientists discovered that fruits and other parts of a plant produce ethylene naturally. Ethylene also promotes the dropping of leaves, flowers, and fruits.

Figure 21-6
a Irises are long-day plants, which bloom in early summer when nights are short and days are long.

b Poinsettias are short-day plants, which bloom in late fall and early winter when nights are long and days are short.

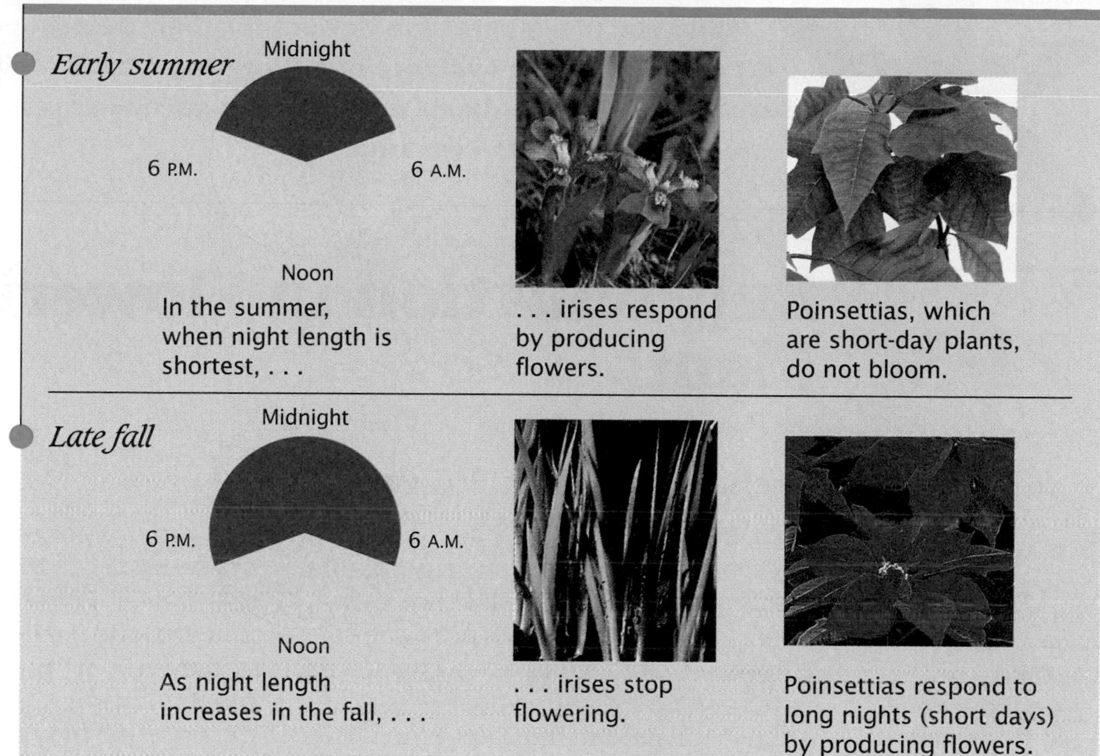

Night Length and Flowering
Early summer
Midnight
6 P.M.
6 A.M.
Noon
In the summer, when night length is shortest, . . .
. . . irises respond by producing flowers.
Poinsettias, which are short-day plants, do not bloom.
Late fall
Midnight
6 P.M.
6 A.M.
Noon
As night length increases in the fall, . . .
. . . irises stop flowering.
Poinsettias respond to long nights (short days) by producing flowers.

Seasonal changes affect plant growth

The growth of many plants varies with the seasons. Different types of plants produce flowers at different times of the year. The seeds of some plants sprout in the spring, while the seeds of others sprout in the fall. Still other plants go through periods of inactivity, or dormancy. Chemicals other than hormones are responsible for many of these growth patterns.

The length of a night (daily period of darkness) affects the production of flowers by some plants, as **Figure 21-6** shows. These plants contain a pigment that senses the change from light to dark. This pigment also enables a plant to sense how long darkness lasts. Long-day (short-night) plants produce flowers only when the daily period of darkness is less than a certain length. Short-day (long-night) plants produce flowers only when the daily period of darkness exceeds a certain length. Some plants, such as tomatoes, are not sensitive to darkness because they evolved in the tropics, where nights are virtually the same length all year.

In many plants, temperature affects flower or fruit production and the release of buds or seeds from a dormant state. For example, tomato plants will not produce fruit if nighttime temperatures are too high. Few varieties of apples are grown in southern regions of the United States because winters there are not cold enough for most varieties to produce fruit. Cold temperatures cause the breakdown of chemicals that cause dormancy in buds and seeds.

SECTION REVIEW

1. **Trace the path of water through a plant.**
2. **How does the movement of water through a plant differ from the movement of organic molecules through a plant?**
3. **List three plant hormones, and describe their effects on plant development.**
4. **How can a change in season affect plant growth?**

Flowering plants are the dominant group of plants on Earth today. While you may appreciate flowers for their decorative role, they are actually the visible evidence of a flowering plant's ability to reproduce sexually. Each flower helps ensure that a particular plant species will produce more of its own kind.

21-3 Reproduction in Flowering Plants

OBJECTIVES

1. **Relate the structure of a flower to the process of sexual reproduction in flowering plants.**
2. **Differentiate between the processes of pollination and fertilization in flowering plants.**
3. **Relate the structure of seeds and fruits to plant dispersal.**
4. **Evaluate tissue culture as a means of plant reproduction.**

Architecture of a Flower

Figure 21-7 shows the sexual reproductive structure of flowering plants, the flower. Most flowers share certain basic features. They consist of four whorls, or circles, of parts. The sepals, which form the outermost whorl and are usually greenish, protect the flower bud. Just inside the sepals is a whorl of petals, which are often brightly colored to attract animal pollinators. The inner whorls of a flower produce the male and female gametophytes. One inner whorl of a flower consists of **stamens** *(STAY mehnz)*, the male parts of a flower. Pollen grains are produced in a structure called the anther, which is the top part of a stamen. The innermost whorl of a flower consists of one or more **pistils**, the female parts of a flower. The base of a pistil is called the ovary. Inside an ovary, eggs are produced within **ovules**. The tip of a pistil is called the stigma, the site where pollen sticks.

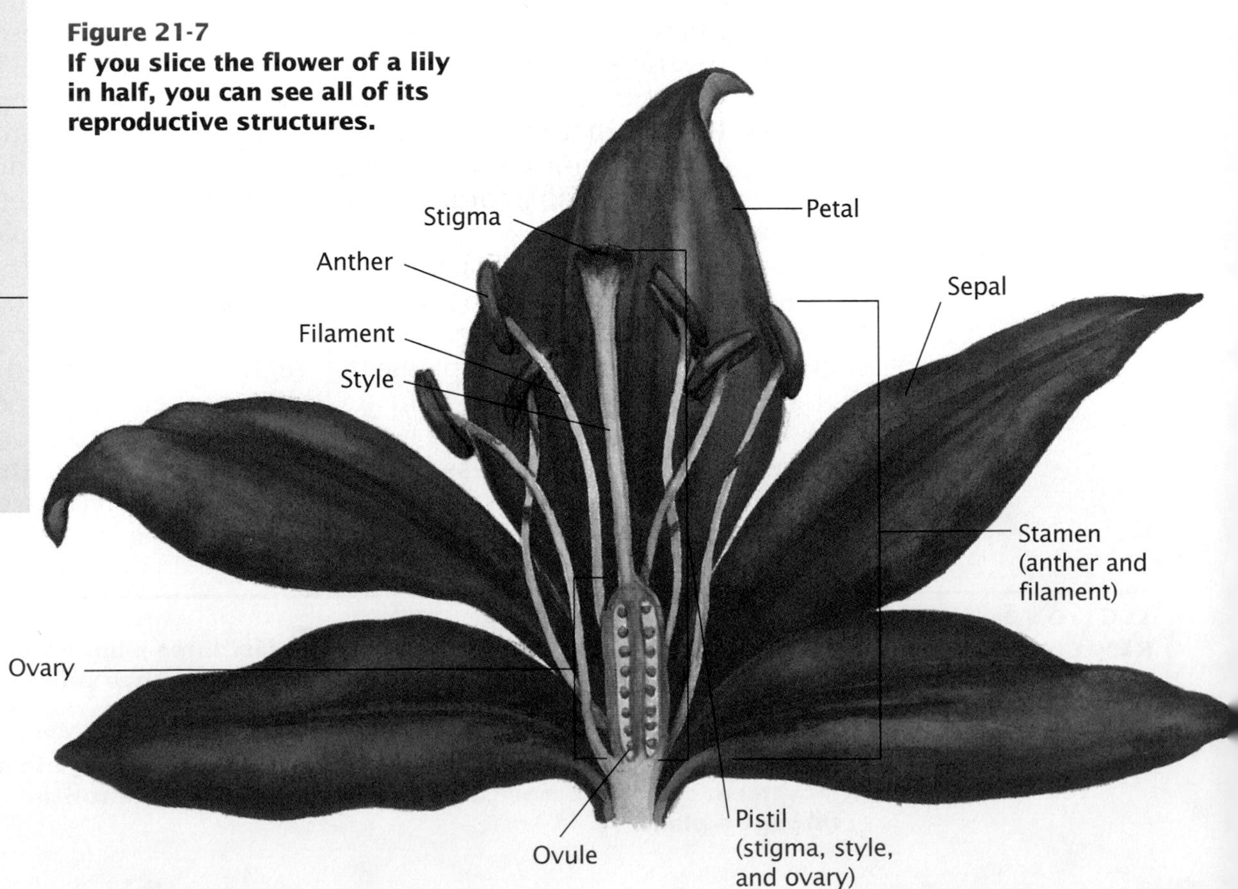

Figure 21-7
If you slice the flower of a lily in half, you can see all of its reproductive structures.

Iris

Iris virginica (Southern blue flag) has violet flowers that resemble cultivated irises. In Roman mythology, Iris was a goddess whose visible sign was a rainbow. The iris is named for its rainbow of colors. Iris flowers occur in many colors, from white to deep violet and yellow to reddish brown.

Iris flowers have parts in multiples of three. Three violet petals curve upward, while three petal-like sepals curve downward. The style branches into three flat, petal-like parts. The flower has three stamens, one under each branch of the style.

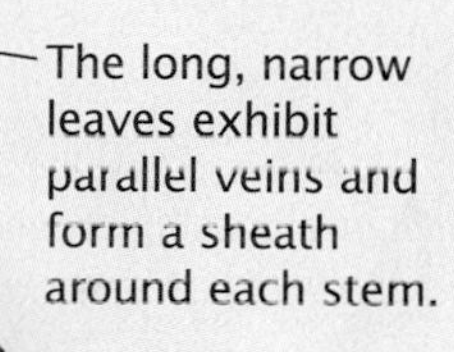

The long, narrow leaves exhibit parallel veins and form a sheath around each stem.

Evolutionary Relationships

Iris virginica is an angiosperm—a flowering plant. Iris flowers are adapted for attracting insect pollinators, in this case bees. While visiting flowers, bees seeking nectar transfer pollen from one flower to another. Irises are also monocots, as the structure of their leaves and flowers indicates.

Habitat

Iris virginica grows in wet meadows, marshes, and other moist places, such as forests and the edges of ponds and streams. Of the many native irises found in the United States, *I. virginica* has one of the widest distributions. It grows throughout the Southeast, as far west as Texas and as far north as Virginia and Missouri.

In addition to forming seeds, irises reproduce asexually from an underground stem called a rhizome. The plants are perennials—old shoots die each winter, and new shoots arise from the rhizome the next spring.

Size

The shoots of *I. virginica* can be up to 1 m (3 ft) tall. Leaves up to 76 cm (2.5 ft) long and up to 3.2 cm (1.2 in.) wide rise from ground level. Flowers with petal-like sepals that are 7.6 cm (3 in.) long crown stems that are 0.6–1.0 m (2–3 ft) tall. Irises often grow in extensive colonies. Some of the clumps in a colony may belong to a single plant. These clumps are connected by a common rhizome.

Pollination and Fertilization

For sexual reproduction to occur, pollen containing the male gametes must reach the stigma. Since flowering plants cannot move about to seek their mates, they must rely on other methods to move pollen. Pollination is the transfer of pollen grains from an anther to the stigma of a flower. Wind causes the pollination of the flowers of many plants, such as oaks and grasses. In many other plants, pollen falls from the anthers onto a stigma of the same flower, resulting in **self-pollination**. However, most flowering plants, from magnolias to orchids to the morning glory seen in **Figure 21-8**, depend on animals for pollination.

The flowers of plants pollinated by animals must attract pollinators. In Chapter 14 you learned that flowering plants and their pollinators have coevolved, or evolved in response to changes in each other. To attract pollinators, some flowers, such as the morning glory in **Figure 21-8b**, have attention-getting "advertisements" such as brightly colored petals or scents. The flowers of many kinds of plants secrete a sugary liquid called nectar, which pollinators use as food. Pollen itself also serves as food for many pollinators.

As an animal explores a flower, pollen may stick to its body. When the animal visits another flower, it may inadvertently transfer some of this pollen to the stigma of that flower. By attracting animals, flowering plants can ensure **cross-pollination**, which is the transfer of pollen to another plant of the same species. Cross-pollination ensures genetic recombination and thus may produce more genetic variation than does self-pollination.

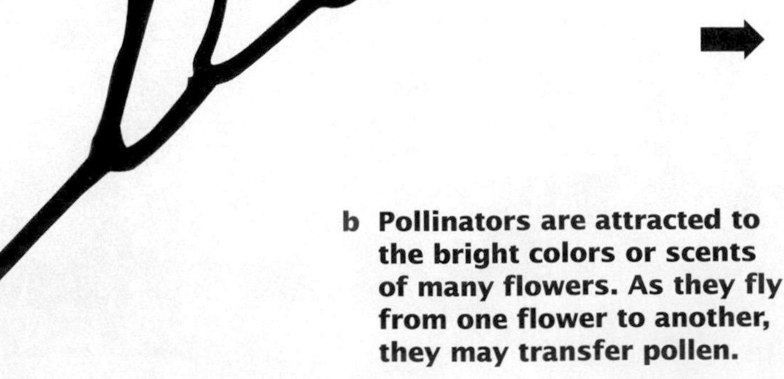

Figure 21-8

a Flowers, the reproductive structures of flowering plants, begin as buds that grow out of a plant's shoots.

b Pollinators are attracted to the bright colors or scents of many flowers. As they fly from one flower to another, they may transfer pollen.

Fertilization in flowers occurs in two stages

After a pollen grain lands on the stigma of a flower, a pollen tube may grow through the style of the pistil and into an ovule, which contains an egg. The egg is one of eight haploid nuclei in an ovule. A pollen grain has two sperm that travel down the pollen tube, as shown in **Figure 21-8c**. One sperm fuses with the egg to form a zygote. The other sperm fuses with two other nuclei in the ovule to form a tissue that will be a source of nutrition for the embryo. This event, in which one sperm fuses with an egg and a second sperm fuses with two nuclei, is called **double fertilization.** Double fertilization has great survival value because it provides each generation with an initial source of nutrition.

Remember that pollination is not the equivalent of fertilization. Fertilization involves the union of egg and sperm and may not occur until weeks or months after pollination has taken place. Sometimes fertilization fails to occur after pollination.

Seeds and fruits form after fertilization

Fertilization causes rapid changes in a flower. The ovule develops into a seed, often with a tough coat protecting the plant embryo and its nutrient supply. As you read in Chapter 20, the development of the seed was a major factor in the success of plants on land. Seeds help plant species to survive in unfavorable environments.

As seeds develop, the ovary grows larger and develops into a fruit. A **fruit** is an enlarged ovary of a flowering plant. You can see the fruit of a morning glory plant in **Figure 21-8d**. Many "vegetables," including squash, green beans, tomatoes, and cucumbers are technically fruits. The development of an ovary into a fruit usually depends on whether fertilization has occurred. If the eggs go unfertilized, the ovary will normally wither and drop, along with the rest of the flower.

Some fruits lack seeds

Have you ever wondered why navel oranges and bananas are seedless? The ability of plants to develop fruits without the fertilization of eggs is called parthenocarpy *(pahr thuh noh KAHR pee).* Seedless varieties of bananas, citrus fruits, and some grapes occur naturally. Parthenocarpy can be induced in plants such as tomatoes and watermelons by applications of auxin at certain stages in development.

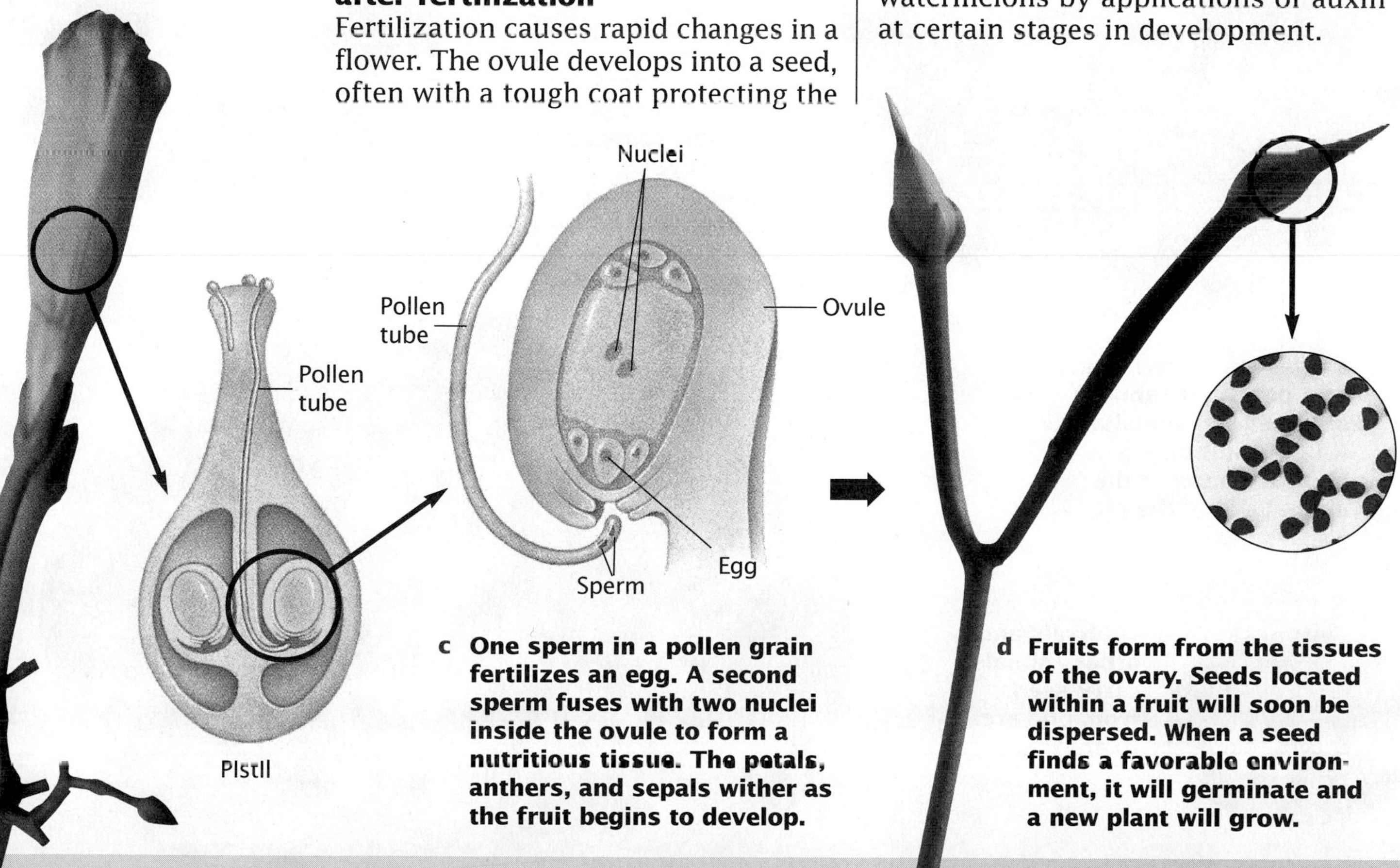

c **One sperm in a pollen grain fertilizes an egg. A second sperm fuses with two nuclei inside the ovule to form a nutritious tissue. The petals, anthers, and sepals wither as the fruit begins to develop.**

d **Fruits form from the tissues of the ovary. Seeds located within a fruit will soon be dispersed. When a seed finds a favorable environment, it will germinate and a new plant will grow.**

How Seeds Are Dispersed

Figure 21-9 Plant seeds are dispersed in a variety of ways. Several examples are shown below.

Once mature, seeds are ready to be dispersed. **Figure 21-9** shows how various types of seeds are dispersed. Some fruits, such as the plump, fleshy peaches, tomatoes, and watermelons you buy at the grocery store, are eaten by animals. When fleshy fruits ripen, their sugar content increases and the fruit becomes soft and juicy. Their colors often change from green to bright red, yellow, or orange. These changes, signs that the seeds are ripe and ready for dispersal, aid in catching the attention of hungry animals.

When fruits are eaten by birds or mammals, the seeds are spread by passing unharmed through the digestive tract. The success of sweet, colorful fruits as a method of seed dispersal is important in the coevolution of animals and flowering plants. Plants such as maple trees, tumbleweeds, and dandelions have extremely light fruits and seeds that can be carried by the wind. The fruits and seeds of many plants growing in or near water, like coconuts, are adapted for floating. Some fruits or seeds, such as burdocks, have hooks that attach them to the feathers, fur, or clothing of passing animals, giving the seed a free ride to a new home.

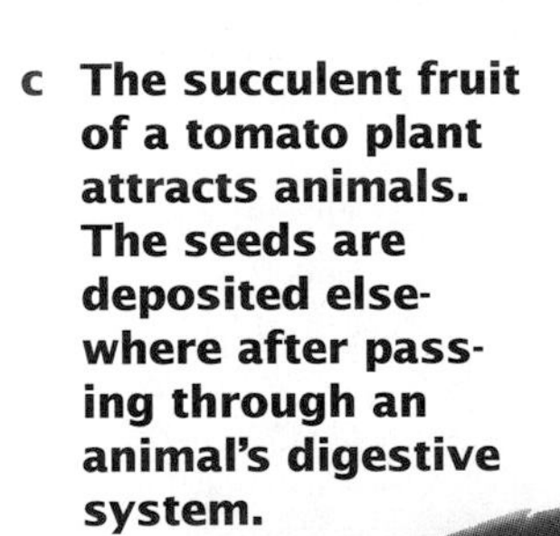

a **The sharp, tough coat of a peach pit cannot be eaten by animals. A new plant may germinate wherever the animal drops the pit.**

b **Squirrels and other animals carry seeds from one area to another.**

c **The succulent fruit of a tomato plant attracts animals. The seeds are deposited elsewhere after passing through an animal's digestive system.**

d **Humans can disperse seeds too. Have you ever spit watermelon seeds on the ground? One of those seeds may have germinated and grown into a new watermelon plant.**

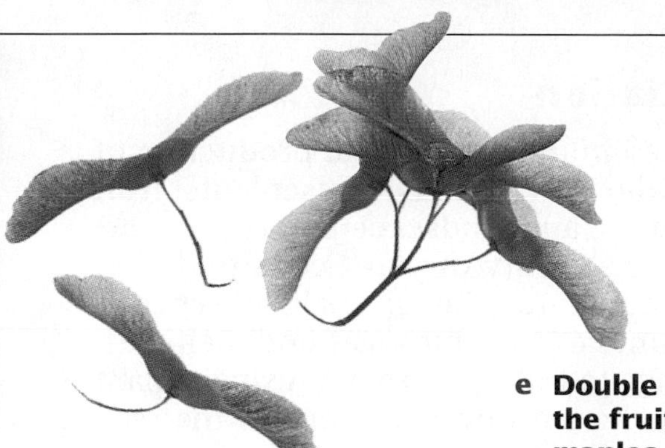

e **Double samaras, the fruits of maples, spin in the wind like helicopter blades.**

f **Wind carries the fruit of dandelions to new locations.**

g **Tumbleweeds can sometimes be seen blowing across roads in the southwestern United States, dispersing their seeds along the way.**

h **The fruit of a coconut palm can float from place to place.**

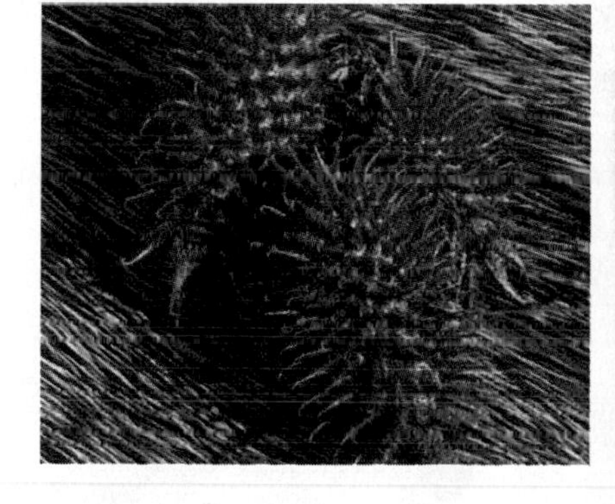

i **Burdocks are transported on the fur of animals such as this deer.**

Seeds resume growth by germinating

After they have formed, many seeds go through a period of decreased activity called dormancy. Their metabolic activity is greatly reduced. The length of time that a seed can be dormant and still grow into a plant varies. The record is held by a seed of the arctic tundra lupine. Seeds from this plant were found in a frozen animal burrow in the Canadian Yukon and were estimated to be 10,000 years old. When these seeds were planted, they sprouted in 48 hours.

In order to germinate, or begin growing into a new plant, seeds need a suitable environment. **Germination**, which is the sprouting of a seed, cannot take place until water and oxygen reach the embryo. Sometimes this involves cracking the seed coat. Some seeds are covered by seed coats that will not crack unless they are exposed to the heat of fire. Other seeds will germinate only after they have passed through the intestines of birds or mammals. After water and oxygen enter a seed, the embryo swells, grows, and breaks through the seed coat. The young plant that grows when a seed germinates is called a seedling.

Plant Cell Growth and Differentiation

When an embryo starts to develop, all of its cells are identical. The cells have the same genes because they came from the same fertilized egg. Over time, hormones act on some cells but not others. The hormones turn some genes on and others off, causing the cells to differentiate, or become different from one another. Some of the cells become vascular tissue, while others become ground tissue or epidermis.

Even after they have differentiated, many kinds of plant cells retain the potential to form other types of cells. As illustrated in **Figure 21-10**, potato plants can grow from pieces of stem tissue. Each cell in the tissue contains all the genetic information needed to grow a new plant. Tissue culture is used in the commercial production of millions of orchids, houseplants, fruit plants, and landscape plants.

The ability of plant cells to differentiate into other specialized types of plant cells enables leaf and stem cuttings to form new roots. A single plant can be cut into many pieces that can grow into identical plants. Bananas and chrysanthemums are produced in this way. The ability of plant cells to differentiate into other cell types also makes it possible to attach, or graft, the stems or buds of one plant to another plant. Grafting and budding are used commercially to produce plants such as hybrid rosebushes and grapevines, and to produce citrus, peach, apple, and pecan trees.

Figure 21-10 The large-scale production of most vegetables by tissue culture is not economically feasible. The procedure is used commercially to grow potatoes that are free of viruses.

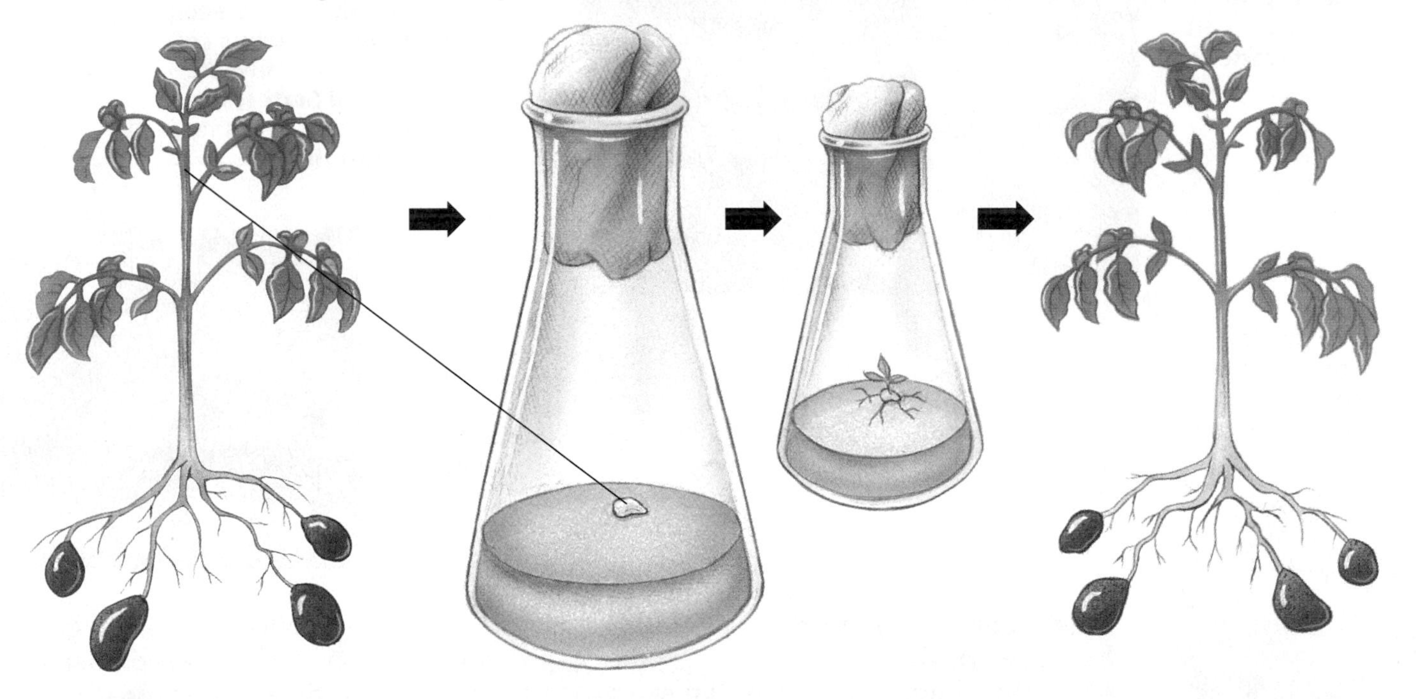

a **To grow a new potato plant by tissue culture, scientists first isolate tissue from the tip of the stem.**

b **The stem tissue is placed in a flask containing nutrients and hormones.**

c **A new plant then begins to grow in the culture flask.**

d **When the plant is large enough, it is planted in soil and allowed to grow to maturity.**

SECTION REVIEW

1. **How does each part of a flower help a plant achieve sexual reproduction?**
2. **How does the process of pollination enable the process of fertilization to occur in plants?**
3. **A certain plant produces seeds within fruits that have many long, straight spines. Predict how this plant's seeds are dispersed.**
4. **What are some possible advantages of using tissue culture to reproduce plants?**

CHAPTER

21 Highlights

Rose hips, the fruit of a rose plant, are sometimes an ingredient of herbal tea.

	Key Terms	Summary
21-1 The Plant Body	root cap (p. 391) root hairs (p. 391) shoot (p. 392) stoma (p. 392) xylem (p. 394) phloem (p. 394) ground tissue (p. 394) epidermis (p. 394) meristem (p. 394) primary growth (p. 394) secondary growth (p. 394)	• Roots absorb water and minerals. Shoots consist of stems and leaves. • Leaves are the main site of photosynthesis. Carbon dioxide enters a leaf, and water and oxygen exit, through the stomata. • Vascular tissue contains xylem and phloem. Xylem transports water and minerals. Phloem transports organic compounds. • Vascular tissue is found within ground tissue, which makes up most of a nonwoody plant. Epidermis covers the outside of a plant. • Regions of cell division are called meristems. Cell division in meristems increases the length and diameter of a plant.
21-2 How Plants Function	capillary action (p. 396) transpiration (p. 397) translocation (p. 397) tropism (p. 398)	• The loss of water through transpiration helps to pull water through a plant. • Stomata in leaves open and close to regulate the amount of water lost through transpiration. • Plant growth and development is regulated by hormones. • Chemicals that sense light and temperature affect the growth and development of many plants.
21-3 Reproduction in Flowering Plants	stamen (p. 400) pistil (p. 400) ovule (p. 400) self-pollination (p. 402) cross-pollination (p. 402) double fertilization (p. 403) fruit (p. 403) germination (p. 405)	• Flowers usually consist of sepals, petals, stamens that form pollen, and pistils that form ovules. • During fertilization in flowering plants, one sperm fuses with an egg to form an embryo, and another sperm fuses with two nuclei to form a nutritious tissue. • After fertilization, seeds develop within ovaries that become fruits. • Germination occurs when the embryo in a seed resumes growth under favorable conditions.

This strawberry plant has roots that anchor it and absorb water and minerals, and shoots that absorb light and perform photosynthesis.

Transpiration drives the uptake of water by a plant, while translocation distributes products of photosynthesis throughout a plant.

Flowers are one of several characteristics that have helped make flowering plants the dominant plants on Earth.

CHAPTER 21

review

Understanding Vocabulary

1. For each set of terms, complete the analogy.
   a. roots : absorb water :: stems : ____
   b. water transport : xylem :: sugar transport : ____
   c. self-pollination : same flower :: cross-pollination : ____
2. Using each set of words below, write one or more sentences summarizing information learned in this chapter.
   a. shoot, leaves, stem, photosynthesis
   b. water, root hairs, transpiration, xylem
   c. stamen, pollen, anther, pistil, ovary, ovule

Understanding Concepts

3. **Relating Concepts** Construct a concept map that describes the types of plant tissues and their functions. Try to include the following terms in your concept map: ground tissue, vascular tissue, xylem, phloem, meristem, minerals, water, organic molecules, epidermis, secondary growth, primary growth.
4. **Comparing Functions** What are the functions of roots? How do they differ from the functions of shoots? Describe three specialized functions of roots.
5. **Summarizing Information** Describe three functions of stems, and explain how stems are able to perform each function.
6. **Analyzing Information** When rings are counted to determine the age of a tree, is primary or secondary growth being measured? Explain your answer.
7. **Summarizing Information** Describe the characteristics of monocots and dicots, and give examples of each.
8. **Analyzing Processes** Describe the major forces behind the movement of water in plants.
9. **Recognizing Relationships** How might the rate of water movement within a plant differ between plants in cool, humid environments and those in hot, dry environments?
10. **Summarizing Information** Describe how plants respond to the hormones auxin, gibberellin, and ethylene.
11. **Reading Graphs** The seeds of certain conifers must experience cold temperatures before they will germinate. Using the graph below, what percent of seeds would germinate after a winter with only two cold months?

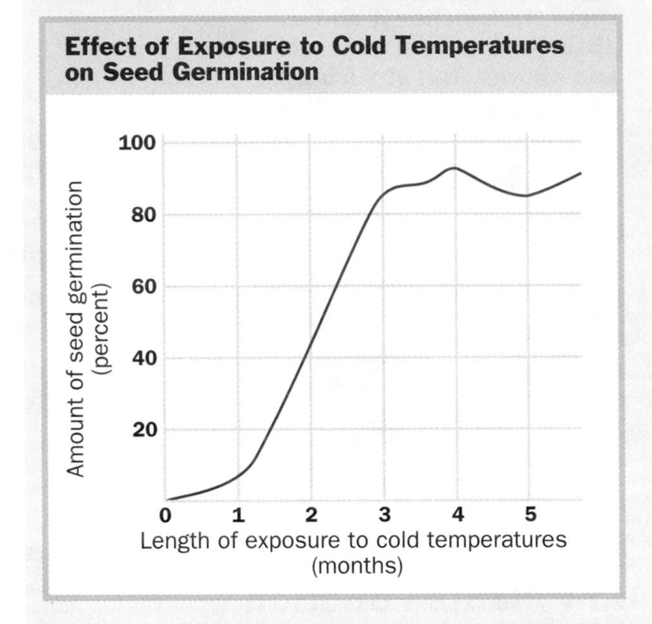

12. **Relating Information** Give an example of a long-night plant and a short-night plant. At what time, or times, of the year would these plants normally produce flowers?
13. **Comparing Functions** Distinguish between pollination and fertilization in angiosperms.
14. **Inferring Relationships** Explain why the process of fertilization in angiosperms is called double fertilization.
15. **Summarizing Information** String beans, cucumbers, and squash are typically called vegetables. Why are they actually fruits?
16. **Inferring Relationships** Explain how the ability of a plant's cells to differentiate can be used to produce new plants.
17. **BUILDING ON WHAT YOU HAVE LEARNED** In Chapters 20 and 21 you read about the symbiotic associations between fungi and plants that are called mycorrhizae. Why do you think plants that develop mycorrhizae have fewer root hairs than plants without mycorrhizae?

Interpreting Graphics

18. Study the photographs below. Identify each structure as belonging to a monocot or a dicot. For each response, write a sentence that explains why the structure belongs to a monocot or dicot.

Reviewing Themes

19. ***Matter, Energy, and Organization***
Explain how the thin, flat structure of a leaf helps increase the efficiency of photosynthesis.

20. ***Evolution***
Describe three features of flowering plants that suggest plants and their animal pollinators have coevolved.

21. ***Heredity***
How does cross-pollination ensure genetic recombination?

22. ***Evolution***
Explain why the development of the seed was a major factor in the success of plants.

Thinking Critically

23. **Evaluating Results** You put several seeds in a moist, airtight container and place the container in the dark. After a few days, you see that the seeds have germinated. After a few more days, you notice that the seedlings are dead. Explain your results.

24. **Evaluating Opinion** Your grandfather told you that if you hammer a nail into the trunk of a young tree, the nail will remain at the same height no matter how tall the tree grows. Do you think your grandfather is right or wrong? Support your answer.

Life/Work Skills

25. **Finding Information** Use library references or search an on-line database to find out about tree-ring dating. Find out what causes the rings that appear in a tree trunk and how differences in the sizes of the rings can indicate climatic conditions at the time the rings were made. What other kind of information about the environment can be determined by studying tree rings? Summarize your findings in a written report.

26. **Finding Information** Use library references or search an on-line database to find out about different kinds of plant movements, such as phototropism, thigmotropism, gravitropism, chemotropism, hydrotropism, and nastic movements. Prepare a poster with illustrations of the different kinds of plant movements to share with your class.

27. **Working Cooperatively** With two or three of your classmates, visit a wholesale plant grower. Orchid growers would be an excellent choice if one is in your area. Find out how tissue culture is used to reproduce different kinds of plants. Try to find out information on two methods of tissue culture that are not described in this chapter. Prepare an illustrated report of your findings to share with the class.

Activities and Projects

28. **Art Connection** Use library references or visit a local art museum or botanical garden to survey the different ways that flowers have been depicted by artists over the years. Look at flowers in paintings, tapestries, stained glass, quilts, sculptures, and porcelains. Compare how flowers are represented by Western artists, such as Claude Monet, Vincent van Gogh, Leonardo da Vinci, and Georgia O'Keeffe, with the ways they are represented by Japanese and Chinese artists.

Discovering Through Reading

29. Read the article "A Garden of Mutants" by Carol Fletcher in *Discover*, August 1995, pages 49–53. Why do botanists study genetics using the *Arabidopsis* plant? What is the MADS-box? What causes the flower mutations in *Arabidopsis*?

CHAPTER

22

REVIEW
- carbohydrates and proteins (Section 2-3)
- leaves, stems, and roots (Section 21-1)
- reproduction in flowering plants (Section 21-3)

Plants in Our Lives

Wheat, shown here in a Colorado field, is one of many important crop plants that provide food for humans.

Plants are an essential part of our lives. You already know that plants produce oxygen. They also provide us with food; almost everything we eat comes from a plant or an animal that ate a plant. And plants provide us with many other useful products, such as wood, medicines, fibers for making cloth and paper, and rubber.

22-1 Uses of Plants

OBJECTIVES

1. **Describe the importance of cereal grains and legumes in the diet.**
2. **Identify the foods that come from plants as fruits, seeds, roots, stems, leaves, or flowers.**
3. **Describe several uses for wood.**
4. **Identify the plant sources of common medicines.**
5. **Name several plants that are natural sources of fibers and rubber.**

Important Grains

Did you know that most of the food people eat comes from fruits? As you read in Chapter 21, fruits are the parts of a flowering plant that contain the plant's seeds. When you think of fruits, you probably think of apples, oranges, and peaches. However, the fruits that provide the most food to humans come from the cereal grasses wheat, rice, corn, and oats. These grain crops occupy more than 70 percent of the world's farmland and provide about 60 percent of the calories eaten by humans, directly or indirectly. The map of the World in **Figure 22-1** shows the regions where wheat, rice, and corn are grown. The fruits of cereal grasses, which are called **grains**, are rich in carbohydrates, protein, vitamins, and dietary fiber. Grains are usually separated from their stalks and then ground into flour or meal and cooked.

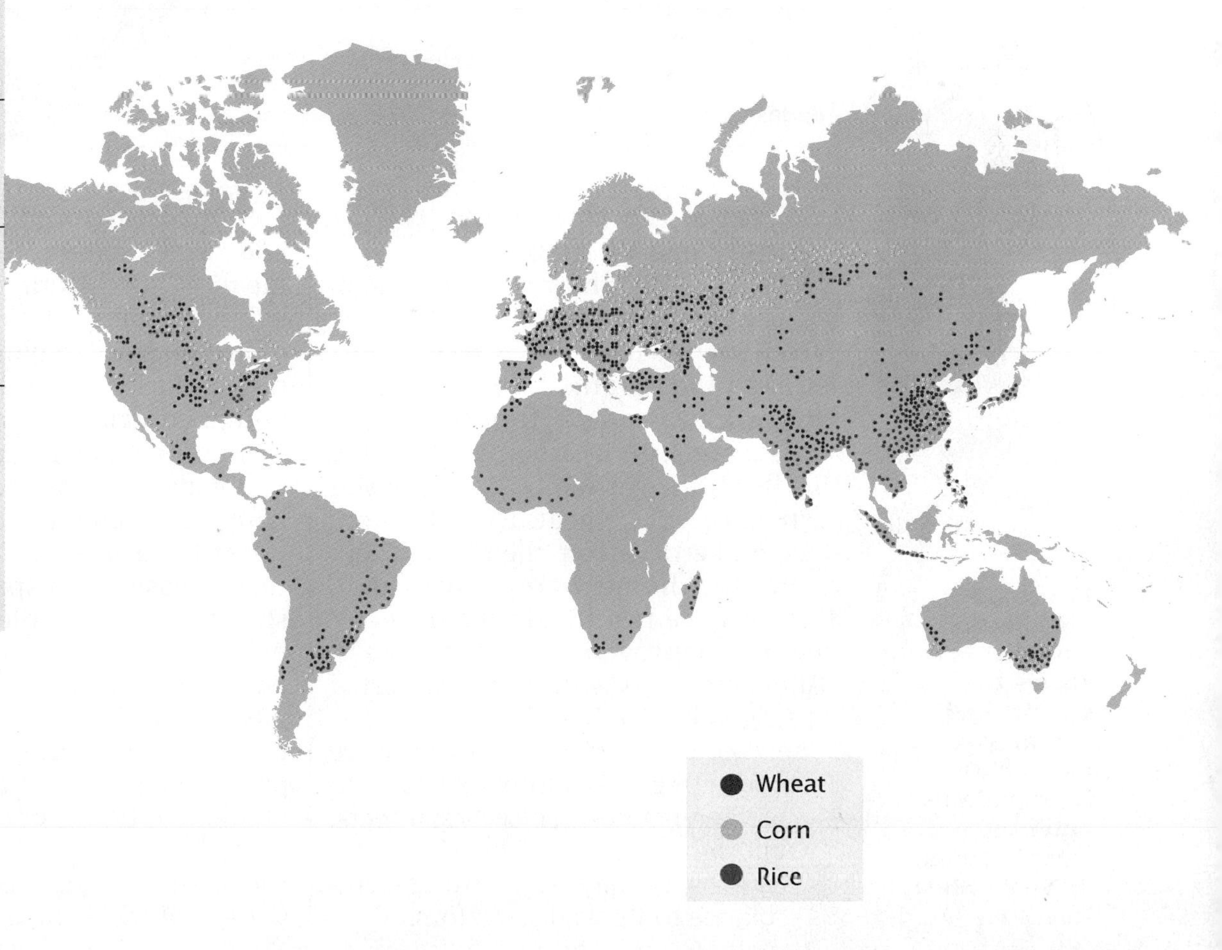

Figure 22-1 Wheat, rice, and corn are grown all over the world.

c Many foods are made from wheat.

b The wheat plant is a grass that can grow up to 1.7 m (5 ft) tall.

Figure 22-2
a This section of a wheat kernel shows the starchy endosperm used in making white flour and the outer bran layer included in whole-wheat flour.

Wheat is the world's most widely grown plant

More than one-third of the people in the world depend on foods made from kernels (grains) of wheat. Wheat stalks and grains are shown in **Figure 22-2**. Long before the beginnings of agriculture, the people of the Middle East gathered and chewed kernels of wild wheat. Early farmers selected kernels from their best wheat plants for the next crop. Over time, this practice resulted in the development of improved varieties of wheat. About a century ago, scientists started using selective breeding to develop new kinds of wheat that are more productive and more resistant to cold temperatures, diseases, insects, and other stresses. The wheat that is grown today bears little resemblance to its wild ancestors.

In many areas of the world, wheat appears in some form at nearly every meal. Wheat flour is excellent for baking because it contains a mixture of proteins called gluten, which makes dough elastic. This elasticity enables dough containing yeast to rise. Yeast-risen dough is used to make a variety of breads and other baked goods. Pasta is made by mixing ground wheat and water to form a thick paste, which is then pressed into spaghetti and other shapes. In the Middle East, wheat kernels are also boiled or soaked and then pounded until they crack. The cracked pieces, which are called bulgur *(BUHL guhr)*, are then dried and used to prepare dishes such as tabbouleh, pilavi, or kisir. Pita, or pocket bread made from wheat flour, may accompany these meals. **Figure 22-2c** shows a variety of foods made from wheat.

Rice is one of the world's most important food crops

More than half of the people in the world eat rice as the main part of their meal. Rice is often grown in standing water, as seen in **Figure 22-3**. Therefore, rice is usually grown in moist regions, such as in the flood plains of rivers. The people of Southeast Asia probably gathered and ate wild rice thousands of years ago, but no one knows exactly when or where domesticated rice originated. China, India, or Indochina is probably where rice was first cultivated. Archaeologists have found evidence that people cultivated rice for food in southern China as early as 5000 B.C. From there, rice spread north to Japan.

Although low in protein, rice is an excellent source of carbohydrates. The nutrient-rich bran layers of rice grains are removed to make white rice, which has few vitamins. Brown rice, however, still contains the nutrient-rich bran layers on the outside of the kernel. **Table 22-1** shows the nutritional content of rice and wheat. In societies where people eat mainly white rice, vitamin-rich sauces such as soy sauce are added to the cooked rice to make meals more nutritious.

Rice is added to many processed foods, including breakfast cereals, soup, baby food, and flour. Breweries use broken rice grains to make mash, an ingredient used to make beer. In Japan, rice grains are used to make an alcoholic drink called sake *(SAHK ee)*, or rice wine. But rice isn't just for eating and drinking. The outer coats of rice grains are used in products such as insulation and cement. Many people in Asia use the dried stalks of rice plants to thatch roofs and to make sandals, hats, and baskets.

Figure 22-3

a The rice plant is a grass that grows to heights of 80 to 180 cm (32 to 72 in.). When the rice is ripe, the plant turns golden yellow.

b Because rice requires a great deal of water, it is often grown in fields of standing water called paddies.

Table 22-1 Protein and Dietary Fiber

Grain (100 g)	Protein (g)	Fiber (g)
Brown rice	7.9	1.3
White rice	7.1	0.3
White wheat flour	10.3	0.3
Whole wheat flour	13.7	2.1

Corn is the most widely grown crop in the United States

Modern corn probably originated from *Zea mexicana*, a wild Mexican annual grass known as teosinte *(tee oh SIHN tee)*. Teosinte produces narrow female flower spikes that mature into "ears" with two rows of grains, as seen in **Figure 22-4a**. The grains of teosinte are covered by a hard case and are difficult to grind until they have been heated and popped like popcorn.

Corn was cultivated by the Aztecs of Mexico, the Mayas of Yucatan and Central America, and the Incas of South America. Columbus and other European explorers introduced corn to many areas of the world. The selective breeding of corn began more than 7,000 years ago in Mexico. American Indian farmers selected plants that produced "ears" with tender grains that could be eaten more easily than the hard teosinte grains. They also selected plants that produced ears with more rows of grains. The high-yield varieties of hybrid corn grown today, such as the one seen in **Figure 22-4b**, are made by crossing inbred strains of corn that each have a desirable trait.

Corn is more suited to the warm climate of the southern United States and Mexico than is wheat. As a result, corn is used to prepare many of the traditional components of the diets in these regions. Foods made from corn are seen in **Figure 22-4c**. Corn bread, corn pone, hominy, and grits are traditional components of the diet in the southeastern United States. Tortillas, which are served with many Mexican dishes, are a flattened, unleavened bread that is often prepared with corn meal. Corn is also used to produce corn syrup, margarine, cornstarch, cooking oil, and many industrial products. In addition, corn is one of the world's chief foods for domestic animals. Seventy percent of the United States corn crop is consumed by livestock.

Figure 22-4
a This ear of teosinte, which is the ancestor of modern corn, has two rows of grains.
b Large ears of tender corn are produced by modern hybrid corn plants.
c Corn tortillas and corn bread are foods made from corn.

b Lentils, another type of legume, are also high in protein. They can be made into dahl, the spicy dish shown at right.

Other Food Plants

Figure 22-5
a Soybeans are very high in protein. They are used to make tofu, soy milk, cheese, and many types of meat substitutes.

Although cereal grains provide most of the calories consumed by humans, they do not contain all of the nutrients you need for a healthy diet. They lack adequate amounts of some of the **essential amino acids**, which are the nine amino acids that are not made by the human body. Foods from other plants are rich sources of these amino acids, and they provide vitamins and minerals as well.

Legumes are high in protein

Legumes *(LEHG yooms)* are members of the pea (or bean) family. As you read in Chapter 9, many legumes host nitrogen-fixing bacteria in their roots. These bacteria provide legumes with extra nitrogen, which is a component of protein. The seeds of legumes have more protein than the seeds of most plants. They are also rich in essential amino acids that are scarce in grains. Many cultures eat combinations of grains and legumes, such as corn and beans, which provide ample amounts of all the essential amino acids.

Interdependence

How do humans benefit from the ability of plants to store the products of photosynthesis?

Soybeans, shown in **Figure 22-5a**, were probably first cultivated in northeast China more than 3,000 years ago. Soybeans can be cooked and pressed into cakes called tofu, also known as bean curd or soybean cakes. Tofu has been the protein staple for people in eastern Asia for 2,000 years and is an important food for more than 1 billion people. Lentils are popular legumes in Egypt and India. They can be ground into flour or made into dahl, which is seen in **Figure 22-5b**.

Sugar comes from sugar cane and sugar beets

Sugar is by far the most important sweetener in the world today. Most of the sugar that you eat comes from sugar beets or sugar cane. Sugar beets have large fleshy roots, like carrots, in which they store sucrose. Sugar cane, seen in **Figure 22-6**, is a tall grass with sturdy stalks that have a large amount of juice containing sucrose. The juice is used to make table sugar and syrup. Sugar cane is thought to have been first used by people in the Far East. The process of producing sugar from sugar cane has been practiced in India since 3000 B.C. The sap was boiled to produce large and small lumps. Large lumps were called *khanda*, which may be the origin of the word *candy*. Small lumps were called *sakkara*, which may be the origin of the word *saccharine*.

Figure 22-6
This man is cutting stalks of sugar cane.

Figure 22-7
Cassava is a starchy root that can be prepared in much the same way as potatoes. In the United States, it is usually made into tapioca.

Other staple foods are roots or stems

Potatoes are the dietary staple (most important food) in many regions of the world. The edible parts of potato plants are underground stems called **tubers**, which store starch. The potato is a native of South America. There is evidence that people in Chile ate wild potatoes 13,000 years ago. When potatoes were introduced in Europe, they became a great success, especially in Ireland. Potatoes grew well in the country's cool, moist climate and provided enough calories for an entire family. As you read in Chapter 19, a disease caused by the water mold *Phytophthora infestans* virtually destroyed the Irish potato crop in the mid-1840s. Yams, which are also tubers, are a food staple in many countries.

Some plants store food for the next season's growth in enlarged roots. Many of these roots are edible. For example, carrots have been grown as food in Europe for at least 2,000 years. Other important root crops include radishes, turnips, beets, sweet potatoes, and cassava, also called manioc. Cassava, shown in **Figure 22-7**, is the staple food of more than 500 million people around the world. It contributes more than one-third of the calories consumed in Africa.

Many vegetables are actually fruits

The term *vegetable* refers to an agricultural commodity, or product, that comes from a plant and is eaten with the main part of a meal. On the other hand, "fruits" are agricultural commodities that are sweet and are usually eaten as desserts. As you learned in Chapter 21, the term *fruit* also refers to a ripened ovary of a flowering plant. All of the "fruits" you eat are also fruits in the botanical sense. However, vegetables such as tomatoes, squash, and green beans are also fruits, as shown in **Figure 22-8**. How can you tell if a vegetable is also a fruit? Vegetables that contain seeds are the fruits of the plants that produce them.

Humans also eat the leaves or flowers of particular plants

The leaves of plants such as cabbage, parsley, spinach, chard, turnips, and lettuce are popular vegetables. Some forms of lettuce may have been cultivated as early as 4500 B.C. Lettuce is even depicted on Egyptian tombs. Other vegetables that you may have eaten are parts of leaves. Onion bulbs and garlic cloves consist of the fleshy bases of leaves, and celery is a petiole (leaf stalk). When you eat Brussels sprouts, you are eating leafy buds. When you eat broccoli, cauliflower, or artichokes, you are eating clusters of immature flowers.

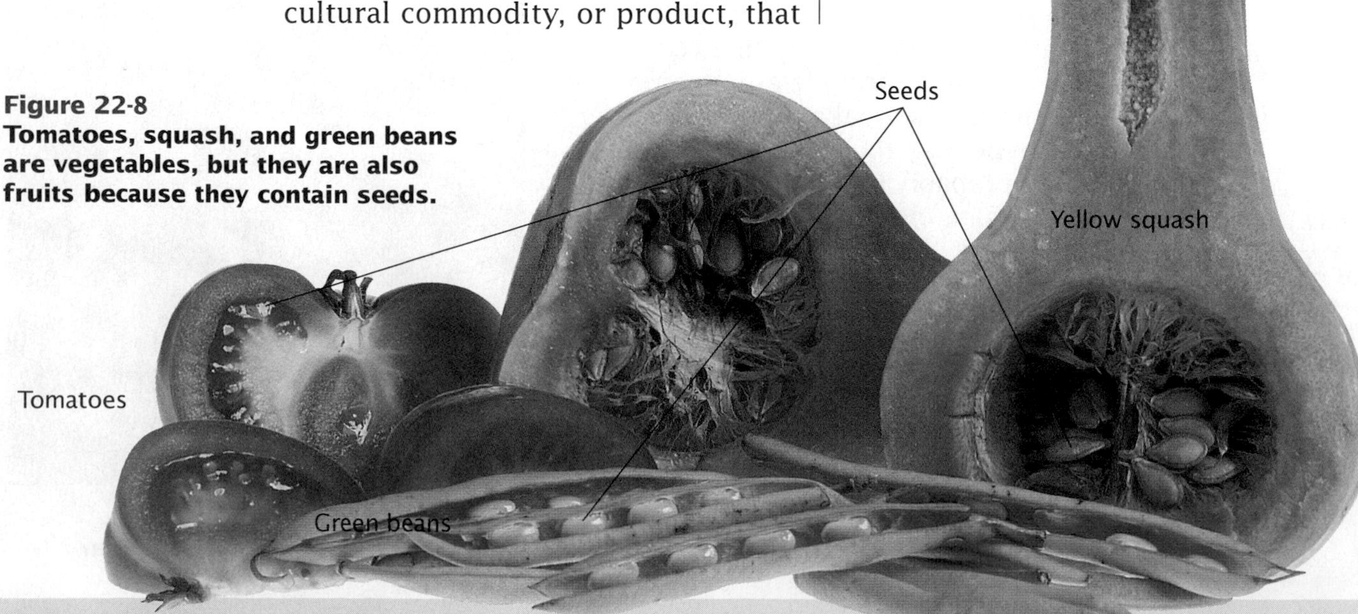

Figure 22-8
Tomatoes, squash, and green beans are vegetables, but they are also fruits because they contain seeds.

Uses of Wood

After food and oxygen, wood is the most valuable resource that people get from plants. As you read in Chapter 21, many plants that live for many years have stems that get thicker each year. The thickening of stems results from the production of secondary xylem and phloem. **Wood** is secondary xylem.

Trees are harvested for lumber

Wood from trees that have been cut down and sawed into boards and planks is called **lumber**. Nearly three-fourths of the lumber in the United States is used for construction. The rest goes to factories that make products such as boxes, crates, toys, railroad cars, boats, and other items. Wood chips and sawdust are treated with chemicals to make wood pulp, which is used to make paper, rayon, and other products.

Some wood is used as fuel

Wood has been used as a fuel since prehistoric times. It is the main fuel in developing countries and for more than half of the people in the world. A family that uses wood for heating and cooking burns about a ton of wood each year. As a result, wood is being burned faster than new trees can grow. In some African countries, wood is so scarce that the average family spends one-third of its income on wood.

Fuel can be made from wood chips through an experimental fermentation process. The wood chips are first treated with sulfuric acid to free cellulose and xylan, another polysaccharide. Additional enzyme treatments break these large compounds down into five- and six-carbon sugars. Finally, an assortment of bacteria and yeasts is added to ferment the sugars into ethanol.

Paper is one of our most important plant products

Cellulose fibers from plants are used to make paper. Papermaking fibers come from many different plants, such as cotton, bamboo, rice, sugar cane, and wheat. However, most paper is made from wood pulp. The process of making paper from wood pulp is summarized in **Figure 22-9**. Paper can be recycled and used to make new paper. But because it is expensive to remove inks and dyes from printed paper, the cost of recycling paper often exceeds the cost of making new paper from wood.

Figure 22-9
The papermaking process involves many steps.

a Logs are ground into wood chips.

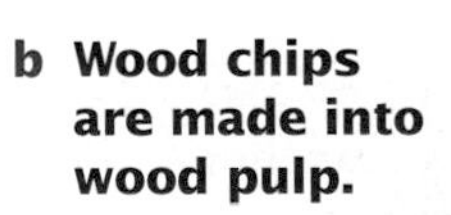

b Wood chips are made into wood pulp.

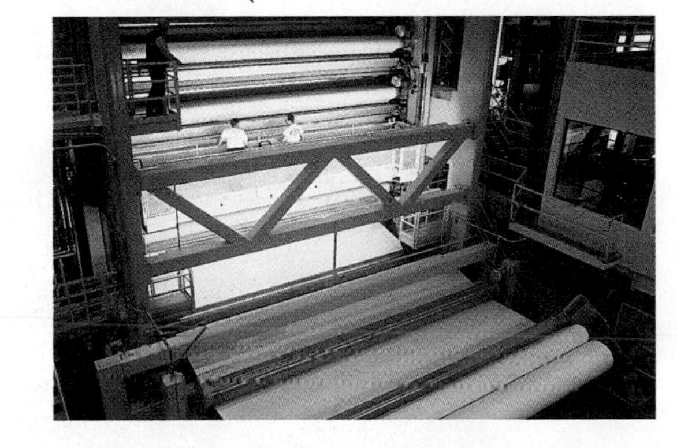

c Huge rollers press the cellulose fibers from wood pulp into paper.

Medicines From Plants

Figure 22-10
An American Indian doctor (left) is teaching a Western doctor (right) about native Amazonian rain-forest plants that can be used to treat disease.

People have always used plants to treat diseases and other ailments. Botany (the study of plants) was considered to be a branch of medicine until the mid-1800s. Doctors, such as the ones in **Figure 22-10**, continue to search for plants that may provide new treatments or cures for diseases such as cancer. Many familiar medicines, their original plant sources, and their uses are listed in **Table 22-2**. Some plants, such as the Pacific yew, nearly faced extinction as demand increased for the medicines made from them. Although the medicinal ingredients of most of these plants have been synthesized, most of these chemicals are still extracted from the plants that form them.

Table 22-2 Medicines Originally Derived From Plants

Name	Original source	Uses
aspirin	willow leaves and bark	relieves pain and reduces fever; is the world's most widely used drug
codeine	poppy fruits	relieves pain
digitalis	foxglove leaves	regulates irregular heartbeat
ephedrine	ephedra stems	relieves high blood pressure and symptoms of asthma and hay fever; acts as a decongestant
hydrocortisone	yam tubers	relieves symptoms of allergies and arthritis
quinine	cinchona tree bark	prevents malaria and relieves symptoms of malaria
reserpine	snakeroot roots	relieves high blood pressure and symptoms of schizophrenia
taxol	yew bark	reduces the size of cancerous tumors
vinblastine, vincristine	rosy periwinkle plants	used to treat cancers such as Hodgkin's disease and acute childhood leukemia

Other Plant Products

Plants are an important source of fibers used to make clothing. Cloth has been made from cotton for centuries. Cotton thread is spun from the strong, fine fibers attached to cotton seeds, shown in **Figure 22-11**. Cotton is still the world's most important plant fiber. The stems of flax plants yield a softer, more durable fiber that is used to make a cloth known as linen. Although synthetic fibers now make up more than 30 percent of the world's fibers, natural fibers are still prized for their durability, comfort, and strength.

Turpentine is a colorless liquid solvent that is used to remove paint. This highly flammable substance is extracted from pine trees that grow in the southeastern United States. A vapor containing turpentine forms when pine trees are made into wood pulp. As the vapor cools, a liquid that contains turpentine forms. Turpentine is also used to make disinfectants, insecticides, medicines, and perfumes.

Rubber is one of the most important raw materials obtained from plants. Natural rubber is made from latex, which is a milky liquid produced by certain plants, such as trees of the genus *Hevea*. American Indians of South America made rubber balls and waterproof shoes from latex, which is shown in **Figure 22-12**. Today most natural rubber comes from plantations of rubber trees in Asia. Natural rubber is also made from the sap of a desert plant called guayule *(gwah YOO lee)*, which is native to Texas and grown as a crop in the southwestern United States. Most of today's rubber, however, is manufactured from petroleum.

Figure 22-11
Fibers that encased seeds in cotton bolls (pods of the cotton plant) are being spun into thread by this Brazilian woman.

Figure 22-12
Latex is the milky sap of certain plants. These men are working with raw latex at a processing plant in Malaysia.

SECTION REVIEW

1. **Why is it important to combine grains with other vegetables, such as legumes?**
2. **Identify each of the following as a fruit, seed, root, stem, leaf, or flower: broccoli, carrot, onion, corn, and potato.**
3. **What are three ways that people use wood?**
4. **Where were aspirin and taxol first obtained? How is each used?**
5. **What are two plant sources of fibers used to make cloth?**

Fossils and art such as this Egyptian wall painting suggest that humans began cultivating plants between 7,000 and 9,000 years ago. Plants that furnish useful products are grown by farmers and horticulturists. Home gardeners grow plants outside and inside of their homes. Successful plant growers understand a plant's needs and provide for those needs.

22-2 Growing Plants

OBJECTIVES

1. **Summarize the importance of soil to the growth of a plant.**
2. **Justify the use of well-adapted plants for landscapes and gardens.**
3. **Relate changes in the appearance of a plant to its health.**
4. **Describe five methods of vegetative propagation.**
5. **Compare and contrast traditional plant breeding with genetic engineering.**

Soils and Plant Growth

Like you, plants require nutrients to grow. Plants obtain their nutrients from inorganic compounds such as carbon dioxide (CO_2) and water (H_2O). The carbon (C), oxygen (O), and hydrogen (H) in carbon dioxide and water are just three of the nutrients plants need. In Chapter 5, you learned that plants use the carbon, oxygen, and hydrogen from carbon dioxide and water in photosynthesis. But plants need more than just air and water to grow. Why do plants grow better in soil than they do in plain water? What is it about soil that helps plants grow?

Plants obtain nutrients from soil
The soil that surrounds a plant's roots, as seen in **Figure 22-13**, provides many of the nutrients that the plant needs. Roots take up these nutrients as they take up water. Plants require relatively large amounts of the nutrients nitrogen (N), phosphorus (P), potassium (K), calcium (Ca), sulfur (S), and magnesium (Mg). The nutrients that plants need in relatively large amounts are called **macronutrients**. Other nutrients, such as iron (Fe), chlorine (Cl), copper (Cu), manganese (Mn), zinc (Zn), and boron (B), are also critical to healthy plant growth, but they are needed only in tiny amounts. The nutrients that plants need in very small amounts are called **micronutrients**.

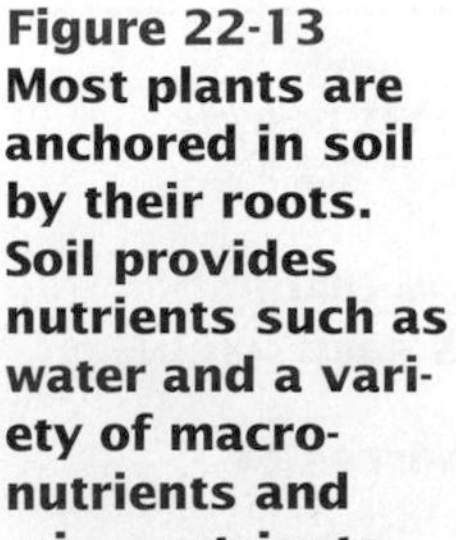

Figure 22-13
Most plants are anchored in soil by their roots. Soil provides nutrients such as water and a variety of macronutrients and micronutrients.

Soil is composed of rock particles and organic matter

Soil is a mixture of pieces of inorganic matter (weathered rock) and decaying organic matter. The rock particles in soil fall into three categories—sand, silt, or clay—based on their size. Sand grains are the largest soil particles, ranging from 2 mm (0.4 in.) to 20 µm (0.004 in.) in diameter. Silt particles range from about 20 µm (0.004 in.) to 2 µm (0.0004 in.) in diameter. Clay particles are the smallest soil particles, less than 2 µm (0.0004 in.) in diameter. If a soil is a mixture of sand, silt, and clay, it is called a **loam**. Most soils are loams. You can determine the amount of sand, silt, and clay in a soil by pouring the soil through a series of sieves with specific mesh sizes. **Figure 22-14** shows another way that soil can be analyzed to determine the types of soil particles it contains.

Soils also contain water, which is held in the air spaces between soil particles. As the size of soil particles decreases, the size of the air spaces between the particles also decreases. Therefore, sandy soils have large spaces between their particles, silty soils have medium-sized spaces, and clay soils have very small spaces.

Have you noticed that relatively few plants grow on sand dunes and on solid clay river banks? What is wrong with these soils? The size of the air spaces in a soil determines its capacity to hold water and air, as the graph in **Figure 22-15** indicates. Sandy soils contain a lot of air but hold very little water. Clay soils hold a large amount of water when all of the spaces are filled but contain very little air as a result. Loams are generally the best soils for plant growth. A loam is packed loosely enough to contain ample air, which provides the oxygen needed by roots for cellular respiration, but tightly enough to hold sufficient water for plant growth.

The organic component of soil is called **humus** *(hyoo muhs)*. It is a rich source of carbon, oxygen, nitrogen, and hydrogen for plants. Humus forms as bacteria and fungi in the soil decompose fallen leaves, dead organisms, and animal wastes. Earthworms and millipedes also consume and break down the remains of organisms. These burrowing animals work humus into a soil and aerate it (make tunnels that enable air to enter) at the same time.

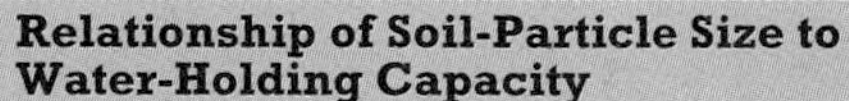

Relationship of Soil-Particle Size to Water-Holding Capacity

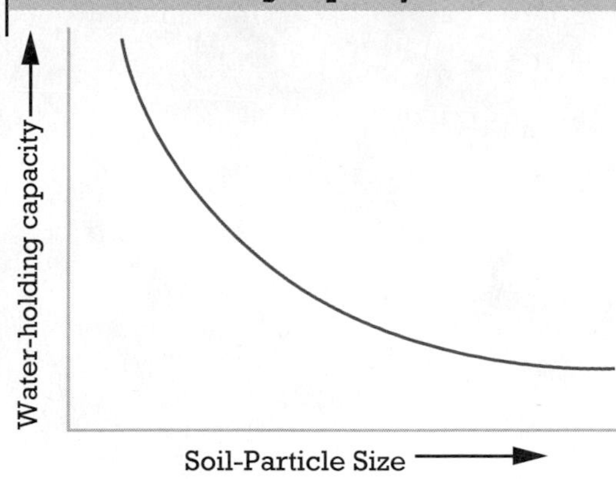

Figure 22-15
This graph shows that the water-holding capacity of a soil increases as the soil-particle size decreases.

Figure 22-14
One way to judge the composition of a soil is to make "soil balls."

a Sandy soils will not make a ball.

b Clay soils make balls that can be worked like modeling clay.

c Loams and silty soils make balls that readily fall apart.

b **Alfalfa plants host nitrogen-fixing bacteria in their roots, like many other legumes. These bacteria add nitrogen to the soil as the crop grows.**

Figure 22-16
a **This farmer is mowing a crop of alfalfa, which is a legume. The plants will be plowed into the soil as green manure, which adds humus to the soil.**

Cultivation practices help maintain soil fertility

Natural processes, such as decomposition and nitrogen fixation, continually add nutrients to the soil. However, plants can remove available nutrients from the soil around their roots more quickly than the nutrients can be replaced. The nutrients that plants take from the soil become part of the plants. These nutrients must be replaced if the soil is to maintain its **fertility**, or its ability to supply plants with nutrients. Many of the nutrients taken from the soil by plants are returned to the soil when the plants die and decay. However, few nutrients are returned to the soil when a crop is harvested. When another crop is planted in the same soil, there may not be enough nutrients in the soil for the crop to grow.

Over the centuries, farmers have developed many methods to help the soil maintain its fertility. For example, a farmer might grow corn in a field one year and soybeans the next year. Both crops remove nutrients from the soil, but the plants have different nutrient requirements, and therefore the soil does not lose the same nutrients two years in a row. Soybean plants even add nitrogen compounds to the soil. Alternating two or more crops in the same field from year to year is called **crop rotation**. Sometimes farmers do not grow a crop in a field for a year or two. This allows natural processes to rebuild the field's store of nutrients.

Other farming practices that help maintain soil fertility involve plowing under plant material left in fields. You can do the same thing in a lawn or garden by not removing grass clippings and dead leaves. Farmers often plow crop residues such as stems, roots, and leaves back into the soil. Occasionally farmers grow a crop not for harvest but strictly for plowing back into the field. Such a crop is called a **green manure**. Plowing under crop residues and green manure, as seen in **Figure 22-16**, adds organic matter to the soil. Decomposers in the soil do the rest, turning the residues and green manure into humus.

Fertilizers are applied to enhance soil fertility

A **fertilizer** is a substance that adds nutrients to soil. Farmers and home gardeners use fertilizers to replace the nutrients that are lost when crops, grass clippings, leaves, and other plant materials are harvested or discarded. The label on a fertilizer bag tells you which nutrients it provides. Look at the fertilizer label in **Figure 22-17**. The series of three numbers (5-10-5) indicates the percentage of available nitrogen (N), phosphorus (P), and potassium (K) that is provided by the fertilizer. This series of numbers is called the NPK rating.

Many different types of fertilizers are available. Complete fertilizers provide all of the nutrients that a plant might need. Special-purpose fertilizers are formulated for particular needs, such as stimulation of root growth or acidification of the soil. These fertilizers contain only one or two of the major nutrients. Chemical fertilizers are made from inorganic chemicals. They are sold in both liquid and dry forms. Some chemical fertilizers release their nutrients as soon as they are applied and are thus called quick-release fertilizers. Fertilizers that release their nutrients over a long period of time are called slow-release fertilizers.

Organic fertilizers contain organic material, such as composted manure (animal droppings), cottonseed meal, or sewage sludge. **Composting** involves mixing organic debris (such as grass clippings and table scraps) and allowing it to begin the process of decomposition. Not all organic materials are good fertilizers. Fresh manure, for example, contains only small amounts of usable nutrients (its NPK rating is typically about 1-1-1). Manure that has not been composted even competes with plants for available nitrogen. However, organic fertilizers also improve the water-holding capacity of soils and promote the growth of bacteria that help plants extract nutrients. Organic fertilizers are slow-release fertilizers because bacterial action releases their nutrients.

Plants can be grown without soil

One method of growing plants without soil is called **hydroponics**. This method involves growing plants in a nutrient solution (water plus nutrients). All of the macronutrients and micronutrients that plants obtain from soil must be included in the nutrient solution surrounding the plant roots. Many greenhouse tomatoes are grown using hydroponics.

Figure 22-17
The three large numbers on the label of this fertilizer bag show its NPK rating, which is the percentage of available nitrogen, phosphorus, and potassium in the fertilizer.

Choosing the Right Plants

Different kinds of plants thrive in different conditions: palm trees grow in the tropics, cacti grow in the desert, and spruce trees grow on cool mountain slopes. Some plants grow better in direct sunlight, while others grow better in shade.

Assess the conditions in your area and select suitable plants

Whether you are planting a field of tomatoes or a single plant, you must consider the conditions under which the plants will grow. For example, plants are adapted to certain ranges of temperature. Unless you grow plants in a greenhouse, you cannot protect plants from temperatures they cannot tolerate. Therefore, if you want to grow plants outside, particularly long-lived trees, shrubs, and perennials, you must plant varieties that are hardy (can survive) in your area. The map in **Figure 22-18** divides the United States into 10 *plant hardiness zones*, which are areas that share a similar climate. Seed packets and plant identification tags usually tell you the hardiness zones where a plant will do well. As a rule, you should grow plants that are rated for your zone.

Plants are also sensitive to the soil in which they grow. Some plants require acidic soils, while others require alkaline soils. Some plants thrive in sandy soils, while others thrive in heavy clay. Again, it is best to grow plants that are suited to the soil in your area, but soils can be altered. Using a soil test kit, you can analyze a soil to determine its water-holding capacity and pH and to identify nutrient deficiencies. The soil can then be altered by adding organic matter or specific nutrients.

Consider the size and shape of mature plants

When planning a garden or a landscape, always plan ahead. An evergreen sapling may look great by the front door now, but what about when it gets to be two stories tall? Whether you plan to grow vegetables, flowers, shrubs, or trees, you must take into account what the full-grown plants will look like. Will the vines you want to plant have something to climb on? Will a tree or shrub crowd out or shade other plants when mature? Be particularly careful when planting trees. The limbs and roots of a large tree that is too close to a house can cause structural damage.

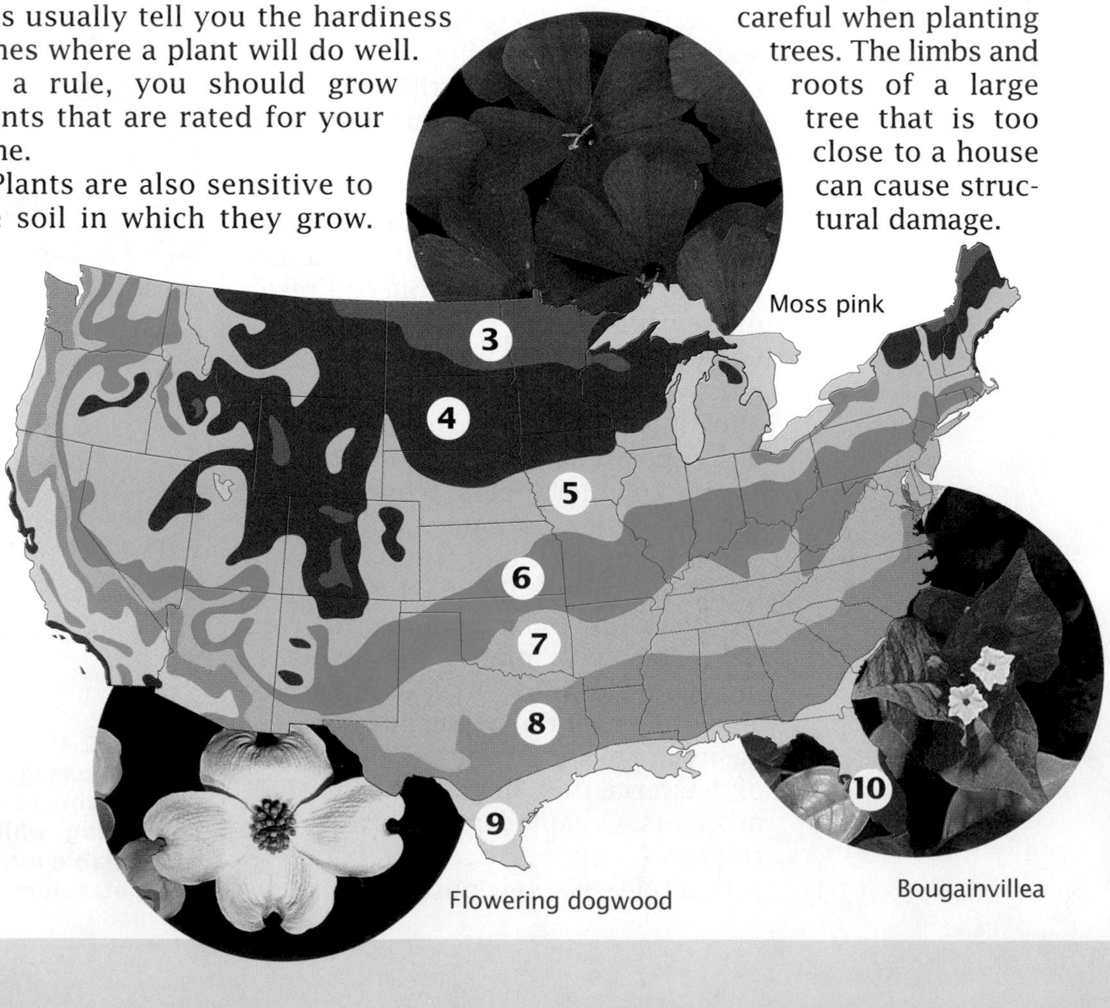

Figure 22-18
A map of plant hardiness zones was prepared by the United States Department of Agriculture (USDA). Moss pinks tolerate cold winters and grow well in zones 3–8. Flowering dogwoods are well adapted to zones 4–8. Bougainvilleas can survive the winters in zones 9 and 10 only.

Average Annual Minimum Temperatures in °C

zone 3	-40 to -30
zone 4	-30 to -20
zone 5	-20 to -10
zone 6	-10 to 0
zone 7	0 to 10
zone 8	10 to 20
zone 9	20 to 30
zone 10	30 to 40

Moss pink

Flowering dogwood

Bougainvillea

Landscape Architecture

What would it be like to . . .

- reclaim, rebuild, and reestablish the natural beauty of neglected, polluted sites in your community?
- plan and design gardens for private and public facilities?

These are just a few of the challenges faced by people working in the field of landscape architecture.

Landscape Architecture: What Is It?

A landscape architect seeks to enhance or restore an environment's natural balance and beauty. Many landscape architects work with architects and urban planners to design buildings, structures, and urban facilities for public use. Landscape architects also play an active role in education and conservation, working with foresters, biologists, and educators to preserve state and national parks, recreational sites, and wilderness areas.

Science/Math Career Preparation

High School
- Art
- Drafting
- Mathematics
- Biology
- Earth science

College
- Drawing and design
- Ecology
- Botany
- Geology

Employment Outlook

The demand for landscape architects will increase dramatically in response to rapid urbanization in all regions of the world.

Career Opportunities

Career Focus: Bob Murase

Landscape Architect in Portland, Oregon

"I would describe myself as an artist and craftsman. I've always been interested in art, especially drawing and sculpting. I had a part-time job helping a local mason build stone walls in San Francisco. This experience not only allowed me to use my imagination and develop my skills as an artist, but also taught me to appreciate the grand and subtle forms of natural beauty that surround us all. Through further, formal studies of the natural sciences, I've developed a keen understanding of the delicate balances in nature."

What I Do: "I went to college to study landscape architecture. Afterward I worked for a firm in San Francisco. When I felt I had enough experience and confidence, I started my own landscape architecture company. That was 15 years ago. Today I have the satisfaction of working with professionals on projects in many different parts of the world. We all have the same goal—to make the environment more pleasant and productive. I'm particularly excited about our work with teachers and students in a local school. We're helping them design real ecological projects that will benefit their communities now and in the future."

To Find Out More

1. For further information, write to:
   American Society of
   Landscape Architects
   4401 Connecticut Avenue NW
   5th Floor
   Washington, DC 20008
2. Visit your school counselor and library to find out which colleges and universities offer programs in landscape architecture. Review the prerequisites and course requirements. Make a poster showing the range of skills involved in this career.
3. Locate resources that contain sketches, photos, or graphics of landscape architecture. Evaluate the architect's design, and prepare a brief critique that explains how you would alter it.

Planting and Caring for Plants

For specific instructions about planting and caring for your plants, ask for advice from someone at the garden store where you purchase the plants. Gardening books and seed packets also contain a wealth of information, such as when to plant, whether to start seeds in pots and then transplant, how plants should be spaced, and how to care for growing plants. Here are some general guidelines you can follow.

Prepare the soil before planting seeds or transplants

Soil preparation begins with plowing, turning, or raking the soil to break up hard clods. These procedures also add air to the soil, or aerate it. Loose soil is easier to dig in, and plant roots need air for healthy growth. You may also need to add nutrients to the soil. A complete fertilizer or a special-purpose fertilizer, such as an acidifying agent, can be applied before planting. Turning under composted organic matter also adds nutrients to the soil and improves its water-holding capacity.

Seeds and transplants should not be planted too deeply. Seeds should be planted no deeper than they are long. Tiny seeds should be barely dusted with soil. Transplants should be planted so that the top of the plant's root ball is at the surface of the soil when it is planted, as seen in **Figure 22-19**.

Pruning and thinning improve the growth of plants

Pruning, or trimming, a plant is more than just giving it a "hair cut" to make its appearance neater. Gardeners prune trees and shrubs to stimulate new growth, to encourage flowering, to remove dead or weak branches, and to shape them. Check with a garden store or a gardening book for advice on how and when to prune specific plants.

When plants are grown from seeds, the seeds are usually planted closer together than mature plants should be spaced to ensure that enough plants sprout. The instructions on a seed packet tell you how far apart the seeds should be planted and how far apart the mature plants should be. After the seeds sprout, thin the seedlings by removing some of them to achieve the proper amount of space between plants. When thinning, leave the stronger plants and remove the weaker ones.

Plants should be watered and fertilized when necessary

Your plants may need regular watering to thrive. As a rule, water plants when they show signs of wilting or when the soil around their roots feels dry 5 to 8 cm (2 to 3 in.) below the surface. A layer of **mulch** (bark chips or other organic matter) helps prevent evaporation of water from the soil. Some plants

Figure 22-19
When transplanting plants grown in pots, like this one, the hole you dig should be no deeper than the plant's root ball.

Table 22-3 Some Nutrient Deficiency Symptoms

Symptoms	Deficient nutrient
all leaves light green; lower leaves yellow	nitrogen (N)
leaves dark green or turning red or purple	phosphorus (P)
small spots of dead tissue at margins and between veins of older leaves	potassium (K)
new leaves of terminal buds hook-shaped; terminal buds die	calcium (Ca)
older leaves mottled, light green, or turning red; leaf margins curled up	magnesium (Mg)
young leaves wilted and light green with light green veins	sulfur (S)
young leaves wilted and light green with dark green veins	iron (Fe)

need applications of specific kinds of fertilizer for healthy growth. But unless you know that a plant has special needs, such as acidic soil, fertilize only when a plant shows signs that it is in need of certain nutrients. Some of these signs are listed in **Table 22-3**.

Watch for warning signs that a plant is unhealthy

Like humans, plants are attacked by viruses, bacteria, fungi, insects, and other parasites. If you notice the shoots of a plant turning yellow, developing spots, wilting, drying, or curling up, this could be an indication of a disease or an insect infestation. Some symptoms of disease and insect problems are shown in **Table 22-4**.

Once a problem has been identified, it can be treated. Plant diseases and insect pests can be controlled through the use of either chemical pesticides or natural techniques. For example, watering in the morning and pruning trees so that lawns and flower beds get more sunlight can help discourage disease-causing fungi. Many "biological controls," such as the one shown in **Figure 22-20**, are also very effective.

Figure 22-20 Ladybugs prey on insects that attack plants in your garden.

Table 22-4 Symptoms of Plant Diseases and Insect Infestations

Symptoms	Possible causes
black, gray, or white fuzzy patches on leaves; rounded target spots	fungi
leaves sticky, developing black patches; misshapen buds; buds that do not open; mottled, yellowed leaves with black specks and webs	sucking insects, mites
waxy bumps or fuzzy lumps	scale insects
young shoots chewed off just above the ground	snails, pill bugs
galls (hard lumps on leaves or swellings of stems)	gall-causing insects or bacteria
holes in leaves	caterpillars, pill bugs
plants wilted or having a blackened, watery area at the base of the stem	soil fungi or bacteria
plants small, misshapen, and yellowed	sucking insects, nematodes
wrinkled or curled leaves; leaves with angular spots or small pits	viruses

Plant Propagation

Plant propagation is the growing of new plants from parent plants. Plants can be propagated by both sexual and asexual means. Growing plants from seeds is a sexual means of propagation because seeds result from sexual reproduction. Asexual propagation methods involve using a variety of vegetative (nonreproductive) plant parts, such as roots, stems, and leaves. Therefore, growing new plants from vegetative parts is called **vegetative propagation**. Several methods of vegetative propagation are shown in **Figure 22-21**.

Underground parts can be used for vegetative propagation

Underground parts of plants that can be used for vegetative propagation include bulbs, corms, rhizomes, roots, and tubers. Bulbs consist of the fleshy bases of leaves, whereas corms, rhizomes, and tubers are modified stems. Bulbs and corms divide naturally as they grow, forming many pieces that can each grow into a new plant. Many rhizomes, roots, and tubers can be cut into pieces with one or more buds that can grow into new shoots. Plants that are typically propagated from underground structures include caladiums, irises, daffodils, lilies, tulips, tuberous begonias, and potatoes.

Aboveground parts can be used for vegetative propagation

Have you ever clipped off a piece of a friend's English ivy and rooted it at home? People are often happy to share cuttings of houseplants such as airplane plants, philodendrons, pothos ivys, and African violets. Cuttings are

Figure 22-21
Plants are vegetatively propagated by many different methods.

Underground Plant Parts

- *Roots*
- *Bulbs*
- *Rhizomes*
- *Tubers*

a These daffodil bulbs have divided into several smaller bulbs, each of which will grow into a new plant.

b Each of the eyes of a potato contains a bud that can develop into a new plant.

leaves or stems that are cut from a plant. Many fruit trees, ornamental trees, and shrubs are grown commercially from stem cuttings. Budding and grafting—which are used extensively to propagate fruit trees, pecan trees, grapevines, and hybrid roses—are specialized techniques that involve attaching the buds or stem cuttings of one plant to the stems of another plant. Tissue culture uses pieces of a plant's stem to produce large numbers of identical offspring. It is particularly important in the propagation of orchids, virus-free potatoes, and genetically engineered plants.

Plants are propagated sexually with seeds

Plants grown from seeds are usually slightly different from their parents because of genetic recombination. Therefore, propagating plants by seeds may not be desirable if a commercial grower or amateur gardener wants to produce large numbers of identical plants. In addition, propagating plants by seeds can be very slow because some plants require two or more growing seasons to produce seeds. Hybrids, which result from crossing genetically different individuals, are usually propagated from seeds.

Aboveground Plant Parts

- *Leaves*
- *Stems*
- *Bulbs*

c Tissue culture, which involves taking a small piece of tissue from a plant and placing it on a nutrient-rich medium, produces large numbers of identical plants.

d This young African violet plant grew from the stalk of the large leaf, a leaf cutting from a mature plant.

e A method of propagation called layering (covering part of a stem with soil) produced the roots on the stem of this dieffenbachia.

Plant Breeding

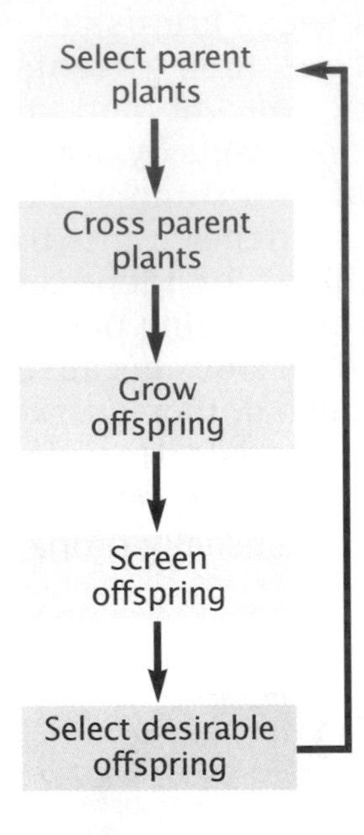

Figure 22-22
a Many steps are involved in traditional plant breeding.

Long before Mendel made his observations of pea plants, people had been observing plants and propagating those with desirable characteristics. As you read in Section 22-1, early farmers collected seeds from the best plants and used them to plant the next crop. By doing this repeatedly, they were practicing artificial selection. The practice of producing new varieties of plants through artificial selection is called **plant breeding**. Plant breeders use their knowledge of genetics and growing plants to produce improved varieties of plants.

Traditional methods of plant breeding, summarized in **Figure 22-22a**, begin with selecting parent plants that have one or more desirable characteristics, such as disease resistance and large fruit. Two desirable parents are then cross-pollinated, and their seeds are grown in a greenhouse or outdoors. Offspring that have the most desirable combination of the parental characteristics are then chosen to be the parents of the next generation. This process is repeated until a variety with all of the desired characteristics is produced.

While traditional plant-breeding methods are still widely used, genetic engineering can accomplish fantastic results in a laboratory in only a few weeks. Many new varieties of plants have been produced through genetic engineering. Some have been genetically engineered to be resistant to insects, diseases, drought, or herbicides. Others have been engineered to grow larger, resist rotting, or taste better. Still others have been engineered to produce proteins or other products that are needed by people with nutritional deficiencies. Genetic engineers have even produced a strain of cotton that has polyester-like qualities. Imagine a 100 percent cotton shirt that you would never have to iron!

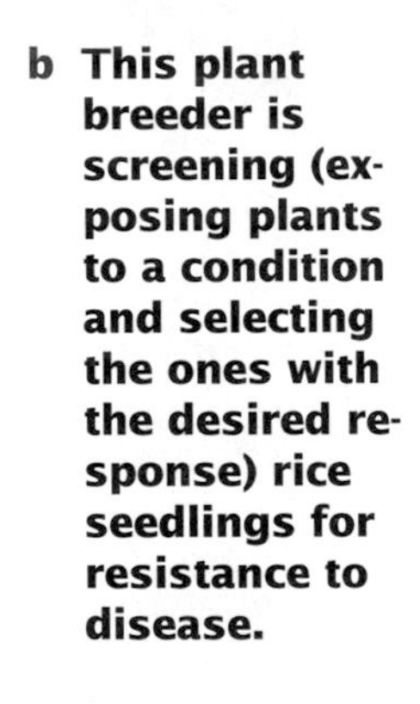

b This plant breeder is screening (exposing plants to a condition and selecting the ones with the desired response) rice seedlings for resistance to disease.

SECTION REVIEW

1. **How is soil important to plants?**
2. **What may be the consequences of trying to grow plants in areas they are poorly adapted to?**
3. **How does a change in appearance indicate the health of a plant?**
4. **How can plants be propagated using vegetative parts from their shoots?**
5. **How are the results of traditional plant breeding different from the results of genetic engineering?**

CHAPTER

22 Highlights

Wheat, rice, and corn are eaten by people all over the world.

	Key Terms	Summary
22-1 Uses of Plants Many important products are made from wood.	grain (p. 411) essential amino acid (p. 415) legume (p. 415) tuber (p. 416) wood (p. 417) lumber (p. 417)	• The most important foods are fruits called grains. • Legumes are high in protein and are often eaten with grains. • Plant parts that provide food include fruits, seeds, stems, roots, leaves, and flowers. • Sugar is produced from sugar cane and sugar beets. • Wood is a valuable source of lumber, fuel, paper, and rayon. • Many familiar drugs, such as aspirin and hydrocortisone, were first isolated from plants. • Important plant products include fibers, turpentine, and rubber.
22-2 Growing Plants People grow plants for a variety of purposes.	macronutrient (p. 420) micronutrient (p. 420) loam (p. 421) humus (p. 421) fertility (p. 422) crop rotation (p. 422) green manure (p. 422) fertilizer (p. 423) composting (p. 423) hydroponics (p. 423) mulch (p. 426) vegetative propagation (p. 428) plant breeding (p. 430)	• Plants need relatively large amounts of macronutrients and relatively small amounts of micronutrients. • The best soils for growing plants are loams, which are a mixture of sand, silt, and clay particles. • Humus provides nutrients and helps soil hold moisture. • Farmers use fertilizers, crop rotation, and green manure to enhance soil fertility. • Knowledge of the local conditions and the needs of plants help people grow plants successfully. • Many plants are produced by vegetative propagation. • New plant varieties are produced by traditional plant breeding and by genetic engineering.

CHAPTER 22

review

Understanding Vocabulary

1. For each set of terms, choose the term that does not fit the pattern, and explain why.
   a. corn, wheat, soy beans, rice
   b. cotton, linen, wood, latex
   c. iron, nitrogen, phosphorus, sulfur

2. For each set of terms, complete the analogy.
   a. carrots and turnips : roots :: yams and potatoes : ____
   b. quinine : prevents malaria :: aspirin : ____
   c. seeds : sexual propagation :: bulbs : ____

Understanding Concepts

3. **Relating Concepts** Construct a concept map that identifies the products people use from plants. Try to include the following terms in your concept map: food, medicines, products, grains, vegetables, digitalis, quinine, paper, fiber, wheat, rice, broccoli, foxglove, cinchona tree, and clothing.

4. **Summarizing Information** What are grains? Which grains provide about 60 percent of the calories eaten by humans? What nutrients do grains provide?

5. **Identifying Structures** Name the part of a plant represented by each of the following foods.

6. **Inferring Relationships** Why is a diet consisting of grains and legumes more nutritious than a diet consisting of grains alone?

7. **Summarizing Information** What are tubers? What two important foods are tubers? What are examples of other foods that are roots?

8. **Summarizing Information** Describe three primary uses of wood.

9. **Describing Processes** Describe the process of making paper.

10. **Summarizing Information** Identify three medicines derived originally from plants. Name the plant that the medicine comes from and the uses of the medicine.

11. **Understanding Functions** What is the function of soil in relation to plant growth?

12. **Inferring Relationships** Why do plants grow better in soil than they do in plain water?

13. **Summarizing Information** What is soil fertility? Describe three cultivation practices that help maintain soil fertility.

14. **Analyzing Methods** Explain why it is advantageous for gardeners and farmers to select and grow plants that are appropriate for their area.

15. **Interpreting Data** After carefully planning and planting your garden, you notice that some plants have waxy bumps, others have black and white patches on their leaves, and still others are small and misshapen with yellow leaves. What are the probable causes of these ailments?

16. **Summarizing Information** What is vegetative propagation? Describe five methods of propagating plants vegetatively.

17. **Recognizing Relationships** What is plant breeding? How do traditional methods of plant breeding compare with plant breeding by genetic engineering?

18. **BUILDING ON WHAT YOU HAVE LEARNED** In Chapter 16 you learned about several environmental problems. What environmental problems are associated with the burning of wood for cooking and heating?

Interpreting Graphics

19. Study the graph shown below.

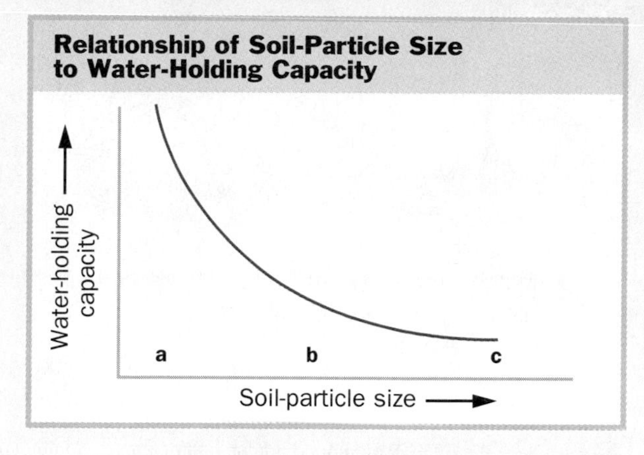

a. What kind of soil is indicated by letter *a*? What is the relationship between this soil's ability to hold water and the size of its particles?
b. What kind of soil is indicated by letter *b*? What is the relationship between this soil's ability to hold water and the size of its particles?
c. What kind of soil is indicated by letter *c*? What is the relationship between this soil's ability to hold water and the size of its particles?

Reviewing Themes

20. ***Heredity***
Modern wheat looks very different from its wild ancestors. What methods have been used to develop the varieties of wheat grown today?

21. ***Stability and Homeostasis***
Describe several ways that farmers and agricultural scientists are striving to maintain adequate food supplies for growing populations.

Thinking Critically

22. **Forming Opinions** Assume that genetic engineers have been able to develop a cereal grass that produces nutritious grains and is also able to fix nitrogen from the soil in much the same way that a legume can. Of what advantage might the nitrogen-fixing characteristic be to a farmer?

23. **Evaluating Results** A gardener planted pole beans (beans that grow on tall, strong vines) too close to corn plants. As both plants grew, the beans intertwined around the corn stalks. The gardener also noticed that the corn plants yielded exceptionally large and numerous ears of corn. How would you explain these results?

Life/Work Skills

24. **Allocating Resources** For three days, record every paper product you use during the day. Include items such as newspaper, paper towels, notebook paper, and product packaging. Can any of these items be recycled? What is the advantage of recycling paper? Identify each product you list as recyclable or nonrecyclable. Write out several recommendations for reducing the amount of paper you use, and share these recommendations with your class.

25. **Finding Information** Use library references or search an on-line database to find out about Xeriscaping. In what parts of the country is this technique most appropriately used? What kinds of plants can be used in Xeriscape gardening? Prepare an illustrated report or informational posters to share with your class.

Activities and Projects

26. **History Connection** Use library references or search an on-line database to find out the ritual and economic importance of corn (maize) to early Mesoamerican cultures such as the Aztecs of central Mexico. How was the corn grown? What gods and rituals were associated with growing corn? How was corn used in the diet? Summarize your findings in an illustrated report to share with your class.

27. **Unit Focus** Use library references or search an on-line database to find out about some of the new varieties of plants that have been produced through genetic engineering. How will these varieties help reduce the amount of environmental pollution that results from agriculture? How might genetically engineered plants change methods of food preservation? Produce a poster that includes several genetically engineered plants and illustrates how these varieties might improve environmental and human health.

Discovering Through Reading

28. Read "A Farming Revolution: Sustainable Agriculture" by Verlyn Klinkenborg in *National Geographic*, December 1995, pages 60–89. What is sustainable agriculture? What do plants such as vetch, peas, and clover provide to farmers? What is a soil's tilth?

UNIT

Five

Animal Kingdom

Herd of reindeer in Russia

PLANS FOR SURVIVAL

Animal body plans first appeared millions of years ago and changed over time to ensure survival. About a million species of animals have since evolved. This wide diversity of animal life is the result of animal adaptations to a changing and challenging environment.

Human Right or Responsibility

The richness of diversity, however, is being threatened by humans striving to provide food, clothing, housing, and medicines for growing populations. Some industries remove animals from an environment. Other actions lead to habitat destruction, endangering or eliminating some species.

You Have a Say

What is lost when an endangered species, such as the spotted owl, becomes extinct? Some biologists believe that biological diversity is vital to long-term survival on Earth. Because of animal and plant diversity, ecosystems are able to recover after change occurs. Will your needs for food, clothing, shelter, and medicines allow ecosystem recovery or threaten it?

LOOKING AHEAD

- In which ways have animal body plans varied? See pages 439–454.
- How have mammals broken the tie to reproduction in water? See pages 472–474.
- What information can be gained by studying an indicator species? See **Science, Technology, and Society: Can One Species Tell All?** pages 498–499.
- How can the study of insects help humans? See **Career Opportunities: Entomology,** page 511.

CHAPTER 23

REVIEW
- natural selection (Section 10-3)
- animal evolution (Sections 11-3 and 11-4)
- classifying living things (Section 17-2)

The Animal Body

23-1	The Origin of Tissues
23-2	Origin of Body Cavities
23-3	Four Innovations in Body Plan

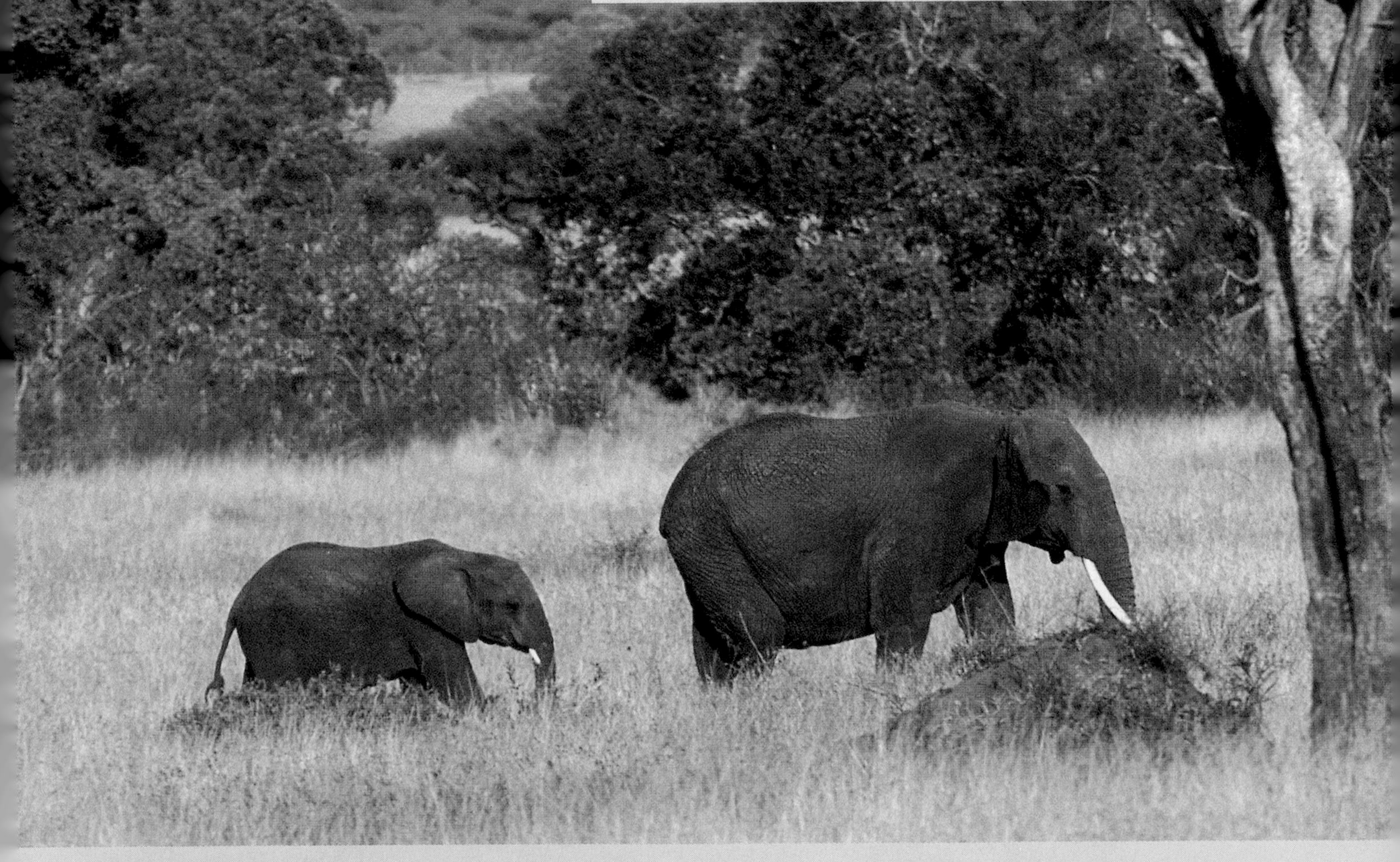

An elephant and her calf cross the Kenyan savanna. This chapter examines some of the milestones in animal evolution.

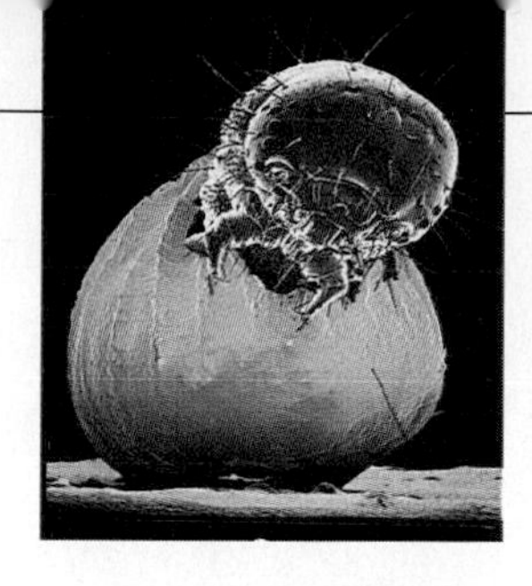

Animals, including the tobacco budworm shown at the left, occur in a great variety of sizes and shapes. This chapter describes the evolution of today's great diversity of animals in terms of a series of key adaptations in body architecture. These adaptations reveal both the cumulative nature of evolution and the importance of body design.

23-1 The Origin of Tissues

OBJECTIVES

1. **List three characteristics of animals.**
2. **Recognize the advantage of multicellularity.**
3. **Identify the difference between tissues and specialized cells.**
4. **List three features found in cnidarians but not in sponges.**

Animal Body Plans

What do the animals in **Figure 23-1** have in common? These animals, and all the other members of the kingdom Animalia, share three characteristics. First, all animals are heterotrophs that consume their food. Second, all animals are multicellular. Third, as you learned in Chapter 3, animal cells do not have cell walls.

Despite these similarities, it is obvious that the animals below differ in shape and structure. The **body plan** of an animal is its overall structure. You learned in Chapter 17 that animals evolved from heterotrophic protists. The first animals probably evolved from multicellular colonies of protists. In this chapter, you will see some of

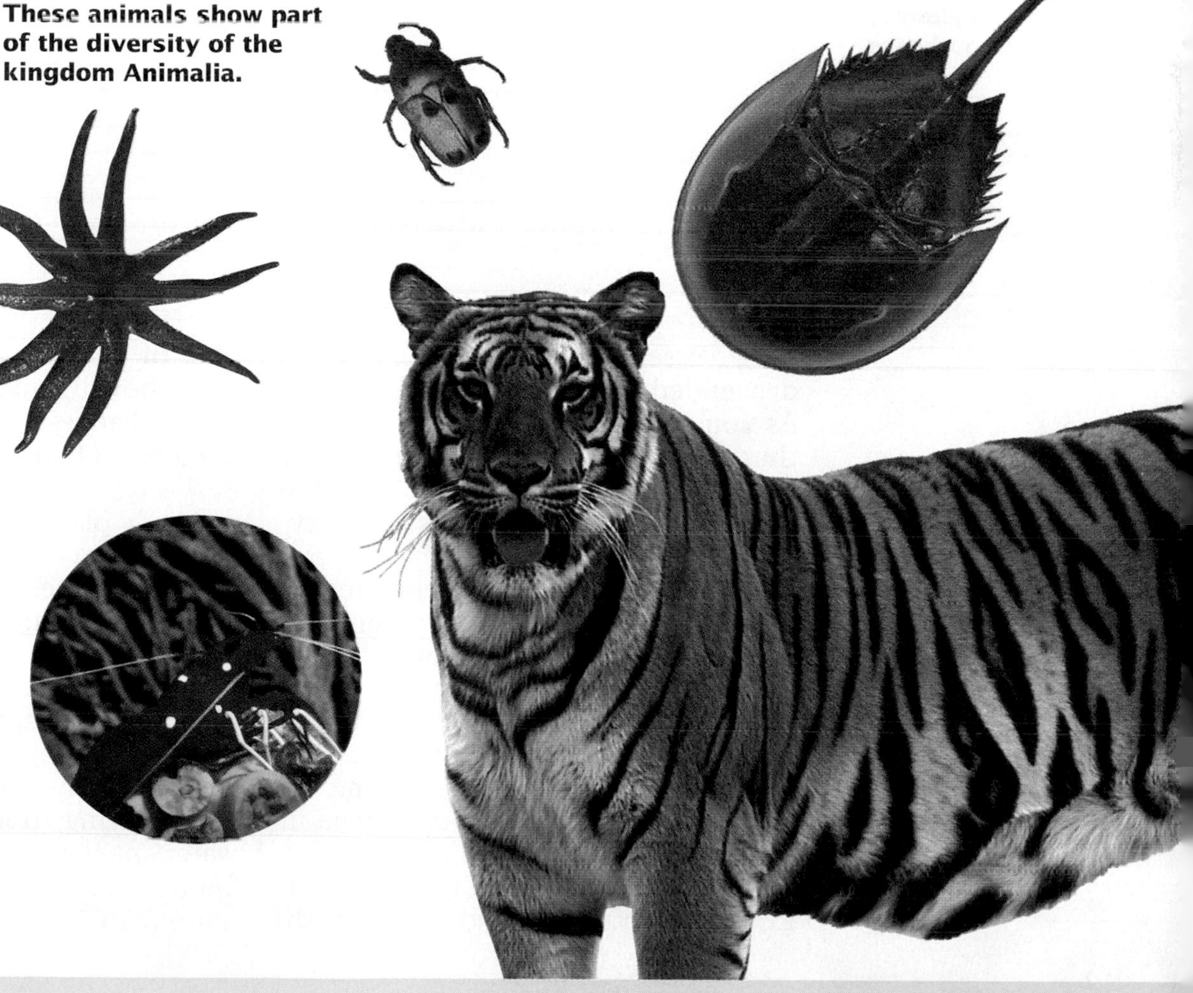

Figure 23-1
These animals show part of the diversity of the kingdom Animalia.

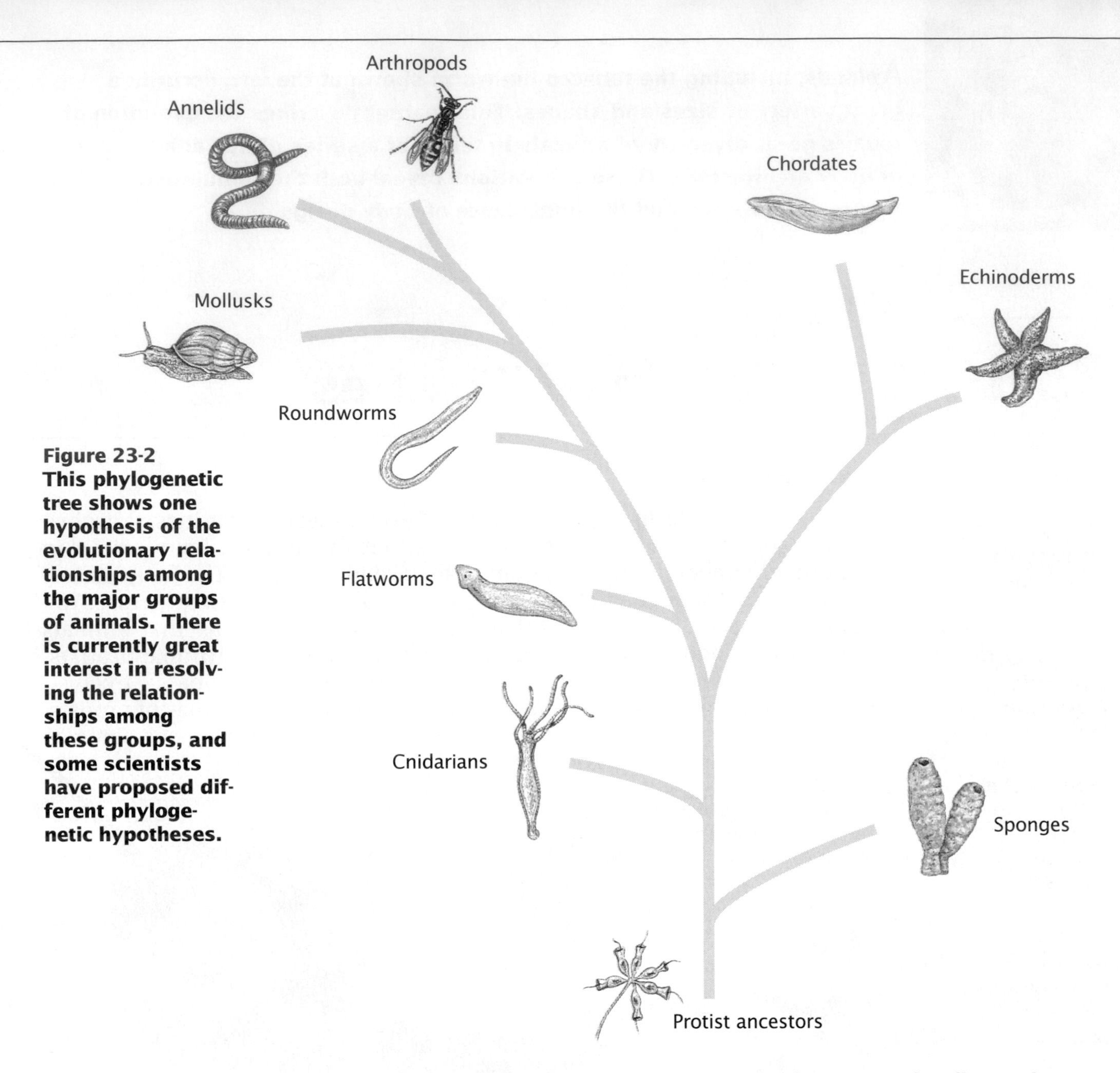

Figure 23-2
This phylogenetic tree shows one hypothesis of the evolutionary relationships among the major groups of animals. There is currently great interest in resolving the relationships among these groups, and some scientists have proposed different phylogenetic hypotheses.

the diverse animal body plans that are descended from this protist ancestor. As you read the chapter, keep in mind that each body plan is the result of many additions and alterations to the body plans of earlier animals. The driving force for this process is natural selection, which favors changes in body plan that increase the likelihood of survival and reproduction.

This chapter traces the major evolutionary changes in animal body plans. The phylogenetic tree in **Figure 23-2** illustrates the relationships among the major groups of animals discussed in this chapter. Each group represents a unique body plan and is profiled in an **Exploration**. Keep in mind that the body plan of an animal reflects the way it lives, how it functions in its environment. Changes in body plan often result in different body functions and entirely new ways of making a living. Each animal body plan is successful for a certain lifestyle. You might hear simple animals such as sponges described as "lower" or "primitive" animals, implying that the are somehow inferior to "higher" animals such as humans. This is not the case. Although sponges cannot fly, run, or think, they are well adapted for their particular way of life—clinging to rocks and filtering food particles from the water passing through their body. Sponges have been living this way for more than 500 million years.

Many Cells Instead of One: Sponges

As you learned in Chapter 11, animals belong to one of the three multicellular kingdoms of life. The most important advantage to multicellularity is the specialization of cells. Individual cells of an animal can specialize on a single task such as digestion or reproduction. Cell specialization enables division of labor among cells in an animal's body. Division of labor is advantageous because a specialized cell can carry out its task more effectively than a cell that must carry out many tasks.

Figure 23-3
Sponges are found in fresh water and salt water. All sponges, such as these tube sponges, have a simple body plan composed of specialized cells.

Sponges are simple, multicellular animals

Sponges, like the one shown in **Figure 23-3**, are members of the phylum Porifera. They demonstrate the advantages of cell specialization. A sponge extracts food and oxygen from the water flowing through its body. The **Exploration** below shows the simple body plan of a representative sponge. The outer layer of cells protects the sponge. Cells called choanocytes *(koh AN oh seyets)* line the internal chamber and produce water currents by beating their flagella. Choanocytes engulf and digest small organisms and small organic particles that are drawn into the sponge.

Although each of the sponge's different cell types is specialized to carry out one or more of the different functions needed for survival, there is very little coordination among cells. The flagella of choanocytes beat independently, for instance. Some sponges can even reassemble into a complete sponge after being forced through a fine silk mesh and separated into individual cells.

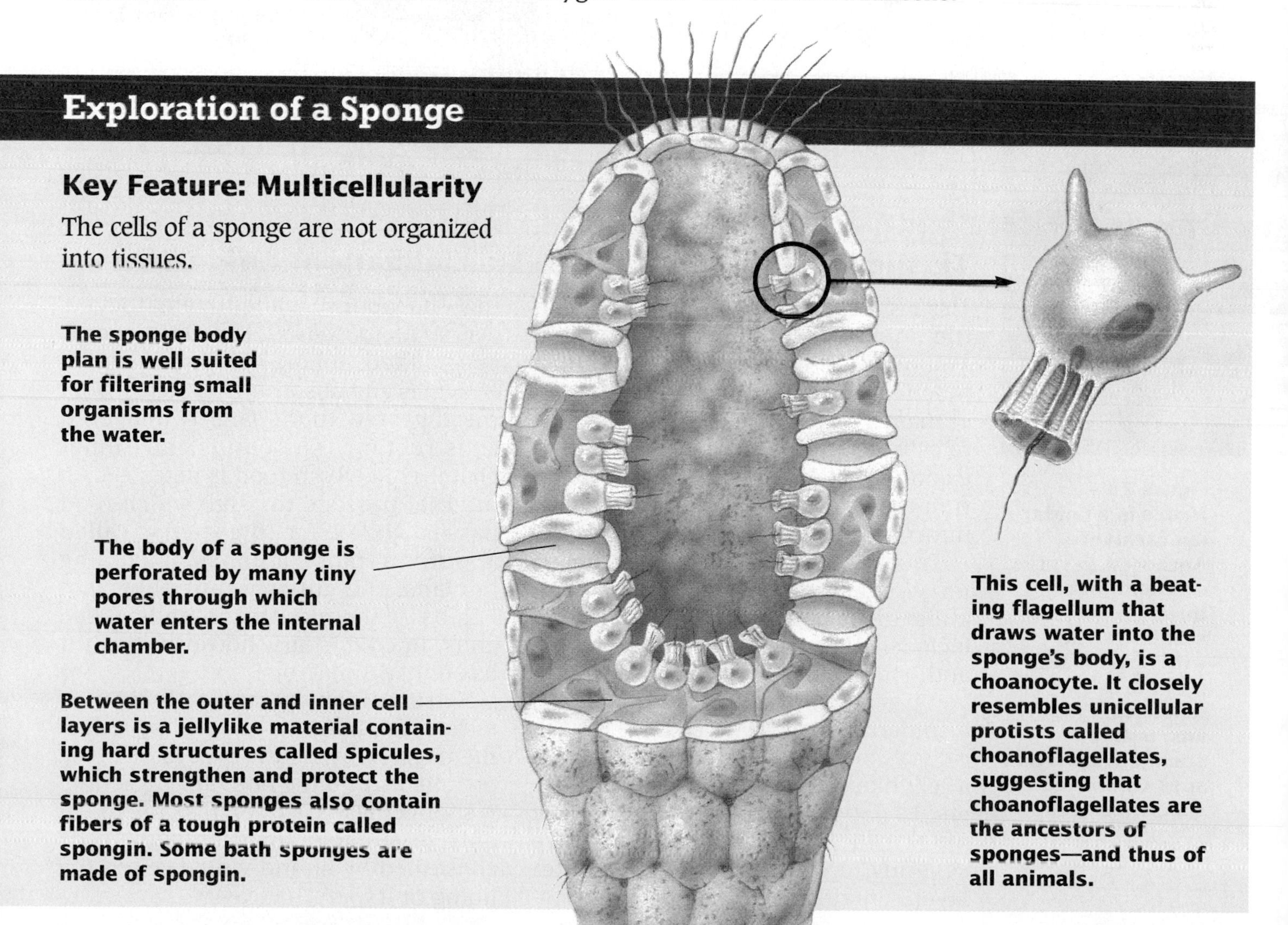

Exploration of a Sponge
Key Feature: Multicellularity
The cells of a sponge are not organized into tissues.
The sponge body plan is well suited for filtering small organisms from the water.
The body of a sponge is perforated by many tiny pores through which water enters the internal chamber.
Between the outer and inner cell layers is a jellylike material containing hard structures called spicules, which strengthen and protect the sponge. Most sponges also contain fibers of a tough protein called spongin. Some bath sponges are made of spongin.
This cell, with a beating flagellum that draws water into the sponge's body, is a choanocyte. It closely resembles unicellular protists called choanoflagellates, suggesting that choanoflagellates are the ancestors of sponges—and thus of all animals.

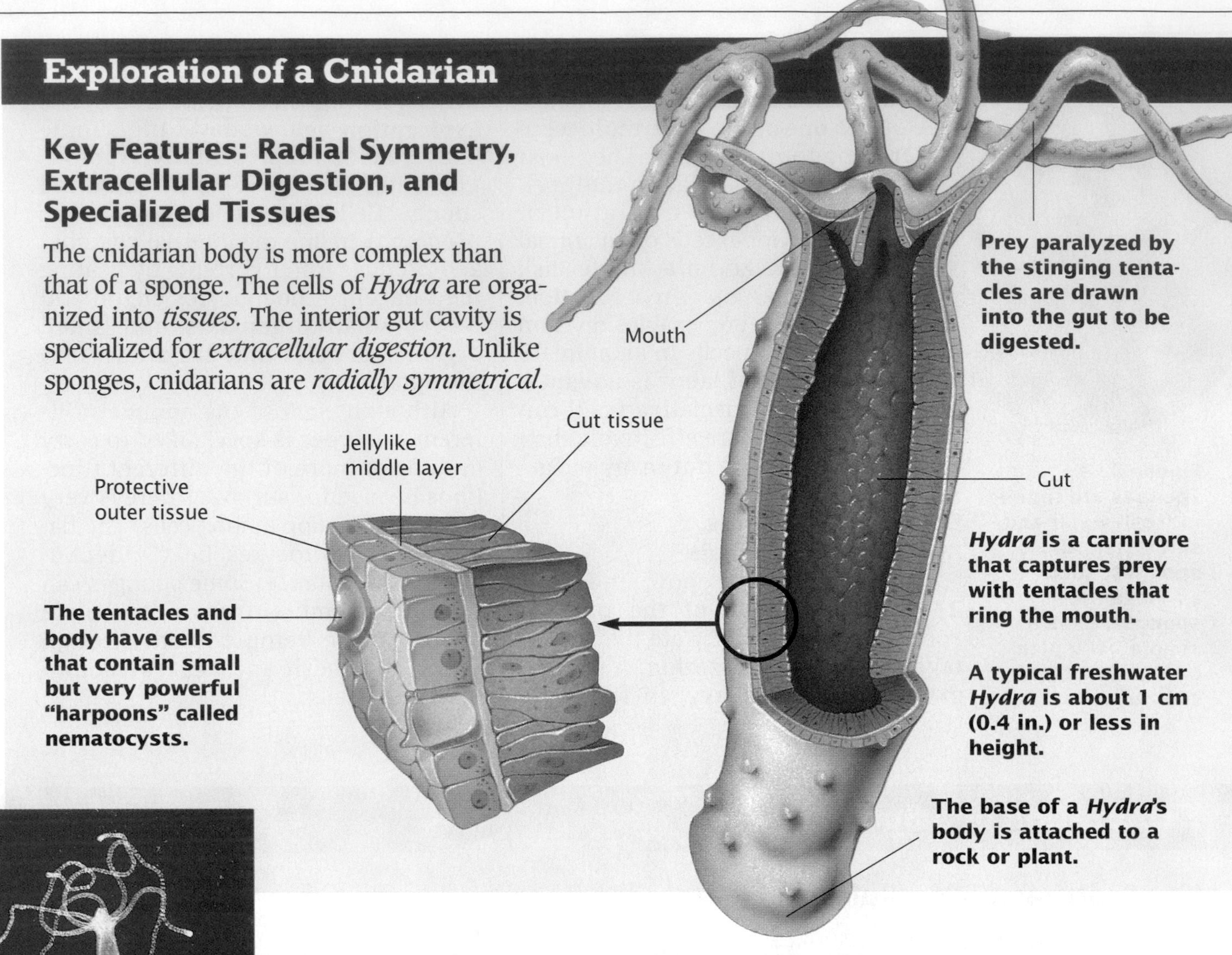

Exploration of a Cnidarian
Key Features: Radial Symmetry, Extracellular Digestion, and Specialized Tissues
The cnidarian body is more complex than that of a sponge. The cells of Hydra are organized into tissues. The interior gut cavity is specialized for extracellular digestion. Unlike sponges, cnidarians are radially symmetrical.
Mouth
Prey paralyzed by the stinging tentacles are drawn into the gut to be digested.
Gut tissue
Jellylike middle layer
Protective outer tissue
Gut
Hydra is a carnivore that captures prey with tentacles that ring the mouth.
The tentacles and body have cells that contain small but very powerful "harpoons" called nematocysts.
A typical freshwater Hydra is about 1 cm (0.4 in.) or less in height.
The base of a Hydra's body is attached to a rock or plant.

Figure 23-4
***Hydra* is a cnidarian carnivore. Anchored to rocks or plants, it captures its prey by stinging them with specialized structures in its tentacles. Cnidarians, unlike sponges, have tissues composed of specialized cells.**

Tissues Enable Greater Cell Specialization: *Cnidarians*

Tissues are the first great innovation in the animal body plan. A **tissue** is a group of cells that are organized into a functional unit. Your body is composed of many types of tissues. The activities of cells in a tissue are coordinated. For example, cells in nervous tissue function together to collect and transmit information.

The phylum Cnidaria *(neye DAIR ee uh)* was one of the first groups of animals to have tissues. This phylum includes jellyfish, corals, sea anemones, and hydras, shown in **Figure 23-4**. As you can see in the **Exploration** above, a cnidarian *(neye DAIR ee uhn)* is basically a double-layered bag of cells with a jellylike substance between the layers. Each layer of cells is a tissue. The outer cell layer forms a protective covering for the cnidarian. This outer layer contains cells armed with stinging "harpoons" called nematocysts *(NEHM uh toh sihsts)*. The inner layer is specialized for releasing digestive enzymes and absorbing nutrients from the digested food. This two-layered construction surrounds an inner chamber in which food is digested. An internal passage through which food passes while being digested is called a **gut**. In most animals, including humans, the gut is a long tube that begins at the mouth and ends at the anus. In cnidarians, however, the gut has only a single opening.

Extracellular digestion allows animals to consume organisms larger than themselves by taking pieces of those organisms into the body for digestion. A sponge has no gut and is able to eat only organisms small enough to be consumed by an individual cell in the lining of its internal cavity.

Figure 23-5
If you were to push on a tennis ball with your thumb, it would simulate what happens during the formation of the endoderm and ectoderm.

Tissue layers form early in the development of the embryo

Figure 23-5 shows how the two-layered construction of a cnidarian is formed during the development of the embryo. This process of forming layers of cells is called **gastrulation** *(gas troo LAY shuhn)*. Gastrulation takes place in the developing embryos of all animals except sponges. In humans, gastrulation occurs when the embryo is a little over one week old. The inner layer of cells formed by gastrulation is **endoderm**, meaning "inner skin." The outer layer of cells is **ectoderm**, meaning "outer skin."

The layers that result from gastrulation will produce all the tissues of the adult body. These layers develop into the same body structures in all animals. Endoderm always gives rise to the lining of the gut, and ectoderm always becomes the outer layer of skin and nervous tissue.

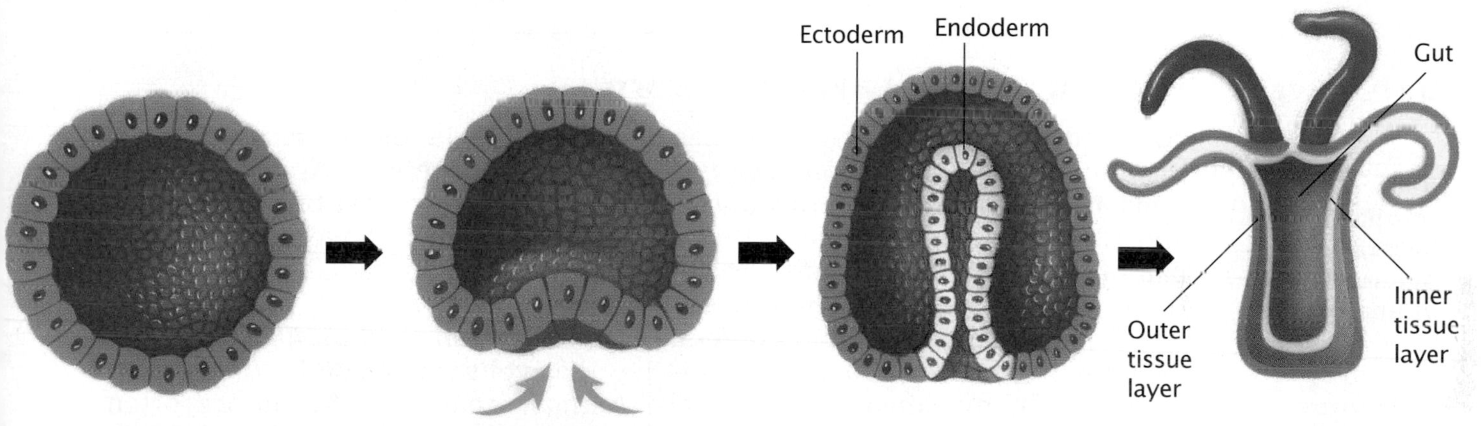

a **A hollow ball of cells forms from continuous divisions of a single fertilized egg.**

b **Gastrulation begins when cell divisions force one surface inward.**

c **The inner layer of cells is the endoderm; the outer layer is the ectoderm.**

d **In *Hydra*, ectoderm forms the outer layer and nervous tissue, and endoderm forms the tissue that lines the gut.**

Regularly Arranged Animals

Compare the shape of the cnidarian shown in the **Exploration** with the shape of the sponge shown in **Figure 23-3**. Most sponges are asymmetrical; they lack a regular arrangement of body parts. In contrast, cnidarians are symmetrical. A hydra's tentacles, for instance, seem to radiate from the body axis (an imaginary line through the center of the body) like spokes radiate from the hub of a wheel.

A test for symmetry is to imagine slicing the body along its axis. If such a slice produces approximate mirror-image body halves, then the animal is symmetrical. **Radial symmetry** is the wheel-like symmetry of hydras, sea anemones, jellyfish and their relatives. You will soon see that radial symmetry is very different from the type of symmetry shown by most other animals, including humans.

SECTION REVIEW

1. **Explain why a protist such as *Amoeba* is not considered an animal, but a sponge is.**
2. **Why can a multicellular organism be more efficient than a unicellular one?**
3. **Why are the choanocytes of a sponge not considered a tissue?**
4. **List three differences between a sponge and a hydra.**

Although cnidarians have a few characteristics of animals you are familiar with, they probably still seem quite unusual to you. How many radially symmetrical animals have you seen today, for instance? Changes in symmetry and other aspects of the body plan eventually resulted in the more complicated animals that you are familiar with.

23-2 Origin of Body Cavities

OBJECTIVES

1. **Contrast the body plans of bilaterally and radially symmetrical animals.**
2. **List three structures derived from mesoderm.**
3. **Describe the advantage of a one-way gut.**
4. **Contrast the three kinds of body cavities.**

A Third Tissue Layer: Flatworms

Your body is neither radially symmetrical nor asymmetrical. You have two arms, two legs, two eyes, two kidneys, and two lungs—one member of each pair on the right side of your body and the other member of the pair on the left. The human body shows **bilateral symmetry**. (The word *bilateral* means "two sides.") Any animal with bilateral symmetry can be separated into nearly mirror-image halves by drawing an imaginary line lengthwise down the middle of the body, as shown in the flatworm in **Figure 23-6**. Like humans, most of the known species of animals are bilaterally symmetrical.

Look again at the hydra in the **Exploration** on page 440. Are you able to locate its head? Most bilaterally symmetrical animals have definite head and tail ends, but radially symmetrical animals such as cnidarians do not. **Cephalization** *(sehf uh lih ZAY shuhn)* is the evolution of a definite head end. Animals that have heads are often active and mobile, moving through their environment headfirst. It is advantageous for sensory organs to be concentrated in the head so that an animal can test for food, danger, hiding places, and mates as it enters new surroundings.

Figure 23-6
All flatworms are bilaterally symmetrical. If you were to cut this flatworm along the dotted line, the result would be two halves that are almost mirror images.

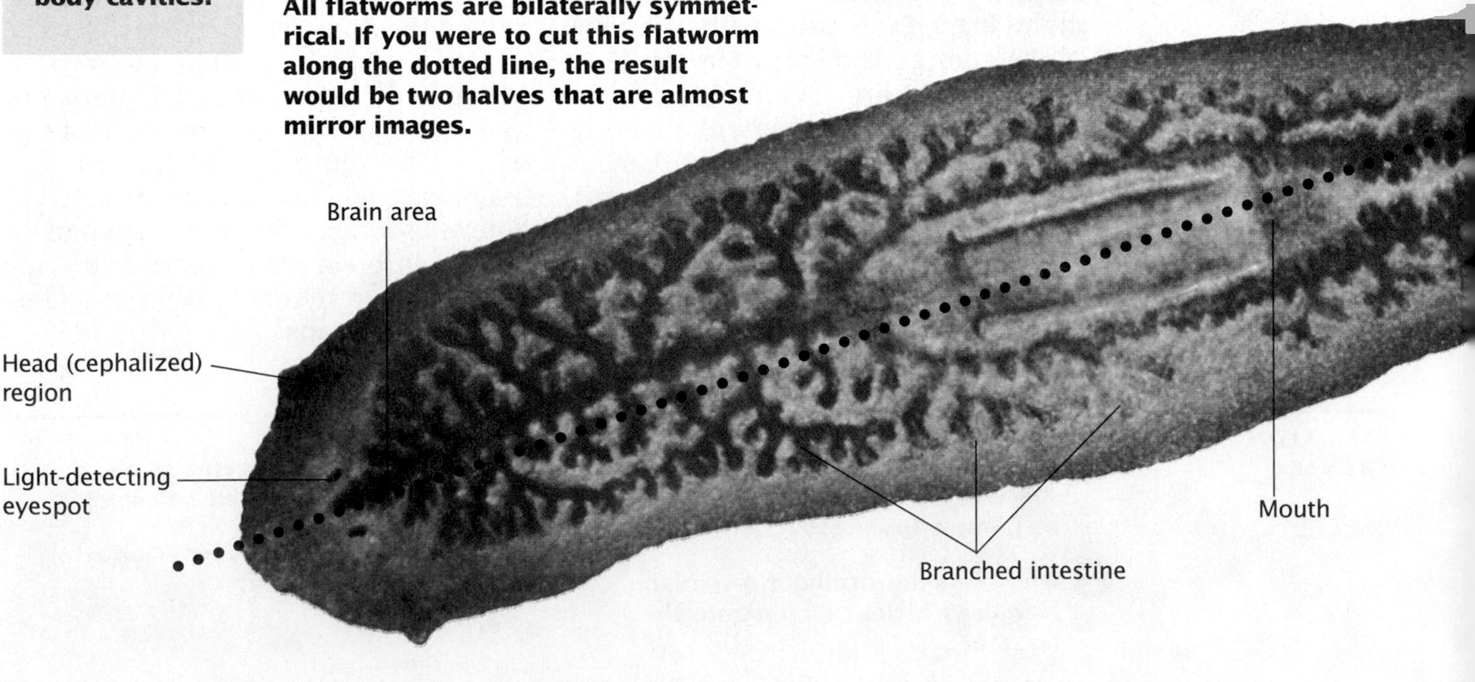

Interpreting and evaluating the information obtained by sensory organs requires a complex processing structure. Cephalization is usually accompanied by the evolution of a collection of nervous tissue at the front end of the animal—a brain or brainlike structure. A brain located near the sensory organs is advantageous because information can rapidly cross the short distance between brain and receptor. This enables an animal to respond quickly to stimulation.

Flatworms are bilaterally symmetrical and cephalized

Flatworms, members of the phylum Platyhelminthes *(plat ih hehl MIHN theez)*, are bilaterally symmetrical and cephalized. The dark spots on the head of the flatworm in **Figure 23-6** are eyespots that can detect light, but they cannot focus an image like your eyes can.

Flatworms have organs

Flatworms also have organs. **Organs** are collections of different kinds of tissue that are dedicated to one function. For instance, the heart of a vertebrate is made up of muscle tissue, connective tissue, and nervous tissue, all of which function together to pump blood. Organs are usually found as units in larger systems known as organ systems. An **organ system** is a group of interrelated organs that carries out one or a few essential body functions. A digestive system, for instance, breaks down food and absorbs nutrients, each organ playing a role in these processes.

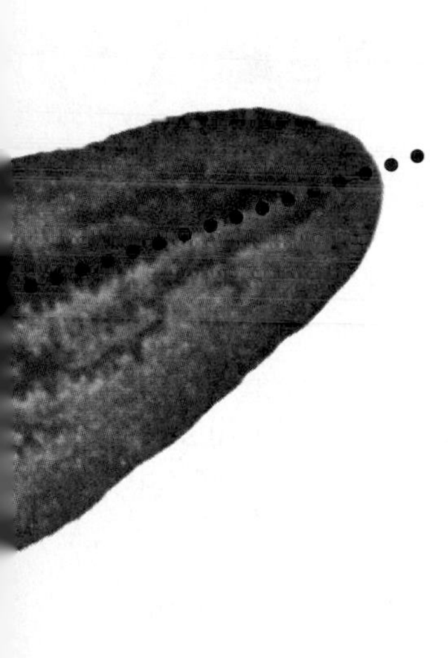

Where do organs come from during development? During gastrulation, endoderm gives rise to the gut lining and ectoderm gives rise to nervous tissue and skin. A flatworm's body is more than these two simple kinds of tissue. In flatworms and all other major groups of animals that you will study in the rest of this chapter, a third tissue layer forms between the ectoderm and endoderm during gastrulation, as illustrated in **Figure 23-7**. This third layer is known as **mesoderm**, meaning "middle skin." Mesoderm develops into muscle, reproductive organs, and circulatory vessels.

Figure 23-7
In flatworms, the process of gastrulation gives rise to three types of tissue layers.

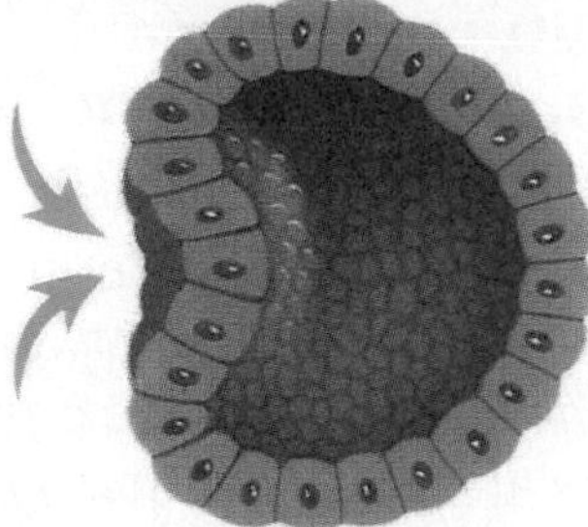

a After gastrulation begins, . . .

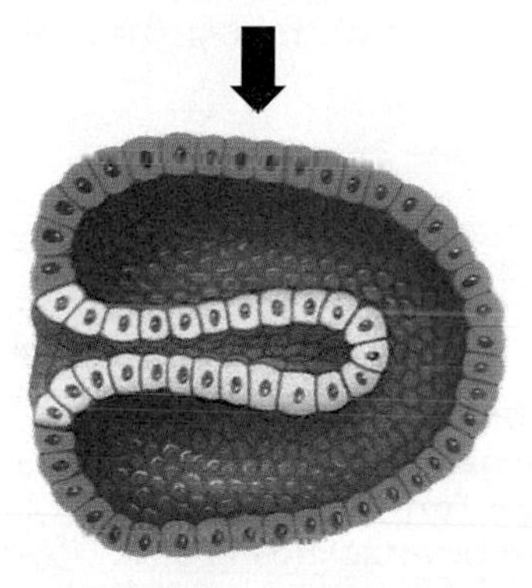

b . . . the endoderm and ectoderm form. Endoderm (yellow) gives rise to gut tissue. Ectoderm (blue) gives rise to nervous tissue and skin.

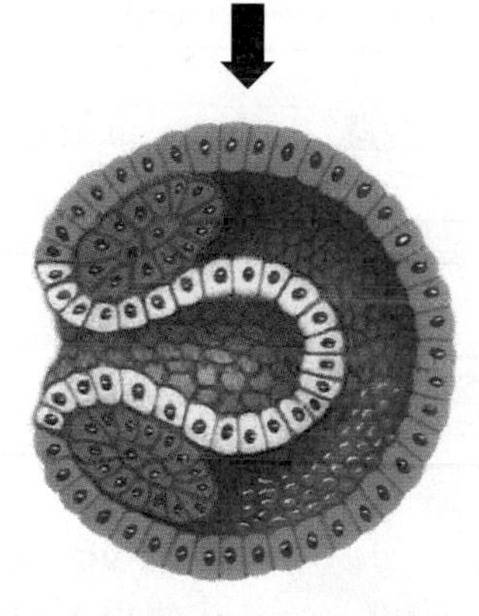

c The mesoderm forms in the space between the endoderm and the ectoderm. Mesoderm (red) gives rise to muscle, reproductive organs, and circulatory vessels.

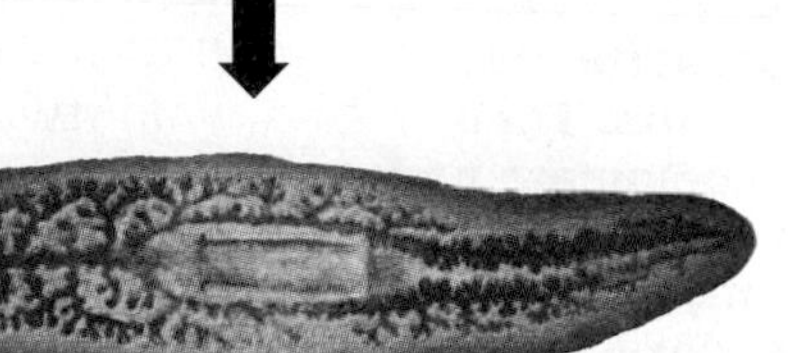

d The embryo eventually develops into an adult flatworm.

Exploration of a Flatworm

Key Features: Internal Organs, Bilateral Symmetry, and Cephalization

The evolution of the mesoderm allowed the formation of *organs.* Flatworms, such as the liver fluke, have organs. Flatworms are *bilaterally symmetrical* and have a distinct *head* end. The body plan of the liver fluke is composed of solid layers of tissues surrounding a branched gut.

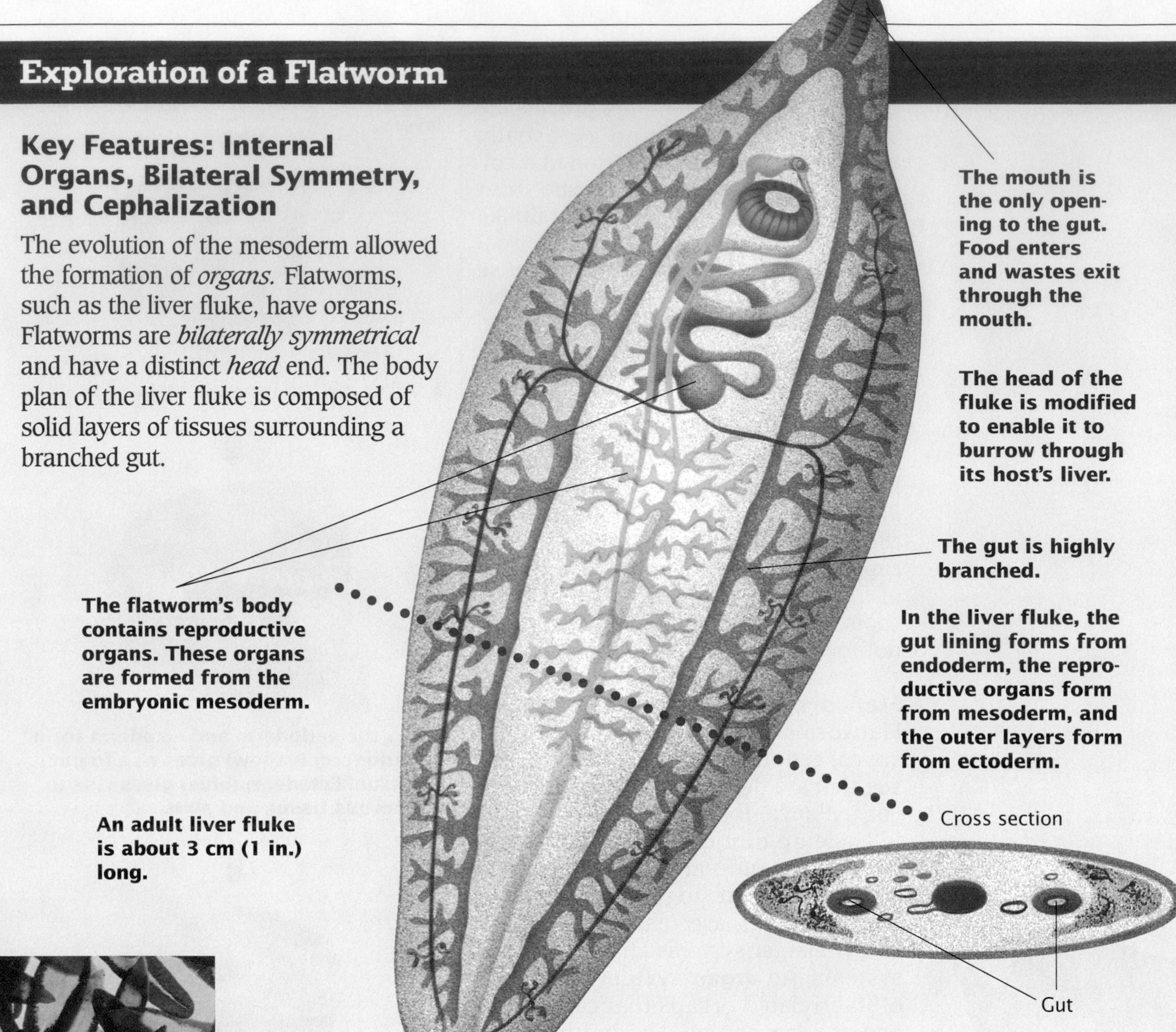

The mouth is the only opening to the gut. Food enters and wastes exit through the mouth.
The head of the fluke is modified to enable it to burrow through its host's liver.
The gut is highly branched.
In the liver fluke, the gut lining forms from endoderm, the reproductive organs form from mesoderm, and the outer layers form from ectoderm.
The flatworm's body contains reproductive organs. These organs are formed from the embryonic mesoderm.
An adult liver fluke is about 3 cm (1 in.) long.
Cross section
Gut

Figure 23-8
Liver flukes are parasitic flatworms. Their acoelomate body plan requires that they be thin in order to allow substances to pass easily to all organs.

Flatworms are solid worms

If you were to cut a flatworm in half across its body, as shown in the **Exploration** above, you would see that the gut is completely surrounded by tissue and organs. This solid body construction is termed **acoelomate** *(ay SEEL oh mayt)*, meaning "without a body cavity."

Flatworms are thin because of their acoelomate body construction. Dissolved substances such as carbon dioxide, oxygen, and nutrients cannot diffuse rapidly through the solid bodies of flatworms, such as the liver flukes shown in **Figure 23-8**. Flatworms are small or thin (or both), which shortens the distance these substances must move. In addition, the gut of a flatworm is often branched so that it runs close to most of the tissues.

The guts of all flatworms have only one opening, the mouth. Because these animals consume food and eliminate wastes through the same opening, two-way movement of material occurs within the gut. The two-way gut is less efficient at extracting nutrients than is a one-way gut such as yours. If animals with an acoelomate body plan eat when food is already in the gut, newly consumed food can mix with partially digested food and wastes.

A One-Way Gut and a Body Cavity: Roundworms

Roundworms (phylum Nematoda), or nematodes, such as *Caenorhabditis elegans* in **Figure 23-9**, have a cavity within the body and a one-way gut with two openings. Food is taken in through the mouth, and wastes are eliminated through an opening at the other end of the gut, the **anus**. This arrangement allows a one-way movement of food through the gut. Also, different regions of the digestive tube can be specialized for different digestive activities. The front part of the gut is adapted for ingesting food. The middle region breaks down food and absorbs nutrients. The last region expels waste products.

Specialization of different regions of the gut brings with it a potential problem, however. Since the absorption of nutrients occurs in only one part of the digestive tract, there must be some effective way of distributing those absorbed nutrients to other parts of the body. Roundworms have a fluid-filled cavity between the gut and the body wall, as illustrated in the **Exploration**. This body cavity is called a **pseudocoelom** *(SOO doh see luhm)*, which means "false body cavity." This term is somewhat misleading. The cavity is real, but it differs from the body cavities of most other animals.

The pseudocoelom permits diffusion of nutrients and other dissolved substances over the short distances between tissues and organs. Diffusion is enhanced by body movements, which cause the fluid within the body cavity to move. Nevertheless, diffusion is a slow process. Pseudocoelomate animals must either be very small—most are less than 2.5 mm (0.1 in.) in length—or have body shapes that maintain short distances between organs and the body surface. For this reason, nematodes are usually thin and threadlike.

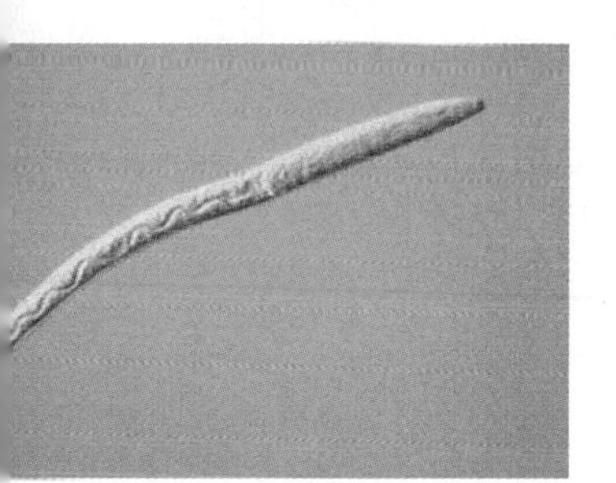

Figure 23-9 Roundworms, such as this nematode, have a body cavity, the pseudocoelom. The presence of the pseudocoelom means that roundworms are not packed with solid tissues as are flatworms. The pseudocoelom allows for more efficient diffusion of nutrients to body organs.

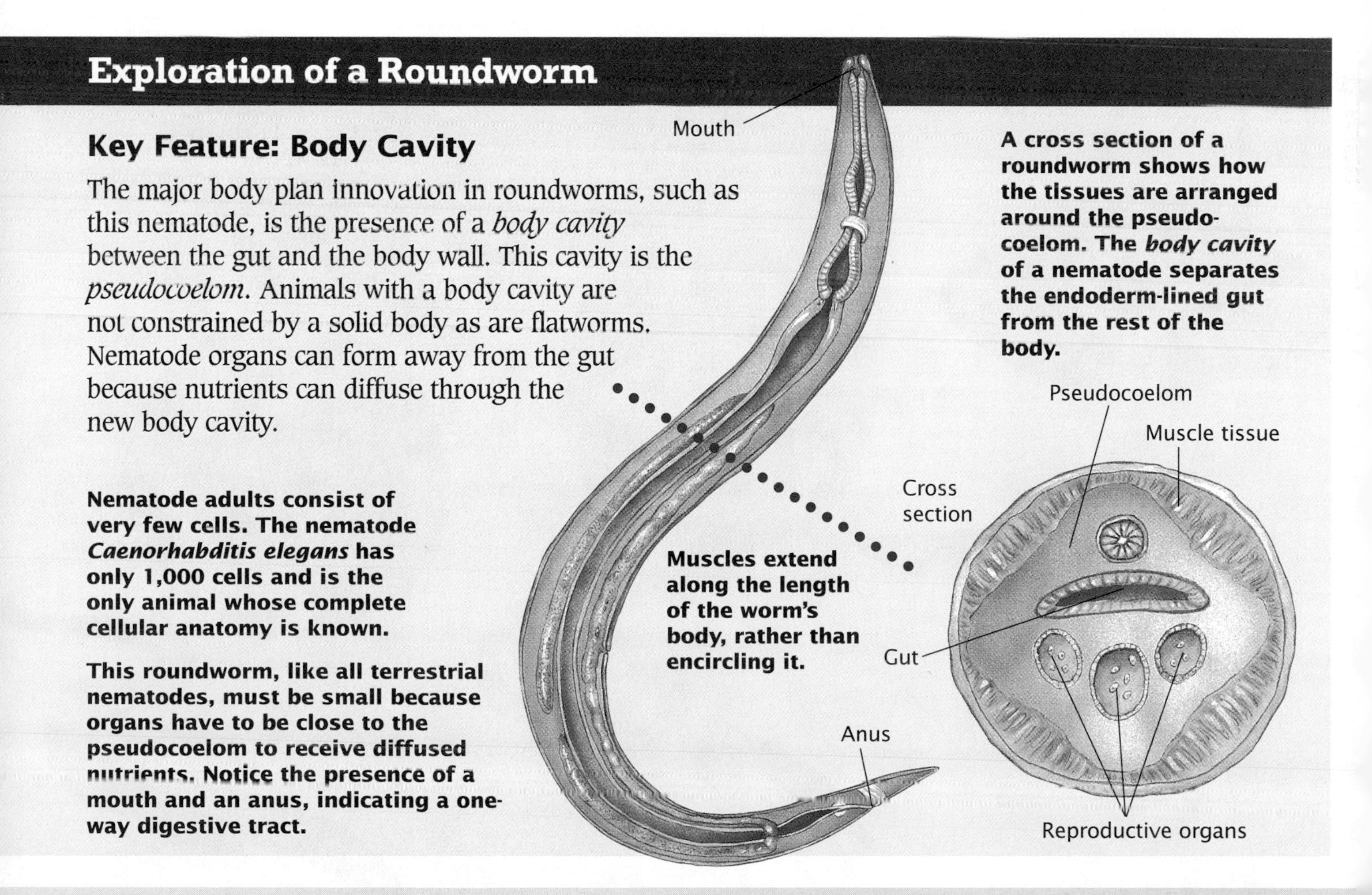

Exploration of a Roundworm
Key Feature: Body Cavity
The major body plan innovation in roundworms, such as this nematode, is the presence of a *body cavity* between the gut and the body wall. This cavity is the *pseudocoelom*. Animals with a body cavity are not constrained by a solid body as are flatworms. Nematode organs can form away from the gut because nutrients can diffuse through the new body cavity.
Nematode adults consist of very few cells. The nematode *Caenorhabditis elegans* has only 1,000 cells and is the only animal whose complete cellular anatomy is known.
This roundworm, like all terrestrial nematodes, must be small because organs have to be close to the pseudocoelom to receive diffused nutrients. Notice the presence of a mouth and an anus, indicating a one-way digestive tract.
Mouth
Muscles extend along the length of the worm's body, rather than encircling it.
Anus
Cross section
Gut
A cross section of a roundworm shows how the tissues are arranged around the pseudocoelom. The *body cavity* of a nematode separates the endoderm-lined gut from the rest of the body.
Pseudocoelom
Muscle tissue
Reproductive organs

A Better Body Cavity: Mollusks

A body cavity is advantageous in two main ways. First, because the organs lie within a fluid-filled cavity, they are insulated from the contractions of the muscles in the body wall. The organs can function independently of the movements of the animal. Food can be digested even while the animal is moving. The second advantage is that the digestive tract can coil up within the body cavity and be much longer than the animal. Your digestive tract is more than 8 m (26 ft) long. A longer digestive system allows food to be more completely digested and increases the efficiency of nutrient absorption.

Most animals with a body cavity have a coelom, or true body cavity. A **coelom** *(SEE luhm)* is a fluid-filled body cavity that lies completely within the mesoderm, as shown in **Figure 23-10**. The coelom separates the muscles of the body wall from the muscles that surround the gut. Animals that have a coelom are called coelomates *(SEE loh maytz)*. Why the coelom evolved is still a mystery. One suggestion is that it allows the mesoderm and endoderm to interact during embryonic development, producing specialized regions of the digestive tract.

In coelomates, as in flatworms, the gut tube is surrounded by solid tissue that is a barrier to rapid diffusion. This arrangement could limit the movement of nutrients, but most coelomates have a **circulatory system**, a network of blood-carrying vessels. The circulatory system brings nutrients and oxygen to the tissues and removes wastes and carbon dioxide. Blood is usually propelled through the circulatory system by contractions of one or more muscular hearts.

Figure 23-10
These cross sections show the differences between acoelomate, pseudocoelomate, and coelomate body constructions.

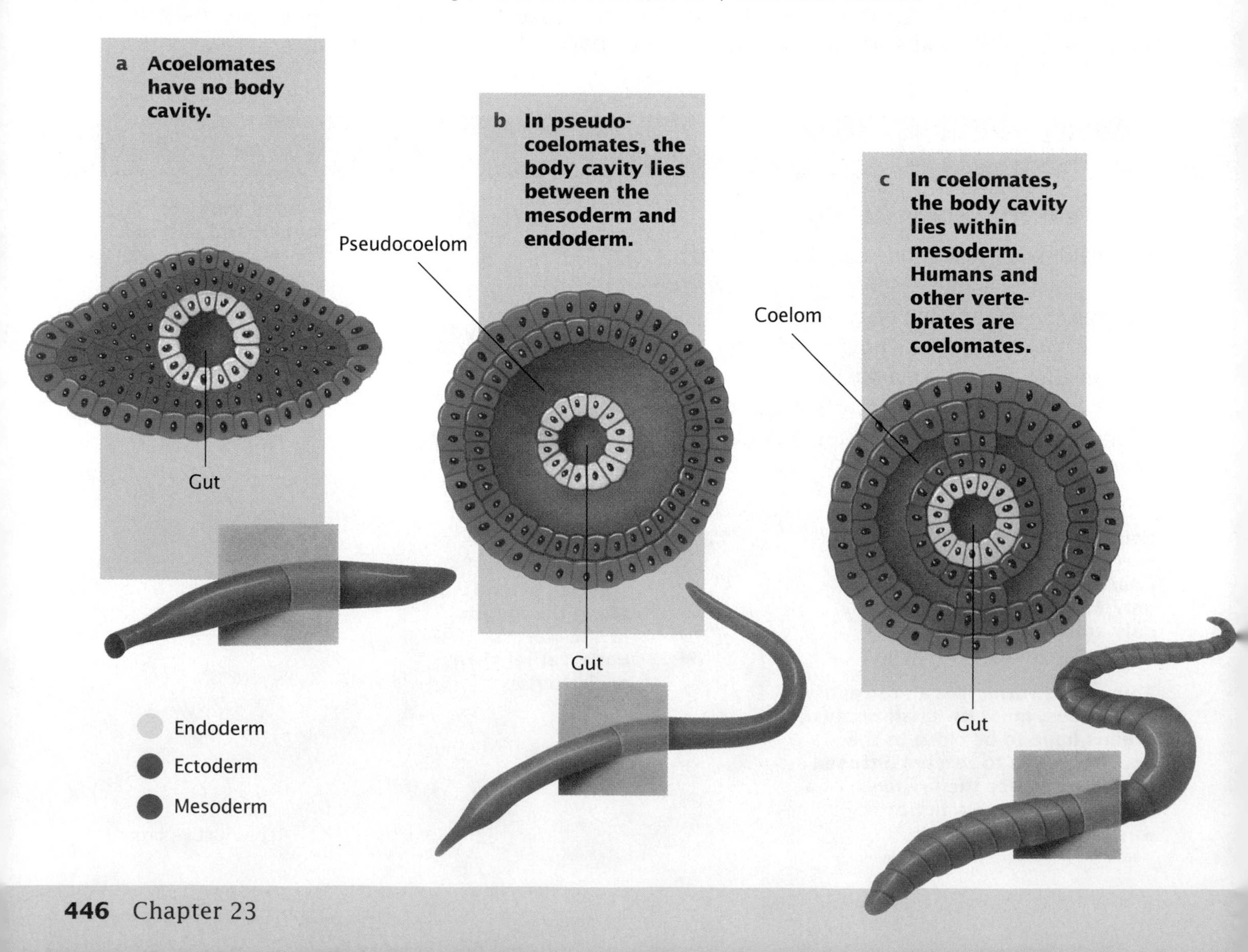

a Acoelomates have no body cavity.
Gut
b In pseudocoelomates, the body cavity lies between the mesoderm and endoderm.
Pseudocoelom
Gut
c In coelomates, the body cavity lies within mesoderm. Humans and other vertebrates are coelomates.
Coelom
Gut
Endoderm
Ectoderm
Mesoderm

Exploration of a Mollusk

Key Feature: Coelom

This snail is a mollusk. Mollusks have a *coelom* and a *circulatory system*. The coelom allows the gut to function independently from the muscles of the body wall and to be longer than the animal.

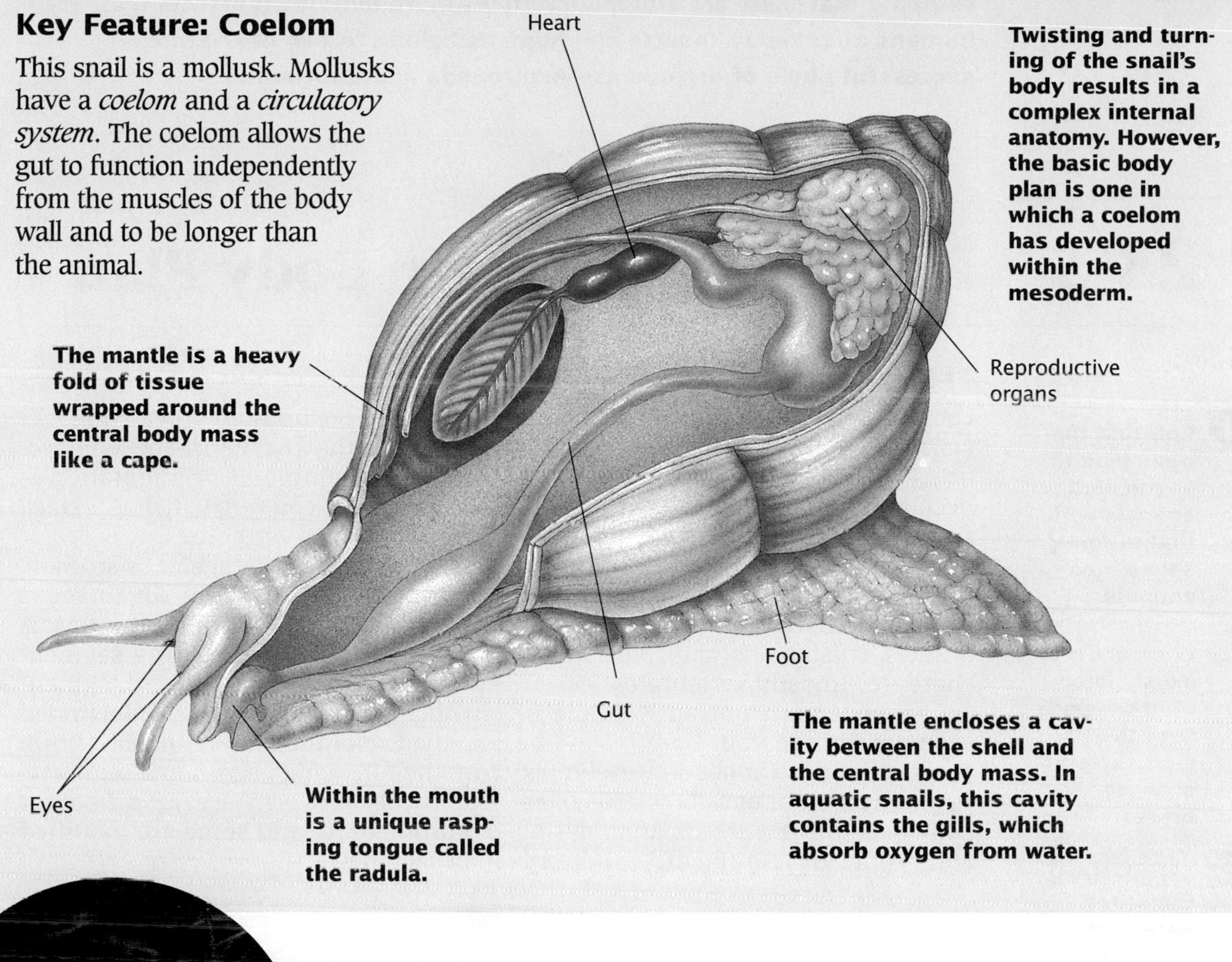

Heart
Twisting and turning of the snail's body results in a complex internal anatomy. However, the basic body plan is one in which a coelom has developed within the mesoderm.
The mantle is a heavy fold of tissue wrapped around the central body mass like a cape.
Reproductive organs
Foot
Gut
Eyes
Within the mouth is a unique rasping tongue called the radula.
The mantle encloses a cavity between the shell and the central body mass. In aquatic snails, this cavity contains the gills, which absorb oxygen from water.

Figure 23-11
This land snail is a mollusk.

Mollusks are coelomates with a circulatory system

Mollusks (phylum Mollusca) have a coelom and a circulatory system. Snails (such as the one in **Figure 23-11**), clams, squids, and mussels belong to this phylum. The mollusk body plan is illustrated in the **Exploration** above.

SECTION REVIEW

1. **Explain the difference between bilateral symmetry and radial symmetry.**
2. **What structures in your body are derived from mesoderm?**
3. **Summarize the advantages of a one-way gut.**
4. **Diagram the two types of body cavities.**

There are more than 5 billion people on Earth. This figure may seem large, but it is tiny compared with the insect population. Scientists estimate that there are 200 million times more individual insects than humans alive today. Insects and humans belong to two of the most successful phyla of organisms—Arthropoda and Chordata.

23-3 Four Innovations in Body Plan

OBJECTIVES

1. **Contrast the body plan of segmented animals with that of non-segmented animals.**
2. **Compare the exoskeleton of arthropods with the endoskeleton of vertebrates.**
3. **Summarize the differences between protostomes and deuterostomes.**
4. **List two traits that reveal the relationship between chordates and echinoderms.**

Segmented Worms: Annelids

Look at the worm in **Figure 23-12**. This worm is made up of many similar units linked together like beads in a necklace. Animals showing **segmentation** are composed of repeated body units. Three very successful animal phyla are segmented: annelids (earthworms and their relatives), arthropods (insects, crustaceans, and spiders), and chordates (mostly vertebrates).

Can you point out an example of segmentation in your body? Don't be surprised if you cannot. In vertebrates, segments are not usually visible externally in adults but are apparent during embryonic development. Vertebrate muscles develop from repeated blocks of tissue that occur in the embryo. Another example of segmentation is the vertebral column, which is a stack of very similar vertebrae.

What great advantage does segmentation provide? Its main advantage is the evolutionary flexibility it offers. A small change in an existing segment can produce a new kind of segment with a different function. As illustrated in the **Exploration of an Annelid** shown on the following page, some segments of the earthworm are modified for reproduction, and some are modified for feeding.

Figure 23-12
The bristle worm is an example of an annelid. Like the earthworm on the next page, its body is segmented. Specialized segments perform specific tasks.

Exploration of an Annelid

Key Feature: Segmentation

Annelids, such as this earthworm, have a body plan that consists of *segments*. Most segments are separated by partitions that cross the coelom. In each segment, parts of the excretory, circulatory, and nervous systems are repeated.

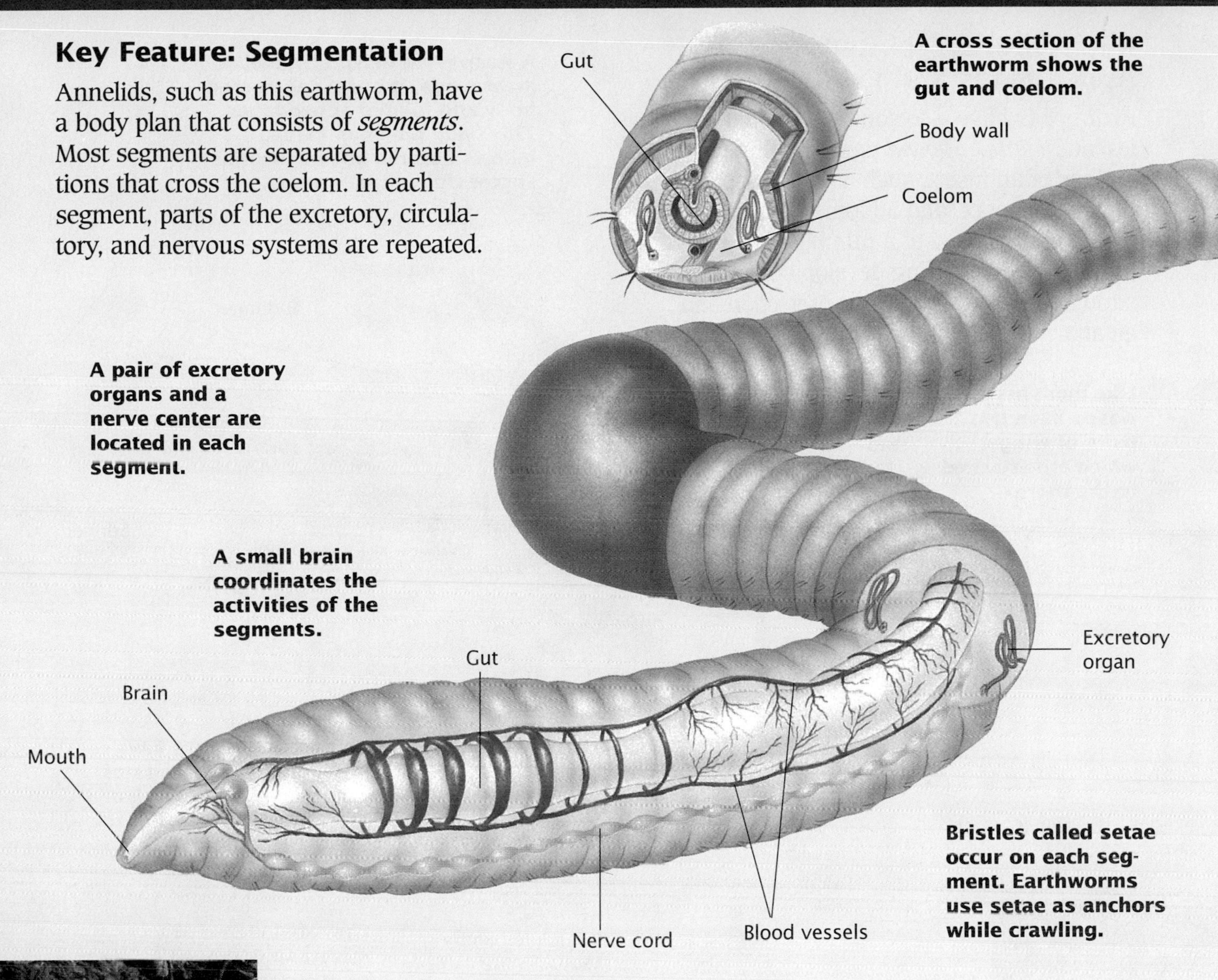

Gut
A cross section of the earthworm shows the gut and coelom.
Body wall
Coelom
A pair of excretory organs and a nerve center are located in each segment.
A small brain coordinates the activities of the segments.
Gut
Excretory organ
Brain
Mouth
Bristles called setae occur on each segment. Earthworms use setae as anchors while crawling.
Nerve cord
Blood vessels

Figure 23-13
The earthworm is an annelid. Members of this phylum have segmented bodies.

Annelids have segments that are specialized

Annelids (phylum Annelida), such as the earthworm in **Figure 23-13**, were among the first segmented animals to evolve. The earthworm body plan is shown in the **Exploration** above. The basic body plan of an annelid is a tube within a tube. The gut tube, which extends from mouth to anus, is suspended within the larger tube of the body wall. The coelom is the space between these tubes. It is divided internally between segments. This partitioning limits diffusion of materials from segment to segment. However, a circulatory system overcomes this limitation by transporting materials between segments.

Segments near the front of an annelid are modified to house a small brain and sense organs. Each segment along the body is controlled by an individual nerve center. A nerve cord running along the underside of the worm connects these nerve centers to the brain so that all of the body's activities can be coordinated.

Exploration of an Arthropod

Key Features: Jointed Appendages and Exoskeleton

Arthropods have a coelom, segmented bodies, and *jointed appendages*. The three body regions of an insect, such as this wasp, are the head, thorax, and abdomen. Each region is composed of a number of segments that fuse during development. The strong *exoskeleton* provides protection and an attachment point for muscles.

A wasp is an insect. Like all arthropods, it has a segmented body and jointed appendages. This body plan has helped insects become one of the most successful animal groups.

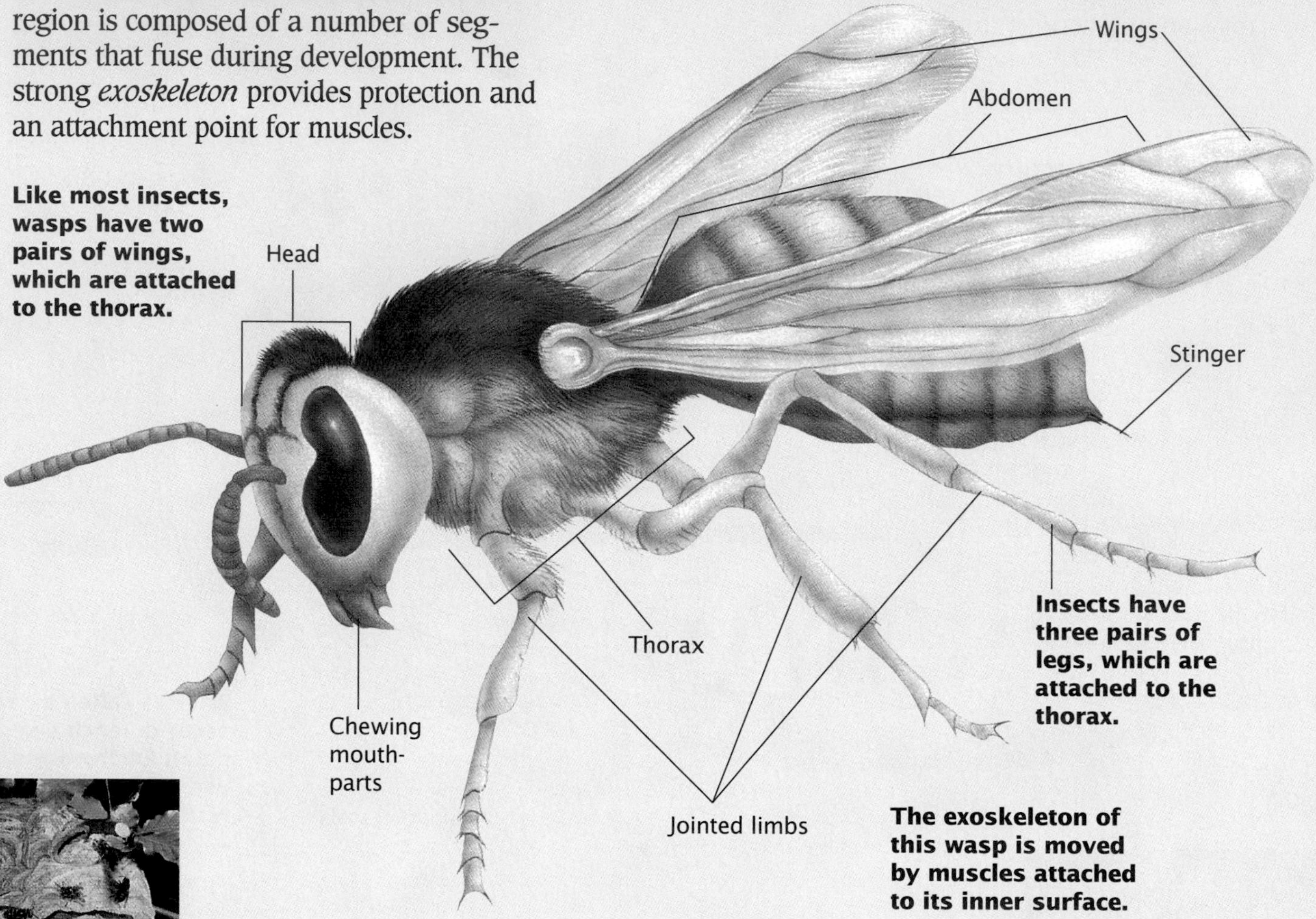

Like most insects, wasps have two pairs of wings, which are attached to the thorax.

Wings
Abdomen
Head
Stinger
Thorax
Chewing mouthparts
Jointed limbs

Insects have three pairs of legs, which are attached to the thorax.

The exoskeleton of this wasp is moved by muscles attached to its inner surface.

Figure 23-14
Wasps are highly social insects that build a nest where the colony lives and breeds. This nest is made of a paperlike material and may house hundreds of wasps.

Limbs and Skeletons: Arthropods

The name *arthropod* comes from the Greek words *arthros*, meaning "jointed," and *podes*, meaning "feet." The great success of arthropods, such as the wasps in **Figure 23-14**, is due largely to their jointed appendages. These appendages have evolved to perform a variety of tasks. Some appendages serve as limbs for walking or grasping. Others function as antennae for sensing the environment or as mouthparts for chewing, sucking, or poisoning prey. For example, a scorpion seizes and tears apart its prey with appendages that have been modified into large pincers capable of grasping and holding tightly.

Like annelids, the basic body plan of arthropods is segmented, as shown in the **Exploration** above. Individual segments of an arthropod often exist only during early development, however, and fuse into functional groups in adults. For example, caterpillars have

Evolution

Arthropods and chordates show segmentation. Many scientists think segmentation evolved independently in each group. What evidence supports this conclusion?

many obvious segments, while butterflies have only three main body units—head, thorax, and abdomen—each composed of several fused segments. Arthropods have an external skeleton, or **exoskeleton**. The arthropod exoskeleton is made of chitin *(KYT uhn)*, a tough polysaccharide, and protein. The muscles are internal to the skeleton and attach to its inner surface. As it grows, an arthropod periodically sheds its rigid exoskeleton. Many arthropods change their body form as they develop. A butterfly begins life as a wormlike caterpillar, later transforming into the flying adult. A silverfish, on the other hand, does not change as it matures; it merely grows larger.

Two Patterns of Development

Humans and the other chordates are more closely related to the echinoderms—the phylum that includes the sea star—than to the arthropods. At first glance, there seems to be little resemblance between humans and sea stars (often called starfish). Sea stars are radially symmetrical, as you can see in the **Exploration** on the next page. However, the embryos of chordates and echinoderms develop similarly.

In animals with mesoderm, except for echinoderms and chordates, the first opening that forms in the embryo during gastrulation eventually becomes the mouth. Animals that develop in this fashion are known as **protostomes**. The word *protostome* means "first mouth." In protostomes, the developmental fate of each cell of the early embryo is usually determined when that cell first appears.

Echinoderms and chordates are **deuterostomes**. The word *deuterostome* means "second mouth" and refers to the fact that the first opening in the embryo does not form the mouth—it becomes the anus of the adult animal. The mouth develops from an opening that appears later in development. In these animals, all the cells of the early embryo are identical. This means that any cell, if isolated early enough in development, can develop into a complete organism. **Figure 23-15** shows some protostomes and deuterostomes.

Figure 23-15
Below are examples of protostomes and deuterostomes. Recent studies using DNA nucleotide sequences support the division of the coelomates into these two groups.

Exploration of an Echinoderm

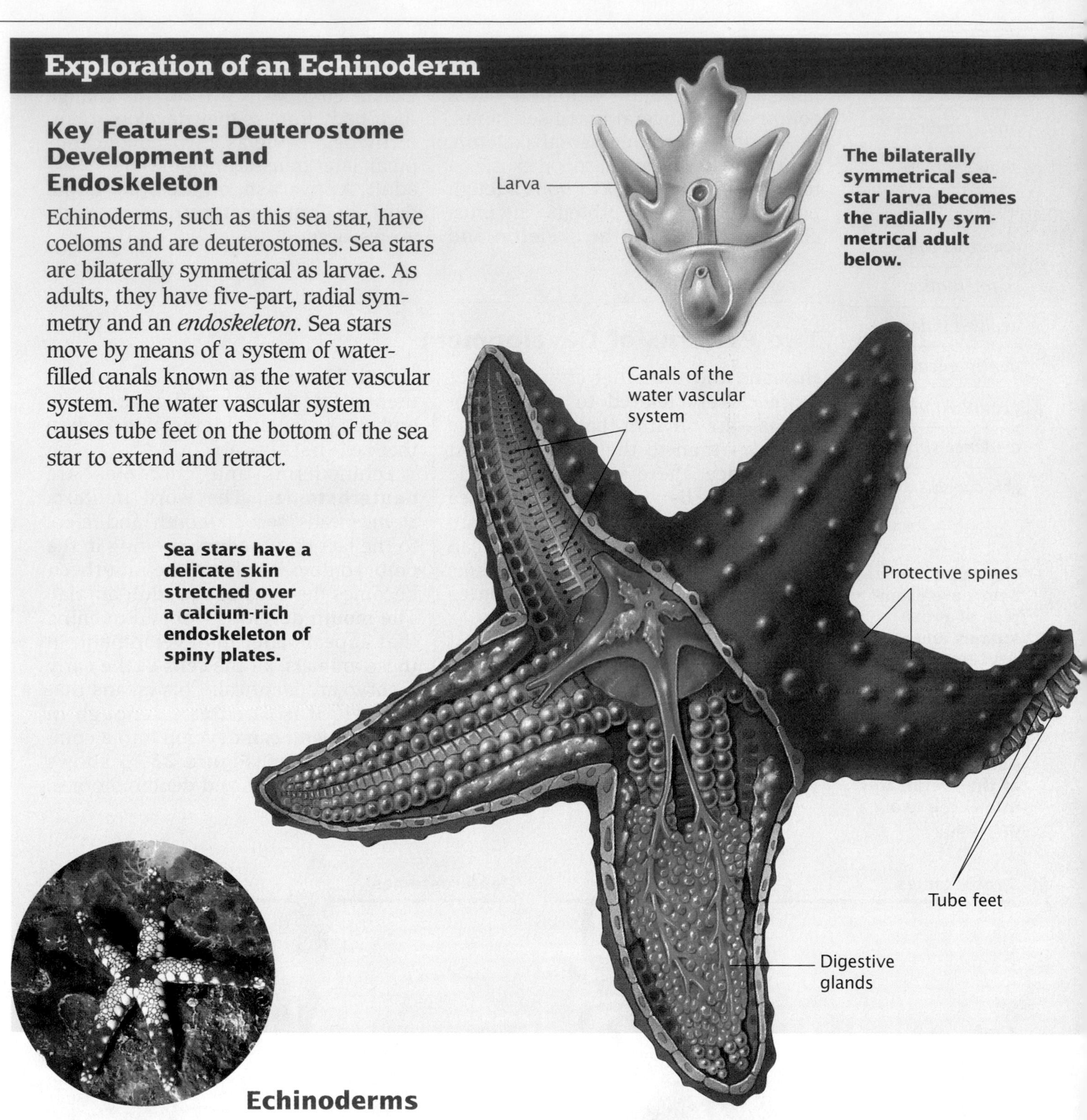

Key Features: Deuterostome Development and Endoskeleton
Echinoderms, such as this sea star, have coeloms and are deuterostomes. Sea stars are bilaterally symmetrical as larvae. As adults, they have five-part, radial symmetry and an *endoskeleton*. Sea stars move by means of a system of water-filled canals known as the water vascular system. The water vascular system causes tube feet on the bottom of the sea star to extend and retract.
Larva
The bilaterally symmetrical sea-star larva becomes the radially symmetrical adult below.
Canals of the water vascular system
Sea stars have a delicate skin stretched over a calcium-rich endoskeleton of spiny plates.
Protective spines
Tube feet
Digestive glands

Figure 23-16 Sea stars, such as this small sea star clinging to a ledge off the Philippine coast, are members of the phylum Echinodermata.

Echinoderms

The larvae of all echinoderms, such as the sea star in **Figure 23-16**, are bilaterally symmetrical and undergo a complex metamorphosis as they mature, transforming into radially symmetrical adults. Unlike arthropods, echinoderms have an internal skeleton composed of many bonelike plates. An internal skeleton is called an **endoskeleton**.

Most adult echinoderms, like the one shown in the **Exploration**, have a five-part body plan. As adults, echinoderms have no head or brain. The body of an echinoderm is controlled by a ring of nerves that branches into each of the arms. Although the arms are capable of complex movements, control of movement is not centralized as it is in bilaterally symmetrical animals.

Figure 23-17
The lancelet is a chordate, a member of the same phylum as human beings. Although it does not resemble a human, the lancelet has characteristics that it shares with humans and all other chordates.

The Most Successful Deuterostomes: Chordates

Fishes, amphibians, reptiles, birds, and mammals are chordates, as is the lancelet shown in **Figure 23-17**. Lancelets do not resemble familiar chordates such as birds, dogs, or humans. Nonetheless, lancelets and all chordates share four key features:

1. dorsal (along the back) hollow nerve cord
2. dorsal supportive rod called the notochord
3. slits in the pharynx (the region of the digestive tract just behind the mouth) that connect the pharynx to the outside
4. a postanal tail—one that extends beyond the anus

This unique combination of characteristics can be seen in the lancelet shown in the **Exploration** below. All chordates have all four of these characteristics at some time during their lives. For example, humans have pharyngeal slits, a nerve cord, a tail, and a notochord as embryos. As adults, humans retain only the nerve cord and one pair of pharyngeal slits (which have become the external ear canals and the eustachian tubes that connect the throat to the middle ear).

In addition to these four principal characteristics, chordates have a number of other distinguishing features, such as a segmented body plan.

Exploration of a Lancelet

Key Feature: Notochord

The lancelet *Branchiostoma* is a chordate. It is a coelomate animal that has a *notochord*, pharyngeal slits, a postanal tail, and a dorsal nerve cord. The notochord persists throughout the life of *Branchiostoma*. In most other chordates, such as birds and humans, the notochord is replaced during embryonic development by the vertebral column.

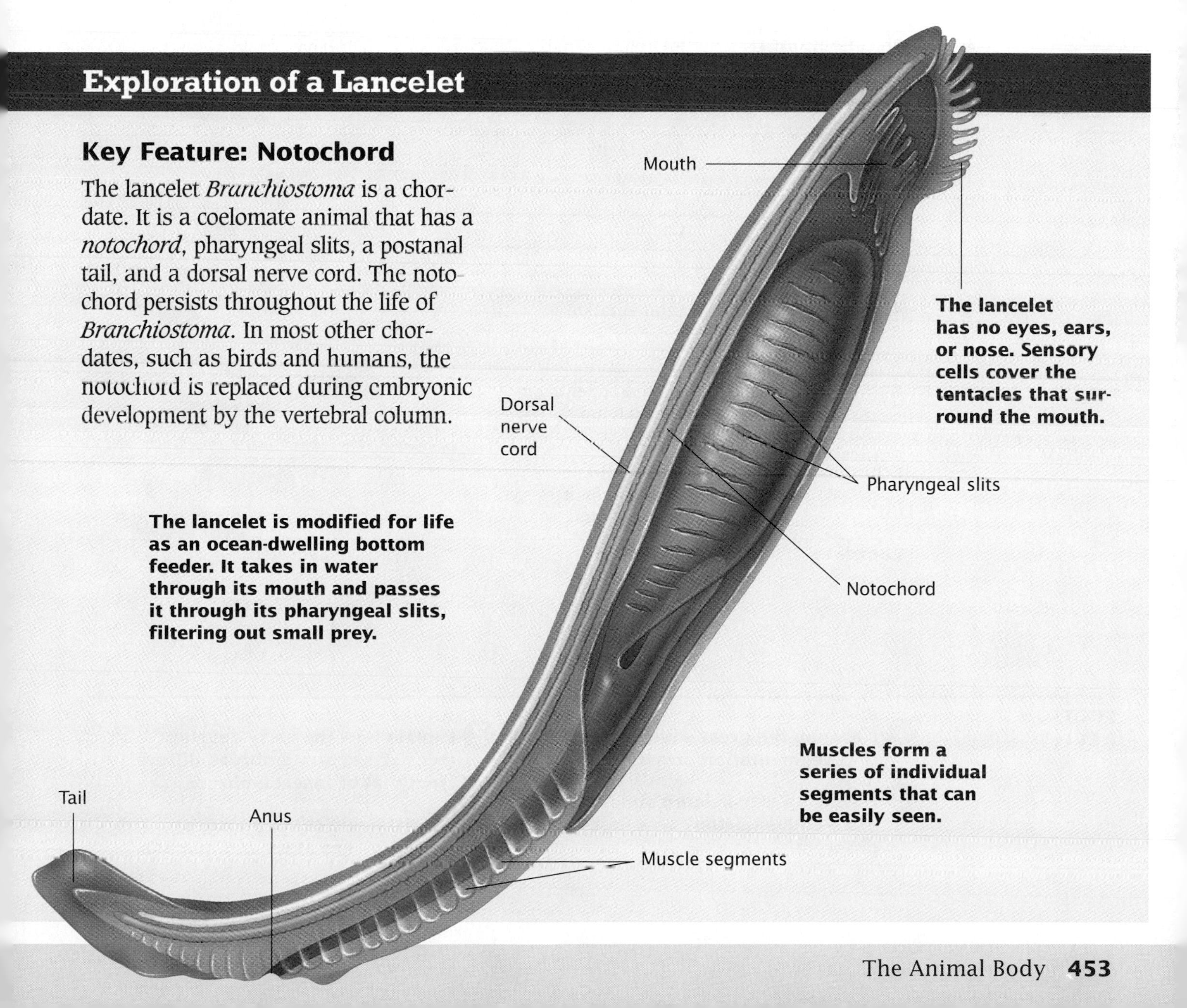

Mouth
The lancelet has no eyes, ears, or nose. Sensory cells cover the tentacles that surround the mouth.
Dorsal nerve cord
Pharyngeal slits
The lancelet is modified for life as an ocean-dwelling bottom feeder. It takes in water through its mouth and passes it through its pharyngeal slits, filtering out small prey.
Notochord
Muscles form a series of individual segments that can be easily seen.
Tail
Anus
Muscle segments

Vertebrates have a sturdy endoskeleton

Most chordates belong to the subphylum Vertebrata, the vertebrates. Fishes, amphibians, reptiles, birds, and mammals are vertebrates, but the lancelet is not. In most vertebrates, the notochord is replaced by the vertebral column during development.

Vertebrates also have an endoskeleton of bone or cartilage against which the muscles work. The vertebrate endoskeleton is lightweight and provides effective support.

Table 23-1 shows the changes in the animal body described in this chapter. You will read more about these groups in Chapters 24–28.

Table 23-1 The Animal Body: An Evolutionary Journey

Phylum	Key features	Typical organism
Porifera	Multicellularity	Sponge
Cnidaria	Radial symmetry Extracellular digestion Specialized tissues	Hydra
Platyhelminthes	Internal organs Bilateral symmetry Cephalization	Liver fluke
Nematoda	Body cavity	Nematode
Mollusca	Coelom	Snail
Annelida	Segmentation	Earthworm
Arthropoda	Jointed appendages Exoskeleton	Wasp
Echinodermata	Deuterostome development Endoskeleton	Sea star
Chordata	Notochord	Lancelet

SECTION REVIEW

1. **Describe the great advantage that segmentation provides.**
2. **How is an exoskeleton similar to an endoskeleton? How is it different?**
3. **Explain how the early development of sea star embryos differs from that of insect embryos.**
4. **Describe two characteristics shared by sea stars and humans.**

CHAPTER

23 Highlights

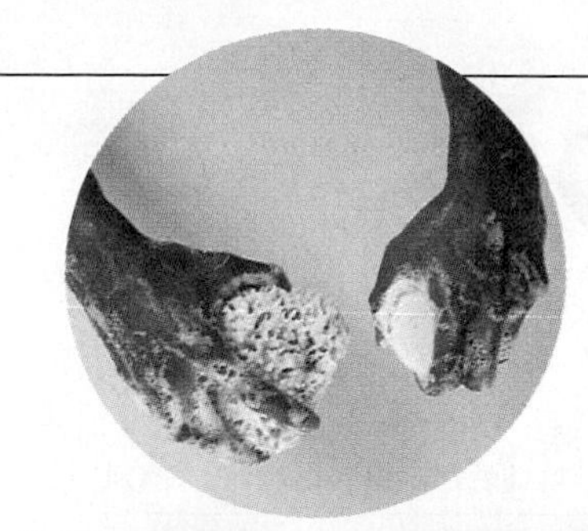
A natural sponge is the skeleton of a marine sponge.

	Key Terms	Summary
23-1 The Origin of Tissues	body plan (p. 437) tissue (p. 440) gut (p. 440) gastrulation (p. 441) endoderm (p. 441) ectoderm (p. 441) radial symmetry (p. 441)	• Multicellularity enables individual cells to specialize on one or a few tasks. • Sponges are multicellular animals without tissues. Sponges filter small food particles from the water. • Cnidarians have two tissue layers and show radial symmetry.
23-2 Origin of Body Cavities	bilateral symmetry (p. 442) cephalization (p. 442) organ (p. 443) organ system (p. 443) mesoderm (p. 443) acoelomate (p. 444) anus (p. 445) pseudocoelom (p. 445) coelom (p. 446) circulatory system (p. 446)	• Flatworms are bilaterally symmetrical and cephalized. • Nematodes have a one-way gut and a pseudocoelom, a cavity lying between the gut and the muscles of the body wall. • A coelom is a fluid-filled body cavity that lies within the mesoderm. Mollusks have a coelom and a circulatory system that distributes nutrients and oxygen to the tissues.
23-3 Four Innovations in Body Plan	segmentation (p. 448) exoskeleton (p. 451) protostome (p. 451) deuterostome (p. 451) endoskeleton (p. 452)	• Annelids have a segmented body. Chordates and arthropods are also segmented. • An exoskeleton of chitin and protein, segmentation, and jointed appendages are characteristics of arthropods. • The embryos of echinoderms and chordates develop similarly. These two groups of animals are closely related.

Sponges have a very simple body plan. They have no tissues, and most are asymmetrical.

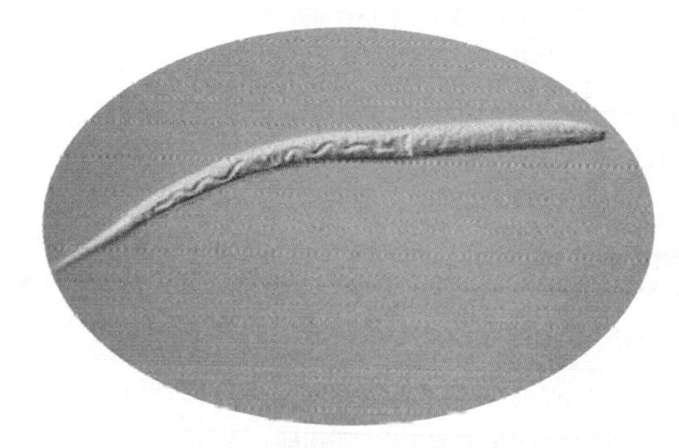
Nematodes have two openings to the gut, allowing for specialization of digestive activities.

Arthropods have jointed appendages, a segmented body, and an exoskeleton.

CHAPTER 23 review

Understanding Vocabulary

1. For each pair of terms, explain the difference in their meanings.
   a. endoderm, ectoderm
   b. radial symmetry, bilateral symmetry
   c. tissue, organ

2. For each set of terms, complete the analogy.
   a. asymmetrical : sponge :: radially symmetrical : ____
   b. acoelomate : flatworm :: pseudocoelomate : ____
   c. jointed appendages, exoskeleton : arthropod :: endoskeleton, radial symmetry as adult : ____

Understanding Concepts

3. **Relating Concepts** Construct a concept map that describes the characteristics of animals. Try to include the following terms in your concept map: heterotrophic, multicellular, cell specialization, tissues, organs, organ system, sponges, arthropods, jointed appendages, exoskeleton, chitin, heart, digestive system.

4. **Identifying Structures** What three characteristics do all animals have in common?

5. **Summarizing Information** What is meant by the phrase "division of labor"? Why is it advantageous for the cells of an animal to exhibit division of labor?

6. **Comparing Structures** Look at the figures below. Which animal has tissues? Which is radially symmetrical?

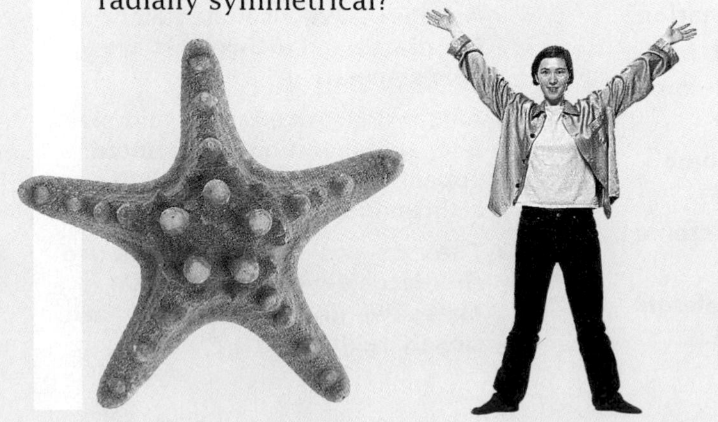

7. **Comparing Structures** What is a tissue? Do sponges have tissues? Explain why or why not.

8. **Comparing Structures** How do the body plans of radially and bilaterally symmetrical animals differ? Give an example of an animal with each kind of symmetry.

9. **Inferring Relationships** The brain and the sense organs in most bilaterally symmetrical animals are located in a head region. Why is this arrangement advantageous?

10. **Summarizing Information** Summarize the steps of gastrulation that lead to the formation of endoderm and ectoderm. What kinds of tissues do endoderm and ectoderm give rise to?

11. **Identifying Structures** What is mesoderm? What structures of an animal's body are derived from the mesoderm?

12. **Analyzing Functions** How does the movement of food in a nematode compare with the movement of food in a flatworm?

13. **Analyzing Functions** What similar problem do coelomates and acoelomates have? How do many coelomates overcome this problem?

14. **Contrasting Structures** Describe and contrast the two kinds of body cavities. Give an example of an animal with each kind of cavity.

15. **Comparing Structures and Functions** What are some of the functions of an exoskeleton? What are some limitations of an exoskeleton?

16. **Comparing Structures** Identify two traits that chordates and echinoderms have in common.

17. **Identifying Structures** What four features do all chordates have at some time during their lives?

18. **BUILDING ON WHAT YOU HAVE LEARNED** In Chapter 17 you learned about how organisms are classified. Combine your knowledge of classification systems with your knowledge of body cavity types to explain why earthworms and sand worms are grouped in the phylum Annelida but flatworms are not.

Interpreting Graphics

19. Look at the diagrams below.

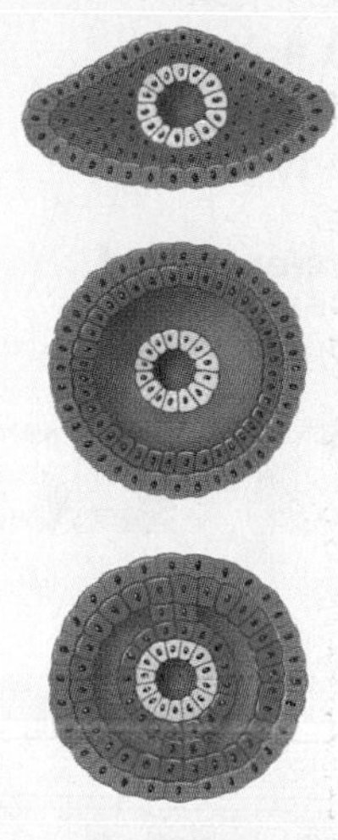

a. Identify each type of body plan shown above. Give an example of an animal with each type of body plan.

b. Describe the limitations and advantages of each type of body cavity.

Reviewing Themes

20. ***Evolution***
What is segmentation? How does segmentation provide for evolutionary flexibility?

21. ***Homeostasis***
Explain how the circulatory system of an animal helps maintain homeostasis.

Thinking Critically

22. **Forming Opinions Based on Sound Reasoning**
As you learned in the chapter, radially symmetrical animals such as echinoderms and cnidarians are not found on land. However, bilaterally symmetrical animals live on land and in water. Propose a hypothesis to explain why radially symmetrical animals are not well suited to life on land.

23. **Recognizing Verifiable Conclusions** A scientist discovers a new species of animal living in a mountain lake. The animal is bilaterally symmetrical and is definitely segmented. It has no eyes and seems to feed on decaying material. The scientist tentatively classifies the animal as a chordate. Is this classification valid? Explain your answer.

Life/Work Skills

24. **Working Cooperatively** Work with two or three of your classmates. Create a chart that you can use in this chapter and in Chapters 24–28 to summarize and organize the characteristics of the different kinds of animals you will study. Include information such as appearance, habitat, type of symmetry, digestion and feeding habits, reproduction, and interactions with humans. On your chart, include room for illustrations of the various animals.

25. **Allocating Resources** Imagine that you are in charge of a large city aquarium. Explain how you would spend a donation given to you by a conservation-minded citizen who wants his money to further the cause of conserving and protecting aquatic and marine wildlife. Decide how much money is being donated. Would you spend some on education programs? Would you purchase new specimens? Make a poster presentation of your choices to convince your benefactor and board of directors.

Activities and Projects

26. **Art Connection** Use three different colors of modeling clay, or any other material, to construct models that show the formation of endoderm and ectoderm as a result of gastrulation and the formation of endoderm, ectoderm, and mesoderm as a result of gastrulation. Also make models that show the three major body plans: acoelomate, pseudocoelomate, and coelomate. Share your models with your class.

27. **Health Connection** Hookworms are parasitic roundworms that infect humans. Look for information in your local library or in an on-line database about the symptoms associated with hookworm infection. What can humans do to prevent becoming infected?

Discovering Through Reading

28. Read the article "Do Lions Purr?" by Doug Stewart in *National Wildlife*, August/September 1995, pages 30–38. Do lions purr? Why do flamingos stand on one leg?

CHAPTER 24

REVIEW
- diffusion (Section 4-2)
- ozone layer (Section 11-3)
- innovations in body plan (Section 23-3)
- the terms *endoskeleton* and *exoskeleton* (Glossary)

Adaptation to Land

These caribou are feeding in Denali National Park in Alaska.

Life arose in the sea and stayed there for more than 3 billion years, until a protective shield of ozone formed in the upper atmosphere. Without this shield, animals like this frog could not live on land. The requirements for living on land are very different from those for living in the sea. In this chapter, you will examine some key problems animals faced in moving onto land.

24-1 Leaving the Sea

OBJECTIVES

1. **List some animal groups that successfully made the transition to life on land.**
2. **Explain the importance of the skeleton to arthropods and vertebrates that live on land.**
3. **Describe the evolution of amphibian limbs from the fins of lobe-finned fishes.**
4. **Relate the structure of the vertebrate ear to the differences between air and water.**

Which Animals Live on Land?

The animal body evolved over many millions of years in the sea. All of the major changes in body plan you read about in Chapter 23 took place in the sea. But the evolutionary journey of animals did not end there.

From the sea, animals invaded the land. Of the major animal phyla, only sponges, cnidarians, and echinoderms —groups of animals that pump water through their bodies—were left behind. Some members of every other major phylum can be found on land. Flatworms, nematodes, and annelids burrow through the soil, and snails creep over damp ground at night. These animals, however, are found mainly in moist habitats. Only two groups of animals—arthropods and vertebrates—have members that are fully adapted to life on dry land. How animals such as the lizard shown in **Figure 24-1** evolved ways to survive the many challenges of living out of water is one of biology's most fascinating stories.

Figure 24-1
The thorny devil is a small lizard that lives in Australia. Its adaptations for water conservation enable it to live in very dry and harsh environments.

Figure 24-2
Several changes in limb structure took place as terrestrial vertebrates evolved from aquatic vertebrates.

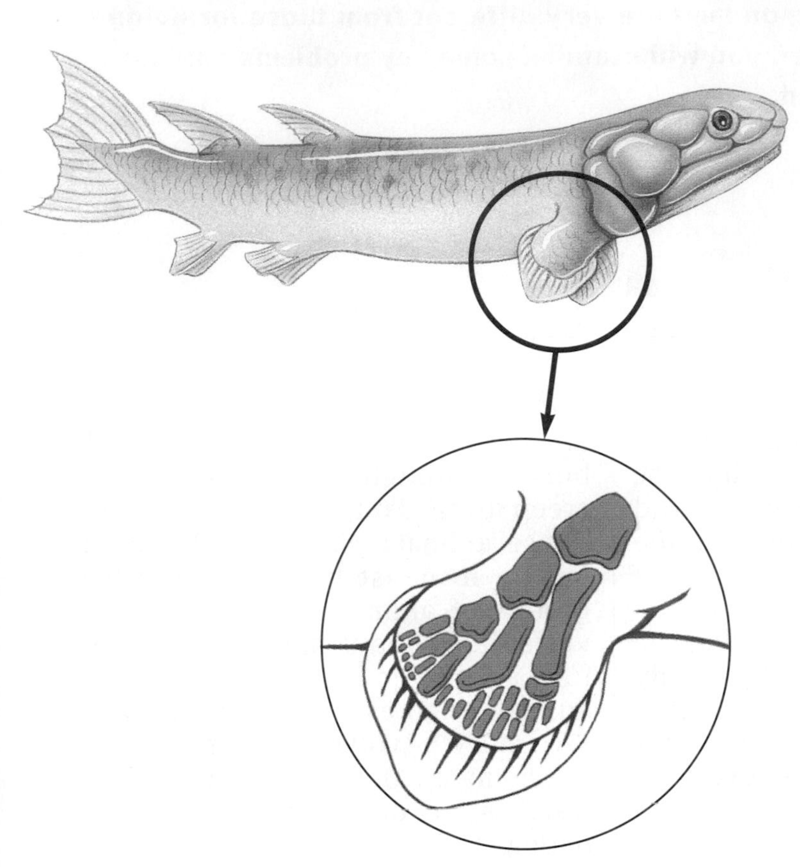

a Scientists think that lobe-finned fishes were the ancestors of the earliest land vertebrates. The drawing above shows the bones in the fin of an extinct lobe-finned fish.

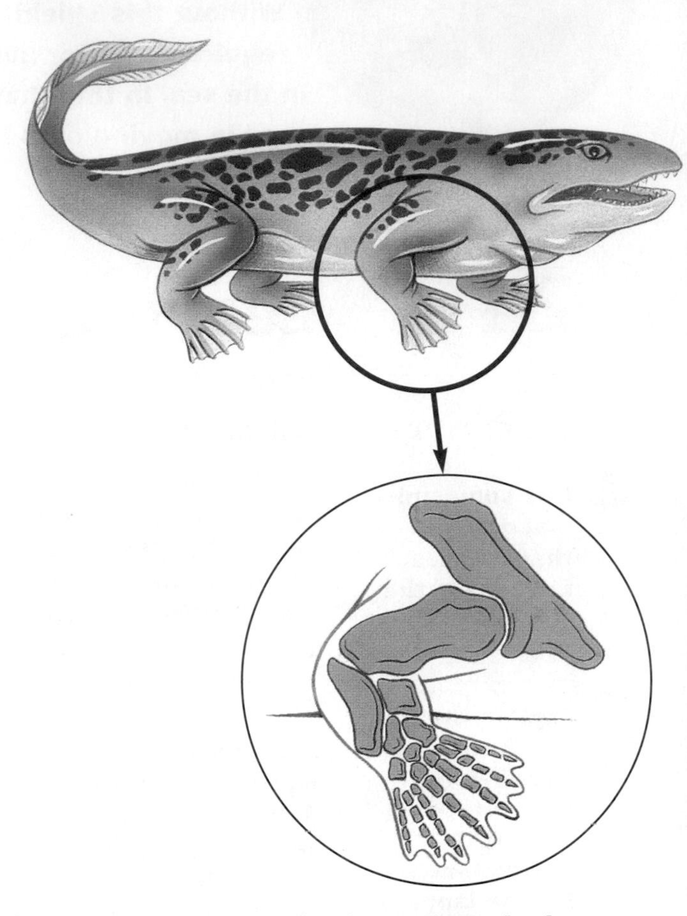

b The earliest known land vertebrate is the amphibian *Ichthyostega*, which lived about 350 million years ago. It had four sturdy legs. Like its fish ancestors, it had a tail fin and some bony scales.

Supporting the Body

Water is about 1,000 times denser than air. Because of water's density, you can float on your back in a pool. For some aquatic animals, water provides much of the support necessary to keep their bodies from collapsing under the pull of gravity. That is why a jellyfish stranded on the beach cannot maintain its shape.

A variety of adaptations were necessary in order for animals that left the sea to overcome the loss of physical support. In nematodes and earthworms, the body cavity is filled with fluid under pressure, which helps to stiffen the body. In arthropods and vertebrates, support of the body is largely taken over by the skeleton. A land animal's skeleton holds up its body against gravity much like beams and girders hold up a skyscraper.

Limbs play an important role in supporting vertebrates on land. When a terrestrial animal is standing, its legs bear the entire weight of its body. In this way, legs function like the pillars that hold up the roof of a building. Unlike pillars, however, animal legs have flexible joints that enable them to move.

In both arthropods and vertebrates, legs evolved from limbs adapted for movement in water. **Figure 24-2a–b** illustrates the evolution of amphibian limbs from the fins of fishes. Recall from Chapter 11 that amphibians, the

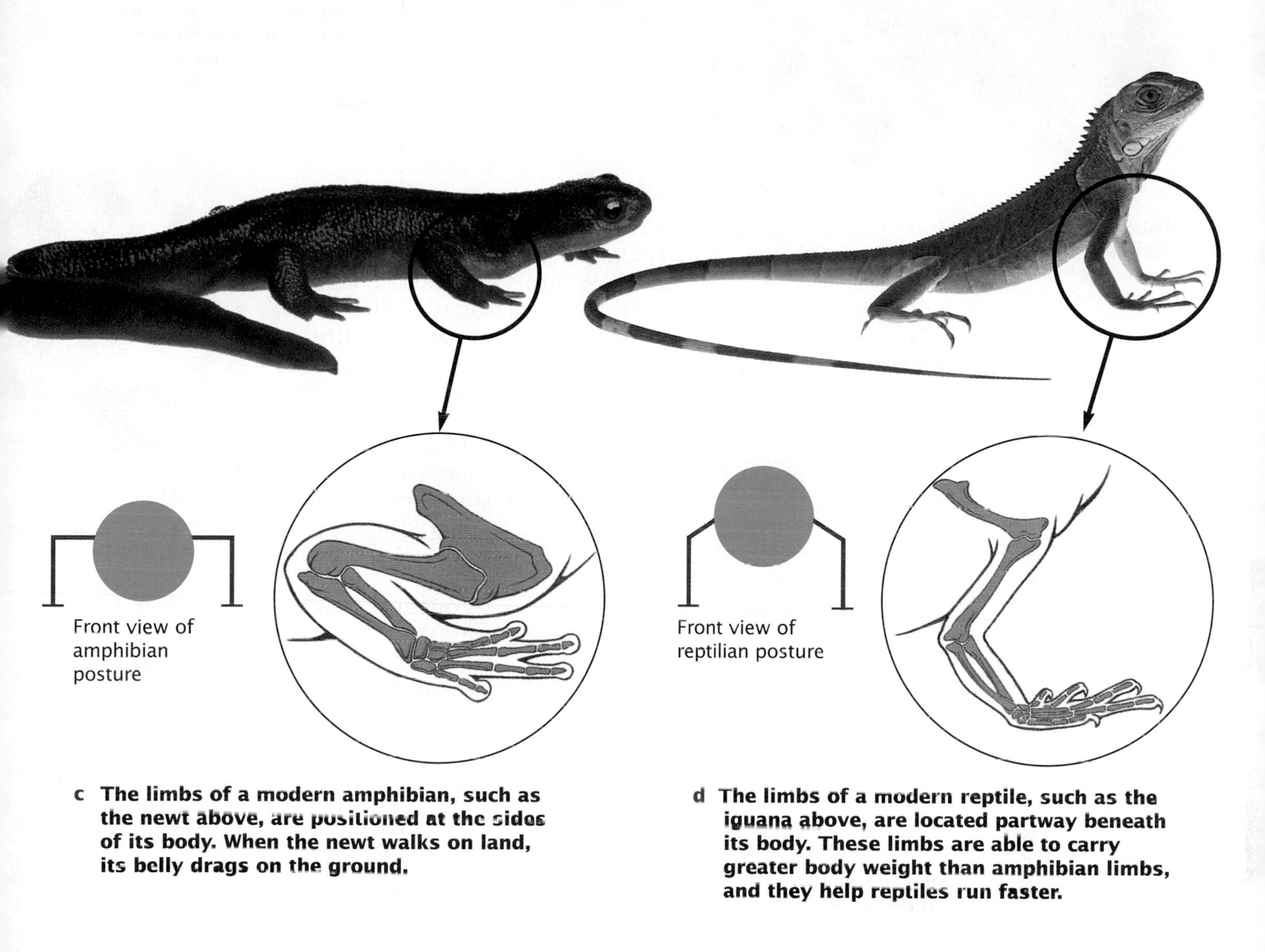

c The limbs of a modern amphibian, such as the newt above, are positioned at the sides of its body. When the newt walks on land, its belly drags on the ground.

d The limbs of a modern reptile, such as the iguana above, are located partway beneath its body. These limbs are able to carry greater body weight than amphibian limbs, and they help reptiles run faster.

first vertebrates on land, evolved from fishes. The limbs of amphibians are homologous to fins of fishes. However, most fishes have thin, paddlelike fins that would be useless for supporting and moving an animal on land. As you can see in **Figure 24-2a–b**, the arrangement and structure of bones in the limb of an early amphibian are similar to those in the fin of a lobe-finned fish. Because of these similarities, scientists think the first amphibians were descendants of the lobe-finned fishes, a group whose modern members include the coelacanth and the lungfishes.

Compare the posture of the amphibian in **Figure 24-2c** with that of the reptile in **Figure 24-2d**. Amphibian limbs are short and join the body horizontally; an amphibian's belly drags on the ground as it walks. The limbs of living reptiles are positioned slightly beneath the body. Limbs positioned in this way support more body weight than do the limbs of amphibians. The limbs of birds and mammals are positioned directly beneath the body, raising the belly well off the ground. Compared with reptilian and amphibian limbs, limbs located directly beneath the body support the animal's weight more effectively. They also enable the animal to run faster and, by elevating the head, help it locate food and be alert to danger.

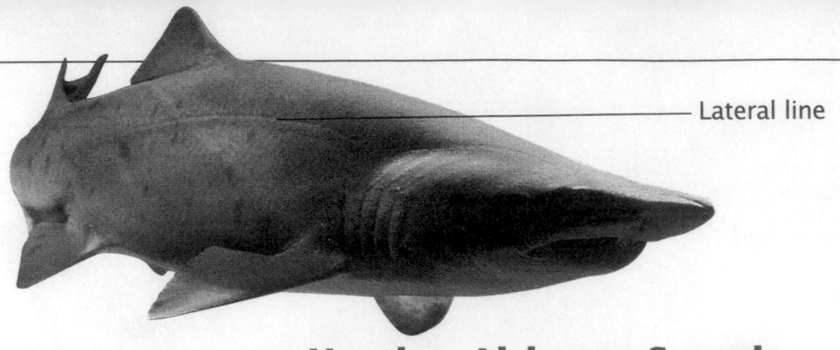

Figure 24-3
Fishes can sense sounds or the movements of other fishes with a row of pressure-sensitive cells called the lateral line.

Figure 24-4
Many mammals, such as this fox, can point their ears in the direction of a sound to pinpoint where the sound is coming from.

Hearing Airborne Sounds

In the ocean, whales are able to communicate with low-pitched "songs" that travel as far as 200 km (125 mi) because water is much denser than air. The denser the medium through which sound travels, the faster and farther the sound can travel.

Water's density makes detecting sound in water easier than doing so on land. Most species of fishes can hear. In addition, fishes are able to sense physical disturbances in the water, such as those caused by the swimming movements of nearby fishes, with their **lateral line** system. The lateral line is a row of pressure-sensitive cells that lies within a fluid-filled canal. One network of canals runs down each side of the fish. The lateral line is shown in **Figure 24-3**.

In air, the pressure changes caused by sound are much smaller because of air's lower density. Terrestrial animals have evolved structures that amplify airborne sounds so that they are detectable.

The ear is the sound-amplifying system of land vertebrates. In mammals, such as the fox in **Figure 24-4**, the eardrum and three small bones amplify sound waves and transmit them to the inner ear, where sounds are detected. In salamanders and snakes, sound is conducted to the inner ear entirely through bones of the jaw and skull. These animals rely more on ground vibrations (which transmit low-frequency sounds even better than water) than on airborne sounds. You will learn more about the workings of the ear in Chapter 30.

Insects can hear well, but they evolved different solutions to the challenges of hearing in air. Some insects detect sounds with sensitive hairs on their legs, body, or head. Others have eardrum-like structures on their body or legs. Crickets, for instance, hear the songs of other crickets through membranes on their front pair of legs, as you can see in **Figure 24-5**.

Figure 24-5
Crickets hear the songs of other crickets through membranes on their front pair of legs.

SECTION REVIEW

1. **Which major groups of animals are not found on land?**
2. **Describe the role of the skeleton in the life of a terrestrial animal.**
3. **What evidence suggests that amphibian limbs evolved from the fins of lobe-finned fishes?**
4. **Compare sound transmission in air with sound transmission in water.**

Conserving body water is one of the greatest challenges of living on land. When animals first moved onto land, they faced the risk of losing excessive amounts of water through evaporation. Many terrestrial arthropods and vertebrates evolved traits that enabled them to conserve body water.

24-2 Staying Moist in a Dry World

OBJECTIVES

1. **Recognize the role of the exoskeleton in preventing water loss.**
2. **Contrast the skin of amphibians with that of reptiles.**
3. **Summarize the evolution of lungs.**
4. **Describe two ways animals conserve water while ridding themselves of wastes.**

Watertight Skin

If you leave a wet towel in the sun, what happens to it? The towel dries through evaporation. Since an animal's body is about 70 percent water, an animal on land is like a wet towel. Its cells are full of water and, like the fibers of the wet towel, are surrounded by water. Like the towel, an animal on land is in danger of drying out through evaporation. Losing even small amounts of body water can be dangerous. Losing 15–20 percent of the water in your body would be fatal. How do terrestrial animals limit their loss of water? In this section you will examine some of the adaptations that enable animals like the desert ants shown in **Figure 24-6** to exist on dry land.

Arthropods and vertebrates have waterproof coatings

What would happen if you sealed a wet towel inside a plastic bag? Because the bag is waterproof, water could no longer escape from the towel into the atmosphere. For the same reason, sandwiches stay moist when they are sealed in sandwich bags. Similarly, the bodies of arthropods and vertebrates are wrapped in a watertight coating. In arthropods, this coating is the exoskeleton. In vertebrates, it is the skin.

Recall from Chapter 23 that the arthropod exoskeleton is a stiff outer coat composed of chitin and protein. In addition to providing protection and physical support, the exoskeleton prevents water from leaving the terrestrial arthropod's body. Especially in spiders and insects, the outer layer of the exoskeleton contains waxes that enhance the waterproofing effect of the exoskeleton.

Figure 24-6
These ants are able to live in the desert because of their many adaptations for preventing water loss. You will read about some of these adaptations in this section.

Figure 24-7
Reptiles have dry, scaly skin that is watertight. The skin of a banded rock rattlesnake is shown above.

Vertebrate skin can be moist and thin or dry and thick

Have you ever touched a frog? What did its skin feel like? Many amphibians secrete a slippery mucus that is responsible for their slimy texture. Like the wax on an insect, this mucous coating helps limit evaporation. Despite this mucous coating, the skins of amphibians are not watertight. As you will see in Chapter 27, amphibians absorb oxygen and release carbon dioxide directly through their skin. To do this, their skin must be moist and thin. Therefore, to keep from drying out, most amphibians must remain in water or in moist environments.

A reptile is not slimy like an amphibian. Reptilian skin is dry and covered with tough scales, as shown in **Figure 24-7**. A watertight skin of scales was a significant evolutionary step, completely freeing reptiles from the necessity of living in a wet or moist environment. Mammals and birds, which evolved from reptiles, also have skin that is dry and relatively watertight. As you might expect, mammals, birds, and reptiles do not exchange gases through their skin.

Gas Exchange

You breathe about 12 times each minute. From the perspective of water conservation, breathing is very costly. About 300 mL (9 oz) of water evaporates from your lungs each day. You breathe, despite the water loss, because breathing is essential for gas exchange. Gas exchange is the process of absorbing oxygen from the environment and ridding the body of carbon dioxide. This "swap" of molecules is necessary for all animals because they require oxygen to carry out cellular respiration and because they make carbon dioxide as a waste product.

Gas exchange occurs by diffusion, which you learned about in Chapter 4. Oxygen diffuses through the cell membrane into the cell, and carbon dioxide diffuses out. Because diffusion works well only over short distances, cells must either be close to the environment or have gases carried to and away from them. Recall from Chapter 23 that flatworms and nematodes cannot be thick because diffusion alone carries oxygen to their innermost tissues. Chordates, arthropods, mollusks, echinoderms, and some annelids can be larger because they have evolved ways to transport oxygen to the cells deep within their body.

Gills are structures for gas exchange in an aquatic environment. Aquatic arthropods, fishes, and amphibian larvae have gills. Amphibian larvae have external gills. The gills of crabs and many of their aquatic arthropod relatives are located in chambers behind the head, as shown in **Figure 24-8**. The gills of fishes also lie directly behind the head, as you can see in **Figure 24-9**.

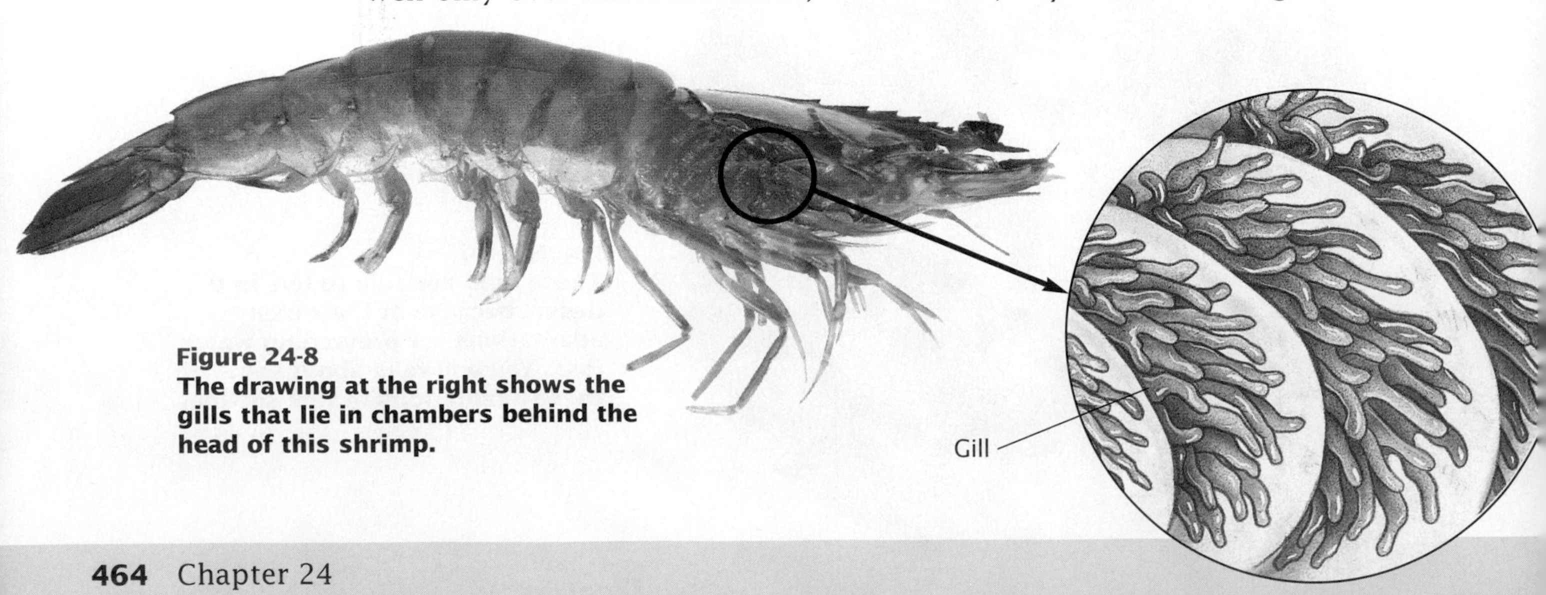

Figure 24-8
The drawing at the right shows the gills that lie in chambers behind the head of this shrimp.

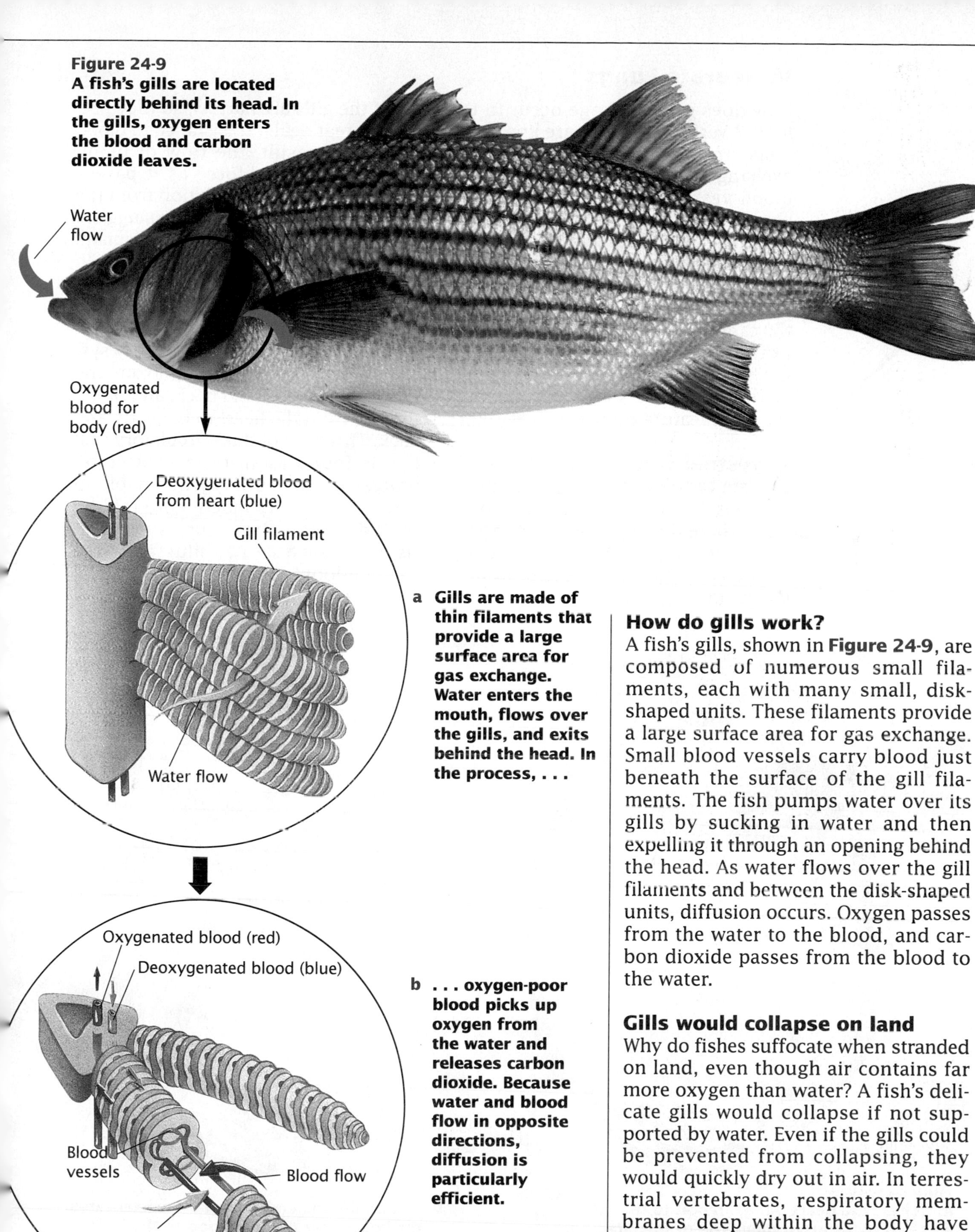

Figure 24-9
A fish's gills are located directly behind its head. In the gills, oxygen enters the blood and carbon dioxide leaves.

a Gills are made of thin filaments that provide a large surface area for gas exchange. Water enters the mouth, flows over the gills, and exits behind the head. In the process, . . .

b . . . oxygen-poor blood picks up oxygen from the water and releases carbon dioxide. Because water and blood flow in opposite directions, diffusion is particularly efficient.

How do gills work?

A fish's gills, shown in **Figure 24-9**, are composed of numerous small filaments, each with many small, disk-shaped units. These filaments provide a large surface area for gas exchange. Small blood vessels carry blood just beneath the surface of the gill filaments. The fish pumps water over its gills by sucking in water and then expelling it through an opening behind the head. As water flows over the gill filaments and between the disk-shaped units, diffusion occurs. Oxygen passes from the water to the blood, and carbon dioxide passes from the blood to the water.

Gills would collapse on land

Why do fishes suffocate when stranded on land, even though air contains far more oxygen than water? A fish's delicate gills would collapse if not supported by water. Even if the gills could be prevented from collapsing, they would quickly dry out in air. In terrestrial vertebrates, respiratory membranes deep within the body have replaced gills. These respiratory membranes line the interior surfaces of sacs called lungs.

Vertebrate Lungs

How does gas exchange occur in the lungs? Vertebrate lungs are lined with moist, thin tissues through which gas exchange occurs. A network of fine blood vessels called capillaries runs close to this lining. Oxygen-rich air is brought into the lungs by inhalation. Oxygen diffuses through the lining of the lungs into the capillaries. At the same time, carbon dioxide moves from the capillaries into the air in the lungs. Exhalation expels the "used" air from the lungs. You can follow the evolution of vertebrate lungs in the *Evolution of the Lung* feature on pages 468–469.

Terrestrial vertebrates have a double-loop circulatory system

In fishes, the heart pumps blood to the gills, where the blood picks up oxygen before flowing to the rest of the body. This creates the "single-loop" system shown in **Figure 24-10a**. The capillaries of the gills are narrow, so they present a great deal of resistance to blood flow. As a result, the flow of blood loses much of its force as it passes through the gills. Circulation from the gills to the rest of the body is sluggish.

The fine capillaries of an amphibian's lung also slow down blood flow. In amphibians, however, the blood returns to the heart for repumping before circulating to the body's tissues. In effect, there are two circulatory systems here: blood flowing from the heart to the lungs and back, and blood flowing from the heart to the body and back. This type of dual circulatory system is found in all terrestrial vertebrates. It carries oxygenated blood through the body much more rapidly than does the circulatory system of fishes. **Figure 24-10b** illustrates the "double-loop" circulation of a typical amphibian.

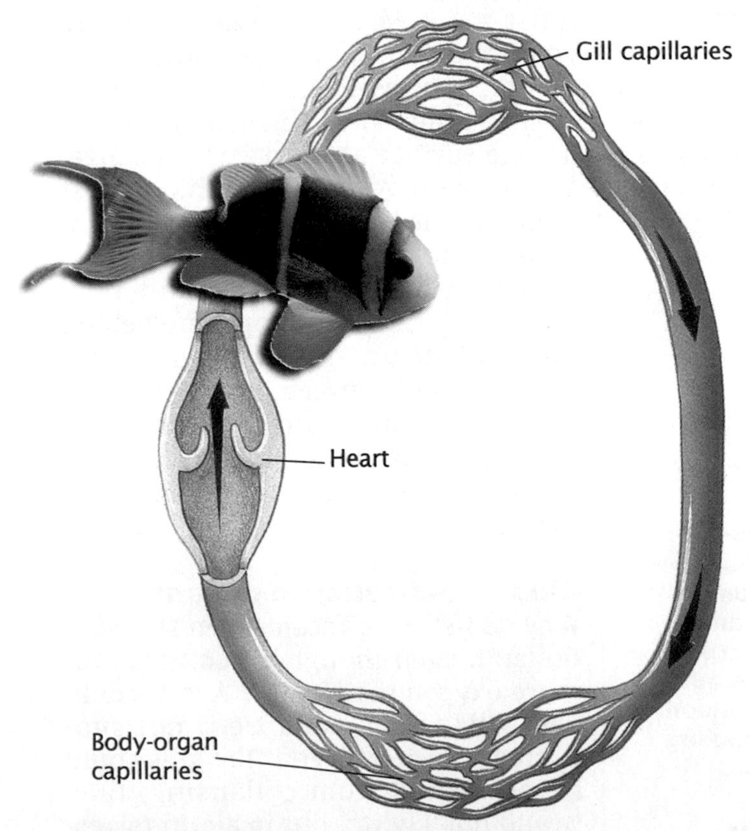

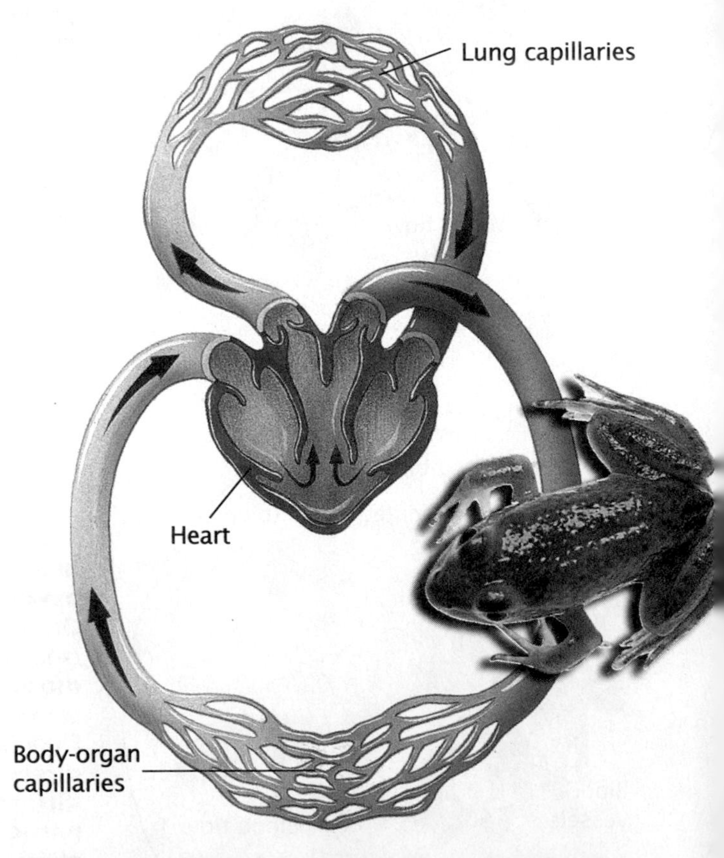

Figure 24-10

a In fishes, the heart pumps blood to the gills, where it picks up oxygen. The oxygenated blood then flows to the rest of the body.

b In amphibians and other land vertebrates, the heart first pumps blood to the lungs, where it picks up oxygen. The oxygenated blood returns to the heart to be pumped to the rest of the body.

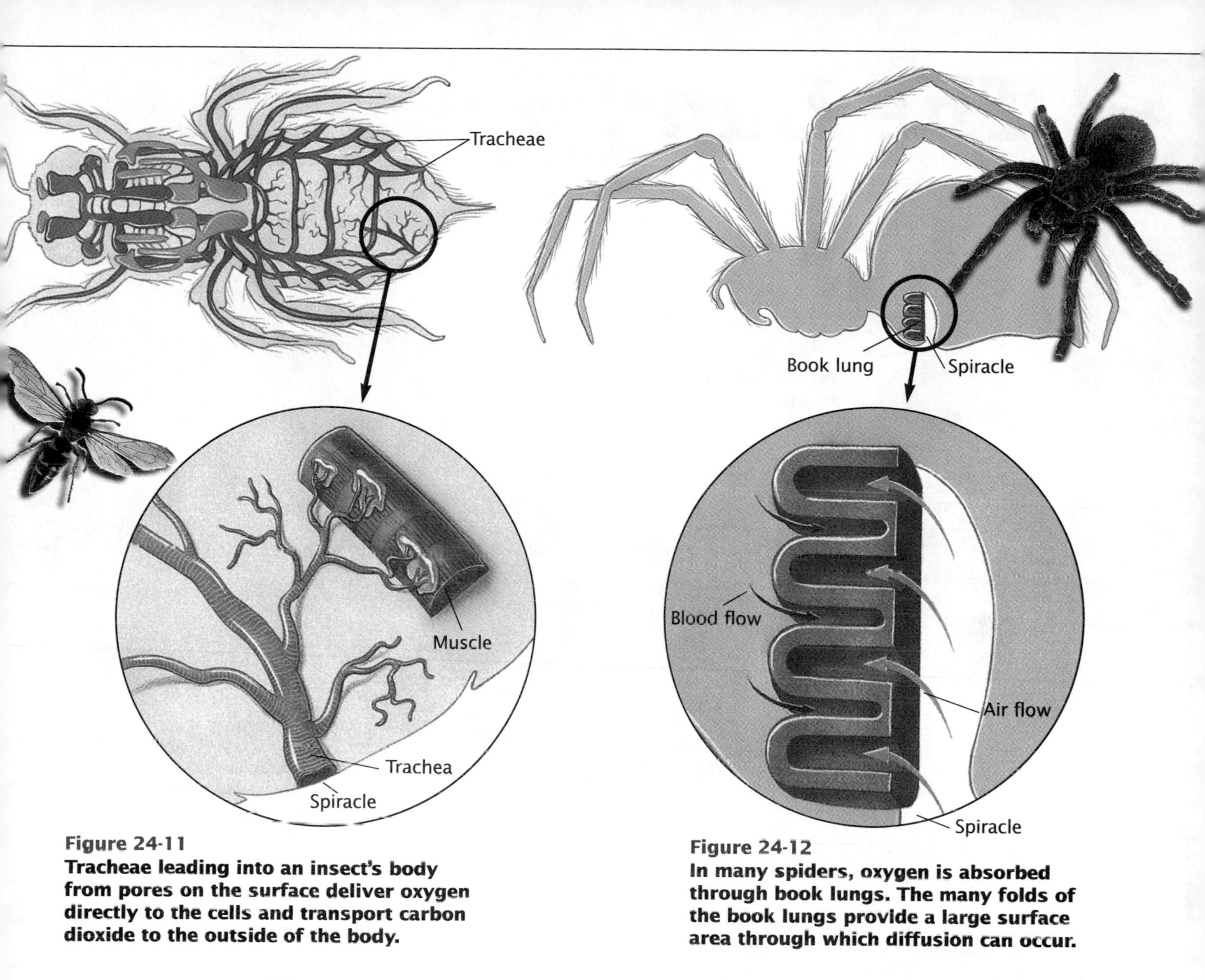

Figure 24-11
Tracheae leading into an insect's body from pores on the surface deliver oxygen directly to the cells and transport carbon dioxide to the outside of the body.

Figure 24-12
In many spiders, oxygen is absorbed through book lungs. The many folds of the book lungs provide a large surface area through which diffusion can occur.

How Terrestrial Arthropods Breathe

Land-dwelling arthropods, which are much smaller than most vertebrates, evolved different gas exchange systems. Insects breathe through **tracheae** *(TRAY kee ee)*. Tracheae (singular, trachea) are tubes that lead into an insect's body cavity from pores called spiracles in the surface of the exoskeleton, as shown in **Figure 24-11**. Each trachea branches and rebranches, penetrating deep into the body. The tips of the tiny branches carry oxygen directly to cells where it is needed. They also allow carbon dioxide to exit. This system is very efficient in small organisms, and it supports the very active lifestyles of most insects. In an insect, oxygen passes directly from the air in the tracheae to the tissues. In your body, oxygen is first transferred to blood cells in the lungs and is then carried to the tissues by blood. Insect blood does not transport oxygen.

Many spiders breathe through **book lungs**. As shown in **Figure 24-12**, book lungs are highly folded sacs that are located inside the body and that open to the outside through a spiracle. Air moves into the sacs, where it is moistened so the oxygen in the air can dissolve and pass into the blood. Having respiratory membranes inside the body allows the surface where gas exchange occurs to remain moist without losing large amounts of water to evaporation.

The Lung

Lungs first evolved in early fishes. Early bony fishes with lungs were the ancestors of amphibians. Lungs provide land vertebrates with a large surface area where gas exchange can occur in a moist environment.

A Simple Sac

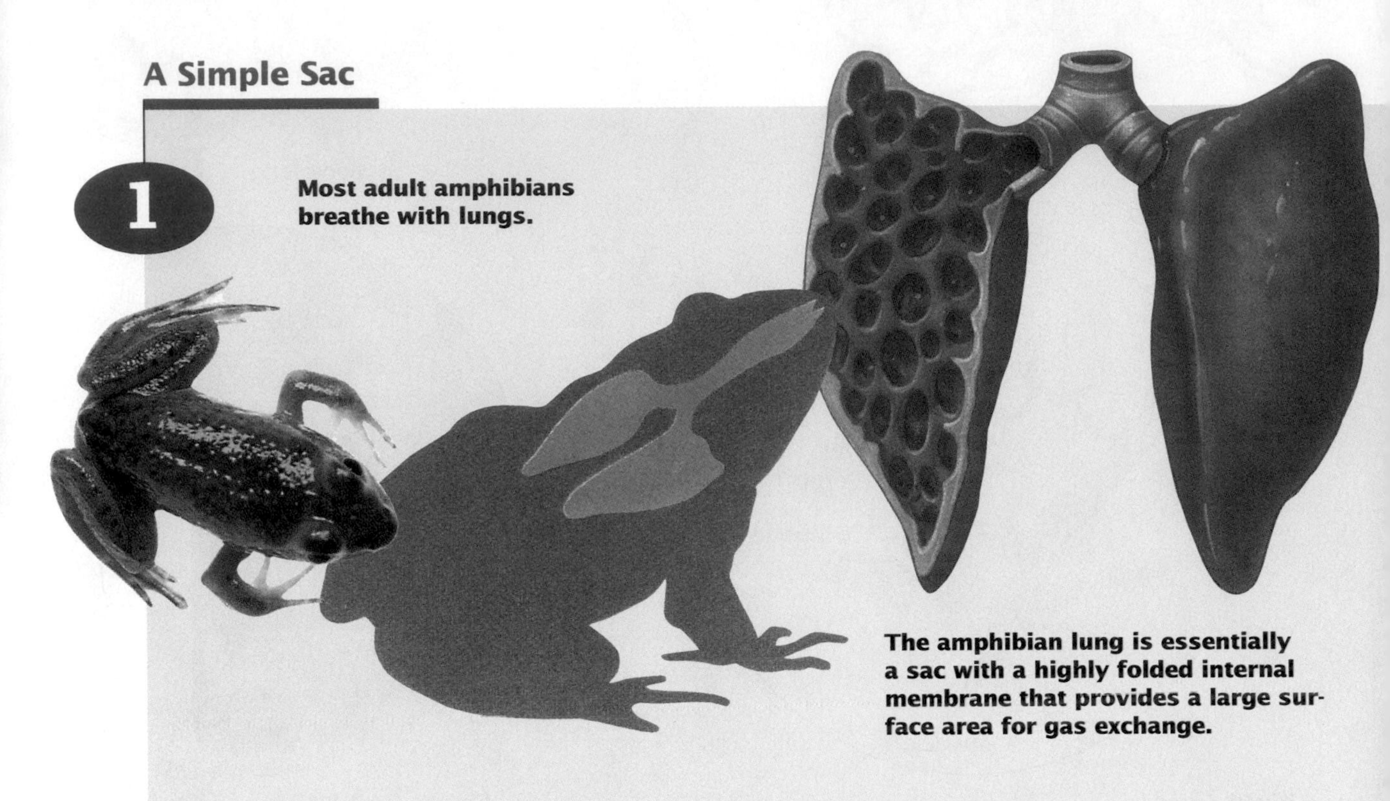

1
Most adult amphibians breathe with lungs.
The amphibian lung is essentially a sac with a highly folded internal membrane that provides a large surface area for gas exchange.

Increased Surface Area

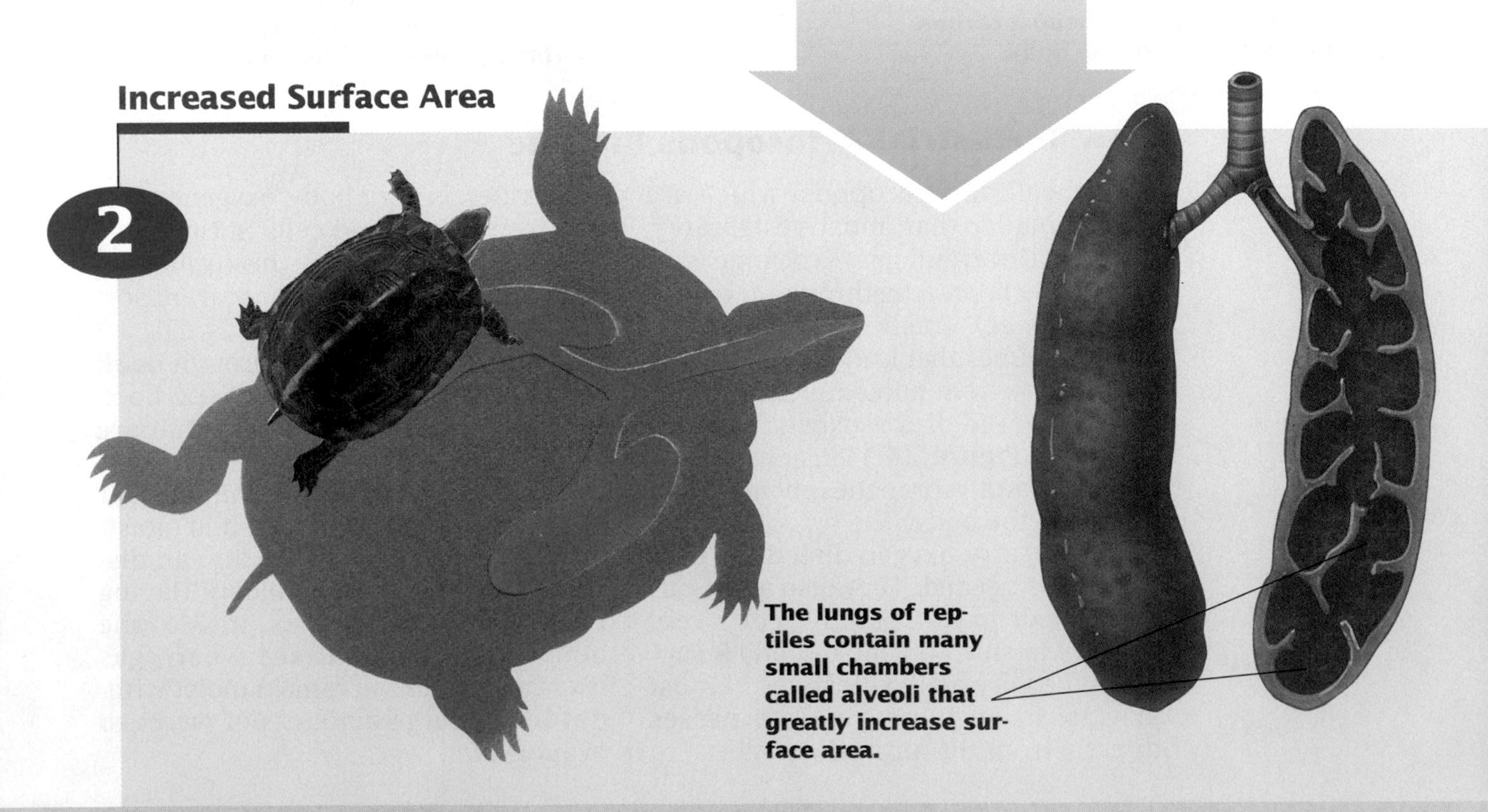

2
The lungs of reptiles contain many small chambers called alveoli that greatly increase surface area.

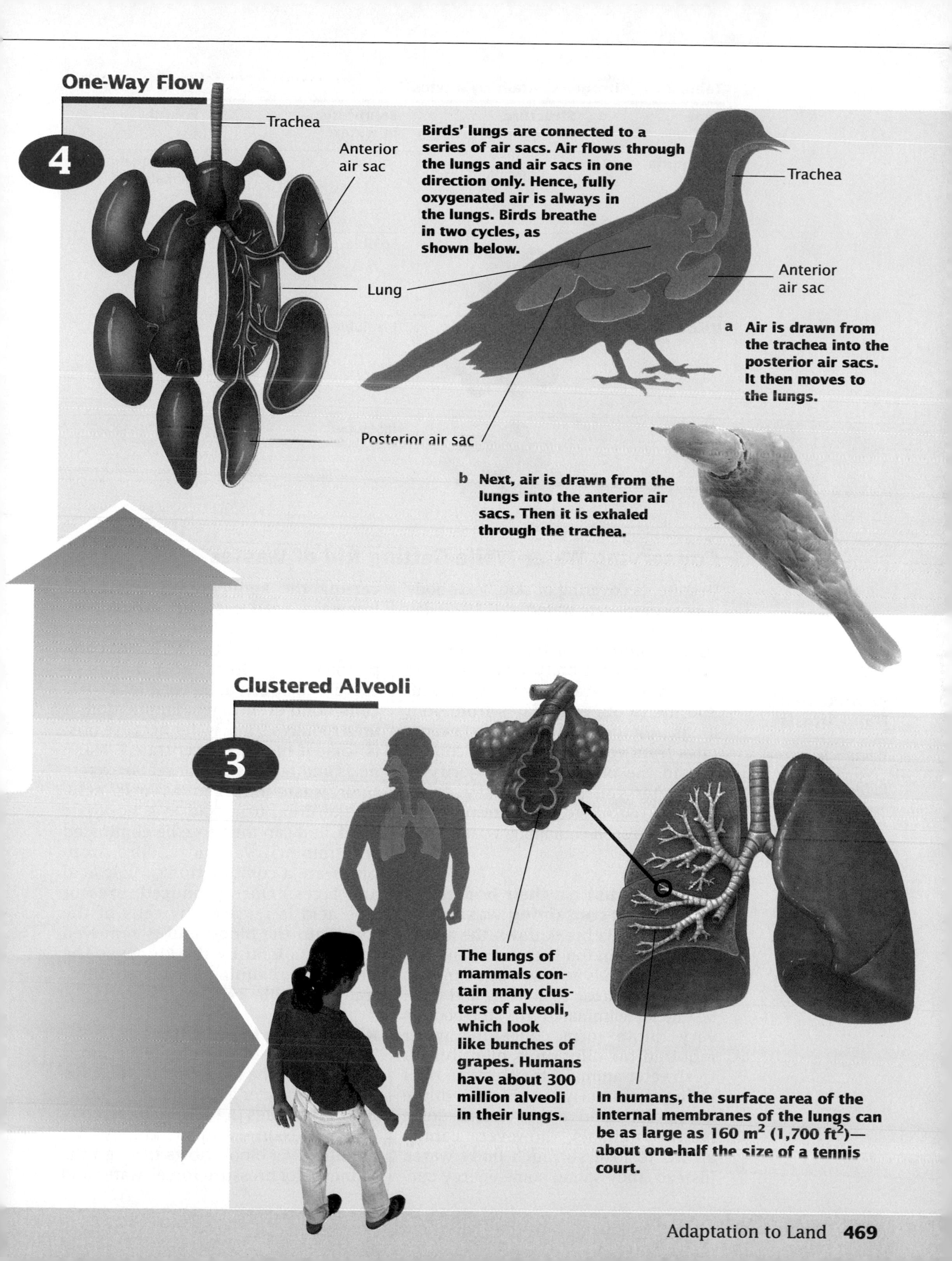
One-Way Flow
4
Trachea
Anterior air sac
Birds' lungs are connected to a series of air sacs. Air flows through the lungs and air sacs in one direction only. Hence, fully oxygenated air is always in the lungs. Birds breathe in two cycles, as shown below.
Trachea
Lung
Anterior air sac
a Air is drawn from the trachea into the posterior air sacs. It then moves to the lungs.
Posterior air sac
b Next, air is drawn from the lungs into the anterior air sacs. Then it is exhaled through the trachea.
Clustered Alveoli
3
The lungs of mammals contain many clusters of alveoli, which look like bunches of grapes. Humans have about 300 million alveoli in their lungs.
In humans, the surface area of the internal membranes of the lungs can be as large as 160 m² (1,700 ft²)—about one-half the size of a tennis court.

Table 24-1 Nitrogen-Containing Wastes

Type	Structure (nitrogen in orange)	Solubility in water	Found in
Ammonia		Soluble	Most aquatic animals
Urea		Soluble	Amphibians, mammals
Uric acid		Insoluble	Insects, reptiles, birds
Guanine		Insoluble	Spiders

Conserving Water While Getting Rid of Wastes

Stability and Homeostasis

What role do the kidneys play in homeostasis?

Despite its covering of skin, your body is not completely watertight. Every day you lose about 2.5 L (2.4 qt) of water—about 2 percent of the total amount of water in your body. How does that much water escape each day? About 300 mL (9 oz) evaporates from your lungs. Another 500 mL (15 oz) evaporates from your skin as sweat, cooling you in the process. The majority of the water you lose each day, about 1.5 L (45 oz), is excreted as urine. Urine carries nitrogen-containing wastes from your body.

Animals must rid their bodies of nitrogen-containing wastes

When animals break down the amino acids found in foods that contain protein, nitrogen is released as ammonia. Ammonia is toxic to animals and must be quickly eliminated from the body in very dilute form. Since the water required for dilution is plentiful for freshwater animals, they excrete their nitrogen directly as highly dilute ammonia. Land animals and some marine animals, however, cannot afford to lose so much body water. Instead, they spend some energy converting the ammonia to less toxic forms that need less water for dilution.

In vertebrates, ammonia is converted into one of two different compounds in the liver. Most mammals excrete nitrogen as **urea**, a less toxic compound of nitrogen. Elimination of urea requires some water because urea is toxic if highly concentrated. Many birds and reptiles (and insects) excrete their waste nitrogen as **uric acid**. Unlike urea, uric acid is a nontoxic solid, and can therefore be eliminated with minimal water loss. (Bird droppings are a combination of uric acid and feces.) Once produced, urea or uric acid is released by cells of the liver into the blood and is removed from the blood by the kidneys. The nitrogen-containing wastes are illustrated in **Table 24-1**.

Kidneys are blood filters

The kidneys of vertebrates remove wastes from the blood and regulate the amount of water and salts in the body. Blood circulates in the vertebrate body under pressure, and the kidneys act as filters. As blood flows through the kidneys, its pressure forces water and

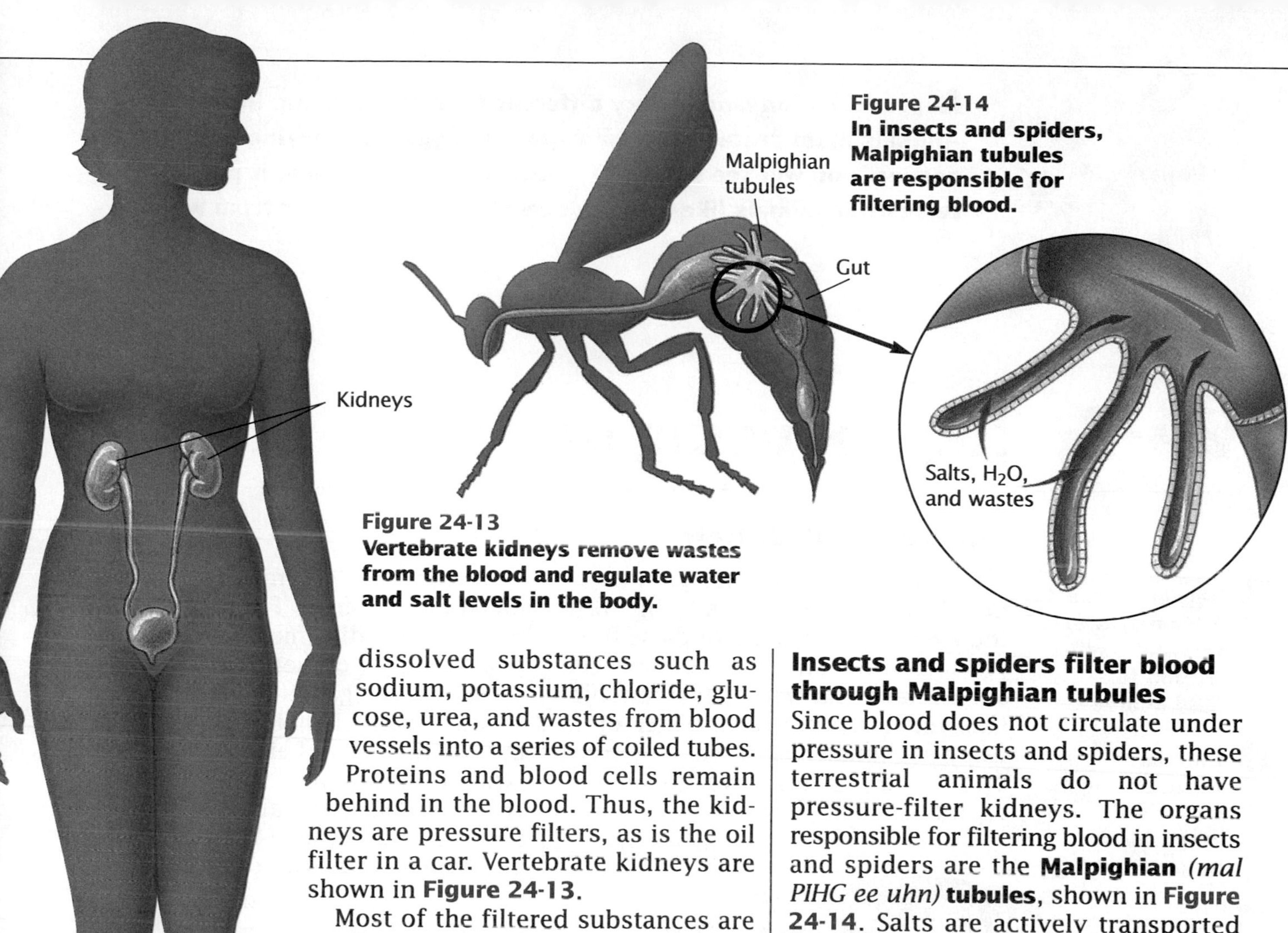

Figure 24-14
In insects and spiders, Malpighian tubules are responsible for filtering blood.

Figure 24-13
Vertebrate kidneys remove wastes from the blood and regulate water and salt levels in the body.

dissolved substances such as sodium, potassium, chloride, glucose, urea, and wastes from blood vessels into a series of coiled tubes. Proteins and blood cells remain behind in the blood. Thus, the kidneys are pressure filters, as is the oil filter in a car. Vertebrate kidneys are shown in **Figure 24-13**.

Most of the filtered substances are useful to the body, and it would be wasteful to discard them. Instead they are reabsorbed back into the bloodstream. Reabsorption is very selective. Urea, wastes, and some water are not reabsorbed; they pass out of the body as urine. The body regulates how much water is reabsorbed and can control how much water leaves the body in the urine. If you drink a large amount of water, your brain senses an increase in blood pressure and releases hormones that cause the kidneys to reabsorb less water. As a result, you produce more urine.

Insects and spiders filter blood through Malpighian tubules

Since blood does not circulate under pressure in insects and spiders, these terrestrial animals do not have pressure-filter kidneys. The organs responsible for filtering blood in insects and spiders are the **Malpighian** *(mal PIHG ee uhn)* **tubules**, shown in **Figure 24-14**. Salts are actively transported from the blood into the Malpighian tubules. The high concentration of salts in the Malpighian tubules *draws* water from the blood. Wastes also move into the tubules. The contents of the Malpighian tubules are released into the gut, where useful salts such as potassium are reabsorbed. Water follows the salts by osmosis. Wastes such as uric acid and guanine are not reabsorbed and remain in the gut as a nearly dry paste that is excreted with the feces. This system enables insects and spiders to conserve water and contributes to their success on land.

SECTION REVIEW

1. **How would the exoskeleton of a land arthropod probably be different from that of an aquatic arthropod? Explain.**
2. **Contrast the amount of evaporation from amphibian skins with that from reptilian skins.**
3. **Describe the differences between amphibian lungs and mammalian lungs.**
4. **List two adaptations that enable land vertebrates to get rid of wastes while still conserving water.**

Reproducing on land is very different from reproducing in water. Animals must protect their gametes and eggs from drying out. In this section, you will see some of the adaptations that make it possible for terrestrial animals like this wildebeest to reproduce far from water.

24-3 Reproducing on Land

OBJECTIVES

1. **Recognize the advantages of internal fertilization for land animals.**
2. **Compare a reptilian egg with an amphibian egg.**
3. **Summarize two advantages of the placental mammals' pattern of development.**

Internal Fertilization

The frogs that you see in **Figure 24-15** may have traveled miles to breed in the water. The behavior of these frogs reflects the strong ties between amphibians and water. Most amphibians must reproduce in water or in damp environments. Insects, mammals, birds, and reptiles have broken the reproductive tie to water.

An egg and a sperm must unite in a moist environment. For many aquatic animals, including most fishes, finding a moist environment is easy: they release their gametes into the surrounding water. Fertilization takes place outside the body of either parent. This type of fertilization is called **external fertilization**. External fertilization is rare on land because both the sperm and the egg could dry out and die.

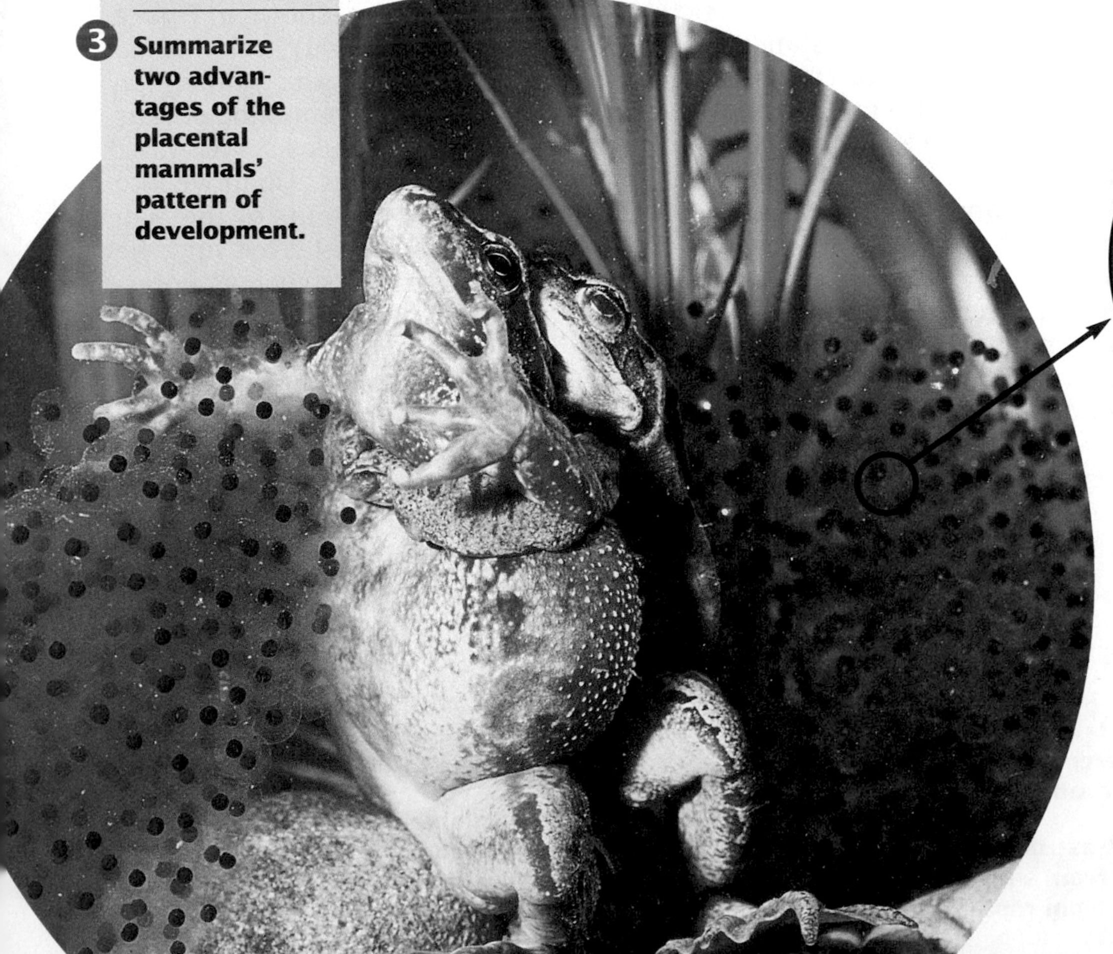

Figure 24-15
The male frog (behind) grasps the female frog and fertilizes her eggs as they are released. Each egg is enclosed in a jellylike coating that is freely permeable to water. These eggs would quickly dry out if laid on land.

Most terrestrial animals reproduce by internal fertilization

How did terrestrial animals overcome the limitations of external fertilization? Recall that each animal carries its own supply of water inside its body. Therefore, if gametes can be transferred directly from the body of one individual to the body of another, the risk that gametes will dry out disappears. In almost all land animals, including flatworms, roundworms, annelids, mollusks, arthropods, reptiles, birds, and mammals, the male deposits his sperm directly inside the female. Fertilization occurs inside the female's body. This type of fertilization is called **internal fertilization**. Most amphibians lack the capability for internal fertilization. They reproduce by external fertilization, as did their fish ancestors. Now you can see one reason why amphibians, such as toads, must return to water to reproduce.

If you go to a pond a few days after toads breed, you will find that the water contains strings of small, round eggs, like those shown in **Figure 24-15**. These eggs are the second reason that most amphibians must reproduce in water or moist places. The embryonic toad develops within the egg, which has a jellylike coating. This coating is freely permeable to carbon dioxide, oxygen, and water. Amphibian eggs removed from moisture soon dry out and die.

The eggs of reptiles and birds are watertight

Unlike amphibian embryos, the embryos of reptiles are surrounded by a watertight protective membrane called the **chorion** *(KAWR ee ahn)*. The chorion is impermeable to water, but it does permit oxygen to enter the egg and carbon dioxide to leave. Lying within the chorion is another membrane, the **amnion** *(AM nee ahn)*. The amnion encloses the embryo within a watery environment, as shown in **Figure 24-16**. This kind of egg is called an **amniotic egg**. Amniotic eggs evolved first in reptiles and are also found in birds and mammals. In the eggs of reptiles and birds, a nutrient-rich yolk supplies nutrition to the developing embryo. A tough shell surrounds and protects the eggs of all birds, most reptiles, and three species of egg-laying mammals.

Like the eggs of birds and reptiles, the eggs of most terrestrial arthropods are watertight and very yolky, providing nutrients for the young during development. Arthropod eggs are usually either thick-walled and resistant to drying or are encased in some sort of waxy protective covering.

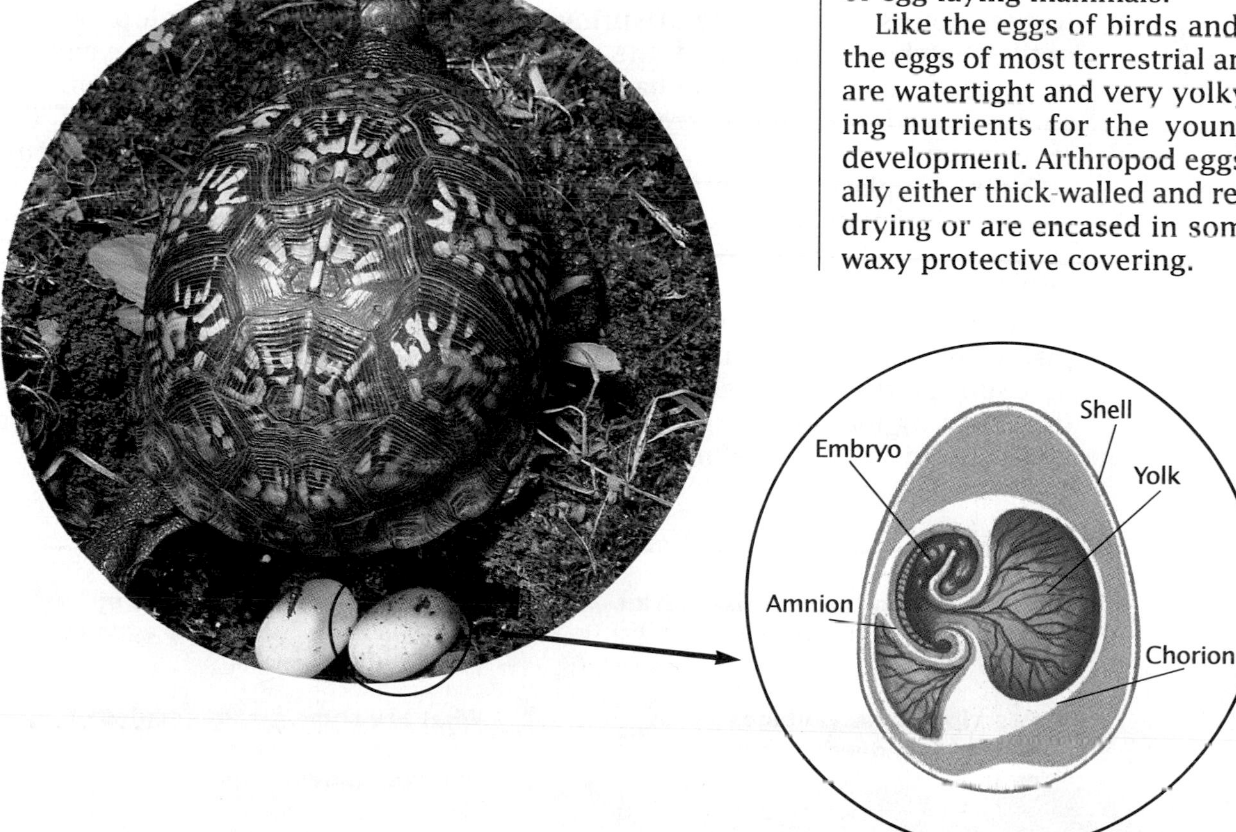

Figure 24-16
The eggs of birds and reptiles, such as the turtle shown below, have a watertight protective coating called the chorion and an internal membrane called the amnion, which encloses the embryo in a watery environment.

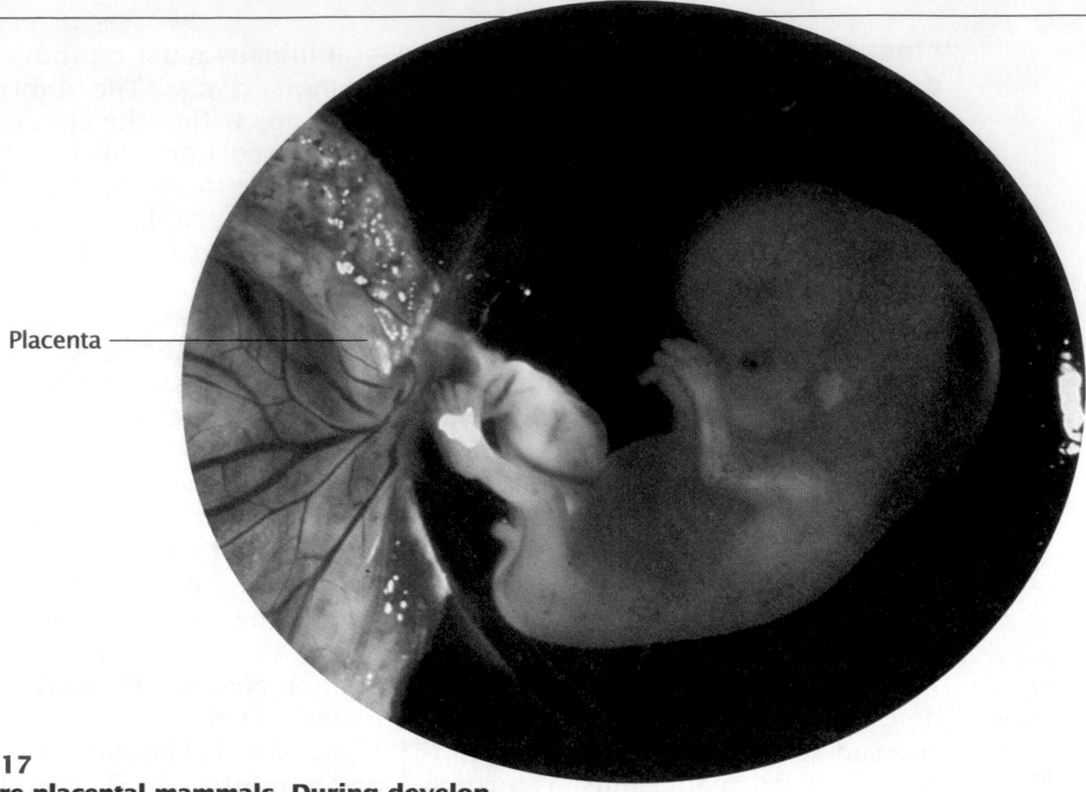

Figure 24-17
Humans are placental mammals. During development, the embryo receives nourishment and eliminates wastes through the placenta.

Eggs Without Shells

Once the egg of a reptile or bird has been laid, the parent cannot provide further nourishment to the offspring until it hatches. Furthermore, the egg is exposed to environmental hazards such as predators and extremes of temperature.

Except for platypuses and echidnas *(ee KIHD nuhz)*, mammals do not lay eggs. Instead, their eggs develop until birth inside the mother's body. No shell forms around the egg. Among marsupial mammals, such as kangaroos and opossums, birth occurs very early, and the offspring matures in the mother's pouch, nourished by suckling her milk. Among placental mammals (mice and humans, for instance), birth occurs much later, when development is essentially complete. A human embryo develops for nine months inside its mother. During its development, the embryo of a placental mammal is nourished by the mother through a unique structure called the **placenta** *(pluh SEHN tuh)*, which is shown in **Figure 24-17**. The placenta is made of embryonic and maternal membranes. Nutrients and oxygen move from mother to embryo through the placenta, while carbon dioxide and other wastes leave the embryo.

SECTION REVIEW

1. **Why isn't external fertilization an effective way to reproduce on dry land?**
2. **Describe the advantages of the shelled egg over the amphibian egg.**
3. **Describe two hazards faced by reptilian eggs but not by eggs of placental mammals.**
4. **What are some disadvantages of the placental mammals' pattern of development?**

CHAPTER

24 Highlights

"Gee, evolution is slow."

24-1 Leaving the Sea

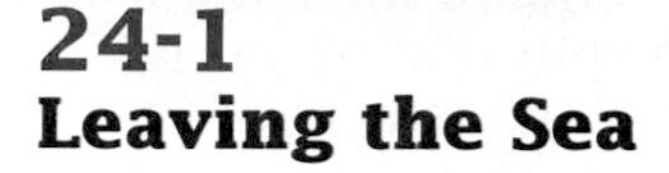

Reptiles have limbs partway beneath their bodies to support their weight more effectively.

Key Terms

lateral line (p. 462)

Summary

- Life on land places different demands on animals than life in water.
- The most successful land groups are the arthropods and the vertebrates. Both of these groups have sturdy skeletons that support their bodies out of water. They also have strong, flexible limbs for moving on land.
- The high density of water makes sound easy to detect. Land animals hear well only with an amplifying system, such as the ear.

24-2 Staying Moist in a Dry World

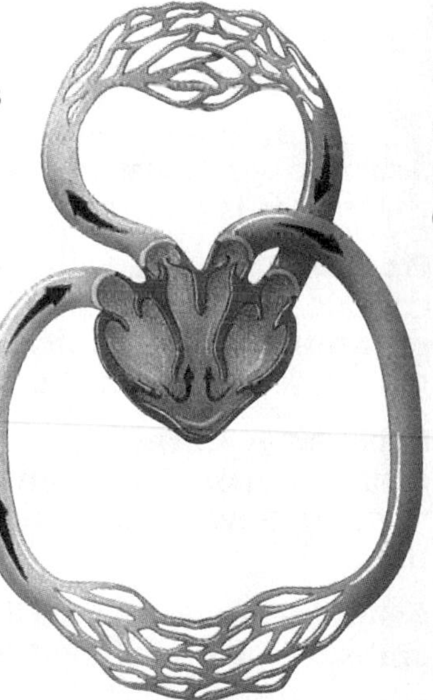

Amphibian blood is pumped through a "double-loop" circulatory system.

Key Terms

gill (p. 464)

trachea (p. 467)

book lung (p. 467)

urea (p. 470)

uric acid (p. 470)

Malpighian tubule (p. 471)

Summary

- The exoskeleton of arthropods serves as a barrier to evaporation. Birds, reptiles, and mammals minimize water loss by means of their watertight skin.
- Animals must obtain oxygen and release carbon dioxide. In land vertebrates, this gas exchange occurs in the lungs. In insects, tracheae carry oxygen.
- The breakdown of amino acids produces the toxic byproduct ammonia. Aquatic animals excrete dilute ammonia. Terrestrial animals transform ammonia into urea or uric acid, which is eliminated with less water loss.

24-3 Reproducing on Land

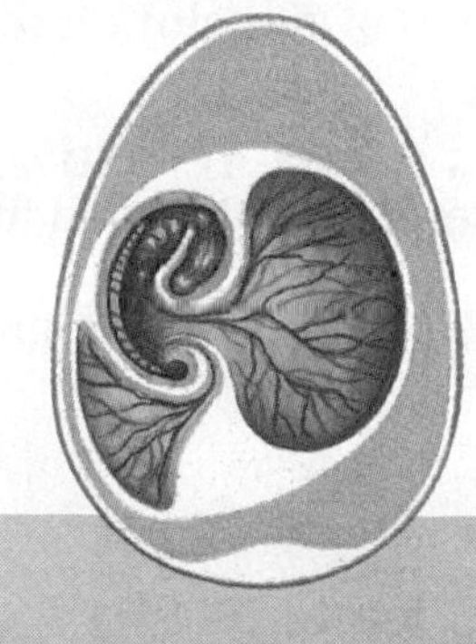

Reptilian eggs have a protective shell and a watery internal environment.

Key Terms

external fertilization (p. 472)

internal fertilization (p. 473)

chorion (p. 473)

amnion (p. 473)

amniotic egg (p. 473)

placenta (p. 474)

Summary

- Most amphibians reproduce in water by external fertilization. Most other land animals reproduce by internal fertilization.
- The eggs of reptiles, birds, and mammals are surrounded by watertight membranes. In most mammals, development is completed within the mother's body.

CHAPTER 24

review

Understanding Vocabulary

1. For each list of terms, identify the one that does not fit the pattern and explain why.
   a. kidneys, lungs, trachea, gills
   b. blood, ammonia, urea, uric acid
   c. chorion, amnion, shell, Malpighian tubules

2. For each pair of terms, explain the differences in their meanings.
   a. lateral line, ear
   b. exoskeleton, scales
   c. trachea, book lungs

Understanding Concepts

3. **Relating Concepts** Construct a concept map that describes the adaptation of animals to life on land and the characteristics that allow them to live on land successfully. Try to include the following terms in your concept map: lobe-finned fishes, amphibians, arthropods, amniotic egg, yolk, birds, reptiles, insects, trachea, oxygen, lizards.

4. **Analyzing Structures** Compare the position of limbs in amphibians with that in reptiles. What advantages result from the arrangement shown by reptiles?

5. **Analyzing Relationships** Explain the relationship between the density of a medium and the speed and distance that sound travels through it. What is the consequence of this relationship for land animals?

6. **Comparing Structures** Arthropods and vertebrates prevent water loss by having waterproof coatings. How do the coatings of arthropods, amphibians, and reptiles differ?

7. **Identifying Processes** What is gas exchange? How does gas exchange occur in the gills of a fish?

8. **Analyzing Structures** Explain why the gills of fishes don't function out of water.

9. **Recognizing Relationships** What features contribute to the high efficiency of a bird's respiratory system?

10. **Comparing Structures** Compare the circulatory system of a fish with that of an amphibian. What is the advantage of the amphibian circulatory system?

11. **Interpreting Graphs** Cheetahs are the fastest land animals. Biologists were interested in learning how the lung size and oxygen consumption of cheetahs differed from those of other cats of similar size. This graph shows the results of their study. Form a hypothesis relating the rate of oxygen consumption in cheetahs to their running performance.

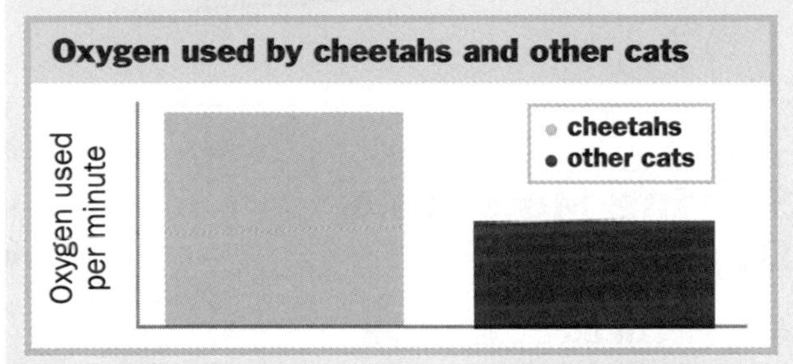

12. **Summarizing Information** Identify three ways that water is lost from the human body.

13. **Predicting Outcomes** Fishes eliminate nitrogen-containing wastes as ammonia. What complications might result if land animals also eliminated waste in this form?

14. **Identifying Functions** Describe how kidneys enable a mammal to remove wastes and to regulate how much water is retained by the body.

15. **Analyzing Processes** Describe how insects and spiders filter blood to remove wastes.

16. **Recognizing Relationships** Explain why most amphibians must reproduce in water or moist environments.

17. **Summarizing Information** Describe two advantages of development in a placental mammal.

18. **BUILDING ON WHAT YOU HAVE LEARNED** In Chapter 4, you learned about diffusion. What role does diffusion play in the exchange of gases in the gills of fishes?

Interpreting Graphics

19. Look at the figure below.

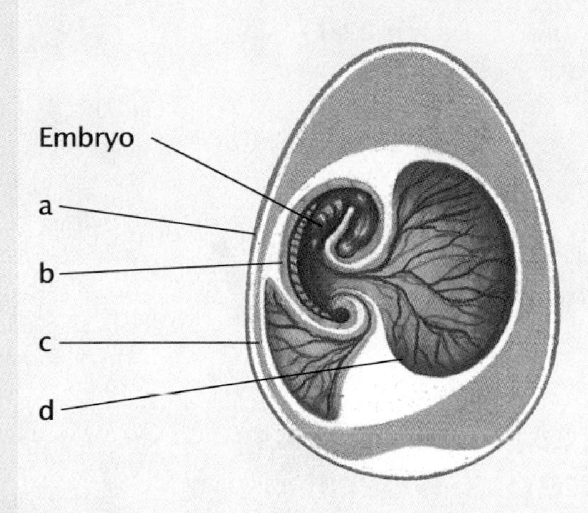

a. What kind of egg is shown in the figure above? Identify the parts of the egg, and describe the function of each part.
b. What animals lay this kind of egg?
c. What evolutionary advantage does an egg of this type have over an amphibian egg?

Reviewing Themes

20. ***Evolution***
What evidence has led scientists to conclude that the earliest land vertebrates evolved from lobe-finned fishes?

21. ***Evolution***
Of what significance was the evolution of a watertight skin to reptiles?

22. ***Matter, Energy, and Organization***
Land animals use energy to convert ammonia into urea or uric acid. What do these animals gain from this conversion?

Thinking Critically

23. **Determining Factual Accuracy** Explain the factual inaccuracies in the following statements: Amphibians evolved from sharks around 350 million years ago. Amphibians lay jellylike eggs in which the embryo is surrounded by an amnion and a chorion. Most amphibians have a moist skin through which they absorb oxygen.

Life/Work Skills

24. **Finding Information** Use library references or search an on-line database to research the fishes called mudskippers (*Boleophthalmus pectinirostris* is one species). Find out the unusual behaviors of male mudskippers that take them out of the water to reproduce and feed. Prepare an illustrated report of your findings to share with your class.

25. **Working Cooperatively** Work together with a partner to make models of a typical amniotic egg and a typical amphibian egg. Use modeling clay, paper, or other materials to make your models. Use library references to find out more detailed information about the structure of these two types of eggs. Write a report about the adaptations exhibited by each kind of egg.

Activities and Projects

26. **Art Connection** Draw a picture of a fictional animal that is adapted for survival in one of the seven biomes studied in Chapter 14. Write a description of how the animal reproduces, breathes, and eliminates wastes.

27. **History Connection** In 1938 a South African fisherman caught a strange type of fish. Scientists later identified it as a kind of lobe-finned fish called a coelacanth. Use library references or search an on-line database to find out the historical significance of this discovery. Write a report on the characteristics of coelacanths that indicate they are close relatives of modern amphibians.

Discovering Through Reading

28. Read the article "Coming Onto the Land" by Carl Zimmer in *Discover*, June 1995, pages 118–127. What is important about the discovery of *Acanthostega*? What evidence about the way the first land animals moved has been revealed by *Acanthostega*?

CHAPTER 25

REVIEW
- animal tissues (Section 23-1)
- origin of body cavities (Section 23-2)
- innovations in body plan (Section 23-3)

Animal Diversity

This colorful nudibranch is a mollusk that lives in the Indian and Pacific Oceans.

Chapter 23 told the story of the evolution of animal body architecture. In this chapter, you will return to the major phyla of animals for a longer visit. You will discover where some of these animals—such as the sea anemone at the left—live, how they feed, and how they reproduce. You will also learn how some of these animals affect humans, both positively and negatively.

25-1 Sponges, Cnidarians, and Simple Worms

OBJECTIVES

1. **Contrast the way of life of a sponge with that of a cnidarian.**
2. **Compare the polyp stage of the cnidarian life cycle with the medusa stage.**
3. **Describe the life cycle of the beef tapeworm.**
4. **List two parasitic nematodes that can live in humans.**

Sponges

Sponges are probably most familiar to you as absorbent pads used for wiping up spills or cleaning dishes. The sponges you buy in the store are usually manufactured, not derived from the animals known as sponges. There are more than 9,000 species of sponges (phylum Porifera), most of which are marine. About 150 species live in fresh water. An adult sponge, like the one in **Figure 25-1**, spends its life attached to a hard surface. For this reason, it was not until 1765 that zoologists realized that sponges are animals, not plants. Most sponges are asymmetrical and grow to conform to the surface on which they live.

A sponge's body is perforated by holes (Porifera means "pore-bearer") that lead to an inner water chamber. Sponges pump water through these pores and expel it through a large opening at the top of the chamber. The water current flowing through the sponge's body carries in food particles and oxygen and carries out wastes.

As you learned in Chapter 23, between the inner and outer layer of many sponges is a jellylike layer. Embedded within this layer is a network of hard structures called **spicules**, a mesh of the tough protein spongin, or both. Spicules and spongin form the skeleton of the sponge. Large sponges with spongin skeletons (and few or no spicules) have been harvested for centuries because their dried skeletons will soak up water and release it when squeezed. Most of the sponges you can buy today are copies of this spongin mesh manufactured from plastic or cellulose.

Figure 25-1 **Sponges vary greatly in size, shape, and color. This purple tube sponge was photographed in the ocean surrounding Bonaire, an island in the Caribbean Sea.**

Sponges reproduce sexually and asexually

Most sponges are able to reproduce asexually. In some sponges, new individuals bud from the parent; in others, the parent sponge breaks into many fragments, and each fragment grows into a new sponge. Sponges can also reproduce sexually, as illustrated in **Figure 25-2**.

Figure 25-2
The sexual reproduction of sponges involves the production and subsequent fusion of eggs and sperm.

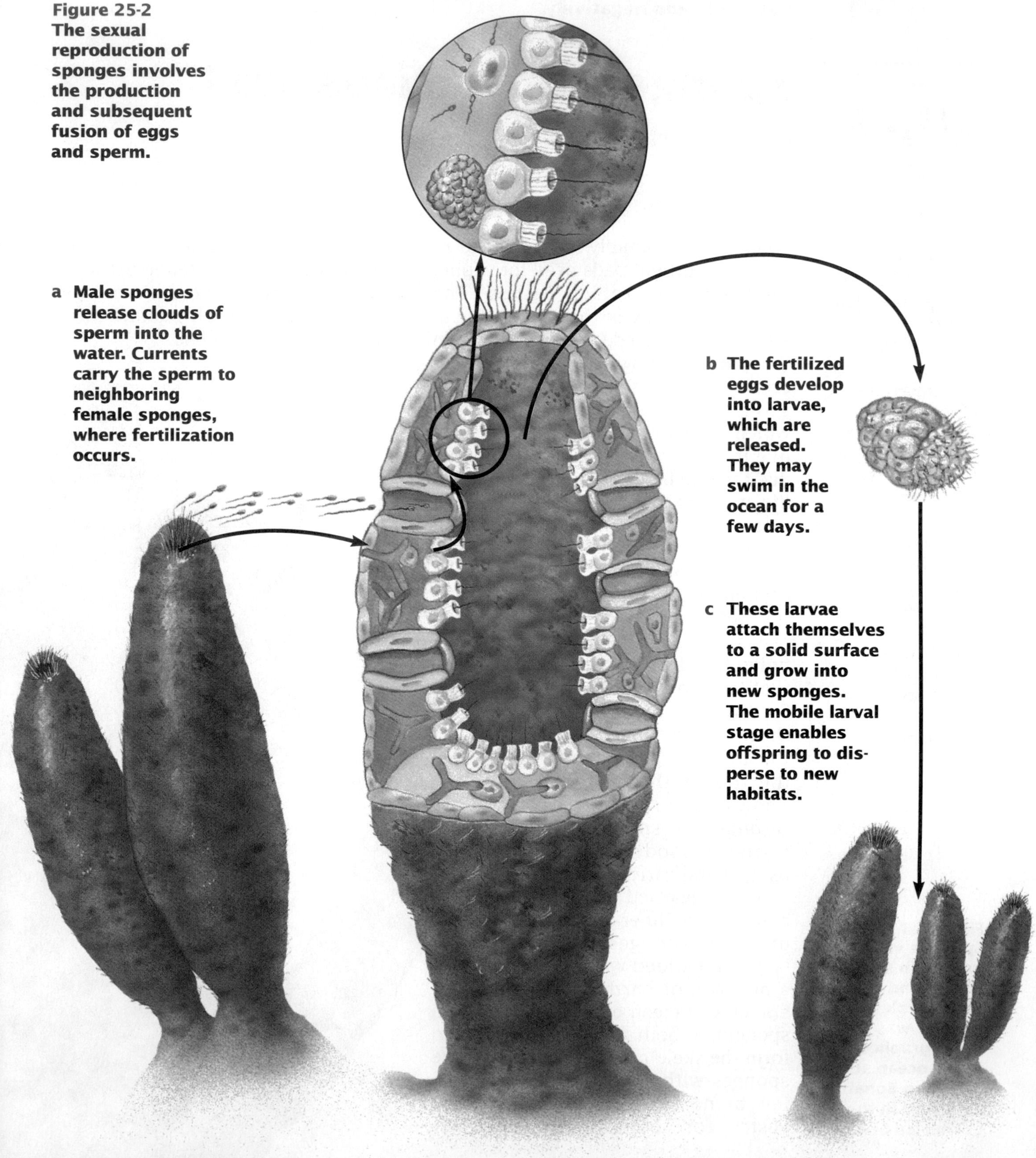

a **Male sponges release clouds of sperm into the water. Currents carry the sperm to neighboring female sponges, where fertilization occurs.**

b **The fertilized eggs develop into larvae, which are released. They may swim in the ocean for a few days.**

c **These larvae attach themselves to a solid surface and grow into new sponges. The mobile larval stage enables offspring to disperse to new habitats.**

Cnidarians: Jellyfish and Relatives

Corals, jellyfish, sea anemones, and hydras are members of the phylum Cnidaria. The name *cnidaria* comes from the Greek word meaning "nettle" and refers to the stinging structures that are characteristic of these animals.

Cnidarians have a gut with only one opening to the outside. This opening is surrounded by a ring of tentacles used to capture food and defend against predators. Cells in the tentacles and outer body surface are armed with stinging, harpoon-like structures called **nematocysts** *(NEHM uh toh sih-sts)*. The projectile fired by a nematocyst contains toxins that can cause paralysis.

Most cnidarians are harmless to people. If you touch the tentacles of a sea anemone, you will feel a sticky sensation as its short projectiles barely pierce your skin. A few jellyfish and corals, however, have very potent toxins that cause a painful, burning rash. The stings of the tropical sea wasp jellyfish can be fatal. To protect against their stings, Australian surfers sometimes wear pantyhose when in the water.

Many cnidarians have two distinct life stages

In Chapter 20 you learned about alternation of generations in plants. Like plants, many cnidarians have two different body forms during their life cycle, as illustrated in **Figure 25-3**. The two forms are called the **polyp** stage and the **medusa** (plural, medusae), or jellyfish, stage. Polyps generally live attached to a hard surface. Sea anemones, hydras, and corals are polyps.

Not all cnidarians go through both polyp and medusa stages. Some remain as either a polyp or medusa throughout their lives. You can read more about cnidarians in the *Tour of a Jellyfish* on page 482.

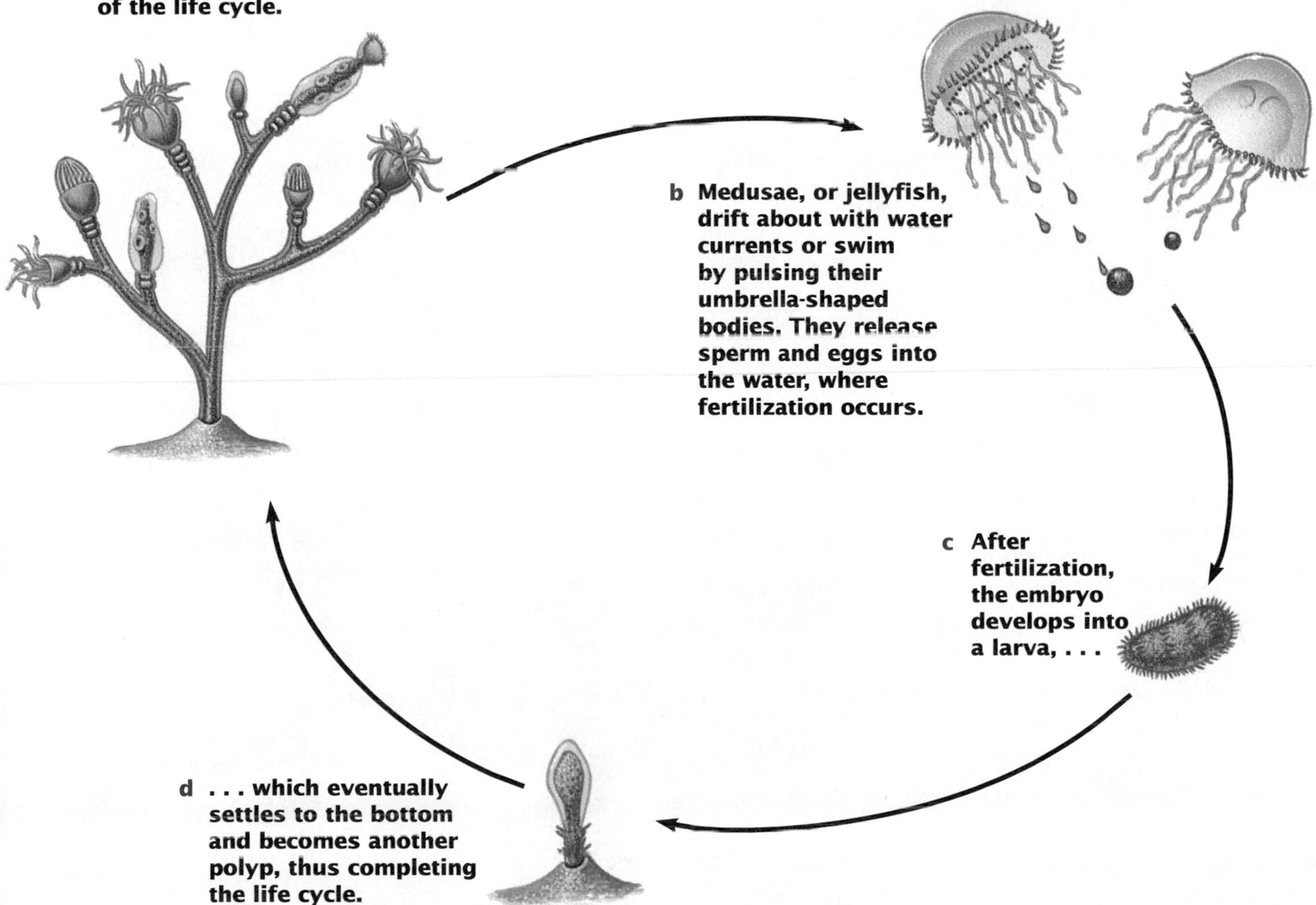

b Medusae, or jellyfish, drift about with water currents or swim by pulsing their umbrella-shaped bodies. They release sperm and eggs into the water, where fertilization occurs.
c After fertilization, the embryo develops into a larva, . . .
d . . . which eventually settles to the bottom and becomes another polyp, thus completing the life cycle.

Figure 25-3
The life cycle of a typical cnidarian involves two different body forms.

a Polyps bud to produce more polyps and, in some cnidarians, to produce the medusa stage of the life cycle.

Tour

Jellyfish

The moon jelly is a common jellyfish. Like all cnidarians, the moon jelly has a body composed of two main layers of cells and a gut with only one opening to the outside.

Habitat
Moon jellies live close to the surface near the shore. In the United States, they are found on the Pacific, Atlantic, and Gulf coasts.

The moon jelly has a bell-shaped body. A muscular ring around the margin of the bell contracts rhythmically to propel the animal through the water.

Size
A large moon jelly can be 41 cm (16 in.) in diameter.

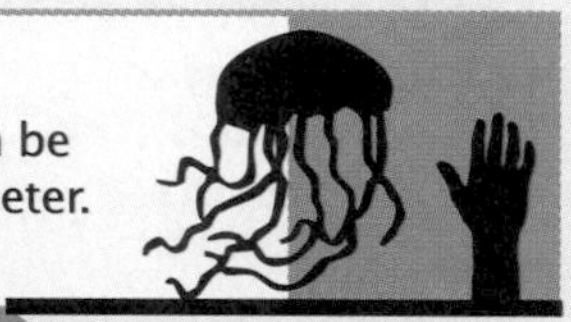

Moon jellies feed on small animals that become trapped in sticky mucus on the underside of the body. Beating cilia carry the prey to the mouth.

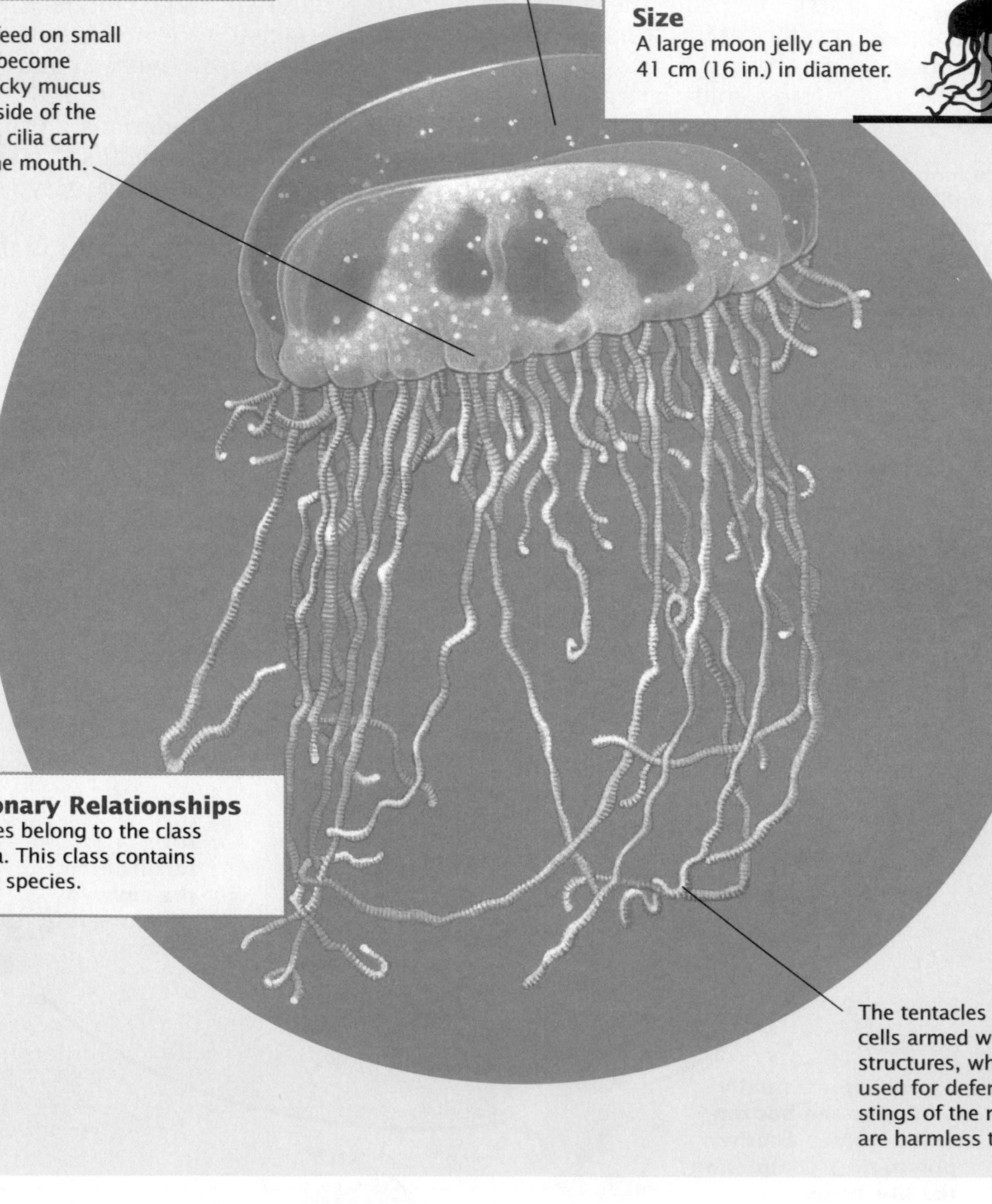

Evolutionary Relationships
Moon jellies belong to the class Scyphozoa. This class contains about 200 species.

The tentacles contain cells armed with stinging structures, which are used for defense. The stings of the moon jelly are harmless to humans.

Flatworms

There are about 20,000 species of flatworms (phylum Platyhelminthes). Most free-living flatworms, such as planarians, are aquatic. They are common in shallow oceans, usually in protected habitats. Freshwater flatworms often reside in gravel or under sunken objects. A few species live on land, but always in very damp areas. The flatworm body is soft and unprotected against predators and dehydration.

Interdependence

In Africa, the construction of man-made lakes has contributed to the spread of schistosomiasis. Form a hypothesis to explain why this has occurred.

Flatworms can reproduce asexually by regeneration

If you cut a planarian in half, either lengthwise or across the body, each half will regrow into a complete worm. This ability to regrow lost parts is called regeneration. Many free-living flatworms can reproduce asexually through regeneration. Scientists are very interested in understanding how flatworms regenerate, and they hope to apply this information to humans. Humans can regenerate some damaged parts, such as growing new skin after a cut or scrape. But human abilities to regenerate are far less dramatic than those of flatworms.

Flatworms also reproduce sexually. Most species are **hermaphrodites**. Hermaphrodites *(huhr MAHF roh deyets)* contain both male and female reproductive systems. During mating, two flatworms exchange sperm so that the eggs of both flatworms are fertilized. A flatworm usually does not fertilize its own eggs.

Most flatworms are parasites

Most species of flatworms are parasitic, infesting a variety of hosts, including humans. Nearly 300 million people are afflicted with the disease schistosomiasis *(shihs tuh soh MEYE uh sihs)*. Each year schistosomiasis kills approximately 800,000 people in Asia, Africa, Latin America, and the Middle East. Microscopic flatworms in the genus *Schistosoma* cause this deadly disease. **Figure 25-4** shows the life cycle of *Schistosoma*.

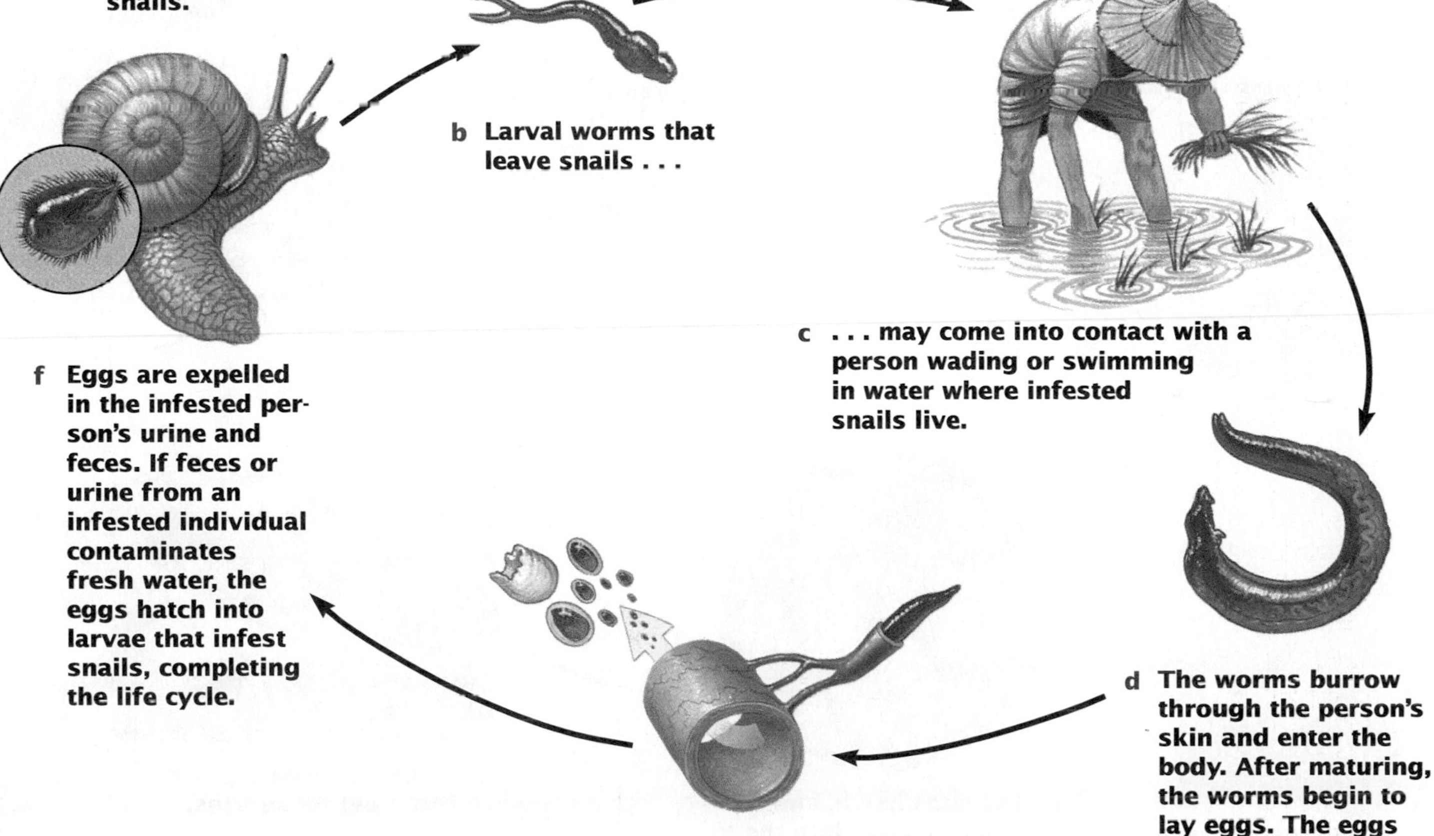

Figure 25-4

a **After hatching, *Schistosoma* larvae invade freshwater snails.**

b **Larval worms that leave snails . . .**

c **. . . may come into contact with a person wading or swimming in water where infested snails live.**

d **The worms burrow through the person's skin and enter the body. After maturing, the worms begin to lay eggs. The eggs are responsible for the symptoms of schistosomiasis.**

e **The eggs can damage or block small blood vessels and often cause internal bleeding in the bladder and intestine.**

f **Eggs are expelled in the infested person's urine and feces. If feces or urine from an infested individual contaminates fresh water, the eggs hatch into larvae that infest snails, completing the life cycle.**

In addition to *Schistosoma*, parasitic flatworms include flukes, which attack the liver or lungs, and tapeworms. The head of the beef tapeworm, which commonly infests humans, is shown in **Figure 25-5**. These parasites can grow to more than 10 m (30 ft) in length. Follow the tapeworm's life cycle in **Figure 25-6**.

Tapeworms are so specialized for a parasitic way of life that they do not have a digestive system. Nutrients from the host's digested food are absorbed directly through the tapeworm's skin. The beef tapeworm often causes no symptoms, but it may cause pain, discomfort, nausea, and abdominal swelling.

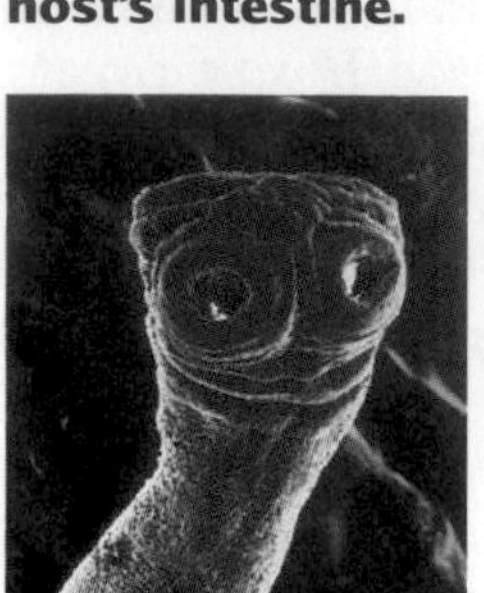

Figure 25-5 The hooks on the head of the beef tapeworm can be thrust out and buried in the wall of its host's intestine.

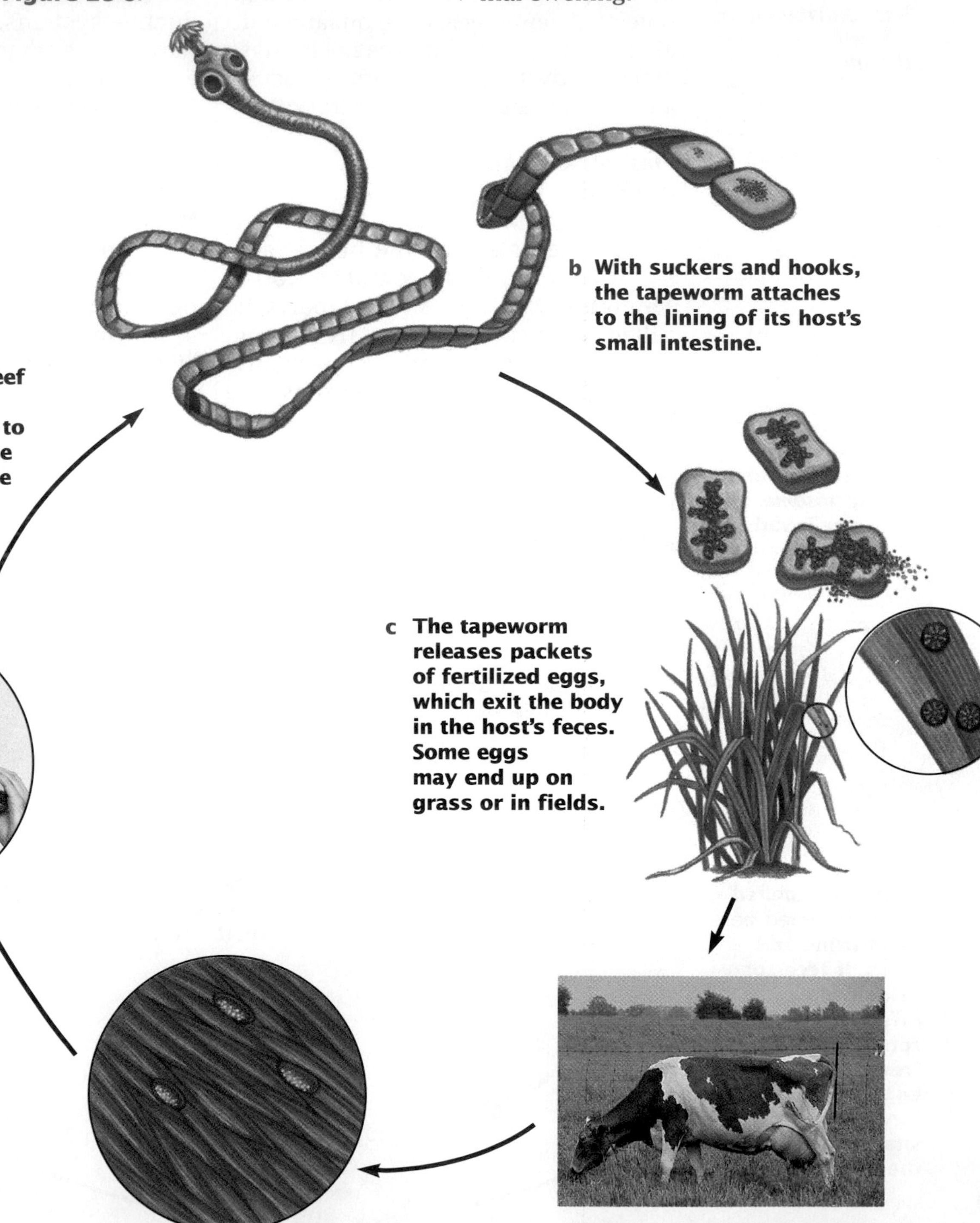

Figure 25-6

a Humans can be infested with tapeworms by eating beef that has not been cooked sufficiently to kill tapeworm larvae lying dormant in the meat.

b With suckers and hooks, the tapeworm attaches to the lining of its host's small intestine.

c The tapeworm releases packets of fertilized eggs, which exit the body in the host's feces. Some eggs may end up on grass or in fields.

d When a cow eats grass, it can eat these eggs.

e The eggs hatch, and the larvae burrow into the cow's muscles, where they become dormant.

Roundworms

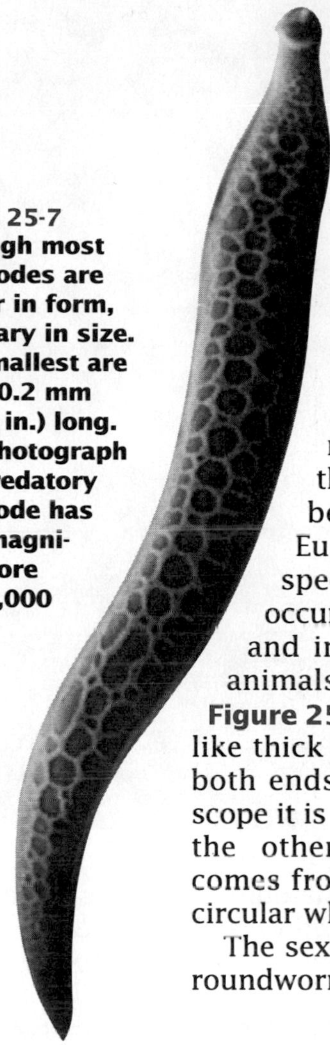

Roundworms, or nematodes (phylum Nematoda), are extremely abundant; a single spadeful of soil can contain more than 1 million individual nematodes. The great abundance of nematodes was illustrated by one very patient zoologist who counted 90,000 nematodes in a single rotting apple. One specimen of nematode is found only in the damp felt coasters under beer mugs in a few eastern European towns. Now *that* is specialization. Nematodes also occur in all aquatic environments and in the bodies of plants and animals as parasites. As shown in **Figure 25-7**, roundworms are shaped like thick threads that are tapered at both ends. In fact, without a microscope it is difficult to tell one end from the other. The name *roundworm* comes from the fact that the body is circular when viewed in cross section.

Figure 25-7
Although most nematodes are similar in form, they vary in size. The smallest are about 0.2 mm (0.008 in.) long. This photograph of a predatory nematode has been magnified more than 2,000 times.

The sexes are separate in nearly all roundworms, and the males are often smaller than the females. After mating, the female secretes a tough case around each fertilized egg and deposits the eggs in the environment, where development takes place. Many parasitic roundworms have complex life cycles, regularly passing from one host to another.

Nematodes play an important role in research on genetics and development. Scientists have learned the complete cellular structure for the nematode *Caenorhabditis elegans*, which has only about 1,000 cells. In addition, scientists are close to identifying the locations of all of the approximately 10,000 genes of this nematode.

Some nematodes cause disease

Nematodes are known to infest virtually all kinds of animals and plants, and some cause millions of dollars of damage each year to livestock and crops. For instance, puppies and adult dogs often must be "wormed," or treated with drugs to kill parasitic nematodes such as the intestinal roundworm *Ascaris*. Heavy infestations of *Ascaris* can be fatal to puppies. *Ascaris* also infests livestock and humans. **Figure 25-8** shows *Ascaris suum*, which infests pigs.

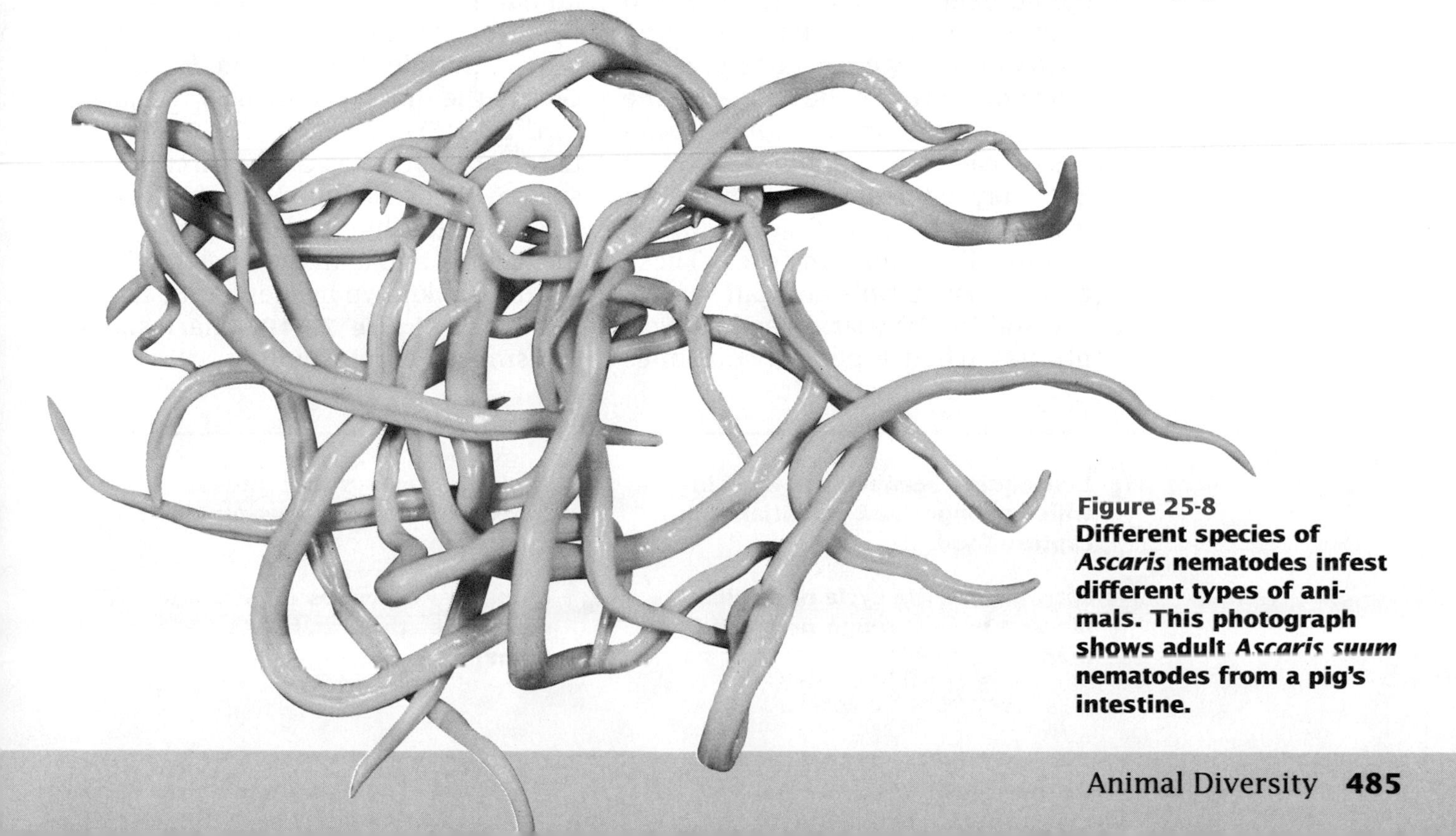

Figure 25-8
Different species of *Ascaris* nematodes infest different types of animals. This photograph shows adult *Ascaris suum* nematodes from a pig's intestine.

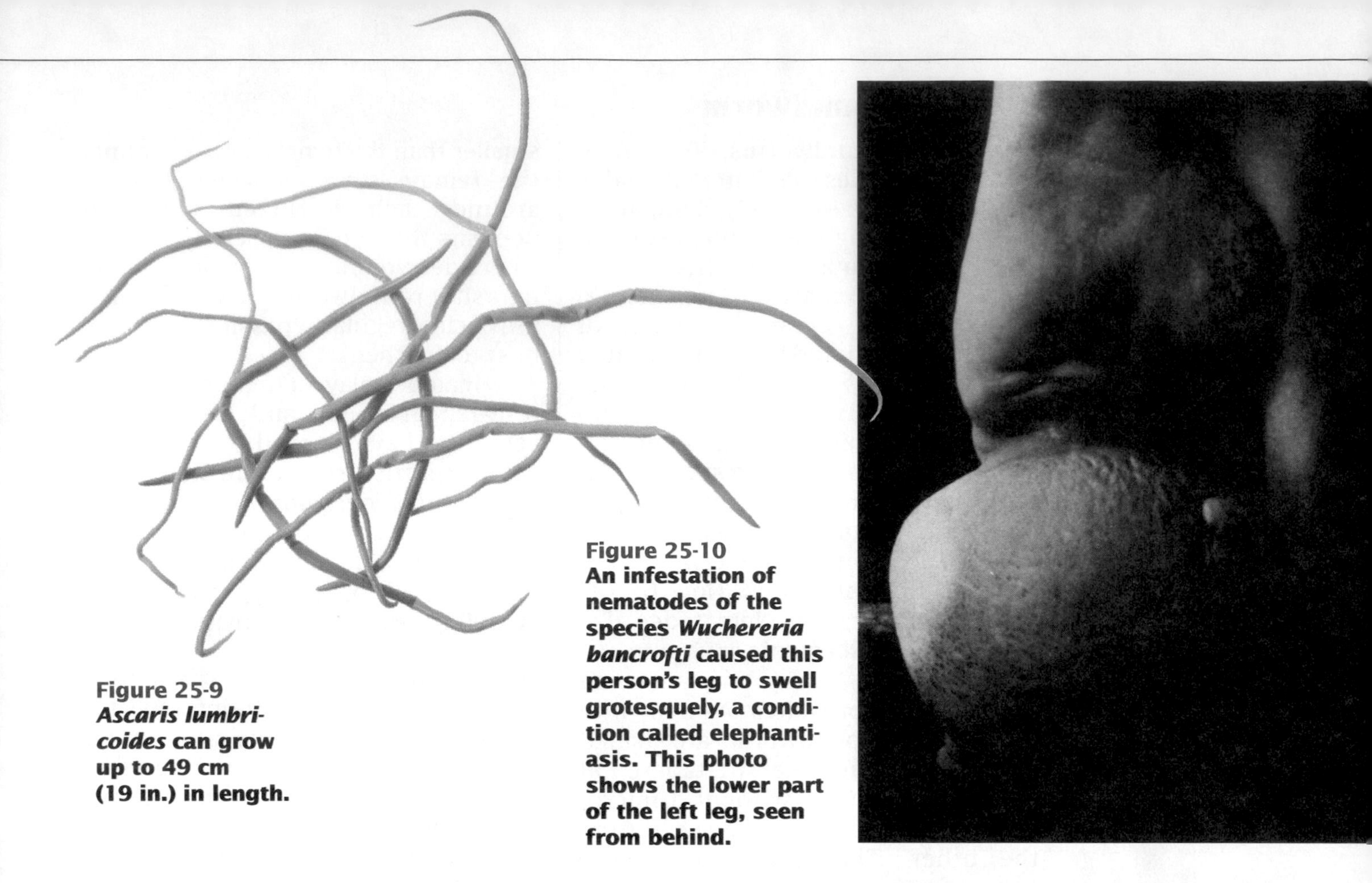

Figure 25-9
***Ascaris lumbricoides* can grow up to 49 cm (19 in.) in length.**

Figure 25-10
An infestation of nematodes of the species *Wuchereria bancrofti* caused this person's leg to swell grotesquely, a condition called elephantiasis. This photo shows the lower part of the left leg, seen from behind.

About 50 species of nematodes parasitize humans. Almost 1 billion people are infested with the nematode *Ascaris lumbricoides*, shown in **Figure 25-9**. This nematode spends its adult life inside the intestine of its host. Large numbers of adult nematodes in the intestine can cause pain and intestinal blockage. Because juvenile worms bore through the tissues of the lungs, an infestation of *A. lumbricoides* can also lead to pneumonia.

Each day a female *A. lumbricoides* lays as many as 200,000 eggs, which pass out of her host in feces. These eggs are very durable and can survive in the soil for 10 years. The life cycle continues when a person consumes the nematode's eggs. This occurs when uncooked vegetables are eaten or when dirty hands are not properly washed before meals. *Ascaris lumbricoides* infections occur mainly in areas where sanitary facilities are poor or where human feces are used as fertilizer.

In tropical regions of Africa and Asia the nematode *Wuchereria bancrofti* causes the disease known as filariasis *(fihl uh REYE uh sihs)*. These nematodes are parasites of the circulatory system. They can clog lymphatic vessels, causing fluid accumulation, skin thickening, and extreme swelling. This condition, known as elephantiasis, is shown in **Figure 25-10**. Filariasis is transmitted by mosquitoes.

SECTION REVIEW

1. **Distinguish between the ways in which sponges and cnidarians capture food.**
2. **Diagram the life cycle of a cnidarian that goes through polyp and medusa stages.**
3. **Explain two ways that infestations of the beef tapeworm could be prevented.**
4. **Name two examples of parasitic nematodes that can infest humans.**

Chances are good that you have never seen a nematode or flatworm. In this section, you will learn about mollusks, annelids, and arthropods, three groups of animals that are probably more familiar to you. You can find members of these groups crawling in a field or garden, burrowing in the soil, flying overhead, or sitting on your plate as part of your evening meal.

25-2 Mollusks, Annelids, and Arthropods

OBJECTIVES

1. **Identify the three main classes of mollusks.**
2. **Summarize the evolutionary advantages of segmentation.**
3. **Describe two ways annelids affect humans.**
4. **List five kinds of arthropods.**

Mollusks

If you visit a shop where fresh seafood is sold, look for some of these mollusks: clams, scallops, mussels, squids, octopuses, and oysters. With more than 50,000 living species, the phylum Mollusca is the second largest animal phylum on Earth, exceeded in size only by the phylum Arthropoda, which includes insects, spiders, and crustaceans. **Figure 25-11** shows two types of mollusks that are often eaten by humans. A shell is a feature of many mollusks. The shell (or shells) provides protection for the body and a solid structure for muscle attachment. The rest of the body is soft; in fact, the name *mollusk* comes from the Latin word for "soft." All mollusk shells are lined with and secreted by a fleshy fold of tissue called the mantle. Another characteristic of mollusks is a muscular foot. Among mollusks, the foot has evolved to perform several functions, including locomotion and burrowing.

Mollusks were one of the earliest groups of animals to evolve efficient excretory organs, the nephridia. **Nephridia** *(nee FRIHD ee uh)* are small tubules that collect wastes from the body fluids and discharge them from the body.

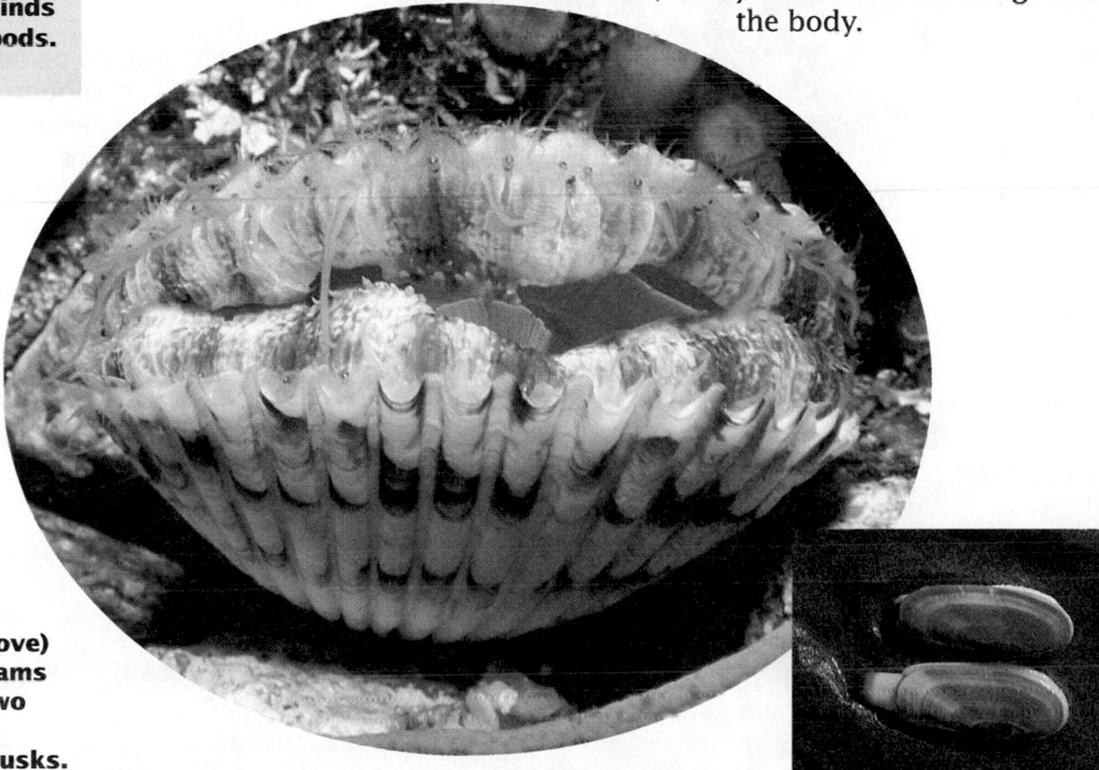

Figure 25-11
Scallops (above) and razor clams (right) are two examples of aquatic mollusks.

Slugs and snails are gastropods

Slugs and snails, which you often see in a garden or aquarium, belong to the class Gastropoda. Gastropods are the most diverse group of mollusks. They are found in the ocean, in fresh water, and on land. Snails and slugs move by creeping along slowly on a muscular foot. The shell, when present, is often brightly colored and is usually coiled.

Just inside the mouth of gastropods and some other mollusks is a unique feeding organ called the **radula** *(RAJ oo lah)*. The radula, shown in **Figure 25-12**, is a flexible structure that is covered by rows of teeth, like a carpenter's file. The radula is used to scrape free small particles of food. In some predatory mollusks, the radula is a sharp fang that is used to stab prey and inject venom.

Figure 25-12
a A snail's shell is made of layers of calcium carbonate. The shell forms a protective barrier against predators.

b This shows an enlargement of the surface of the radula. The radula consists primarily of chitin and is covered with rows of pointed, backward-curving teeth.

Oysters, mussels, clams, and scallops are bivalves

If you are a fan of oysters, mussels, scallops, or clam chowder, you have enjoyed the delicate taste of bivalves (class Bivalvia). Some of these animals are of great value, supporting commercial harvesting industries around the world.

Bivalves have two shells. Unlike other mollusks, bivalves feed by filtering small particles from the water. They lack a radula. Most bivalves are sedentary, and some, such as oysters and mussels, permanently attach themselves to hard surfaces as adults. The foot is wedge-shaped and is used for digging in sand or mud or for secreting tough attachment threads.

Squids, octopuses, and nautiluses are cephalopods

In contrast to the bivalves and gastropods, the cephalopods (class Cephalopoda) are active and can be fast swimmers. Squids, octopuses, and nautiluses are examples of cephalopods. *Tour of a Mollusk* on page 489 profiles one kind of octopus.

Cephalopods are mobile marine predators that are known for their well-developed nervous system. Their body form reflects their active habits. Except for nautiluses, cephalopods have reduced shells or no shell at all. The body of a cephalopod includes the head, tentacles or "arms," and a large, fleshy mantle. **Figure 25-13** shows an example of a cephalopod. The mantle

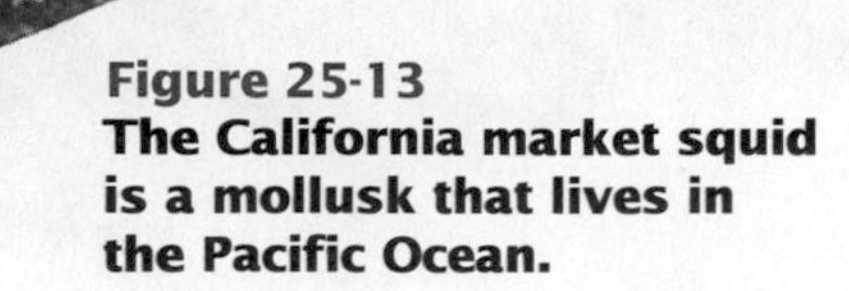

Figure 25-13
The California market squid is a mollusk that lives in the Pacific Ocean.

Mollusk

The white-spotted octopus is a bottom-dwelling mollusk found in tropical oceans. Unlike snails, clams, mussels, and most other familiar mollusks, octopuses have no external shell. All octopuses have eight arms.

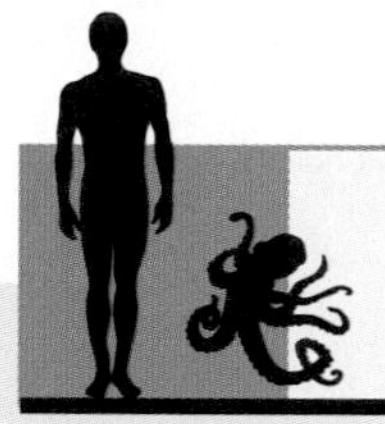

Size
A large adult white-spotted octopus can be 1 m (39 in.) in total length, including the arms.

An ink sac produces dark ink that can be expelled to confuse enemies.

Octopuses swim by forcefully expelling water from the mantle cavity—a sort of jet propulsion.

Most mollusks have a large muscular foot that is used for burrowing or crawling. In cephalopods, the foot has been modified into numerous arms. The eight arms of an octopus have sucking disks for seizing prey such as crabs or other mollusks.

Habitat
White-spotted octopuses live on coral reefs, where they can find abundant food and places to hide. They are found in tropical waters worldwide.

Evolutionary Relationships
Octopuses belong to the class Cephalopoda. This group of about 600 species also includes squids and nautiluses. All cephalopods have many arms, which they use to capture prey.

is thick and muscular. It covers the internal organs and a large mantle cavity. A tubelike siphon derived from the foot serves as an outlet for water leaving the mantle cavity. When the mantle muscles contract, water is rapidly forced out of the siphon, allowing these animals to move by what is essentially jet propulsion. This is the usual manner of movement in squids. Octopuses use this method of movement to escape predators.

Annelids: Segmented Worms

Figure 25-14
Each body segment of this clam worm, found on Long Island, New York, has lateral flaps that function as gills.

The 15,000 or so species of segmented worms (phylum Annelida) include earthworms, leeches, and a host of marine species such as tube worms, feather dusters, and the clam worm shown in **Figure 25-14**.

The segmented body of annelids has allowed the evolution of tremendous diversity in the phylum. Since the segments or groups of segments can be operated somewhat independently, different regions of the body can be specialized for different functions. Such specializations include the suckers of leeches, localized reproductive organs in earthworms, and tube-secreting segments in certain marine worms. **Figure 25-15** describes some of the functions of the specialized segments in an earthworm.

Earthworms feed by sucking in soil and decaying matter. Organic material is digested, and wastes are eliminated as castings, smooth blobs of soil that you can see at the entrances to earthworm burrows. Charles Darwin calculated that a single earthworm could eat its own weight in soil every day. The action of earthworms is beneficial to plants because it breaks up and aerates the soil and because castings are rich in nutrients.

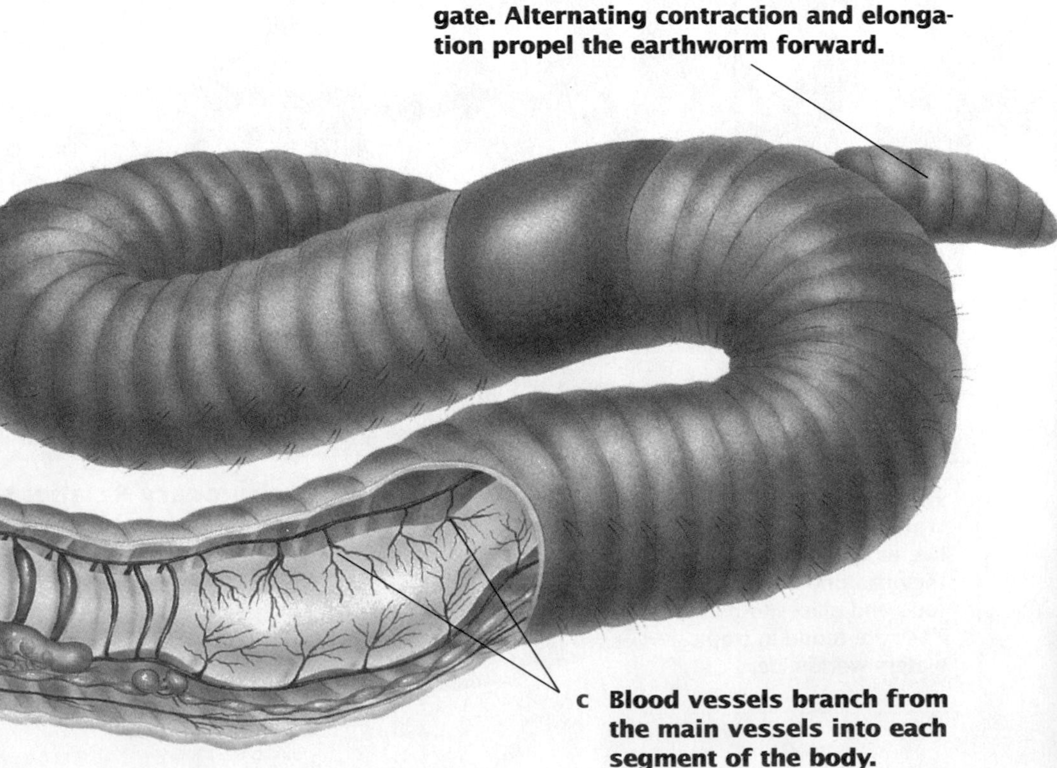

a Segments at the front of the earthworm's body are specialized for burrowing in the soil. This region contains sensory nerve endings.
b Segments in one part of the worm can contract while those in another part elongate. Alternating contraction and elongation propel the earthworm forward.
c Blood vessels branch from the main vessels into each segment of the body.

Figure 25-15
Segmentation is a feature of annelids. The body of an earthworm can have more than 100 segments.

Figure 25-16
The body of this aquatic leech is segmented, just like the body of an earthworm.

Some leeches are parasitic annelids

Externally, leeches are distinct from the other annelids. The body is somewhat flattened and not quite as wormlike as an earthworm's body, as you can see in **Figure 25-16**. Some leeches are scavengers or predators. Many leech species are at least part-time parasites. They feed on other animals, usually vertebrates. Suckers on the front and back of the leech enable it to attach to its host.

Parasitic leeches use enzymes or three sharp, bladelike teeth to make an opening in the host's skin through which blood is sucked. Some leeches release proteins into the host that prevent blood from clotting and ending their meal. Scientists have isolated one of the proteins produced by leeches. This protein is being used in research on blood clotting. Another leech protein has been effective in treating lung cancer. Some leeches secrete a local anesthetic, so the host is not even aware that it is providing a meal for the leech.

For centuries, leeches were used to "bleed" sick patients because people believed that excess blood caused disease. This practice fell out of favor when scientists showed that viruses, bacteria, and other microorganisms cause disease. Recently leeches have made a medical comeback. The large blood-sucking leech *Hirudo medicinalis* is used by modern physicians to liquefy blood clots and to remove blood from bruises and severely damaged tissue.

Arthropods: The Most Abundant Animals

Arthropods (phylum Arthropoda) dominate virtually all habitats, both in numbers of individuals and in numbers of species. About 1 million species of arthropods have been described. Some specialists believe that as many as 30 million species still remain to be discovered.

As you learned in Chapter 23, arthropods have a segmented body that is covered by an exoskeleton of chitin. Most important, they have jointed appendages. The phylum Arthropoda includes a variety of familiar animals, including those shown in **Figure 25-17**. Spiders, insects, centipedes, scorpions, shrimp, crabs, and lobsters all belong to the phylum

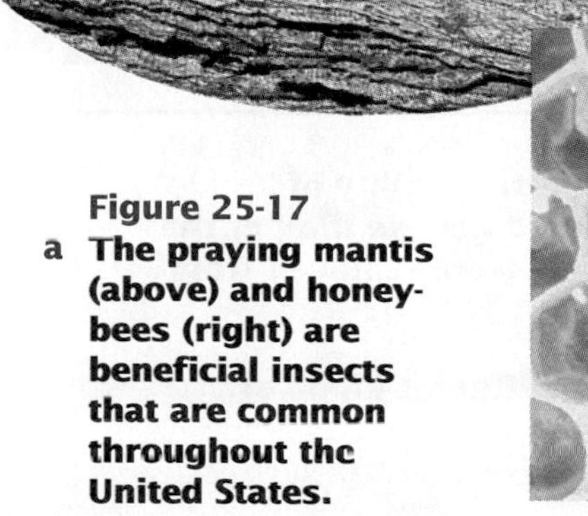

Figure 25-17
a The praying mantis (above) and honeybees (right) are beneficial insects that are common throughout the United States.

b This tarantula is native to Asia.

Figure 25-18 Beetles, butterflies, ticks, scorpions, and centipedes are all arthropods.

Arthropoda. Stop for a moment to consider the many ways in which arthropods affect our lives. Lobsters, crabs, and shrimp are sources of food. Some insects are serious pests of food crops and cause millions of dollars in damage each year. Other insects are extremely beneficial as pollinators of crop plants. Still others are parasites. Malaria and African sleeping sickness, two of the world's most serious diseases, are transmitted by insects.

Look at the arthropods shown in **Figure 25-18**. These photos show just a tiny sample of the variety within this phylum. You will read much more about arthropods and their importance to humans in Chapter 26.

a The lady beetle kills pests that attack crop plants and flowers such as this buttercup.

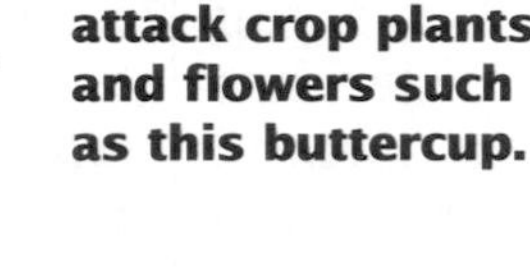

b The giant centipede is found in Texas.

c The scorpion *Pandinus imperoitor* is found in Africa.

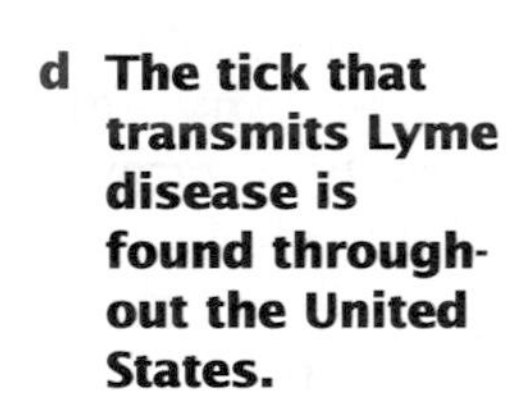

d The tick that transmits Lyme disease is found throughout the United States.

e Most monarch butterflies live in the United States during the summer and migrate to Mexico or South America in the winter.

SECTION REVIEW

1. **Name the major classes of mollusks. Give an example of each.**
2. **What are the advantages of a segmented body?**
3. **Why would the fertility of a farmer's field decline if all of the earthworms were removed from the soil?**
4. **Name five different kinds of arthropods.**

Echinoderms and chordates belong to a fundamentally different evolutionary line from the annelids, mollusks, and arthropods. Echinoderms and chordates are deuterostomes, which are distinguished from the groups you studied in the other sections of this chapter by differences in their embryonic development.

25-3 Echinoderms and Chordates

OBJECTIVES

1. **Identify three kinds of echinoderms.**
2. **List two characteristics of animals in the phylum Echinodermata.**
3. **Contrast the three subphyla of chordates.**

Echinoderms: Sea Stars and Their Relatives

Sea stars, sea cucumbers, sand dollars, and sea urchins are echinoderms (phylum Echinodermata). As adults, nearly all echinoderms show a five-part radial symmetry: parts of the body radiate from its center like the arms of a five-pointed star, as shown in **Figure 25-19**. The sand dollar shown in **Figure 25-20** also has five-part body symmetry, but it is less obvious.

In addition to radial symmetry, echinoderms have a **water vascular system**. As you learned in Chapter 23, the water vascular system is a complex arrangement of fluid-filled tubes that operates the tube feet on the animal's lower surface. The water vascular system functions in locomotion, feeding, and gas exchange.

Echinoderms have no excretory organs and thus have no efficient means of regulating water balance. That is one reason echinoderms are restricted to marine habitats. Most of these animals obtain oxygen by diffusion through thin parts of the body surface or through the walls of the tube feet. Internal distribution of oxygen and nutrients occurs mainly by circulation of fluid within the coelom.

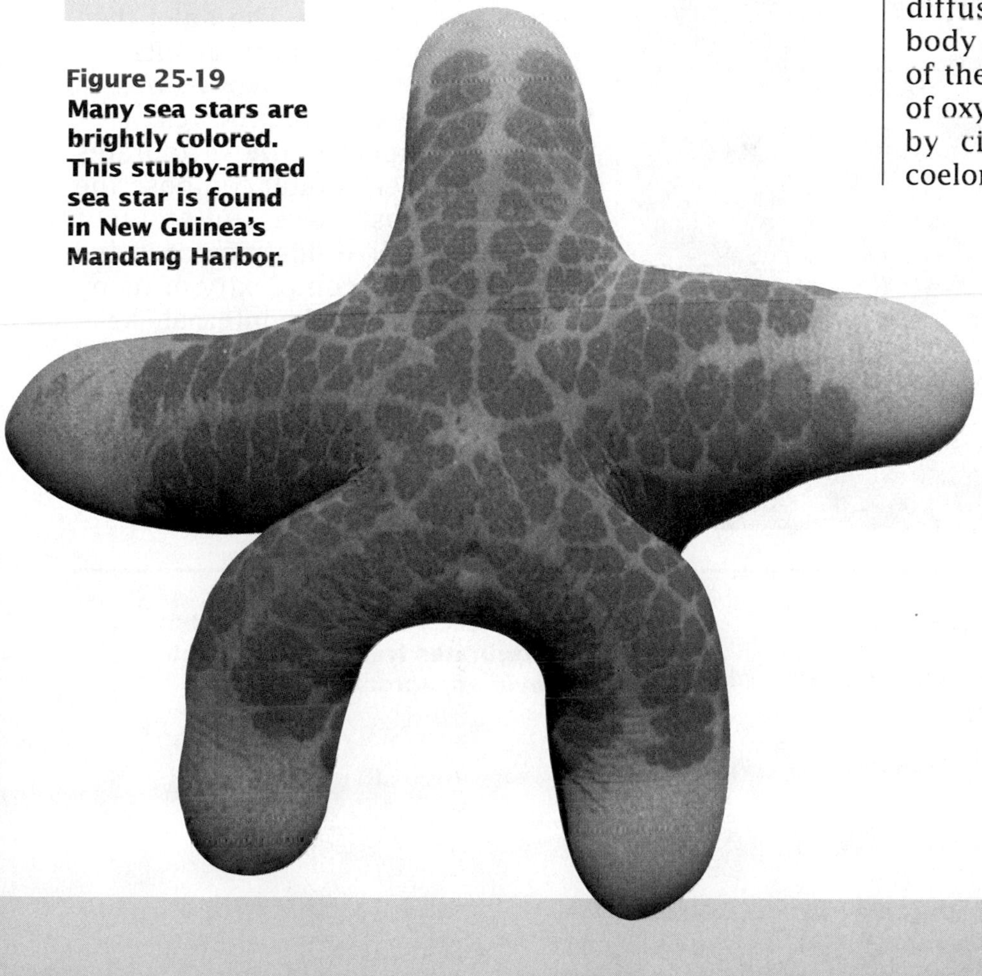

Figure 25-19
Many sea stars are brightly colored. This stubby-armed sea star is found in New Guinea's Mandang Harbor.

Figure 25-20
This endoskeleton of a sand dollar shows the five-part radial symmetry found in all echinoderms.

Chordates

Like echinoderms, chordates (members of the phylum Chordata) are deuterostomes. Recall that all chordates share four features: a notochord, a dorsal nerve cord, a post-anal tail, and pharyngeal slits. Three subphyla make up the phylum Chordata.

Tunicates lose some chordate characteristics

Most adult tunicates, or sea squirts (subphylum Urochordata), live attached to rocks or the sea floor, as shown in **Figure 25-21**. Like sponges, these tunicates obtain food by filtering particles of food from the water. They secrete a leathery or jellylike covering composed of a substance similar to cellulose. This outer body layer is known as a tunic, hence the name of the group. During development, urochordates go through a free-swimming larval stage. During this stage the four chordate features are apparent. Adult tunicates retain only the pharyngeal slits.

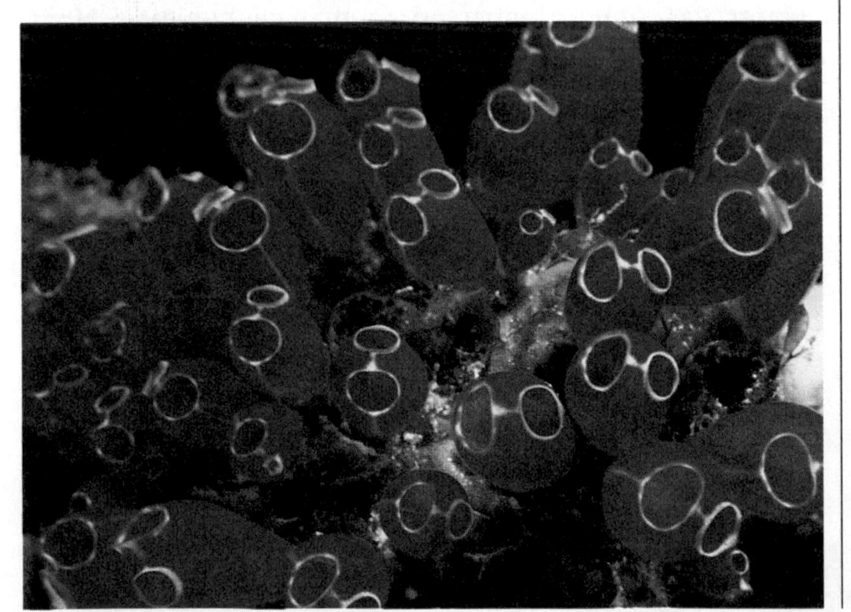

Figure 25-21
Unlike their larval form, these adult tunicates are not free-swimming.

Figure 25-22
Unlike tunicates, lancelets retain all of their chordate characteristics throughout their lives.

Lancelets retain their chordate features throughout their lives

Lancelets (subphylum Cephalochordata) are usually less than 5 cm (2 in.) long. **Figure 25-22** shows the lancelet *Branchiostoma*. Lancelets bury their tails in sand or mud and keep their heads exposed in the overlying water. As water enters the mouth, tentacles around the mouth exclude large particles. The water passes into the pharynx, where food is trapped in the mucus-covered lining. The water then exits through the pharyngeal slits.

Vertebrates have backbones

With 42,500 species, the vertebrates (subphylum Vertebrata) compose the largest and most successful group of chordates. Much of this success is due to the increased complexity of many organ systems and the internal skeleton. A distinguishing feature of the vertebrates is the vertebral column, for which the group was named. Fishes, amphibians, reptiles, birds, and mammals (including humans) are members of this very diverse subphylum.

SECTION REVIEW

1. **List three echinoderms.**
2. **Describe two characteristics of echinoderms.**
3. **What feature distinguishes the vertebrates from the other subphyla of chordates?**

CHAPTER 25

Highlights

Many bivalves are important sources of food for humans.

25-1 Sponges, Cnidarians, and Simple Worms

Structures on the head of this tapeworm allow it to attach to the intestinal wall of its host.

Key Terms

spicule (p. 479)

nematocyst (p. 481)

polyp (p. 481)

medusa (p. 481)

hermaphrodite (p. 483)

Summary

- Sponges live attached to rocks. They filter food from the water.
- Jellyfish, hydras, corals, and sea anemones are cnidarians. Cnidarians are characterized by stinging nematocysts.
- Many flatworms are parasites. Parasitic flatworms of humans include the beef tapeworm and *Schistosoma*, which causes the disease schistosomiasis.
- Roundworms, or nematodes, are among the most abundant organisms. Some nematodes are parasites.

25-2 Mollusks, Annelids, and Arthropods

Snails belong to the class Gastropoda, one of the three major groups of mollusks.

Key Terms

nephridium (p. 487)

radula (p. 488)

Summary

- Clams, mussels, snails, slugs, octopuses, and squids are mollusks.
- Most mollusks have a protective shell and feed by using a rasping structure called a radula.
- The three main classes of mollusks are Gastropoda, Bivalvia, and Cephalopoda.
- Earthworms, leeches, and certain marine worms belong to the phylum Annelida, the segmented worms.
- Most of the known species of animals are arthropods.

25-3 Echinoderms and Chordates

This sea star is an echinoderm. Like you, sea stars are deuterostomes.

Key Terms

water vascular system (p. 493)

Summary

- Echinoderms and chordates are deuterostomes.
- Echinoderms have a five-part radial symmetry. They also have a water vascular system that operates their tube feet.
- The three main groups of chordates are tunicates, lancelets, and vertebrates. You are a vertebrate.

CHAPTER 25

review

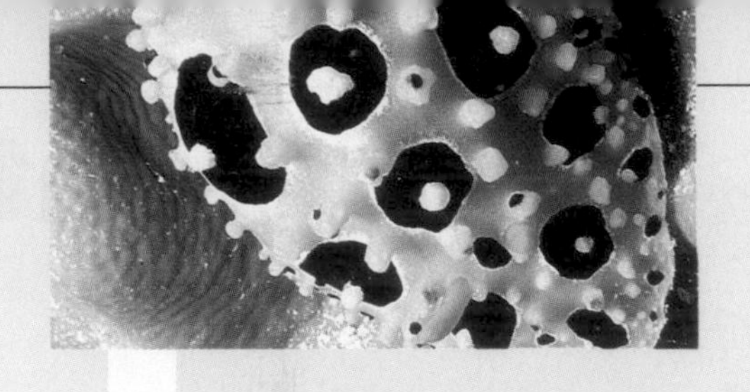

Understanding Vocabulary

1. For each set of terms, complete the analogy.
   a. medusa : free-floating :: polyp : _____
   b. schistosomiasis : *Schistosoma* :: filariasis : _____
   c. flatworms : unsegmented :: annelids : _____

2. For each list of terms, identify the one that does not fit the pattern and explain why.
   a. coral, jellyfish, sea anemone, sponge
   b. octopus, planarian, scallop, snail
   c. leech, *Ascaris*, *Schistosoma*, lancelet

Understanding Concepts

3. **Relating Concepts** Construct a concept map that identifies and describes the characteristics of the three main classes of mollusks. Try to include the following terms in your concept map: phylum Mollusca, bivalves, cephalopods, gastropods, radula, oysters, octopuses, snails, and filtration.

4. **Inferring Conclusions** Considering that an adult sponge does not move from place to place, how can a population of sponges move to a different area?

5. **Comparing Structures** How are the adult stages of sponges and sea anemones similar? How are they different?

6. **Comparing Structures** How do the polyp stage and the medusa stage of the cnidarian life cycle differ?

7. **Identifying Structures** When a swimmer is stung by a jellyfish, what causes the pain felt by the swimmer?

8. **Recognizing Relationships** Explain how eating an inadequately cooked hamburger or steak puts you at risk of infestation by the beef tapeworm.

9. **Identifying Structures** Give an example of segmentation in an earthworm. How does the earthworm benefit from segmentation?

10. **Recognizing Relationships** Suggest two points at which the life cycle of *Schistosoma* could be interrupted, thus preventing infestations in humans. Describe the actions necessary to interrupt the life cycle at each point.

11. **Summarizing Information** Copy the following table onto a separate sheet of paper. Complete the table by filling in the blanks.

	Shell present?	Habitat	Example
Cephalopods		Aquatic	
Gastropods	Usually		
Bivalves			Razor clam

12. **Recognizing Relationships** How does an earthworm affect the soil? How do humans benefit from these activities?

13. **Summarizing Information** Describe two chemicals produced by leeches. What are the functions of these chemicals? How do doctors use leeches in modern medicine.

14. **Identifying Structures** Which of the following are not characteristics of arthropods: segmentation, endoskeleton of chitin, notochord, radial symmetry, jointed appendages?

15. **Identifying Structures** Describe the main characteristics of animals in the phylum Echinodermata.

16. **Organizing Information** What are the three subphyla in the phylum Chordata? What are the characteristics of each subphylum?

17. **Summarizing Information** If only adult forms were considered, would sea squirts be classified as chordates?

18. **BUILDING ON WHAT YOU HAVE LEARNED** In Chapter 20, you learned about alternation of generations in plants. Alternation of generations is also a characteristic of many members of the animal kingdom. Name one animal that exhibits alternation of generations, and describe the body form for each stage of its life cycle.

Interpreting Graphics

19. Look at the graph below, which shows the number of cases of malaria and schistosomiasis in four areas.

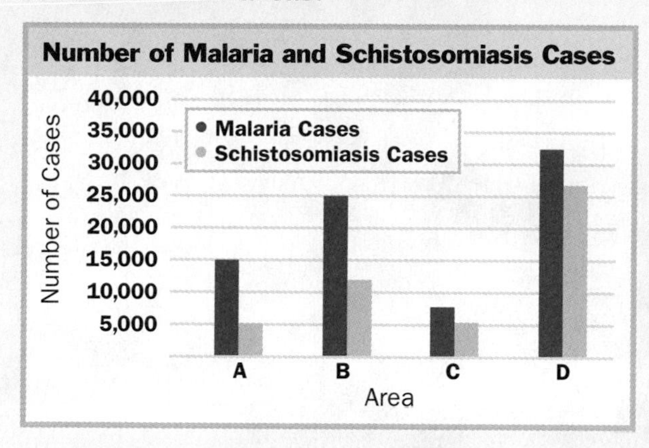

Which of the following statements is best supported by the data?

a. Malaria and schistosomiasis are caused by the same organism.
b. There are more cases of malaria in areas that have few cases of schistosomiasis.
c. Malaria cases outnumber schistosomiasis cases in each area.
d. Spraying for mosquitoes would reduce the number of cases of schistosomiasis.

Reviewing Themes

20. ***Interdependence***
Explain how the life cycle of *Schistosoma* is dependent on snails.

21. ***Homeostasis and Stability***
What are nephridia? What function do they perform? How have they contributed to the ability of some mollusks to live on land?

Thinking Critically

22. **Distinguishing Relevant Information** All cephalopods are aquatic, while some gastropods are terrestrial. Which of the following statements could help explain this difference?

a. Cephalopods move by expelling water from their mantle cavity.
b. Both groups have nephridia
c. Gastropods often have a shell.
d. Cephalopods have much larger brains than do gastropods.

Explain your answer.

Life/Work Skills

23. **Communicating** Many professionals, such as surgeons and cosmetologists, prefer to use natural sponges rather than manufactured ones. Find out how natural sponges are collected and prepared for use. Present an oral report of your findings to your class.

24. **Working Cooperatively** The class Turbellaria in the phylum Platyhelminthes contains about 3,000 different species. Form a group with three or four of your classmates. Each of you should research a different species of turbellarian. Draw accurate, color pictures of the species. Find out the animal's common and scientific names, its habitat, the part of the world in which it lives, how it feeds and reproduces, and how it protects itself from predators. Present your work to your class.

Activities and Projects

25. **Economics Connection** Use library references to find out about the economic importance of mollusks. Research the importance of mollusks as food. Also investigate the economic impact of plant destruction caused by snails. Investigate the uses of shells and the economic importance of shell collecting. Prepare a written report of your findings.

26. **Art Connection** Many people collect shells as a hobby. Ask friends and relatives if they have a shell collection that you can see. Many natural history museums and aquariums have extensive shell collections. If you cannot see a shell collection, consult a field guide to shells or a reference such as *The LaRousse Guide to Shells of the World*, by A.P.H. Oliver. Select five of your favorite shells. Make life-sized color drawings of each shell. Under each drawing, give the common and scientific name of the mollusk, its habitat, and in what part of the world the mollusk lives. Share your drawings with your class.

Discovering Through Reading

27. Read the article "The Mussels' Message" by William Stolzenburg in *Nature Conservancy*, November/December 1992, pages 17–23. What effect did building a dam on the Tennessee River have on mussel populations in Muscle Shoals?

Science, Technology, and Society

Can One Species Tell All?

Researchers sometimes use a single species to determine the health of an entire ecosystem. Studying a single species is convenient and relatively inexpensive, and it can give scientists valuable information about what is happening in an ecosystem.

The golden-cheeked warbler is just one of many endangered species whose existence is threatened by habitat destruction.

Using Indicator Species

A single species, called an indicator species, can provide an enormous amount of information about an ecosystem. Careful observation of the species' characteristics, including population size, appearance, behavior, reproductive health, and feeding habits, can help scientists evaluate the overall health of the ecosystem.

Ecosystems are sometimes exposed to stressful changes, such as climate shifts, immigration of new species, pollution, and deforestation. The response of the indicator species to these changes can be used to predict how other species will react and what long-term effects may result.

Predictors of Environmental Health

Consider Buchman's warbler, a migratory songbird that once made its summer home throughout the southeastern United States. The warblers used the forests in Cuba as their winter nesting site. Several decades ago, these forests were destroyed in order to plant sugar cane. No Buchman's warbler has been seen since the 1960s, and the species is assumed to be extinct. Is the fate of the Buchman's warbler an early predictor of the fate of other songbirds?

From the 1960s to the 1980s, the number of migratory songbirds in the United States decreased by 50 percent due to loss of habitat. These birds spend the winter months in tropical forests extending from Mexico to Argentina. If tropical forests continue to be logged, some experts think these forests may vanish in the next 50 years. With this habitat destruction, are the more than 250 species of songbirds facing the same fate as the Buchman's warbler?

Bioindicators Point to Pollution

Species that are used to predict changes for an entire ecosystem are called bioindicators. Samples taken from these species can be analyzed for levels of pollution, diversity, reproductive health, and disease.

Pollutants appear to be harming vertebrates worldwide. A growing concern among scientists is the presence of chemicals in the environment that mimic hormones. A large number of substances have been suggested as possible sources of these chemicals, including certain adhesives, plastics, insecticides, and solvents. Animals that have been exposed to these agents—alligators in Florida, rainbow trout in Europe, and many others—show disrupted reproductive systems.

Human health may be affected by these pollutants too. Some studies suggest that sperm counts are declining worldwide and that incidences of endometriosis and testicular cancer are rising.

Technique

Electronic Identification for Animals

Scientists often track specific animals to study migration patterns or mating habits. This requires that the animals be tagged for identification. Identification tags, radio collars, metal bands, and tattoos are methods of tagging animals.

A new method of tagging involves implanting a small electronic device under an animal's skin. This device, called a transponder, is as small as a grain of rice, can't be broken or lost, and never needs new batteries. A computer chip inside the transponder contains an identification number that can be easily scanned, just like barcodes on products in a supermarket. Computers store information about the animal, allowing scientists to track its movements and behavior. Transponders have been used to track rattlesnakes in the American desert, hammerhead sharks in the Gulf of Mexico, and animals such as wolves and big cats that are reintroduced into environments where they once flourished.

Applications: tracking animal migrations in the wild, identifying lost pets, locating individual animals in wildlife preserves

Users: environmental scientists, wildlife trackers, and veterinarians

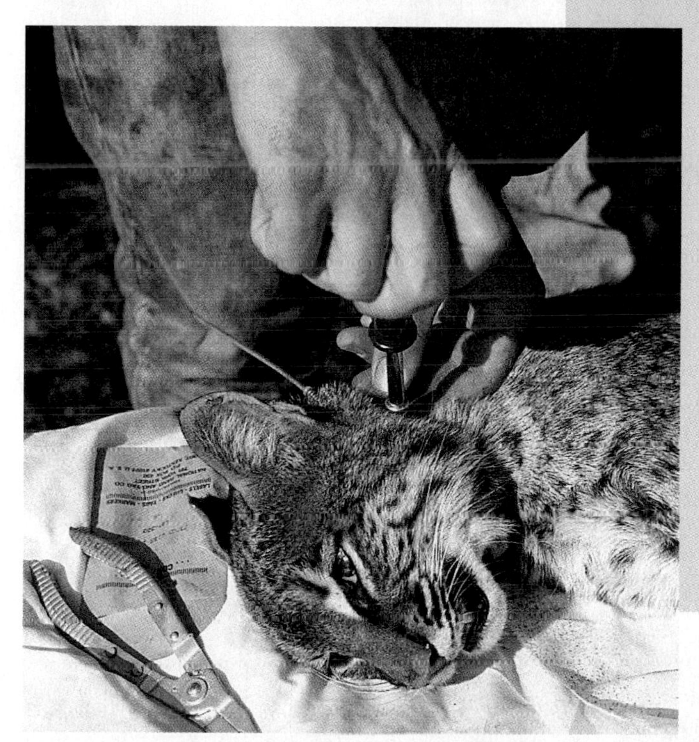

This sedated bobcat in Cumberland Islands, Georgia, is being fitted with a radio-tracking collar.

Analyzing the Issues

What makes a species a good indicator?

1. List the characteristics that are desirable in an indicator species. Compare your ideas with those of other students in the class. Does everyone agree? Could some characteristics be desirable in some indicator species but detrimental in others?

Can information about an indicator species be applied to an entire ecosystem?

2. Read "The Puzzle of Declining Amphibian Populations" by Andrew Blaustein in the April 1995 issue of *Scientific American*, pages 52–57. What possible effect on humans could the disappearance of frogs predict?

How can your community use an indicator species?

3. Consider the area in which you live. Do any potential environmental problems exist? Is your community recovering from an environmental threat? How could a local species provide information about the health of your ecosystem?

Can indicator species predict short-term damage?

4. Read "Red Tides" by Donald M. Anderson in the August 1994 issue of *Scientific American*, pages 62–68. What species would be a good indicator of the presence of organisms that cause red tide? Is it possible to stop red tides or prevent humans from being affected?

CHAPTER 26

REVIEW

- key features of the insect body plan (Section 23-3)
- gas exchange (Section 24-2)
- the terms *exoskeleton*, *gill*, *trachea*, *book lungs*, and *Malpighian tubules* (Glossary)

Arthropods

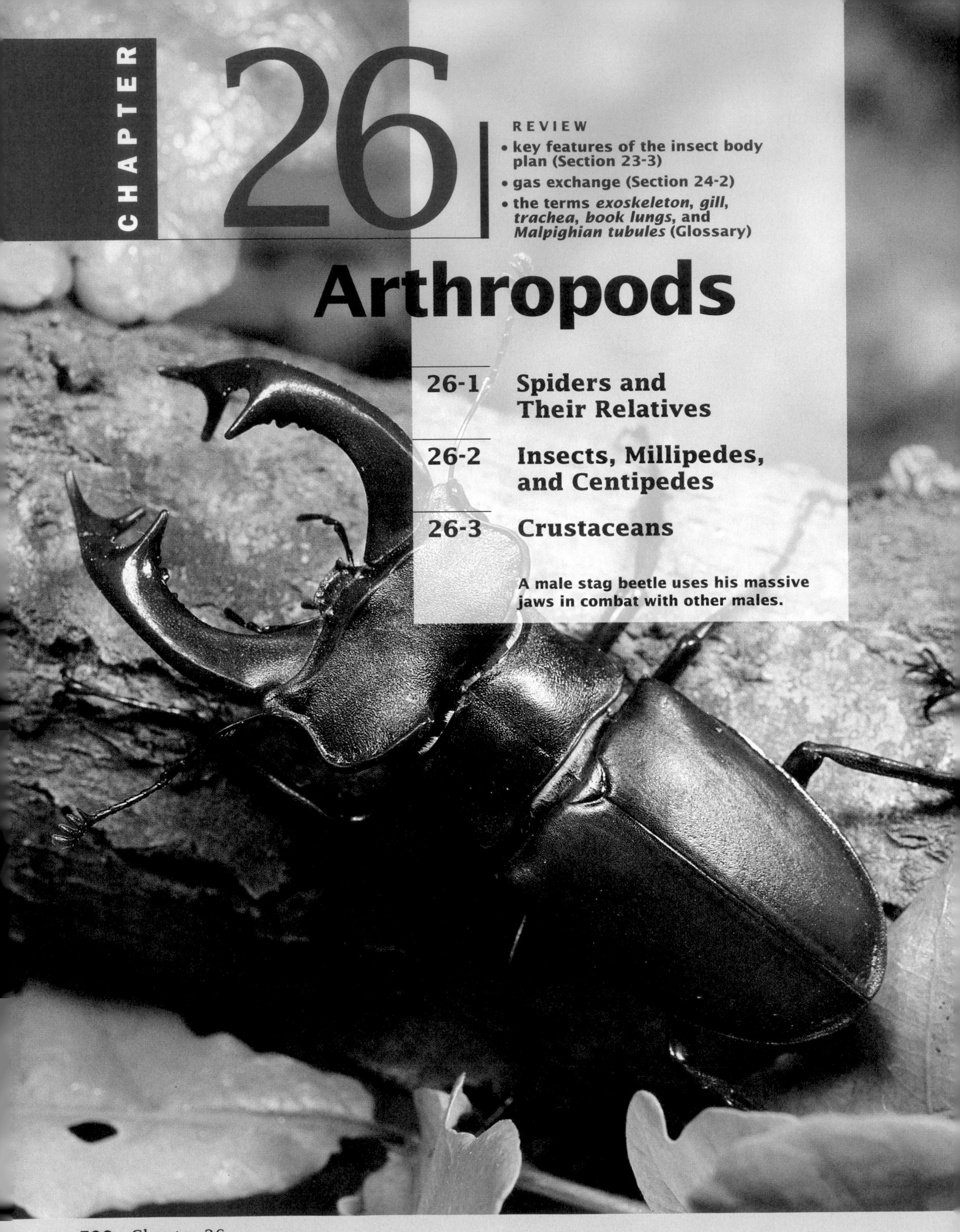

A male stag beetle uses his massive jaws in combat with other males.

This spider is an arthropod. There are about 1 million named species of arthropods, far more species than belong to any other phylum. Arthropods have a great impact on humans and on world ecology. There are three major groups of living arthropods—spiders and their relatives, insects and their relatives, and crustaceans.

26-1 Spiders and Their Relatives

OBJECTIVES

1. **Describe the functions of chelicerae and pedipalps in spiders.**
2. **Explain three ways spiders use silk.**
3. **List two arachnids that directly affect people.**

Characteristics of Arachnids

What do you see when you look closely at a spider like the tarantula shown in **Figure 26-1**? One of the spider's key features is the pair of appendages called **chelicerae** *(kuh LIHS uh ree).* Chelicerae, which are located at the front of a spider's body, are the hallmark of spiders and their relatives, including scorpions, mites, and ticks. Chelicerae are used for feeding. The chelicerae of spiders are poison-delivering fangs. A second pair of appendages, known as **pedipalps**, lies just behind the chelicerae. Pedipalps are used to capture and manipulate prey and for courtship displays and mating.

Spiders, scorpions, mites, and ticks are **arachnids** *(uh RAK nihdz)*, members of the class Arachnida. Remember from Chapter 23 that segmentation is a characteristic of the arthropods. In arachnids, the body segments are fused to form two main body regions: the cephalothorax *(sehf uh luh THAWR aks)* and the abdomen. Arachnids have four pairs of legs, which are attached to the cephalothorax.

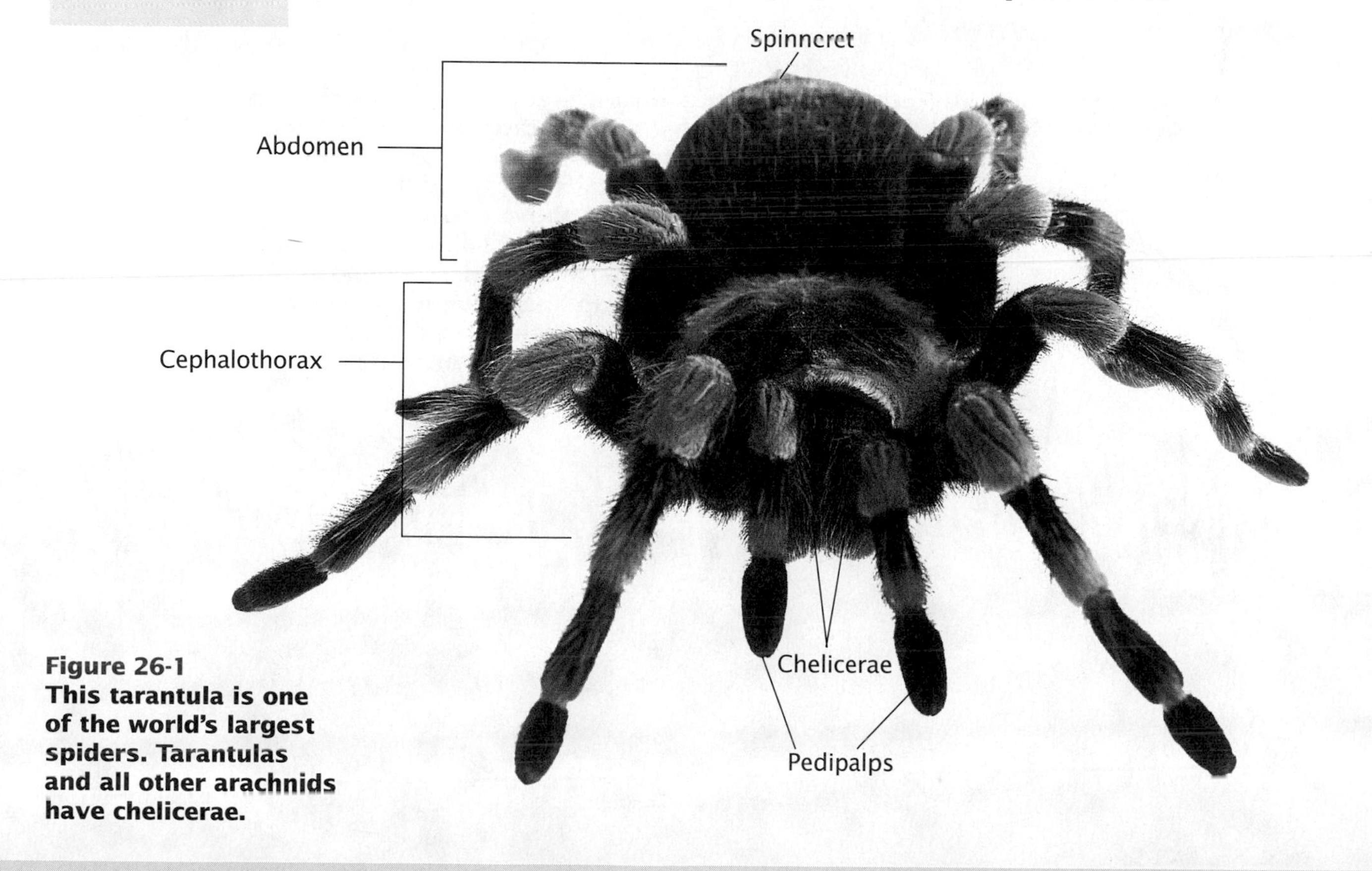

Figure 26-1
This tarantula is one of the world's largest spiders. Tarantulas and all other arachnids have chelicerae.

Spiders

There are about 35,000 named species of spiders. While a few live in fresh water, the majority are land dwellers. Indeed, spiders are thought to have been among the first arthropods to successfully colonize the land. Spiders are predators, feeding on nearly any other animal of manageable size.

Spiders have several adaptations that contribute to their success. First, they have efficient organs of excretion and water balance, so they can rid their bodies of wastes while still conserving precious water. Second, many spiders are able to breathe air by using book lungs. Recall from Chapter 24 that book lungs allow air to come close to the spider's blood in a humid chamber where gas exchange occurs. Third, nearly all spiders produce poison. The chelicerae of spiders are fangs equipped with poison glands. Fourth, and perhaps most important, spiders have the ability to produce silk threads, which they use in a great variety of ways.

Silk and fangs are a deadly combination

You may have watched a spider building its web. At the end of a spider's abdomen are small nozzle-like structures called **spinnerets**. Spinnerets direct the flow of silk from silk-producing glands in the abdomen. Spiders can produce different kinds of silk threads in various diameters.

Spider silk is composed of a complex structural protein that is produced in the glands as a liquid. As the liquid silk flows out through holes in the spinnerets, it solidifies into the threads that make up the web. As illustrated in **Figure 26-2**, spiders spin many different types of webs.

Figure 26-2
Spiders have many different kinds of specialized webs that enable them to catch their prey. Some spiders build webs in high tree branches, while others build horizontal sheet webs that catch insects that drop from above or are carried by the wind. Some spiders build silk-lined burrows beneath the ground. These spiders extend threads that act as trip lines around the burrow entrance. Then they sit inside, waiting for an unsuspecting insect to become entangled.

Spider silk is elastic and extremely strong, about as strong as a nylon thread of the same diameter. Spiders use silk for many purposes. Threads are used as safety lines when spiders dangle from branches or drop to the ground. Many spiders also use silk to line their nests or burrows, to wrap prey, or to fashion cocoons for their young. Some newly hatched spiderlings spin long, thin threads with which they ride the winds over great distances.

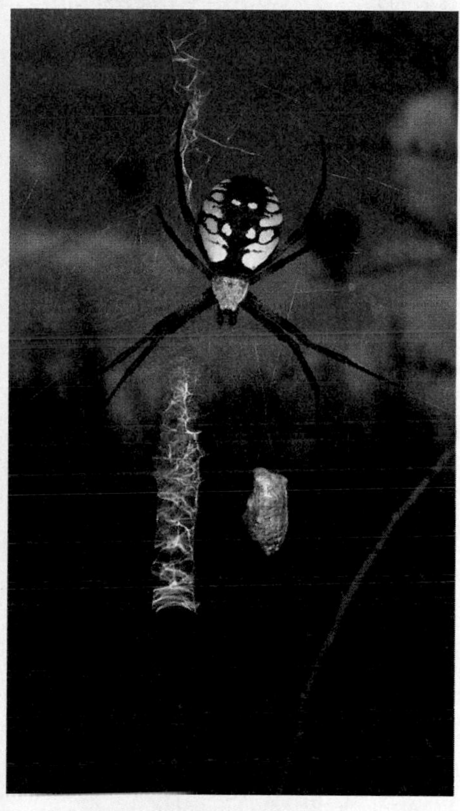

Figure 26-3
This *Argiope* spider has captured a bee, injected it with digestive enzymes, and wrapped it in silk. Later, the spider will suck out the liquefied tissues of the bee and discard the carcass.

A spider's web is a sticky trap to ensnare prey. Constructing a web is a complex architectural feat that requires several steps. When an insect strikes the web and struggles, the spider is alerted by the threads' vibrations. The spider then rushes to the victim, bites it, and injects poison and digestive enzymes. The ability to stun or kill the prey quickly with poison from the chelicerae is a great advantage. It saves the spider from having to wrestle with large or potentially dangerous prey, such as bees and wasps. The spider shown in **Figure 26-3** has captured a bee in its web.

Although nearly all spiders produce poison, only about 20 kinds are considered dangerous to people. Two such species are found in the United States: the American black widow spider and the brown recluse spider, both shown in **Figure 26-4**. Most spiders are more beneficial than harmful to humans. They are the world's champion pest controllers, consuming millions of insects in their daily activities. Many of these insects are agricultural pests.

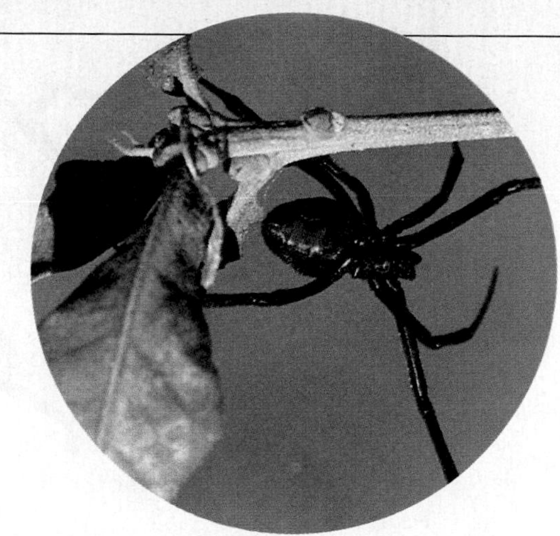

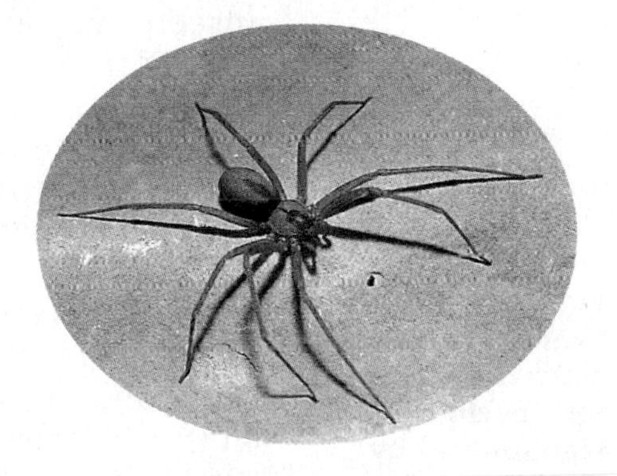

Figure 26-4
a The American black widow spider is most common in the southern United States. The bite of the female injects a powerful venom that acts on the victim's nervous system.

b The brown recluse spider can be recognized by the violin shape on its upper cephalothorax. Brown recluse spiders live in dark places such as basements and woodpiles. Their venom causes tissue damage in the area of the bite.

Spiders have courtship rituals

Spiders are usually solitary animals. When they do seek a mate, it is important that males and females of the same species be able to recognize each other. The male is usually smaller than the female and could easily be mistaken for a meal if some signal did not identify him to his prospective partner. Spiders have evolved complex behaviors that ensure successful mating and fertilization of the female's eggs.

Mating behavior in the black widow spider provides an example of the dangers a male spider faces. As the male approaches the web of a female, he drums a specific rhythm on the strands of the web to alert the female that he is a potential mate and not a meal. If he vibrates the female's web inappropriately, she will kill and eat him. After mating, the female sometimes kills the male anyway, wrapping and saving him for her offspring's first dinner.

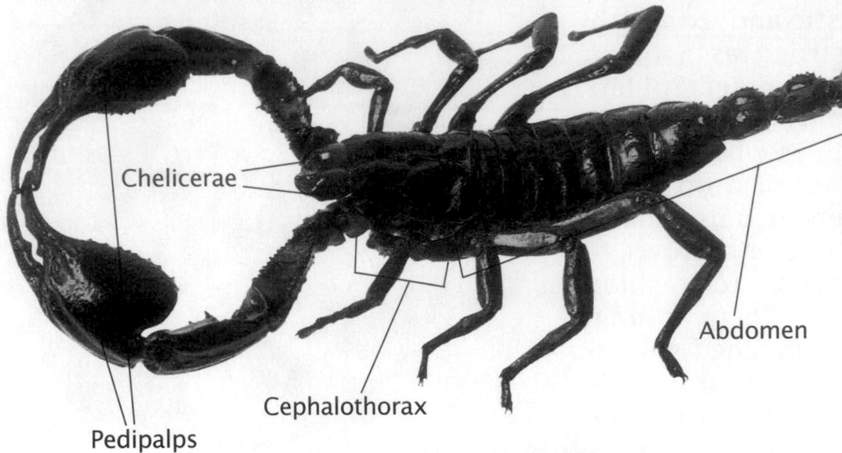

Figure 26-5
This African scorpion is about 18 cm (7 in.) long. Scorpions found in the United States range from 5 to 13 cm (2 to 5 in.) in length.

Other Arachnids

Relatives of scorpions were the first arthropods to invade land, more than 400 million years ago. Today's scorpions share many of the adaptations that enable spiders to survive on land, including book lungs and efficient excretory and water-conserving organs. Unlike the poison-injecting chelicerae of spiders, scorpions' chelicerae are ripping claws. Their pedipalps have evolved into enlarged pincers used to capture prey. Scorpions feed mostly on insects. As you can see in **Figure 26-5**, the scorpion's stinger is found at the tip of its abdomen. While the scorpion holds its victim in its pincers, it brings its abdomen forward to jab into the prey and then tears the prey into pieces.

Scorpions live in the warmer regions of the world. About 20 scorpion species are found in the United States, mostly in the Southwest. Scorpions hide during the day and hunt at night. Although a scorpion's sting is painful, it is rarely lethal to humans. Like spiders, scorpions perform elaborate mating rituals.

Figure 26-6
Lyme disease is caused by a bacterium that is transmitted by the bite of the deer tick. About one-half of the people infected with Lyme disease develop this "bull's-eye" rash within a few days of being bitten.

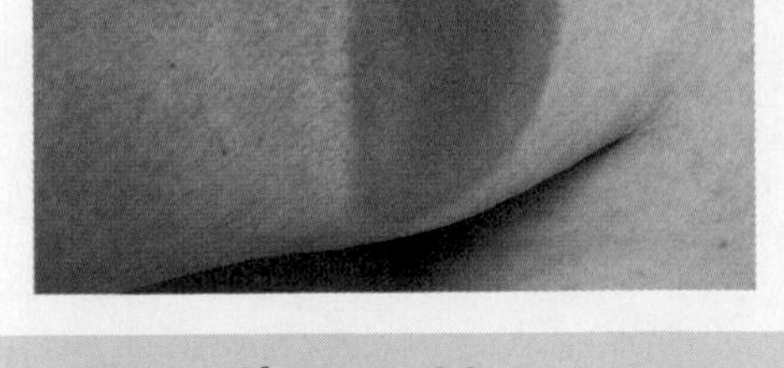

Most mites and ticks are parasites

If dust makes you sneeze, you may not be reacting to the dust itself, but to the tiny mites found in the dust. Over 30,000 species of mites and ticks have been described. A number of these species are free-living on land or in water, but many are parasites on the bodies of other animals. The free-living forms are mostly scavengers or predators on other tiny creatures.

If you have done much camping or hiking in brushy terrain, chances are you have spent some time removing ticks from your body. Most ticks are bloodsuckers that are parasites on vertebrates. Their sharp chelicerae are specialized for slicing skin. Although normally quite small, some ticks can swell to 3 cm (just over 1 in.) in length after a full meal of blood. Ticks transmit Lyme disease and Rocky Mountain spotted fever to humans, and they transmit Texas cattle fever to livestock. **Figure 26-6** shows the rash that usually occurs a few days after a person has been bitten by a tick that carried Lyme disease.

Mites, such as the red water mite in **Figure 26-7**, are much more diverse than ticks. They parasitize virtually all groups of animals and many plants. The incredible specialization of mites is shown by two species, *Demodex folliculorum* and *D. brevis. Demodex folliculorum* lives only in hair follicles,

Figure 26-7
The red water mite is an active swimmer. The larval stages of water mites may be parasites on insects or fishes.

and *D. brevis* lives only in oil glands of the human forehead. Mites have serious direct and indirect effects on humans. Many, such as chiggers, punch holes in the skin and cause severe itching. Mites also carry viruses that plague food crops, such as mosaic viruses of rye and wheat. The red-legged mite and the winter grain mite both feed on stored crops, destroying many tons of grain each year. Other mites cause feather loss in birds, decreased wool production in sheep, and mange in dogs. If you find yourself itching a bit after reading this section, don't be alarmed; we are all hosts to some of these parasites.

Horseshoe Crabs

Horseshoe crabs, such as the one shown in **Figure 26-8**, are arthropods that have changed little in 200 million years. There are only five species of horseshoe crabs today. They live in shallow parts of the oceans, where they plow through sandy bottoms and use their clawlike legs to grasp small animals on which they feed. They also eat dead animals.

Horseshoe crabs have chelicerae but are not arachnids. They are closely related to the extinct giant "water scorpions," or eurypterids, such as the one shown in **Figure 26-9**. Eurypterids lived from about 500 million years ago until about 300 million years ago.

Figure 26-8
Horseshoe crabs are not really crabs. They are more closely related to spiders and scorpions than to crabs.

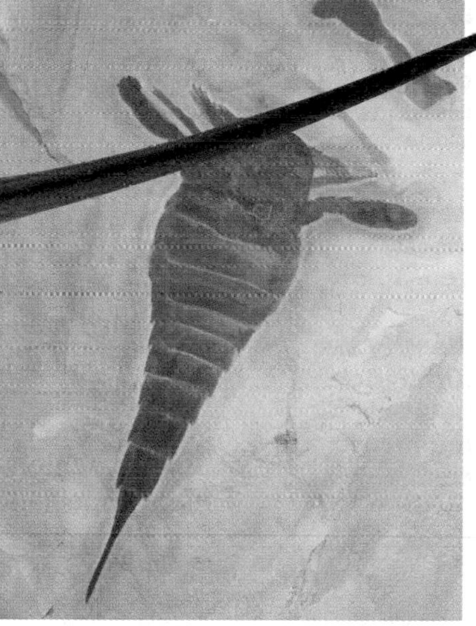

Figure 26-9
Some eurypterids, ancient relatives of the horseshoe crab, reached 3 m (10 ft) in length. This eurypterid lived in a shallow marine environment.

SECTION REVIEW

1. **What functions do pedipalps serve in spiders?**
2. **Describe two ways spiders use silk, other than for building webs.**
3. **Describe one way your life is affected positively by arachnids.**
4. **How do arachnids negatively affect humans?**

Tour

Insect

Almost one-half of the named species of insects are beetles. The lady beetle is often called ladybug or ladybird beetle. There are more than 400 species of lady beetles in the United States. This species is the two-spotted lady beetle, which is common in northern parts of the United States.

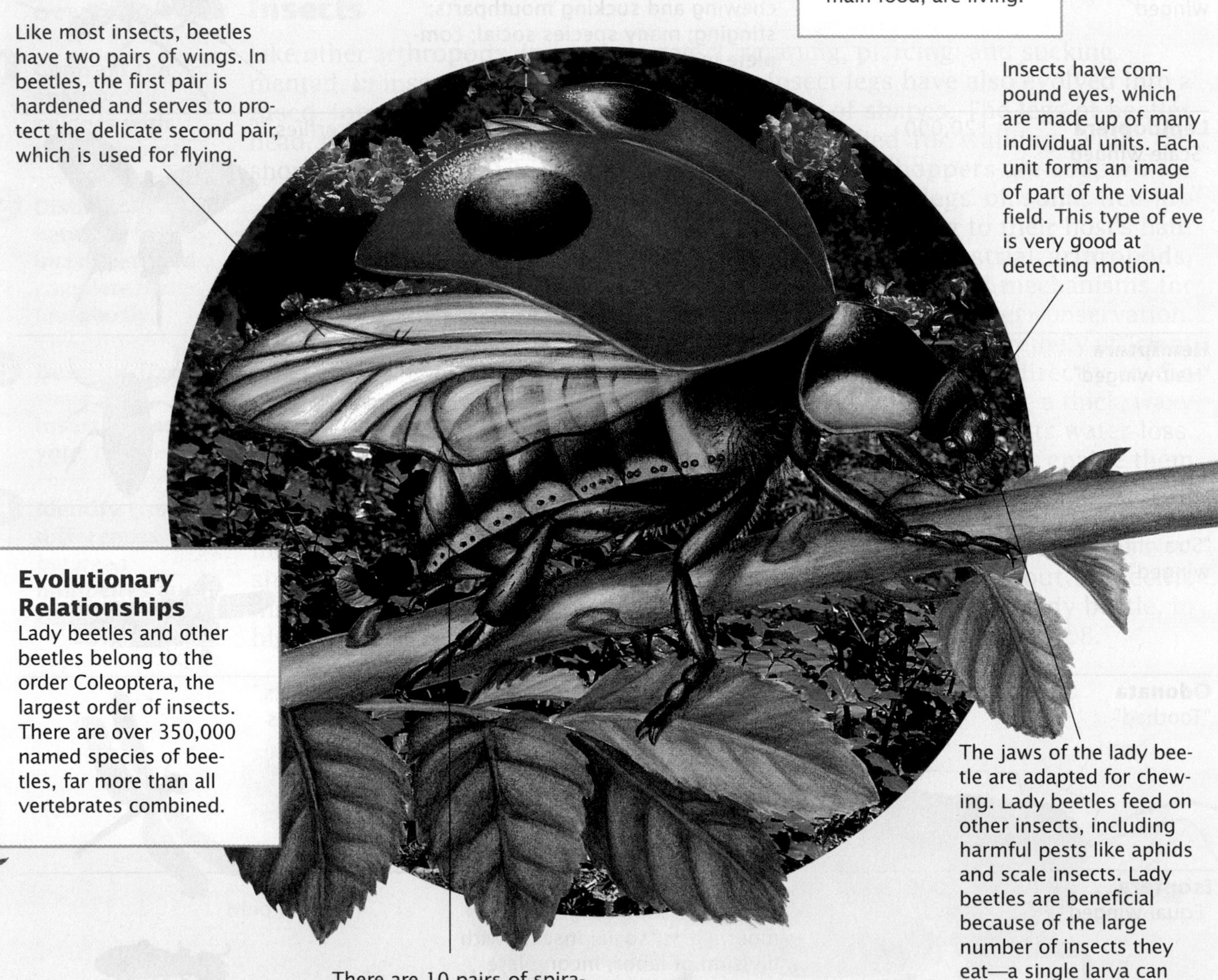

Beetles, as well as most other insects, are characterized by three distinct body regions: head, thorax (to which legs and wings attach), and abdomen.

Habitat
The lady beetle is often found on plants on which aphids, its main food, are living.

Like most insects, beetles have two pairs of wings. In beetles, the first pair is hardened and serves to protect the delicate second pair, which is used for flying.

Insects have compound eyes, which are made up of many individual units. Each unit forms an image of part of the visual field. This type of eye is very good at detecting motion.

Evolutionary Relationships
Lady beetles and other beetles belong to the order Coleoptera, the largest order of insects. There are over 350,000 named species of beetles, far more than all vertebrates combined.

The jaws of the lady beetle are adapted for chewing. Lady beetles feed on other insects, including harmful pests like aphids and scale insects. Lady beetles are beneficial because of the large number of insects they eat—a single larva can eat 400 aphids.

There are 10 pairs of spiracles along each side of the abdomen. Spiracles allow air into a network of tubules that carry oxygen throughout the body.

Size
The two-spotted lady beetle is about 5 mm (0.2 in.) long.

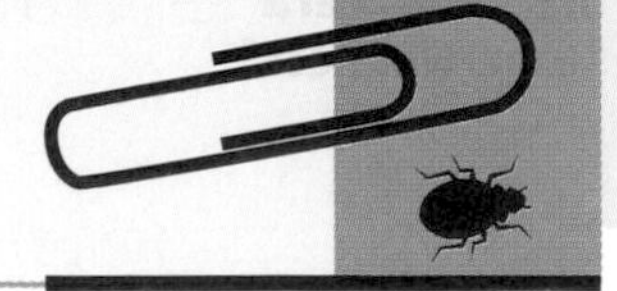

Most insects change their body form as they mature

Spiders, scorpions, mites, and ticks all undergo **direct development**. In this kind of development, a miniature copy of the adult form hatches from the egg. Some species of insects also undergo direct development.

Most insects, however, hatch as a form that is not like the adult. As it grows, the young insect changes to become more and more like its parents. Each of these changes is called a **metamorphosis** *(meht uh MAWR fuh sihs)*. If the metamorphosis into the adult form involves a series of gradual changes, it is called **incomplete metamorphosis**. Insects that undergo incomplete metamorphosis include grasshoppers, dragonflies, mayflies, and cockroaches. In this type of development, the insect that emerges from the egg is called a **nymph** *(NIHMF)*. The nymph is a smaller version of the adult insect, similar in structure but without wings or mature reproductive organs.

In **complete metamorphosis**, illustrated in **Figure 26-11**, the immature form usually does not resemble the adult. When the egg of an insect hatches, an immature form called a larva emerges. The larva cannot fly. It feeds and grows until ready to transform into an adult. When the larval stage is complete, the insect enters a stage called the pupa *(PYOO puh)* or, in butterflies, the chrysalis *(KRIHS uh lihs)*. Many pupae form a covering around themselves. This covering may develop from the larva's "skin," or it may be a cocoon of silk. During this stage, most insects remain immobile while larval tissues and organs are replaced with new tissues and organs. At the end of the pupal stage, a fully formed, mature adult emerges. About 90 percent of insect species

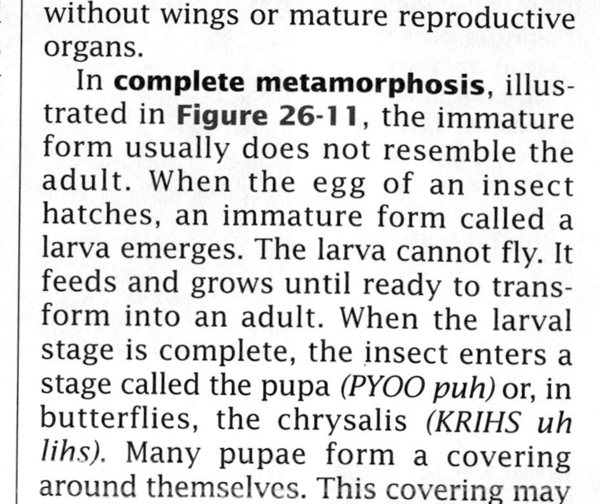

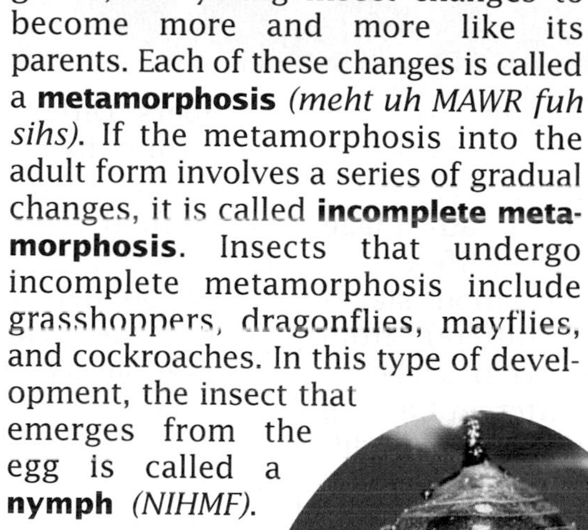

a Egg

b Larva (caterpillar)

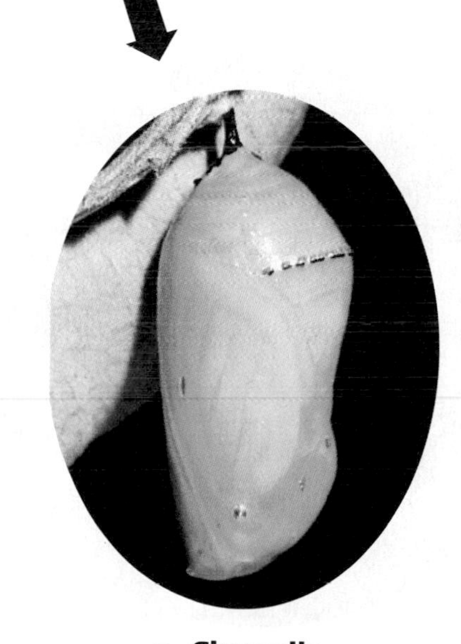

c Chrysalis

d Adult butterfly about to emerge

e Adult butterfly

Figure 26-11
The life cycle of the monarch butterfly illustrates complete metamorphosis.

Figure 26-
a A termite
usually ha
single que
that lays a
the eggs.
nest of ter
may conta
as many a
5 million
individual

b Worker ter
mites sea
for food,
which the
supply to
queen, kir
and soldi

Figure 26-19
Large numbers of *Daphnia* can be found in almost any sample of water taken from a nutrient-rich lake or pond. There are nearly 400 species of *Daphnia* in freshwater environments. They range from 0.2 to 3 mm (0.008 to 0.1 in.) in length.

Among the most important crustaceans in freshwater environments are the so-called water fleas, or cladocerans *(kluh DAHS uhr uhns)*, like *Daphnia*, shown in **Figure 26-19**. *Daphnia* feed on phytoplankton (photosynthetic plankton). In high concentrations, water fleas consume huge quantities of photosynthetic organisms and in turn are fed upon by small predators such as young fishes. In many freshwater habitats, these little crustaceans are a vital link between the producers and the rest of the food web.

Look at the photograph in **Figure 26-20**. This tiny crustacean is very important in marine ecosystems. It is a **copepod** *(KOH puh pahd)*. While some kinds of copepods live on the sea bottom, and a few are parasitic, the vast majority are part of the zooplankton, the heterotrophic organisms that feed on phytoplankton. Copepods occur both in the sea and in fresh water, often in incredibly high concentrations. Their abundance follows the seasonal changes in concentrations of phytoplankton. Like water fleas in lakes, copepods link the ocean's photosynthetic life to the rest of the ocean's food web. Copepods are consumed by a variety of small predators, which are eaten by larger predators, and so on. Virtually all animal life in the open sea depends on the copepods, either directly or indirectly. Although humans do not eat copepods directly, our sources of food from the ocean would disappear without the copepods.

Figure 26-20
Although most copepods are pale and transparent, some species are brilliant red, orange, purple, blue, or black. Most copepods are less than 10 mm (0.4 in.) long.

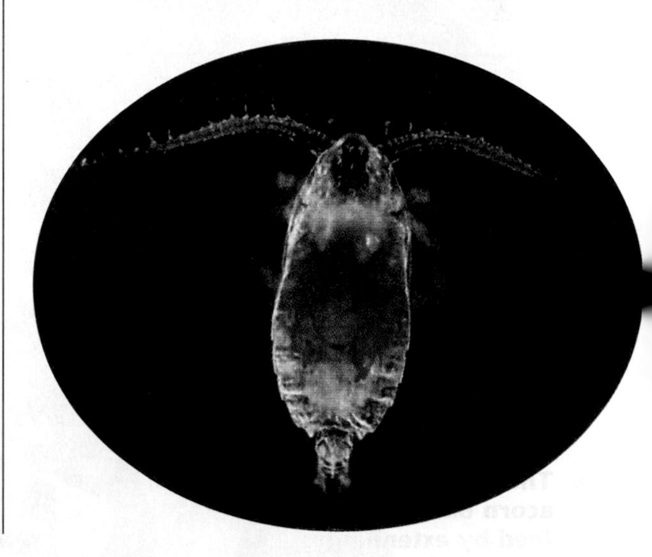

SECTION REVIEW

1. **Name three kinds of crustaceans.**
2. **Describe three differences between insects and crustaceans.**
3. **How would the ecology of the sea and the human food supply be affected if all crustaceans died?**
4. **Explain the advantages and disadvantages of an exoskeleton.**

CHAPTER 26 Highlights

This praying mantis is lying in wait for flying insects.

	Key Terms	Summary
26-1 Spiders and Their Relatives	chelicera (p. 501) arachnid (p. 501) pedipalp (p. 501) spinneret (p. 502)	• Spiders, scorpions, mites, ticks, and horseshoe crabs have chelicerae, a specialized pair of appendages at the front of the body. They have two main body units: the cephalothorax and the abdomen. Spiders, scorpions, mites, and ticks are arachnids. • The chelicerae of spiders are poison delivering fangs. Spiders use silk to spin safety lines, to line burrows, to protect the young, and to trap food. • Scorpions' chelicerae are adapted for tearing prey. The pedipalps of scorpions are pincers for grasping prey. • Many ticks and mites are parasites. Some transmit diseases to humans and domestic animals.
26-2 Insects, Millipedes, and Centipedes	mandible (p. 506) direct development (p. 509) metamorphosis (p. 509) incomplete metamorphosis (p. 509) nymph (p. 509) complete metamorphosis (p. 509)	• Insects have three main body units: the head (with three sets of mouthparts), the thorax (with three sets of legs), and the abdomen. • Insect development usually involves either complete or incomplete metamorphosis. • Millipedes have two pairs of legs per segment and feed on decaying matter. Centipedes have only one pair of legs per segment and are predators.
26-3 Crustaceans	decapod (p. 517) copepod (p. 518)	• Crabs, lobsters, shrimp, pill bugs, and barnacles are crustaceans, subphylum Crustacea. Most crustaceans are aquatic. • The crustaceans known as copepods are extremely important links in marine food webs.

Spiders and their relatives are characterized by chelicerae.

There are more species of insects on Earth than any other group of animals.

Most crustaceans live in or close to the ocean.

CHAPTER 26

review

Understanding Vocabulary

1. For each set of terms, complete the analogy.
   a. first appendages : chelicerae :: second appendages : ____
   b. insects : tracheae :: crustaceans : ____
   c. Lyme disease : ticks :: bubonic plague : ____
2. For each pair of terms, explain the differences in their meanings.
   a. spider, insect
   b. incomplete metamorphosis, complete metamorphosis
   c. crustacean, arachnid

Understanding Concepts

3. **Relating Concepts** Construct a concept map that identifies the three major groups of arthropods and gives characteristics of each. Try to include the following terms in your concept map: arachnids, insects, crustaceans, spiders, silk, bees, ants, social, lobsters, ocean, pill bugs, land.
4. **Comparing Structures** How do the chelicerae and pedipalps of spiders compare with those of scorpions? How do scorpions capture their prey? How are their pedipalps and chelicerae adapted for this task?
5. **Identifying Structures** Identify three physical characteristics of animals classified as spiders. Name two kinds of arachnids that are not spiders. Describe two ways that the arachnids you chose affect humans
6. **Summarizing Information** Spiders are predators but have no teeth or jaws. How do they capture their prey? How do they obtain nutrients from their prey?
7. **Inferring Conclusions** An unknown arthropod is found. It has three body segments, one pair of antennae, and three pairs of walking legs. What kind of arthropod is it? Explain how you arrived at this conclusion.
8. **Forming Hypotheses** A scientist knew that a particular fungus was known to kill fire ants. Using 500 ants, he set up an experimental mound with a glass window, through which he could watch the ants inside the mound. On Day 1 he introduced the fungus. By Day 6 how many ants were left? Propose a hypothesis to explain this result. How would you test your hypothesis?

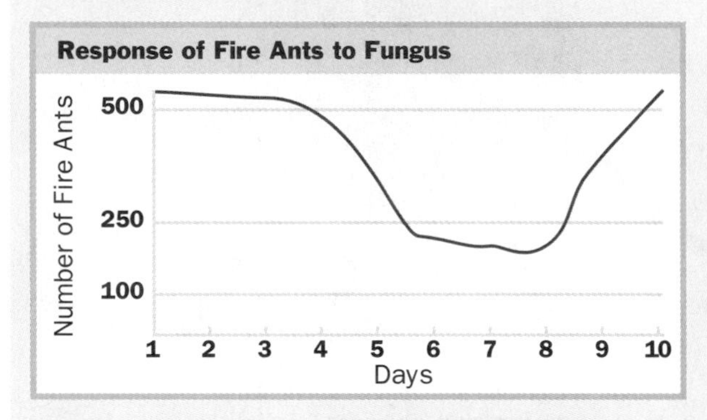

9. **Recognizing Relationships** The different stages of an insect that undergoes complete metamorphosis are both anatomically and ecologically different. How is this beneficial for the insect? How does this pose a problem for the farmer on whose crops the insect feeds?
10. **Analyzing Information** Identify the four castes in a termite colony and describe the role of each.
11. **Recognizing Relationships** What are two ways that insects benefit humans?
12. **Summarizing Information** Which of the following are not crustaceans: shrimp, crayfish, horseshoe crab, scorpion, wasp, water flea, tick?
13. **Identifying Functions** A lobster's heavy exoskeleton cannot grow with the animal. How does a lobster grow?
14. **BUILDING ON WHAT YOU HAVE LEARNED** In Chapter 24, you learned that terrestrial animals have many adaptations that enable them to live on land. How are the eggs of many arthropods well suited for survival on land?

Interpreting Graphics

15. Look at the photograph of the arthropod below.

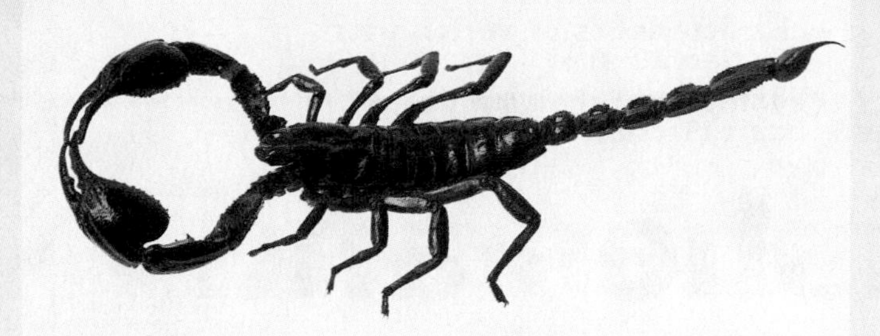

a. What kind of arthropod is this animal?
b. Explain the adaptations that enable this arthropod to conserve water, capture prey, and defend itself.

Reviewing Themes

16. ***Interdepence***
How do humans depend on copepods?

17. ***Evolution***
Some species of insects can produce a new generation every few weeks. How does such rapid reproduction affect their ability to evolve resistance to pesticides?

Thinking Critically

18. **Recognizing Logical Inconsistencies** A neighbor was recently complaining about the increased number of insects around her house. At the same time, she was removing every spider web and killing every spider she saw. What is wrong with the neighbor's attempt to control insects?

19. **Evaluating Proposals** To help control a species of scale insect that feeds on orange trees, a scientist proposes introducing a parasitic wasp from Australia. You are asked to help evaluate the proposal. What are some of the biological issues that you should consider before making your decision? What non-biological issues might you consider?

Life/Work Skills

20. **Allocating Resources** Working as a group, plan how you would spend a budget of $5,000 to eradicate fire ants on the grounds of a school in the South. This budget must last for three years, so you will have to think about long-term effectiveness. How well have chemical pesticides worked to kill fire ants in other places? What other methods have been tried, and how successful were they? Investigate these questions before you decide how to spend your budget. Prepare a report or poster presentation to inform the school board of the positive and negative aspects and the costs of the methods you have researched.

Activities and Projects

21. **Health Connection** Head lice are sometimes a problem in schools. Do library research or search an on-line database to discover the symptoms of lice infestation and to learn what can be done to stop the spread of head lice. Summarize your findings in a written report.

22. **History Connection** Ancient Egyptians worshiped the sun and held sacred the scarab beetle, *Scarabaeus saber*, as a symbol of immortality. Use library references or search an on-line database to find out about scarab beetles and their historical importance to Egyptians. Also research the importance of scarab beetles in Chinese history and folklore. Prepare a written report of your findings.

Discovering Through Reading

23. The entire March 1995 issue of *Natural History* is devoted to arachnids. Read the article "Webs of Deceit" by Catherine L. Craig on pages 33–35. How do the number and kinds of silk produced by a tarantula compare with those produced by spiders in the Araneoidea superfamily? How might the color of the golden orb spider's web be an advantage in capturing prey?

CHAPTER 27

REVIEW

- characteristics of chordates (Section 23-3)
- characteristics of vertebrates (Section 23-3)
- evolution of the amphibian heart (Section 24-2)

Fishes and Amphibians

Amphibians, such as these red-eyed tree frogs, thrive in the moist conditions of tropical rain forests.

You and most of the animals you commonly see are vertebrates. As you learned in Chapter 11, vertebrates are chordates that have a vertebral column, or spine. The first vertebrates evolved more than 500 million years ago. There are approximately 40,000 species of vertebrates today, and about half of them are fishes, including the great white shark shown at the left.

27-1 Jawless Fishes and Cartilaginous Fishes

OBJECTIVES

1. **List three characteristics of agnathans.**
2. **Describe how a lamprey feeds.**
3. **Describe how jaws are thought to have evolved.**
4. **Contrast sharks and rays with agnathans.**

The First Fishes: Class Agnatha

The first vertebrates evolved about 500 million years ago. They were jawless fishes belonging to the class Agnatha (*Agnatha* means "without jaws"). Thick bony plates covered the bodies of these ancient agnathans *(AG na thuns)*, as you can see in **Figure 27-1**. Bone and cartilage first evolved in this group of fishes. Agnathans were an abundant and diverse group for about 150 million years, until they were largely replaced by jawed fishes about 360 million years ago. Only 81 species of agnathans exist today.

The living agnathans are the lampreys and hagfishes, which are shown on the next page. These eel-like creatures have scaleless, slimy skin and lack paired fins. Their skeleton is mostly composed of cartilage, and there is no well-developed vertebral column. The notochord, which is replaced by vertebrae in most vertebrates, functions as the major support structure in adult lampreys and hagfishes. The gills of agnathans lie within pouches that branch from the pharynx. Water exits these pouches through several openings behind the head.

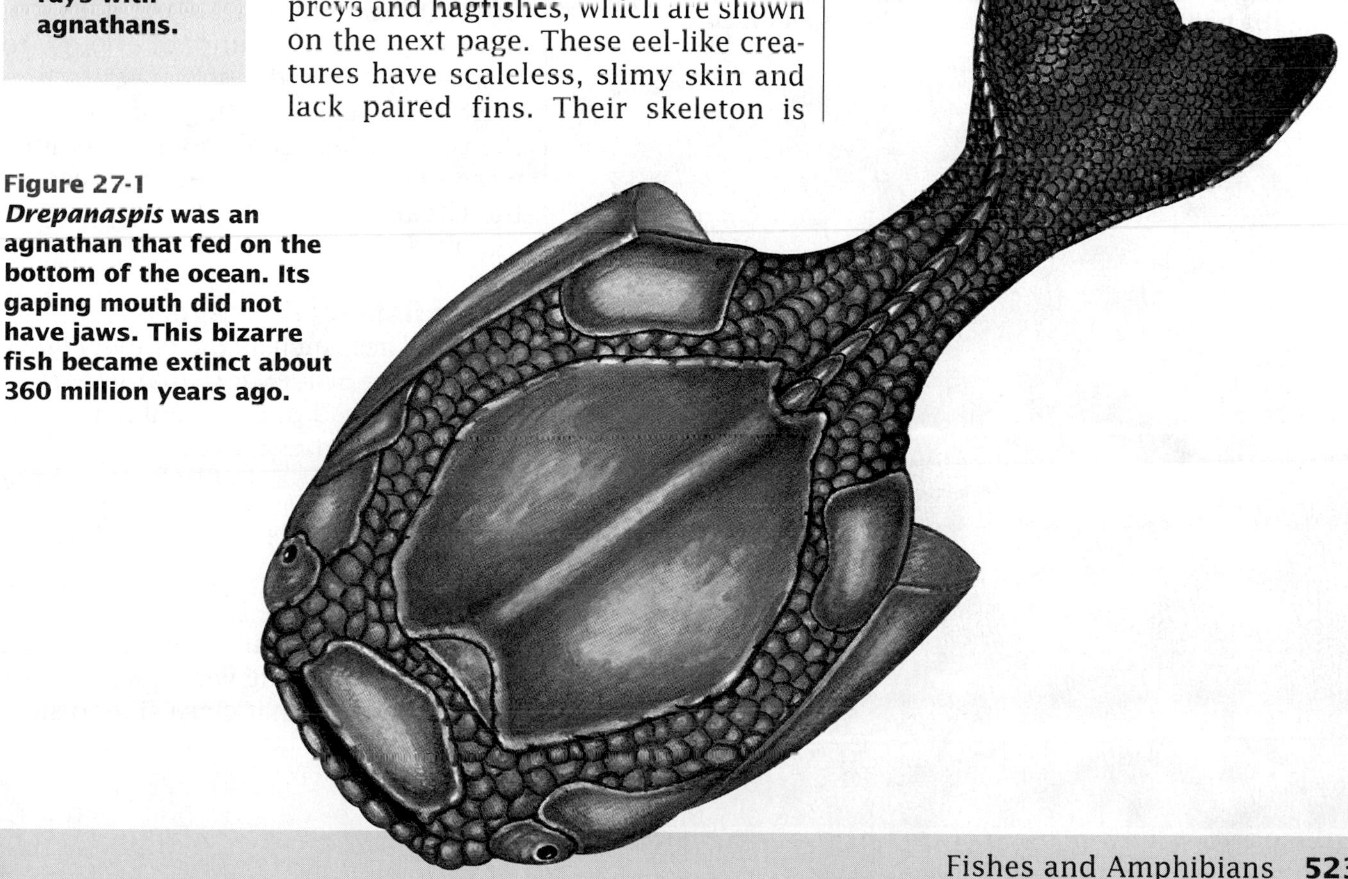

Figure 27-1
***Drepanaspis* was an agnathan that fed on the bottom of the ocean. Its gaping mouth did not have jaws. This bizarre fish became extinct about 360 million years ago.**

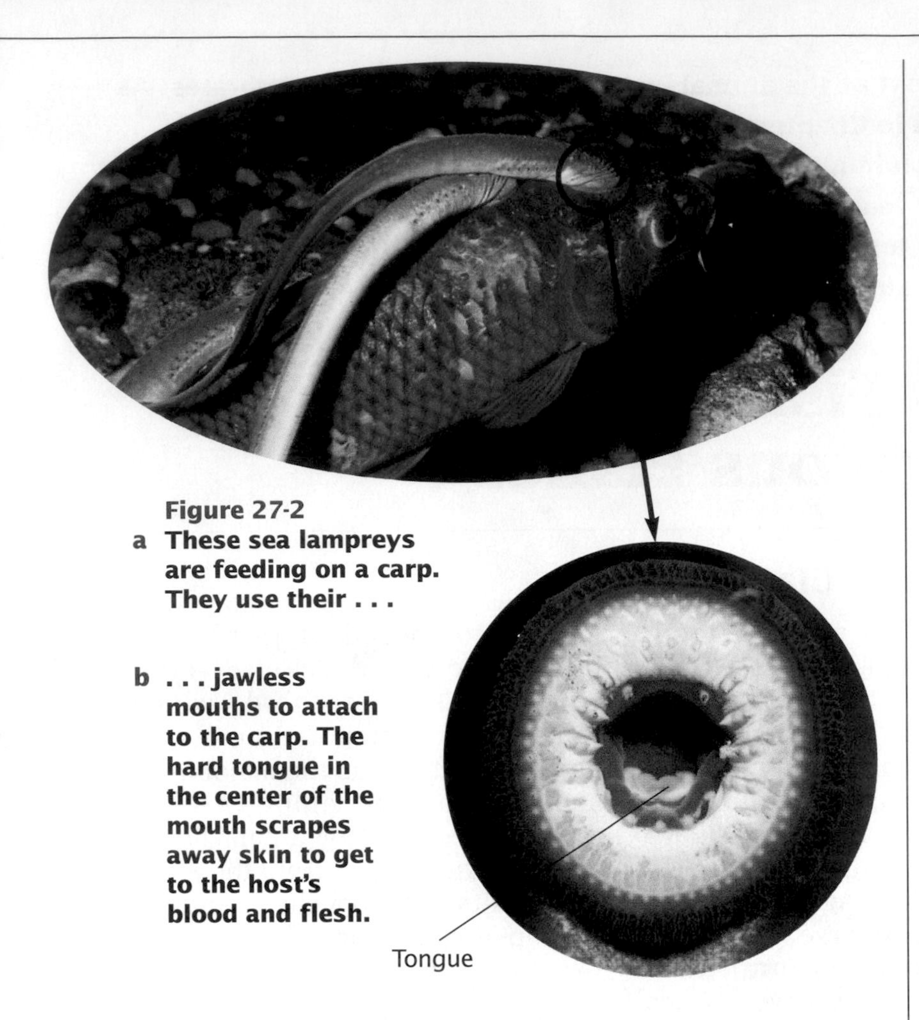

Figure 27-2
a These sea lampreys are feeding on a carp. They use their . . .

b . . . jawless mouths to attach to the carp. The hard tongue in the center of the mouth scrapes away skin to get to the host's blood and flesh.

Figure 27-3
Hagfishes are jawless scavengers. They invade the bodies of dying or dead animals and feed on the internal organs.

Many lampreys are parasites

Most kinds of lampreys spend their entire lives in fresh water. A few species live in the sea as adults but migrate into fresh water to breed. Some lampreys, like the sea lampreys shown in **Figure 27-2**, are external parasites that feed on other fishes. A parasitic lamprey's mouth is recessed within a funnel-like structure. Sharp toothlike hooks in the funnel help the lamprey attach to its host. The rim of the funnel functions as a suction cup. A rough tongue scrapes off small particles of the host's skin and flesh. The lamprey sucks in these particles along with the host's blood. Like some leeches, the lamprey secretes a substance that prevents its host's blood from clotting.

After a lamprey has fed, it drops off of its host. Damage to the host can be severe since the wound left by the lamprey may become infected or cause the host to bleed to death. For this reason, large lamprey populations can cause great damage to populations of other fishes. For instance, when a canal that allowed ships to bypass Niagara Falls was deepened in the early 1900s, the ocean lamprey was able to move from the Atlantic Ocean into the upper Great Lakes. By the 1940s and 1950s, this lamprey was abundant enough to cause a serious decline in the commercial and sport fishing industries of the Great Lakes. Lamprey populations were eventually reduced by treating the Great Lakes with poisons toxic to lamprey larvae.

Hagfishes are scavengers

Hagfishes, such as the one in **Figure 27-3**, are scavengers that generally feed on dead or dying animals, such as large invertebrates or other fishes. When a hagfish locates a potential meal, it enters the body of the other animal by squirming in through the gill openings, the mouth, or the anus. Once inside, it feeds on the internal organs of the animal, biting with jawlike folds of muscle that close side to side.

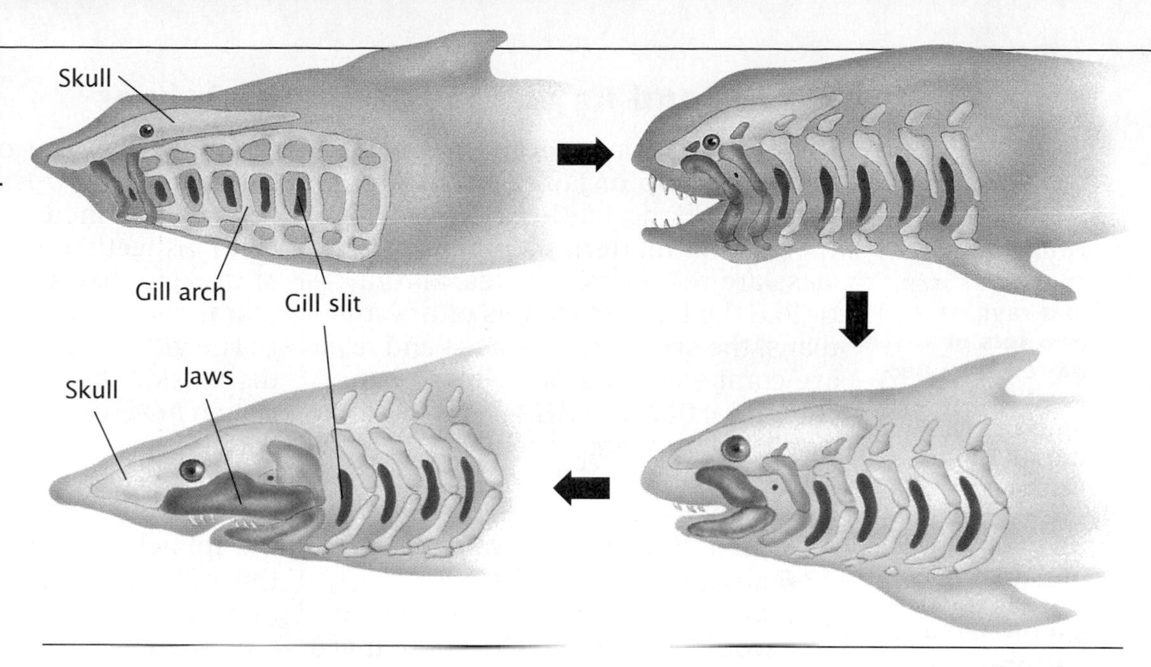

Figure 27-4
The jaw is an evolutionary modification of the first pair of gill arches that were present in ancient jawless fishes.

Evolution of Jaws

Because they lack jaws, agnathans can eat only food that can be sucked into their mouths. Apart from the agnathans, all vertebrates have movable jaws. Animals with jaws can exploit a much wider range of foods than can jawless animals. Scientists think that jaws evolved from one or more of the gill arches that support the pharynx in agnathans, possibly in the way shown in **Figure 27-4**.

The earliest jawed fishes were acanthodians *(uh KAN thoh dee uhns)*, members of the class Acanthodii. Although superficially resembling sharks, the acanthodians, or "spiny fishes," had a bony internal skeleton and were definitely not sharks. Acanthodians evolved about 435 million years ago. Another group of early jawed fishes was the placoderms *(PLAK uh durms)*, class Placodermi, which evolved about 400 million years ago. Placoderms (meaning "plate skin") were armored with heavy, bony plates and had strong jaws, as in the predator *Dunkleosteus*, illustrated in **Figure 27-5**. Both placoderms and acanthodians show a characteristic found in most other jawed vertebrates—paired appendages, in this case paired fins. Placoderms and acanthodians diversified rapidly. Perhaps because of jaws and paired fins, these fishes largely replaced the agnathans. Lampreys and hagfishes may have survived because of their specialized feeding habits. The placoderms became extinct about 345 million years ago, and the acanthodians died out about 270 million years ago. Scientists do not know why these once-diverse groups became extinct.

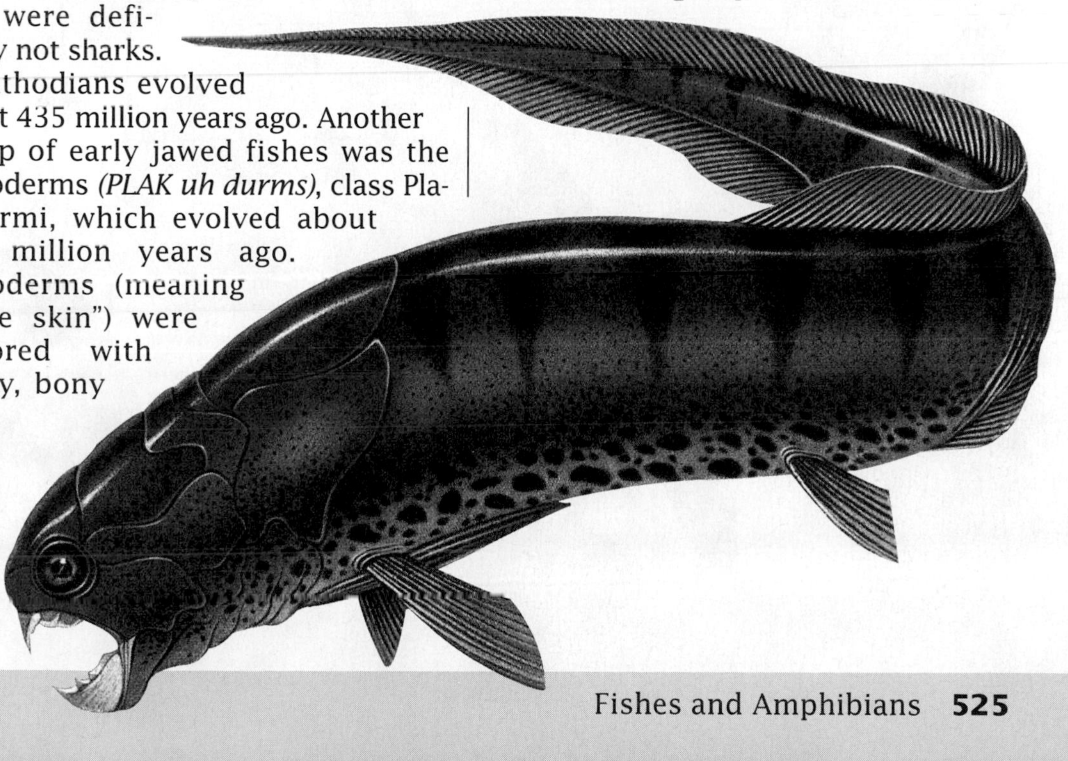

Figure 27-5
***Dunkleosteus*, a placoderm that was a predator in the ancient seas, might have looked like this drawing. Its skull and jaws were over 65 cm (2 ft) long.**

Sharks and Rays: Class Chondrichthyes

Soon after placoderms appeared, another group of jawed fishes arose, the class Chondrichthyes *(kahn DRIHK thees)*. The modern members of this class are the sharks, skates, and rays. Unlike the bony skeletons of acanthodians, the skeletons of sharks and rays are composed of a flexible substance called cartilage *(KAHRT'l ihj)*. Chondrichthyes means "cartilage fishes." There are about 850 living species of cartilaginous fishes, the vast majority of which live in salt water. **Figure 27-6** shows a shark and a ray.

If you were to touch a shark or ray, you would notice rough, sandpaper-like skin. This texture results from the many small scales that are embedded in the skin. As shown in **Figure 27-6b**, shark scales are very similar to shark teeth. The mouth of a cartilaginous fish includes upper and lower jaws, generally armed with rows of hard teeth. As teeth in the outer row are lost or broken, they are replaced by others that move up from the row behind.

The gills of sharks and rays open to the outside through a series of slits, as shown in **Figure 27-6c**. One gill slit, the spiracle, opens directly to the outside on the side or top of the head. Water can be brought to the pharynx through the spiracle and then expelled through the gill slits. This arrangement not only frees the mouth for feeding but also enables sharks and rays to lie on the bottom of the ocean and draw water in through the spiracle rather than the mouth, which would also bring in sediment.

Figure 27-6
a Sharks, skates, and rays have two sets of paired fins: pectoral fins and pelvic fins. The dorsal fin (on the back) and the caudal fin (the tail) are unpaired. This ray has no dorsal fin, but its pectoral fins are greatly enlarged into a pair of large, winglike fins.

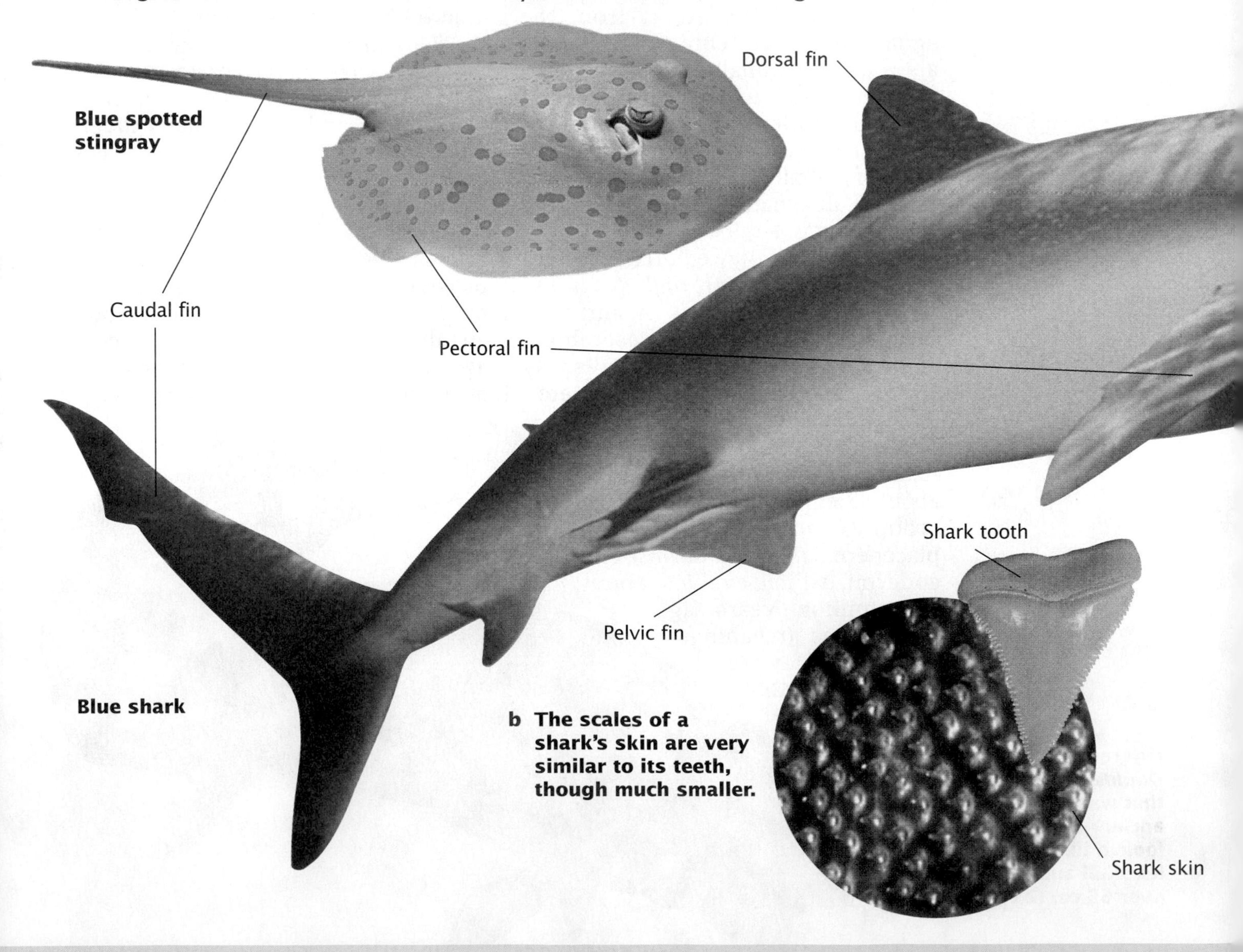

b The scales of a shark's skin are very similar to its teeth, though much smaller.

How sharks detect and capture prey

Most sharks and rays are carnivores. Although sharks have a reputation for being man-eaters, only a few species of sharks are dangerous to humans. Three well-developed senses enable sharks and rays to detect their prey. First, they have an acute sense of smell. Second, they have the ability to sense electric currents in water. This sense is particularly useful in detecting the small electric currents generated by the muscle movements of animals. The electric ray *Torpedo* is also able to produce a powerful electric current, which it uses to deter predators and to stun prey. Third, sharks and rays have a lateral line system. As you learned in Chapter 24, the lateral line system is composed of pressure-sensitive cells that lie within canals along the sides of a fish. Changes in pressure caused by a fish or other animal swimming nearby can be detected by the cells in the lateral line.

Given the variety of shapes and sizes of cartilaginous fishes, it is not surprising that they exhibit a variety of feeding methods. Like the largest whales, the largest shark (the whale shark) and the largest ray (the manta ray) feed on plankton. The whale shark may exceed 13 m (45 ft) in length, and the manta ray can have a "fin span" of nearly 6 m (20 ft). In both the whale shark and the manta ray, the mouth is filled with a hard mesh that is used to trap tiny organisms. Both of these large animals cruise the oceans and filter small animals from the water.

Evolution

Why is it advantageous for a shark to be able to replace lost teeth throughout its life?

c The gill openings of a shark open directly to the outside. Water is taken in through the mouth or spiracle, passes over the gills, and then exits through the gill slits.

Sharks and rays have internal fertilization

The pelvic fins of male cartilaginous fishes are modified into a pair of claspers, which are used to transfer sperm to the female during mating. Thus, fertilization in these animals is internal. Skates, rays, and some sharks lay eggs. The extremely yolky fertilized eggs are usually housed in an elaborate leathery case. Many species of sharks do not lay eggs. Instead, the female keeps the yolky eggs inside her body until they hatch, and the young sharks are born alive. In some sharks, the female provides nutrients to her developing young through a membranous sac somewhat like the placenta.

SECTION REVIEW

1. **Identify two characteristics of agnathans.**
2. **Explain two adaptations that enable some lampreys to be external parasites of fishes.**
3. **Explain how the absence of jaws limits what a jawless fish can eat.**
4. **Describe two differences between sharks and lampreys.**

Of the nearly 22,000 living species of fishes, 21,000 species are bony fishes. Bony fishes evolved about 400 million years ago and are one of evolution's greatest success stories. The bony fishes alive today include familiar animals such as trout, bass, perch, tuna, swordfish, guppies, catfish, bluefish, mackerel, and cod.

27-2 Bony Fishes

OBJECTIVES

1. **Identify three differences between bony fishes and sharks.**
2. **Describe the importance of the swim bladder for bony fishes.**
3. **Summarize the importance of lobe-finned fishes in the evolution of land vertebrates.**

Structure of a Bony Fish

Bony fishes belong to the class Osteichthyes *(ahs tee IHK thees)*, which means "bone fish." What are the differences between sharks and bony fishes? As shown in **Figure 27-7**, most bony fishes have a skeleton composed largely of bone and are covered by scales formed of bone. These scales are very different from the toothlike scales of sharks and rays, as you can see by comparing **Figure 27-7a** with **Figure 27-6b**. The gills of bony fishes are housed in a common chamber on each side of the head. Each chamber is covered by a hard plate called the operculum *(oh PUR kyoo luhm)*. The

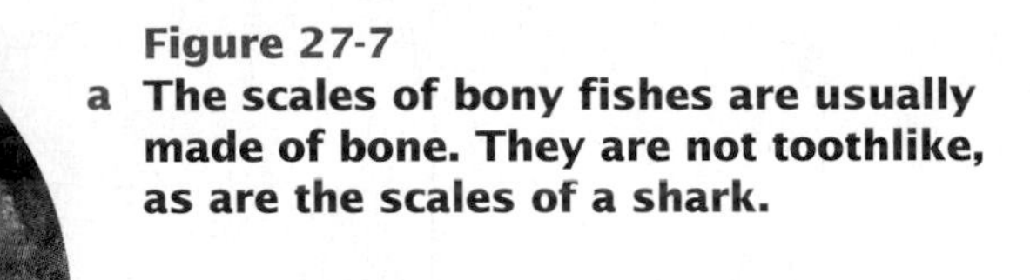

Figure 27-7

a The scales of bony fishes are usually made of bone. They are not toothlike, as are the scales of a shark.

b The bones of this bony fish are evident in this 40-million-year-old fossil.

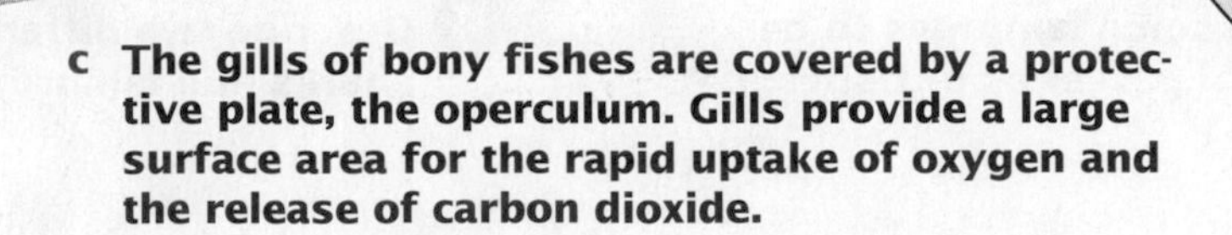

c The gills of bony fishes are covered by a protective plate, the operculum. Gills provide a large surface area for the rapid uptake of oxygen and the release of carbon dioxide.

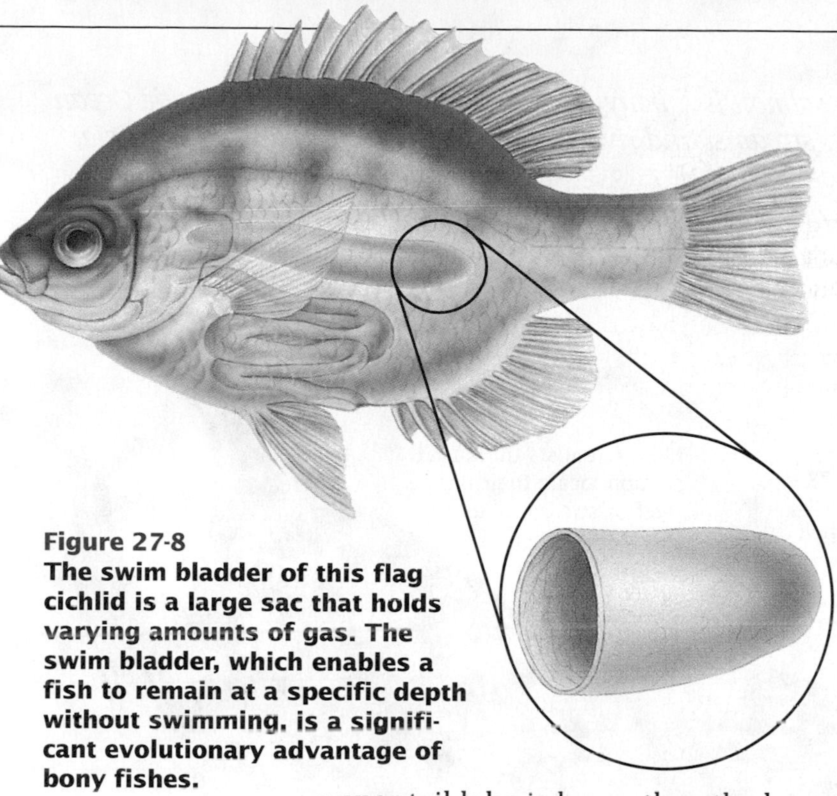

Figure 27-8
The swim bladder of this flag cichlid is a large sac that holds varying amounts of gas. The swim bladder, which enables a fish to remain at a specific depth without swimming, is a significant evolutionary advantage of bony fishes.

upper tail lobe is larger than the lower lobe in sharks, but the lobes are more symmetrical in bony fishes.

A critical difference between bony fishes and sharks is the presence of a **swim bladder** in most bony fishes. This gas-filled sac, shown in the cutaway view in **Figure 27-8**, gives the fish buoyancy. By regulating the amount of gas in the swim bladder, a fish can adjust its buoyant density. Thus, the fish can remain at a particular depth without expending energy through swimming. Although a shark's oil-rich liver gives it some buoyancy, a shark is denser than water and will sink if it stops swimming.

Like sharks, bony fishes sense pressure changes in water with a lateral line system. Vision is also an important sense for most bony fishes. Most bony fishes have color vision. The bright colors of many fishes, such as those in **Figure 27-9**, serve as signals to potential mates and rivals. Fishes living in different environments are sensitive to different colors. For instance, fishes that live at moderate depths often are more sensitive to blue light than to red light; red light is absorbed by the ocean, so little red light penetrates to the depths inhabited by these fishes.

Many species of bony fishes can detect electric fields. The knife fishes of South America produce their own weak electric fields. Knife fishes detect disturbances in their fields, which can be caused by rocks or other objects, by prey or predators, or by other fishes of the same species. Knife fishes live in silty streams where vision is limited. They use their electric fields to navigate, to find prey, to avoid predators, and to signal to other members of their species.

Most bony fishes lay eggs

In almost all bony fishes, fertilization and development occur outside the female's body. The eggs and sperm of bony fishes are typically released into the water or into a nest fashioned by the parents. In live-bearing fishes, such as guppies, mollies, and swordtails, the male transfers his sperm directly to the female via a modified fin near his tail. Fertilization is internal, and the young develop inside the female. You can read more about fishes in the *Tour of a Fish* on the next page.

Figure 27-9
The coloration of these glassy sweepers allows members of the species to easily recognize one another, which is especially important for schooling and mating.

Fish

The sockeye salmon is a bony fish. It lives in the northern Pacific Ocean and in lakes, streams, and rivers in North America and northern Asia. Like other bony fishes, it has a skeleton made of bone. The sockeye salmon also has a swim bladder, a gas-filled sac that allows the fish to adjust its buoyancy.

Habitat
Sockeye salmon live in the ocean and in fresh water. Salmon hatch in fresh water. The young migrate to the sea, where they mature. Adult salmon live in the ocean but return to the stream in which they hatched to reproduce.

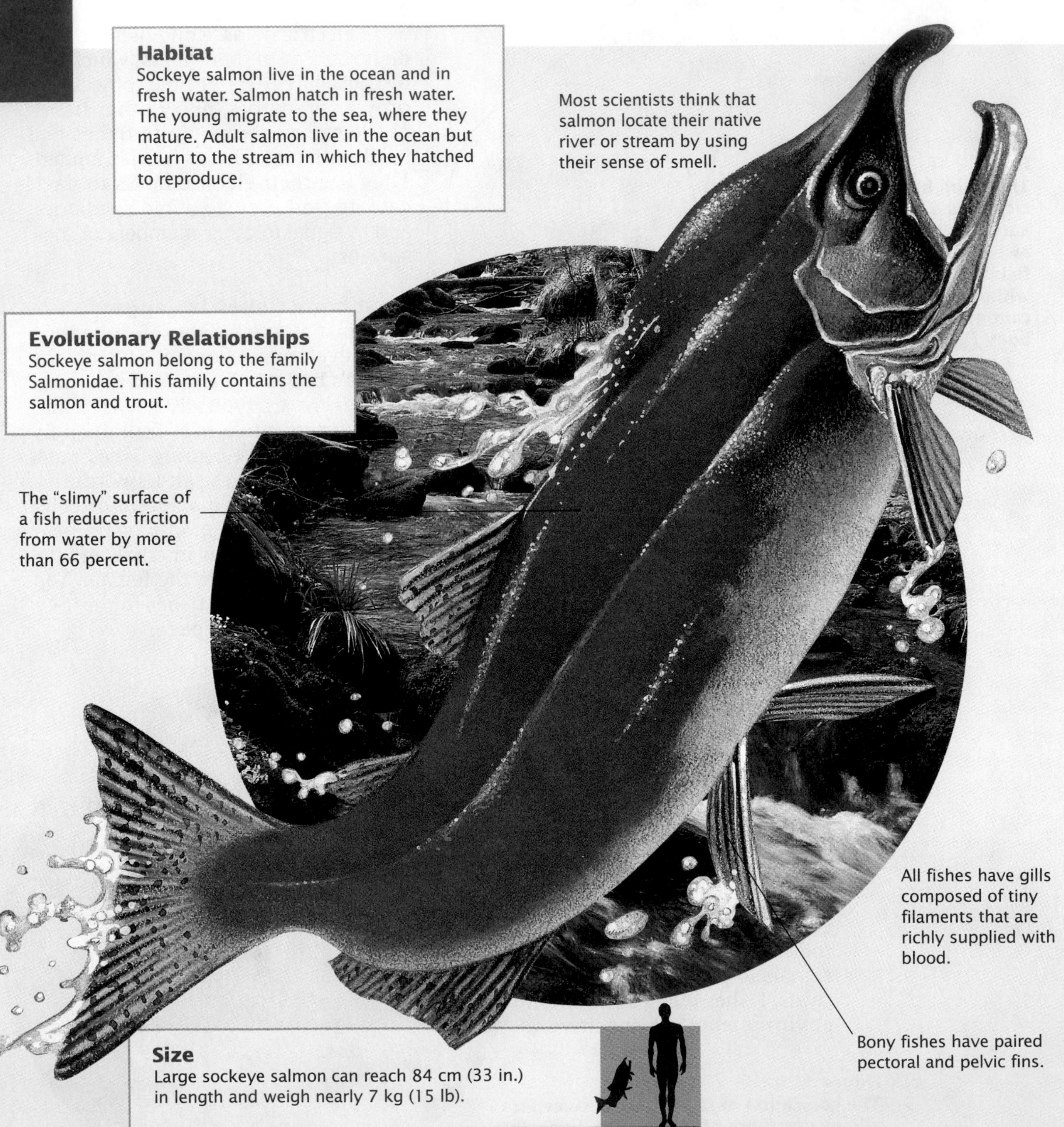

Most scientists think that salmon locate their native river or stream by using their sense of smell.
The "slimy" surface of a fish reduces friction from water by more than 66 percent.
All fishes have gills composed of tiny filaments that are richly supplied with blood.
Bony fishes have paired pectoral and pelvic fins.

Evolutionary Relationships
Sockeye salmon belong to the family Salmonidae. This family contains the salmon and trout.

Size
Large sockeye salmon can reach 84 cm (33 in.) in length and weigh nearly 7 kg (15 lb).

Major Groups of Bony Fishes

Look carefully at the fins of the salmon shown in the *Tour of a Fish* on page 530. Salmon belong to the group of bony fishes known as the ray-finned fishes. All but seven species of modern-day bony fishes are **ray-finned fishes**. In these fishes, the fins are fan-shaped and are supported by thin, bony rays. Ray-finned fishes are the most successful and diverse group of vertebrates.

The other group of bony fishes is the **lobe-finned fishes**. In this group, the fins are fleshy and are supported by central bones. The existing lobe-finned fishes are six species of lungfishes and one species of **coelacanth** *(SEE luh kanth)*. **Table 27-1** lists the major groups of fishes.

As their name suggests, lungfishes have functional lungs. These fishes live in Africa, South America, and Australia. A lungfish from Africa is shown in **Figure 27-10**. Some lungfishes

Figure 27-10
Lungfishes, like this African lungfish, have fleshy, lobed fins. Lungs enable these fishes to live in water that is low in oxygen.

Table 27-1 Major Groups of Fishes

Class	Number of modern species	Description	Examples
Agnatha	81	No jaws; no paired appendages	Lampreys, hagfishes
Acanthodii	Extinct	Jaws; spiny, paired fins	Acanthodians
Placodermi	Extinct	Jaws; paired fins, bony armor	Placoderms
Chondrichthyes	850	Jaws; paired fins; skeleton of cartilage; no swim bladder; spiracle; internal fertilization	Sharks, skates, rays
Osteichthyes	21,000	Jaws; paired fins supported by bony rays; bony skeleton; most have swim bladder	Ray-finned fishes
	7	Jaws; paired lobed fins; bony skeleton; extinct forms are ancestors of amphibians	Lobe-finned fishes

Figure 27-11 ***Latimeria chalumnae*, the living species of coelacanth, was named to honor Marjorie Courtney-Latimer, a museum curator in South Africa. She recognized and preserved the first specimen of this species in 1938.**

inhabit stagnant water that is low in oxygen. Other lungfishes live in ponds that dry up annually. Lungfishes survive these dry periods by burying themselves in mud at the bottom of the pond. Lungs enable these fishes to supplement their oxygen intake by breathing air.

The coelacanth, by contrast, inhabits deep ocean environments. Fossils of coelacanths from the time of the dinosaurs are common, but no coelacanth fossils younger than 80 million years old have been found. Coelacanths were thought to be extinct until a specimen was netted off the eastern coast of South Africa in 1938. **Figure 27-11** shows a coelacanth that was captured in deep water off the east coast of Africa.

Lobe-finned fishes are the ancestors of amphibians

Zoologists are convinced that land vertebrates evolved from lobe-finned fishes that are now extinct. Until recently, scientists thought that coelacanths, which have sturdier fins than lungfishes, were the closest living relatives of the amphibians. However, comparisons of mitochondrial DNA reported in 1990 suggest that lungfishes, not coelacanths, are the closest living relatives of modern amphibians.

SECTION REVIEW

1. **How could you tell whether a fish was a shark or a bony fish?**
2. **Explain how a swim bladder enables a fish to conserve energy.**
3. **Describe two differences between ray-finned fishes and lobe-finned fishes.**
4. **Could legs have evolved from the fins of ray-finned fishes? Explain your answer.**

The name *amphibian* is derived from the Greek words meaning "double" and "life." This name reflects the fact that although most amphibians are terrestrial, they must reproduce in water or in moist environments. Amphibians were the first vertebrates to walk on land. They evolved from lobe-finned fishes about 350 million years ago.

27-3 Amphibians

OBJECTIVES

1. **Identify two characteristics that enabled amphibians to invade the land.**
2. **Describe the life cycle of a frog.**
3. **Describe the characteristics of the three orders of living amphibians.**

The First Land Vertebrates

More than 4,300 species of amphibians exist today. Amphibians include frogs, toads, salamanders, and the wormlike, legless caecilians *(see SIHL ee uhns)*. Two examples of amphibians are shown in **Figure 27-12** and **Figure 27-13**. As adults, amphibians are clearly different from fishes, showing many adaptations for life on land. Except for caecilians, all amphibians have legs. Most amphibians have lungs for gas exchange, although some salamanders lack lungs. Gas exchange also occurs across the thin, moist skin of most amphibians. The lack of a skin that is resistant to drying is one feature that limits most amphibians to moist environments.

Another difference between fishes and amphibians is evident in the amphibian circulatory system. Amphibians achieve more efficient circulation than fishes because of their double-loop circulatory system. As you saw in **Figure 24-10b** on page 466, a fish's heart pumps only deoxygenated blood,

Figure 27-13
This spotted newt is an amphibian. Limbs enable the newt to move on land.

Figure 27-12
The fire-bellied toad lives near slow-flowing rivers in the lowlands of central and eastern Europe. Although adapted to land, amphibians such as this toad must have access to a watery environment to reproduce.

sending it to the gills, where it picks up oxygen. When the blood reaches the gills, it slows down while passing through narrow capillaries. The amphibian heart, in contrast, pumps both deoxygenated and oxygenated blood. Deoxygenated blood returning from the body is sent back out by the heart to the lungs to absorb oxygen. This oxygen-rich blood, which also slows as it passes through the narrow capillaries of the lungs, returns to the heart to be pumped to the rest of the body. Thus, in amphibians blood is pumped to the body at higher pressures and faster rates of flow than in fishes. There is a new problem, however. Oxygenated blood partially mixes with deoxygenated blood in the single, undivided ventricle of the amphibian heart.

The Tie to Water

As you learned in Chapter 24, amphibians do not lay watertight eggs and so must reproduce in water or in moist environments. Many species of frogs and toads, for example, congregate in ponds during the spring in preparation for mating. The males generally arrive first and begin noisy mating calls that attract nearby females. You have probably heard the raucous symphony of male frogs on warm spring evenings. Once mating begins, the male grasps a female and holds her. While she releases eggs, he simultaneously releases sperm. Fertilization takes place externally. Follow the development of a frog in **Figure 27-14**. You can read more about frogs in *Tour of a Frog* on the next page.

Mating in some salamanders resembles that in frogs and toads. In most salamanders, however, fertilization occurs internally. After performing complex courtship behaviors to attract a female, the male deposits a packet of sperm on the ground or in the water. The female then draws the sperm packet into her reproductive opening. She later deposits the fertilized eggs in water or a moist area. Salamander larvae retain external gills until the time of metamorphosis. Other than that, they resemble their parents. All salamander larvae, like the adults, are carnivorous.

Figure 27-14

a The life cycle of a frog involves large-scale changes in body form. First a mass of eggs is laid in a wet or moist environment.

b The young tadpole emerges from the egg with external gills, which are later replaced by internal gills. After feeding and growing, the tadpole begins to transform into an adult frog.

c Dramatic changes occur in the tadpole. The tail and gills recede. Lungs and front and hind limbs grow. Feeding habits may also change. Herbivorous tadpoles change into carnivorous adults.

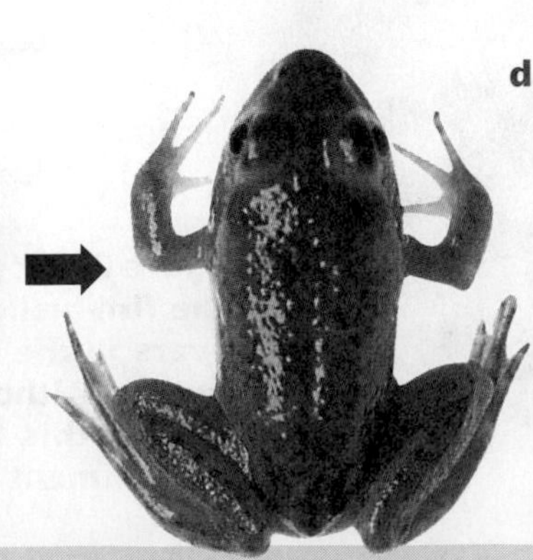

d By the time a tadpole becomes an adult frog, it has completely lost its tail and gills. Its lungs enable it to breathe air. The changes that transform a tadpole into an adult frog are called metamorphosis.

Frog

The red-eyed tree frog is an amphibian that lives in Central America and northern South America. Like most amphibians, the red-eyed tree frog has a moist, scaleless skin and passes through an aquatic tadpole phase during its development.

Tour

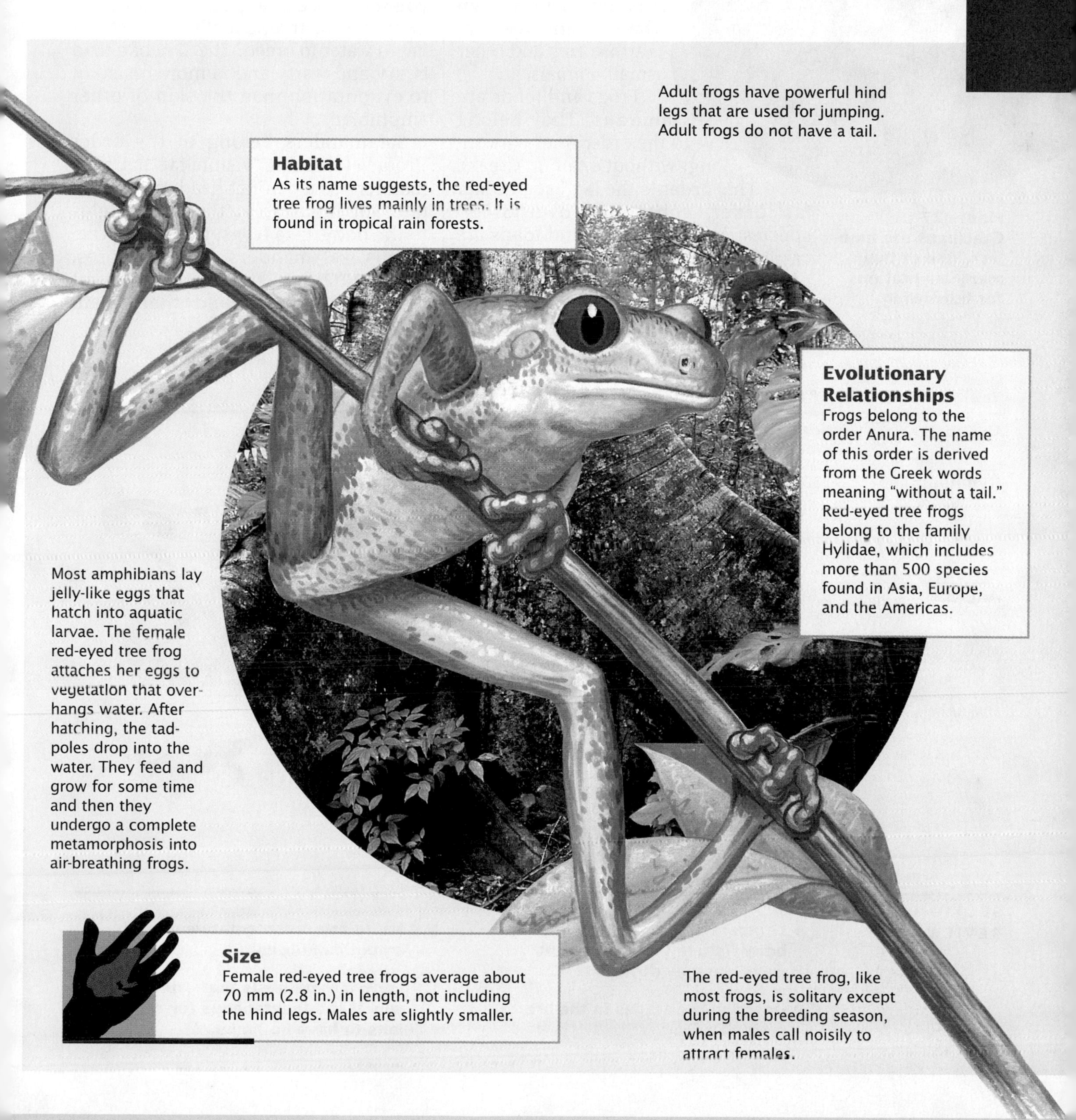

Adult frogs have powerful hind legs that are used for jumping. Adult frogs do not have a tail.

Habitat
As its name suggests, the red-eyed tree frog lives mainly in trees. It is found in tropical rain forests.

Evolutionary Relationships
Frogs belong to the order Anura. The name of this order is derived from the Greek words meaning "without a tail." Red-eyed tree frogs belong to the family Hylidae, which includes more than 500 species found in Asia, Europe, and the Americas.

Most amphibians lay jelly-like eggs that hatch into aquatic larvae. The female red-eyed tree frog attaches her eggs to vegetation that overhangs water. After hatching, the tadpoles drop into the water. They feed and grow for some time and then they undergo a complete metamorphosis into air-breathing frogs.

Size
Female red-eyed tree frogs average about 70 mm (2.8 in.) in length, not including the hind legs. Males are slightly smaller.

The red-eyed tree frog, like most frogs, is solitary except during the breeding season, when males call noisily to attract females.

Kinds of Amphibians

Figure 27-15 shows a tropical burrowing amphibian called a caecilian. Caecilians are members of the order Gymnophiona (from the ancient Greek words for "naked" and "snakelike"). Caecilians burrow through the soil and feed on earthworms and other small animals.

Frogs and toads are **anurans**. They belong to the order Anura (meaning "without a tail" in Greek). This order is the largest amphibian order, containing over 3,800 species. As adults, frogs and toads are carnivores. They have large mouths, often with sticky tongues. The body form of adult frogs and toads is distinct—the hind legs are specialized for jumping, and there is no tail. Frogs and toads are found in a variety of habitats. Some species are completely aquatic, while others spend some time on land. Some species of frogs spend most of their adult lives in trees. Toads, which you might see in a garden or park, are mainly terrestrial, usually only returning to water to breed. The skin of a toad is dry and warty and is more resistant to evaporation than the skin of other amphibians.

Figure 27-15 Caecilians are limbless, one of their many adaptations for burrowing.

Salamanders belong to the order Urodela (meaning "visible tail"). A salamander has a distinct head, trunk, and tail. The limbs are set at right angles to the body. Like frogs, toads, and caecilians, salamanders are carnivores. **Table 27-2** lists the three orders of living amphibians and their main characteristics.

Table 27-2 Orders of Amphibians

Order	Number of modern species	Description	Examples
Gymnophiona	160	Wormlike body with no limbs; tail short or absent; restricted to tropics	Caecilians 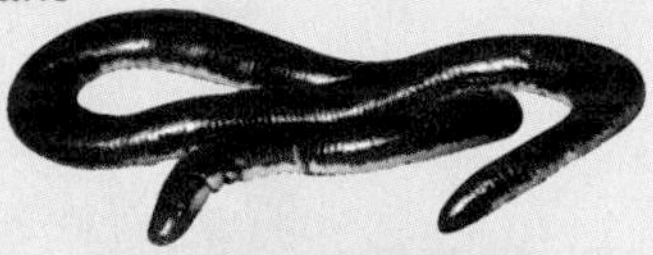
Anura	3,800+	No tail; hind limbs specialized for jumping; worldwide distribution	Frogs, toads 
Urodela	390	Body has distinct head, trunk, and tail; limbs set at right angles to body; distributed worldwide except Australia	Salamanders, newts 

SECTION REVIEW

1. **Compare an amphibian with a bony fish. In what important ways are they different?**
2. **Describe the stages in the life cycle of a frog.**
3. **Name one representative of each amphibian order.**
4. **Form a hypothesis that explains why it is advantageous for caecilians to have no limbs.**

CHAPTER

27 Highlights

A male Darwin's frog carries developing offspring in his vocal sacs. Here, a young frog has just been released.

	Key Terms	Summary
27-1 Jawless Fishes and Cartilaginous Fishes		• Jawless fishes were the first vertebrates to evolve. • The existing jawless fishes include lampreys, many of which are external parasites on other fishes, and hagfishes, which are scavengers. • Acanthodians were the first vertebrates with jaws. • Sharks and rays have skeletons of cartilage.
27-2 Bony Fishes	swim bladder (p. 529) ray-finned fishes (p. 531) lobe-finned fishes (p. 531) coelacanth (p. 531)	• Unlike sharks and rays, most bony fishes have skeletons of bone. • Bony fishes have a swim bladder, a gas-filled sac that helps them to maintain their level in the water. • Most bony fishes are ray-finned fishes. • A group of bony fishes known as the lobe-finned fishes are thought to be the closest relatives of the amphibians.
27-3 Amphibians	anuran (p. 536)	• Amphibians were the first terrestrial vertebrates. • Frogs, toads, salamanders, and caecilians are amphibians. • Because their skins are not watertight, most amphibians live in water or in damp environments. • Amphibian eggs are not watertight and will dry out if not kept wet or moist. • Caecilians are legless, tropical amphibians. They burrow in moist soils and eat worms and other small animals. • Frogs and toads, the anurans, are adapted for jumping. • Salamanders have four limbs and a tail. They are carnivorous and have internal fertilization.

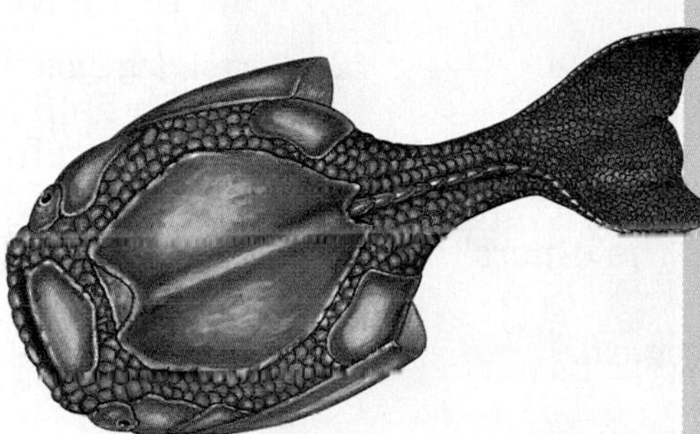

Jawless fishes were the only vertebrates for over 150 million years.

Most bony fishes are ray-finned fishes.

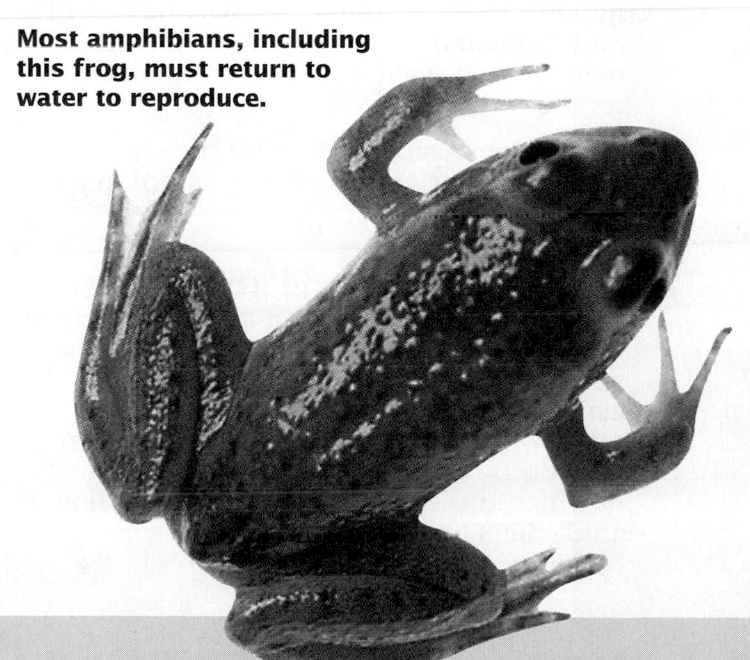

Most amphibians, including this frog, must return to water to reproduce.

CHAPTER 27

review

Understanding Vocabulary

1. For each set of terms, complete the analogy.
   a. fishes with a swim bladder : Osteichthyes :: jawless fishes : ____
   b. skeleton of cartilage : Chondrichthyes :: skeleton of bone : ____
   c. frog : jumping :: caecilian : ____
2. For each list of terms, select the one that does not fit the others and explain why.
   a. agnathan, lamprey, hagfish, acanthodian
   b. internal fertilization, slimy skin, skeleton of cartilage, jaws
   c. frog, toad, coelacanth, newt, caecilian

Understanding Concepts

3. **Relating Concepts** Construct a concept map that identifies members of the class Osteichthyes and describes their characteristics. Try to include the following terms in your concept map: jaws, swim bladders, buoyancy, endoskeleton, bone, lobe-finned fishes, ray-finned fishes, amphibians, salmon, frogs.
4. **Identifying Structures** What does the term *agnathan* mean? What are three characteristics of agnathans? What are the two main groups of modern agnathans?
5. **Comparing Structures and Functions** What is the lateral line system? What function does it serve? What structure in terrestrial vertebrates most closely resembles the lateral line in function?
6. **Summarizing Information** Explain how parasitic lampreys obtain food.
7. **Comparing Functions** What structures in agnathans are thought to have given rise to jaws?
8. **Identifying Structures** Describe three senses that enable sharks to detect and capture their prey.
9. **Comparing Structures** Name one similarity and one difference between sharks and agnathans.
10. **Comparing Structures** How does fertilization in sharks differ from fertilization among most bony fishes?
11. **Comparing Functions** Explain why a shark will sink if it stops swimming. Why doesn't a bony fish have this same problem?
12. **Comparing Functions** How do the lungs of a lungfish help it survive? Why wouldn't lungs be useful for the living species of coelacanth?
13. **Summarizing Information** Copy the following table onto a separate sheet of paper. Complete the table by filling in the blanks.

	Agnathans	Sharks and rays	Bony fishes
Composition of skeleton		Cartilage	
Paired fins present?			Yes
Example	Hagfish		

14. **Recognizing Relationships** What characteristics enable amphibians to live on land? What characteristics require that most amphibians live in or near the water throughout their lives?
15. **Inferring Relationships** The populations of some species of amphibians have decreased dramatically in the last few decades. One hypothesized cause for this decline is acid rain. Explain one reason why amphibians might be more vulnerable to acid rain than reptiles are.
16. **Summarizing Information** Identify the characteristics of the three orders of amphibians. Give an example of an amphibian in each order.
17. **BUILDING ON WHAT YOU HAVE LEARNED** In Chapter 15 you learned about the evolutionary effects of competition. Using what you learned in that chapter, form a hypothesis to explain why the larvae of some amphibians feed on different foods from the adults. Name another group of animals in which this difference between larval and adult diets occurs.

Interpreting Graphics

18. Look at the photograph of the animal below.

a. What kind of animal is shown in the photo? What characteristics enabled you to identify the animal?
b. Describe the adaptations that enable this animal to live on land.

Reviewing Themes

19. ***Evolution***
Analysis of mitochondrial DNA indicates that lungfishes are the closest living relatives of amphibians. Identify one similarity in structure between lungfishes and amphibians.

Thinking Critically

20. **Determining Factual Accuracy** "The first fishes evolved about 200 million years ago. These fishes had no jaws and no paired fins and belonged to the class Chondrichthyes. Sharks, which appeared later, have paired fins, jaws, and a swim bladder, which they use to breathe air. Bony fishes are the descendants of sharks and the ancestors of amphibians. The first amphibians appeared about 360 million years ago. The three groups of amphibians alive today are the frogs and toads, the caecilians, and the lizards." Identify and correct any factual errors in this passage.

Life/Work Skills

21. **Finding Information** Use library references or search an on-line database to find out about some of the different species of sharks, such as hammerhead sharks, nurse sharks, basking sharks, great white sharks, gray reef sharks, "cookie cutter" sharks, bull sharks, tiger sharks, and whale sharks. Create a table that includes information about the scientific and common names of the sharks, their size, habitats, feeding habits, and unusual characteristics. Include illustrations of the sharks in your table. Share your findings with the class.

22. **Allocating Resources** Fish have been a primary source of food for humans around the world for thousands of years. Use library references or search an on-line database to research where the world's most productive fisheries are located and what kinds of fishes are harvested. Also explore the various techniques by which fish are collected. Investigate how changes in environmental conditions and increased demands by humans have affected the major fisheries of the world. Summarize your findings in a written report to your class.

Activities and Projects

23. **Language Arts Connection** Use library references to find out how amphibians are portrayed in literature. In Shakespeare's play *Macbeth*, the witches call for "eye of newt" and "toe of frog" as part of their mysterious concoctions. Children's fairy tales frequently mention kissing a frog and having it turn into a handsome prince. Collect several literary references to amphibians. Share your findings in an oral report to your class.

Discovering Through Reading

24. Read the article "Hammerhead City" by A. Peter Klimley in *Natural History*, October 1995, pages 33–38. How does a female hammerhead establish dominance within a school of sharks? Why do hammerhead sharks school near underwater mountains?

CHAPTER 28

REVIEW
- placenta (Section 24-3)
- structure of the amniotic egg (Section 24-3)
- the terms *ectotherm* and *endotherm* (Glossary)

Reptiles, Birds, and Mammals

The ringtail is one of more than 4,000 living species of mammals. It rests by day and hunts insects and other small animals by night.

Reptiles (class Reptilia) were the first fully terrestrial vertebrates. Their freedom from aquatic environments was made possible by several adaptations that made them and their eggs essentially watertight. Modern reptiles include snakes; lizards, like the iguana shown at left; turtles; crocodiles; and alligators. They represent the living members of a group that once dominated the land.

28-1 Reptiles

OBJECTIVES

1. **Identify three adaptations that make reptiles well suited to terrestrial life.**
2. **Contrast ectothermy with endothermy.**
3. **Describe one hypothesis that explains the disappearance of the dinosaurs.**
4. **List the four orders of living reptiles.**

Reptilian Adaptations to Terrestrial Life

The skin of a reptile, contrary to what many people think, is not wet and slimy. Instead, reptilian skin is dry and covered with tough, hard, platelike scales, as shown in **Figure 28-1**. This dry skin forms a barrier to water loss in land environments. Reptilian skin is resistant to water loss because it contains large amounts of lipids and the protein keratin. Keratin is the tough, wear-resistant material that also composes your hair and fingernails and the feathers of birds. Unlike the scales of fishes, the scales of reptiles do not contain bone. In some reptiles, however, thick bony plates develop beneath the scales, such as those that form the shells of turtles.

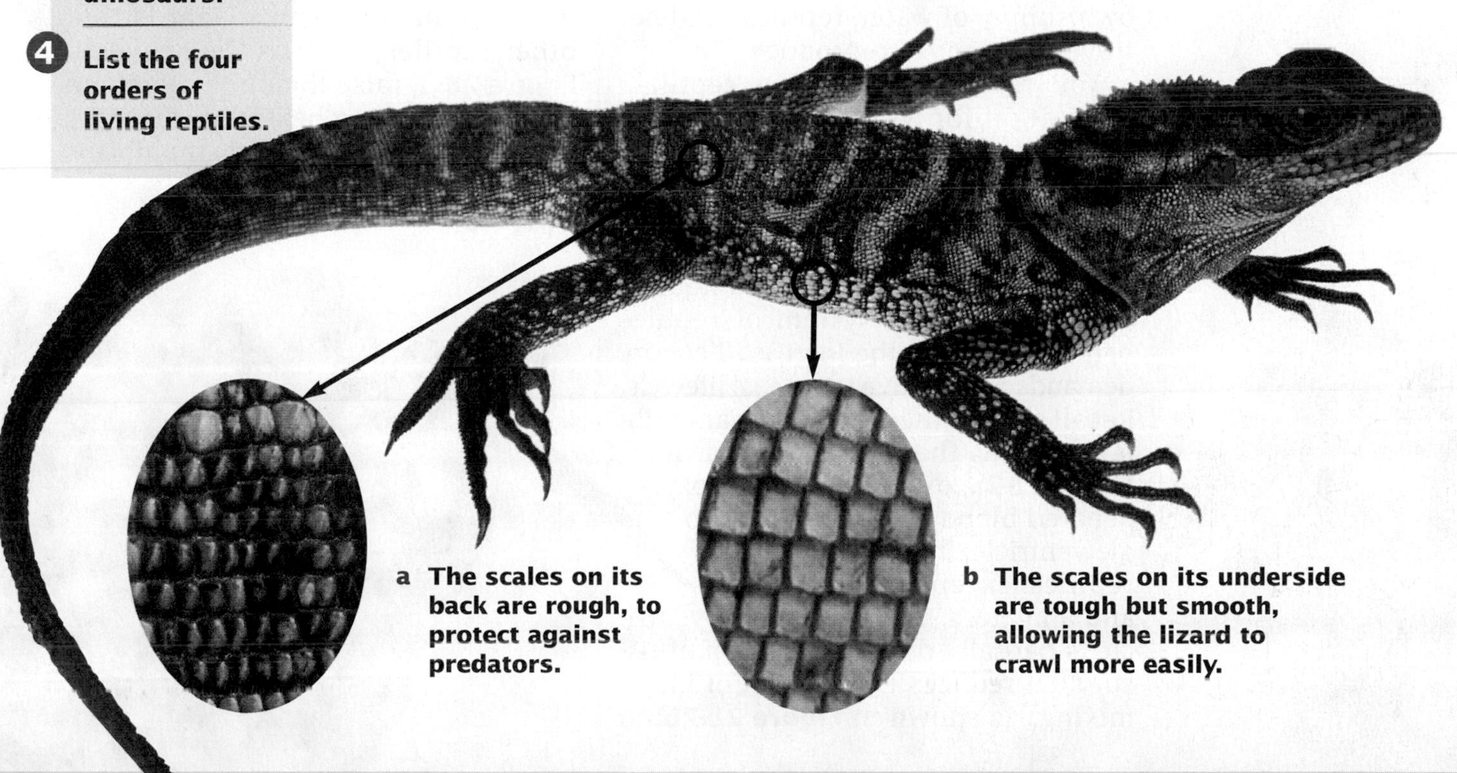

Figure 28-1
This Philippine sail-finned lizard has dry, scaly skin that protects its body from drying out and from being cut or scratched while moving over the ground.

a **The scales on its back are rough, to protect against predators.**

b **The scales on its underside are tough but smooth, allowing the lizard to crawl more easily.**

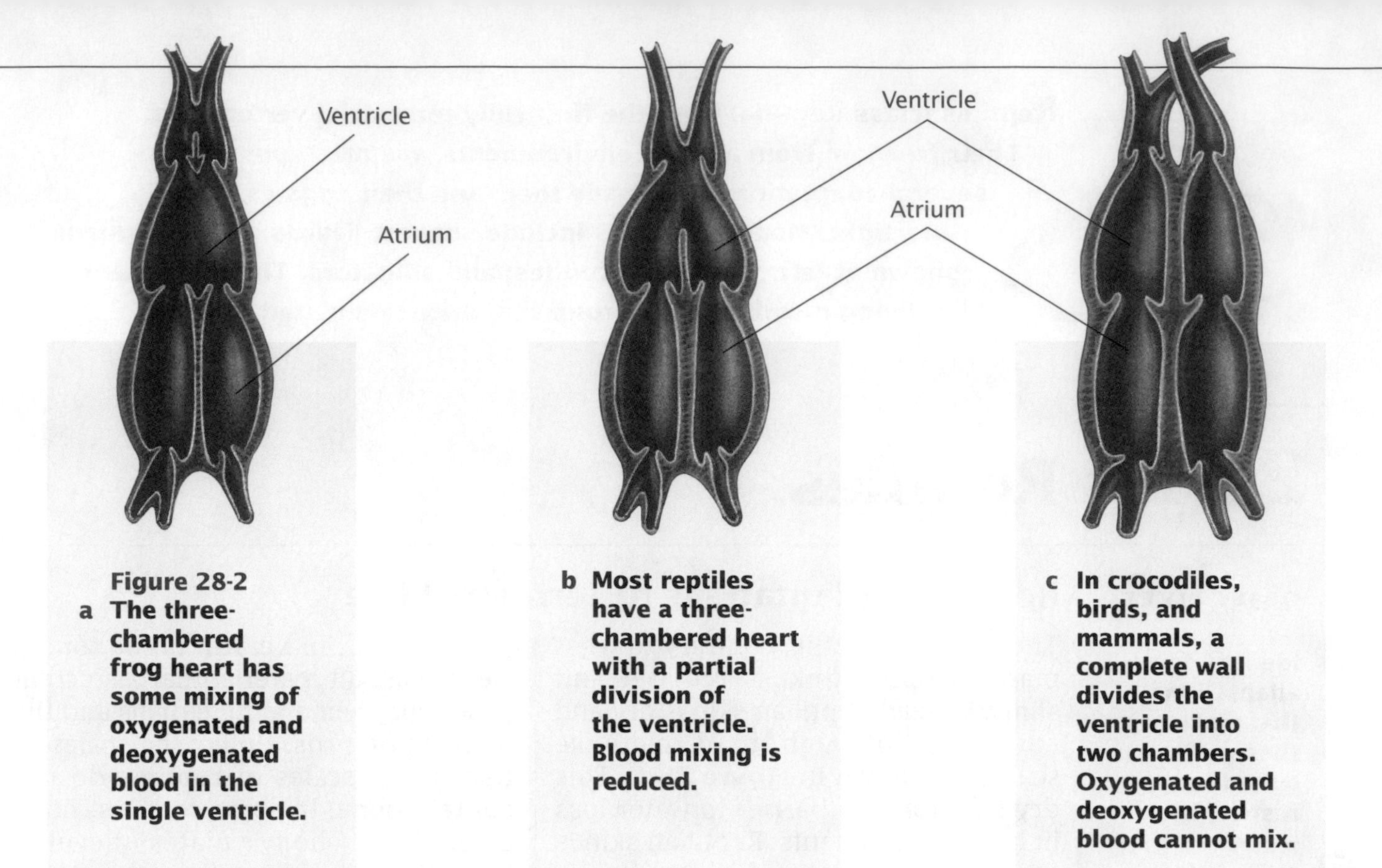

Figure 28-2
a The three-chambered frog heart has some mixing of oxygenated and deoxygenated blood in the single ventricle.

b Most reptiles have a three-chambered heart with a partial division of the ventricle. Blood mixing is reduced.

c In crocodiles, birds, and mammals, a complete wall divides the ventricle into two chambers. Oxygenated and deoxygenated blood cannot mix.

An important adaptation of reptiles to life on dry land is the amniotic egg, which is resistant to water loss. As you learned in Chapter 24, the reptilian egg protects the vulnerable embryo and provides for its needs as it develops. Since the amniotic egg contains its own supply of water, reptiles need not travel to water to reproduce.

You read about another reptilian adaptation for water conservation in Chapter 24. Reptiles excrete nitrogenous waste as uric acid, a form that requires very little water for dilution. Reptile urine contains so little water that it is a paste rather than a liquid.

The circulatory system of reptiles helps them meet the increased energy demands of an active terrestrial lifestyle. Recall that in the heart of an amphibian, such as the frog heart shown in **Figure 28-2a**, oxygenated and deoxygenated blood can mix in the single ventricle. In the hearts of all reptiles except crocodiles and alligators, by contrast, the ventricle is partially divided by a wall of tissue that reduces the amount of blood mixing, as shown in **Figure 28-2b**. In crocodiles, alligators, birds, and mammals, this partition is complete, as you can see in **Figure 28-2c**.

Are reptiles "coldblooded"?

Drive along a desert road on a warm morning and you may see lizards basking in the sun. Lizards and other reptiles, such as the snakes in **Figure 28-3**, raise their body temperature by absorbing heat from their surroundings. Like fishes and amphibians,

reptiles are ectotherms. The metabolism of an ectotherm is too slow to produce enough heat to warm its body. Instead, ectotherms must absorb heat from their surroundings. Birds and mammals, in contrast, are endotherms. Their rapid metabolism produces large amounts of heat, some of which is used to maintain a high, relatively constant body temperature.

Throughout the day, the body temperature of an ectotherm often follows the temperature of its surroundings. Body temperature falls at night, when the air is cool. In the morning, many reptiles seek sunny places, letting the sun's rays warm their bodies. Once warmed, many reptiles maintain a high and relatively constant body temperature by moving into and out of the sunshine, their bodies warming and cooling, as described in **Figure 28-4**. You might hear ectotherms called "coldblooded," but this description is inaccurate. On warm days, some desert lizards have body temperatures higher than yours.

An ectotherm's slow metabolism limits where it can live and what activities it can perform. For instance, endotherms such as birds and mammals can sustain activity for longer periods of time than can reptiles and amphibians. Birds and mammals can also be active on colder days and live in colder climates than can reptiles and amphibians. But there is a high cost to endothermy—it requires large amounts of food. A mouse must eat about 10 times as much food as a similarly sized lizard. You can read more about reptiles in *Tour of a Lizard* on page 544.

Figure 28-3
These red-sided garter snakes are emerging from their den, where they have spent the winter. By clustering together, they are able to conserve heat.

Figure 28-4
a Heat exchange is very important for ectotherms. This spiny lizard is warmed by heat from the rock on which it sits and by sunlight.

b During the hottest part of the day, the lizard may retreat into the shade to lower its body temperature.

Tour

Lizard

The largest group of reptiles today is the lizards. This day gecko is a lizard that lives on the islands of Mauritius and Madagascar in the Indian Ocean. Day geckos are unusual among lizards because they feed on nectar from flowers as well as on insects.

Size
Adult day geckos range from 10 to 15 cm (4 to 7 in.) in total length.

All reptiles have dry, tough, scaly skin. The outer layer of scales is shed periodically.

The day gecko has adhesive pads on its toes that allow it to climb vertical surfaces and even to walk on glass.

Habitat
Day geckos are tree dwellers.

The day gecko, like most reptiles, lays eggs with a protective shell. Reptilian eggs do not dry out when exposed to air.

Many lizards drop their tails when seized by a predator. The tail regrows in several weeks.

Evolutionary Relationships
Lizards belong to the suborder Sauria, which contains about 3,000 species. Day geckos belong to the family Gekkonidae. Most geckos are active at night, but nearly all day geckos are active during the day.

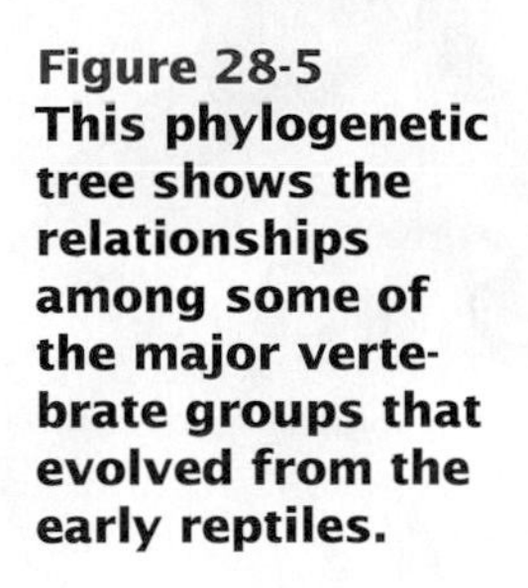

Figure 28-5
This phylogenetic tree shows the relationships among some of the major vertebrate groups that evolved from the early reptiles.

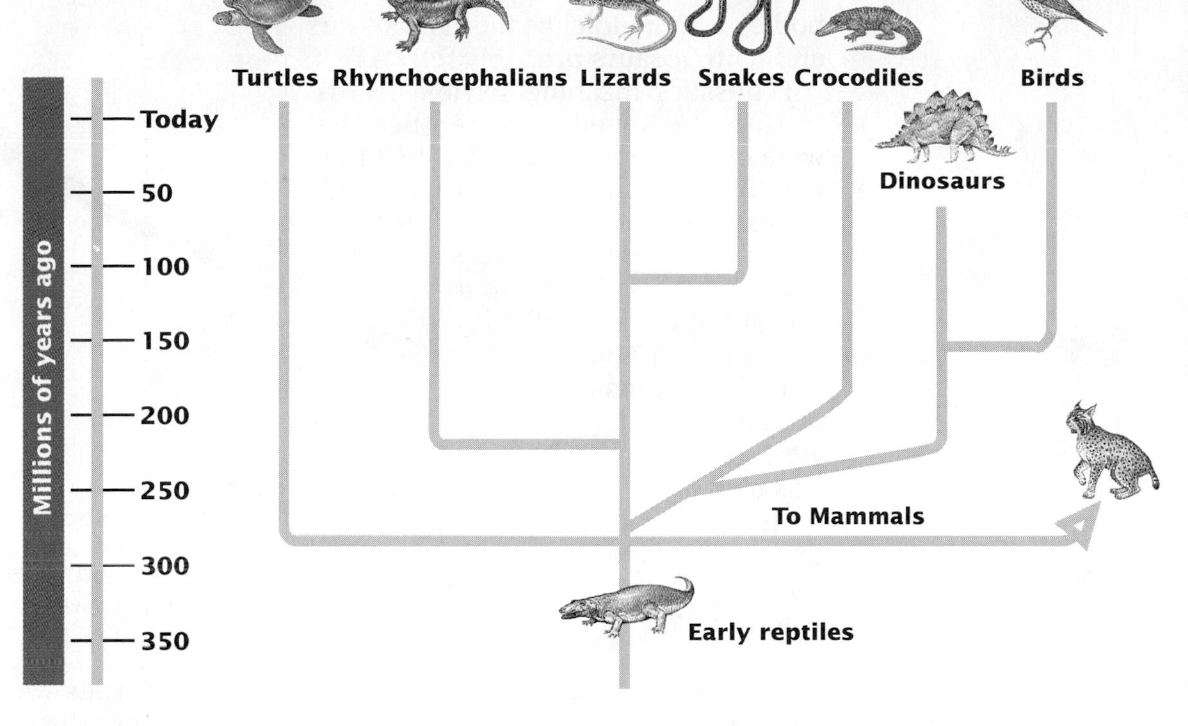

The Age of Reptiles

The first reptiles evolved about 320 million years ago, when the world was entering a dry period. Well suited to dry conditions, reptiles diversified rapidly after their initial appearance, giving rise not only to the ancestors of modern reptiles but also to the wide variety of dinosaurs and other reptiles that were the dominant animals on Earth for over 170 million years. The period of reptilian dominance, which lasted from about 250 million to 65 million years ago, is called the Age of Reptiles. The evolutionary relationships of the reptiles and their descendants are shown in **Figure 28-5**.

Two groups of reptiles, shown in **Figure 28-6**, inhabited the oceans while the dinosaurs lived on land. Ichthyosaurs *(IHK thee oh sawrs)* were fully adapted for an aquatic existence. Ichthyosaurs ("fish lizards" in Greek) resembled dolphins, having a pointed snout, streamlined body, paddlelike fins, and a flattened tail. Plesiosaurs *(PLEE see oh sawrs)* had barrel-shaped

Figure 28-6
The 2 m (6 ft 6 in.) *Ichthyosaurus* gave birth to live young. The 2.3 m (7 ft 6 in.) body of *Plesiosaurus* was adapted for maneuverability to catch fish. Both existed during the Jurassic period.

bodies with paddlelike fins. Plesiosaurs and ichthyosaurs probably fed on fish.

Dinosaurs, meaning "terrible lizards" in Greek, evolved from reptiles that were only 0.5 m to 1 m (1.5 ft to 3 ft) in length. Among the dinosaurs were the largest land animals, such as the herbivores *Apatosaurus*, *Diplodocus*, and *Brachiosaurus*. *Brachiosaurus* was about 23 m (75 ft) long and about 17 m (56 ft) tall. The dinosaurs also included the largest land carnivore, *Tyrannosaurus rex*. This great predator was over 5 m (16 ft) tall when standing on its hind legs. Dinosaurs were very diverse in size and shape. You can see a sampling of this diversity in **Figure 28-7**.

Pterosaurs were the first vertebrates to fly

No reptiles can fly today, but one group of flying reptiles existed alongside the dinosaurs—the pterosaurs *(TEHR oh sawrs)*, such as *Rhamphorhynchus* shown in **Figure 28-8**. These reptiles were flying 75 million years before the first birds. During the 160 million years they existed, pterosaurs were a diverse group. Some were as small as a sparrow. Others had wingspans of 11 m (35 ft), greater than the wingspan of many small airplanes.

Figure 28-8
Pterosaurs, such as *Rhamphorhynchus*, were the first group of vertebrates to evolve the ability to fly. They evolved 75 million years before the first birds.

Figure 28-7
The popular conception of dinosaurs is of huge lumbering animals like *Stegosaurus*, which was 6 m (20 ft) long and weighed up to 1,500 kg (2 tons). But many dinosaurs, such as *Struthiomimus*, were small and fast. Although the dinosaurs shown here did not all exist at the same time, this illustration shows some of the great diversity among dinosaurs.

How did the dinosaurs become extinct?

Dinosaurs, pterosaurs, plesiosaurs, and virtually all other land and sea animals larger than a small dog abruptly became extinct about 65 million years ago. Although this mass extinction is clearly recorded in the fossil record, its cause is not fully understood. One formerly popular hypothesis is that the Earth entered a time of significant cooling 65 million years ago. According to this hypothesis, dinosaurs were unable to maintain sufficiently high body temperatures to survive.

The most widely accepted hypothesis today was proposed in 1980 by the Nobel Prize–winning physicist Luis Alvarez, his son Walter, and several other scientists. These scientists proposed that the Earth was struck by a very large meteorite about the time the dinosaurs disappeared. In rocks of that age, Alvarez and other scientists discovered a layer of iridium (an element very rare on Earth but common in meteorites) and bits of melted rock that exhibit stress fractures characteristic of blast damage. The impact point of the meteorite appears to have been in northern Yucatan, Mexico. A collision between the Earth and a large meteorite would have raised thick dust clouds that could have blocked out sunlight for a time, perhaps for months. A reduction in the amount of sunlight reaching Earth's surface would have adversely affected plant life and might have caused large-scale climatic changes. Though evidence for a meteorite impact is strong, whether such an impact caused this mass extinction is still not established. The controversy over the end of the Age of Reptiles is still far from settled.

The Survivors

The mass extinction 65 million years ago spared four groups of reptiles: lizards and snakes (order Squamata), turtles and tortoises (order Testudines), crocodiles and alligators (order Crocodylia), and the tuataras (order Rhynchocephalia).

Of these four groups, alligators, such as the one in **Figure 28-9**, and crocodiles are the most closely related to dinosaurs. Twenty-two species of crocodiles and alligators live in tropical and subtropical regions of the world. Two species, the American alligator and the American crocodile, occur in the United States. Crocodiles and alligators lead a largely aquatic life, feeding on aquatic animals such as fishes and turtles and on terrestrial animals that come to drink or feed in the water. Crocodiles and alligators have been hunted intensely for their hides, which are used to make handbags, shoes, and other leather products. Because of overhunting and habitat destruction, 12 species of crocodiles are now endangered.

All turtles, such as the box turtle shown in **Figure 28-10a**, have a shell. A turtle's shell is composed of bony plates that are fused together. The vertebrae and ribs are fused to the interior of the shell, as shown in **Figure 28-10b**. Turtles lack teeth but have a sharp beak. Most turtles spend some of their

Figure 28-9
The American alligator and other crocodilians differ from other reptiles in having a four-chambered heart. All crocodilians are carnivores, and all lay eggs.

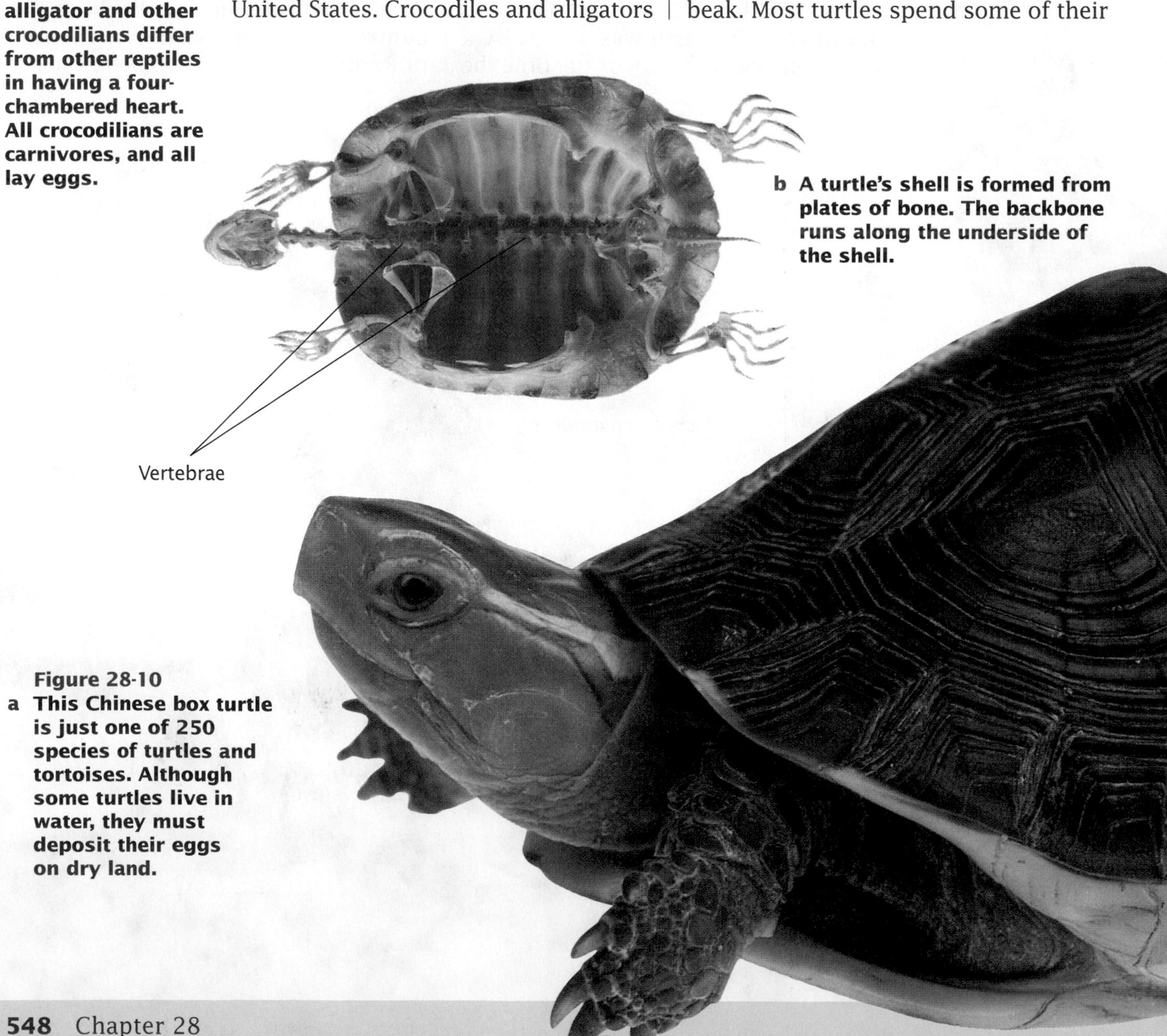

b A turtle's shell is formed from plates of bone. The backbone runs along the underside of the shell.

Figure 28-10
a This Chinese box turtle is just one of 250 species of turtles and tortoises. Although some turtles live in water, they must deposit their eggs on dry land.

Figure 28-11
This is one of the two species of tuataras. The name *tuatara* means "spiny crest."

time in water. Extreme examples are the sea turtles, which leave the ocean only to lay their eggs. The limbs of sea turtles are flattened and paddlelike for steering and propulsion in the ocean. Tortoises, on the other hand, are almost completely terrestrial and have heavy limbs for walking.

The tuatara shown in **Figure 28-11** is a member of one of the two surviving species in the order Rhynchocephalia. The lizardlike tuataras closely resemble their relatives that were common 150 million years ago. Tuataras live only on a few of the smaller islands of New Zealand.

Snakes and lizards belong to the largest order of reptiles, order Squamata. There are about 6,000 species of snakes and lizards, and they live on every continent except Antarctica. All snakes lack limbs, movable eyelids, and external ears. Snakes are carnivores and can feed on animals much larger than themselves. A snake's jaw has five joints (your jaw has only one), which give the jaw great flexibility. This flexibility enables the snake to engulf prey several times its own diameter, as shown in **Figure 28-12**. Most snakes are nonpoisonous, but a few species can deliver a poisonous bite. Rattlesnakes, coral snakes, copperheads, and cobras are examples of poisonous snakes.

Figure 28-12
The structure of the snake's jaw and the flexible nature of its skin and skeleton allow the snake to eat animals larger in diameter than its own body. It may take this snake an hour or more to swallow its prey.

a After suffocating its prey by constriction, . . .

b . . . the snake begins the long process of swallowing.

Unlike snakes, most lizards have four limbs and external ears. Lizards are most abundant in the tropics and in deserts. Most species are carnivorous, feeding on insects, small mammals, or other lizards. A few species are herbivorous. Only two species of lizards are poisonous: the Gila monster of the southwestern United States and northern Mexico, shown in **Figure 28-13**, and the Mexican beaded lizard of southern Mexico and Guatemala. **Table 28-1** summarizes the major orders of reptiles.

Figure 28-13
The glands that produce the Gila monster's poison are located in the lower jaw. The lizards chew it into their victims.

Table 28-1 Orders of Reptiles

Order	Number of living species	Description	Examples
Ornithischia	Extinct	Mostly plant-eating dinosaurs with two pelvic bones facing backward, as in a bird's pelvis; hole in the skull in front of eye socket; legs positioned beneath the body; over 150 genera	*Triceratops*, *Ankylosaurus*, *Iguanodon* *Stegosaurus*
Saurischia	Extinct	Flesh-eating and plant-eating dinosaurs with one pelvic bone facing forward and the other facing backward, as in a lizard's pelvis; terrestrial; three or sometimes five toes; hole in skull in front of eye socket; legs positioned beneath body; over 200 genera	*Allosaurus*, *Apatosaurus*, *Brachiosaurus* *Tyrannosaurus rex*
Pterosauria	Extinct	Flying reptiles with wings of skin between fourth finger and body; wingspan of early (Jurassic) *Rhamphorhynchus* was typically 60 cm (2 ft), and wingspan of later (Cretaceous) *Pteranodon* was over 7.5 m (25 ft)	*Pteranodon*, *Pterodactylus*, *Quetzalcoatlus* *Rhamphorynchus*
Plesiosauria	Extinct	Marine reptiles with very large paddle-shaped fins, a barrel shaped body, and long jaws with sharp teeth; some had a snakelike neck twice as long as the body; others had a short neck and elongated skull	*Peloneustes*, *Elasmosaurus*, *Kronosaurus* *Plesiosaurus*

Ichthyosauria	Extinct	Marine reptiles with streamlined bodies up to 3 m (10 ft) in length; the four legs modified into balancing fins; apparently fast swimmers, with many body similarities to modern fishes such as tuna or mackerel	*Ichthyosaurus*
Squamata suborder Sauria	3,000	Lizards; limbs set at right angles to body; dry skin of scales; socketless teeth; heart with partially divided ventricle; most species are terrestrial, but a few are at least partially aquatic	Anoles, geckos, horned lizards
Squamata suborder Serpentes	2,500	Snakes; no legs; scaly skin is shed periodically; socketless teeth; heart with partially divided ventricle; no external ear openings; most species are terrestrial, but a few are aquatic	Rattlesnakes, garter snakes
Testudines	250	Body encased in shell of bony plates; sharp, horny jaw edges without teeth; vertebrae and ribs fused to shell; terrestrial and aquatic species	Turtles, tortoises, terrapins
Crocodylia	22	Four-chambered heart; extended jaw with socketed teeth, five digits on forelimbs, four digits on hind limbs; live near or in water	Crocodiles, alligators
Rhynchocephalia	2	Sole survivors of a group that largely disappeared about 100 million years ago. Skull like those of early Permian reptiles; fused, wedgelike, socketless teeth; terrestrial	Tuataras

SECTION REVIEW

1. **Describe two adaptations to land shown by reptiles.**
2. **What is one disadvantage of endothermy?**
3. **What evidence indicates that a meteorite collided with Earth about the time the dinosaurs became extinct?**
4. **List a representative of each of the four living orders of reptiles.**

Although the dinosaurs died out 65 million years ago at the end of the Cretaceous period, their descendants, the birds, survived. Today birds are the most diverse group of land vertebrates, with nearly 10,000 species. All birds have feathers, and almost all birds are capable of powered, sustained flight.

28-2 Birds

OBJECTIVES

1. **List two similarities between birds and reptiles.**
2. **Identify two differences between birds and reptiles.**
3. **Identify two functions of feathers.**
4. **Describe two bird adaptations, other than feathers, for flight.**

Birds Evolved From Reptiles

Birds (class Aves) evolved more than 150 million years ago. The oldest known fossils of birds were found in fine limestone in Bavaria. These specimens were named *Archaeopteryx*, which means "ancient wing." *Archaeopteryx* shared many characteristics with the small dinosaurs from which it evolved, including teeth in sockets and a long, bony tail. But as you can easily see in **Figure 28-14**, *Archaeopteryx* had feathers, a characteristic unique to birds.

Modern birds lack teeth and have only a vestigial tail, but they still retain many reptilian characteristics. For instance, birds lay amniotic eggs, although the shell of a bird's egg is hard rather than leathery. Scales are also present on the feet and lower legs of birds.

Feathers are unique to birds

Feathers have replaced scales as the body covering of birds. Feathers are flexible, strong structures that can be regrown—they make an excellent wing for flying. Feathers provide most of the surface area of a bird's wing. A bird can change the surface area of its wing and alter its flight patterns by spreading or collapsing its wing feathers and the wing itself.

Like wing feathers, a bird's tail feathers can be spread or collapsed, changing their effective surface area. Tail feathers can also be used for braking and steering during flight. Watch the movement of the tail of a bird in flight and during landing to see some of the actions of these feathers.

Figure 28-14
***Archaeopteryx lithographica* was about the size of a pigeon.**

As shown in **Figure 28-15**, birds have two major types of feathers. Most of a bird's feathers are **contour feathers**. These feathers cover the body of the bird and give the wings and tail their shape. Contour feathers also insulate against heat loss. Fine **down feathers** growing underneath or among the contour feathers are specialized for insulation. The down feathers of geese and eider ducks are used in sleeping bags because they are a lightweight, effective insulation.

Bird skeletons are lightweight

The skeleton of a bird is adapted for flight. The bones are thin and hollow. Many are reinforced by internal struts, like the wings of an airplane. The breastbone, or sternum, is large and has a keel, providing solid anchorage for some of the large flight muscles.

Bird wings are modified forelimbs. The bones of the forelimbs fully support and move the wings. The finger bones are very tiny, but the arm and hand bones are long, providing strength and enabling complex movements of the wings. As you learned in Chapter 10, the bones of a bird's wing are homologous to the bones of your arm and hand.

Birds are endothermic and active

Flight is an energy-demanding activity. Birds, like mammals, are endothermic. Their rapid metabolism provides the energy needed for flight. In addition, birds have a four-chambered heart with separate circulatory loops to the lungs and to the body. Therefore, oxygen-rich blood is rapidly delivered to tissues where it is needed, without mixing with deoxygenated blood. Respiration in birds is very efficient because their system of air sacs permits air to flow in only one direction through the lungs, as explained in Chapter 24. You can read more about birds in the *Tour of a Bird* on page 555.

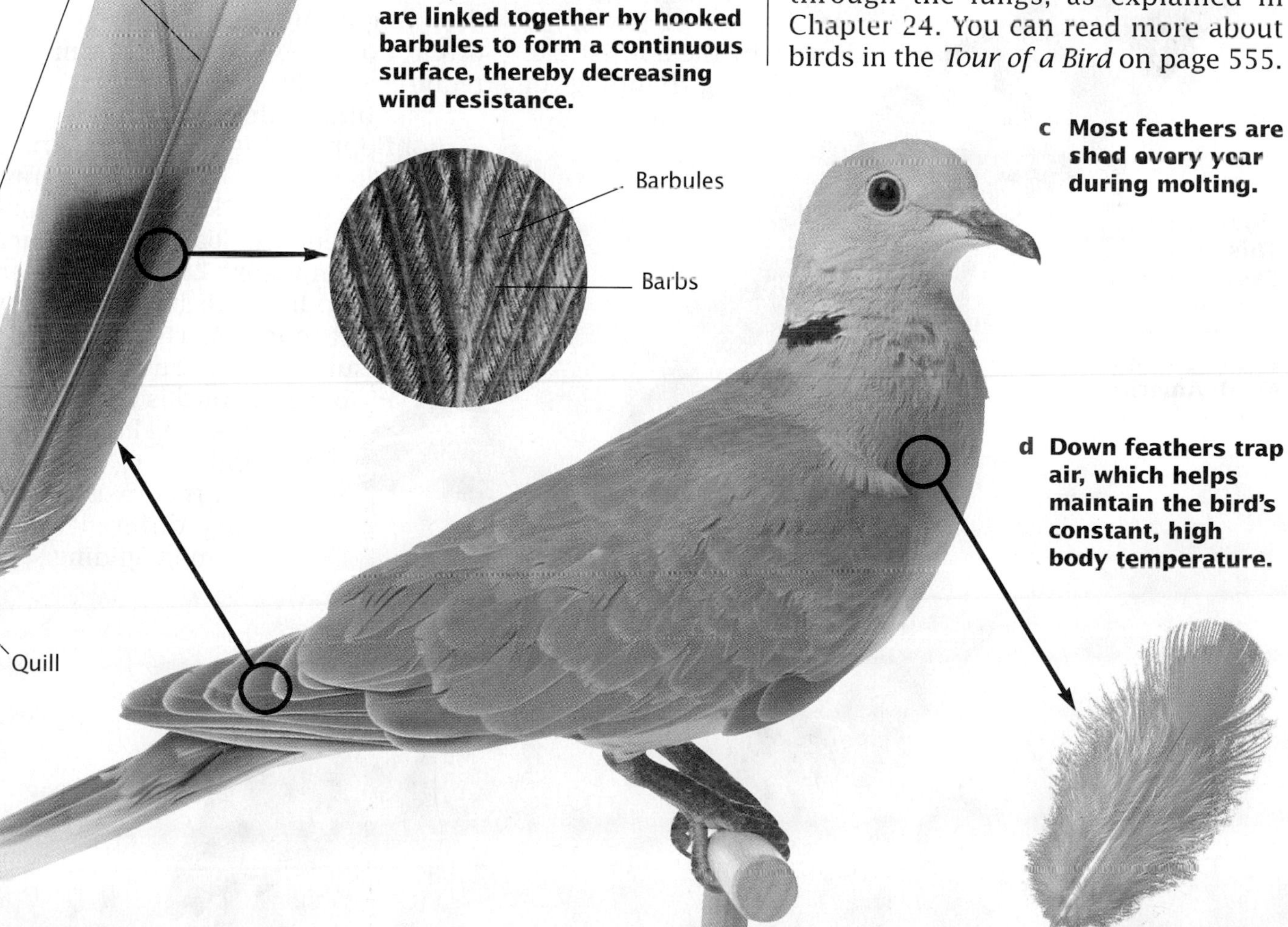

Figure 28-15

a Contour feathers and down feathers have different functions. Down feathers serve primarily as insulation. Contour feathers provide insulation and help the bird with steering and balance.

b The individual filaments, or barbs, of a contour feather are linked together by hooked barbules to form a continuous surface, thereby decreasing wind resistance.

c Most feathers are shed every year during molting.

d Down feathers trap air, which helps maintain the bird's constant, high body temperature.

How Birds Fly

Birds must overcome gravity in order to achieve and maintain flight. The wing is the feature that enables birds (and airplanes) to fly. An upward force known as lift is generated when air passes across the surface of a wing. Lift is produced when the pressure of the air passing over the top of the wing is less than that of the air passing under the wing. This pressure difference can be created in two ways. First, the wing can be arched. Second, the front edge of the wing can be held higher than the back edge, increasing what is called the angle of attack.

Lift is produced only when air is moving across the wing, either when the wind blows toward the bird or when the bird travels forward. Since wind direction and speed are unpredictable, efficient flight demands that birds generate force to propel themselves forward. This forward force is known as thrust. Birds typically move forward by beating their wings. Thrust is produced largely by the downstroke during flapping of the wings.

Active flight requires strong muscles to raise and lower the wings. The muscles that power the wings of birds during flight attach to the sternum, or breastbone. The sternum of a bird that flies is large and keeled, like the hull of a boat, to accommodate these large muscles.

There are different kinds of flight among birds

Birds fly in a variety of ways. The most common method is called flapping flight. It is also the most energy-demanding form of flight. The bird must actively lower and raise the wings in a complex twisting pattern that produces both lift and forward thrust.

If you watch birds in the air, you will notice that not all of them flap their wings all of the time. The simplest form of nonflapping flight is gliding. A bird always loses altitude as it glides because gravity overcomes the lift provided by its forward descent. Birds dropping from treetops to the ground often do so by simple gliding; it requires very little energy.

Some birds exhibit a different kind of nonflapping flight called soaring, which allows them to overcome gravity and stay aloft for long periods while using little energy. Soaring birds, such as the tern in **Figure 28-16**, are generally large-bodied and have relatively small flight muscles. The large body results in forward momentum once the bird is flying, thus maintaining lift. It is this ability to maintain lift that makes soaring different from simple gliding.

Figure 28-16 This soaring Caspian tern is a member of the gull family and is the largest North American tern.

Bird

The cedar waxwing is a common bird throughout North America. All modern birds have forelimbs modified into wings, are covered with feathers, feed with a horny beak, and lay hard-shelled eggs.

Tour

Size
Cedar waxwings reach a maximum length of 18 cm (7 in.).

Birds are endotherms. The body temperatures of birds range from 40°C to 42°C (104°F to 108°F), higher than the lethal limit for many mammals.

The bones of birds are laced with air cavities, making them light and strong at the same time. The skeleton of a bird typically weighs less than its feathers.

Feathers evolved from reptilian scales. Most birds molt, or shed their feathers, at least once a year. In most birds, however, only a few feathers are shed at a time.

The cedar waxwing, like all flying birds, has a keeled sternum to which powerful flight muscles are attached.

Like many birds, the cedar waxwing migrates seasonally. In the spring, cedar waxwings fly north. Some birds spend the summer as far north as Alaska and northern Canada. In the fall, cedar waxwings migrate south and spend the winter in warmer areas, sometimes as far south as Panama.

Habitat
Cedar waxwings prefer wooded areas in which berries, their favorite food, are plentiful. They will live in areas disturbed by human activities, such as fields and orchards.

Evolutionary Relationships
Birds are the descendants of dinosaurs. The close relationship between birds and dinosaurs prompted Thomas Huxley, a colleague of Darwin, to call birds "glorified reptiles." Cedar waxwings belong to the order Passeriformes, the largest order of birds. This order, commonly called "songbirds," contains more than 5,000 species.

Major Orders of Birds

You can often tell a great deal about the habits and food of a bird by examining its beak and feet. For instance, carnivorous birds, such as eagles, have curved talons for seizing prey and a sharp beak for tearing apart their meal. The beaks of seed-eating birds, such as finches, are short, thick, seed crushers. What birds are common in your neighborhood? What sorts of adaptations do they show? Twenty-eight orders of living birds have been described. **Table 28-2** lists 16 of the most common orders of birds and shows an example of each. Pay particular attention to the feet and beaks of these birds.

Table 28-2 Major Orders of Birds

Order	Number of living species	Description	Examples
Passeriformes	5,276	Songbirds; perching feet; well-developed vocal organs; dependent young; largest bird order, containing 60 percent of all bird species	Sparrows, robins, warblers, crows, starlings, mockingbirds
Apodiformes	428	Small bodies, short legs, rapid wing beat; hummingbirds are the smallest birds	Hummingbirds, swifts
Piciformes	383	Sharp, chisel-like bills for pounding through wood in search of insects; grasping feet	Woodpeckers, toucans, honeyguides
Psittaciformes	340	Well-developed vocal organs; large powerful bills for crushing seeds	Parrots, cockatoos
Charadriiformes	331	Shorebirds; typically with long, slender, probing bills and long, stiltlike legs	Gulls, terns, plovers, auks, sandpipers
Columbiformes	303	Stout bodies; perching feet	Pigeons, doves
Falconiformes	288	Birds of prey; day-active carnivores; sharp pointed beaks for tearing flesh; keen vision; strong fliers	Eagles, hawks, falcons, vultures
Galliformes	268	Rounded bodies; often limited flying ability	Chickens, quail, grouse, pheasants

Gruiformes	209	Marsh dwellers; diverse body shapes; long, stiltlike legs	Rails, coots, bitterns, cranes
Anseriformes	150	Waterfowl; webbed toes; broad bill with filtering ridges at margins	Swans, geese, ducks
Ciconiiformes	114	Long-legged waders; often with large bodies	Storks, herons, ibises
Strigiformes	146	Nocturnal birds of prey; large eyes; powerful beaks and feet	Owls
Procellariiformes	104	Sea birds; tube-shaped bills; many can fly for long periods of time	Albatrosses, petrels
Sphenisciformes	18	Marine; flightless; confined to Southern Hemisphere; thick coat of insulating feathers; wings modified as paddles for swift swimming	Penguins
Apterygiformes	3	Small and flightless; long probing bill; keen sense of smell; found only in New Zealand	Kiwis
Struthioniformes	1	Large; flightless; only two toes; long, strong running legs	Ostrich

SECTION REVIEW

1. **What evidence indicates that birds evolved from reptiles?**
2. **Name two differences between a lizard and a bird.**
3. **Ostriches have feathers but cannot fly. What functions do feathers perform for an ostrich?**
4. **List two features of the bird's skeleton that make it well suited for flight.**

You are a mammal. So are many of the animals that humans eat, use as work animals, and keep as pets. All mammals have hair or fur on their bodies, a characteristic not found in any other kind of modern animal. Mammals are also endothermic, like birds. Female mammals produce milk with which they nurse their offspring.

28-3 Introduction to Mammals

OBJECTIVES

1. **List the unique characteristics of mammals.**
2. **Identify two features of monotremes.**
3. **Contrast the pattern of development of marsupials with that of placental mammals.**

Origin of Mammals

Mammals (class Mammalia) arose from early reptiles called therapsids *(thuh RAP sihdz)*, one of which is illustrated in **Figure 28-17a**. The fossil record provides a well-documented transition between these reptiles and mammals, with fossil forms ranging from reptiles with a few mammalian characteristics to true mammals. The first mammals appeared about 220 million years ago, just as the first dinosaurs were evolving. Early mammals, such as the *Eozostrodon* shown in **Figure 28-17b**, were small (about the size of mice), fed on insects, and lived in trees. Their relatively large eye sockets suggest that they were active at night. These early mammals had a jaw composed of only one bone, a characteristic that distinguishes them from the therapsids, which had a jaw composed of several bones. Two of the bones in the therapsid jaw became part of the chain of bones that transmits sound in the middle ear of mammals.

Figure 28-17

a ***Cynognathus*** **is a therapsid that lived about 250 million years ago. It was about as large as a medium-sized dog. Some therapsids lived in cold climates and may have had hair. The jaw of a therapsid is composed of several bones, while a mammal's jaw is a single bone.**

b ***Eozostrodon*** **was a typical early mammal. It was about 18 cm (5 in.) long and lived about 215 million years ago.**

For over 150 million years, while the dinosaurs flourished, mammals were a minor group that changed little. Mammalian fossils from this time are very rare, indicating that mammals were not abundant. At the end of the Cretaceous period, 65 million years ago, the dinosaurs and many other kinds of aquatic and land-dwelling animals became extinct, but the mammals survived. In the time since the dinosaurs disappeared, mammals have diversified and have taken over many of the ecological roles once occupied by dinosaurs. Mammals reached their maximum diversity about 15 million years ago. At that time, tropical conditions existed over much of the world. During the last 15 million years, world climates have cooled and the area covered by tropical rain forests has decreased, causing a decline in the number of mammalian species. Today, human activities such as overhunting and habitat modification threaten many species of mammals.

Mammalian Characteristics

Mammals are distinguished by two characteristics: the presence of hair or fur on the body and the ability to produce milk. Even the apparently naked whales and dolphins grow sensitive bristles on their snouts. Evolution of fur and the ability to regulate body temperature through metabolism enabled mammals to inhabit colder environments than could be tolerated by ectothermic reptiles and amphibians.

Like birds, mammals have a four-chambered heart. There is no mixing of deoxygenated and oxygenated blood in the heart.

Female mammals have **mammary glands** that secrete milk. Newborn mammals, which are born without teeth, suckle this rich milk until they are able to feed on their own. **Figure 28-18** summarizes some of the key characteristics of mammals.

Figure 28-18 This cheetah shows the main characteristics of mammals.

Four-chambered heart
The cheetah's heart has four chambers—two atria and two ventricles.
Endothermic metabolism
The cheetah maintains a high and constant body temperature.
Fur
The cheetah is covered with a coat of fur. Its coloration and pattern help conceal her from predators and potential prey.
Mammary glands
The cheetah feeds her young nutrient-rich milk produced by her mammary glands.

Egg-Laying Mammals: The Monotremes

Imagine an egg-laying, furry animal with a large bill and webbed feet. Such an animal exists in Australia—the duck-bill platypus. The platypus and two species of echidnas *(ee KIHD nuhs)*, or spiny anteaters, are the only living **monotremes**. A platypus and an echidna are shown in **Figure 28-19a** and **b**. Although they are mammals, monotremes have some reptilian characteristics. They lay shelled eggs, and the structure of their pelvis is very similar to that of the early reptiles. Also like reptiles, monotremes have a single opening through which feces, urine, and reproductive products leave the body. Scientists think monotremes resemble the early mammals more closely than do any other living mammals.

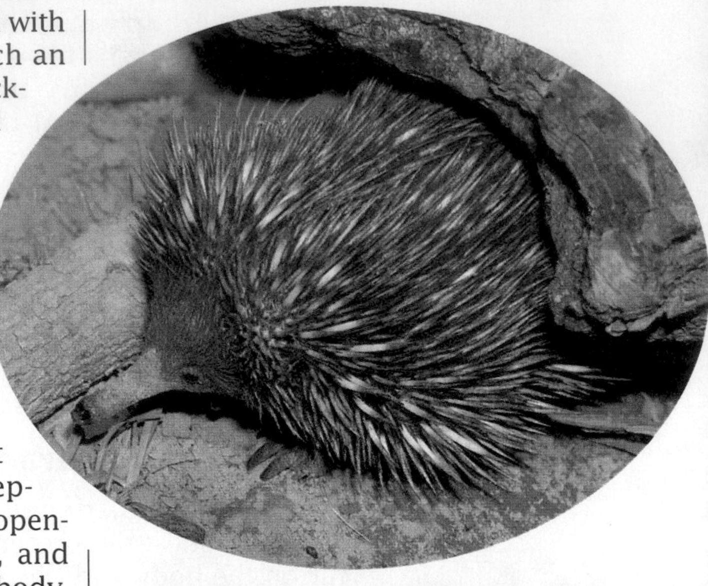

b This species of echidna can be found in New Guinea and Australia. It uses its long nose to root for insects and other small animals.

In addition to their reptilian features, monotremes have both defining mammalian features: fur and functional mammary glands. Young monotremes drink their mother's milk after they hatch from eggs. Because of their strange mouths and the absence of well-developed nipples on the females, the young cannot suckle. Instead, the milk oozes onto the mother's fur and the young lap it up with their tongues.

Figure 28-19

a The platypus, about half the size of a house cat, is a mammal that lays eggs. The young are nourished by milk from the mother's mammary glands.

The platypus is a good swimmer. It uses its bill much as a duck does, rooting in the mud for worms and other soft-bodied animals. Echidnas have very strong, sharp claws, which they use for burrowing and digging for insects. The echidna probes for insects, especially ants and termites, with its long snout.

Among living mammals, only monotremes lay eggs. In the rest of the mammals, the offspring are born after developing for some time inside the mother. **Marsupials**, such as kangaroos, koalas, and opossums, carry the offspring internally for only a short time. The young are born at an early developmental stage and finish developing in a pouch on the mother's abdomen. In contrast, the offspring of **placental mammals** remain inside their mother until development is essentially complete. They are nourished through a structure known as the placenta, which was described in Chapter 24. Dogs, cats, cows, horses, humans, and most other mammals are placental mammals.

Pouched Mammals: The Marsupials

The major difference between marsupials and placental mammals is their pattern of embryonic development. In marsupials, the fertilized egg is surrounded by a shell membrane, but no shell forms around the egg. During most of its early development the marsupial embryo is nourished by the abundant yolk within the egg. Shortly before birth, the shell membrane is lost and a short-lived placenta forms. Still early in its development, the baby marsupial is born. Newborn marsupials emerge hairless and tiny, as you can see in **Figure 28-20**. Smaller than your thumbnail, they must crawl to the pouch, where they can nurse.

Marsupials evolved not long before placental mammals, about 100 million years ago. Today, all but about 20 of the species of marsupials live in New Guinea and Australia, where there are relatively few species of placental mammals. Marsupials in Australia and New Guinea have diversified to fill ecological positions occupied by placental mammals elsewhere in the world. For example, some species of large kangaroos are the Australian grazers, playing the ecological role that antelope, horses, and buffalo perform elsewhere. The only marsupial native to the United States is the Virginia opossum.

Humans have had a negative impact on the unique marsupials of Australia and New Guinea. Humans introduced placental mammals, including dogs, cats, sheep, foxes, and rabbits, that prey on native marsupials or compete with them for food. The negative impact of humans has intensified since European settlers arrived in Australia about 200 years ago.

Figure 28-20
a This newborn opossum has crawled to its mother's pouch and is nursing.

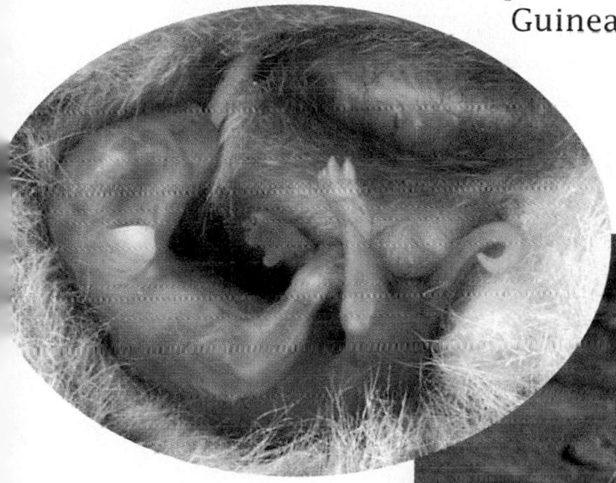

b After several months, the young opossums are able to grasp the hair on their mother's back.

True Placental Mammals

Mammals that produce a true placenta, through which the embryos are nourished for their entire development, are called placental mammals.The majority of mammal species, including humans, are placental mammals. Placental mammals are very diverse, ranging in size from 1.5 g (0.05 oz) bats to 100,000 kg (110 ton) whales. Today, placental mammals are found on all continents except Antarctica. They live in a variety of habitats, from hot, moist rain forests to the frigid tundra. They have also invaded the air and returned to the sea. There are more than 4,000 species of placental mammals. **Table 28-3** describes 12 of the 19 orders of modern-day placental mammals.

Table 28-3 Major Orders of Placental Mammals

Order	Number of living species	Description	Examples
Rodentia	1,814	Herbivores with chisel-like incisor teeth that grow continuously; typically small	Squirrels, rats, mice, beavers, porcupines
Chiroptera	986	The only flying mammals; elongated fingers support a thin wing membrane; mainly fruit or insect eaters; many fly at night, navigating by sonar	Bats
Insectivora	390	Small, chiefly night-active mammals; feed on insects; sharp-snouted; spend most of their time underground; the most primitive placental mammals	Moles, shrews, hedgehogs
Carnivora	240	Land-living predators; teeth adapted for seizing prey and shearing flesh; no native families in Australia	Dogs, bears, cats, wolves, otters, weasels
Primates	233	Largely tree dwellers; binocular vision and an opposable thumb; large brains; the end product of a line that branched off early from other mammals; retains many primitive characteristics	Prosimians, apes, monkeys, humans

Artiodactyla	211	Hoofed mammals with two or four toes; large herbivores; most are grass eaters	Sheep, pigs, cattle, deer, giraffes
Cetacea	79	Aquatic; streamlined bodies; front limbs modified into broad flippers; no hind limbs; nostrils are blowholes on top of head; hairless except on muzzle	Whales, dolphins, porpoises
Lagomorpha	69	Rodentlike mammals with four upper incisors, rather than the two seen in rodents; hind legs often longer than forelegs, an adaptation for jumping	Rabbits, hares, pikas
Pinnipedia	34	Marine carnivores with limbs modified for swimming; feed mainly on fish	Seals, sea lions, walruses
Edentata	30	Mostly insect eaters; many are toothless, but some have degenerate, peglike teeth	Sloths, anteaters, armadillos
Perissodactyla	17	Hoofed mammals with one or three toes; herbivores with teeth adapted for chewing	Horses, zebras, rhinoceroses, tapirs
Proboscidea	2	Enormous herbivores with long trunks; two upper incisors elongated as tusks; the largest living land animals	Elephants

SECTION REVIEW

1. **Why are monotremes classified as mammals even though they lay eggs?**
2. **List two differences between monotremes and other mammals.**
3. **Describe two differences between reproduction in marsupials and that in placental mammals.**
4. **What are two advantages of the placental pattern of development over the marsupial pattern?**

Although there are fewer species of mammals than reptiles, mammals thrive in a far greater variety of habitats and show a greater variety of sizes and shapes than do reptiles. In this section, you will learn about some of the unique adaptations of mammals that suit them to their wide range of habitats and ways of life.

28-4 Mammalian Adaptations

OBJECTIVES

1. **Identify three functions of hair.**
2. **List two mammalian structures that contain keratin.**
3. **Contrast the teeth of herbivores with those of carnivores.**
4. **Describe how bats navigate in the dark.**

Hair Has Many Functions

All mammals have hair, including the seemingly hairless whales and dolphins. A hair extends from a bulblike structure known as a hair follicle, which lies below the surface of the skin. Hair is composed mainly of air spaces and dead cells filled with the protein keratin.

Most mammals are covered by a coat of hair, shown in **Figure 28-21**. Two different types of hair make up this coat. **Guard hairs** are the long, thick outer hairs responsible for the coat's color. Between the guard hairs grows a dense coat of **underhair**, consisting of thinner, shorter hairs. You can see the underhair of a dog or cat by brushing back the guard hairs with your hand.

What are the functions of hair? One of its functions is insulating against heat loss. Endothermic animals such as mammals often maintain body temperatures higher than the temperature of their surroundings and so tend to lose body heat. The dense underhair coat of many mammals reduces the amount of body heat that escapes, performing the same function as a sweater that you wear on cold days.

Another function of hair is camouflage. The coloration and pattern of a mammal's coat usually match its environment. Hairs also function as sensory structures. The whiskers of cats and dogs are stiff hairs that are sensitive to touch. Mammals that are active at night or that live underground often rely on their whiskers to locate prey or to avoid colliding with objects. Hair can also serve as a defensive weapon. For instance, porcupines and hedgehogs protect themselves with long, sharp, stiff hairs.

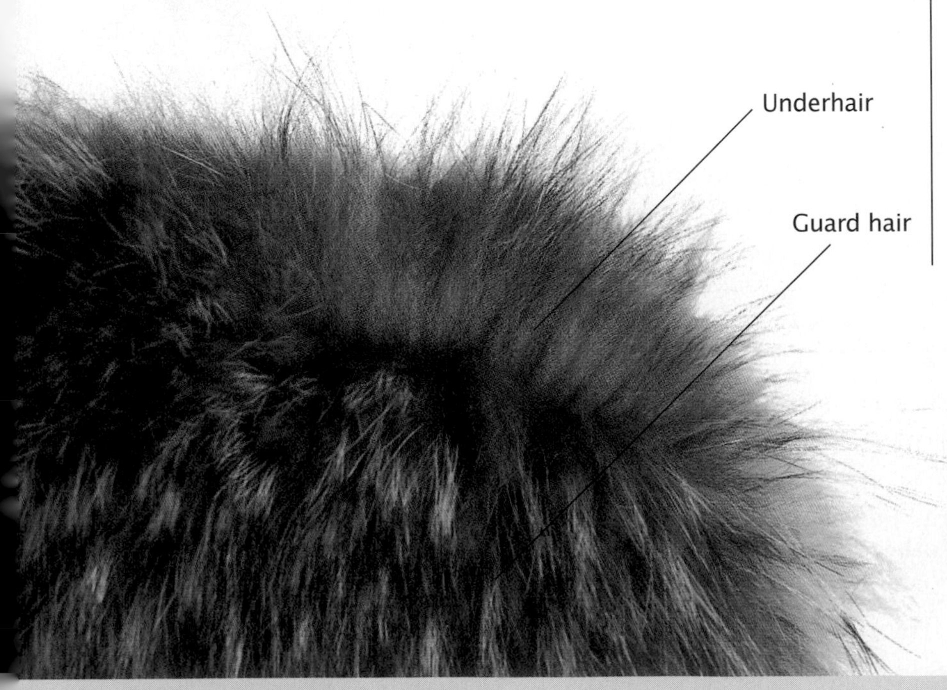

Figure 28-21
All mammals have fur, which functions primarily as insulation. Underhair and guard hair aid in this function.

Claws, Hooves, Horns, and Antlers

Keratin is a versatile protein that is a component of many mammalian structures. For example, it is found in claws, fingernails, and hooves. Hooves, such as the zebra's hoof shown in **Figure 28-22**, are specialized pads on the toes of horses, cattle, sheep, antelopes, and other running mammals. In addition, the horns of cattle, sheep, and antelopes are composed of a core of bone surrounded by a sheath of keratin. The bony core is attached to the skull, and the horn is not shed. The horns of rhinoceroses are made of hairlike fibers of keratin that are compacted into a very hard structure.

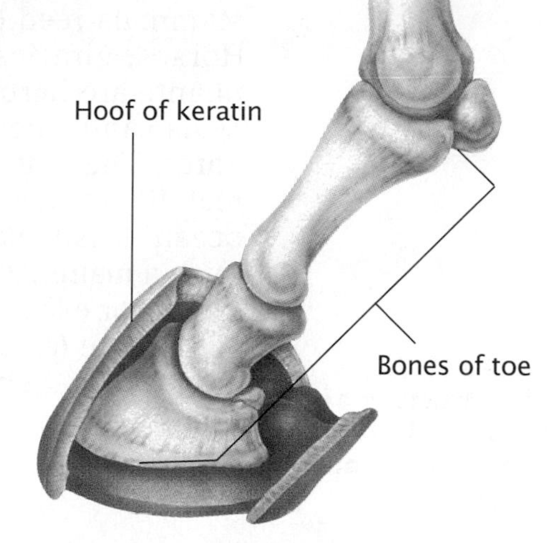

Figure 28-22
A zebra's foot is modified so that the zebra actually walks on a single toe that is covered with a hoof made of keratin.

Deer grow and shed a set of antlers each year. Deer antlers, which are grown only by males, are made of bone and have no keratin sheath. While growing, antlers are covered by a thin layer of skin known as velvet. The velvet dies and is scraped off when the antlers are fully grown. The male deer uses his antlers to attract females and to combat other males. After the breeding season, the male sheds his antlers. Some examples of horns and antlers are shown in **Figure 28-23**.

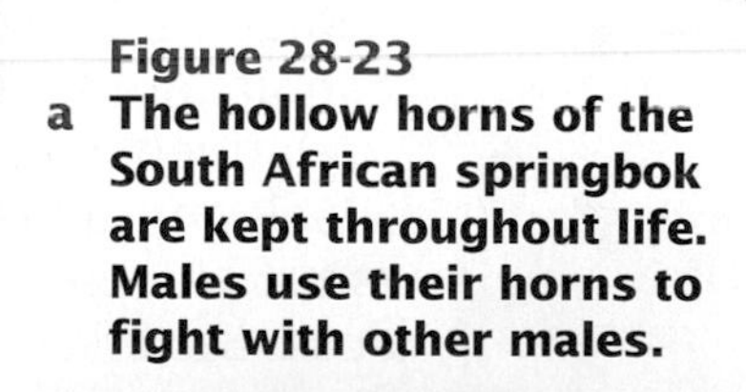

Figure 28-23
a The hollow horns of the South African springbok are kept throughout life. Males use their horns to fight with other males.

b The antlers of the mule deer are used for combat and to attract females. They drop off every winter and regrow during the spring and summer.

Food and Feeding

Mammals feed on a variety of foods. Horses, giraffes, antelopes, and elephants are herbivores. Lions, wolves, seals, and sperm whales are carnivores. The blue whale, the largest animal, filters small crustaceans from the ocean. It is usually possible to determine a mammal's diet by examining its teeth. For example, look at the skull of the coyote (a carnivore) and the deer (a herbivore) shown in **Figure 28-24**. The coyote's long canine teeth are suited for biting and holding prey. Its premolar and molar teeth are triangular and sharp for shearing off chunks of flesh. The deer's canines, in contrast, are small. It clips off mouthfuls of plants with its flat incisors. The deer's molars are large, and their surfaces are covered with ridges that form an effective grinding surface for breaking up tough plant tissues.

Figure 28-24 The structure of a mammal's jaw and teeth usually reveals its diet.

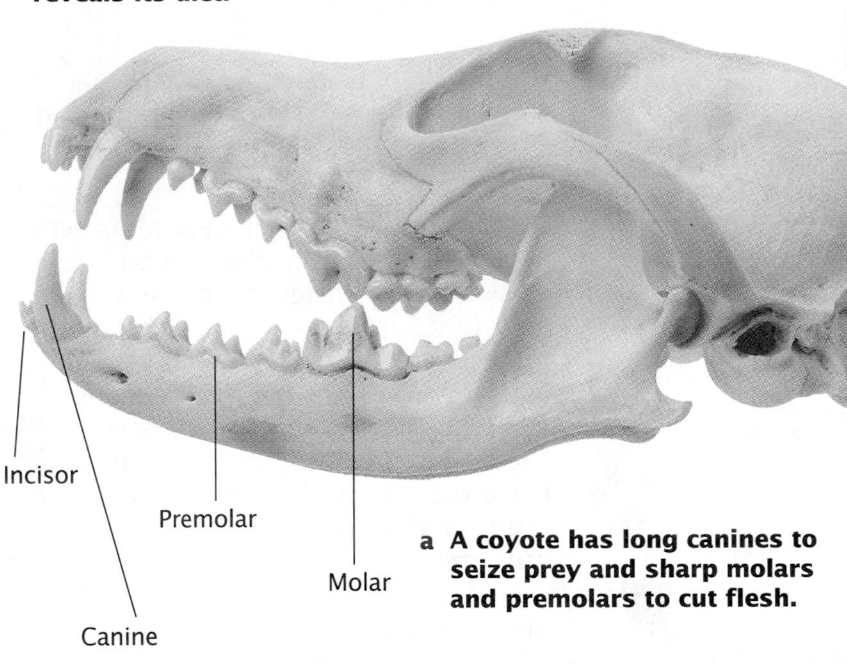

a A coyote has long canines to seize prey and sharp molars and premolars to cut flesh.

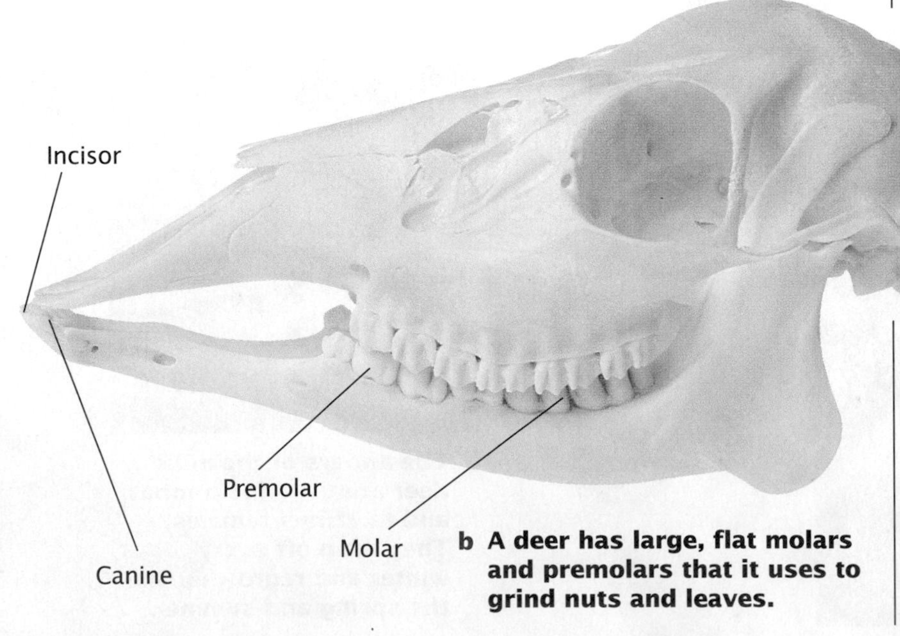

b A deer has large, flat molars and premolars that it uses to grind nuts and leaves.

Cellulose is the major component of plant cell walls and is thus a major constituent of the plant body. Mammals do not have enzymes that can digest cellulose. Herbivorous mammals rely on a mutualistic partnership with bacteria and protists that can produce cellulose-splitting enzymes. Mammals such as cows, buffaloes, antelopes, goats, deer, and giraffes have huge four-chambered stomachs that function as storage and fermentation vats. The first chamber is the largest and holds a large population of bacteria and protists. When the animal swallows, chewed plant material passes into this chamber. Bacteria and protists partly digest the material, which is then regurgitated and chewed again. A cow chewing its cud is rechewing this partly digested food. After another thorough grinding, the cud is swallowed and further digested in the stomach. It then passes from the stomach into the intestines.

Rodents, horses, elephants, and rabbits are herbivores but have relatively small stomachs that lack mutualistic microorganisms. These animals do not chew a cud. Bacteria that aid in digestion live in a pouch that branches off the large intestine.

Even with these complex adaptations for breaking down cellulose, a mouthful of plants is less nutritious than a mouthful of flesh. Herbivores must consume large amounts of plant material to gain sufficient nutrition. An elephant eats 135–150 kg (300–400 lb) of food per day. You can read more about mammalian adaptations in the *Tour of a Mammal* on page 567.

Mammal

The deer mouse is one of the most common and widespread mammals in the United States. It is found in most of the continental United States. Like all mammals, deer mice have hair, and the females feed their young with milk.

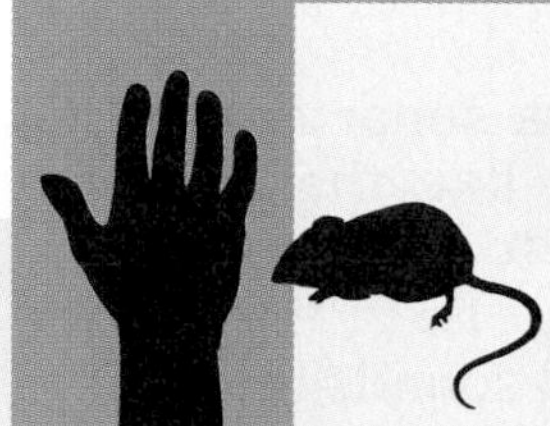

Size
The deer mouse ranges from 120 to 230 mm (about 5 in. to 9 in.) in length, including its tail. It weighs from 18 to 35 g (0.67 to 1.25 oz).

Whiskers are sensory hairs that enable the mouse to detect nearby objects in the dark.

All mammals are endotherms. Per gram of body mass, a mouse consumes five times as much food as a dog. Smaller mammals, such as the deer mouse, must eat more to remain active because they lose much more heat due to their greater surface-area-to-volume ratio.

The chisel like incisors of rodents grow continually, an adaptation for gnawing. A mammal's teeth are replaced only once during its lifetime, unlike the teeth of a reptile, which are replaced continually throughout its life.

Habitat
Because it is such a widespread species, the deer mouse lives in many habitats. It prefers dry areas, such as forests and grasslands. It nests in burrows, fallen logs, and rock piles and in buildings in rural areas.

Evolutionary Relationships
Deer mice are rodents, members of the order Rodentia, which is the largest order of mammals. Rats, squirrels, porcupines, beavers, and chipmunks are rodents. Rodents are the most numerous and widely distributed mammals.

Flying Mammals

Bats are the only mammals capable of powered flight. Like the wings of birds, the wings of bats are modified forelimbs. The bat wing is a leathery membrane of skin and muscle supported by the bones of four fingers. The membrane attaches to the side of the body and to the hind leg (and to the tail in some bats). When resting, most bats prefer to hang by their feet, as shown in **Figure 28-25**.

Contrary to popular belief, not all bats emerge only at night. The so-called flying foxes, for instance, feed on fruit during the day. Most bats, however, are active at dusk or at night. Some eat fish and frogs, which they pluck from the water's surface. Others feed on nectar from flowers or prey on small mammals, including other bats. Vampire bats of Central America and South America drink blood from large mammals. Flying insects are the main food source for most nocturnal bats. Since few birds fly at night, bats have almost exclusive access to this rich food supply.

Figure 28-25
Many bats, such as this California leaf-nosed bat, are social animals. They roost together in groups of thousands, and some species hunt cooperatively.

How do bats navigate in the dark?

How can bats fly in dark caves and hunt their food at dusk or even at night? Do bats have particularly sensitive vision? Or do they use another sense to find their way around? Late in the eighteenth century, the Italian scientist Lazzaro Spallanzani showed that a blinded bat could fly and capture flying insects. However, when Spallanzani plugged the ears of the bat, it was unable to navigate and collided with objects. Spallanzani concluded that bats "hear" their way through the world. It was not until the late 1930s that bats' ability to "fly blind" was explained.

Bats have evolved a sonar system that functions much like the sonar devices used by ships to locate underwater objects. As a bat flies, it emits extremely high pitched sounds that are outside our range of hearing. These high-frequency pulses are emitted through the mouth or, in some cases, through the nose. The sound waves reflect from obstacles or flying insects, and the bat hears the echo. Through sophisticated processing of this echo within its brain, a bat can determine not only the direction of an object but also the distance to the object. This sonar system enables bats to navigate and to capture their prey at night, as shown in **Figure 28-26**.

Figure 28-26
This bat used its sonar to catch a moth in midair in the dark.

SECTION REVIEW

1. **List two functions of hair.**
2. **Name two keratin-containing structures on your body.**
3. **Contrast the teeth of a coyote with those of a deer.**
4. **Explain why a blinded bat can catch prey but a deafened one cannot.**

CHAPTER 28 Highlights

"Well, of course I did it in cold blood, you idiot! ... I'm a reptile!"

	Key Terms	Summary
28-1 Reptiles  Reptiles control their body temperature through their behavior.		• Reptiles have dry, largely watertight skin and are ectotherms. Most species lay watertight eggs. • Reptiles evolved about 320 million years ago. • Pterosaurs were the only reptiles that evolved the ability to fly. • Dinosaurs became extinct about 65 million years ago. • The surviving reptiles include crocodiles and alligators, turtles, tuataras, lizards, and snakes.
28-2 Birds  Some birds may have more than 25,000 feathers on their body.	contour feather (p. 553) down feather (p. 553)	• The first birds evolved more than 150 million years ago. • All birds have feathers, are endothermic, and lay eggs. • Birds have down feathers and contour feathers. • The skeletons of birds are lightweight. The bones are hollow and thin.
28-3 Introduction to Mammals Mammals, such as these cheetahs, receive nourishment from their mother's mammary glands while young. 	mammary gland (p. 559) monotreme (p. 560) marsupial (p. 560) placental mammal (p. 560)	• Mammals evolved from reptiles about 220 million years ago. • Mammals have hair and are endothermic. Female mammals have mammary glands. • Monotremes are mammals that lay eggs. • Marsupial mammals are born early in development and complete their development in the mother's pouch. • Placental mammals nourish their young through the placenta.
28-4 Mammalian Adaptations The skull of a coyote has long canines and strong jaws. 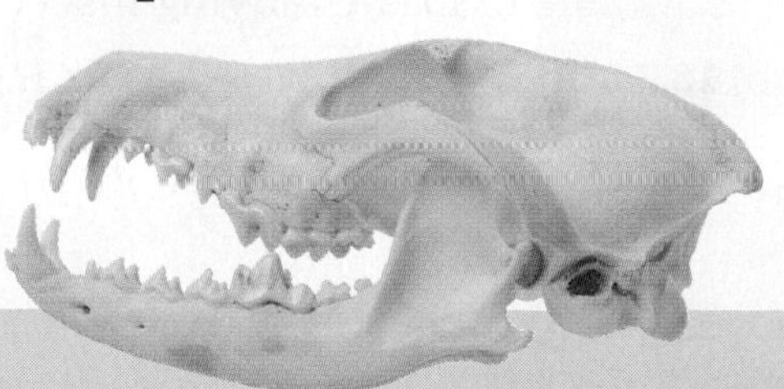	guard hair (p. 564) underhair (p. 564)	• Hair serves as insulation, as camouflage, as a sensory structure, and as a defense. • Bats are the only flying mammals. Bats are able to navigate and capture prey in the dark by means of a sonar system.

CHAPTER 28

review

Understanding Vocabulary

1. For each pair of terms, explain the difference in their meanings.
   a. endotherm, ectotherm
   b. contour feathers, down feathers
   c. monotremes, marsupials

2. Using each set of words below, write one or more sentences summarizing information learned in this chapter.
   a. ichthyosaur, plesiosaur, dinosaur, pterosaur
   b. flapping flight, gliding, soaring
   c. hair, mammary glands, endothermic, four-chambered heart

Understanding Concepts

3. **Relating Concepts** Construct a concept map that describes the characteristics of the three vertebrate groups described in this chapter. Try to include the following terms in your concept map: vertebrate, reptiles, birds, mammals, ectothermic, endothermic, feathers, hair, mammary glands, dinosaurs, contour, down, and insulation.

4. **Comparing Structures** How does the skin of a reptile differ from the skin of an amphibian? Describe the differences between the eggs of amphibians and the eggs of reptiles.

5. **Analyzing Information** Suppose that some new dinosaur fossils are discovered. Radioactive dating shows that these fossils are 60 million years old. Do these fossils support or contradict the hypothesis that the dinosaurs became extinct because a meteorite struck the Earth? Explain your answer.

6. **Summarizing Information** List the four orders of living reptiles and give the main characteristics of each. Give an example of an animal in each order.

7. **Analyzing Data** The graph below shows how the body temperatures of two animals change over the day. Which animal is probably an ectotherm? Explain your answer.

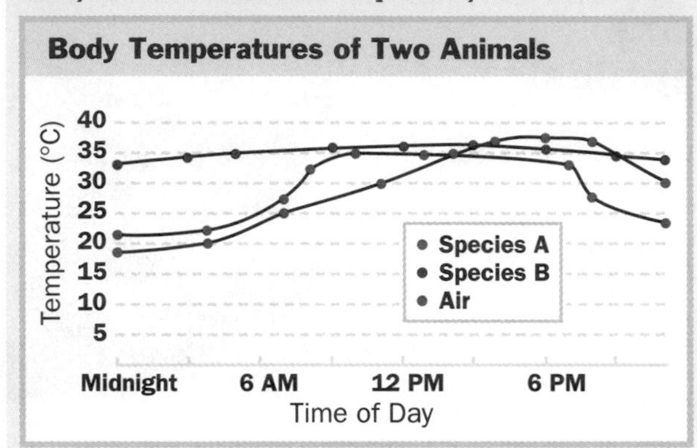

8. **Recognizing Relationships** Early specimens of *Archaeopteryx* were misidentified as reptiles. What characteristics does *Archaeopteryx* share with reptiles? What features does it share with modern birds?

9. **Comparing Structures** Describe two ways the skeleton of a bird is modified for flight.

10. **Comparing Structures** How were the early mammals different from therapsids?

11. **Identifying Structures** Identify the two characteristics that make mammals unique.

12. **Comparing Functions** Contrast the pattern of development of marsupials with that of placental mammals. What makes monotremes unique among living mammals?

13. **Identifying Functions** Describe three functions of hair.

14. **Comparing Functions and Structures** How are underhair and down feathers alike? How are they different?

15. **Identifying Structures** Name three structures of your body that contain keratin.

16. **Comparing Structures** How could you tell if a mammalian skull was that of a herbivore or a carnivore?

17. **Recognizing Relationships** Explain how a bat is able to locate a flying insect on a dark night.

18. **BUILDING ON WHAT YOU HAVE LEARNED** In Chapter 18 you learned about bacteria. How are bacteria beneficial to herbivorous mammals?

Interpreting Graphics

19. Examine the animals pictured below.

a. Identify the class to which each animal belongs.
b. Which animal is ectothermic?
c. In which animal is the ventricle of the heart incompletely divided?

Reviewing Themes

20. ***Evolution***
Pterosaurs had a large, keeled breastbone. What might this feature indicate about how pterosaurs flew? Explain your answer.

21. ***Evolution***
What evidence suggests that reptiles and birds have a common ancestor?

22. ***Interdependence***
Explain the impact that humans have had on the survival of marsupials in Australia and New Guinea.

Thinking Critically

23. **Justifying Conclusions** When a female leatherback turtle comes up on the beach to lay her eggs, she first digs a deep hole, lays her eggs, and covers them with sand. Then she crawls about 100 m and digs another hole. This time she lays no eggs but covers the hole with sand anyway. Suggest a hypothesis to explain this behavior. Describe how you could test your hypothesis.

Life/Work Skills

24. **Finding Information** Use library references or search an on-line database to find out about the different ways that snakes move. Prepare an illustrated written report to share your findings with your class.

25. **Working Cooperatively** Working with two or three of your classmates, use library references and field guides to find out what kinds of birds are common in your area. Find out what the birds eat and what kind of shelter they require. Design and build a birdhouse appropriate for a particular bird in your area. Also design and build a bird feeder for the same bird. Place the house and feeder where they can be easily observed. Keep a journal of your observations to share with your class.

Activities and Projects

26. **Literature Connection** Throughout history, reptiles have been portrayed in literature as evil and dangerous monsters. Read fairy tales, folk tales, and other stories, and compile a list of the ways that reptiles are portrayed. Then use library references or search an on-line database to provide evidence to support or disprove these portrayals.

27. **Language Arts Connection** Use library references to look up common phrases that refer to characteristics of animals, such as "happy as a lark," "sings like a bird," "eats like a bird," "blind as a bat," and "dog tired." Are these phrases based on fact? Compile a list of phrases and information about their accuracy to share with the class.

Discovering Through Reading

28. Read the article "The Private Lives of Pit Vipers" by Michael Lipske in *National Wildlife*, August/September 1995, pages 14–20. Describe the function of the pits located behind the nostrils of all pit vipers. How frequently do blacktail rattlesnakes eat during a year? Describe one way that pit vipers have benefited medicine.

UNIT

Six

Human Life

A white blood cell devours invading bacteria

VIRUSES AND YOU

When you get a cold or the flu, a virus is trying to take over your cells as it spreads throughout your body. A few days after the virus invades, you begin to feel ill. Though colds and flu are uncomfortable, few healthy people die from them because the body's immune system battles viruses and other pathogens that cause disease.

A Deadly Virus

The virus that causes AIDS is another story. It can hide in the body for months or years before it produces any signs of illness. But all that time the virus is destroying the ability of the immune system to protect the body from other pathogens. That is why this virus is known as the human immunodeficiency virus (HIV).

A Profound Challenge

HIV mutates rapidly, disguising itself as a slightly different virus with each mutation. This rapid mutation makes the development of vaccines or treatments a challenge. Because the virus is now spreading most rapidly through sexual contact between men and women, it will be very difficult to eliminate.

LOOKING AHEAD

- How does the immune system usually fight off invasions by a virus? How does HIV disable the immune system? See pages 683–691.
- How frequently is AIDS affecting people aged 15–24? See page 693.
- What other pathogens are transmitted in the same ways as HIV? How can the risk of HIV infection be reduced? See pages 737–740.
- Why should health-care costs for the elderly be of concern to you? See **Science, Technology, and Society: Who Pays the Bill for Growing Old?** pages 716–717.

CHAPTER 29

REVIEW
- cell division (Section 6-2)
- tissues (Section 23-1)
- innovations in body plan (Section 23-3)

The Human Body

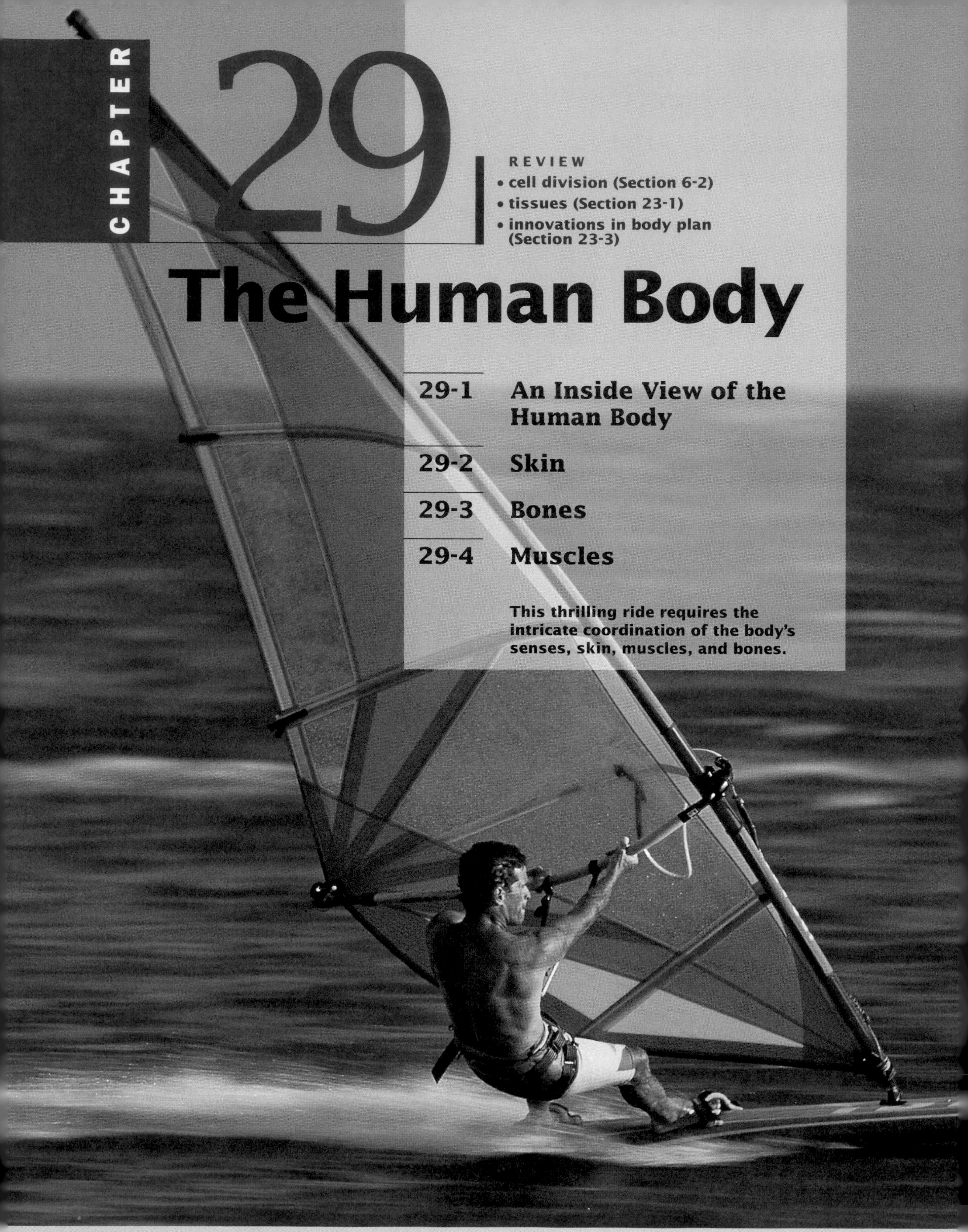

This thrilling ride requires the intricate coordination of the body's senses, skin, muscles, and bones.

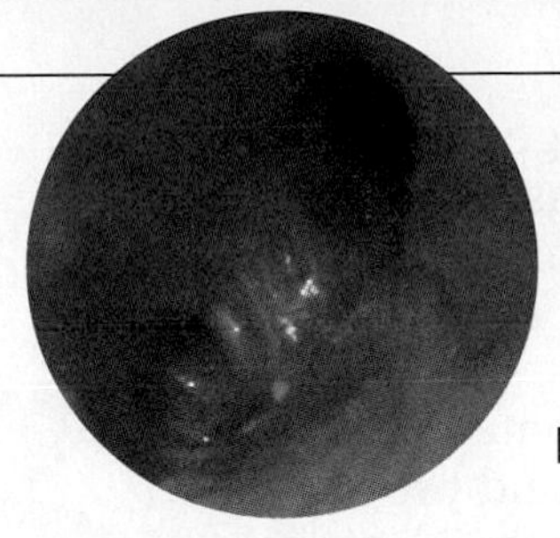

Your body consists of many specialized tissues, such as those seen in this view of the body's airway to the left lung. These tissues, each made of specialized cells, are arranged differently in each part of your body. This process of specialization and arrangement began before you were born, when a single cell began to divide.

29-1 An Inside View of the Human Body

OBJECTIVES

1. **Explain the roles of the four kinds of tissues in the human body.**
2. **Classify connective tissue into three types, and name common examples.**
3. **Identify three organ systems that protect or support the body.**

Similar Cells Form Tissues

All of the trillions of cells in the human body arise from a single cell—the fertilized egg. The first few cells that form by cell division after fertilization look alike. Soon, however, the new cells become specialized to carry out particular tasks.

As cells specialize, they begin to group together to form tissues. A **tissue** is a group of similar cells that work together to perform a specific function, such as movement or protection. Your body contains four types of tissue—epithelial tissue, muscle tissue, nervous tissue, and connective tissue, as shown in **Figure 29-1**.

Epithelial tissue covers and protects

Epithelial *(ehp uh THEE lee uhl)* **tissue** is made of tightly connected cells that are arranged in flat sheets, often only a few cells thick. Just as canvas protects machinery that must be stored outdoors, epithelial tissue prevents damage to the cells that lie beneath it. Skin is an example of epithelial tissue. Epithelial tissue also lines spaces within the body and covers the inner and outer surfaces of your internal organs.

Some epithelial tissue has an entirely different function. This epithelial tissue forms exocrine glands. An exocrine gland is a cell or group of cells that produces and releases secretions onto a body surface. One type of exocrine gland in your skin secretes sweat, which helps cool your body. Exocrine glands in the lining of your digestive system produce enzymes that digest food.

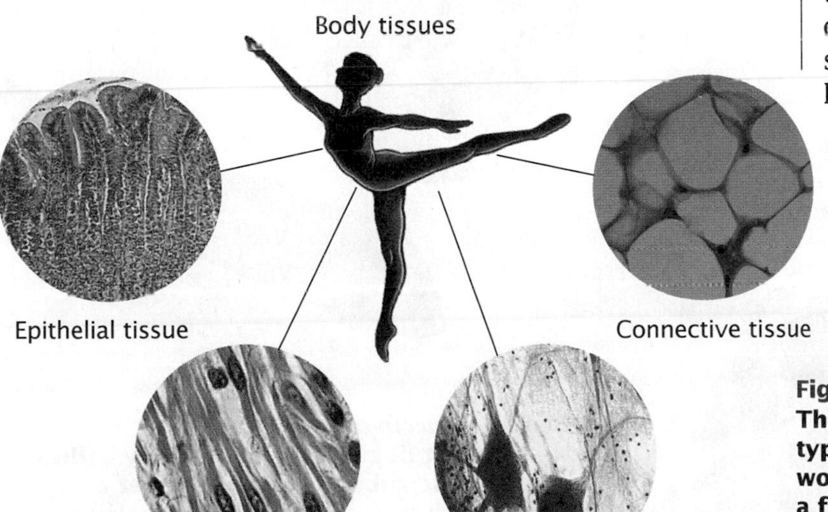

Figure 29-1
The human body is made of four types of tissue, shown at left. All work together to make the body a functional unit capable of all life activities.

Muscle tissue moves the body

Have you ever awakened with sore muscles? If so, you probably became aware of how often you use those particular muscles. **Muscle tissue** moves the parts of your body. Even when you are sitting still, muscles are at work maintaining your posture, moving food through your digestive system, and moving blood through your heart.

Muscle tissue is made of cells that contract and then relax to return to their usual length. Another characteristic of muscle tissue is that it responds to electrical stimulation. Electrical signals control when a muscle will contract or relax.

Nervous tissue sends electrical signals through the body

As you walk up a flight of stairs, the electrical signals that cause the muscles in your legs to contract are produced by nerve tissue. **Nervous tissue** is found in your brain, nerves, and sense organs. It contains cells called neurons that generate electrical impulses and transfer the impulses to other cells. Neurons have long extensions that carry electrical impulses long distances. Nervous tissue also contains other cells that nourish and protect the neurons that carry the electrical signals.

Connective tissue joins, supports, and transports

Connective-tissue cells are spaced farther apart than cells of most other tissues. **Connective tissue** actually contains few cells. Some connective tissues are fluids; others are more like crystals. There are three types of connective tissue: defensive, structural, and sequestering. Sequestering tissues store, or "sequester," particular compounds. Defensive tissue defends the body; structural tissue supports it. **Figure 29-2** describes these types of connective tissue and gives examples of each.

Figure 29-2
The graceful leap of a dancer depends on many types of tissue, including muscle, nervous, and epithelial tissues. Examples of these tissues and all three types of connective tissue are shown below and to the right.

a ***Defensive connective tissue***
Defensive connective tissue defends the body from invaders. This kind of tissue circulates in the blood, protecting the body from harm. White blood cells are a common example.

b ***Structural connective tissue***
This tissue gives the body its support. Bone and softer connective tissues like cartilage, ligaments, tendons, and fibroblasts are examples. Fibroblasts secrete the protein strands that help wounds heal.

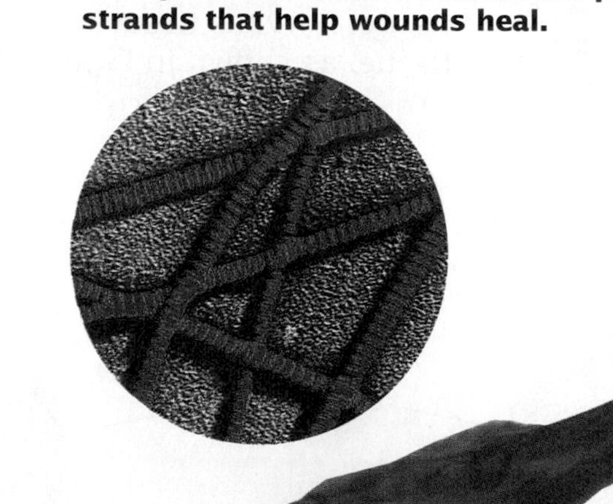

c ***Sequestering connective tissue***
These tissues act like storehouses. They collect and store specific substances such as fat or melanin in the skin. Some transport particular molecules. For example, red blood cells store hemoglobin for transporting oxygen throughout the body.

Tissues Form Organs

How can only four kinds of tissue make up all the different organs in your body? The answer is simple: different tissues are combined in different ways to form each organ. An **organ** is a structure composed of a number of tissues that work together to perform a specific job in the body.

For example, the largest organ of your body—your skin—contains all four types of tissue. Your skin is made of many watertight sheets of epithelial cells covering a cushioning layer of connective tissue. Nervous tissue in your skin makes it sensitive to touch. Tiny muscles can make the fine hairs that cover most of your body stand on end when you are cold. All of these tissues cooperate to protect your body.

Organs that work together form an **organ system**. An example of an organ system is the circulatory system. The heart and the blood vessels are separate organs that work together to carry substances through the body. **Figure 29-3** shows the three organ systems that protect and support the body or produce movements. You will learn more about these organ systems in this chapter. Later chapters will explain other systems.

Figure 29-3
Each organ in this dancer's body performs a specific task to help other organs in its system. For instance, her muscular and skeletal systems support her; her muscles move her bones; and her skin protects all the other systems and keeps her from overheating.

SECTION REVIEW

1. **How does each of the four types of tissue function in the human body?**
2. **How might a disease that slowly destroys nerve cells affect muscle tissue in the body?**
3. **In what ways does connective tissue help the body?**
4. **How do organ systems protect and support the human body?**

Skin is a lot more than just an outer covering. It is the largest—and one of the most important—organs of the body. It cushions and protects internal organs and prevents the loss of the fluids that bathe all your cells. Small cuts, abrasions, and burns are repaired with remarkable speed. Old skin cells are continually replaced by new, healthy cells.

29-2 Skin

OBJECTIVES

1. **Compare and contrast the dermis and the epidermis.**
2. **Describe the functions of the different components of the dermis.**
3. **Explain why acne is a common problem for adolescents.**
4. **Recommend three ways to reduce the risk of skin cancer.**

The Dermis

Your skin is composed of two main parts: an outer layer called the **epidermis** and an underlying, thicker layer. This inner part of your skin is called the **dermis**. Shown in **Figure 29-4**, the dermis is composed mainly of connective tissue. Fibers in the dermis make your skin tough, flexible, and elastic. The dermis also contains nerves, muscles, blood vessels, hair follicles and glands.

Nerves and muscles run throughout your skin

When a friend taps you on the shoulder or you accidentally prick yourself with a pin, you immediately react. Nerves in the dermis make it possible for you to respond appropriately with a "hello" or an "ouch." These nerves enable you to sense pressure, temperature, and, of course, pain.

Figure 29-4
On the surface, skin looks uniform. But underneath, it is a complex organ made of blood vessels, nerve fibers, glands, and muscles.

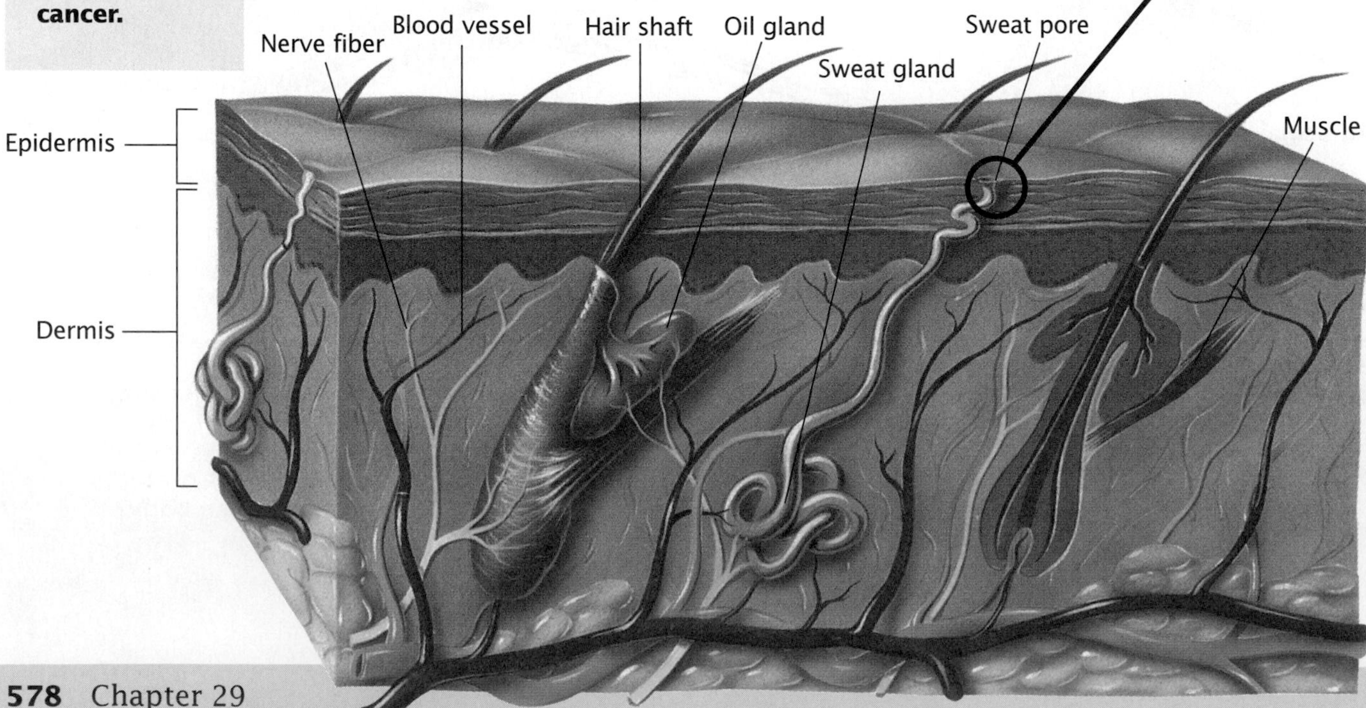

Figure 29-5
The evaporation of sweat from the surface of your skin helps cool your body during and after a tough workout.

Your dermis also has tiny muscles that are attached to the hairs in your skin. When you are cold or afraid, the muscles contract, pulling the hairs upright. This same process happens in the skin of other mammals. The fur of a cat, for instance, will stand up when the cat is threatened by a dog, making the cat look larger and more dangerous. When the cat is cold, its fur fluffs up and traps more air near its body. Because trapped air is a good insulator, this helps the cat stay warm. The muscles in your skin behave just like those of a cat. However, because you don't have fur, you just get goose bumps—a leftover from our evolutionary past.

Stability and Homeostasis

How does sweating enable the human body to maintain homeostasis?

Blood vessels carry nourishment, remove wastes, and cool the body

Your skin, like all other living parts of your body, requires nourishment to live. This nourishment is supplied by blood that courses through tiny blood vessels in the dermis. In addition to carrying nutrients, the blood in these vessels carries away waste products and helps regulate body temperature. Blood radiates heat into the air as it passes near the surface of the skin. If your body becomes too hot, the tiny blood vessels enlarge, allowing more blood to flow through the dermis near the body surface. This increased flow of blood is easy to see in light-skinned people as their skin becomes reddish during strenuous exercise.

Sweat glands assist in cooling

Sweat is another way that your body removes excess heat. Your skin contains about 100 sweat glands per square centimeter. The evaporation of sweat from the surface of your skin removes heat much more efficiently than simply radiating heat from the blood into the air. Without sweat, you would have great difficulty cooling your body on a hot day or after exercising like the person in **Figure 29-5**.

Not all sweat is the same. Most sweat is about 99 percent water, mixed with small amounts of salts, acids, and waste products. However, the sweat from certain sweat glands, called apocrine glands, also contains proteins and fatty acids. These substances provide a rich source of food for bacteria that live on the skin. The waste products of these bacteria give sweat an unpleasant odor, particularly when a person's diet is high in protein and fats from animal sources. Most apocrine glands are located in the armpits and groin area. You have probably noticed that sweat glands are activated by nervousness or stress, even in cool temperatures.

The Epidermis

Not all of your skin is alive. The outermost layer of your skin is made of layers of dead cells. These cells were formed two to four weeks earlier as new epithelial cells in the innermost layer of the epidermis, a region called the basal layer. After the cells are formed, they are pushed outward by new cells forming underneath. As they move outward, many cells become flattened and begin to make a water-insoluble protein called keratin. The epithelial cells also move farther from the nutrients that diffuse from the blood flowing through the dermis. When the cells are pushed too far from their nutrient supply and are filled with keratin, they die. The dead epithelial cells become part of the skin's "surface armor," absorbing the wear and tear of life until each cell is eventually shed to make way for its replacement. The dead epidermal cells, shown in **Figure 29-6a**, are packed with the waterproof protein called keratin and are ready to be shed. Keratin works with the oil glands in the dermis to waterproof your skin. It also prevents the loss of water by evaporation from your skin and body.

Hair and nails are dead cells

A strand of hair is made of dead, keratin-filled cells that overlap like roof shingles, as shown in **Figure 29-6b**. Hairs grow from specialized epidermal structures called hair follicles. Each hair on your head grows for several years before its follicle enters a resting phase for a few months. While the follicle is dormant, the hair falls out. The length of time that a hair follicle is active before the hair falls out is a genetic trait. If you have inherited a short cycle of hair growth, you will never be able to grow very long hair, no matter how carefully you care for it.

Like hair, nails are also produced by specialized epidermal cells, as described in **Figure 29-6c**.

Inner epidermal cells are very alive

The inner layer of living epidermal cells is one of the most active regions of cell division in the body, even though the epidermis is the thinnest layer of skin on your body. Actually, there are many layers of cells and many shapes. As the cells migrate to the upper surface, some change shape. Some cells assume new roles. Other cells adapt for protection.

Most of the living epidermal cells sit on the basement membrane of the epithelium and produce all of the new epidermal cells. These are called basal

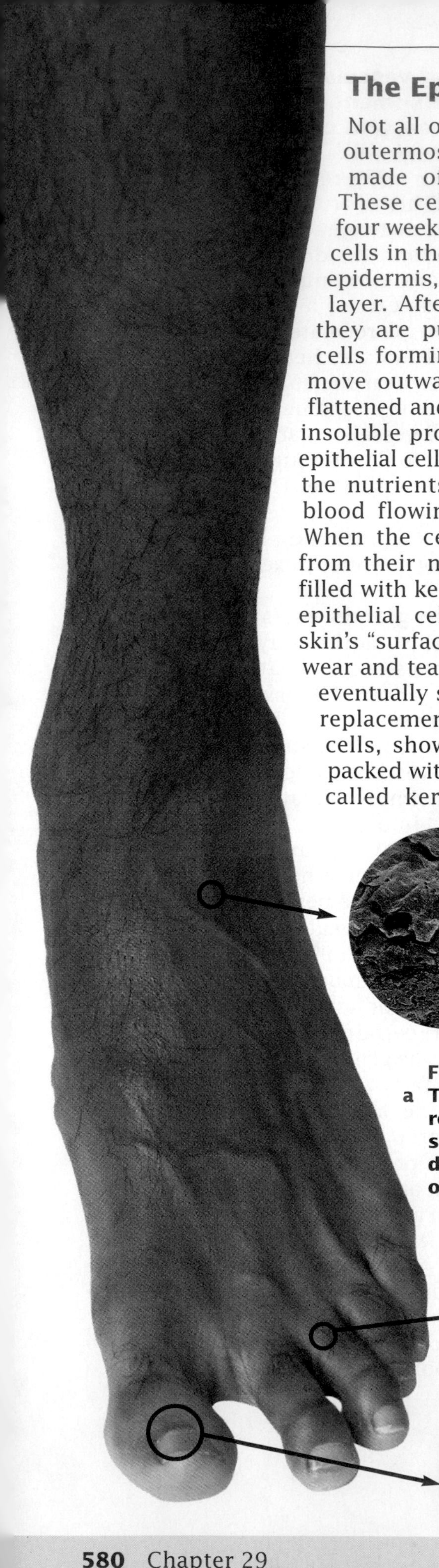

Figure 29-6

a This SEM of the epidermis reveals that by the time a skin cell is shed, it is a dead, flattened scale made of keratin.

b The center of each hair is filled with melanin and air bubbles. Dark hair contains more melanin than blond hair.

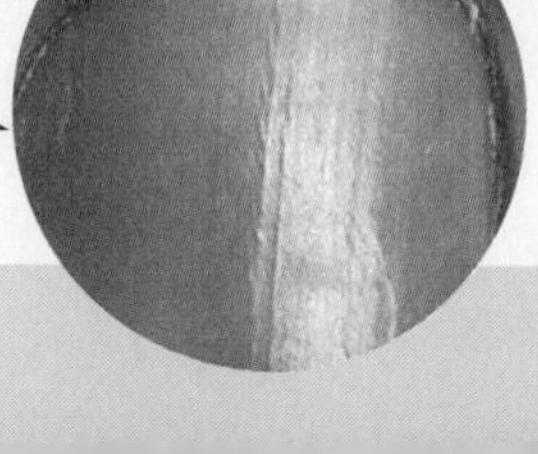

c In nails, new cells are produced in the white half-moons. These cells fill with keratin as they push cells that were produced earlier toward the free edge of the nail.

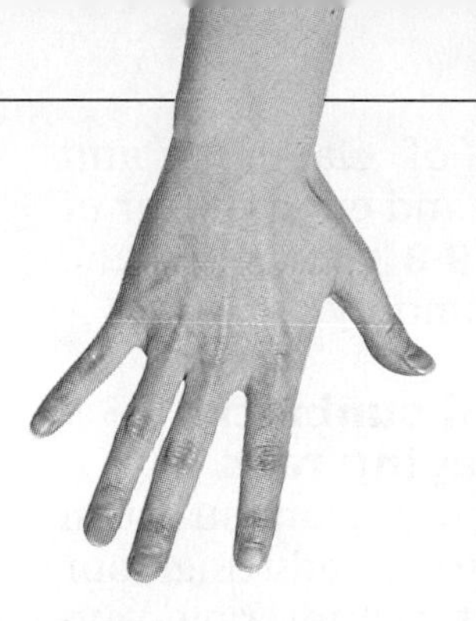
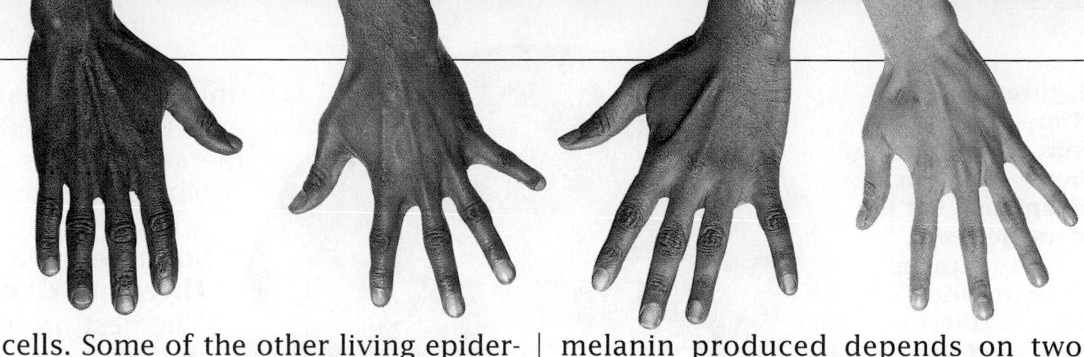

Figure 29-7 Variety in skin colors is caused by the pigment melanin. The amount of melanin varies from person to person.

cells. Some of the other living epidermal cells store chemicals known as provitamins, which are activated by sunlight to form vitamin D. Fifteen minutes of sun exposure two to three times a week activates this vitamin, which helps to calcify and harden your bones. Still other living epidermal cells in the innermost layer produce the pigments that give skin its color.

Pigments color the skin and help protect it

Human skin color ranges from pale pink to almost blue-black. Skin is mainly colored by cells that produce a brown pigment called **melanin** *(MEHL uh nihn)*. These cells are found in the lowest epidermal layer. The amount of melanin produced depends on two things. First, melanin concentration is a genetic trait that varies from person to person. For example, melanin is missing in the skin of albinos but is abundant in black skin. Second, the amount of melanin in the skin depends on the length of time one is exposed to ultraviolet radiation. Ultraviolet radiation activates melanin to produce a darker coloration, commonly called a "tan." The longer you are exposed to the sun, the darker your melanin becomes. This darkening prevents you from further burning. Melanin, then, provides some protection from the sun's rays. **Figure 29-7** shows a variety of people with different levels of melanin.

Skin Disorders

Your skin is the most exposed part of your body. It serves as a protective barrier. Nevertheless, skin can be, and often is, damaged. The damage may be minor—a blister, an insect bite, or a small cut. Other damage is more serious, as in the case of skin cancer. Your skin may also be affected by changes within your body, by microbes, and by physical and chemical hazards in the environment.

Oil glands, heredity, and bacteria interact to produce acne

Oil glands in your skin produce oils and release them into nearby hair follicles. Normally the amount of oil produced is just enough to waterproof your hair and to seal moisture in the skin. However, high levels of sex hormones produced during adolescence increase oil production in these glands. The oil may combine with cellular debris and bacteria to clog follicles, resulting in the skin eruptions that are so common during the teen years.

Not all skin eruptions are diagnosed as acne. Some eruptions, known as whiteheads, are merely the build-up of oil. If a whitehead is invaded by bacteria and then exposed to air, it darkens and forms a blackhead. If the oil builds up or if the area is squeezed to the point that the oil gland actually bursts, the area around the gland becomes red and inflamed. The pimple that forms is evidence that the body is fighting the tissue damage and the invading bacteria.

Heredity determines the amount of oil and hormones produced. Most acne cures itself over a few months or a couple of years, but proper cleansing and daily skin care will certainly reduce the conditions leading to bacterial infections, whether or not acne is diagnosed.

Figure 29-8
Time spent in the sun or in a tanning bed can produce deep tans. But the damage may appear decades later as wrinkles, leathery skin, and skin cancers.

Skin changes are linked to ultraviolet radiation

Many changes occur in the skin after exposure to ultraviolet radiation. This form of radiation, often called black light, is invisible to the human eye and is produced by the sun's rays or by lamps that produce certain wavelengths of light. Radiation of this kind affects only the layers of the epidermis because the rays have low penetrating power.

Ultraviolet rays are classed as ultraviolet A rays (UVA) and ultraviolet B rays (UVB). UVB rays burn the skin. This is one of the first signs that one has been exposed to the sun too long. UVA rays, on the other hand, result in deeper "tanning" without burning; there is no warning that overexposure has occurred.

Overexposure to sunlight or tanning devices also results in a thickening of the skin, more rapid skin aging (including a loss of elasticity and greater wrinkling), and even cancer of the skin. **Figure 29-8** shows a scene where skin is commonly damaged.

Sunscreens and sunblocks filter out damaging rays

The best protection from sunburn and overexposure is reducing your exposure to ultraviolet rays and remaining covered when you are in the sun. You can reduce your exposure to ultraviolet radiation by doing the following:

- Stay out of the sun between 10 A.M. and 3 P.M.
- Cover up with hats, UV-filtering eyeglasses, and clothes when outdoors.
- Take care even on cloudy days. You can still burn.
- Avoid tanning beds and lamps.

Protection from overexposure can be achieved by using sunscreens and sunblocks. A product's ability to block ultraviolet light is rated by its SPF value. SPF stands for sun protection factor. Sunscreens block the shorter UVB rays, which can burn the skin. Sunblocks also block the UVA rays, which damage the deeper layers of the epidermis without burning.

Information on a product's label will indicate if the product blocks both UVA and UVB rays. Look for chemicals containing a mineral like titanium or zinc. These chemical combinations shield the skin from UVA rays and are found in products called broad-spectrum blocking agents. **Figure 29-9** illustrates the different SPF values found in lotions and creams and describes what each value means. A sunblock with an SPF of at least 15 is recommended. Reapply often because sweating and swimming wash the product away.

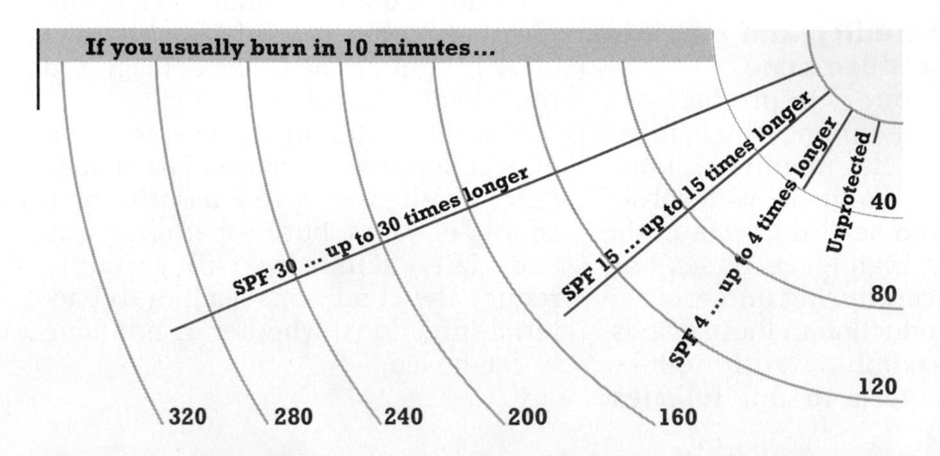

Figure 29-9
SPF stands for sun protection factor. The SPF number indicates the product's ability to block UVB rays only. It tells you how much longer you can stay in the sun without burning in relation to when you would burn without protection. Broad-spectrum sunblocks of SPF 15 or greater may give the best protection from both UVA and UVB rays.

Figure 29-10
Moles exhibit characteristics that can be monitored by using the ABCD rule for dangerous changes that may signal skin cancer.
A—Asymmetry
B—Border irregularities
C—Color changes and differences
D—Diameter increases

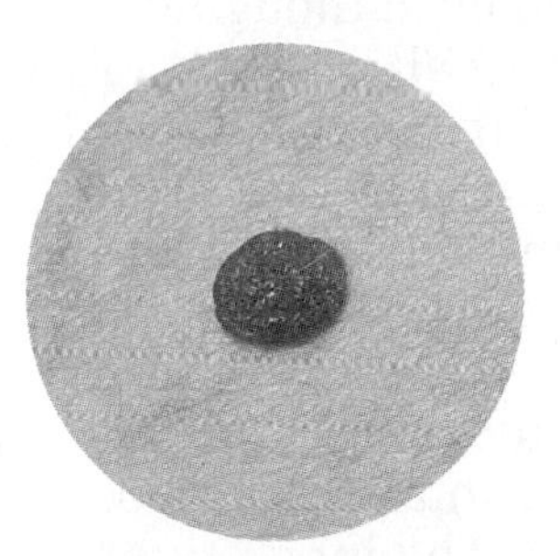

a This is a normal mole. Each half matches the other.

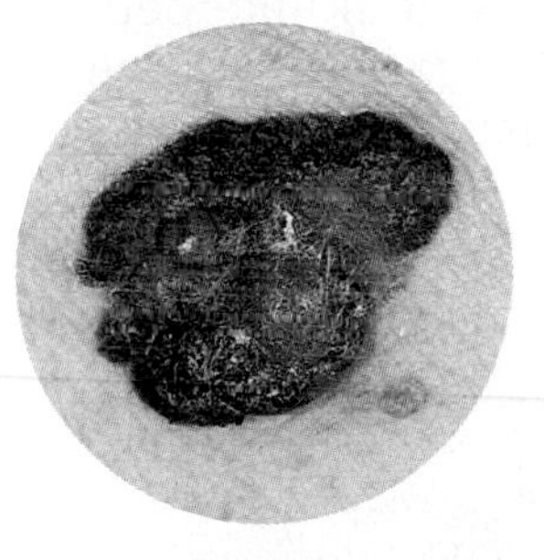

b This mole has two halves that do not match, a characteristic called asymmetry.

Changes in epithelial cells give rise to skin cancers

Continued exposure to ultraviolet light can trigger cellular mutations in DNA, which can then cause cells to lose the ability to stop dividing. This uncontrolled cell division may produce tumors, which can spread to other parts of the body and are then called cancers.

The danger of a particular skin cancer depends on the type of epidermal cell that becomes cancerous. The most common skin cancers have a very high cure rate when detected early. These include cancers that usually form around the ears or on the face in the basal cells of the epidermis. Once thought of as a disease of middle-aged or older people, basal cell cancer is being seen more often among younger people. This rise in the rate of basal cell cancer is believed to be primarily due to sunbathing, especially as more of the Earth's ozone layer is destroyed.

A second type of cancer begins in the flattened cells of the epidermis on sun-exposed areas or within skin lesions like scars or burns. Some of these cancers begin as flat, scaly, reddish patches on the skin, an indication that the skin has been damaged. These cancers are more invasive than the previous kind and are more likely to spread, although they are not as common as the first kind.

About 1 percent of all cancers begin in the melanin-producing cells. Called malignant **melanomas**, these cancers account for 75 percent of the deaths from skin cancers. Certain types of moles make it more likely that a person will develop a melanoma. **Figure 29-10** describes the difference between a melanoma and an ordinary mole. Checking moles, spots, and birthmarks monthly is recommended. If these areas change or if patches of red, scaly skin are found, a doctor should be consulted.

Sunburn before age 18 increases risk of skin cancer later in life

What you do now as a teen affects how your skin may change in the future. Your risk for skin cancer and melanoma is higher if you

- experienced a severe, blistering sunburn before age 18,
- have used or now use tanning beds or sunlamps,
- sunbathe or work in the sun,
- have naturally blond, red, or light-brown hair,
- have freckled or fair skin, or
- wear sunscreen less than half of the time outdoors.

Knowing how ultraviolet radiation affects skin and how to reduce exposure may help you to maintain your skin's homeostasis.

A faulty gene may allow skin cancer to develop

The body cells contain a gene called p53 that acts as a tumor suppressor. This gene is believed to be activated in normal skin cells when UV light damages the cells' DNA. Following the damage, the p53 gene directs the injured cells to "commit suicide." As the cells die, they are shed from the skin. It appears, though, that repeated exposure to the sun damages the p53 gene. When a damaged p53 gene gets zapped with UV light, it does not direct cells to die as it normally would. Instead, the cells with damaged DNA continue to survive and divide. Eventually they may become pre-malignant skin patches that can later form cancer.

SECTION REVIEW

1. **Which part of the skin has the simpler structure, the dermis or the epidermis? Explain your answer.**
2. **What do nerves, blood vessels, glands, and connective tissue contribute to the skin?**
3. **How could keeping your skin very clean help minimize the effects of acne?**
4. **What factors increase a person's risk of developing skin cancer?**

You may think of your skeleton as just a rigid framework of bone that enables you to sit, stand, or run. Bone itself may appear as lifeless as rock. However, bone is in fact a dynamic connective tissue made of living cells. Throughout your life, these cells continue to produce the many fibers and minerals that fill the spaces between the cells.

29-3 Bones

OBJECTIVES

1. **Draw a diagram of a typical long bone, and identify changes in bone from infancy to old age.**
2. **Differentiate a fracture from a sprain.**
3. **Discuss the causes and effects of osteoporosis.**
4. **Compare the actions of five types of joints in your body, and explain the movement that each permits.**

Bone Structure and Growth

Throughout your childhood and adolescence, bone cells build more and more bone as your body grows. Compare the size and shape of the infant's skull with that of the adult's skull in **Figure 29-11**. Even in adults, specialized bone cells continue to break down and rebuild bone tissue. For example, the bone tissue at the end of the thighbone where it joins the knee is completely replaced every six months.

Four minerals are built into the bone cells to make bones strong. These minerals—calcium, phosphorus, magnesium, and manganese—also regulate an amazing variety of activities in your body. For example, nerves and muscles cannot work without the proper balance of calcium and magnesium. Bones act as a warehouse for storing calcium and phosphorus. And even though the body stores a very small amount of manganese, some of the highest concentrations of manganese are stored in bone. The storage and release of minerals, especially calcium, in your bones help maintain a precise level of minerals inside your body.

The human body contains bones of all shapes and sizes. The long bones of your arms and legs are shaped like cylinders. Curved, flat plates of bone form the part of the skull that protects your brain. Wrists and ankles contain many small bones that look like pebbles, and bones of the face and spine have unusual, irregular shapes.

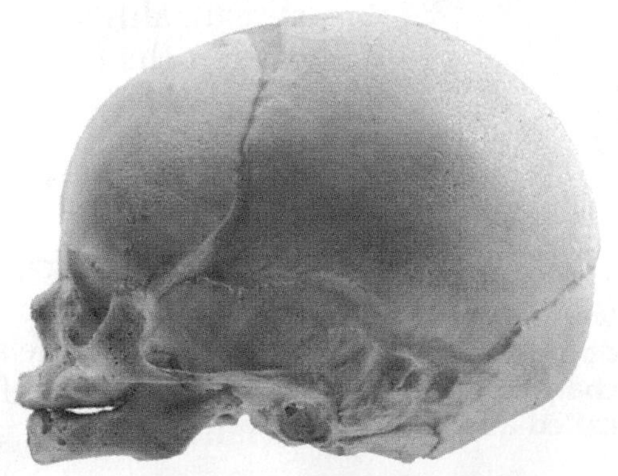

b Infant skull

a Adult human skull

Figure 29-11
The skull of an infant has more separate bones than that of an adult. As the child grows, many of these bones will fuse together. The "channels" you see in the infant skull will disappear by the age of four.

Bone growth begins with cartilage

Bone growth begins long before birth. The basic shape of a long bone, such as an arm bone, is first formed in cartilage. Later, the cartilage cells begin to be replaced by cells that form bone. While you are growing, long bones still have a region of cartilage near each end that allows bones to grow longer. When you reach your adult size, these regions, too, are converted to bone.

Figure 29-12 shows a cross section of the femur in the thigh, a typical long bone. The outer layer of a long bone consists mainly of minerals and mature bone cells "trapped" by the minerals that they have deposited. To form this dense outer layer, bone cells deposit minerals in concentric rings, leaving a canal in the center of each group of rings. Each canal contains a small blood vessel that carries nutrients to the bone cells.

Inside the ends of a long bone are bone cells and minerals with large spaces in between, like a sponge. These spaces, and the entire center of the long middle part of the bone, are filled with **marrow**. The marrow inside long bones produces blood cells in newborns. As you get older, this marrow gradually changes its job to storing fat. In adults, most blood cells are produced in the marrow of flat bones like the sternum, or breastbone.

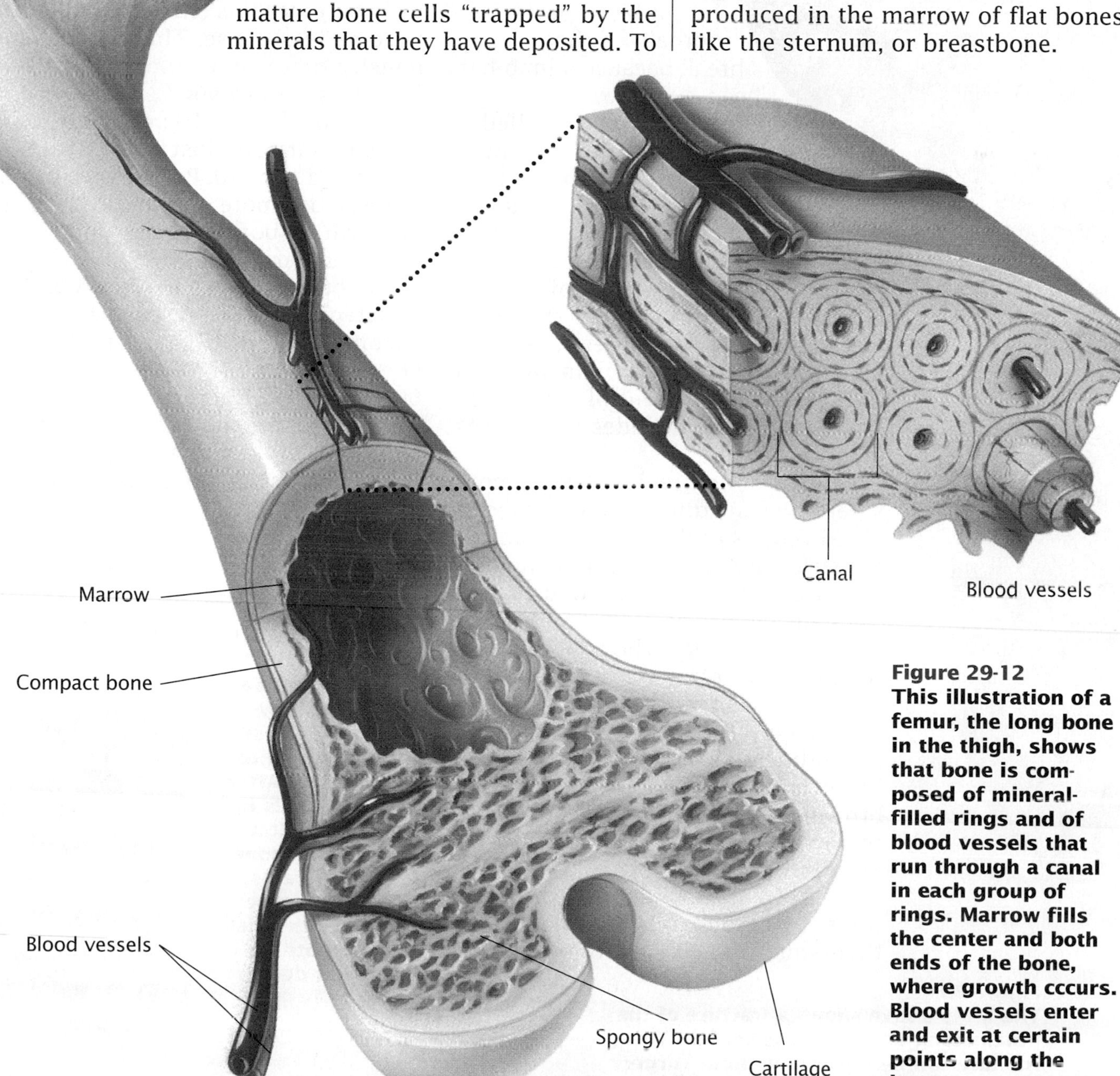

Figure 29-12
This illustration of a femur, the long bone in the thigh, shows that bone is composed of mineral-filled rings and of blood vessels that run through a canal in each group of rings. Marrow fills the center and both ends of the bone, where growth cccurs. Blood vessels enter and exit at certain points along the bone.

Bones break under stress

The minerals deposited in bone tissue make bones hard and rigid. These characteristics enable bone to protect and support the body. Although the fibers in bones make them much less brittle than a piece of chalk, severe stress placed on a bone may cause it to break. For example, certain types of falls while snow skiing can cause the bones in the lower leg to break at the top of the ski boot. In contact sports like football, a bone may break because a limb has been twisted.

A broken bone is called a **fracture**. A fracture may be a simple crack, or the bone may actually break into two or more pieces, as shown in **Figure 29-13**. The most serious type of fracture is a compound fracture, in which pieces of broken bone protrude through the skin. One reason compound fractures are so serious is that they may result in infection of the bone. Treatment of compound fractures usually includes large doses of antibiotics.

When a bone breaks, there is considerable bleeding from damaged blood vessels in the bone itself and in the surrounding tissues. Healing of a fracture begins as the blood in this swollen region around the broken bone begins to clot. The bone tissue is then rebuilt between the two broken ends of bone in much the same way as bone tissue forms before birth. The rebuilding is not always perfect, however. Many people who have broken a bone can still feel a thicker region of bone where the fracture occurred.

Bone fractures heal at different rates. Large bones heal more slowly than small bones, and the bones of young people heal more quickly than those of older people. Holding the broken ends close to each other and keeping them completely still speeds healing of bones. This is why bone fractures are often treated by encasing the fractured bone in a cast.

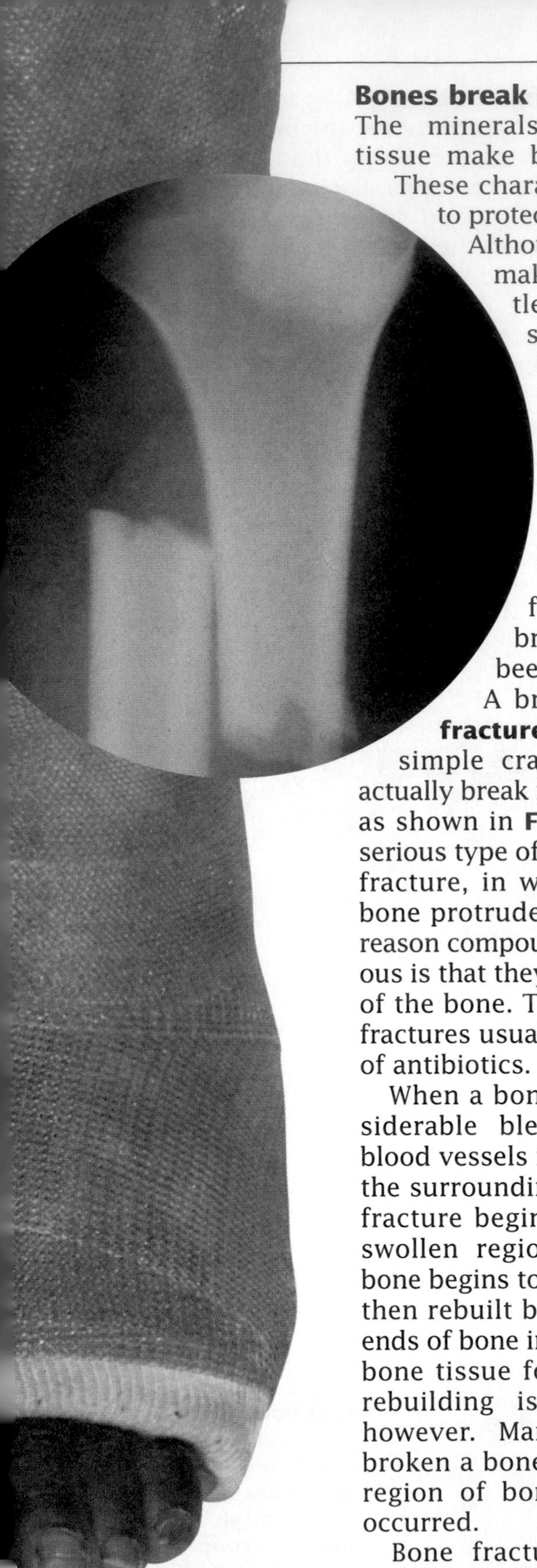

Figure 29-13
The X ray above shows a fracture of the tibia, a long bone in the lower leg. This fracture can be repaired without surgery by setting the fracture with a cast.

Osteoporosis causes bones to become brittle

As bones grow longer, they also grow thicker and denser. In young adults, bone density stays relatively constant as bone tissue is broken down and replaced at a steady rate. From about the age of 40, bone replacement gradually becomes less efficient, and bones become less dense. The loss of bone density that may eventually result is called **osteoporosis** *(ahs tee oh puh ROH sihs)*. About 5–10 percent of bone mass is normally lost every 10 years after the age of 40. People with osteoporosis lose more. Compare a healthy bone with a bone that has undergone severe mineral loss after the onset of osteoporosis, shown in **Figure 29-14**.

Osteoporosis can cause bones to become light, brittle, and easily broken. In the United States, more than 600,000 bone fractures a year result from osteoporosis. Severe osteoporosis in the bones of the spine often changes the posture of very old

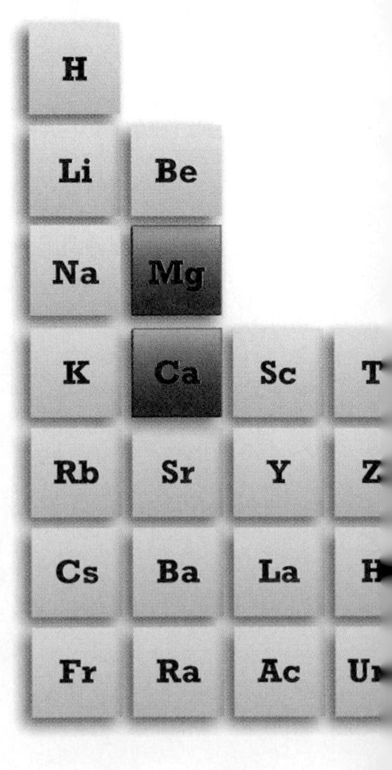

Figure 29-14
You may think that osteoporosis is something that you won't have to think about for a long time. However, bone density increases only to the mid-30s. After that, increases in bone mass slow down, while bone loss accelerates. Regular exercise and a healthy diet will promote the growth of denser bone.

people. Although both men and women lose bone as they age, women are at a greater risk for osteoporosis for two reasons. First, women's bones are usually smaller and lighter than men's bones. Therefore, the loss of the same amount of bone in a woman and a man results in thin, fragile bones in the woman, while the man's bones might still be quite strong. Second, the production of female sex hormones declines rapidly during menopause. Because sex hormones help to maintain bone density, this decline in hormone production increases the rate of bone loss.

Researchers have found that exercise can increase the amount of minerals deposited in bone. For people in their teens and twenties, regular exercise and a nutritious diet that includes a balance of all minerals, including calcium and magnesium, can actually increase bone density. For older people, regular exercise can slow the bone loss that can lead to osteoporosis. That's good news for the tennis player in **Figure 29-14**. Other factors that increase the amount of calcium available to the body and that increase bone density include adequate dietary protein, vitamin D, and digestive acids in the stomach.

a This woman is over 60 years old and is still very active. Regular exercise at any age can slow the bone loss that leads to osteoporosis.

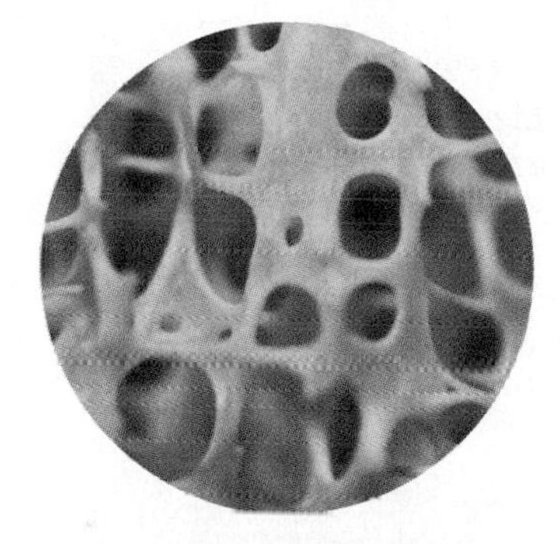

c In healthy bone tissue (magnified 10× in this photograph) minerals are continuously replaced, so the bone remains strong.

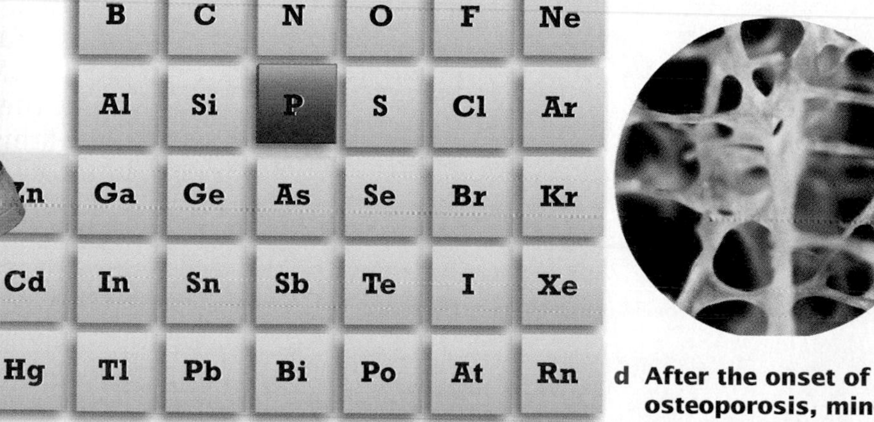

b The elements highlighted in the periodic table represent the four primary minerals built into strong, healthy bones. Eating foods that provide these minerals will help to prevent osteoporosis.

d After the onset of osteoporosis, minerals are not replaced as rapidly. As a result, bones become brittle and can break easily.

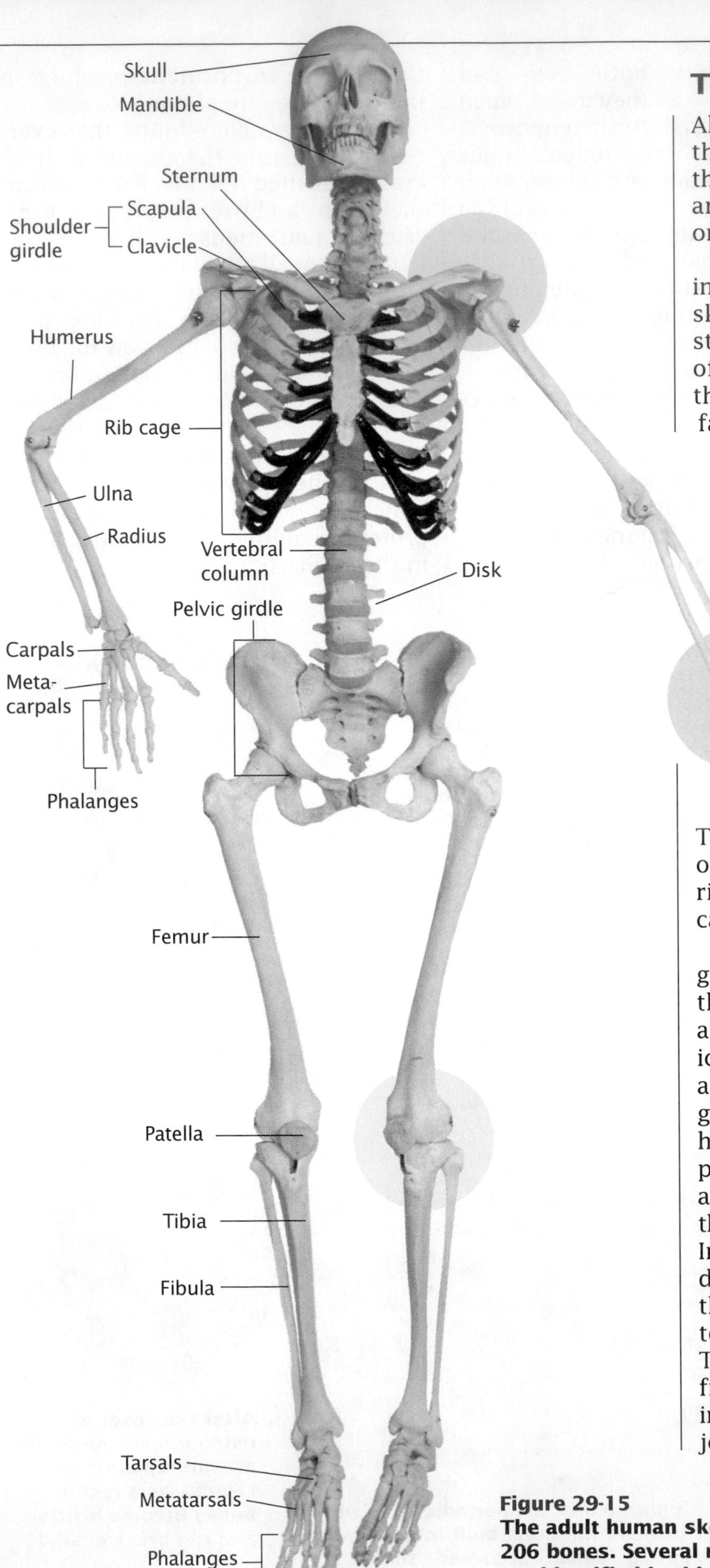

The Skeleton

All of the bones in the body make up the skeleton. The skeleton supports the body's weight, enables it to move, and protects many of its internal organs.

A typical human skeleton is shown in **Figure 29-15**. The central part of the skeleton consists of the skull, spine, sternum, and ribs. The skull consists of many fused bones that protect the brain and form the shape of the face. The lower jaw, the mandible, is the only bone of the skull that moves easily and visibly. The spine is made up of vertebrae that support the trunk and allow flexibility. In between the vertebrae are fluid-filled disks. These disks act as shock absorbers and prevent the vertebrae from rubbing against one another. The vertebrae surround and protect the spinal cord. Attached to the spine are 12 sets of ribs. The ribs protect the heart, lungs, and other organs in the chest cavity. Some ribs are attached to the sternum by cartilage.

Two frameworks of bone, called girdles, connect the arms and legs to the central skeleton. The legs are attached to the pelvic girdle. The clavicle and scapula (shoulder blade) move as a unit and are called the shoulder girdle. The attachment between the humerus and the clavicle is the only point of attachment between the arms and the main skeleton. This permits the arms and shoulders to move freely. In contrast, hipbones in the pelvic girdle attach directly to the lower part of the spine, enabling the hips and legs to bear the full weight of the body. This is why the legs cannot move as freely as the arms. The colored circles in **Figure 29-15** match those of specific joints in **Figure 29-16**.

Figure 29-15
The adult human skeleton has 206 bones. Several major bones are identified in this skeleton.

Figure 29-16
A joint is where two or more bones connect. Some examples of joints are shown here, along with the kinds of movement they allow. The wires, screws, and bolts holding these bones together represent the ligaments in the body.

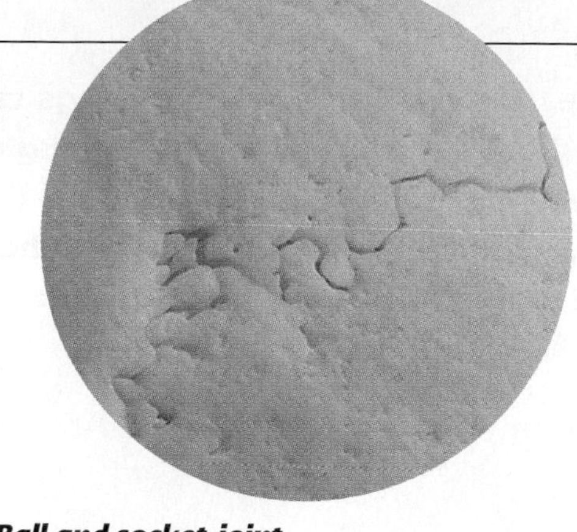

a *Suture joint*
The skull is virtually immovable, shifting unseen in its own unique rhythm. Fusion of its bones prohibits other movement.

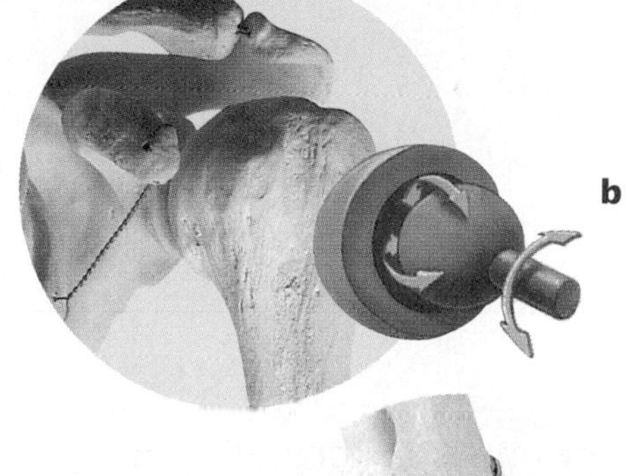

b *Ball-and-socket joint*
The shoulder enables your arm to move freely in all directions.

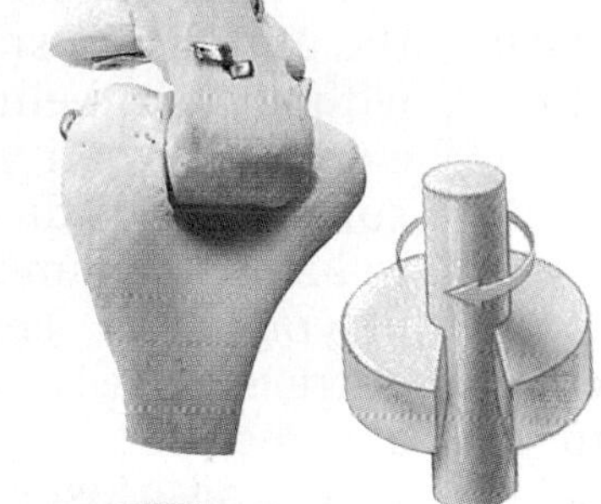

c *Pivot joint*
A forearm bone pivots at the elbow and enables your hand to turn over.

d *Plane joint*
The lower arm has bones that glide over those of the hand.

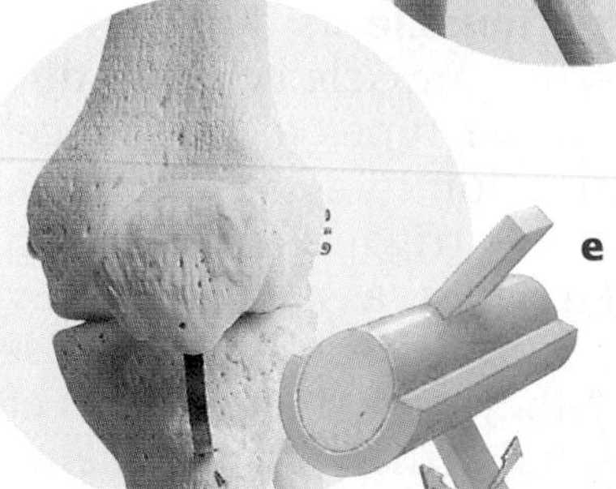

e *Hinge joint*
The knee enables you to flex and extend your lower leg.

Bones fit together with joints

The place where two or more bones connect is called a **joint**. Joints vary greatly in their flexibility. For example, the shoulder and hip can be moved in many directions, but the joints in your spinal column can move only slightly. Some examples of joints are shown in **Figure 29-16**.

Bones are joined to each other by strong elastic bands of connective tissue called **ligaments**. Although ligaments are very elastic, there is a limit to how far they can stretch. When ligaments are stretched too far, the injury that results is called a **sprain**. If the ligament is actually torn by overstretching, it will not heal and must be repaired surgically.

The hip is a particularly complex joint where the leg is attached to the pelvic girdle. Because of the upright posture of humans, the hip must support the body's full weight and allow leg movement in all directions. Many ligaments are needed to connect the pelvic and leg bones to one another at this joint and to make the joint as stable as possible. Proper placement of the pelvis and knees plus strong ligaments and muscles are necessary to prevent injury to the back and knees when lifting heavy objects.

SECTION REVIEW

1. Draw a diagram of a typical long bone, and label the dense bone, spongy bone, marrow, and blood vessels.
2. How do fractures and sprains differ?
3. What actions can you take now to prevent osteoporosis later in life?
4. Identify five different types of joints in your body, and give an example of each.

Large, powerful muscles in the legs can propel an athlete high enough to slam-dunk a basketball or fast enough to run 100 m (328 ft) in less than 10 seconds. Smaller muscles enable your eyes to read the words on this page. A smile or a frown would not be possible without the action of more than a dozen muscles beneath the skin of your face.

29-4 Muscles

OBJECTIVES

1. **Compare and contrast the three types of muscles.**
2. **Explain how muscles work to move a bone.**
3. **Describe how different forms of exercise affect muscles.**
4. **Discuss the negative effects of taking anabolic steroids.**

The Actions of Muscles

Your body has more than 600 muscles, each containing thousands of muscle cells. A muscle can move a part of the body when electrical signals cause the muscle cells to contract and change in length. For most of the muscles in your body, these electrical signals are provided by nerves. After contracting, a muscle relaxes and returns to its original length until the next signal.

Skeletal muscle is attached to the bones of the skeleton. When skeletal muscles like those in **Figure 29-17** contract, bones move, and actions like walking, grasping, or bending result. Skeletal muscles are sometimes called voluntary muscles because you can control their actions. You also use skeletal muscles for activities that happen without your thinking about them, such as blinking or maintaining your posture. However, you can consciously control these actions at any time.

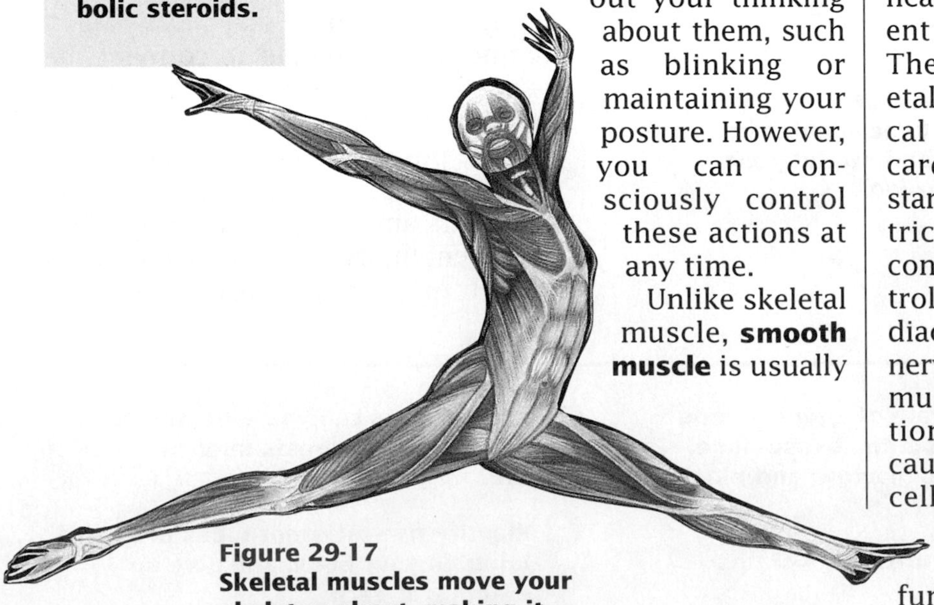

Unlike skeletal muscle, **smooth muscle** is usually not under conscious control. Most of the actions of smooth muscle occur without your being aware of them. For example, when you go outside on a cold, sunny day, smooth muscles reduce the diameter of blood vessels in your skin. This automatic action decreases the flow of blood to the skin and reduces the amount of heat that your body loses. At the same time, smooth muscles in your eye cause your pupils to become smaller so that you will not be blinded by the bright sunshine. Smooth muscles are also found in the walls of internal organs such as those in your digestive system, where they move food through your body.

Cardiac muscle is found in the heart. Cardiac muscle cells are different from other types of muscle cells. The contraction of a smooth or skeletal muscle is triggered by an electrical signal from a nerve. However, cardiac muscle cells are "self-starters"—they generate their own electrical signals that cause them to contract. Nerves to the heart can control how fast the heart beats, but cardiac muscle will contract even if these nerves are cut. Connections among the muscle cells coordinate the contractions of the individual cells. This causes the entire heart to beat as the cells contract together. **Figure 29-18** on the next page shows the three types of muscle tissue and their functions in the human body.

Figure 29-17
Skeletal muscles move your skeleton about, making it possible to do leaps like this.

Making Your Skeleton Move

A skeletal muscle can move a bone like the one in your thigh by pulling on it. When the muscle contracts, it pulls on the bone to which it is attached. No muscle can push a bone; muscles can only pull on bones.

Muscle pairs work together

Moving a bone requires the cooperation of at least two sets of muscles. Pairs of muscles run parallel to each other on either side of a bone in the skeleton. Why are two sets of muscles needed to move a bone? When a muscle pulls on a bone, it moves the bone in one direction. Because the muscle cannot push on the bone, another muscle is needed to pull the bone in the opposite direction. Contracting one set of muscles while relaxing the other set can cause a particular bone to move and decreases the joint angle. Contracting the other set of muscles while relaxing the first set causes the joint angle to increase once again. **Figure 29-18d–e** shows how the main pair of muscles that control the human elbow works.

What would happen if both sets of muscles contracted and shortened at the same time? Under normal circumstances, the bone would not move because the muscles would be pulling in opposite directions at the same time. However, large muscles in your legs are so powerful that if all of the muscles contracted strongly at the same time, they could break a bone. This is prevented by a nerve message sent from one set of muscles to the other. When one set contracts, the other set is signaled to relax. An intricately coordinated sequence of muscle contractions and relaxations alternately bends and straightens the hips, knees, and ankles, allowing you to walk or run smoothly.

Figure 29-18
The human body contains three main types of muscle tissue that move different parts of the body.

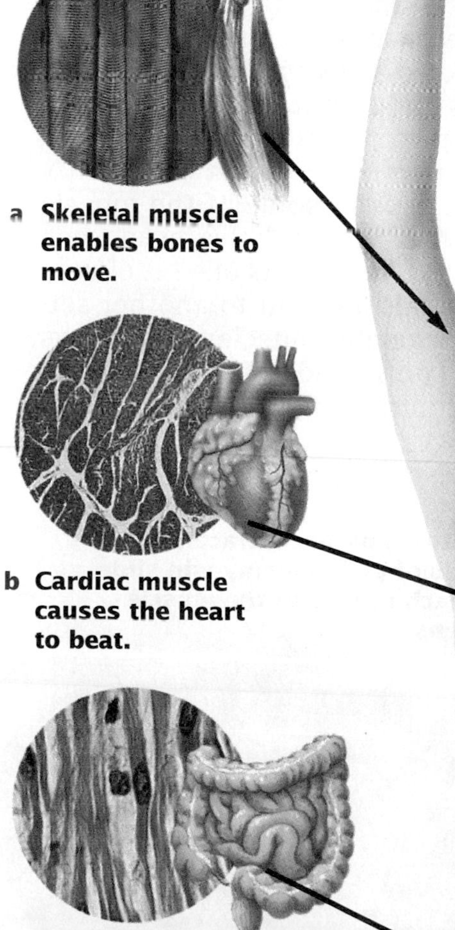

a Skeletal muscle enables bones to move.

b Cardiac muscle causes the heart to beat.

c Smooth muscle moves food through digestive organs and makes other internal organs "work."

d When you bend your elbow, the biceps—a skeletal muscle in the front of your upper arm—contracts and shortens. The triceps muscle in the back of your upper arm relaxes.

e When you straighten your arm, the triceps muscle contracts and shortens while the biceps relaxes.

Tendons attach muscles to bones

Muscles, are connected to bones by **tendons**. For example, the biceps muscle in your arm is attached by tendons to a bone in your shoulder and to another bone in your forearm. Tendons are made of the same tough, elastic connective tissue as ligaments. The difference between tendons and ligaments is that ligaments connect a bone to another bone, while tendons connect a muscle to a bone.

Sometimes the distance between a muscle and the bone to which it is attached is quite long. When you bend your fingers, you are actually contracting muscles in your forearm. There are no muscles in your fingers—only long tendons that extend from muscles in your forearm to the bones in your fingers. Move your fingers and watch the muscles in your forearm move.

Skeletal muscle cells have light and dark bands

Each of your skeletal muscles contains many bundles of long, thin muscle cells, as shown in **Figure 29-19**. When stained and viewed with a microscope, muscle cells show alternating dark and light bands in repeating units called sarcomeres *(SAR kuh mihrz)*.

Two types of protein threads—actin and myosin—make up the dark and light bands within each sarcomere. These protein threads are arranged so that they can slide past each other. When a muscle cell receives an electrical signal from a neuron, the sets of actin and myosin threads slide past one another, and the entire muscle cell shortens. You can read more about actin and myosin in *A Closer Look at Skeletal Muscle* on page 593.

Nerves send messages to muscles to generate force

If the muscles in your arms contracted with the same force when picking up a pencil as when picking up a bowling ball, the pencil would probably go flying to the ceiling. How does a muscle create just enough force to perform a particular job?

The contraction of each muscle cell generates a certain amount of force that pulls on the bone. But not every muscle cell contracts at once. Signals from the nervous system make a muscle generate more force by contracting more muscle cells. When you pick up a pencil, a small percentage of muscle cells in each muscle contract. But when you pick up a bowling ball, more muscle cells contract in order to provide enough force to lift the ball.

Your nervous system constantly sends signals to some of the muscle cells in all of your muscles, even when you are not moving. As one set of cells tires, signals are sent to another set, so your skeletal muscles are always partly contracted. This constant, partial contraction of your muscles is known as muscle tone. Muscle tone maintains your posture.

Figure 29-19 These photos show the relaxation and contraction of the frontalis muscle of the forehead.

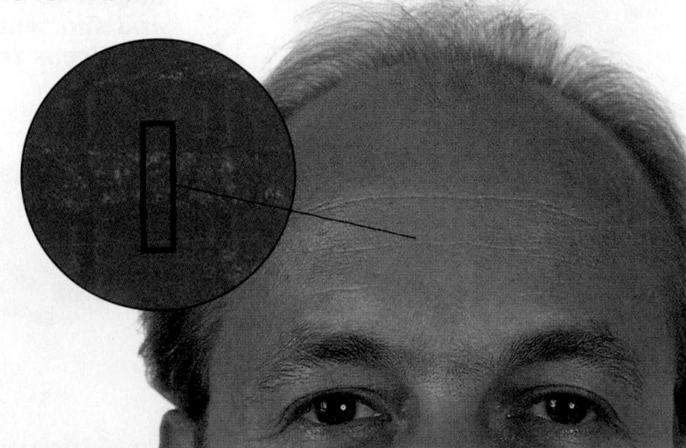

a When the muscle is relaxed, threads of actin and myosin are at rest, side by side in parallel stacks.

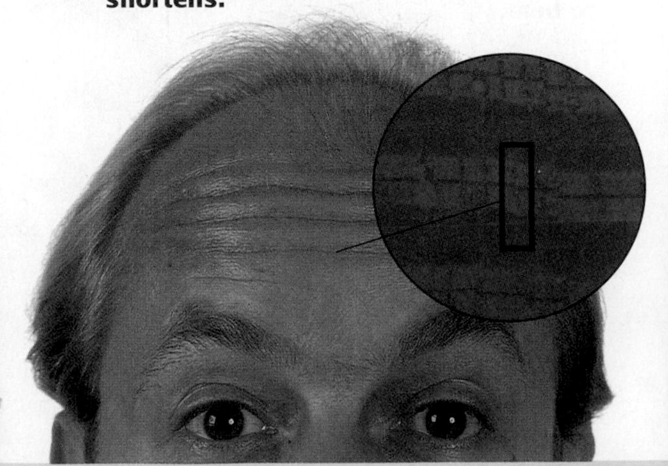

b When the muscle contracts, the threads of actin and myosin slide past each other and the muscle shortens.

Skeletal Muscle

Vertebrates have three kinds of muscle cells: smooth muscle cells that are organized into sheets, cardiac muscle cells that form a lattice, and striated muscle cells (also called skeletal muscle cells) that appear to be striped.

A Closer Look

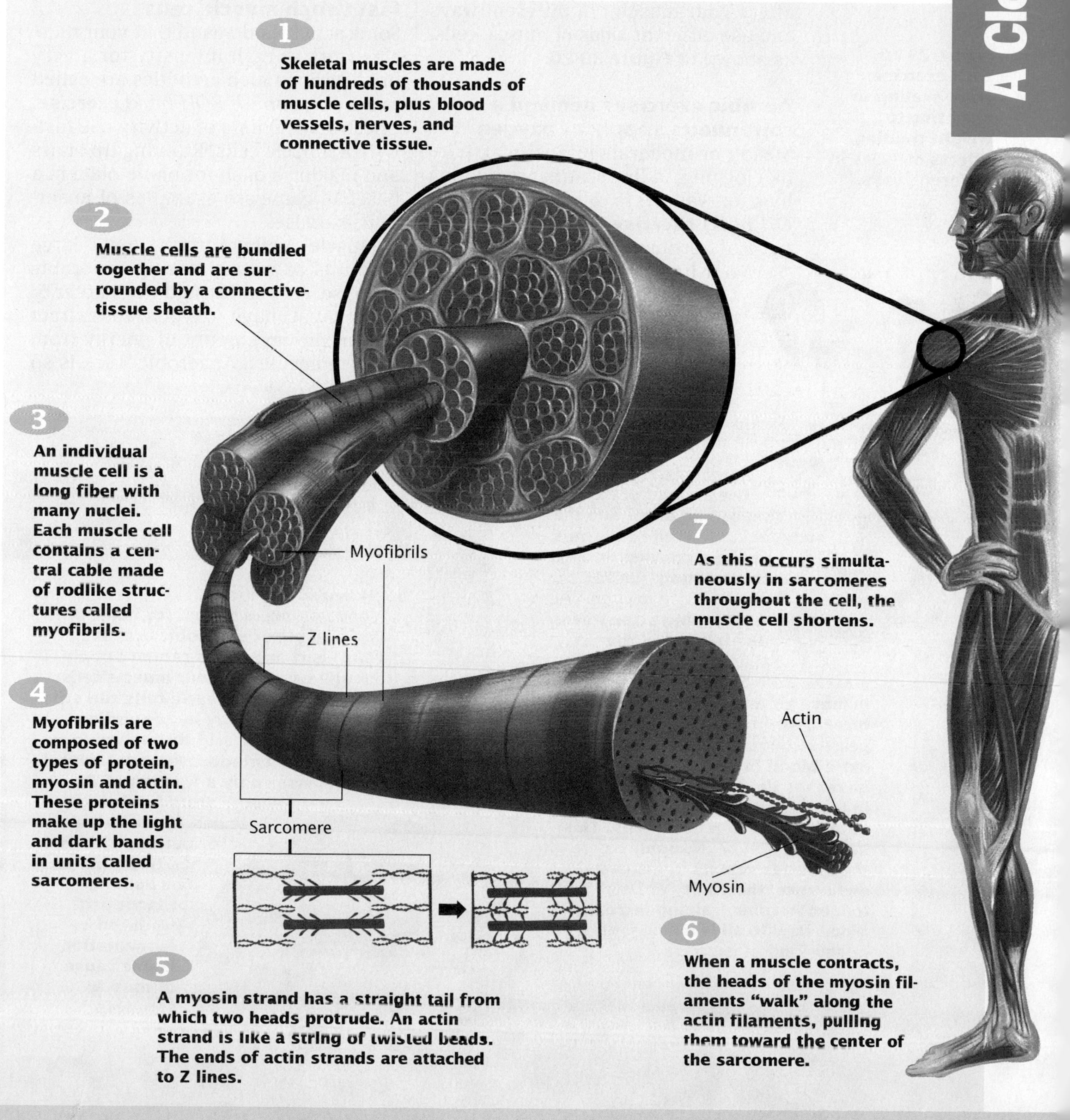

1
Skeletal muscles are made of hundreds of thousands of muscle cells, plus blood vessels, nerves, and connective tissue.
2
Muscle cells are bundled together and are surrounded by a connective-tissue sheath.
3
An individual muscle cell is a long fiber with many nuclei. Each muscle cell contains a central cable made of rodlike structures called myofibrils.
Myofibrils
Z lines
4
Myofibrils are composed of two types of protein, myosin and actin. These proteins make up the light and dark bands in units called sarcomeres.
Sarcomere
Actin
Myosin
5
A myosin strand has a straight tail from which two heads protrude. An actin strand is like a string of twisted beads. The ends of actin strands are attached to Z lines.
6
When a muscle contracts, the heads of the myosin filaments "walk" along the actin filaments, pulling them toward the center of the sarcomere.
7
As this occurs simultaneously in sarcomeres throughout the cell, the muscle cell shortens.

What Exercise Does for Muscles

Muscles are structured to move and to do work. When muscles are not used, they fatigue easily, and over time they harden, become smaller, and lose strength. Most people can increase their strength, flexibility, and endurance through physical activity and exercise. Different kinds of exercise affect your muscles in different ways and use different kinds of muscle cells, as shown in **Figure 29-20**.

Aerobic exercises demand a continuous supply of oxygen

Steady or moderate-intensity activity like jogging, in-line skating, swimming laps, or walking is called **aerobic** *(ehr OH bihk)* **exercise.** The slow, steady pace of aerobic activity ensures that your lungs and heart can deliver oxygen to your muscles at the same rate at which the muscle cells are using it. Your muscles can use this continuous supply of oxygen to extract the maximum amount of energy from food molecules and to continue the activity for as long as 20 minutes or more. **Figure 29-20a** describes some of the benefits of aerobic exercise.

Anaerobic exercises use fast-twitch muscle cells

Some activities demand that your muscles work at high intensity for a very brief period. Such activities are called **anaerobic** *(an uh ROH bihk)* **exercise**. These short bursts of activity use fast-twitch muscle cells. Running up stairs and making a dash for home plate in a baseball game are examples of anaerobic exercise.

Muscle cells rapidly use large amounts of energy during anaerobic exercise. As you learned in Chapter 5, cells must have oxygen to extract the maximum amount of energy from food molecules. Anaerobic work is so

Figure 29-20
Each exercise, from skating to sprinting to weight training, affects muscles in different ways.

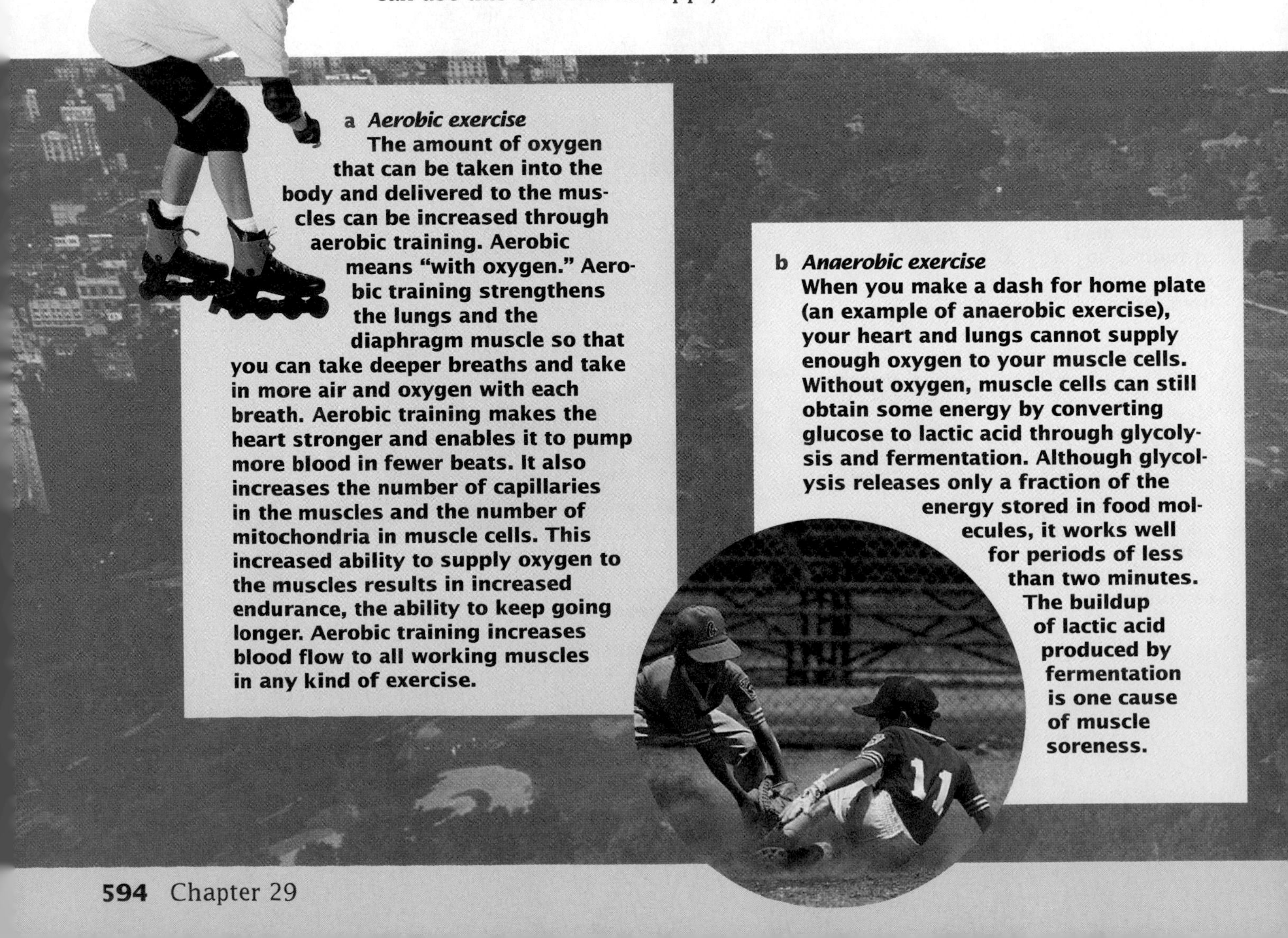

a ***Aerobic exercise***
The amount of oxygen that can be taken into the body and delivered to the muscles can be increased through aerobic training. Aerobic means "with oxygen." Aerobic training strengthens the lungs and the diaphragm muscle so that you can take deeper breaths and take in more air and oxygen with each breath. Aerobic training makes the heart stronger and enables it to pump more blood in fewer beats. It also increases the number of capillaries in the muscles and the number of mitochondria in muscle cells. This increased ability to supply oxygen to the muscles results in increased endurance, the ability to keep going longer. Aerobic training increases blood flow to all working muscles in any kind of exercise.

b ***Anaerobic exercise***
When you make a dash for home plate (an example of anaerobic exercise), your heart and lungs cannot supply enough oxygen to your muscle cells. Without oxygen, muscle cells can still obtain some energy by converting glucose to lactic acid through glycolysis and fermentation. Although glycolysis releases only a fraction of the energy stored in food molecules, it works well for periods of less than two minutes. The buildup of lactic acid produced by fermentation is one cause of muscle soreness.

intense that there's no time to take in enough oxygen in the short time that the work lasts. See **Figure 29-20b** for an explanation of how the needed energy is supplied for this form of activity.

After you stop anaerobic exercise, your body needs extra oxygen to burn up the excess lactic acid and return your energy reserves to normal. This need for extra oxygen is called oxygen debt. Oxygen debt is the reason you must breathe rapidly and deeply for a few minutes after a hard run.

Resistance exercises increase muscle size and strength

Aerobic exercise will increase the size and strength of your skeletal muscles somewhat, but it mostly strengthens the heart and lungs. To significantly increase the size or the strength of your skeletal muscles, resistance exercises are the most effective form of exercise. **Resistance exercises** are those that require muscles to overcome the resistance (weight) of another object. Overcoming your own body weight by doing chin-ups or push-ups and overcoming the resistance provided by free weights, weight machines, or rubber tubing are examples of resistance training. You can read more about resistance training in **Figure 29-20c**.

Skeletal muscle cells can be fast-twitch or slow-twitch

Every skeletal muscle in your body contains two types of muscle cells. Fast-twitch muscle cells are called into action when a person, such as a pole-vaulter, requires speed and quick movements over a short period. Slow-twitch muscle cells respond more slowly but do not tire as quickly as fast-twitch muscle cells. Slow-twitch muscles are used for activities, such as cycling, that require endurance. You can read more about fast- and slow-twitch muscles in **Figure 29-20d**.

c ***Resistance exercise***
Resistance training increases both muscle strength and muscle endurance. Unlike free weights, rubber tubing places the muscle under constant resistance throughout the range of motion and is especially useful for rehabilitation after injury. Regular resistance training helps to increase bone mass and also strengthens the abdominal muscles necessary for a healthy back. As you get older, it becomes more important to include resistance training in your exercise routine. Your daily tasks become easier as you increase or maintain your strength.

d ***Fast- and slow-twitch muscles***
The number of fast- and slow-twitch muscle cells that a person has is primarily an inherited trait that does not change with moderate exercise. Researchers measured the numbers of fast- and slow-twitch muscle cells in different athletes. Sprinters, jumpers, and weight lifters had a high number of fast-twitch muscle cells. Long distance runners and cyclists had more slow-twitch muscle cells.

Anabolic steroids are dangerous

Some people are tempted to experiment with anabolic steroids to increase the size of their muscles. Anabolic steroids are powerful synthetic compounds that chemically resemble the male sex hormone, testosterone. The use of steroids may produce serious side effects and health problems, as summarized in **Figure 29-21**.

When anabolic steroids are used before the skeleton matures completely, as in a teenager, the steroids stop the bones from growing. Remember that a bone grows from its soft ends, the regions called growth plates. Steroids cause premature closure of these growth plates. When the plates close, the once-soft cartilage begins to turn into bone earlier than it should. The bones stop growing, while the muscles continue to get larger. The body never reaches adult height and may look distorted. These and other symptoms of steroid use may not be reversible.

Some males who use anabolic steroids develop female-like breasts and shriveled testes. Females who use these chemicals may develop facial hair, deepening of the voice, and male-pattern baldness. The long-term risks to health are often greater than any benefits that might be gotten from the use of steroids.

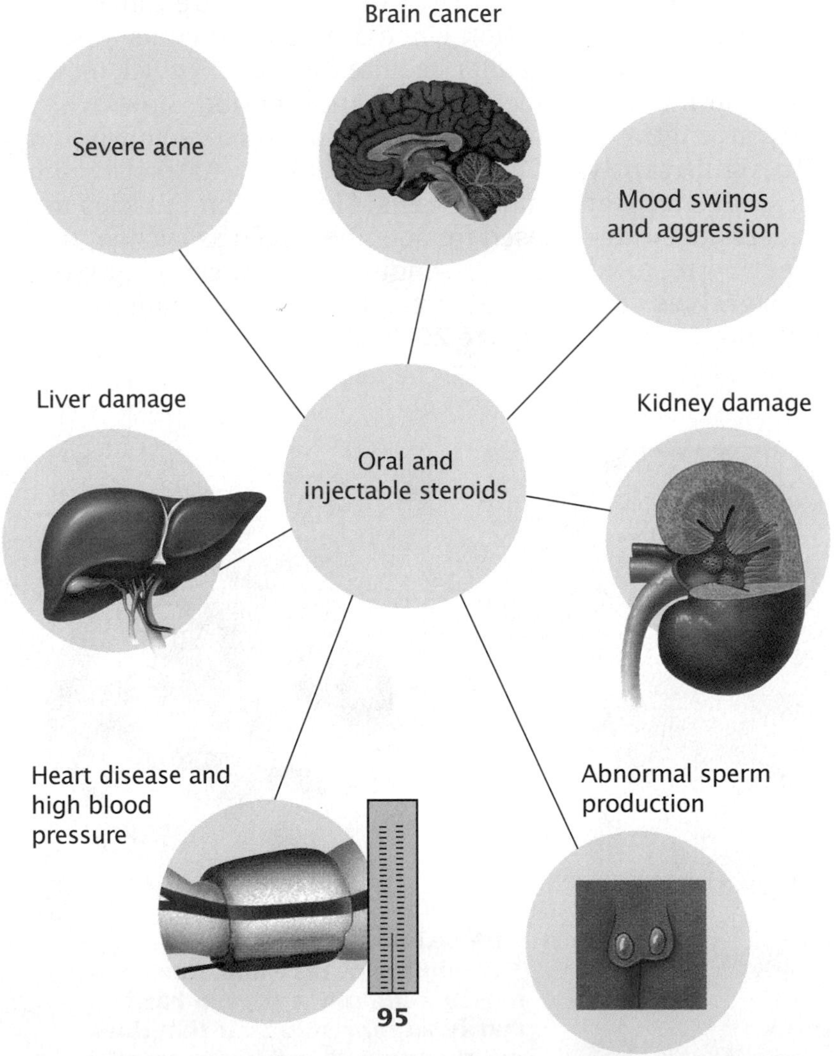

Figure 29-21 The use of anabolic steroids poses a threat to numerous body systems. Both physical and psychological health may be affected.

Overuse causes muscle injuries

Overusing muscles by exercising too much or without proper conditioning can lead to muscle injury. A muscle strain, commonly called a "pulled muscle," is the overstretching or even tearing of a muscle. Muscle strain may occur when a muscle is overused or not properly warmed up or when proper form is not used. Muscles are wrapped and supported by collagen and elastin fibers. Elastin stretches; collagen does not. When a muscle heals after a tear, its elastin is replaced by collagen. Then the muscle doesn't stretch as well as it once did. Tendons, as well as muscles, can get injured from overuse. Tendinitis is the painful inflammation of a tendon caused by too much friction or stress on the tendon.

SECTION REVIEW

1. **Describe the three main types of muscle.**
2. **How might your ability to move your arm be affected by an injury to the biceps muscle? to the triceps muscle?**
3. **If you wanted to increase your endurance, which kind of exercise would be most effective?**
4. **Describe how anabolic steroids damage bones, and identify three other disadvantages of steroid use.**

CHAPTER 29 # Highlights

Aerobic exercise, such as cross-country skiing, increases oxygen flow to the muscles.

	Key Terms	Summary
29-1 An Inside Look at the Human Body	tissue (p. 575) epithelial tissue (p. 575) muscle tissue (p. 576) nerve tissue (p. 576) connective tissue (p. 576) organ (p. 577) organ system (p. 577)	• Specialized cells are organized into groups called tissues. • The four types of tissue in the body are epithelial (for protection), muscle (for movement), nervous (for communication), and connective (for support and transportation). • Tissues work together to form organs. • Organs work together to form organ systems.
29-2 Skin	epidermis (p. 578) dermis (p. 578) melanin (p. 581) melanomas (p. 583)	• Skin protects the body and aids in elimination. • Several skin layers contain a variety of cells, living and dead.
29-3 Bones	marrow (p. 585) fracture (p. 586) osteoporosis (p. 586) joint (p. 589) ligament (p. 589) sprain (p. 589)	• Bone is a living tissue made of cells that deposit minerals. • Marrow is important in blood cell production and fat storage. • The skeleton's joints are held in place by ligaments, which connect bone to bone.
29-4 Muscles	skeletal muscle (p. 590) smooth muscle (p. 590) cardiac muscle (p. 590) tendons (p. 592) aerobic exercise (p. 594) anaerobic exercise (p. 594) resistance exercise (p. 595)	• The three basic types of muscle tissue are skeletal muscle, smooth muscle, and cardiac muscle. • Moving a bone requires two sets of muscles working in opposition. • Muscles attach to bones directly by means of tendons. • Muscle action depends on the sliding action of protein threads (actin and myosin). • Three types of exercise (aerobic, anaerobic, and resistance) make different demands on muscles.

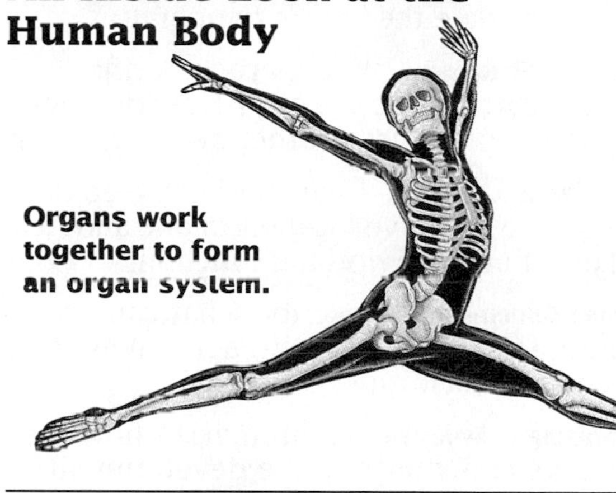
Organs work together to form an organ system.

Dead skin cells flake off and are replaced.

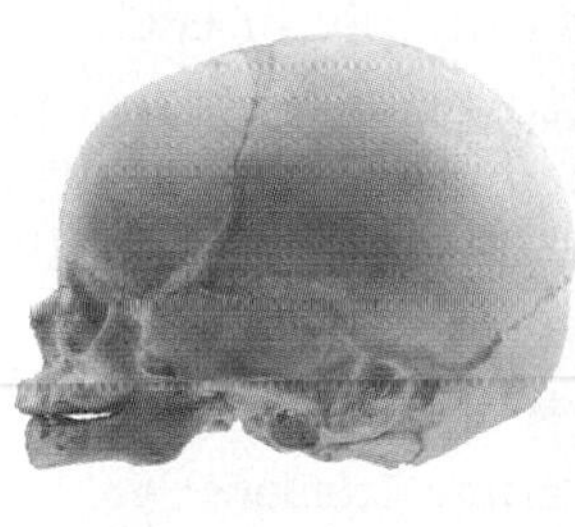
As you grow, bone cells build more bone tissue.

Exercise helps increase endurance, muscle strength, agility, and joint flexibility.

CHAPTER 29

review

Understanding Vocabulary

1. For each set of terms, complete the analogy.
   a. cartilage : connective tissue :: sense organs : ____
   b. contains blood vessels : dermis :: contains melanin : ____
   c. broken bone : fracture :: ligament injury : ____
2. Using each set of words below, write one or more sentences summarizing information learned in this chapter.
   a. tissue, organ, circulatory system
   b. epidermis, keratin, melanin, hair follicles
   c. skeletal muscle, smooth muscle, cardiac muscle

Understanding Concepts

3. **Relating Concepts** Construct a concept map that identifies the three types of muscle tissue and their characteristics. Include these terms: cardiac, smooth, skeletal, heart, digestive system, bones, unconsciously, voluntarily, electrical impulses.
4. **Identifying Functions** Describe the functions of the three types of connective tissue, and give examples of each.
5. **Organizing Information** Describe how the different types of tissue work together in the skin, bones, and muscles. Give examples of each.
6. **Inferring Relationships** According to the graph below, under what outdoor conditions is it safe to exercise? When is outdoor exercise dangerous or to be avoided?

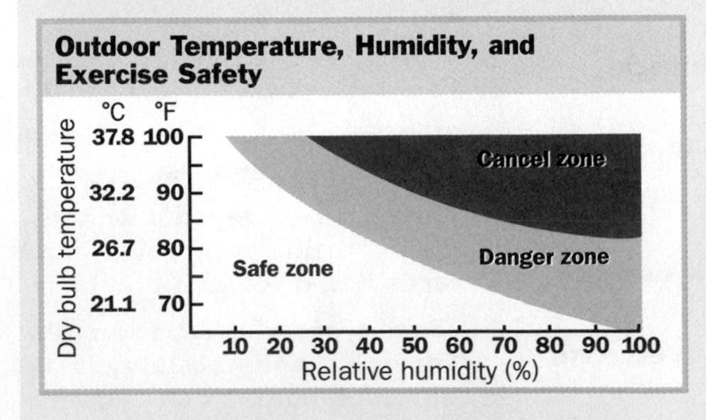

7. **Identifying Functions** Identify the organ systems that work together to protect, support, and move the body.
8. **Identifying Relationships** Describe the functions of blood vessels within the dermis, and explain how they work with other structures there to regulate the body's temperature.
9. **Identifying Structures** Where are apocrine glands located in your body? How do apocrine glands differ from other sweat glands?
10. **Organizing Information** Characterize the epidermal layers by relative location to one another, level of cellular activity, and function.
11. **Inferring Conclusions** Describe what causes the pimples associated with acne. Why is this condition common in teens?
12. **Recognizing Relationships** Summarize how the sun's rays contribute to the development of skin cancer, and explain how you can protect yourself from sun damage.
13. **Inferring Relationships** Describe how a bone grows, ages, breaks, and heals.
14. **Summarizing Information** Identify five kinds of joints in the human skeleton. Describe the movement allowed by each joint, and give an example of each.
15. **Identifying Structures** Classify the muscles of the body into three types, and describe each type's primary function.
16. **Analyzing Information** How are sprains and fractures alike? How are they different?
17. **Inferring Relationships** How does the term *osteoporosis* describe changes in mature bones? What causes these changes?
18. **Summarizing Information** Compare the effects of anaerobic exercise, aerobic exercise, and resistance exercise on the body.
19. **Inferring Relationships** Describe the results of using anabolic steroids before the skeleton matures completely.
20. **BUILDING ON WHAT YOU HAVE LEARNED** As you learned in Chapter 16, scientists have noted a reduction in the Earth's ozone layer. How is the destruction of this layer related to an increase in the number of warnings to use sunscreens and to avoid the midday sun?

Interpreting Graphics

21. Look at the photo of the athlete below.

a. How can you tell that this person has just finished exercising?
b. Which sweat glands are responsible for the pattern of sweat shown here?

Reviewing Themes

22. ***Evolution***
What do goose bumps reveal about our evolutionary past?

23. ***Heredity***
Explain why some people are never able to grow long hair.

Thinking Critically

24. **Justifying Conclusions** In some countries, potential athletes have been selected for particular sports at a very young age based on the percentages of fast-twitch and slow-twitch muscle cells in their bodies. If an examination of muscle tissue from the leg reveals that athlete A has fewer fast-twitch muscle cells than athlete B, which athlete would be chosen for training as a marathon runner? Explain your answer.

25. **Inferring Relationships** Why are two sets of muscles required to move a bone?

Life/Work Skills

26. **Using Technology** Search the Internet to find information about the Visible Human Project, which was conducted to create a detailed, computerized atlas of the human body. The project produced computer-generated, three-dimensional models of a human male (the "Visible Man") and a human female (the "Visible Woman"). Find out how the images were created and how they might be used. Information about the Visible Human Project is also available on CD-ROM.

Activities and Projects

27. **Reading Connection** Study the beautifully illustrated book *The Incredible Machine*, by the National Geographic Society (1986). Read a chapter of your choice, such as "The Powerful River," which is about the circulatory system, and prepare a written report on information in the chapter that interested you.

28. **Health Connection** Use library references or search an on-line database to find out about muscular disorders such as muscular dystrophy and myasthenia gravis. What are the symptoms of these conditions? What are the causes? Obtain photos of some famous people who have suffered from these disorders, and share their stories with your classmates.

Discovering Through Reading

29. Read the article "Saving Face: Yes, You Can Prevent Wrinkles and Even Get Rid of Some You've Got" by Patricia Long in *Health*, November/December 1994, pages 56–63. When do men get facial wrinkles? When do women get them? Why?

30. Read the article, "Saving Burn Victims" from *Discover*, Volume 15, Issue 9. September 1994, pages 16–18. Why are third degree burns so dangerous? From what kind of tissue do fibroblasts make new skin?

CHAPTER 30

REVIEW
- ion (Section 2-2)
- receptor proteins (Section 4-1)
- sodium-potassium pump (Section 4-2)
- diffusion (Section 4-2)
- nerve tissue (Section 29-1)

The Nervous System

Balance, strength, coordination, and memory—this gymnast's successful performance is controlled by the body's nervous system.

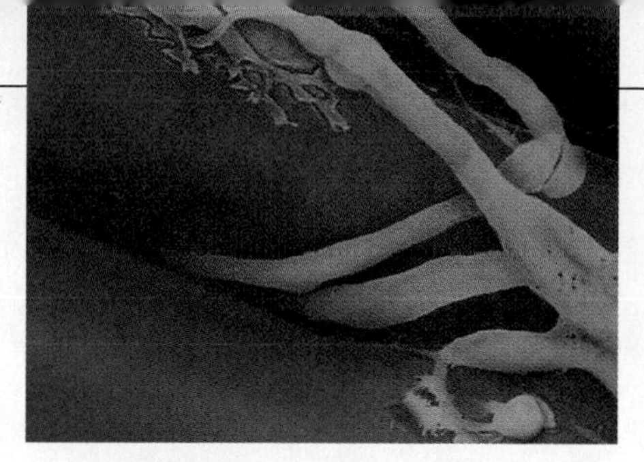

When you touch a hot stove, your hand instantly jerks back because of a message from your spinal cord. When you swat a fly, the muscles of your arm contract quickly because of a message from your brain. These messages travel along nerves, which are bundles of cells that carry electrical signals throughout your body.

30-1 Nerve Impulses

OBJECTIVES

1. **Describe the basic structure of a neuron.**
2. **Summarize the changes that occur in a neuron during a nerve impulse.**
3. **Explain how a nerve impulse is carried across a synapse.**
4. **Describe addiction in terms of a neuron's membrane.**

Structure of a Neuron

Nerve cells, or **neurons** *(NOO rahnz)*, are specialized for conducting information. Bundles of neurons form thin cables called **nerves**. Like electrical cables running through a town, nerves form a communications network that extends throughout your body, including organs like your skin, eyes, and stomach.

There are many types of neurons, all with structures that reflect their particular roles. However, all neurons have the same basic architecture. A typical neuron, shown in **Figure 30-1**, consists of an enlarged region called the cell body, which contains the nucleus and organelles. The cell body generally receives information at short, branched extensions called **dendrites** *(DEHN dryts)*. Typically, a neuron has many dendrites, which enables it to receive input from thousands of other cells. From the cell body, information is transmitted to other cells by long fibers called **axons**. In contrast to its high number of dendrites, a neuron usually has only one axon. The tip of the axon often branches into short knobs so that a single neuron can pass information to many other cells.

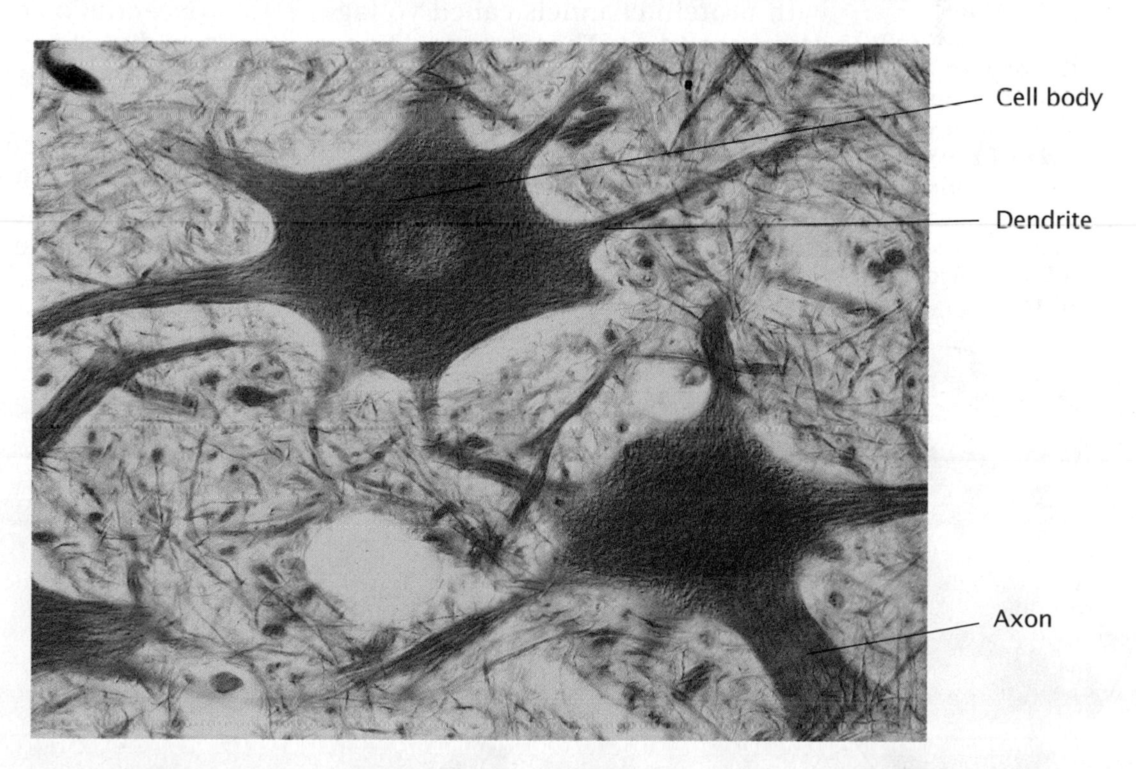

Figure 30-1 Neurons, or nerve cells, are specialized to receive information and send it to other cells in other parts of the body.

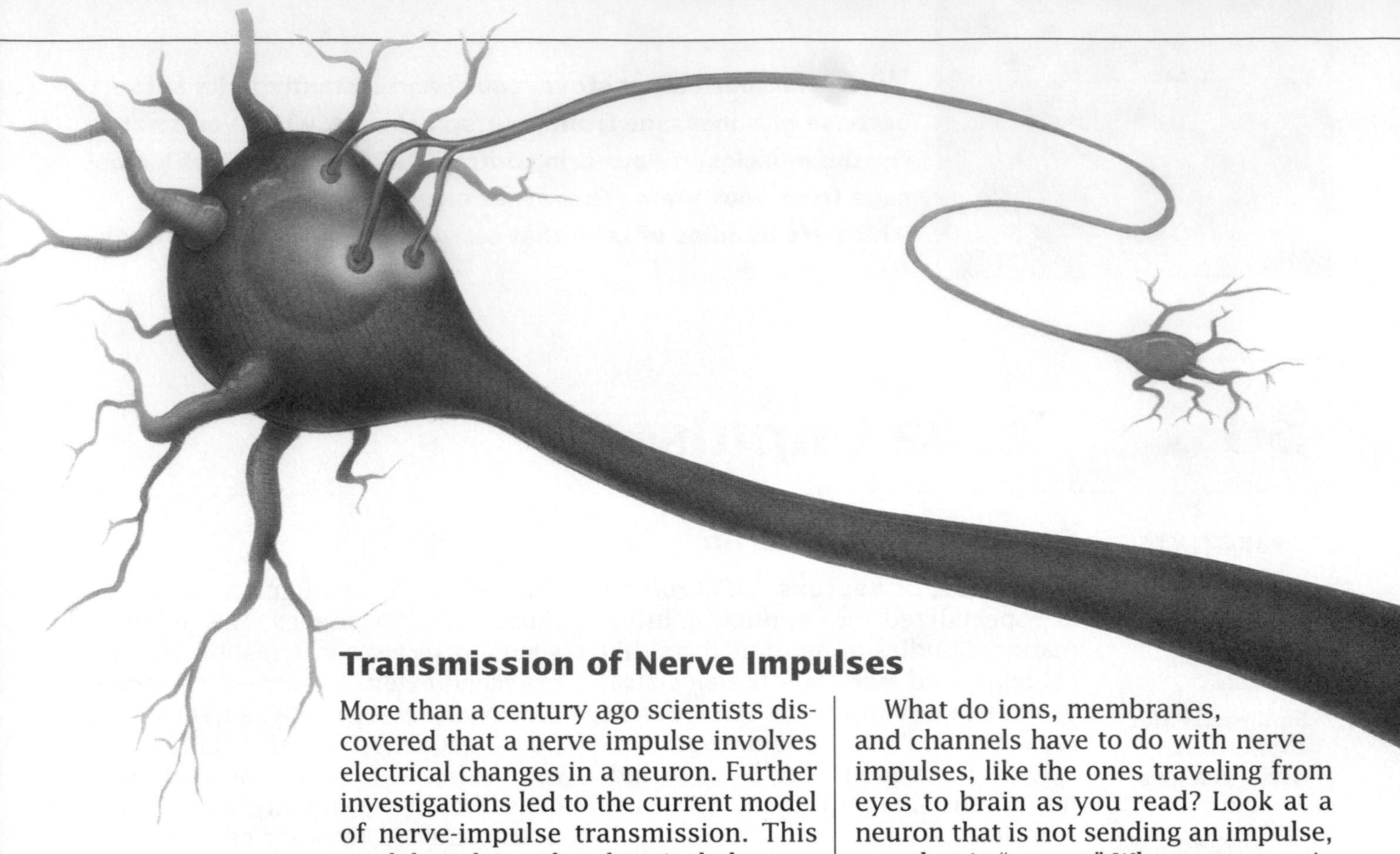

Transmission of Nerve Impulses

More than a century ago scientists discovered that a nerve impulse involves electrical changes in a neuron. Further investigations led to the current model of nerve-impulse transmission. This model explains the electrical changes as differences in concentrations of two ions—sodium (Na^+) and potassium (K^+).

The key to the transmission of nerve impulses lies in a neuron's membrane. The membrane of a neuron is studded with protein channels called voltage-gated channels. These channels open and close in response to changes in voltage (electrical charge) across the membrane. They are ion specific—some allow only Na^+ ions to pass through the membrane, while others allow only K^+ ions to pass.

What do ions, membranes, and channels have to do with nerve impulses, like the ones traveling from eyes to brain as you read? Look at a neuron that is not sending an impulse, one that is "at rest." When a neuron is at rest, its voltage-gated channels are closed. Inside the concentration of Na^+ ions is low, and the concentration of K^+ ions is high. Outside the neuron, the concentration of Na^+ ions is high, and the concentration of K^+ ions is low. The unequal distribution of ions, illustrated in **Figure 30-2a**, results in a difference in voltage across the cell membrane called the resting potential. During resting potential, the inside of the cell has a negative charge (about −70 millivolts) compared with the outside.

c Sodium-potassium pumps maintain the proper balance of ions inside and outside of the neuron, enabling it to function properly over a long period of time.

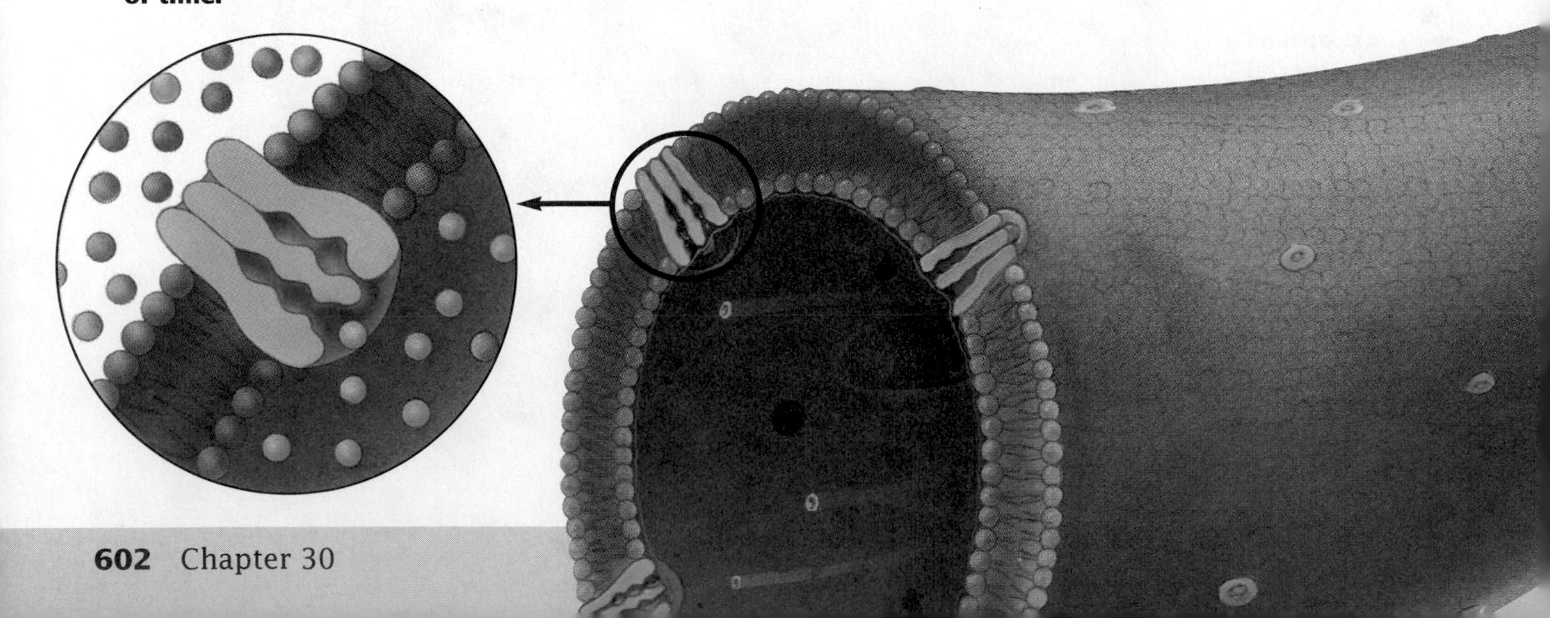

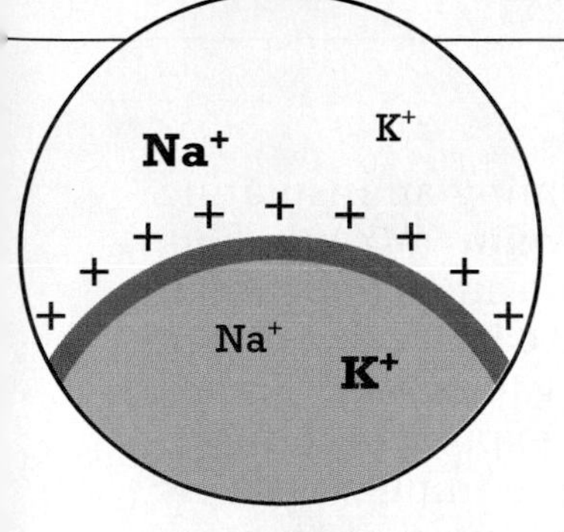

Figure 30-2
a *Resting potential* **The concentration of sodium ions is much greater outside the cell, and the concentration of potassium ions is greater inside the cell.**

A nerve impulse begins when membrane voltage is reversed

When a neuron is stimulated, such as by light, odor, a pinch, or a mild electric shock, its membrane undergoes a change. For a fraction of a second, the voltage-gated channels in the membrane open, allowing Na^+ ions to pass through. Powered by diffusion and attracted by negative charges inside the cell, Na^+ ions rush across the membrane into the cell. The inward rush of Na^+ ions is so great that, for a moment, the inside of the neuron becomes positively charged (+35 millivolts) compared with the outside. This sudden reversal of electrical charge across the neuron membrane is called an action potential, illustrated in **Figure 30-2b**.

The inward rush of Na^+ ions lasts for only half of a millisecond before the voltage-gated channels close. At the same moment, other voltage-gated channels open, allowing K^+ ions to pass through. As a result, K^+ ions rush out of the cell, powered by diffusion and repelled by the momentarily high positive charge inside the cell. This exit of K^+ ions restores the cell to resting potential.

One action potential triggers another action potential

During an action potential, the voltage across a small patch of membrane is reversed. This change in voltage causes nearby voltage-gated Na^+ ion channels to open, triggering an action potential in another region of the membrane. Likewise, this action potential triggers another one farther along the axon and continues like a chain of falling dominoes. In short, a nerve impulse is a series of action potentials flowing along a neuron.

A neuron must maintain the balance of ions

If a nerve impulse involves an inward flow of Na^+ ions and an outward flow of K^+ ions, how does a neuron restore its original balance of ions? After an impulse passes, diffusion restores the balance between Na^+ ions outside of the cell and K^+ ions inside. This balance is maintained by sodium-potassium pumps, which actively transport sodium ions out of the cell and potassium ions into the cell, as shown in **Figure 30-2c**. Without the action of the sodium-potassium pump, a neuron would lose its ability to conduct impulses.

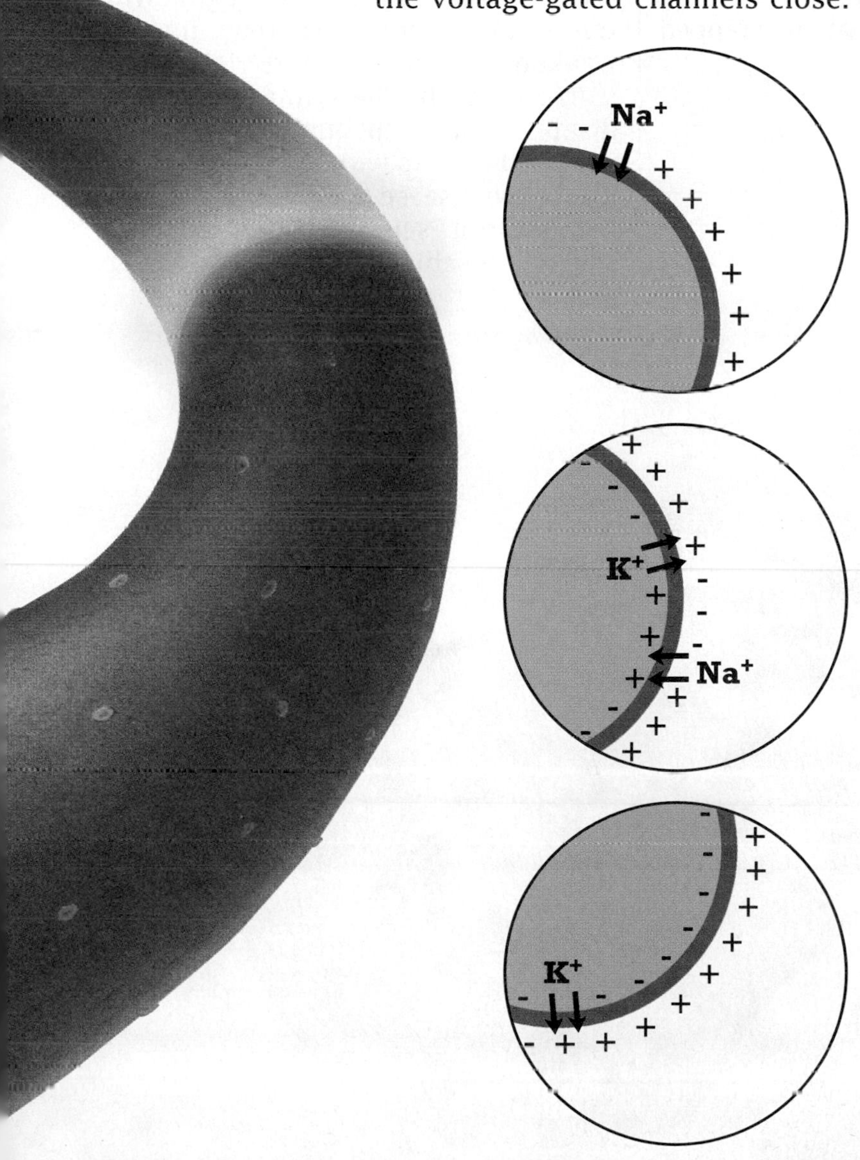

b *Action potential* **Sodium ions rush into the cell through voltage-gated channels, upsetting the resting potential. Potassium ions rush out of the cell through other channels, restoring the voltage.**

Speeding Up Nerve Impulses

Evolution

Why are myelin sheaths considered an evolutionary advancement?

Humans and other vertebrates have neurons that run from the brain and spinal cord directly to muscles. Such neurons have very long axons—imagine a single axon stretching 5 m (16 ft) from a giraffe's spine to its foot! It takes time for a nerve impulse to travel the length of a long axon; this can have dangerous consequences. For example, a giraffe would be in trouble if the information traveling from its brain to the muscles in its legs were slower than a hungry, charging lion. Similarly, you could severely burn your hand on a hot stove if the message to remove it were slow in reaching your spinal cord.

Humans, giraffes, and other vertebrates have specialized nerve cells that speed up nerve impulses. These cells have fatty membranes that are wrapped around axons, forming an insulating layer called a **myelin** *(MY uh lihn)* **sheath**. A myelin sheath is not continuous, as you can see in **Figure 30-3**. It is interrupted by gaps called nodes, which are sites of exposed cell membrane. When a nerve impulse travels along a myelinated axon, it "jumps" from node to node. Because fewer action potentials need to be produced, a nerve impulse travels faster along an axon with a myelin sheath.

The importance of myelin sheaths becomes evident in people who suffer from a disease called multiple sclerosis. Multiple sclerosis destroys large patches of myelin around neurons in the brain and spinal cord. Without the insulating layer, normal nerve impulses are impaired. The disease begins with temporary symptoms that include weakened limbs, blurred vision, and slurred speech. The symptoms usually disappear spontaneously but return in repeated attacks with increasing severity. As the disease progresses, a person with multiple sclerosis is left weak and severely disabled.

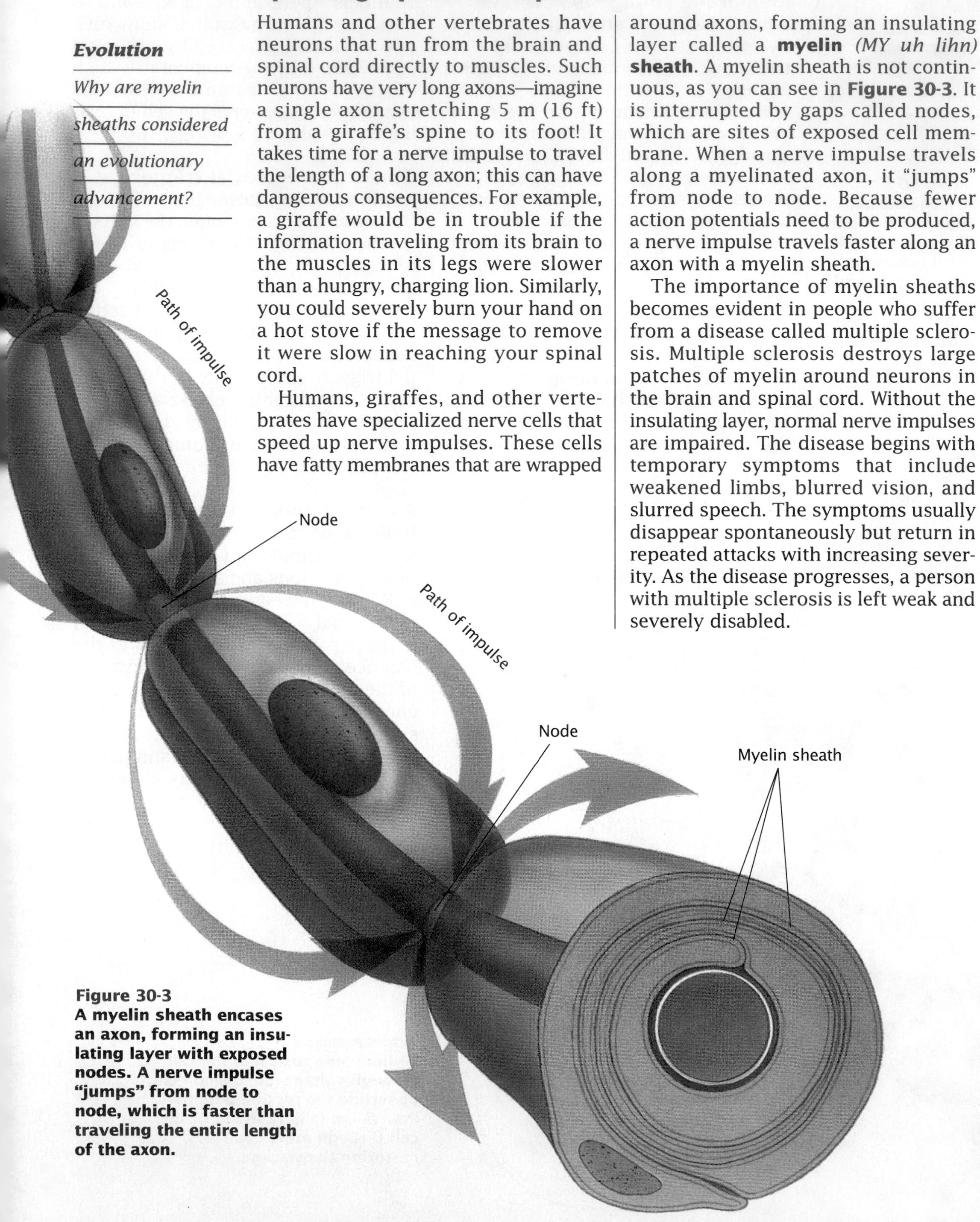

Figure 30-3
A myelin sheath encases an axon, forming an insulating layer with exposed nodes. A nerve impulse "jumps" from node to node, which is faster than traveling the entire length of the axon.

Transmission Across Synapses

A nerve impulse can travel only so far along a cell membrane before it reaches the end of the axon. In most cases, neurons do not touch each other directly. They are separated by a tiny gap called a **synapse** *(SIHN aps)*, shown in **Figure 30-4**. Usually, the axon of one neuron forms a synapse with the dendrites or cell body of another neuron. Since the end portion of an axon may be branched, a single axon may form synapses with many other neurons.

When a nerve impulse reaches the end of an axon, it must cross the synapse if the message is to continue. In most cases, an action potential is unable to "jump" from one neuron to another. Instead, the impulse is carried across the synapse by chemical messengers called **neurotransmitters**. Neurotransmitters are contained in tiny sacs in the end of the axon. When a nerve impulse reaches the end of the axon, the sacs empty the neurotransmitters into the synapse, as shown in **Figure 30-4**. The neurotransmitters diffuse across the synapse and bind to receptors in the membrane of the adjacent neuron. This binding opens the gates of channel proteins in the neuron membrane, which allows specific ions to pass through the membrane in either direction. Ionic movement across the membrane triggers (or, in some cases, inhibits) an action potential.

After a neurotransmitter has carried an impulse across a synapse, it must be removed. Otherwise, it would remain in the synapse, stimulating a neuron indefinitely. How are spent neurotransmitters removed from a synapse? Some simply diffuse away, while others are absorbed by the axon and reused later. Still others are destroyed by specific enzymes.

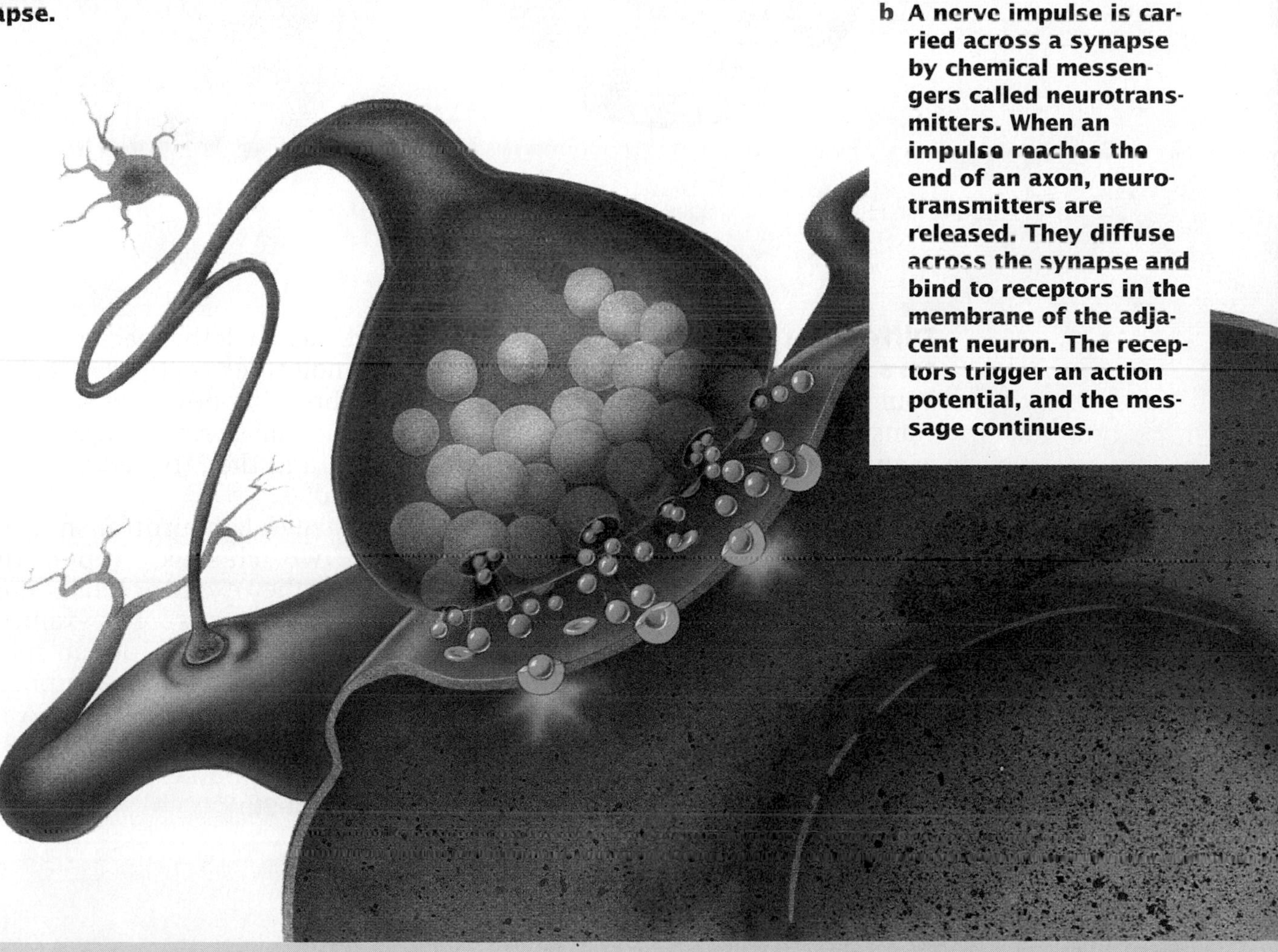

Figure 30-4
a Neurons come very close to each other but usually never touch. They are separated by a tiny gap called a synapse.

b A nerve impulse is carried across a synapse by chemical messengers called neurotransmitters. When an impulse reaches the end of an axon, neurotransmitters are released. They diffuse across the synapse and bind to receptors in the membrane of the adjacent neuron. The receptors trigger an action potential, and the message continues.

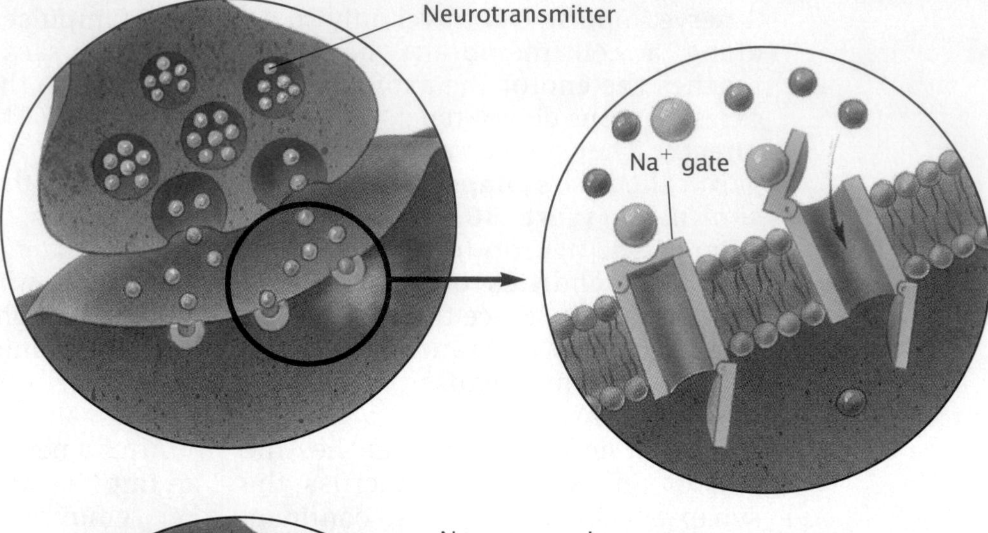

Figure 30-5

a Dopamine is an example of a neurotransmitter that excites nerve impulses. When it crosses a synapse, it opens sodium gates.

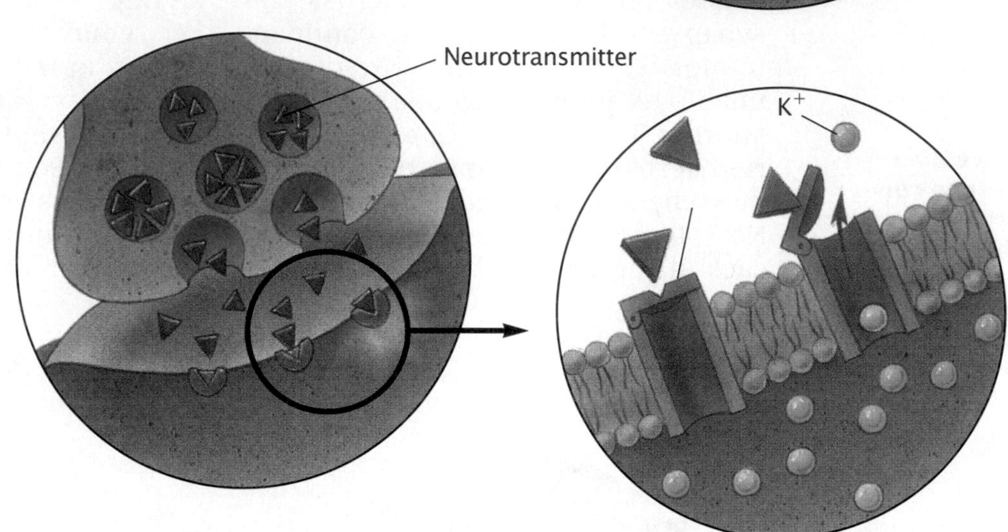

b GABA is an example of a neurotransmitter that inhibits nerve impulses. When it crosses a synapse, it opens potassium gates.

Different types of neurotransmitters produce different effects

Your body produces a variety of neurotransmitters. A neurotransmitter called acetylcholine binds to receptors in the membranes of muscle cells. This triggers an action potential, making the muscle fiber contract. Many drugs that cause paralysis work by affecting acetylcholine receptors. For example, molecules of a drug called curare, used to poison arrow tips, have the same shape as acetylcholine. Curare binds to the acetylcholine receptors, permanently blocking all nerve impulses, which leads to death. Insecticides and nerve gas are lethal because they contain molecules that prevent enzymes from breaking down acetylcholine. As a result, muscles become permanently active, and the affected animal eventually dies.

In general, neurotransmitters fall into two classes, depending on whether they excite or inhibit impulses across the synapse. For example, the neurotransmitters serotonin and dopamine both act to excite impulses, as illustrated in **Figure 30-5a**. A neurotransmitter abbreviated as GABA acts to inhibit impulses, as illustrated in **Figure 30-5b**.

Matter, Energy, and Organization

Why is the term resting potential misleading when describing a neuron's cell membrane?

Drugs and the Nervous System

Recent investigations have found that certain disorders, such as schizophrenia and severe clinical depression, are triggered by malfunctions of neurotransmitters and receptors. This research has yielded new ways to use chemicals to treat certain emotional disorders. It has also revealed new, intriguing information about how drugs affect the nervous system, reinforcing the dangers of their use.

In the broadest sense, a drug is a chemical that can alter biological structures and functions of tissues in your body. A class of drugs called psychoactive drugs alters tissues in the nervous system, which often makes these drugs addictive. Addiction to psychoactive drugs is a biological response that involves drug molecules and receptors in neuron membranes. In effect, addiction is the body's attempt to cope with the chemical disruption that a drug causes in a synapse.

To understand the nature of drug addiction, keep in mind how neurons communicate with one another. Recall that when a nerve impulse reaches the end of an axon, neurotransmitters are released, cross the synapse, and bind to receptors on the adjacent neuron, triggering an impulse. Psychoactive drugs interfere with this activity in the synapse. For example, certain drugs have chemical structures similar to those of a particular neurotransmitter. When a molecule of one of these drugs reaches a neuron, it may bind to receptors for that neurotransmitter. The result is that the neuron reacts as if the neurotransmitter were present. This is how drugs such as morphine and heroin work. Other drugs prevent a neurotransmitter from being destroyed or recycled, as shown in **Figure 30-6**.

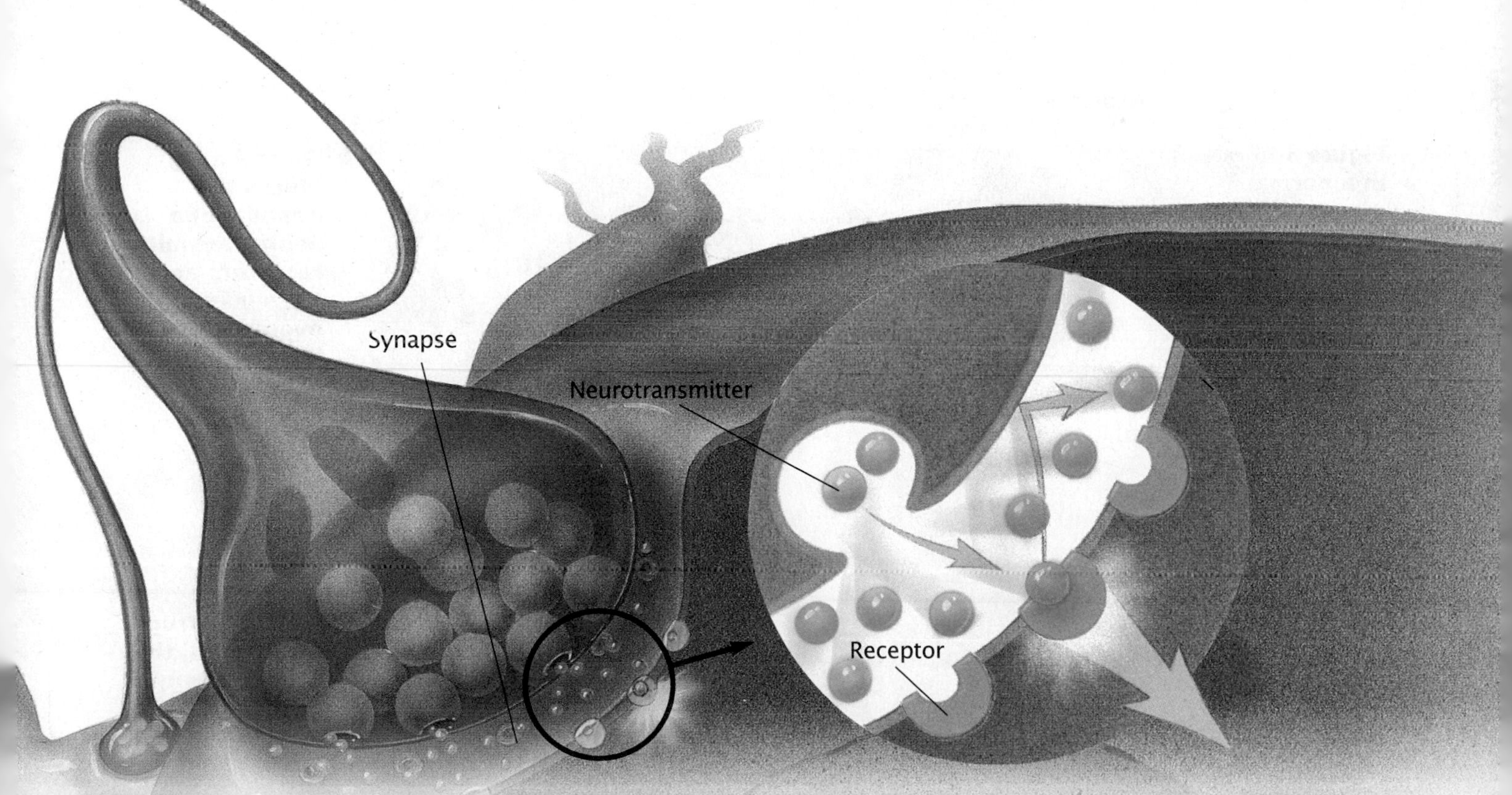

Figure 30-6

a **Depression can result from a shortage of the neurotransmitter serotonin (green spheres).**

b **The drug Prozac® works as an antidepressant because it blocks reabsorption of serotonin from the synapse, thus increasing the level of the neurotransmitter.**

Addiction is caused by a change in the number of receptors

When a neuron is exposed to a signal for a prolonged period of time, it tends to lose its ability to respond to the stimulus with its original intensity. For example, when you put on a wristwatch, you soon become used to wearing it and forget that it is there. Neurons are particularly affected by prolonged exposure to a signal. When neurotransmitters remain in a synapse for a prolonged period of time, a neuron will respond by building fewer receptors for its membrane. This feedback mechanism enables a cell to adjust its number of receptors to match levels of a neurotransmitter.

This feedback mechanism is illustrated in **Figure 30-7**. In this figure, a drug that blocks the removal of neurotransmitter molecules is present in a synapse. As a result, large amounts of the neurotransmitter remain in the synapse for a long time. In response, the neuron produces fewer receptors, creating a less sensitive neuron.

This change in the number of receptors is one of the physiological changes that occur in chemical dependence. As long as the use of the same dosage of the drug continues, large amounts of neurotransmitter molecules remain in the synapse. But when the drug is absent, neurotransmitter molecules can again be removed, and the level of neurotransmitters in the synapse declines. However, the neuron now has too few receptors to receive information. In the absence of the drug, the neuron is unable to function normally until it restores its original number of receptors, which takes time. In drug-treatment programs, a drug is sometimes withdrawn slowly to allow the number of receptors to adjust gradually, thus minimizing withdrawal symptoms such as anxiety, depression, and strong cravings for the drug. As you can see, addiction is not simply a psychological state that can be overcome by willpower; addiction is a physiological dependency caused by a change in the number of receptors.

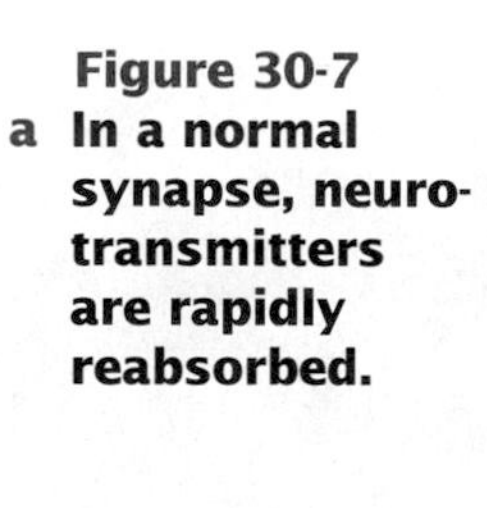

Figure 30-7
a In a normal synapse, neurotransmitters are rapidly reabsorbed.

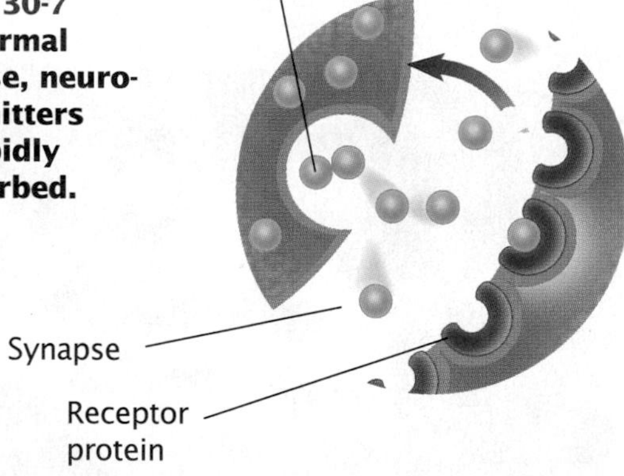

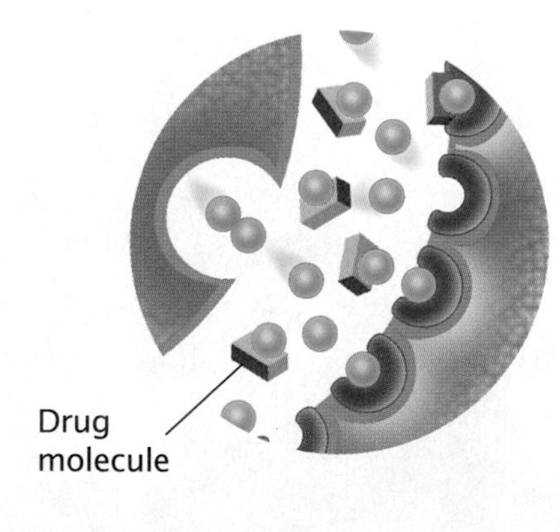

b When a drug blocks the removal of a neurotransmitter, receptors across the synapse are overstimulated by the excess neurotransmitters.

c The neuron responds by reducing the number of receptors.

d When the drug is removed, the excess supply of neurotransmitters dwindles. Now there are too few receptors in the synapse to receive the message.

Psychoactive drugs affect neurons in a variety of ways

Most psychoactive drugs affect the nervous system by regulating activity at the synapse. For example, narcotics are powerful drugs derived from a species of poppy called *Papaver somniferum*. When this substance is prepared in the form of drugs such as opium, morphine, or heroin, it can be used to relieve pain or induce sleep. Narcotics function by imitating enkephalins, a natural pain reliever produced in your body. Enkephalins open potassium channels in spinal neurons, which blocks pain messages from traveling to the brain. Cocaine is another highly addictive psychoactive drug. It is extracted from the leaves of coca plants that grow in the mountains of South America. In the body, cocaine blocks the reabsorption of dopamine, a neurotransmitter that helps send pleasure messages to the brain. The dopamine trapped in the synapse continually stimulates neurons, producing an intense euphoria. Cocaine can also cause agitation, paranoia and severe depression. The mode of action of these and several other drugs are summarized in **Table 30-1**.

Table 30-1 Drug Actions and Effects

Drug	Mode of action	Effects
Nicotine (found in tobacco products such as cigarettes, chewing tobacco, and snuff)	Stimulants; mimic the effects of acetylcholine, a neurotransmitter that stimulates skeletal muscles	Contains highly mutagenic substances; highly addictive
Cocaine, crack (made from leaves of coca plants)	Stimulants; prevent reabsorption of dopamine, causing overstimulation of nerve pathways leading to regions of the brain that interpret pleasure	Highly addictive; can be fatal
Amphetamines, caffeine	Stimulants; stimulate the release of noradrenaline, a neurotransmitter that constricts blood vessels and elevates levels of blood glucose	Irregular heartbeat, high blood pressure, stomach disorders, exhaustion, violent behavior
Barbiturates, tranquilizers, phencyclidine hydrochloride (PCP)	Depressants; interact with GABA, a neurotransmitter that inhibits nerve impulses	Drowsiness, depression, emotional instability; especially dangerous when mixed with alcohol use
Narcotics (opium, morphine, heroin)	Mimic the effects of enkephalins, chemicals that act as natural pain relievers in your body	Extreme anxiety, tremors, drowsiness
Marijuana	At different dose levels it can produce sedative or hallucinatory effects.	Lung damage, loss of motivation, short-term memory loss
Hallucinogens	Inhibit the action of the neurotransmitter serotonin	Dangerous hallucinations, unpredictable behavior
Inhalants	Depressants; depress the central nervous system and tend to slow down the function of the brain and spinal cord	Hallucinations; permanent brain, kidney, and liver damage; can be fatal

Alcohol changes the shape of receptors

Of all the psychoactive drugs, alcohol (ethanol) is one of the most widely used and abused. Alcohol is a product of fermentation; wine is made from fermented grapes, and beer from fermented grain. Both have been consumed for centuries.

Alcohol is known for its ability to reduce inhibitions and produce a sense of well-being. Unfortunately, consuming alcohol can alter both judgment and reaction time, a condition described as being "drunk." Every year, many high school students are killed in automobile accidents caused by drunken driving, making alcohol a dangerous drug in more ways than one.

Figure 30-8
This young woman, like many other Americans, is seeking help to overcome her addiction to alcohol.

Alcohol is an unusual psychoactive drug. Unlike narcotics and cocaine, it does not bind to receptors. Instead, research has shown that alcohol is able to alter the structure of the cell membrane, producing changes in the shapes of receptors. An altered receptor may become more sensitive to a stimulus or may be, in effect, switched off. As you can imagine, a change in a receptor can affect normal brain function in complicated ways. Alcohol also blocks nerves that repress the region of the brain responsible for feelings of pleasure and exhilaration.

Addiction to alcohol, or alcoholism, is the major drug-abuse problem in the United States. People who drink excessively over long periods of time can develop serious health problems. For example, many alcoholics suffer from a vitamin deficiency because they do not eat properly when drinking heavily. This can lead to malnutrition, abnormalities in the circulatory system, and inflammation of the stomach lining. In addition, alcohol is readily converted to energy, preventing the body from breaking down other nutrients, such as sugars, amino acids, and fatty acids. These nutrients are stored as fat in the liver. After years of drinking, liver cells become clogged with fat and die. If heavy drinking continues, a liver condition called cirrhosis may develop. In cirrhosis, liver cells are replaced with useless scar tissue, and the liver gradually shrinks into a small, hard mass, and can no longer function normally. Overcoming an addiction to alcohol can require physical, mental and emotional support, which counseling, as seen in **Figure 30-8,** can help to provide.

SECTION REVIEW

1. **How does the structure of a neuron suit its function?**
2. **Devise a table that summarizes the events of a nerve impulse.**
3. **What role do neurotransmitters play in nerve-impulse transmission?**
4. **How do psychoactive drugs cause addiction?**

Signals from nerve cells enable you to play a piano, throw a baseball, write in your notebook, or just sit and think. A highly developed brain coordinates activities and enables you to learn, imagine, remember, and reason. The complex system that coordinates and controls body functions is the nervous system.

30-2 The Nervous System

OBJECTIVES

1. **Describe the functions of the two main parts of the nervous system.**
2. **Discuss the functions of four parts of the brain.**
3. **Describe two types of technology used to study the brain.**
4. **Explain the roles of the peripheral nervous system.**

Organization of the Human Nervous System

You can think of your nervous system as having two main parts. Your **central nervous system** consists of your brain and spinal cord. It is your control center, organizing all incoming and outgoing information. Your brain and spinal cord are connected to the rest of your body by nerves that make up the **peripheral nervous system**. The sensory neurons and the motor neurons that compose the peripheral nervous system act as independent communications pathways in that each monitors a different part of the body. The organization of the human nervous system is summarized in **Figure 30-9**.

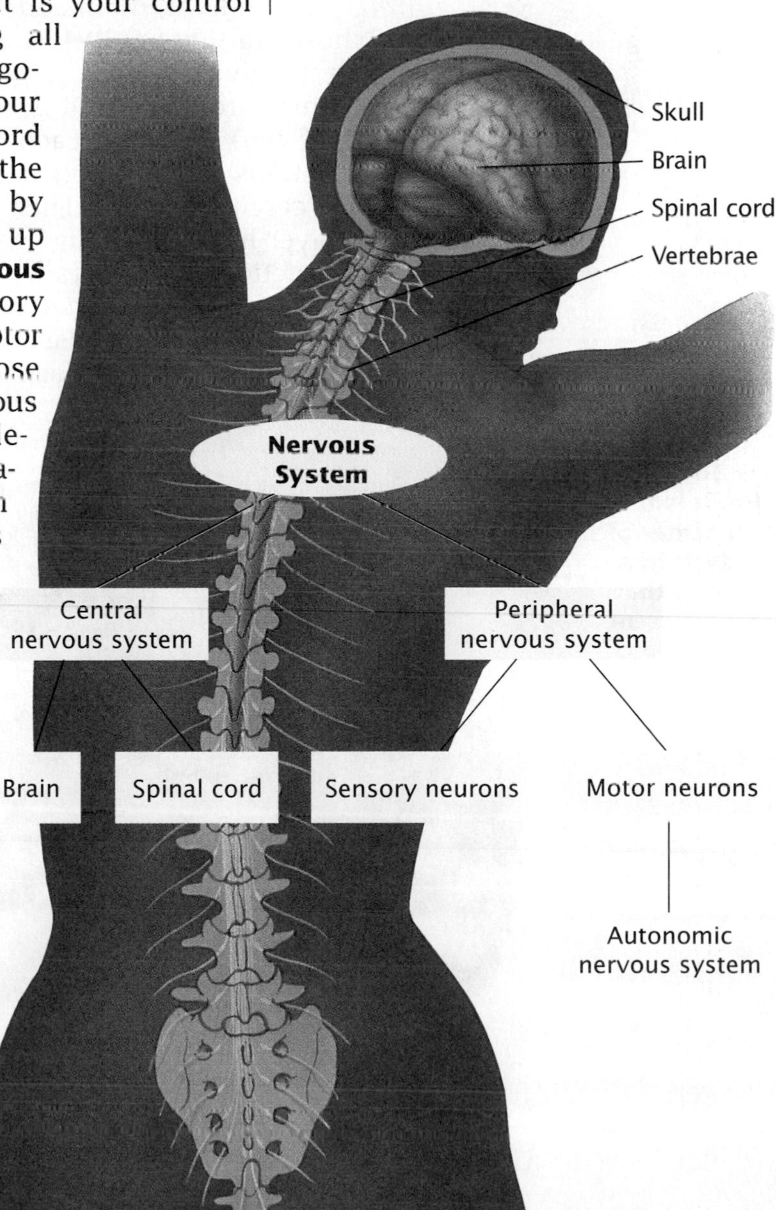

Figure 30-9
The human nervous system can be organized into different categories based on functions. The fragile organs of the central nervous system are protected by the vertebrae and the skull.

The Central Nervous System

At any given time of the day, your brain and spinal cord are being bombarded with information from sensory organs and receptors located throughout your body. After sorting this vast amount of information, they issue commands to your muscles, glands, and organs. Here is a brief look at some important structures found in the central nervous system.

The cerebrum is the control center of the brain

In the average adult, the brain is one of the largest organs of the body, weighing about 1,500 g (approximately 3 lb). About 85 percent of the weight of the brain is made up of the **cerebrum** *(suh REE bruhm)*, the familiar, wrinkled outer layer shown in **Figure 30-10a**. The cerebrum is divided into two hemispheres, which are connected by a large band of nerve fibers. Each hemisphere contains an outer gray layer called the cerebral cortex. This 2–4 mm thick layer handles the most sophisticated functions of the brain, from processing visual images to thinking and planning. Because it is easily accessible, the cerebral cortex has been extensively studied, and some parts have been mapped in great detail. Underlying the cerebral cortex is a region of white matter that relays information to the cortex.

The cerebellum coordinates muscle movements

Tucked beneath the rear of the cerebrum is the cauliflower-shaped structure called the **cerebellum** *(sehr uh BEHL uhm)*, shown in **Figure 30-10c**. The cerebellum is essential for coordination of movement. Damage to the cerebellum causes uncoordination and tremors. During the course of vertebrate evolution, the cerebellum has increased in size and complexity in many animals. While well developed in mammals, it is even more developed in birds. Birds maneuver through the air, which requires complicated feats of balance that most mammals never have to face. Imagine the coordination, balance, and control needed to land on a branch or fence post at precisely the right moment. You can learn more about the differences among vertebrate brains in the *Evolution of the Brain* on pages 616 and 617.

Figure 30-10 The brain is the control center of the body. It has many parts that control different body functions.

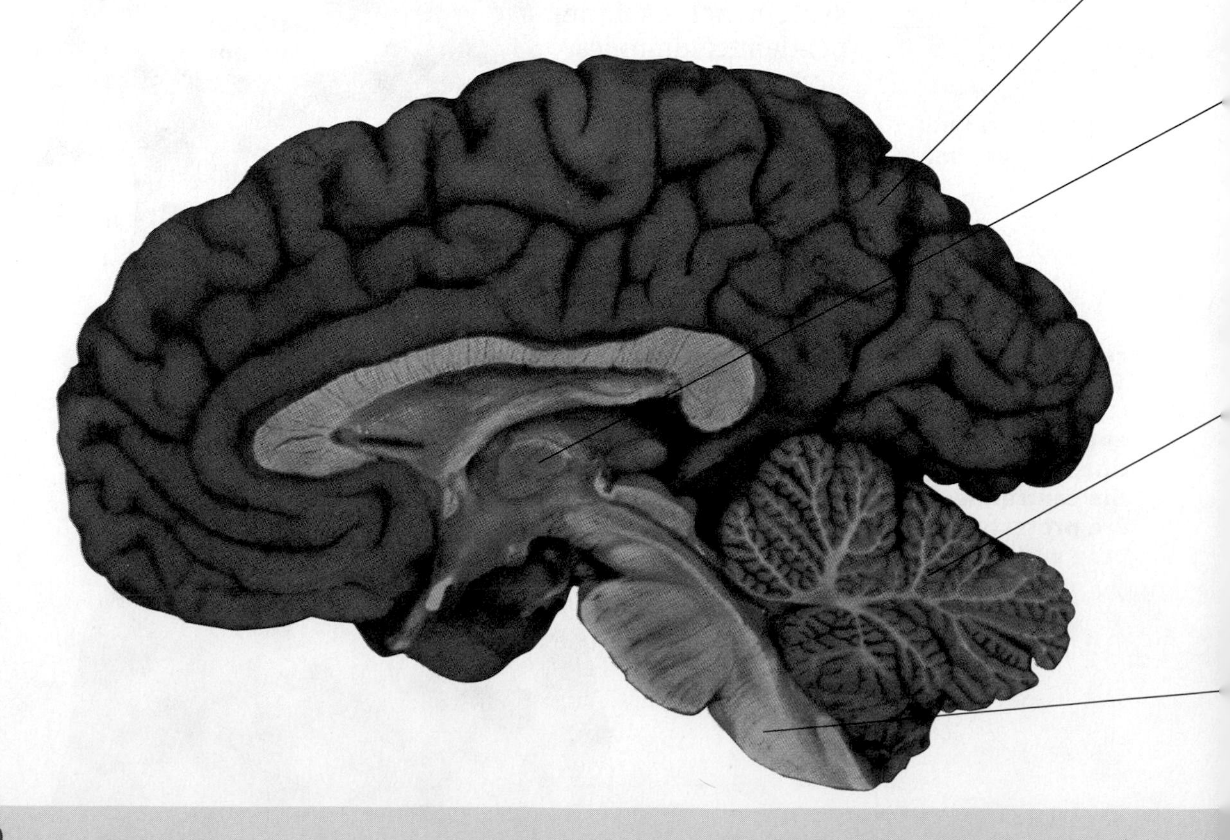

The brain stem controls essential body processes

Underneath the cerebrum lies the brain stem, a stalklike structure buried deep in the brain. The brain stem, shown in **Figure 30-10d**, contains nerves that control vital life processes such as breathing, swallowing, and digestion as well as heart rate and blood pressure. Many nerves that control muscles in the body cross in the brain stem. Thus, damage to one side of the brain above the crossover causes weakness or paralysis on the opposite side of the body.

An extremely important network of nerves called the reticular formation also runs through the brain stem. The reticular formation is the brain's arousal system; it activates other parts of the brain with appropriate information. Consisting of nerves running to and from the brain, the reticular formation "listens to" the messages entering and leaving. It then selects which messages to enhance and which to suppress. In a very real way, it "decides" what you will be aware of. Your reticular formation enables you to sleep through repetitive noises such as traffic yet awaken when an alarm clock rings.

The brain stem also contains the **hypothalamus**, shown in **Figure 30-10b**, the body's center for emotions and instincts. Scientists have identified areas in the hypothalamus that control hunger, thirst, body temperature, water balance, and blood pressure as well as sexual desire, pleasure, hostility, and pain.

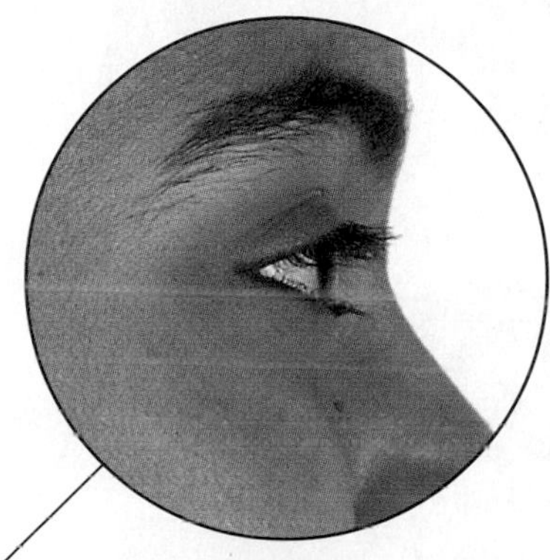

a The cerebrum is the center of intellect, memory, language, and consciousness. It receives information from this girl's eye and translates it into a meaningful image.

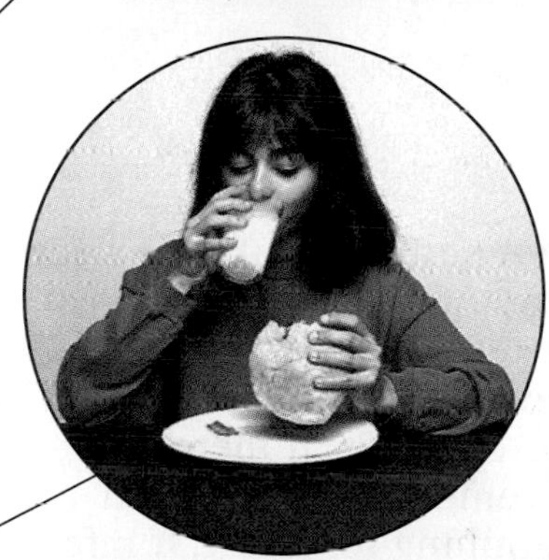

b The hypothalamus controls your body's homeostasis, maintaining functions such as body temperature and water balance. It also regulates hunger and thirst and tells this girl it is time to eat lunch.

c The cerebellum is responsible for smooth, coordinated body movements. It helps this gymnast maintain posture, muscle tone, and equilibrium.

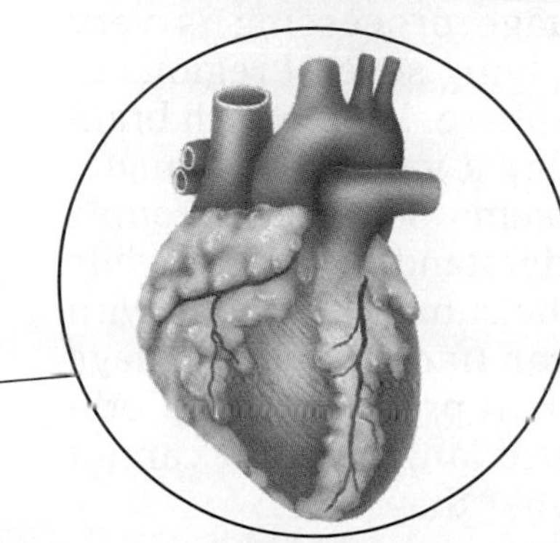

d The brain stem regulates vital body processes that you do not consciously control, such as heartbeat, respiration, and blood pressure.

The spinal cord shuttles information between the body and brain

Messages from the body and brain run up and down the spinal cord. The spinal cord is a cable of nerve tissue that extends from the brain stem. Like the brain, the spinal cord contains gray matter, which forms a column down the center of the cord. A layer of white matter surrounds the column. The spinal cord is surrounded and protected by the backbone, a tunnel of bone formed by the rings of the vertebrae. Injuries to the spinal cord can result in pain, numbness, or paralysis in limbs and extremities.

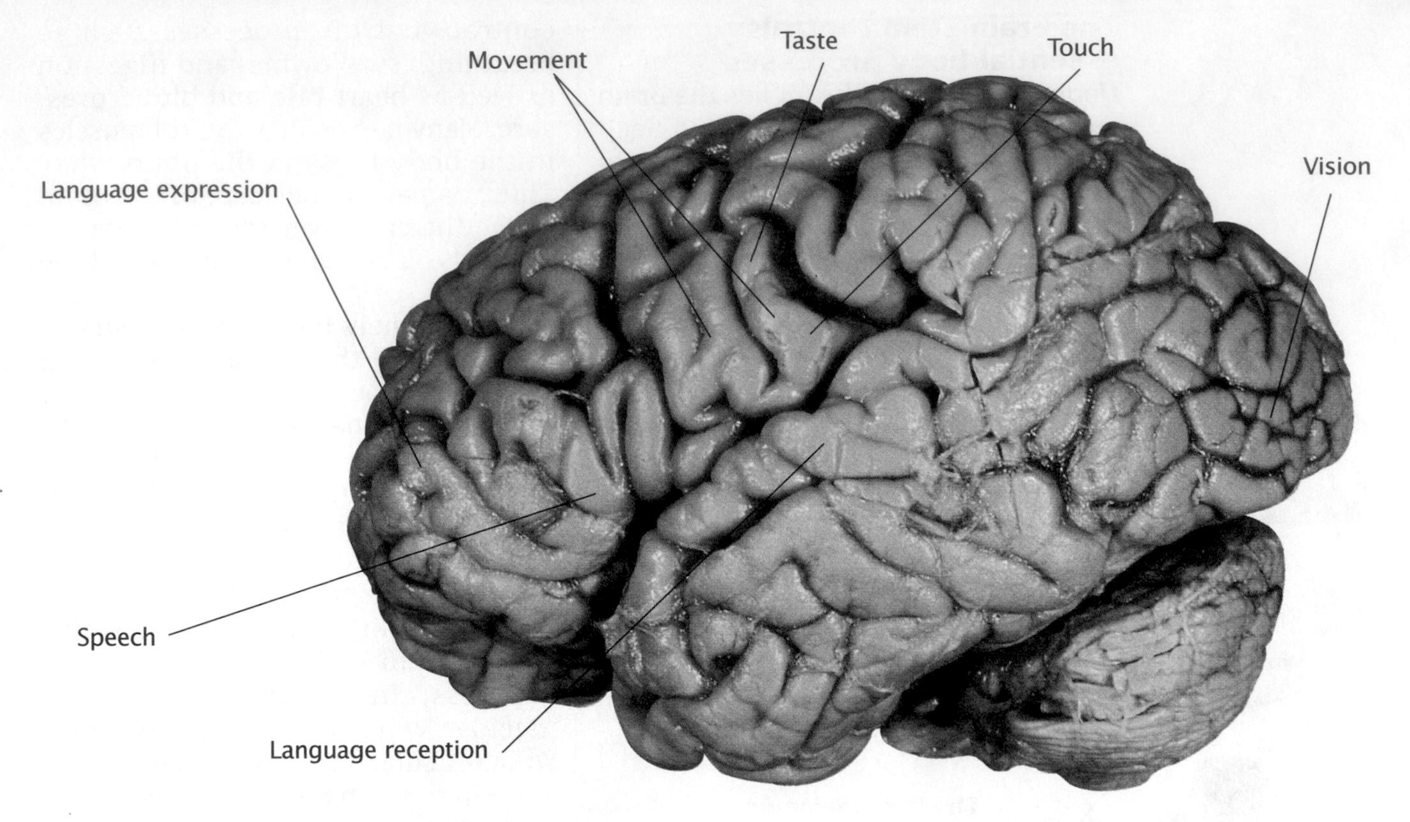

Figure 30-11 Advances in technologies are giving scientists new information about the exact position of many sites controlling the brain's processing.

Mapping the Brain

For centuries, people have been intrigued by the mysteries of the brain. In recent years science has taken great strides toward unraveling its secrets. Advances in technologies are now giving scientists the opportunity to probe deeply into the living tissues of the brain and to construct a blueprint of its functions, such as the map shown in **Figure 30-11**.

Learning and memory are related brain functions

Learning and memory are brain functions that are especially well developed in humans. Learning occurs when two or more pieces of information are linked into a pattern of connections between neurons. The repeated stimulation of this pattern of neurons results in memory. Almost every memory you have is made of different patterns of connections. Many overlapping patterns are stored together, so a single stimulus can cause a flood of memories, which you may have noticed if you've ever caught yourself "daydreaming."

Memory lasts for different lengths of time. When you take notes in class, you often remember what you hear only long enough to write it down. This is short-term memory, which can last from a few seconds to a few hours. You use your short-term memory for things like remembering to run an errand or cramming for a test. If information is stored for any length of time, it is transferred to long-term memory. Here it may remain for life.

Language is processed in the left hemisphere

Although scientists have learned a lot about language processing in the brain, much remains a mystery. Research has indicated that the human brain is already wired at birth with the nerve pathways necessary for language. Language processing is very complex, involving several regions in the left hemisphere. People with brain damage display a variety of unusual language problems. Some have trouble using and understanding nouns, while others have the same difficulties with verbs. Some can understand language but are unable to produce it. Still others can produce language but cannot understand what they hear.

How scientists study the brain

Scientists and researchers are able to study the brain by using technology that has become increasingly sophisticated over the years. The invention of the computerized axial tomography (CAT) in the early 1960s marked an important era in brain research. When undergoing a CAT scan, a patient is placed on a table that slides into a large tube. Inside the tube, a weak X-ray beam rotates around a person's head. Electronic sensors pick up the emerging rays and feed them into a computer, where they are integrated into a single image. This image can be viewed on a video screen and photographed. CAT scans are used to locate brain tumors, blood clots, and areas of cerebral damage.

Other new technologies used to study the brain and other organs have yielded diagnostic and investigative advances. In magnetic resonance imaging (MRI), a patient is placed in an intense magnetic field that stimulates the protons in the brain to resonate or vibrate and emit radio frequency pulses. This information is transformed into images such as the one shown in **Figure 30-12a**. These images reveal different tissue characteristics of the brain.

Positron emission tomography (PET) reveals tissue structure as well as biochemical activity in the brain. In a PET scan, a patient is injected with glucose (the main energy source for brain activity) that has been prepared with radioactive labels. The labels emit radioactive particles called positrons, which are detected by the PET scanner. Computer processing produces colored images, as shown in **Figure 30-12b**, that indicate a region of activity in the brain. PET scans can be used to diagnose brain tumors and strokes as well as mental illnesses such as Alzheimer's disease and schizophrenia. It is also being used to study how the brain processes language and the ways it constructs images from both memory and imagination.

All of these techniques have advantages and disadvantages, depending on which region and activity of the brain is being studied. Such noninvasive methods for studying the living brain has enabled researchers to probe the brain in ways never before possible.

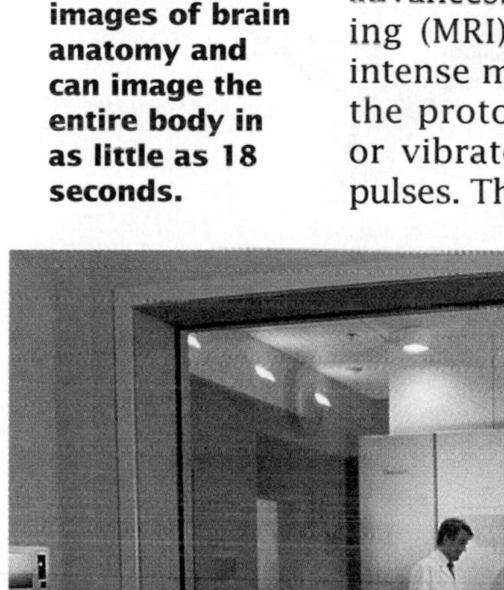

Figure 30-12
a Magnetic resonance imaging (MRI) creates images of brain anatomy and can image the entire body in as little as 18 seconds.

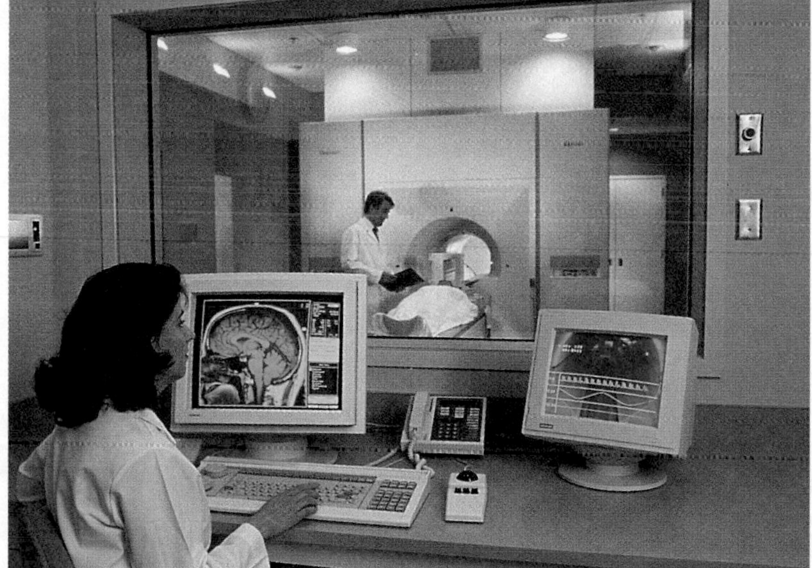

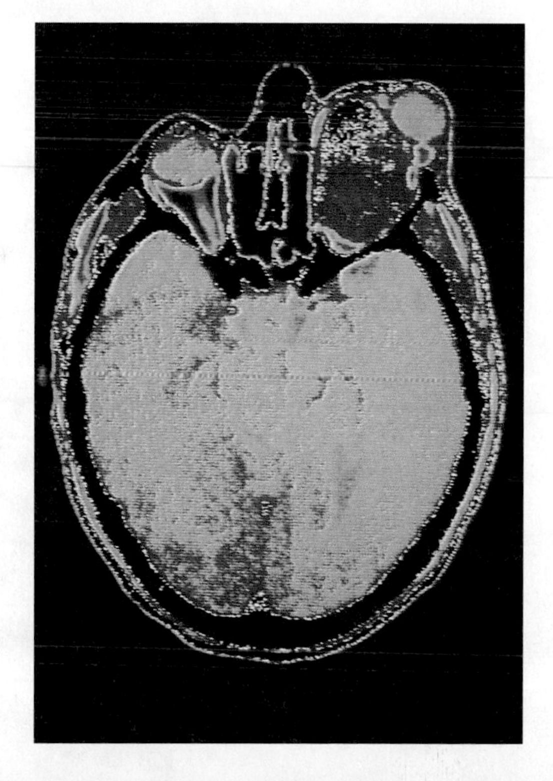

b Computer processing produces colored images such as this brain scan, which enable us to learn more than ever before. For example, this brain scan reveals the presence of a tumor, shown here in red.

Evolutionary Journey

The Brain

Every vertebrate brain has a forebrain, a midbrain, and a hindbrain. As land vertebrates have evolved, the forebrain has become increasingly dominant.

Cerebrum | Midbrain | Hindbrain

A Dominant Hindbrain

1 **In sharks and other fishes, the principal part of the brain is the hindbrain, which is devoted largely to coordinating motor reflexes.**

Cerebrum | Optic lobe | Cerebellum | Spinal cord | Medulla

The rest of the brain serves mainly to process sensory information.

Shark

The Forebrain Enlarges

2 Cerebrum | Medulla | Optic lobe | Cerebellum | Spinal cord

In frogs and other amphibians, and in reptiles, the forebrain is far larger.

A larger cerebrum is devoted to associative activity.

Frog

A Distinct Cerebrum

3 Cerebrum | Cerebellum | Optic lobe | Medulla | Spinal cord

In birds, which evolved from reptiles, the cerebrum is more pronounced.

Bird

A Powerful Cerebrum

5

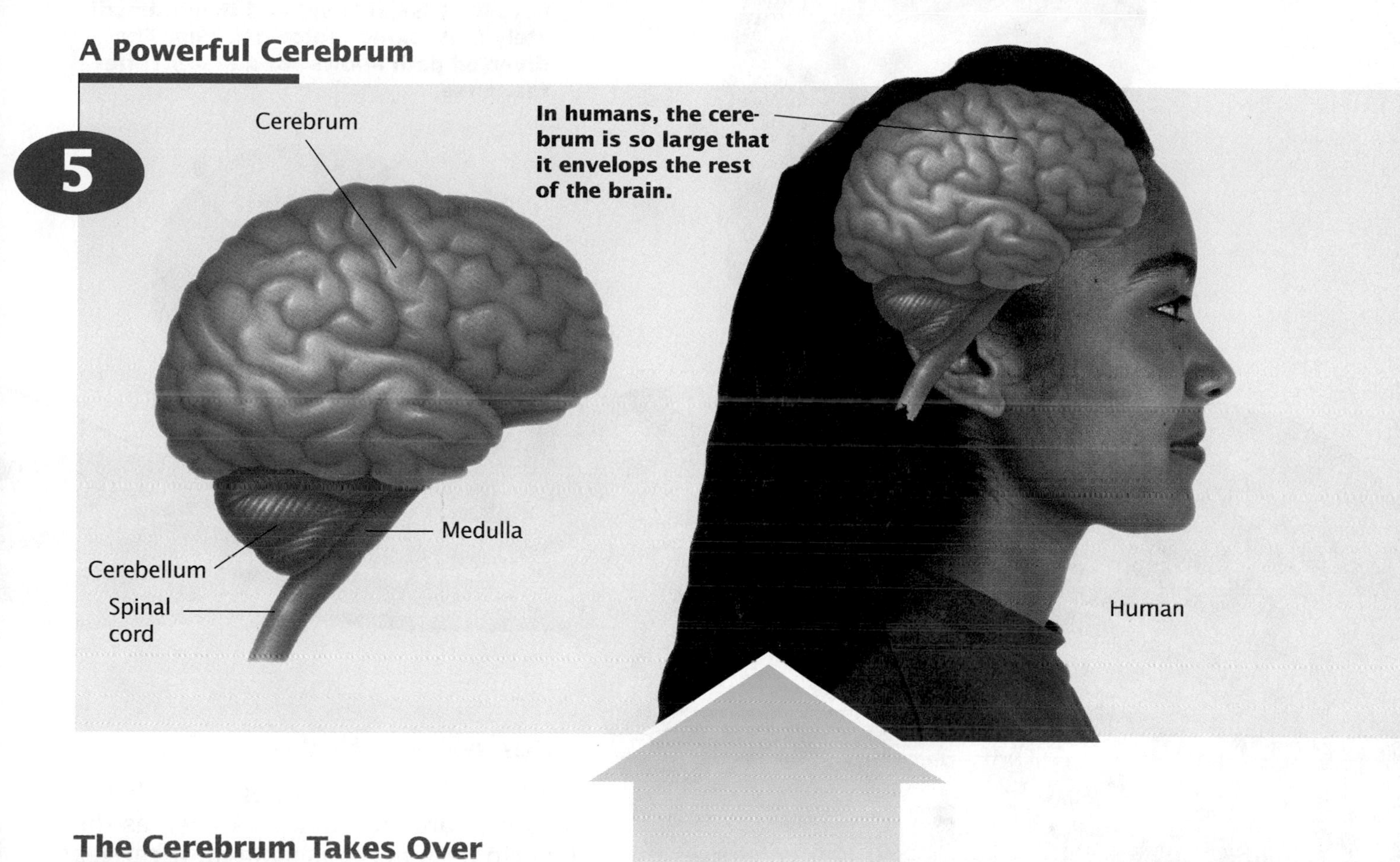

The Cerebrum Takes Over

4

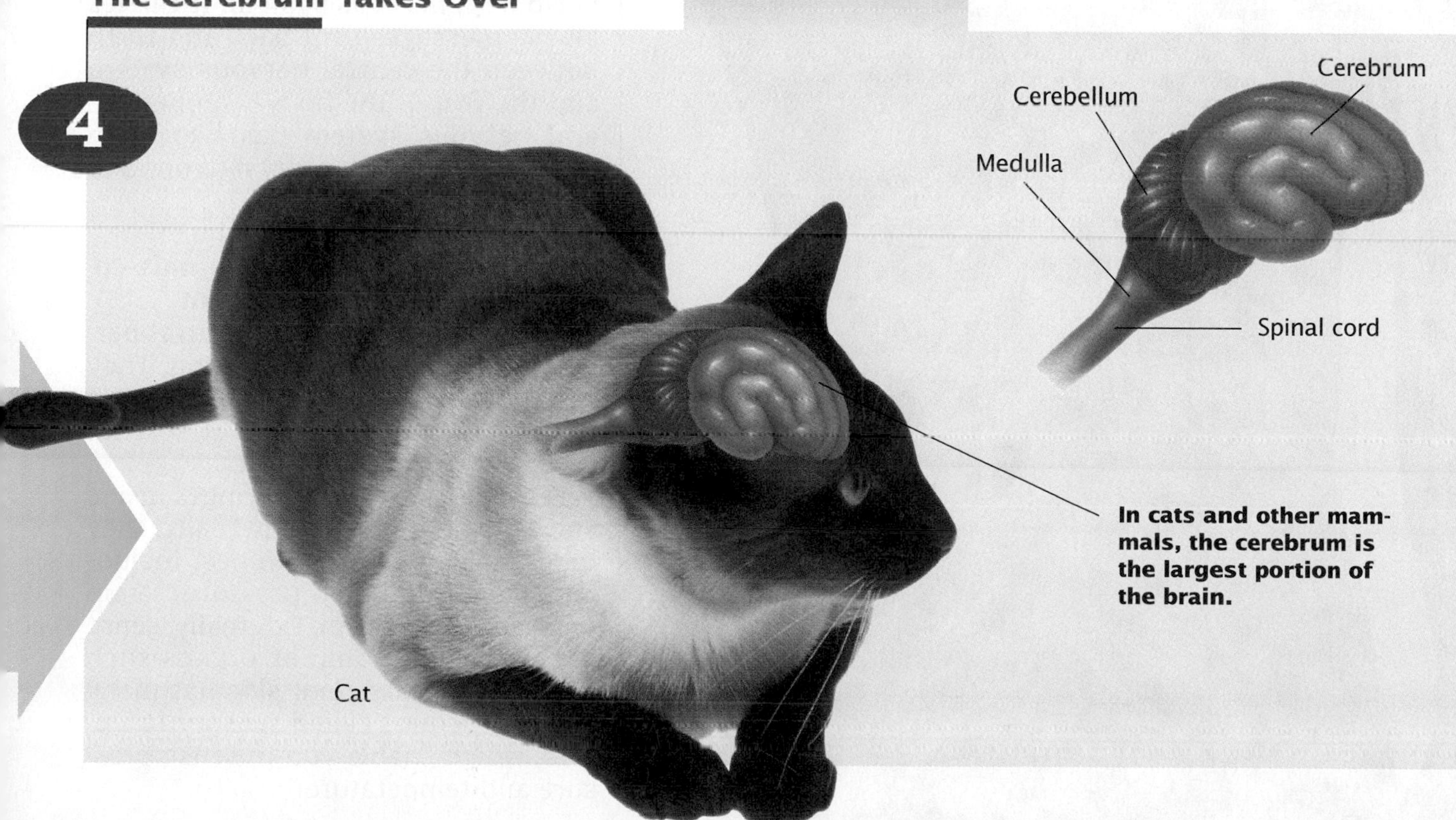

Cerebrum
Cerebellum
Medulla
Spinal cord
In cats and other mammals, the cerebrum is the largest portion of the brain.
Cat

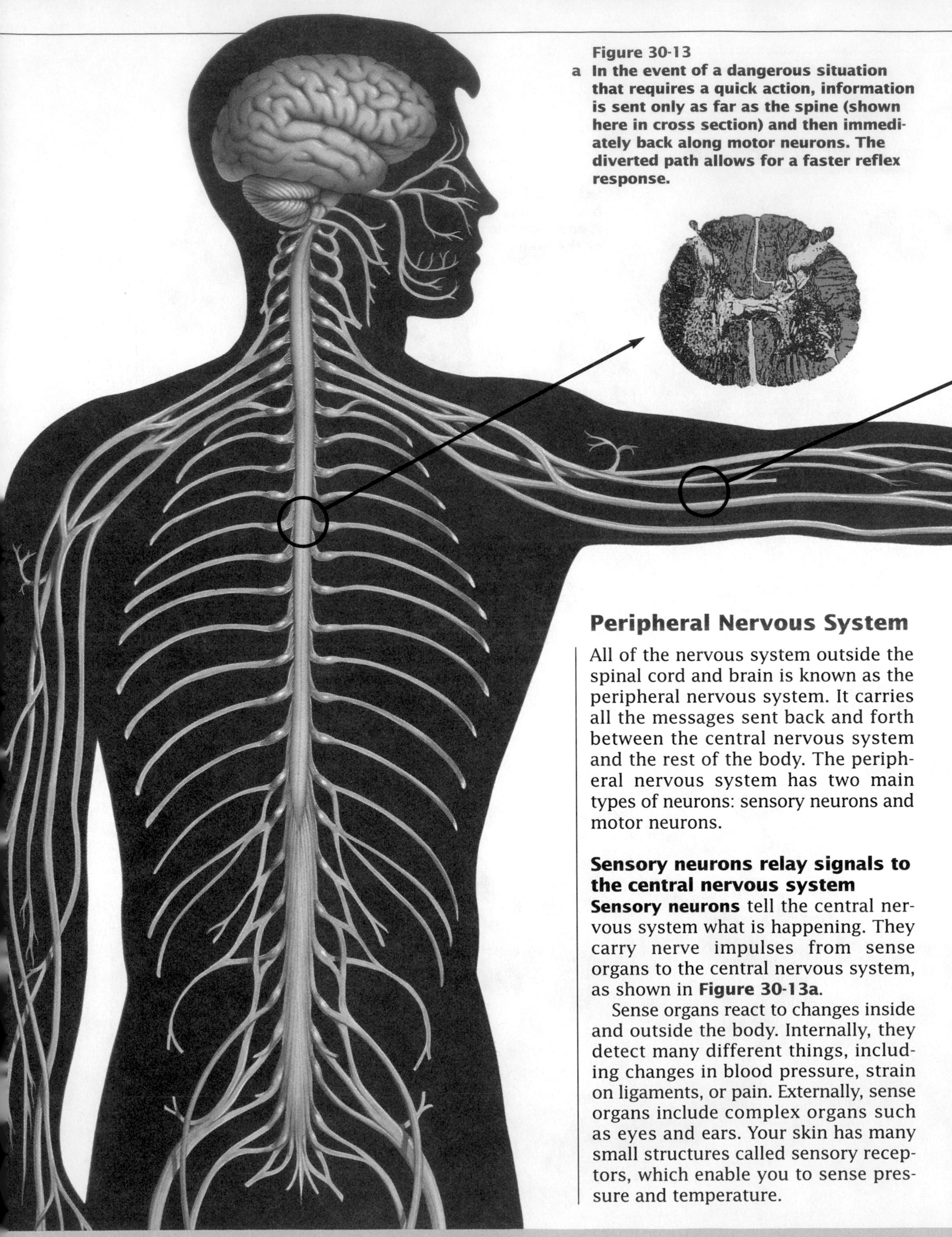

Figure 30-13

a **In the event of a dangerous situation that requires a quick action, information is sent only as far as the spine (shown here in cross section) and then immediately back along motor neurons. The diverted path allows for a faster reflex response.**

Peripheral Nervous System

All of the nervous system outside the spinal cord and brain is known as the peripheral nervous system. It carries all the messages sent back and forth between the central nervous system and the rest of the body. The peripheral nervous system has two main types of neurons: sensory neurons and motor neurons.

Sensory neurons relay signals to the central nervous system

Sensory neurons tell the central nervous system what is happening. They carry nerve impulses from sense organs to the central nervous system, as shown in **Figure 30-13a**.

Sense organs react to changes inside and outside the body. Internally, they detect many different things, including changes in blood pressure, strain on ligaments, or pain. Externally, sense organs include complex organs such as eyes and ears. Your skin has many small structures called sensory receptors, which enable you to sense pressure and temperature.

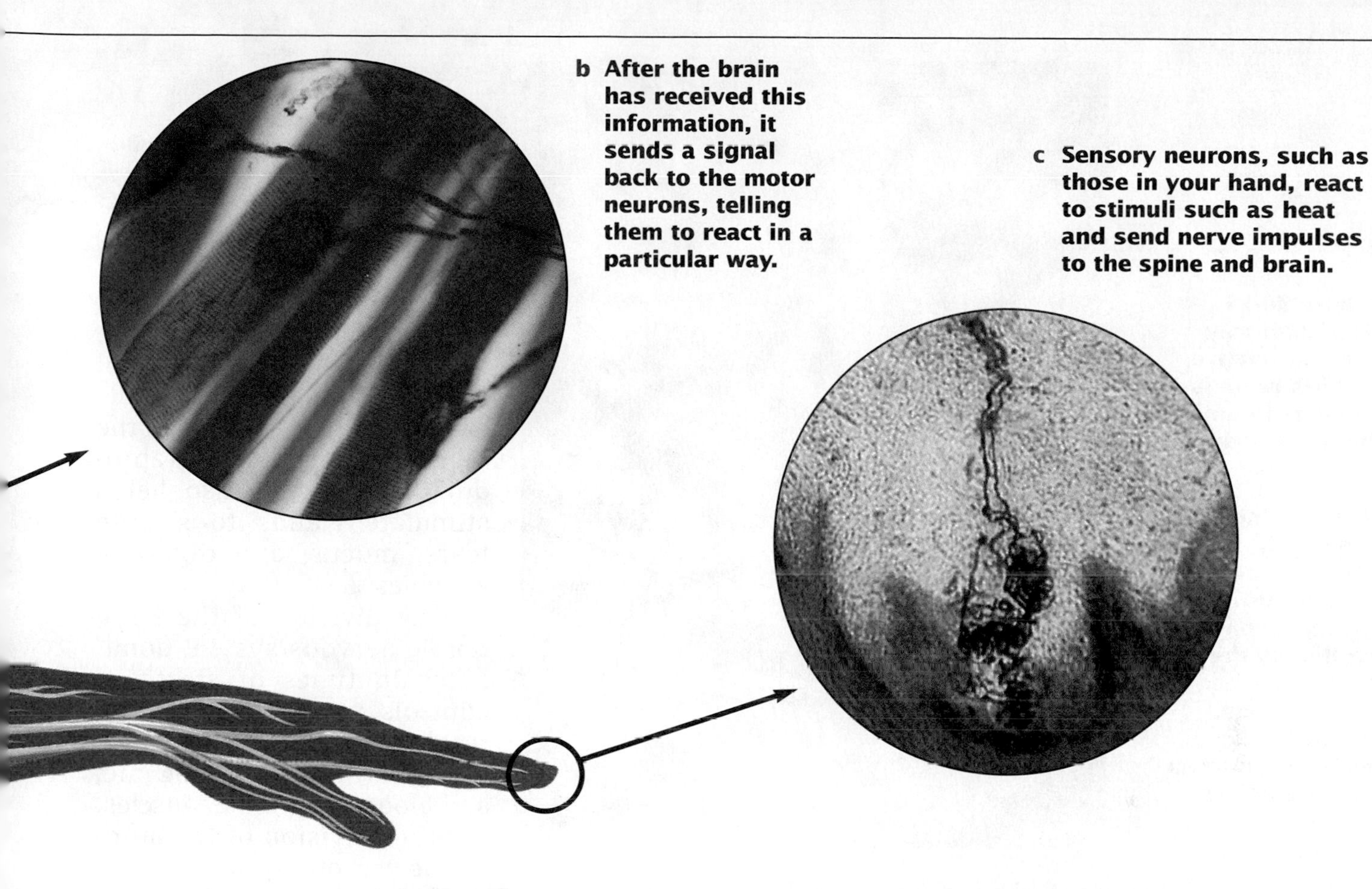

b After the brain has received this information, it sends a signal back to the motor neurons, telling them to react in a particular way.

c Sensory neurons, such as those in your hand, react to stimuli such as heat and send nerve impulses to the spine and brain.

Motor neurons deliver information to muscles and glands

Motor neurons are partners of sensory neurons. Motor neurons carry information from the central nervous system to a muscle or gland, as shown in **Figure 30-13**. They act on the information delivered by the sensory neurons. If your eyes see a runaway truck speeding toward you, the central nervous system sends messages through motor neurons to glands that secrete adrenaline. The adrenaline increases your heartbeat and breathing rate. The central nervous system also sends messages through motor neurons to many muscles, which contract and get the body out of there—fast!

In each segment of the spine, sensory nerves go into the cord and motor nerves come out of it. The motor nerves control most of the muscles below the head. This is why injuries to the spinal cord often paralyze the lower part of the body. A muscle is paralyzed and cannot move if its motor neurons are damaged.

Reflexes enable you to act quickly

All animals have the ability to act particularly quickly at times of danger. These sudden, involuntary movements are reflexes. A **reflex** produces a rapid motor response to a stimulus because the sensory neuron synapses directly with a motor neuron. The escape reaction of a cockroach is a reflex. A good example in humans is a reflex that protects the eye. If anything, such as an insect or a cloud of dust, approaches the eye, the eyelid blinks closed even before we realize what has happened. The reflex occurs before the cerebrum is aware that the eye is in danger.

Reflexes involve few neurons and are therefore very fast. Many reflexes never reach the brain. The nerve impulse travels only as far as the spinal cord, as indicated in **Figure 30-13**. If you step on something sharp, your leg jerks away from the danger. The prick causes nerve impulses in sensory neurons. These neurons synapse in the spinal cord with motor neurons that cause the leg to pull away.

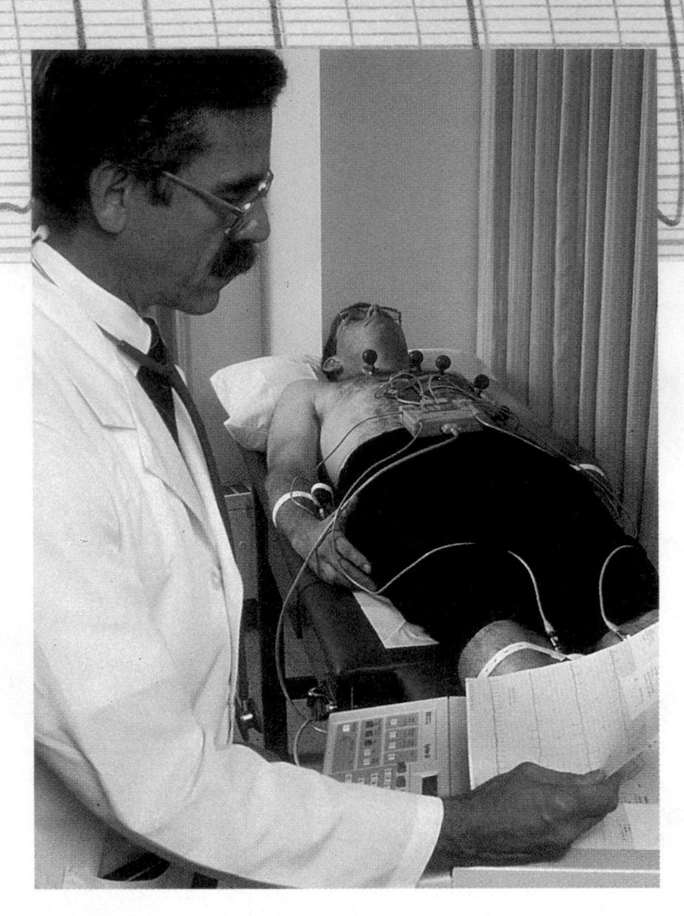

Figure 30-14
a This man may appear inactive, but his brain is busy maintaining his vital body functions.

b The beating of his heart, recorded on an electrocardiogram, is controlled by the autonomic nervous system, which is still active even when you are asleep.

The autonomic nervous system controls internal organs

Some motor neurons are active all the time, even when the body is asleep. These neurons carry messages from the central nervous system that keep the body going even when it is not active. These are the neurons of the **autonomic nervous system**. The autonomic nervous system carries messages to muscles and glands that usually work without our notice. For example, these muscles and glands control the blood pressure and the movement of food through the digestive system even during sleep, as shown in **Figure 30-14**.

The autonomic nervous system enables the central nervous system to govern most of the body's homeostasis. It helps regulate heartbeat and helps control muscle contraction in the walls of the blood vessels and in the digestive, urinary, and reproductive tracts. It also helps stimulate glands to secrete tears, mucus, and digestive enzymes.

One division of the autonomic nervous system dominates in times of stress. It controls the "fight-or-flight" reaction, increasing blood pressure, heart rate, breathing rate, and blood flow to the muscles. Another division of the autonomic nervous system has the opposite effect. It conserves energy by slowing the heartbeat and breathing rate and by promoting digestion and elimination.

Although the autonomic nervous system can carry out its tasks automatically, it is not completely independent of voluntary control. For instance, breathing is controlled by the autonomic nervous system, but one can decide to stop breathing for a short time. However, any voluntary control of the autonomic nervous system that endangers life disturbs homeostasis of the brain tissue, causing unconsciousness. The autonomic nervous system then takes over again and restores normal functions. This is why you cannot hold your breath indefinitely.

SECTION REVIEW

1. **What are the functions of the two main parts of the nervous system?**
2. **Draw a sketch of a human brain and label four main parts. Explain the major functions of each part.**
3. **How is technology enabling scientists to learn more about the brain?**
4. **Compare two different roles of the peripheral nervous system.**

Vivid reports about your environment stream into your brain through your eyes, ears, skin, nose, and mouth. The fact is, there are more than a dozen different types of sensory cells that detect changes outside and inside your body. Together, they help you interact with the world that surrounds you.

30-3 The Sense Organs

OBJECTIVES

1. **Describe four different receptors that are found throughout your body.**
2. **Explain two functions of the inner ear.**
3. **Explain how humans perceive vision.**
4. **Explain how taste and smell are sensed.**

Sensory Receptors

All information about your surrounding environment and the conditions within your body enters your brain through specialized cells called sensory receptors. There are many different kinds of sensory receptors in your body. There are receptors that detect and respond to internal changes, such as when your skin feels temperature changes or pain. Sensory receptors in your blood vessels respond to changes in blood pressure. Sensory receptors in your muscles are responsible for reflexes such as the knee jerk. **Table 30-2** lists several kinds of sensory receptors that your body uses to keep its internal environment constant. To provide information about the external environment, you also have sensory receptors associated with specialized cells and tissues. These form **sensory organs** such as your ears, eyes, nose, and tongue. The sensory receptors in your ears and eyes respond to sound and light to provide you with hearing and vision. The sensory receptors in your tongue and nose respond to particular types of molecules to produce the sensations of smell and taste.

Table 30-2 Sensory Receptors

Respond to internal stimuli	Stimulus	Location	Process
Heat receptor and cold receptor	Temperature changes	Skin and hypothalamus	Changes in temperature alter activity of ion channels
Nociceptor	Pain	In all tissues and organs except brain	Changes in temperature and pressure open membrane channels
Mechanoreceptor	Pressure	Skin surface	Changes in pressure deform membrane
Stretch receptor	Pressure; muscle contractions	Wrapped around muscle fibers	Motion of muscle fibers changes membrane
Respond to external stimuli			
Specialized structures of inner ear	Position and motion; sound	Inner ear	Cilia on cell membranes respond
Taste bud cell	Dissolved chemicals	Tongue	Bind to specific receptors in cell membranes
Olfactory neuron	Scents	Nasal passages	Bind to specific receptors in cell membranes
Rod and cone cell	Light	Retina of eye	Initiates process of focusing and visual processing

Receptors in the Ear

Your ear is really two sense organs in one. It not only detects sound waves but also establishes equilibrium—your sense of position.

The receptor cells that establish equilibrium are located in the fluid-filled semicircular canals and chambers in your inner ear, shown in **Figure 30-15**. Each canal and chamber contains a bed of receptor cells. Each receptor cell has a protruding bundle of hairs. The hairs are covered with a jellylike layer, shown in **Figure 30-15c**, that contains tiny embedded mineral crystals. This layer has the ability to move over the bed of receptor cells. When you change the position of your head or your speed of motion, like the girl in **Figure 30-15a**, the jellylike layer, along with the minerals, slides over the hairs and bends them. This sends signals to the brain, which enables you to determine and adjust, if necessary, your position in space.

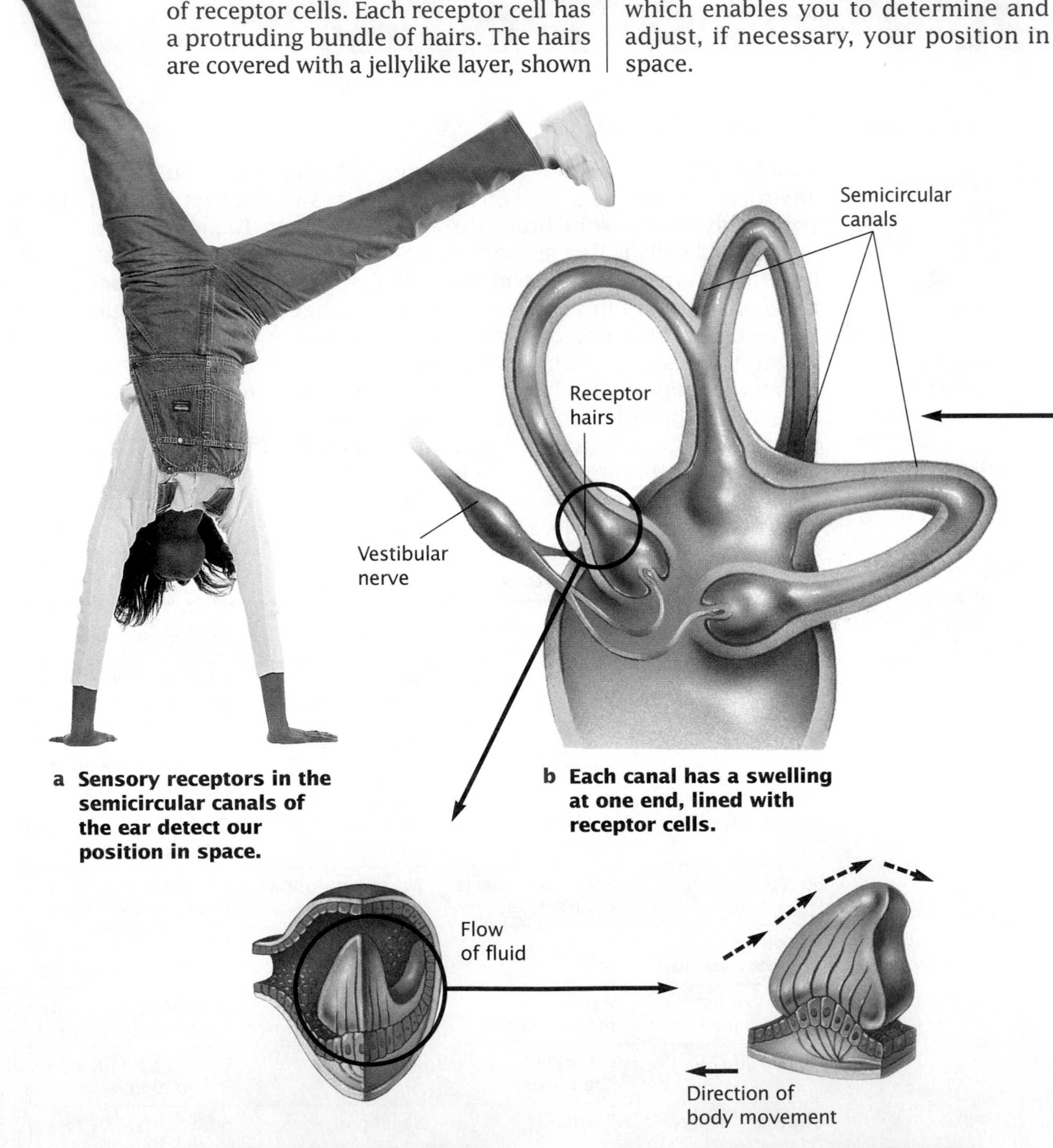

a Sensory receptors in the semicircular canals of the ear detect our position in space.

b Each canal has a swelling at one end, lined with receptor cells.

c When your head is upright, the fluid is still and the hairs are straight.

d Any movement causes the fluid to slide over the hairs and bend them in the opposite direction.

The inner ear also senses sound waves

Look again at the structure of the inner ear in **Figure 30-15**. The cuplike outer ear helps capture sound waves in the air and funnel them into the ear canal. The waves strike the eardrum, causing it to vibrate. These vibrations pass through three small bones that stretch across a cavity inside the ear. One of these bones presses against a snail-shaped structure called the cochlea, shown in **Figure 30-15g**. The cochlea is the organ of hearing. It contains fluid-filled canals that are lined with hair cells. The hair cells project into an overhanging membrane. When vibrations are carried through the cochlea, the hair cells vibrate, moving the hairs up and down against the membrane. As the hairs strike the overhanging membrane, they are bent. This sends signals through the cochlea, as shown in **Figure 30-15g**, and then to the brain, where they are interpreted as sound.

The hairs of the receptor cells are delicate and can be sheared off from forceful vibrations such as those from loud music. Because the hairs cannot grow back, prolonged exposure to loud noise can result in permanent hearing loss. Also, as humans get older, they tend to lose more and more hair cells, which is why some older adults may not hear very well.

Figure 30-15
In the inner ear, there are two kinds of sensory receptors that enable us to hear and maintain our balance.

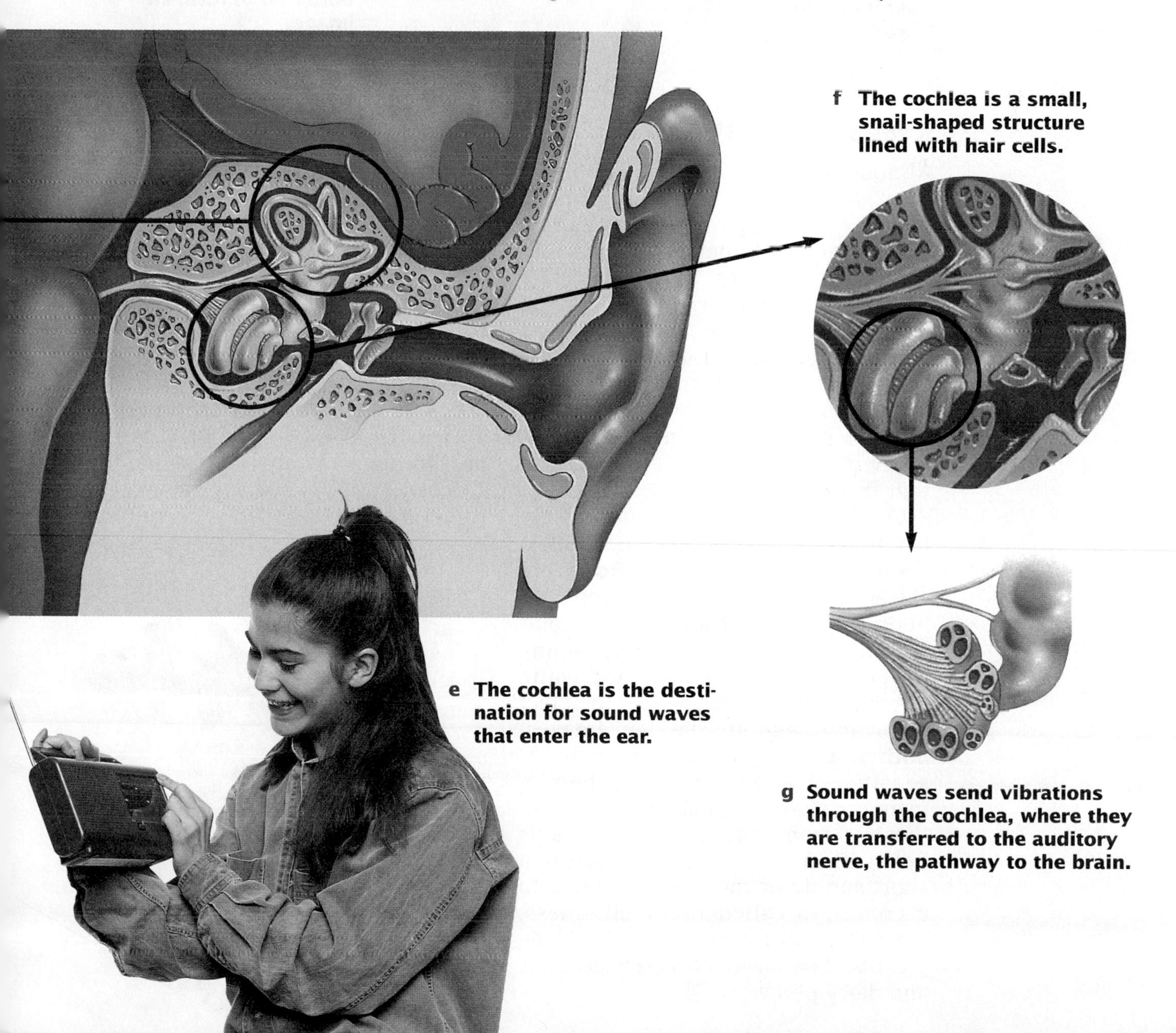

f The cochlea is a small, snail-shaped structure lined with hair cells.
e The cochlea is the destination for sound waves that enter the ear.
g Sound waves send vibrations through the cochlea, where they are transferred to the auditory nerve, the pathway to the brain.

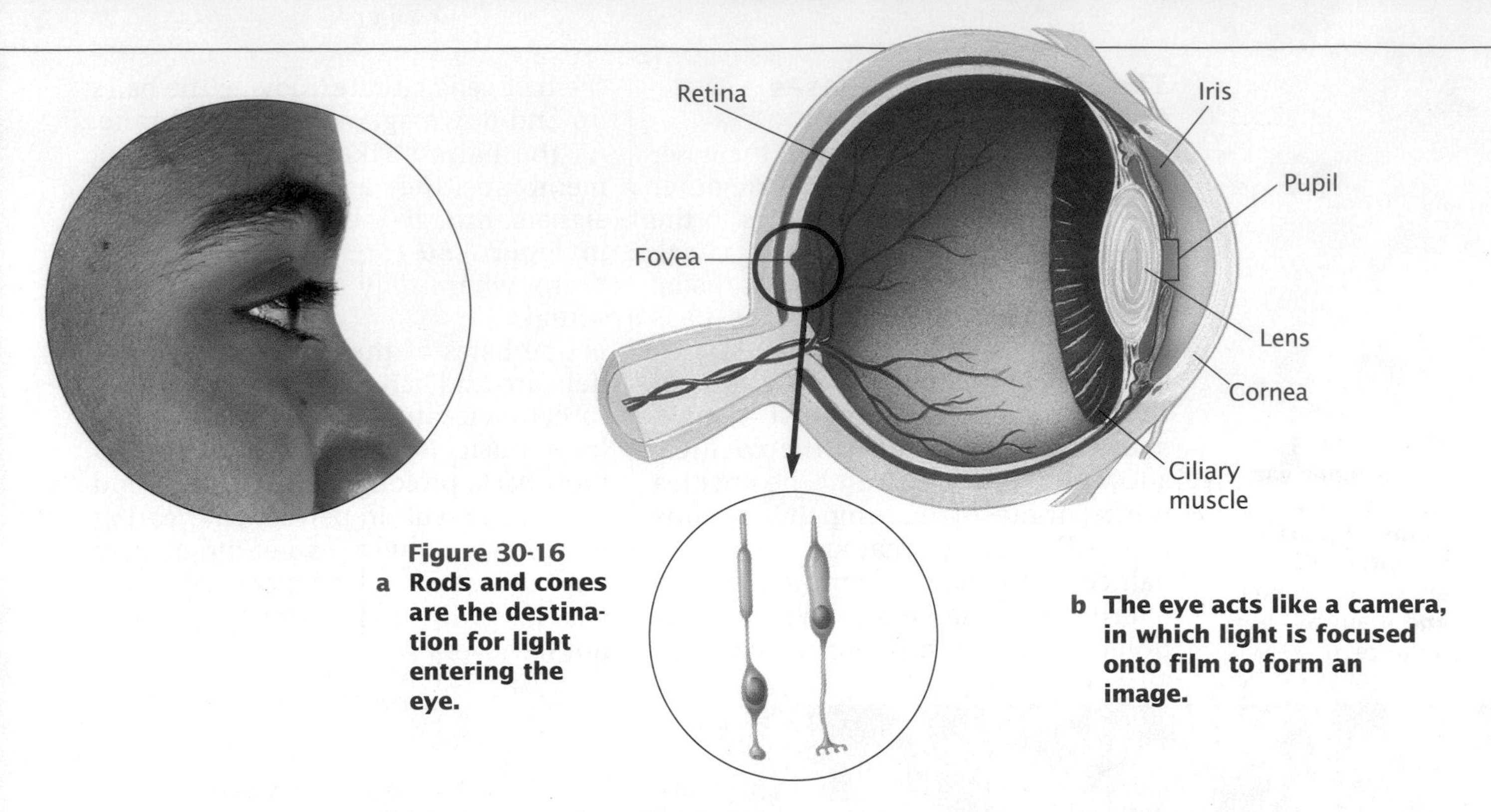

Figure 30-16
a Rods and cones are the destination for light entering the eye.

b The eye acts like a camera, in which light is focused onto film to form an image.

Receptors in the Eye

Almost all animals respond to light. Even some single-celled organisms have a special region that serves as a light detector. Humans and other animals have two eyes, each of which is a complex sense organ that is the starting point of vision. Both eyes contain specialized receptor cells, called photoreceptors, that contain light-sensitive pigments. The energy in light produces changes in these pigments, triggering a nerve impulse that travels to the brain.

You have two types of photoreceptor in your eyes—rods and cones, shown in **Figure 30-16a**. **Rods** are extremely sensitive and function only in dim light. They cannot detect color and therefore generate coarse, grainy images like those of black-and-white photographs. **Cones** are able to detect color and generate detailed images. Both rods and cones contain light-sensitive pigments. In rods, the pigment is called rhodopsin, which is synthesized from vitamin A. A diet that is deficient in vitamin A can result in a short supply of rhodopsin, leading to a condition called night blindness. Important sources of vitamin A include green and yellow fruits and vegetables and dairy products.

Light waves must be focused onto the retina

Rods and cones are packed close together in a layer called the **retina**, which covers the surface of the back of the eye. Before light waves can strike rods and cones, they must travel through a series of transparent materials that bend and focus them onto the retina. As you can see in **Figure 30-16b**, light waves are bent as they pass through an outer transparent film. They enter the eye through an opening called the pupil, which is surrounded by a ring of muscle fibers. The ring of

muscles fibers is able to contract and relax in different configurations, thus regulating the amount of light that enters the eye.

Suspended behind the pupil is an elastic, almond-shaped disk called the lens. Its elasticity is necessary for bending and focusing different angles of light waves that enter the eye. For example, when you are looking at a distant object (one more than 6 m away), light waves are traveling parallel to one another. They require little bending, if any, in order to strike the retina. As a result, the lens stretches, exerting little influence on the angle of incoming light waves. However, when you are viewing an object at close range, light waves are traveling at sharp angles to one another. In order to strike the retina, they must be bent and focused. Your lens accomplishes this by becoming thicker and rounder, which bends the incoming light rays. Finally the light waves strike the retina, stimulating the rods and cones. Nerve impulses travel along the optic nerve to the brain, where they are interpreted as meaningful images, as shown in **Figure 30-16c**.

Perception varies among animals

Humans and other primates have excellent color vision, which is unusual for mammals. Most mammals evolved as nocturnal animals and thus have eyes that are adapted for vision in dim light. Their retinas contain only rods, making them extremely sensitive to shades of gray. Humans are not the only animals that can see color, however. Apparently many fishes and reptiles, as well as most birds, have color vision. Many insects also have color vision, an important adaptation for those that feed on flowers. In addition, many insects and other animals can detect ultraviolet light, but humans cannot.

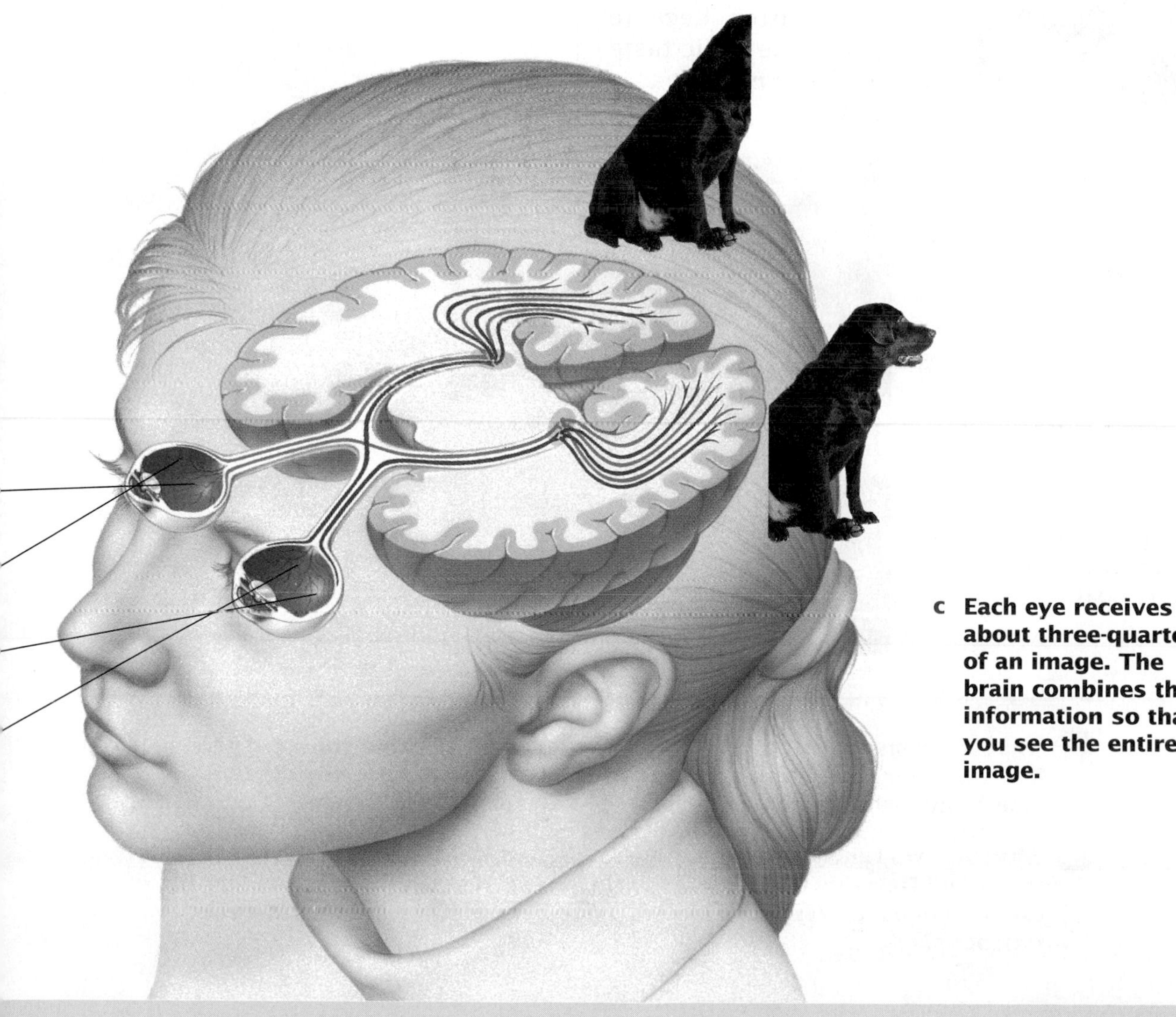

c Each eye receives about three-quarters of an image. The brain combines this information so that you see the entire image.

Receptors for Taste and Smell

Have you ever tried to describe the flavors and aromas of your favorite meal? If so, you have probably realized that your senses of taste and smell are very much alike. The receptor cells for taste and smell are both stimulated by particular chemicals. Your receptor cells for taste are located in taste buds that line the upper surface of your tongue, as shown in **Figure 30-17**. Receptor cells that detect smell are located in the wall of the upper part of your nasal cavity. Before chemicals can stimulate either kind of receptor cell, they must pass through a film of liquid coating the membranes in your mouth and nose. One reason you cannot taste food when you have a cold is that your nasal passages are inflamed and coated with mucus, rendering your smell receptors inoperable.

Humans have four basic taste senses: sweet, sour, salty, and bitter. Receptor cells for these basic tastes are in different areas of the tongue. Most chemicals stimulate two or more types to varying degrees. The sensation described as taste is produced by the blending of four basic senses in different intensities. Incidentally, the "hot" sensations of foods such as chili peppers is detected by pain receptors, not taste receptors.

While there appear to be four basic types of taste receptors, research shows that there may be more than 1,000 different receptors for smell. Most animals depend on their sense of smell to a far greater extent than humans. Certain insects, invertebrates, and mammals release scents that contain information about their species and sex. Their sense of smell can also be used to locate food, recognize family members, and mark territory, which you may have encountered if you've ever walked a dog.

Figure 30-17

a Taste buds in your tongue enable you to sense taste, like the sour bite of a lemon.

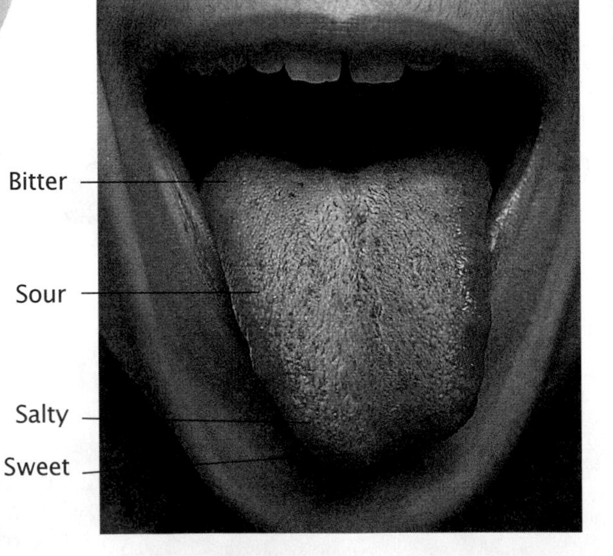

b Your tongue contains receptors that detect four basic tastes.

c Receptors in your taste buds react to chemicals in food. These reactions send nerve impulses to the brain.

SECTION REVIEW

1. **What is a sensory receptor? Describe four receptors that help your body maintain homeostasis.**
2. **What are two functions of the inner ear? Describe the structures responsible for these functions.**
3. **What roles do rods and cones play in vision? How is light focused onto them?**
4. **In what ways are the sensations of taste and smell similar? In what ways are they different?**

CHAPTER 30

Highlights

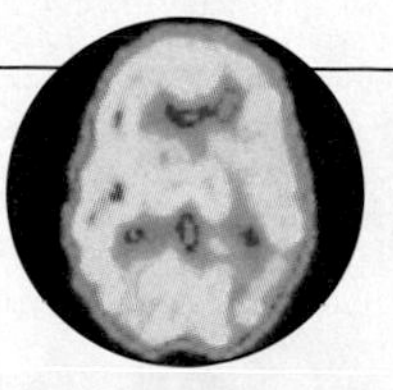
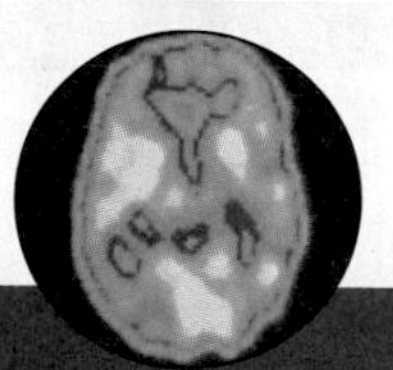

Positron emission tomography (PET) scans are useful for detecting activity in the brain. These scans compare deoxyglucose levels in a normal (top) and a clinically depressed (bottom) person.

	Key Terms	Summary
30-1 Nerve Impulses	neuron (p. 601) nerve (p. 601) dendrite (p. 601) axon (p. 601) myelin sheath (p. 604) synapse (p. 605) neurotransmitter (p. 605)	• A neuron is a cell specialized for conducting electrical information. • A nerve impulse is an action potential traveling along a neuron. During an action potential the balance of sodium and potassium on both sides of the membrane is briefly changed. • Myelin sheaths speed up nerve impulses, which jump from node to node instead of traveling the full length of the axon. • Neurons communicate with each other by sending neurotransmitters across synapses. • Drugs affect a neuron's membrane and cause addiction by altering the number of receptors.
30-2 The Nervous System	central nervous system (p. 611) peripheral nervous system (p. 611) cerebrum (p. 612) cerebellum (p. 612) hypothalamus (p. 613) sensory neuron (p. 618) reflex (p. 619) autonomic nervous system (p. 620)	• The central nervous system consists of the brain and spinal cord. Its main functions are to sort incoming information and to send messages to the body. • The brain is the subject of intense research using new technologies. • The peripheral nervous system consists of nerves that run throughout the body. • The autonomic nervous system regulates involuntary functions and is always active.
30-3 The Sense Organs	sensory organ (p. 621) rod (p. 624) cone (p. 624) retina (p. 624)	• Different receptors in the ear establish equilibrium and detect sound waves. • Receptors in the eye detect light, enabling you to see.

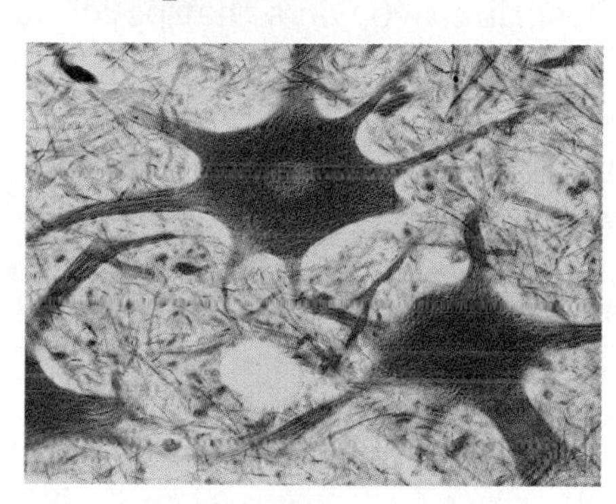

Neurons are the basic units of the nervous system. They conduct electrical information throughout your body.

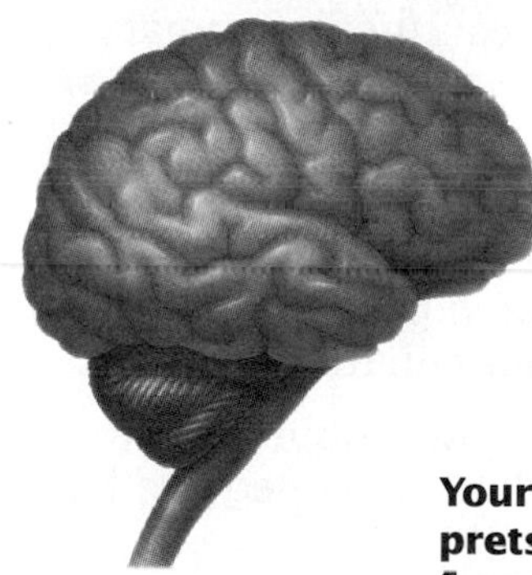

Your brain interprets messages from sensory neurons into meaningful information.

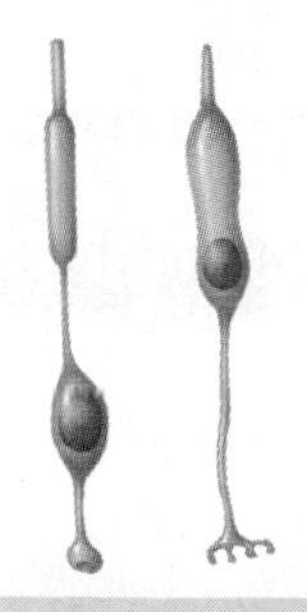

Your eyes contain light-sensitive cells called rods and cones.

CHAPTER 30

review

Understanding Vocabulary

1. For each list of terms, identify the one that does not fit the pattern and describe why it does not fit.
   a. cerebellum, cerebrum, reflex, brain stem
   b. pupil, retina, cochlea, iris
   c. chemical, sound, light, neuron

2. For each set of terms, complete the analogy.
   a. cell composed of dendrites and axons : neuron :: bundles of neurons that form network : ____
   b. incoming message : dendrite :: outgoing message : ____
   c. control center : cerebrum :: muscle coordination : ____

Understanding Concepts

3. **Relating Concepts** Construct a concept map that describes the structure and functions of the two main parts of the nervous system. Try to include the following terms in your concept map: central nervous system, peripheral nervous system, spinal cord, brain, sensory neurons, motor neurons, gray matter, autonomic nervous system, reflexes.

4. **Identifying Structures and Functions** Describe the parts and functions of a neuron.

5. **Identifying Structures** What is a voltage-gated channel? How does it function in the transmission of a nerve impulse?

6. **Reading Graphs** Look at the graph below. The curve was obtained when a neuron was stimulated by an electrode. Explain the shape of the curve in terms of changes in the concentration of ions inside and outside the cell.

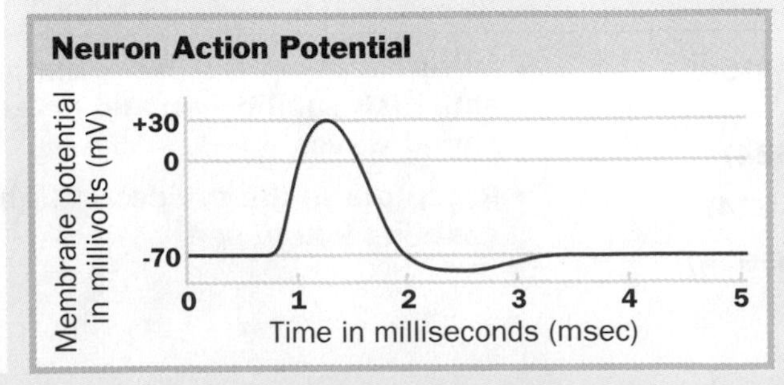

7. **Comparing Functions** Compare the concentration of sodium (Na^{+}) and potassium (K^{+}) ions both inside and outside the cell during the resting potential and during the action potential.

8. **Inferring Relationships** What is a psychoactive drug? Describe two ways that psychoactive drugs affect the way that neurons communicate with each other. Give examples.

9. **Analyzing Processes** Describe the feedback mechanism that results in addiction.

10. **Recognizing Relationships** Describe some of the long-term health problems associated with addiction to alcohol.

11. **Comparing Functions** Identify and compare the functions of the three main parts of the brain.

12. **Predicting Outcomes** Why is a blow to the back of the skull near the brain stem extremely dangerous?

13. **Summarizing Information** Describe two types of technology used to study the brain.

14. **Inferring Relationships** What is a reflex? How does a reflex enable you to act quickly?

15. **Identifying Functions** What is the autonomic nervous system? Describe its primary function.

16. **Inferring Conclusions** Sufferers of vertigo feel dizzy and disoriented in certain situations, such as when they are on a roller coaster. What is the relationship between vertigo and the semicircular canal?

17. **Predicting Outcomes** How would a person be affected if an accident damaged the hairs inside the cochlea?

18. **Comparing Functions** Differentiate between the two types of photoreceptors.

19. **Analyzing Information** How are the senses of taste and smell similar? How are they different?

20. **BUILDING ON WHAT YOU HAVE LEARNED** You learned about sodium-potassium pumps in Chapter 4. How do sodium-potassium pumps in a neuron's cell membrane achieve resting potential?

Interpreting Graphics

21. Study the figure below and answer the questions that follow.

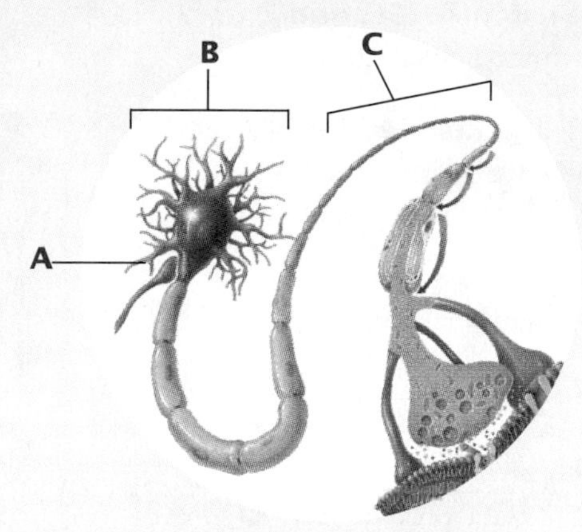

a. What kind of cell is shown in the figure? What evidence do you have for your decision?
b. What part of the cell is labeled *A*? What is its function?
c. What part of the cell is labeled *B*? What is its function?
d. What part of the cell is labeled *C*? What is its function?
e. Which part of the cell would be damaged by multiple sclerosis? What is the name of this structure?

Reviewing Themes

22. ***Evolution***
What are the three main parts of every vertebrate brain? How do these parts compare in a shark, a frog, a bird, and a cat?

23. ***Stability and Homeostasis***
A child may try to hold his breath during a tantrum. How does the autonomic nervous system prevent damage to the body in this situation?

Thinking Critically

24. **Recognizing Verifiable Conclusions** Epilepsy is a condition that affects one out of every 200 people in the United States. In a healthy human, brain cells produce small bursts of impulses in varying patterns. In a person with epilepsy, large numbers of brain cells occasionally send rapid bursts of impulses simultaneously, causing the person to have convulsions. How would you explain these symptoms? Why do you think the symptoms go away after only a short period of time?

Life/Work Skills

25. **Finding Information** Use library references or search an on-line database to find out about the different types of drugs and how they affect the nervous system. Investigate specific drugs that are classified as stimulants (amphetamines), depressants (barbiturates), and hallucinogens. Summarize your findings in an oral report to the class.

26. **Using Technology** Contact a local library or search an on-line database to observe sample images of the brain produced by one of the technologies described in this chapter, such as MRI or PET. Invite a physician or technician to class to describe the components of the images and what they reveal about the brain. Also ask about how the procedure is done and about the time and costs involved.

Activities and Projects

27. **Health Connection** Depression is a mental disorder that affects approximately 8 million people in the United States. Use library references or search an on-line database to find out about the symptoms, causes, and treatments of depression. Summarize your findings in a written report.

28. **Career Connection: Optometrist** Optometrists are professionals who provide and adjust eyeglasses and contact lenses for people with vision problems. Use library references or search an on-line database to find out what kind of education and specialized training is required to become an optometrist. Where do most optometrists work? For additional information, write to The American Optometry Association, 243 North Lindbergh Boulevard, St. Louis, Missouri 63141-7851.

Discovering Through Reading

29. Read the article "Scientists Explain Eight Wonders of the Human Brain" by Trina Chang in *American Health*, June 1995, pages 68–71. Why is handwriting unique? Why do some people have perfect pitch?

CHAPTER 31

REVIEW
- protein structure (Section 2-3)
- cell membrane architecture (Section 3-2)
- chemical signals (Section 4-1)
- autonomic nervous system (Section 30-2)

Hormones

Islet cells of the pancreas secrete hormones that regulate blood sugar.

After fighting a blazing fire for many hours, a firefighter nears exhaustion. Suddenly a wall collapses! The firefighter rushes to control the fiery debris. How? Chemical signals carried by the bloodstream ensure that the firefighter's body gets the extra oxygen and energy-supplying sugars needed to sustain peak performance until the crisis passes.

31-1 What Hormones Do

OBJECTIVES

1. **Describe the major components of the endocrine system.**
2. **Summarize how the hypothalamus and pituitary gland work together to regulate body systems.**
3. **Explain the importance of feedback.**
4. **Analyze the relationship among chemicals involved in a negative-feedback loop.**

Hormones: Chemical Signals

In the emergency described above, the signals traveling through the firefighter's blood are small molecules called hormones. A **hormone** is a chemical signal, made in one place and delivered (usually through the blood) to another, that regulates the body's activities. Various organs scattered throughout the body are sources of hormones. Organs that produce most of the hormones in your body are called **endocrine** *(EN duh krihn)* **glands**. The endocrine system is made up of the ductless endocrine glands and other organs, shown in **Figure 31-1**, that produce hormones. These other organs are part of other body systems as well.

Endocrine glands secrete hormones directly into the bloodstream. The hormones travel to a specific tissue or organ called a **target**. Once a hormone arrives at its target, the hormone will elicit a specific response. For example, some hormones speed up the heart. Others help digest food and enable the body to use the food as fuel. Without hormones, humans would not grow or mature. Like the brain and nerves, hormones are essential to maintaining homeostasis. Often, two or more hormones work together to produce their effects.

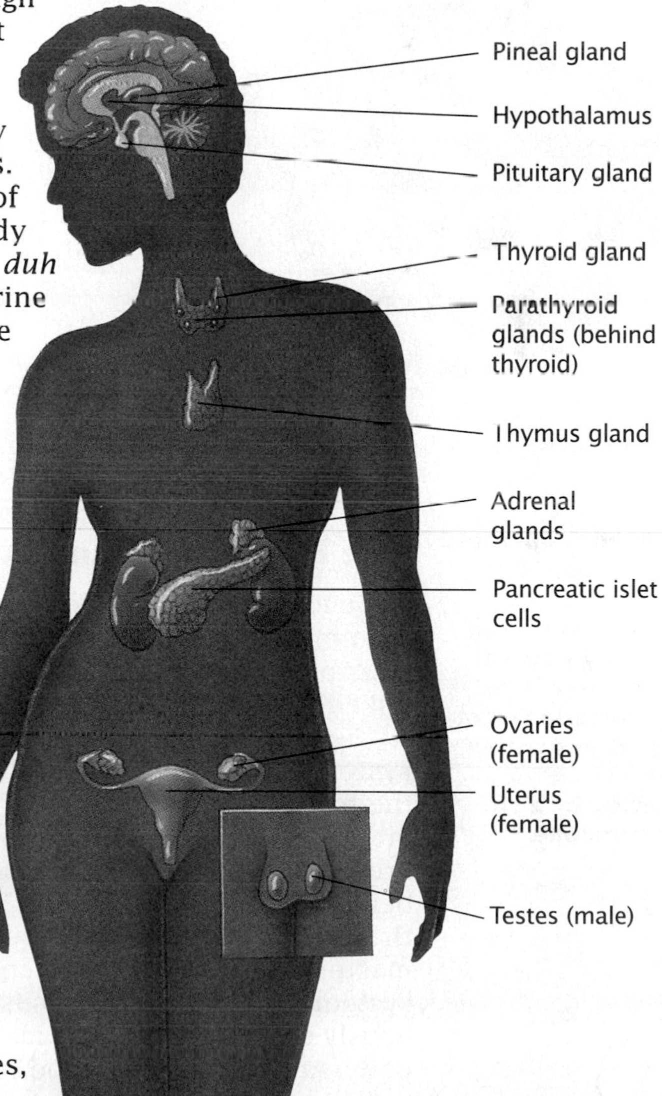

Figure 31-1 The endocrine system regulates overall metabolism, growth, reproduction, and maintenance of homeostasis.

The effects of hormones last longer than those of nerve impulses

Imagine that you start to cross a street. Suddenly a car speeds around the corner. Instantly, nerve impulses deliver the message "Danger!" from your eyes to your brain. The effect is immediate nervous control of one target cell at a time, spurring your body to action.

It takes your body just a moment to respond, but that brief moment is long compared with the nerves' rapid firing. Still, in a matter of seconds, hormones from your adrenal glands excite different target cells in several locations. Your heart pounds, and your blood pressure rises. Your skin sweats. Your muscles tense, and you sprint back to the curb—safe.

But, unlike the nerve impulses that have now ceased, some of the hormones that spurred you to action will affect you for 10 or 20 minutes. And when emergencies last for hours or days, other hormones are released. For reactions that occur over longer periods of time, like the changes of puberty leading to adulthood, hormones exert their influence over days, weeks, months, or even years.

Notice the shapes and colors used in **Figure 31-2**. You will see them used throughout this chapter to illustrate the nervous system, endocrine glands, target cells, and changes in tissue activity that occur as a result of hormones.

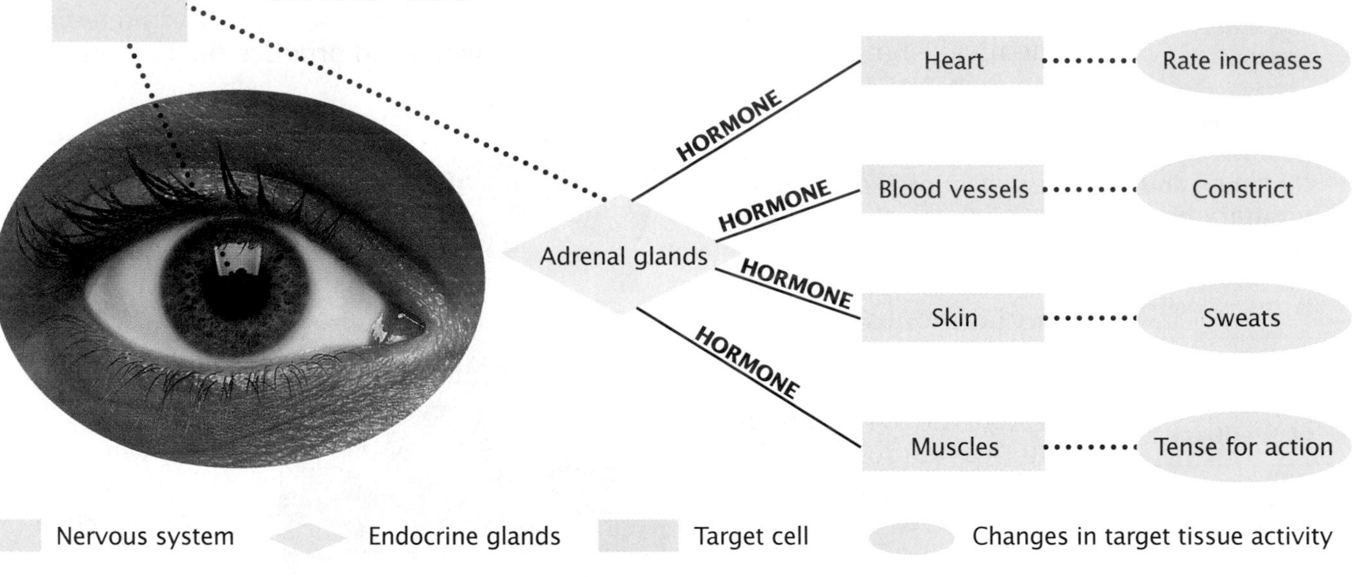

Figure 31-2 Hormones operate more slowly than the nervous system in controlling the body's operations, but notice how several target organs and activities may be affected by a single hormone.

The Hypothalamus-Pituitary Connection

The endocrine system and the nervous system are the two main control systems of the body. They are so closely linked that they often are considered a single system—the **neuroendocrine system**. In Chapter 30 you learned that the hypothalamus is the part of your brain that regulates body temperature, breathing, hunger, and thirst. Besides being a part of your nervous system, the hypothalamus also serves as the master switchboard of the endocrine system. Your hypothalamus is continuously checking conditions inside your body. Are you too hot or too cold? Are you running out of fuel? Are you in pain? If your internal environment starts to get out of balance, your hypothalamus can set things right again. In some cases, it acts like an endocrine gland itself, sending out commands in the form of hormones. In other cases, the hypothalamus sends messages instantaneously to other parts of the brain and to other glands in the body through direct neuron links. Together with hormones, these unique nerve transmissions are the hypothalamus's way of setting up a chain of command that allows it to control much of the activity of the endocrine system.

The pituitary gland secretes and stores hormones

All of the hormones produced by the hypothalamus move through a slender thread of tissue to the **pituitary** *(puh TOO uh tehr ee)* **gland**. The pituitary gland produces at least six different hormones in response to the hormones released from the hypothalamus. As described in **Figure 31-3a–b**, two of these hormones, growth hormone and prolactin, have direct effects on tissues in the body. Growth hormone, for instance, stimulates protein synthesis and cell division in target cells. It also profoundly influences the growth of cartilage and bone. Because these pituitary hormones are controlled by releasing hormones from the hypothalamus, the brain exercises direct control over the endocrine system.

Four of the hormones secreted by the pituitary gland control the activities of a number of other endocrine glands, such as the thyroid gland and the adrenal glands. Some of these glands are discussed in more detail later in this chapter.

The pituitary gland also stores hormones made in the hypothalamus. Two of these hormones, oxytocin *(ahk see TOHS ihn)* and antidiuretic hormone (ADH), discussed in **Figure 31-3c–d**, are released from the pituitary gland when needed by the body.

Figure 31-3
The pituitary gland was formerly called the "master gland" because so many of its hormones regulate other endocrine functions. Now the hypothalamus is considered to be the "master gland."

a *Growth hormone* stimulates general growth of the body, particularly of the skeleton. When too little growth hormone is produced, a condition called dwarfism can result. With too much growth hormone, a condition called gigantism can occur.

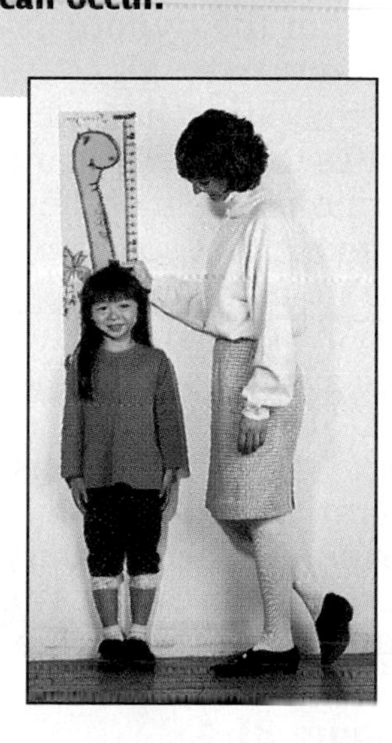

b *Prolactin* causes glands in the mother's breasts to produce milk. Then the baby's nursing stimulates the release of oxytocin.

c *Oxytocin* aids in the release of milk from the mammary glands and also causes contractions of the uterus during and shortly after labor.

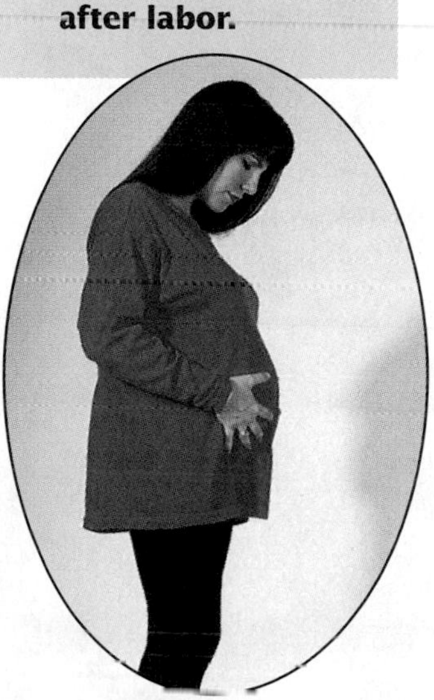

d *ADH* causes the kidneys to form more concentrated urine, thereby conserving water in the body.

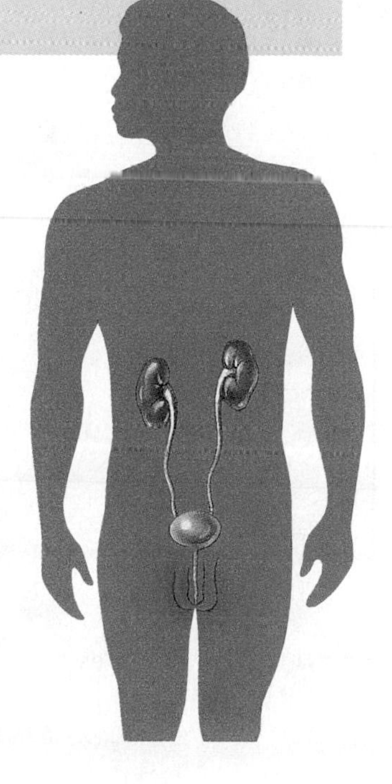

Regulating Hormone Release

The human body produces more than 40 hormones and must be able to regulate the release of these hormones. Although nerve impulses alone can increase or decrease hormone production, in many cases chemical signals, including hormones, regulate the activity of the endocrine system. These chemical signals are used as feedback by the endocrine system to regulate itself.

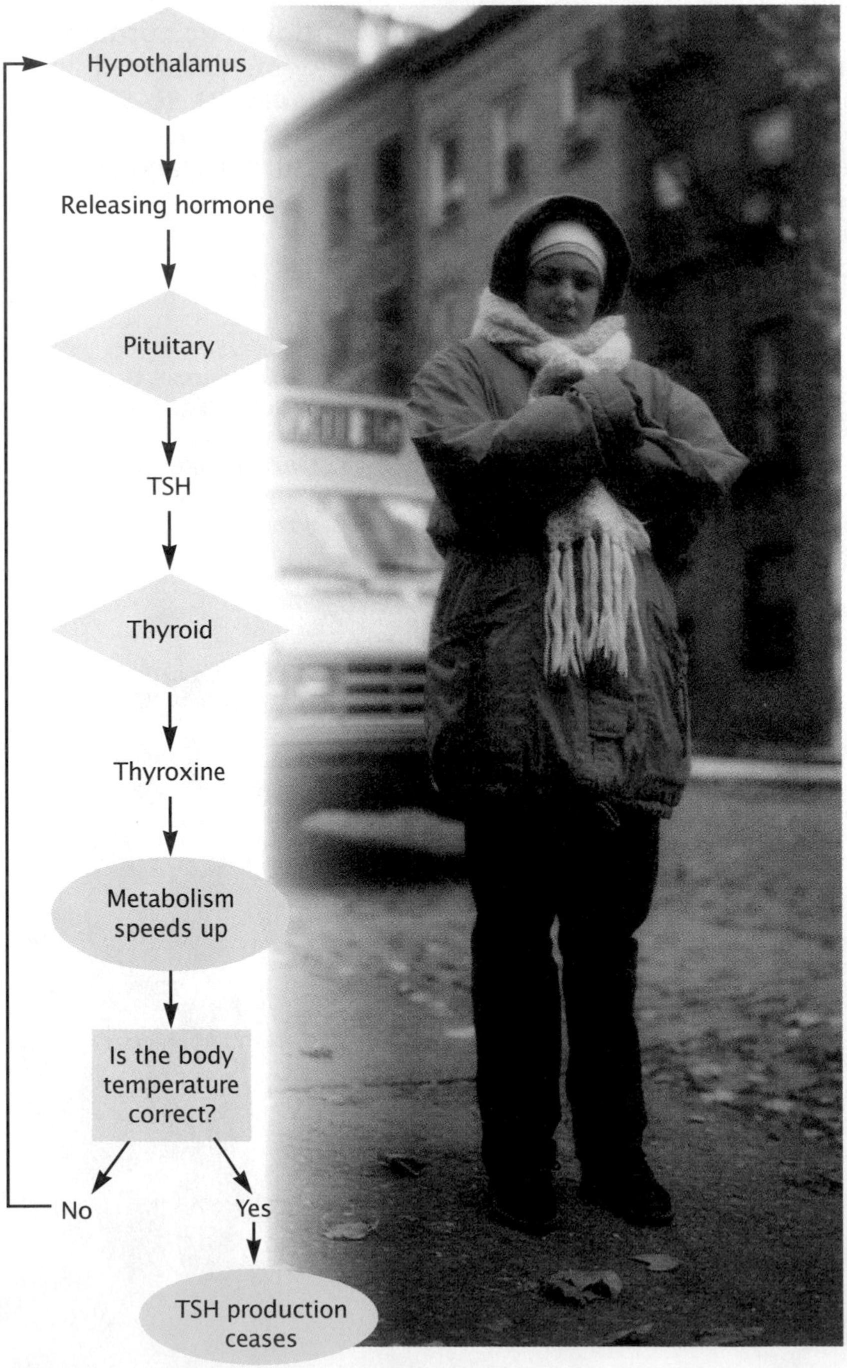

Figure 31-4
When you are cold, the hypothalamus sends signals to the pituitary gland, which signals the thyroid to produce thyroxine. Thyroxine speeds up metabolism until your body temperature rises slightly. Then the system shuts off.

Hypothalamus
Releasing hormone
Pituitary
TSH
Thyroid
Thyroxine
Metabolism speeds up
Is the body temperature correct?
No
Yes
TSH production ceases

Feedback is a system for self-regulation. A feedback mechanism detects the amount of hormones already in circulation or the amount of other chemicals produced because of hormone action. After receiving feedback, the endocrine system adjusts the amount of hormones being made or released. If feedback stimulates the output of more hormone, it is called positive feedback. In contrast, a **negative feedback** system inhibits the production of more hormone. Most homeostatic control is accomplished by negative feedback, which is a fairly complex mechanism because hormones are never turned completely on or off. Sometimes more than one negative-feedback loop operates to control the rate of secretion and thus the levels of hormones in the blood. At other times, nervous stimulation, like that produced by the nursing infant on the previous page, regulates hormone secretion.

Thyroid hormones maintain body temperature

Imagine that you are waiting for a bus in cold, rainy weather. Your hypothalamus uses negative feedback to monitor your body temperature. Are you cold? If you are, neural adjustments will be made. Are you still cold? If you are, three hormones are released to help restore your body temperature. First, a hormone from the hypothalamus causes your pituitary gland to release thyroid stimulating hormone (TSH). TSH stimulates the thyroid gland to secrete two hormones. One of these is thyroxine, a hormone that speeds up your body's rate of metabolism. This is like turning up your internal furnace. When the feedback system senses that your temperature has returned to normal, it slows endocrine operations. The hypothalamus stops production of the releasing hormone. The pituitary reduces TSH secretion. In turn, the thyroid stops secreting thyroxine. This negative-feedback process is summarized in **Figure 31-4**.

Negative feedback also regulates blood calcium

You will remember that calcium is vital to bones, muscles, and nerves. Your bones are more than just support beams for your body; they are mineral reservoirs. More than 98 percent of your body's calcium is stored in your bones. Hormones control the level of calcium in the blood by controlling calcium's movement into and out of bone and other tissues.

Calcitonin from the thyroid gland lowers the level of calcium in the blood. Parathyroid hormone (PTH) from the parathyroid glands raises the level of calcium in the blood. **Figure 31-5** illustrates the negative-feedback loops that control the release of these hormones. If the level of calcium is too high, the thyroid gland secretes calcitonin, and blood calcium is lowered. When the level of calcium in the blood drops even slightly below normal, the parathyroid gland receives a message to secrete PTH. PTH causes the release of calcium from bone into the blood. What happens to calcium in the blood now? When the level of calcium in the blood rises to a normal level, the parathyroid glands halt production of PTH.

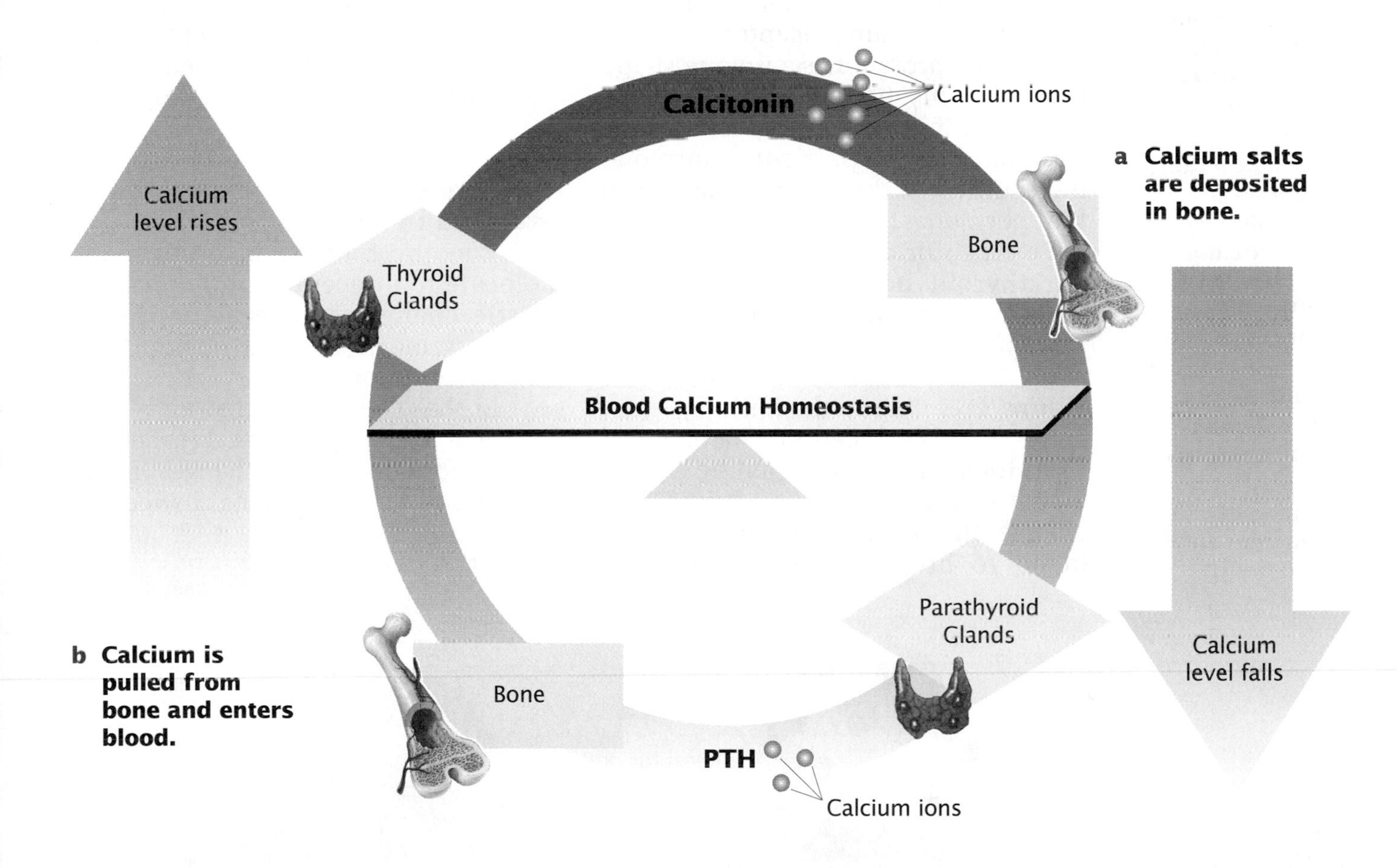

Figure 31-5 Blood calcium level is regulated by moving calcium into and out of storage. The major storage site for calcium is bone.

SECTION REVIEW

1. **Define the term *hormone*, and name four endocrine glands in the body.**
2. **Why are the nervous system and endocrine system often called the neuroendocrine system?**
3. **How does the body regulate the release of hormones?**
4. **What roles do the hormones calcitonin, PTH, and thyroxine play in homeostasis?**

Hormones are carried throughout the human body in the bloodstream. Only target cells actually respond to the message that has been sent. After a hormone arrives at its target, the hormone causes certain actions to occur. Two types of hormones cause changes in cells in different ways, as you will learn in this section.

31-2 How Hormones Work

OBJECTIVES

1. **Describe how a target cell responds to a hormone.**
2. **Compare the actions of steroid and peptide hormones in affecting a target cell.**
3. **Explain how fewer than 50 hormones cause thousands of changes in the body.**

Hormone Receptor Proteins

As you read in Chapter 4, the cell membrane contains receptor proteins. A receptor protein has a unique shape that will only hold a certain type of molecule. If a cell has a receptor protein that holds a particular hormone molecule, then the cell will respond to that hormone. For example, bone cells have receptor proteins that will hold parathyroid hormone molecules. A parathyroid hormone molecule will fit into one of these receptor proteins like a key fitting into a lock, as shown in **Figure 31-6**. Fitting the hormone molecule into the receptor changes the receptor's shape, like opening a lock, which causes the cell's activities to change. In this case, the bone cell will begin to break down the minerals stored in the bone so that calcium can be released. While many receptor proteins are located in the cell membrane, others are found inside the cytoplasm or in the nucleus of the cell.

Hormones affect enzymes in cells

How can the binding of a hormone molecule to a receptor protein cause drastic changes in the activities of a cell? Some hormones change the activity or amounts of enzymes present in that cell. Recall that enzymes initiate or speed up chemical reactions. When a hormone causes changes in the enzymes in a cell, it causes changes in the chemical reactions that occur inside the cell.

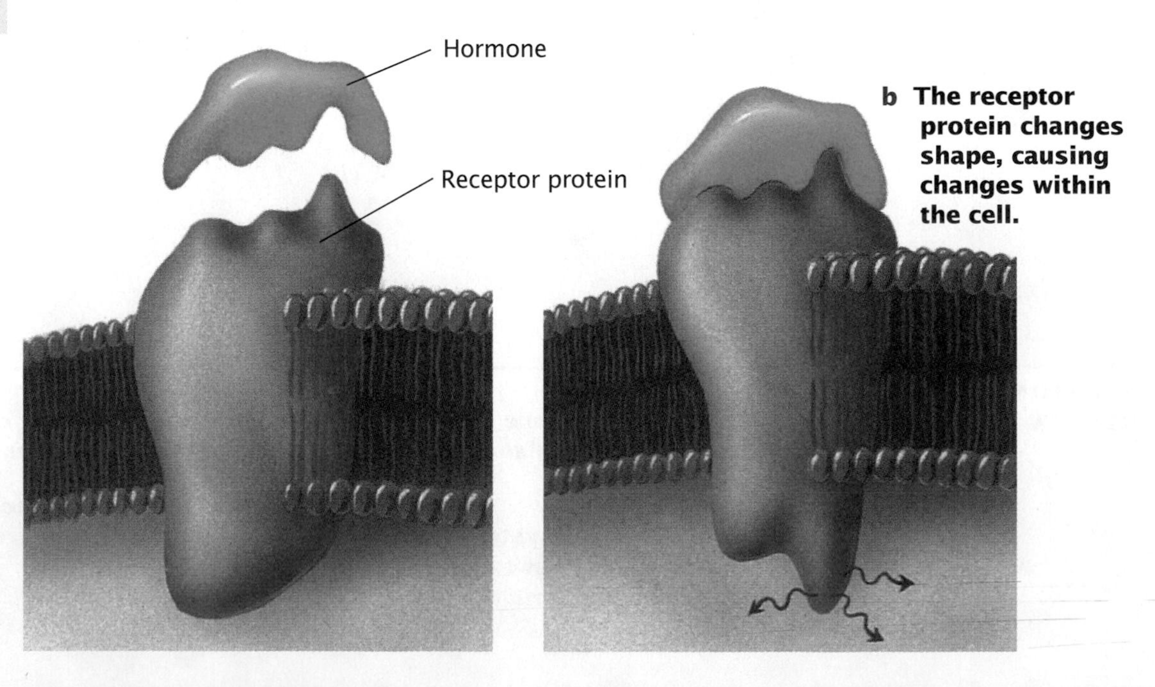

Figure 31-6
a When a hormone comes into contact with a cell membrane receptor protein that will hold it, the hormone binds to the receptor protein like a key fitting into a lock.

b The receptor protein changes shape, causing changes within the cell.

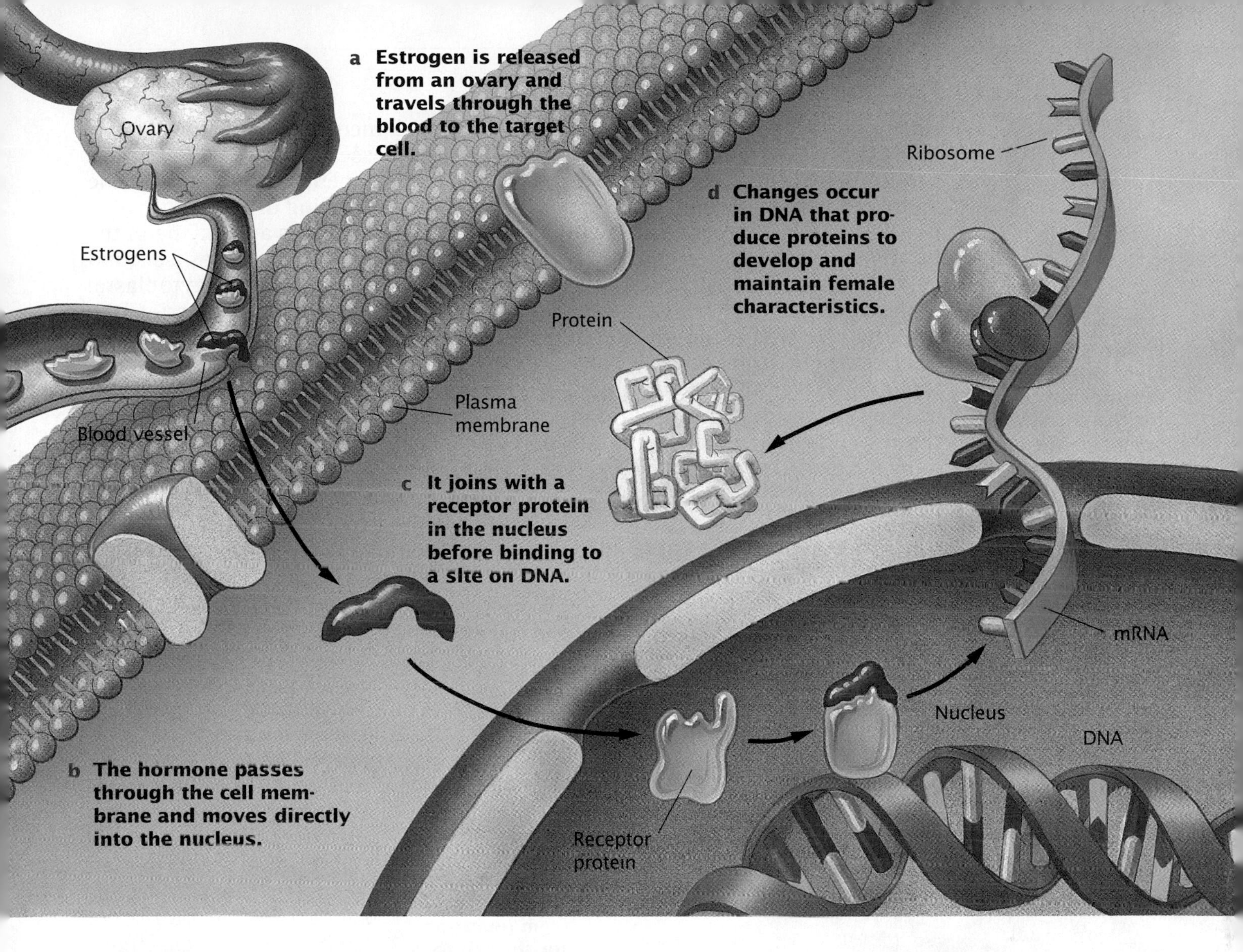
a Estrogen is released from an ovary and travels through the blood to the target cell.
Ovary
Estrogens
Blood vessel
Plasma membrane
Protein
Ribosome
d Changes occur in DNA that produce proteins to develop and maintain female characteristics.
c It joins with a receptor protein in the nucleus before binding to a site on DNA.
mRNA
Nucleus
DNA
b The hormone passes through the cell membrane and moves directly into the nucleus.
Receptor protein

Figure 31-7 **The group of steroid hormones known as estrogens influence development of female sexual traits. The way these steroid hormones cause changes in cellular activity is described above.**

Steroid Hormones

Hormones assembled from cholesterol are **steroid hormones**. Steroid hormones include the sex hormones and adrenal cortex hormones. All steroid hormones do their work inside the cell's nucleus, where they activate genes and trigger changes in chromosomes.

Steroid hormones pass through the cell membrane without help and enter the cytoplasm. Steroid hormones made in the adrenal cortex have receptors in the cytoplasm. Once bound to a receptor, the steroid hormone enters the cell's nucleus, where it binds to DNA and influences genes.

Receptors for most of the sex hormones reside in the cell's nucleus. After crossing the cell membrane, the sex hormone goes directly to the nucleus, where a receptor awaits. In the nucleus, the hormone and receptor bind first to one another, then to a site on DNA, as shown in **Figure 31-7**.

Steroid hormones switch genes on and off

Recall that every cell in your body (except sperm or egg cells) contains all of the genes that you inherited from your parents. However, different sets of genes are switched on or off in different types of cells. Steroid hormones operate some of these switches. Switching genes on or off alters a cell's enzyme activity. When the switch is on, enzyme activity, and thus protein synthesis, is in full swing. New proteins may then produce other receptors, proteins, or hormones, having far-reaching effects on the cell's metabolism.

Peptide Hormones

Hormones made of chains of amino acids are called **peptide hormones**. Amino acids can be joined to form short and simple or long and complex peptides. Most hormones are peptide hormones. Recall that amino acids dissolve in water because both are polar molecules. However, the positive and negative charges on peptide hormones prevent them from passing through the lipid bilayer of a cell membrane to the cell's interior. Thus, peptide hormones and other polar hormones or neurotransmitters must send messages from outside the target cell.

Second messengers carry information into the cytoplasm

Because peptide hormones cannot enter the cell, they bind to receptors on the cell membrane. The hormone is the first messenger, carrying the message from the endocrine gland to the cell surface. Binding of the hormone to its receptor activates an enzyme in the cell membrane, which causes a second messenger to appear in the cytoplasm. The **second messenger** carries the message into the cytoplasm.

One of the most common second messengers is cyclic AMP. Cyclic AMP is made from ATP by an enzyme that removes two phosphate groups, forming AMP. The ends of the AMP join to form a circle. Second messengers such as cyclic AMP generally activate enzymes in the cytoplasm of target cells. **Figure 31-8** shows how the peptide hormone glucagon *(GLOO kuh gahn)* signals liver cells to release glucose into the blood.

Figure 31-8 Liver cells have receptor proteins for the peptide hormone glucagon, which is made by the pancreas. Glucagon uses cyclic AMP as a second messenger to release glucose into the blood.

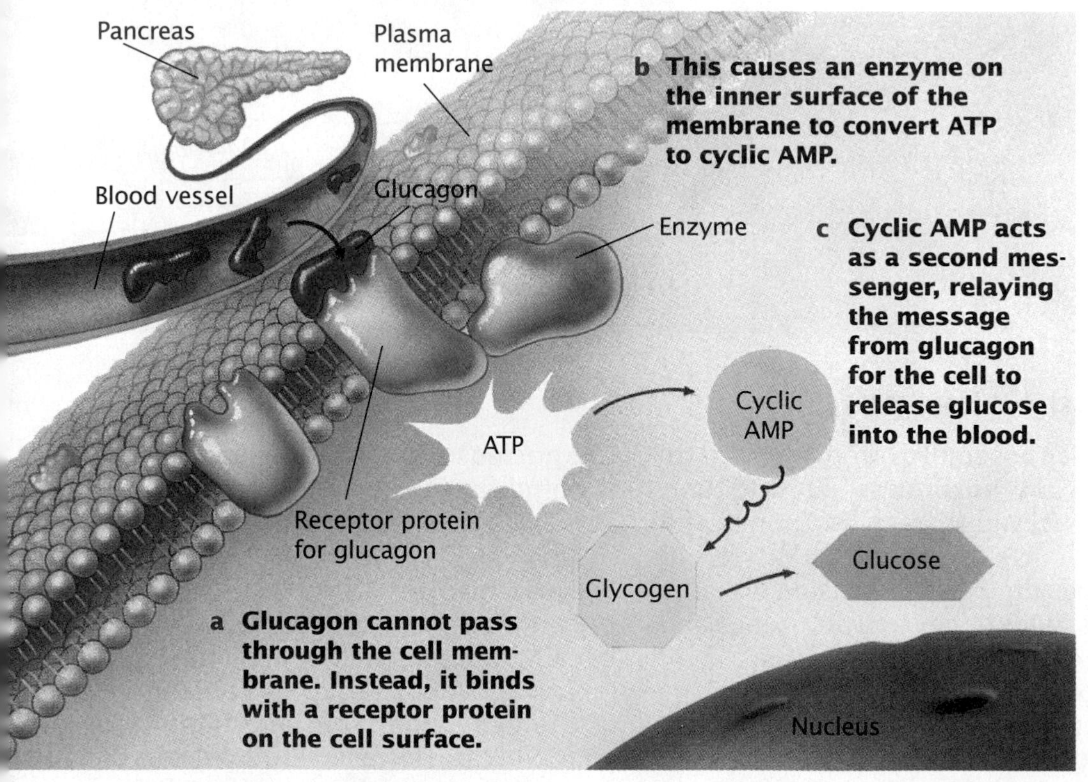
Pancreas
Plasma membrane
Blood vessel
Glucagon
Enzyme
b This causes an enzyme on the inner surface of the membrane to convert ATP to cyclic AMP.
c Cyclic AMP acts as a second messenger, relaying the message from glucagon for the cell to release glucose into the blood.
Cyclic AMP
ATP
Receptor protein for glucagon
Glycogen
Glucose
a Glucagon cannot pass through the cell membrane. Instead, it binds with a receptor protein on the cell surface.
Nucleus

Second messengers are hormone amplifiers

A single hormone molecule binding to a receptor in the cell membrane can result in the formation of many second messengers in the cytoplasm. Each second messenger, in turn, can activate many enzymes. Sometimes these enzymes activate other enzymes. This cascade of activity greatly amplifies the original signal, enabling each hormone molecule to have a tremendous effect inside a cell, even though it never actually enters the cell.

SECTION REVIEW

1. **How does a receptor protein respond to a hormone molecule?**
2. **Name the two ways that hormones affect enzymes.**
3. **How do the estrogens deliver a message to a target cell?**
4. **How does the hormone glucagon deliver a message to a target cell?**

Your body functions depend on the delicate interaction between your endocrine system and your body tissues. The endocrine system orchestrates several processes used by the body to cope with stress, to regulate metabolism, to control blood sugar levels, to carry on respiration, and even to reproduce.

31-3 Glands and Their Functions

OBJECTIVES

1. **Locate three major endocrine glands or organs, and identify the hormone(s) each produces.**
2. **Explain the effects of producing or administering too much or too little of a particular hormone.**
3. **Predict the effect of long-term stress on homeostasis.**
4. **Evaluate the advantages and disadvantages of inhibiting prostaglandin synthesis.**

The Adrenal Glands

Fewer than 10 major endocrine glands and several other organs in the body produce dozens of different hormones. Six major hormones are produced by the **adrenal glands** alone. The adrenal glands are two almond-sized glands located on top of the kidneys. Each adrenal gland is two endocrine glands in one. The inner part of each adrenal gland is called the adrenal medulla, and the outer part is called the adrenal cortex.

The adrenal medulla helps the body react to a sudden crisis

The adrenal medulla is different from other glands because of the signal that activates it. Rather than being activated by hormones, the adrenal medulla is stimulated to release its hormones by nerves that run directly to it from the hypothalamus.

The adrenal medulla produces epinephrine *(ep ih NEF rihn)*, or adrenaline, and norepinephrine, or noradrenalin. These hormones work with the autonomic nervous system to produce the **"fight-or-flight" reaction** and are secreted in response to sudden stresses such as fear, anger, pain, or physical exertion. In a fraction of a second, the adrenal medulla can respond to any emergency that may arise, such as the scene in **Figure 31-9**. The heart beats faster, and blood flow increases to the heart and muscles. At the same time, respiration increases, and more oxygen is delivered throughout the body. In addition, blood pressure rises, sweating increases, and muscles become tense, ready for action. Because these changes help us to act quickly in times of sudden stress, it is also known as the alarm reaction phase of the "General Adaptation Syndrome," or GAS.

Figure 31-9
This is a famous movie scene. But if it were a real, life-threatening emergency, the adrenal medulla would secrete the hormone epinephrine. This hormone speeds up the heart rate and increases blood flow to the muscles.

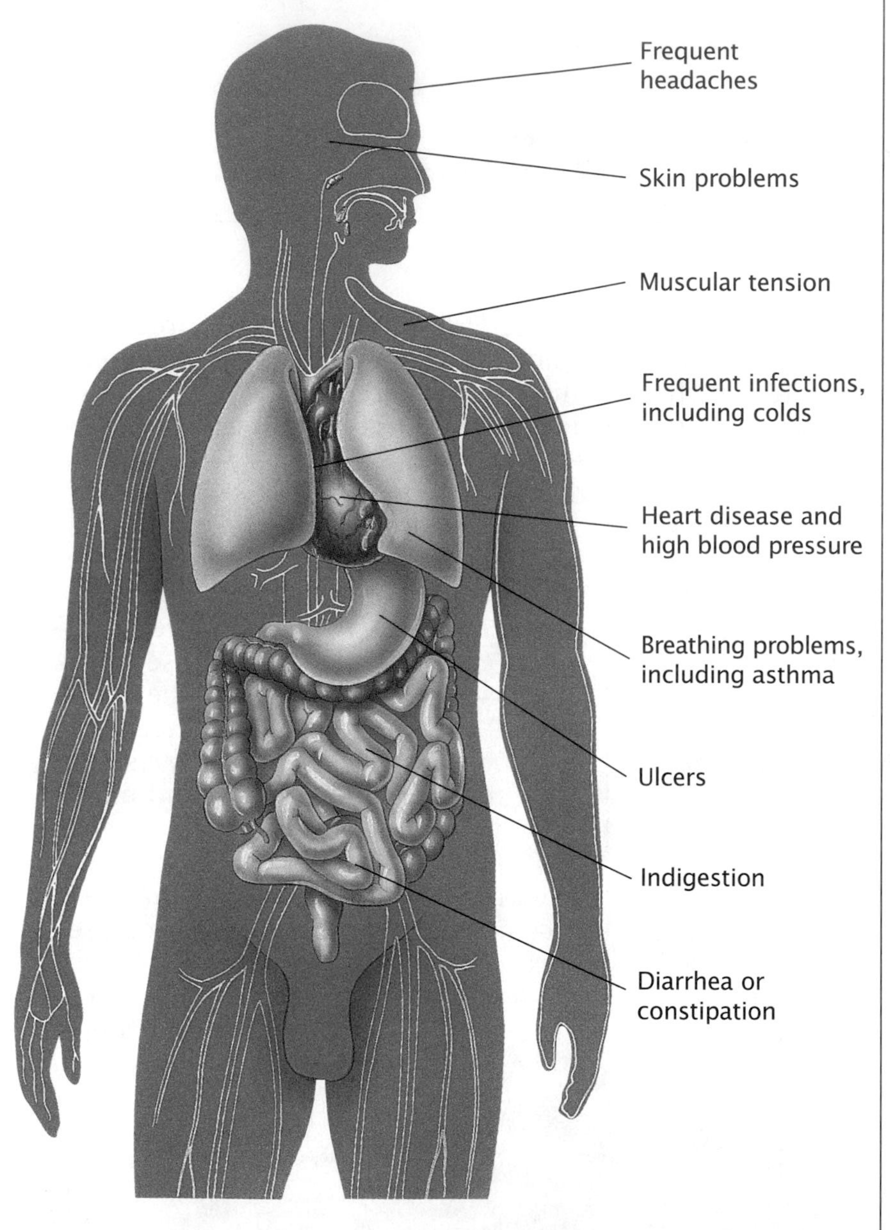

Figure 31-10
Long-term stress and continued release of cortisol gradually threaten the homeostasis of the body. Long-term stress can be severe enough to cause death, but it is more likely to result in long-term adaptation or illness.

The adrenal cortex helps the body deal with long-term stress

Once the alarm stage is over, the effects of epinephrine wear off within a few minutes. If the crisis continues for more than a few hours, however, other hormones produced by the adrenal cortex enable the body to adapt to stress for hours, days, or even weeks. This new period of adaptation is called the resistance stage of adaptation.

In this second stage, cortisol is the steroid hormone that makes more energy available to the body, especially the brain. Cortisol causes the body to burn lean tissue (muscle) for energy. Its release increases the level of blood glucose so that the body increases in strength, endurance, and sensory response. In the alarm and resistance stages of adaptation, this effect helps you to adapt more efficiently. When stressful situations continue without relief, however, the body enters a state of chronic fatigue. Organs malfunction, and physical as well as mental burnout may result, signaling the third and final phase of the adaptation syndrome, the exhaustion stage.

Long-term exposure to steroid hormones weakens the immune system

When you are under stress for long periods of time, the continued release of cortisol has negative effects on your body's ability to defend itself. The immune system becomes suppressed, or weakened, and loses its ability to fight disease effectively. Immune suppression may play a role in the development of certain types of cancers and other health problems. For example, ulcers, bowel disorders, heart disease, high blood pressure, asthma, skin disorders, and even autoimmune diseases are linked to stress. Other side effects of continued release of cortisol include the loss of resistance to infections, loss of bone mass and muscle protein, and poor healing of wounds. Cortisone is similar to cortisol but is made in much smaller quantities by the body. Synthetically produced cortisone is injected, given orally, or used in cream form as an anti-inflammatory drug in the treatment of arthritis, joint injuries, and severe insect bites and other allergic responses. The prolonged use of cortisone or similar drugs can have the same negative effects as the continued release of cortisol. **Figure 31-10** illustrates a range of ways in which the body may adapt to long-term stress.

The Thyroid Gland

The thyroid gland is located in the neck, just below the Adam's apple. The thyroid gland releases thyroxine and other hormones containing iodine, which regulate the body's metabolic rate. Thyroid hormones are necessary for normal growth and development during childhood and for maintaining mental alertness in adults. Thyroid hormones also affect reproductive functions.

Figure 31-11
In 1988, American sprinter Gail Devers experienced severe weight loss, uncontrollable shaking fits, and loss of vision in her left eye. Nearly two years later she was diagnosed with Graves' disease, a condition in which too much thyroxine is produced by the thyroid gland. With proper treatment, however, Devers went on to win the gold medal in the women's 100 m dash in the 1992 and 1996 Summer Olympics.

Too much thyroxine causes Graves' disease

Some common disorders of the endocrine system are caused by problems with the thyroid. If too much thyroxine is produced, the condition that results is called **hyperthyroidism**. One form of hyperthyroidism is Graves' disease. People with Graves' disease have a rapid, irregular heart rate, feel very nervous, and lose weight. Olympic gold medalist Gail Devers, pictured in **Figure 31-11**, began to experience the symptoms of Graves' disease just prior to the 1988 Summer Olympics. Most people with Graves' disease can lead normal, productive lives once they are treated.

Graves' disease seems to be caused by a malfunction of the immune system. Recall that hormones work by binding to receptor proteins, which causes changes in the activity of the cell. The binding of a hormone to receptor proteins in thyroid cells causes the thyroid gland to release thyroxine. In a person with Graves' disease, however, antibodies made by the immune system bind to these receptor proteins instead. As a result, the thyroid gland releases more hormone in the body even though too much is already present.

Graves' disease is usually treated with a dose of radioactive iodine. The iodine collects in the thyroid gland and, over the course of several months, slowly destroys thyroid tissue. It is hard to control just how much of the gland is destroyed, however, so the patient often makes too little thyroxine after the treatment. Thyroxine levels are then brought up to normal by taking pills containing the hormone.

Low levels of thyroxine cause hypothyroidism

The production of too little thyroxine is called **hypothyroidism**. Thyroid cells make thyroid hormones by attaching three or four iodine atoms to a large protein. Iodine comes from the diet, and the large protein is made and stored in the thyroid. When the diet contains little or no iodine, there is little to attach to the large protein. The negative-feedback loop that controls thyroxine production fails. More and more of the large protein is stored. The thyroid stays stimulated, trying unsuccessfully to make hormone. The thyroid gland swells with the growing store of protein, producing a **goiter** *(GOY tuhr)*, shown in **Figure 31-12a**.

Iodine is found in foods like the ones shown in **Figure 31-12b**. People living near an ocean usually include seafood in their diets and rarely have thyroid problems caused by a lack of iodine. However, goiters used to be common in inland regions of the United States. Today, small amounts of iodine are added to plain table salt. But goiters are still common in developing countries, where they are considered a serious problem because of their effect on growth and brain development in children and on mental functions in adults.

Some people who get enough dietary iodine still suffer from hypothyroidism. Like those with Graves' disease, these people suffer from a misdirected attack by antibodies from their immune system. In this case, the damaged thyroid gland cannot produce adequate thyroxine.

Because the thyroid gland controls metabolic rate, people with too much thyroid hormone are thin. Sometimes being overweight is attributed to an "underactive thyroid." However, even very low thyroxine levels cause a weight gain of only 5–10 lbs, so being seriously overweight is not usually caused by low thyroxine levels.

Figure 31-12

a The lack of iodine in this woman's diet has caused her thyroid gland to swell, producing a goiter.

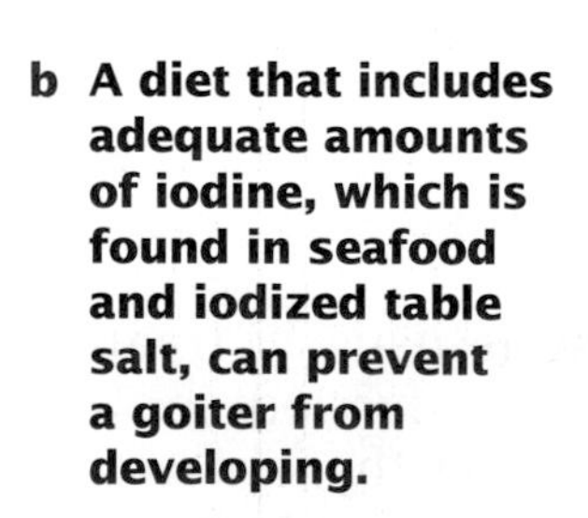

b A diet that includes adequate amounts of iodine, which is found in seafood and iodized table salt, can prevent a goiter from developing.

Stability and Homeostasis

How do hormones assist in maintaining homeostasis?

Other Glands and Hormones

The hormones that are responsible for the development and maintenance of sexual characteristics are **sex hormones**. In males, a group of sex hormones called the androgens are produced by the testes. The most important of these is testosterone *(tehs TAHS tuh rohn)*. Anabolic steroids are synthetic androgens. In females, the ovaries produce two groups of sex hormones known as estrogens and progestins. The most abundant of these are estrogen *(EHS truh jehn)* and progesterone *(proh JEHS tuh rohn)*, respectively. Sex hormones affect sexual behavior.

The release of sex hormones is controlled by the hypothalamus and the pituitary gland. Males and females produce the same hormones in these glands, but the level of these hormones varies in males and in females. You will learn more about the sex hormones in Chapter 35.

Small glands produce hormones

Other endocrine glands release hormones. Specialized white blood cells called T-cells mature in the thymus gland because of thymosin, which stimulates T-cell development. Although the gland gets smaller after puberty, its role in adult immunity is still debated. You will learn more about T-cells in Chapter 33.

The pineal gland contains light-sensitive cells that mark changes in the length of days and nights. In some way not yet understood, the gland translates light information and then relates it to biological rhythms associated with reproduction. There is some reason to believe that melatonin may help to regulate the onset and ending of puberty. This gland, too, gets smaller with age, but it continues to function to a small degree in healthy adults. The functions of these glands and others are summarized in **Table 31-1**.

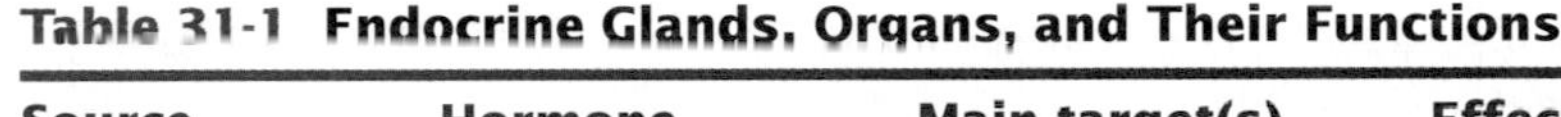

Table 31-1 Endocrine Glands, Organs, and Their Functions

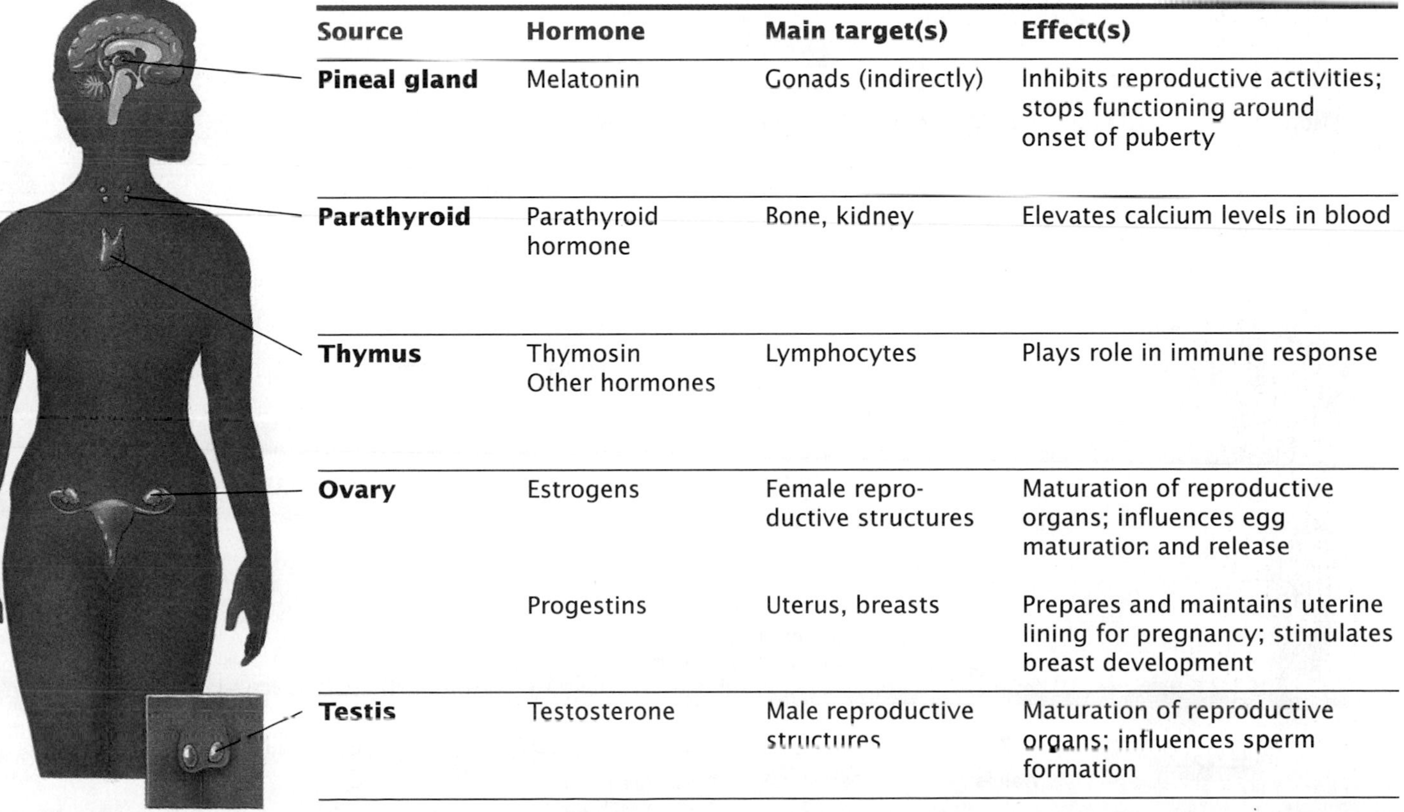

Source	Hormone	Main target(s)	Effect(s)
Pineal gland	Melatonin	Gonads (indirectly)	Inhibits reproductive activities; stops functioning around onset of puberty
Parathyroid	Parathyroid hormone	Bone, kidney	Elevates calcium levels in blood
Thymus	Thymosin Other hormones	Lymphocytes	Plays role in immune response
Ovary	Estrogens	Female reproductive structures	Maturation of reproductive organs; influences egg maturation and release
	Progestins	Uterus, breasts	Prepares and maintains uterine lining for pregnancy; stimulates breast development
Testis	Testosterone	Male reproductive structures	Maturation of reproductive organs; influences sperm formation

The Pancreas

The pancreas contains small clusters of hormone-secreting cells that help regulate how much glucose is dissolved in the blood. This **blood sugar**, as it is commonly called, circulates to all cells in the body, providing the energy needed to fuel all kinds of cellular activity. Two hormones from the pancreas work opposite one another to control blood sugar.

After you eat, enzymes break down the food you have eaten into amino acids, glucose, other simple sugars, and numerous other molecules that the body uses for fuel. The sugars pass through the walls of the small intestine and enter your bloodstream. Blood sugar level rises. The glucose molecules are then transported to all of the body's cells. The rise in blood sugar triggers release of insulin.

At the cell membrane, **insulin**, a peptide hormone, interacts with a receptor and helps glucose cross the membrane and enter the cytoplasm. Getting glucose into cells has the effect of lowering blood sugar. This is extremely important because high blood sugar can seriously upset homeostasis and even threaten life. After you have eaten, extra sugar not used by the cells is stored in the liver in the form of carbohydrate called glycogen or under the skin as fat.

If you haven't eaten for a while, your blood sugar level begins to fall. A different hormone, also made by the pancreas, helps you in this case. Glucagon raises the level of glucose in the blood. It does this by converting glycogen that has been stored in the liver into glucose, which the cells can use. This action, too, is important because low blood sugar can also upset homeostasis. Normal brain function requires a certain amount of glucose. If the level of sugar in the blood falls, the brain cannot get what it needs. Symptoms of lowered blood sugar may include nervousness, irritable mood, difficulty concentrating, weak or shaking muscles, and, frequently, headaches. The homeostatic control of blood sugar is shown in **Figure 31-13**.

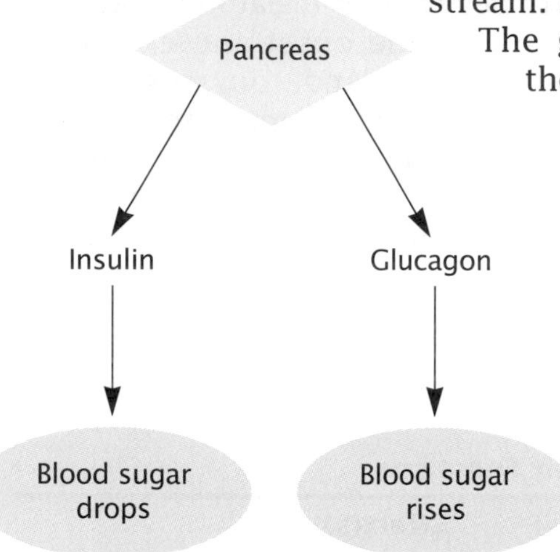

Figure 31-13 **The hormone insulin lowers blood sugar levels after meals. The hormone glucagon raises blood sugar levels between meals. A balanced breakfast like #1 stabilizes blood sugar for the entire morning. A breakfast high in sugar like #2 produces the symptoms of low blood sugar soon after it's eaten.**

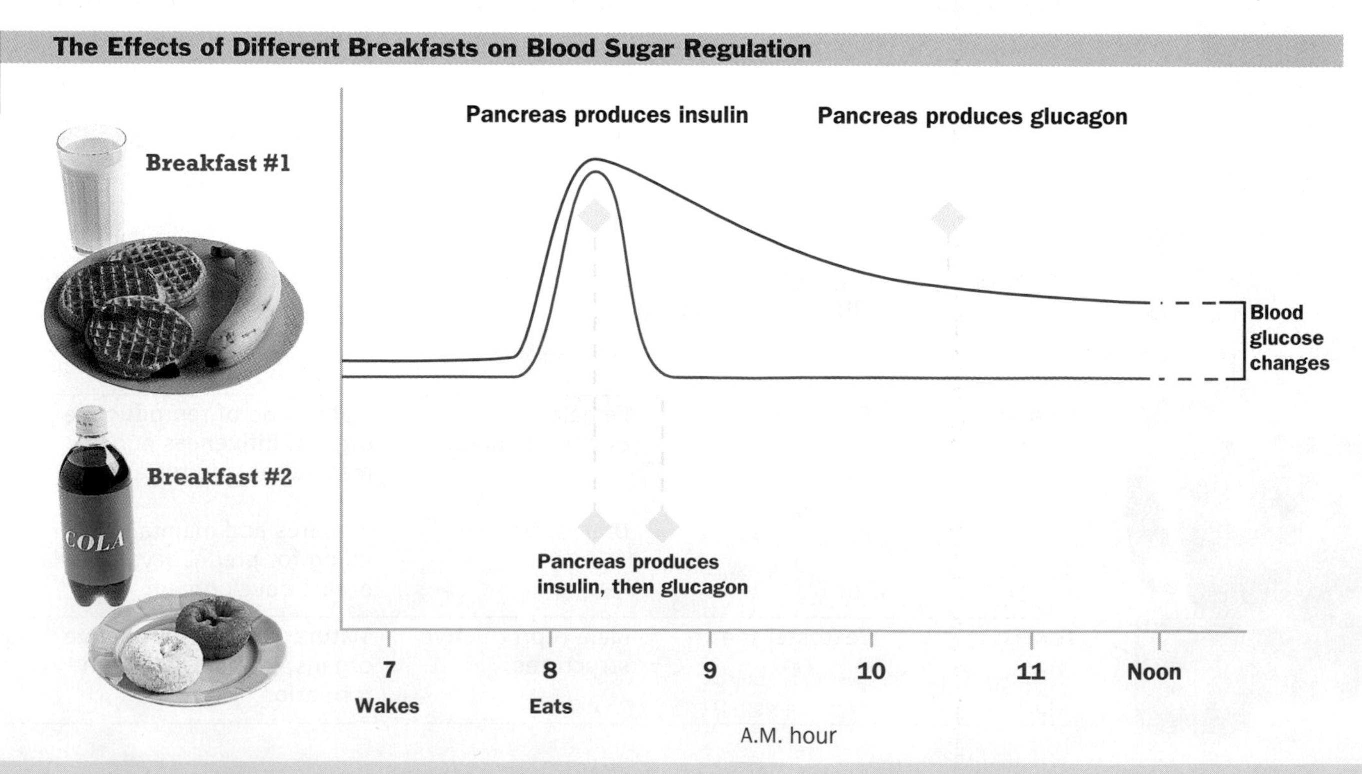

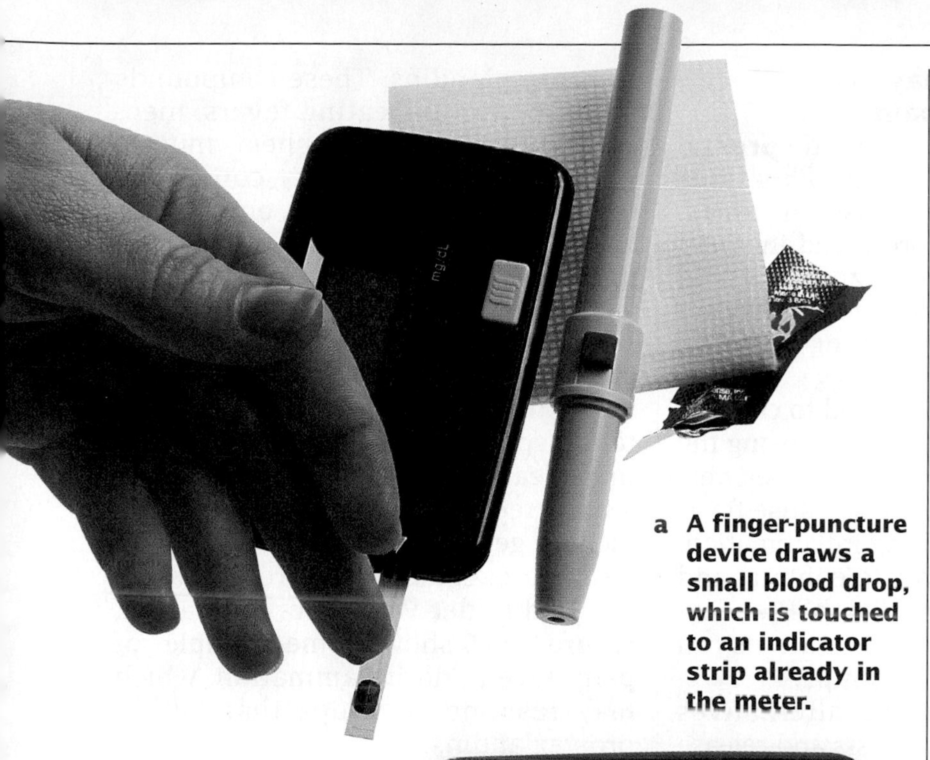

a **A finger-puncture device draws a small blood drop, which is touched to an indicator strip already in the meter.**

Figure 31-14
A glucose meter is used to monitor blood sugar level.

b **The meter reads the strip and gives a glucose reading in 20–45 seconds.**

Insulin plays a role in diabetes

Diabetes mellitus *(deye uh BEET eez muh LYT uhs)* is a disease that results from abnormally high blood sugar. Heredity, hormones, diet, and stress interact to produce different types of diabetes.

When a body makes little or no insulin, insulin-dependent diabetes mellitus (IDDM) is diagnosed. IDDM, also called Type I diabetes, usually develops in childhood or early adolescence. This form of diabetes requires daily injections of insulin. The amount of insulin given is determined by the calories taken in at each meal or snack.

When insulin is produced but glucose cannot enter cells, non-insulin dependent diabetes (NIDDM), or Type II diabetes, is diagnosed. NIDDM occurs when there are not enough insulin receptors on the cells, when the receptors are defective, or when too little insulin is produced. This form of diabetes accounts for about 95 percent of diabetics, and it usually occurs in adults over the age of 40. Obesity and inactivity contribute to the development of NIDDM, which can be prevented. Proper diet and aerobic exercise, which improves the cell's uptake of glucose, are used to manage the disease, although oral medicines may be used to lower blood sugar.

Uncontrolled blood sugar is dangerous for diabetics. When blood sugar rises but cannot enter cells, proteins and fats are burned for energy. Fat breakdown releases fatty acids and ketones into the bloodstream. Fatty acids release energy, but ketones are toxic and can disrupt heart rate, breathing, and brain activity if they are not excreted by the kidneys. Coma and death can follow. High blood sugar can result if diabetics eat or drink too much or fail to take their medication. When blood sugar is low, the brain cannot receive the glucose it needs. Dizziness, confusion, and fainting may result. Shock results if blood sugar drops too low. Blood sugar can get too low if a diabetic takes too much medication or does not eat regularly.

To prevent diabetic emergencies such as these, diabetics monitor their blood sugar several times a day. Glucose meters measure blood glucose and indicate if blood sugar is within the desired range. If needed, diabetics can adjust their caloric intake, their exercise, or their medication so that blood sugar stays within the normal range. **Figure 31-14** shows a glucose meter being used to monitor blood sugar, and **Table 31-2** summarizes the symptoms of IDDM and NIDDM.

Table 31-2 Warning Signs of Diabetes

Type I (IDDM)	Type II (NIDDM)
Frequent urination	Any of the Type I symptoms
Unusual thirst	Frequent infections
Extreme hunger	Blurred vision
Unusual weight loss	Cuts or bruises that are slow to heal
Extreme fatigue	Tingling or numbness in the hands or feet
Irritability	Recurring skin, gum, or bladder infections

Prostaglandins may cause fever and pain

A group of lipids called **prostaglandins** *(prahs tuh GLAN dihnz)* also function in the body as hormones. Prostaglandins are produced in small quantities by specialized cells in almost every part of the body. They act locally rather than traveling by blood to a distant site.

Prostaglandins help blood to clot and can raise blood pressure by causing the smooth muscle surrounding blood vessels to constrict. They also cause fever and increase pain during inflammation. Some cause the uterus to contract and start labor during childbirth. High levels of prostaglandins in the uterus can also cause menstrual cramps.

Aspirin and aspirin alternatives interfere with the synthesis and release of prostaglandins. These compounds are effective in treating fevers, menstrual cramps, headaches, muscle pains, and inflammatory conditions such as arthritis. However, only aspirin alternatives are recommended for children and teens. The use of aspirin is associated with Reye's syndrome, a potentially fatal disease of the liver and brain. Although many people are aware that young children should not take aspirin, they do not know that most deaths from Reye's syndrome are among teenagers. Aspirin alternatives include generic drugs such as ibuprofen, naproxen, and fenoprofen, which are sold under various brand names. **Figure 31-15** shows some examples of pain, fever, and inflammation, which may respond to drugs that inhibit prostaglandins.

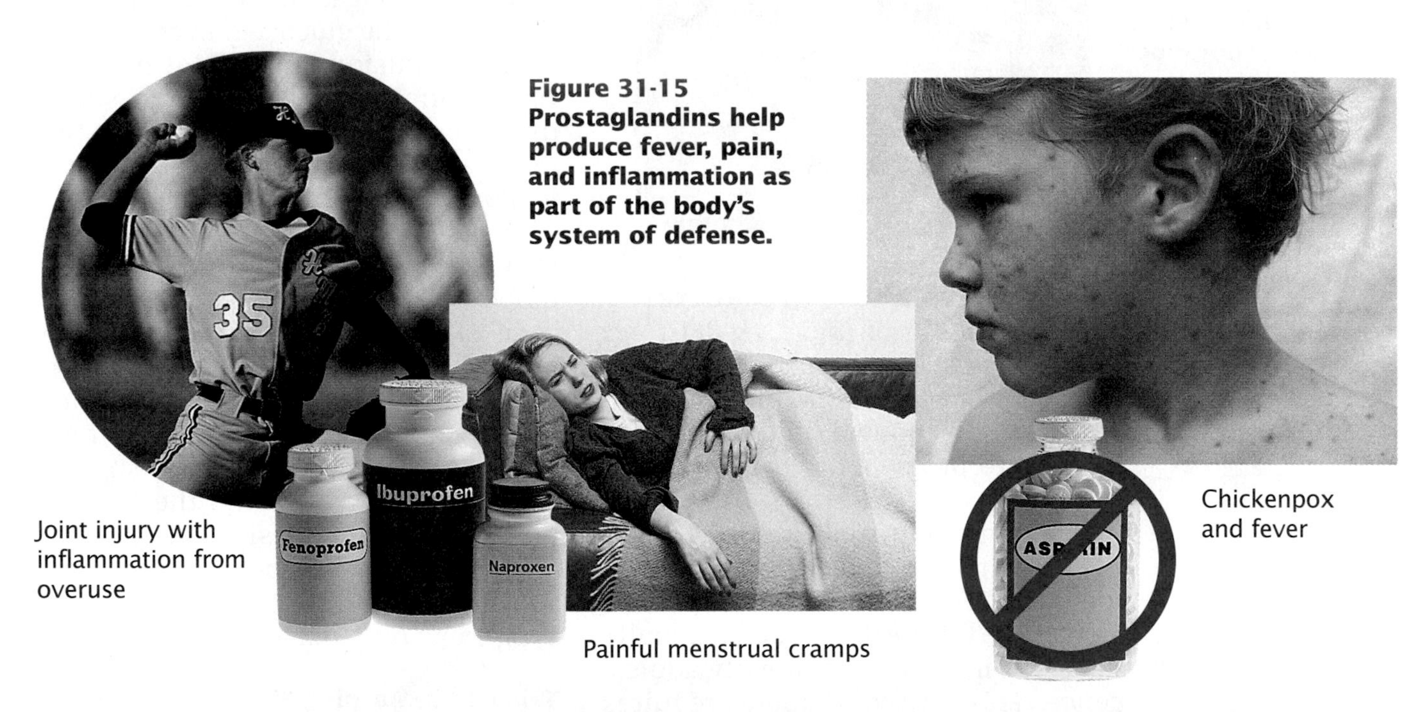

Figure 31-15 Prostaglandins help produce fever, pain, and inflammation as part of the body's system of defense.

a Drugs that inhibit the production and release of prostaglandins can help reduce these symptoms.

b Aspirin alternatives are considered safer than aspirin for children and teens with viral infections, although many fevers respond well to ample fluids, calories, and rest, without medication.

SECTION REVIEW

1. **Name three major endocrine glands in your body and the hormones they produce.**
2. **What occurs if the thyroid gland or the pancreas produces too much or too little hormone?**
3. **How does the fight-or-flight (alarm) reaction differ from the later stages of adaptation?**
4. **Why are aspirin and ibuprofen effective in treating headaches?**

CHAPTER

31 Highlights

When the body produces too little growth hormone, dwarfism can result. When too much growth hormone is produced, gigantism can occur.

31-1 What Hormones Do

In a negative-feedback process, thyroxine speeds up metabolism, keeping body temperature constant.

Key Terms

- hormone (p. 631)
- endocrine gland (p. 631)
- target (p. 631)
- neuroendocrine system (p. 632)
- pituitary gland (p. 633)
- feedback (p. 634)
- negative feedback (p. 634)

Summary

- Endocrine glands and other organs secrete hormones into the bloodstream.
- The effects of the endocrine system are similar to those of the nervous system, but these effects usually occur more slowly and last longer.
- The brain is connected directly to the endocrine system by the hypothalamus, which helps regulate the body's internal environment through the pituitary gland.
- Feedback systems allow for endocrine self-regulation.

31-2 How Hormones Work

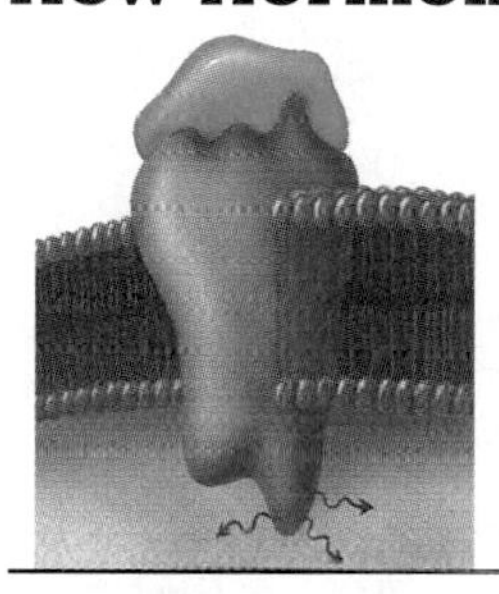

Hormones bind with receptors like keys fitting into locks.

Key Terms

- steroid hormone (p. 637)
- peptide hormone (p. 638)
- second messenger (p. 638)

Summary

- Hormones travel throughout the body in the bloodstream. They affect only the appropriate target cells.
- Steroid hormones pass through the cell membrane. Peptide hormones act from outside the cell by means of second messengers.
- Hormones act by affecting the kinds or amounts of enzymes produced.

31-3 Glands and Their Functions

A treatable condition called Graves' disease causes the thyroid gland to produce too much thyroxine.

Key Terms

- adrenal glands (p. 639)
- "fight or flight" reaction (p. 639)
- hyperthyroidism (p. 641)
- hypothyroidism (p. 642)
- goiter (p. 642)
- sex hormone (p. 643)
- blood sugar (p. 644)
- insulin (p. 644)
- diabetes mellitus (p. 645)
- prostaglandin (p. 646)

Summary

- Fewer than 10 endocrine glands produce dozens of hormones in the human body.
- The adrenal medulla helps the body react quickly to crises. The adrenal cortex makes steroids that regulate energy levels and adaptation to long-term stress.
- The thyroid gland regulates metabolism, while the pancreas regulates blood sugar.
- Other hormones, made by smaller endocrine glands or by organs, also have powerful effects on their respective targets.
- Hormone-like substances called prostaglandins act locally to regulate changes in the body.

CHAPTER 31

review

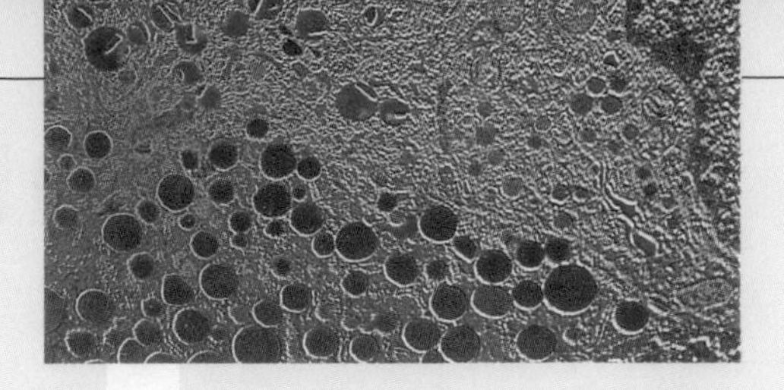

Understanding Vocabulary

1. For each set of terms, complete the analogy.
   a. short-term effects : nerve impulses :: long-term effects : ____
   b. chains of amino acids : peptide hormones :: assembled from cholesterol : ____
   c. epinephrine : adrenal medulla :: cortisol : ____
2. Using each set of words below, write one or more sentences summarizing information learned in this chapter.
   a. hormone, endocrine gland, endocrine system, target
   b. pituitary gland, prolactin, oxytocin, ADH
   c. peptide hormone, target, first messenger, second messenger

Understanding Concepts

3. **Relating Concepts** Construct a concept map that describes the operation of the pituitary gland and its relationship to the rest of the endocrine system. Include the following terms in your concept map: hypothalamus, nervous system, endocrine system, neuroendocrine system, thyroid gland, thyroxine, hormones, adrenal gland, medulla, cortex, stress, and hypothyroidism.
4. **Summarizing Information** What is the function of the endocrine system? What components make up this system?
5. **Identifying Functions** Describe how thyroid cells produce thyroid hormones. What happens when iodine is in short supply?
6. **Recognizing Relationships** Describe the relationship between the hypothalamus and the pituitary gland.
7. **Recognizing Patterns** Explain the role of feedback systems. How does negative-feedback maintain homeostasis?
8. **Inferring Relationships** Describe how the binding of a hormone to a receptor protein changes the activities of a cell.
9. **Organizing Information** Identify the locations where steroid and peptide hormones are activated and ultimately exert their influence.
10. **Comparing Functions** Compare what happens to the body during the alarm-reaction phase of the general adaptation syndrome with what happens during the resistance phase. Identify which endocrine glands and hormones are involved in each phase.
11. **Summarizing Information** Identify three major endocrine glands. Describe where each gland is located and the major hormones produced by each.
12. **Reading Graphs** Look at the graph below. In males, the hormone testosterone regulates aggression and increases the rate at which glucose gets to the muscles. What is happening with baboon A and B? Can you relate this graph with any behavior?

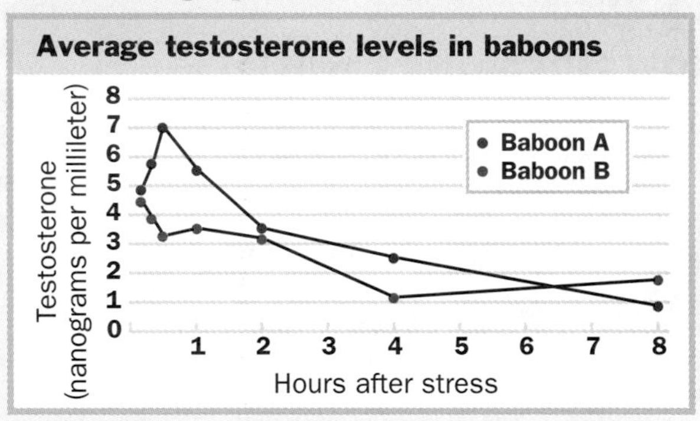

13. **Recognizing Processes** Describe the relationship between the hormones glucagon and insulin in the regulation of blood sugar.
14. **Recognizing Relationships** Differentiate between the two types of diabetes mellitus. Describe the symptoms of each and how each is treated.
15. **Inferring Relationships** Why might prostaglandins be called "local" hormones?
16. **Analyzing Information** When might inhibiting prostaglandin synthesis be helpful? When might it be unwanted?
17. **BUILDING ON WHAT YOU HAVE LEARNED** In Chapter 30, you learned about neurotransmitters. How are the actions of hormones and neurotransmitters alike? How are they different?

Interpreting Graphics

18. A scientist wanted to know the effects of thyroxine on tadpole growth. She placed three tadpoles in three different beakers of water. In one beaker she added 5 mL of thyroxine; in another beaker she added 10 mL of thyroxine. Nothing was added to the third beaker. She measured the length of the tadpoles every three days for three weeks. The results of her experiment are shown in the graph below.

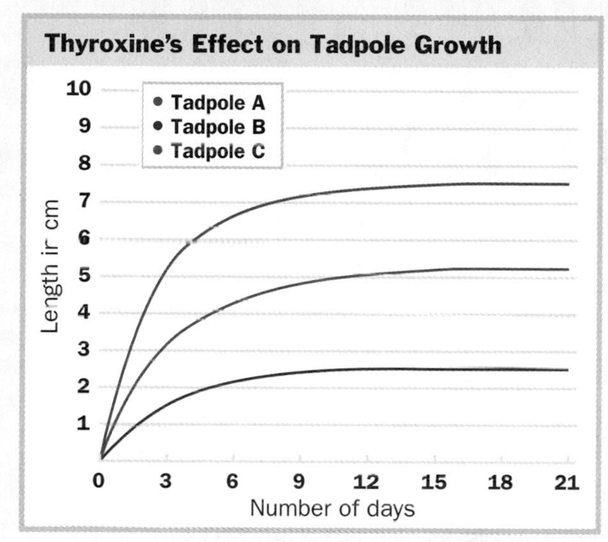

a. Which line on the graph represents the tadpoles exposed to 10 mL of thyroxine? to no thyroxine? Explain your answer.
b. How would you verbally summarize the results of this experiment?

Reviewing Themes

19. ***Stability and Homeostasis***
What are two ways that the hypothalamus helps the body maintain homeostasis?

20. ***Matter, Energy, and Organization***
Describe the importance of "fit" between a receptor protein and a hormone.

Thinking Critically

21. **Distinguishing Relevant Information** After a medical examination, a person is found to be unable to move glucose, stored as glycogen, from the liver into the blood. Further tests show that glucagon levels are normal and so is the hormone itself. Why do you think glucagon is unable to carry out its function in this case?

Life/Work Skills

22. **Finding Information** Use library references or search an on-line database to find out about the hormonal causes of giantism and dwarfism. Research the tallest and shortest humans known. Prepare an illustrated report to share your findings with the class.

23. **Using Technology** Working with a group of your classmates, contact a reference librarian to find out about computer software programs such as SUPERDOC and PUFF. How are these programs used and by whom? What advantage is there for using such programs? Summarize your findings in a written report.

Activities and Projects

24. **Health Connection** Use library references or search an on-line database to find out about disorders of the endocrine system such as Addison's disease and Cushing's syndrome. Describe the symptoms and treatments for these and other disorders of the endocrine system. Summarize your findings in a written report to your class.

Discovering Through Reading

25. Read the article "Melatonin Mania" in *Newsweek*, November 1995, pages 60–63. What is melatonin? How do people think melatonin can affect their lives? What are the dangers of high doses of melatonin?

26. Read the article "A Chance to Be Taller," in *TIME*, January 8, 1990, page 70. As more and more synthetic hormones become available through technology, their use in healthy individuals raises questions. When a young person's height varies from that of most others of the same age, social and emotional problems may result. The synthesis in 1985 of human growth hormone (HGH), makes it possible to increase the height of normal, but short, children. With one or several friends, separate the issues into categories such as physical, financial, social, mental, and emotional concerns and present them to your class. After your presentation, develop a series of 8–12 questions to survey your classmates' opinions about the use of HGH.

CHAPTER

32

REVIEW
- homeostasis (Section 1-3)
- gas exchange (Section 24-2)
- evolution of the lung (Section 24-2)

Circulation and Respiration

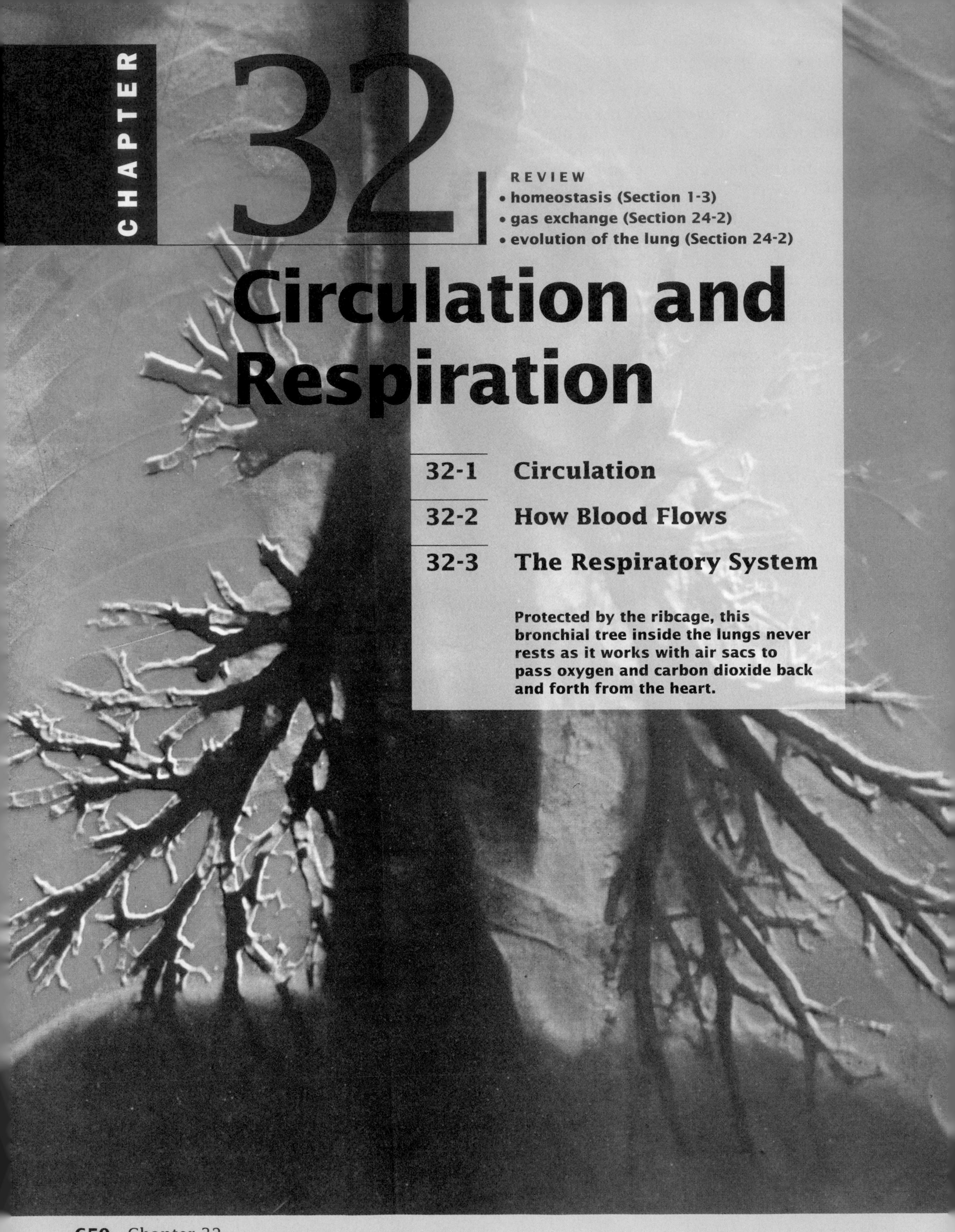

Protected by the ribcage, this bronchial tree inside the lungs never rests as it works with air sacs to pass oxygen and carbon dioxide back and forth from the heart.

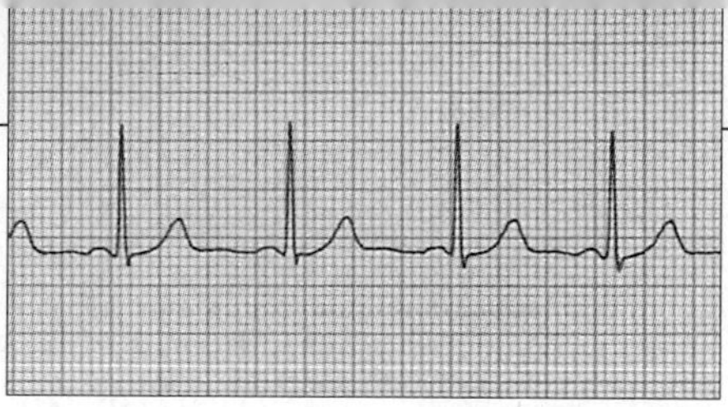

Each lubb-dup of your beating heart can be seen as an up and down pair of spikes on an ekg strip. Inside your body, blood flows along a vast 60,000 mile highway, long enough to circle Earth two and a half times. With every beat, blood nourishes and cleanses the entire body, bringing life to all it touches.

32-1 Circulation

OBJECTIVES

1. **Explain the three functions of circulation.**
2. **Describe the components of blood and identify their roles.**
3. **Explain the relationship between the three major types of blood vessels.**
4. **Contrast the role of the lymphatic system with that of general circulation.**

Transporting Materials Through the Body

All living things must capture materials from their environment that enable them to carry on their life processes. Many bacteria, single-celled protists, and simpler multicellular animals live within a liquid environment, enabling materials to diffuse directly into and out of the organism through plasma membranes. However, larger animals with many cells stacked in layers, such as earthworms, armadillos, and humans, cannot rely solely on diffusion to supply needed materials and carry away wastes. In these organisms, a circulatory system transports oxygen, carbon dioxide, food molecules, hormones, and other materials to and from the cells of the body. In addition, the circulatory systems of mammals and birds help maintain their constant body temperatures. The circulatory system also carries cells that help protect the body from disease. All of these functions help the body maintain homeostasis. All are essential to survival.

Our modern understanding of the human circulatory system began with the work of the seventeenth-century English physician William Harvey, who demonstrated that blood circulates in one direction in a closed circuit throughout the body. An illustration of his experiment is shown in **Figure 32-1**.

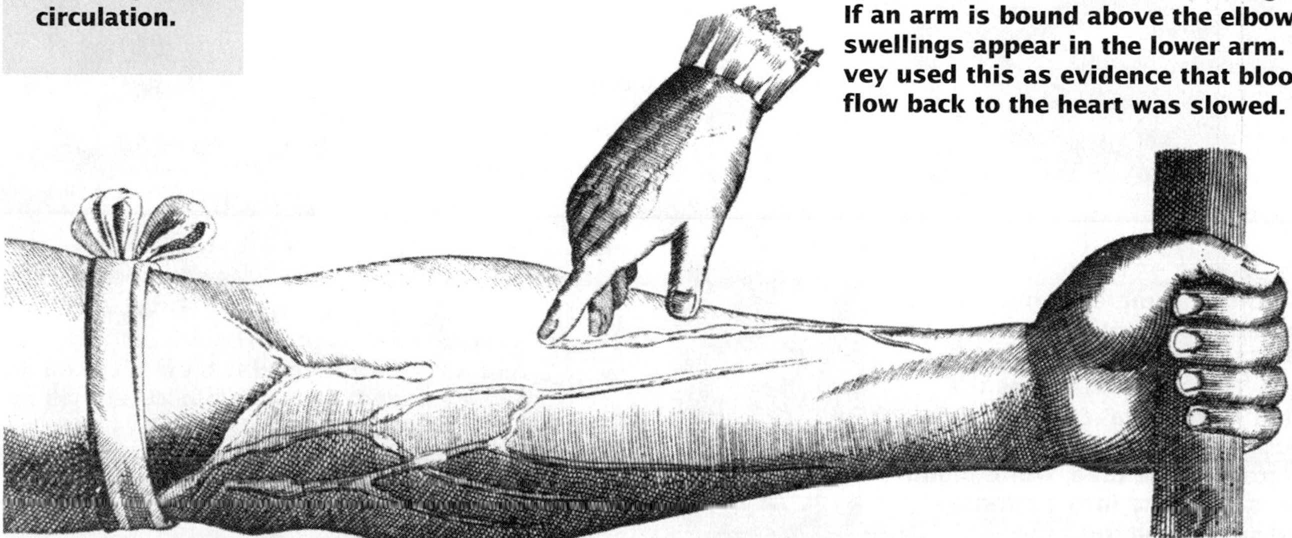

Figure 32-1
In 1628, William Harvey demonstrated that blood travels from the heart to the body's limbs and then back again. If an arm is bound above the elbow, swellings appear in the lower arm. Harvey used this as evidence that blood flow back to the heart was slowed.

Blood: A Liquid Tissue

Blood has been called the river of life. It is the tissue responsible for transporting nearly everything within the body. Among all the body's tissues, blood is the only liquid tissue. The material that makes blood a liquid is a protein-rich substance called **plasma** *(PLAZ muh)*. Plasma is about 90 percent water and makes up approximately 55 percent of blood. The other 45 percent is made mostly of cells—red blood cells, white blood cells, and platelets.

Red blood cells carry oxygen to other body tissues

Most of the cells that make up blood are red blood cells. There are approximately 5 million red blood cells per cubic millimeter of blood! The main function of a **red blood cell** is to carry oxygen from the lungs to all cells of the body. Red blood cells also carry carbon dioxide from cells back to the lungs to be exhaled. The structure of a red blood cell is well suited to this function. As shown in **Figure 32-2a**, normal red blood cells are shaped like round cushions, squashed in the center. Red blood cells are filled with a protein called **hemoglobin**, an iron-containing molecule that gives blood its red color. Oxygen binds easily to the iron in hemoglobin, making red blood cells efficient oxygen carriers. A single red blood cell contains about 250 million hemoglobin molecules. Each red blood cell can carry about 1 billion molecules of oxygen.

Anemia *(uh NEE mee uh)* is a blood condition that results when there is an oxygen shortage in the cells of the body. Blood loss, too little hemoglobin, and not enough iron or certain vitamins can cause abnormalities in the shape, size, or color of red blood cells. Iron-deficiency anemia is common in young girls and women of childbearing age who lack adequate dietary or stored iron. Common symptoms include fatigue, stomach problems, broken and brittle nails, and paleness.

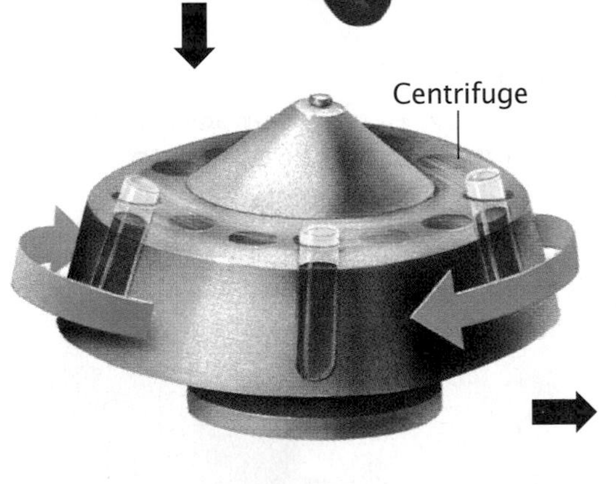

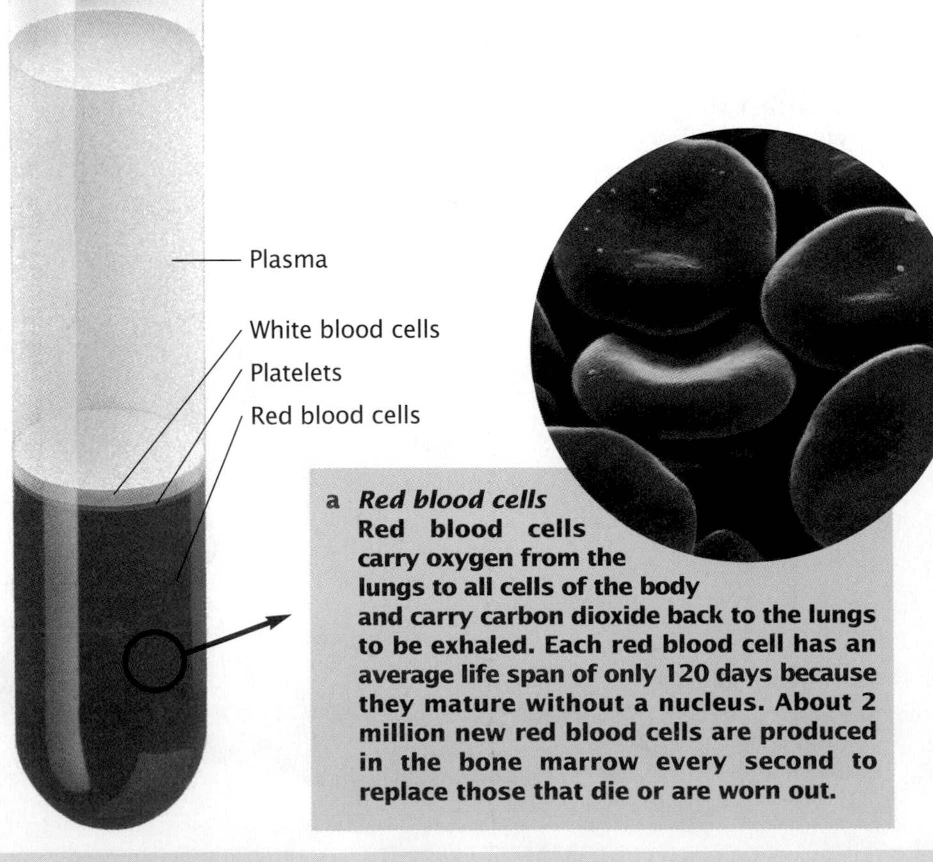

Figure 32-2
After a blood sample is collected in a vacuum syringe, it is spun rapidly in a device called a centrifuge (above). This rotation forces the blood to settle in layers, with the heaviest cells (red blood cells) at the bottom of the tube, white blood cells and platelets in the middle, and plasma at the top.

a *Red blood cells*
Red blood cells carry oxygen from the lungs to all cells of the body and carry carbon dioxide back to the lungs to be exhaled. Each red blood cell has an average life span of only 120 days because they mature without a nucleus. About 2 million new red blood cells are produced in the bone marrow every second to replace those that die or are worn out.

White blood cells defend the body

White blood cells form a mobile army that protects the body against invading bacteria, viruses, or other foreign cells. White blood cells called macrophages *(MAK roh fayj iz)* act as a clean-up crew, scavenging worn-out or dead cells. While red blood cells are confined to the bloodstream, white blood cells are able to squeeze in and out of blood vessels. They move toward areas of tissue damage and infection in response to chemicals released by damaged cells. By following the chemical trail, macrophages are able to migrate to infected areas in the body, where they engulf and digest foreign cells.

As shown in **Figure 32-2b**, white blood cells are colorless and irregularly shaped. Most of them are manufactured and stored in bone marrow until needed by the body. Although they are larger than red blood cells, they are considerably less numerous; normally there are 4,000 to 11,000 white blood cells per cubic millimeter of blood. Whenever white blood cells are mobilized for action, the body speeds up their production so that twice the normal number may appear within a few hours. A white blood cell count over 11,000 generally indicates the presence of a bacterial or viral infection in the body.

Platelets help repair damaged blood vessels

Platelets are not really cells; they are fragments of large white blood cells formed within bone marrow. Each cubic millimeter of blood contains between 250,000 and 500,000 tiny platelets. Each platelet circulates for only three to eight days.

Platelets are essential for the clotting process that occurs when blood vessels are injured, as shown in **Figure 32-2c**. When you cut your finger, platelets make contact with the ends of the broken blood vessels. There, the platelets swell and stick to the injured surfaces as well as to each other. Certain platelets rupture and release a protein that triggers a series of reactions, resulting in the formation of a clot. The clot stops the bleeding and hardens into a patch over the injured area. In time, new cells grow and replace those that were damaged.

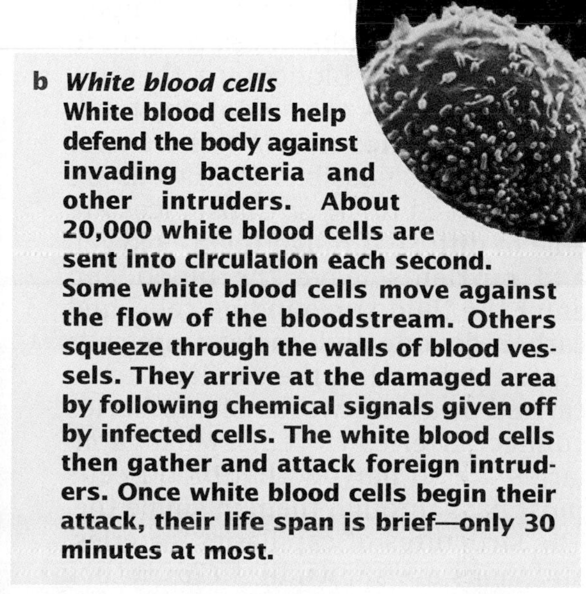

b *White blood cells*
White blood cells help defend the body against invading bacteria and other intruders. About 20,000 white blood cells are sent into circulation each second. Some white blood cells move against the flow of the bloodstream. Others squeeze through the walls of blood vessels. They arrive at the damaged area by following chemical signals given off by infected cells. The white blood cells then gather and attack foreign intruders. Once white blood cells begin their attack, their life span is brief—only 30 minutes at most.

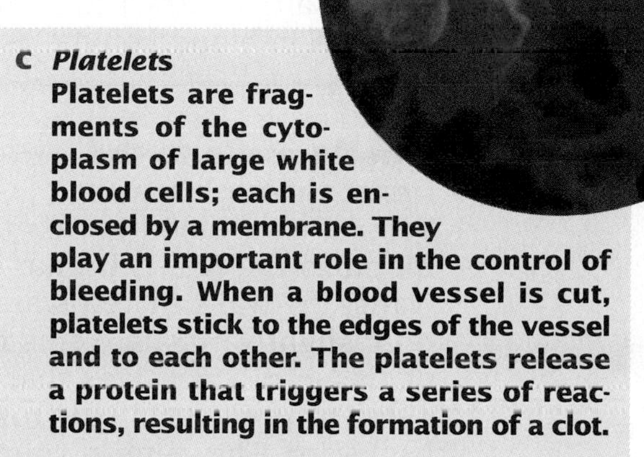

c *Platelets*
Platelets are fragments of the cytoplasm of large white blood cells; each is enclosed by a membrane. They play an important role in the control of bleeding. When a blood vessel is cut, platelets stick to the edges of the vessel and to each other. The platelets release a protein that triggers a series of reactions, resulting in the formation of a clot.

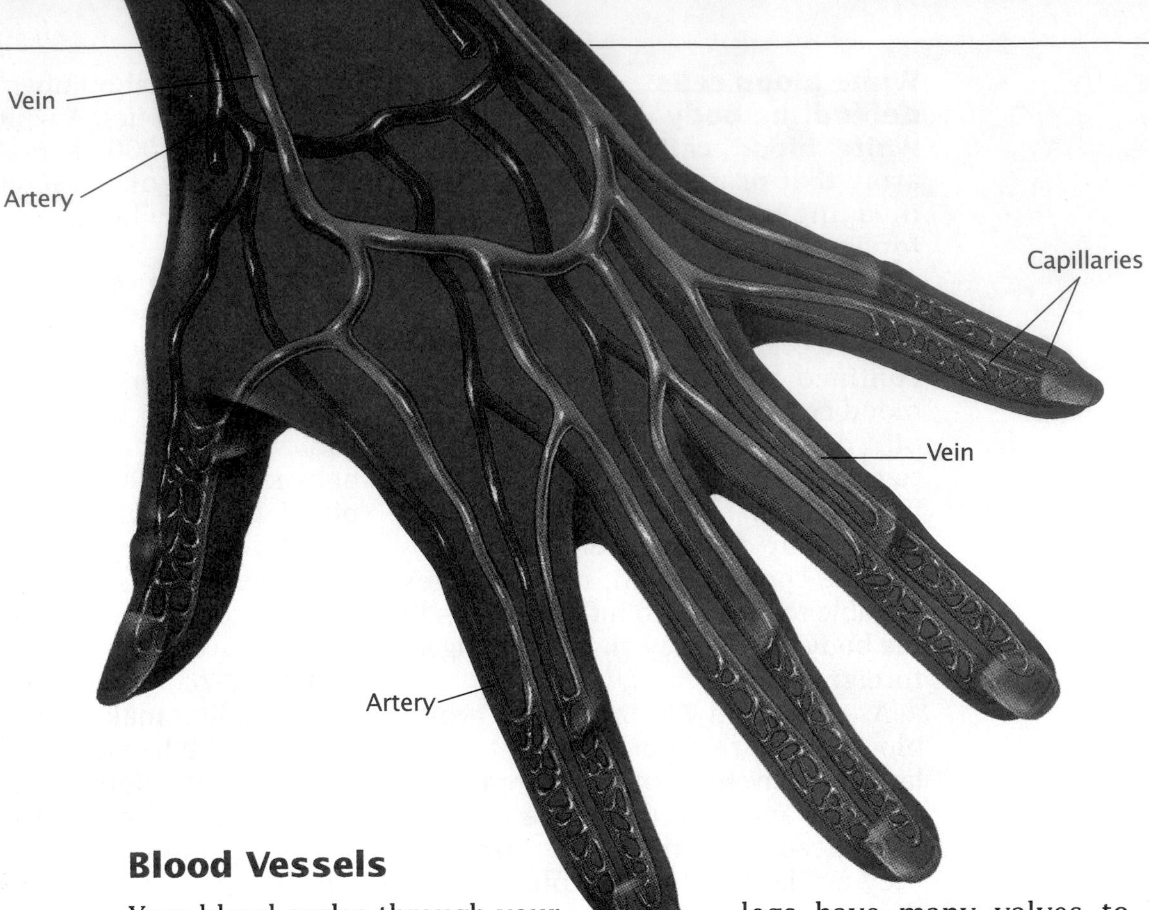

Figure 32-3 Blood travels through all parts of the body (such as the hand above) in blood vessels. Arteries (red) carry blood away from the heart to the body's organs and tissues. Veins (blue) return blood to the heart. Exchange of oxygen and cellular wastes occurs in tiny tubes called capillaries (purple). These tubes form a network that connects arteries to veins.

Blood Vessels

Your blood cycles through your body 1,440 times a day, traveling through tubes called **blood vessels**. Blood vessels carry blood from the heart to the organs in the body and back again to the heart. They have flexible walls so that they can change diameter when more blood is needed. The blood vessels of the hand are shown in **Figure 32-3**.

Blood travels away from the heart through arteries

Arteries are the blood vessels that carry blood away from the heart. In most instances, arteries carry oxygen-rich blood. Arteries have highly elastic, muscular walls, enabling them to expand and contract to accommodate changes in blood volume as the heart pumps.

Blood travels to the heart through veins

Veins are the blood vessels that return blood to the heart. Unlike the artery walls, the walls of veins are thin and only slightly elastic. Veins have flaps of tissue called valves that open and close to prevent blood from flowing backward. The veins in your arms and legs have many valves to help return blood to the heart against the force of gravity. When valves do not work properly, blood tends to build up within the vein. The walls of the vein become stretched and twisted, causing a condition called varicose veins.

Materials are exchanged in the capillaries

Since the purpose of blood circulation is to transport substances, it is important that these substances be able to move from the blood into the surrounding cells. A network of tiny tubes called **capillaries** makes this transfer of substances possible. Their walls are only one cell thick, so substances are able to diffuse through them—oxygen and nutrients leave the blood and enter the fluid surrounding cells, and carbon dioxide and wastes from cells enter the blood in the capillary to be carried away. Networks of capillaries connect arteries to veins. Most capillaries are so narrow that blood cells must pass through them in single file. The structures of capillaries, arteries, and veins are shown in *A Closer Look at the Blood Vessels* on page 655.

Blood Vessels

Arteries, veins, and capillaries are the three main types of blood vessels in the human body. The walls of these vessels are well constructed for the passage of blood throughout the body.

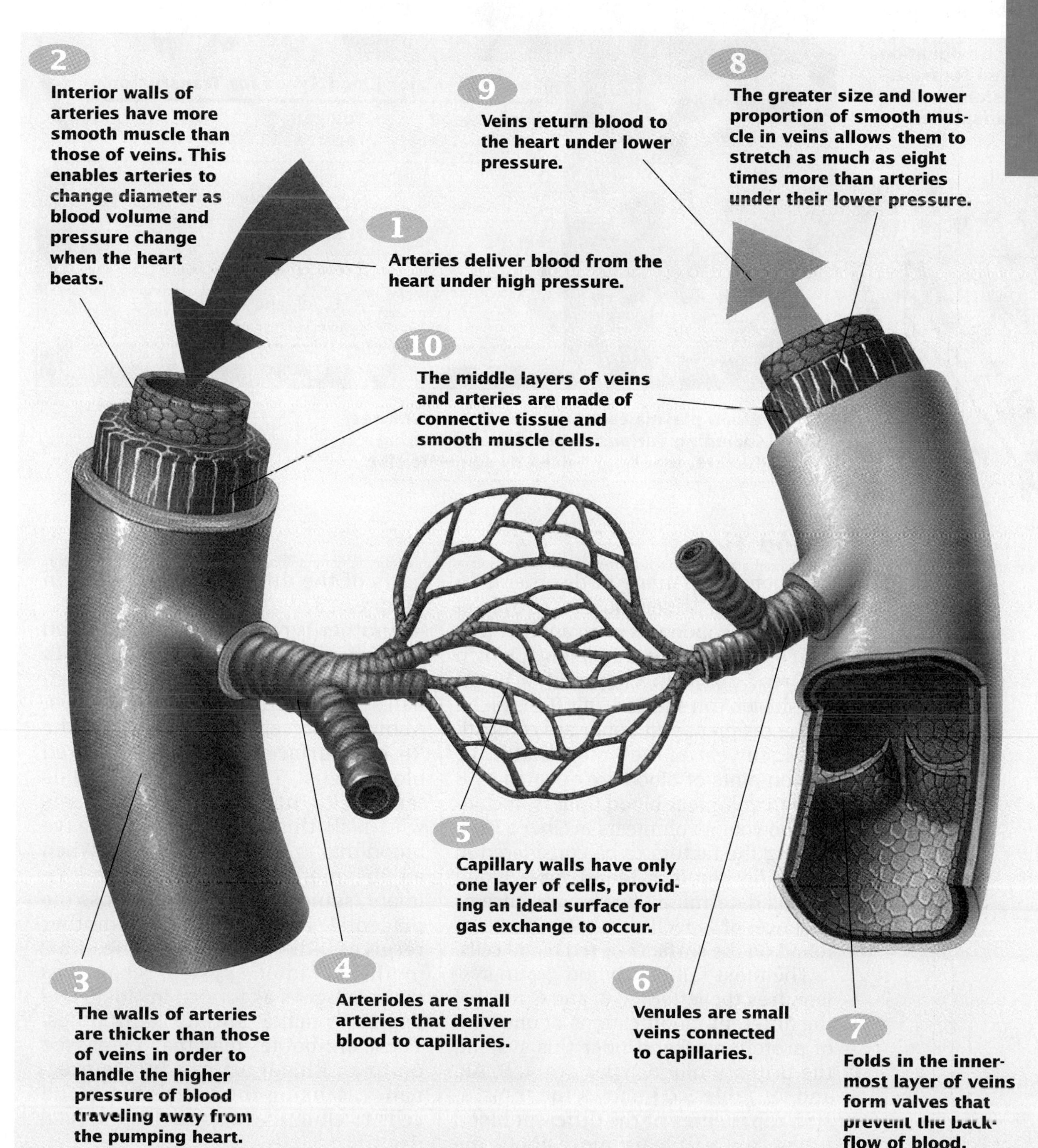

1
Arteries deliver blood from the heart under high pressure.
2
Interior walls of arteries have more smooth muscle than those of veins. This enables arteries to change diameter as blood volume and pressure change when the heart beats.
3
The walls of arteries are thicker than those of veins in order to handle the higher pressure of blood traveling away from the pumping heart.
4
Arterioles are small arteries that deliver blood to capillaries.
5
Capillary walls have only one layer of cells, providing an ideal surface for gas exchange to occur.
6
Venules are small veins connected to capillaries.
7
Folds in the innermost layer of veins form valves that prevent the backflow of blood.
8
The greater size and lower proportion of smooth muscle in veins allows them to stretch as much as eight times more than arteries under their lower pressure.
9
Veins return blood to the heart under lower pressure.
10
The middle layers of veins and arteries are made of connective tissue and smooth muscle cells.

Figure 32-4
A blood transfusion can succeed only if the blood of the recipient is compatible with that of the donor.

a In this mobile unit, medical laboratory technicians collect blood donations used for transfusions and transplants.

Table 32-1 Major Blood Types for Transfusion

If your blood type is . . .	You can receive . . .
O	O only (universal donor)
A	A and O
B	B and O
AB	A, B, AB, and O (universal recipient)

b Blood plasma contains over 150 substances, including various antibodies and clotting factors, that can be used by someone else.

Blood Types

Occasionally an injury or disorder is so serious that a person must receive blood or blood components from another person. As with organ transplants, which you read about in Chapter 4, a blood transfusion can succeed only if the blood of the recipient and donor are compatible. Each year in the United States, 12 million pints of blood are donated by 8 million volunteer blood donors, including the young volunteers in **Figure 32-4**. Among the factors to be considered in matching blood is **blood type**. Blood type is determined by the presence or absence of specific marker proteins found on the surfaces of red blood cells.

The most familiar blood group system uses the letters A, B, and O to label the different modifications of one kind of protein marker. Under this system, the primary blood types are A, B, AB, and O. **Table 32-1** shows the transfusion capabilities of the different blood types. You will learn more about the basis of the different blood types in Chapter 33.

Another type of marker protein on the surface of red blood cells is the Rh factor, so named because it was originally identified in rhesus monkeys. About 85 percent of humans have the Rh cell surface marker on their red blood cells. This type of blood is termed Rh positive (Rh+). Humans who lack this surface marker have blood that is Rh negative (Rh−). When an Rh− mother gives birth to an Rh+ infant, some fetal blood cells cross the placental barrier, and the mother receives Rh+ antigens. The Rh− mother's immune system recognizes these antigens as foreign invaders and begins to make "anti-Rh" antibodies. These antibodies may then be passed on to an Rh+ fetus in a future pregnancy, causing the fetus's red blood cells to clump, which can lead to fetal death.

The Lymphatic System

Stability and Homeostasis

Explain why swollen lymph nodes are often considered a symptom of illness.

All the cells in your body are bathed in a clear, watery fluid that moves materials between the capillaries and body cells. This fluid is formed from parts of the blood that diffuse out of capillaries—water, proteins, and other nutrients. If this fluid had no way of returning to the circulatory system, body tissues would soon become flooded and would swell up. At the same time, the constant loss of fluid from the blood would eventually drain the circulatory system. These excess fluids and proteins are returned to the blood by a system of vessels called the **lymphatic system**. Lymph nodes are located at various places along the lymphatic vessels, shown in **Figure 32-5**. Each lymph node contains many white blood cells that help defend the body against cancerous cells or disease-causing organisms. Lymph nodes also return water, proteins, and other nutrients to the blood.

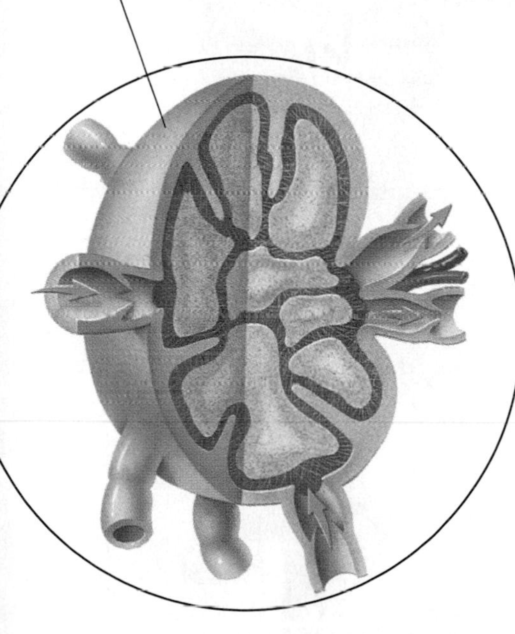

Figure 32-5

a **Lymph nodes about the size of small kidney beans are found in groups near veins.**

b **Enlarged lymph nodes in three sites (shown in white) can be felt by a health care professional.**

c **Lymph nodes filter foreign matter from the body's fluid and prevent cancerous cells and disease-causing agents from entering the bloodstream.**

SECTION REVIEW

1. **List three functions of the circulatory system.**
2. **Summarize the functions of the three cellular components of blood.**
3. **Distinguish between arteries and veins.**
4. **How does the lymphatic system help the body fight infection?**

Blood could not meet the body's needs if it did not flow. Blood flows because the circulatory system includes a pump—the heart—that forces blood to move throughout the body. Keeping fit, not smoking, and eating foods low in saturated fats will reduce your risk of diseases affecting the heart and blood vessels.

32-2 How Blood Flows

OBJECTIVES

1. **Describe the structure of the heart.**
2. **Trace the two routes blood can take through the body.**
3. **Explain the significance of blood pressure, and explain how it is measured.**
4. **Summarize the value of a blood analysis in describing homeostasis.**
5. **Describe the causes and symptoms of two cardiovascular diseases.**

The Heart

The heart is a muscular organ that pumps blood throughout the body. When you are sitting still, your heart pumps about 5 L (5.3 qt) of blood each minute. If you are riding a bike, like the person in **Figure 32-6**, your heart may have to pump up to seven times that amount per minute.

Your heart is divided into left and right halves. The right half pumps blood to the lungs to pick up oxygen and release carbon dioxide. The left half of the heart pumps the oxygen-rich blood to the rest of the body—your head, arms, legs, and all the tissues and organs in between.

As shown in **Figure 32-7**, each half of the heart has an upper and a lower chamber. The upper chamber, called the **atrium**, receives blood coming into the heart. The lower chamber, called the **ventricle**, pumps blood out of the heart. The upper and lower chambers are separated from each other by flap-like valves that control the direction of the blood flow inside the heart. Blood flows into both atria at the same time, and the atria contract together. Similarly, the ventricles contract together. All this activity causes the heartbeat, a sound that is usually described as "lubb dup." The "lubb" sound relates to the closing of the valves that lead from the atrium to the ventricle. The "dup" comes very shortly afterward and is related to the closing of the valves between the ventricles and the arteries that lead to the lungs and the rest of the body.

Figure 32-6
When you are active, your heart pumps up to 35 L (37 qt) of blood each minute. That's seven times the amount your heart pumps when you are at rest.

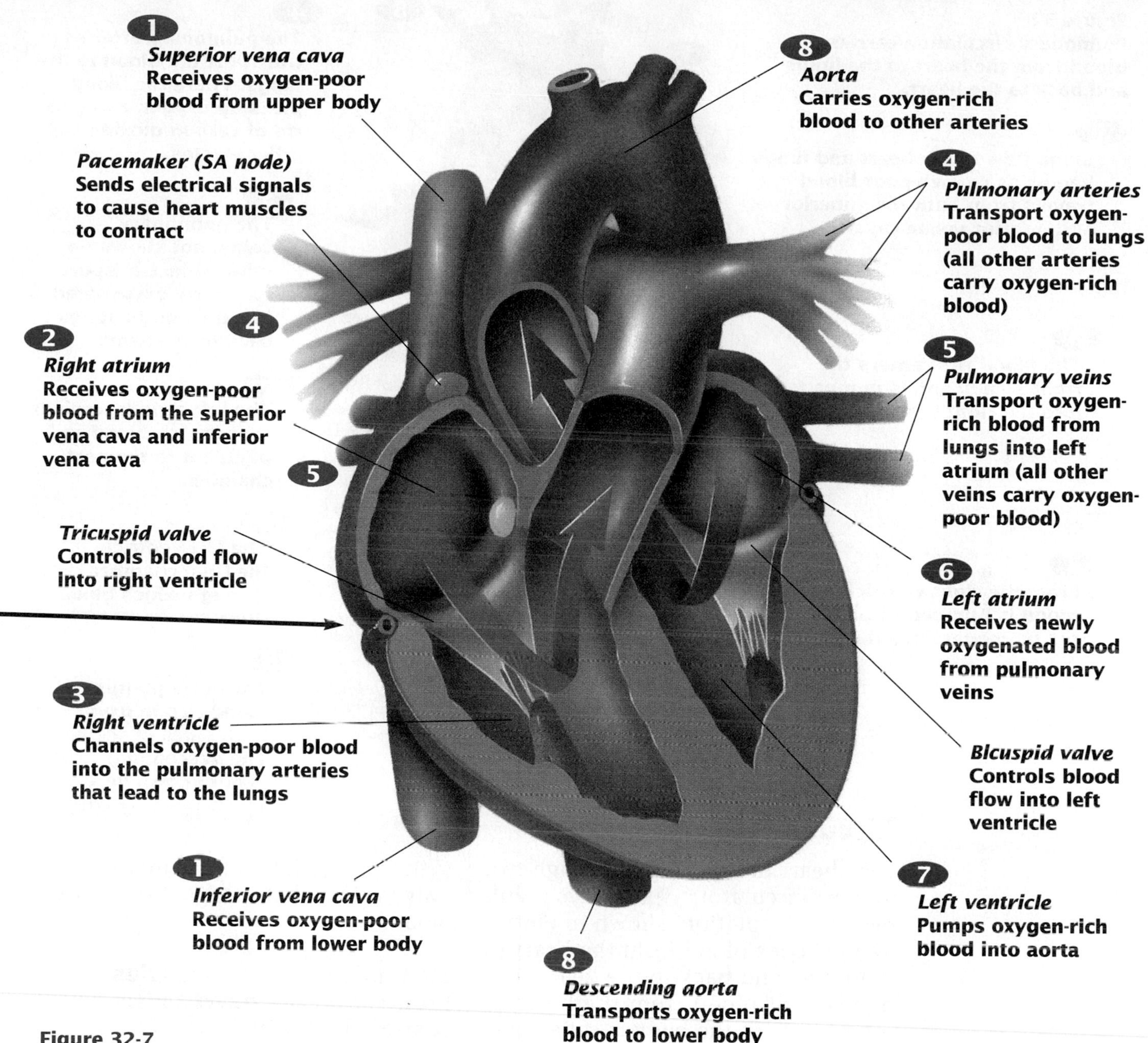
1
Superior vena cava
Receives oxygen-poor blood from upper body
Pacemaker (SA node)
Sends electrical signals to cause heart muscles to contract
2
Right atrium
Receives oxygen-poor blood from the superior vena cava and inferior vena cava
4
5
Tricuspid valve
Controls blood flow into right ventricle
3
Right ventricle
Channels oxygen-poor blood into the pulmonary arteries that lead to the lungs
1
Inferior vena cava
Receives oxygen-poor blood from lower body
8
Descending aorta
Transports oxygen-rich blood to lower body
8
Aorta
Carries oxygen-rich blood to other arteries
4
Pulmonary arteries
Transport oxygen-poor blood to lungs (all other arteries carry oxygen-rich blood)
5
Pulmonary veins
Transport oxygen-rich blood from lungs into left atrium (all other veins carry oxygen-poor blood)
6
Left atrium
Receives newly oxygenated blood from pulmonary veins
Bicuspid valve
Controls blood flow into left ventricle
7
Left ventricle
Pumps oxygen-rich blood into aorta

Figure 32-7
The human heart pumps oxygen-poor blood to the lungs and oxygen-rich blood to the rest of the body. The numbered steps above allow you to follow the flow of blood from its entry into the heart (1) to its exit from the heart (8).

The four-chambered human heart is shown in **Figure 32-7**. The *Evolution of the Heart* on pages 662–663 summarizes the evolution of the vertebrate heart.

What makes the heart beat?

The heartbeat originates in heart tissue. It begins to beat in the embryo, before any nerves connect it to the brain. It can continue to beat during transplant surgery, after all nerves have been cut. How is this possible?

Each heartbeat is started by the **pacemaker**, a small bundle of cells at the entrance to the right atrium. An electrical signal from the pacemaker, shown in **Figure 32-7**, travels through the heart muscles in the right and left atria, causing them to tighten, or contract. When the signal reaches the right and left ventricles, they also contract. These contractions cause the chambers to squeeze the blood through the heart and push it to other parts of the body.

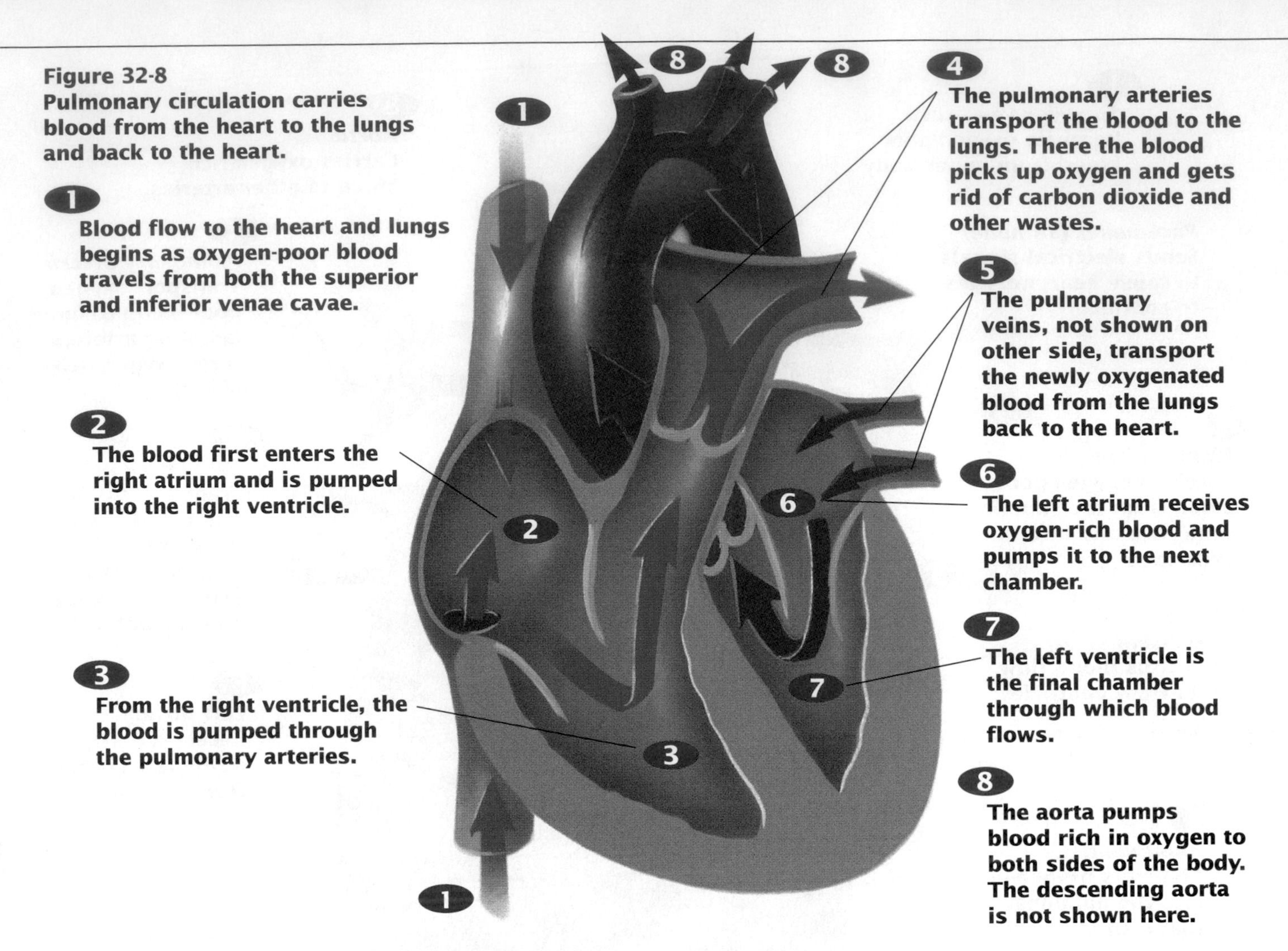

1
Blood flow to the heart and lungs begins as oxygen-poor blood travels from both the superior and inferior venae cavae.
2
The blood first enters the right atrium and is pumped into the right ventricle.
3
From the right ventricle, the blood is pumped through the pulmonary arteries.
4
The pulmonary arteries transport the blood to the lungs. There the blood picks up oxygen and gets rid of carbon dioxide and other wastes.
5
The pulmonary veins, not shown on other side, transport the newly oxygenated blood from the lungs back to the heart.
6
The left atrium receives oxygen-rich blood and pumps it to the next chamber.
7
The left ventricle is the final chamber through which blood flows.
8
The aorta pumps blood rich in oxygen to both sides of the body. The descending aorta is not shown here.
1
8
8
2
3
6
7
1

Figure 32-8
Pulmonary circulation carries blood from the heart to the lungs and back to the heart.

Circulatory Pathways

Your heart pumps blood through two major circulatory pathways. Pulmonary circulation, shown in **Figure 32-8**, carries blood from the heart to the lungs and back to the heart. The pathway of blood from the heart to other parts of the body is called systemic circulation.

Blood picks up oxygen and releases carbon dioxide in the lungs

Blood returning to the heart from the body is low in oxygen and high in carbon dioxide. This blood is pumped through the right side of the heart into arteries that lead to the lungs. These arteries, the pulmonary arteries, are the only arteries that carry oxygen-poor blood. As blood flows through the capillaries in the lungs, it picks up oxygen and gets rid of carbon dioxide. Leaving the lungs, the blood travels through veins, back to the heart. These veins, the pulmonary veins, are the only veins that carry oxygen-rich blood.

Systemic circulation carries blood from the heart to the rest of the body

Blood returning to the heart from the lungs is ready to deliver oxygen to the rest of the body. Notice in **Figure 32-8** that oxygen-rich blood is pumped through the left side of the heart and out the aorta, the largest artery in the body. From the aorta, blood flows to all parts of the body through a system of increasingly smaller arteries. Systemic circulation has three branches of special importance: the branch that carries blood to the heart, the branch that carries blood to the digestive tract and liver, and the branch that carries blood to and from the kidneys. Several of the major veins and arteries of systemic circulation are shown in **Figure 32-9**.

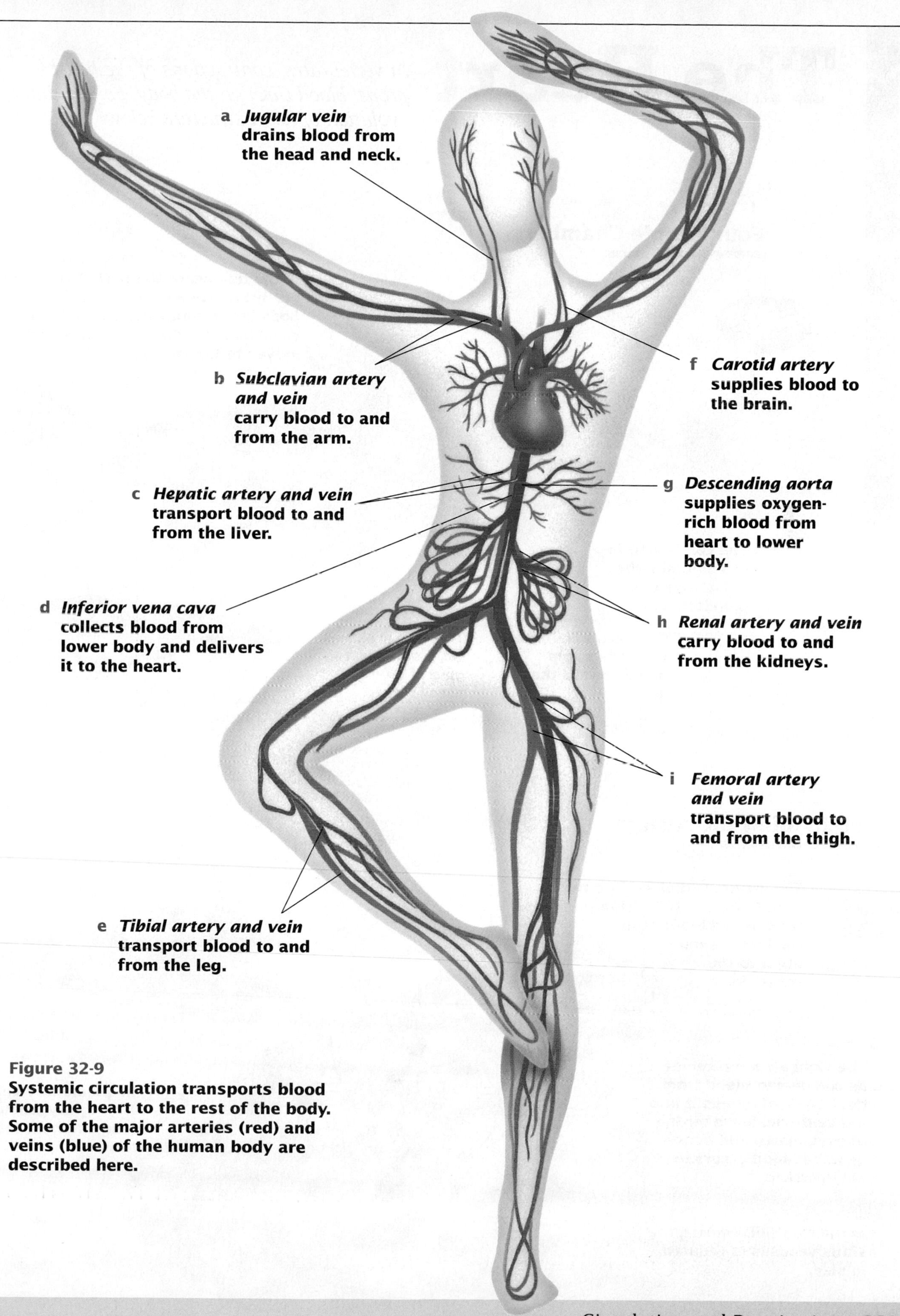
a *Jugular vein*
drains blood from the head and neck.
b *Subclavian artery and vein*
carry blood to and from the arm.
c *Hepatic artery and vein*
transport blood to and from the liver.
d *Inferior vena cava*
collects blood from lower body and delivers it to the heart.
e *Tibial artery and vein*
transport blood to and from the leg.
f *Carotid artery*
supplies blood to the brain.
g *Descending aorta*
supplies oxygen-rich blood from heart to lower body.
h *Renal artery and vein*
carry blood to and from the kidneys.
i *Femoral artery and vein*
transport blood to and from the thigh.

Figure 32-9
Systemic circulation transports blood from the heart to the rest of the body. Some of the major arteries (red) and veins (blue) of the human body are described here.

Evolutionary Journey

The Heart

In vertebrates, contractions of the heart propel blood through the body. Follow the evolution of heart structure below.

Four Simple Chambers

1

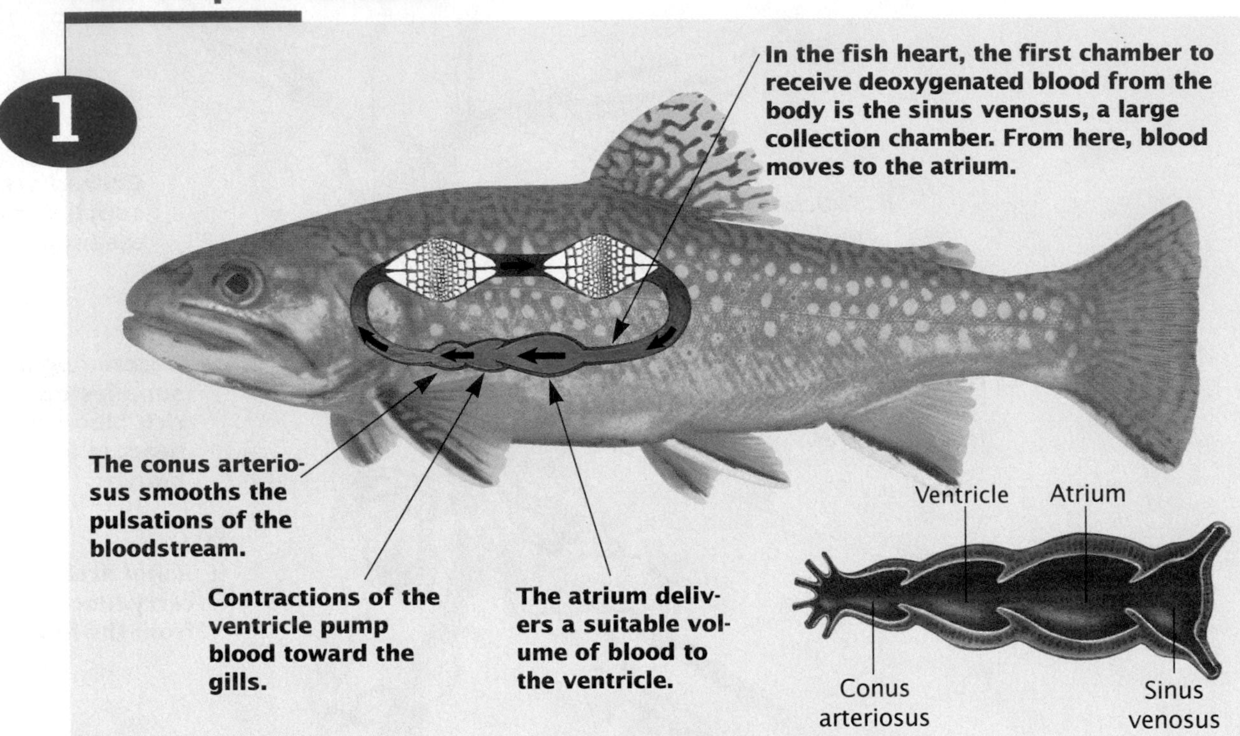

In the fish heart, the first chamber to receive deoxygenated blood from the body is the sinus venosus, a large collection chamber. From here, blood moves to the atrium.
The conus arteriosus smooths the pulsations of the bloodstream.
Contractions of the ventricle pump blood toward the gills.
The atrium delivers a suitable volume of blood to the ventricle.
Ventricle
Atrium
Conus arteriosus
Sinus venosus

Subdividing the Atrium

2

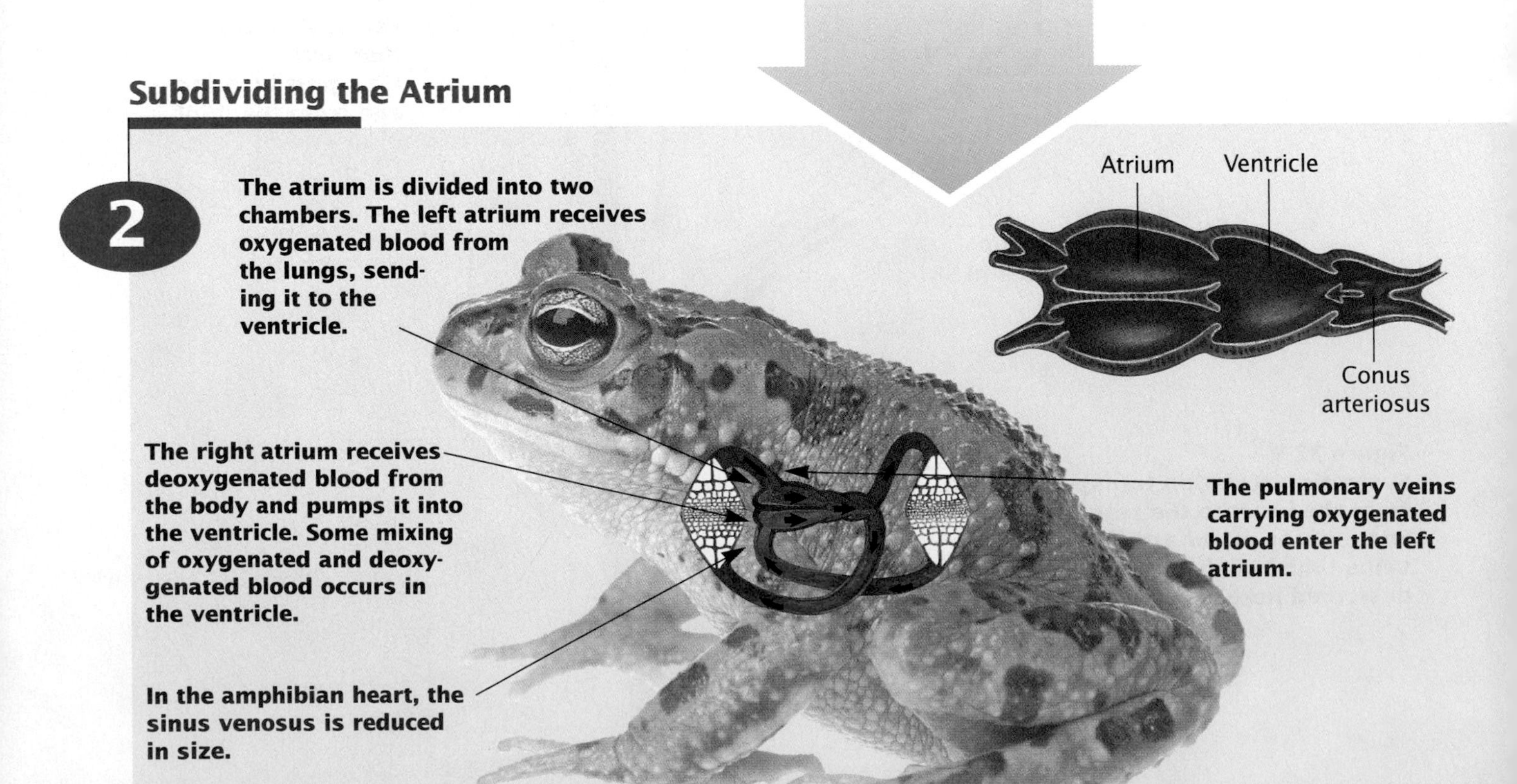

The atrium is divided into two chambers. The left atrium receives oxygenated blood from the lungs, sending it to the ventricle.
Atrium
Ventricle
Conus arteriosus
The right atrium receives deoxygenated blood from the body and pumps it into the ventricle. Some mixing of oxygenated and deoxygenated blood occurs in the ventricle.
The pulmonary veins carrying oxygenated blood enter the left atrium.
In the amphibian heart, the sinus venosus is reduced in size.

Completing the Septum

4

In birds, crocodiles, and mammals, the septum, or wall, dividing the ventricle is complete. There is no mixing of oxygenated and deoxygenated blood.

The SA node is a remnant of the sinus venosus. It controls the heartbeat.

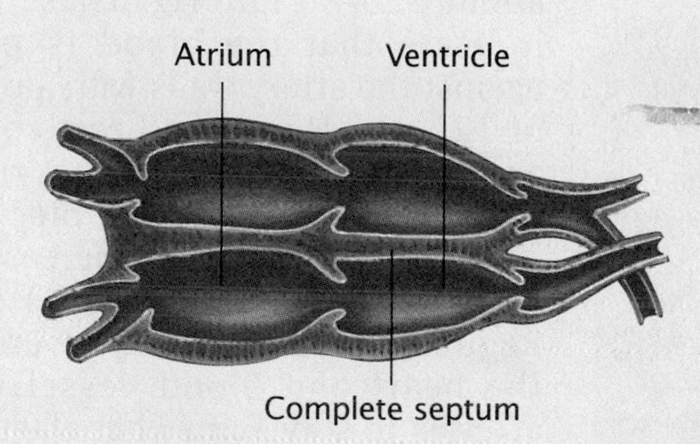

Partially Dividing the Ventricle

3

Atrium

Ventricle

Partial septum

In the reptilian heart, the conus arteriosus is replaced by arteries that direct oxygenated blood into the aorta and deoxygenated blood to the lungs.

The septum dividing the atrium extends partway into the ventricle.

Blood Pressure

When the ventricles contract, blood is forced into the arteries, which exerts pressure on the walls of the blood vessel. This force is called **blood pressure**. When ventricles relax, the pressure decreases. The muscular, elastic walls of the arteries are able to adjust to the changing pressure. The elasticity helps maintain the pressure between heartbeats. This way blood is kept flowing through the body continuously. When you take your pulse, you are feeling the expansion and relaxation of an artery with each heartbeat.

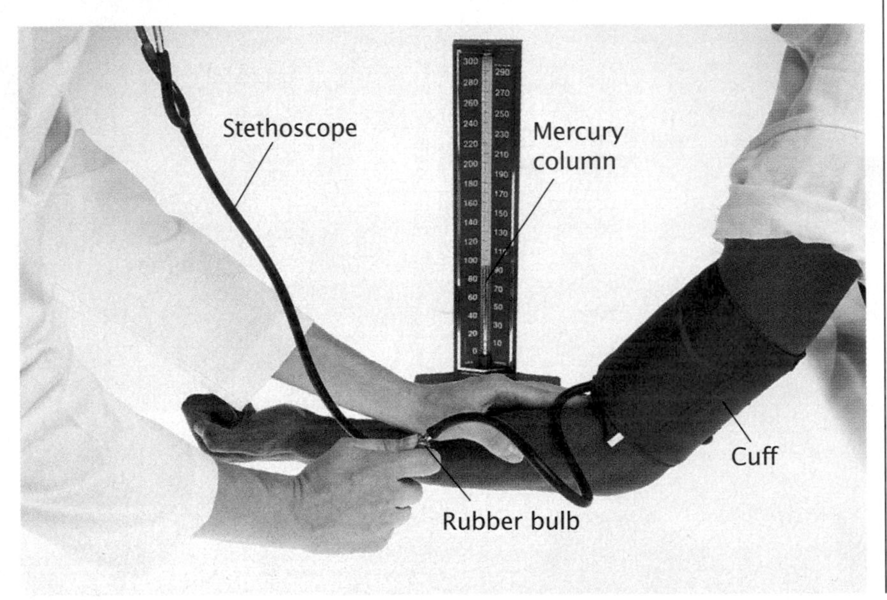

Figure 32-10
A kit for measuring blood pressure includes a stethoscope, a cuff that can be filled with air, a hollow rubber bulb that pumps air into the cuff, and a gauge or hollow glass tube containing a column of mercury.

How is blood pressure measured?

As described in **Figure 32-10**, two numbers are used to measure blood pressure. The first number, called the systolic pressure, tells how much pressure is exerted when the heart contracts and blood flows through the arteries. The second number is the diastolic pressure, which tells how much pressure is exerted when the heart relaxes. Blood pressure is expressed in terms of millimeters of mercury (mm Hg). Blood pressure in a healthy adult is typically about 120/80 mm Hg. These figures indicate that the blood is pushing against the artery walls with a pressure of 120 mm Hg as the heart contracts and 80 mm Hg as the heart rests. In industrialized countries, blood pressure steadily rises with increasing age as the arteries become less elastic. Regular aerobic activity strengthens the heart and blood vessels, helps arteries to stay somewhat elastic even with aging, and can minimize this expected rise in blood pressure. Blood pressure figures provide information about the conditions of the arteries and are useful in diagnosing high blood pressure and hardening of the arteries.

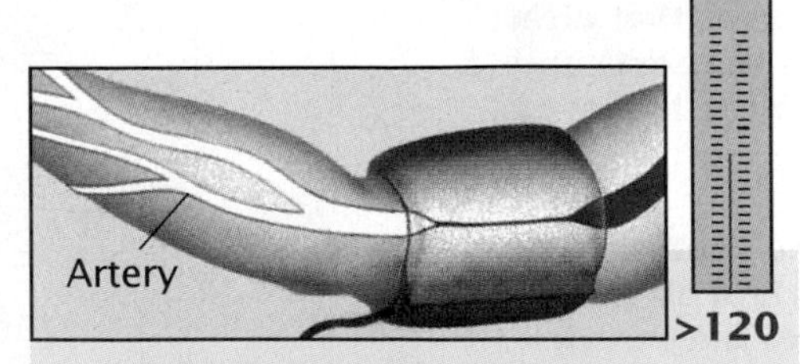

a The cuff is wrapped around the patient's upper arm and positioned over the brachial artery. When the rubber bulb is pumped, air inflates the cuff. This squeezes the artery until no blood passes through. The stethoscope is used to listen for the sound of blood flow to ensure that the artery is closed.

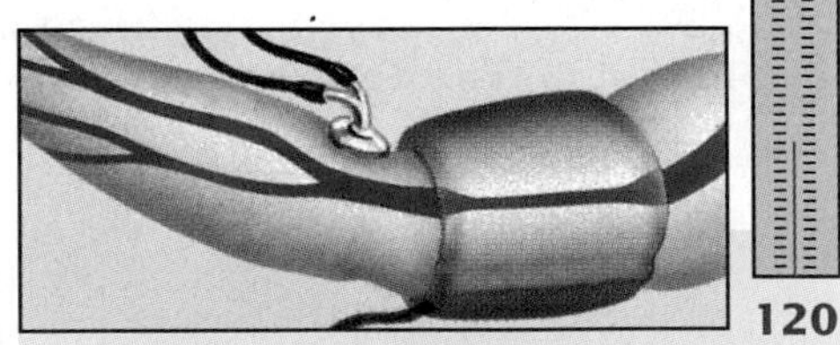

b The cuff is gradually deflated until blood begins to flow into the arm. A sound of blood pulsing can be heard, indicating that the blood pressure is greater than pressure exerted by the cuff. The pressure at this point (120 mm Hg) is the systolic pressure, which is read from the scale. Systolic pressure is exerted by contracting ventricles.

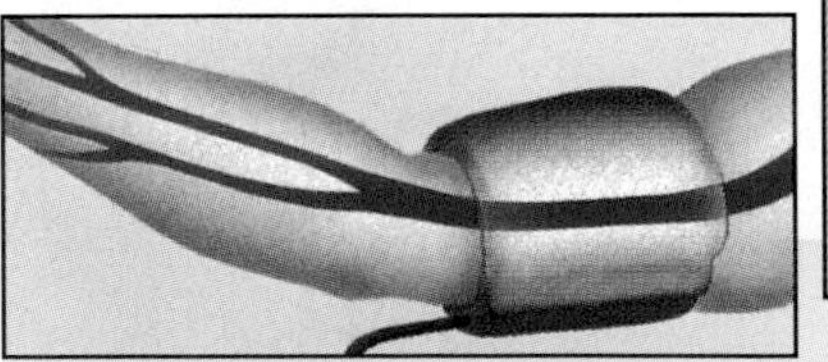

c The cuff is loosened until blood flows freely through the artery and the sounds below the cuff disappear. The pressure at this point (80 mm Hg) is the diastolic pressure, the pressure between heart contractions.

Pharmaceutical Industry

What would it be like to . . .

- help develop a drug that prevents or cures a disease?
- use experiences in business or health care to contribute to the health of people around the world?

These are just a few of the opportunities open to people who work in the pharmaceutical industry.

Pharmaceutical Industry: What Is It?

The pharmaceutical industry is made up of hundreds of companies that discover, develop, produce, and market drugs used for medicinal purposes. Since the widespread availability of antibiotics in the 1940s, the pharmaceutical industry has discovered thousands of new drugs that help cure diseases, prevent epidemics, and increase life expectancy. Most of the work in this industry focuses on researching and developing new drugs or new uses for existing drugs.

Science/Math Career Preparation

High School
- Biology
- Chemistry
- Mathematics

College
- Botany
- Zoology
- Anatomy and physiology
- Chemistry
- Biochemistry
- Biostatistics

Employment Outlook

Accelerated growth in the pharmaceutical industry is predicted as advances in technology increase.

Career Focus: Dora Duncan

Clinical Research Associate for Pharmaco International, Inc.

"When I was growing up, a woman had few career choices—she became either a teacher or a nurse. I chose nursing because I had always been interested in human biology and wanted to have an impact on people's lives. I was in nursing for 26 years, and I'm amazed how a nurse's image and role have changed. Nurses play a more active role in caring for patients, and have the ability to specialize in a diversity of areas in health care."

What I Do: "Today I'm using my experience in a different area—the pharmaceutical industry. I work for a contract research organization in the pharmaceutical industry as a clinical research associate. Before a new drug can be approved for sale and use, it must undergo a long series of tests to determine its safety and effectiveness. My job is to monitor these tests to ensure that they are conducted according to strict guidelines."

To Find Out More

1. For further information, write to:
   Department of Health and Human Services
   Public Health Service
   Food and Drug Administration
   5600 Fishers Lane
   Rockville, MD 20857
2. Visit your school counselor and library to find out which colleges and universities offer training for careers in the pharmaceutical industry. Review the prerequisites and course requirements. Make a poster showing the types of careers that relate to this industry.
3. Find articles listing the procedures necessary before a new drug is approved by the Food and Drug Administration.

Blood Chemistry

Blood is the fluid of life for the human body. As you have learned, the blood transports raw materials to cells so that they can carry on their necessary activities. Blood also removes wastes from cells so that they can continue their activities without interruption. Blood then transports these wastes to the lungs, skin, or kidneys for excretion so that the body does not experience a toxic buildup. In addition, blood maintains a fluid environment for cells. Water moves from the blood to spaces between the cells. The solids (salts, sugars, etc.) dissolved in that water must be kept at a proper concentration if the cells are to function properly. The blood regulates the amount of water by moving dissolved solids into and out of the spaces between the cells and into and out of the cells. Finally, blood contains weapons to protect the body from disease, from extreme changes in acidity, and from blood loss. The blood's ability to perform these vital functions depends on the kinds and concentration of substances in water, or depends on the blood's chemistry.

Changes in blood composition may indicate disease

Recall the composition of blood. There are three kinds of blood cells suspended in plasma. Most of your blood (about 85 percent) is circulating throughout your blood vessels at any given time. The rest of your blood is in your heart or lungs. The content of circulating blood is important to homeostasis. Medical science has established healthy ranges for the concentrations of solids and gases that are circulating in your blood. When these concentrations move outside the healthy range, it is an indication that the body is out of balance. Severe imbalances are termed "disease." How do we know when we are out of balance? One way is to perform a blood analysis on a sample of blood.

Blood drawn from an individual becomes a blood sample

A blood analysis is performed on a sample of blood drawn from the vein of an individual. Specialized collection tubes make it possible to keep the cells in suspension or to allow the cells to clot, yielding packed red cells. When the blood is allowed to clot at room temperature, the blood-clotting components form a netlike trap around the red cells, and a clot forms at the bottom of the tube. For certain types of blood analyses and for blood transfusions, it is necessary to prevent the blood from clotting, so an anticoagulant is added to the collection tube. The anticoagulant keeps the whole cells in suspension and yields a sample of whole blood for analysis.

A blood analysis can be simple or detailed

Performing a blood analysis is like looking inside the body without actually going inside. Sometimes the analysis is as simple as a complete blood count, or CBC, which is a count of the number of blood cells in a given volume of whole blood. Sometimes, when more detail is needed, the plasma is also analyzed.

In a CBC, a sample of whole blood is run through a specialized machine. The red blood cells are counted and analyzed for size, shape, and color. In addition, the different classes of white blood cells are identified, and the number of each class is counted. Finally, the platelets are counted. The general information learned from these counts is indicated in **Figure 32-11a–c**.

Remember that centrifuging the blood forces its components to settle in layers with a liquid portion called plasma at the top. The blood's plasma contains over 150 substances. The amount of each specific substance in circulation is vital to homeostasis. When the blood plasma is analyzed, a blood analysis provides more detail about the body's system of complex interdependences that are required to maintain homeostasis. **Figure 32-11d** lists some of the substances that can be analyzed in plasma.

Analysis of blood lipids can be used to indicate disease risk

An analysis of blood lipids is becoming a routine part of annual health examinations. You have probably heard people talk about having their cholesterol checked. Cholesterol is a lipid that is an essential part of cell membranes and a starting compound for the manufacture of many substances, including sex hormones and vitamin D. But while the body needs cholesterol, it doesn't need some of the cholesterol compounds that collect in the blood vessels and the heart. These cholesterol compounds block the blood vessels and reduce blood flow. A blood analysis called a cardiac-risk profile reveals the presence of specific cholesterol compounds in the blood and the amounts of each. The results of the cardiac-risk profile can be used to assess one's risk of developing diseases of the circulatory system. Cholesterol measurements are recommended for people 20 years of age and older.

The blood contains combination carrier molecules for cholesterol

Recall that plasma is primarily water and that lipids like cholesterol are soluble in oil, but not in water. How, then, does the body transport cholesterol in blood plasma, which is mostly water? The body uses a combination of lipid and protein to carry water-insoluble compounds like cholesterol. These lipid-protein carriers are called lipoproteins. When carrying cholesterol, a lipoprotein forms a lipoprotein-cholesterol complex. Sophisticated blood analyses can now measure the different kinds of lipoprotein-cholesterol complexes in the blood.

Figure 32-11
The most common blood analysis counts each of the blood cells in a sample of whole blood. Greater detail about homeostasis or disease can be learned by further analyzing the blood plasma.

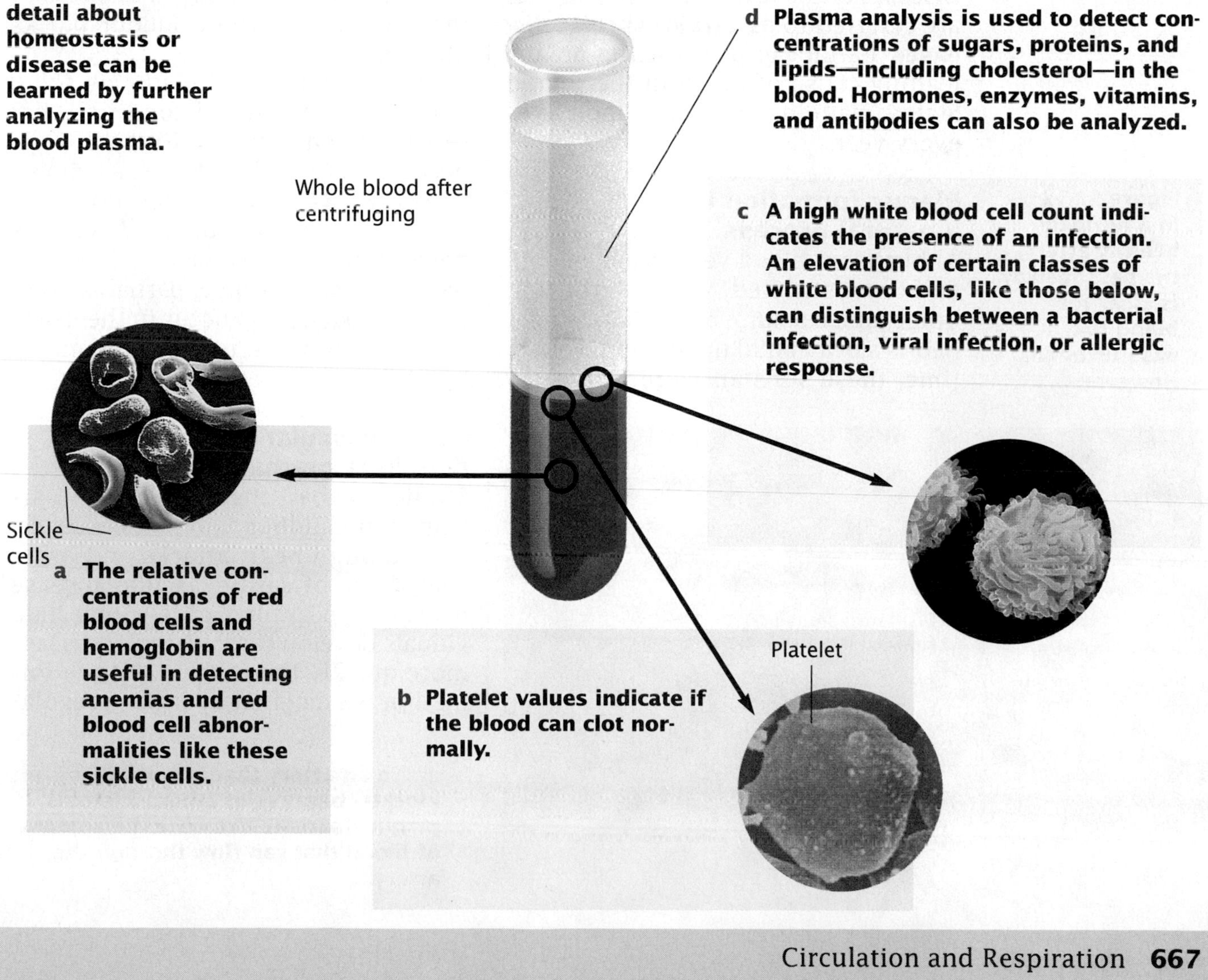

d Plasma analysis is used to detect concentrations of sugars, proteins, and lipids—including cholesterol—in the blood. Hormones, enzymes, vitamins, and antibodies can also be analyzed.

c A high white blood cell count indicates the presence of an infection. An elevation of certain classes of white blood cells, like those below, can distinguish between a bacterial infection, viral infection, or allergic response.

a The relative concentrations of red blood cells and hemoglobin are useful in detecting anemias and red blood cell abnormalities like these sickle cells.

b Platelet values indicate if the blood can clot normally.

Blood analysis distinguishes good and bad cholesterol

Low-density lipoprotein (LDL) is the main cholesterol carrier in the blood, but cholesterol carried on LDL is readily deposited on the walls of arteries. As a result, LDL-cholesterol is often called "bad cholesterol."

On the other hand, cholesterol transported on high-density lipoprotein (HDL) is carried back to the liver for removal from the body. Because HDL-cholesterol is removed from body tissues, it is often referred to as "good cholesterol."

A cardiac-risk profile measures the amount of total cholesterol and HDL-cholesterol in the blood. In addition, it measures the amount of freely circulating lipids called triglycerides. From these measurements, the amount of LDL can then be calculated. **Table 32-2** gives the healthy ranges established by medical scientists for these blood lipids.

Stability and Homeostasis

How could calcium deposits in blood vessels affect blood pressure?

Table 32-2
Healthy Ranges for Blood Lipids

Lipid compound	mg/100 mL
Total cholesterol	130–200
HDL-cholesterol	
males	30–75
females	40–90
LDL-cholesterol	less than 130
Triglycerides	30–150

Diseases of the Heart and Blood Vessels

Diseases of the heart and blood vessels are referred to as **cardiovascular diseases**. Cardiovascular diseases are the leading cause of death in the United States, claiming about 1 million lives every year.

Plaque formation is a natural process

At birth, our blood vessels are smooth and unobstructed. Cholesterol and other fats, calcium, and other cellular debris slowly build up in arteries. Over time, these substances form a pasty mass called plaque *(plak)* that sticks to the artery walls. Plaque buildup blocks blood flow.

Atherosclerosis *(ATH uhr oh skluh ROH sihs)* is a medical condition that results when the blood-vessel walls become coated with plaque, as shown in **Figure 32-12b**. Tissues normally nourished by the blocked vessels become oxygen starved. A few cells begin to die. Over time, partial or complete blockages can occur in the heart, the brain, or the arteries that connect the heart to other tissues.

Cardiovascular disease develops gradually

Studies show that atherosclerosis begins in childhood and adolescence, even though heart attacks and other symptoms of cardiovascular disease may not appear for decades. Some individuals develop cardiovascular disease more quickly than others. Plaque formation is hastened by lack of regular

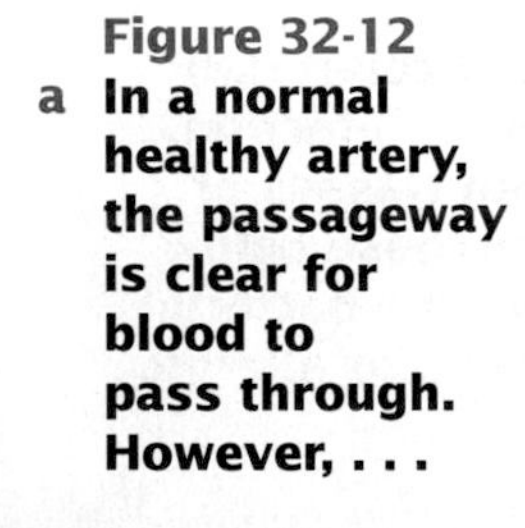

Figure 32-12
a In a normal healthy artery, the passageway is clear for blood to pass through. However, . . .

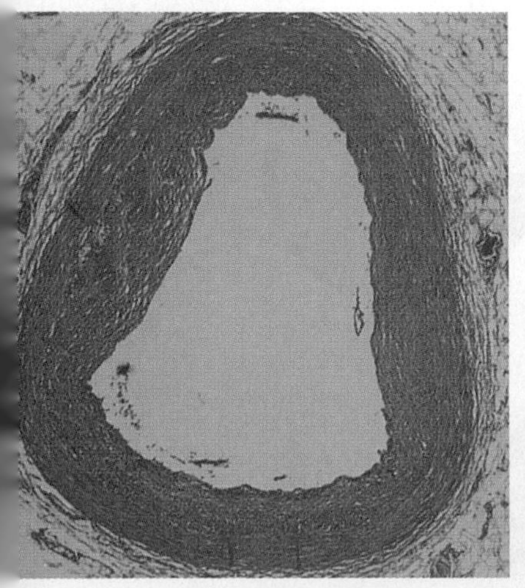

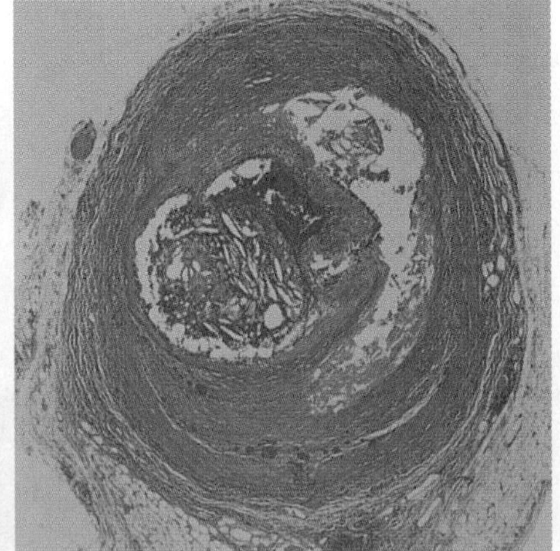

b . . . in an artery that is partially blocked because of atherosclerosis, plaque deposits decrease the amount of blood that can flow through the artery.

exercise, a diet high in saturated fats (dairy and meat fats, eggs, and hydrogenated vegetable oils), and cigarette smoke. High LDL and low HDL levels as well as high blood pressure also hasten plaque formation. Triglycerides are the most common fatty substance in the blood. About 95 percent of the lipids in our diets consist of triglycerides. Combined with cholesterol, triglycerides hasten plaque formation. Family history of cardiovascular disease is a critical factor for some people.

Blocked vessels starve tissues

When blood vessels are blocked, the pressure within them builds. When blood pressure against an artery wall remains higher than normal, a condition called **hypertension**, or high blood pressure, results. Because it has no warning symptoms, hypertension is often called the silent killer. High blood pressure is a key indicator of cardiovascular disease.

More obvious than hypertension, though, are the sudden emergencies that result when a heart stops or when part of the brain ceases to direct the body's functions. A heart attack results from a blockage that reduces the supply of blood to an area of heart muscle, causing heart tissue to die. A stroke results from a blockage that reduces the supply of blood to an area of the brain, causing brain tissue to die. The severity of disability depends on how much of the heart muscle or the brain is damaged.

Blood lipids may indicate cardiovascular disease

More than half of the people who have heart attacks have normal total cholesterol levels. How can that be? Recall that there are good and bad cholesterol complexes in the blood. The *ratio* of total cholesterol to good cholesterol (total cholesterol divided by HDL) is more important than total cholesterol alone. Higher levels of good cholesterol lower the ratio. Lower ratios mean greater protection against cardiovascular disease.

Look at the picture and the lipid profile of the teen in **Figure 32-13**. How does her cardiac-risk profile shown in the figure compare with the healthy values for blood lipids in **Table 32-2**? Given her family history of high cholesterol and cardiovascular disease, what can she do to protect her heart and blood vessels?

Blood lipid	Amount in mg/100 mL
Total cholesterol	210
HDL-cholesterol	40
Triglycerides	90
Calculated LDL	152

Figure 32-13
This teen appears lean and healthy, but her cardiac-risk profile indicates factors that may hasten the development of plaque.

SECTION REVIEW

1. **Describe the roles played by the left and right halves of the heart.**
2. **Compare and contrast systemic and pulmonary circulation.**
3. **What does a blood pressure of 120/80 mean?**
4. **What steps could you take to reduce your chances of developing cardiovascular disease?**

How gases are exchanged in animals depends on the size of the animal. Very small animals can obtain oxygen and give off carbon dioxide directly through the plasma membranes of their outermost layer of cells. In contrast, larger animals have evolved special respiratory systems that provide a large surface area to exchange gases.

32-3 The Respiratory System

OBJECTIVES

1. **Explain how the diaphragm and rib muscles work to move air into and out of the lungs.**
2. **Describe the pathway by which oxygen from the air travels to a cell in the body.**
3. **Explain how your body regulates your breathing rate.**
4. **Describe three diseases of the respiratory system.**

Lungs and Breathing

Throughout this book, you have learned that most living organisms, like the soccer players in **Figure 32-14**, need to obtain oxygen from their environment and to remove carbon dioxide from their bodies. As you read in Chapter 5, oxygen is needed for cellular respiration, the chemical process that allows organisms to obtain energy from substances such as glucose. During aerobic respiration, carbon dioxide is produced as a waste product and must be removed from the cell. As you learned in Chapter 24, the evolution of lungs enables terrestrial animals to obtain oxygen from the atmosphere and release excess carbon dioxide.

Figure 32-14
On or off the soccer field, humans must obtain oxygen from the atmosphere and release excess carbon dioxide in the process of gas exchange.

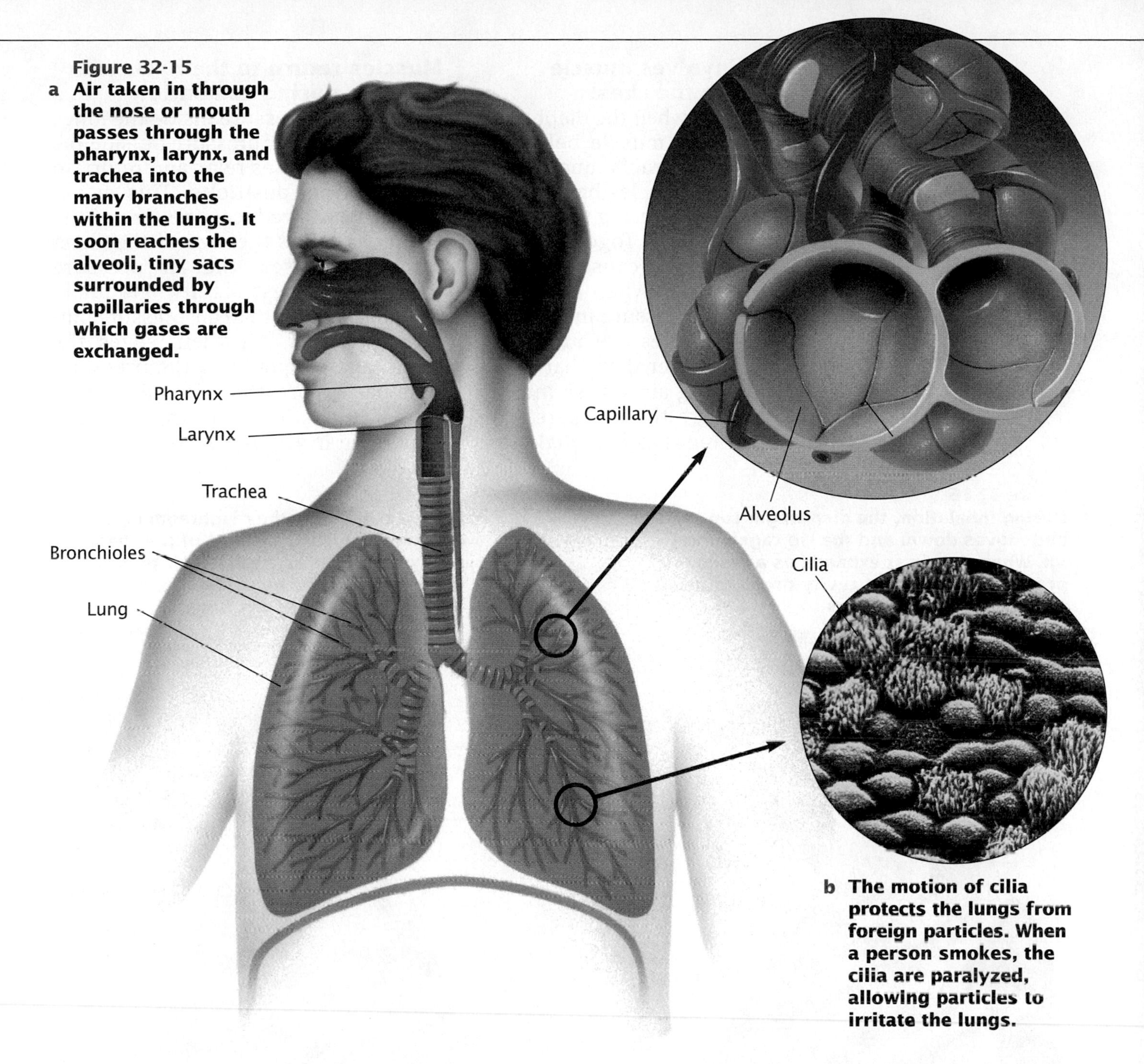

Figure 32-15

a Air taken in through the nose or mouth passes through the pharynx, larynx, and trachea into the many branches within the lungs. It soon reaches the alveoli, tiny sacs surrounded by capillaries through which gases are exchanged.

b The motion of cilia protects the lungs from foreign particles. When a person smokes, the cilia are paralyzed, allowing particles to irritate the lungs.

Lungs contain branched tubes

Follow the passage of air through the respiratory system in **Figure 32-15**. Air enters the body through the nose or mouth and travels to the pharynx *(FAIR ihnks)*, a tube at the back of the nose and mouth. From the pharynx, air enters the voice box, or larynx *(LAR ihnks)*, and the windpipe, or trachea *(TRAY kee uh)*. These passageways are lined with tissues that warm and moisten incoming air. The lower end of the trachea divides into two branches, which divide many times into smaller branches of tubes called bronchioles *(BRAHNG kee ohlz)*. This branching network of tubes is lined with cilia, shown in **Figure 32-15b**, and a layer of protective mucus. The smallest of these tubes lead into **alveoli** *(al VEE uh leye)*, clusters of tiny air sacs. Gases enter and leave the circulatory system through the alveoli. Each alveolus is surrounded by a network of capillaries. Blood in these capillaries picks up oxygen from the alveoli and releases carbon dioxide to be exhaled. Each lung contains about 150 million alveoli, providing a surface area larger than that of a small house for gas exchange.

Inhalation involves muscle contraction in the chest

Breathing begins when the diaphragm, the dome-shaped muscle below the chest cavity, contracts and moves downward. The muscles between the ribs also contract, causing the rib cage to move up and out. Together, these muscle contractions cause the chest cavity to enlarge. When the chest expands, the air pressure in the chest cavity drops. Air pressure outside the body is then greater than that inside the chest, causing air to rush into the lungs to equalize the pressure. This part of breathing is called **inhalation**.

Muscles return to their relaxed position during exhalation

When the air pressure inside the lungs is equal to the air pressure outside the lungs, the muscles relax and return to their original positions. This movement, in turn, reduces the size of the chest cavity. As the size of the chest cavity decreases, the air pressure inside the chest cavity gradually becomes greater than air pressure outside the body. Air then leaves the lungs, again equalizing the pressure. This part of breathing is called **exhalation**. Inhalation and exhalation are illustrated in **Figure 32-16**.

Figure 32-16

a During inhalation, the diaphragm contracts and moves down, and the rib cage moves up. When the chest expands as a result, air pressure in the chest cavity drops, causing air to rush into the lungs.

b During exhalation, the diaphragm relaxes and moves up, and the size of the chest cavity decreases as a result. Air pressure in the chest cavity increases, forcing air to be exhaled out of the lungs.

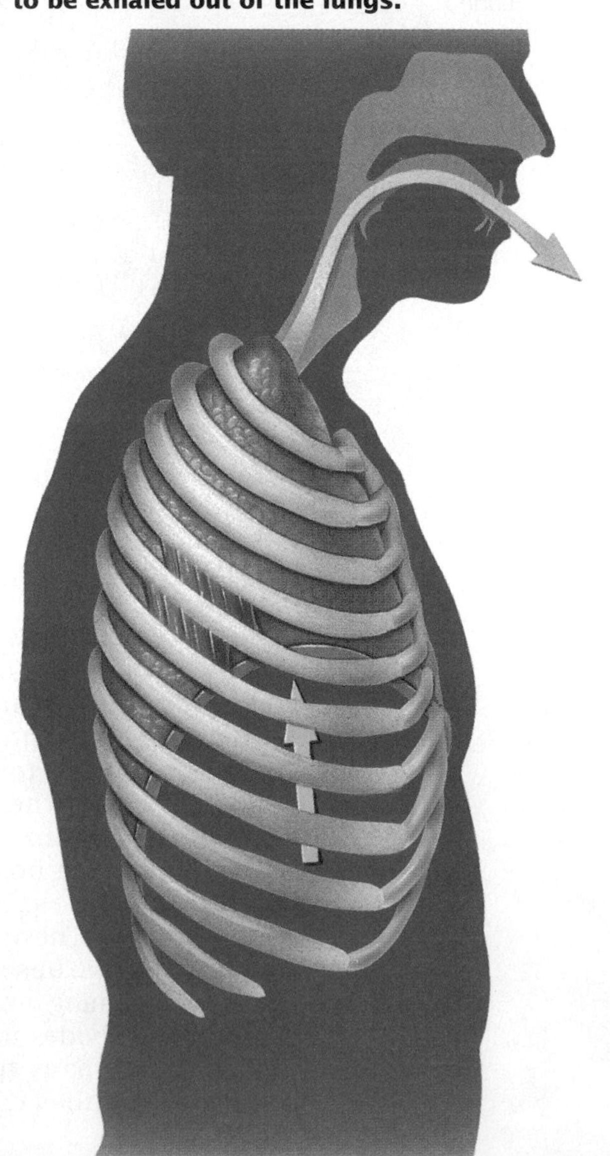

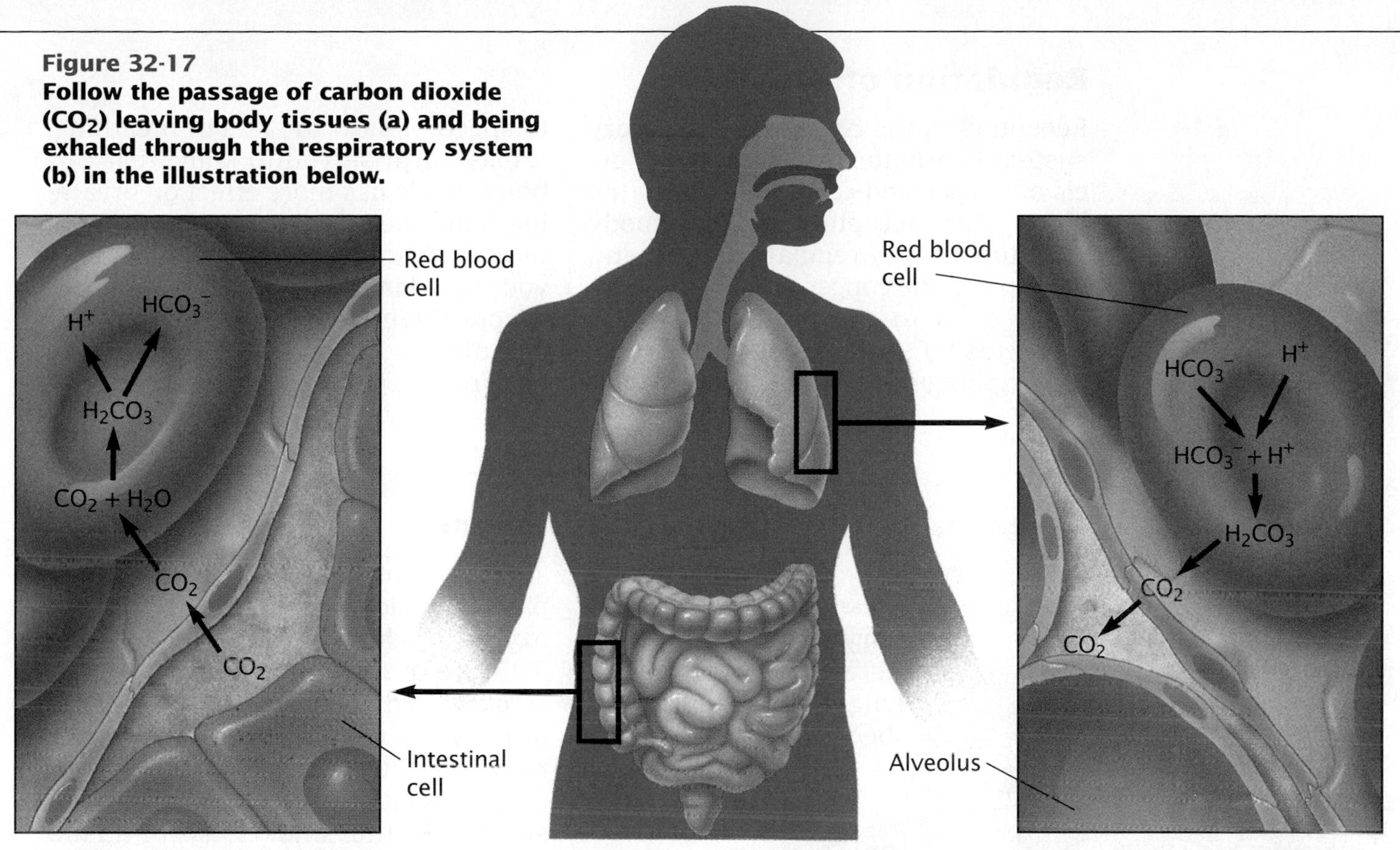

Figure 32-17
Follow the passage of carbon dioxide (CO_2) leaving body tissues (a) and being exhaled through the respiratory system (b) in the illustration below.

a **When CO_2 leaves cells of the intestines, it can enter red blood cells and combine with hemoglobin. Or it may combine with water to form carbonic acid, H_2CO_3, which breaks down to form hydrogen ions and bicarbonate ions.**

b **When blood reaches the lungs, the process is reversed, and CO_2 is released. The CO_2 diffuses from the blood into the alveoli in the lungs. From there, it is exhaled along with water vapor.**

Gas Exchange

The destination of the inhaled air is the alveoli, which are surrounded by capillaries. The tissue forming the walls of the alveoli and capillaries is only one cell thick. Gas exchange occurs when oxygen in the alveoli diffuses into the blood in the capillaries. In turn, the carbon dioxide in the blood diffuses into the air of the alveoli.

Oxygen binds to hemoglobin

In the blood, oxygen quickly binds with hemoglobin, the protein in red blood cells. Hemoglobin soaks up oxygen extremely effectively, which causes still more oxygen to enter red blood cells. The red blood cells then give up their oxygen to the cells of body tissues, where it is used in metabolism, the chemical activities of cells. As a result of metabolism, oxygen concentration in the body's cells is low, but carbon dioxide concentration is high.

Carbon dioxide is transported to the lungs to be exhaled

Carbon dioxide is a waste product that must be eliminated from cells. It is transported in the blood in three ways. About 5 percent of the carbon dioxide in the body dissolves in the plasma in blood. Another 25 percent enters the red blood cells and combines with hemoglobin. With help of an enzyme, the remaining 70 percent combines with water in the red blood cells to form carbonic acid, H_2CO_3. Because carbonic acid is unstable, hydrogen ions and bicarbonate ions quickly form, as shown in **Figure 32-17a**.

When blood reaches the lungs, chemical reactions occur that reverse the process, releasing carbon dioxide. As shown in **Figure 32-17b**, the carbon dioxide diffuses from the blood into the alveoli in the lungs. The carbon dioxide is exhaled with water vapor.

Regulation of Breathing

Receptors in the brain and circulatory system continuously monitor the levels of oxygen and carbon dioxide in the blood. These receptors enable the body to automatically regulate oxygen and carbon dioxide concentrations by sending signals to the brain. The brain responds by sending nerve signals to the diaphragm and rib muscles in order to speed or slow the rate of breathing. It may surprise you to know that carbon dioxide has more effect on breathing than does oxygen. For example, if the concentration of carbon dioxide in your blood increases, you breathe more deeply, ridding your body of excess carbon dioxide. When the carbon dioxide level drops, your breathing slows.

Diseases of the Respiratory System

Respiratory diseases affect millions of Americans. **Asthma** *(AZ muh)* is a respiratory disease in which certain airways in the lungs become constricted because of sensitivity to certain stimuli. The narrowing of the airways reduces the efficiency of respiration, which decreases the amount of oxygen reaching body cells.

Figure 32-18 These posters, warning about some of the life-threatening illnesses associated with cigarette smoking, were designed by high school students.

Cigarette smoking is harmful to both smokers and nonsmokers and is linked to emphysema and lung cancer, two respiratory diseases that claim millions of lives annually. Direct inhalation of smoke by the smoker, secondhand inhalation of the smoker's exhaled smoke by the nonsmoker, and inhalation of sidestream smoke from the smoker's burning cigarette by the nonsmoker—all increase the risk of developing emphysema and other serious respiratory diseases. In people who have **emphysema** *(em fuh SEE muh)*, the lung tissue loses its elasticity, greatly reducing the efficiency of gas exchange. In **lung cancer**, carcinogens present in tobacco smoke trigger the growth of cancerous cells in lung tissue. More than 90 percent of lung cancer patients are smokers, as suggested by the antismoking posters shown in **Figure 32-18**. Lung cancer has an extremely low cure rate. Fewer than 10 percent of its victims live more than five years after diagnosis.

SECTION REVIEW

1. **What role do the diaphragm and rib muscles play in the processes of inhalation and exhalation?**
2. **How does the exchange of gases occur in the lungs?**
3. **When you exercise, you automatically begin to breathe faster. Explain how and why this occurs.**
4. **How do asthma and emphysema affect respiration?**

CHAPTER

32 Highlights

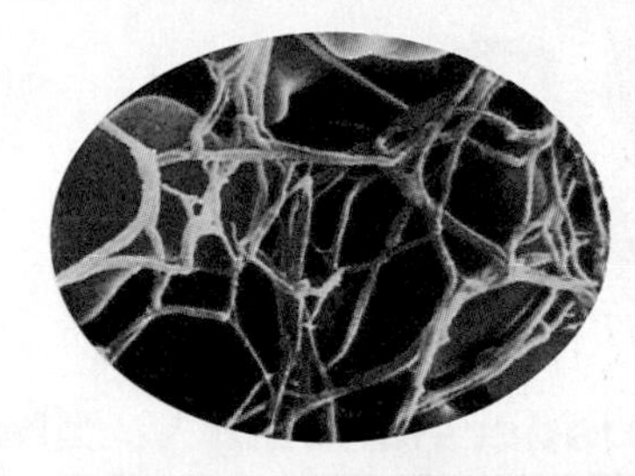

Strands of a protein called fibrin trap red and white blood cells. Soon a blood clot will form.

	Key Terms	Summary
32-1 Circulation	plasma (p. 652) red blood cell (p. 652) hemoglobin (p. 652) white blood cell (p. 653) platelet (p. 653) blood vessel (p. 654) artery (p. 654) vein (p. 654) capillary (p. 654) blood type (p. 656) lymphatic system (p. 657)	• The circulatory system transports materials to and from cells. • Blood consists of plasma, red blood cells that transport oxygen, white blood cells that protect against infection, and platelets that help clotting. • Blood types are defined by proteins on the surface of red blood cells. • Blood vessels include arteries that carry blood away from the heart, veins that carry blood back to the heart, and capillaries that connect the arteries to the veins. • The lymphatic system returns fluids back to the blood vessels.
32-2 How Blood Flows	atrium (p. 658) ventricle (p. 658) pacemaker (p. 659) blood pressure (p. 664) cardiovascular disease (p. 668) hypertension (p. 669)	• The heart has two sides: the right side moves blood to the lungs, and the left side moves blood to the rest of the body. • Blood pressure consists of a high value (systolic) when the heart contracts and a lower value (diastolic) when the heart rests. • Hypertension, or high blood pressure, is an example of a cardiovascular disease.
32-3 The Respiratory System	alveoli (p. 671) inhalation (p. 672) exhalation (p. 672) asthma (p. 674) emphysema (p. 674) lung cancer (p. 674)	• When the diaphragm and the rib muscles contract, enlarging the chest cavity, inhalation occurs. When the muscles relax, air is forced out and exhalation occurs. • Gas exchange occurs when oxygen in the alveoli diffuses into the blood in the capillaries. Carbon dioxide in the blood diffuses into the air of the alveoli. • Breathing is regulated mainly by the level of carbon dioxide detected in the blood. • Cigarette smoking is linked to emphysema and lung cancer.

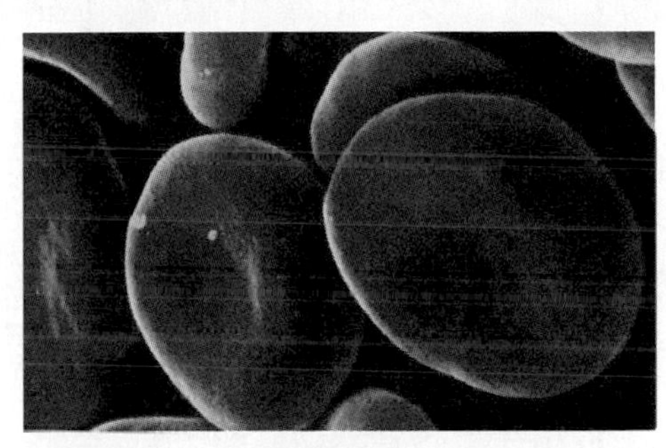

Red blood cells carry oxygen from the lungs to all cells of the body and carry carbon dioxide back to the lungs to be exhaled.

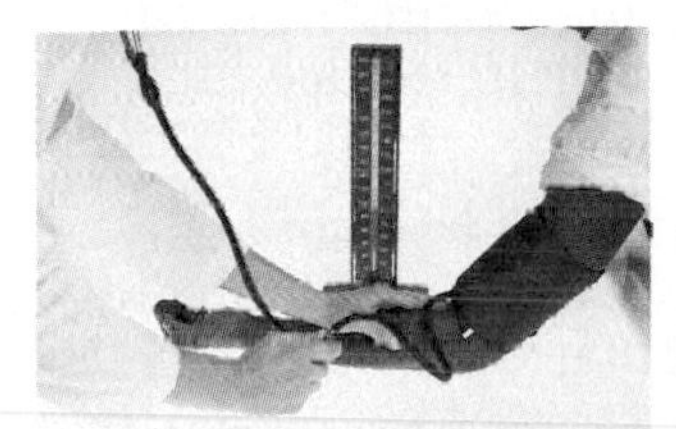

Blood pressure in a healthy adult is typically about 120/80 mm Hg.

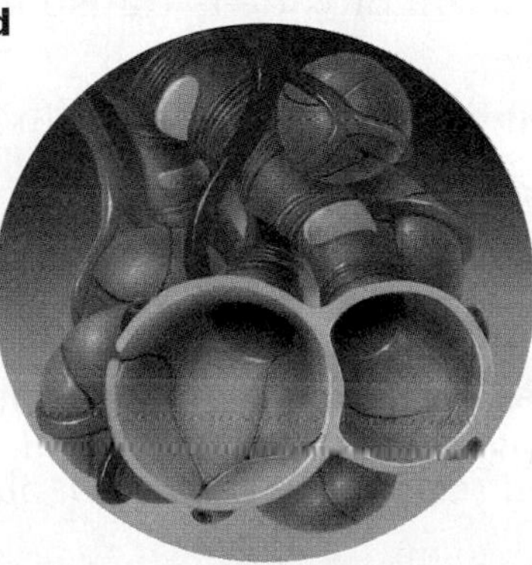

The walls of the alveoli and the capillaries around them provide a large surface area for gas exchange.

CHAPTER 32 review

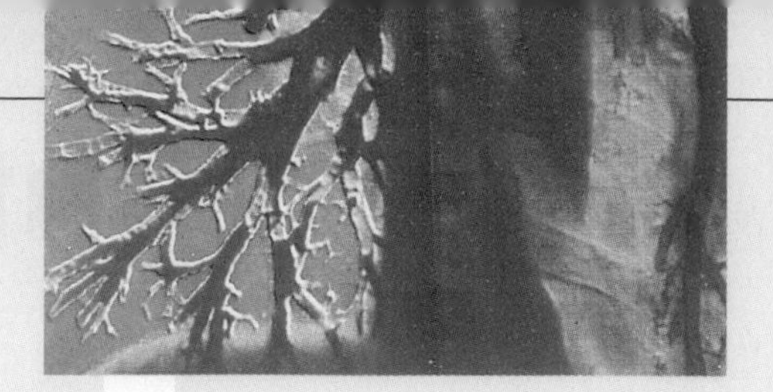

Understanding Vocabulary

1. For each pair of terms, explain the differences in their meanings.
   a. red blood cell, white blood cell
   b. heart attack, stroke
   c. systolic pressure, diastolic pressure

2. For each set of terms, complete the analogy.
   a. material that makes blood a liquid : plasma :: material that helps repair damaged blood vessels : ____
   b. confined to the bloodstream : red blood cells :: able to squeeze in and out of blood vessels : ____
   c. carries oxygen-poor blood from upper body : superior vena cava :: carries oxygen-rich blood to other arteries : ____

Understanding Concepts

3. **Relating Concepts** Construct a concept map that describes the structure and function of blood. Try to include the following terms in your concept map: vessels, plasma, cell platelets, white blood cells, red blood cells, hemoglobin, blood clotting, defense, oxygen, vessels, carbon dioxide.

4. **Identifying Functions** How do bacteria, single-celled protists, and simple multicellular animals transport materials from the environment throughout their bodies?

5. **Summarizing Information** The circulatory system transports oxygen, carbon dioxide, and hormones. What else does it do?

6. **Summarizing Information** Identify the main components of blood, and describe the function of each.

7. **Comparing Functions and Structures** How do red blood cells and white blood cells differ in terms of function, size, and number?

8. **Comparing Functions** What are the three main types of blood vessels and their functions? What characteristics of blood vessels allow them to adjust to different volumes of blood flow?

9. **Reading Graphs** Using the graph below, describe the relationship between the amount of table salt consumed daily and systolic blood pressure.

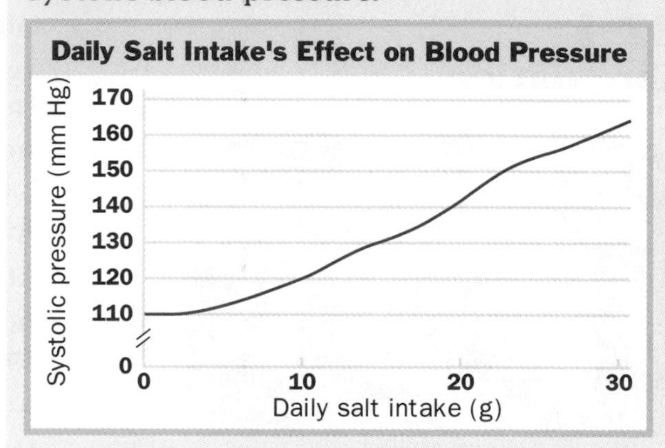

10. **Recognizing Patterns** Trace the path of blood during pulmonary circulation. Begin with oxygen-poor blood entering the heart.

11. **Comparing Functions** How does the lymphatic system assist the circulatory system? How do the operations of these two systems differ?

12. **Inferring Relationships** What kinds of information can be gained by measuring blood pressure? How can you help reduce high blood pressure?

13. **Inferring Relationships** Explain how an analysis of blood lipids can be used to indicate the risk of cardiovascular disease.

14. **Recognizing Relationships** Describe the factors that contribute to cardiovascular diseases.

15. **Summarizing Information** Describe the pathway of air as it travels through the respiratory system to a cell in the body.

16. **Analyzing Information** Describe the three ways that carbon dioxide is transported in the blood.

17. **Recognizing Relationships** Explain how your body regulates your breathing rate.

18. **Summarizing Information** Describe three diseases of the respiratory system.

19. **BUILDING ON WHAT YOU HAVE LEARNED** In Chapter 4 you learned about organ transplants. What role does blood type play in receiving organ transplants?

Interpreting Graphics

20. Look at the cross sections of the two coronary arteries below and answer the questions that follow.

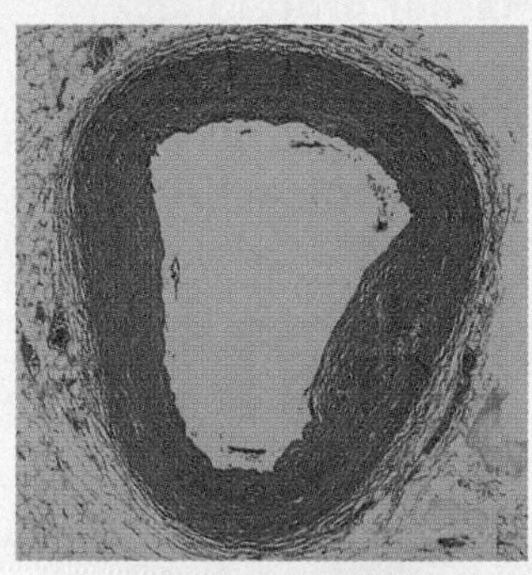

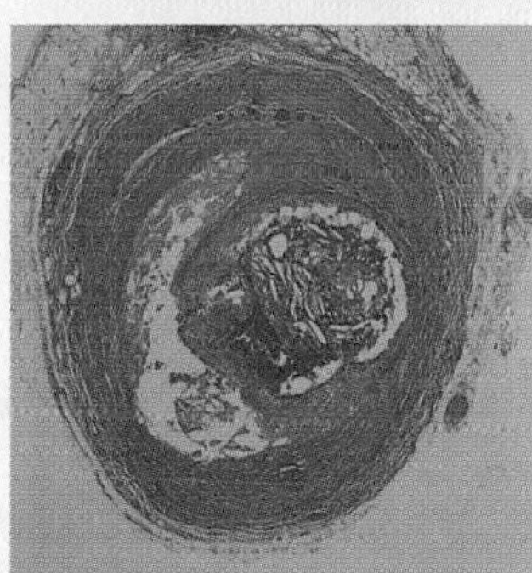

a. How are the two arteries different?
b. How can you explain the differences observed? What are the effects of the observed differences?

Reviewing Themes

21. ***Cell Structure and Function***
Why can the four-chambered heart of mammals be thought of as two hearts in one?

22. ***Homeostasis***
How can a blood analysis determine if the body is maintaining homeostasis?

23. ***Evolution***
Explain why the vertebrate lung is considered an adaptation to terrestrial life.

Thinking Critically

24. **Recognizing Verifiable Conclusions** Endurance athletes, such as long-distance runners, often have larger hearts that can pump a greater volume of blood with each heart beat. How would these modifications help the heart respond to the demands placed on it by endurance activities?

Life/Work Skills

25. **Working Cooperatively** Work with a partner. Contact the local chapter of the American Lung Association to find out if certain diseases are more common in your area than in other parts of the country. Also contact the local chapter of the American Heart Association. Find out the warning signs of a heart attack and how to care for a heart attack victim until help arrives. Share your findings with your class.

26. **Using Technology** An electrocardiogram is a measurement and recording of the electrical activity of the heart. Contact a cardiac rehabilitation center or a cardiologist to find out about electrocardiograms. What kinds of information can be learned from this type of measurement? How are electrocardiograms taken?

Activities and Projects

27. **Mathematics Connection** Assume that the average heart beats 70 times per minute. How many times does the heart beat in a person who lives 75 years? Assume that the heart of an overweight person beats 10 additional times each minute. Explain why being overweight can strain the heart.

28. **Career Connection: Respiratory Therapist** A respiratory therapist, sometimes called an inhalation therapist, assists people with breathing problems by giving them respiratory treatment and life support. These patients might be suffering from asthma, bronchitis, emphysema, pneumonia, or other disease. Use library references or search an on-line database to find out about becoming a respiratory therapist. What are the educational and training requirements? For additional information, write to the American Association for Respiratory Therapy, 1720 Regal Row, Dallas, TX 75235.

Discovering Through Reading

29. Read the article "A Town With a Golden Gene" by Patricia Long in *Health*, January/February 1994, pages 60–66. What is unique about the people in Limone, Italy, who are called *portatori*? What are the benefits of high levels of HDL? Identify three things you can do to improve your various cholesterol levels.

CHAPTER 33

REVIEW

- membrane marker protein (Section 3-2)
- viral reproduction (Section 18-3)
- the term *pathogen* (Glossary)

The Immune System

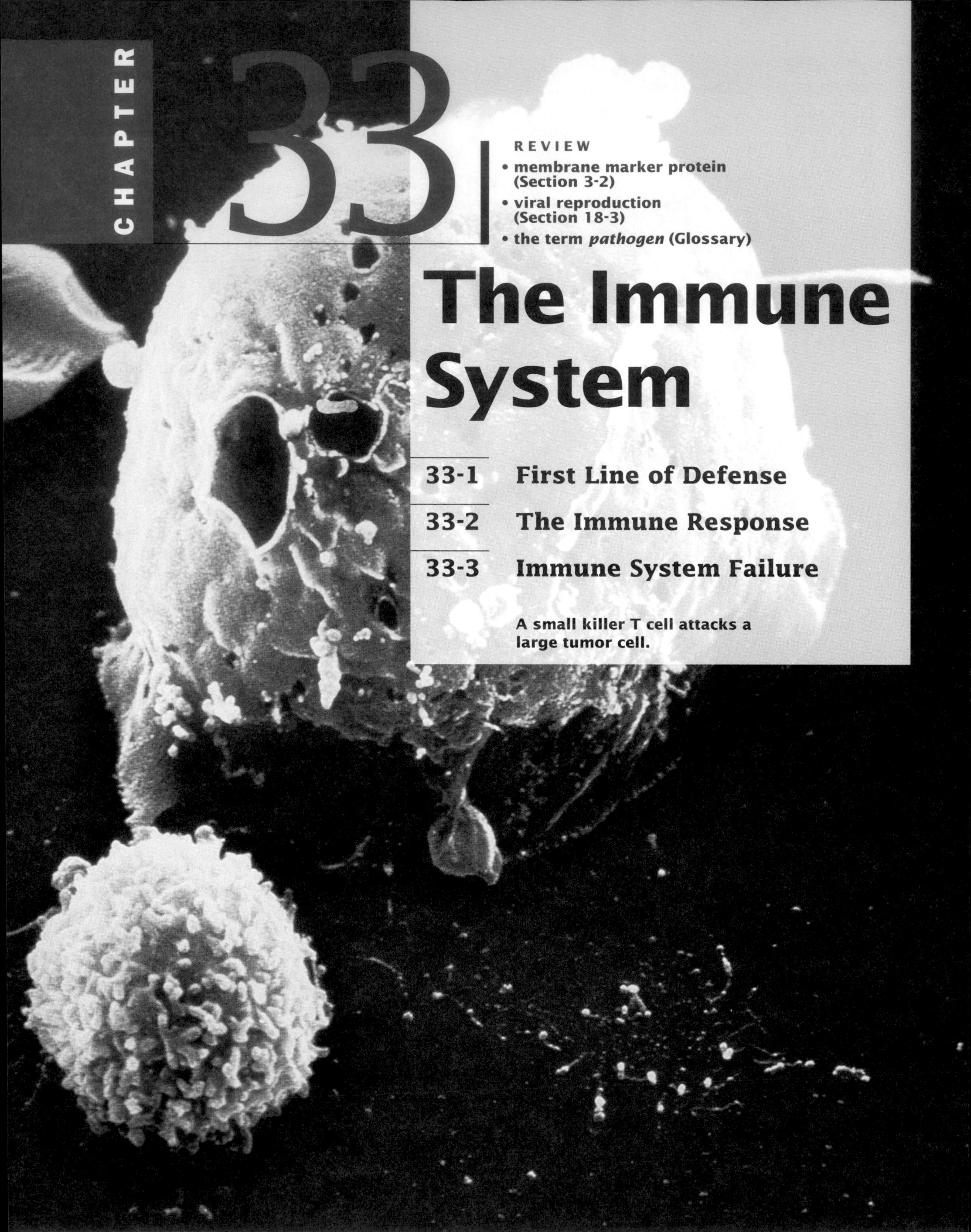

A small killer T cell attacks a large tumor cell.

Your body is routinely surrounded by legions of would-be invaders: bacteria, viruses, protists, parasitic animals, and parasitic fungi. Yet you are usually well, because your body mobilizes cells like the yellow one at left to repel these pathogens. It defeats the few pathogens that manage to enter your body, using a powerful defense called the immune system.

33-1 First Line of Defense

OBJECTIVES

1. **Identify two types of defensive epithelial tissue, and describe how each works.**
2. **Summarize the reactions that make up the inflammatory response.**
3. **Describe how your body distinguishes its own cells from invading pathogens.**

Keeping Pathogens Out

Most pathogens must enter the body to cause disease. Your skin functions as a wall to keep out foreign organisms, including viruses. As you learned in Chapter 29, skin is the body's dry, flexible covering. Cells of the outer layers of epidermis are continually being worn away and replaced by cells moving up from the lower layers of epidermis. In only 40 minutes, your body loses and replaces approximately 1 million skin cells. Such rapid replacement of body cells enables punctures or cuts to be sealed very quickly.

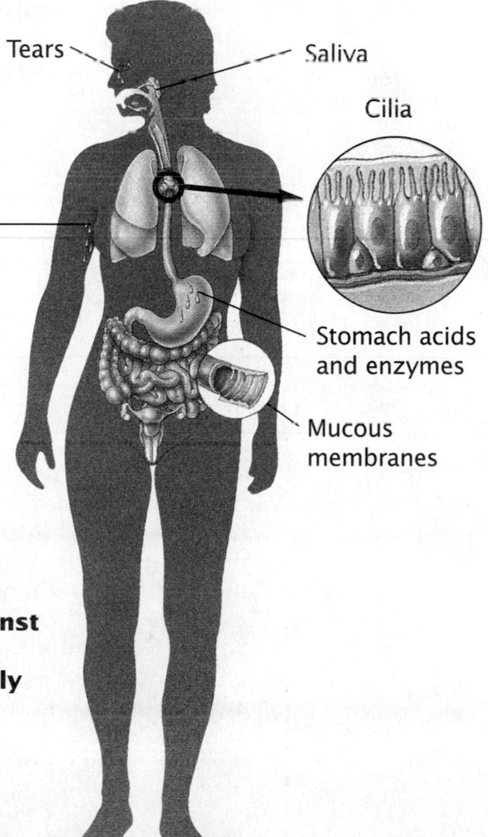

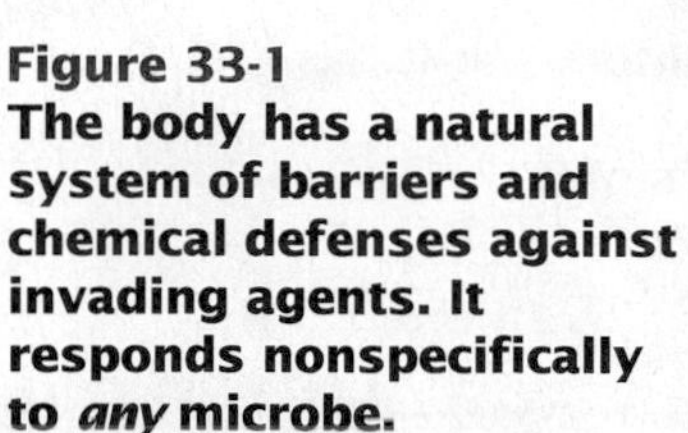

Figure 33-1
The body has a natural system of barriers and chemical defenses against invading agents. It responds nonspecifically to *any* microbe.

Skin not only acts as a barrier to pathogens but also engages in chemical warfare against them. Oils and sweat secreted by glands in the skin acidify its surface, creating an unfavorable environment for bacterial growth. In addition, sweat contains enzymes that digest the cell walls of certain bacteria.

Your body uses a different series of nonspecific defenses to guard the natural openings in your skin, such as the mouth, nostrils, eyes, vagina, urethra, and anus. These internal body surfaces that come into contact with the environment are covered with **mucous membranes**.

A mucous membrane is a moist epithelial layer that is impermeable to most pathogens. Mucous membranes line the nasal passages, mouth, lungs, digestive tract, urethra, and vagina. Mucous membranes contain glands that secrete mucus, a sticky fluid that traps pathogens. Pathogens inhaled into the respiratory tract become lodged in mucus and are swept up to the pharynx by cilia. They are then swallowed and pass into the stomach. Digestive enzymes and strong acids in the stomach destroy most pathogens that are swallowed.

Similar antibacterial enzymes are found in saliva and tears. **Figure 33-1** illustrates several nonspecific natural defenses used by the body to protect against invasion by disease-causing agents.

Organizing the Defense

The outer defenses of the body may not always be successful in preventing pathogens or other foreign agents from entering. When the body's defensive barriers are broken, the body uses a variety of specialized white blood cells as defenders. White blood cells (**leukocytes**) are the "soldiers" of the immune system. Some white blood cells are best suited for short, minor invasions. These cells are known as **phagocytes**. Other white blood cells, known as **lymphocytes**, are specialized for longer, more serious invasions.

Phagocytes handle small-scale infections

A splinter in your fingertip is only a minor injury, even though pathogens can enter the puncture in your skin. Your body quickly activates a set of chemical and cellular defenses to destroy any invaders. Your finger gets red, swells, hurts, and feels hot. These are signs that your body is attacking pathogens. Together these signs make up the **inflammatory response** to infection or injury, as shown in **Figure 33-2**. The inflammatory response is sufficiently powerful to repel most small-scale infections.

Phagocytes are crucial to the inflammatory response because they rapidly trap, consume, and completely destroy foreign material by phagocytosis. Some white blood cells, called natural killer cells, destroy your own infected cells. Phagocytes, dead cells, and pathogens are the components of pus. The presence of pus indicates that phagocytes are combating pathogens.

Several kinds of white blood cells use phagocytosis, but they cannot continue to act repeatedly. This line of defense peaks within a day or two of the infection. If homeostasis is not restored, the body activates its main line of defense, a more powerful and specific response called the immune response.

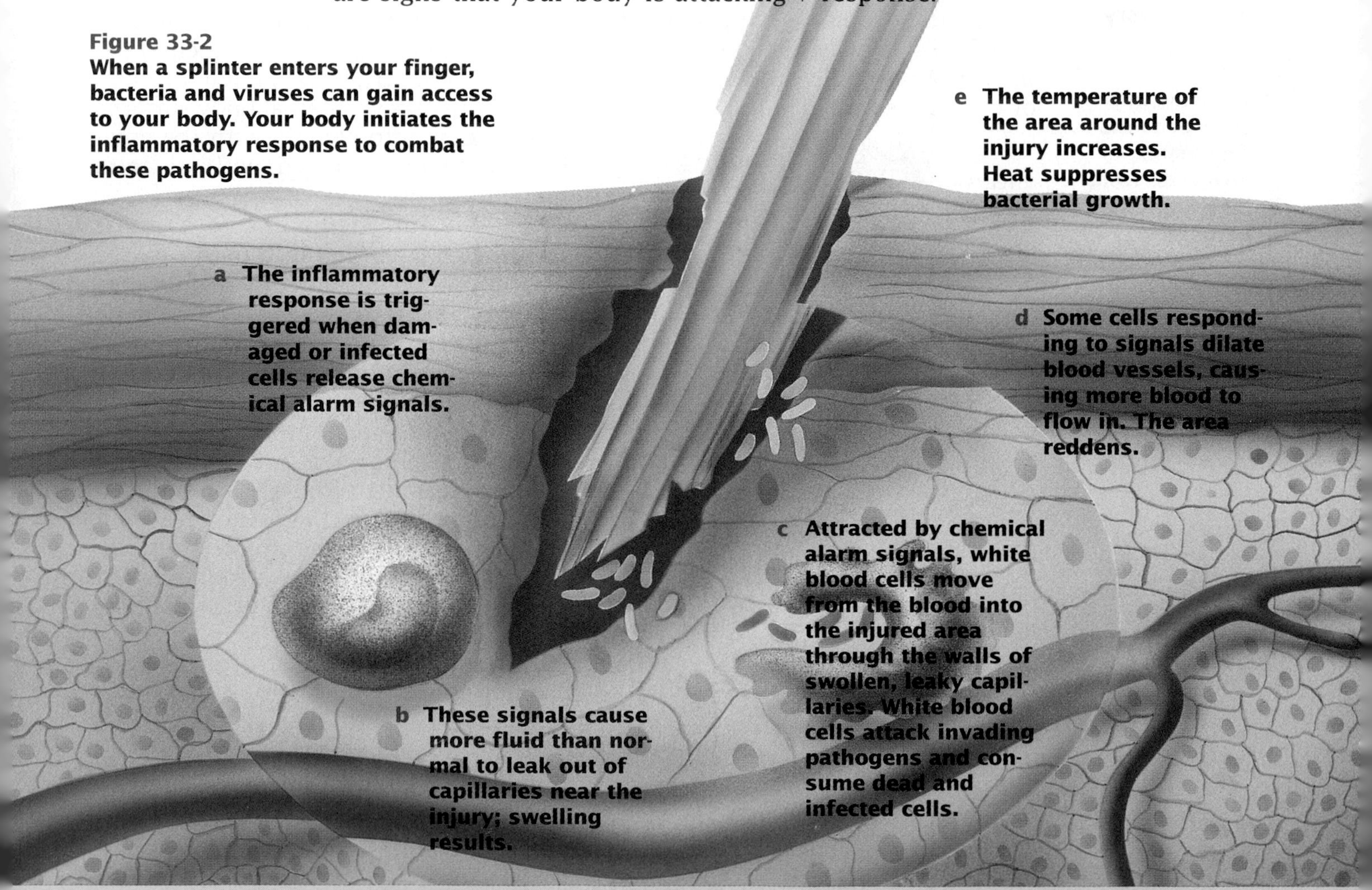

Figure 33-2
When a splinter enters your finger, bacteria and viruses can gain access to your body. Your body initiates the inflammatory response to combat these pathogens.

a The inflammatory response is triggered when damaged or infected cells release chemical alarm signals.

b These signals cause more fluid than normal to leak out of capillaries near the injury; swelling results.

c Attracted by chemical alarm signals, white blood cells move from the blood into the injured area through the walls of swollen, leaky capillaries. White blood cells attack invading pathogens and consume dead and infected cells.

d Some cells responding to signals dilate blood vessels, causing more blood to flow in. The area reddens.

e The temperature of the area around the injury increases. Heat suppresses bacterial growth.

Figure 33-3
a The brief inflammatory response buys time so that the main line of defense can be recruited.

b The main immune response provides both short- and long-term defenses.

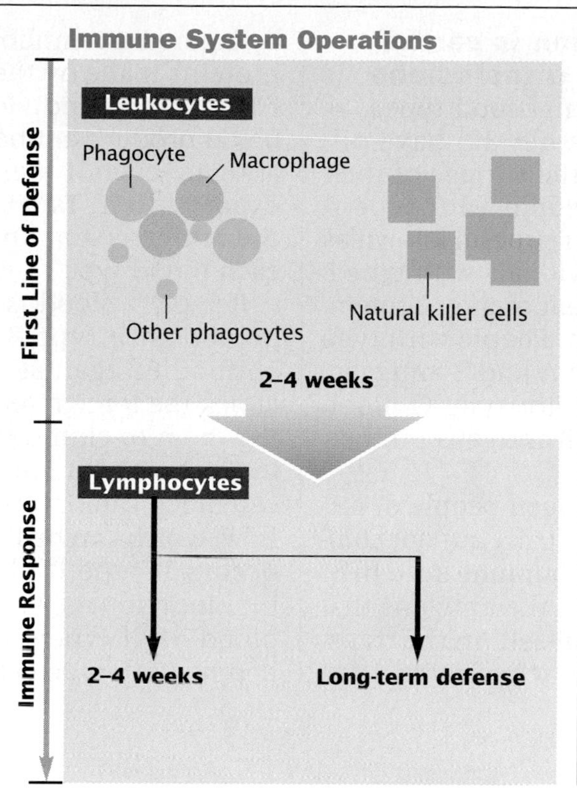

A variety of other white blood cells help to defend the body

Phagocytes do well in short-term skirmishes like the splinter incident. But phagocytes alone cannot wage a long-term defense against specific agents that cause disease. For specific protection, the body uses a main line of defense, called the immune response, that involves recognition and then destruction. White blood cells called **macrophages** *(MAHK roh fayj ehz)* are used in this defense because they live longer than other leukocytes and can repeat their destructive actions many times. Macrophages and other phagocytes are joined in this longer battle by two other kinds of white blood cells called lymphocytes. Each of these white blood cells is a specialist in a particular field of defense. **Figure 33-3** shows the general classes of white blood cells involved in defending the body.

Antigen Recognition

Your white blood cells need to distinguish your body's cells from those of the enemy in order to defend your body successfully. Recall from Chapter 3 that there are marker proteins and receptor proteins in the cell membrane. Your marker proteins have a unique shape, which makes them different from marker proteins on cells of other humans and from marker proteins on the surface of disease-causing agents. Thus, marker proteins serve as identification tags. **Figure 33-4** illustrates marker proteins.

Marker proteins that differ from your own are identified as "non-self" by your white blood cells. Once they identify "non-self" cells, macrophages sound an alarm. How do they do that? When a macrophage battles an enemy, it uses phagocytosis, as other leukocytes do. But instead of destroying the enemy completely, the macrophage saves some of the invader's marker protein and moves it to its own surface as a "marker alert" for other white blood cells. Once alerted, receptor proteins of other white cells also recognize the material as "non-self."

The display and recognition of foreign marker proteins trigger the immune response. Marker proteins that trigger the immune response are called **antigens** *(AN tuh juhns)*. Once antigens are displayed by macrophages and are identified as the enemy, the cells in your body bearing that antigen are marked as targets for elimination.

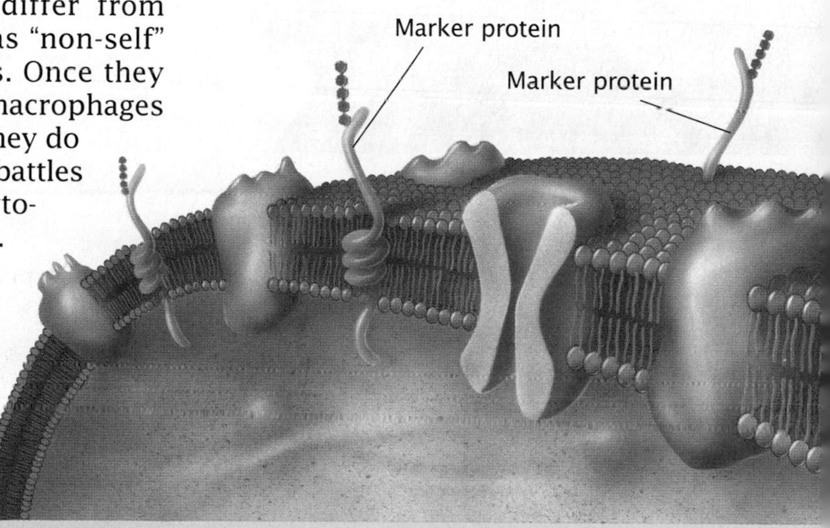

Figure 33-4
Because no two organisms have the same marker proteins, cells of the immune system are able to identify and destroy invading pathogens.

Antigen recognition is part of successful blood transfusions

People with different blood types, as you learned in Chapter 32, have different antigens on the surface of their red blood cells. People with type A blood have a marker protein known as the A antigen. Individuals with type B blood have a different marker protein, called the B antigen. People with type AB blood have both A and B antigens, while individuals with type O blood have neither A nor B antigens on their red blood cells.

Transfusions between people of different blood types usually are not compatible because the immune system of the recipient attacks the antigens that it recognizes as "non-self" in the transfused blood. The attack is carried out by antibodies. Antibodies are defensive proteins made by the immune system. For instance, individuals with type A blood produce antibodies against the B antigen, even if they have never been exposed to it. **Table 33-1** summarizes the antigens and antibodies found in each blood type.

If type A blood is transfused into a person with type B or type O blood, antibodies against the A antigen will attack the foreign red blood cells, causing them to clump together, as shown in **Figure 33-5**. Clumps of just 2–3 cells can block capillaries and cut off blood flow, which can be fatal. Clumping also occurs if type B blood is transfused into individuals with type A or type O blood, or if type AB blood is given to people with type O blood.

Figure 33-5
Some blood types are not compatible. When incompatible blood types are mixed, they form clumps (shown below) that prevent blood from circulating properly.

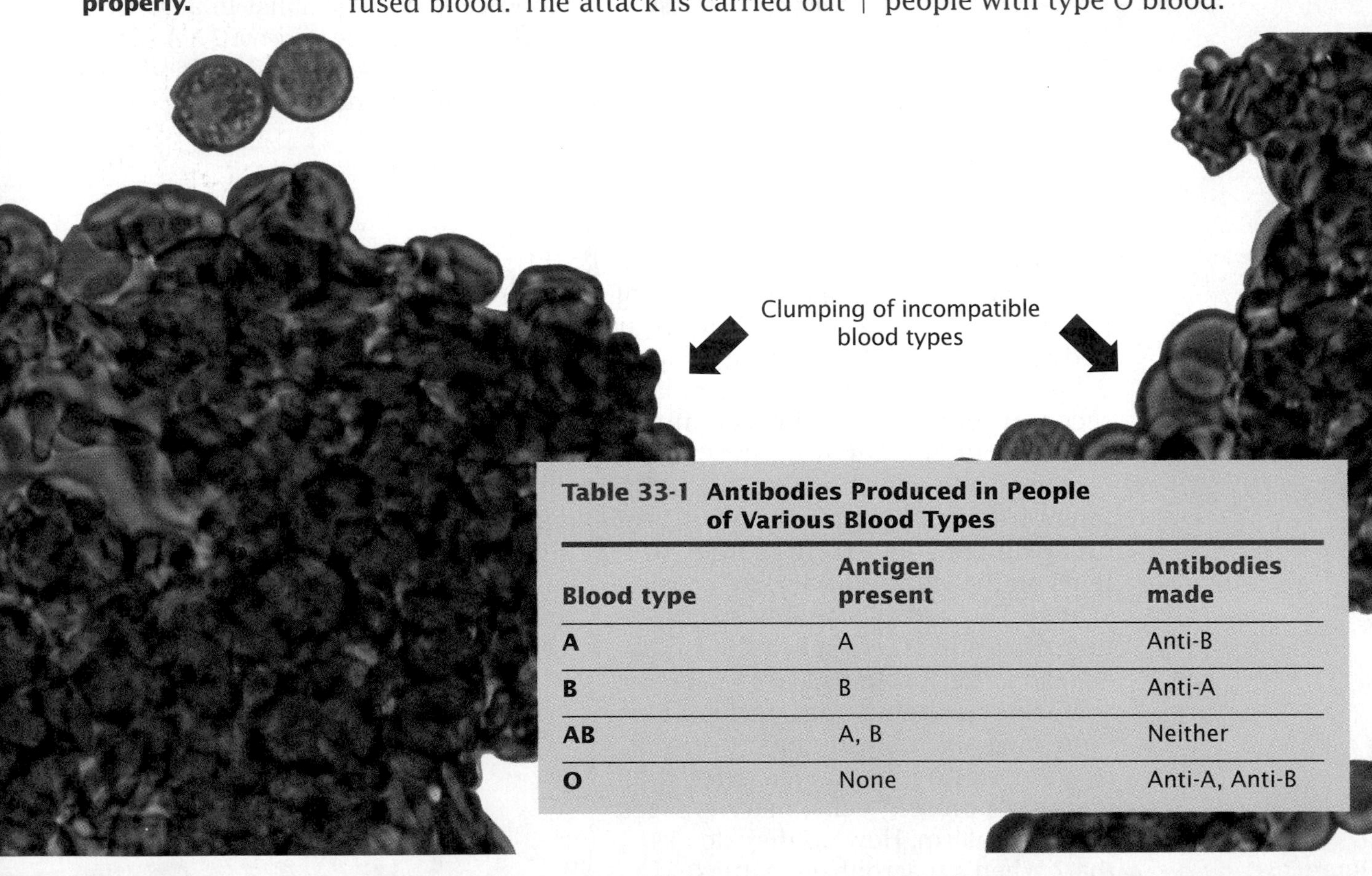

Table 33-1 Antibodies Produced in People of Various Blood Types

Blood type	Antigen present	Antibodies made
A	A	Anti-B
B	B	Anti-A
AB	A, B	Neither
O	None	Anti-A, Anti-B

SECTION REVIEW

1. **Describe the chemical and mechanical defenses of skin and mucous membranes.**
2. **Describe the defensive events that occur after a splinter enters your finger.**
3. **Contrast the first line of defense with the main line of defense.**
4. **Explain how white blood cells can distinguish your cells from those of invading pathogens or of incompatible blood.**

The pain of a splinter does not compare to the suffering you endure when you have the flu. The symptoms of flu are more severe because influenza viruses break through your first line of defense. As you will see in this section, your body activates the main line of defense, called the immune response, that combats and defeats specific pathogens, including viruses.

33-2 The Immune Response

OBJECTIVES

1. **Identify the functions of helper T cells, killer T cells, and B cells.**
2. **Explain how fever helps to defeat pathogens.**
3. **Describe how the immune system protects against a pathogen's second attack.**
4. **Relate vaccines to the functioning of the immune system.**

Main Line of Defense

The immune system consists of the defenses that your body uses to attack pathogens or other foreign agents that get past your first line of defense. For example, consider what happens when you get the flu. Influenza viruses are transmitted in small water droplets that are expelled when an infected person sneezes or coughs. If you inhale some of these droplets, influenza viruses can enter cells of the mucous membrane lining your respiratory tract. As you recall from Chapter 18, viruses seize control of their host cells and transform them into virus-producing factories. Influenza viruses take over and infect mucous-membrane cells. You feel sick because infected cells lining your respiratory tract are being destroyed by your own natural killer cells.

At this point during a case of flu, the viruses have the upper hand, but your body's defenses are beginning to fight back. As mentioned earlier, macrophages can engulf and consume the viruses. Macrophages then break down the viruses' surface proteins, capture fragments of the viral antigens, and display the antigens on their cell surface. This display serves as a victory banner, but, more important, it acts as an alert to other white blood cells, and it stimulates the immune system to carry out a full-blown immune response. The **immune response** is the immune system's main attack on a *specific* pathogen. Bacteria like the *E. coli* shown in **Figure 33-6** and other specific pathogens are also attacked by macrophages.

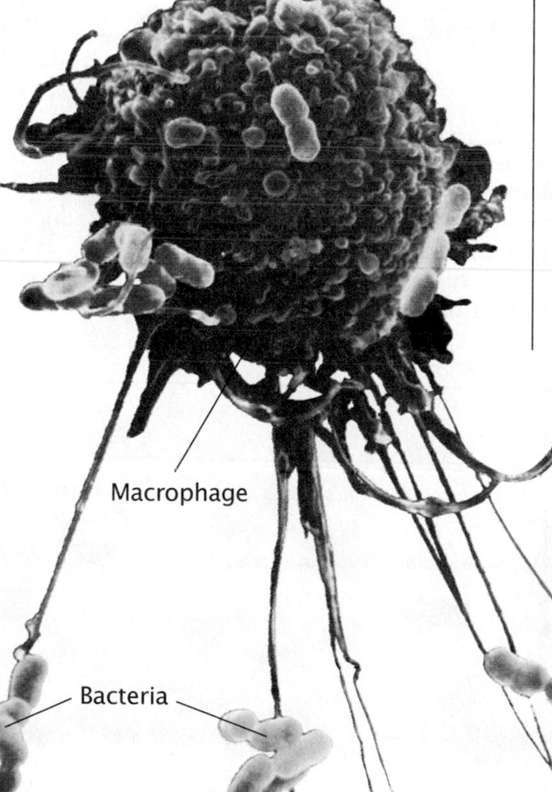

Figure 33-6 Macrophages attack foreign pathogens, such as viruses or the *E. coli* bacteria shown here. This macrophage's pseudopod is reaching out to retrieve the enemy for destruction.

T Cells: Command and Attack

When a macrophage parades its victory banner of captured antigens, it sends an alarm to the immune system. This alarm activates **T-cell lymphocytes**. Some, called helper T cells, circulate in your blood and lymph. Helper T cells control the immune response and direct two kinds of lymphocytes at the same time—killer T cells and suppressor T cells. Suppressor T cells regulate the immune response, and some can even change into killer cells, which are the main "soldiers" of the immune response army.

Killer T cells carry specifically shaped receptor proteins on their cell membrane. These enable each killer T cell to recognize and bind to one specific antigen. Your body can respond to millions of different antigens because it manufactures millions of different types of T cells, each type bearing uniquely shaped proteins. When you have the flu, T cells that match the antigens of the influenza viruses in your body will be "called up" to fight.

Killer T cells destroy only infected cells. **Figure 33-7** describes the role of helper T cells and suppressor T cells.

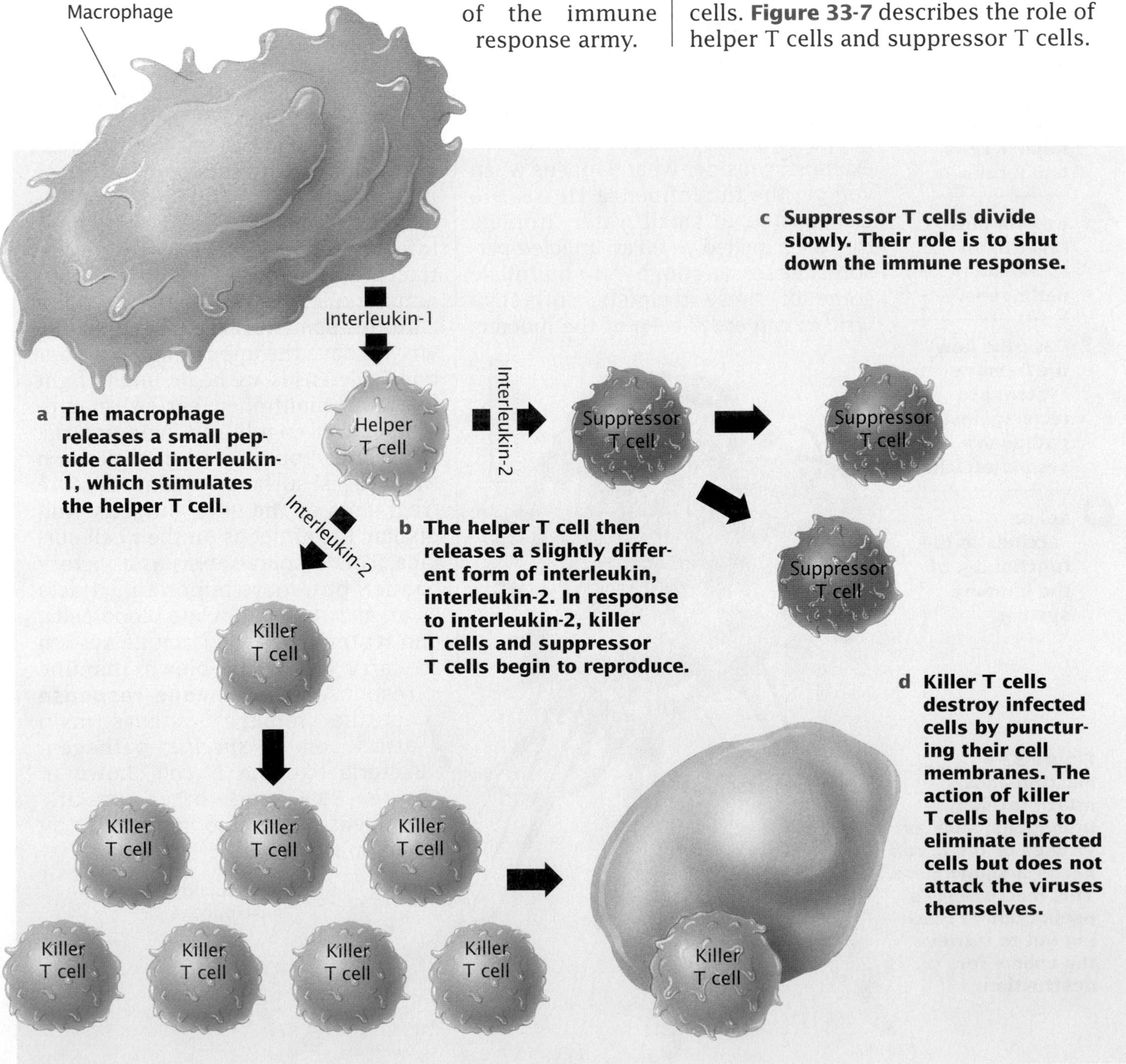
Macrophage
Interleukin-1
Helper T cell
Interleukin-2
Suppressor T cell
Suppressor T cell
Suppressor T cell
Interleukin-2
Killer T cell
Killer T cell
Killer T cell
Killer T cell
Killer T cell
Killer T cell
Killer T cell
Killer T cell
Killer T cell
a The macrophage releases a small peptide called interleukin-1, which stimulates the helper T cell.
b The helper T cell then releases a slightly different form of interleukin, interleukin-2. In response to interleukin-2, killer T cells and suppressor T cells begin to reproduce.
c Suppressor T cells divide slowly. Their role is to shut down the immune response.
d Killer T cells destroy infected cells by puncturing their cell membranes. The action of killer T cells helps to eliminate infected cells but does not attack the viruses themselves.

Figure 33-7 The immune response is activated when a macrophage that has consumed a pathogen, such as an influenza virus, comes into contact with a helper T cell and presents the captured antigens.

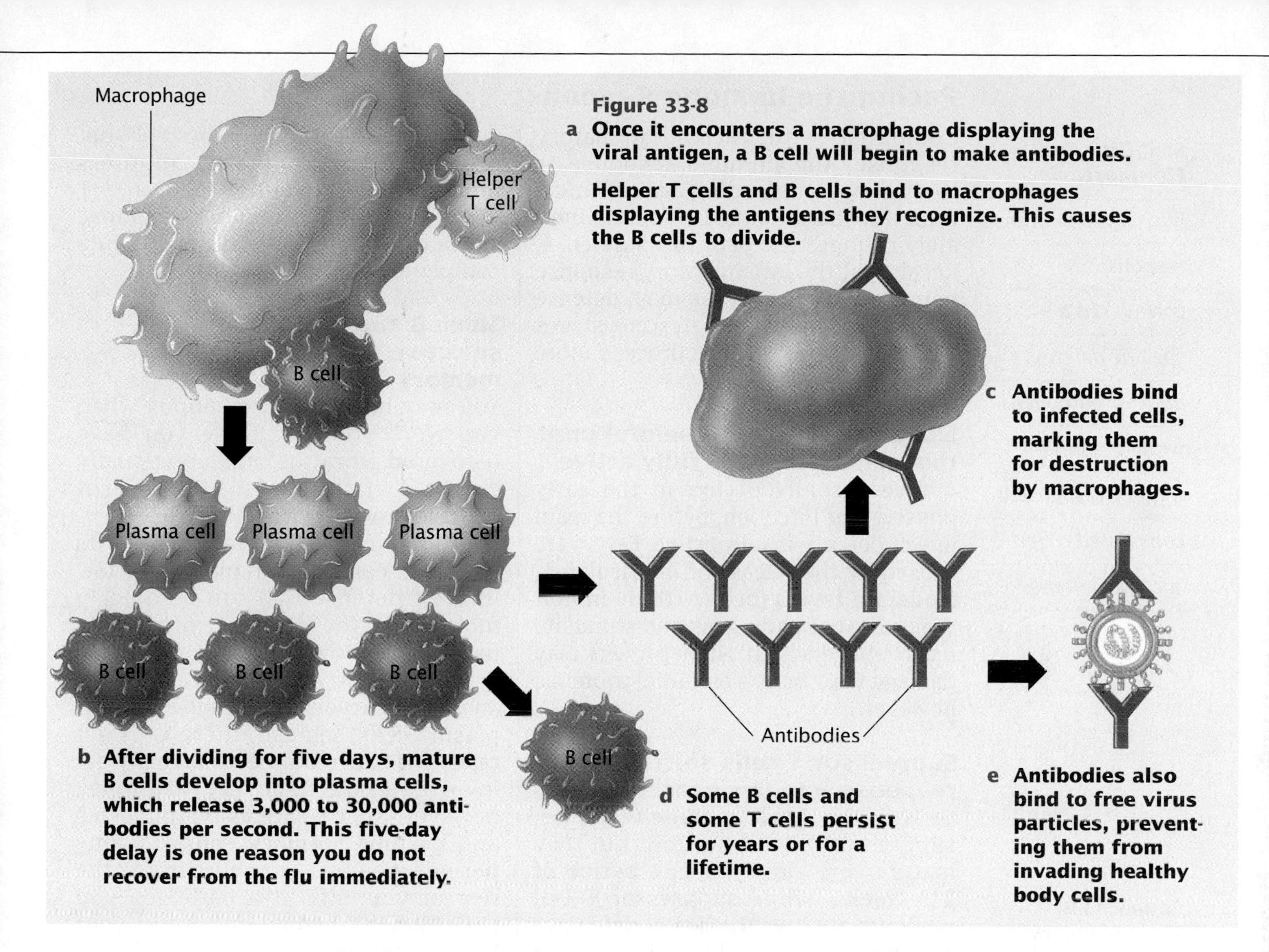

Figure 33-8
a Once it encounters a macrophage displaying the viral antigen, a B cell will begin to make antibodies.

Helper T cells and B cells bind to macrophages displaying the antigens they recognize. This causes the B cells to divide.

b After dividing for five days, mature B cells develop into plasma cells, which release 3,000 to 30,000 antibodies per second. This five-day delay is one reason you do not recover from the flu immediately.

c Antibodies bind to infected cells, marking them for destruction by macrophages.

d Some B cells and some T cells persist for years or for a lifetime.

e Antibodies also bind to free virus particles, preventing them from invading healthy body cells.

B Cells: Biological Warfare

Recall that killer T cells kill only infected cells. What happens to the invading pathogen or foreign agent? Other lymphocytes, called B cells, are deployed to attack the invaders. An assault by B cells is the other half of the attack commanded by helper T cells. Recall that antibodies are defensive proteins made by the immune system. B cells develop into plasma cells, which produce and release the **antibodies**. Antibodies recognize and bind to antigens on the surface of free-floating pathogens or to antigen fragments displayed by macrophages. Binding by antibodies does not kill a pathogen directly. Rather, it marks the pathogen for destruction by macrophages and phagocytes that are guided to enemy target cells coated with antibodies. Receptors on the surface of these macrophages and phagocytes react with the tail of specific antibodies, allowing them to secrete chemicals that destroy the enemy targets.

Your body makes millions of different types of **B cells**. Each type has a unique receptor protein on its membrane that will bind to one particular antigen. At any one time, millions of B cells are circulating in your blood and lymph but are not producing antibodies. Antibody production is turned on when B cells interact with antigen-displaying macrophages and helper T cells, as illustrated in **Figure 33-8**. Your body is capable of making millions of different antibodies. B cells become specialized plasma cells to manufacture only one chosen antibody of the millions possible. In this way, your body prepares defenses against a variety of potential invaders. Many will never be encountered.

Pacing the Immune Response

Stability and Homeostasis

Why is the immune response considered a specific defense against pathogens, while the inflammatory response is considered a nonspecific defense?

Compared with the inflammatory response, the immune response is fairly slow in defending against infection. But the immune system deliberately changes the pace of its defense. Recall that the inflammatory response bought time to allow the main defense to kick in. Even after that, some stages in the immune response proceed more slowly than others.

Moderate fevers are helpful until the main defense is fully active

A fever usually occurs in the early phase of an infection, before the main line of defense is fully active. Fevers are caused by the release of interleukin-1. Moderate fevers (below 103°F) inhibit the growth of pathogens and stimulate macrophage action. Higher fevers may damage your body's essential proteins, however.

Suppressor T cells shift immune response into low gear

Early in the T-cell immune response, suppressor T cells multiply, but they mature very slowly, over a period of 2–3 weeks. While suppressor T cell numbers are low, the heavy defensive action of T cells and B cells can take place. After two or more weeks, however, there are large numbers of suppressor T cells, and they suppress (inhibit) the production of B and T killer cells. The immune response slows down, although it never shuts completely off.

Figure 33-9 The secondary immune response to a pathogen occurs much more quickly and produces more antibodies than the primary immune response.

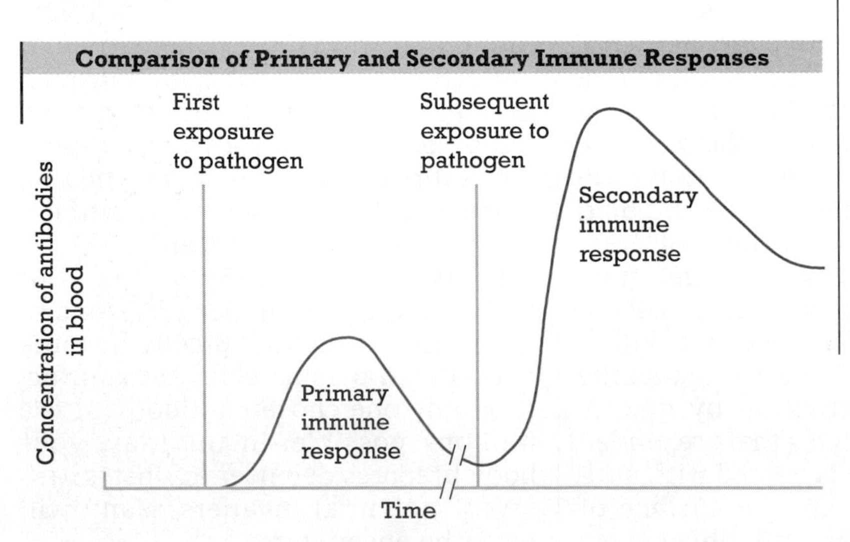

Some B and T cells have selective, yet vitally useful, memory

You may have had chickenpox when you were younger. Once you have recovered from the pox, you usually cannot catch this illness again, even if exposed to the virus that causes it. Your immune system "remembers" its battles. It continues to maintain a low level of defense that can be quickly mobilized to destroy previously defeated pathogens, if they return.

Some helper T cells, killer T cells, and B cells never become killer cells or plasma cells. Instead, these cells get ready for battle but never actually get involved in the fight. They continue, however, to circulate as lymphocytes and become **memory cells** for long periods of time—sometimes for the rest of your life. If a pathogen you have already encountered enters your body, memory cells recognize it and start to divide, quickly forming a new generation of antibody-producing cells. The antibodies generated by the new cells destroy the pathogen before you become ill; you are not even aware of the battle going on within your body. An immune response to a previously encountered pathogen is much faster than an initial immune response and produces many more antibodies than the first response, as shown in **Figure 33-9**. This later immune response is known as a **secondary immune response** because it occurs after the first exposure to a pathogen. The first exposure is known as the **primary immune response**.

Vaccination makes use of the immune system's "memory"

Many serious diseases can be prevented through vaccination, which you read about in Chapter 18. Vaccination triggers an immune response against a particular pathogen without causing the disease itself. Vaccines work because they contain antigens that have been stripped of their disease-causing abilities. For instance, vaccines for bacterial diseases usually contain bacteria killed by heat or chemical treatment. Mature B (plasma) cells respond to the antigens on the dead bacteria by producing antibodies. Similarly, viral vaccines stimulate the immune response with viruses made harmless by chemicals or genetic engineering.

The production of antibodies brings about **immunity**, or resistance to disease.

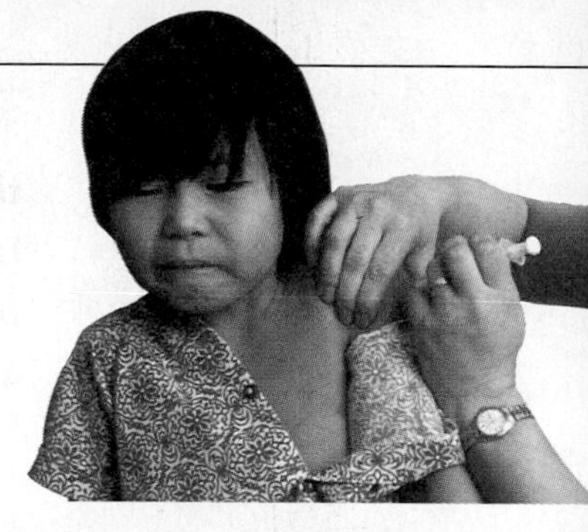

Figure 33-10 This child is receiving the vaccinations required before entering school.

Immunity as a result of vaccination is artificial compared with natural immunity, which results from having had the disease. Both types of immunity make use of the memory of T and B cells.

In many countries, children, like the girl in **Figure 33-10**, are routinely immunized against common childhood diseases, as identified in **Table 33-2**. Each year, more of the United States are requiring that newly developed vaccines for hepatitis B and chickenpox be given to children before they enter school. The immunity achieved from some vaccines lasts a lifetime, as with mumps, for example. Other vaccines, like the tetanus vaccine, require a "booster" shot from time to time to stimulate additional antibody production.

Table 33-2 Single Vaccines and Vaccine Combinations Used in Immunization

■	●	▮	▲	⬮
Diphtheria **T**etanus **P**ertussis	**O**ral **P**oliovirus **V**accine	**M**easles **M**umps **R**ubella	**H**aemophilus **I**nfluenzae **B**	**H**epatitis **B**

Immunization Schedule for Children and Adults†

	1–2 mo	2 mo	2–4 mo	4 mo	6 mo	6–18 mo	12–15 mo	15 mo	4–6 yr	11–12 yr	+10 yr intervals
⬮	⬮		⬮			⬮					
■		■		■	■			■	■		■
●		●		●		●					
▮							▮		▮	▮	
▲		▲		▲	▲		▲				

†Immunization schedule as recommended by the American Academy of Pediatrics and the USPHS

Vaccines aren't effective against rapidly evolving viruses

Genes that carry codes for the surface proteins of flu viruses mutate, or change, rapidly. Thus, the shapes of these surface proteins change frequently. Your body does not recognize viruses with altered surface proteins as the same viruses it has already successfully defeated or been vaccinated against. When these viruses invade your body, they provoke a new primary immune response, and you get sick again. This explains why you can catch the flu more than once.

SECTION REVIEW

1. **Predict how the destruction of helper T cells would affect the immune response.**
2. **Contrast the roles of B cells and T cells in the immune response.**
3. **Explain the role of memory cells in the immune system.**
4. **What can you conclude about the surface proteins of viruses for which effective vaccines exist?**

The immune system is a powerful weapon for self protection. But there are times when the immune system works in an unwanted fashion. Sometimes, the system may be overly sensitive and overreact to even mild stimulation. At other times, the body wages an immune assault against its own tissues.

33-3 Immune System Failure

OBJECTIVES

1. **Describe the events of an allergic reaction.**
2. **Contrast cancer cells with normal cells.**
3. **Relate the symptoms of AIDS to the action of HIV.**
4. **Describe how HIV is transmitted.**

Immune Overreaction

Petting a cat, smelling a flower, or walking into a dusty room can make some people miserable. In these individuals, these activities lead to sneezing, itchy nose and eyes, nasal congestion, and even difficulty in breathing. These are symptoms of an allergy. An **allergy** is an immune system response against a non-pathogenic antigen. A variety of substances, including pollen, certain foods, insect stings or bites, and dust, can trigger allergies. Dust allergies are linked to tiny mites, such as the one in **Figure 33-11**, that live on dust.

If you inhale pollen that you are allergic to, your body produces many more antibodies to the pollen antigen than it needs. These antibodies bind to the antigen and trigger a strong inflammatory response. This strong response causes the cells in the nasal passages to release a set of chemical messengers that includes **histamine** *(HIHS tuh meen)*, as shown in **Figure 33-12** on page 689. Histamine and the other messengers stimulate nearby capillaries to swell and release fluid. Histamines also increase mucus production from the mucous membranes, resulting in a runny nose, nasal congestion, or watery eyes. Many allergy medicines relieve these symptoms with **antihistamines**, chemicals that block histamine action.

Asthma is a form of allergic response that can be life threatening. In people who have asthma, histamines also cause the narrowing of air passages in the lungs, making breathing difficult.

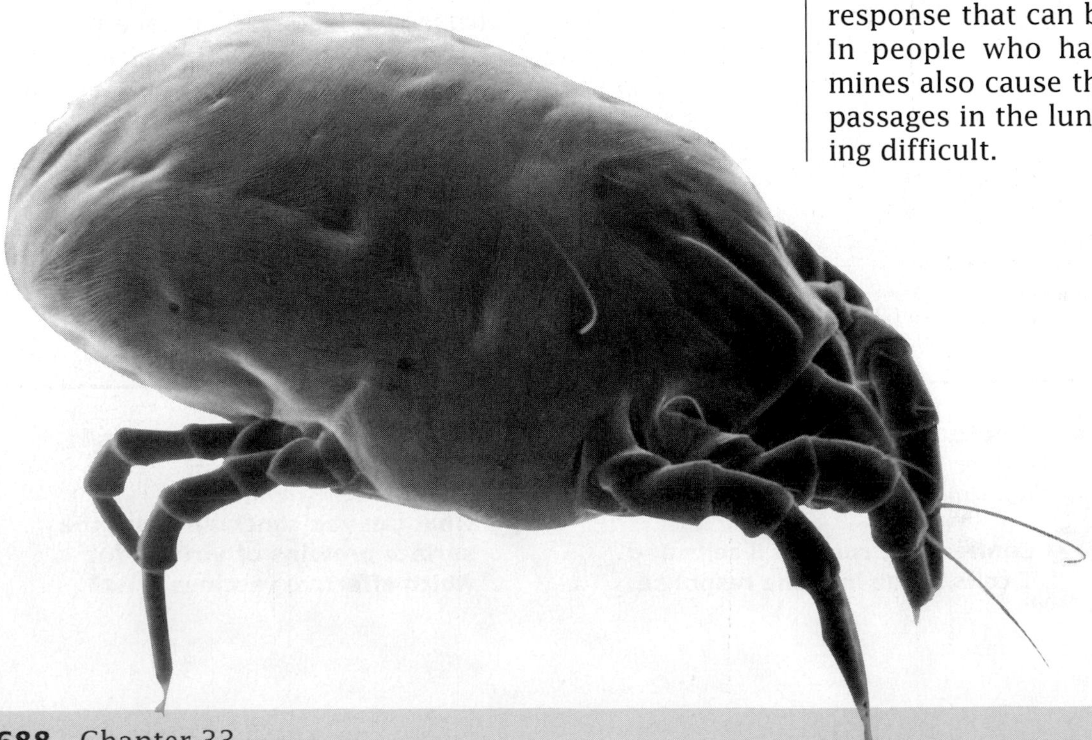

Figure 33-11
People with allergies to dust are actually allergic to the feces of the house-dust mite, which lives on dust particles. This photograph is magnified 300 times.

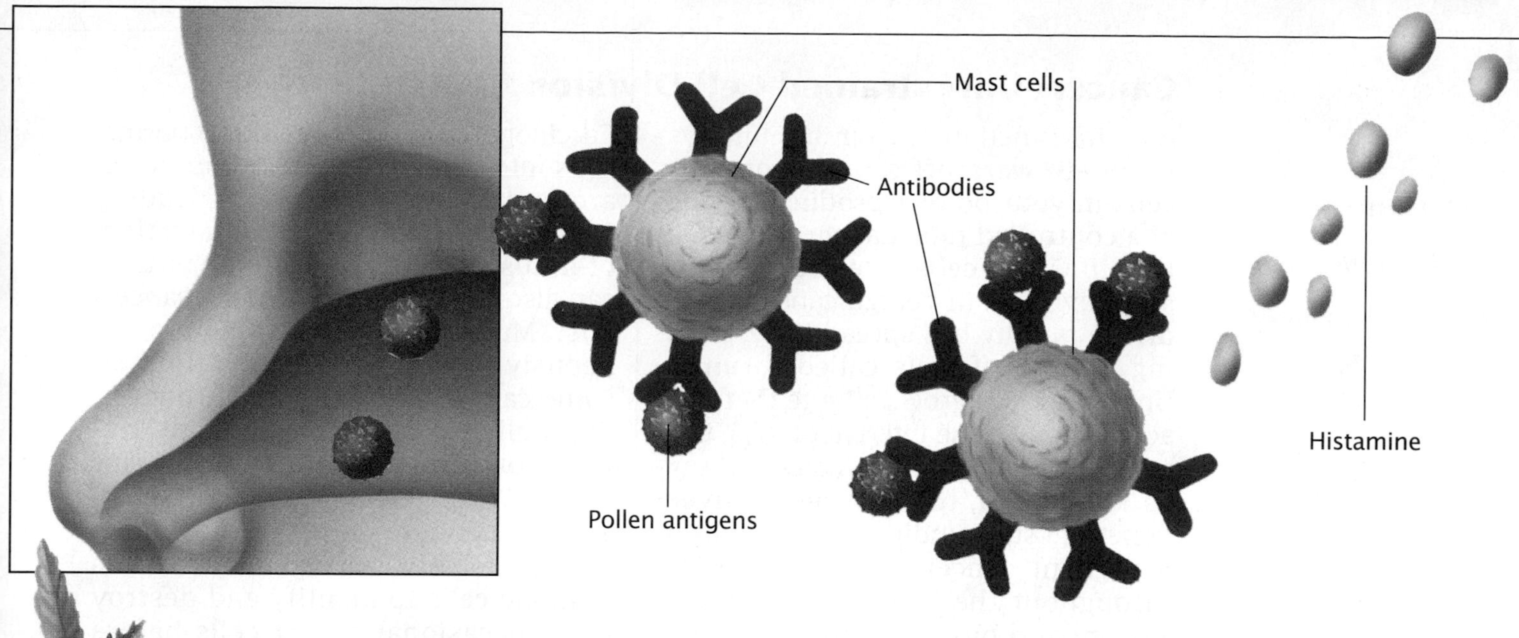

Figure 33-12
Phase 1
When pollen from a plant enters the nose of a person who is allergic to it, pollen antigens attach to antibodies on the surface of the mast cells lining the nasal passages.

Phase 2
The binding of antibodies stimulates cells to release histamine and other chemicals that cause symptoms such as sneezing, runny nose, and itchy eyes.

Autoimmune diseases result when the body manufactures "anti-self" antibodies

Distinguishing self from non-self is the key ability of the cells of the immune system. In certain diseases, this ability breaks down, and the body attacks its own tissues. Such diseases are called **autoimmune diseases**. Multiple sclerosis is an autoimmune disease that usually strikes people between the ages of 20 and 40. In multiple sclerosis, the immune system attacks and destroys the insulating myelin sheath that covers neurons. Degeneration of the myelin sheath interferes with transmission of nerve impulses, which eventually cannot travel at all. Voluntary functions, such as movement of the limbs, and involuntary functions, such as bladder control, are lost. Multiple sclerosis usually leads to paralysis and death. Scientists do not know what stimulates the immune system to attack myelin. Some other autoimmune diseases, the tissues or organs they affect, and their symptoms are listed in **Table 33-3**.

Table 33-3 Some Autoimmune Diseases

Disease	Areas affected	Symptoms
Systemic lupus erythematosus	Connective tissue, joints, kidney	Facial skin rash, painful joints, fever, fatigue, kidney problems, weight loss
Type 1 diabetes	Insulin-producing cells in pancreas	Excessive urine production, blurred vision, weight loss, fatigue, irritability
Graves' disease	Thyroid	Weakness, irritability, heat intolerance, increased sweating, weight loss, insomnia
Rheumatoid arthritis	Joints	Crippling inflammation of the joints

Cancer: Unrestrained Cell Division

Evolution

The protists that cause African sleeping sickness can rapidly change the antigens on their cell surfaces. Why is this an effective way to evade the immune system?

A major function of your immune system is to ward off cancer. Normally, cells in your body reproduce (divide) at a controlled rate. Cancer is a condition in which cells lose the ability to stop dividing. In benign cancers, cells divide rapidly but spread little, forming a mass of cells called a **tumor**. Unless tumors grow so large that they actually crush the internal organs, they are not usually fatal. In malignant cancers, however, the cells aggressively penetrate surrounding tissues. Cells of malignant cancers can even spread throughout the body via the lymph system and bloodstream. Because of the damage they cause to the tissues that they invade, malignant cancers are often fatal unless detected early.

Scientists think that the mutation of genes that regulate the cell division cycle initiates cancer. As you read in Chapter 14, a mutagenic agent that causes cancer is known as a carcinogen. Carcinogens that can transform normal cells into cancer cells are found in cigarette smoke (smoking is a leading cause of lung cancer) and in a variety of industrial chemicals. Some viruses can also cause cells to become cancerous. Mutation also occurs spontaneously in your body all the time, so some cancer cells are being produced "naturally" each day. The immune system is on constant patrol, watching for surface changes on the body's cells. This constant watch, known as **immunological surveillance**, allows immune cells to identify and destroy these occasional cancer cells before they spread. **Figure 33-13** shows a T cell killing a cancer cell. Surveillance by the immune system is your body's primary defense against cancer. Among AIDS patients, who lack effective immune systems, cancer is one of the leading causes of death.

Figure 33-13

a When a killer T cell comes into contact with the cells of a tumor, it is able to recognize the tumor cells as enemies. Below you can see a killer T cell (upper right) that has bound to a tumor cell.

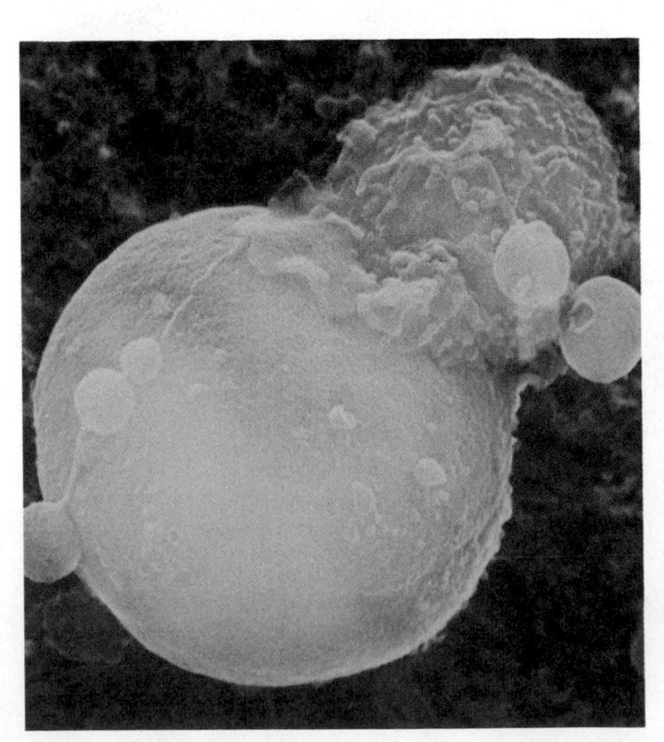

b The killer T cell releases proteins that disrupt the tumor cell's membrane. The damage results in osmotic imbalances within the cell's interior. The tumor cell below is "blebbing" prior to death.

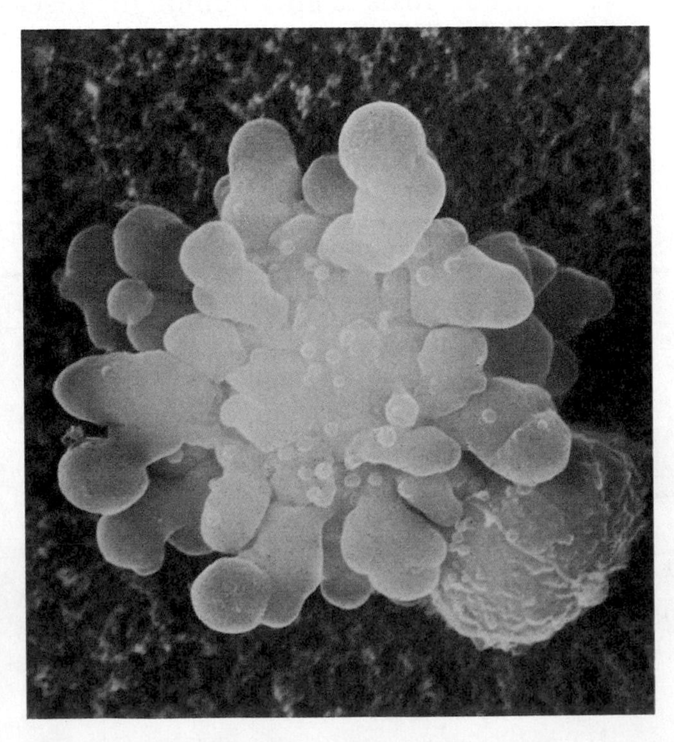

AIDS: Immune System Collapse

When this book was published, more than 300,000 Americans had died of AIDS (acquired immune deficiency syndrome) and more than 440,000 had the disease. Most are young. It is estimated that one out of every 92 men ages 27 to 39 and one out of every 250 people in general is infected.

HIV attacks and cripples the immune system. It invades macrophages and helper T cells, like the one shown in **Figure 33-14**. HIV transforms these cells into virus factories, killing large numbers of helper T cells. When the number of helper T cells in the blood has fallen to very low levels, a person is said to have AIDS. Without helper T cells to stimulate and direct B cells and killer T cells, the immune response cannot occur. The body is overwhelmed by pathogens and cancers that it normally could defeat. Scientists think that practically everyone infected with HIV will eventually develop AIDS. The time between infection and onset of AIDS can be 10 years or longer. During this time, an infected person with or without symptoms can still transmit the virus to others.

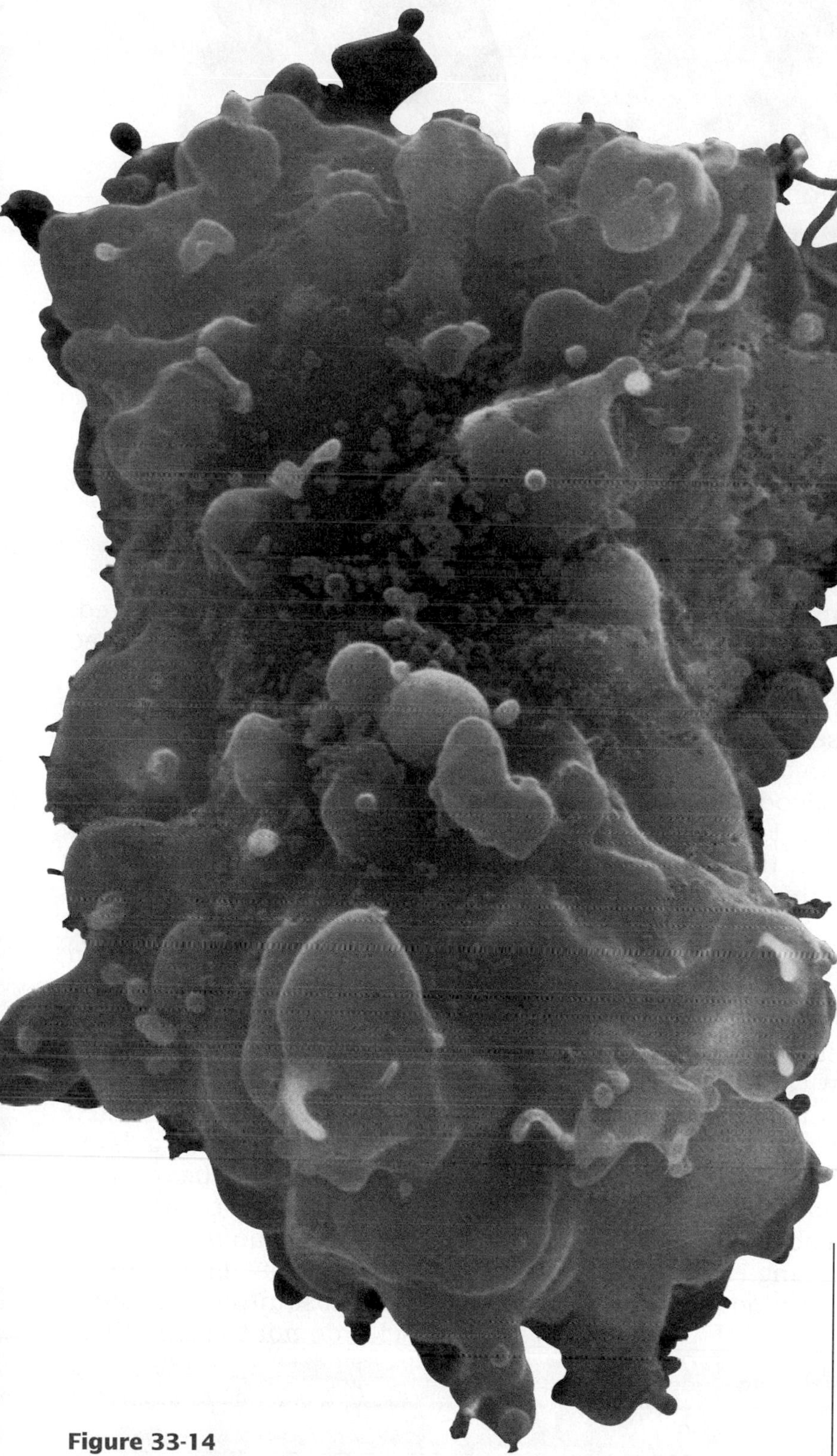

Figure 33-14
The blue particles in this scanning electron micrograph are particles of HIV. They have been produced in this infected helper T cell and are leaving the cell to infect other cells.

HIV is transmitted in body fluids

There is no cure for AIDS. You can protect yourself from AIDS by avoiding exposure to HIV. HIV is a fragile virus that cannot exist for long outside cells. You can get HIV only by contacting the HIV-infected blood cells or body fluids of infected individuals. Because HIV is found in both semen and vaginal secretions, you can contract HIV through sexual intercourse with an infected person; worldwide, most HIV infections are spread this way. The virus can be transmitted to either person during intercourse. Newer strains of HIV are more easily spread in this way. Use of a condom during sex reduces but does not eliminate the risk of getting HIV.

Figure 33-15
Anyone can be infected with HIV, regardless of age, sex, ethnic background, or sexual orientation. A person can be infected with the virus for 10 years or longer without knowing it.

Ryan White died of AIDS at the age of 18. His appearances before the United States Senate helped gain support and understanding for people with the disease. A major government funding act bears his name—the Ryan White Act.

Arthur Ashe, a professional tennis player, died in 1994 from an AIDS-related infection.

HIV can be transmitted when blood from an infected person is transferred to an uninfected person. People who inject drugs can be infected with HIV if they share or reuse needles or hypodermic syringes, both of which can be contaminated with blood carrying the virus. The majority of HIV infections in the United States in the late 1980s were transmitted in this manner.

In the late 1970s and early 1980s, many people contracted HIV by receiving blood transfusions from infected individuals. Many hemophiliacs, including Ryan White (one of the HIV-infected people shown in **Figure 33-15**), were infected by receiving blood clotting factor that had been isolated from blood of infected individuals. Since 1985, donated blood has been tested for the presence of HIV, so the likelihood of contracting HIV through blood transfusions or blood products is very low.

If a pregnant woman is infected, her fetus has about a one-in-three chance of being infected during the pregnancy. HIV can also be transmitted from mother to child in breast milk. **Table 33-4** summarizes the ways that HIV can be transmitted.

HIV is not transmitted through the air or on toilet seats. It cannot be contracted through casual contact, such as shaking hands, sharing food, or drinking from the same water fountain as an infected person. Although HIV is found in saliva, tears, and urine, it occurs there in very low concentrations. One drop of infected blood contains as many viruses as one quart of infected saliva. Scientists think that a large quantity of the virus must be received for infection to occur. You cannot catch HIV from the small amount of saliva exchanged when kissing. Biting arthropods, such as mosquitoes, bedbugs, fleas, and ticks, do not transmit HIV.

Table 33-4 Known Routes of HIV Transmission
• Vaginal, oral, or anal intercourse with an infected person
• Injecting drugs or other substances with hypodermic syringes or needles used by an infected individual
• Use of skin piercing equipment, such as tattooing needles, that has been used by an infected person
• From infected mother to fetus through the placenta; from infected mother to baby in breast milk
• Transfusions or injections of blood or blood products or organs donated from an infected person (Transmission rarely occurs by this route in the United States and other developed nations because blood is tested for the presence of HIV. It is still a common transmission route in the less developed countries, where such blood tests are often unavailable.)

After the death of her eight-year-old daughter in 1988, **Elizabeth Glaser**, herself infected, cofounded the Pediatric AIDS Foundation. Elizabeth died in 1994, leaving her husband and their second infected child, age 10.

Actor **Pedro Zamora** played one of the roommates on MTV's *Real World*. He died in 1994 at age 22.

AIDS has killed more than 300,000 Americans

As you can see in **Figure 33-16**, the number of AIDS cases and deaths have risen dramatically since 1981, when the disease was first described. American teenagers, such as those in **Figure 33-17**, are increasingly at risk of contracting HIV. More than 20 percent of all reported AIDS cases occur among people in their 20s. Given the average 10-year period before symptoms appear, the majority of these people were probably infected during their teenage years. In 1992, AIDS was the sixth leading cause of death among Americans between the ages of 15 and 24. In 1993, AIDS became the number one killer of people aged 25–44 in the United States. And AIDS is killing greater proportions of minority women and children than individuals from other groups. In 1993, AIDS became the fourth leading cause of death among women aged 25–44. Sexual intercourse of any kind—anal, oral, or vaginal—between males and females accounts for the greatest increase in AIDS cases in women in recent years.

Figure 33-16
This graph shows the cumulative numbers of AIDS cases and deaths in the United States since 1981.

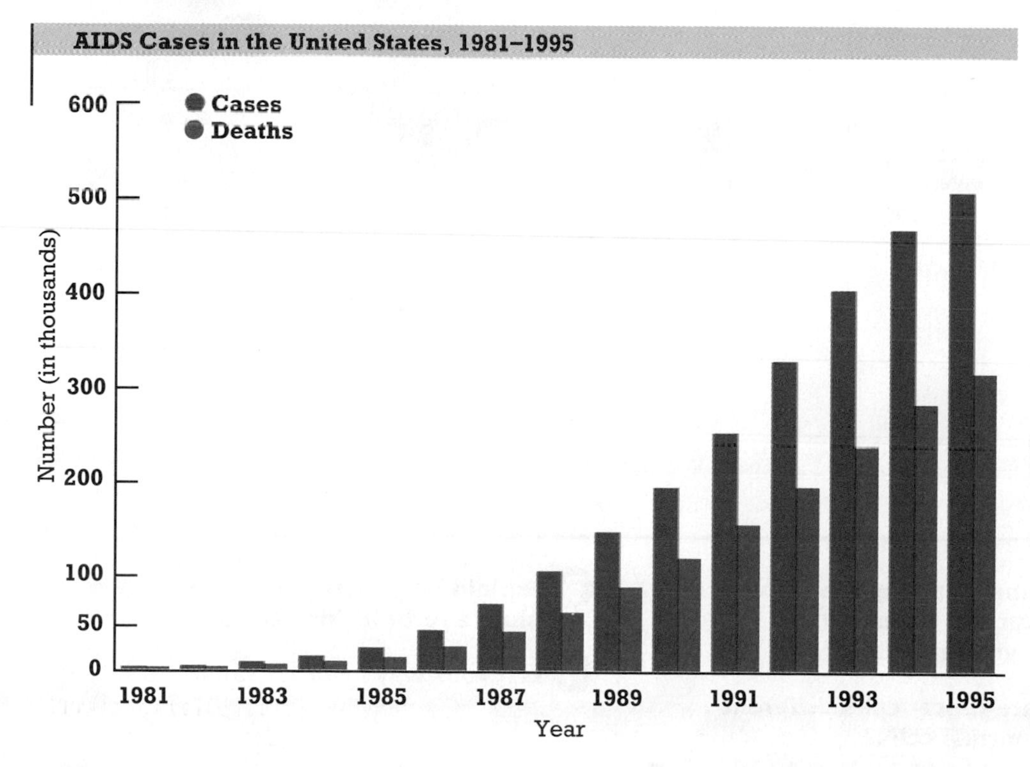

Figure 33-17
Anyone can get AIDS—even you.

AIDS Is a Worldwide Disease

AIDS was first recognized as a disease in the United States. It has since been found in 163 countries. (The origin of the disease remains undetermined.) The World Health Organization estimates that by the year 2000 there will be 10 million cases of AIDS and that about 30 million people will be infected with HIV. Most of these cases will occur in the less-developed countries, where AIDS is now spreading rapidly. Each day, 6,000 people become infected with the virus, and the number of AIDS cases doubles every nine months in Nigeria alone. Currently, AIDS is one of the two biggest killers of women in Africa. **Figure 33-18** shows the projected increase in the number of people infected with HIV worldwide. AIDS is the final stage of HIV infection.

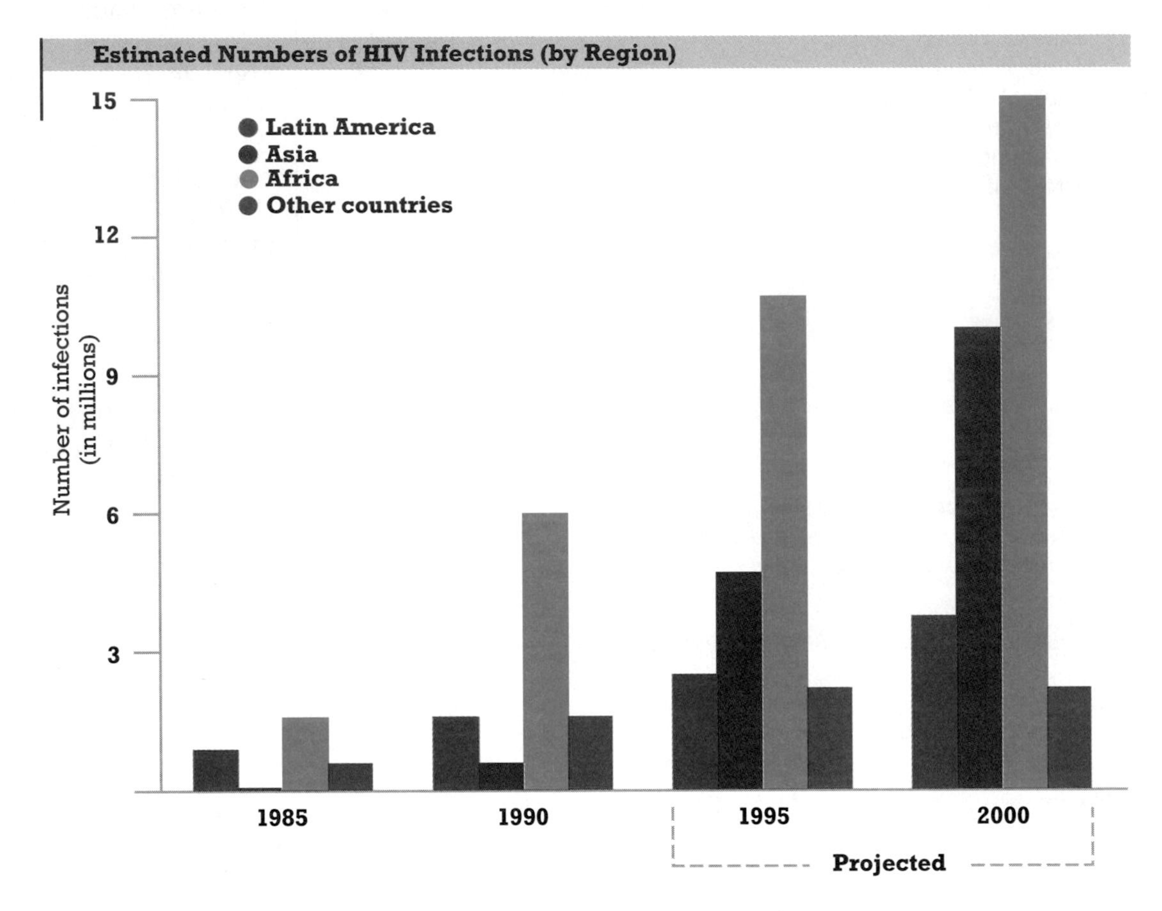

Figure 33-18 The World Health Organization (WHO) projects that 30 million people will be infected with HIV by the year 2000. WHO predicts that Africa and Asia will have the most infections and that the industrialized countries, the category that includes the United States, will have the fewest infections.

SECTION REVIEW

1. **Describe what happens when you are exposed to something to which you are allergic.**
2. **How are cancer cells different from normal cells?**
3. **Explain why AIDS patients are unable to fight infections.**
4. **List four ways HIV is transmitted.**

CHAPTER

33 Highlights

Killer T cells protect the body by destroying tumors, like the one in the lower left corner of the photograph.

	Key Terms	Summary
33-1 First Line of Defense	mucous membrane (p. 679) leukocyte (p. 680) phagocyte (p. 680) lymphocyte (p. 680) inflammatory response (p. 680) macrophage (p. 681) antigen (p. 681)	• The body has several natural defenses for disabling pathogens. • Several kinds of white blood cells work together to defend the body. • Defensive cells and responses are suited for either short or long battles. • The display and recognition of marker proteins are critical to successful defense against invaders.
33-2 The Immune Response	immune response (p. 683) T-cell lymphocytes (p. 684) antibody (p. 685) B cell (p. 685) memory cell (p. 686) secondary immune response (p. 686) primary immune response (p. 686) immunity (p. 687)	• The immune response is used when pathogens get past the first line of defense. • It takes several days after invasion before the immune response is activated. • T cells command and regulate the immune response. Some T cells kill their own body's infected cells. • B cells are the main killers, attacking the enemy most directly. Some B cells become plasma cells and manufacture antibodies. • Long-term immunity results from the memories of enemies held by certain T cells and B cells.
33-3 Immune System Failure	allergy (p. 688) histamine (p. 688) antihistamine (p. 688) autoimmune disease (p. 689) tumor (p. 690) immunological surveillance (p. 690)	• An allergy is a response to a nonpathogenic antigen. • Cancer is uncontrolled cell division. The immune system normally destroys cancer cells before they spread. • HIV causes the immune system to fail by invading and destroying helper T cells and macrophages.

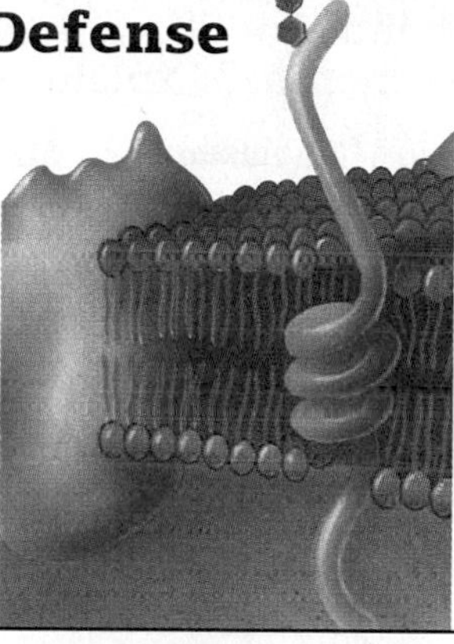

The immune system depends on marker proteins to distinguish self from foreign invaders.

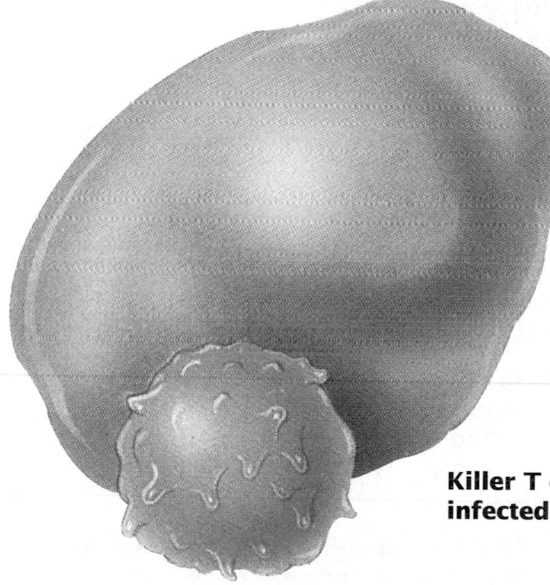

Killer T cells attack infected body cells.

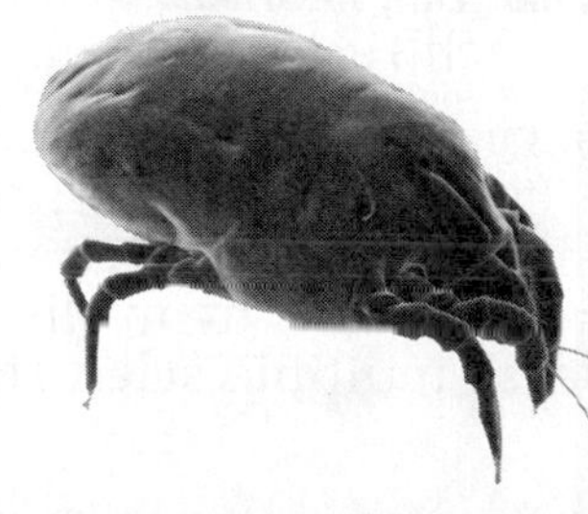

This house-dust mite is a source of misery for thousands of people who are allergic to dust.

CHAPTER 33

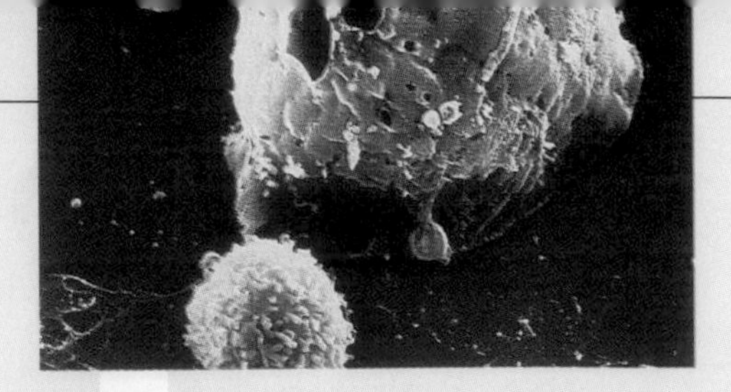

review

Understanding Vocabulary

1. For each pair of terms, explain the differences in their meanings.
   a. inflammatory response, immune response
   b. antigens, antibodies
   c. T cells, B cells
2. Using each set of words below, write one or more sentences summarizing information learned in this chapter.
   a. mucous membranes, sweat, tears, saliva
   b. leukocytes, lymphocytes, macrophages
   c. immune response, interleukin-1, pathogens, macrophage

Understanding Concepts

3. **Relating Concepts** Construct a concept map that describes an immune response. Include the following terms: macrophages, bacteria, dead cells, viruses, T cells, killer T cells, B cells, pathogens, HIV, phagocyte, antibodies.
4. **Identifying Structures and Functions** Describe how your skin defends your body.
5. **Identify Structures and Functions** Describe three nonspecific natural defenses that protect the body from invading microbes and pathogens.
6. **Summarizing Information** Describe the events that occur after a splinter enters your finger.
7. **Comparing Functions** Differentiate between the inflammatory response and the immune response in terms of duration of response and kinds of white cells involved.
8. **Inferring Relationships** Explain how white blood cells can distinguish your cells from those of invading pathogens.
9. **Inferring Relationships** What is the function of antibodies in the immune response?
10. **Summarizing Information** Summarize the steps leading to the production of antibodies by plasma cells that encounter a macrophage showing a viral antigen.
11. **Inferring Relationships** Identify the cells responsible for long-term immunity, and explain how they achieve it.
12. **Inferring Conclusions** What would be your chances of contracting German measles for a second time if memory cells lived only three months? Explain your answer.
13. **Inferring Conclusions** Explain why vaccines are not effective against viral diseases such as influenza and AIDS.
14. **Interpreting Graphs** Over time, people with AIDS become sicker and sicker, unable to fight infection. Look at the graph and explain why this occurs.

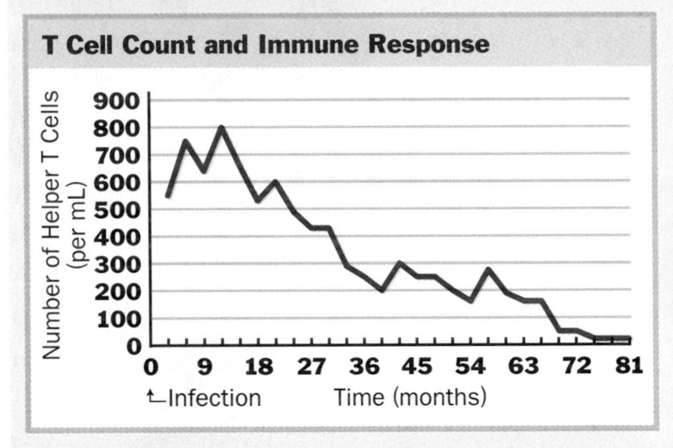

15. **Summarizing Information** Describe the events of an allergic reaction.
16. **Recognizing Relationships** What happens to the immune system's ability to distinguish self from non-self cells in an autoimmune disease? Identify two autoimmune diseases and their symptoms.
17. **Inferring Relationships** What causes normal cells to become cancer cells?
18. **Recognizing Relationships** What is AIDS? Why do people with AIDS die from bacterial and viral infections?
19. **Summarizing Information** Identify four routes by which HIV is transmitted.
20. **BUILDING ON WHAT YOU HAVE LEARNED** In Chapter 30 you learned about the function of myelin in the transmission of nerve impulses. How is myelin affected by the disease multiple sclerosis?

Interpreting Graphics

21. Study the diagram below.

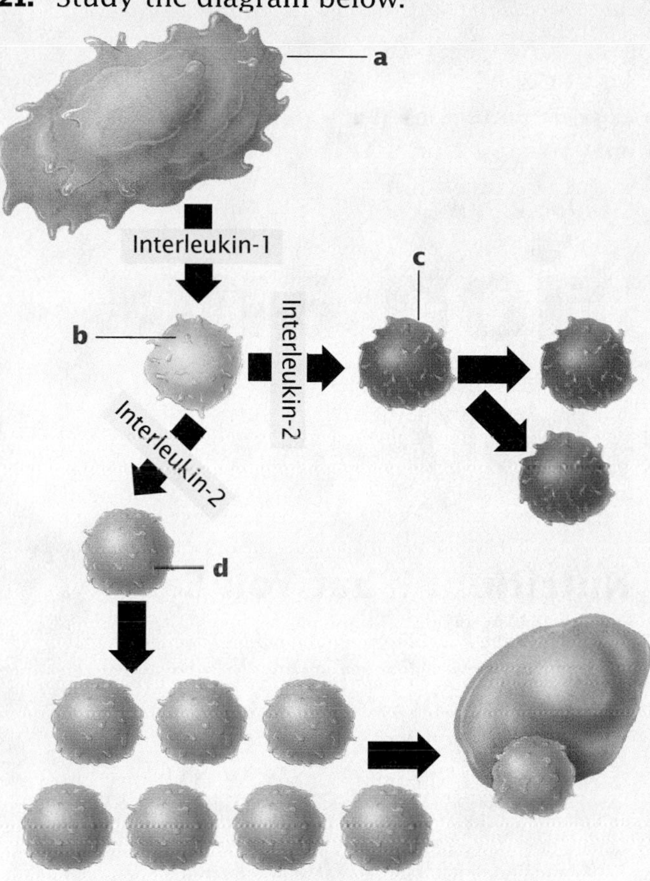

a. Identify the cell labeled *a*. What is its function in the immune response?
b. What is the function of the cells labeled *d*?
c. Describe how HIV infection affects the process illustrated above.

Reviewing Themes

22. ***Cell Structure and Function***
Long-term smoking causes paralysis of cilia lining the trachea. How would this affect the body's ability to repel pathogens?

23. ***Stability and Homeostasis***
What symptoms are associated with an asthma attack? What effect would taking medicines that contain antihistamines have on the symptoms?

Thinking Critically

24. **Distinguishing Fact From Opinion** A newspaper columnist wrote "The ongoing fight against AIDS should be limited to only those people who are in high-risk groups, such as those people who abuse drugs." Comment on the effectiveness of this approach.

Life/Work Skills

25. **Finding Information** Use library references or search an on-line database to find out about the work of German physician Robert Koch. What did Koch contribute to the knowledge about the spread of disease? What are Koch's postulates? Summarize your findings in a written report.

26. **Allocating Resources** Every year billions of dollars are spent on research into the causes and treatment of tuberculosis. Use library references or search an on-line database to find out how much money is spent and by what agencies. How has the number of TB cases changed? What percentage of the money has been spent on testing, treatment, and education for prevention? Where is the bulk of the money currently being spent? Why? How has the amount of money spent changed over the last 10, 15, and 20 years? Summarize your findings in a series of graphs or charts.

Activities and Projects

27. **Health Connection** For decades, polio, or poliomyelitis, affected people around the world. This incurable disease causes pain, stiffness, paralysis, and eventually death. Use library references or search an on-line database to find out about Jonas Salk and the polio vaccine. How was the vaccine developed? How effective was it? What contribution did Albert Sabin make in the fight against polio? How common is polio today? Summarize your findings in a written report to your class.

28. **Geography Connection** According to the World Health Organization (WHO), the majority of future AIDS cases will occur in Africa, Asia, and South America. Look for information in your library about living conditions, medical care, and education in these areas and about their projected population growth. How might those conditions support WHO's projection?

Discovering Through Reading

29. Read the article "Our Immune System: The Wars Within" by Peter Jaret in *National Geographic*, June 1986, pages 702–735. Why is the AIDS virus so deadly? Explain why the analogy of the Trojan horse might be appropriate in describing HIV.

CHAPTER 34

REVIEW
- organic molecules (Section 2-3)
- enzymes (Section 5-1)
- water conservation (Section 24-2)

Digestion and Excretion

The large intestine in this X-ray photo is an important last stop for food in the digestive system.

Animals cannot make their own food as plants do, so they must eat other organisms. Some animals, called carnivores, eat only other animals; others, called herbivores, eat only plants. But humans, like bears, are omnivores, eating both plants and animals.

34-1 Nutrition: What You Eat and Why

OBJECTIVES

1. **Analyze the relationship between the six essential nutrients and the food pyramid.**
2. **List two ways the body uses lipids, proteins, carbohydrates, water, minerals, and vitamins.**
3. **Discuss two factors that lead to nutritional problems in the United States.**
4. **Discuss the effects of the eating disorders anorexia nervosa and bulimia.**

Nutrition and Health

Historically, protein, mineral, and vitamin deficiencies were significant nutritional problems in both developed and developing countries. Today, too much food (overconsumption) and too little physical activity (sedentary lifestyle) are the two major factors leading to nutritional problems in the United States, although nutrient deficiencies still exist. Excess calories and fat lead to **obesity** and heart disease. Excess dietary fat has also been shown to increase the risk of breast and colon cancer. Too much protein harms the kidneys, and too much sodium is linked to high blood pressure. Too little fiber results in poor intestinal elimination and is linked to cancer of the large intestine (colon).

Nutrition is the science that studies how food affects the functions of living organisms. Proper nourishment is essential for healthy biological functioning. When the body does not receive the substances it needs, imbalances occur, biological processes are impaired, and diseases or death can result. Many of today's diets lead not only to inadequate nutrition but also to problems with digestion and elimination.

Essential nutrients are required for life

Food contains nutrients. You need several types of nutrients in your diet. To obtain energy, your body uses carbohydrates, lipids, and proteins as fuel nutrients. To regulate how your body uses energy from the fuel nutrients, your body needs water, minerals, and vitamins, which do not provide energy. From these six essential nutrients, your body gets all it needs in order to grow and to maintain and repair itself. Most of the food and beverages you consume provide a combination of the six nutrients. **Table 34-1** lists the nutrients required in the diet.

Table 34-1 Essential Nutrients

Fuel nutrients	Regulatory nutrients
Carbohydrates	Water
Protein	Minerals
Fats	Vitamins

Figure 34-1

a The USDA Food Guide Pyramid serves as a guide for healthy eating. The placement of each food group reminds you which groups are needed in abundance and which ones should be eaten in smaller amounts.

Fats, oils, and sweets group
Saturated and hydrogenated fats cause the greatest damage to health.

Key
These symbols show fats, oils, and added sugar:
● **Fat (naturally occurring or added)**
▼ **Sugars (added)**

Use sparingly

Milk, yogurt, and cheese group

Meat, poultry, fish, dry beans, eggs, and nuts group
These two groups, side by side, are the major sources of protein.

2–3 servings daily

Vegetable group
The expression "five-a-day," often seen in food markets, reminds you to eat at least five servings of highly colored fruits and vegetables daily.

3–5 servings daily

2–4 servings daily

Fruit group

Bread, cereal, rice, and pasta group
This group is the foundation of a healthy diet.

6–11 servings daily

Food Energy

Energy stored in food is trapped in chemical bonds. When these bonds are broken in the body, the energy is released and then captured again in a form that the body can use. The amount of energy provided by the energy nutrients is measured in calories. A calorie is the unit of measurement used to describe the amount of heat energy that is released from food or that is used during physical activity. Three types of nutrients provide energy bonds that are useful to the human body.

b Each group of fuel nutrients should provide a specific percentage of the day's total calories.

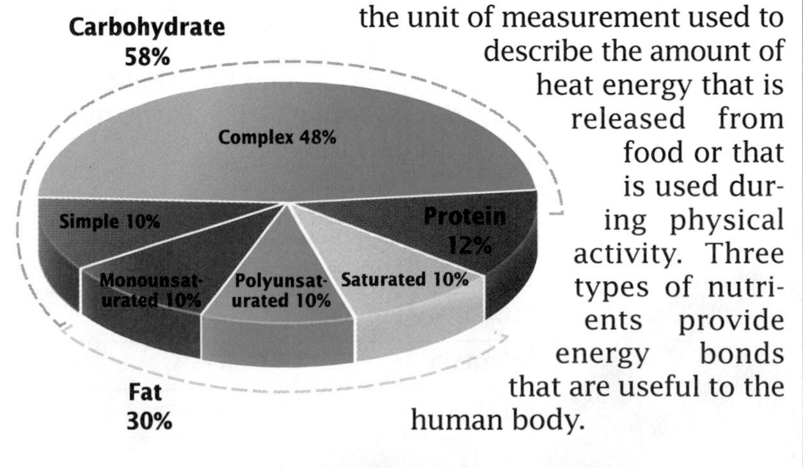

Carbohydrates are the body's main source of fuel

Carbohydrates provide the most readily available energy for your body. More than half of your body's energy needs should come from those carbohydrates classified as starches and sugars. Recall that glucose, a simple sugar, is the molecule that all of your cells use for energy. Glucose and other naturally occurring sugars are found in many foods, especially fruits, vegetables, and dairy foods. Simple sugars, such as glucose, are combined into longer chains to make complex carbohydrates called starches. Cereals, grains, seeds, and beans are examples of foods high in complex carbohydrates. **Figure 34-1** shows the USDA Food Guide Pyramid, a model for creating a healthful diet. Notice that carbohydrate foods are the foundation of a healthy diet. Other complex carbohydrates, such as cellulose, are components of plant tissues but are not digested by humans. These indigestible carbohydrates provide roughage, or fiber, in our diets. Fiber does not provide energy, but it is so important in the body that it is sometimes called the seventh essential nutrient. Whole grains, legumes, fruits, and vegetables are high in fiber.

Lipids store highly concentrated energy

Lipids such as fats (solid at room temperature) and oils (liquid at room temperature) are very concentrated energy sources. One gram of fat provides more than twice as much energy as one gram of carbohydrate or protein. Lipids in the body form the bilayer foundation of cell membranes and organelles; thus, they are vital to the life of cells. Fats in the diet act as a solvent for fat-soluble vitamins and essential fatty acids. Dietary fats also give a sense of satisfaction after eating, because they take a long time to digest.

Fats in the diet are classified according to their chemical makeup. Recall that saturated fats and too many calories from fat in the diet are linked to cardiovascular disease, obesity, and certain kinds of cancer. Saturated fats are found in animal fats and animal protein, coconut and palm oils, and vegetable oils that have been processed to be a solid at room temperature (hydrogenated). Monounsaturated fats are found in canola, olive, and peanut oils. Polyunsaturated fats are found in safflower, corn, soybean, and cottonseed oils, among others. Monounsaturated and polyunsaturated fats should each provide about 10 percent of the day's calories. Saturated fats should provide 10 percent or less of the day's calories. **Figure 34-2** shows a food label that provides information about calories and nutrients found in food.

Figure 34-2
"Nutrition Facts" is the title for the new food label required by law. The food label below helps consumers make food choices based on the nutrients that are most important to health.

❶ *Serving size* is based on the amount of the food people usually eat. Label calculations are for one serving of the food.

❷ *Total fat* displays the amount of all kinds of fat in the food.

❸ *Total carbohydrate* shows the amounts of complex carbohydrates, fiber, and sugar.

Nutrition Facts
Serving Size ¾ cup (30g)
Servings Per Container About 14

Amount Per Serving	Corn Crunch	with ½ cup skim milk
Calories	120	160
Calories from Fat	15	20
		% Daily Value**
Total Fat 2g*	**3%**	**3%**
Saturated Fat 0g	**0%**	**0%**
Cholesterol 0mg	**0%**	**1%**
Sodium 160mg	**7%**	**9%**
Potassium 65mg	**2%**	**8%**
Total Carbohydrate 25g	**8%**	**10%**
Dietary Fiber 3g		
Sugars 3g		
Other Carbohydrate 11g		
Protein 2g		
Vitamin A	15%	20%
Vitamin C	25%	25%
Calcium	2%	15%
Iron	25%	25%
Vitamin D	0%	10%
Thiamin	25%	30%
Riboflavin	25%	35%
Niacin	25%	25%
Viamin B6	25%	25%
Folic Acid	25%	25%
Phosphorus	6%	20%
Zinc	25%	30%

*Amount in Cereal. A serving of cereal plus skim milk provides 2g fat, less 5mg cholesterol, 220mg sodium, 270mg potassium, 31g carbohydrate (19g sugars) and 6g protein.

**Percent Daily Values are based on a 2,000 calorie diet. Your daily values may be higher or lower depending on your calorie needs:

	Calories	2,000	2,500
Total Fat	Less than	65g	80g
Sat Fat	Less than	20g	25g
Cholesterol	Less than	300mg	300mg
Sodium	Less than	2,400mg	2,400mg
Potassium		3,500mg	3,500mg
Total Carbohydrate		300g	375g
Dietary Fiber		25g	30g

❹ *Calories from fat* tells how many calories in the food come from fat.

❺ *Cholesterol* and *sodium* amounts are also required on the label. A teaspoon of table salt contains about 2 g of sodium.

❻ *Daily values* footnote lists the current nutrition recommendations for selected essential nutrients.

❼ Some labels tell the approximate number of calories per gram of carbohydrate, fat, and protein.

Corn: calcium

Proteins build body tissue

Proteins in the diet are the body's only source of nitrogen. The body needs nitrogen to make amino acids that are then built into muscles, organs, enzymes, hormones, and most carrier molecules. The body makes most, but not all, of the amino acids that it needs. Essential amino acids are those that the body cannot make; they must come from the diet. If a food contains all of the essential amino acids, its protein is said to be "complete." Some foods, like eggs, milk, seafood, and meat, have high levels of the essential amino acids. Other foods, such as legumes, grains, nuts, and seeds, have high levels of some essential amino acids and low levels of others. Vegetarians must plan their diets to get a balanced level of the essential amino acids. Foods can be combined to improve their amino acid balance. For instance, combinations like beans and rice or macaroni and cheese provide a more complete variety of protein than the individual foods would provide if eaten separately. Nutritious food combinations are staples in many cultures and complement each other's protein.

Tomatoes: potassium

Regulatory Nutrients

The biological processes that release food energy are regulated by three kinds of nutrients that do not provide energy.

Water is the most vital nutrient

Water is the most vital nutrient, but its importance is often overlooked and its intake is often low. You can live for weeks without food, but you can live only a few days without water. Water acts as a lubricant, a solvent, and a coolant. Drinking one-half of an ounce of water daily for each pound of body weight will provide an adequate daily intake. You can monitor to see if you are drinking enough water throughout the day. When your water intake is adequate, your urine will be pale to almost clear in color.

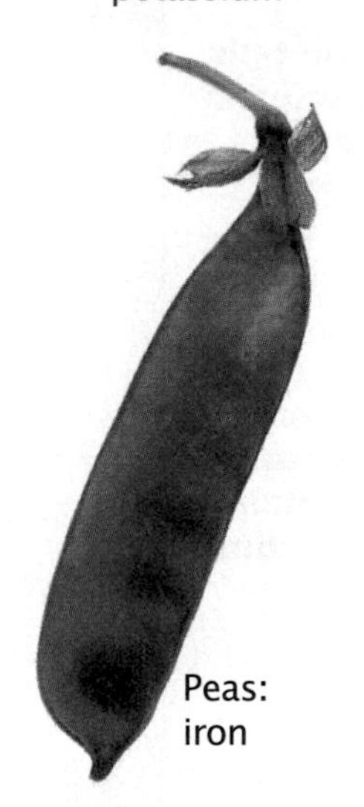
Peas: iron

Minerals are built into bone, essential compounds, and enzymes

Unlike other nutrients, **minerals** are inorganic compounds—they are not made by living things and must be obtained elsewhere. Plants absorb minerals from the water or soil, and animals get minerals by eating plants or plant-eating animals. Minerals are not broken down within the body. Some minerals, such as iron, copper, iodine, and zinc, are called trace elements because they are needed in very small amounts. Besides their use in body structure, specific minerals are required for the normal action of enzymes. **Table 34-2** lists some minerals, their sources, and their functions.

Table 34-2 Important Minerals and Their Functions

Mineral	Function	Sources
Calcium	Needed for building bones and teeth; nerve and muscle function	Dairy products, legumes, dark green leafy vegetables
Iodine	Essential for production of thyroid hormone	Seafood, seaweed, spinach, iodized salt
Iron	Found in hemoglobin; needed for some enzymes	Meats, whole grains, legumes, dark green leafy vegetables
Potassium	Needed for normal metabolism; nerve and muscle function	Meats, poultry, fish, fruits, vegetables
Sodium	Essential for proper water balance in cells and tissues; bio-electrical conduction	Widely distributed in nature

Celery: sodium

Vitamins are needed in small amounts for regulation

Vitamins are organic substances needed by the body in very small amounts. Vitamins activate enzymes and regulate the release of energy in the body.

There are two classes of vitamins: those that are fat-soluble and those that are water-soluble. Fat-soluble vitamins, including vitamins A, D, E, and K, can accumulate in body fat and in the liver. Excessive amounts of these vitamins can be harmful. Plants provide substances called carotenoids that form vitamin A in the human body. These substances are not toxic to the body, as excessive vitamin A can be.

Water-soluble vitamins, like vitamins B and C, are not stored in body tissue. Because these vitamins easily dissolve in water, any excess is excreted in urine. Therefore, regular, balanced sources of water-soluble vitamins must be available to prevent deficiencies. A varied diet containing foods from each of the groups in the food pyramid is recommended in order to provide adequate amounts of vitamins and minerals. **Table 34-3** lists some common vitamins and their sources, and explains their various functions.

Whole-grain bread: B-complex vitamins, vitamins E, K

Spinach: carotenoids, vitamins C, E, K

Tangerine: vitamin C

Yellow pepper: carotenoids

Cheese: vitamins A, B_2

Table 34-3 Some Important Vitamins and Their Functions

Vitamin	Function	Sources	Deficiency
Vitamin A (fat-soluble) Carotenoids	Maintains healthy eyes, skin, bones, teeth; keeps lining of digestive tract resistant to infection	Milk products, liver, bright fruit and vegetables and dark green leafy vegetables	Night blindness, impaired growth
Vitamin B_1 Thiamine (water-soluble)	Assists with conversion of carbohydrates to energy; normal appetite and digestion; nervous system function	Pork products, liver, legumes, enriched and whole-grain breads, cereals, nuts	Beriberi (inflamed nerves, muscle weakness, heart problems)
Vitamin B_2 Riboflavin (water-soluble)	Assists with nerve cell function; healthy appetite; release of energy from carbohydrates, protein, and fats	Dairy foods, eggs, whole-grain products, meat, dried beans, enriched breads, cereals, and pasta	Cheilosis (skin sores on nose and lips, sensitive eyes)
Vitamin B_3 Niacin (water-soluble)	Maintenance of normal metabolism, digestion, nerve function, energy release	Red meats, organ meats, fish, enriched breads and cereals, green vegetables	Pellagra (soreness of tongue and lips, diarrhea, irritability, and depression)
Vitamin B_{12} (water-soluble)	Necessary for formation of red blood cells; normal cell function	Lean meats, liver, egg products, milk, cheese	Pernicious anemia, stunted growth
Vitamin C (water-soluble)	Needed for normal wound healing and the development of connective tissues, including those holding teeth	Citrus fruits, melons, green vegetables, potatoes	Scurvy (slow healing of wounds, bleeding gums, and loose teeth)
Vitamin D (fat-soluble)	Promotes normal growth; assists with calcium and phosphorus use in building bones and teeth	Fish-liver oils, fortified milk, liver, egg yolk, salmon, tuna	Rickets (inadequate growth of bones and teeth)
Vitamin E (fat-soluble)	Prevents destruction of red blood cells; needed for certain enzymes	Wheat germ, vegetable oils, legumes, nuts, dark green vegetables	Red blood cell rupture, causing anemia
Vitamin K (fat soluble)	Assists with blood clotting	Leafy green vegetables and vegetable oils, tomatoes, potatoes	Slow blood clotting, hemorrhage

Eating Disorders

Millions of teens, afraid of gaining weight, are on weight-reducing diets. Unfortunately, obsessive dieting sometimes leads to eating disorders. Anorexia nervosa *(an oh RECK see ah ner VOH sah)* is an eating disorder in which afflicted people starve themselves. In contrast, bulimia *(byoo LEE mee uh)* is a disorder characterized by bingeing (consuming a large amount of food in a short time) and purging, either by self-induced vomiting or by taking laxatives, enemas, or diuretics (fluid pills).

Individuals with anorexia (such as the young woman in **Figure 34-3**) and bulimia have a distorted perception of their body and believe they are too fat. They see their eating behavior as a way to control their weight. The anorexic or bulimic is very secretive about eating. Those around them often do not know there is a problem. In addition, some anorexics and bulimics exercise obsessively. About 1 in 100 females between the ages of 12 and 18 are estimated to be anorexic. Between five to fifteen percent of people with eating disorders are male.

The metabolic changes caused by anorexia or bulimia are serious. Females may stop menstruating because they lose too much body fat and the level of sex hormones falls. Many anorexics never mature sexually. In addition, muscle mass and bone mass are lost, hair and skin become dry, and nails are brittle. Although bulimics may appear healthier than individuals with anorexia, bulimia leads to serious trauma to the lining of the mouth, stomach, and esophagus due to repeated vomiting. Tooth enamel may erode because of long-term exposure to stomach acids, and gums pull away from the teeth. Fatigue and muscle cramps are common because minerals are lost during purging. The binge-purge routine is characteristic of bulimia whether it occurs once a month or several times a day. At its most extreme, bulimia can lead to death by heart failure, stomach rupture, or kidney failure. Both anorexia and bulimia, however, can be managed with a combination of medical treatment, counseling, and family support.

"I got to be very thin, but I thought I looked normal. The disease persists to this day as a constant battle between my perception of my body and my common sense."

–Roxanne A.

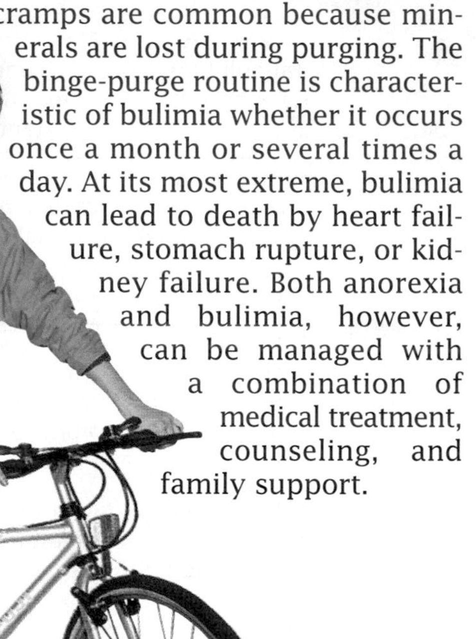

Figure 34-3 Roxanne is a recovering anorexic. Today, she looks and feels healthy.

SECTION REVIEW

1. **What are the six nutrients that the body requires for good health?**
2. **What roles do each of the six nutrients play in the body?**
3. **Which diseases are caused by nutrient deficiencies? by excesses?**
4. **What kind of damage can an eating disorder do to the body?**

Pediatrics

Career Opportunities

What would it be like to . . .

- diagnose and treat ailments and diseases in children?
- teach children about preventive medicine?
- contribute to the health of children in your community?

These are just a few of the daily adventures experienced by people working in the field of pediatrics.

Pediatrics: What Is It?

Pediatricians generally provide medical care for children from infancy through adolescence. In infants, a pediatrician is likely to encounter birth defects, infectious diseases, and problems arising from prematurity. In older children, a pediatrician is more likely to treat infectious diseases and injuries due to accidents. A regular task of a pediatrician is teaching children how to prevent, identify, and deal with health problems.

Science/Math Career Preparation

High school
- English
- Science

College
- Biology
- Chemistry
- Psychology

Medical school
- Biochemistry
- Physiology
- Anatomy
- Pathology

Employment Outlook

As the world population climbs, the need for pediatricians throughout the world will increase dramatically.

Career Focus: Dr. Cheryl Coldwater

Pediatrician

"As a child I had to spend a lot of time in the hospital because I was born with a birth defect—a broken leg that healed improperly in the womb. I wore a brace until I was six, at which point my leg was amputated. That way I could wear a prosthesis and have a greater range of movement.

"In high school, I volunteered in pediatrics in a hospital, weighing babies and helping nurses. I won't deny that medical school was hard work—long hours of intense study. However, working with patients was easy. That's what I like best about being a doctor—working with children."

What I Do: "As a pediatrician, I concentrate on preventive medicine. I discuss the importance of proper nutrition with young children. It's important for them to understand that learning to eat healthfully can prevent many kinds of health problems later. Other subjects I address include exercise, safety, and drugs and alcohol. Talking directly and frankly with children lets them know that there are adults other than their parents who care about them."

To Find Out More

1. For further information, write to:
   American Academy of Pediatrics
   141 N.W. Point Blvd.
   P. O. Box 927
   Elk Grove Village, IL 60009
2. Visit your school counselor and library to find out what colleges and universities offer programs in pediatrics. Review the prerequisites. Make a poster showing the range of job opportunities in this field.
3. Find articles that describe efforts of the Centers for Disease Control and the World Health Organization to develop immunization services worldwide. Prepare a report that convinces readers of the importance of immunization.

Supplies of energy and building materials for the body exist only in potential forms in food. Whatever we eat must be processed into smaller pieces before it can be used by the body. Food undergoes this transformation in the digestive system, a hollow muscular tube where food is broken down and either absorbed or eliminated.

34-2 The Digestive System

OBJECTIVES

1. **List the main organs of the digestive system.**
2. **Identify the sites of digestion for each of the three fuel nutrients.**
3. **Describe the absorption of food from the digestive tract.**
4. **Explain the roles of the pancreas and liver in the digestive system.**

Digestion

Suppose you eat a taco for lunch. The taco is rich in the three kinds of fuel nutrients found in food. The tortilla and beans are made of complex carbohydrates. Proteins occur in the meat, cheese, corn, and beans. Fats are found in the meat and cheese. The taco also contains essential vitamins, minerals, water, and some fiber. When you eat the taco, however, the nutrient-containing substances found within it are too large for your body to use immediately. As the taco travels through your digestive system, a journey of more than 8 m (26 ft), it is crushed and churned, and the nutrients in it are broken down by various chemicals in order to release its store of energy and building materials. In short, it is digested.

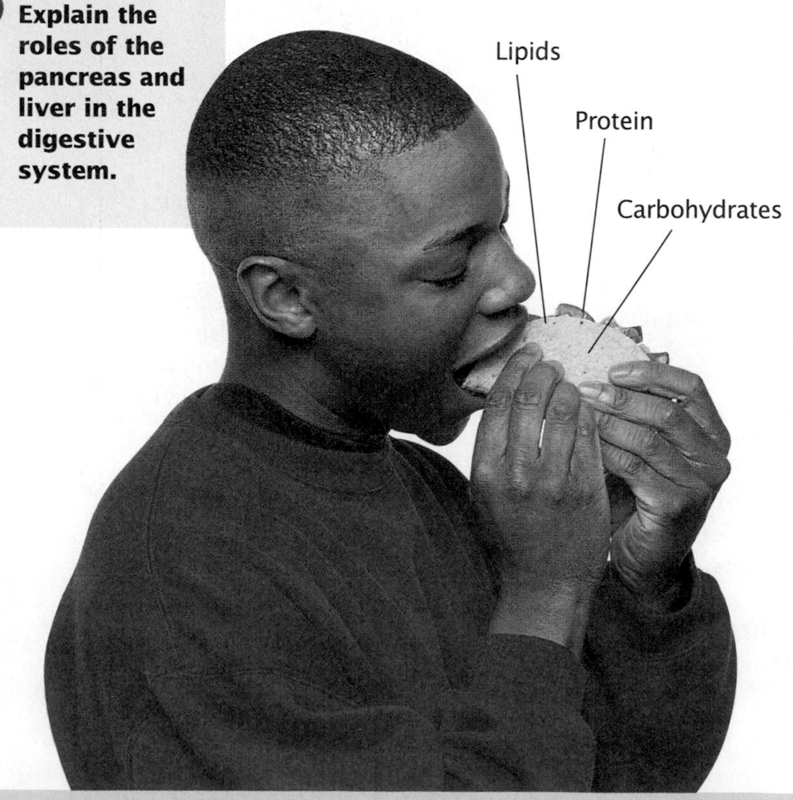

Digestion begins in the mouth

Taking a bite of the taco, as shown in **Figure 34-4**, begins the process of digestion. As the food is chewed, it is broken apart so that it will be accessible to digestive enzymes. You have probably noticed that your mouth "waters" when you are hungry and you smell or see food (or sometimes just think of food). Salivary glands lying in and near the mouth increase their production of saliva before you eat, and saliva plays an important role in digestion. It moistens and lubricates food, making it easier to swallow. Saliva also contains an enzyme that begins to break down starches present in the food.

After being chewed, food is forced into the rest of the digestive tract by swallowing. The tongue and floor of the mouth rise, squeezing food into a muscular tube called the **esophagus**

Figure 34-4
Chewing not only makes a taco easier to swallow but also increases the surface area of the food so that digestive enzymes can come into contact with more of the food.

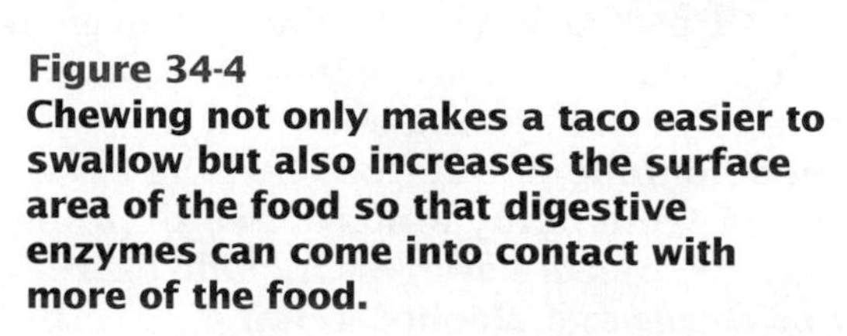

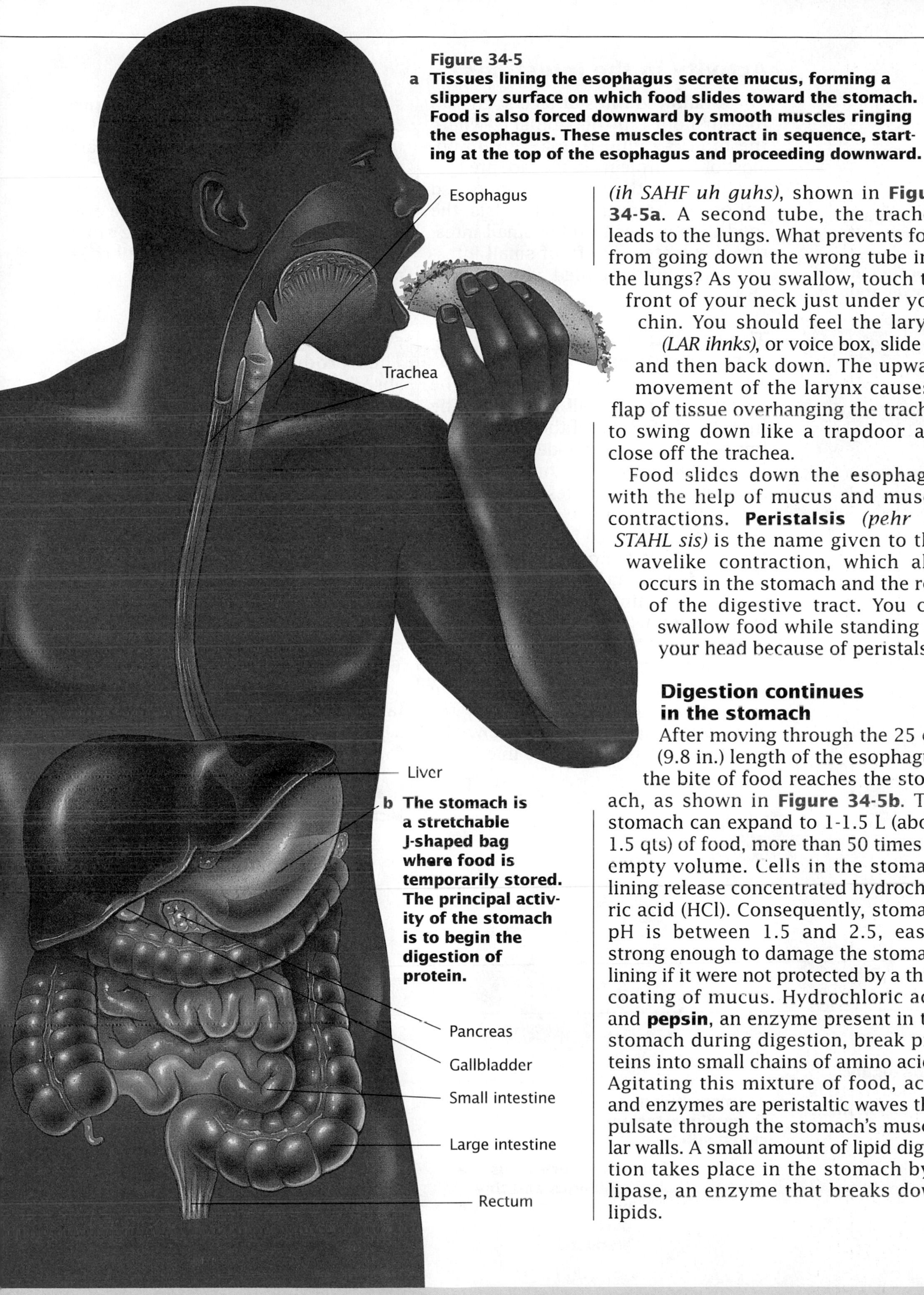

Figure 34-5

a Tissues lining the esophagus secrete mucus, forming a slippery surface on which food slides toward the stomach. Food is also forced downward by smooth muscles ringing the esophagus. These muscles contract in sequence, starting at the top of the esophagus and proceeding downward.

b The stomach is a stretchable J-shaped bag where food is temporarily stored. The principal activity of the stomach is to begin the digestion of protein.

(ih SAHF uh guhs), shown in **Figure 34-5a**. A second tube, the trachea, leads to the lungs. What prevents food from going down the wrong tube into the lungs? As you swallow, touch the front of your neck just under your chin. You should feel the larynx *(LAR ihnks)*, or voice box, slide up and then back down. The upward movement of the larynx causes a flap of tissue overhanging the trachea to swing down like a trapdoor and close off the trachea.

Food slides down the esophagus with the help of mucus and muscle contractions. **Peristalsis** *(pehr uh STAHL sis)* is the name given to this wavelike contraction, which also occurs in the stomach and the rest of the digestive tract. You can swallow food while standing on your head because of peristalsis.

Digestion continues in the stomach

After moving through the 25 cm (9.8 in.) length of the esophagus, the bite of food reaches the stomach, as shown in **Figure 34-5b**. The stomach can expand to 1-1.5 L (about 1.5 qts) of food, more than 50 times its empty volume. Cells in the stomach lining release concentrated hydrochloric acid (HCl). Consequently, stomach pH is between 1.5 and 2.5, easily strong enough to damage the stomach lining if it were not protected by a thick coating of mucus. Hydrochloric acid and **pepsin**, an enzyme present in the stomach during digestion, break proteins into small chains of amino acids. Agitating this mixture of food, acid, and enzymes are peristaltic waves that pulsate through the stomach's muscular walls. A small amount of lipid digestion takes place in the stomach by a lipase, an enzyme that breaks down lipids.

Activity in the Intestines

The swallowed bites of taco spend up to four hours in the stomach, where the food is reduced to a thin liquid. Most of the proteins in the taco are broken down in the stomach. Squeezed out of the stomach by peristalsis, the food then moves into the small intestine. Six meters (20 ft) of small intestine are tightly coiled within your abdomen.

The small intestine absorbs nutrients

The rest of digestion occurs in the entrance to the **small intestine**. The first 25 cm (9.8 in.) of the small intestine is known as the duodenum *(doo uh DEE nuhm)*, and it is here that the remaining carbohydrates and proteins are broken down. In adults, almost all lipids are digested in the small intestine. As in the stomach, peristalsis in the duodenum churns its contents.

The remaining 90 percent of the small intestine has a critical function: it absorbs the nutrients released by digestion and transfers them to the blood. Water is also absorbed. The small intestine is well adapted for absorption. In a cutaway view, the intestine looks as if it were lined with shag carpeting, as shown in **Figure 34-6c**. The projections of the intestinal lining are **villi** (singular, villus). Each villus is coated with hairlike projections known as **microvilli**. With villi and microvilli, the total surface area available for absorption in the small intestine is about 300 m^2 (3,229 ft^2)—an area larger than a tennis court.

The pancreas and liver secrete digestive enzymes

The products of digestion are absorbed into the bloodstream during the three to six hours that food spends in the small intestine. Enzymes and chemicals secreted by the upper end of the small intestine cause additional food breakdown. Most of these secretions come from two organs, the **liver** and the **pancreas**. In Chapter 31 you studied the role of the pancreas in regulating blood sugar. The pancreas also secretes a variety of digestive enzymes. These enzymes break down carbohydrates into simple sugars. They also split proteins into amino acids, and fats into short chains of fatty acids. In addition, the pancreas secretes bicarbonate (the chemical found in antacids and baking soda),

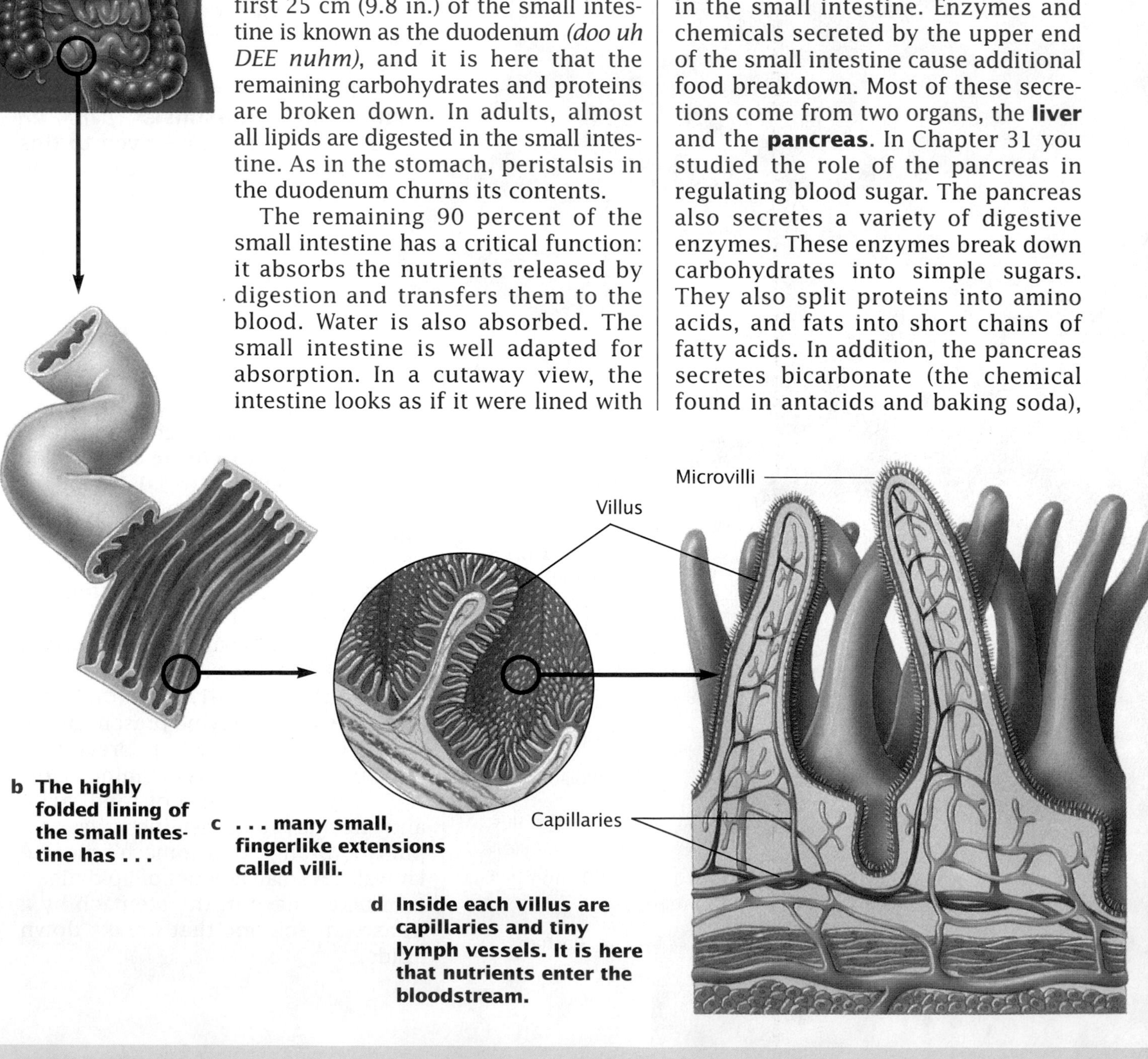

Figure 34-6
a The small intestine is located in the lower abdomen.

b The highly folded lining of the small intestine has . . .

c . . . many small, fingerlike extensions called villi.

d Inside each villus are capillaries and tiny lymph vessels. It is here that nutrients enter the bloodstream.

which helps neutralize the stomach acids so that lipid digestion can proceed and so that the acids do not digest the walls of the small intestine. When too much stomach acid is produced, the pancreas cannot form enough bicarbonate to compensate and acid begins to eat away the lining of the small intestine or stomach. Stress and certain foods (alcohol and coffee) or medications (aspirin) can cause the stomach to secrete excess acid.

Recall from Chapter 3 that fats are not soluble in water. Fat molecules avoid water and form small clusters or globules in the stomach. Your liver secretes bile, a greenish fluid that emulsifies fat globules so that they can be broken down and absorbed. The gall-bladder is a green muscular sac, attached to the liver, that stores bile until it is needed in the small intestine.

Figure 34-7

a The liver, a football-sized organ, is located above and to the right of the stomach, just below the diaphragm.

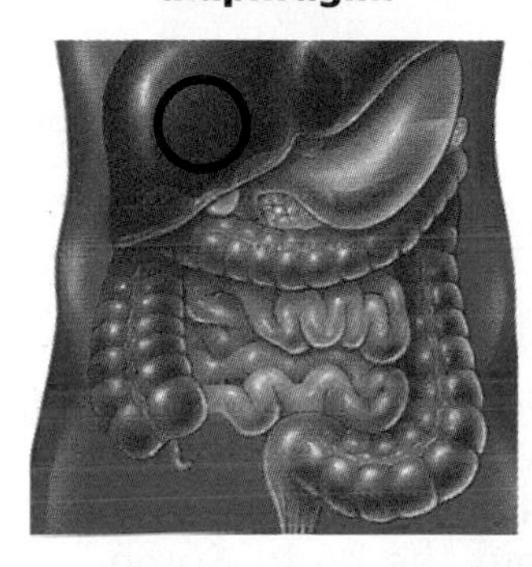

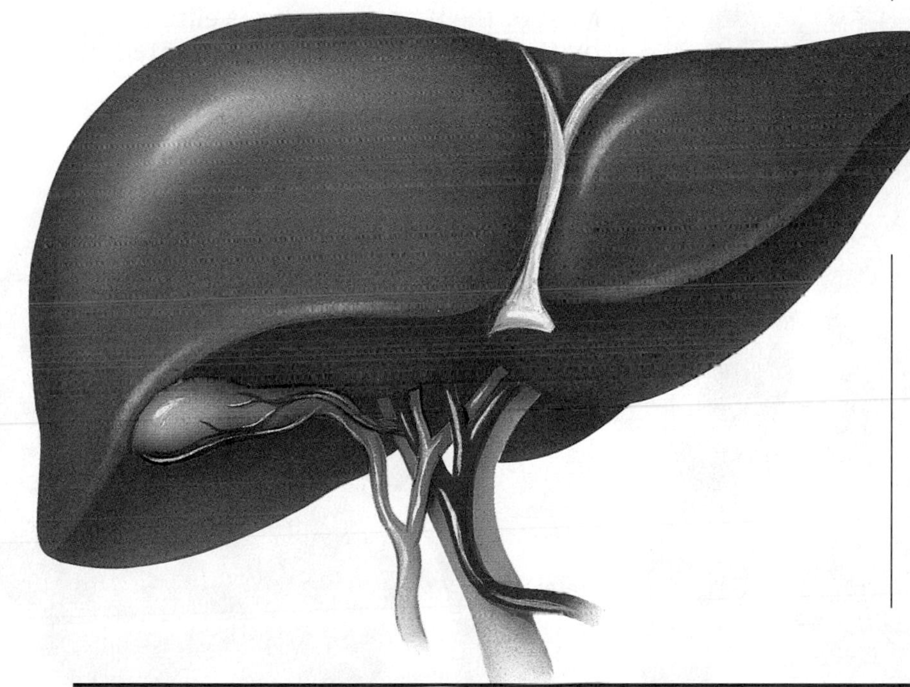

b The liver produces bile, stores nutrients, regulates the level of blood sugar, and filters the blood.

Nutrients are collected in the liver

After digested food molecules are absorbed into the bloodstream, they go first to the liver. The liver, shown in **Figure 34-7**, then sends these molecules into the rest of the bloodstream as needed. For example, when you eat a meal, the liver removes excess glucose from the blood and stores it as glycogen. When the level of glucose in the blood falls, glucagon causes the liver to release some of this glucose back into the blood. The liver also detoxifies many substances. For instance, it produces enzymes that break down drugs and alcohol.

The large intestine compacts wastes

Your lunch-time taco now is largely undigestible material such as cellulose. From the small intestine, what is left of the food now moves into the **large intestine.** The large intestine's role is to briefly store and compact this undigestible material and then, with the help of adequate dietary fiber and water, eliminate it. Some nutrients, including water, sodium, and vitamin K, are absorbed, and a number of B-vitamins and vitamin K are made in the large intestine. These vitamins are produced by the billions of beneficial bacteria living in the large intestine. In fact, the bacteria in the large intestine may play an important role in nourishing their host and in resisting disease.

After traveling the 1.5 m (5 ft) length of the large intestine, what remains of the taco passes out through the rectum and anus as feces. Feces contain undigested food, such as cellulose, and bacteria.

SECTION REVIEW

1. **What organs are needed to digest food?**
2. **Which nutrient is digested mainly in the stomach? in the small intestine?**
3. **How is the small intestine specialized for absorption?**
4. **How would digestion be affected if the pancreas or liver were damaged?**

Regulating the concentrations of substances in your body's fluids is essential. A slight rise in potassium levels in the blood can cause heart failure. But this rarely happens because the kidneys maintain the level of potassium and other substances within narrow limits. Kidneys also serve as filters to remove nitrogen-containing wastes from the blood.

34-3 The Excretory System

OBJECTIVES

1. **Explain how the kidneys determine what leaves the blood and what remains.**
2. **Describe the structure of a kidney nephron.**
3. **Compare and contrast the two stages of urine formation.**
4. **Explain how medical technology has helped people with kidney disorders.**

Kidney Form and Function

When the body breaks down amino acids, metabolic wastes—especially nitrogen compounds in the form of ammonia—are released. In Chapter 24 you read that land animals must rid their bodies of ammonia and other nitrogen-containing wastes. The body is able to make ammonia less poisonous by combining it with carbon dioxide in the liver to form a less toxic compound called urea. Urea enters the bloodstream and circulates throughout the body. Some urea is eliminated from the body through the skin as perspiration, which is a mixture of water, minerals, and urea. Most of the urea, however, is eliminated by the excretory system, illustrated in **Figure 34-8**.

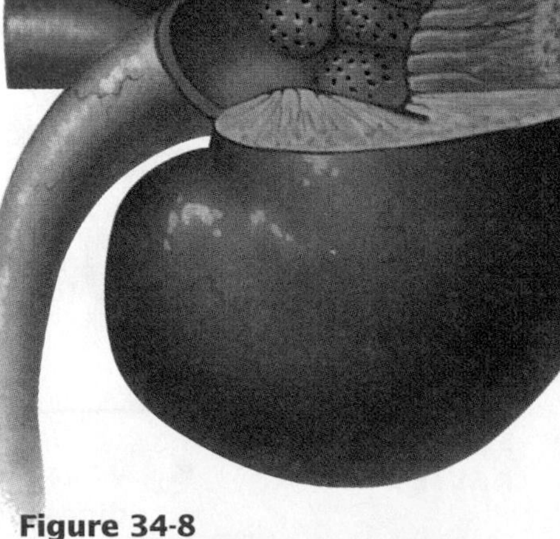

Figure 34-8
The kidneys filter out toxins, urea, water, and mineral salts from the blood. These substances are then stored in the bladder as a liquid called urine until they are eliminated.

The kidneys clean the blood

Your **kidneys** are fist-sized organs located toward the back of your abdomen, about even with the bottom of the rib cage. The kidneys play a vital role in maintaining homeostasis; they remove urea and other wastes, regulate the amount of water in the blood, and adjust the concentrations of various substances in the blood. Because of their crucial function, the kidneys receive large amounts of blood. About one-fourth of the blood that your heart pumps every minute travels to the kidneys for cleaning.

Nephrons carry out the processes that form urine

Each kidney contains over 1 million blood-cleaning units called **nephrons** *(NEHF rahns)*. A nephron, shown in **Figure 34-9**, is a tiny tube with a cup-shaped capsule at one end. The capsule surrounds a tight ball of capillaries. Substances in the blood inside the capillaries filter into the capsule. The substances then travel through the twists and loops in the nephron, forming urine, an amber-colored fluid that contains nitrogenous waste products and excess amounts of water and solutes.

Urine Formation

The first stage of urine formation is called **filtration**, as shown in **Figure 34-9a**. During filtration, the blood cells, proteins, and other large solutes remain in the blood, while smaller solutes (such as glucose, salts, amino acids, and urea) and water are forced out of the blood.

Each day, about 180 L of fluid filter out of the blood and into the nephrons. Most of this fluid is absorbed back into the blood, leaving about 1 L of urine every day. The amount and content of urine varies, depending on what has been taken into the body.

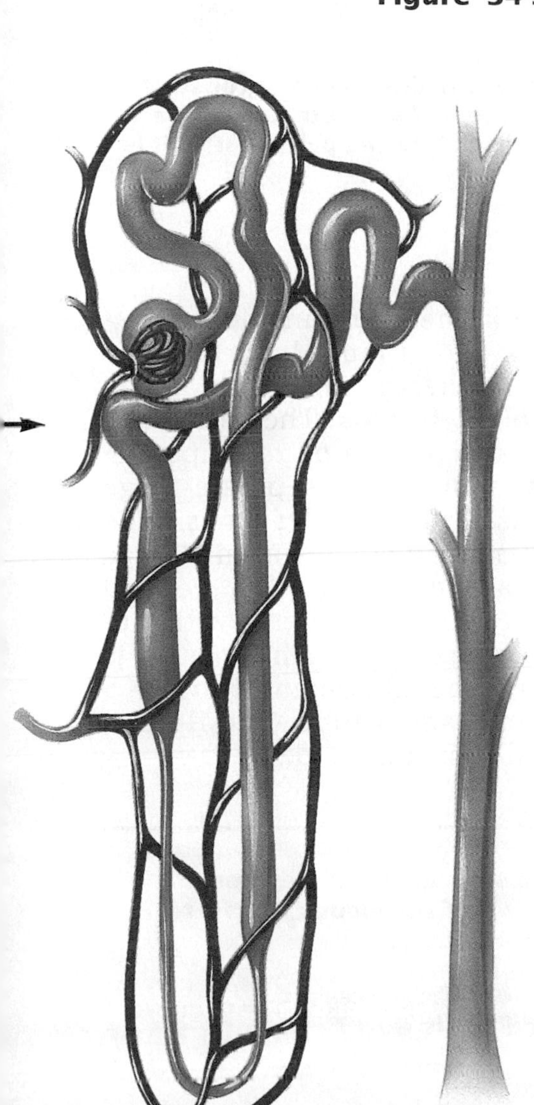

Figure 34-9
a During filtration, blood pressure in the capillaries forces water, glucose, amino acids, and various salts through small pores in the capillaries and into the capsule of the nephron.

b During reabsorption, water and solutes move out of the nephron and into the capillaries.

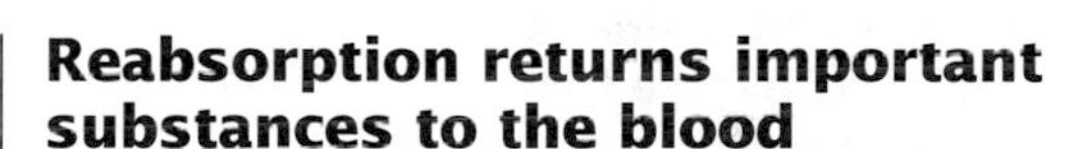

Reabsorption returns important substances to the blood

From the nephron capsule, the filtered fluid passes into the nephron, beginning **reabsorption**, the second process of urine formation, shown in **Figure 34-9b**. About 99 percent of the water, all of the glucose and amino acids, and many of the salts are reclaimed and sent back into your bloodstream. Water reabsorption is controlled by your hypothalamus. When your body needs to conserve water, the hypothalamus triggers the secretion of a hormone called ADH (antidiuretic hormone). ADH makes the ends of nephrons more permeable to water. As a result, more water is reabsorbed, and the urine leaving your body is more concentrated. When your body must get rid of excess water, the secretion of ADH is inhibited, and your urine is more dilute.

After reabsorption, the fluid remaining in the nephron tubes is urine. It is made up of water, urea, and various salts. Urine flows from each kidney into a slender tube called a ureter. It then flows into the urinary bladder, where it is stored. It leaves the body through a tube called the urethra, which leads to the outside of your body.

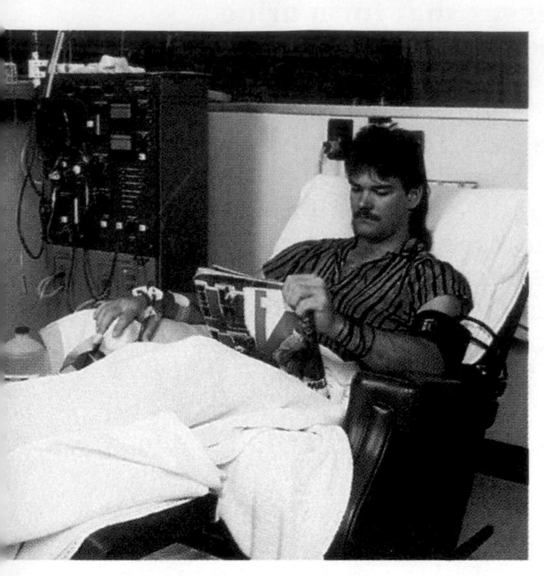

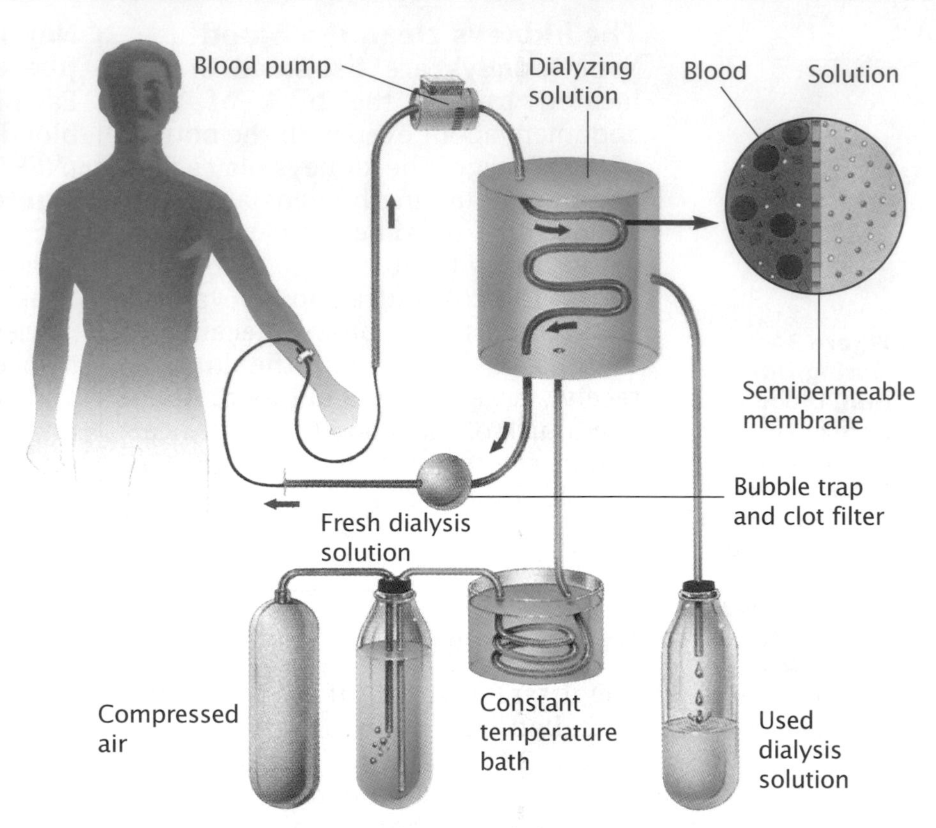

Figure 34-10
The kidney machine is an efficient device for filtering blood.

a Blood in a solution is passed through a membrane, where the wastes are removed.

b The blood is maintained at a constant temperature and pH so that it can be pumped back into the patient after it is filtered.

Kidney Disorders and Treatment

Stability and Homeostasis

Why might high-protein weight-loss diets contribute to the development of kidney stones?

An estimated 13 million people in the United States suffer from kidney disorders. One common disorder is the development of kidney stones. Kidney stones are deposits of uric or oxalic acid, calcium salts, and other substances that collect inside the kidney. The stones may become lodged in the urethra, the tube through which urine leaves the body, where they interfere with urine flow and cause pain. Kidney stones usually pass naturally from the body, but they can be eliminated by medical or surgical procedures.

When kidneys are unable to function, the blood must be artificially filtered. Filtering of the blood is called **hemodialysis**. The term *dialysis* refers to the use of a semipermeable membrane to separate large particles from smaller ones. The kidney machine shown in **Figure 34-10** can be used twice a week to filter the blood of patients with kidney disorders. A dialyzing solution is fed into the abdomen through a catheter, a small tube. The fluid is changed four times a day.

SECTION REVIEW

1. **How do the kidneys help maintain homeostasis?**
2. **How are nephrons specialized for urine formation?**
3. **Which substances do nephrons filter out of the blood? Which are reabsorbed?**
4. **How does the process of hemodialysis work?**

CHAPTER

34 Highlights

Microvilli can be seen with the aid of a scanning electron microscope.

34-1 Nutrition: What You Eat and Why

Eating healthful foods is important for maintaining a healthy body.

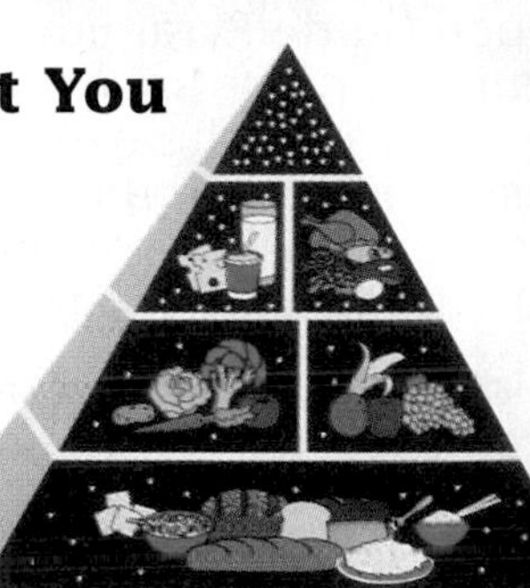

Key Terms

- obesity (p. 699)
- nutrition (p. 699)
- mineral (p. 702)
- vitamin (p. 703)
- anorexia nervosa (p. 704)
- bulimia (p. 704)

Summary

- A healthful diet includes carbohydrates and lipids (for fuel), proteins (for building tissues), vitamins, minerals, and water.
- Too much food and too little activity lead to nutritional problems.
- Anorexia nervosa and bulimia are eating disorders that can cause serious health problems.

34-2 The Digestive System

Digestion is the process of breaking down food so that the body can absorb its nutrients.

Key Terms

- esophagus (p. 706)
- peristalsis (p. 707)
- pepsin (p. 707)
- small intestine (p. 708)
- villi (p. 708)
- microvilli (p. 708)
- liver (p. 708)
- pancreas (p. 708)
- large intestine (p. 709)

Summary

- In the first stage of digestion, the teeth and saliva break down food and begin carbohydrate digestion.
- In the stomach, proteins are broken down by pepsin.
- Digestion is completed in the small intestine, where nutrients are absorbed.
- The large intestine reabsorbs most of the water from the mass of undigested material left after the removal of nutrients in the small intestine.
- The liver stores and regulates the levels of food molecules in the blood.

34-3 The Excretory System

The kidneys are the filtering organs in the body, removing urea and other wastes from the blood.

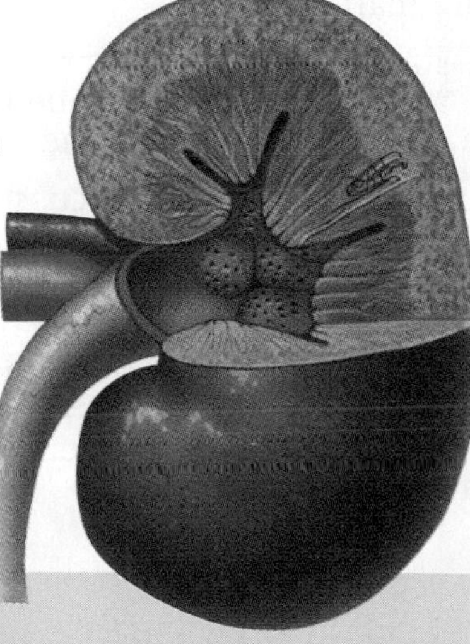

Key Terms

- kidney (p. 711)
- nephron (p. 711)
- filtration (p. 711)
- reabsorption (p. 711)
- hemodialysis (p. 712)

Summary

- The functional unit of the kidney is the nephron.
- During filtration, water, glucose, amino acids, salts, and urea move out of the capillary and into the nephron. During reabsorption, water, glucose, amino acids, and many salts move out of the nephron and back into the bloodstream.
- Some kidney disorders can be managed with hemodialysis, a process that uses a semipermeable membrane to simulate the filtering action of the kidney.

CHAPTER 34

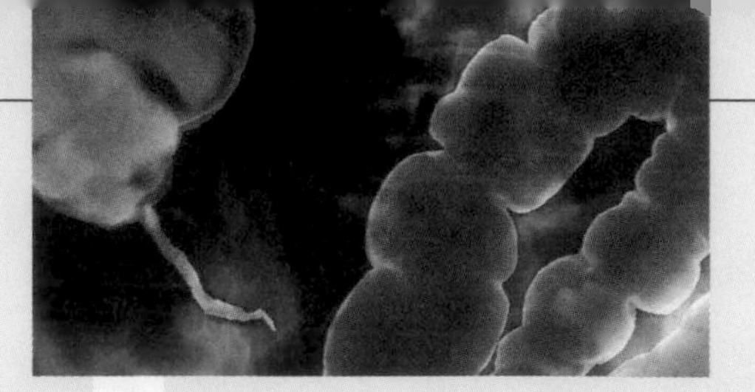

review

Understanding Vocabulary

1. For each set of terms, complete the analogy.
   a. inorganic compounds : minerals :: organic substances : ____
   b. bicarbonate : pancreas :: bile : ____
   c. stomach and small intestine : digestion :: kidneys and large intestine : ____
2. For each pair of terms, explain the difference in their meanings.
   a. fuel nutrients, regulatory nutrients
   b. fat-soluble vitamins, water-soluble vitamins
   c. filtration, reabsorption

Understanding Concepts

3. **Relating Concepts** Construct a concept map that illustrates how the essential fuel nutrients are digested by the body. Include the following terms in your concept map: carbohydrates, proteins, fats, fiber, enzymes, saliva, mouth, pancreas, pepsin, HCl, stomach, mucus, bile, liver, small intestine, and large intestine.
4. **Identifying Relationships** What are the two major factors leading to nutritional problems today in the United States? What health problems are caused by these factors?
5. **Reading Graphs** Look at the pie chart below, showing the percentages of fuel nutrients recommended for our daily diet. What is incorrect in the chart?

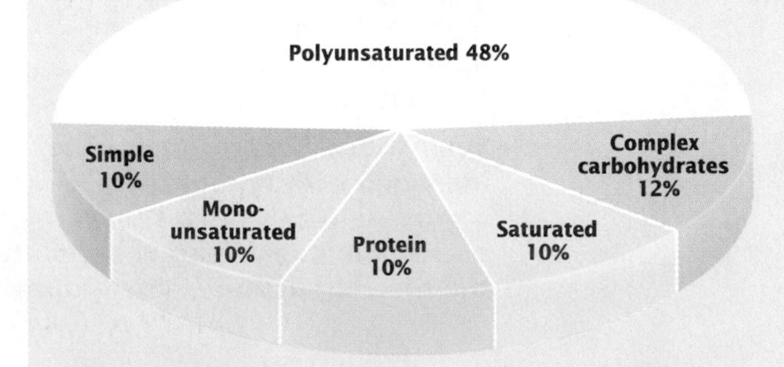

6. **Summarizing Information** List two ways that the body uses water, carbohydrates, fats, proteins, minerals, and vitamins.
7. **Summarizing Information** Draw the USDA Food Guide Pyramid, which serves as a guide for healthy eating. List the food groups at each level of the pyramid, the essential nutrients provided, and how many servings from each food group should be eaten daily.
8. **Distinguishing Relevant Information** How could you use the information from a food label to help you meet the food pyramid's recommendations?
9. **Inferring Relationships** Why must vegetarians plan their diets carefully?
10. **Inferring Relationships** Describe the physiological and metabolic changes that are caused by anorexia and bulimia.
11. **Identifying Functions** Describe the mechanical and chemical aspects of digestion that take place in the mouth.
12. **Summarizing Information** Describe the main organs of the digestive system, their location, and their function.
13. **Understanding Structures** Astronauts in space are often seen eating upside down. How is it possible for them to swallow and digest food in this inverted position?
14. **Identifying Structures and Functions** Explain how the structure of the small intestine enables it to absorb water and nutrients.
15. **Summarizing Information** How do the kidneys help to maintain homeostasis within the body? How does sweating contribute to elimination?
16. **Identifying Structures** Describe the structure of a nephron, and describe its involvement in the production of urine.
17. **Inferring Relationships** Discuss the similarities between natural and artificial dialysis.
18. **BUILDING ON WHAT YOU HAVE LEARNED** In Chapter 24 you learned about the different forms of nitrogen-containing wastes excreted by organisms. Why would it be a problem for humans to excrete ammonia rather than urea?

Interpreting Graphics

19. Name the identified parts of the digestive tract, and describe the function of each.

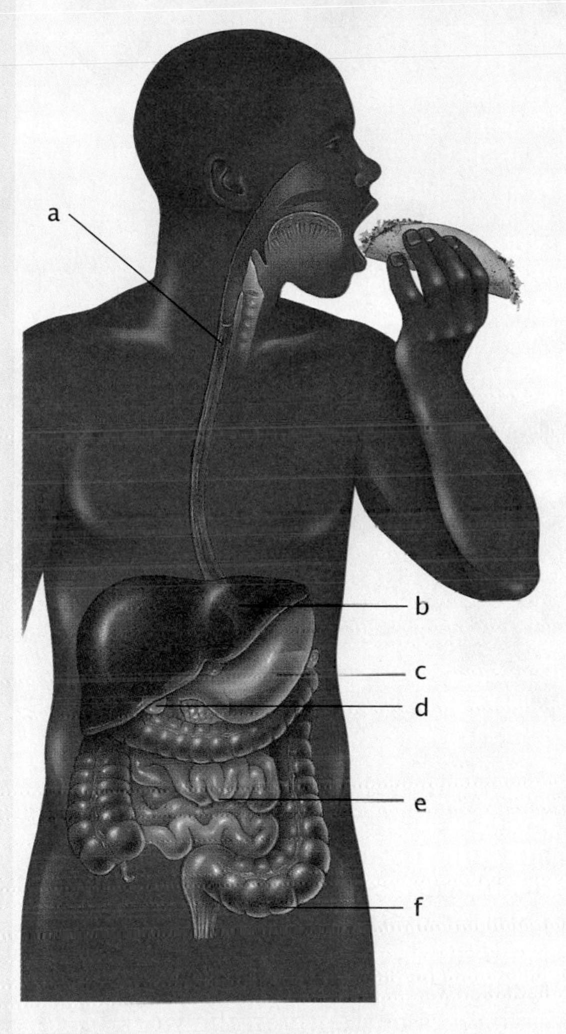

Reviewing Themes

20. ***Cell Structure and Function***
Both chemical and mechanical processes are involved in digestion. Identify a chemical and a mechanical process that occur in one location, and describe how each aids digestion.

21. ***Stability and Homeostasis***
How is water balance regulated in the human body?

Thinking Critically

22. **Evaluating Opinions** A friend trying to lose weight for several months now has just declared that he is now a "fruitarian"—one who eats nothing but fruits and fruit juices. Your friend is excited about this "healthy way to lose weight." How would you react to your friend's decision?

Life/Work Skills

23. **Working Cooperatively** Work with a partner to collect labels from packaged and prepared foods. Study the nutritional information provided on the labels. Refer to the USDA Food Guide Pyramid on page 700. Collect enough appropriate labels to represent the proper number of servings from each group for one day. Arrange the labels as meals on a poster, and share your information with your class.

24. **Finding Information** Use library references or search an on-line database to find out about common disorders of the digestive system. Research the causes and symptoms of disorders such as ulcers, gastroenteritis, gallstones, pancreatitis, diverticulosis, and acid reflux. Summarize your findings in a written report.

Activities and Projects

25. **Career Connection: Dietitian** Dietitians are professionals who are trained in nutrition and who use dietary management to help patients maintain good health and prevent or manage disease. They may specialize in research, education, food service, or nutritional care of people of different age groups. Use library references or search an on-line database to find out about the requirements to become a clinical dietitian, a clinical nutritionist, a registered dietitian, and a dietetic assistant. Where do most dietitians work? To find out more, see the *Occupational Outlook Handbook* published by the U.S. Department of Labor and the *Health Careers Guidebook* published by the U.S. Department of Labor and the U.S. Department of Health and Human Services. You might also want to write to the American Dietetic Association at 216 West Jackson Boulevard, Suite 800, Chicago, IL 60606.

Discovering Through Reading

26. Read the article "Children of the Corn" in *Newsweek*, August 28, 1995, pages 60–62. What is the main reason that many teenagers become vegetarians? What is the leading problem facing teenage vegetarians? Could becoming a vegetarian be a warning sign of an eating disorder?

Science, Technology, and Society

Who Pays the Bill for Growing Old?

The population of the United States is growing older, and health-care bills for the elderly increase each year. How can we deal with this aging society?

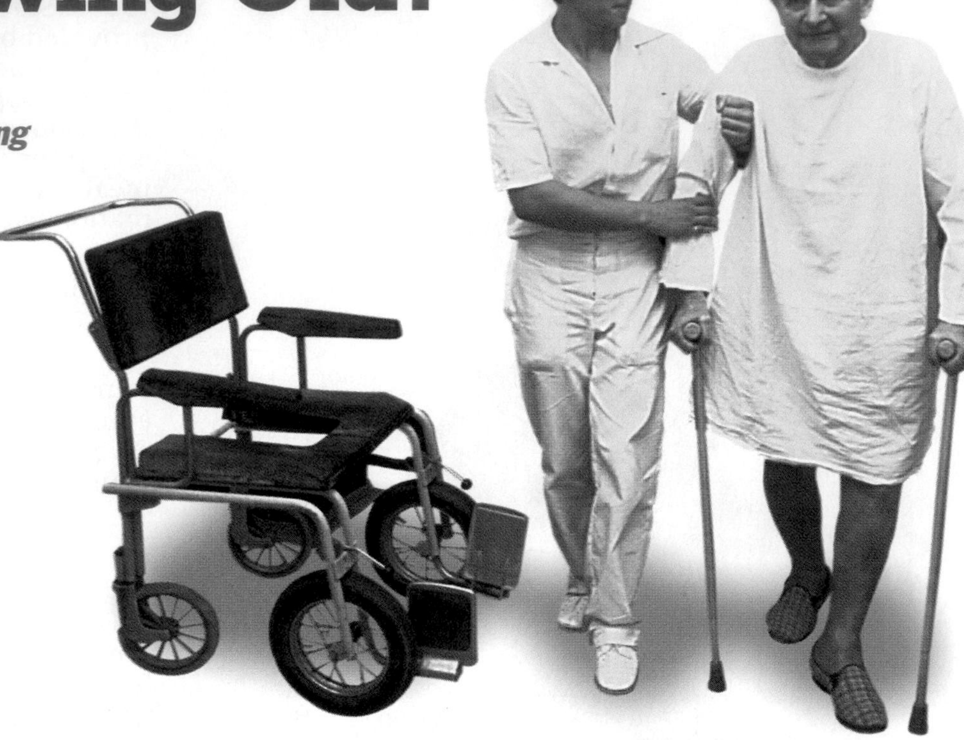

A Mature Society

In 1900, less than 1 percent of our society was over the age of 65. By 1994, that figure had increased to 13 percent, and experts predict that by the year 2050, 25 percent of the population will be over 65. Why is our society aging so rapidly?

In 1900, one-third to one-half of all children died before their first birthday. Life expectancy was only 47 years. In a span of less than 100 years, many of the diseases that plagued humans have been controlled. Advances in medicine as well as improvements in housing, nutrition, and hygiene have increased life spans. Today, life expectancy is 82 years.

Caring for the Elderly

The elderly tend to have more health problems than younger people. Many of these problems are chronic, or long-term, illnesses, such as heart disease, stroke, cancer, and Alzheimer's disease. Many elderly people need professional nursing help or require expensive nursing-home care.

Although the elderly represent only 13 percent of the population, they account for about one-third of all health-care spending. Much of this is paid for by the federal government from funds such as Medicare and Medicaid. This money is supplied by taxpayers currently in the work force.

The High Cost of Health Care

Health-care costs have risen dramatically in the past few years. One reason is the use of high-tech procedures for diagnosing and treating illnesses. These procedures are perceived as the best in medical care, but they are not always cost effective (that is, they are not significantly better than less expensive alternatives). Today more than ever, people view high-quality health care as a right.

Quantity Versus Quality

The common use of life-extending procedures is an indication of how far medical discoveries have come. Even though doctors can help a person live longer, the quality of that life may be poor; a cancer patient may live two years longer but be in constant pain. Some feel we need to improve the quality of life rather than just extend it. Other recent studies seem to indicate that disease and disability rates are declining and that every year there is a smaller percentage of people unable to care for themselves.

Cutting the Costs

Many solutions have been proposed for reducing the cost of health care, especially for the elderly. Ideas include the following:

- providing more funds for in-home care
- preventing illnesses rather than just curing them
- using alternative, low-tech (and hence low cost) treatments

Technique

Computer Tomography (CT)
Many sophisticated, high-tech procedures are used for the diagnosis of illnesses rather than for treatment. One of these techniques, called computer tomography (CT), can find tumors in the body too small to be detected by other means. CT was such a major development in diagnostics that its inventors were awarded the Nobel Prize in 1979.

CT uses a combination of dyes, X rays, and computer images. The X-ray camera rotates completely around the patient to take pictures at all angles. A computer analyzes these pictures and creates a three-dimensional image of the region. If a tumor is found, the computer can measure its exact size, shape, and location.

The cost of CT rose by more than 400 percent in just a decade, making CT scans very expensive.

Applications: detecting small tumors in various organs; diagnosing tissue damage from head injuries, strokes, and arthritis; monitoring the effectiveness of treatments

Users: surgeons, oncologists (doctors who treat cancer patients), and X-ray technicians

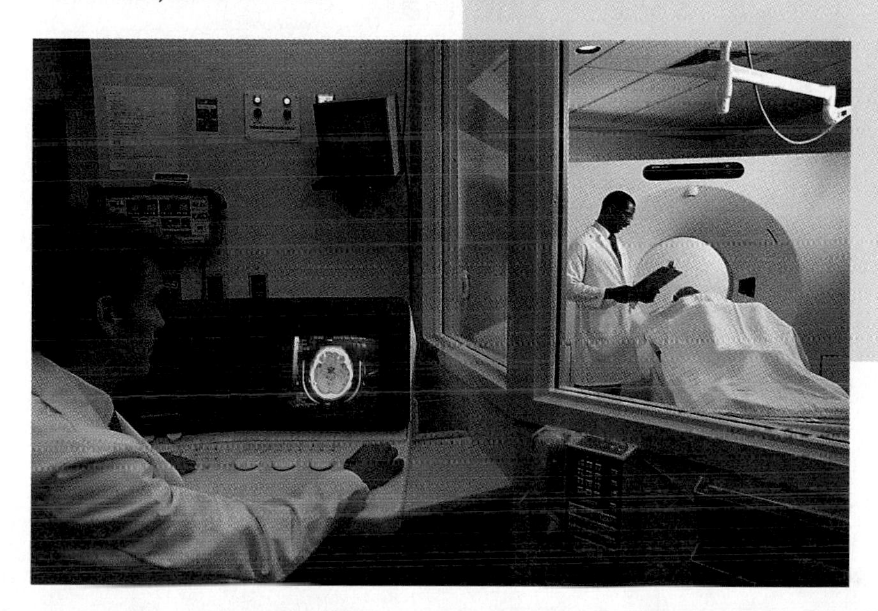

Analyzing the Issues

What is a hospice?

1 Research the definition and history of the hospice movement. What services are offered by a hospice? What is the philosophy behind hospices? Present your findings to the rest of the class.

What solutions are there for funding health care?

2 The largest costs for the elderly are due to long-term care. Research the different solutions proposed for financing these health-care costs. What are some of the options? Describe the benefits and shortcomings of each option.

Does growing older mean growing sicker?

3 Read "The Oldest Old" by Thomas T. Perls in the January 1995 issue of *Scientific American*, pages 70–75. What does the author mean by "the oldest old," and what does he say about their health? How might the increasing numbers of the oldest old affect health-care costs in the future?

How can you increase your chances of staying healthy?

4 Doctors estimate that 70 percent of the illnesses they treat, such as lung cancer and heart disease, are preventable. Research lifestyles and habits that can help extend life span. Summarize your findings in a written report.

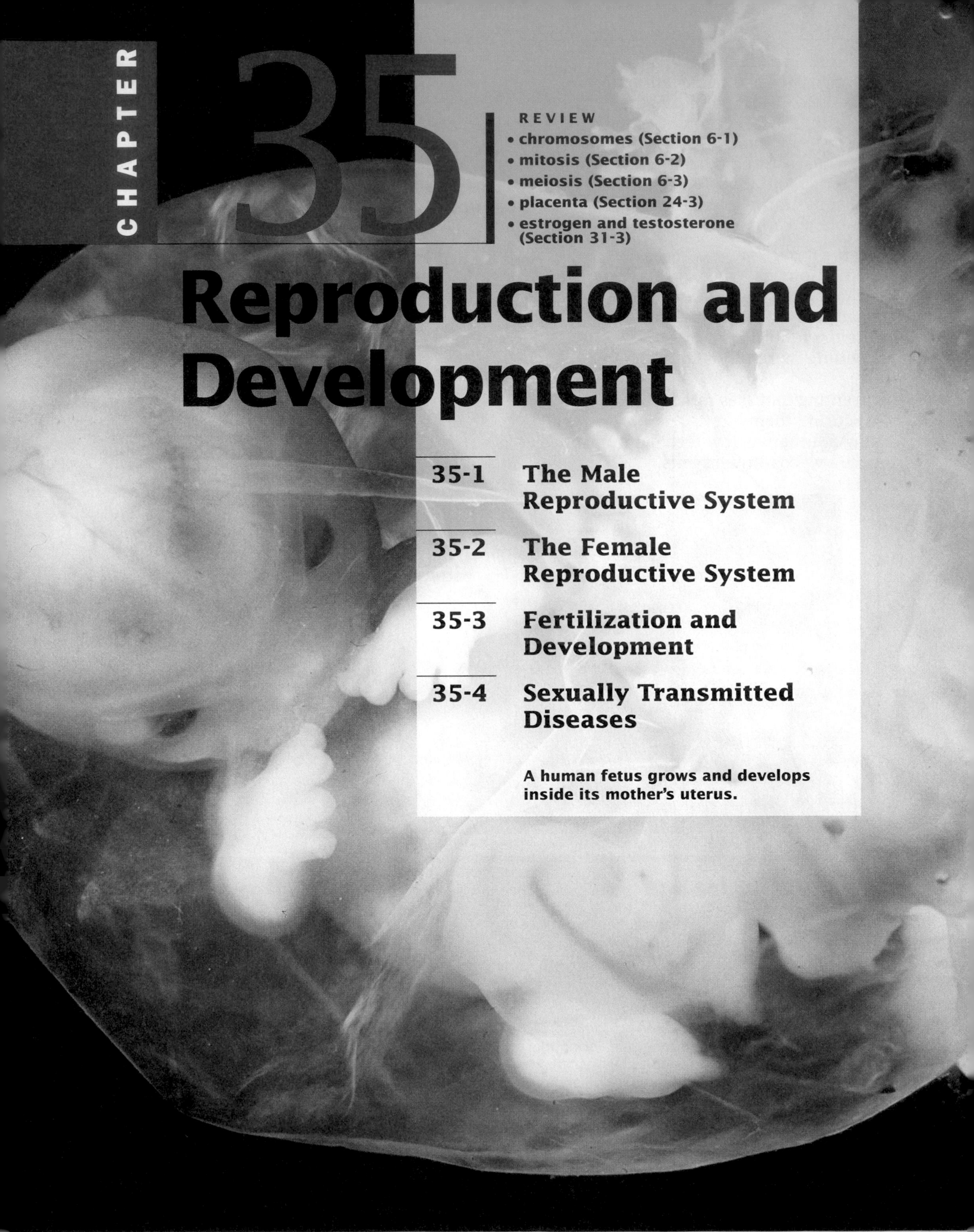

CHAPTER 35

REVIEW
- chromosomes (Section 6-1)
- mitosis (Section 6-2)
- meiosis (Section 6-3)
- placenta (Section 24-3)
- estrogen and testosterone (Section 31-3)

Reproduction and Development

A human fetus grows and develops inside its mother's uterus.

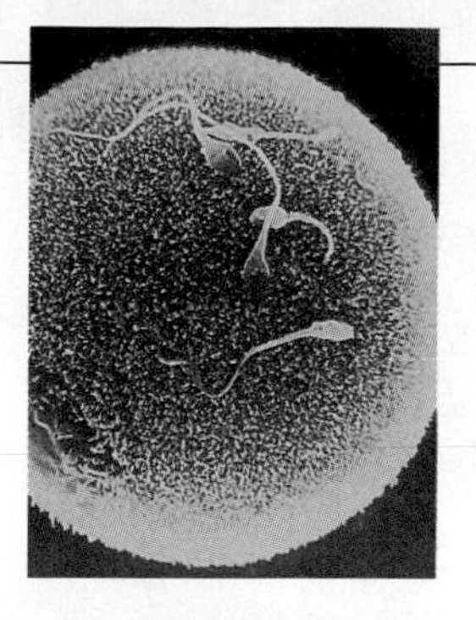

Both males and females produce specialized sex cells called gametes, which contain one-half of the genetic information needed to form a new individual. The role of the male reproductive system is to produce sperm, the male gametes, and to deliver them to the female gamete, the egg, so that fertilization can occur.

35-1 The Male Reproductive System

OBJECTIVES

1. **Identify two features of sperm cells.**
2. **Trace the pathway of sperm from the testes to the outside of the body.**
3. **Describe how the hypothalamus regulates the functioning of the testes.**

Structure of Sperm

A sexually mature human male produces millions of sperm each day. How are sperm cells different from other cells of the body? Recall from Chapter 6 that sperm are haploid. They have only one set of chromosomes instead of the two sets found in the body's other cells. The cells that give rise to sperm are initially diploid. These cells undergo meiosis, which reduces the number of chromosomes in the developing sperm cell by half.

After meiosis, a sperm matures for several weeks, becoming a highly specialized DNA delivery cell. As you can see in **Figure 35-1**, a sperm has a long tail and has discarded most of its cytoplasm and organelles. Mitochondria remain in the sperm cell, clustering around the top part of the tail. Mitochondria are "motors" for the sperm, providing energy for the tail to whip back and forth. A cap of digestive enzymes forms at the tip of the sperm's head. These enzymes enable the sperm to penetrate the egg.

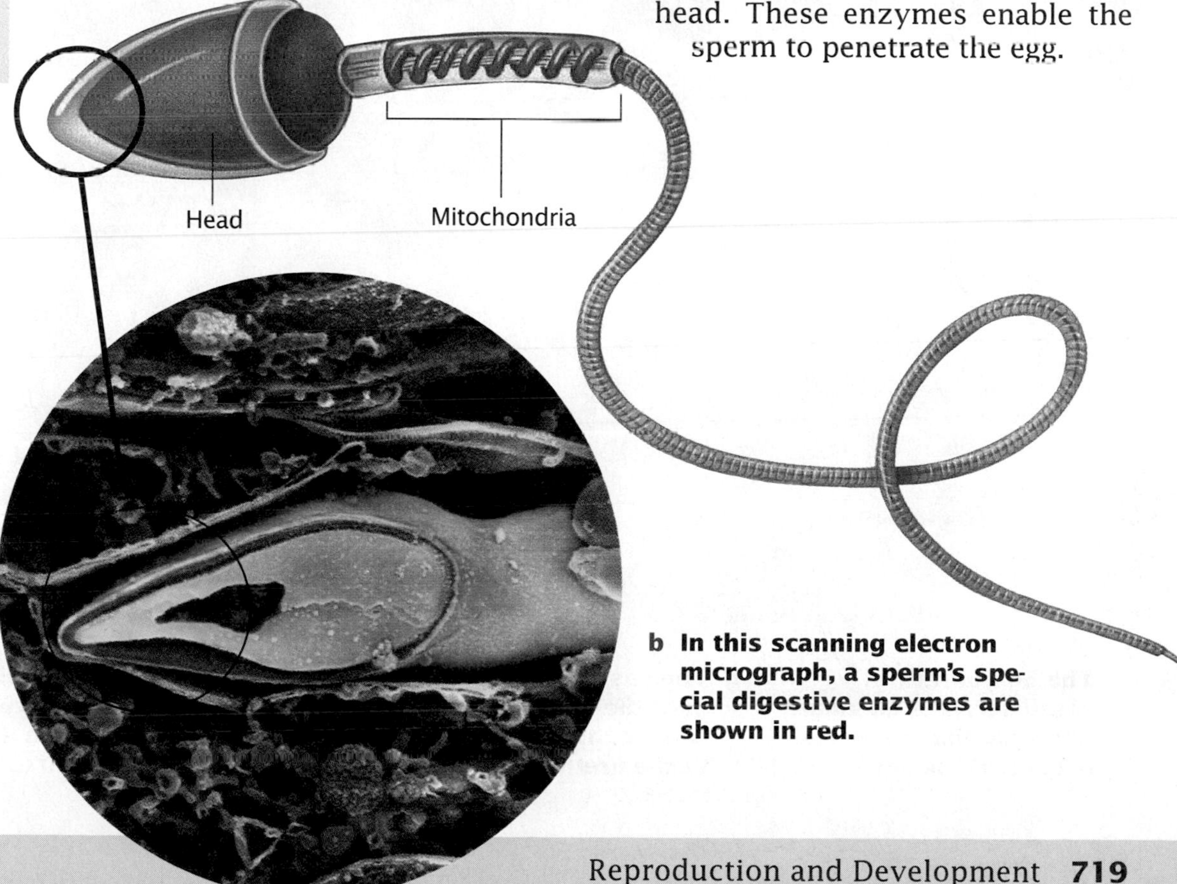

Figure 35-1

a A sperm cell consists of a tail used for locomotion and a head that contains DNA.

b In this scanning electron micrograph, a sperm's special digestive enzymes are shown in red.

The Path Traveled by Sperm

Sperm are produced in two oval-shaped organs called **testes** *(TEHS tees)*, or testicles. The testes are located in the scrotum, a loose sac hanging below the base of the penis. After sperm form, they mature and are stored temporarily in the **epididymis** *(ehp uh DIHD ih mihs)*, shown in **Figure 35-2**. The lower end of each epididymis gradually unwinds to become the **vas deferens** *(vas DEHF uh REHNZ)*.

Figure 35-2
The male reproductive system is a set of organs that produces and delivers sperm.

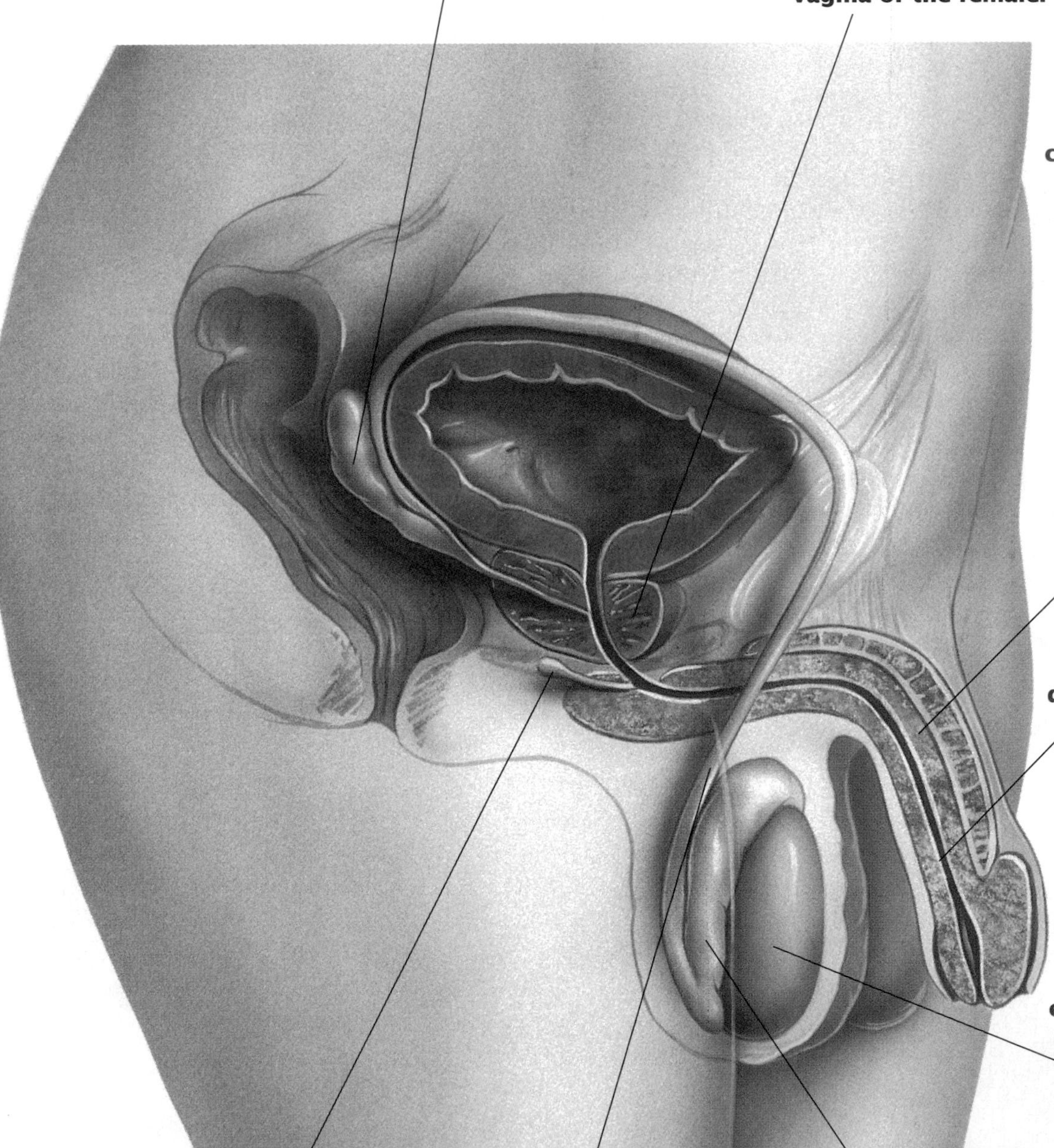

a The seminal vesicles produce fluid that nourishes sperm.
b The prostate gland secretes an alkaline fluid that counteracts the acids found in the male urethra and in the vagina of the female.
c When a male becomes sexually excited, blood accumulates in the penis, causing it to become firm and erect. The firmness of the erect penis makes it possible for the penis to be inserted into the vagina of the female.
d The urethra is the tube that carries urine from the bladder (center of picture) to the outside of the body. It also carries semen during ejaculation.
e Sperm production begins in each of the two testes at puberty and continues throughout life.
f The epididymis is a coiled tube attached to the surface of each testis. If uncoiled, an epididymis would be about 6 m (20 ft) long.
g The vas deferens passes over the urinary bladder, connecting the epididymis to the urethra.
h The bulbourethral gland secretes an alkaline fluid that becomes part of the semen.

For fertilization to occur, sperm first must be transferred from a male to a female. The penis is the male reproductive organ that delivers sperm to the body of the female. **Ejaculation** *(ee jak yoo LAY shuhn)* is the process by which sperm leave the male's body. The walls of the vas deferens contain a thick layer of smooth muscle. During ejaculation, rhythmic contractions of this smooth muscle propel sperm through the vas deferens. Sperm move from the vas deferens into a tube that leads into the urethra. During ejaculation, the tube from the bladder is closed off by a small valve. This valve keeps urine and semen separated and prevents urination during ejaculation. The path that sperm travel from the time they are produced to the time they are expelled during ejaculation is outlined in **Figure 35-3**.

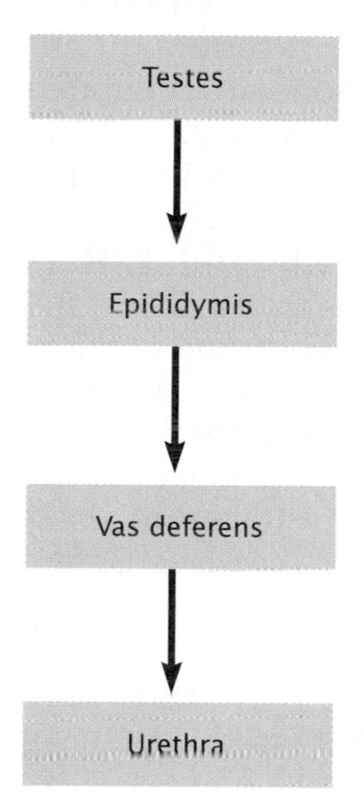

Figure 35-3
After sperm are produced in the testes, they are stored in the epididymis and vas deferens. As they pass out of a male's body during ejaculation, they pass through the vas deferens and urethra.

Several glands add fluids to the sperm

As they travel through the vas deferens and urethra, sperm are bathed in fluids that are produced by glands along the route of travel. The seminal vesicles *(SEHM uh nuhl VEHS ih kuhlz)* are two small glands located near the bladder. The fluid produced by these glands nourishes the sperm. The bulbourethral *(buhl boh yoo REE thruhl)* gland is located just past the prostate *(PRAH stayt)* gland. Both glands secrete a thick, clear, alkaline fluid. These secretions mix with sperm to form **semen**. Semen is expelled from the penis during ejaculation. **Figure 35-4** shows sperm that have reached the uterus of a female.

If sperm cannot reach the egg, fertilization cannot occur. The surest way to prevent a pregnancy is to not have sexual intercourse. Blocking the path traveled by sperm can also prevent fertilization. One way to accomplish this is for a male to wear a latex rubber sheath called a condom whenever the penis is hard. If the condom is used properly and no semen leaks out, fertilization cannot occur because no sperm are deposited in the body of the female. If a male chooses to have no more children, he can have an operation called a vasectomy. During a vasectomy, a small incision is made in the scrotum. Each vas deferens is then cut and the ends are tied. A male who has had a vasectomy is sterile because the sperm's path to the outside of the body is blocked. Sperm are still produced, but they disintegrate in the epididymis.

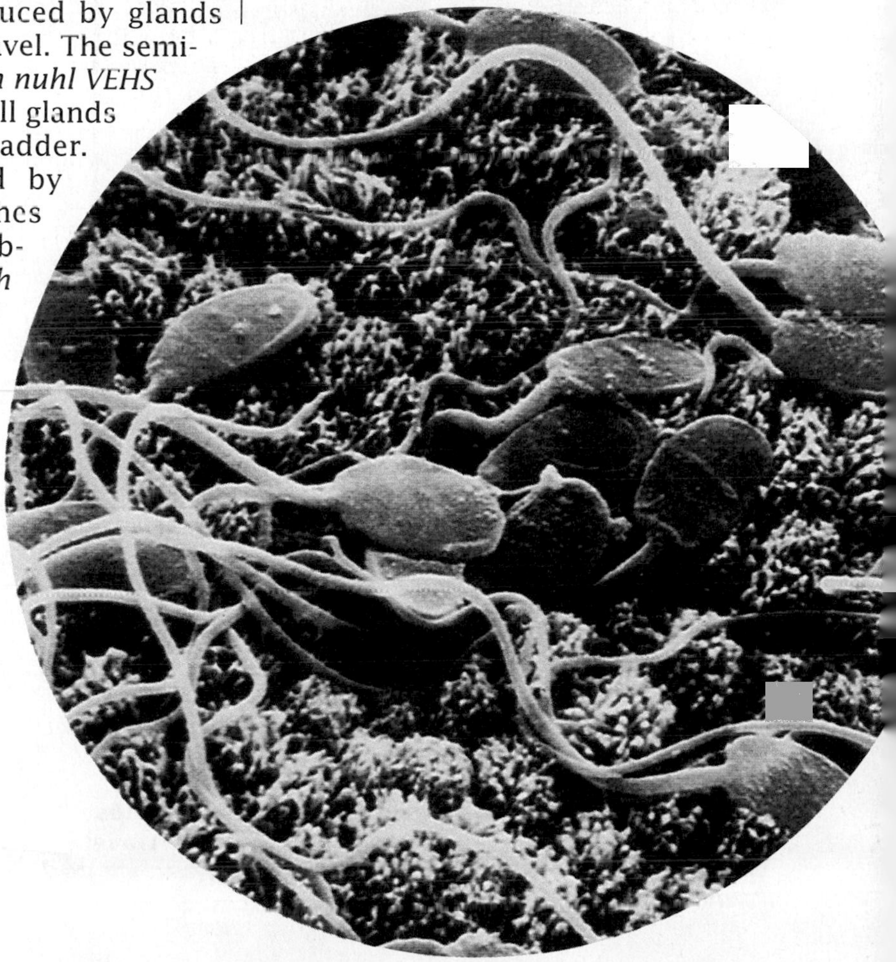

Figure 35-4
As many as 350 million sperm, shown here in the uterus, are released into the vagina of the female during sexual intercourse. Sperm can live for as long as 3–5 days after being deposited within the female.

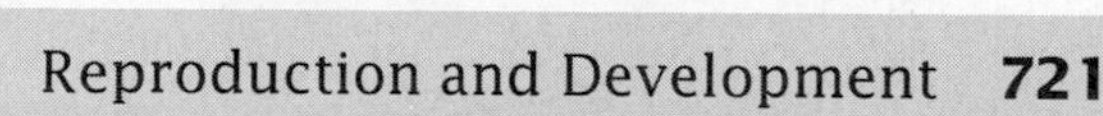

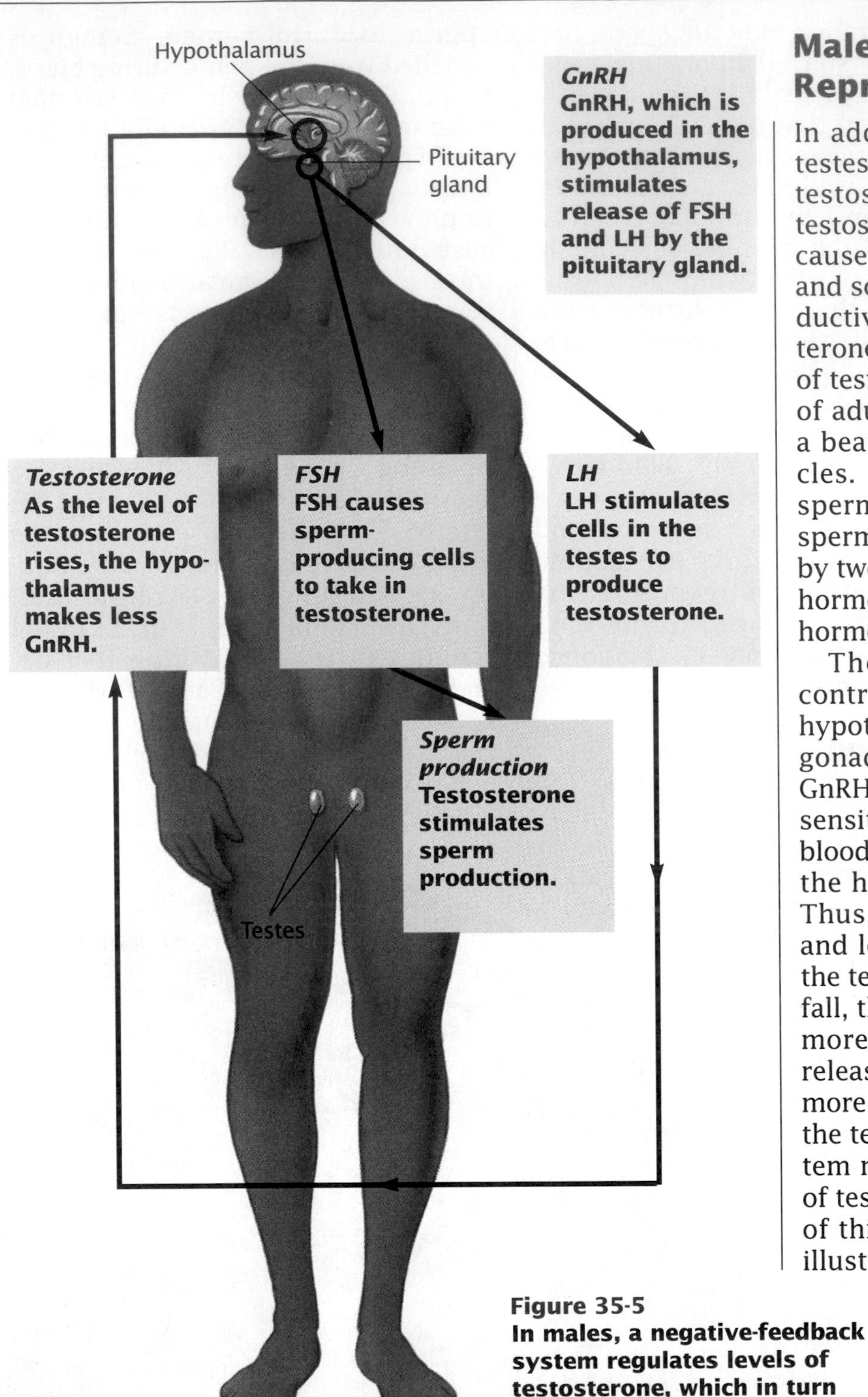

Hypothalamus
Pituitary gland
GnRH
GnRH, which is produced in the hypothalamus, stimulates release of FSH and LH by the pituitary gland.
Testosterone
As the level of testosterone rises, the hypothalamus makes less GnRH.
FSH
FSH causes sperm-producing cells to take in testosterone.
LH
LH stimulates cells in the testes to produce testosterone.
Sperm production
Testosterone stimulates sperm production.
Testes

Figure 35-5
In males, a negative-feedback system regulates levels of testosterone, which in turn controls sperm production.

Male Hormones and Reproduction

In addition to producing sperm, the testes also produce the male hormone testosterone. A male begins to make testosterone before birth. Testosterone causes an embryo to develop a penis and scrotum rather than female reproductive structures. At puberty, testosterone levels rise. These higher levels of testosterone cause the development of adult male characteristics, such as a beard, a low voice, and large muscles. Testosterone also stimulates sperm production. Production of sperm and testosterone are regulated by two pituitary hormones: luteinizing hormone (LH) and follicle-stimulating hormone (FSH).

The release of LH and FSH is controlled by a hormone from the hypothalamus. This hormone is called gonadotropin-releasing hormone, or GnRH for short. The hypothalamus is sensitive to testosterone levels in the blood. If the level of testosterone rises, the hypothalamus makes less GnRH. Thus, less LH and FSH are released, and less testosterone is produced by the testes. As the levels of testosterone fall, the hypothalamus begins to make more GnRH. More LH and FSH are released from the pituitary gland, and more testosterone is then produced by the testes. This negative feedback system maintains a nearly constant level of testosterone in the blood. The steps of this negative-feedback system are illustrated in **Figure 35-5**.

SECTION REVIEW

1. **In what ways are sperm different from the body's other cells?**
2. **What is the function of fluids added to sperm as they travel from the epididymis?**
3. **Describe two functions of the testes.**
4. **What role does the hypothalamus play in sperm production?**

The male contributes one set of chromosomes when the female's egg is fertilized. Once fertilization has occurred, the male's biological role is finished. In contrast, the female carries the developing fetus for about nine months, until birth. The female reproductive system thus must play two biological roles: making gametes and nourishing the fetus.

35-2 The Female Reproductive System

OBJECTIVES

1. **Identify two differences between the male and female reproductive systems.**
2. **Compare and contrast sperm production with egg production.**
3. **Relate the events of the ovarian cycle to the levels of LH, FSH, and estrogens.**
4. **Describe the events of the uterine cycle.**

Structure of the Female Reproductive System

Female gametes are called **ova** (singular, ovum), or eggs. The **ovaries** are the organs responsible for producing eggs. Ovaries are located in the lower part of the abdomen. An egg released from an ovary moves into a **fallopian tube**, which carries the egg into the uterus, as you can see in **Figure 35-6**. The **uterus** is a hollow, muscular organ about the shape and size of a small pear. During pregnancy, the uterus expands to many times its usual size.

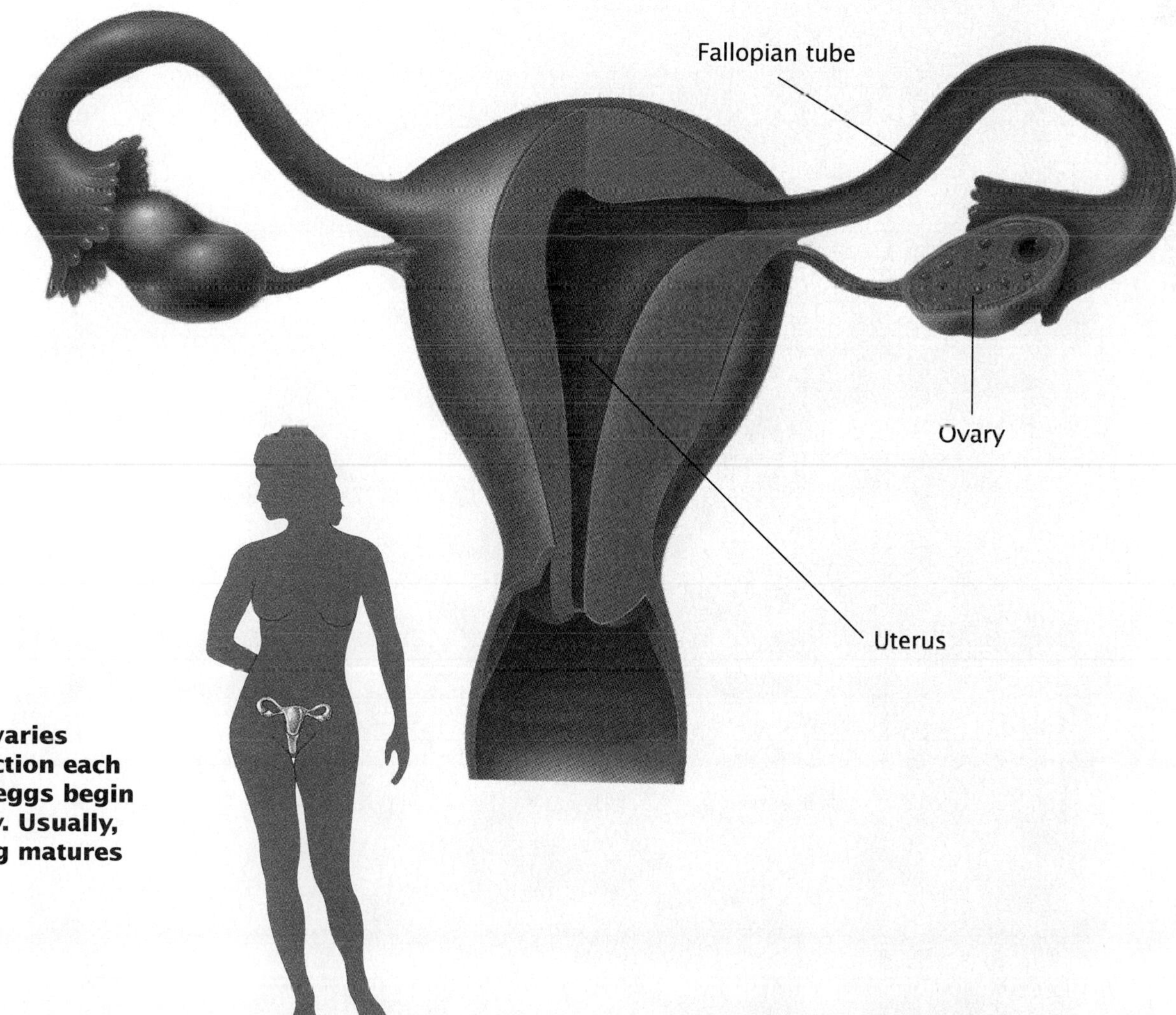

Figure 35-6
After puberty, the ovaries alternate egg production each month, and several eggs begin to grow in one ovary. Usually, though, only one egg matures each month.

The lower part of the uterus is a tubular ring of strong muscles known as the **cervix** *(SER vihks)*. As shown in **Figure 35-7**, the **vagina** *(vuh JEYE nuh)* is a muscular tube that receives the penis and any sperm ejaculated by a male during intercourse. The lower end of the vagina opens to the outside of the body, and the upper end connects to the cervix. During birth, the baby passes through the cervix and the vagina, also called the birth canal.

Pregnancy can be prevented by not allowing sperm to enter the opening in the cervix. A diaphragm is a shallow, round cup made of rubber that is placed inside the vagina to block the entrance to the cervix. A cervical cap is similar to a diaphragm but fits tightly over the cervix. A diaphragm or a cervical cap must be obtained from a doctor or health-care professional. Spermicidal creams, spermicidal jellies, vaginal suppositories, and spermicidal foams work by killing sperm in the vagina so that, in theory, no living sperm can enter the opening in the cervix (*spermicidal* means "sperm-killing"). Usually spermicides are used with a barrier such as a condom, diaphragm, or cervical cap. A barrier and spermicide used together are much more effective in preventing fertilization than either used by itself.

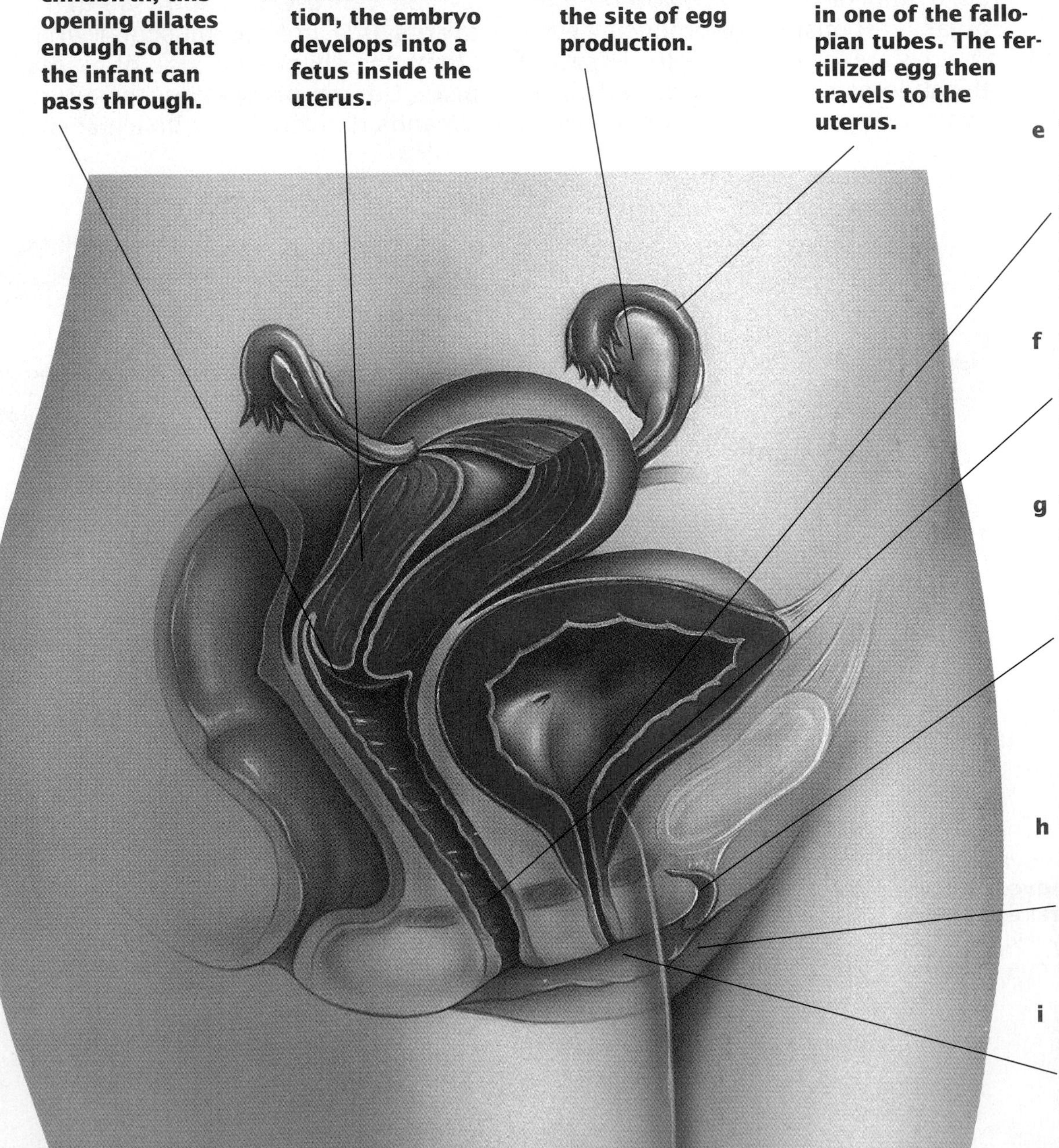

a This end of the cervix has a small opening through which sperm can enter. During childbirth, this opening dilates enough so that the infant can pass through.
b After fertilization, the embryo develops into a fetus inside the uterus.
c Ovaries are the site of egg production.
d Fertilization occurs in one of the fallopian tubes. The fertilized egg then travels to the uterus.
e The urethra carries urine from the bladder to the outside of the body but plays no role in reproduction.
f The vagina connects the uterus to the outside of the body.
g The clitoris is similar to the penis in that it also has the capacity to become erect. It is covered by a hood of skin that is connected to the labia.
h The two larger folds of skin between the thighs are called the labia majora.
i Inside the labia majora are two smaller folds called the labia minora.

Figure 35-7
The female reproductive system releases eggs and provides nourishment to a developing fetus.

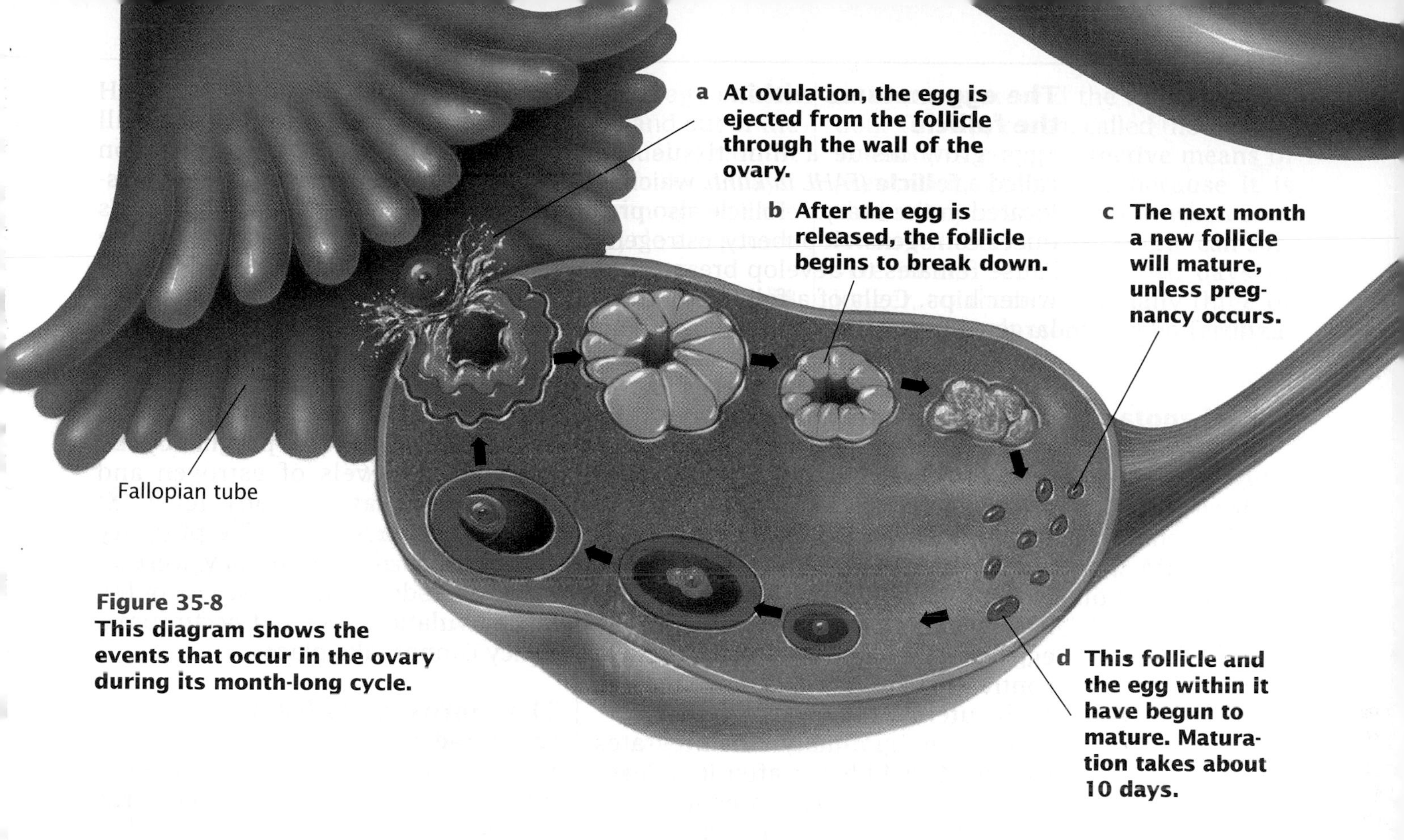

a At ovulation, the egg is ejected from the follicle through the wall of the ovary.
b After the egg is released, the follicle begins to break down.
c The next month a new follicle will mature, unless pregnancy occurs.
Fallopian tube
d This follicle and the egg within it have begun to mature. Maturation takes about 10 days.

Figure 35-8
This diagram shows the events that occur in the ovary during its month-long cycle.

Matter, Energy, and Organization

How are mature sperm and eggs different in size and structure?

The Female Reproductive Cycle

The two reproductive tasks of the female, making gametes and nourishing the fetus, are accomplished in two overlapping cycles—the ovarian cycle and the uterine cycle. Together these cycles allow for the maturation and release of female gametes (eggs) and for accommodation of the egg, whether fertilized or not. A delicate mix of hormones produces these changes in a sequence of events that is known as the **menstrual** *(MEHN struhl)* **cycle**.

The ovarian cycle prepares a mature egg for fertilization

Once a male has reached puberty, his testes continuously produce and release millions of sperm. In a female, however, egg release occurs only at a specific time once a month. The release of an egg from the ovary is called **ovulation** *(ahv yoo LAY shun)*. A female usually releases only one egg each month. If this egg is not fertilized, the ovaries prepare another egg for release the next month. This cycle of activity in the ovaries is called the ovarian cycle.

Eggs begin to mature in the ovaries of a female while she is still in her mother's uterus. Thousands of her eggs grow and begin to undergo meiosis. Growth and meiosis stop before the eggs are completely mature, however. All of the eggs in the ovaries remain in this immature state until puberty. Several immature eggs are shown in **Figure 35-8**.

After puberty, follicle-stimulating hormone (FSH), released by the pituitary, signals several eggs to resume the process of maturation each month. The maturing eggs become large, highly complex cells, growing to be nearly 200,000 times larger than a sperm. Although many eggs begin to mature each month, usually only one egg completes the process and is released, as shown above. If two eggs mature, fraternal, or nonidentical, twins may result.

Before fertilization, a sperm and an egg are separate cells, each containing one-half of the information needed for the development of an embryo. During fertilization, gametes fuse into one cell containing all of the instructions needed to form a new individual. After fertilization, the embryo will grow and develop for about nine months inside its mother's uterus.

35-3 Fertilization and Development

OBJECTIVES

1. **Describe the process of fertilization.**
2. **Explain the process of implantation.**
3. **List three substances that can pass through the placenta.**
4. **Describe three changes that occur as the fertilized egg develops into a new individual.**

Fertilization: Fusion of Gametes

During ejaculation, 150 million to 350 million sperm are deposited just a few inches away from the fallopian tubes, in which **fertilization** can occur. Then a long and difficult journey begins. To reach the egg, a sperm must first pass through the small opening in the cervix. Then it must cross the uterus and swim up one of the fallopian tubes. The vast majority of ejaculated sperm never leave the vagina. Of the sperm that find the small opening to the cervix, many will be trapped in mucus that covers this opening. Over the next two days, sperm will gradually free themselves from this mucus and start their journey across the uterus. Even if a mature egg is waiting in one of the fallopian tubes, half of the sperm will swim up the wrong tube.

Although only one sperm can fuse with the egg, many sperm can arrive at the egg at about the same time. The egg is surrounded by a protective layer of smaller cells. Enzymes in the heads of the sperm cells help loosen this layer of cells. Once a sperm reaches the surface of the egg, as shown in **Figure 35-11**, it leaves its tail behind. The genetic information inside the head of the sperm enters the egg and fuses with the genetic information inside the egg. A protective shield then forms around the egg to prevent other sperm from entering.

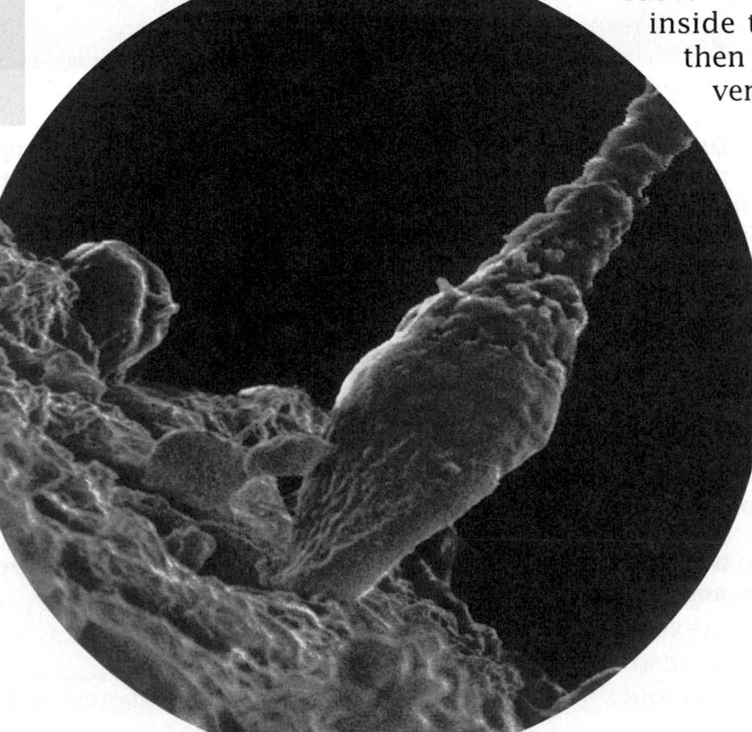

Figure 35-11 This sperm is penetrating an egg.

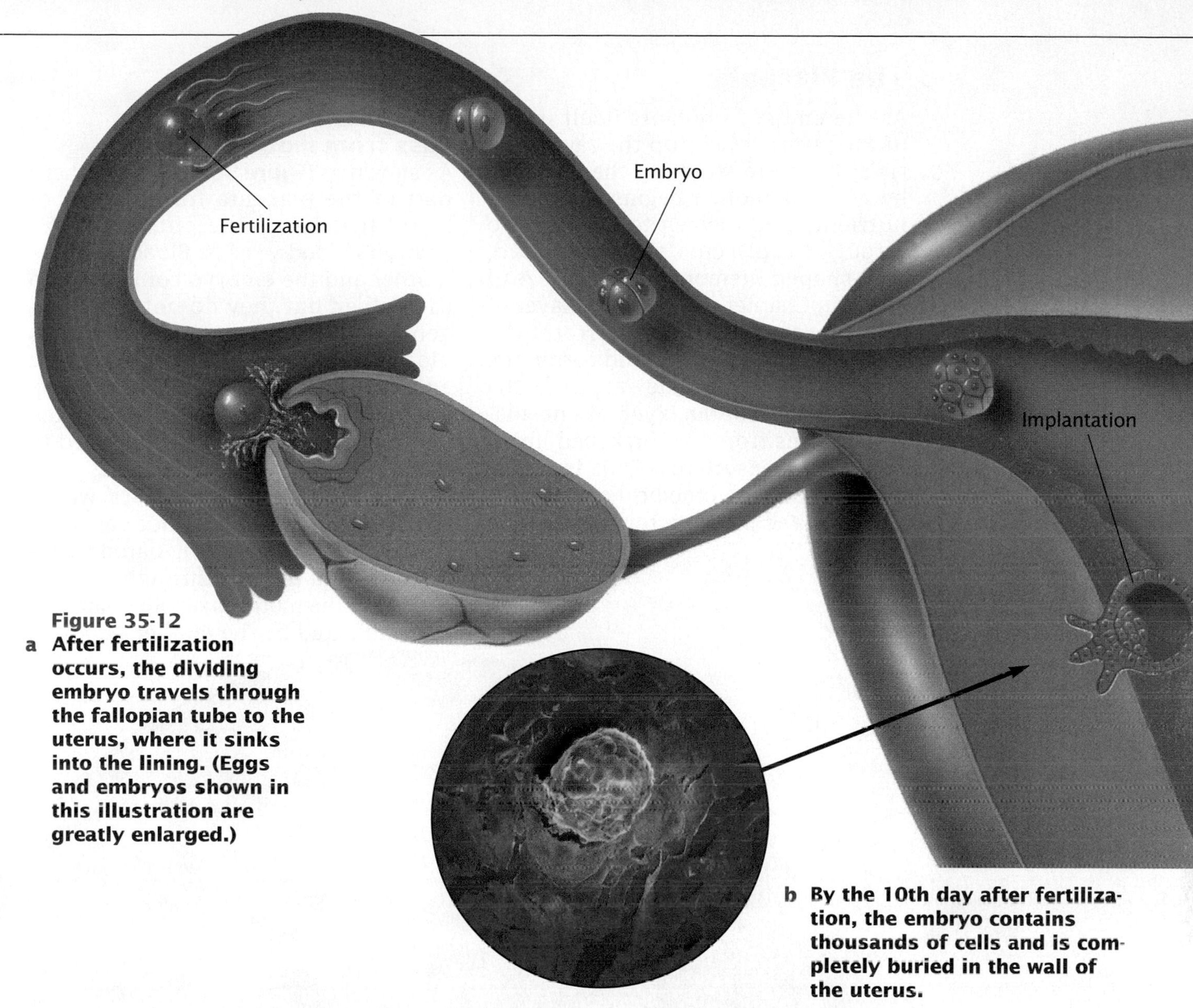

Figure 35-12
a After fertilization occurs, the dividing embryo travels through the fallopian tube to the uterus, where it sinks into the lining. (Eggs and embryos shown in this illustration are greatly enlarged.)

b By the 10th day after fertilization, the embryo contains thousands of cells and is completely buried in the wall of the uterus.

The Embryo Enters the Lining of the Uterus

Cell Structure and Function

Explain why a young girl can become pregnant even if she has not had her first menstrual period.

A newly fertilized egg is called a **zygote**. After the egg is fertilized, it continues to travel through the fallopian tube toward the uterus, as shown in **Figure 35-12**. After about 30 hours the zygote divides into two cells and is called an embryo. The cells of the embryo continue to divide by mitosis. (If the young embryo splits completely in two, identical twins will result.)

By the time the embryo reaches the uterus, about 4–5 days after fertilization, it is made up of a hundred or more cells. The embryo embeds itself in the soft, thickened uterine lining and then releases hormones to ensure that the thick lining of the uterus remains healthy and is not washed away by a menstrual period. The embryo's entry into the uterine wall is known as **implantation**. From the ninth week of development until birth, the developing embryo is called a **fetus**.

There are several ways for a female to determine whether she is pregnant. A missed menstrual period is one possible sign of pregnancy. A more accurate way of determining pregnancy is to test her blood or urine for the presence of the hormone produced by the embryo. By three weeks after fertilization, a pregnant woman will probably have enough of this hormone in her blood and urine to produce a positive test result. Doctors and health clinics are equipped to administer these tests, and they perform more reliable tests than those performed at home.

The Placenta

As the embryo implants itself in the uterine wall, it uses up the raw materials that were stored in the egg. The body of the mother begins to transfer nutrients and oxygen to the embryo through the placenta, the two-layered, disk-shaped membrane that you read about in Chapter 24. The inner layer of the placenta is derived from the chorion *(KAWR ee ahn)* and forms the outermost membrane around the embryo. The outer layer of the placenta forms from the thickened lining of the mother's uterus. Cells from the embryo form the umbilical cord, which connects the placenta to the embryo.

Nutrients and oxygen pass from mother to child

As shown in **Figure 35-13**, the mother's part of the placenta forms pools of blood that bathe the outside of the embryo's blood vessels. Blood from the mother and the embryo come close to each other, but they do not mix. Oxygen and nutrients from the mother's blood pass through the walls of the embryo's blood vessels. The oxygen and nutrients then travel through the blood vessels in the umbilical cord to the embryo.

Since the embryo is enclosed within the mother's body, it cannot get rid of waste products on its own. Carbon dioxide and urea diffuse through the placenta into the mother's bloodstream and are eliminated by the mother's body.

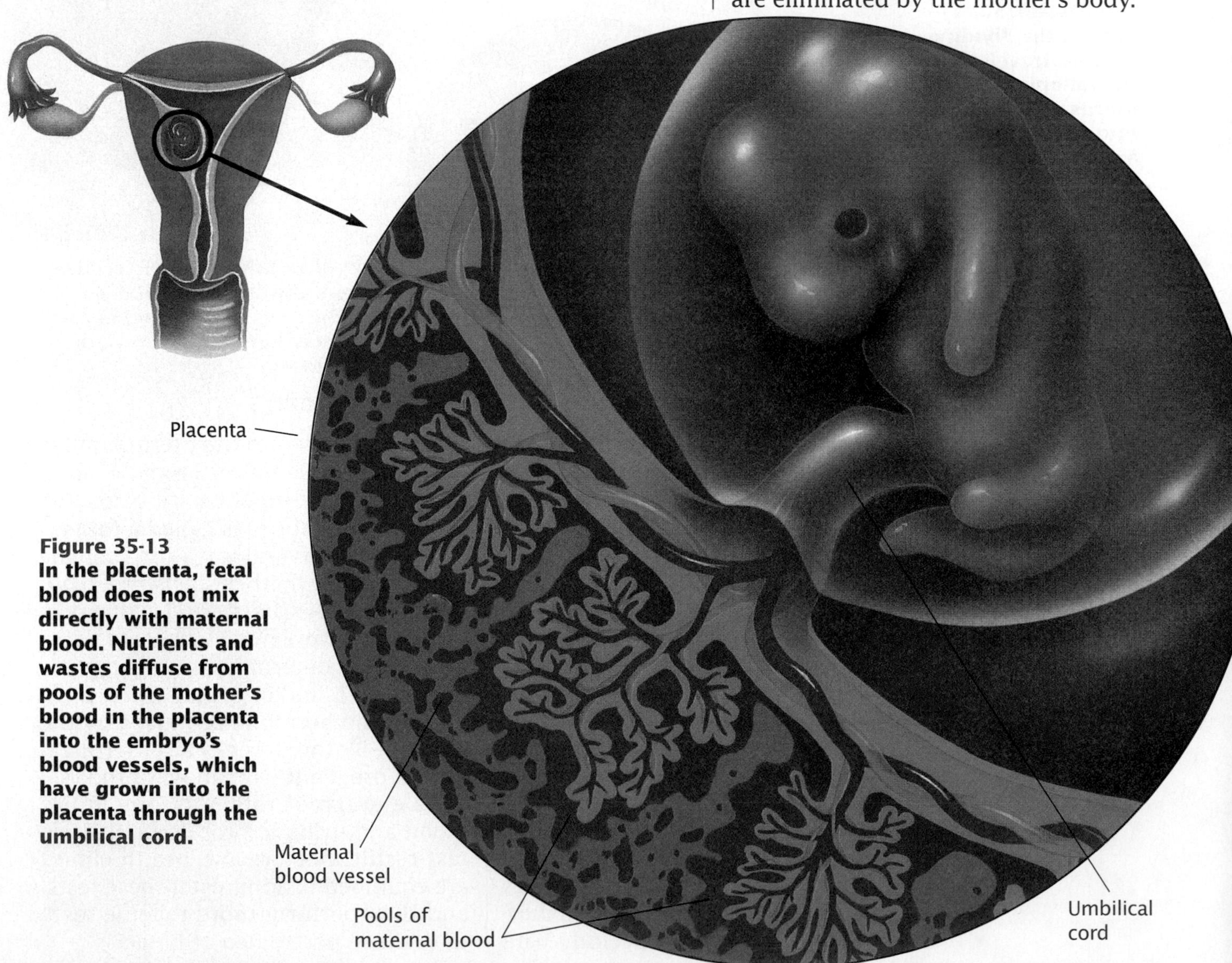

Figure 35-13
In the placenta, fetal blood does not mix directly with maternal blood. Nutrients and wastes diffuse from pools of the mother's blood in the placenta into the embryo's blood vessels, which have grown into the placenta through the umbilical cord.

Drugs
Babies of mothers who use drugs during pregnancy are often born underweight. These babies can face a variety of medical complications. For this reason, no pregnant woman should take any drug without checking with her doctor first.

Nutrition and Exercise
Nutrition and exercise are important during pregnancy. A mother must provide all the nutrients that a fetus needs to develop normally. An undernourished mother will give birth to a smaller, less healthy baby.

Smoking
Smoking cigarettes decreases the amount of oxygen available to the growing fetus. Babies born to mothers who smoke heavily tend to be much smaller than the babies of nonsmokers. Babies of heavy smokers also face a much higher risk of complications after birth than the babies of nonsmokers.

Alcohol
Women who drink alcohol during pregnancy may have babies with severe facial deformities and mental retardation. Most doctors think that even a small amount of alcohol may damage a fetus and therefore recommend that women do not drink at all during pregnancy.

Figure 35-14
This baby was born healthy because his mother was careful during pregnancy. She ate healthy foods, exercised regularly, and avoided drugs, including alcohol.

Maternal Health

The health and activities of its mother influence whether an embryo develops normally. From the instant of fertilization, an embryo is vulnerable to damage. In the first three months of pregnancy, the major systems and organs are forming. In the last six months of pregnancy, a fetus grows and matures. As **Figure 35-14** indicates, the diet and behavior of a woman during her pregnancy can affect the mental and physical development of the child, before and even years after birth. Even when it is too early for a woman to suspect she is pregnant, her use of alcohol, nicotine, and other drugs—prescription drugs, over-the-counter drugs, or street drugs—can damage the embryo.

Radiation and certain diseases may damage the fetus

Chemicals and other agents that cause malformation to a fetus are called **teratogens** *(TUH rat oh jinz)*. All types of ionizing radiation, including X rays, and ultraviolet rays, are teratogens. Radiation is most damaging to cells undergoing rapid division, as fetal cells are. For this reason, a woman who suspects or knows she is pregnant should not undergo X rays.

Certain diseases can also harm a fetus. Rubella (German measles) causes birth defects in early pregnancy. Toxoplasmosis is caused by a protozoan that is transmitted by undercooked meat or by taking in food, water, or dust contaminated by an infected cat. Adults show no symptoms or only mild ones, but fetal infection is more severe and may result in brain or eye damage and sometimes death. Both rubella and toxoplasmosis pathogens can pass from the mother's blood, through the placenta, and into the blood of the fetus.

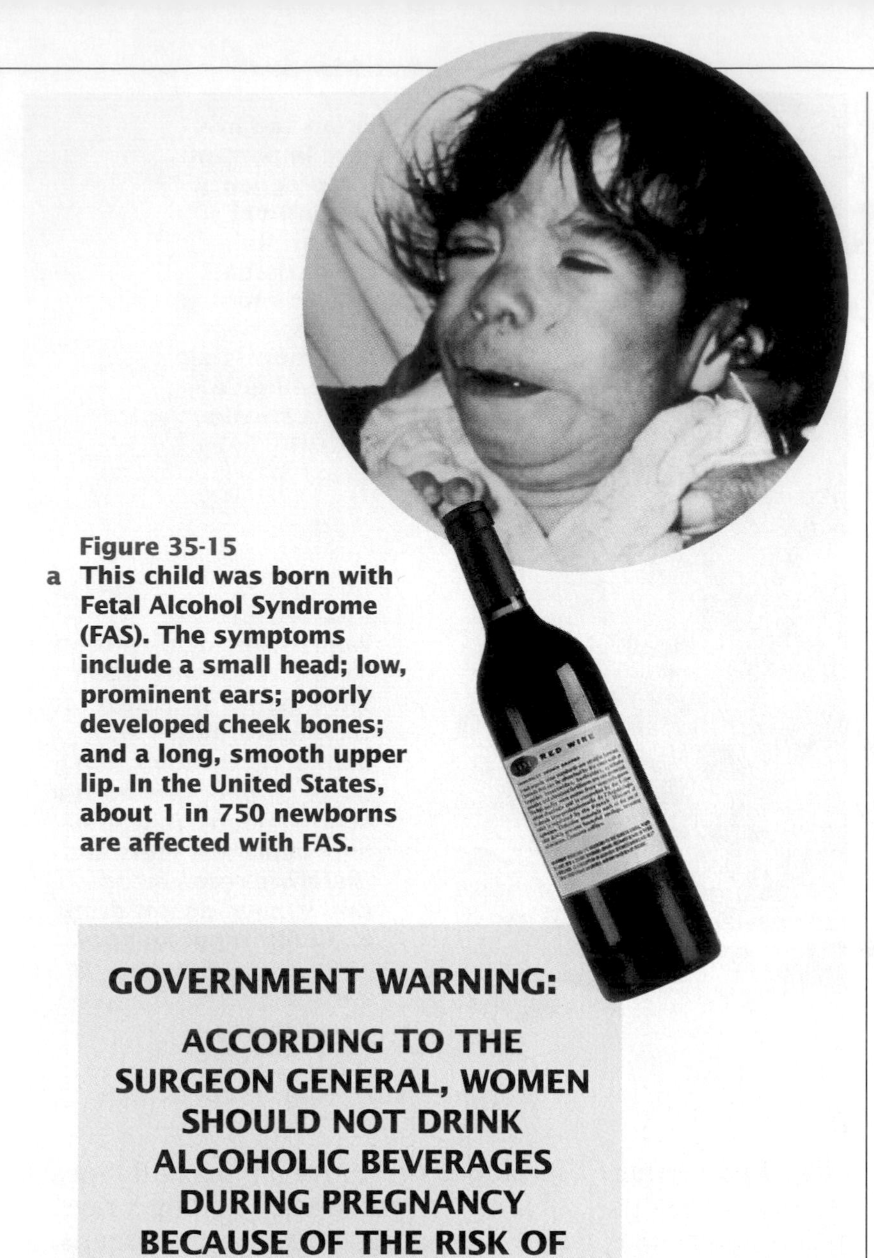

GOVERNMENT WARNING:
ACCORDING TO THE SURGEON GENERAL, WOMEN SHOULD NOT DRINK ALCOHOLIC BEVERAGES DURING PREGNANCY BECAUSE OF THE RISK OF BIRTH DEFECTS.

Figure 35-15
a This child was born with Fetal Alcohol Syndrome (FAS). The symptoms include a small head; low, prominent ears; poorly developed cheek bones; and a long, smooth upper lip. In the United States, about 1 in 750 newborns are affected with FAS.

b Warning labels now appear on all alcoholic beverage containers.

Alcohol can cause severe birth defects

Alcohol use is considered to be the main cause of preventable birth defects in the United States today. Heavy drinking can cause a variety of defects that are collectively called Fetal Alcohol Syndrome (FAS). FAS babies often have deformed faces like that of the infant shown in **Figure 35-15a**. They often are physically as well as mentally retarded, having trouble with simple functions such as telling time or counting. The average IQ of children with FAS is 30–40 points below normal. The damage is non-treatable and permanent.

Drinking alcohol during the first three months of pregnancy can result in severe physical damage—malformed organs, misshapen spine, and impaired brain development. Drinking later in pregnancy often results in stunted physical growth, hyperactivity, brain damage, and behavioral problems long after the child is born. Because the brain develops throughout the nine months of pregnancy, fetal brain damage due to a pregnant woman's alcohol consumption can occur at any time. **Figure 35-15b** shows a warning label that appears on alcoholic beverage containers.

Many toxic chemicals and drugs cross the placenta and can damage the embryo

The stimulants nicotine and caffeine cross the placenta too. In addition to nicotine, many of the other chemicals in cigarette smoke cross the placenta. Women who smoke have more miscarriages than women who do not smoke. The more a woman smokes, the more problems she and her baby are likely to have. Compared with nonsmoking mothers, smoking mothers have smaller infants and more babies who die at or shortly after birth. Caffeine, on the other hand, damages chromosomes in animal cells before conception. Although the extent of the damage that caffeine causes to the human fetus is not yet clear, women are advised to reduce caffeine intake or to avoid caffeine completely, beginning early in pregnancy.

A growing problem in the United States is the increasing number of babies born to mothers who use crack cocaine during pregnancy. These "crack babies" have a very high risk of mental retardation, learning disabilities, and personality disorders. Other drugs, legal and illegal, also affect the fetus. Babies of mothers who are addicted to heroin or cocaine may themselves be addicted when born.

Growth and Development

Once the embryo implants in the uterus, it grows rapidly. It continues to grow for a period of about 40 weeks, or 240–300 days. The due date is calculated from the expectant mother's last menstrual period, but the fetus actually spends about 260 days in the uterus from fertilization to birth. Because only 5 percent of women deliver on their due dates, pregnancy is divided into three-month segments called **trimesters** that are used as guideposts to key events in the development of the embryo and fetus.

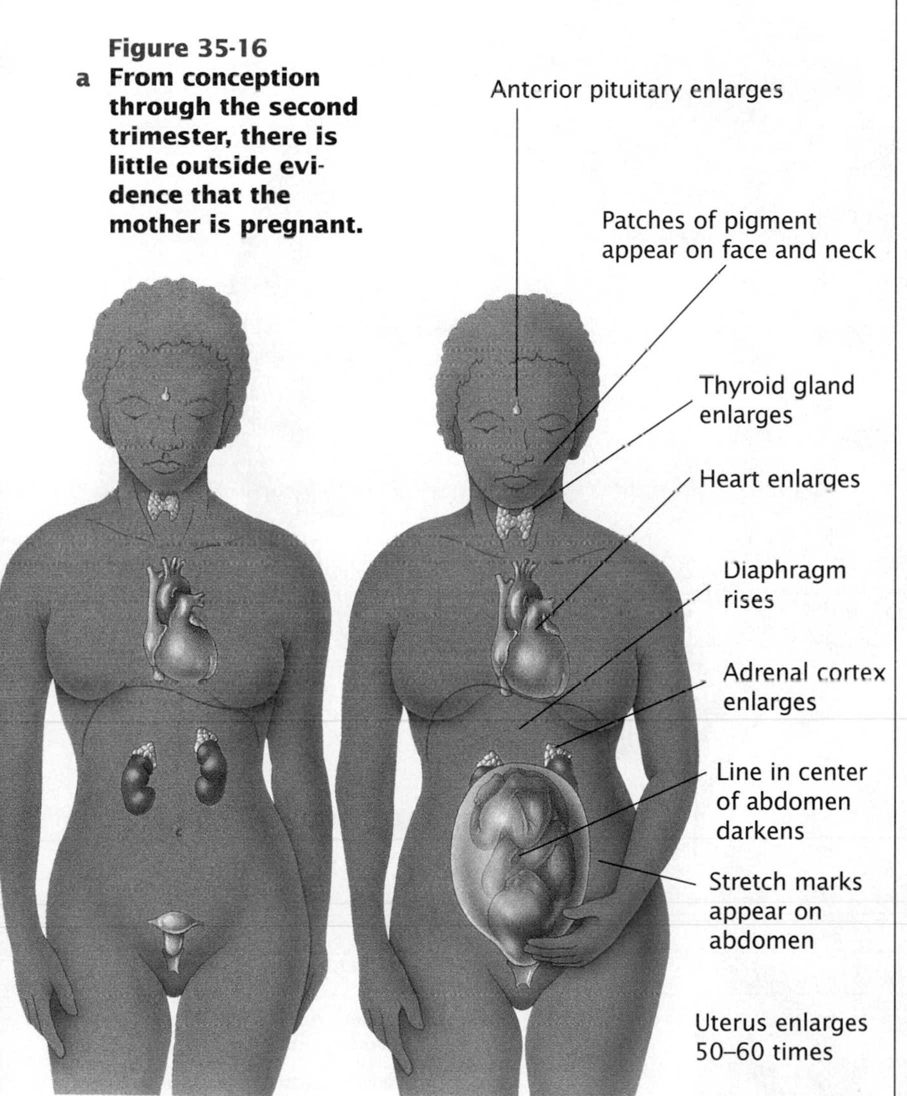

Figure 35-16
a From conception through the second trimester, there is little outside evidence that the mother is pregnant.
b By the third trimester the fetus is growing rapidly. The mother's glands and organs have enlarged to take care of her and her fetus, and her pregnancy is obvious.

Key events in fetal development and growth occur in each trimester

During the first trimester, very few changes occur in the mother's body, while the embryo is rapidly developing all of its major organ systems. By the end of the first trimester, all of the fetus's organ systems are in place. If damage occurs during the first trimester, significant birth defects may result, for this is the time that the embryo is most vulnerable.

At the beginning of the second trimester, changes in the mother's body are more noticeable as the fetus begins to make greater demands on the mother. Fetal organ systems are growing larger, the placenta is well established, and the circulatory systems for both mother and child are in full operation. **Figure 35-16** shows some of the changes that the mother's body experiences during pregnancy. Notice how certain organs and glands enlarge to handle a greater volume of blood, increased hormone secretion, and an increase in overall metabolism that supports rapid development and growth of cells, tissues, and organs. By the end of the second trimester, most of the fetus's development is complete.

Now growth of the fetus begins in earnest. The fetus gains the most weight in the last trimester. The fetus is refining respiratory and digestive organs, storing a fat layer, and gaining immunity from certain communicable diseases. Full-term infants (infants who stay within the mother for the full course of pregnancy) are much better equipped to survive outside the mother. Premature infants (those born before the 37th week of pregnancy) have problems related to immature organ systems, such as the lungs, kidneys, central nervous system, digestive organs, or liver.

Profound changes take place during the nine months of development

Four weeks after fertilization, the embryo is only about one-fourth of an inch long. The inner and outer cells that developed after fertilization have continued to expand and have formed a membrane that surrounds the embryo and is filled with fluid. In this liquid environment, the climate is controlled, and the embryo can grow freely without nearby tissues pressing on it. The fluid also cushions the embryo or fetus from bumping or jarring. To prevent the fetus from being overwhelmed by the fluid, a mucous plug fills its nose and part of its throat. At the end of 40 weeks, the fetus has grown to an average length of about 20 in. and a weight of about 7.25 lb. **Figure 35-17** will help you to follow the changes that occur during prenatal (before birth) development.

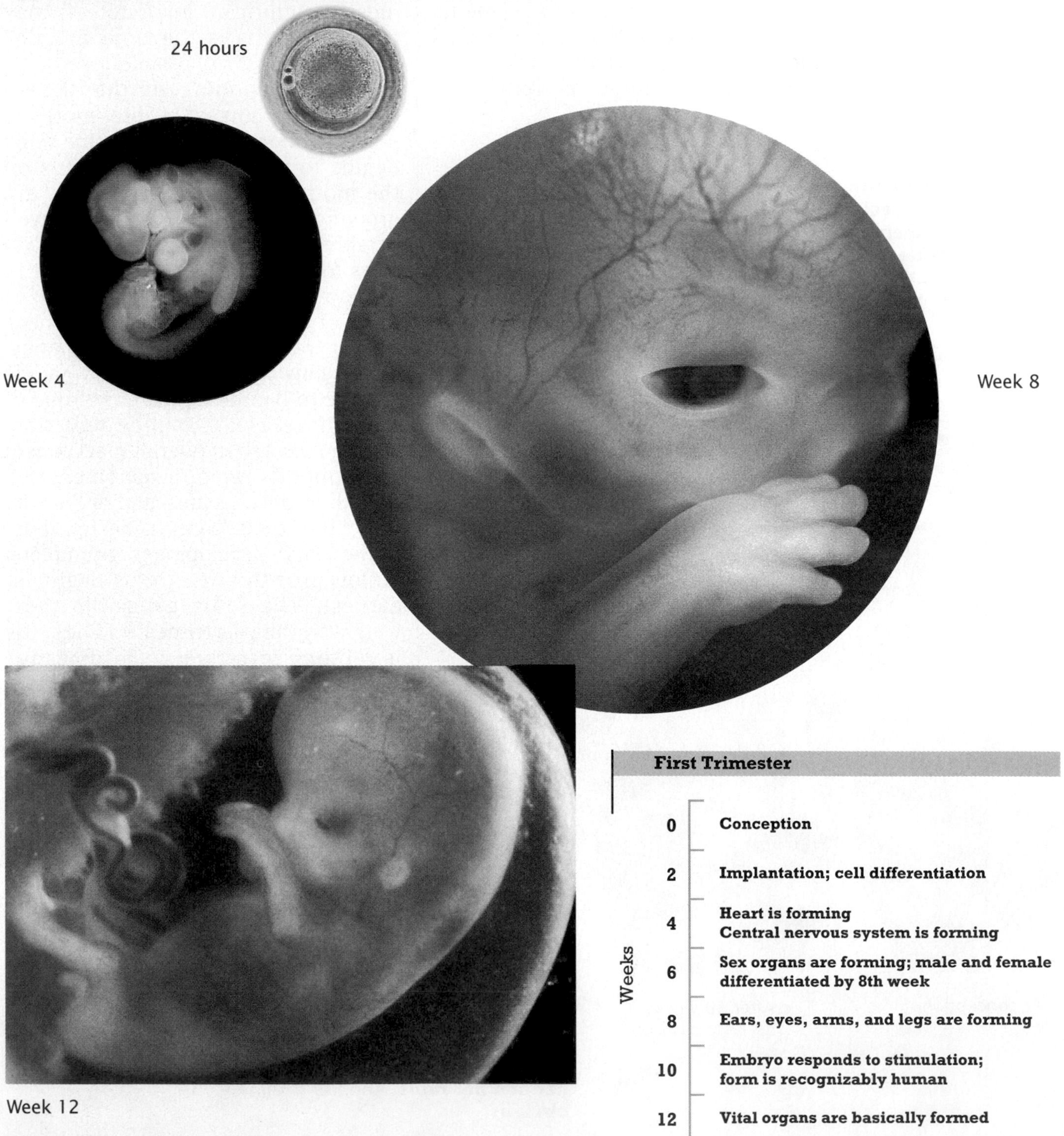

24 hours
Week 4
Week 8
Week 12
First Trimester
Weeks
0 Conception
2 Implantation; cell differentiation
4 Heart is forming
Central nervous system is forming
6 Sex organs are forming; male and female differentiated by 8th week
8 Ears, eyes, arms, and legs are forming
10 Embryo responds to stimulation; form is recognizably human
12 Vital organs are basically formed

Figure 35-17 Key developmental events occur in each trimester of pregnancy. Follow the development and growth of zygote, embryo, and fetus over the course of nine months.

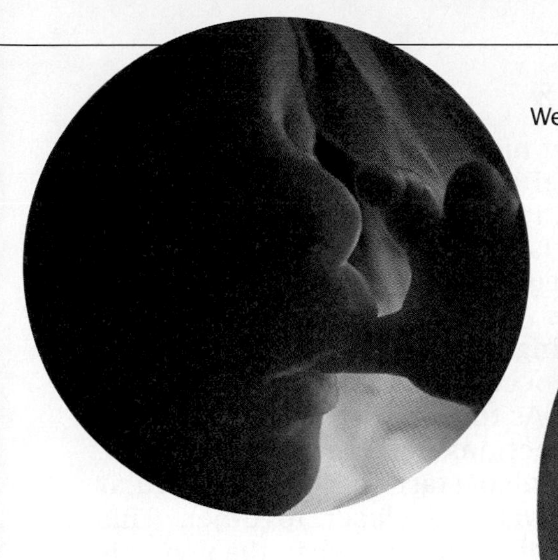

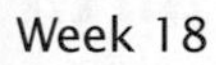

Week 18

Week 21

Week 22

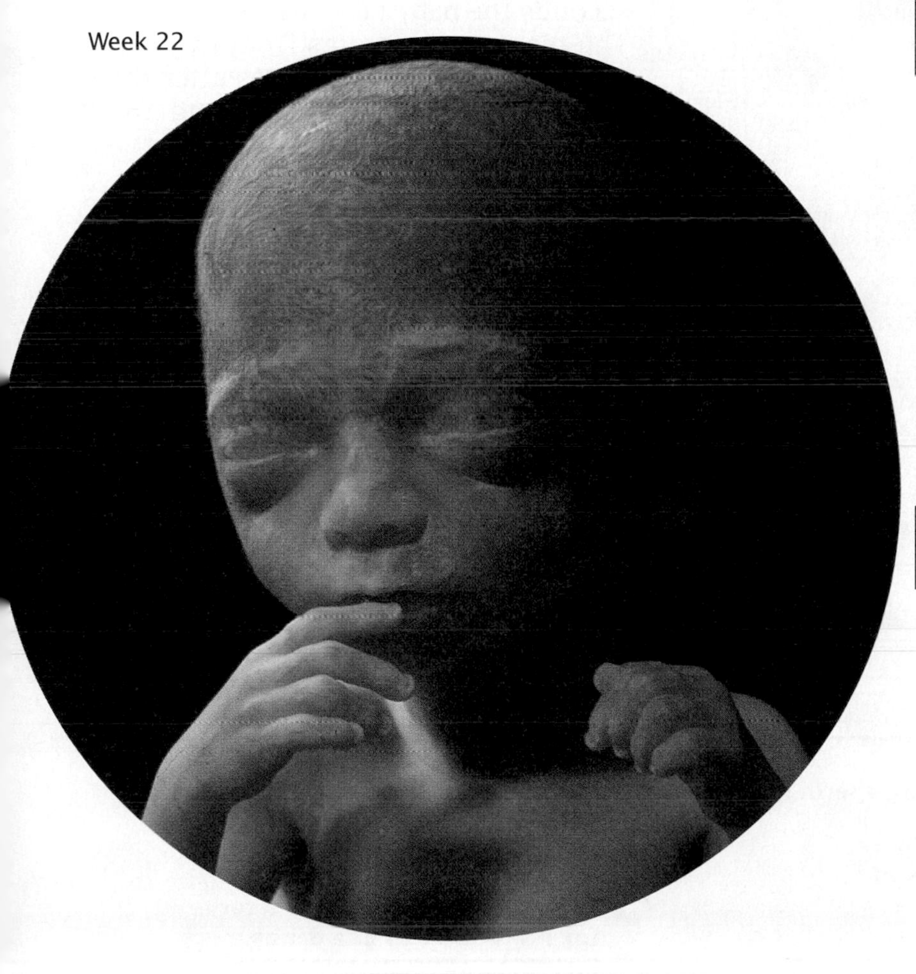

	Second Trimester
14	Circulatory system is operating
16	Mother can feel movements Fetus's sex organs are distinct
18	
20	Hair is forming on body and head
22	
24	Sucking movements; vigorous body movements; eyelashes and eyebrows appear; skin is thin,wrinkled, translucent

Weeks

	Third Trimester
26	Eyes open Survival outside womb is possible
28	
30	Layer of fat starts forming beneath skin
32	
	Survival outside of womb is probable
34	
36	
38	Birth

Weeks

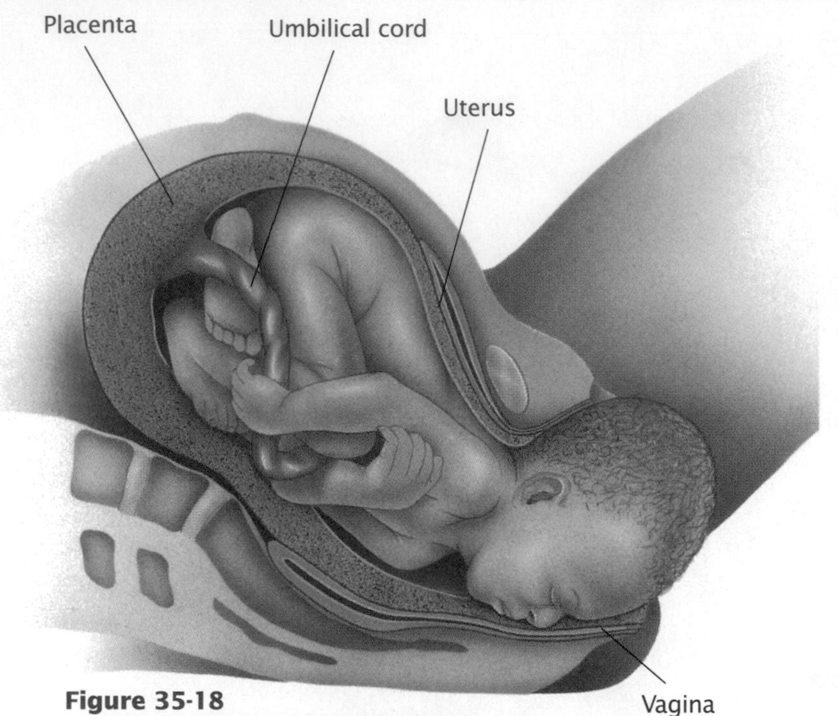

Figure 35-18

a During birth, the cervix is forced open and the baby is pushed out through the vagina.

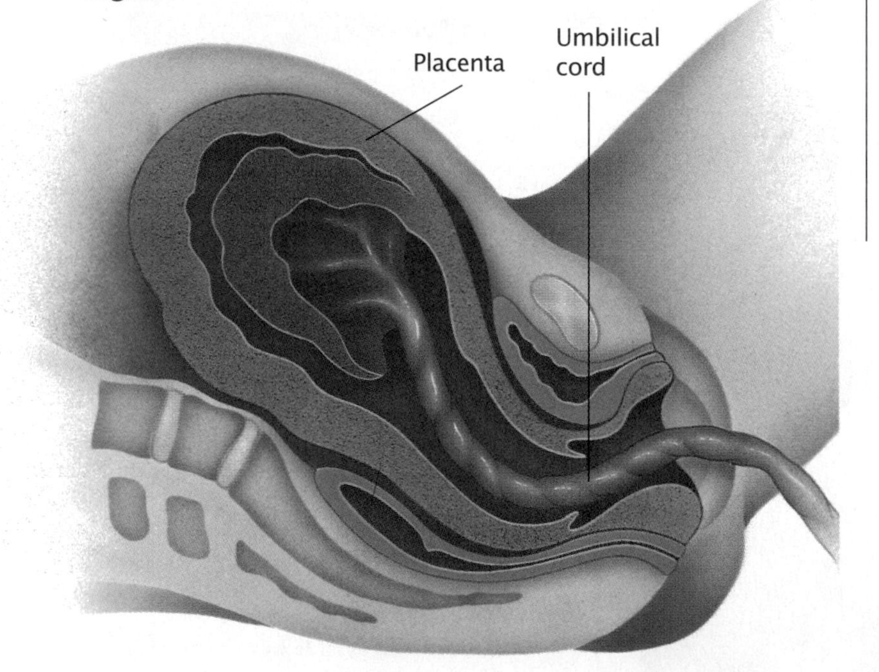

b Shortly after a child is born, fluid, blood, and the placenta are expelled from the uterus. This material is the afterbirth.

Birth

About nine months after fertilization, the fetus is ready to be born. By this time, it has usually moved so that its head is against the cervix. Near the end of pregnancy, hormones cause muscles in the walls of the uterus to begin contracting. Weak, irregular contractions may occur for several weeks before birth. As the time for birth approaches, contractions become much stronger. They also start to come at regular intervals and closer together. This stage of pregnancy, when the uterus is contracting strongly enough to push the baby out of the mother's body, is called **labor**. Labor usually takes between 5 and 20 hours.

Contractions of the uterus press the fetus's head against the cervix, forcing it to open enough for the baby to pass through, as shown in **Figure 35-18**. The baby then passes through the vagina and out into the world. Within a few seconds the baby takes its first breath. The placenta separates from the wall of the uterus. A few minutes after the baby is born, the placenta and the umbilical cord are pushed out of the uterus, marking the final stage of birth. The process of childbirth is complete.

SECTION REVIEW

1. **Explain why only one sperm fertilizes an egg.**
2. **Describe the events of implantation.**
3. **Describe two changes that occur in the fertilized egg before it is implanted in the uterus.**
4. **Why can it be harmful to a fetus for its mother to use drugs, including alcohol, during pregnancy?**

Disease-causing bacteria and viruses travel from one host to another in characteristic ways. Some are carried in water, in air, or by insects, while others are transmitted by sexual contact. Diseases that can be spread from one person to another by sexual contact are called sexually transmitted diseases, or STDs.

35-4 Sexually Transmitted Diseases

OBJECTIVES

1. **Identify three ways HIV is transmitted.**
2. **Describe two sexually transmitted diseases that are caused by bacteria and two caused by viruses.**
3. **Identify three ways to avoid HIV or HBV infection.**
4. **Describe the causes and consequences of pelvic inflammatory disease.**

Sexually Transmitted Viral Diseases

As you learned in Chapter 18, viruses cause many human diseases, including measles and influenza. In addition, viruses cause some sexually transmitted diseases, including AIDS, hepatitis B infection, papilloma virus infection, and genital herpes.

AIDS is a fatal viral disease

As you learned in Chapter 33, acquired immune deficiency syndrome (AIDS) is caused by the human immunodeficiency virus (HIV). AIDS patients, such as the man being examined in **Figure 35-19**, die from infections and cancers that their immune systems cannot fight.

HIV is present in blood, semen, and vaginal fluids of infected individuals and can be transmitted from one person to another during sexual intercourse. Anal intercourse, oral sex, and vaginal intercourse are each capable of transmitting HIV. Both females and males can get the virus from an infected partner during intercourse. Transmission of the virus can also occur anytime unsterilized needles are used. Sharing needles and syringes used for injecting drugs is the most common route of infection; however, tattoo needles and needles for piercing ears can also transmit the virus if not sterilized. In Africa, many cases of HIV have been transmitted through the reuse of needles in medical clinics. HIV can be transmitted from mother to infant across the placenta or in breast milk. About one-third of babies born to HIV-infected mothers are infected with the virus.

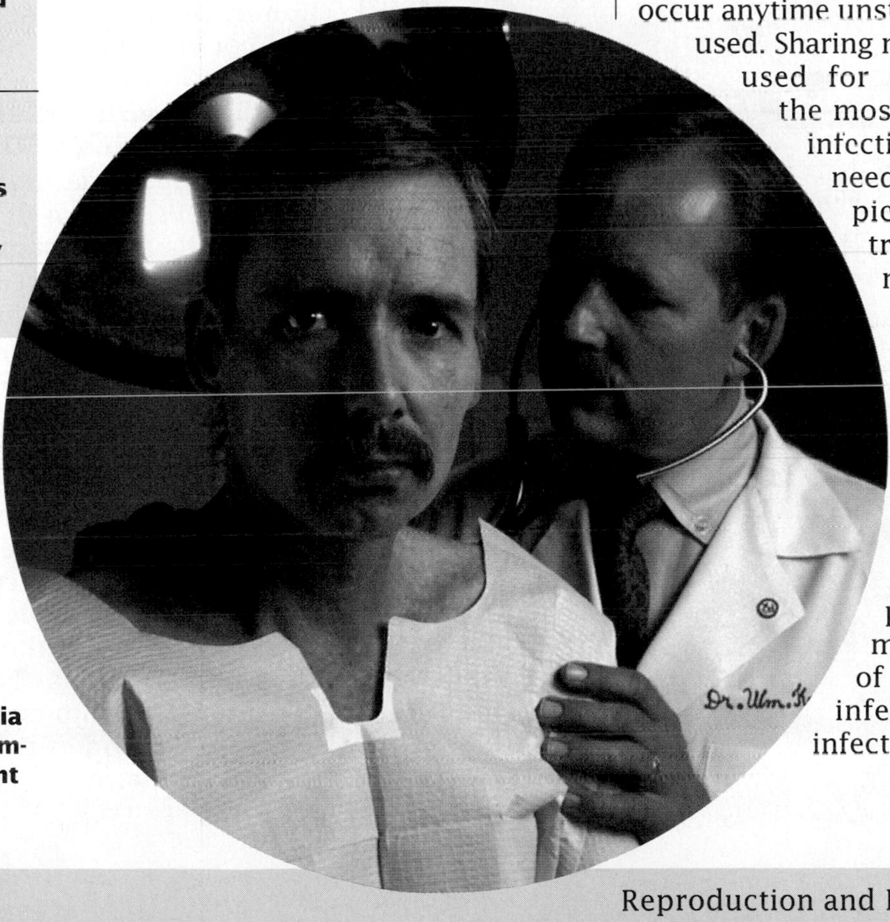

Figure 35-19 This California doctor is examining a patient with AIDS.

A person can be infected with HIV for many years without knowing it. During this time, the infected person can spread the disease through sexual contact or by sharing needles and syringes. Although there is no cure for HIV or AIDS to date, HIV infection can be prevented. Abstaining from sexual intercourse and from the use of drugs of any kind are ways to eliminate exposure to HIV. Using a condom during sexual intercourse reduces but does not eliminate the risk of being exposed to HIV.

HBV infection is more common than HIV

Hepatitis B infection, caused by hepatitis B virus (HBV), can lead to serious liver damage, including cancer, in both infants and adults. Sexual contact is a major route of transmission, but drug users who share needles are also at risk for getting HBV. HBV can be transmitted from mother to child during pregnancy. Testing pregnant women for the HBV antigen is recommended, and infants born to mothers who test positive for the antigen should be immunized. The most commonly used vaccine in the United States is made using recombinant DNA technology.

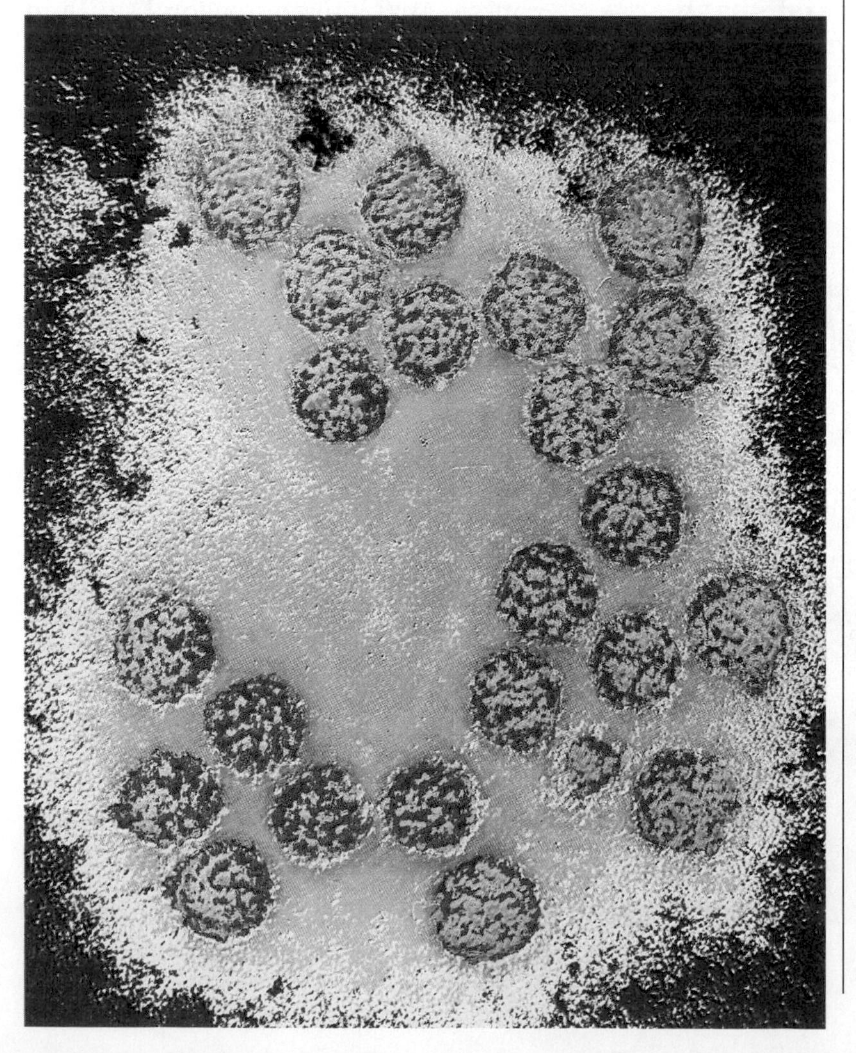

Figure 35-20
The human papilloma virus is the cause of today's fastest growing sexually transmitted disease, genital warts.

Genital herpes is also caused by a virus

Genital herpes is a sexually transmitted disease caused by herpes simplex virus (HSV). Two types of HSV can cause genital herpes: HSV-1 and HSV-2. About 80 percent of genital herpes is caused by HSV-2, which causes painful, recurring blisters and sores on the genitals, similar to the cold sores and fever blisters most commonly caused by HSV-1. Antiviral drugs can temporarily eliminate the blisters, but these drugs cannot eliminate the virus from the body. The virus is never completely destroyed by treatment; it remains dormant in the body and may recur at any time. The blisters transmit the infection to new hosts, including sex partners and infants delivered by vaginal birth from an infected mother. Infants infected in this manner may develop brain fever or infection of the membranes covering the brain or spinal cord.

Papilloma virus infection is the fastest growing STD

Human papilloma *(pap ih LOH mah)* virus (HPV), shown in **Figure 35-20**, causes today's fastest growing sexually transmitted disease, genital warts. HPV appears to play a role in cancers of the cervix, vagina, vulva, and penis as well. Even when warts and lesions are not seen or felt, viral particles are shed and transmitted through sexual contact. Having intercourse at an early age and having many sex partners over a lifetime greatly increase the risk of getting HPV. Some evidence suggests that up to 80% of the people exposed to the virus pick it up, although most will have no symptoms. There are many treatments, but no cures, for genital warts. A person carrying HPV may have it for life.

Sexually Transmitted Bacterial Diseases

Evolution

Strains of penicillin-resistant gonorrhea bacteria are becoming more common. How does penicillin resistance represent an adaptation by these bacteria?

Like viruses, bacteria cause many human diseases. Among these are sexually transmitted diseases such as syphilis, gonorrhea, and chlamydia. These sexually transmitted diseases, unlike those caused by viruses, can be treated with antibiotics.

Syphilis often goes unnoticed

Syphilis *(SIHF uh lihs)* is a serious disease that eventually afflicts the entire body if untreated. Although it can be cured with antibiotics, syphilis often is not diagnosed or treated because its early and later symptoms go unnoticed or ignored.

Two weeks to three months after infection, a small painless ulcer called a chancre *(SHAHNG kuhr)* appears on the genitals at the exact site of infection. Often the chancres are inside the vagina or rectum and are not seen. Obvious chancres on the penis or labia may be ignored because they are painless. Weeks or months later, a rash that does not itch appears on the hands or feet along with fever, sore throat, or joint pains. Large, moist sores on the sex organs or mouth may develop. Like chancres, these sores are highly contagious, and the symptoms disappear without treatment. Sometimes unnoticed for years, the latent infection causes serious damage to the brain, spinal cord, and circulatory organs. The bacteria can cross the placenta and infect a fetus, usually after the fifth month of pregnancy. **Figure 35-21** shows the incidence of syphilis and gonorrhea among 15- to 19-year-olds. As cases of syphilis and gonorrhea have dropped over the years, cases of other sexually transmitted diseases have risen dramatically.

Gonorrhea can damage the reproductive organs

Gonorrhea *(gahn uh REE uh)* is another sexually transmitted disease that often goes untreated or ignored. Although antibiotics can cure the disease, some strains of gonorrhea are resistant to the more commonly used antibiotics, such as penicillin. Different antibiotics are required in the treatment of these resistant bacterial strains.

In males, gonorrhea bacteria usually cause painful urination and pus discharge from the penis. Untreated

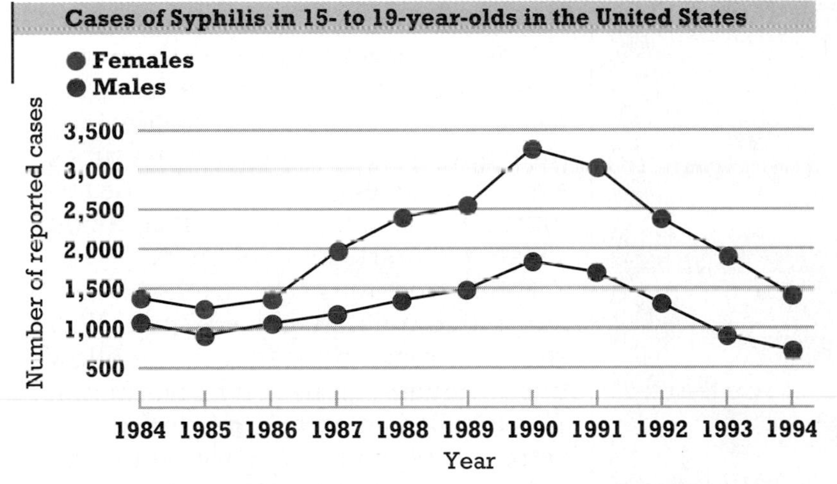

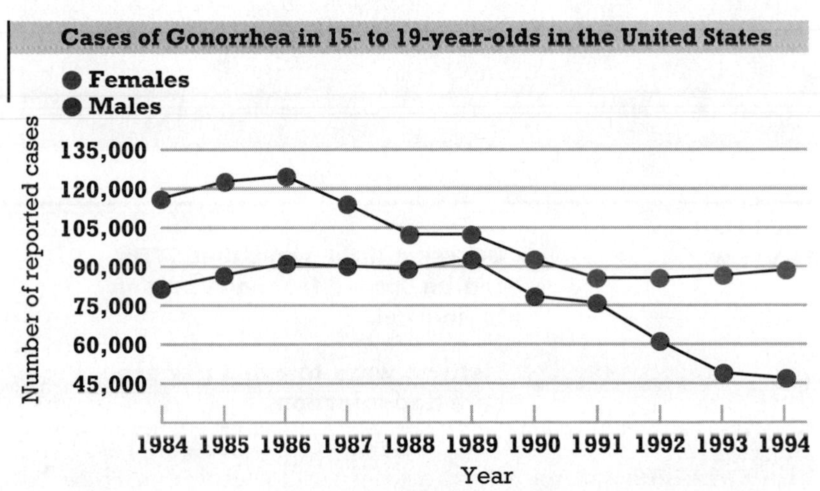

Figure 35-21
The incidence of syphilis and gonorrhea appears to be dropping in the teen population. This is not, however, a reliable picture. Many cases of both diseases are not reported, and many infected people never seek treatment.

gonorrhea can spread to the vas deferens, prostate, or testes. In females, the pathogen causes vaginal discharge but more often causes no symptoms. The disease can spread to the fallopian tubes and can affect the rectum, throat, or eyes as well. Infection of the male and female reproductive structures can cause scarring that results in infertility. Infants can be infected during birth, resulting in genital infections or eye infections that can lead to blindness.

Chlamydia is the most common sexually transmitted disease

Chlamydia *(kluh MIHD ee ah)* is the most common sexually transmitted disease in the United States, with up to 10 million new cases each year. The *Chlamydia* bacterium shown in **Figure 35-22** is usually transmitted through vaginal intercourse. The symptoms of chlamydia are similar to those of a mild case of gonorrhea: painful urination and a urethral discharge in males, and vaginal discharge and bleeding between menstrual periods in females. Chlamydia often produces no symptoms and goes unnoticed and undiagnosed. Untreated chlamydia can spread to the vas deferens and fallopian tubes and cause infertility. A baby born to a mother with chlamydia can be infected as it passes through the birth canal and may suffer eye or lung damage. Although gonorrhea and chlamydia often have no symptoms, routine laboratory tests available at hospitals and clinics can detect infections by either bacterium, and both partners should be treated simultaneously.

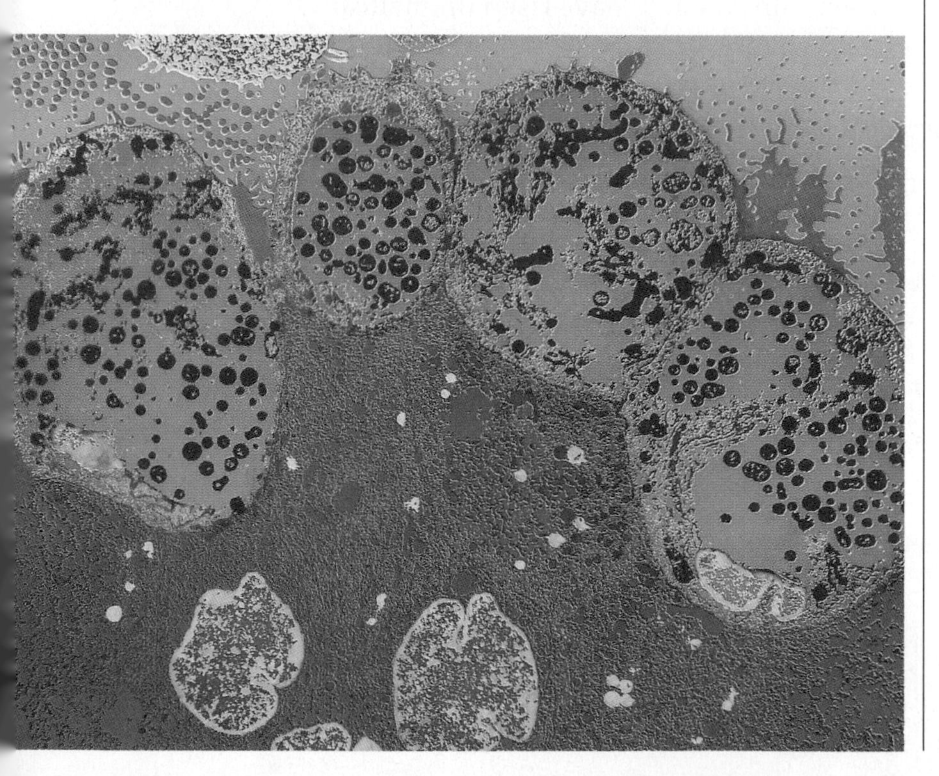

Figure 35-22
The bacterium that causes chlamydia, shown here in dark red, grows within cells of its human host. These bacteria are growing inside an infected woman's fallopian tubes.

Untreated gonorrhea or chlamydia can cause pelvic inflammatory disease

Bacterial infection of the cervix, uterus, ovaries, and fallopian tubes is called pelvic inflammatory disease, or PID. PID is usually caused by gonorrhea or chlamydia bacteria that attach themselves to sperm and then move into the pelvic region. When the fallopian tubes are infected, scar tissue may form and close the tubes. If the tubes are closed, an egg can no longer be fertilized. PID is one of the most common causes of infertility in women.

Fallopian tubes that are damaged, but not completely closed, may allow sperm to reach the egg, but not allow the fertilized egg to reach the uterus. This is called an **ectopic pregnancy**. An ectopic pregnancy occurs when the embryo grows somewhere other than the uterus. An ectopic pregnancy in a fallopian tube can be life-threatening to a woman. As the embryo grows, it can cause the fallopian tube to rupture.

SECTION REVIEW

1. **Describe three ways that STDs can be spread from one person to another.**
2. **List two ways to avoid HIV exposure and infection.**
3. **Describe three sexually transmitted diseases that are caused by bacteria and three caused by viruses.**
4. **Discuss the possible consequences of untreated gonorrhea or chlamydia.**

CHAPTER 35

Highlights

This young lady, not yet two years old, has much to learn before she can wear daddy's hat.

	Key Terms	Summary
35-1 The Male Reproductive System	testes (p. 720) epididymis (p. 720) vas deferens (p. 720) ejaculation (p. 721) semen (p. 721)	• Sperm are produced in the testes, stored in the epididymis, and expelled during ejaculation. • Several hormones are responsible for sexual changes, including sperm production and release.
35-2 The Female Reproductive System	ova (p. 723) ovaries (p. 723) fallopian tube (p. 723) uterus (p. 723) cervix (p. 724) vagina (p. 724) menstrual cycle (p. 725) ovulation (p. 725) follicle (p. 726) embryo (p. 726) menopause (p. 727)	• The release of an egg from an ovary is called ovulation. • Ovulation occurs once a month, about 14 days before the next menstrual period starts. • The ovarian cycle explains what happens in the ovary prior to and at ovulation. • The uterine cycle explains what happens in the uterus after ovulation and during menstruation.
35-3 Fertilization and Development	fertilization (p. 728) zygote (p. 729) implantation (p. 729) fetus (p. 729) teratogens (p. 731) trimester (p. 733) labor (p. 736)	• Fertilization of an egg by sperm usually takes place in a fallopian tube. • Fertilization marks the start of pregnancy as the zygote moves to the uterus. • After implantation, the placenta is formed for fetal circulation. • Pregnancy is divided into trimesters. • Birth occurs about nine months after fertilization.
35-4 Sexually Transmitted Diseases	chlamydia (p. 740) ectopic pregnancy (p. 740)	• Some STDs are treatable, some are curable, and some are deadly. • Untreated STDs can cause serious problems in both males and females.

Millions of sperm are expelled by the male during ejaculation.

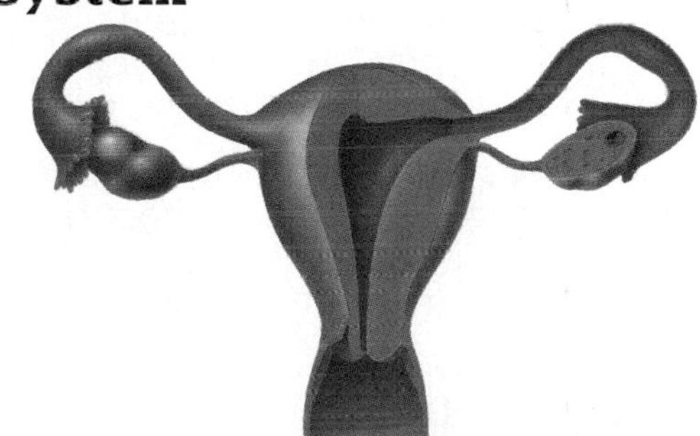
Usually, one egg is released from an ovary each month. If it is fertilized by a sperm, an embryo will develop.

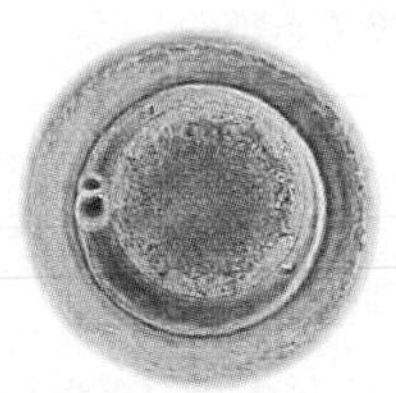
The fetus will grow and develop for about nine months, until it is ready to be born.

In the United States, more than 500,000 new cases of genital herpes are reported each year.

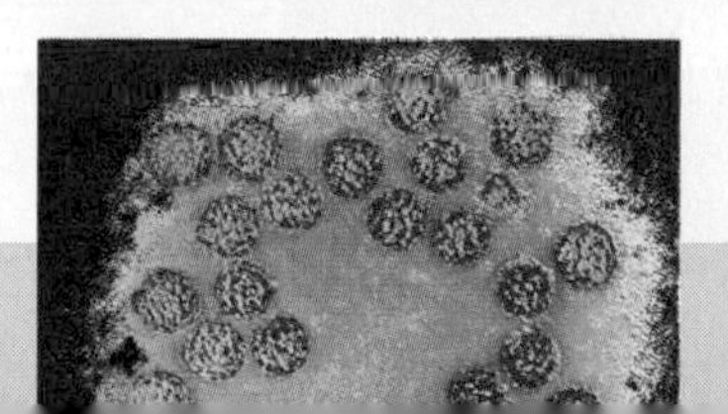

CHAPTER 35

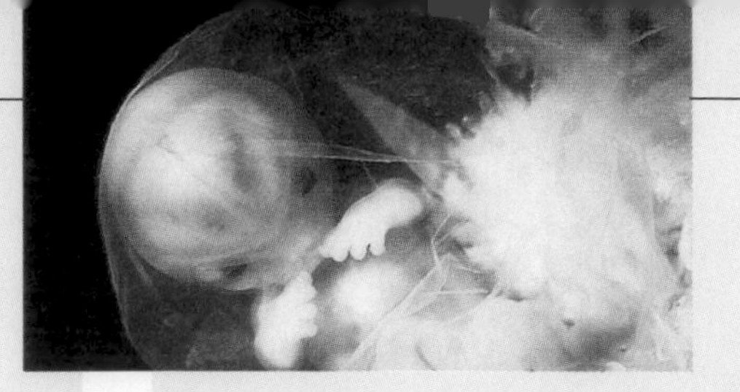

review

Understanding Vocabulary

1. For each set of terms, choose the term that does not fit the pattern, and explain why.
   a. seminal vesicles, testes, urethra, fallopian tubes
   b. uterus, ovaries, epididymis, cervix

2. For each set of terms, complete the analogy.
   a. body cells of sexually mature male : diploid :: sperm cells : ____
   b. organs that produce sperm : testes :: organs that produce ova : ____
   c. the joining of an egg and sperm : fertilization :: an embryo's attachment to the uterine wall : ____

Understanding Concepts

3. **Relating Concepts** Construct a concept map that illustrates how the male hypothalamus regulates sperm production. Include the following terms: male hypothalamus, GnRH, pituitary, testes, LH, FSH, testosterone, sperm production, male characteristics, facial hair, muscles, and low voice.

4. **Identifying Functions** Trace the pathway of sperm from their production to their exit during ejaculation. Identify all important organs along the pathway.

5. **Inferring Relationships** Urine is acidic. After urine passes through the urethra, the urethra becomes acidic. How does the male body prevent sperm from being killed as they pass through the urethra?

6. **Comparing Structures and Functions** How are the male and female reproductive systems similar? In what two ways are they different?

7. **Organizing Information** Relate the rise, peak, and decline of four major hormones released during the menstrual cycle to the physiological events of ovulation, fertilization, implantation, and the onset of the menstrual period.

8. **Distinguishing Relevant Information** Explain why the menstrual cycle is better explained by an ovarian cycle and a uterine cycle.

9. **Predicting Outcomes** What happens to an egg if it is not fertilized within 24 hours after its release?

10. **Recognizing Relationships** Describe the role of progesterone in the uterine cycle.

11. **Summarizing Information** Describe the path a sperm must take to reach an egg. Also describe the events that occur once a sperm reaches the egg.

12. **Summarizing Information** Describe what happens to the egg between fertilization and implantation.

13. **Identifying Functions** List the substances that pass through the placenta and the direction in which they move.

14. **Predicting Outcomes** What are teratogens? What are some examples? How do they affect a developing fetus?

15. **Inferring Relationships** How can alcohol consumption by a pregnant woman affect the fetus she carries? How can smoking cigarettes during pregnancy affect a developing fetus?

16. **Organizing Information** Describe at least two major developmental events that occur in the fetus during each trimester of pregnancy.

17. **Inferring Relationships** What is AIDS? Identify three ways that the human immunodeficiency virus (HIV) is transmitted.

18. **Summarizing Information** What is a sexually transmitted disease? Describe two sexually transmitted diseases that are caused by viruses and two that are caused by bacteria.

19. **Inferring Relationships** What is pelvic inflammatory disease, or PID? How is it caused? What are the results of this disease?

20. **BUILDING ON WHAT YOU HAVE LEARNED** In Chapter 31 you learned about oxytocin and prolactin. What is the function of oxytocin-induced contractions during labor? after labor?

Interpreting Graphics

21. Study the illustration below.

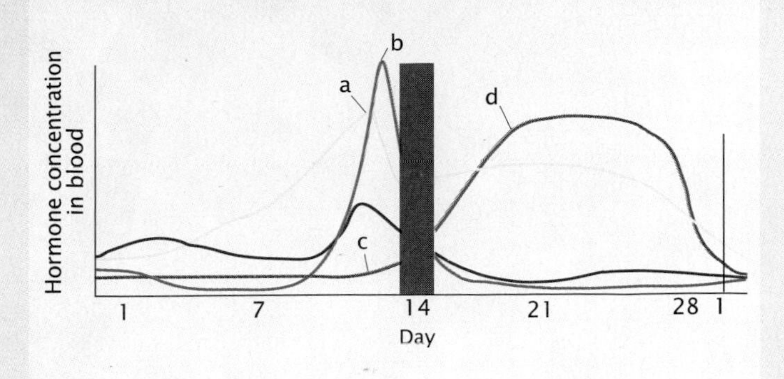

If day 1 indicates the start of the menstrual period,

a. What hormone is represented by the blue line? Support your answer. What primary event accompanies point *b*?
b. At which point, *a* or *c*, is the uterus stimulated to prepare for a fertilized egg? Support your answer. A gradual rise in which hormone is represented at this point?
c. What is point *a*'s relationship to point *b*?
d. If the egg were fertilized during this cycle, what would point *d* represent? Which hormone is primarily responsible for sustaining pregnancy, according to this graph?
e. What does the shaded area represent?

Reviewing Themes

22. ***Cell Structure and Function***
Why does a woman who has undergone tubal ligation not get pregnant?

23. ***Cell Structure and Function***
How are a vasectomy and a tubal ligation similar?

Thinking Critically

24. **Evaluating Results** In the early 1960s, many pregnant women took a tranquilizer called thalidomide. Women who took the tranquilizer during the first trimester of pregnancy gave birth to babies with serious limb deformities. Women who took the tranquilizer later in their pregnancies gave birth to normal babies. What does this information reveal about the sequence of development of the fetus?

Life/Work Skills

25. **Using Technology** Use library references or search an on-line database to find out about the technology used in the process of in vitro fertilization. Describe the process of in vitro fertilization. Who was the first baby born as a result of this process? Summarize your findings in a written report.

Activities and Projects

26. **Health Connection** Do library research on the use of ultrasound technology during pregnancy. What information can be learned about the fetus from this technology? Gather information from a community health-care provider about the cost of this technology in your area and when its use is recommended.

27. **Language Arts Connection** Do library research on folk tales and myths about the birth of human babies. One familiar myth is that babies are brought by the stork. Include modern myths. Summarize your findings in an oral report to share with your class.

Discovering Through Reading

28. Read the article "Prevent Sexually Transmitted Diseases" by Lauren Picker in *American Health*, October 1995, pages 62–66. Identify three sexually transmitted viral diseases that are incurable and whose numbers are on the rise. How do sexually transmitted diseases facilitate HIV's entry into the body? Describe two reasons women have a greater susceptibility than men to contract an STD. Identify five techniques to prevent sexually transmitted diseases.

Lab Program Introduction

Nutrient
Indicator
Positive test (standard)
Negative test
Food sample
Sugar
Benedict's solution
Turns orange or brick red
Stays blue-green
Starch
Lugol's iodine
Protein
Biuret solution
Lipid
Appearance on brown paper
esc
tab
caps lock
shift
control
alt
option

Lab Program Explorations

Using the Explorations

Each Exploration gives you an opportunity to practice techniques that are used in biological research by providing a step-by-step procedure for you to follow. You will use many of these techniques later in one or more Investigations.

Each Exploration Contains the Following Parts:

Skill identifies the technique you will learn.

Objectives tell you what you will be expected to accomplish.

Situation is the setting for the lab.

Background provides additional information you will need for the lab.

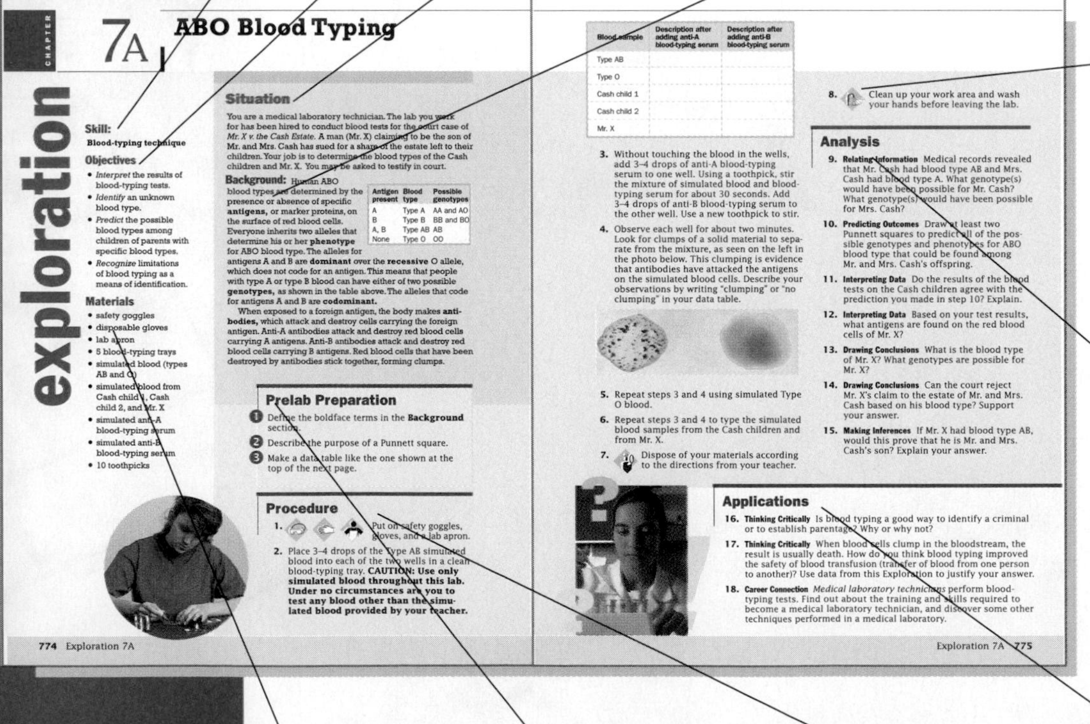

CHAPTER 7A exploration

ABO Blood Typing

Skill:
Blood-typing technique

Objectives
- *Interpret* the results of blood-typing tests.
- *Identify* an unknown blood type.
- *Predict* the possible blood types among children of parents with specific blood types.
- *Recognize* limitations of blood typing as a means of identification.

Materials
- safety goggles
- disposable gloves
- lab apron
- 5 blood-typing trays
- simulated blood (types AB and O)
- simulated blood from Cash child 1, Cash child 2, and Mr. X
- simulated anti-A blood-typing serum
- simulated anti-B blood-typing serum
- 10 toothpicks

Situation

You are a medical laboratory technician. The lab you work for has been hired to conduct blood tests for the court case of *Mr. X v. the Cash Estate*. A man (Mr. X) claiming to be the son of Mr. and Mrs. Cash has sued for a share of the estate left to their children. Your job is to determine the blood types of the Cash children and Mr. X. You may be asked to testify in court.

Background: Human ABO blood types are determined by the presence or absence of specific **antigens,** or marker proteins, on the surface of red blood cells. Everyone inherits two alleles that determine his or her **phenotype** for ABO blood type. The alleles for antigens A and B are **dominant** over the **recessive** O allele, which does not code for an antigen. This means that people with type A or type B blood can have either of two possible **genotypes,** as shown in the table above. The alleles that code for antigens A and B are **codominant.**

Antigen present	Blood type	Possible genotypes
A	Type A	AA and AO
B	Type B	BB and BO
A, B	Type AB	AB
None	Type O	OO

When exposed to a foreign antigen, the body makes **antibodies,** which attack and destroy cells carrying the foreign antigen. Anti-A antibodies attack and destroy red blood cells carrying A antigens. Anti-B antibodies attack and destroy red blood cells carrying B antigens. Red blood cells that have been destroyed by antibodies stick together, forming clumps.

Prelab Preparation

1. Define the boldface terms in the **Background** section.
2. Describe the purpose of a Punnett square.
3. Make a data table like the one shown at the top of the next page.

Procedure

1. Put on safety goggles, gloves, and a lab apron.
2. Place 3–4 drops of the Type AB simulated blood into each of the two wells in a clean blood-typing tray. **CAUTION: Use only simulated blood throughout this lab. Under no circumstances are you to test any blood other than the simulated blood provided by your teacher.**

774 Exploration 7A

Blood sample	Description after adding anti-A blood-typing serum	Description after adding anti-B blood-typing serum
Type AB		
Type O		
Cash child 1		
Cash child 2		
Mr. X		

3. Without touching the blood in the wells, add 3–4 drops of anti-A blood-typing serum to one well. Using a toothpick, stir the mixture of simulated blood and blood-typing serum for about 30 seconds. Add 3–4 drops of anti-B blood-typing serum to the other well. Use a new toothpick to stir.
4. Observe each well for about two minutes. Look for clumps of a solid material to separate from the mixture, as seen on the left in the photo below. This clumping is evidence that antibodies have attacked the antigens on the simulated blood cells. Describe your observations by writing "clumping" or "no clumping" in your data table.
5. Repeat steps 3 and 4 using simulated Type O blood.
6. Repeat steps 3 and 4 to type the simulated blood samples from the Cash children and from Mr. X.
7. Dispose of your materials according to the directions from your teacher.
8. Clean up your work area and wash your hands before leaving the lab.

Analysis

9. **Relating Information** Medical records revealed that Mr. Cash had blood type AB and Mrs. Cash had blood type A. What genotype(s) would have been possible for Mr. Cash? What genotype(s) would have been possible for Mrs. Cash?
10. **Predicting Outcomes** Draw at least two Punnett squares to predict all of the possible genotypes and phenotypes for ABO blood type that could be found among Mr. and Mrs. Cash's offspring.
11. **Interpreting Data** Do the results of the blood tests on the Cash children agree with the prediction you made in step 10? Explain.
12. **Interpreting Data** Based on your test results, what antigens are found on the red blood cells of Mr. X?
13. **Drawing Conclusions** What is the blood type of Mr. X? What genotypes are possible for Mr. X?
14. **Drawing Conclusions** Can the court reject Mr. X's claim to the estate of Mr. and Mrs. Cash based on his blood type? Support your answer.
15. **Making Inferences** If Mr. X had blood type AB, would this prove that he is Mr. and Mrs. Cash's son? Explain your answer.

Applications

16. **Thinking Critically** Is blood typing a good way to identify a criminal or to establish parentage? Why or why not?
17. **Thinking Critically** When blood cells clump in the bloodstream, the result is usually death. How do you think blood typing improved the safety of blood transfusion (transfer of blood from one person to another)? Use data from this Exploration to justify your answer.
18. **Career Connection** *Medical laboratory technicians* perform blood-typing tests. Find out about the training and skills required to become a medical laboratory technician, and discover some other techniques performed in a medical laboratory.

Exploration 7A 775

Procedure includes safety precautions and cleanup and disposal instructions.

Analysis items help you analyze the lab's techniques and your data to come to a conclusion.

Applications questions require you to apply what you have learned and direct you to research a career related to the lab's skills.

Materials are the items you will need to do the lab.

Prelab Preparation items are things you should do before beginning the procedure.

Procedure is a step-by-step explanation for doing the lab.

What You Should Do Before an Exploration

Preparation helps you work safely and efficiently. The day before a lab, be sure to complete the following steps.

- **Read the *Situation, Background,* and *Procedure*** to make sure you understand what you will do.
- **Read the safety information** that begins on page 752, as well as that provided in the lab procedure.
- **Complete the *Prelab Preparation* section** by defining the terms, answering the questions, and preparing data tables in your lab notebook.

What You Should Do After an Exploration

Most teachers require a lab report as a way of making sure that you understood what you were doing in the lab. Your teacher will give you specific details about how to organize the lab report. Most lab reports include the following information.

- **title** for the lab
- **summary paragraph** describing the purpose and procedure
- **data tables and observations** that are organized and comprehensive
- **answers** for the items in the Analysis and Applications sections

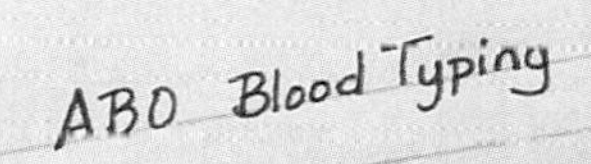

Tips for Success in the Lab

Whether you are doing an Exploration or an Investigation, you can take the following steps to help ensure success.

- **Read each lab twice** before you come to class.
- **Read and follow the safety precautions** in the lab and on pages 752–755.
- **Prepare a data table** before you come to class.
- **Record all data and observations immediately** in your data table or lab notebook.
- **Use appropriate units** whenever you record data.
- **Keep your lab table organized** and free of clutter.

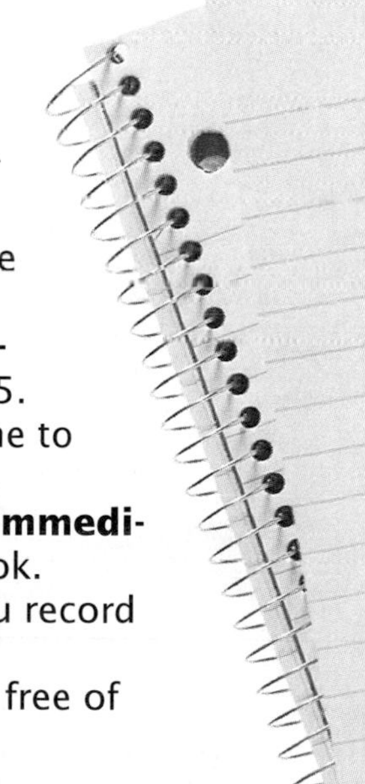

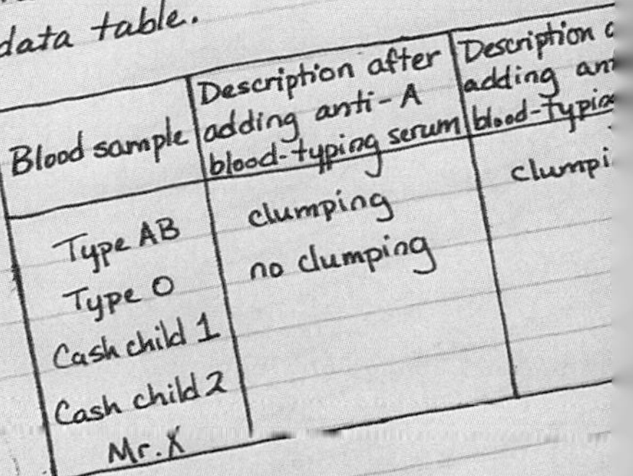

Using the Investigations

Each Investigation requires you to assume the role of an employee in a biological consulting firm. You will develop your own procedure to solve a problem that has been presented to your company by a client. The procedures you develop will be based on the procedures and techniques you learned in previous Explorations. You must also decide what equipment to use for a project and manage expenses to earn a profit.

Each Investigation Contains the Following Parts:

Prerequisites tell you which Explorations contain procedures that apply to the Investigation and which concepts you will need to understand to complete the lab.

The **Letter** contains a request from a client to solve a problem or to do a project.

The **Memorandum** is a note from your supervisor directing you to perform the work requested by the client. The memorandum contains clues or directions that will help you design a successful experiment or project.

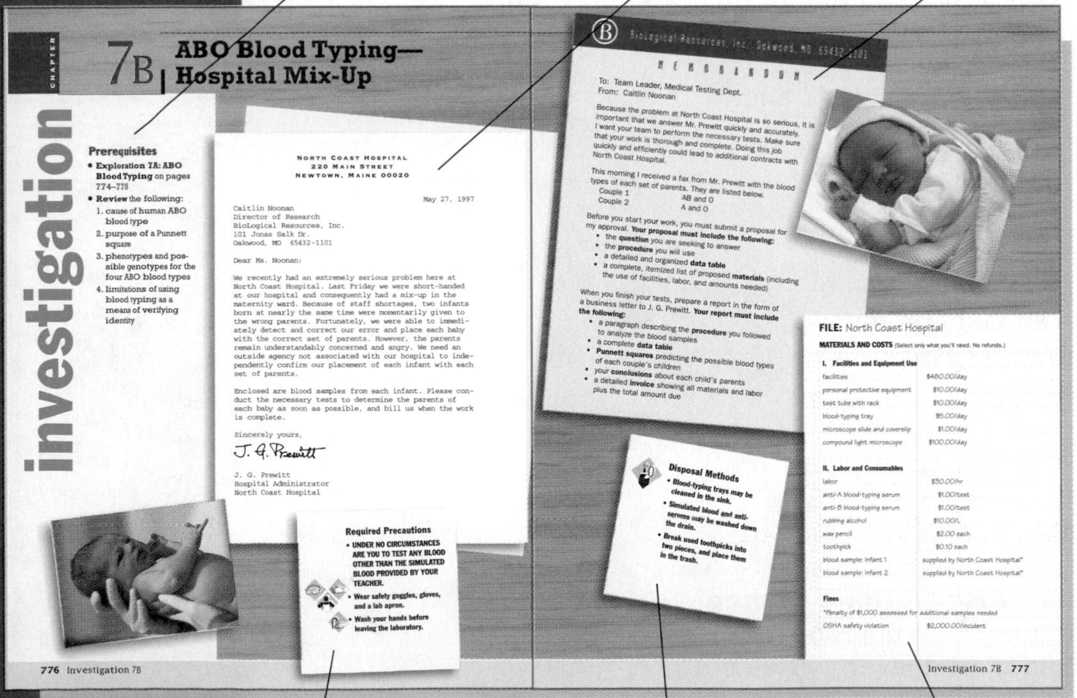

Required Precautions are the safety procedures you must follow during the lab.

Disposal Methods tell you how to dispose of the materials used.

Materials and Costs is a list of the services and items that might be needed to complete the work and the unit cost of each item. Select from this list only the services and items you will need to complete the project within your budget.

What You Should Do Before an Investigation

Before you will be allowed to work on an Investigation, you must turn in a proposal that includes the question to be answered, the procedure you will use, a complete and organized data table, and a list of all the required services and materials with their costs. Before you begin writing your proposal, follow these steps.

- **Read the Investigation thoroughly,** and jot down any clues you find that will help you successfully complete the lab.
- **Consider what you must measure or observe** to solve the problem.
- **Think about Explorations** you have done that required similar measurements and observations.
- **Imagine working through a procedure,** keeping track of each step and what equipment you will need.

What You Should Do After an Investigation

After you finish, organize a report as described in the Memorandum. The report can be in the form of a one- or two-page letter to the client plus an invoice showing how much you charged for each phase of the work and the total cost of the work. (Your teacher may have additional requirements for your report.) Carefully consider how to convey the information the client needs to know. In some cases, graphs and diagrams may communicate information better than words.

BioLogical Resources, Inc. Oakwood, MO 65

June 4, 1997

J. G. Prewitt
Hospital Administrator
North Coast Hospital
220 Main Street
Newtown, Maine 00020

Dear Mr. Prewitt,

We have completed the blood-typing tests as you reque
following procedures: 1) place 3-4 drops of Infant 1's
two wells of a blood-typing tray; 2) add 3-4 drops of ar
serum to one well and 3-4 drops of anti-B blood-typin
well; 3) stir the blood and typing serum in each well
toothpick for 30 seconds; 4) observe each well for 2
observations in a data table; 5) repeat step 1–4 usi
6) repeat steps 1–5 two more times; and 7) compa
Punnett squares completed for each couple. The P
below.

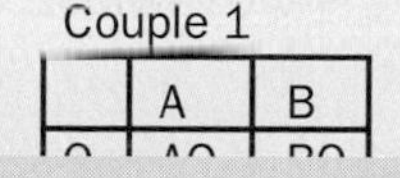

Couple 1

	A	B
O	AO	BO

Couple 2

	A	A
O	AO	AO

or

BILL TO: North Coast Hospital

MATERIALS AND COSTS

I. Facilities and Equipment Use

Item	Rate	Number	Total
facilities	$480.00/day	2	$960.00
personal protective equipment	$10.00/day	2	$20.00
blood-typing tray	$5.00/day	2	$10.00

II. Labor and Consumables

Item	Rate	Number	Total
labor	$30.00/hour	2	$60.00
anti-A blood-typing serum	$1.00/test	3	$3.00
anti-B blood-typing serum	$1.00/test	3	$3.00
toothpicks	$0.10 each	12	
blood sample: Infant 1			

Using the Interactive Explorations

Imagine that you could watch what happens within a plant cell during photosynthesis, or that you could observe how drug use changes the nervous system. You can't do these things in a conventional laboratory exercise, but you can in an Interactive Exploration.

Using the Disk *Interactive Explorations in Biology: Human Biology*

1. Obtain the CD-ROM *Interactive Explorations in Biology: Human Biology* from your teacher. Load the disk into the CD-ROM drive according to the instructions provided by your teacher.

2. An icon for the CD-ROM will appear on the screen. Click it, and then click the icon labeled "Start Explorations."

3. After a few moments, the title screen will appear. Click the button labeled "Menu" at the bottom right of the screen.

4. The computer will then display the menu screen, which lists all the Interactive Explorations on the disk. Select the Interactive Exploration you wish to run and click on its title.

5. After a few moments, the computer will display a screen containing a column of buttons along the right edge, as shown in **Figure 1**. This column is the Navigation Palette, and the buttons it contains allow you to access the various features found in each Interactive Exploration. To begin, click the top button on the column, labeled "Interactive Investigation."

6. If you have questions about an Interactive Exploration, you can play the instructions for operating the Interactive Exploration. Click the word *File* at the top of the screen, and select "Interactive Exploration Help." You may also refer to the User's Guide for additional information.

Using the Disk *Interactive Explorations in Biology: Cell Biology and Genetics*

1. Obtain the CD-ROM *Interactive Explorations in Biology: Cell Biology and Genetics* from your teacher. Load the disk into the CD-ROM drive according to the instructions provided by your teacher.

2. An icon for the CD-ROM will appear on the screen. Click it, and then click the icon labeled "Start Explorations."

3. After a few moments, the title screen will appear. Click the button labeled "Menu" at the bottom right of the screen.

4. The computer will then display the menu screen, which lists all the Interactive Explorations on the disk. Select the Interactive Exploration you wish to run and click on its title.

5. After a few moments, the computer will display the Interactive Exploration screen and play a short narration that describes the problem you will be studying. (To shut off this narration, select "Sound Off" from the Sound menu at the top right of the screen.) After this brief narration has played, the Interactive Exploration is ready to run.

6. If you have questions about an Interactive Exploration, you can play the instructions for operating the Interactive Exploration. Click the word *Help* at the top of the screen, and select "How to Use This Exploration" from the menu that appears. Clicking the Help button on the Navigation Palette will call up instructions on how to use the features of the Navigation Palette. You may also refer to the User's Guide.

Figure 1
Clicking the Navigation button at the bottom right of the screen calls up the Navigation Palette, a ribbon of 10 buttons, each with its own icon. The title of each button appears when you position the pointer on the icon. The buttons of the Navigation Palette allow you to navigate among features within the Interactive Exploration, to move to a different Interactive Exploration, or to quit entirely.

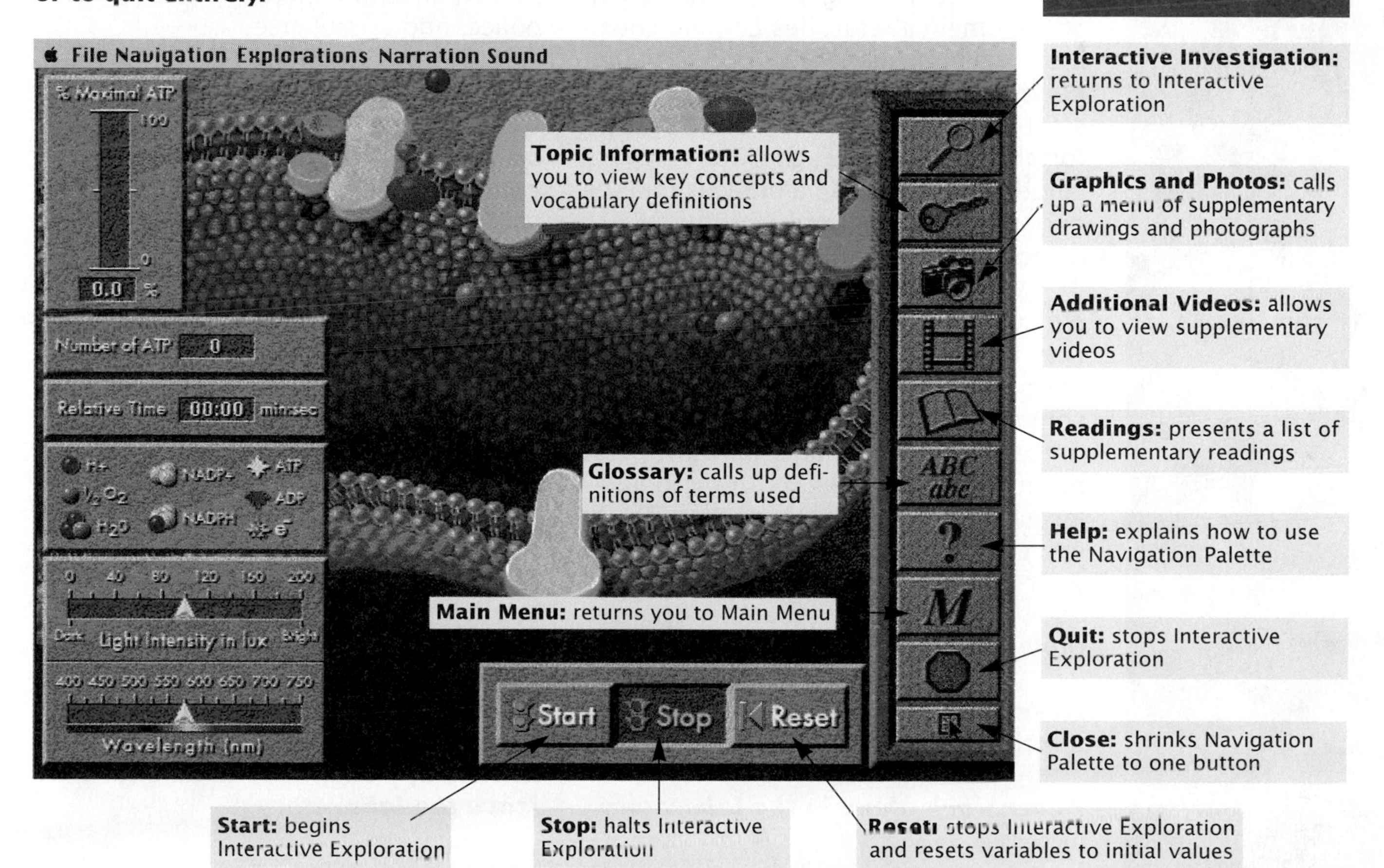

Lab Program Safety

Safe Laboratory Practices

It is your responsibility to protect yourself and other students by conducting yourself in a safe manner while in the laboratory. Familiarize yourself with the printed safety symbols—they indicate additional measures that you must take.

While in the Laboratory

Familiarize yourself with the investigation—especially safety issues—before entering the lab. Know the potential hazards of the materials, equipment, and procedures required. Ask the teacher to explain any parts you do not understand before you start.

Always wear a lab apron and safety goggles. Wear this equipment even if you are not working on an experiment at the time. Laboratories contain chemicals that can damage your clothing. Keep the apron strings tied. If your safety goggles cloud up, or are uncomfortable, ask your teacher for help. Try lengthening the strap a bit, washing the goggles with soap and warm water, or using an anti-fog spray.

No contact lenses are allowed in the lab. Even while you are wearing safety goggles, chemicals could get between contact lenses and your eyes and cause irreparable eye damage. If your doctor requires that you wear contact lenses instead of glasses, then you should wear eye-cup safety goggles in the lab. Ask your doctor or your teacher how to use this very important and special eye protection.

Never perform any experiment not specifically assigned by your teacher.

Never work alone in the laboratory.

Know the location of all safety and emergency equipment used in the laboratory. Examples include eyewash stations, safety blankets, safety shower, fire extinguisher, first aid kit, and chemical spill kit.

Know the location of the closest telephone. Be sure there is a posted list of emergency phone numbers, including poison control center, fire department, police, and ambulance.

Before beginning work: tie back long hair, roll up loose sleeves, and put on any required personal protective equipment as directed by your teacher. Avoid or confine loose clothing that could knock things over, catch on fire, or soak up chemical solutions. Remember that nylon and polyester fabrics burn and melt more readily than does cotton. Open-toed shoes or sandals offer no protection, and even cloth shoes are not a good idea in the laboratory.

Report any accident, incident, or hazard—no matter how trivial—to your teacher immediately. Any incident involving bleeding, burns, fainting, chemical exposure, or ingestion should also be reported immediately to your school nurse or school physician.

In case of fire, alert the teacher and leave the laboratory.

Horseplay and fooling around in the lab are very dangerous. NEVER be a clown in the laboratory.

Never consume food or drink, or apply cosmetics, in the laboratory. Never store food in the laboratory. Keep your hands away from your face. Wash your hands at the conclusion of each laboratory investigation and before leaving the laboratory. Remember that some hair-care products are highly flammable, even after application.

Keep your work area neat and uncluttered. Bring only books and other materials that are needed to conduct the experiment.

Clean your work area at the conclusion of the experiment as your teacher directs.

When Called for, Use These Safety Procedures:

Eye Safety

Wear approved chemical safety goggles as directed. Goggles should always be worn in the laboratory, especially when you are working with a chemical or solution, heat source, mechanical device, or physical process.

In case of eye contact:
Go to an eyewash station immediately and flush your eyes (including under the eyelids) with running water for at least 15 minutes. Hold your eyelids open with your thumb and fingers, and roll your eyeball around. After starting to flush the eye, have another student notify your teacher or another adult in charge.

Wearing contact lenses for cosmetic reasons is prohibited in the laboratory. Liquids or gases can be drawn up under the contact lens and into the eye. If you must wear contacts prescribed by a physician, inform your teacher. You must wear approved eye-cup safety goggles.

Never look directly at the sun through any optical device or lens system, or reflect direct sunlight to illuminate a microscope. Such actions will concentrate light rays that will severely burn your retina, possibly causing blindness!

Electrical Supply

Never use equipment with frayed cords.

Make sure that electrical cords are taped to work surfaces. This will prevent tripping and will ensure that equipment can't be pulled off the table.

Never use electrical equipment around water, or with wet hands or clothing.

Clothing Protection

Wear an apron or laboratory coat when working in the laboratory to prevent chemicals or chemical solutions from contacting skin or street clothes.

Animal Care

Do not touch or approach any animal in the wild. Be aware of poisonous or dangerous animals in any area where you will be doing outside fieldwork.

Always obtain your teacher's permission before bringing any animal (including pets) into the school building.

Handle animals only as your teacher directs. Mishandling or abuse of any animal will not be tolerated!

continued

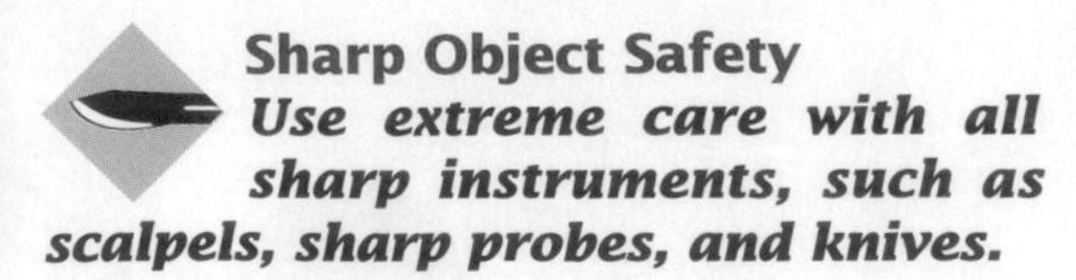

Sharp Object Safety

Use extreme care with all sharp instruments, such as scalpels, sharp probes, and knives.

Never use double-edged razor blades in the laboratory.

Never cut objects while holding them in your hand. Place objects on a suitable work surface.

Hygienic Care

Keep your hands away from your face and mouth.

Wash your hands thoroughly before leaving the laboratory.

Remove contaminated clothing immediately. If the spill is on your pants or somewhere else that will not fit under the sink faucet, use the safety shower. Remove the pants or other affected clothing while under the shower, and call your teacher. (It may be temporarily embarrassing to remove clothing in front of your classmates, but failing to flush that chemical off your skin could cause permanent damage.)

Launder contaminated clothing separately.

Use the proper technique demonstrated by your teacher when handling bacteria or similar microorganisms. Examine microorganism cultures (such as petri dishes) without opening them.

Return all stock and experimental cultures to your teacher for proper disposal.

Chemical Safety

Always wear appropriate personal protective equipment. Eye goggles, gloves, and an apron or lab coat should always be worn when working with any chemical or chemical solution.

Never taste, touch, or smell any substance, or bring it close to your eyes, unless specifically told to do so by your teacher. If you are directed by your teacher to note the odor of a substance, do so by waving the fumes toward you with your hand. Never pipette any substance by mouth; use a suction bulb as directed by your teacher.

Always handle chemicals or chemical solutions with care. Check the label on the bottle and observe safety procedures. Never return unused chemicals or solutions to their containers. Return unused reagent bottles or containers to your teacher.

Never mix any chemical unless specifically told to do so by your teacher.

Never pour water into a strong acid or base. The mixture can produce heat and splatter. Remember this rhyme:

Do as you oughta—
Add acid (or base) to water.

Report any spill immediately to your teacher. Handle spills only if your teacher instructs you to do so.

Check for the presence of any source of flames, sparks, or heat (open flames, electric heating coils, etc.) before working with flammable liquids or gases.

Proper Waste Disposal

Clean and decontaminate all work surfaces and personal protective equipment as directed by your teacher.

Dispose of all sharp objects (such as broken glass) and other contaminated materials (biological or chemical) in special containers as directed by your teacher.

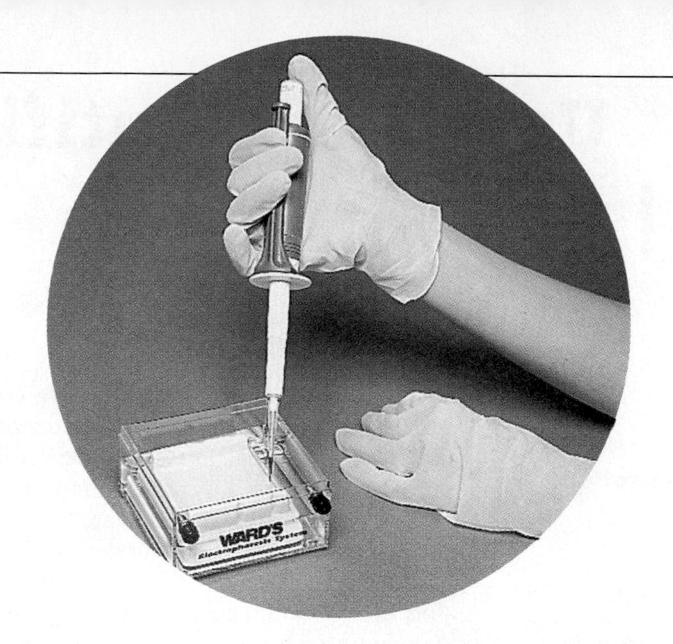

Heating Safety

When heating chemicals or reagents in a test tube, never point the test tube toward anyone.

Use hot plates, not open flames. Be sure hot plates have an "On-Off" switch and indicator light. Never leave hot plates unattended, even for a minute. Never use alcohol lamps.

Know the location of laboratory fire extinguishers and fire blankets. Have ice readily available in case of burns or scalds.

Use tongs or appropriate insulated holders when heating objects. Heated objects often do not have the appearance of being hot. Never pick up an object with your hand unless you are certain it is cool.

Keep combustibles away from heat and other ignition sources.

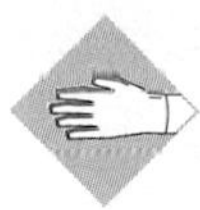

Hand Safety

Never cut objects while holding them in your hand.

Wear protective gloves when working with stains, chemicals, chemical solutions, or wild (unknown) plants.

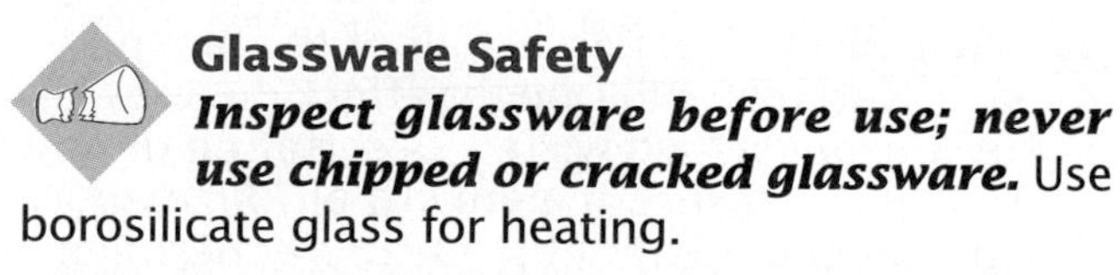

Glassware Safety

Inspect glassware before use; never use chipped or cracked glassware. Use borosilicate glass for heating.

Do not attempt to insert glass tubing into a rubber stopper without specific instruction from your teacher.

Always clean up broken glass by using tongs and/or a brush and dustpan. Discard the pieces in an appropriately labeled container.

Safety With Gases

Never inhale any gas or vapor unless directed by your teacher.

Handle materials prone to emit vapors or gases in a well-ventilated area. This work should be done in an approved chemical fume hood.

Plant Safety

Do not ingest any plant part used in the laboratory (especially seeds sold commercially). Do not rub any sap or plant juice in your eyes, on your skin, or on mucous membranes.

Wear protective gloves (disposable polyethylene gloves) when handling any wild plant.

Wash hands thoroughly after handling any plant or plant part (particularly seeds). Avoid touching your face and eyes.

Do not inhale or expose yourself to the smoke of any burning plant.

Do not pick wildflowers or other plants unless directed to do so by your teacher.

CHAPTER 1A

Using a Scientific Method

exploration

Skill:

Conducting a controlled experiment

Objectives

- *Identify* independent variables, dependent variables, and constants.
- *Measure* changes in a dependent variable.
- *Write* a hypothesis.
- *Conduct* an experiment to test a hypothesis.

Materials

- safety goggles
- lab apron
- 6 plastic petri dishes
- wax pencil
- paper towels
- scissors
- plastic knife
- day-old bread
- water
- 10 mL graduated cylinder
- growth chamber
- thermometer
- metric ruler

Situation

You have just started working as a laboratory technician for a pharmaceutical company. The company is interested in developing new antibiotics that will combat antibiotic-resistant strains of bacteria. Your supervisor is hoping to discover whether *Rhizopus*, which is a common mold that grows on bread, produces any compounds that kill disease-causing bacteria. Your research team has been given the task of determining the optimum conditions for growing the *Rhizopus* needed for testing. The first factor your team decides to test is light.

Background: In designing an experiment, you must identify and define all of the factors that can possibly change. Factors that can change are called **variables.** There are two types of variables. **Independent variables** are factors, such as temperature and amount of moisture, that you can change, or manipulate. In a simple experiment, only one independent variable should be allowed to change, and all other independent variables should be **constants** (factors that are not allowed to change). **Dependent variables** are factors, such as an increase or decrease in size, that change in response to an independent variable. Every experiment must have a **control,** a standard against which test groups can be compared.

Prelab Preparation

1. Define the following terms: observation, hypothesis, prediction, experiment, scientific method.
2. Define the boldface terms in the **Background** section.
3. Make a data table like the one at the top of the next page.

Procedure

1. On a sheet of paper, list all of the independent variables that might affect the growth of *Rhizopus*. Also write a description of how you will measure the growth of *Rhizopus*, the dependent variable for this experiment.
2. The question you will answer in this experiment is, How does light affect the growth of *Rhizopus*? Look back at your list of independent variables, and label "light" as the "independent variable to be tested." Label all other variables as "constant."

Day	Experimental group	Control group
1		
2		
3		
4		
5		

3. On your data sheet, write a hypothesis that can be tested with an experiment to determine the effect of light on the growth of *Rhizopus*. Write an "if . . . then" statement that restates your hypothesis in the form of a prediction.
4. Put on safety goggles and a lab apron.
5. Prepare six petri dishes as follows: Label the bottom of three dishes Control 1, 2, and 3, respectively. Label the other dishes Experimental 1, 2, and 3, respectively. Cut a circle of paper towel to fit into the bottom of each petri dish. Add 1 mL of water to each petri dish.
6. Using a plastic knife, cut six 5 cm × 5 cm squares of day-old bread. Place one square on the paper towel in each petri dish. Place the tops on the petri dishes, but do not seal the dishes.
7. Put the petri dishes marked "Experimental" in a lighted area. Do not put them in direct sunlight or in a place where they will get additional heat. Put the petri dishes marked "Control" in a darkened growth chamber. *Note: The temperature in the growth chamber and in the lighted area should be the same.*
8. Observe each petri dish in the "Experimental" group and in the "Control" group for 10 days. **CAUTION: Do not remove the tops of the petri dishes.** Use a metric ruler to measure the size of any colonies that develop. Record any changes you see in your data table.
9. Dispose of your materials according to the directions from your teacher.
10. Clean up your work area and wash your hands before leaving the lab.

Analysis

11. **Relating Concepts** Why is light an independent variable in this experiment? Why is mold growth a dependent variable?
12. **Identifying Functions** What is the function of the bread in this experiment?
13. **Summarizing Data** Write a brief description of what happened in each group of petri dishes.
14. **Analyzing Data** Do your data support your hypothesis? Explain.
15. **Drawing Conclusions** How does light affect the growth of *Rhizopus*?
16. **Making Predictions** What would happen if you placed the "Experimental" group in a darkened growth chamber?

Applications

17. **Thinking Critically** A team decides to test the hypothesis that *Rhizopus* will grow better if more water is available. They place bread samples in containers of different sizes, add 10 mL of water to each container, and then place the containers in a darkened growth chamber. What can you say about this experimental design?
18. **Personal Connection** Where should you store bread at home if you want to prevent the growth of *Rhizopus*?
19. **Career Connection** *Pharmacologists* develop new medicinal drugs. Find out about the training and skills required to become a pharmacologist, and discover the source of some common antibiotics.

CHAPTER

1B | Using a Scientific Method—Tropical-Fish Dealer

investigation

Prerequisites

- **Exploration 1A: Using a Scientific Method** on pages 756–757
- **Review** the following:
  1. independent variable
  2. dependent variable
  3. control
  4. constant

Farmaquatics, Inc.

August 30, 1997

Caitlin Noonan
Director of Research
BioLogical Resources, Inc.
101 Jonas Salk Dr.
Oakwood, MO 65432-1101

Dear Ms. Noonan:

I have recently started a tropical-fish breeding company in Denver. I have extensive experience in raising and breeding fish but have always relied on others for food stocks, including brine shrimp for tropical fish. Because of the size of the operation, my staff will need to raise the brine shrimp for feeding.

I've heard that it is possible to increase the number of brine shrimp eggs that hatch by controlling the salt content of the water in the hatching containers, but I do not know the specifications. I need to know what salt concentration will enable the most brine shrimp to hatch.

Dr. Lamberton at the university told me of your new company and assured me that this is just the sort of problem you could handle. I wish to begin operating our facility as soon as possible. Please conduct the appropriate experiments to determine the optimal salt concentration for hatching brine shrimp. Bill us when you send your report.

Sincerely,

Arthur Beauchamp

Arthur Beauchamp
President
Farmaquatics, Inc.

Required Precautions

- Wear safety goggles and a lab apron.
- Do not touch or taste any solution or chemical used in the laboratory.

- Wash your hands before leaving the laboratory.

BioLogical Resources, Inc., Oakwood, MO 65432-1101

MEMORANDUM

To: Team Leader, Zoology Dept.
From: Caitlin Noonan

This could be a great first client for us! Good references from Mr. Beauchamp might bring us more business. Farmaquatics is relying on us to provide quick and accurate results about the optimal salt concentration for hatching brine shrimp. Test at least 5 different salt concentrations. Make sure that your work is thorough and accurate.

Last night I did some research on brine shrimp and found some potentially useful information:

- Brine shrimp live in tidal pools, estuaries, and salt lakes, where salt concentration may vary from 10 ppt to 60 ppt.
- Hatched brine shrimp and unhatched eggs can be collected with a clean cotton swab.

Before you start your work, you must submit a proposal for my approval. **Your proposal must include the following:**

- the **question** you are seeking to answer
- the **procedure** you will use
- a detailed **data table**
- a complete, itemized list of the proposed **materials** and **costs** (including the use of facilities, labor, and amounts needed)

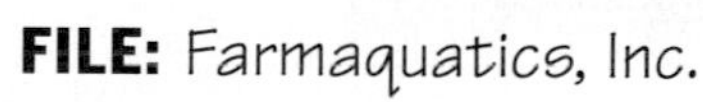

When you finish your tests, prepare a report in the form of a business letter to Arthur Beauchamp. **Your report must include the following:**

- a paragraph describing the **procedure** you followed to determine the optimal salt concentration for hatching brine shrimp
- a complete **data table**
- your **conclusions** about the salt concentration that Mr. Beauchamp should use for hatching brine shrimp
- a detailed **invoice** showing all materials and labor plus the total amount due

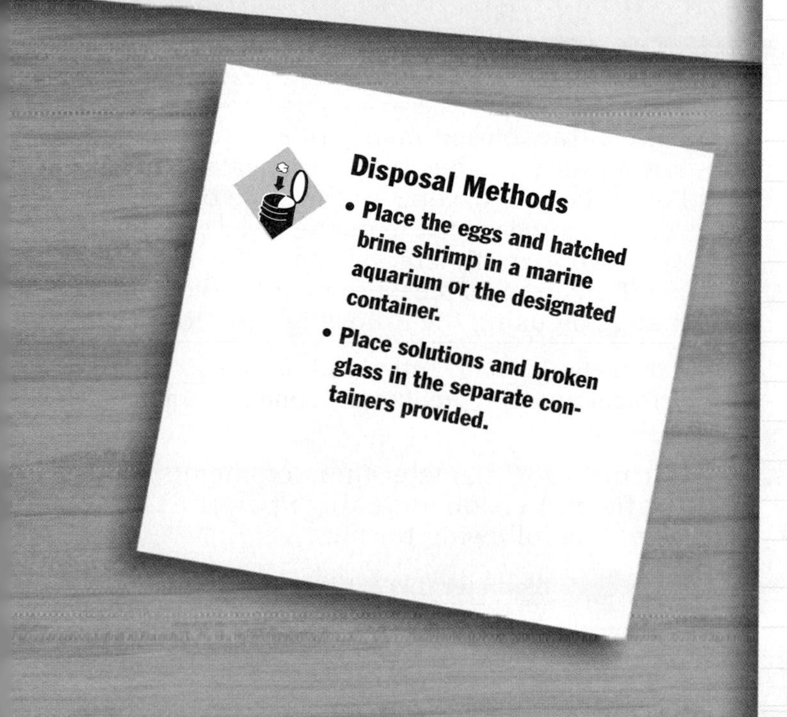

Disposal Methods
• Place the eggs and hatched brine shrimp in a marine aquarium or the designated container.
• Place solutions and broken glass in the separate containers provided.

FILE: Farmaquatics, Inc.

MATERIALS AND COSTS (Select only what you'll need. No refunds.)

I. Facilities and Equipment Use	
facilities	$480.00/day
personal protective equipment	$10.00/day
stereomicroscope or hand lens	$100.00/day
microscope slide and coverslip	$1.00/day
forceps	$5.00/day
dropper	$5.00/day
10 mL graduated cylinder	$10.00/day
balance	$10.00/day
beaker	$10.00/day
II. Labor and Consumables	
labor	$30.00/hr
brine-shrimp eggs	$25.00/25 g package
dried yeast (brine-shrimp food)	$5.00/package
petri dishes	$5.00 each
wax pencil	$2.00 each
aluminum foil	$0.25/m
synthetic sea salts	$5.00/kg
dechlorinated water	$10.00/L
cotton swabs	$0.10 each
plastic foam cups	$1.00 each
Fines	
OSHA safety violation	$2,000.00/incident

CHAPTER

3A Observing Cells With a Microscope

exploration

Skills:
- **Using a compound microscope**
- **Making wet mounts**

Objectives
- *Make* wet mounts of cells.
- *Estimate* the size of cells.
- *Stain* cells.
- *Identify* cell structures.
- *Identify* a cell as eukaryotic or prokaryotic.

Materials
- safety goggles
- lab apron
- compound microscope
- lens paper
- transparent metric ruler
- depression slide
- microscope slide
- 2 coverslips
- culture of organism A
- medicine dropper
- DETAIN™, protist-slowing agent
- toothpick
- Lugol's solution (stain)
- paper towel
- leaf from organism B
- prepared slide of organism C
- prepared slide of organism D

Situation

You are a member of a biological research team that is studying the life in a polluted pond. You have been given samples of the organisms found in the pond. Your job is to observe the organisms with a compound microscope, determine the size of their cells, and categorize them as prokaryotic or eukaryotic.

Background: Cells are measured in micrometers (μm). There are 1000 μm in 1 mm. **Organelles** and other cell structures that can be seen with a compound microscope may include the **cell membranes, cell walls, nuclei,** and **chloroplasts.** Chemicals called **stains,** which are taken up by certain parts of cells, make cell structures more visible under a microscope.

Prelab Preparation

1. Define the boldface terms in the **Background** section.
2. List the characteristics of prokaryotic cells and eukaryotic cells.
3. Review the information about the care and use of a compound microscope in the Appendix.
4. Review the procedures for making a wet-mount slide in the Appendix.
5. Prepare a data table like the one at the top of the next page.

Procedure

1. Place a transparent metric ruler under the low power (LP) objective of a microscope. Focus the microscope on the scale of the ruler, and measure the diameter of the field of vision in millimeters. Record this number. Convert this measurement to micrometers by using the following equation:

$$\text{diameter in micrometers } (\mu m) = \text{diameter in millimeters (mm)} \times \frac{1000\ \mu m}{mm}$$

2. Calculate the diameter (in micrometers) of the field of vision under high power (HP) using the following formula:

$$\text{diameter (HP)} = \frac{\text{diameter (LP)} \times \text{mag. of LP objective}}{\text{mag. of HP objective}}$$

Record this number.

Characteristics	Organisms			
	A	B	C	D
General observations				
Size (in µm)				
Cell membrane				
Cytoplasm				
Nucleus				
Chloroplast				
Cell wall				

3. To estimate the size of an object seen with a microscope, first estimate what fraction of the diameter of the field of vision that the object occupies (e.g., one-half, one-tenth). Then multiply the diameter you calculated in micrometers by that fraction. For example, if the field of vision's diameter is 400 µm and the object's estimated length is about one-tenth of that diameter, multiply the diameter by one-tenth to find the object's length (400 µm × 1/10 = 40 µm).
4. Using a medicine dropper, place some of the solution from the container marked "Organism A" into the well of a depression slide. Add one drop of DETAIN™ to this solution and stir with a toothpick. Place a coverslip over the well.
5. View the slide with the low-power objective of a microscope. Find one of the organisms and observe it.
6. Put on safety goggles and a lab apron. Place a drop of Lugol's solution at one edge of the coverslip. **CAUTION: Lugol's solution will stain your skin and clothing. Promptly wash off spills to minimize staining.** Place a piece of paper towel at the opposite edge of the coverslip to draw the stain under the coverslip. Observe the organism.
7. Switch to high power, and continue observing the organism. Use the procedure in step 3 to estimate the size of the organism's cells. Draw what you see on a piece of paper. Label the visible cell structures. Describe this organism by completing the column for it in your data table.
8. Using a clean slide and coverslip, make a wet mount of a leaf from organism B. Repeat steps 5, 6, and 7 for this organism.
9. View a prepared slide of organism C under low power, and then repeat step 7.
10. View a prepared slide of organism D under low power, and then repeat step 7.
11. Dispose of your materials according to the directions from your teacher.
12. Clean up your work area and wash your hands before leaving the lab.

Analysis

13. **Making Inferences** What can you *not* infer about the organisms you observed on prepared slides?
14. **Drawing Conclusions** Your research team must summarize their findings in a report. Write a paragraph in which you describe organisms A, B, C, and D and explain why their cells are prokaryotic or eukaryotic.

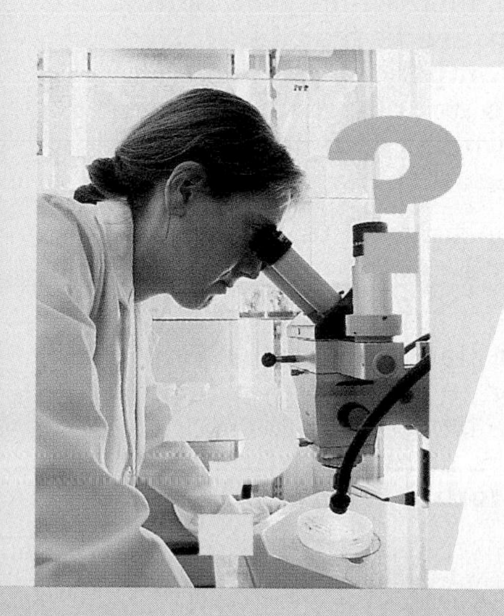

Applications

15. **Thinking Critically** What was the purpose of adding a stain to the slides of organisms A and B?
16. **Thinking Critically** Biologists once classified all organisms into two kingdoms. Evaluate the accuracy of the two-kingdom system of classification on the basis of your observations.
17. **Career Connection** *Microscopists* prepare microscope slides of living cells and tissues. Find out about the training and skills required to become a microscopist, and discover some of the special techniques they must use.

CHAPTER

4A Modeling the Action of a Cell Membrane

exploration

Skill:
Using indicators to detect starch and sugar

Objectives

- *Demonstrate* the movement of molecules through a selectively permeable membrane.
- *Perform* the chemical tests that detect starch and sugar.
- *Collect* and *interpret* data.

Materials

- safety goggles
- lab apron
- wax pencil
- eight 18 × 150 mm test tubes
- test-tube rack
- 10 mL graduated cylinder
- starch solution
- glucose solution
- Lugol's iodine solution
- Benedict's solution
- test-tube holder
- boiling-water bath
- clock or timer
- two 250 mL beakers
- water
- two 15 cm pieces of dialysis tubing
- heavy-duty thread or string

Situation

A manufacturing company has asked you to test a new type of dialysis tubing. The company is concerned that the small gaps in the tubing membrane may be too large for the tubing to be selectively permeable. If it is not selectively permeable, the company will not be able to sell the tubing. Your job is to determine whether this tubing is selectively permeable.

Background: A **selectively permeable membrane** allows only certain particles to pass through it. **Dialysis tubing** is made of a membrane that is selectively permeable, like a cell membrane. It is used in **dialysis,** which is a process that filters soluble materials from blood. Small molecules, such as glucose, pass through dialysis-tubing membrane, while large molecules, such as starch, cannot. Starch and glucose can be detected with **indicators,** chemicals that change color in the presence of particular substances. **Lugol's iodine solution** indicates starch by turning blue-black. **Benedict's solution** indicates glucose by turning rusty red.

Prelab Preparation

1. Define the boldface terms in the **Background** section.
2. What is a selectively permeable membrane?
3. Prepare a data table like the one on the next page.

Procedure

1.   Put on safety goggles and a lab apron.
2. Using a wax pencil, label two clean test tubes "Glucose" and two "Starch." **CAUTION: Glassware is fragile. Notify your teacher promptly of any broken glass or cuts. Do not clean up broken glass or spills unless your teacher tells you to do so.**
3. Add 5 mL of starch solution to each test tube labeled "Starch." Add 5 mL of glucose solution to each test tube labeled "Glucose."
4. Perform a test for starch by adding 2 drops of Lugol's solution to one test tube labeled "Glucose" and one labeled "Starch." **CAUTION: Lugol's solution will stain your skin and clothing. Promptly wash**

	Solution	Color after Lugol's test for starch	Color after Benedict's test for sugar
Initial tests	Glucose solution		
	Starch solution		
	Beaker 1 water		
	Beaker 2 water		
Final tests	Beaker 1 water		
	Beaker 2 water		

off spills to minimize staining. Record any color change in your data table.

5. Perform a test for glucose by adding 10 drops of Benedict's solution to each of the remaining test tubes. Place the test tubes in a boiling-water bath for 5 minutes. **CAUTION: Do not touch hot objects with your bare hands. Use tongs, test-tube holders, and padded gloves, as appropriate.** Remove the test tubes from the water bath.

6. Using a wax pencil, label a 250 mL beaker "Beaker 1—Glucose" and another 250 mL beaker "Beaker 2—Starch." Fill each beaker half full with water.

7. Label two test tubes "Beaker 1 water," and two test tubes "Beaker 2 water." Add 5 mL of water from the designated beaker to each test tube. Set the test tubes in a test tube rack.

8. Obtain two 15 cm pieces of dialysis tubing from your teacher. Securely tie one end of a piece of dialysis tubing as shown in the photograph at right. Open the opposite end of the tubing, and pour glucose solution into the tubing until it is about two-thirds full. Using another piece of thread, tightly tie off the other end of the tubing to make a bag. Rinse the outside of the bag, and place it in the beaker labeled "Beaker 1—Glucose."

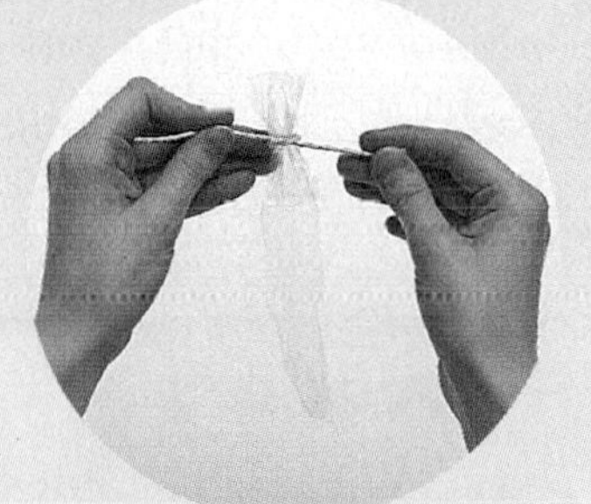

9. Make a second dialysis-tubing bag with starch solution. Rinse the outside of the bag, and place it in the beaker labeled "Beaker 2—Starch." Set the beakers aside.

10. Using the test tubes you labeled in step 7, repeat steps 4 and 5 to test the water from each beaker for the presence of starch and glucose. Record your results. Empty and clean the test tubes.

11. After 15 minutes, again test the water in the beakers for the presence of glucose and starch. Record your results.

12. Dispose of your materials according to the directions from your teacher.

13. Clean up your work area and wash your hands before leaving the lab.

Analysis

14. **Interpreting Data** What do your results suggest about the contents of the water in each beaker at the end of the experiment?

15. **Inferring Conclusions** What would you report to the manufacturing company about the permeability of their dialysis tubing? What evidence supports your conclusion?

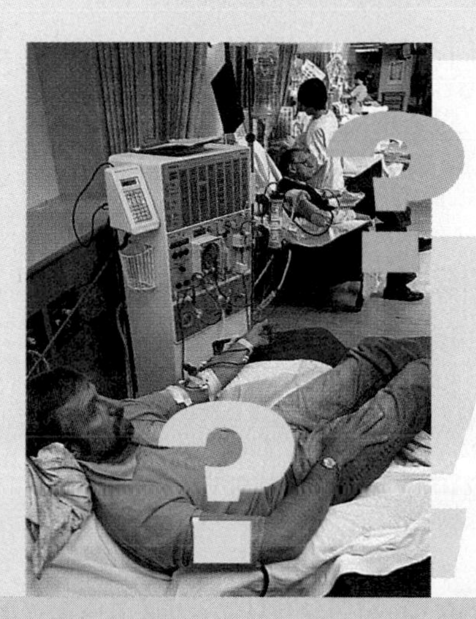

Applications

16. **Thinking Critically** Why can dialysis tubing be used as a model for the cell membrane?

17. **Thinking Critically** How could you determine whether water from the beakers moved into the dialysis-tubing bags?

18. **Career Connection** *Hemodialysis technicians* perform kidney dialysis. Find out about the training and skills required to operate a hemodialysis machine, and discover how semipermeable membranes are used in these machines.

CHAPTER

5A Measuring the Rate of a Process

exploration

Skill:
Measuring the production of a gas

Objectives

- *Measure* the rate of photosynthesis.
- *Recognize* the relationship between color of light and the rate of photosynthesis.
- *Relate* a change in pH to evidence of photosynthesis.

Materials

- safety goggles
- lab apron
- 5 *Elodea* sprigs
- 5 large test tubes
- 5 one-hole stoppers with glass tubing
- water with food coloring
- fine-point marker
- test-tube rack
- 250 mL graduated cylinder
- 400 mL beaker
- 125 mL of water
- 25 mL of bromothymol blue solution
- soda straw
- colored cellophane (red, blue, green, yellow, and colorless)

Situation

You are an environmental engineer. A client wants to start a small retail garden center that will feature a greenhouse. Your client wants to use a bright color of plastic to cover the greenhouse for two reasons. First, bright colors are appealing to customers. Second, colored plastic will reduce the amount of light entering a greenhouse and will therefore keep the greenhouse cooler. Your client knows that the color of light affects the rate of photosynthesis. Your job is to determine which color of plastic will promote the highest rate of photosynthesis.

Background: Colored plastic acts as a **filter,** which permits a certain color of light to pass through it and absorbs the other colors. The color of a filter is due to the color of light it transmits (allows to pass through it). When carbon dioxide gas mixes with water, it reacts with the water to form a weak acid. Bromothymol blue is a **pH indicator** that is yellow in an acid solution and blue in a basic solution.

Prelab Preparation

1. Define the boldface terms in the **Background** section.
2. Write a chemical equation that summarizes photosynthesis. Underline the symbols for carbon dioxide gas and oxygen gas.
3. Describe how you will measure the rate of photosynthesis in this Exploration.
4. Prepare a data table like the one shown at the top of the next page.

Procedure

1. Put on safety goggles and a lab apron.
2.  Prepare a solution containing 125 mL of water and 25 mL of bromothymol blue solution. Use a soda straw to blow your breath into the solution until the solution turns yellow. Move the straw through the solution as you blow. Fill five large test tubes, to within 3 cm of the top, with this solution. Place the test tubes in a test-tube rack. **CAUTION: Glassware is fragile. Notify your teacher promptly of any broken glass or cuts. Do not clean up broken glass or spills unless your teacher tells you to do so.**

Color of light	Distance	Starting color/pH	Ending color/pH
Blue			
Red			
Yellow			
Green			
White			

3. Obtain 5 stoppers into which glass tubing has been inserted. Place a tiny drop of water with food coloring in each glass tube, and position each drop just above the top of the stopper, as seen in the photo at right. Mark the top of each drop with a fine-point marker.
4. Select 5 healthy, green sprigs of *Elodea* that have been growing in bright light. Each sprig should be about 8 cm long. Immerse one sprig in each test tube. Close each test tube with a stopper and glass tube, as seen in the photo at right.
5. Wrap each test tube in a different color of cellophane. *Note: Use the same number of layers of cellophane for each test tube.* Put the test tubes in the test-tube rack, and place the rack under bright light. After 30 minutes, measure how far each drop of colored water moved in the glass tubes. Record your data.
6. Remove the cellophane from each test tube, and record the color of the solutions in the test tubes.
7. Dispose of your materials according to the directions from your teacher.
8.  Clean up your work area and wash your hands before leaving the lab.

Analysis

9. **Relating Information** What substance in your breath caused bromothymol blue to turn yellow in step 1? Explain why the color change occurred.
10. **Relating Concepts** What is the control in this experiment?
11. **Interpreting Data** How would you explain any color change that occurred in the test tubes? If there were no color change in a test tube, how would you explain this result?
12. **Interpreting Data** How does a color change, or change in pH, indicate that photosynthesis has occurred?
13. **Summarizing Data** List the colors of the cellophane-covered test tubes in order from the fastest rate of photosynthesis to the slowest rate of photosynthesis.
14. **Making Inferences** What is indicated by the bubbles produced in the five test tubes?
15. **Drawing Conclusions** Which color of plastic (light) will enable the fastest rate of photosynthesis?
16. **Making Predictions** How will using colored plastic on the greenhouse affect the growth of the plants?

Applications

17. **Thinking Critically** What other experiments might you conduct to better advise your customer?
18. **Personal Connection** Incandescent light has a high percentage of red light, while fluorescent light has a high percentage of blue light. Which type of light would promote the highest rate of photosynthesis in houseplants? Explain.
19. **Career Connection** *Environmental engineers* study the effect of the environment on living things. Find out about the training and skills required to become an environmental engineer, and discover some of the types of work that they do.

CHAPTER 5B

Measuring the Rate of a Process—Falling Bread

investigation

Prerequisites

- **Exploration 5A: Measuring the Rate of a Process** on pages 764–765
- **Review** the following:
  1. measuring the rate of a gas-producing process
  2. the reactants and products of fermentation

Windy Hill Bakery 

Boston, MA

December 30, 1997

John Williams, Director
Food Technology Division
BioLogical Resources, Inc.
101 Jonas Salk Dr.
Oakwood, MO 65432-1101

Dear Mr. Williams,

We are having a problem here at the bakery, and I hope you will be able to help us. Last week two batches of our bread failed to rise. Yesterday the problem occurred again.

As part of our continuous bread-making procedure, we mix together all of our ingredients except the flour. This mixture is called a broth. The yeast in the broth causes it to ferment. The fermentation process begins at room temperature. Flour is added at the last step, and then the dough is shaped into loaves and baked. Carbon dioxide gas produced by the yeast in the broth causes the bread to rise.

We suspect that our yeast supply may be bad. Dead yeast cells release an amino acid that affects the gluten in flour and prevents bread from rising. I am enclosing a sample of our yeast for you to test. It is very important that we know your results quickly. Please bill us when you send your report.

Very truly yours,

Simone Prevan

Simone Prevan
Windy Hill Bakery
Boston, MA

Required Precautions

- **Wear safety goggles and a lab apron.**
- **Glassware is fragile. Notify your teacher promptly of any broken glass or cuts.**
- **If you get a chemical on your skin or clothing, wash it off while calling to your teacher.**

- **Wash your hands before leaving the laboratory.**

Biological Resources, Inc., Oakwood, MO 65432-1101

MEMORANDUM

To: Team Leader, Food Testing Dept.
From: John Williams

This is our first big job in food technology. I phoned Ms. Prevan to assure her that we will deliver the results right away. She said that the bakery had recently tried to cut utility bills by lowering the temperature in the production area. This, not the yeast, may be the cause of their problem.

Please conduct the appropriate tests to compare the rate of fermentation by the bakery's yeast (Yeast A) with that of another sample of yeast (Yeast B). I also want you to test the rate of fermentation by the bakery's yeast at four different temperatures: 10°C, 20°C, 30°C, and 40°C.

Prepare yeast samples for testing by adding a measured amount of yeast to warm (not hot) water and stirring gently with a glass rod.

Before you start your work, you must submit a proposal for my approval. **Your proposal must include the following:**

- the **question** you are seeking to answer
- the **procedure** you will use
- a detailed **data table**
- a complete, itemized list of proposed **materials** and **costs** (including the use of facilities, labor, and amounts needed)

When you finish your tests, prepare a report in the form of a business letter to Simone Prevan. **Your report must include the following:**

- paragraph describing the **procedure** you followed to determine the rate of fermentation by the yeast
- a complete **data table**
- your **conclusions** about the bakery's yeast and how temperature affects the rate of fermentation by yeast
- a detailed **invoice** showing all materials and labor plus the total amount due

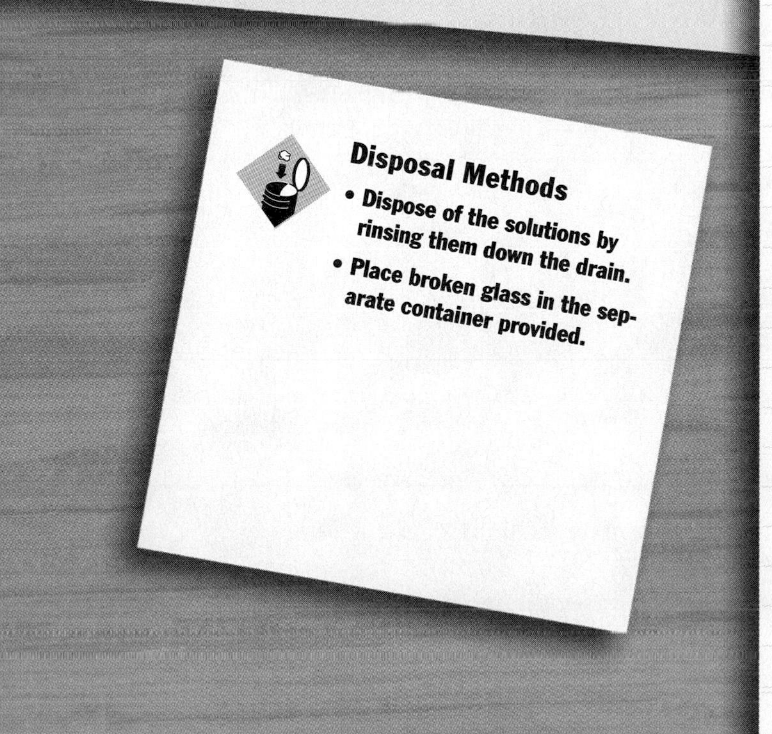

Disposal Methods

- **Dispose of the solutions by rinsing them down the drain.**
- **Place broken glass in the separate container provided.**

FILE: Windy Hill Bakery

MATERIALS AND COSTS (Select only what you'll need. No refunds.)

I. Facilities and Equipment Use	
facilities	$480.00/day
personal protective equipment	$10.00/day
glass stirring rod	$1.00/day
stopper with glass tubing	$5.00/day
large test tube	$5.00/day
test-tube rack	$5.00/day
balance	$10.00/day
graduated cylinder	$10.00/day
beaker	$10.00/day
thermometer	$10.00/day
II. Labor and Consumables	
labor	$30.00/hr
10% glucose solution	$0.10/mL
fine-point marker	$3.00 each
hot and cold tap water	$1.00/L
ice	$1.00/cup
bromothymol blue solution	$5.00/mL
food coloring	$2.00/bottle
Yeast A	no charge
Yeast B	$5.00/package
Fines	
OSHA safety violation	$2,000.00/incident

CHAPTER

5C Photosynthesis

interactive exploration

Background: Photosynthesis is the process in which plants, some bacteria, and some protists capture energy contained in sunlight and store it in carbohydrates. Recall that in photosynthesis, solar energy is first trapped and used to synthesize ATP and NADPH. ATP and NADPH are then used in making carbohydrates. This interactive exploration allows you to explore the effects that varying the intensity and wavelength of light have on the rate of ATP synthesis. The more ATP produced, the greater the amount of carbohydrate that can be synthesized.

Objectives

- *Simulate* how the rate of photosynthesis is affected by variations in the intensity and wavelength of light.
- *Relate* the rate of photosynthesis to the production of ATP in a chloroplast.

Materials

- computer with CD-ROM drive
- CD-ROM *Interactive Explorations in Biology: Cell Biology and Genetics*
- graph paper

Prelab Preparation

1. Load and start the program Photosynthesis. You will see an animated diagram of a chloroplast like the one below. Click the Navigation button, and then click the Topic Information button. Read the focus questions and review these concepts: Photons, Pigments, and The Action Spectrum of Photosynthesis.
2. Click the word *Help* at the top left of the screen, and select How to Use This Exploration. Listen to the instructions that explain the operation of the exploration. Click the Interactive Exploration button on the Navigation Palette to begin the exploration.

Feedback Meters

a **% Maximal ATP:** ratio of number of ATP molecules produced to the maximum number that can be produced

b **Number of ATP:** the number of ATP molecules that have been synthesized

c **Relative Time:** how long the investigation has been running

Variables

d **Light Intensity:** varies the amount of light

e **Wavelength of Light:** varies the wavelength of light

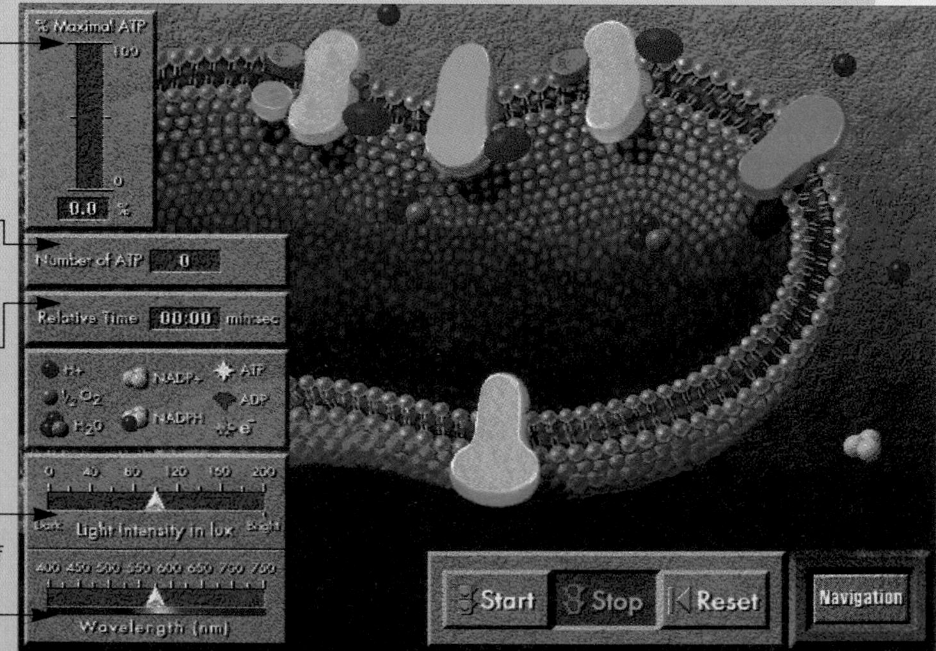

Procedure

Part A: The Effect of Light Intensity on the Rate of Photosynthesis

You will first investigate how varying light intensity affects the rate of photosynthesis. Prepare a three-column table with the headings Light Intensity, % Maximal ATP, and Number of ATP. In the first column, list the following light intensities: 0, 40, 80, 120, 160, and 200 lux. Note on your table that the wavelength of light will be kept constant at 650 nm.

1. Click the word *Speed* at the top of the screen. Select Slow.
2. Click and slide the wavelength indicator to 650. Click and slide the light intensity indicator to 120 lux.
3. Click the Start button. Observe the protein channel at the bottom of the screen. What molecule is synthesized near this protein channel?
4. Allow the simulation to run for 30 seconds, as indicated on the relative time meter, then click the Stop button. In your table, record the % Maximal ATP and the Number of ATP molecules that were produced.
5. Click the Reset button, and then click and slide the light intensity indicator to 160 lux.
6. Click the Start button and allow the simulation to run for 30 seconds, as indicated on the relative time meter.
7. Click the Stop button and record the % Maximal ATP and Number of ATP in your table.
8. Continue testing until you have evaluated all the light intensities listed on your table.
9. Explain why the wavelength of light is kept constant in this series of steps.

Part B: Examining the Effect of Wavelength of Light on the Rate of Photosynthesis

10. Click the Reset button. Prepare another three-column table with the headings Wavelength, % Maximal ATP, and Number of ATP. In the first column, list the following wavelengths of light that you will test in this investigation: 400, 450, 500, 550, 600, 650, 700, and 750 nm.
11. Click and slide the light intensity indicator to 200 lux. Click and slide the wavelength of light indicator to 400 nanometers.
12. Click the Start button and allow the simulation to run for 30 seconds, as indicated on the relative time meter.
13. Click the Stop button and record the % Maximal ATP and Number of ATP in your data table.
14. Continue testing until you have evaluated all the wavelengths of light listed on your table.
15. Explain why you didn't vary the intensity of light in this part of the investigation.

Analysis

16. Prepare a bar graph that shows the % Maximal ATP on the *y*-axis and the intensity of light on the *x*-axis. Above each bar, indicate the number of ATP molecules produced.
17. What can you conclude about the effect of light intensity on the rate of photosynthesis? What assumption are you making when drawing your conclusion?
18. Prepare another bar graph, this time plotting the % Maximal ATP versus the wavelength of light (in nanometers). Again, above each bar indicate the number of ATP molecules synthesized.
19. Refer to your second graph. What wavelengths of light are most effective for photosynthesis? What wavelengths are least effective?
20. Which combination of wavelength and intensity of light would produce the highest rate of photosynthesis?

CHAPTER

6A Modeling Mitosis and the Cell Cycle

exploration

Skill:

Modeling a biological process

Objectives

- *Relate* cell reproduction to the cell cycle.
- *Model* the stages of mitosis in an animal cell.

Materials

- 1 piece of string (1 m long)
- 2 pieces of string (each 40 cm long)
- 4 strips (1 cm × 6 cm) of colored paper, 1 strip each of four different colors
- scissors
- 4 paper clips
- 8 pieces of string (each 10 cm long)
- transparent tape

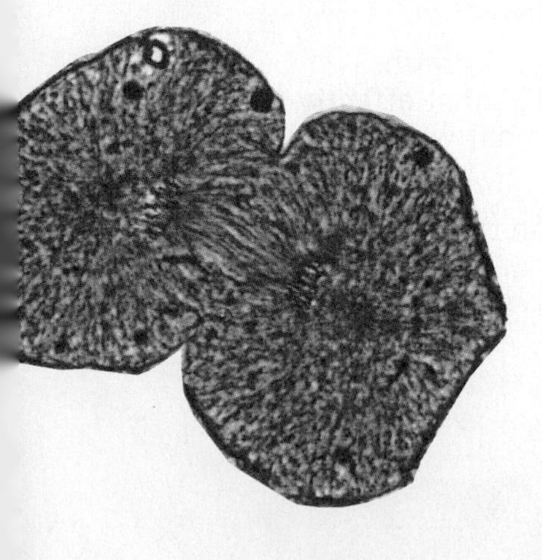

Situation

You are a science teacher who has been asked to test an activity that models mitosis and the cell cycle in animal cells. This activity will be part of a packet of instructional materials on cancer, which results when the cell cycle is disrupted. Your job is to evaluate whether the activity will help someone to visualize the stages of mitosis and the cell cycle.

Background: In science, a model is a hypothetical representation of an object or a process that can be used to study how the object or process works. Biologists use models to study many biological processes. You can use a model of an animal cell to study how mitosis occurs and to discover how errors in mitosis and the cell cycle might lead to cancer.

Prelab Preparation

1. List the five stages of the cell cycle, and describe the events that occur in each stage.
2. List the four stages of mitosis, and describe the events that occur in each stage.
3. Describe how mitosis and the cell cycle relate to cell reproduction.
4. Make a data table like the one at the top of the next page.

Procedure

1. Using a 1 m piece of string, make a circle on your lab table to represent the cell membrane of a cell. Using a 40 cm piece of string, make another circle inside of the "cell" to represent the nuclear membrane. Place 4 strips of paper (one of each color) inside the "nucleus" to represent chromosomes. The model you have built represents a cell in the G_1 phase. Record the number of "chromosomes" inside the "nucleus" in your data table.
2. Use scissors to cut each paper strip in half lengthwise. The two pieces that result from each strip represent the two chromatids that make up a chromosome after replication. Use a paper clip to join the two strips of paper so that they form an X. Each paper clip represents a centromere.

Name of phase	Chromosome/ chromatid number	Events
G_1 phase		
S phase		
G_2 phase		
M phase		
Prophase		
Metaphase		
Anaphase		
Telophase		
C phase		

3. You modeled the events of the S phase in step 2. Count the total number of "chromatids" (half-strips of paper) in the "nucleus," and record this number in your data table. The cell then enters the G_2 phase. Mitosis (the M phase) follows the G_2 phase.

4. **Prophase** Remove the "nuclear membrane." Place the 10 cm strings, which represent spindle fibers, inside the "cell" so that they look similar to those shown in the photo at right.

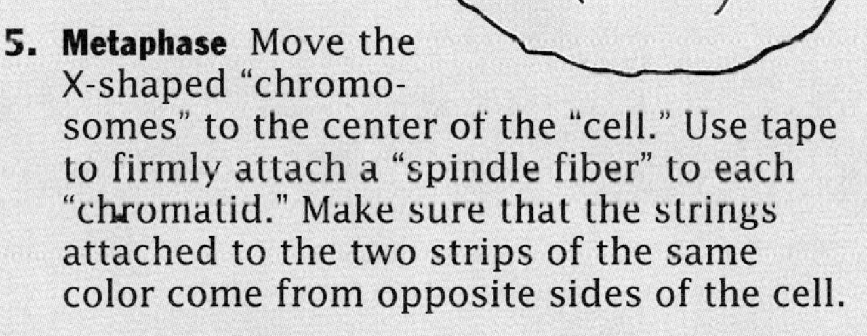

5. **Metaphase** Move the X-shaped "chromosomes" to the center of the "cell." Use tape to firmly attach a "spindle fiber" to each "chromatid." Make sure that the strings attached to the two strips of the same color come from opposite sides of the cell.

6. **Anaphase** Gather the loose ends of the strings on both sides of the cell, and pull the strings in opposite directions.

7. **Telophase** Untape the string from the paper strips. These half-strips of paper, which formerly represented chromatids, now represent chromosomes. Place one 40 cm string around each group of "chromosomes," forming two nuclei. Count the number of "chromosomes" in one of the new nuclei, and record this number in your data table. The C phase follows telophase.

8. Use the model to complete your data table.

9. Dispose of your materials according to the directions from your teacher.

Analysis

10. **Relating Information** What step in mitosis did you model when you removed the "nuclear membrane" in step 4?

11. **Interpreting Results** In what way are the two new cells that result from cell reproduction different from the original?

12. **Predicting Outcomes** Draw a line on one of the model chromatids. This line represents a mistake in the replication of a chromosome's DNA. How might such a mistake affect the outcome of cell reproduction?

13. **Drawing Conclusions** Evaluate this model's ability to help someone visualize animal cell reproduction.

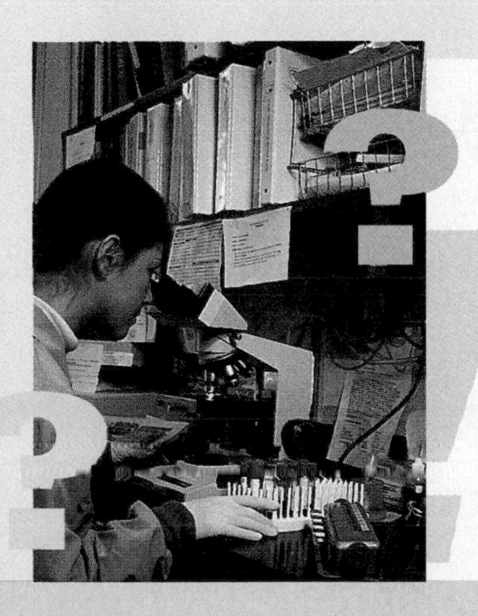

Applications

14. **Thinking Critically** How can a model help someone understand a biological process?

15. **Thinking Critically** The G_1 phase is normally the longest phase of the cell cycle. A "tumor suppressor" gene holds a cell in the G_1 phase. Cancer results when this gene is defective. How do you think this defective gene affects the process of cell reproduction?

16. **Career Connection** *Cytologists* are scientists who study cells. Many cytologists are involved in cancer research. Find out about the training and skills required to become a cytologist, and discover where cytologists do their work.

CHAPTER

6B Meiosis: Down Syndrome

interactive exploration

Objectives

- *Compare* the events of a normal meiotic division with those resulting in Down syndrome.
- *Determine* the relationship between parental age and the likelihood that a child will have Down syndrome.

Materials

- computer with CD-ROM drive
- CD-ROM *Interactive Explorations in Biology: Cell Biology and Genetics*
- graph paper

Background: This interactive exploration allows you to explore the cause of Down syndrome, an inherited condition that results in mental retardation, short stature, stubby hands and feet, and a characteristic heavy eyefold. This condition is a consequence of nondisjunction, the failure of chromosomes to separate normally during meiosis. In this exploration, you will have the opportunity to examine how the age of each parent affects the events of meiosis.

Prelab Preparation

1. Load and start the program Meiosis: Down Syndrome. You will see an animated diagram like the one below. Click the Navigation button, and then click the Topic Information button. Read the focus questions and review these concepts: Ploidy, Meiosis, Aneuploidy, and Nondisjunction.
2. Now click the word *Help* at the top left of the screen, and select How to Use This Exploration. Listen to the instructions that explain the operation of the exploration. Click the Interactive Exploration button on the Navigation Palette to begin the exploration.

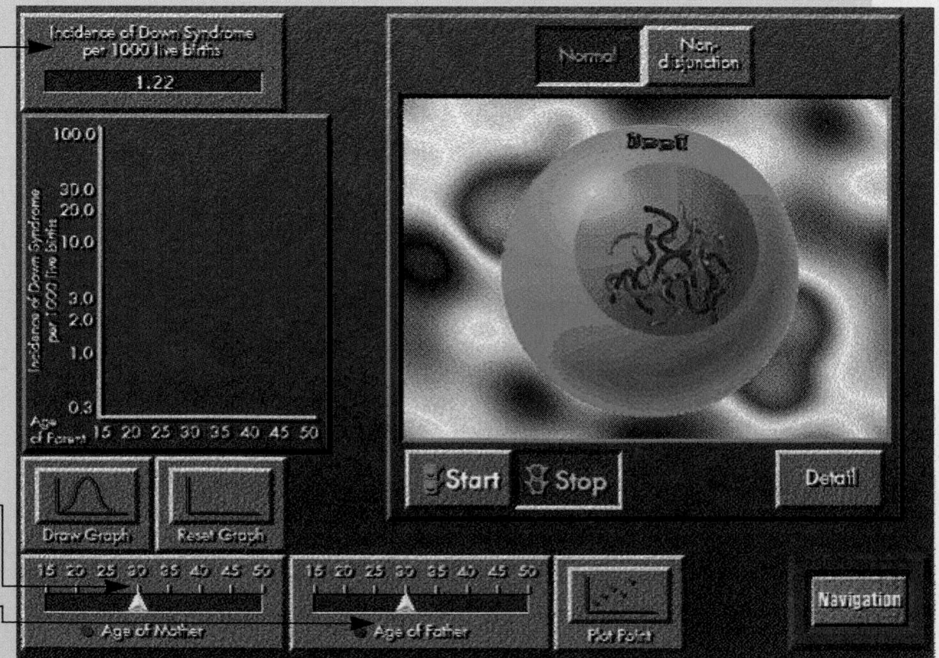

Feedback Meter

a **Incidence of Down Syndrome:** displays the frequency of Down syndrome births for the combination of parents chosen

Variables

b **Age of Mother:** allows you to set the mother's age

c **Age of Father:** allows you to set the father's age

Procedure

Part A: Normal Meiosis

First you will observe an animation showing the events of normal meiosis. For simplicity's sake, this animation shows only two pairs of chromosomes.

1. Click the Normal button to observe a simulation of normal meiosis.
2. Click the Detail button. Then click the Forward button (the one with the right-facing arrow) to progress through the eight steps of a normal meiotic division. As you view each step, read the caption and study the illustration.
3. Click the Close button.
4. Click the Start button to observe the same meiosis in continuous motion. How many gametes have formed when meiosis is complete?

Part B: Nondisjunction

Now you will observe an animation showing a meiotic division in which nondisjunction occurs.

5. Click the Nondisjunction button.
6. Click the Detail button. Then click the Forward button to observe nondisjunction. As you view each step, read the caption and study the illustration. What differences between normal meiosis and nondisjunction do you observe in frame 4?
7. Click the Close button.
8. Click the Start button to observe nondisjunction in continuous motion.

Part C: Determining the Effect of the Father's Age

Make a table like the one shown below.

Parent Being Evaluated: ______	
Age of parent	**Incidence of Down syndrome**

9. Click and slide the left-hand indicator so that the age of the mother is 25.
10. Click and slide the other indicator so that the age of the father is 25.
11. Click the Plot Point button. In your table, record the probability that these parents will have a child with Down syndrome, shown by the meter labeled Incidence of Down Syndrome per 1000 Live Births.
12. Repeat steps 10–11 five times, each time increasing the age of the father by 5 years. Make a graph of the age of the father versus the incidence of Down syndrome. What can you conclude about the effect of the father's age on the likelihood of having a child with Down syndrome?

Part D: Determining the Effect of the Mother's Age

Make a table like the one you made in Part C.

13. Click the Reset Graph button.
14. Click and slide the indicator so that the age of the father is 25.
15. Click and slide the indicator so that the age of the mother is 25.
16. Click the Plot Point button. In your table, record the Incidence of Down Syndrome per 1000 Live Births.
17. Repeat steps 15–16 five times, each time increasing the age of the mother by 5 years. Be sure to record the incidence of Down syndrome for each age. Make a graph of the mother's age versus the incidence of Down syndrome. What can you conclude about the effect of the mother's age on the likelihood of having a child with Down syndrome?

Analysis

18. In step 12, why did you hold the mother's age constant while varying the father's age?
19. In the United States, the age at which women have their first child has been increasing. How might this affect the incidence of Down syndrome?

CHAPTER 7A

ABO Blood Typing

exploration

Skill:
Blood-typing technique

Objectives

- *Interpret* the results of blood-typing tests.
- *Identify* an unknown blood type.
- *Predict* the possible blood types among children of parents with specific blood types.
- *Recognize* limitations of blood typing as a means of identification.

Materials

- safety goggles
- disposable gloves
- lab apron
- 5 blood-typing trays
- simulated blood (types AB and O)
- simulated blood from Cash child 1, Cash child 2, and Mr. X
- simulated anti-A blood-typing serum
- simulated anti-B blood-typing serum
- 10 toothpicks

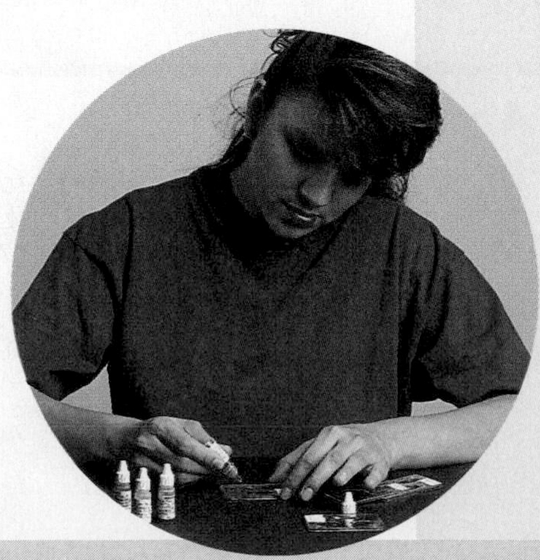

Situation

You are a medical laboratory technician. The lab you work for has been hired to conduct blood tests for the court case of *Mr. X v. the Cash Estate.* A man (Mr. X) claiming to be the son of Mr. and Mrs. Cash has sued for a share of the estate left to their children. Your job is to determine the blood types of the Cash children and Mr. X. You may be asked to testify in court.

Background: Human ABO blood types are determined by the presence or absence of specific **antigens,** or marker proteins, on the surface of red blood cells. Everyone inherits two alleles that determine his or her **phenotype** for ABO blood type. The alleles for antigens A and B are **dominant** over the **recessive** O allele, which does not code for an antigen. This means that people with type A or type B blood can have either of two possible **genotypes,** as shown in the table above. The alleles that code for antigens A and B are **codominant.**

Antigen present	Blood type	Possible genotypes
A	Type A	AA and AO
B	Type B	BB and BO
A, B	Type AB	AB
None	Type O	OO

When exposed to a foreign antigen, the body makes **antibodies,** which attack and destroy cells carrying the foreign antigen. Anti-A antibodies attack and destroy red blood cells carrying A antigens. Anti-B antibodies attack and destroy red blood cells carrying B antigens. Red blood cells that have been destroyed by antibodies stick together, forming clumps.

Prelab Preparation

1. Define the boldface terms in the **Background** section.
2. Describe the purpose of a Punnett square.
3. Make a data table like the one shown at the top of the next page.

Procedure

1. 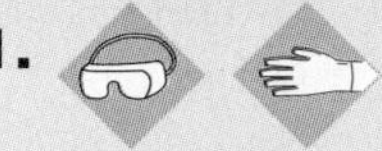  Put on safety goggles, gloves, and a lab apron.
2. Place 3–4 drops of the Type AB simulated blood into each of the two wells in a clean blood-typing tray. **CAUTION: Use only simulated blood throughout this lab. Under no circumstances are you to test any blood other than the simulated blood provided by your teacher.**

Blood sample	Description after adding anti-A blood-typing serum	Description after adding anti-B blood-typing serum
Type AB		
Type O		
Cash child 1		
Cash child 2		
Mr. X		

3. Without touching the blood in the wells, add 3–4 drops of anti-A blood-typing serum to one well. Using a toothpick, stir the mixture of simulated blood and blood-typing serum for about 30 seconds. Add 3–4 drops of anti-B blood-typing serum to the other well. Use a new toothpick to stir.

4. Observe each well for about two minutes. Look for clumps of a solid material to separate from the mixture, as seen on the left in the photo below. This clumping is evidence that antibodies have attacked the antigens on the simulated blood cells. Describe your observations by writing "clumping" or "no clumping" in your data table.

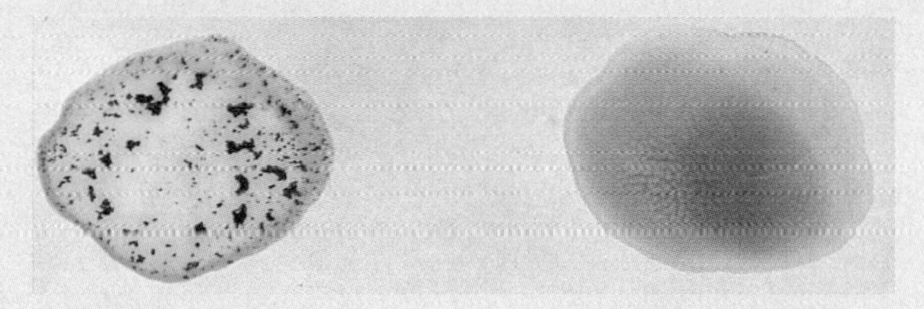

5. Repeat steps 3 and 4 using simulated Type O blood.

6. Repeat steps 3 and 4 to type the simulated blood samples from the Cash children and from Mr. X.

7. Dispose of your materials according to the directions from your teacher.

8.  Clean up your work area and wash your hands before leaving the lab.

Analysis

9. **Relating Information** Medical records revealed that Mr. Cash had blood type AB and Mrs. Cash had blood type A. What genotype(s) would have been possible for Mr. Cash? What genotype(s) would have been possible for Mrs. Cash?

10. **Predicting Outcomes** Draw at least two Punnett squares to predict all of the possible genotypes and phenotypes for ABO blood type that could be found among Mr. and Mrs. Cash's offspring.

11. **Interpreting Data** Do the results of the blood tests on the Cash children agree with the prediction you made in step 10? Explain.

12. **Interpreting Data** Based on your test results, what antigens are found on the red blood cells of Mr. X?

13. **Drawing Conclusions** What is the blood type of Mr. X? What genotypes are possible for Mr. X?

14. **Drawing Conclusions** Can the court reject Mr. X's claim to the estate of Mr. and Mrs. Cash based on his blood type? Support your answer.

15. **Making Inferences** If Mr. X had blood type AB, would this prove that he is Mr. and Mrs. Cash's son? Explain your answer.

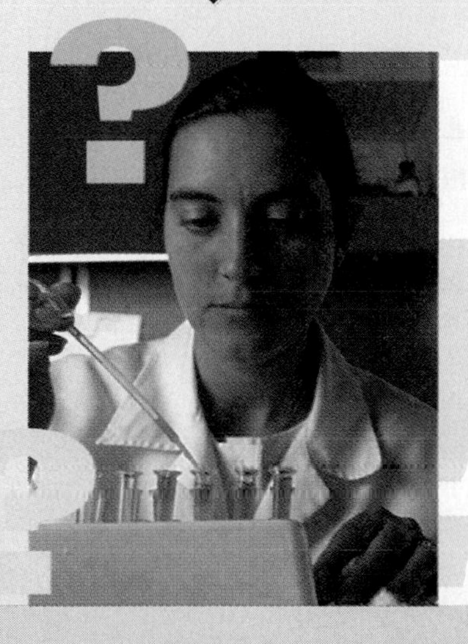

Applications

16. **Thinking Critically** Is blood typing a good way to identify a criminal or to establish parentage? Why or why not?

17. **Thinking Critically** When blood cells clump in the bloodstream, the result is usually death. How do you think blood typing improved the safety of blood transfusion (transfer of blood from one person to another)? Use data from this Exploration to justify your answer.

18. **Career Connection** *Medical laboratory technicians* perform blood-typing tests. Find out about the training and skills required to become a medical laboratory technician, and discover some other techniques performed in a medical laboratory.

CHAPTER

7B | ABO Blood Typing—Hospital Mix-Up

investigation

Prerequisites

- **Exploration 7A: ABO Blood Typing** on pages 774–775
- **Review** the following:
  1. cause of human ABO blood type
  2. purpose of a Punnett square
  3. phenotypes and possible genotypes for the four ABO blood types
  4. limitations of using blood typing as a means of verifying identity

NORTH COAST HOSPITAL
220 MAIN STREET
NEWTOWN, MAINE 00020

May 27, 1997

Caitlin Noonan
Director of Research
BioLogical Resources, Inc.
101 Jonas Salk Dr.
Oakwood, MO 65432-1101

Dear Ms. Noonan:

We recently had an extremely serious problem here at North Coast Hospital. Last Friday we were short-handed at our hospital and consequently had a mix-up in the maternity ward. Because of staff shortages, two infants born at nearly the same time were momentarily given to the wrong parents. Fortunately, we were able to immediately detect and correct our error and place each baby with the correct set of parents. However, the parents remain understandably concerned and angry. We need an outside agency not associated with our hospital to independently confirm our placement of each infant with each set of parents.

Enclosed are blood samples from each infant. Please conduct the necessary tests to determine the parents of each baby as soon as possible, and bill us when the work is complete.

Sincerely yours,

J. G. Prewitt

J. G. Prewitt
Hospital Administrator
North Coast Hospital

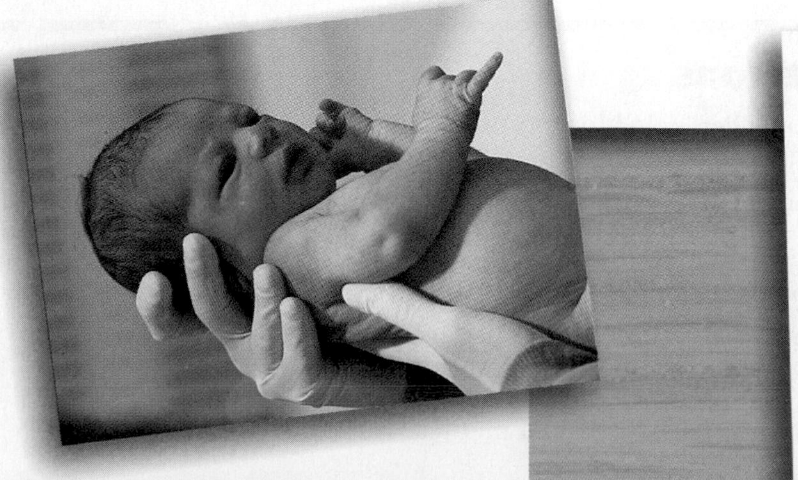

Required Precautions

- **UNDER NO CIRCUMSTANCES ARE YOU TO TEST ANY BLOOD OTHER THAN THE SIMULATED BLOOD PROVIDED BY YOUR TEACHER.**
- **Wear safety goggles, gloves, and a lab apron.**
- **Wash your hands before leaving the laboratory.**

BioLogical Resources, Inc., Oakwood, MO 65432-1101

MEMORANDUM

To: Team Leader, Medical Testing Dept.
From: Caitlin Noonan

Because the problem at North Coast Hospital is so serious, it is important that we answer Mr. Prewitt quickly and accurately. I want your team to perform the necessary tests. Make sure that your work is thorough and complete. Doing this job quickly and efficiently could lead to additional contracts with North Coast Hospital.

This morning I received a fax from Mr. Prewitt with the blood types of each set of parents. They are listed below.

Couple 1	AB and O
Couple 2	A and O

Before you start your work, you must submit a proposal for my approval. **Your proposal must include the following:**

- the **question** you are seeking to answer
- the **procedure** you will use
- a detailed and organized **data table**
- a complete, itemized list of proposed **materials** (including the use of facilities, labor, and amounts needed)

When you finish your tests, prepare a report in the form of a business letter to J. G. Prewitt. **Your report must include the following:**

- a paragraph describing the **procedure** you followed to analyze the blood samples
- a complete **data table**
- **Punnett squares** predicting the possible blood types of each couple's children
- your **conclusions** about each child's parents
- a detailed **invoice** showing all materials and labor plus the total amount due

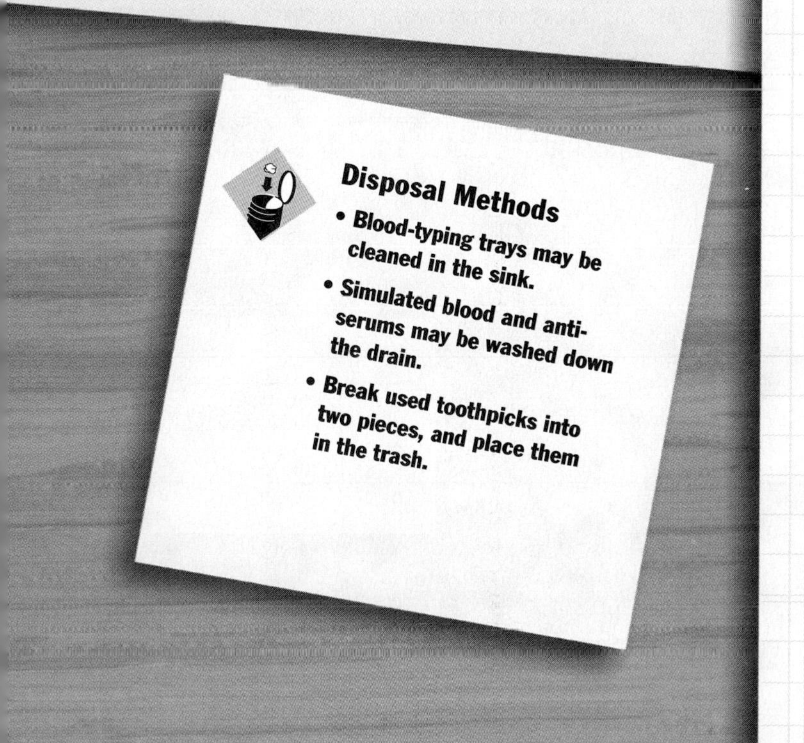

Disposal Methods

- **Blood-typing trays may be cleaned in the sink.**
- **Simulated blood and anti-serums may be washed down the drain.**
- **Break used toothpicks into two pieces, and place them in the trash.**

FILE: North Coast Hospital

MATERIALS AND COSTS (Select only what you'll need. No refunds.)

I. Facilities and Equipment Use	
facilities	$480.00/day
personal protective equipment	$10.00/day
test tube with rack	$10.00/day
blood typing tray	$5.00/day
microscope slide and coverslip	$1.00/day
compound light microscope	$100.00/day
II. Labor and Consumables	
labor	$30.00/hr
anti-A blood-typing serum	$1.00/test
anti-B blood-typing serum	$1.00/test
rubbing alcohol	$10.00/L
wax pencil	$2.00 each
toothpick	$0.10 each
blood sample: Infant 1	supplied by North Coast Hospital*
blood sample: Infant 2	supplied by North Coast Hospital*
Fines	
*Penalty of $1,000 assessed for additional samples needed	
OSHA safety violation	$2,000.00/incident

CHAPTER

7C Cystic Fibrosis

interactive exploration

Objectives

- *Simulate* the effects of three different genotypes on the movement of water molecules and chloride ions through the cell membrane.
- *Relate* the symptoms of cystic fibrosis to the ability of water molecules and chloride ions to pass through the cell membrane.

Materials

- computer with CD-ROM drive
- CD-ROM *Interactive Explorations in Biology: Human Biology*

Background: Cystic fibrosis is a fatal genetic disease that results from the failure of chloride ions (Cl^-) to pass through the cell membrane. As a result, chloride ions accumulate inside the cells of the body. When this happens in the lungs, pancreas, and liver, mucus builds up on the lining of these organs. In this interactive exploration, you will explore how an individual's genotype affects the membrane proteins involved in transporting Cl^- ions through the cell membrane.

Prelab Preparation

1. Load and start the program Cystic Fibrosis. Click the Topic Information button on the Navigation palette. Read the focus questions and review these concepts: Transport Channels, Osmosis, Mutation, and Cystic Fibrosis.
2. Click the word *File* at the top left of the screen, and select Interactive Exploration Help. Listen to the instructions that explain the operation of the exploration. Click the Exploration button at the bottom right of the screen to begin the exploration. You will see an animated drawing like the one below.

Feedback Meters

a. **Cl^- Ion Transfer:** the rate at which chloride ions move across the membrane

b. **Mucus Buildup:** how much mucus has accumulated on the cell

c. **Exterior/Interior Cl^- Concentration:** ratio of chloride ion concentration outside the cell to concentration inside the cell

d. **Exterior/Interior Water Concentration:** ratio of number of water molecules outside the cell to number inside the cell

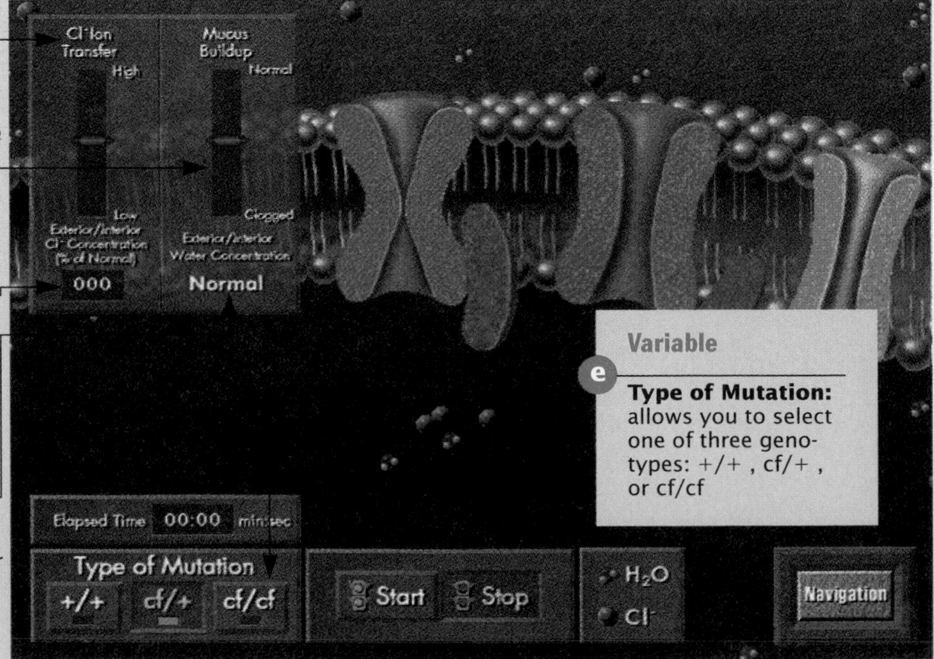

Procedure

Create a table like the one below for recording your data.

Genotype	Exterior/ interior Cl^- concentration	Exterior/ interior water concentration	Mucus buildup
+/+			
cf/+			
cf/cf			

Part A: Two Normal Alleles (+/+)
First you will investigate how water molecules and Cl^- ions pass through the membranes of cells in individuals having two normal alleles.

1. Move the pointer to the Type of Mutation box and click the +/+ button. What is the phenotype of a person with this genotype? Note that water molecules and Cl^- ions are located outside the cell (the portion of the screen above the cell membrane) and inside the cell (the portion of the screen below the cell membrane). Also note the position of the two CF proteins in the cell membrane.

2. Click the Start button to begin the simulation. Allow the simulation to run for 3 to 4 minutes. Describe how water molecules pass through the cell membrane. How does this differ from the passage of Cl^- ions through the membrane?

3. Click the Stop button. In your table, record the Exterior/Interior Cl^- Concentration, the Exterior/Interior Water Concentration, and the amount of mucus buildup outside the cell membrane.

Part B: One Mutant Allele and One Normal Allele (cf/+)

4. Move the pointer to the Type of Mutation box and click the cf/+ button. What is the phenotype of a person with this genotype? Describe what happened to the CF protein shown on the left and how this has affected the Cl^- ion channel.

5. Click the Start button and allow the simulation to run for 3 to 4 minutes. Describe the changes in the way water molecules and Cl^- ions now move through the membrane. How will the change in the movement of water molecules and Cl^- ions affect the external surface of the cell?

6. Click the Stop button. In your table, record the same three variables as you did in step 3. On which side of the membrane are chloride ions at the higher concentration? On which side is water at the higher concentration? Compare the results to those you obtained in Part A, and explain any differences.

Part C: Two Mutant Alleles (cf/cf)

7. Move the pointer to the Type of Mutation box and click the cf/cf button. What is the phenotype of a person with this genotype? Describe what has happened to both CF membrane proteins and how this has affected the Cl^- ion channels.

8. Click the Start button and allow the simulation to run for 3 to 4 minutes. Describe the changes in the movement of water molecules and Cl^- ions through the membrane. How will the external surface of the cell be affected?

9. Click the Stop button. Describe the difference between the external and internal environments of the cell. In your table, record the same variables as you did in Steps 3 and 6. On which side of the membrane are chloride ions at the higher concentration? Compare the results with those you obtained in Parts A and B, and explain any differences.

Analysis

10. Why are the Cl^- ion channels, but not the water channels, called gated channels?

11. Explain why people with one normal allele and one mutant allele for cystic fibrosis have a normal life expectancy.

CHAPTER

8A Modeling the Effect of a Mutation

exploration

Skill:
Modeling a biological process: protein synthesis

Objectives

- *Compare* and *contrast* the structures and functions of DNA and RNA.
- *Use* a model to demonstrate how DNA directs protein synthesis.
- *Explain* how a mutation can affect a protein.

Materials

- masking tape
- metric ruler
- oval-shaped card marked "ribosome"
- 24 plastic soda-straw pieces of the same color
- 13 plastic soda-straw pieces of a different color
- 12 red (A) pushpins
- 10 green (G) pushpins
- 8 yellow (C) pushpins
- 6 clear (T) pushpins
- 1 blue (U) pushpin
- 37 paper clips
- set of 8 "tRNA/amino acid" cards
- scissors
- transparent tape
- piece of paper

Problem

You are a genetic counselor who has been asked to evaluate a hands-on activity that demonstrates the cause of sickle cell anemia. The activity models protein synthesis and the effect of a mutation on protein production. First you must perform the activity to see how it works.

Background: The process that uses the instructions in DNA to make proteins is called **protein synthesis.** The result of protein synthesis is a chain of **amino acids** that will become part of a protein. If a change, or **mutation,** occurs in a cell's DNA, a resulting protein may be defective. One such mutation causes a genetic blood disorder called **sickle cell anemia.** This disorder results from the production of abnormal **hemoglobin** molecules, which impair the blood's ability to carry oxygen.

Prelab Preparation

1. Define the boldface terms in the **Background** section.
2. Describe the events that occur during transcription and translation.
3. Describe the process of base pairing as it occurs in DNA and RNA.

Procedure

PART 1—DNA Structure

1. Using masking tape, mark off a 45 cm × 60 cm rectangle on your desk or lab table to represent the cell membrane of a cell. Within this rectangle, mark off a 20 cm × 45 cm rectangle to serve as the cell's nucleus. Place the clear plastic oval (ribosome) and the 8 tRNA/amino acid cards in the "cytoplasm" of your "cell."
2. Assemble the 24 soda-straw pieces of the same color, 24 paper clips, and 6 each of the red, clear, green, and yellow pushpins as shown in the photo at the top of the next page. To do this, insert a pushpin midway along the length of each soda-straw piece. **CAUTION: Pointed objects can cause injury.** Then push a paper clip into one end of each soda-straw piece until it touches the pushpin. The resulting structures represent DNA nucleotides. Each pushpin color represents a base.

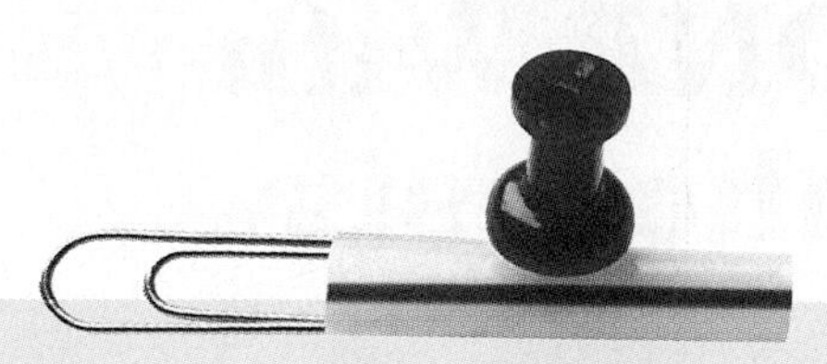

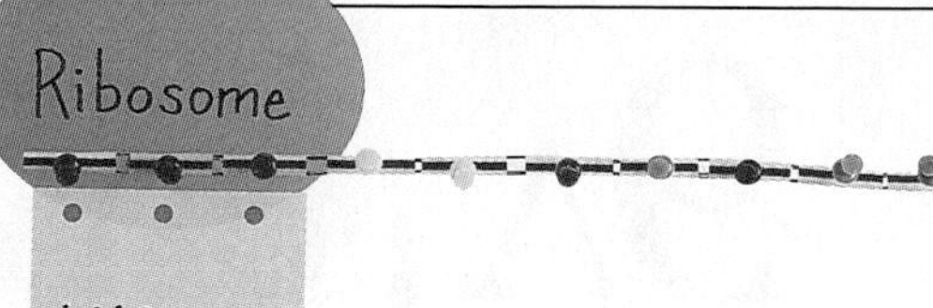

3. Use the 24 model DNA nucleotides to construct a strand of DNA inside your cell's "nucleus." First join two T "nucleotides" by inserting the paper clip of one into the open end of the other. Then, keeping the pushpins in a straight line, add 10 more "nucleotides" to make a strand of 12 "nucleotides" in the following order:

   T—T—T—G—G—T—C—T—C—C—T—C

4. Using your DNA strand as a template, make the complementary DNA strand with the remaining "nucleotides." First pair the base in each "nucleotide" to a complementary base in the DNA strand, and then join the "nucleotides" as you did before.

PART 2—Transcription

5. Assemble the remaining soda-straw pieces, paper clips, and pushpins as before. These structures represent RNA nucleotides.

6. Using your first DNA strand as a template, construct a strand of mRNA. First separate the two DNA strands. Then pair the base in each mRNA "nucleotide" with a complementary base in your first DNA strand. *Note: Uracil (U) replaces thymine (T) in RNA and pairs with adenine (A).* Now join the RNA "nucleotides" to make a strand of mRNA.

7. Carefully move the mRNA strand from the nucleus to the ribosome in the cytoplasm.

PART 3—Translation

8. Starting at the red (A) end of the mRNA strand, place the "ribosome" above the first 3 bases. The 3 bases (AAA) make up a *codon*.

9. Match a tRNA/amino acid card to the codon, as shown in the photo above. To do this, pair the bases in the mRNA codon to the complementary set of 3 bases on a tRNA/amino acid card. These 3 tRNA bases (UUU) make up an *anticodon*. Move the "ribosome" to the next mRNA codon.

10. Repeat step 9 until all tRNA/amino acid cards have been matched to mRNA codons. Connect the amino-acid cards with tape. Then cut off the resulting amino-acid chain, tape it to a piece of paper, and label it "Fragment of Normal Human Hemoglobin."

PART 4—Effect of a Mutation

11. Repeat steps 3–10, using the following sequence of DNA nucleotides:

    T—T—T—G—G—T—C—A—C—C—T—C

    Label the resulting amino-acid chain "Fragment of Human Sickle Cell Hemoglobin."

12. Dispose of your materials according to the directions from your teacher.

Analysis

13. **Comparing Structures** Compare and contrast the structures of DNA and RNA.

14. **Inferring Relationships** How do the protein fragments you made differ? Explain this difference.

15. **Drawing Conclusions** Evaluate the effectiveness of this model in explaining what a mutation is.

Applications

16. **Critical Thinking** Can this model be used to explain the relationship between genes and proteins? Explain your answer.

17. **Critical Thinking** A person has been referred to you for genetic counseling. This person is concerned about contracting sickle cell anemia from a friend who has just been diagnosed with the disease. What would you tell your client about sickle cell anemia?

18. **Personal Connection** Research the effects of mutagens on DNA. Explain how you might protect yourself from the possible effects of a mutagen.

CHAPTER

9A Separating DNA With Gel Electrophoresis

exploration

Skill:

Gel electrophoresis

Objectives

- *Use* restriction enzymes to cut DNA.
- *Separate* DNA fragments of different sizes.
- *Identify* the restriction enzyme used to cut a sample of DNA.
- *Determine* the R_f of different DNA fragments.

Materials

- personal protective gear
- crushed ice
- ice bucket
- microtube rack
- microtube A—Uncut DNA
- microtube B—*Hind*III
- microtube C—*Bam*HI
- microtube D—*Eco*RI
- microtube E—Unknown
- permanent marker
- 0.5–10 µL micropipetter
- micropipetter tips
- 10× restriction buffer for each restriction enzyme
- 34 µL Lambda virus DNA
- 37°C water bath
- gel-casting tray
- 6-well gel comb
- 0.8% agarose
- hot mitt
- zipper-lock plastic bag
- 10 mL graduated cylinder
- 1× TBE buffer
- freezer
- 5 µL loading dye
- gel chamber and power supply
- WARD's DNA stain
- staining trays
- distilled water
- metric ruler

Situation

You have just started working as a forensic technician for a genetics lab that does **forensic analysis,** which is the analysis of materials for legal proceedings. You will be analyzing DNA for the lab. Your supervisor has given you a sample of DNA from the Lambda virus to use while you practice your technique. The Lambda virus is a **bacteriophage,** which is a virus that infects only bacteria. The DNA in the sample has been cut into pieces with an enzyme. Unfortunately, the sample did not get labeled. The DNA sample must be identified before your supervisor can use it in research. Your job is to determine which of several possible enzymes has cut the DNA.

Background: Cutting DNA into fragments is the first step in certain types of DNA analysis. The DNA is cut into fragments with a **restriction enzyme.** A restriction enzyme cuts DNA where a specific sequence of base pairs occurs. Each type of restriction enzyme cuts DNA at a different base-pair sequence. After a sample of DNA is cut by a restriction enzyme, the resulting DNA fragments are separated according to size by **gel electrophoresis** *(ee lehk troh foh REE sihs)*. This process involves passing an electric current through a jellylike substance called a **gel.** The electric current causes one end of the gel to become negative and the other end to become positive.

A sample of cut DNA is placed in a **well** (depression) at the negative end of the gel. DNA has a negative charge and therefore flows toward the positive end of the gel during electrophoresis. Small DNA fragments move faster than larger fragments. After electrophoresis, the largest fragments are found closest to their wells, while the smallest fragments are found the farthest away. The distance that each fragment moves is used to calculate the R_f, or relative mobility, of a fragment. The R_f is used to calculate the number of base pairs in the fragment.

Prelab Preparation

1. Define the boldface terms in the **Situation** and **Background** sections.
2. Describe how gel electrophoresis moves DNA.

Procedure

Part 1—Cutting DNA

1.   Wear safety goggles, gloves, and a lab apron.

2. Fill an ice bucket with ice. Obtain one each of the following microtubes: A—Uncut DNA, B—*Hin*dIII, C—*Bam*HI, D—*Eco*RI, and E—Unknown. Microtubes B–D contain 1 μL of the indicated restriction enzyme. Place all microtubes in the ice. *Note: Restriction enzymes MUST be kept on ice until step 6.*

3. With a permanent marker, write the initials for everyone in your group ON THE TOP of microtubes A–E.

4. Set a micropipetter to 1 μL, and put a tip on the end of the micropipetter, as seen in the photo at right. Using a new tip for each microtube, add 1 μL of the corresponding 10× restriction buffer to each of microtubes B–D. Place the buffer on the side of the tube, as seen in the photo at right. *Note: DO NOT touch the micropipetter tips to the solutions in the microtubes.* **CAUTION: Report all spills to your teacher.**

5. Reset the micropipetter to 8 μL. Using the micropipetter and a new tip for each microtube, add 8 μL of the Lambda virus DNA to the side of each of microtubes B–D. Gently tap each microtube on your lab table until the solutions are thoroughly mixed. *Note: Do not shake the microtubes!* Reset the micropipetter to 10 μL, and add 10 μL of Lambda DNA to microtube A.

6. Place all of the microtubes into a 37°C water bath. After 50–60 minutes, remove the microtubes from the water bath, and immediately put them into a freezer. If the class period ends before 50 minutes has passed, your teacher will give you further directions. While the restriction enzymes are working, prepare an agarose gel according to the directions in Part 2.

Part 2—Preparing an Agarose Gel

7. Set up a gel-casting tray, as seen in the photo at right, according to directions from your teacher. Place a gel comb in the grooves of the gel-casting tray. Make sure that the comb does not touch the bottom of the tray. If it does, get another gel comb and inform your teacher.

8. Write the names of the members of your group on a paper towel. Carry your tray to the table with the melted agarose, and place your tray on the paper towel.

9. Using a hot mitt, pour melted 0.8% agarose into your gel-casting tray until the agarose reaches a depth of 3 mm. Make sure that the agarose spreads evenly throughout the tray. *Note: Do not move your gel tray before the agarose solidifies.*

10. Let the gel cool (about 20–30 minutes) until the agarose solidifies.

11. While the gel is cooling, write your name, the date, and your class period on a zipper-lock plastic bag. Pour 5 mL of 1× TBE buffer into the bag.

12. When the gel has solidified, carefully remove the gel comb by pulling it straight up, as shown in the photo at right. *Note: Do not wiggle the comb. This may tear the gel.* If the comb does not come up easily, pour a little 1× TBE buffer on the comb area. After removing the gel comb, open the plastic bag and carefully slide the gel tray into the bag. *Note: Do not remove the gel from the gel-casting tray.* Store the gel according to your teacher's instructions.

Part 3—Loading and Running a Gel

13. Retrieve your microtubes (A–E) and your gel. If the materials in the microtubes are frozen, hold each tube in your hand until the solutions thaw.

14. Set a micropipetter to 1 µL, and place a tip on the end. Add 1 µL, of loading dye to each microtube. *Note: Use a new tip for each microtube. Do not touch the bottom of the tube.* Gently tap each microtube on your lab table to thoroughly mix the solutions. *Do shake the microtubes.*

15. Remove your gel (still in the gel-casting tray) from the plastic bag, and place it in a gel chamber, as seen in the photo at right. Orient the gel so that the wells are closest to the BLACK wire, or anode.

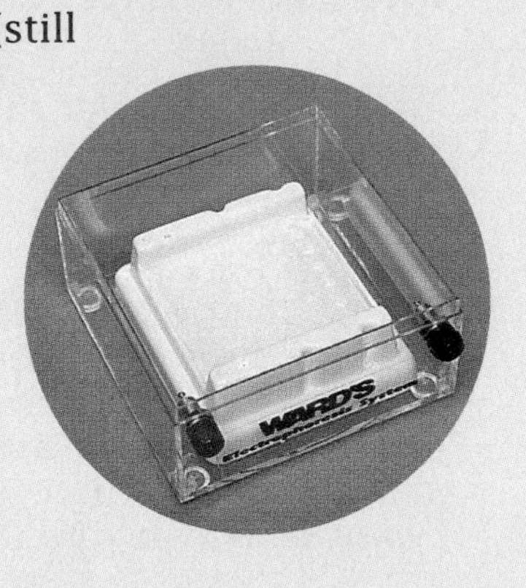

16. Set your micropipetter to 10 µL, and place a new tip on the end. Open microtube A, and remove 10 µL of solution. Carefully place the solution into the well in lane 1, as shown in the diagram below. To do this, place both elbows on the lab table, lean over the gel, and slowly lower the micropipetter tip into the opening of the well before depressing the plunger, as seen in the photo at right. *Note: Do not jab the micropipetter tip through the bottom of the well.*

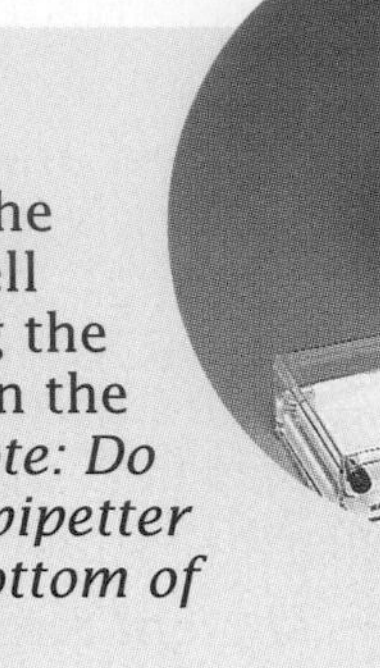

17. Using a new micropipetter tip for each tube, repeat step 16 for each of the remaining microtubes. Use lane 2 for microtube B, lane 3 for microtube C, lane 4 for microtube D, and lane 5 for microtube E.

18. Very slowly fill the gel chamber with 1× TBE buffer until the level of the buffer is approximately 1–2 mm above the surface of the gel.

19. Close the gel chamber and connect it to a power supply according to your teacher's instructions. **CAUTION: Follow all of the manufacturer's precautions regarding the use of this equipment.**

20. Allow an electric current to flow through the gel. You will see a blue line moving away from the wells. When the blue line is approximately 5 mm from the end of the gel, disconnect the power supply and remove the gel. Store the gel overnight in the plastic bag.

Part 4—Staining and Analyzing a Gel

21. To stain a gel, carefully place the gel (wells up) into a staining tray, as seen in the photo at right. Pour WARD's DNA stain into the staining tray until the gel is completely covered. Cover the staining tray, and label it with your initials. Allow the stain to sit for at least 2 hours. Next, carefully pour the stain into the sink drain, and flush it down the drain with water. *Note: Be careful. Do not let the stained gel break or slip out of the staining tray.*

Lambda DNA Gel Analysis Data Table

Well 1 No enzyme		Well 2 *Hind*III		Well 3 *Bam*HI		Well 4 *Eco*RI		Well 5 Unknown	
Distance (in mm)	R_f	Distance (in mm)	R_f	Distance (in mm)	R_f	Distance (in mm)	R_f	Distance (in mm)	R_f

22. To destain a gel, cover the gel with distilled water. *Note: Do not pour water directly over the gel. Pour water to one side of the gel.* Let the gel sit overnight (or at least 8–12 hours). After destaining the gel, the bands of DNA will appear as purple lines against a light background, as seen in the photo below.
23. Prepare a data table like the one on this page. Record your data in the table as you complete step 24.
24. Using a metric ruler, measure in millimeters the distance that each DNA fragment traveled from a well. Calculate the R_f for each fragment using the following equation:

$$R_f = \frac{\text{distance in mm that DNA fragment migrated}}{\text{distance in mm from well to the dye}}$$

25. Draw and label your stained gel on a piece of paper or in your lab notebook. Instead of drawing your gel, you may tape a photo of the gel or the dried gel to your report.
26. Dispose of your materials according to the directions from your teacher.
27. Clean up your work area and wash your hands before leaving the lab.

Analysis

28. **Relating Concepts** What is the purpose of testing the DNA in microtube A?
29. **Analyzing Data** Where on your gel were the fragments with the largest R_f values?
30. **Analyzing Results** Was the DNA in any of the bands darker than in others? What could account for this?
31. **Interpreting Data** Which two samples appear to have the same pattern of DNA bands?
32. **Drawing Conclusions** What restriction enzyme cut the DNA in the unknown sample? Justify your answer.
33. **Predicting Outcomes** In this lab, DNA from the *same* type of virus was cut with *different* restriction enzymes. What results would you expect from the gel electrophoresis of DNA from *different* organisms but cut by the *same* restriction enzyme?

Applications

34. **Thinking Critically** Would you expect different individuals to have the same or different patterns of DNA bands? Justify your answer.
35. **Thinking Critically** What are some measures that you took to prevent contamination of DNA samples during this lab? Why is it important to prevent contamination while performing DNA analysis?
36. **Career Connection** *Forensic technicians* perform DNA analysis on blood and hair samples taken from crime scenes. Find out about the training and skills required to become a forensic technician, and discover some of the types of tests that they perform.

CHAPTER 10A

Analyzing Variation Within a Species

exploration

Skills:

- **Taking numerical data**
- **Determining mean, median, and range**
- **Graphing data**

Objectives

- *Collect* numerical data on three characteristics of leaves.
- *Graph* the data for each characteristic.
- *Determine* the mean, median, and range of the data for each characteristic.
- *Compare* your data with the class data.
- *Infer* a possible relationship between two plants.

Materials

- 1 bag of leaves marked "Plant A"
- 1 bag of leaves marked "Plant B"
- metric ruler
- graph paper
- graphing calculator (optional)

Situation

You are a botanist who has been hired to study the plant life in a new park. To complete your study, you must identify (find the name of) a plant you found in the park. Normally you would use a *key* to identify an unknown plant. However, a plant key requires you to examine flowers, and the plant has no flowers at this time. The plant looks like one that you identified for the local school. You think the two plants might belong to the same species, so you take sample leaves from both plants to find out.

Background: The individuals of a **species** differ slightly in appearance. Many of these differences occur because of **genetic variation** (the different versions of the genes within a species). Some of these differences are caused by environmental factors, such as light, temperature, and moisture.

Numerical data (numbers obtained by counting and measuring) can be analyzed to determine how similar two individuals are. Graphing numerical data enables you to visualize the **range** (difference between largest and smallest value) and **frequency** (number of occurrences) of the numbers. To judge how similar the numbers are, their **distribution** (range and frequency) is compared with their **mean** (arithmetic average) and **median** (middle number in the range).

Prelab Preparation

1. Define the boldface terms in the **Background** section.
2. Prepare two data tables similar to the one at the top of the next page. Title one "Group Data" and the other "Class Data."

Procedure

1. Measure each leaf in the bags for Plant A and Plant B to determine its length and width (at the widest part) to the nearest tenth of a centimeter. *Note: Handle leaves carefully so that they do not break apart.* Record your measurements (numerical data) in your group's data table.

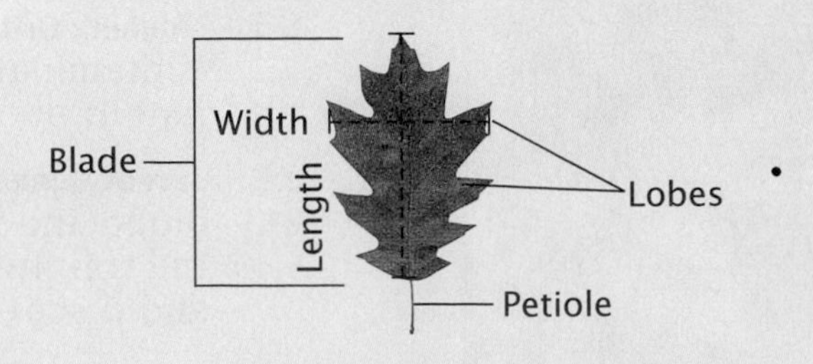

Leaf/ Group	PLANT A			PLANT B		
	Leaf length	Leaf width	Other	Leaf length	Leaf width	Other
1						
2						
3						
Mean						

2. Collect numerical data for one other leaf characteristic, such as number of lobes or length of petiole (leaf stalk). Record these numbers in your group's data table.
3. For each characteristic observed, make a line graph or a bar graph showing the distribution of the data for that characteristic. Label the *x*-axis of each graph with the name of a leaf characteristic, and label the *y*-axis of each graph "Frequency." Graph the data for Plants A and B separately.
4. Calculate the mean of each characteristic by adding all of the numbers in the column for a characteristic and then dividing by the number of rows (leaves measured). Record each mean in your group's data table.
5. Compile the data collected by your class in the class data table. In this table, record the mean of each lab group's data on each characteristic.
6. Repeat step 3 using the means reported by each lab group.
7. Find the mean, median, and range of the numbers in each column of the class data table. On the graph of the class data for each leaf characteristic, draw vertical lines to indicate the mean and median of the data.

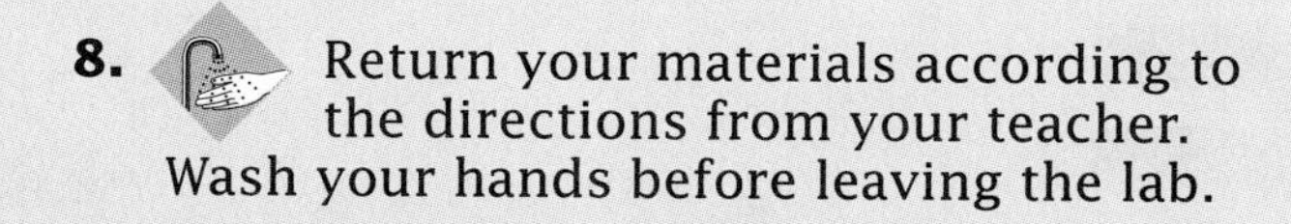

8. Return your materials according to the directions from your teacher. Wash your hands before leaving the lab.

Analysis

9. **Interpreting Graphs** How does your graph for each characteristic of Plant A compare with your graph for that characteristic of Plant B?
10. **Interpreting Data** How does the mean, median, and range of each characteristic of Plant A compare with the mean, median, and range of that characteristic of Plant B?
11. **Interpreting Data** How do the mean, median, and range of each leaf characteristic compare with one another?
12. **Recognizing Relationships** How does the mean, median, and range of each leaf characteristic relate to the graph of the class data for that characteristic?
13. **Recognizing Patterns** How do the graphs of your group's data compare with the graphs of the class data?
14. **Inferring Conclusions** Based on your group's data, what does the relationship between the two plants seem to be?
15. **Predicting Patterns** How might two graphs of leaf length compare if the leaves came from two different species?
16. **Making Comparisons** How does the number of leaves from each plant examined by the class compare with the number of leaves from each plant examined by your group?

Applications

17. **Critical Thinking** Do you think that the differences between the leaves of Plants A and B were caused by genetic variation or by environmental factors? Justify your answer.
18. **Critical Thinking** After comparing your group's data with the class data, are you more certain of your conclusion about the relationship of the two plants? Why or why not?
19. **Career Connection** *Statisticians* analyze numerical data. Find out about the training and skills required to become a statistician, and discover how statisticians help biologists interpret their data.

CHAPTER 14A

Modeling an Ecosystem

exploration

Skill:
Modeling an ecosystem

Objectives

- *Construct* a semiclosed, microscale ecosystem.
- *Identify* successful and unsuccessful organisms in a model ecosystem.
- *Recognize* the limitations of environmental microscale modeling.

Materials

- three 2 L clear plastic soda bottles
- wax pencil
- scissors
- 2 pieces of plastic, 5 cm square
- nail or awl
- dissecting needle
- candle or Bunsen burner
- clear waterproof tape
- algae
- *Elodea*
- *Vallisneria*
- *Myriophyllum*
- *Paramecium*
- *Daphnia*
- 2 or 3 aquatic snails
- soil
- 20–30 Rapid radish seeds
- dead organic matter (leaves, sticks, rotten fruit)
- terrestrial animals (flies, spiders, pill bugs)
- compound microscope
- slides
- coverslips
- dropping pipet

Situation

You are an ecologist who has been hired by a manufacturing company that is planning to build a new factory. The company wants to build the factory on a piece of land that is adjacent to a wilderness park. A group of citizens is trying to block construction of the factory because they believe that the factory will destroy the delicate ecosystem of the park. Your job is to perform tests that will assure the citizens that the wastes produced by the factory will not affect the park's ecosystem. First you must build a model of the park's ecosystem.

Background: An ecosystem can be studied with the aid of a **microscale model,** which is a small-scale model. A good model should include both **biotic** (living) and **abiotic** (nonliving) components of the ecosystem to be studied.

Prelab Preparation

1. Define the boldface terms in the **Background** section.
2. Define the following terms: food chain, producer, consumer, scavenger, decomposer, herbivore, omnivore, carnivore.
3. Prepare a data table with four columns and the following headings: Date, Section 1 observations, Section 2 observations, and Section 3 observations.

Procedure

1. Remove the labels from three 2 L clear plastic soda bottles. Remove the bases from two of the bottles. Using a wax pencil, mark the bottles to be cut, as shown in the diagram below. Using scissors, cut the bottles along the wax-pencil lines.

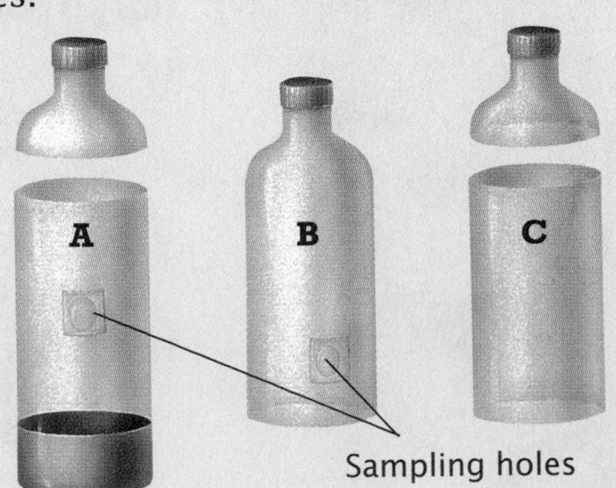

2. Cut sampling holes 4 cm in diameter in Bottles A and B, as shown in the diagram on the previous page. Make a cover for each sampling hole by placing a 5 cm square piece of plastic over each hole and attaching one side of the plastic square to the bottle with tape.

3. Using a nail or an awl, make drain holes in each of the caps from the three bottles.

4. Heat the point of a dissecting needle in the flame of a candle or Bunsen burner. Use the hot needle to make several holes in the top of Bottle A and in the sides of Bottles B and C. **CAUTION: Keep combustibles away from flames.**

5. Assemble the soda bottle pieces into a microscale ecosystem with an aquatic community (Section 1), a lowland community (Section 2), and an upland community (Section 3), as seen in the diagram at right.

6. To make Section 1, place 5 cm of sand in the base of Bottle A, and then fill with water to within 3 cm of the sampling hole. Place aquarium plants, algae, *Daphnia*, *Paramecium*, and aquatic snails in the water.

7. To make Section 2, invert Bottle B, place it on top of Section 1, and then tape the bottles together. Fill this section one-third full with soil. Place 20 radish seeds, some dead organic matter, and two or three animals on the soil.

8. To make Section 3, place the top of Bottle A (cap up) on top of Section 2. Then invert the bottom of Bottle C and place it on top of Bottle A. Tape the bottles together. Place soil in Section 3, but do not cover the cap of Bottle A. Place seeds, dead organic matter, and two or three animals on top of the soil.

9. Invert the top of Bottle C, place it on top of Section 3, and tape these pieces together. Add water to your ecosystem, when necessary, through the opening at the top.

10. Observe your ecosystem for 2 to 3 weeks, and record any changes that you observe in your data table. Count the organisms you see. To count the microorganisms in the "pond," remove some of the water with a pipet, make a wet mount with it, and observe it under a microscope.

11. Clean up your work area and wash your hands before leaving the lab.

Analysis

12. **Interpreting Data** As your model ecosystem changed over time, which organisms increased in number? Which organisms declined in number?

13. **Drawing Conclusions** In your ecosystem, which trophic level had the most biomass?

14. **Making Inferences** Which was the most successful organism in your model ecosystem?

Applications

15. **Relating Information** Describe one food chain that you could study using your model ecosystem.

16. **Thinking Critically** How is your model ecosystem similar to a real ecosystem? How is it different?

17. **Thinking Critically** How might a microscale model of an ecosystem be useful? What are some of its limitations?

18. **Career Connection** *Wildlife biologists* study ecosystems that can't be modeled on a microscale. Find out about the training and skills required to be a wildlife biologist, and discover how they do their work.

CHAPTER 15A

Modeling an Ecosystem—Fertilizer Dilemma

investigation

Prerequisites

- **Exploration 14A: Modeling an Ecosystem** on pages 788–789
- **Review** the following:
  1. independent variable
  2. dependent variable
  3. control
  4. constant

CITY ENVIRONMENTAL PLANNING COMMISSION

August 30, 1997

Rosalinda Gonzales
Director of Environmental Analysis
BioLogical Resources, Inc.
101 Jonas Salk Dr.
Oakwood, MO 65432-1101

Dear Ms. Gonzales,

I am the chairman of the City Environmental Planning Commission, which evaluates all proposed development in the local area. Our mission is to preserve the environment by intelligently managing economic development.

A large wholesale nursery is planning to relocate on the vacant farmland southeast of town. The land is very close to City Lake, which is used for recreation. The nursery will be applying a water-soluble 15-30-15 fertilizer to their plants on a regular basis. We need to know how these nutrients might affect City Lake. The city council needs our recommendation as soon as possible. Please conduct the appropriate experiments to determine any impact that the nursery may have on City Lake. Bill us when you send your report.

Sincerely,

Donald Crosby

Donald Crosby, Ph.D.
Chairman
City Environmental
Planning Commission

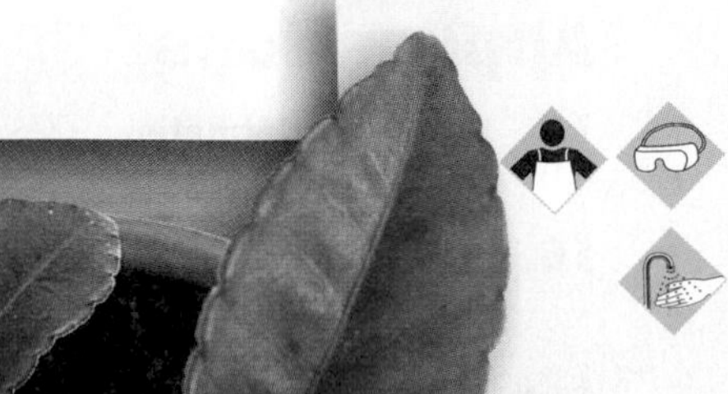

Safety Precautions

- **Wear safety goggles and a lab apron.**
- **Wash your hands before leaving the laboratory.**

Biological Resources, Inc., Oakwood, MO 65432-1101

MEMORANDUM

To: Team Leader, Ecology Dept.
From: Rosalinda Gonzales

Dr. Crosby is relying on us to provide quick and accurate results. His decision depends on our correctly identifying the effect that increased nutrients could have on City Lake. Make sure that your work is thorough and accurate.

The following information may be useful:

- A model ecosystem can be used to determine how excess nutrients from fertilizer might affect the ecology of a lake.
- An excess of nutrients can cause an increase in the organic matter in a lake.
- A 15-30-15 fertilizer is 15 percent nitrogen, 30 percent phosphorus, and 15 percent potassium. Different fertilizers supply different percentages of these nutrients.

Before you start your work, you must submit a proposal for my approval. **Your proposal must include the following:**

- the **question** you are seeking to answer
- the **procedure** you will use
- a detailed **data table**
- a complete, itemized list of proposed **materials** and **costs** (including use of facilities, labor, and amounts needed)

When you finish your tests, prepare a report in the form of a business letter to Donald Crosby, Ph.D. **Your report must include the following:**

- a paragraph describing the **procedure** you followed to determine how excess fertilizer might affect the lake
- a complete **data table**
- your **conclusions** about how runoff containing fertilizer nutrients will affect the lake
- a detailed **invoice** showing all materials and labor plus the total amount due

FILE: City Environmental Planning Commission

MATERIALS AND COSTS (Select only what you'll need. No refunds.)

I. Facilities and Equipment Use	
facilities	$480.00/day
personal protective equipment	$10.00/day
microscope	$100.00/day
microscope slide and coverslip	$1.00/day
forceps	$5.00/day
dropping pipet	$5.00/day
graduated cylinder	$10.00/day
beaker	$10.00/day
thermometer	$10.00/day
II. Labor and Consumables	
labor	$30.00/hr
detergent	$2.00/bottle
15-30-15 fertilizer	$0.25/g
10-0-0 water-soluble fertilizer	$0.25/g
0-10-0 water-soluble fertilizer	$0.25/g
0-0-10 water-soluble fertilizer	$0.25/g
paper cups	$0.10 each
model ecosystem (from Exploration 14A)	$500.00 each
Fines	
OSHA safety violation	$2,000.00/incident

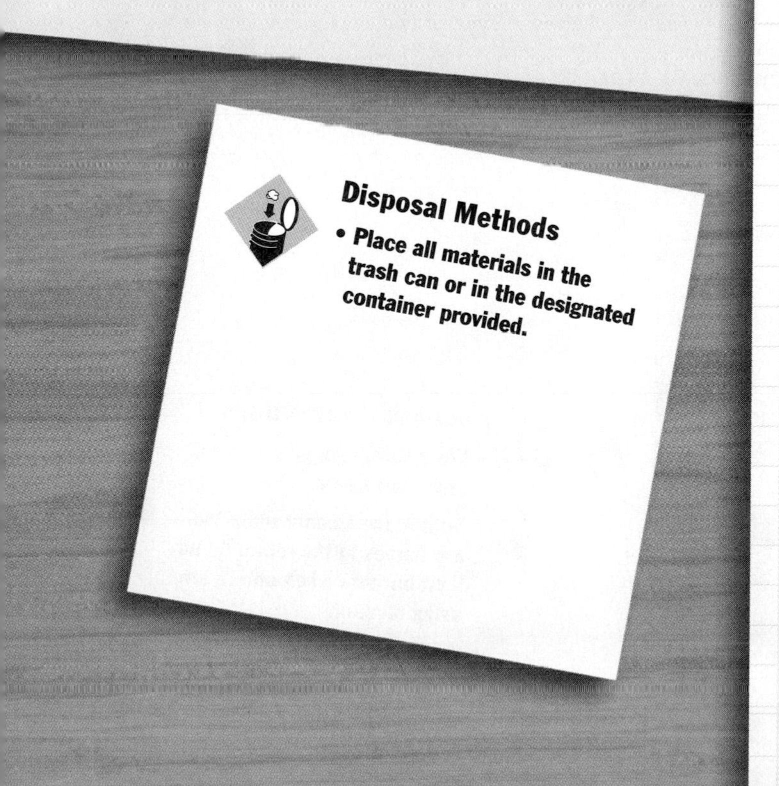

Disposal Methods

- **Place all materials in the trash can or in the designated container provided.**

CHAPTER 18A

Culturing Bacteria—New Antibacterial Soap

investigation

Prerequisites

- **Laboratory Techniques and Experimental Design: Using Aseptic Technique** in the Holt BioSources Lab Program
- **Review** the following:
  1. aseptic technique
  2. how to establish a bacterial lawn
  3. how to identify a positive result

Advertising Associates
New York, New York

February 14, 1997

Caitlin Noonan
Director of Research
BioLogical Resources, Inc.
101 Jonas Salk Dr.
Oakwood, MO 65432-1101

Dear Ms. Noonan,

I am heading up an advertising campaign for a new product. The product is an antibacterial, liquid soap that our client wants to market for use as a surface cleaner for the kitchen. As you are aware, foods can be contaminated with bacteria such as *Salmonella*, *Staphylococcus*, and *Escherichia coli*. These bacteria can live on kitchen counters and contaminate other foods if the counters are not cleaned before and after food preparation. Each of these bacteria can cause serious food poisoning.

All claims made in our advertisements must have research data to back them up. We want to know how the effectiveness of this new product compares with that of other antibacterial cleaning agents, such as bleach, household disinfectant, hand soap, and dishwashing soap.

Please perform the necessary tests. Bill us for your work when you send your report. Your quick action on this project will enable us to finalize our advertising campaign as soon as possible.

Sincerely,

Anda Chan

Anda Chan
Account Manager

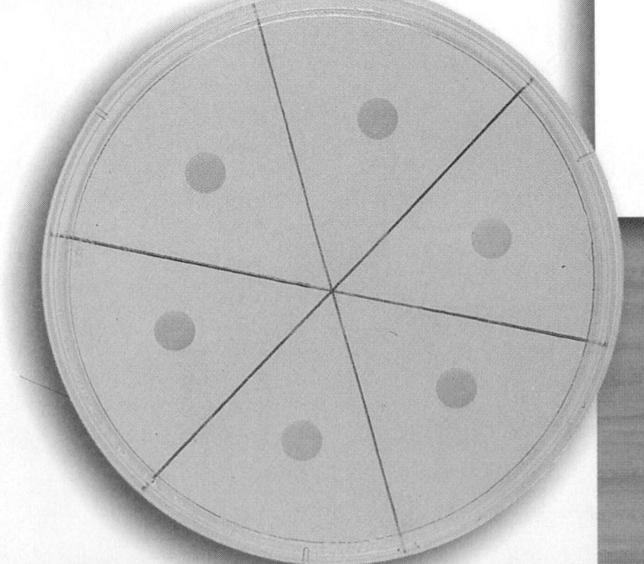

Required Precautions

- Wear safety goggles, gloves, and a lab apron.
- Do not use alcohol when there are flames in the room. Do not light burners when others are using alcohol.
- Do not open sealed petri dishes.

- Wash your hands before leaving the lab.

B Biological Resources, Inc., Oakwood, MO 65432-1101

MEMORANDUM

To: Team Leader, Microbiology Dept.
From: Caitlin Noonan

Please get your team on this job right away. You should develop a procedure to determine whether the soap prevents new bacterial growth. Let's compare a 10 percent solution of the test soap with 10 percent solutions of four types of cleaning agents—bleach, hand soap, dishwashing detergent, and a household disinfectant. Small disks of filter paper can be used to apply the cleaning agents to cultures of *E. coli*. One petri dish of *E. coli* should accommodate a disk for each test solution, plus one for the control. Please have each of your teams perform one test with each of the cleaning agents. Compile the data from all tests into one report.

Before you start your work, you must submit a proposal for my approval. **Your proposal must include the following:**

- the **question** you are seeking to answer
- the **procedure** you will use
- a detailed **data table**
- a complete, itemized list of proposed **materials** and **costs** (including use of facilities, labor, and amounts needed)

When you finish your tests, prepare a report in the form of a business letter to Anda Chan. **Your report must include the following:**

- a paragraph describing the **procedure** you followed to test the effectiveness of each cleaning agent
- a complete **data table**
- your **conclusions** about the effectiveness of the new product
- a detailed **invoice** showing all materials and labor plus the total amount due

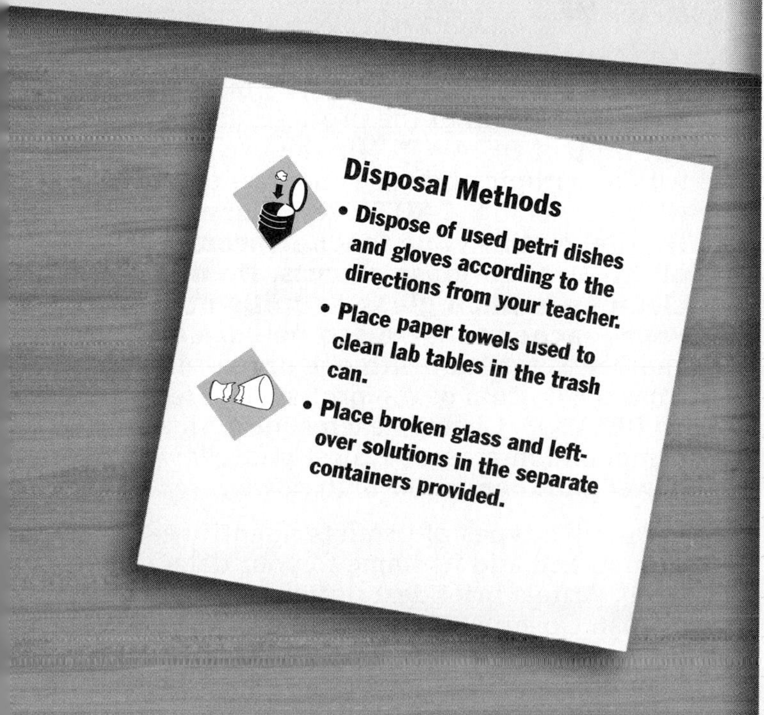

Disposal Methods

- **Dispose of used petri dishes and gloves according to the directions from your teacher.**
- **Place paper towels used to clean lab tables in the trash can.**
- **Place broken glass and left-over solutions in the separate containers provided.**

FILE: A to Z Advertising

MATERIALS AND COSTS (Select only what you'll need. No refunds.)

I. Facilities and Equipment Use	
facilities	$480.00/day
personal protective equipment	$10.00/day
microscope	$100.00/day
incubator	$50.00/day
forceps	$5.00/day
metric ruler	$5.00/day
Bunsen burner	$10.00/day
graduated cylinder	$10.00/day
beaker	$10.00/day
II. Labor and Consumables	
labor	$30.00/hr
petri dishes of *E. coli*	$10.00 each
fine-point marker	$3.00 each
tape	$0.25/m
distilled water	$1.00/L
cleaning agents	$0.50/mL
alcohol	$0.10/mL
saline solution	$0.50/mL
filter-paper disks	$0.10 each
solution of test soap	provided
Fines	
OSHA safety violation	$2,000.00/incident

CHAPTER

19A Observing the Behavior of Protists

exploration

Skill:

Using a compound microscope

Objectives

- *Make* wet mounts of aquatic protists.
- *Observe* protists with a microscope.
- *Describe* the structure and behavior of aquatic protists.
- *Draw* and *label* the parts of protists.

Materials

- compound microscope
- 3 microscope slides
- 2 coverslips
- DETAIN™ (protist-slowing agent)
- aquatic protist culture
- toothpick
- 40 × 20 mm piece of black paper
- hole punch
- white paper
- light source
- forceps
- 2 capillary tubes
- WARD's silicone culture gum
- hot water
- ice water

Situation

You work for a company that installs backyard ponds. The company is developing a kit that will contain cultures of the protists necessary for a healthy aquatic ecosystem in their ponds. The kit will also have some educational literature about the protists so that customers will know what features a pond must have to ensure the survival of the protists. Your job is to observe the protists that will be included in the kit, describe the protists, and write recommendations for establishing a pond that will provide environments favorable to all of the protists.

Background: Protists are an important part of an ecosystem. They are a part of many food chains, acting as producers, consumers, decomposers, and food for other organisms.

Prelab Preparation

1. List the general characteristics of the kingdom Protista.
2. Review the procedures for using the microscope and for making a wet mount found in the Laboratory Skills section of your textbook.
3. Prepare a data table like the one at the top of the next page.

Procedure

1. Using a microscope slide, make a wet mount with one drop of liquid from the bottom of the protist culture and one drop of DETAIN™. Mix the two drops with a toothpick before you cover them with a coverslip. **CAUTION: Glassware is fragile. Notify your teacher promptly of any broken glass or cuts. Do not clean up broken glass or spills unless your teacher tells you to do so.** Make another wet mount with one drop of liquid from the bottom of the protist culture and no DETAIN™. Observe each slide with a compound microscope, first under low power and then under high power.
2. Locate five types of protists. Identify each protist, and add its name to your data table. Write a brief description of each protist. Observe the distribution of the protists. Predict whether each protist is heterotrophic or autotrophic. Record your

Observations	Name of protist				
Description					
Distribution					
Autotroph or heterotroph					
Response to light					
Response to cold					
Response to warmth					

descriptions and predictions in your data table.

3. On a sheet of paper, draw and label an example of each type of protist listed in your data table.

4. Punch a hole in the center of a 40 × 20 mm piece of black construction paper. Wind the paper around your finger so that it has a slight curl to it.

5. Position the piece of black paper on top of the slide *without* DETAIN™ so that the hole is over the center of the coverslip. Place the slide on white paper, and leave it in a lighted area for about 10 minutes.

6. Gently place the slide on the stage of the compound microscope. **Be careful:** *Do not disturb the black paper.* Using low power, focus on the hole in the black paper. Notice which types of protists are visible through the hole. Using forceps, carefully remove the black paper. Continue observing for 2 to 3 minutes. Record what happens.

7. Hold your finger over one end of a capillary tube, and lower the open end of the tube to the bottom of the protist culture. Remove your finger from the end of the tube. Liquid from the culture will fill the tube. Plug the tube by pushing each end into silicone culture gel.

8. Place the filled and plugged capillary tube on a clean microscope slide. Roll a ball of silicone culture gum on your lab table to make a thin tube about 6 cm long, and make a well around one end of the capillary tube, as shown in the diagram above. Place the micro-observation tube you have made on the stage of your microscope.

9. Place several drops of hot tap water into the well on the microscope slide. Observe the movement of the protists in the tube first under low power and then under high power. Record your observations.

10. Repeat the procedure in step 9 using ice water instead of hot water.

11. Dispose of your materials according to the directions from your teacher.

12. Clean up your work area and wash your hands before leaving the lab.

Analysis

13. **Summarizing Data** How did the protists adapt to changes in their environment?

14. **Analyzing Data** Which of the protists preferred shade? Which preferred sunlight?

15. **Drawing Conclusions** What physical features should a backyard pond have in order to provide favorable environments for many different types of protists? Explain.

Applications

16. **Thinking Critically** Where in a pond are protists likely to be found during very hot weather? Justify your answer.

17. **Thinking Critically** Why is it important to know how protists behave when planning an aquatic ecosystem?

18. **Career Connection** *Landscape architects* design gardens and landscapes for all types of buildings. Find out about the training and skills required to become a landscape architect, and discover how landscape architects combine a variety of organisms in the landscapes they design.

CHAPTER 20A

Making a Cladogram

exploration

Skill:
Classifying plants by making a cladogram

Objectives
- *Identify* major characteristics of plants.
- *Determine* derived characteristics among four groups of plants.
- *Identify* the out-group among several groups of plants.
- *Construct* a cladogram that shows the evolutionary relationships of four groups of plants.

Materials
- paper
- pencil
- metric ruler
- gardening magazines (optional)
- scissors (optional)
- glue (optional)

Situation

You are the botanist for the Parks and Recreation Department of a small city. The city council recently voted to assemble an interpretive display of the plants found in the area for the city's nature preserve. The display will include a cladogram that shows the evolutionary relationships among common plants found in the local area. Your job is to make a cladogram using the following plants: spruce, horsetail, liverwort, and rose.

Background: Cladistics is used to determine the order in which evolutionary lines diverged, or branched. Among several groups of organisms, a group that lacks a characteristic seen in the others is called an **out-group.** The out-group is the most distantly related to the other groups because it has the fewest characteristics in common with the other groups. Any deviations from the characteristics of the out-group are considered **derived characteristics,** or evolutionary changes. Derived characteristics are used to build a **cladogram,** which shows the order in which major lines of organisms diverged from a **common ancestor.** To construct a cladogram showing the evolution of major plant groups, you can use the following derived characteristics: **flowers, seeds,** and **vascular tissue.**

Prelab Preparation

1. Define the boldface terms in the **Background** section.
2. Describe the process of classification.
3. Prepare a data table like the one at the top of the next page.

Procedure

1. On a separate sheet of paper, draw a diagonal line that rises from left to right and is about 18 cm long. Draw four lines, about 3 cm long and 6 cm apart, that are perpendicular to the diagonal line, as seen in the diagram at right. Your diagram will become a cladogram with four branches.

Plants	Derived Characteristics		
	Flowers	Seeds	Vascular system
Spruce			
Horsetail			
Liverwort			
Rose			
Total number of plants with characteristic (+)			

2. Using your textbook, determine whether each of the plants in your data table has a vascular system, seeds, or flowers. If the plant has a particular characteristic, mark a plus sign (+) in the column for that characteristic. If the plant does not have a particular characteristic, mark a minus sign (−) in the column for that characteristic.
3. Determine the number of groups that share each derived characteristic. To do this, count the number of plus signs (+) in each column of your data table. Write these numbers at the bottom of each column.
4. Look at your data table. The column with the greatest number of plus signs (+) indicates the derived characteristic that is shared by the most plants. In the same column, the plant with the minus sign (−) is considered the out-group. Write the name of this out-group at the end of the branch to the far left side of your cladogram.
5. To the right of the diagonal line and midway between the first two branches, write the name of the derived characteristic shared by the remaining plant groups.
6. Look at your data table again. Go to the column with the next largest number of plus signs. Determine the out-group among the remaining groups. Write the name of this out-group at the end of the second branch from the left of your cladogram. To the right of the diagonal line and midway between the second and third branches, write the name of the derived characteristic shared by the remaining groups.
7. Determine the out-group among the remaining groups of plants. Write the name of the third out-group at the end of the third branch from the left of your cladogram. To the right of the diagonal line and midway between the third and fourth branches, write the name of the derived characteristic possessed by the last group.
8. Write the name of the last group at the end of the last branch of your cladogram.
9. Illustrate the plant groups in your cladogram with drawings, photographs, or specimens of the plants.

Analysis

10. **Analyzing Diagrams** According to your cladogram, which plants are in the group that lacks seeds? Which plants are in the group that lacks flowers?
11. **Summarizing Information** Using your cladogram, describe the sequence of events that occurred in the evolution of the four types of plants in this Exploration.
12. **Evaluating Methods** What are the drawbacks of using cladistics to classify organisms?

Applications

13. **Thinking Critically** A city council member wants you to add a favorite garden plant—the bearded iris—to your cladogram. What would be your response to this request?
14. **Thinking Critically** Can a cladogram be used to identify an unknown plant? Explain your answer.
15. **Career Connection** *Botanists* study all aspects of plants, including their structure, function, and classification. Find out about the training and skills required to become a botanist, and discover some places that employ botanists.

CHAPTER

21A Altering Plant Growth With a Hormone

exploration

Skills:
- **Working with living plants**
- **Graphing results**

Objectives
- *Apply* plant hormones to bean plants.
- *Compare* and *contrast* the effects of different concentrations of an auxin on root growth.
- *Evaluate* the effectiveness of two methods of applying an auxin to bean plants.

Materials
- safety goggles
- lab apron
- 5 flasks (150 mL)
- wax pencil
- 100 mL graduated cylinder
- 300 mL of water
- 5 bean seedlings
- scalpel
- cotton
- 105 mL of indoleacetic acid solution (IAA), 1:10,000
- 105 mL of indoleacetic acid solution (IAA), 1:1,000,000
- 2 small paintbrushes
- metric ruler

Situation

You have just been hired by the United States Department of Agriculture (USDA) as a lab technician for a plant physiologist. As part of your job, you will be conducting experiments to determine how the plant hormone auxin can be used to stimulate root growth. Your first assignment is to compare the effects of auxin concentration and method of application on root growth in bean plant cuttings.

Background: The auxin **indole-3-acetic acid (IAA)** is produced in the new leaves of a plant and then transported to the stem. The IAA initiates the formation of **adventitious roots,** which are roots that grow from a stem.

Prelab Preparation

1. Define the boldface terms in the **Background** section.
2. Predict which method of applying IAA will produce the best root growth.
3. List this experiment's independent variable, dependent variables, constants, and control.
4. Make a data table with six main columns and the following headings: Day, Treatment 1, Treatment 2, Treatment 3, Treatment 4, and Treatment 5. Divide each of the Treatment columns into two columns with the following headings: Number of roots, Length of roots.

Procedure

1.   Put on safety goggles and a lab apron.
2. Using a wax pencil, put the initials of everyone in your group on each of the five flasks. **CAUTION: Glassware is fragile. Notify your teacher promptly of any broken glass or cuts. Do not clean up broken glass or spills unless your teacher tells you to do so.** Label the flasks as follows:

   Treatment 1—water (leaves and stem)

   Treatment 2—1:10,000 IAA (leaves)

   Treatment 3—1:1,000,000 IAA (leaves)

   Treatment 4—1:10,000 IAA (stem)

   Treatment 5—1:1,000,000 IAA (stem)

3. Place 100 mL of distilled water in the flasks for Treatments 1, 2, and 3. Place 100 mL of the 1:10,000 IAA in the flask for Treatment 4. Place 100 mL of the 1:1,000,000 IAA in the flask for Treatment 5.

4. Using a scalpel, carefully cut the stems of five bean seedlings near soil level. Also remove the cotyledons from each plant. **CAUTION: Scalpels are very sharp. Be careful not to cut yourself.**

5. Place one plant in each of the five flasks. Make sure the stems are submerged in the liquid. Pack cotton at the top of each flask to keep the plants upright, as seen in the photo above. *Note: Be careful not to crush or break the stems of the plants.*

6. Using a small paintbrush, coat the leaves of the plant in the Treatment 1 flask with distilled water. Coat the leaves of the plant in the Treatment 2 flask with the 1:10,000 IAA solution. **Using a different paintbrush,** coat the leaves of the plant in the Treatment 3 flask with the 1:1,000,000 IAA solution. Repeat each day for 10 days.

7. Set the flasks in a place where the plants will receive the same light. Observe the plants daily for 10 days. Record the number of roots that appear.

8. Every day after roots appear, carefully remove each plant from its flask, and measure the length (from stem to root tip) of three typical roots. Measure the same three roots each day. Carefully replace each plant in its original flask. Find the average length of the roots for each plant, and record this average in your data table.

9. Dispose of your materials according to the directions from your teacher.

10. Clean up your work area and wash your hands before leaving the lab.

11. Make two graphs of your data. Show "Days" on the x-axis of both graphs. Show "Number of roots" on the y-axis of one graph and "Average length of roots" on the y-axis of the other graph. Use a different color or symbol to plot the data for each treatment.

Analysis

12. **Summarizing Data** What happened to each bean plant over the observation period?

13. **Interpreting Data** Does IAA concentration affect root growth? Explain your answer.

14. **Drawing Conclusions** Which method of applying IAA was the best? What criteria did you use in deciding which was the best method?

15. **Predicting Outcomes** What would happen if you increased the IAA concentration? What would happen if you decreased it?

Applications

16. **Thinking Critically** Why is it important to handle living plants carefully during experiments?

17. **Designing an Experiment** Auxin produced in the buds at the tips of some plants prevents the growth of buds lower down on the stems. Design an experiment to determine the effect of removing the bud at the tip of a plant.

18. **Career Connection** *Plant physiologists* study how plant hormones affect plants. Find out about the training and skills required to become a plant physiologist, and research some of the ways that hormones are used to control the growth of plants used for food.

CHAPTER

22A Identifying Nutrients in Foods

exploration

Skill:

Qualitative analysis for food nutrients

Objectives

- *Prepare* standards to use in testing a food for the presence of sugar, starch, protein, and lipid.
- *Compare* tests for sugar, starch, protein, and lipid in a food to standards.
- *Identify* the nutrients present in food.

Materials

- safety goggles
- lab apron
- wax pencil
- 9 test tubes
- test-tube rack
- test-tube holder
- distilled water
- 2 mL of glucose solution
- dropper bottle of Benedict's solution
- hot-water bath
- 2 mL of 1% starch solution
- dropper bottle of Lugol's solution
- 2 mL of gelatin solution
- dropper bottle of biuret solution
- 5 in. square of brown paper
- 2 mL of vegetable oil
- food sample for four tests
- wooden splint or spatula

Peanut plant

Situation

You have applied for a job as a food technologist with a food manufacturer. If you get the job, you will be analyzing the company's new line of vegetarian foods for nutrient content. All applicants must demonstrate the ability to test a food for the presence of carbohydrates, proteins, and lipids. Obtaining accurate results, and demonstrating good lab techniques and the ability to follow instructions, will mean that you will be asked to return for the final round of interviews.

Background: When you test for *the presence of* a substance, you are performing a **qualitative analysis.** Nutrients such as carbohydrates (sugars and starch) and proteins can be detected by an **indicator,** which is a chemical that produces a characteristic color when a particular substance is present. Lipids can be detected by their ability to make paper **translucent** (able to let light shine through). Tests on substances with an unknown makeup are compared to **standards,** which are the results of tests that show a positive response for a known substance.

Prelab Preparation

1. Define the boldface terms in the **Background** section.
2. Explain how standards are used in research.
3. Prepare a data table like the one at the top of the next page.

Procedure

1. 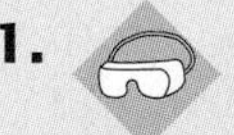  Put on safety goggles and a lab apron.
2. Using a wax pencil, label nine test tubes as follows: sugar—positive; sugar—negative; sugar—sample; starch—positive; starch—negative; starch—sample; protein—positive; protein—negative; protein—sample. Group the tubes in a test-tube rack according to the substance being tested for. Add 2 mL of distilled water to each test tube labeled "negative."
3. Pour 2 mL of glucose solution into the "sugar—positive" test tube. To test for sugar, add a dropperful of Benedict's solution to both the positive and negative test tubes. **CAUTION: If you get a chemical on your skin or clothing, wash it off while calling to your teacher.**

Nutrient	Indicator	Positive test (standard)	Negative test	Food sample (________)
sugar	Benedict's solution			
starch	Lugol's solution			
protein	biuret solution			
lipid	appearance on paper			

4. Place the test tubes in a hot-water bath using a test-tube holder. **CAUTION: Do not touch hot objects with your bare hands. Use tongs, test-tube holders, and padded gloves, as appropriate.** Heat for about 5 minutes. Record any color changes in the test tubes after heating. Using a test-tube holder, set the test tubes in the rack to cool.

5. Pour 2 mL of starch solution into the "starch—positive" test tube. To test for starch, add a dropperful of Lugol's solution to both the positive and negative test tubes. Record the colors you observe in your data table. **CAUTION: Lugol's solution will stain your skin and clothing. Promptly wash off spills to minimize staining.**

6. Pour 2 mL of gelatin solution into the "protein—positive" test tube. To test for protein, add a dropperful of biuret solution to both the positive and negative test tubes. Record your observations. **CAUTION: If you get a chemical on your skin or clothing, wash it off while calling to your teacher.**

7. Place a square of brown paper on a dry area of your lab table. Smear a drop or two of vegetable oil on the paper, and label the area "positive test" below the spot. Place a few drops of water in another area away from the oil spot. Label the area "negative test" below the water spot. Let the spots dry, and then record your observations.

8. Obtain enough food from your teacher to test for each of the four nutrients. Using a wooden splint or spatula, place a small amount of food in each of the three test tubes labeled "sample." Use the proper indicator to test for the nutrient marked on each test tube. Smear the fourth sample on an unused area of the brown paper. Compare your results with the standards you prepared earlier. Record your results in your data table.

9. Dispose of your materials according to the directions from your teacher.

10. Clean up your work area and wash your hands before leaving the lab.

Analysis

11. **Recognizing Patterns** Which test tubes contained standards?

12. **Analyzing Results** What indicator detects sugar? starch? protein?

13. **Analyzing Results** What color resulted when sugar was present? starch? protein?

14. **Analyzing Methods** How did the test for lipids differ from the other tests?

15. **Drawing Conclusions** What nutrients were present in the food sample?

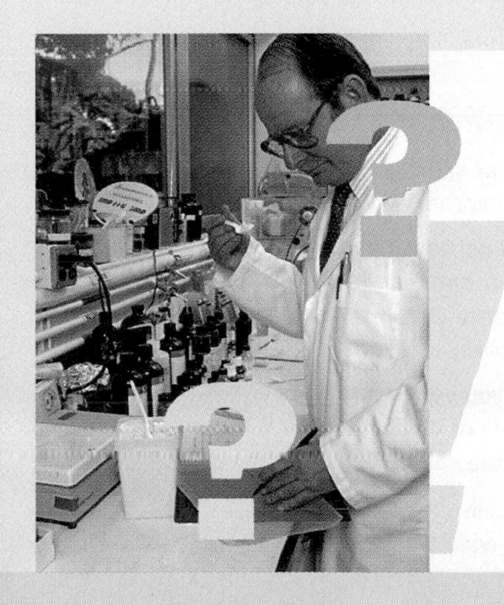

Applications

16. **Thinking Critically** If *qualitative analysis* tests for the presence of specific substances, what do you think *quantitative analysis* tests for?

17. **Thinking Critically** Evaluate the use of qualitative tests such as these to help you plan a diet with fewer than 50 g of fat.

18. **Career Connection** *Food technologists* perform a variety of qualitative and quantitative tests on foods. Find out about the training and skills required to become a food technologist, and discover which industries employ food technologists.

CHAPTER 22B

Identifying Nutrients in Foods—School Menu

Investigation

Prerequisites

- **Exploration 22A: Identifying Nutrients in Foods** on pages 800–801
- **Review** the following:
  1. indicators
  2. the use of standards

Mt. Pleasant Private Academy

March 1, 1997

Mr. John Williams, Director
Food Technology Division
BioLogical Resources, Inc.
101 Jonas Salk Dr.
Oakwood, MO 65432-1101

Dear Mr. Williams:

Our elementary students (284 in all) are busy getting ready to celebrate National Nutrition Week in April. I would love for our teachers to give them an exciting lesson on foods from plants. Young children just love colors, and I have heard that there are bright color tests for the good things in foods. Here is a list of some foods that will be served in the cafeteria during that week: potatoes, corn, white beans, whole-wheat rolls, potato chips, bananas, and peanut butter. Could you run some of those tests and let me know what good things are in these foods? Tell me how to do the tests, too, so that the teachers in our lower grades can perform the tests for their students. If the students can see some bright colors, I just know they won't forget which foods have those good carbohydrates, lipids, and proteins. The board is willing to pay your normal fee. Thank you so much.

Sincerely,

Amanda Brown

Ms. Amanda Brown, Chairperson
Education Advisory Board
Mt. Pleasant Private Academy

Required Precautions

- Wear safety goggles and a lab apron.
- Do not touch hot objects with your bare hands. Use tongs, test-tube holders, and padded gloves, as appropriate.

- If you get a chemical on your skin or clothing, wash it off while calling to your teacher.
- Wash your hands before leaving the laboratory.

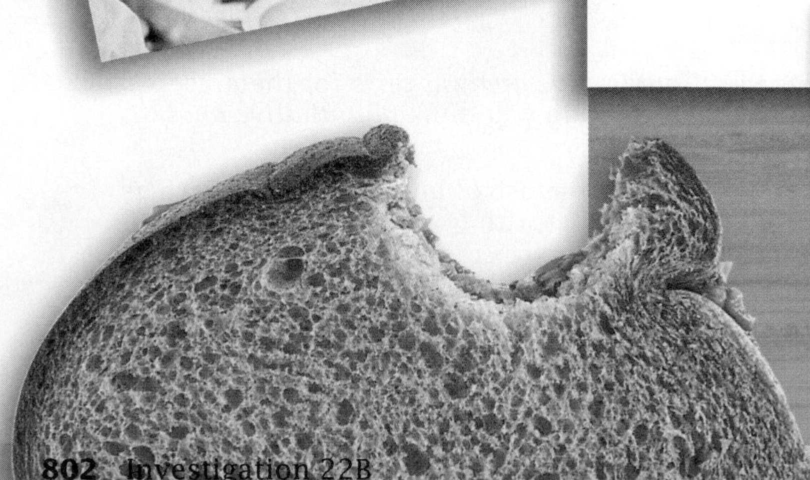

BioLogical Resources, Inc., Oakwood, MO 65432-1101

MEMORANDUM

To: Team Leader, Food Testing Dept.
From: John Williams

An education board member wants an exciting lesson for elementary students who are studying nutrition. Please analyze the foods she has listed using indicator tests, when possible, that teachers can repeat. Ms. Brown will need a list of nutrients, including lipids, in these foods and directions on repeating the tests so that teachers can show their students the colors that result when indicators are used to test for nutrients in foods. Thanks for doing this promptly.

Before you begin, you must submit a proposal for my approval. **Your proposal must include the following:**

- the **question** you are seeking to answer
- the **procedure** you will use
- a detailed **data table**
- a complete, itemized list of proposed **materials** and **costs** (including the use of facilities, labor, and amounts needed)

When you finish your tests, prepare a report in the form of a business letter to Ms. Amanda Brown. **Your report must include the following:**

- a paragraph describing the **procedure** you followed to test the food samples
- a complete **data table**
- your **conclusion** about the nutrients in the cafeteria foods being served that week
- a detailed **invoice** showing all materials and labor plus the total amount due

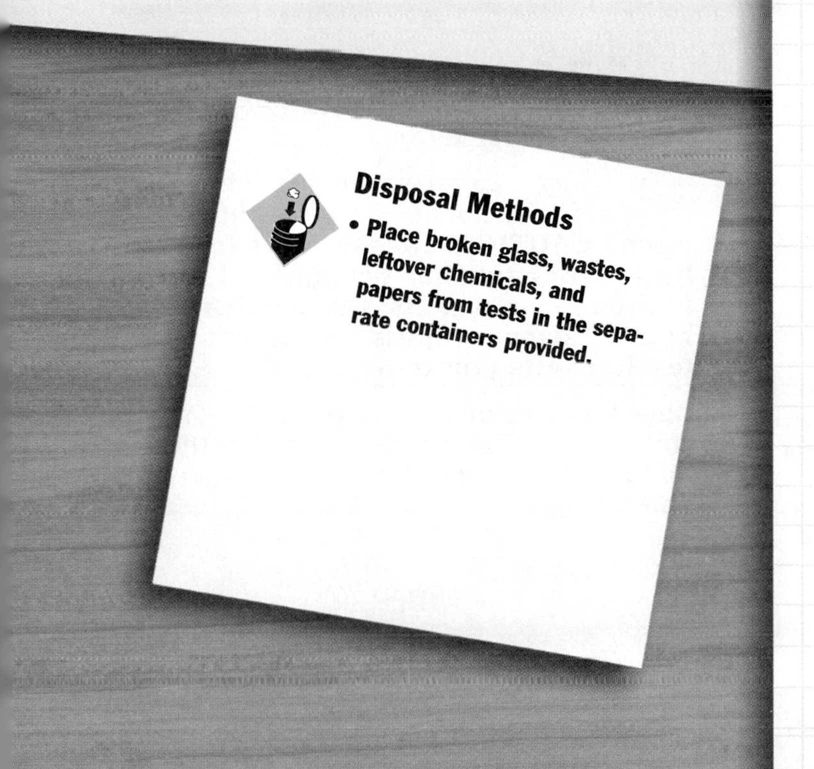
Disposal Methods

- **Place broken glass, wastes, leftover chemicals, and papers from tests in the separate containers provided.**

FILE: Mt. Pleasant Private Academy

MATERIALS AND COSTS (Select only what you'll need. No refunds.)

I. Facilities and Equipment Use	
facilities	\$480.00/day
compound light microscope	\$100.00/day
personal protective equipment	\$20.00/day
hot-water bath	\$15.00/day
test-tube rack	\$5.00/day
test-tube holder	\$5.00/day
test tubes	\$2.00 each
II. Labor and Consumables	
labor	\$30.00/hour
food samples	\$5.00 each
Benedict's solution	\$20.00/dropper bottle
Lugol's solution	\$20.00/dropper bottle
biuret reagent	\$20.00/dropper bottle
Sudan III	\$20.00/dropper bottle
wax pencil	\$2.00 each
wax paper	\$0.50/sheet
brown paper	\$0.50/square
wooden splints	\$0.10 each
Fines	
OSHA safety violation	\$2,000.00/incident

CHAPTER 24A

Modeling the Action of a Kidney

exploration

Skill:
Modeling a biological process

Objectives

- *Construct* a model of the action of a kidney.
- *Predict* the outcome of an experiment.
- *Relate* changes in color and turgor to evidence of osmosis.
- *Evaluate* the function of the kidney as an adaptation for living on land.

Materials

- safety goggles
- lab apron
- 250 mL beaker
- graduated cylinder
- distilled water
- kidney dialysis tubing
- "unfiltered blood"
- masking tape, adhesive label, or wax pencil
- 2 pieces of string

Situation

You are an assistant to the zoologist for a natural-science museum. Part of your job is to set up demonstrations that teach museum visitors about the structure, function, and behavior of animals. The museum is building a large display that features adaptations that enabled animals such as the Iguana lizard seen below to live successfully on land. Today you will be setting up a demonstration that shows how a kidney removes nitrogen-containing wastes from the blood with little loss of water.

Background: Within a kidney, blood passes through a series of filtering units called nephrons. A nephron contains capillaries, which carry blood to be filtered, and ducts (tubes), which collect small particles of substances such as salt, water, glucose, and urea. Some of these particles are reabsorbed into the blood along with much of the water. Some of the water and the urea remain in the ducts and are excreted from the body as urine.

Prelab Preparation

1. Define the following terms: nephron, capillary, duct, diffusion, osmosis, urea.
2. Describe the function of a kidney.
3. Make a data table like the one at the top of the next page.

Procedure

1.   Put on safety goggles and a lab apron.
2. Using a 25 mL graduated cylinder, measure about 20 mL of "unfiltered blood." **CAUTION: Glassware is fragile. Notify your teacher promptly of any broken glass or cuts. Do not clean up broken glass or spills unless your teacher tells you to do so.**
3. Obtain a piece of wet dialysis tubing from your teacher. Tightly tie one end of the tubing with a piece of string, as seen in the photo at right.

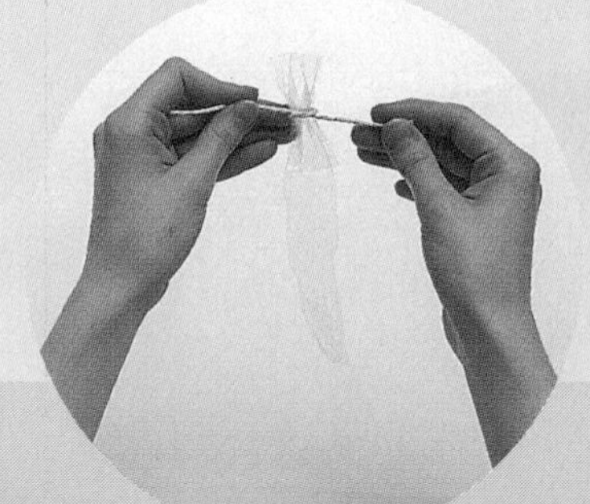

Observations	Inside dialysis tubing	In the beaker
Initial color		
Color after 30 minutes		
Prediction		
Color after 24 hours		

4. Wet your fingers with tap water, and open the tubing by rubbing the untied end of the tubing between your fingers. Pour the 20 mL of "unfiltered blood" into the open end of the tubing. Tightly tie this end of the tubing with a piece of string. **CAUTION: Dyes in the "unfiltered blood" will stain your skin and clothing. Promptly wash off spills to minimize staining.** The dialysis-tubing bag you have made represents the blood vessels that enter a kidney.
5. Rinse the dialysis-tubing bag with tap water in the sink. Place the dialysis-tubing bag into a 250 mL beaker. Using a wax pencil, write your initials on the beaker. Fill the beaker to within 2–3 cm (about 1 in.) of the top with cold tap water. Record the beginning colors of the solutions inside the bag and in the beaker.
6. After 30 minutes, record the color of the solution in the beaker. While observing the contents of the tubing and of the beaker for the remainder of the lab period, predict what will happen in the next 24 hours. Write your prediction in your data table.
7. Clean up your work area and wash your hands before leaving the lab.
8. After 24 hours, observe any changes that have occurred in the beaker and in the dialysis-tubing bag. Record these observations in your data table.
9.  Dispose of your materials according to the directions from your teacher.
10.  Clean up your work area and wash your hands before leaving the lab.

Analysis

11. **Summarizing Results** Describe any changes that occurred inside the dialysis-tubing bag and in the beaker.
12. **Inferring Relationships** In what part of a kidney do you find the liquid that the water in the beaker represents? Write a label for the water in the beaker to use in your display.
13. **Relating Information** On the audiotape that will accompany your display, you have no more than 15 seconds to explain the changes you observed in your model. Write a script for what you would say, and record the time required to read the script.
14. **Drawing Conclusions** Provide four multiple-choice answers for each of the following questions, which you will include in your display: What substance remained in the dialysis-tubing bag? What substance entered the water in the beaker? Indicate the correct answer for each question.
15. **Making Inferences** For your audiotape program, write a script that explains why the "blood" remains in the dialysis-tubing bag while the "urine" passes into the beaker.

Applications

16. **Thinking Critically** What circumstances would cause water to move out of the dialysis-tubing bag (bloodstream) and into the beaker (collecting ducts of a nephron)?
17. **Thinking Critically** Complete the script for your audiotape program by writing an explanation of why the kidney is an important evolutionary advance for land animals.
18. **Career Connection** *Museum-exhibit technicians and education specialists* design and build museum exhibits. Find out about the training and skills necessary to become either an exhibit technician or an education specialist for a natural-science museum.

CHAPTER 26A

Classifying Insects

investigation

Prerequisite

- **Review** the following:
  1. characteristics of insects
  2. definition and use of a dichotomous key

Aldous Huxley Elementary School
Oakwood, MO 65432-1984

March 28, 1997

Shing Chun
Director of Development
BioLogical Resources, Inc.
101 Jonas Salk Dr.
Oakwood, MO 65432-1101

Dear Mr. Chun,

I can't tell you how much we have enjoyed the beautiful garden that your company donated to our school. Our students have learned a lot about gardening and plants. The garden has also attracted many insects. I want our students to be able to identify some of these insects and to understand that many garden insects are beneficial. Our PTA suggested that we add a display on insects to the garden. The following is a list of the things we would like the display to do.

- give the common name and scientific name of each insect
- indicate whether insects are harmful or beneficial to plants
- teach students how to use a key to identify an organism

I thought you might be interested in helping us provide this educational opportunity for our students. Could your staff construct an insect display like the one described above? I would appreciate any help that you can give us.

Sincerely,

Danielle Andrews

Danielle Andrews
Principal
Aldous Huxley
Elementary School

Required Precautions

- **Wear gloves when handling stinging insects.**

- **Do not attempt to collect stinging insects if you are allergic to insect stings.**

- **Wash your hands before leaving the laboratory.**

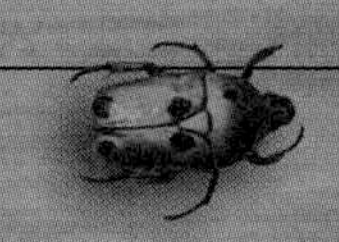

Biological Resources, Inc., Oakwood, MO 65432-1101

MEMORANDUM

To: Team Leader, Entomology Dept.
From: Shing Chun

Our board of directors has given me the go-ahead for this project. I want your team to put together a display that will be donated to the school. The display should include at least 20 insect specimens. Please collect a variety of insects from many different habitats.

I have attached a dichotomous key to the major insect orders. Use this key in constructing your display. Not all of the orders in this key are found in our area. You may also include other orders of insects that you find.

Before you start your work, you must submit a proposal for my approval. **Your proposal should include the following:**

- the **procedure** you will use
- a complete, itemized list of proposed **materials** and **costs** (including use of facilities, labor, and amounts needed)

When you finish, write a business letter to Ms. Andrews describing the procedure you used to construct your display. Your letter will accompany the completed project. Keep a careful record of the cost of all materials and labor. Because our donations are tax deductible, we will need this for our records.

Dichotomous Key to Major Orders of Insects

1a	Wings lacking	Go to 2
1b	Wings present	Go to 3
2a	Abdomen narrowly joined to thorax	Hymenoptera (ants, bees, wasps)
2b	Antlike body; abdomen broadly joined to thorax	Isoptera (termites)
3a	Two wings; mouthparts adapted for sucking	Diptera (flies)
3b	Four wings	Go to 4
4a	Forewings hard, shield-like, and meeting in a straight line down the back	Coleoptera (beetles)
4b	Wings membranous, at least in part	Go to 5
5a	Wings broad and covered with scales; tubelike, sucking mouthparts	Lepidoptera (moths, butterflies)
5b	Not so	Go to 6
6a	Hindwings smaller than forewings; chewing mouthparts	Hymenoptera (wasps, ants, bees)
6b	Not so	Go to 7
7a	Hindwings as large as or larger than forewings; wings held straight	Odonata (dragonflies, damselflies)
7b	Not so	Go to 8
8a	First pair of wings thickened at base; upper thorax shield shaped	Hemiptera (true bugs)
8b	Not so	Go to 9
9a	Wings held sloping at sides when at rest; piercing and sucking mouthparts	Homoptera (cicadas, leafhoppers)
9b	Forewings straight and leathery; hindwings membranous and folded under forewings	Orthoptera (cockroaches, crickets, grasshoppers)

Disposal Methods

- **Live insects can be released.**
- **Place dead insects in the trash.**

FILE: Aldous Huxley Elementary School

MATERIALS AND COSTS (Select only what you'll need. No refunds.)

I. Facilities and Equipment Use	
facilities	$480.00/day
personal protective equipment	$10.00/day
stereomicroscope	$100.00/day
hand lens	$50.00/day
insect net	$2.00/day
small jar with lid	$2.00/day
forceps	$5.00/day
field guide for insects	$5.00/day
II. Labor and Consumables	
labor	$30.00/hr
markers	$2.00 each
colored pencils	$2.00 each
glue	$2.00/bottle
poster board	$0.50/piece
paper	$0.25/piece
colored string or yarn	$0.25/m
Fines	
OSHA safety violation	$2,000.00/incident

CHAPTER 28A

Making a Cladogram—Wildlife Zoo

investigation

Prerequisites

- **Exploration 20A: Making a Cladogram** on pages 796–797
- **Review** the following:
  1. characteristics of vertebrates
  2. definition and preparation of a cladogram

Wildlife Zoo, Inc.
Valleyview, VT 10201-2111

January 15, 1997

Shing Chun
Director of Development
BioLogical Resources, Inc.
101 Jonas Salk Dr.
Oakwood, MO 65432-1101

Dear Mr. Chun,

I am the director of a small zoo. We also do wildlife rehabilitation. Each year, hundreds of school children from the surrounding area tour our zoo. Our enthusiastic volunteers conduct educational programs for all age groups.

Recently, we received a grant from a wildlife conservation organization. The grant is to be used to provide more educational opportunities for the zoo's visitors. Our board of directors wants to use some of the grant money to build a new exhibit that shows the evolutionary relationships among the animals in the exhibit.

An architect will produce the blueprints for the exhibit. However, I need someone to "do the science" for the project. I want your company to submit a plan for displaying zoo animals in a way that shows their evolutionary relationships. We are willing to pay your normal fees. Please bill us when you finish the work.

Sincerely,

Jerome Franklin
Director,
Wildlife Zoo, Inc.

Required Precautions

- **Handle scissors carefully.**

B Biological Resources, Inc., Oakwood, MO 65432-1101

MEMORANDUM

To: Team Leader, Zoology Dept.
From: Shing Chun

This sounds like an exciting and worthwhile project. I want your team to develop a plan for the zoo's exhibit on vertebrate evolution. Then build a model that shows what the exhibit might look like. Use preserved specimens or photographs of the animals for the model, or have one of our artists draw the animals.

I suggest that you base the exhibit on a cladogram showing the relationships among the major groups of vertebrates: jawless fish, cartilaginous fish, bony fish, amphibians, reptiles, birds, and mammals. This could be done in many ways. Be creative! Don't forget to write the description to be placed by each animal in the exhibit. Finally, produce an interpretive brochure that will guide visitors through the exhibit.

Before you build your model, you must submit a proposal for my approval. **Your proposal must include the following:**

- the **plan** you propose for the zoo's new exhibit
- the **procedure** you will use to build your model
- a detailed **data table**
- a **cladogram** showing the evolutionary relationships among vertebrates
- a complete, itemized list of proposed **materials** and **costs** (including use of facilities, labor, and amounts needed)

When you finish, prepare a report in the form of a business letter for Mr. Franklin. **Your report must include the following:**

- a paragraph describing the **plan** you propose for the zoo's new exhibit
- your **interpretive brochure**
- a detailed **invoice** showing all materials and labor plus the total amount due

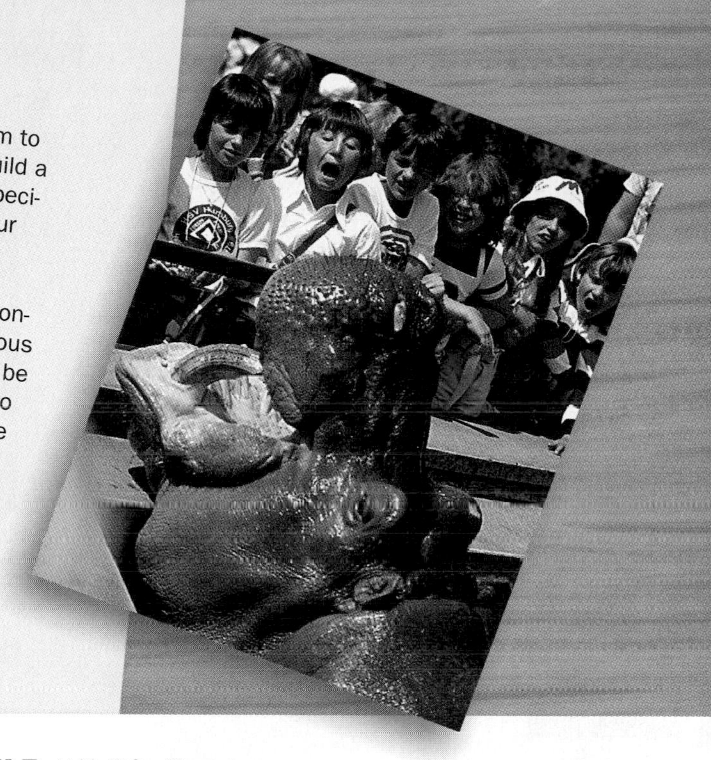

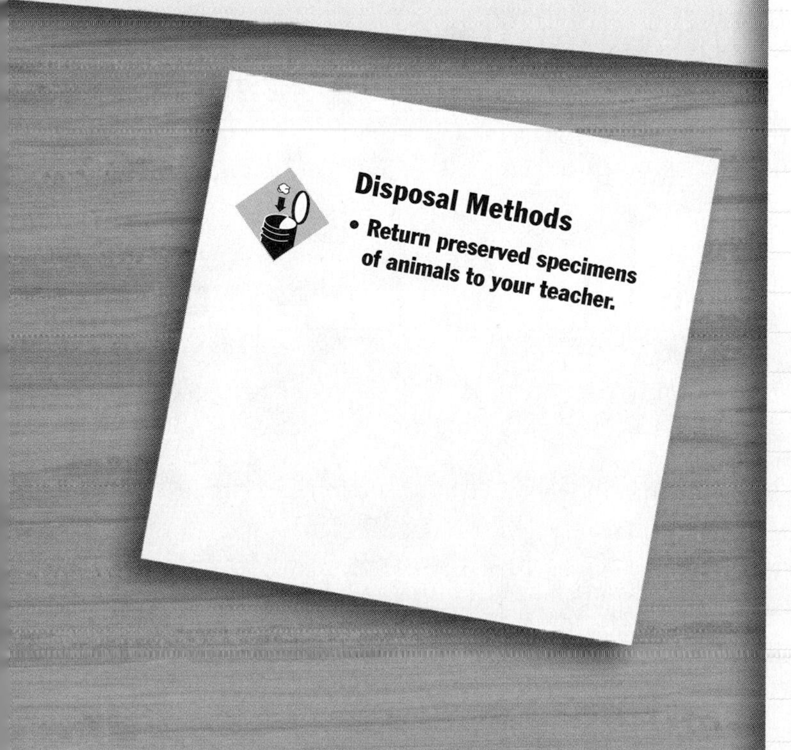

Disposal Methods

- **Return preserved specimens of animals to your teacher.**

FILE: Wildlife Zoo, Inc.

MATERIALS AND COSTS (Select only what you'll need. No refunds.)

I. Facilities and Equipment Use	
facilities	$480.00/day
personal protective equipment	$10.00/day
hand lens	$50.00/day
scissors	$2.00/day
II. Labor and Consumables	
labor	$30.00/hr
markers	$2.00 each
colored pencils	$2.00 each
glue	$2.00/bottle
poster board	$0.50/piece
paper	$0.25/piece
colored string or yarn	$0.25/m
photographs of vertebrates	$5.00 each
particle board	$10.00/piece
preserved specimens of vertebrates	$50.00 each
Fines	
OSHA safety violation	$2,000.00/incident

CHAPTER 29A

Life Span and Lifestyle

interactive exploration

Background: How long will you live? Although genes play a role in longevity, lifestyle and diet are also very important. Some important factors include how much exercise you get, how much you eat, and what you eat. This interactive exploration allows you to see how these and other factors affect life span.

Objectives

- *Simulate* how weight, smoking, amount of exercise, and diet affect life span.
- *Determine* your projected life span based on your current lifestyle.

Materials

- computer with CD-ROM drive
- CD-ROM *Interactive Explorations in Biology: Human Biology*

Prelab Preparation

1. You will need to calculate the energy content of your diet and your daily energy use. Follow your teacher's instructions.
2. Load and start the program Life Span and Lifestyle. Click the Topic Information button on the Navigation Palette. Read the focus questions and review these concepts: Life Expectancy, Smoking and Health, Diet and Health, and Exercise and Health.
3. Click the word *File* at the top left of the screen, and select Interactive Exploration Help. Listen to the instructions that explain the operation of the exploration. Click the Exploration button at the bottom right of the screen to begin the investigation. You will see an animated diagram like the one below.

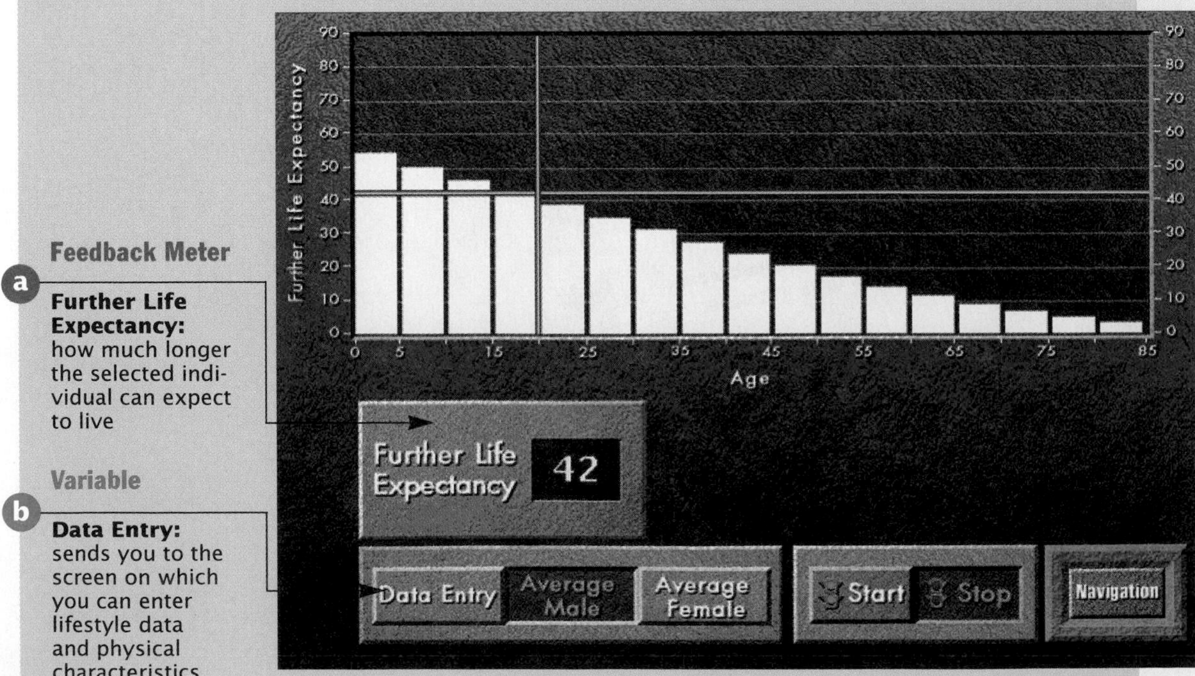

Procedure

Part A: Comparing Different Lifestyles

1. Make a table like the one below in which to record your data. Now compare two 15-year-olds of the same sex who follow different lifestyles. Begin by clicking the Data Entry button.

Age	Sex	Ht.	Wt.	Exercise	Animal fat in diet	Smoking	Calories in diet	Expected life span

2. One 15-year-old is a female athlete who eats a low-fat diet and doesn't drink or smoke. Slide the indicator in each scale that appears on the screen to reflect her characteristics and lifestyle: height, 70 inches; weight, 130 pounds; exercise, 1500 Cals/day; animal fat in diet, 10 percent; smoking, 0 cigarettes/day; alcohol, 0 drinks/day; and diet, 2500 kcal/day. Set the age to 15 years, and click female on the Sex button. Record these values in your table, and click the Close button.
3. Click the Start button. The graph that appears is called a life expectancy histogram and indicates how many additional years this person is expected to live. What is the life expectancy for this 15-year-old athlete? Record this value in your table.
4. Compare this value with that of a female who is the same height, weight, and age but who exercises less. Click the Data Entry button. Slide the exercise scale to 750 Cals/day. Do not change any of the other variables. Record the value of each variable in your table. Click the Close button and then the Start button.
5. Record the life span in your table. Compared with the life expectancy of the 15-year-old athlete, how many years less is this 15-year-old female expected to live?

Part B: Determining the Impact of Smoking

6. Enter the following data for an 18-year-old male: height, 70 inches; weight, 165 pounds; exercise, 1000 Cals/day; animal fat in diet, 20 percent; smoking, 0 cigarettes/day; alcohol, 0 drinks/day; and diet, 2200 kcal/day. When you have entered all the data, record the values in your table, and click the Close button.
7. Click the Start button. What is the life expectancy for this 18-year-old?
8. Click the Data Entry button. Slide the smoking indicator to 5 cigarettes/day and record all values in your table. How does this change in lifestyle affect his life expectancy?
9. Repeat step 8, but change the number of cigarettes smoked to 20 per day. How much is life expectancy reduced by smoking 20 cigarettes per day?

Part C: Calculating Your Life Expectancy

10. Click the Data Entry button. Refer to the summary of your energy use and energy intake that you prepared earlier. Select the values for your age, sex, height, weight, and lifestyle. Determine your life expectancy, and record it in your table. How could you change your lifestyle to lengthen your life span?

Analysis

11. Assume that the 18-year-old described in step 6 is smoking 10 cigarettes per day. Concerned about his life expectancy, he decides to increase his exercise program to 1200 Cals per day, thinking that the additional exercise will extend his life span. Enter the data and determine if he is correct in his thinking.
12. Assume that the 18-year-old male described in the previous question is thinking about either quitting smoking or reducing the amount of animal fat in his diet to 10 percent. Enter the data for each decision, and determine which factor would have the greater impact on his life expectancy.

CHAPTER 30A

exploration

Determining the Effect of Drugs on Heart Rate

Skills:
- **Measuring heart rate**
- **Graphing and analyzing numerical data**

Objectives
- *Observe* the effect of drugs on heart rate.
- *Categorize* several drugs as stimulants or depressants.

Materials
- safety goggles
- disposable gloves
- lab apron
- 5 medicine droppers
- 5 *Daphnia*
- petroleum jelly
- depression slide
- coverslip
- compound microscope
- paper towel cut into five 5 cm squares
- clock with second hand
- 100 mL beaker
- 50 mL of culture water
- Solution 1—tobacco
- Solution 2—coffee
- Solution 3—orange pekoe tea
- Solution 4—diet pill containing phenylpropanolamineHCL
- Solution 5—ethyl alcohol

Situation

As a pharmacist's assistant, you work for a clinic that helps patients recover after heart surgery. Some of these patients need to lose weight. Others need to avoid substances that make their heart beat too fast or too slowly. Your job is to develop a list of beverages and over-the-counter medications that do not cause a change in heart rate of more than 10 percent and that doctors could recommend to patients. For your first round of tests, you decide to use the invertebrate *Daphnia*.

Background: Drugs known as **stimulants** increase the body's metabolism. Others, known as **depressants,** slow the rate of metabolism. **Heart rate** (HR) is one indicator of increased or decreased metabolism. Stimulants and depressants affect the heart rate of *Daphnia* and the heart rate of humans in similar ways.

Prelab Preparation

1. Define the boldface terms in the **Background** section.
2. What is a drug?
3. Prepare a data table like the one at the top of the next page. Complete your data table by adding a row for each of the five solutions to be tested.

Procedure

1. 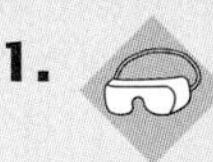   Put on safety goggles, gloves, and a lab apron.
2. Using a clean medicine dropper, collect one *Daphnia* in fluid from the classroom stock culture. Place the *Daphnia* into the well of a clean depression slide. Place a small dab of petroleum jelly in the well to slow the *Daphnia*'s movement. Add a coverslip, and observe under low power.
3. Collect data to serve as a control by performing step 4.
4. Count the *Daphnia*'s heartbeats for 10 seconds. Record this number under Trial 1 in your data table. Turn off the microscope light, and wait 20 seconds. Repeat the count for Trials 2 and 3, always turning off the light and waiting 20 seconds between counts.

Solution	10-second Heart Rate			Formulas for calculations			
	Trial 1 (A)	Trial 2 (B)	Trial 3 (C)	Average 10 s HR $(A + B + C)/3$ (D)	Average 60 s HR $(D \times 6)$ (E)	Change in HR $(E_{sol} - E_{control})$ (F)	Percent difference $(E_{sol} \div E_{control}) \times 100$ (G)
Control						________	________
1							

5. Use a clean medicine dropper to place a drop of Solution 1 on the depression slide so that it touches one edge of the coverslip. Place a piece of paper towel along the opposite edge of the coverslip to draw the solution into the fluid around the *Daphnia.*
6. Wait one minute for the solution to take effect. Repeat step 4.
7. Place the *Daphnia* in the container provided. Clean the slide and coverslip with soap and water. Rinse thoroughly. Prepare a slide with a fresh *Daphnia* by repeating step 2.
8. Repeat steps 5–7 for test solutions 2–4. Repeat steps 5 and 6 for Solution 5. Use a clean dropper for each solution.
9. Complete the calculations indicated in columns D, E, F, and G of your data table.
10. Make two graphs similar to the one below. Plot the data in column E on the *y*-axis of graph 1. Plot the data in column F on the *y*-axis of graph 2. Give each graph a title and a label for the *y*-axis. *Note: Allow for negative changes in graph 2. A bar graph would work well in this instance.*

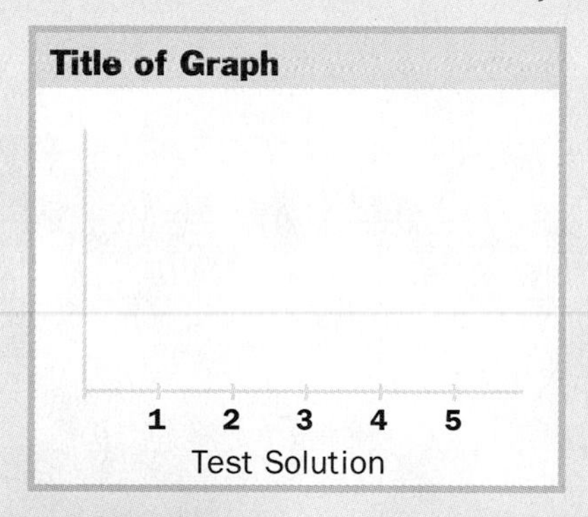

11. Dispose of your materials according to the directions from your teacher.
12. Clean up your work area and wash your hands before leaving the lab.

Analysis

13. **Recognizing Relationships** Which solutions increased the heart rate of *Daphnia*? decreased the heart rate of *Daphnia*? What drug does each solution contain?
14. **Analyzing Data** Which solutions caused the heart rate of *Daphnia* to differ by more than 10 percent?
15. **Interpreting Data** Based on your data, which solutions contain drugs classified as stimulants? as depressants?
16. **Analyzing Methods** Why did you need to use a fresh *Daphnia* for each solution?
17. **Drawing Conclusions** Which of the substances tested would you recommend to patients whose heart rate should not be allowed to vary by more than 10 percent?

Applications

18. **Designing an Experiment** Design an experiment that investigates the effects of different drug concentrations on heart rate.
19. **Thinking Critically** Why use the percent difference in heart rate, instead of the actual change in heart rate, when comparing the effects of drugs on the heart rate of different organisms?
20. **Career Connection** *Pharmacist's assistants* help pharmacists fill prescriptions and educate patients about their medications and how to take them properly. Find out about the training and skills required to become a pharmacist's assistant, and discover what other tasks they perform.

CHAPTER 30B

Drug Addiction

interactive exploration

Objectives

- *Simulate* the effects of cocaine use on a synapse in the brain.
- *Observe* the effect of cocaine use on the number of receptors in a synapse.

Materials

- computer with CD-ROM drive
- CD-ROM *Interactive Explorations in Biology: Human Biology*

Background: This interactive exploration allows you to learn about the physical basis of drug addiction by exploring the direct consequences of using a particular addictive drug, cocaine, on a nerve. The exploration presents an animated diagram of a single nerve synapse within the limbic system of the human brain. You will explore the consequences of introducing cocaine into the synapse.

Prelab Preparation

1. Load and start the program Drug Addiction. Click the Topic Information button on the Navigation Palette. Read the focus questions and review these concepts: The Synapse, Neurotransmitters, Transporters, Neuromodulators, Desensitization, and Addiction.
2. Now click the word *File* at the top left of the screen, and select Interactive Exploration Help. Listen to the instructions that explain the operation of the exploration. Click the Exploration button at the bottom right of the screen to begin the exploration. You will see an animated diagram like the one below.

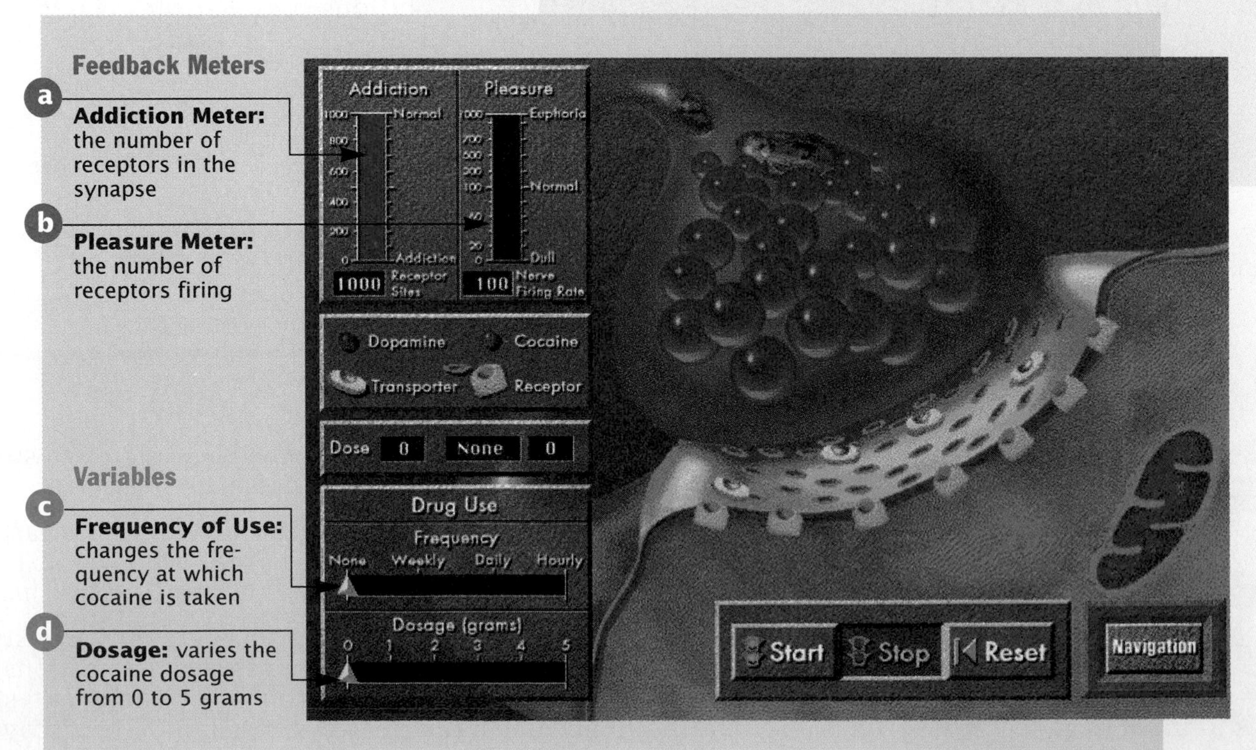

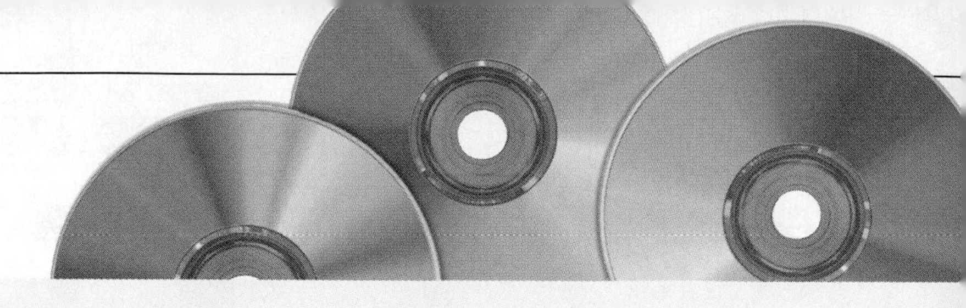

Procedure

Part A: Normal Synapse

First you will investigate how the synapse works when no cocaine is present.

1. Before beginning the simulation, record the number of receptors in the synapse. As the simulation runs, observe what happens to the Pleasure Meter and the Addiction Meter. Record the highest and lowest levels reached on the Pleasure Meter.
2. Click the Start button to begin the simulation. Observe what happens when neurotransmitters (blue spheres) are released into the synapse. Each contact with a receptor on the other side of the synapse "fires" the post-synaptic neuron, producing a sensation of pleasure. Note carefully the function of the little yellow "boats," which are transporters that recycle the neurotransmitter molecules so that they don't accumulate in the synapse.
3. Allow the simulation to run for 3 to 4 minutes, noting any changes in the number of receptors (recorded by the Addiction Meter) and in the post-synaptic neuron's firing rate (recorded by the Pleasure Meter). End the simulation by clicking the Stop button.
4. Record the number of receptors in the synapse at the end of the simulation. Did the number of receptors change during the simulation? Explain why or why not.

Part B: Data Collection

Now you will explore the effects of cocaine on the synapse.

5. Click the Reset button to clear the simulation.
6. Slide the Dosage indicator to 1 gram and the Frequency of Use indicator to Weekly.
7. On a separate sheet of paper, create a four-column table like the one shown below.

Addiction meter		Pleasure meter	
Start	End	Start	End

8. Record the number of receptors in the synapse in your table in the column labeled Addiction Meter: Start. Record the post-synaptic neuron firing rate in the column labeled Pleasure Meter: Start.
9. Click the Start button. For each dose of cocaine, you will see a burst of red spheres, representing cocaine, enter the synapse. What happens to the red spheres after they enter the synapse?
10. Based on what happens to the red spheres, explain why cocaine induces pleasurable sensations.
11. The simulation will run until five doses have been taken, which should require about 5 minutes. At the end of the simulation, record in your table the number of receptors and the post-synaptic neuron's firing rate.
12. How does the number of receptors in the synapse change with repeated cocaine doses?
13. Is this dose large and frequent enough to cause addiction? Explain your answer.
14. What was the highest level reached by the Pleasure Meter? Compare this level to the highest level reached without cocaine use, and account for the difference.
15. Explain how the simulation you carried out in Part A serves as a control for this simulation.
16. Click the Reset button to clear the simulation. Set the Frequency of Use Indicator at weekly, but slide the Dosage indicator to 2 grams. Repeat steps 8–14. How do the results of this simulation differ from those obtained with a 1-gram dose?

Analysis

17. Can you conclude, based on your results, that there is a safe dose of cocaine at which addiction does not occur? Explain your answer.
18. Explain why people who are very addicted to cocaine require higher and higher doses to produce pleasure.

CHAPTER

31A Determining the Effect of Hormones—Polluted Well

investigation

Prerequisites

- **Exploration 30A: Determining the Effect of Drugs on Heart Rate** on pages 812–813
- **Review** the following:
  1. monitoring heart rate of *Daphnia*
  2. using *Daphnia* as a test organism
  3. how epinephrine affects the body

Beatrice, SD 40007

Rosalinda Gonzales
Director of Environmental Analysis
BioLogical Resources, Inc.
101 Jonas Salk Dr.
Oakwood, MO 65432-1101

Dear Ms. Gonzales,

I am a licensed water-quality engineer who owns a consulting and testing service in the rural Midwest. A rancher has hired us to test water samples from his land. His cattle have been acting strangely, and a veterinarian advised him that the cattle show signs of stress. In some of the samples, we found a chemical that appears to be similar to the hormone epinephrine, which causes the stress response. We have three samples that are different concentrations of this chemical. We want to know how these samples affect animals; specifically, we want to know the threshold concentration for a response.

We don't do any testing on animals, but I have heard that your company does. Could you please test the enclosed samples on a suitable animal? A rapid turn-around time is needed. FAX the results to me at 1-999-AOK-AGUA (265-2482), and invoice my company at the address above. Thank you for your help.

Urgently,

Henry Perales

Henry Perales

enclosures:
Three samples and their concentrations:
Sample A (0.000001%)
Sample B (0.00001%)
Sample C (0.0001%)

Required Precautions

- Wear safety goggles, gloves, and a lab apron.
- Glassware is fragile. Notify your teacher promptly of any broken glass or cuts.
- Epinephrine is toxic and is absorbed through the skin. If you get a chemical on your skin or clothing, wash it off while calling to your teacher.
- Wash your hands before leaving the lab.

B BioLogical Resources, Inc., Oakwood, MO 65432-1101

MEMORANDUM

To: Team Leader, Water Quality Dept.
From: Rosalinda Gonzales

This request takes priority! The courier with this memo has samples of three different concentrations of an unknown chemical that mimics epinephrine. Start testing with the least concentrated sample. Use *Daphnia* as your test animal and, due to urgency, check only change in heart rate—the most obvious response to epinephrine. We can test for other responses later. Identify the lowest concentration of the chemical that stimulates a response. This concentration is the threshold concentration. Please send me a copy of your report. This may be a serious environmental contaminant and a danger to the livestock and people in that area.

Before you start your work, you must submit a proposal for my approval. **Your proposal must include the following:**

- the **question** you are seeking to answer
- the **procedure** you will use
- a detailed **data table**
- a complete, itemized list of proposed **materials** and **costs** (including the use of facilities, labor, and amounts needed)

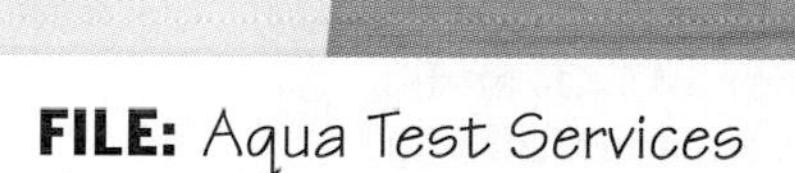

When you finish your tests, prepare a report in the form of a business letter to Mr. Henry Perales. **Your report must include the following:**

- a paragraph describing the **procedure** you followed to determine the threshold concentration for the stress response
- a complete **data table**
- your **conclusions** about the threshold concentration
- a detailed **invoice** showing all materials and labor plus the total amount due

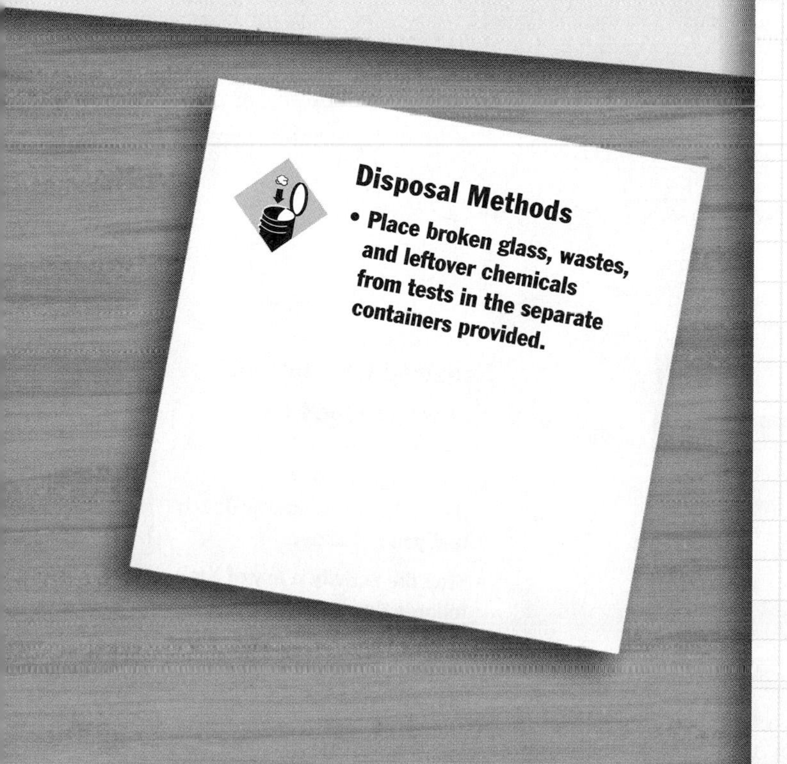

Disposal Methods

- **Place broken glass, wastes, and leftover chemicals from tests in the separate containers provided.**

FILE: Aqua Test Services

MATERIALS AND COSTS (Select only what you'll need. No refunds.)

I. Facilities and Equipment Use	
facilities	$480.00/day
personal protective equipment	$10.00/day
compound light microscope	$100.00/day
dropper and bottle	$10.00/set
depression slide and coverslip	$2.00/day
clock with second hand	$15.00/day
beaker	$10.00/day
forceps	$5.00/day
II. Labor and Consumables	
labor	$30.00/hour
Daphnia	$5.00/organism
cotton swabs	$0.10/swab
paper towels	$0.10/sheet
hot and cold water	$1.00/liter
biuret reagent	$10.00/bottle
petroleum jelly	$0.10/mL
test solutions (provided)	no charge
Fines	
OSHA safety violation	$2,000.00/incident

CHAPTER 32A

Observing the Effects of Exercise—Dance Team

investigation

Prerequisites

- **Exploration 5A: Measuring the Rate of a Process** on pages 764–765
- **Exploration 30A: Determining the Effect of Drugs on Heart Rate** on pages 812–813
- **Review** the following:
  - cellular respiration and gas exchange
  - the influence of CO_2 concentration on solvent pH
  - expiration of CO_2 and indirect measurements of CO_2 concentration
  - monitoring heart rate
  - methods for measuring pH and pulse

RHS

Roosevelt High School

Caitlan Noonan
Director of Research
Biological Resources, Inc.
101 Jonas Salk Dr.
Oakwood, MO 65432-1101

Dear Ms. Noonan:

I am an exercise physiologist working with your former high school's dance team, which is now training for a national high-kick competition. The prize money ($20,000) will be used to upgrade the training equipment used by students. Our team has a good chance to win if the girls stick to a training schedule. The team ultimately has to perform a three-minute kick routine three times in 15 minutes. I recently read about BioLogical's lab at the Olympic Training Center. Data from that lab would allow our girls to compare their physical fitness with that of other athletes. If you can provide us with a simple procedure for testing the effects of exercise on heart rate and respiration, as well as data on your athletes for comparison, we can track our progress over the next three months. Each team member can then see how her training efforts are paying off.

You know our financial position. I am hoping that you can donate your company's services for this project. We will publicize your company's name at each opportunity we get. Thank you for your consideration.

Sincerely,

Alexandra Hoeppner

Alexandra Hoeppner
Dance Team Trainer

Required Precautions

- Wear safety goggles and a lab apron.
- Avoid hyperventilating (breathing so rapidly or deeply that you get dizzy).
- Stop the activity if any of the following occurs: chest pain, breathing difficulty, nausea, dizziness, weakness, or muscle cramping.

- Wash your hands before leaving the lab.

B Biological Resources, Inc., Oakwood, MO 65432-1101

MEMORANDUM

To: Physiology Lab Director, Olympic Training Center
From: Caitlan Noonan

Please develop a simple procedure that Ms. Hoeppner can use to test the effect of exercise on the dance team, and collect sample data. Include the following in your procedure:

1. measure changes in heart rate and exhaled CO_2
2. collect data for a total of 9 minutes of intense activity using knee lifts or 12-inch step-ups
3. mimic what the dance team must do in competition—follow an interval of intense activity with an interval of light activity

Before you start your work, you must submit a proposal for my approval. **Your proposal must include the following:**

- the **question** you are seeking to answer
- the **procedure** you will use
- a detailed **data table**
- a complete, itemized list of proposed **materials** and **costs** (including the use of facilities, labor, and amounts needed)

When you finish your tests, prepare a report in the form of a business letter to Alexandra Hoeppner. **Your report must include the following:**

- a paragraph describing the **procedure** you followed to study the effects of exercise on heart rate and respiration
- a complete **data table**
- your **conclusions** about the effect of exercise on heart rate and respiration
- a detailed **invoice** showing all materials and labor plus the total amount due

Note: Even though our services will be donated, we need to track costs for inventory and tax records.

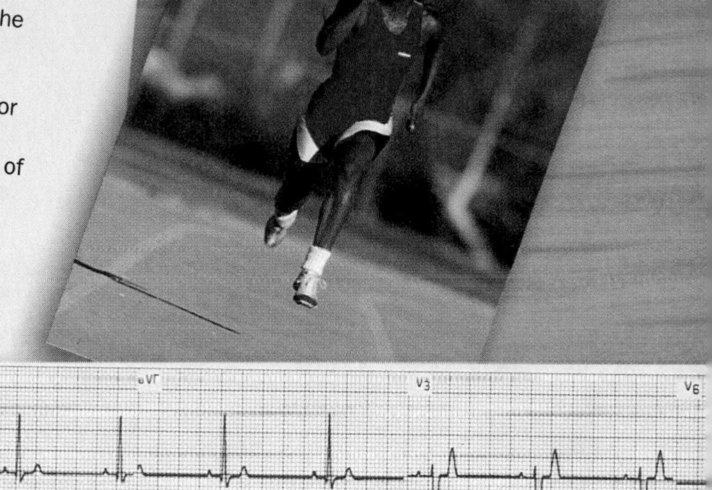

FILE: Roosevelt High School

MATERIALS AND COSTS (Select only what you'll need. No refunds.)

I. Facilities and Equipment Use	
facilities	$480.00/day
personal protective equipment	$10.00/day
clock with a second hand	$15.00/day
150 mL beaker	$10.00/day
MBL probe (optional)	$50.00/day
thermometer	$10.00/day
metronome	$25.00/day
II. Labor and Consumables	
labor	$30.00/hour
distilled water	$10.00/L
soda straw	$0.10 each
bromothymol blue solution	$5.00/mL
wax pencil	$1.00 each
plastic disposable pipet	$1.50 each
plastic wrap	$0.10 each
Fines	
OSHA safety violation	$2,000.00/incident

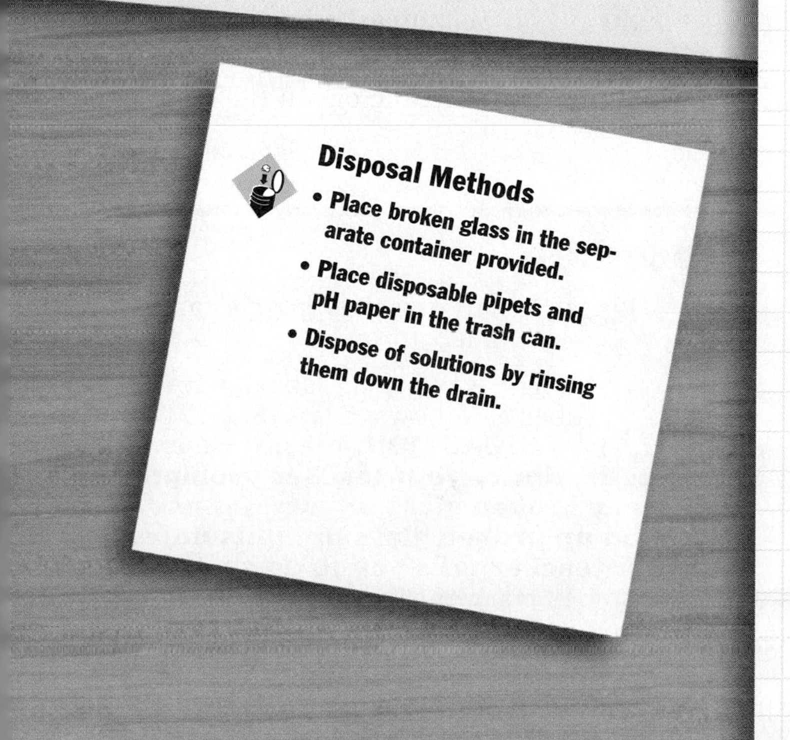

Disposal Methods

- **Place broken glass in the separate container provided.**
- **Place disposable pipets and pH paper in the trash can.**
- **Dispose of solutions by rinsing them down the drain.**

CHAPTER 34A

Determining the Effect of pH on Protein Digestion

exploration

Skill:

Testing the effect of pH on enzyme activity

Objectives

- *Observe* the effect of digestive enzymes on the size of a food particle.
- *Compare* the efficiency of pepsin and trypsin at two different pHs.
- *Determine* the pH at which a protease breaks down protein most completely.

Materials

- safety goggles
- lab apron
- wax pencil
- 6 test tubes
- test-tube rack
- 18 cubes of boiled egg white (protein)
- metric ruler
- 10 mL graduated cylinder
- 10 mL of distilled water
- 20 mL of pepsin
- 20 mL of trypsin
- 15 mL of dilute hydrochloric acid (HCl)
- 15 mL of dilute sodium hydroxide (NaOH)
- test-tube holder
- water bath or incubator at 37°C

Situation

As a biochemist, you are well prepared to work in pharmaceutical research. Your company is developing a digestive aid for use by individuals whose stomach has been surgically removed or reduced in size. Two enzymes have been suggested. You must select one to be marketed as a digestive aid to these individuals.

Background: During digestion, food is broken down in stages by a series of chemical reactions. These chemical reactions are catalyzed by **digestive enzymes.** In the early stages of digestion, several different enzymes break down large food molecules into molecules of intermediate size. In the later stages of digestion, other enzymes break down the intermediate molecules into the final products of digestion—forms such as amino acids and simple sugars that can be used by the body. Different types of digestive enzymes work on different types of food molecules. For example, a **protease** is an enzyme that breaks down proteins. Two proteases, pepsin and trypsin, are secreted at different stages and at different sites during digestion. Each digestive enzyme works best at an **optimum pH,** the pH at which it breaks down a specific nutrient most completely.

Prelab Preparation

1. Define the boldface terms in the **Background** section.
2. Where are proteins digested in the body?
3. What is the pH of the stomach and small intestine during digestion?
4. Prepare a data table like the one at the top of the next page.

Procedure

1. 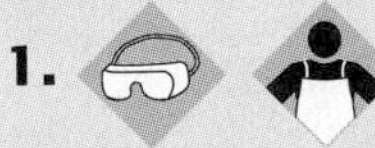 Put on safety goggles and a lab apron.
2. Using a wax pencil, label six test tubes as follows: C-2, C-8, P-2, P-8, T-2, and T-8. **CAUTION: Glassware is fragile. Notify your teacher promptly of any broken glass or cuts. Do not clean up broken glass or spills unless your teacher tells you to do so.** Use the test tubes marked *C* for your controls. The letter *P* indicates pepsin, and the letter *T* indicates trypsin. The number will indicate the pH of the test tube's contents.

HCl (pH2)

NaOH (pH8)

Test tube	Before		After	
	Food size	Food appearance	Food size	Food appearance
C-2				
C-8				
P-2				
P-8				
T-2				
T-8				

3. Place three cubes of boiled egg white in each test tube. Observe the size and appearance of the food cubes in each tube. Record your observations in your data table.

4. Add 5mL of distilled water to each test tube labeled *C*. Add 5 mL of pepsin to each test tube labeled *P*. Add 5 mL of trypsin to each test tube labeled *T*. **CAUTION: If you get a chemical on your skin or clothing, wash it off at the sink while calling to your teacher.**

5. Add 5 mL of dilute hydrochloric acid (HCl) to each tube labeled *2*. Add 5 mL of dilute sodium hydroxide (NaOH) to each tube labeled *8*. **CAUTION: If you get an acid or a base on your skin or clothing, wash it off at the sink while calling to your teacher. If you get an acid or a base in your eyes, promptly flush it out at the eye-wash station while calling to your teacher. Notify your teacher in the event of an acid or base spill.**

6. Place the test tubes in a water bath or incubator at 37°C, and leave them there overnight). Make sure the temperature remains between 35°C and 39°C.

7. After the required time, observe the tubes and note any changes in the size and appearance of the cubes of egg in each tube.

8. Dispose of your materials according to the directions from your teacher.

9. 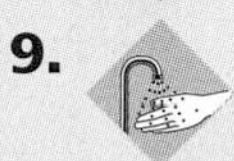 Clean up your work area and wash your hands before leaving the lab.

Analysis

10. **Analyzing Results** How did the contents of the tubes with a protease added differ from the tubes with no protease?

11. **Comparing Functions** At what pH did pepsin break down protein most completely? At what pH did trypsin break down protein most completely?

12. **Relating Concepts** How do your findings about pH and enzyme activity relate to what you know about the sites and the conditions of protein digestion?

13. **Inferring Relationships** According to today's results, what specific protease would break down protein in the small intestine? Justify your answer.

14. **Recognizing Patterns** What is the significance of using a water bath at 37°C?

15. **Relating Conclusions** What enzyme would you select as the digestive aid most useful to people with little or no stomach? Why?

Applications

16. **Thinking Critically** If a person were taking large amounts of antacids because of excess stomach acidity, how might this affect the early stages of protein digestion?

17. **Designing an Experiment** Design an experiment that would enable you to test whether taking antacids affects protein digestion. Before conducting the experiment, discuss it with your teacher.

18. **Career Connection** *Biochemists* develop new medicines and work in many other fields of biology. Find out about the training and skills required to become a biochemist, and discover the kinds of jobs that biochemists perform.

CHAPTER 34B

Conducting a Urinalysis

exploration

Skill:

Urinalysis testing

Objectives

- *Test* artificial urine for protein, glucose, and ketones.
- *Interpret* the results of a quantitative test.

Materials

- safety goggles
- lab apron
- wax pencil
- 5 test tubes
- test-tube rack
- 10 mL graduated cylinder
- 5 mL of artificial urine with glucose
- 5 mL of artificial urine with ketones
- 5 mL of artificial urine with albumin
- 5 mL of artificial urine (control)
- forceps
- 3–4 glucose test strips
- 3–4 ketone test strips
- 3–4 protein test strips
- paper towel
- 5 disposable pipets
- 5 mL of 3% sulfosalicylic acid
- 3 rubber stoppers
- 10 mL of artificial urine (unknown)

Situation

Your main task as a medical-laboratory assistant at a health clinic is to perform routine tests on urine specimens. You follow industry testing procedures, initially using test strips to detect the presence of specific substances that may be found in urine, such as glucose, ketones, and protein. When you obtain a positive result for protein in a urine sample, you perform a quantitative test to find out how much protein is present.

Background: A **urinalysis** (urine analysis) gives clues about how the body is working. Organic and inorganic chemicals that are filtered from the blood are found in urine. In the presence of disease or trauma, compounds that are not usually present in the urine may appear. For example, when the kidneys can no longer reabsorb glucose because the blood-glucose level is too high, the kidneys excrete glucose in urine. **Ketones** are organic compounds that are produced when the body uses its own tissues for energy. **Albumin** is a protein that is found in blood plasma. The presence of proteins in urine may be a sign of an abnormality because proteins are too large to pass through the filtration system of the kidney. The **turbidity** (cloudiness) of urine indicates how much protein is present.

Prelab Preparation

1. Define the boldface terms in the **Background** section.
2. Make a data table with six columns and five rows. Write the following headings for the columns: Test, Urine with glucose, Urine with ketones, Urine with protein, Control urine, and Unknown urine. Under the Test heading, write the following headings for the rows: Glucose test strips, Ketone test strips, Protein test strips, and Sulfosalicylic acid.

Procedure

1. 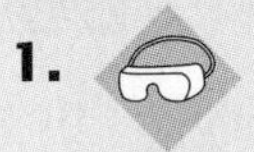 Wear safety goggles and a lab apron.
2.  Using a wax pencil, label five test tubes as follows: P+, G+, K+, Control (P−, G−, K−), and Unknown. **CAUTION: Glassware is fragile. Notify your teacher promptly of any broken glass or cuts. Do not clean up broken glass or spills unless your teacher tells you to do so.** Place the test tubes in a test-tube rack.

Turbidity Grading Scale

Grade	Appearance	Protein in urine (g/100 mL)
negative	no turbidity	<0.005
trace	barely perceptible turbidity	0.010
11	distinct turbidity but no distinct granulation*	0.050
21	turbidity with granulation but no flocculation†	0.2
31	turbidity with granulation and flocculation	0.5
41	clumps of precipitated protein or solid precipitate††	1.0

*Granulation — appearance of small granules (Sugar and salt are granular.)
†Flocculation — small, individual masses resembling tufts of wool
††Precipitate — sediment at bottom

3. Using a 10 mL graduated cylinder, add 5 mL of artificial urine with albumin to the "P+" test tube. Add 5 mL of artificial urine with glucose to the "G+" test tube. Add 5 mL of artificial urine with ketones to the "K+" test tube. Add 5 mL of the artificial urine labeled Control to the "Control" test tube. Rinse the graduated cylinder well between artificial urine samples.
4. Using forceps, obtain two glucose test strips. Lay the strips on a paper towel, and label one "+" and one "−." *Note: Do not touch the strips with your fingers.* Using a pipet, place a drop of liquid from the test tube marked G+ onto the "+" strip. Use a clean pipet to place a drop of the "Control" sample on the "−" strip. Wait 60 seconds. Compare the resulting color to the color guide. Record the results in your data table.
5. Repeat step 4 using ketone test strips and the test tube marked "K+."
6. Repeat step 4 using protein test strips and the test tube marked "P+."
7. Using a clean graduated cylinder, add 5 mL of 3% sulfosalicylic acid to the "P+" and "Control" test tubes. Close each tube with a stopper, and then invert the tubes several times. Let the tubes sit for 10 minutes. Invert the tubes once more, and observe them for turbidity (cloudiness). Compare the degree of turbidity to the grading scale on this page. Record your results in your data table.
8. Obtain 5 mL of the artificial urine labeled "Unknown," and place it in the test tube marked "Unknown." Using test strips, test for glucose, ketones, and protein.
9. If you obtain a positive result for protein in step 8, repeat step 7 using the artificial urine labeled "Unknown." Record your results in your data table.
10. Dispose of your materials according to the directions from your teacher.
11.  Clean up your work area and wash your hands before leaving the lab.

Analysis

12. **Analyzing Methods** How are the glucose, ketone, and protein test strips like the indicators you used in Exploration 22A?
13. **Analyzing Results** What did turbidity of the artificial urine indicate in this lab? Is this a qualitative or quantitative test? Explain.
14. **Interpreting Data** What abnormal components were present in the unknown sample?

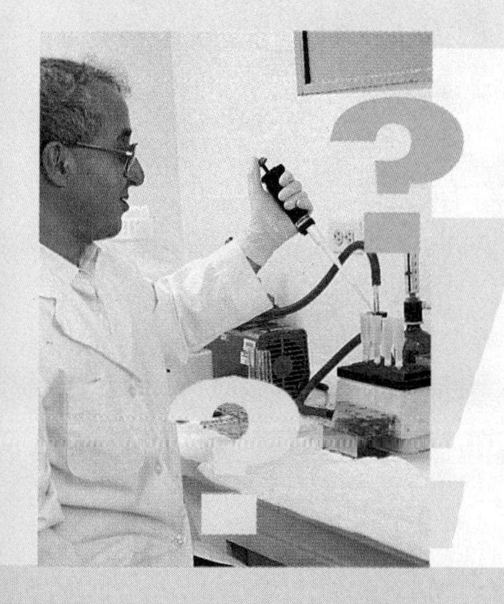

Applications

15. **Thinking Critically** Diabetics may produce urine with ketones if their blood-glucose level is too low and their body burns proteins for energy. What body tissues are broken down in this situation?
16. **Thinking Critically** The medical testing industry recommends performing additional urinalysis tests to confirm positive results obtained with test strips. Why do you think this practice is recommended?
17. **Career Connection** *Medical-laboratory assistants* help medical technologists by performing routine tests, such as a urinalysis. Find out about the training and skills required to become a medical-laboratory assistant, and discover some other tests that they perform.

CHAPTER

34C Thermodynamics

interactive exploration

Objectives

- *Simulate* the effect of pH, temperature, and enzyme concentration on the activity of two enzymes.
- *Determine* the optimal conditions for each enzyme.

Materials

- computer with CD-ROM drive
- CD-ROM *Interactive Explorations in Biology: Cell Biology and Genetics*
- graph paper

Background: All living things rely on enzymes to catalyze chemical reactions. Recall that an enzyme is a protein molecule that increases the rate of a chemical reaction by lowering the activation energy necessary to start the reaction. Enzyme effectiveness depends on several factors. This interactive investigation allows you to explore how three of these factors—enzyme concentration, temperature, and pH—affect the activity of the enzymes carbonic anhydrase and lysozyme.

Prelab Preparation

1. Load and start the program Thermodynamics. You will see an animated diagram like the one below. Click the Navigation button, and then click the Topic Information button. Read the focus questions and review these concepts: Reaction Rate, Enzyme Catalysis, Temperature and Reaction Rates, and pH Can Influence Enzyme Shape.
2. Now click the word *Help* at the top left of the screen, and select How to Use This Exploration. Listen to the instructions that explain the operation of the exploration. Click the Interactive Exploration button on the Navigation Palette to begin the exploration.

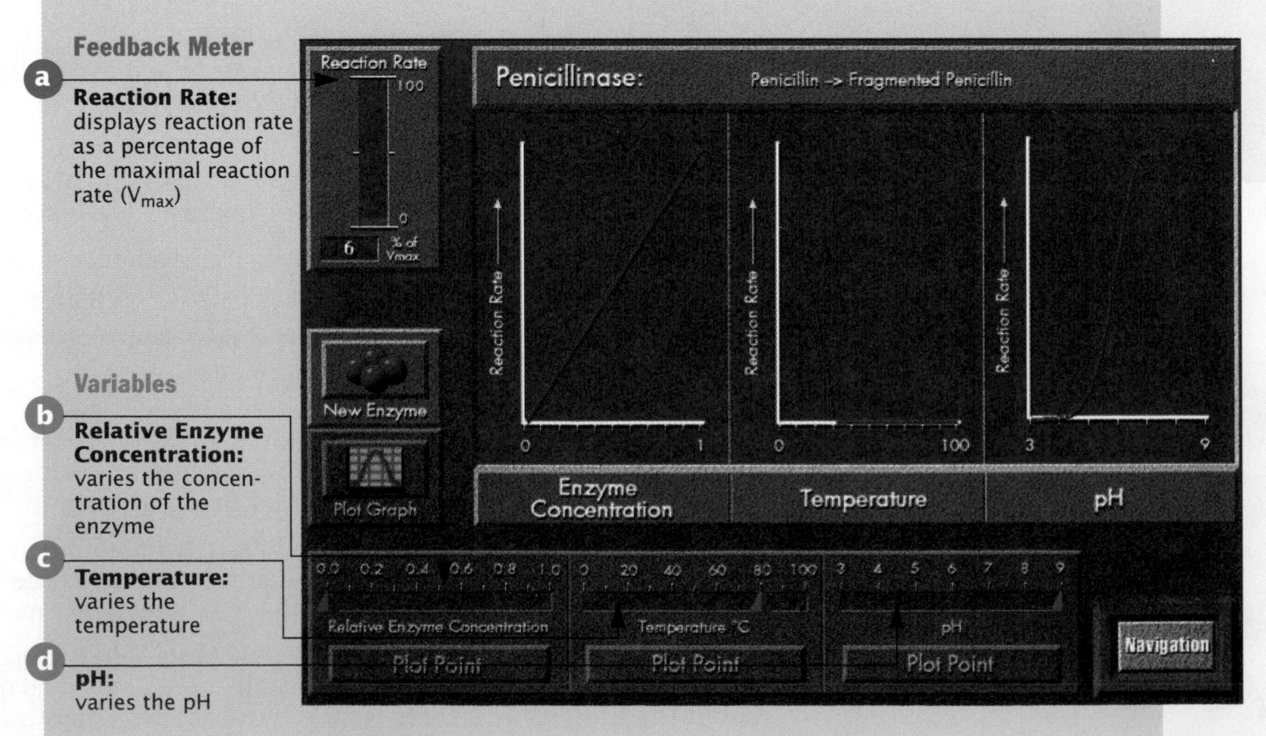

Procedure

Part A: Carbonic Anhydrase

1. Click the New Enzyme button until "Carbonic Anhydrase" appears at the top of the screen. Carbonic anhydrase is found in red blood cells, where it helps regulate the concentration of carbon dioxide.
2. Click and slide the indicator so that the Relative Enzyme Concentration is 0.2.
3. Click the Plot Point button and observe the point that appears on the graph. This point indicates the rate of reaction for this enzyme concentration. The rate of reaction is expressed as a percentage of V_{max}, the highest possible reaction rate. What is the reaction rate indicated by the Reaction Rate meter in the upper-left corner of the screen?
4. Repeat steps 2 and 3 until you have tested the following Relative Enzyme Concentrations: 0.4, 0.6, 0.8, and 1.0.
5. Click the Plot Graph button. How does increasing the enzyme concentration affect the rate of reaction?
6. Make a table like the one shown below.

Enzyme: ______ Temperature	Reaction rate (% of V_{max})

7. Click and slide the indicator so that the Temperature meter indicates 0°C. Click the Plot Point button and observe the point that appears on the graph. In your table, record the reaction rate.
8. Repeat step 7 until you have tested the effects of the following temperatures: 10°C, 20°C, 30°C, 40°C, and 50°C.
9. Use graph paper to make a graph of temperature versus reaction rate. Draw a curve that connects your six points. What can you conclude about the effect of temperature on the activity of carbonic anhydrase?
10. Click the Plot Graph button. Compare your graph with the one shown on the screen. Explain any differences between the two graphs.
11. Make a two-column table similar to the one you made in step 6. Label the left column "pH" and the right column "Reaction rate (% of V_{max})."
12. Click and slide the indicator so that the pH meter indicates a pH of 3. Click the Plot Point button and observe the point that appears on the graph. In your table, record the reaction rate.
13. Repeat step 12 until you have tested the following pH values: 4, 5, 6, 7, and 8.
14. After testing all the values above, make a graph of pH versus reaction rate. Draw a curve that connects your points. What can you conclude about the effect of pH on the activity of carbonic anhydrase?
15. Click the Plot Graph button. How does your graph compare with the graph on the screen? Explain any differences.

Part B: Lysozyme

16. Click the New Enzyme button until "Lysozyme" appears at the top of the screen. Lysozyme catalyzes a reaction in which polysaccharides are broken down into simple sugars.
17. Repeat steps 2–15 for the enzyme lysozyme.

Analysis

18. What is the optimal temperature for each enzyme? What is the optimal pH for each enzyme?
19. The optimal pH of pepsin, a stomach enzyme, is about 2, while the optimal pH of trypsin, an enzyme of the small intestine, is about 8. Give a functional explanation for this difference.

Reference and Skills Appendix
Introduction

Laboratory Skills

Reading and Study Skills

Math and Problem-Solving Skills

Reference

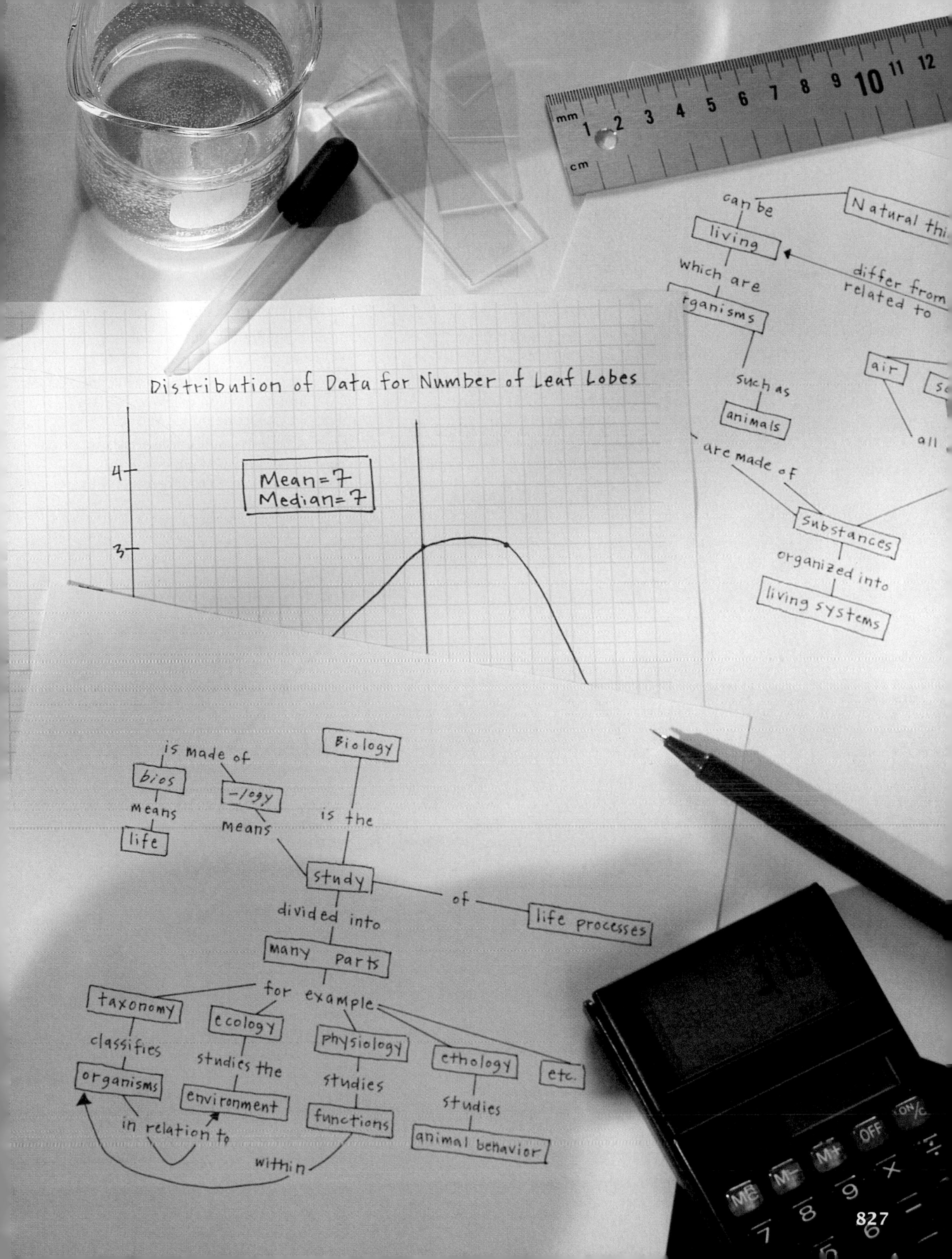

Distribution of Data for Number of Leaf Lobes
Mean=7
Median=7
can be
living
which are
differ from
related to
such as
animals
air
are made of
substances
organized into
living systems
Biology
is made of
bios
-logy
means
life
means
is the
study
of
life processes
divided into
many parts
for example
taxonomy
ecology
physiology
ethology
etc.
classifies
organisms
studies the
environment
studies
functions
studies
animal behavior
in relation to
within

Using a Compound Light Microscope

Parts of the Compound Light Microscope

- The *eyepiece* magnifies the image, usually 10X.
- The *low-power objective* magnifies the image more, often 10X.
- The *high-power objective* magnifies the image even more, such as 43X.
- The *revolving nosepiece* holds the objectives and can be turned to change from one magnification to another.
- The *body tube* maintains the correct distance between the eyepiece and objectives. This is usually about 25 cm, the normal distance for reading and viewing objects with the naked eye.
- The *coarse adjustment* moves the body tube up and down in large increments to allow gross positioning and focusing of the low-power objective lens.
- The *fine adjustment* moves the body tube slightly to bring the image into sharp focus.
- The *stage* supports a slide.
- *Stage clips* secure the slide in position for viewing.
- The *diaphragm* (or *iris diaphragm*) controls the amount of light allowed to pass through the object being viewed.
- The *light source* provides light for viewing the image. It can be either a light reflected with a mirror or incandescent light from a small lamp.

  NEVER use direct sunlight as a light source.
- The *arm* supports the body tube.
- The *base* supports the microscope.

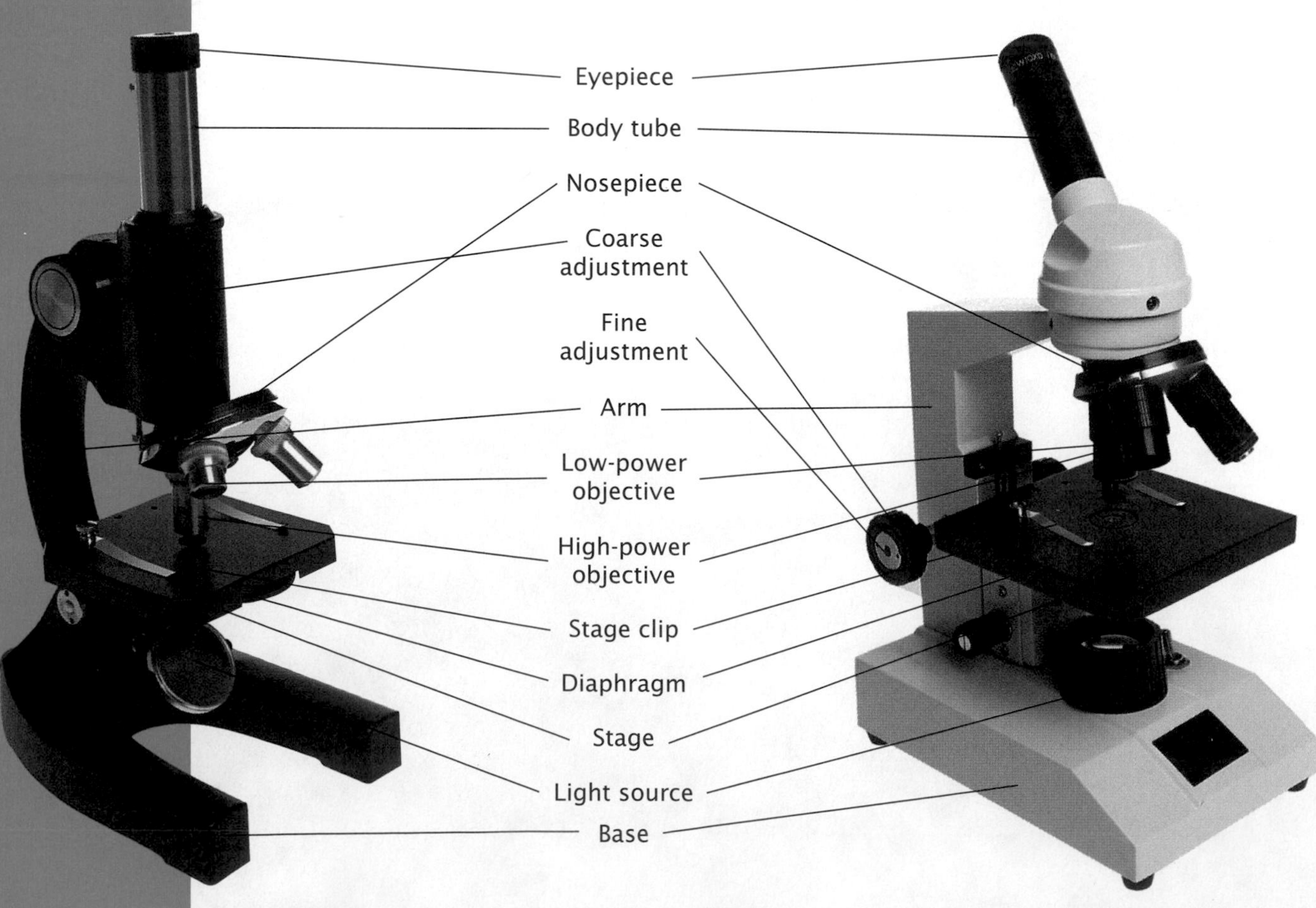

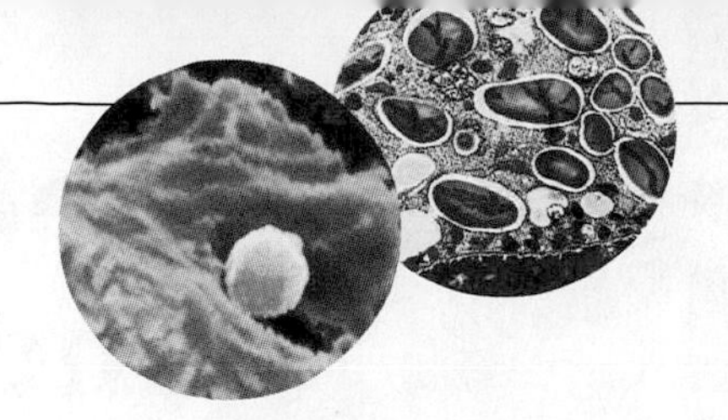
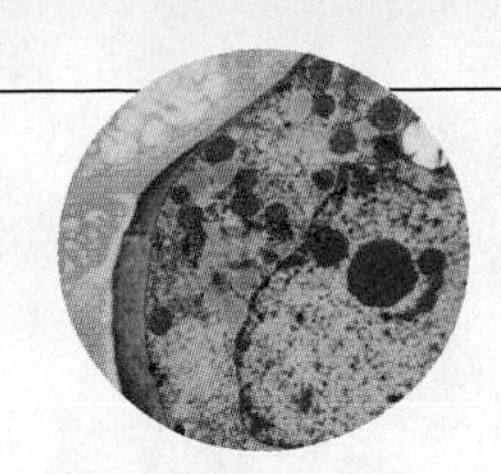

Proper Handling and Use of the Compound Light Microscope

1. Carry the microscope to your lab table using both hands, one beneath the base and the other holding the arm of the microscope. Hold the microscope close to your body.
2. Place the microscope on the lab table, at least 5 cm (2 in.) from the edge of the table.
3. Check the type of light source for the microscope. If the microscope has a lamp, plug it in, making sure that the cord is out of the way. If the microscope has a mirror, adjust it to reflect light through the hole in the stage.

   CAUTION: ***If your microscope has a mirror, do not use direct sunlight as a light source. Direct sunlight can damage your eyes.***
4. Adjust the revolving nosepiece so that the low-power objective is in line with the body tube.
5. Place a prepared slide over the hole in the stage and secure the slide with stage clips.
6. Look through the eyepiece and move the diaphragm to adjust the amount of light coming through the specimen.
7. Now look at the stage at eye level, and slowly turn the coarse adjustment to lower the objective until it almost touches the slide. Do not allow the objective to touch the slide.
8. While looking through the eyepiece, turn the coarse adjustment to raise the objective until the image is in focus. Never focus the objectives by moving them downward. Use the fine adjustment to raise the objective until you achieve a sharply focused image. Keep both eyes open while viewing a slide.
9. Make sure that the image is exactly in the center of your field of vision. Then, switch to the high-power objective. If necessary, increase the power of the objective to the next level. Focus the image with the fine adjustment. Never use the coarse adjustment at high power.
10. When you are finished using the microscope, remove the slide. Clean the eyepiece and objectives with lens paper and return the microscope to its storage area.

Making a Wet Mount

1. Use lens paper to clean a glass slide and coverslip.
2. Place the specimen you wish to observe in the center of the slide.
3. Using a medicine dropper, place one drop of water on the specimen.
4. Position the coverslip so that it is at the edge of the drop of water and at a 45-degree angle to the slide. Make sure that the water runs along the edge of the coverslip.
5. Lower the coverslip slowly to avoid trapping air bubbles.

6. If a stain or solution is to be added to a wet mount, place a drop of the staining solution on the microscope slide along one side of the coverslip. Place a small piece of paper towel on the opposite side of the coverslip.

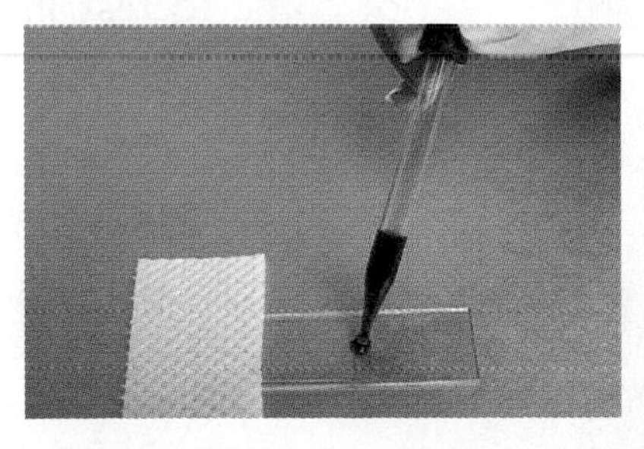

7. As the water evaporates from the slide, add another drop of water by placing the tip of the medicine dropper next to the edge of the coverslip, just as you would do when adding stains or solutions to a wet mount. If you have added too much water, remove the excess by using the corner of a paper towel as a blotter. Do not lift the coverslip to add or remove water.

Determining Mass and Temperature

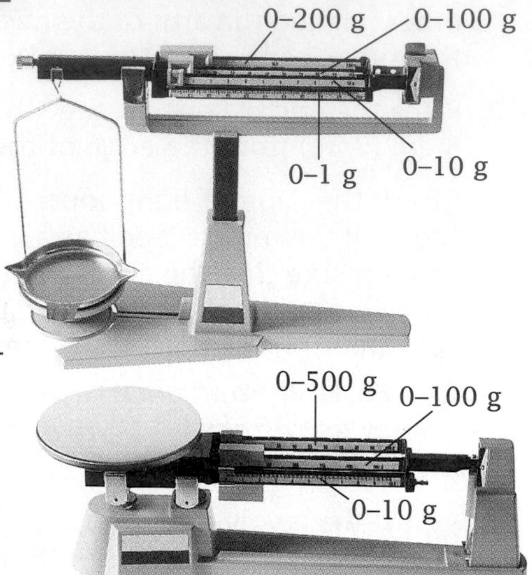

Reading a Balance

A single-pan balance has one pan and three or four beams. The scale of measure for each beam depends on the model of the balance. You can see these differences by looking at the two balances shown. When an object is placed on the pan, the riders are moved to the right as needed until the mass on the beams equals the mass of the object.

Measuring Mass

CAUTION: ***Never place a hot object or chemical directly on a balance pan. For example, when determining the mass of a chemical or powder, use weighing or filter paper. Determine the mass of the paper and subtract that mass from the final mass.***

1. Make sure the balance is on a level surface and allow the pan to move freely. Position all the riders at zero. Use the adjustment knob (usually located under and to the left of the pan) if the pointer does not come to rest in the middle of the scale.
2. Place the mass on the pan.
3. Move the largest rider along the beam to the right until it is at the last notch that does not move the pointer below the zero point in the center of the scale.
4. Follow the same procedure with the next largest rider.
5. Move the smallest rider until the pointer rests at zero in the middle of the scale.
6. Total the readings on all the beams to determine the mass of the object.

Practice Exercises

1. Use a single-pan balance to determine the mass of each of the following items.
   a. an empty 250 mL beaker
   b. 250 mL beaker filled with 100 mL of water
   c. the same beaker filled with 100 mL of cooking oil
   d. a house key
   e. a small book
   f. a paper clip or small safety pin
2. Determine the mass of each object represented by the balance readings shown.

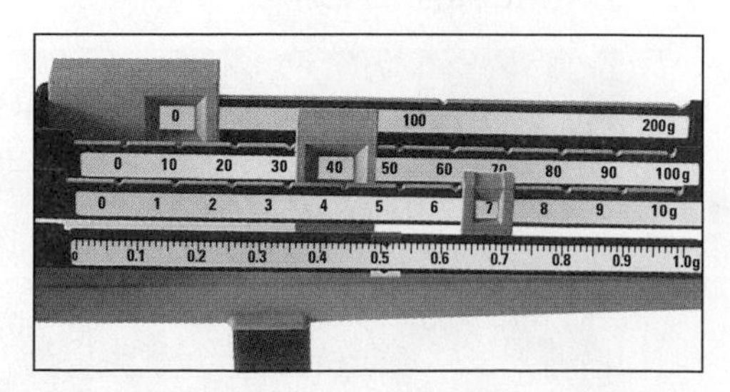

a.

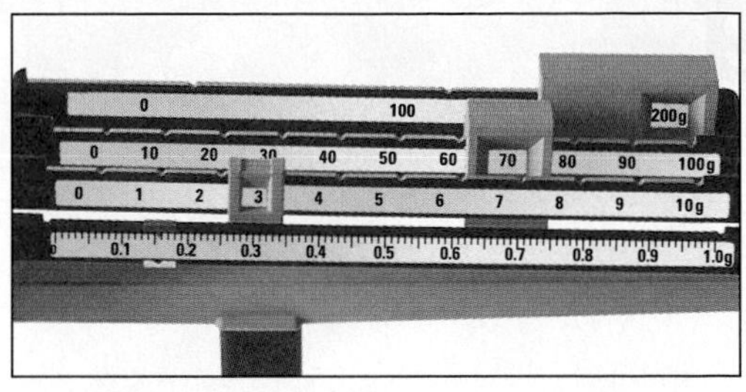

b.

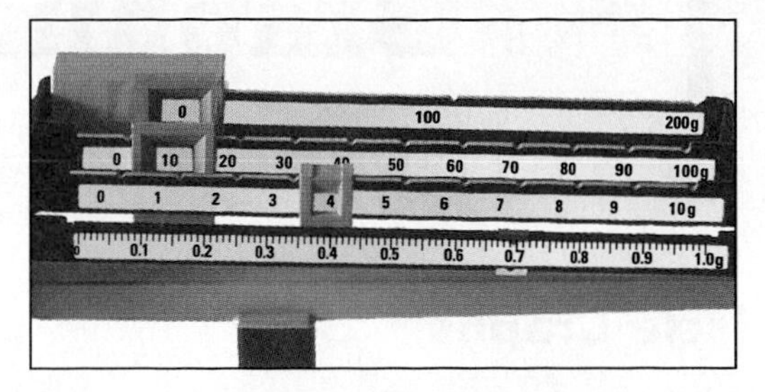

c.

Reading a Thermometer

Most laboratory thermometers are of the bulb-type illustrated at the right. The sensing bulb of the thermometer is filled with a colored liquid that expands predictably when heated. The liquid expands into a capillary tube. Most laboratory thermometers are marked in small increments in degrees Celsius.

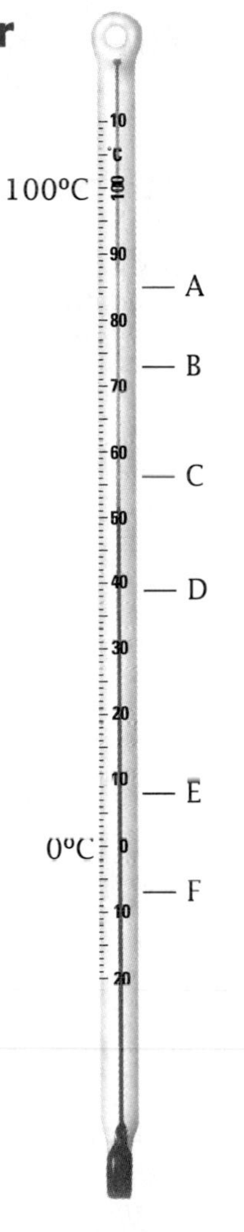

Figure 1

Measuring Temperature

1. The sensing bulb of the thermometer must be carefully lowered into the material. The thermometer may rest against the side of the container but the bulb should never rest on the bottom of a container where heat is being applied. If the thermometer has an adjustable clip for the side of the container, the thermometer can be suspended in the liquid. For safety, never hold a thermometer in your hand while measuring the temperature of a hot substance.
2. Gently rotate the thermometer in the clip. Watch the rising colored liquid in the capillary tube. When the liquid in the capillary tube stops rising, note the whole degree increment nearest the top of the liquid column. If your thermometer is marked in tenths of a degree, count the increments up or down from the nearest whole degree to obtain your reading. For example, if the top of the colored liquid column is nearest the 52°C mark but somewhat below it, as shown, what is the accurate temperature reading?

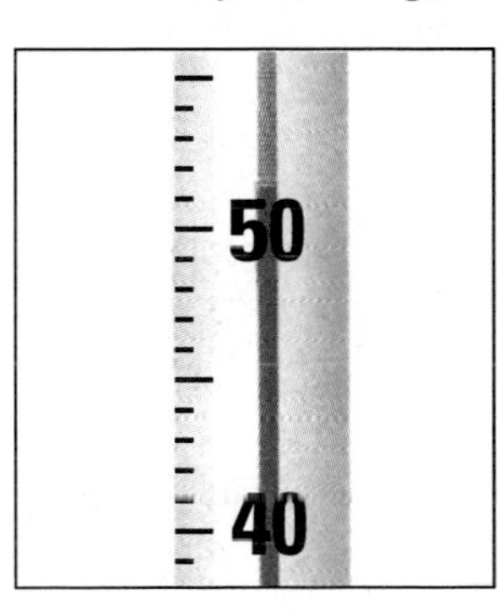

Because the top of the column is half of a degree below the 52°C mark, the temperature is 51.5°C.

Practice Exercises

1. Examine the thermometer in **Figure 1**. Note that only 0°C and 100°C are labeled.
   a. What scale is used to mark this thermometer—Celsius or Fahrenheit?
   b. Is this thermometer marked only in whole degrees or in tenths of degrees?
   c. What is the temperature reading on this thermometer?
2. What would be the temperature reading of the thermometer shown in **Figure 1** if the top of the column were resting at
   a. point A?______
   b. point B?______
   c. point C?______
   d. point D?______
   e. point E?______
   f. point F?______

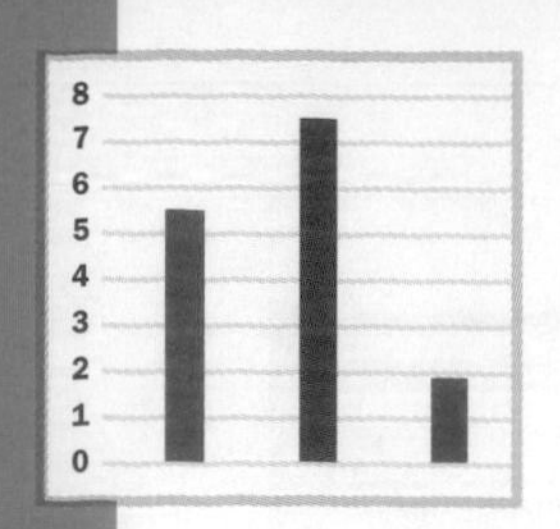

Graphing

Circle Graphs

A circle graph, sometimes called a pie chart, shows groups of data in relation to all of the data. Each part of the circle represents a category of the data. The entire circle represents all of the data. For example, a biologist studying a plot of land in a hardwood forest in Wisconsin found that there were five different types of trees in the plot. The data table below summarizes the biologist's findings.

WISCONSIN HARDWOOD TREES

Type of tree	Number found	Percentage of total (rounded to nearest whole-number percent)
Oak	539	17
Maple	758	24
Beech	319	10
Birch	1327	42
Hickory	222	7
Total	3,165	100

To aid in the interpretation of the data and to compare it with other sets of data, the biologist constructed a circle graph of the data.

Characteristics of a Circle Graph

1. The sections are accurately calculated and drawn. The number of each type of tree is divided by the total number of trees and multiplied by 100 to get the tree percentages.
2. The sections are clearly labeled.
3. Labels, colors, or shading represent each section of the circle.
4. The graph is easy to interpret.

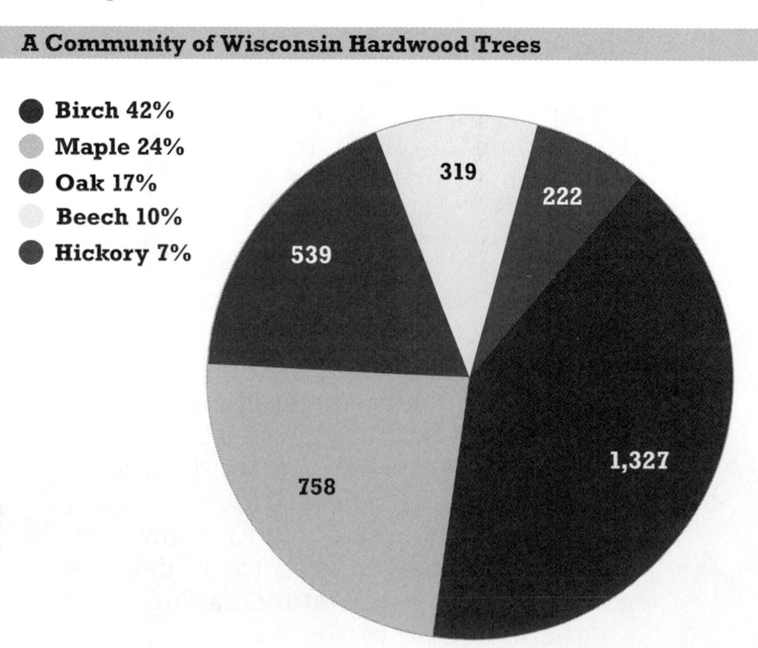

Practice Exercise

1. Construct a circle graph from the following data.

Animal	Wildlife refuge population
Baboon	20
African elephant	35
Leopard	12
Tiger	16
Giraffe	10

Line Graphs

Line graphs are most often used to demonstrate continuous change. Examine the following data.

POPULATION OF THE UNITED STATES 1880 – 1990	
Year	**Population (in millions)**
1881	50.2
1890	62.9
1900	76
1910	92
1920	105.7
1930	122.8
1940	131.7
1950	151.3
1960	179.2
1970	203.2
1980	226.5
1990	251.4

Both the year and the population are variables. The variable to which values are assigned, in this case the year, is called the **independent variable**. The population is determined by the year. The population is called the **dependent variable**. Each set of values, the independent and dependent variables, is called a **data pair**. To prepare a line graph, data pairs must first be organized in a table like that shown above.

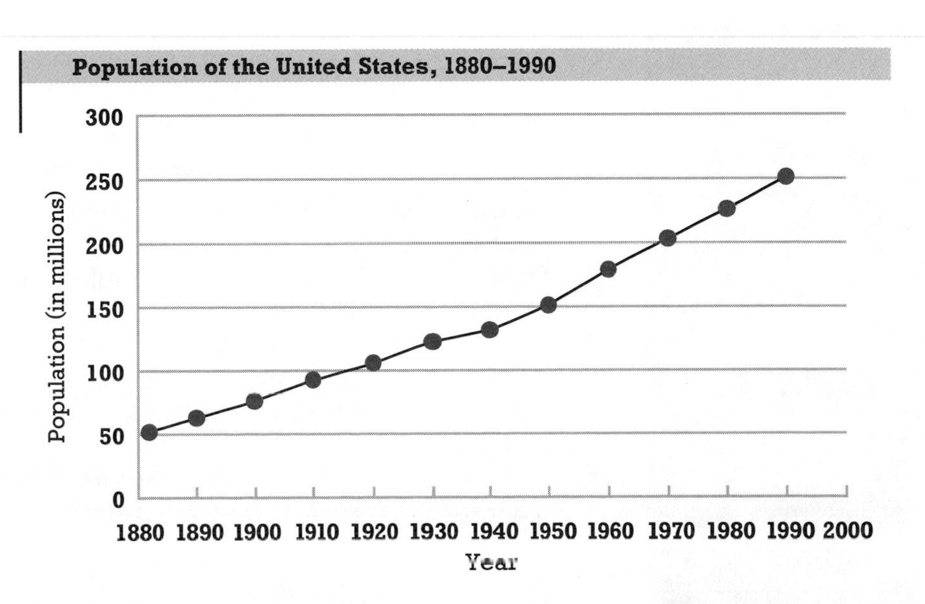

continued

Characteristics of a Line Graph

1. The independent variable is put on the horizontal (x) axis. The dependent variable is put on the vertical (y) axis.
2. The axes are labeled and an appropriate scale and interval are used on each axis. The same scale and interval must be used for the total length of the axis. For example, if 1 cm represents an interval of 10 years, 30 years must be represented by 3 cm.
3. Reasonable starting points are used for each axis. Examine the population graph. The horizontal (x) axis goes from 1880 to 2000 and the vertical (y) axis goes from 0 to 300 million.
4. The data pairs are plotted as accurately as possible.
5. The title of the graph accurately reflects the data presented.
6. If more than one set of data is presented on a graph, a key must accompany the graph. For example, if the populations of Canada, Mexico, and the United States from 1880 to 1990 were presented on the same graph, it would be important to use a different color or type of line (solid, dotted, dashes, etc.) for each country. The key indicates what each color or type of line represents.
7. The graph is easy to interpret.

Using Graphs to Make Predictions

Examine the graph of the population of the United States from 1880 to 1990 on page 833. Do you notice any trends? Do you think you can predict the population of the United States in the year 2000? The process of going beyond the data points in a graph is called extrapolation. Scientists use extrapolation to make predictions about future data based on data patterns from the past.

We can also use graphs to find information between the data points on a data table. For example, we can use the graph to determine that the population of the United States in 1905 was approximately 84 million people. We find the point on the line that represents 1905 and then determine the position of that point on the y axis. The reading is about 84 million. Finding information between two sets of data pairs is called interpolating.

Practice Exercises

Use the population graph on page 833 to answer the following.

2. Predict what the population of the United States will be in the year 2010.
3. Determine the approximate population of the United States in 1935, 1945, and 1985.
4. What was the approximate population of the United States in 1870?
5. What will be the approximate population of the United States in 2020?
6. How certain are you of your answer? Why?

Bar Graphs

Consider the following information:

PRECIPITATION IN HARTFORD, CONNECTICUT APRIL 1 – 15, 1996			
Date	**Precipitation (in centimeters)**	**Date**	**Precipitation (in centimeters)**
April 1	0.5	April 9	0.25
April 2	1.25	April 10	0.0
April 3	0.0	April 11	1.0
April 4	0.0	April 12	0.0
April 5	0.0	April 13	0.25
April 6	0.0	April 14	0.0
April 7	0.0	April 15	6.50
April 8	1.75		

A bar graph is appropriate for data that are not continuous. A bar graph is a good indicator of trends if the data are taken over a sufficiently long period of time. Precipitation records require years of data, and even then predictions are difficult to make with reliability. The bar graph shown at the right depicts the data presented in the above table.

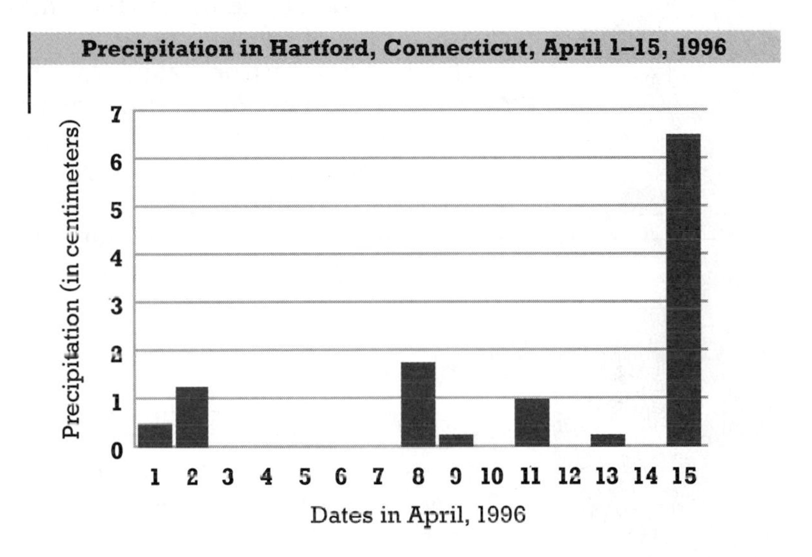

Characteristics of a Bar Graph

1. An appropriate scale and reasonable starting points are used for each axis.
2. The axes are labeled and data are accurately plotted.
3. The title of the graph accurately reflects the data presented.
4. The graph is easy to interpret.

Practice Exercises

Use the bar graph above to answer the following.

7. How many centimeters of rain fell in Hartford on April 11?
8. Can a bar graph show a trend, even if the data are not continuous?
9. Can the bar graph be used to predict precipitation in Hartford on April 20?
10. Could a line graph of these data be used to make a prediction about the rainfall on April 20? Why or why not?

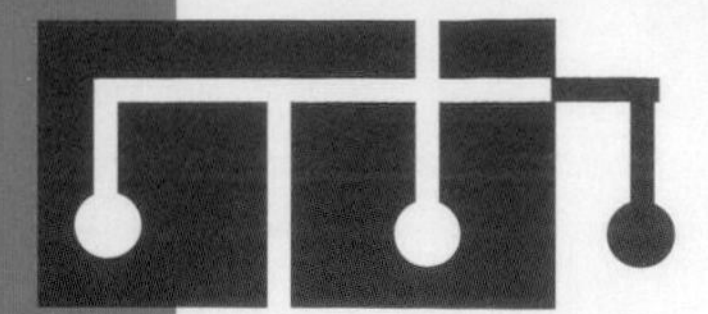

Concept Mapping

What's the Best Way to Study?

If you want to learn information that you can remember longer and use more easily, you need to move that information into your long-term memory.

Concept mapping may help you understand the meanings of ideas by showing you their connections to other ideas. If you define soccer as a team sport, you have already related the idea or concept of "soccer" to the ideas of "team" and "sport." Making an idea map may help you see the information more clearly.

This is different from taking notes or making an outline. Your map not only identifies the major ideas of interest from a chapter or your class notes, but also shows you the relationships among the ideas in much the same way that a road map illustrates how highways and other roads link cities.

Examine these words:

pool	**water**	**biking**
grass	**sky**	**raining**
tree	**playing**	**thinking**

Every one of these words is a concept. Concepts usually form a picture in your mind.

Look at the following words. Are they concepts?

the	**when**	**with**
to	**was**	**can**
has	**be**	

No. They don't cause a picture to form in your mind. They are linking words. They connect ideas or concepts.

Organizing Concepts

From the concept map in **Figure A**, on the next page, you can see what a concept map looks like, but we will continue to discuss the fine points of mapping and you will do some maps on your own.

Another feature of mapping ideas is that some concepts or ideas are more central, or general, and include other concepts. What is the main idea of this list of concepts: Tokyo, Mexico City, Seoul, New York, Bombay? Cities, of course. What is the main idea of this list: car, bus, train, bicycle, truck? Each of these concepts is an example of a vehicle.

Try to remember the words below after reading them once or twice:

racket	**net**
ellipse	**helmet**
starfish	**mitt**
tuna	**bat**
square	**diamond**
seahorse	**star**

You probably couldn't remember all of them. Even if you did remember almost all of them, it's a good bet that you were already categorizing them.

You shouldn't need as much time to memorize the list of words below.

square	**touchdown**	**necklace**
circle	**goal**	**bracelet**
triangle	**basket**	**earring**

Easier, wasn't it? This second list was already categorized. If you learn in a way that carries more meaning, it's easier to remember! That's how concept mapping works, too.

Making Learning Faster

You can see in **Figure A** that every concept map should have at its top the most general and all-inclusive concept. The smaller, more specific concepts and examples should go below. It might help you to capitalize the first letter of the main concept or idea. You can write the other, more specific, concepts using all lowercase letters. Your maps should show a pattern from most general to most specific, from the top of the map to the bottom of the map. Try to understand how the concept map in **Figure B** shown on the following page was made before you try one yourself. First read the following paragraph and make a list of any important words that you think should be learned.

What is life? What is the difference between living and nonliving things? If you were in a natural wilderness area, it would be easy for you to pick out the living things and nonliving things. The animals and plants are the living organisms. Organisms are made up of many substances organized into living systems. The rocks, air, water, and soil are nonliving. They contribute substances to the living organisms.

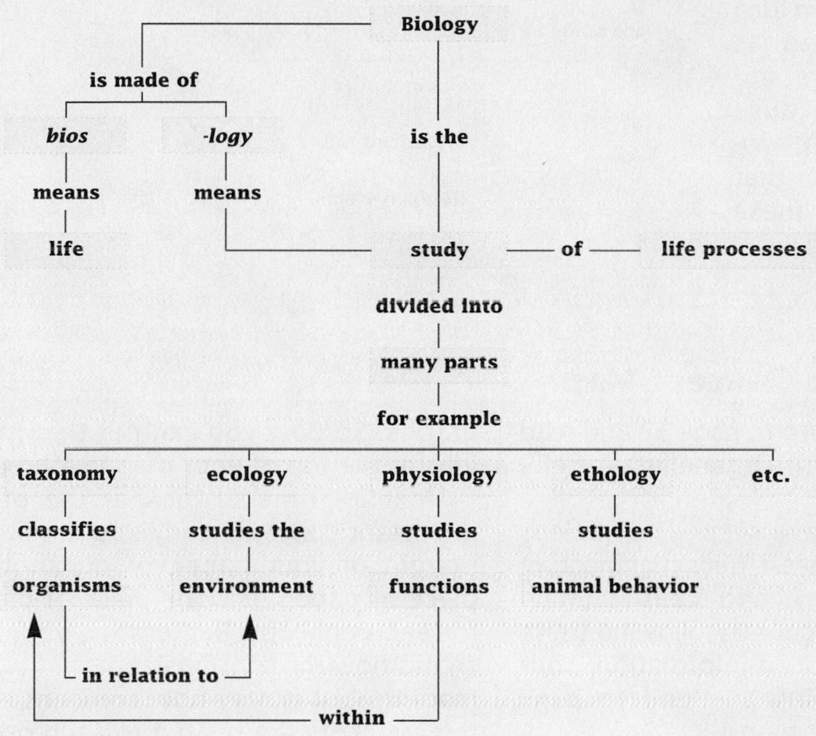

Biology
is made of
bios
-logy
means
means
life
is the
study
of
life processes
divided into
many parts
for example
taxonomy
ecology
physiology
ethology
etc.
classifies
studies the
studies
studies
organisms
environment
functions
animal behavior
in relation to
within

Figure A

Connecting Ideas

In the preceding paragraph, there seem to be two general ideas: living and nonliving. You can make either two separate maps, the chief concept of one being "living" and of the other being "nonliving," or one map, using "natural things" as the main idea or main concept. Capitalize the first letter of the main concept and put it at the top of the map. In this case, the main concept "natural things" did not actually appear in the paragraph, but frequently it does. Since natural things can be living or nonliving, you can make these subconcepts parts of the main one. As you read the paragraph, you learned that living things are organisms, such as plants and animals, and that they are made up of substances that are organized into living systems.

The paragraph shows that living things are different from nonliving things, but that they are related to nonliving things. You can make that connection near the top of the map. The rest of the map should explain the connection, so you can follow what you learned in the paragraph. Nonliving things such as soil, water, and rocks all contribute substances that are organized into living systems.

continued

Now you have the entire map shown in **Figure B**. If you had a choice of studying the paragraph or looking at this map, you might agree that the map shows the concepts more clearly. This map gives you the main idea more quickly, and it's easier to understand everything because you've put it in a two-dimensional form that shows how all of these things are related.

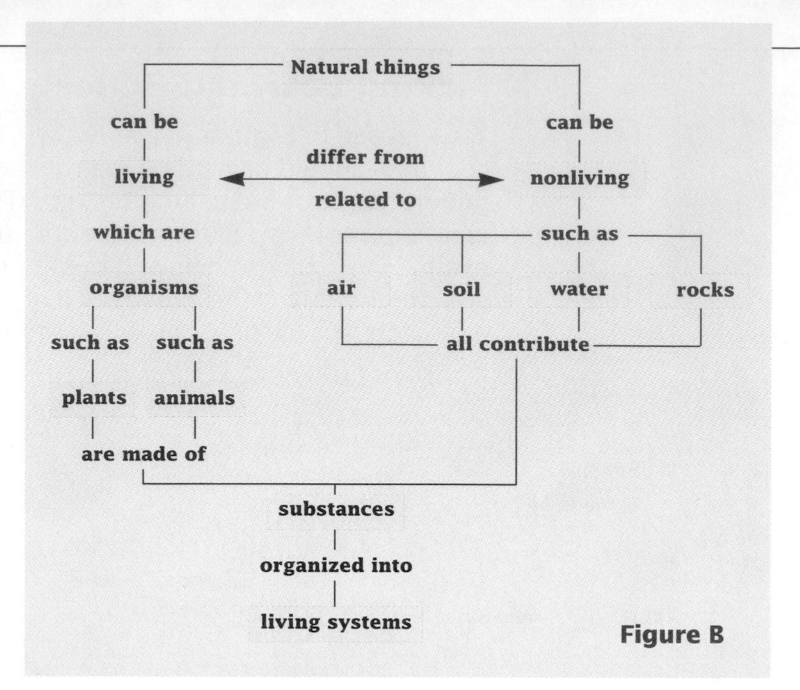

Figure B

Planning the Concept Map

Now you try it. Turn to page 52 and read the major titles in "Organelles: A Cell's Laborers."

You should have seen "The nucleus directs cell activities," "Cells manufacture and release energy," and "Cells maintain homeostasis." Of course, following those titles is quite a lot of information. Your job is to turn the information from page 52 into a concept map.

Identifying the main idea is easy—it's in the main title. It could be "Basic cell functions" or "Life functions of cells," or a similar shortened version of the title. This book is organized and written around concepts. In some other texts, the main idea may be buried in the paragraphs. So now you can put that main idea at the top of your map in a box.

Next it's time to look at the other important concepts and list them in the margin of your paper, or on another piece of paper. Use one piece of paper for each concept. Remember, your list shows how the concepts appeared in the reading, but it may not represent how the concepts are related to each other. Making those connections is your job.

Some of the concepts that should be in your list include the following: reproduction, energy manufacture, energy release, mitochondria, chloroplasts, homeostasis, endoplasmic reticulum, and Golgi apparatus.

Building the Concept Map

The next step is to put the concepts in order, from the most general to the most specific. Some concepts will share the same rank and be equally specific. Remember that the examples are the most specific and will be at the bottom of the map.

Now begin to rearrange the concepts you've written on the pieces of paper. Start with the most general; get the main idea. Then, if there are two or more concepts that are equally specific, you're going to place them on the same level.

You probably figured out that the nucleus, mitochondria, chloroplasts, and the endoplasmic reticulum are some of the organelles described by the title. They should be placed below the main idea so that they are equally spread out across the map.

Now continue to lay out all of the other concepts under the subconcepts in the first row or layer until you have used them all. You can rearrange your pieces of paper at any time, so keep pushing them around as if they were the pieces of a jigsaw puzzle until you've arranged them the way you think they belong.

Remember not to use only the words given here. You're going to have to read the text material and add concepts to explain the major ones. For instance, your map may need to include the functions performed by the various cell organelles. Thus, under "homeostasis" you need to have "endoplasmic reticulum" and "Golgi apparatus." Under "energy manufacture" you will have "chloroplasts" and "photosynthesis," "food," "sugar," and "green plants."

Concept mapping takes some time at first; it will take about six times longer the first few times than it will after a week or two of practice.

Now you're going to make the connections among the concepts. Use lines to connect the concepts, and write linking words on the lines to show or tell why they are connected. Do this for all the lines connecting all the concepts. Glue or tape down your concept papers if you want to make the map permanent, or use a separate piece of paper to draw a sketch showing the way you've arranged the concepts.

Features of Good Concept Maps

Remember that practice is the key to good concept mapping and that you'll get better at it as you go along. Here are some things to remember as you get started.

- **A concept map does not have to be symmetrical. It can have more concepts on one side than on the other.**
- **There are no perfectly correct concept maps, only maps that come closer to the meanings you have for those concepts. As the mapmaker, you must make it correct for you.**
- **Do not put more than three words in a concept box.**
- **Do not string out more than three boxes in a row or line without branching out.**
- **Write linking words to connect possible concepts using as few words as possible.**

If the relationships you have made between any two concepts are wrong, your teacher will help you sort out the misconception. Even if you are absolutely correct in your relationships, you may see another student's map that has differences. Both maps could be equally correct even though they may look nothing alike. Everyone thinks a little bit differently, and, as a result, everyone sees different relationships between certain concepts.

As you get practice in making concept maps, your teacher will examine your linking statements more closely. Since these links represent the relationships between concepts, whatever you write will tell whether you really understand how those two concepts are connected. A good concept map should always have the following characteristics:

- **It is two dimensional, not just a list of concepts. It should go from top to bottom and from general to specific, with lines that show how the concepts are related.**
- **It shows which concepts are more important by their placement on the map and by what concepts branch from them.**
- **It has many branches with no more than three concept boxes in a row and no more than three words in a concept box.**
- **It has only concepts in the boxes and only linking words on the lines.**

Word Roots

Analyzing Word Parts

A knowledge of common prefixes, roots, and suffixes can give you clues to the meaning of an unfamiliar word and help make learning new words easier. For example, each of the word parts in the chart below can be combined with the root *derm* to form another word.

Prefix	Root	Suffix
	derm	-otology
	derm	-otologist
hypo-	derm	-ic
	derm	-atitis
pachy		-derm

Example: dermatologist

derm (root): skin
-logy (suffix): study of
-ist (suffix): someone who practices or deals with something
dermatologist—someone who studies or deals with skin

Following is a chart of prefixes, suffixes, and roots commonly found in the study of biology. Each is followed by its usual meaning, an example of a word in which it is used, and a definition of that word. You can use the chart to determine the meaning of a word like dermatology.

Prefix, root, or suffix	Definition	Example
a-	not, without	• asymmetrical, not symmetrical
ab-	away, apart	• abduct, move away from the middle
-able	able	• viable, able to live
ad-	to, toward	• adduct, move toward the middle
amphi-	both	• amphibian, type of vertebrate, like a frog, that lives both on land and in the water
ante-	before	• anterior, front of an organism
anti-	against	• antibiotic, substance, such as penicillin, capable of killing some disease-causing organisms
arche-	ancient	• *Archaeopteryx*, earliest known fossil bird
arthro-	joint	• arthropod, jointed-limbed organism belonging to the phylum Arthropoda
auto-	self, same, automatic	• autotrophic, able to manufacture its own food; autonomic, self-controlling, functionally independent, as in the autonomic nervous system
bi-	two	• bivalve, mollusk with two shells
bio-	life	• biology, the study of life
blast-	embryo	• blastula, hollow ball stage in the development of an embryo
carcin-	cancer	• carcinogenic, causing cancer
cereb-	brain	• cerebrum, part of the vertebrate brain

Prefix, root, or suffix	Definition	Example
chloro-	green	• chlorophyll, green pigment in plants necessary for photosynthesis
chromo-	color	• chromosome, structure found in eukaryotic cells that contains DNA
chondro-	cartilage	• Chondrichthyes, cartilaginous fish
circ-	around	• circulatory, system for moving fluids through the body
-cide	kill	• insecticide, a substance that kills insects
co-, con-	with, together	• conjoined twins, identical twins physically joined by a shared portion of anatomy at birth
-cycle	circle	• pericycle, layer of plant cells
cyt-	cell	• cytology, the study of cells
de-	remove	• dehydration, removal of water
derm-	skin	• dermatology, study of the skin
di-	two	• diploid, full set of chromosomes found in somatic cells
dia-	through	• dialysis, separation of molecules using a membrane
ecol-	dwelling, house	• ecology, the study of living things and their environments
ecto-	outer, outside	• ectoderm, outer germ layer of developing embryo
-ectomy	removal	• appendectomy, removal of the appendix
endo-	inner, inside	• endoplasm, cytoplasm within the plasma membrane
epi-	upon, over	• epiphyte, plant growing upon another plant
ex-, exo-	outside of	• exobiology, the search for life elsewhere in the universe
gastro-	stomach	• gastropod, type of mollusk
-gen	type	• genotype, genes in an organism
-gram	write or record	• climatogram, record depicting the annual precipitation and temperature for an area
hemi-	half	• hemisphere, half of a sphere
hetero-	different	• heterozygous, different alleles inherited from parents
hist-	tissue	• histology, the study of tissues
homeo-	the same	• homeostasis, maintenance of the same conditions in an organism as the environment changes
hydro-	water	• hydroponics, growing plants in water instead of soil
hyper-	above, over	• hypertension, blood pressure consistently higher than normal
hypo-	below, under	• hypothalamus, part of the brain
-ic	of or pertaining to	• hypodermic, under or below the skin
inter-	between, among	• interbreed, breed within a family or strain
intra-	within	• intracellular, inside a cell
iso-	equal	• isogenic, having an identical set of genes
-ist	someone who practices or deals with something	• biologist, someone who studies life
-logy	study of	• biology, the study of life
macro-	large	• macromolecule, large molecule such as DNA or protein

continued

Prefix, root, or suffix	Definition	Example
mal-	bad	• malnourished, poorly nourished
mega-	large	• megaspore, larger of two types of spores produced by some ferns and flowering plants
meso-	in the middle	• mesoglea, jellylike material found between outer and inner layer of coelenterates
meta-	change	• metamorphosis, change in form in insects
micro-	small	• microscopic, too small to be seen with the unaided eye
mono-	one, single	• monoploid, one set of alleles
morph-	form	• morphology, study of form of organisms
neo-	new	• neonatal, newborn
nephr-	kidney	• nephron, functional unit of the kidney
neur-	nerve	• neurotransmitter, chemical released by a neuron to stimulate or inhibit other neurons
oo-	egg	• oogenesis, gamete formation in female diploid organisms
org-	living	• organism, living thing
-oma	swelling	• carcinoma, cancerous tumor
orth-	straight	• orthodontist, dentist who straightens teeth
para-	near, on	• parasite, organism that lives on and gets nutrients from another organism
path-	disease	• pathogen, disease-causing organism
peri-	around	• pericardium, membrane around the heart
photo-	light	• phototropism, bending toward light by plants
phyto-	plants	• phytoplankton, part of plankton consisting of plant life
poly-	many	• polypeptide, sequence of amino acids joined together to form a protein
-pod	foot	• pseudopod, false foot that projects from the main part of an amoeboid cell
pre-	before	• prediction, a forecast of events before they take place; the outcome, if a hypothesis is accepted
-scope	instrument to see something	• microscope, instrument to see very small objects; telescope, instrument to see very distant objects
semi-	partially	• semipermeable, allowing some particles to move through
-some	body	• chromosome, structure found in eukaryotic cells that contains DNA
sub-	under	• substrate, molecule on which an enzyme acts
super-, supra-	above	• superficial, on or near the surface of a tissue or organ
syn-	with	• synapse, junction of one neuron with another
-tomy	to cut	• appendectomy, surgery to cut out the appendix
trans-	across	• transformation, the transfer of genetic material from one organism to another
visc-	organ	• viscera, internal organs of the body cavity

SI Measurement

Using Conversion Factors

Length

The metric system is used for making measurements in science. The official name of this system is the Système International d'Unités, or International System of Measurements (SI). SI is a decimal system. This means that all relationships between units of measurement are based on powers of 10. There are several prefixes that can be combined with an SI unit. These prefixes change the unit to mean some other power of 10. For example, the SI base unit of length is the meter. When the prefix *kilo* is added to the base unit *meter*, the word *kilometer* is formed (a derived unit meaning 1000 meters). When the prefix *centi* is combined with the base unit *meter,* the word *centimeter* is formed (a derived unit meaning 1/100 of a meter).

Table 1 lists some commonly used prefixes and their meanings. The prefixes modify the value of the base to describe very small or very large measurements, like the diameter of an animal cell or distances between cities. You can easily convert from one SI derived unit to another. For example, there are 100 centimeters in a meter and there are 10 decimeters in a meter. Therefore, there are 10 centimeters in 1 decimeter. **Table 2** shows you the relationship of SI units to commonly used English units.

Table 1 SI Prefixes

Prefix	Symbol	Exponential	Rational	Meaning
tera	T	10^{12}	1,000,000,000,000	trillion
giga	G	10^{9}	1,000,000,000	billion
mega	M	10^{6}	1,000,000	million
kilo	k	10^{3}	1,000	thousand
hecto	h	10^{2}	100	hundred
deka	da	10^{1}	10	ten
deci	d	10^{-1}	1/10	one-tenth
centi	c	10^{-2}	1/100	one-hundredth
milli	m	10^{-3}	1/1,000	one-thousandth
micro	m	10^{-6}	1/1,000,000	one-millionth
nano	n	10^{-9}	1/1,000,000,000	one-billionth
pico	p	10^{-12}	1/1,000,000,000,000	one-trillionth

continued

Table 2 SI Conversion Table

SI Units	From SI to English	From English to SI
Length		
kilometer (km) = 1,000 m	1 km = 0.62 mile	1 mile = 1.609 km
meter (m) = 100 cm	1 m = 3.28 feet	1 foot = 0.305 m
centimeter (cm) = 0.01 m	1 cm = 0.394 inch	1 inch = 2.54 cm
millimeter (mm) = 0.001 m	1 mm = 0.039 inch	
micrometer (µm) = 0.000 001 m		
nanometer (nm) = 0.000 000 001 m		
Area		
square kilometer (km^2) = 100 hectares	1 km^2 = 0.3861 square mile	1 square mile = 2.590 km^2
hectare (ha) = 10,000 m^2	1 ha = 2.471 acres	1 acre = 0.4047 ha
square meter (m^2) = 10,000 cm^2	1 m^2 = 10.765 square feet	1 square foot = 0.0929 m^2
square centimeter (cm^2) = 100 mm^2	1 cm^2 = 0.155 square inch	1 square inch = 6.4516 cm^2
Volume		
liter (L) = 1,000 mL = 1 dm^3	1 L = 1.06 fluid quarts	1 fluid quart = 0.946 L
milliliter (mL) = 0.001 L = 1 cm^3	1 mL = 0.034 fluid ounce	1 fluid ounce = 29.57 mL
microliter (µL) = 0.000 001 L		
Mass		
kilogram (kg) = 1,000 g	1 kg = 2.205 pounds	1 pound = 0.4536 kg
gram (g) = 1,000 mg	1 g = 0.0353 ounce	1 ounce = 28.35 g
milligram (mg) = 0.001 g		
microgram (µg) = 0.000 001 g		

Temperature

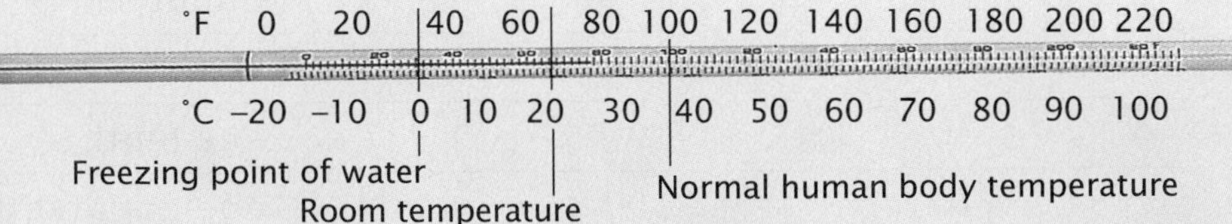

The top of the thermometer is marked off in degrees Fahrenheit (°F). To read the corresponding temperature in degrees Celsius (°C), look at the bottom half of the thermometer. For example, 50°F is the same temperature as 10°C. You may also use the formulas given at the right for conversions.

Conversion of Fahrenheit to Celsius:
°C = 5/9 (°F – 32)

Conversion of Celsius to Fahrenheit:
°F = (9/5 °C) + 32

Choosing the Correct Unit

Remember that every measurement may be expressed using more than one prefix. Usually, the simplest expression is best. For example, it is better to state the diameter of a cell as 5 μm rather than 0.000,0005 m, even though the two measurements are equivalent. The distance between cities is more appropriately stated in km rather than in cm.

Practice Exercises

1. What prefix and unit represents 10 meters?

2. What derived unit represents 1/100 of a meter?

3. There are 1000 millimeters in 1 meter (1 millimeter = 1/1000 of a meter). Express 500 millimeters in meters.

4. How many gigameters is 145 million kilometers?

5. The length of line segment **A** in **Figure 1** can be represented in millimeters, centimeters, and meters. All of the measures are equivalent, and all represent the length of line segment A. Estimate the length of segment A, and then measure it. Express the measurement in each of the following units.

a. millimeters______
b. centimeters______
c. meters______

6. Animal and plant cells are studied under a microscope. Some cells are so small that their parts are measured in micrometers. What part of a meter is a micrometer?

7. A millimeter is the smallest distance pictured on the metric ruler in **Figure 2**. Use the ruler to answer each of the following.

a. A points to _______ mm or _______cm
b. B points to _______ mm or _______cm
c. C points to _______ mm or _______cm
d. D points to _______ cm or _______mm
e. E points to _______ cm or _______mm

8. Choose the most appropriate metric unit for measuring each of the following:

a. Area of a sheet of notebook paper
b. Area of a desktop
c. Area of an airport runway

9. A rectangular plot of land is to be seeded with grass. The plot to be seeded is 22 m by 28 m. If 1 kg of seed is needed for each 85 m^2 of land, how many kg of grass seed are needed?

A

Figure 1

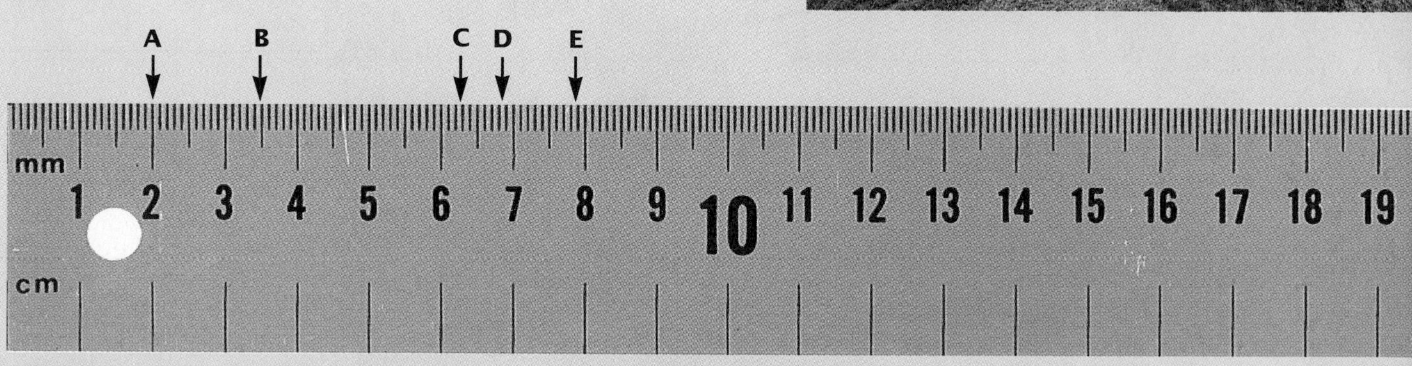

Figure 2

Interpreting Food Labels

Nutrition Facts Labels

The government now requires uniform nutrition labeling of packaged foods. The Nutrition Facts label on similar packaged foods indicates nutrition information based on similar serving sizes. The label at the right shows the nutrition information on a carton of whole milk.

The Nutrition Facts label indicates the serving size, the actual amount of a nutrient per serving, and the % Daily Value of that nutrient in one serving. We can see that one cup (236 mL) of whole milk contains 8 g of protein, which is 16 percent of the recommended daily amount (daily value) of protein we should consume. How would you determine the total amount, in grams, of protein needed each day?

The Nutrition Facts label lists *some* of the nutrients important to your health. For example, only two vitamins, vitamin A and vitamin C, and two minerals, calcium and iron, are listed.

Since vitamin D is added to milk, it is also listed on containers of milk. The Nutrition Facts label also indicates the maximum or minimum recommended daily consumption of some nutrients. We can see that on a daily basis, we should consume a maximum of 300 mg of cholesterol and a minimum of 25g of fiber.

Nutrition Facts
Serving Size 1 cup (236mL)
Servings Per Container 2

Amount Per Serving	
Calories 150	Calories form Fat 70
	% **Daily Value***
Total Fat 8g	**12**%
Saturated Fat 5g	**25**%
Cholesterol 35mg	**11**%
Sodium 125mg	**5**%
Total Carbohydrate 12g	**4**%
Dietary Fiber 0g	**0**%
Sugars 12g	
Protein 8g	**16**%

Vitamin A 6% • Vitamin C 4%
Calcium 30% • Iron 0% • Vitamin D 25%

*Percent Daily Values are based on a 2000 calorie diet. Your daily values may be higher or lower depending on your calorie needs:

	Calories:	2,000	2,500
Total Fat	Less than	65g	80g
Sat Fat	Less than	20g	25g
Cholesterol	Less than	300mg	300mg
Sodium	Less than	2,400mg	2,400mg
Total Carbohydrate		300g	375g
Dietary Fiber		25g	30g
Protein		50	65

Practice Exercises

Use the Nutrition Facts label from a can of peas to answer the questions below.

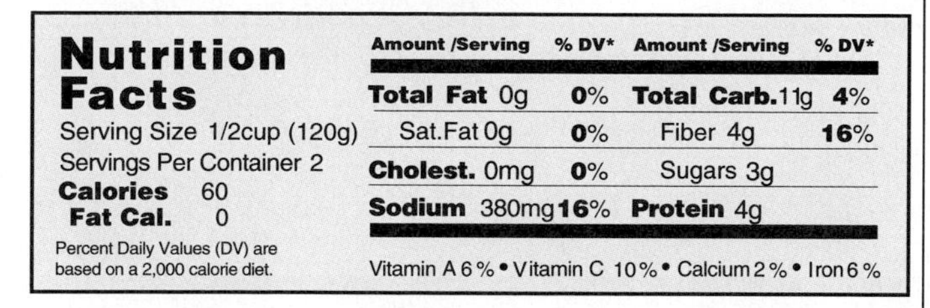

Nutrition Facts
Serving Size 1/2cup (120g)
Servings Per Container 2
Calories 60
Fat Cal. 0
Percent Daily Values (DV) are based on a 2,000 calorie diet.

Amount /Serving	% DV*	Amount /Serving	% DV*
Total Fat 0g	**0%**	**Total Carb.** 11g	**4%**
Sat.Fat 0g	**0%**	Fiber 4g	**16%**
Cholest. 0mg	**0%**	Sugars 3g	
Sodium 380mg	**16%**	**Protein** 4g	

Vitamin A 6% • Vitamin C 10% • Calcium 2% • Iron 6%

1. What is the size of one serving of peas?
2. What percent of the recommended daily amount of carbohydrate consumption is in one serving of peas?
3. How many servings of these canned peas would you have to eat in order to get 50 percent of the vitamin C that you should have on a daily basis?
4. How many grams of fiber does one serving of peas provide?

Calories and Equivalents

One important piece of information listed on the Nutrition Facts label is the number of calories per serving. The calorie content of food depends on its components. Notice that the calories per gram on the label shows that fats contain more than twice as many calories as equal masses of carbohydrates and proteins.

Nutrition Facts
Serving Size 1oz. (28g/about 22 chips)
Servings Per Container 6

Amount Per Serving	
Calories 150	Calories form Fat 90
	% Daily Value*
Total Fat 10g	**15%**
Saturated Fat 1.5g	**8%**
Cholesterol 0mg	**0%**
Sodium 140mg	**6%**
Total Carbohydrate 14g	**5%**
Dietary Fiber 1g	**4%**
Sugars 0g	
Protein 2g	

Vitamin A 0% • Vitamin C 10%
Calcium 0% • Iron 2%

*Percent Daily Values are based on a 2000 calorie diet. Your daily values may be higher or lower depending on your calorie needs:

		Calories: 2,000	2,500
Total Fat	Less than	65g	80g
Sat Fat	Less than	20g	25g
Cholesterol	Less than	300mg	300mg
Sodium	Less than	2,400mg	2,400mg
Total Carbohydrate		300g	375g
Dietary Fiber		25g	30g

Calories per a gram:
Fat 9 • Carbohydrate 4 • Protein 4

You probably have heard the term *junk food*. Junk foods are foods that have a high caloric content and contain few or no nutrients. The amount of a specific nutrient that a food provides in relation to its calories is known as its nutrient density. A food that is high in calories and low in a specific nutrient is called nutrient-limited for the nutrient. Foods with greater amounts of nutrients in relation to calories are called nutrient-plus.

Practice Exercises

5. How many calories are in one serving of whole milk?
6. How many calories are in three servings of canned peas?
7. Why are the calories in junk foods often referred to as empty calories?
8. Use the Nutrition Facts label from a bag of potato chips to answer the following.
   a. Which nutrient group makes up the greatest percentage of the potato chips?
   b. How many calories are there in six servings of potato chips?
   c. Are potato chips a good source of iron? Why or why not? Are potato chips nutrient-plus or nutrient-limited for iron?
9. Keep a record of your total food intake for an entire day. Write down the name of the food, the approximate serving size, and the number of servings of the food you eat. Use the Nutrition Facts label on the packaged foods and a calorie chart to answer the following questions.
   a. What was your approximate caloric intake for the day?
   b. How did your consumption of total fat, vitamin C, fiber, and sodium compare with the recommended amounts for these nutrients?
   c. Which foods were nutrient-plus for vitamin A?
10. A type of food contains 200 calories per serving, 120 of which come from fat. It contains less than 2 percent of the % Daily Values of vitamins A and C, no calcium or iron, and 3 percent of the % Daily Value for fiber. Is it a junk food? Why or why not?
11. One serving of a food contains 20 g of fat, 14 g of protein, and 3 g of carbohydrate. Approximately how many calories are in one serving?

Reference and Skills Appendix
Reference

Oak tree

The Six Kingdoms

Kingdom	Number of Named Species*	Description	Examples
Archaebacteria	100	Prokaryotic; unicellular; lack nucleus or other internal compartments; most primitive form of life; existed for more than 2 billion years before protists appeared; usually occupy hostile environments (extremes of temperature, salinity, no oxygen)	*Methanobacterium, Halobacterium, Thermoplasma*
Eubacteria	4,000+	Prokaryotic; unicellular; lack nucleus or other internal compartments; most primitive form of life; existed for more than 2 billion years before protists appeared; most-studied of the bacteria	*Escherichia coli, Vibrio cholerae, Salmonella typhimurium*
Protista	43,000	Eukaryotic; all but algae are unicellular; aquatic; all have nucleus and internal compartments; almost all have mitochondria and many also have chloroplasts; most have cell walls; aquatic; most diverse kingdom	*Volvox, Paramecium,* red algae
Fungi	77,000	Eukaryotic; multicellular; almost all terrestrial; cell walls of chitin; nonmotile cells; exclusively heterotrophic with nutrition by absorption; mitosis occurs within nuclear envelope	Mushrooms, yeasts
Plantae	265,350	Eukaryotic; multicellular; terrestrial; cell walls of cellulose; nonmotile cells (except gametes of a few forms); photosynthetic, with chloroplasts	Oak trees, mosses, roses, ferns
Animalia	1,203,000	Eukaryotic; multicellular; both aquatic and terrestrial; no cell walls or chloroplasts; heterotrophic with nutrition by ingestion; reproduction is predominately sexual	Humans, sea stars, earthworms, flies

* Approximate number identified

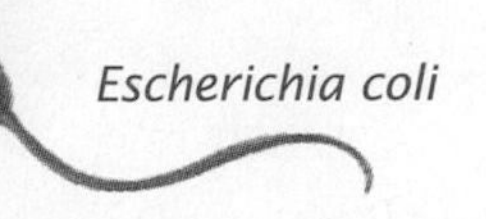
Escherichia coli

Volvox

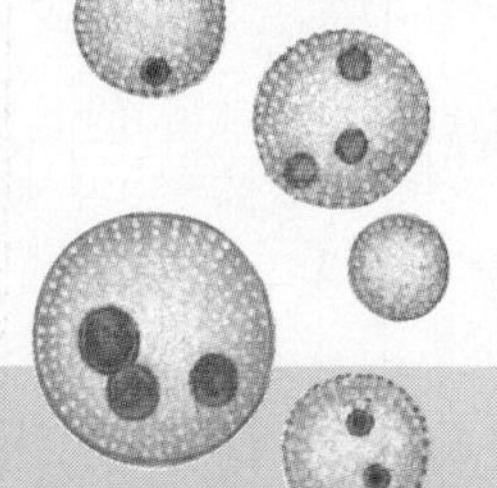

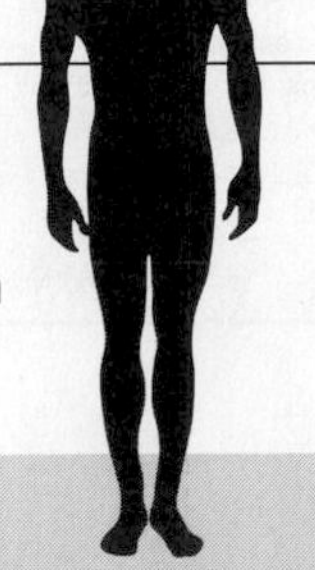
Human

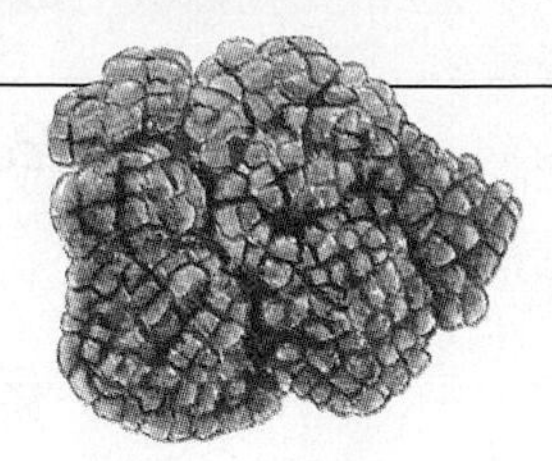
Methanosarcina

Kingdom Archaebacteria

Group*	Main Characteristics	Examples
Methanogens	Strict anaerobes that produce methane from carbon dioxide and hydrogen; die in the presence of oxygen; found in swamps and marshes	*Methanosarcina, Methanobacterium*
Thermoacidophiles	Many grow in hot, acidic environments (hot springs); use hydrogen sulfide as energy source	*Acidianus, Thermoproteus*
Halophiles	Require high concentrations of salt to survive; live in the Dead Sea and the Great Salt Lake	*Halococcus, Natronococcus*

Kingdom Eubacteria

Group*	Main Characteristics	Examples
Cyanobacteria	Photosynthetic bacteria; common on land and in the ocean; responsible for "blooms" in polluted waters	*Anabaena, Spirulina, oscillatoria*
Chemoautotrophs	Ancient group of bacteria that can grow without sunlight or other organisms; play critical role in Earth's nitrogen cycle	*Nitrobacteria, Nitrosomonas,* sulfur bacteria
Enterobacteria	Typically rigid, rod-shaped, heterotrophic bacteria; responsible for many diseases of animals and plants	*Salmonella typhimurium, Escherichia coli, Vibrio cholerae*
Pseudomonads	Straight or curved rods with flagella at one end; very common in soil; many are serious plant pathogens	*Pseudomonas aeruginosa*
Spirochaetes	Long spiral-shaped cells; responsible for several serious diseases	*Borrelia* (Lyme disease), *Treponema* (syphilis)
Actinomycetes	Filamentous (and thus sometimes mistaken for fungi); spore-producing; sources of several antibiotics	Mycobacteria, dental plaques
Rickettsias	Obligatory parasites within the cells of vertebrates and arthropods; responsible for Rocky Mountain spotted fever	*Rickettsia rickettsii*
Gliding and budding bacteria	Long rods; often aggregate into gliding masses; some members form upright spore-bearing structures of unusual complexity for bacteria	*Myxobacteria*

* The bacteria are the least studied kingdoms. Since thousands of species are yet to be discovered and named, the number of species identified is not provided in this table.

Anabaena
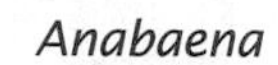
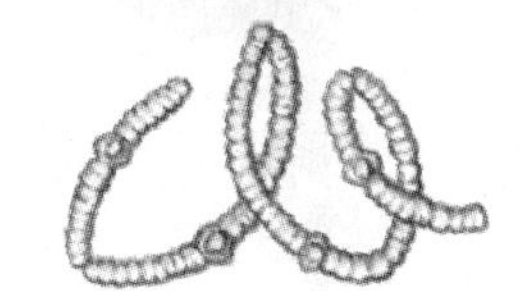

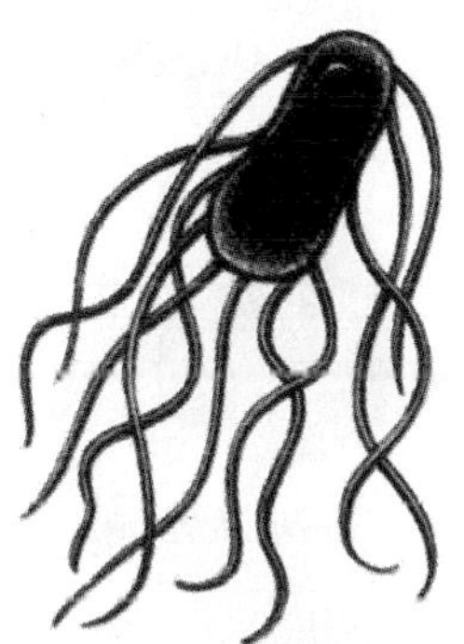
Salmonella typhimurium

Borrelia

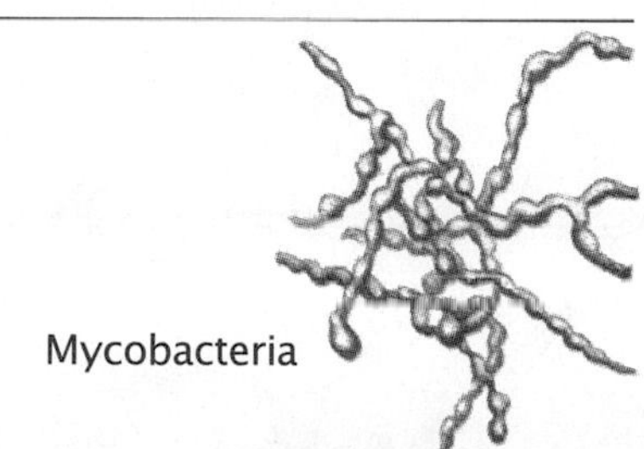
Mycobacteria

continued

Kingdom Protista

Diatom

Vorticella

Chlamydomonas

Kelp

Phylum	Number of Species*	Characteristics	Examples
Bacillariophyta "Diatoms"	11,500+	Unicellular; photosynthetic; cells within a unique shell made of opaline silica—resembles a small box with a lid; some radially symmetrical; chloroplasts resemble those of brown algae; shells often form very thick deposits called diatomaceous earth	Diatoms
Ciliophyta "Ciliates"	8,000	Very complex single cells; heterotrophic; possess rows of cilia and two types of cell nuclei; genetic code differs in several respects from that of other organisms; some scientists argue that ciliates should be assigned to a separate kingdom of their own	*Vorticella, Paramecium*
Chlorophyta "Green algae"	7,000	Unicellular, colonial, and multicellular species; all photosynthetic; chloroplasts very similar to those of green plants; most scientists believe that plants evolved from green algae	*Chlamydomonas, Chlorella, Volvox, Ulva, Spirogyra*
Rhodophyta "Red algae"	4,000	Almost all multicellular; photosynthetic; chloroplasts evolved symbiotic cyanobacteria; no cilia or centrioles; one of the most ancient groups of protists	Nori (edible algae)
Sporozoa "Sporozoans"	3,900	Unicellular; heterotrophic; nonmotile; spore-forming parasites of animals; complex life cycles, often involving alternation between different hosts; *Plasmodium* is responsible for malaria, which kills more than 1 million people each year	*Plasmodium*
Zoomastigina "Zoomastigotes"	3,000	Mostly unicellular; heterotrophic; flagellated; free-living or parasitic; a highly diverse phylum; similar cells exist in sponges, suggesting that animals evolved from a zoomastigote	Choanoflagellates
Phaeophyta "Brown algae"	1,500	Multicellular; photosynthetic; marine; some are the longest, fastest growing, and most photosynthetically productive species on Earth	Kelps, *Sargassum*

* Approximate number identified

Phylum	Number of Species*	Characteristics	Examples
Dinoflagellata "Dinoflagellates"	1,000+	Unicellular; photosynthetic; body enclosed within cellulose plates; two unequal flagella beating in grooves that encircle the body; often symbiotic with corals	*Gonyaulax*, "Red tides"
Euglenophyta "Euglenoids"	800	Unicellular; both photosynthetic and heterotrophic species; asexual; most live in fresh water; very similar to zoomastigotes; chloroplasts similar to those of green algae; thought to have evolved from the same symbiotic bacteria as algae	*Euglena*
Oomycota "Water molds"	475	Unicellular parasites on living or dead tissue; cell walls of cellulose (not chitin, like those of fungi); spores have two unequal flagella, one directed forward, the other backward; responsible for Irish potato famine of 1845–1847	*Saprolegnia*, rusts, mildews
Myxomycota "Plasmodial slime molds"	450	Heterotrophic; individuals stream along as a nonwalled multinucleate mass of "slime" cytoplasm; no centrioles; all nuclei undergo mitosis synchronously; when dry or starving, spores form to start a new individual in a more favorable environment	*Arcyria*
Foraminifera "Forams"	300	Unicellular; heterotrophic; marine; have shells of organic material with many pores through which thin cytoplasmic threads project; shells are reinforced with grains of calcium carbonate or sand; the White Cliffs of Dover and other limestone deposits are made almost entirely of foram shells	Forams
Rhizopoda "Amoebas"	300	Unicellular; heterotrophic; amorphously shaped cells that move using cytoplasmic extensions called pseudopods; asexual reproduction (fission) only	*Amoeba*
Acrasiomycota "Cellular slime molds"	65	Heterotrophic; amoeba-shaped cells that aggregate into a moving mass called a slug when deprived of food; the slug produces spores that form new amoebas elsewhere	*Dictyostelium*
Caryoblastea "Pelomyxa"	1	Unicellular; lack mitochondria and chloroplasts; have two kinds of bacterial symbionts; do not undergo mitosis; possibly an early stage in the evolution of eukaryotes	*Pelomyxa palustris*

Gonyaulax

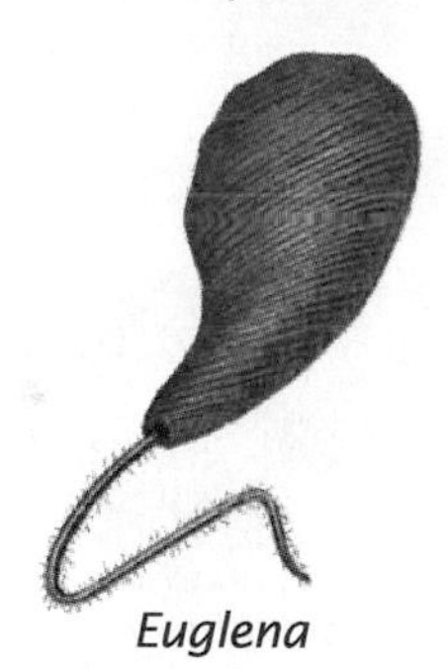

Euglena

Foram

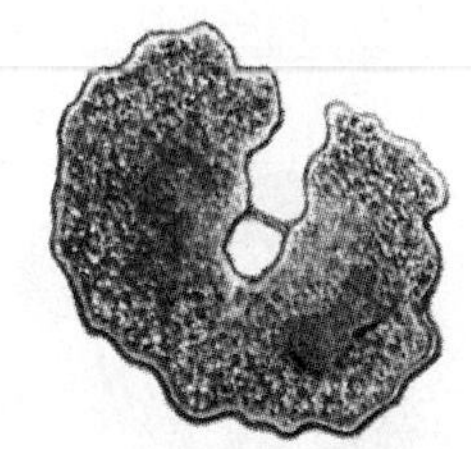

Amoeba

continued

Kingdom Fungi

Division	Number of Species*	Main Characteristics	Examples
Ascomycota "Sac fungi"	30,000	Fusion of hyphae leads to formation of reproductive structures of densely interwoven hyphae (visible part) that contain characteristic microscopic structures called asci (singular, ascus); two nuclei fuse within the ascus, forming a diploid zygote that undergoes meiosis; asexual reproduction is also common	Morels, yeasts, truffles, cup fungi
Deuteromycetes "Fungi imperfecti"	17,000	Sexual reproduction not observed; mostly ascomycetes that have lost the ability to reproduce sexually	*Penicillium, Aspergillus*
Basidiomycota "Club fungi"	16,000	Fusion of hyphae leading to formation of a densely interwoven reproductive structure (mushroom) with characteristic microscopic, club-shaped structures called basidia (singular, basidium); only the terminal cell of a basidium produces spores (all cells of an ascus do); reproduction is usually sexual	Puffballs, mushrooms, toadstools, rusts
Zygomycota "Molds"	665	Multinucleate hyphae without septa except for reproductive structures; fusion of hyphae leads directly to formation of a zygote, which divides by meiosis when it germinates; asexual reproduction is also common	*Rhizopus* (black bread mold)

Morel

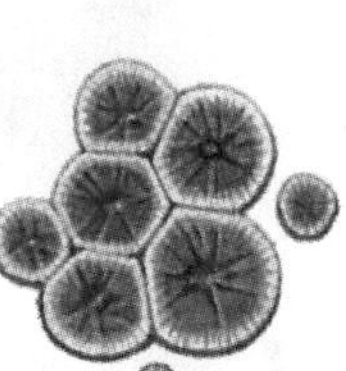

Penicillium

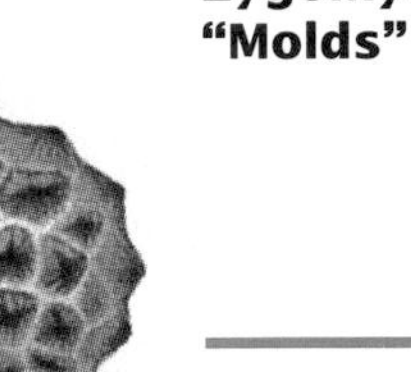

Puffball

Symbiotic Associations**

Lichens	15,000	Fungi (almost always ascomycetes) that have cyanobacteria, green algae, or both, living in their cells; derive energy from their photosynthetic partners and cannot survive without them; able to survive freezing or drying out; can invade the harshest of habitats; break down rocks and set the stage for invasion by other organisms	
Mycorrhizae	5,000	Symbiotic associations between fungi and plants; 80 percent of all plants have mycorrhizae in their roots; the most common by far are endomycorrhizae, in which zygomycete fungal hyphae penetrate the cells of the plant root; some plants of temperate regions have ectomycorrhizae, in which basidiomycete (or, rarely, ascomycete) hyphae surround but do not penetrate the root	

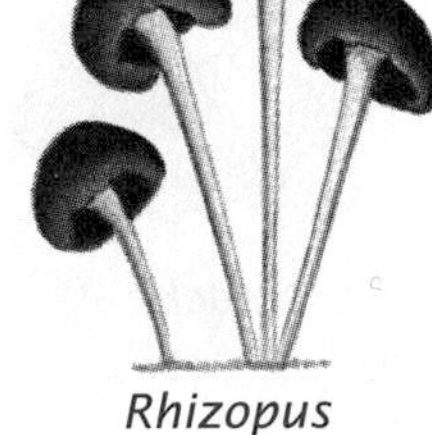

Rhizopus

* Approximate number identified

** Fungi can form mutualistic associations with algae to form lichens or with plants to form mycorrhizae.

Kingdom Plantae

Division	Number of Species*	Main Characteristics	Examples
Anthophyta "Flowering plants"	235,000	Angiosperms; ovules fully enclosed by carpel; after fertilization, carpel seeds mature to become fruit; flowers are reproductive structures; fertilization involves two sperm nuclei—one forms the gamete, the other fuses with polar bodies to form endosperm for the seed	Roses, oak trees, wheat
Bryophyta "Mosses"	16,600	No vascular tissue; lack true roots, stem, and leaves; obtain nutrients by osmosis and diffusion; found chiefly in moist habitats; seedless; sperm must swim to eggs; two other divisions, Hepaticophyta (liverworts) and Antherocerophyta (hornworts), are also considered bryophytes and make up 40 percent of the bryophyte species	*Polytrichum*, *Sphagnum* (peat moss)
Pterophyta "Ferns"	12,000	Seedless, vascular plants; motile sperm germinate into free-living haploid individuals; two minor divisions, Sphenophyta (horsetails) and Psilophyta (whisk ferns), contain 21 additional species	*Sphaeropteris* (tree ferns), Azolla
Lycophyta "Lycopods"	1,000	Seedless vascular plants; motile sperm; similar in appearance to mosses, but are diploid; found chiefly in moist habitats	*Lycopodium*, club mosses
Coniferophyta "Conifers"	550	Gymnosperms; ovules exposed at time of pollination; pollen dispersed by wind; sperm lack flagella; needle-like or scale-like leaves; most species are evergreens	Fir, spruce, pine, redwood trees
Cycadophyta "Cycads"	100	Gymnosperms; sperm flagellated and motile, but reach egg by a pollen tube; palmlike trees; very slow growing	Sago palms, cycads
Gnetophyta "Shrub teas"	70	Gymnosperms; sperm nonmotile; shrubs and vines	*Welwitschia*, Mormon teas
Ginkgophyta "Ginkgo"	1	Gymnosperms; motile sperm reach egg by a pollen tube; deciduous (drop leaves in winter); fanlike leaves; fleshy and illscented seeds	Ginkgos

* Approximate number identified

Rose bush

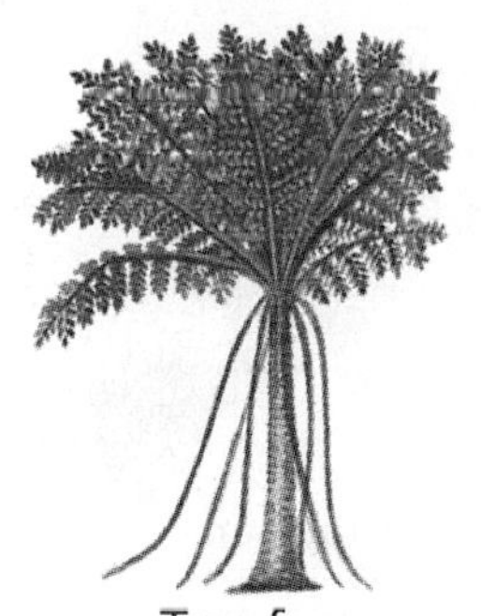
Tree fern

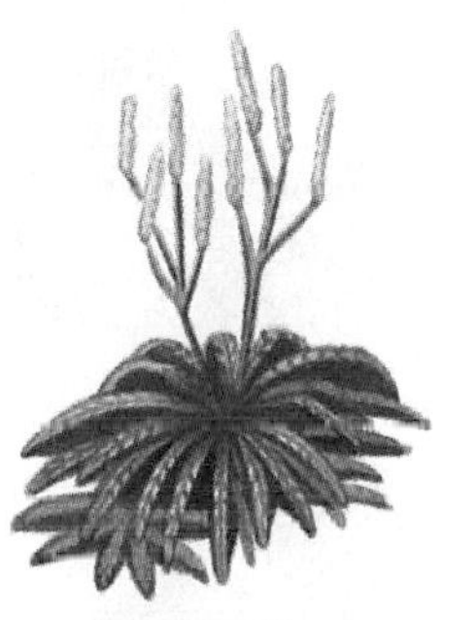
Lycopodium

Fir tree

continued

Kingdom Animalia

Beetle

Oyster

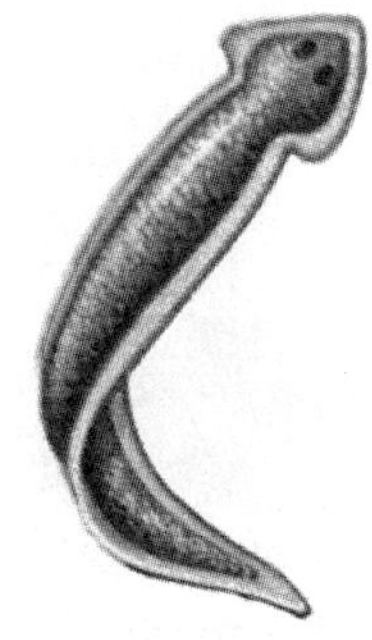
Planarian

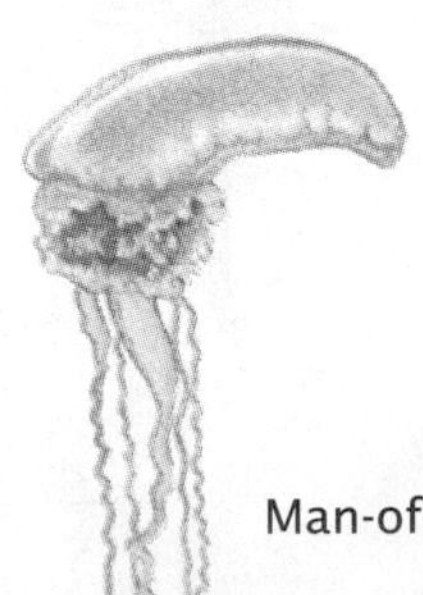
Man-of-war

Phylum	Number of Species*	Main Characteristics	Examples
Arthropoda "Arthropods"	1,000,000	Segmented bodies with paired, jointed appendages; chitinous exoskeletons; open circulatory system; many groups possess wings; most successful of all animal phyla	Beetles, spiders, flies, lobsters
Mollusca "Mollusks"	110,000	Soft-bodied animals with a true coelom; three-part body plan consisting of head-foot, visceral mass, and mantle; many have shells; 35,000 species are terrestrial; almost all have a unique rasping tongue called a radula	Oysters, snails, squids, octopuses
Chordata "Chordates"	42,500	Possess a notochord, a dorsal nerve cord, pharyngeal slits, and a tail at some stage of life; notochord is replaced during development by the spinal column in vertebrates; pharyngeal slits are lost during development of terrestrial vertebrates; 20,000 species are terrestrial	Fishes, dinosaurs, mammals
Platyhelminthes "Flatworms"	12,000	Solid worms; bilaterally symmetrical; three germ layers; digestive cavity has only one opening; no coelom or pseudocoelom; simplest animals in which organs occur	Planaria, liver flukes, tapeworms
Nematoda "Roundworms"	12,000	Tiny, parasitic, unsegmented worms; tubular; bilaterally symmetrical; complete digestive tract with mouth and anus, pseudocoelom; no cilia	*Ascaris*, pinworms, *Filaria*, hookworms
Annelida "Segmented worms"	12,000	Serially segmented worms; complete digestive tract with coelom; some have chitinous bristles (setae) on each segment that anchor them while crawling	Leeches earthworms, polychaetes
Cnidaria "Jellyfish"	10,100	Radially symmetrical; gelatinous; digestive cavity with only one opening; tentacles with stinging cells called nematocysts	Hydras, man-of-wars
Porifera "Sponges"	5,000	Asymmetrical animals; lack distinct tissues and organs; sessile; mostly marine; body consists of two layers that are supported by a stiff skeleton with many pores; internal cavity lined with unique food-filtering cells called choanocytes	Barrel sponges

Phylum	Number of Species*	Main Characteristics	Examples
Echinodermata "Echinoderms"	6,000	Deuterostomes; adults radially symmetrical; endoskeleton of calcium plates; water vascular system with tube feet; five-part body plan; able to regenerate lost body parts	Sea stars, sand dollars, sea urchins
Bryozoa "Moss animals"	4,000	Deuterostomes; microscopic, aquatic organisms that form branching colonies; U-shaped row of ciliated tentacles for feeding (the lophophore)	Plumatella, *Bowerbankia*
Rotifera "Wheel animals"	2,000	Small, wormlike or spherical animals; have digestive tract; have crown of cilia around mouth resembling a wheel; pseudocoelomates; almost all live in fresh water	Rotifers
Minor worms	980	**Rhynchocoela** (ribbon worms): bilaterally symmetrical; acoelomates; marine; long, extendible proboscis **Pognophora** (tube worms): deep-sea worms that live within chitinous tubes in ocean floor; sessile; long, beardlike tentacles **Hemichordata** (acorn worms): marine worms with dorsal and ventral nerve cords **Onychophora** (velvet worms): protostomes; chitinous exoskeleton; evolutionary relicts **Chaetognatha** (arrow worms): deuterostomes; coelomates; bilaterally symmetrical; marine worms with large eyes and powerful jaws	Acorn worms, giant tube worms, velvet worms
Brachiopoda "Lamp shells"	250	Have a lophophore like bryozoans, but within two clamlike shells; more than 30,000 fossilized species known	*Lingula*
Ctenophora "Comb jellies"	100	Transparent, gelatinous marine animals; eight bands of cilia; largest animals that use cilia for locomotion; often bioluminescent; digestive tract with anal pore	Comb jellies, sea walnuts
Phoronida "Lophophores"	12	Lophophorate tube worm; U-shaped gut instead of straight digestive tube of other tube worms; protostomes; often live in dense populations	*Phoronis*
Loricifera "Sand animals"	6	Minute; bilaterally symmetrical; pseudocoelomates; live in spaces between grains of sand; mouthparts exhibit a unique flexible tube; discovered in 1983	*Nanoloricus mysticus*

* Approximate number identified

Sea star

Bowerbankia

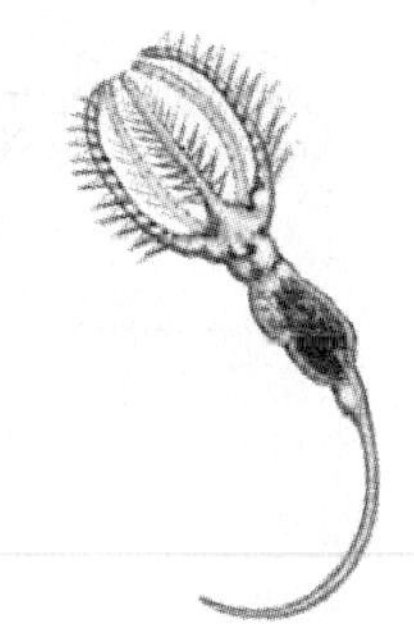
Rotifer

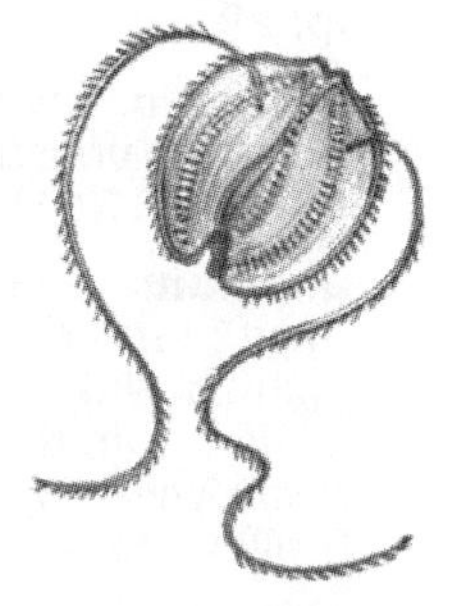
Comb jelly

Glossary

Pronunciation Key

Sound	As In	Phonetic Respelling
a	bat	(BAT)
ay	face	(FAYS)
ah	lock argue	(LAHK) (AHR gyoo)
ow	out	(OWT)
ch	chapel	(CHAP uhl)
eh	test	(TEHST)
ai	rare	(RAIR)
ee	eat feet ski	(EET) (FEET) (SKEE)
ih	bit	(BIHT)
eye *y*	idea ripe	(eye DEE uh) (RYP)
ihng	going	(GOH ihng)
k	card kite	(KAHRD) (KYT)
ng	anger	(ANG guhr)
oh	over	(OH vuhr)
aw	dog horn	(DAWG) (HAWRN)
oy	foil	(FOYL)
u	pull	(PUL)
oo	pool	(POOL)
s	cell sit	(SEHL) (SIHT)
sh	sheep	(SHEEP)
th	that thin	(THAT) (THIHN)
uh	cut	(CUHT)
ur	fern	(FURN)
y	yes	(YEHS)
yoo *yu*	globule cure	(GLAHB yool) (KYUR)
z	bags	(BAGZ)
zh	treasure	(TREHZH uhr)
uh	medal talent pencil onion playful dull	(MEHD uhl) (TAHL uhnt) (PEHN suhl) (UHN yuhn) (PLAY ful) (DUHL)
uhr	paper	(PAY puhr)

acid: compound capable of donating a hydrogen ion to a water molecule; from the Latin *acere,* meaning "to be sour" (p. 291)

acid rain: any precipitation with unusually high acidity (p. 291)

acoelomate (ay SEEL oh mayt): term describing animals without a body cavity; from the Greek *a* + *koilia,* meaning "not" + "hollow" (p. 444)

activation energy: amount of energy needed to start a chemical reaction (p. 78)

active site: pocket formed in the folds of an enzyme and to which a substrate or substrates binds; site of the reaction catalyzed by the enzyme (p. 80)

active transport: one-way transport of substances across a membrane; requires energy (p. 70)

adaptation: process by which a species becomes better suited to its environment (p. 182)

adrenal gland: either of two almond-sized endocrine organs lying atop the kidney and producing more than two dozen hormones; from the Latin *ad* + *renes,* meaning "at" + "kidneys" (p. 639)

aerobic (ehr OH bihk) **exercise**: continuous, low- or moderate-intensity exercise; from the Greek *aēr* and *bias,* meaning "air" and "life" (p. 594)

AIDS: **A**cquired **I**mmune **D**eficiency **S**yndrome, the final stages of a disease in which the immune system has been disabled; caused by HIV (p. 345)

allele (uh LEEL): one of two or more alternate forms of a gene; from the Greek *allēlon* meaning "of one another" (p. 119)

allergy: immune system response to non-pathogenic antigens such as pollen; from the Greek *all* + *ergon,* meaning "different" + "work" (p. 688)

alternation of generations: describes a life cycle in which a haploid form alternates with a diploid form (p. 381)

altruism (AHL troo ihz uhm): self-sacrificing behavior; from the Latin *alter,* meaning "another" (p. 245)

alveoli (al VEE oh leye): tiny air sacs within the lungs, through which gases enter and leave the circulatory system; from the Latin *alveus,* meaning "hollow cavity" (p. 671)

amnion (AM nee ahn): thin membrane forming a closed sac around the embryo; from the Greek *amnos,* meaning "lamb" (p. 473)

amniotic egg: the egg of reptiles, birds, and mammals; the embryo develops within two protective membranes, the chorion and amnion (p. 473)

anaerobic (an uh ROH bihk) **exercise**: exercise that requires muscles to work at high intensity for a very brief period, creating a shortage of oxygen; from the Greek *an* + *aēr* and *bios,* meaning "not" + "air" and "life" (p. 594)

analogous structures: features of organisms that are similar but which have evolved independently; from the Greek *analogos,* meaning "proportionate" (p. 320)

angiosperm (AN jee oh spurm): seed plant that has flowers and produces seeds within an ovary; from the Greek *angeion* and *speirein,* meaning "case" and "to sow" (p. 386)

anorexia nervosa (an oh RECK see ah ner VOH sah): eating disorder in which sufferers starve themselves; from the Greek *an* + *oregein,* meaning "without" + "appetite" (p. 704)

anthropoid (AN thruh poyd): member of the primate suborder that contains monkeys, apes, and humans; from the Greek *anthrōpos,* meaning "man" (p. 215)

antibiotic: chemical substance with capacity to destroy or inhibit growth of bacteria; from the Greek *anti* + *bios,* meaning "against" + "life" (p. 343)

antibiotic resistance: arises when a few bacteria with mutant genes are able to survive antibiotic treatment and reproduce, passing that resistance on to the next generation (p. 344)

antibody: defensive protein produced by white blood cells in response to a foreign substance, released by B cells (pp. 163, 685)

anticodon: three-nucleotide sequence of transfer RNA (tRNA) molecule, which is complementary to an mRNA codon; from the Latin *anti-* + *code,* meaning "against" + "book" (p. 144)

antigen (AN tuh juhn): molecule that can be recognized by white blood cells and that can trigger a defensive response (p. 681)

antihistamine: chemical that blocks the action of histamines thus reducing the body's reactivity to an allergen (p. 688)

anuran (uh NUR uhn): member of the order Anura, which contains frogs and toads; from the Greek *an* + *oura,* meaning "without" + "tail" (p. 536)

anus: posterior opening of the digestive tract; from the Latin *anus,* meaning "ring" (p. 445)

arachnid (uh RAK nihd): any of the animals of the class Arachnida, such as spiders, scorpions, mites, and ticks; from the Greek *arachnē,* meaning "spider" (p. 501)

archaebacteria (AHR kee bahk THR ee ah): distinctive group of bacteria thought to have given rise to eukaryotes; from the Greek *archaikos* + *baktērion,* meaning "ancient" + "small staff" (p. 195)

artery: blood vessel that carries blood away from the heart (p. 654)

asthma (AZ muh): respiratory disease in which airways in the lungs become restricted because of sensitivity to stimuli; from the Greek *asthma,* meaning "panting" (p. 674)

atom: smallest unit of matter that cannot be broken down by chemical means; from the Greek *a* + *tomos,* meaning "not" + "cut" (p. 26)

ATP: adenosine triphosphate; molecule used by cells to deliver energy for chemical reactions (p. 83)

atrium (AY tree uhm): chamber of the heart that receives blood from the body; from the Latin *atrium,* meaning "large room" (p. 658)

autoimmune disease: disease in which the ability of the immune system to distinguish self from non-self breaks down and the body attacks its own cells (p. 689)

autonomic nervous system: division of the nervous system that carries messages to muscles and glands, and that usually works without our noticing (p. 620)

autotroph (AWT oh trahf): organism that makes its own food; from the Greek *auto-* + *trophē,* meaning "self" + "food" (p. 257)

axon (AK sahn): long extension of a nerve cell that carries impulses away from the cell body; from the Greek *axōn,* meaning "axis" (p. 601)

balancing selection: situation in which natural selection for an allele is balanced by selection against it, and the allele remains at the same frequency (p. 183)

B cell: white blood cell that produces and releases antibodies (p. 685)

behavior: an action or series of actions performed by an animal in response to a stimulus (p. 233)

bilateral symmetry: characteristic of an animal that can be separated into nearly mirror-image halves by being cut lengthwise along the midline of the body; from the Latin *bi* + *latus* and *symmetria,* meaning "two" + "side" and "measured together" (p. 442)

binocular vision (beye NAHK yuh luhr VIHZH uhn): vision in which both eyes simultaneously view the same object; from the Latin *bi-* + *ocularis* and *videre,* meaning "double" + "of the eye" and "to see" (p. 214)

biology: study of living things; from the Greek *bios* and *logos,* meaning "life" and "study of" (p. 5)

biome (BEYE ohm): major type of terrestrial ecological community; from the Greek *bi-* and the Latin *oma,* meaning "life" and "mass" (p. 264)

bipedal (beye PEHD uhl): walking on two legs; from the Latin *bi-* + *pedalis,* meaning "double" + "of the foot" (p. 220)

blood pressure: force exerted on the inner walls of arteries (p. 664)

blood sugar: glucose dissolved in the blood, providing energy to the cells (p. 644)

blood type: classification of blood according to the presence or absence of specific marker proteins (p. 656)

blood vessel: any of the tubes that carry blood from the heart to the organs and cells of the body and back again; from the Latin *vas,* meaning "vessel" (p. 654)

body plan: overall organization of an animal's body (p. 437)

book lung: in arachnids, highly folded sacs, used for gas exchange, that are located inside the body and that open to the outside through a spiracle (p. 467)

bulimia (buh LEEM ee uh): eating disorder in which sufferers engage in binge-purge cycles of eating followed by vomiting or intestinal elimination; from the Greek *boulimia,* meaning "ox hunger" (p. 704)

cancer: a condition in which normal restraints on cell reproduction break down, resulting in rapid, uncontrolled cell division; from the Greek *karkinōma,* meaning "crab" or "cancer" (p. 109)

capillary: one of the network of tiny blood vessels from which substances in the blood move into surrounding cells; from the Latin *capillus,* meaning "hair" (p. 654)

capillary action: movement of a liquid up a narrow tube caused by the forces of attraction between the liquid molecules and the walls of the tube (p. 396)

carbohydrate (KAHR boh HEYE drayt): compound composed of carbon, hydrogen, and oxygen in the ratio of 1:2:1; used by living things as a source of energy; from the Latin *carbon* and Greek *hydōr,* meaning "coal" and "water" (p. 30)

carcinogen (kahr SIN uh juhn): any substance that causes cancer; from the Greek *karkinōma,* meaning "crab" or "cancer" (p. 289)

cardiac muscle: muscle found in the heart; from the Greek *kardia* and the Latin *musculus,* meaning "heart" and "little mouse (muscle)" (p. 590)

cardiovascular disease: disease of the heart or blood vessels (p. 668)

carnivore (KAHR nuh vawr): flesh-eating organism; from the Latin *caro* and *vorare,* meaning "flesh" and "to devour" (p. 257)

catalyst (KAT uh lihst): material that speeds up a chemical reaction without being used itself; from the Greek *katalyein,* meaning "to dissolve" (p. 79)

cell: smallest unit that can perform all life processes; from the Latin *cella,* meaning "small room" (pp. 16, 41)

cell cycle: 5-phase sequence of eukaryotic cell growth and division (p. 107)

cell membrane: bilipid layer that forms the cells outer surface, essential to the cell's metabolism; also called the plasma membrane; from the Latin *membrana,* meaning "skin" (p. 41)

cell surface marker: membrane protein that distinguishes a body's cells from foreign matter (p. 66)

cellular respiration: process by which living things obtain energy from the bonds of food molecules; from the Latin *respirare,* meaning "to breathe" (p. 84)

cell wall: thick, outer layer that encloses and supports most plant, fungal, and bacterial cells and some protist cells (p. 54)

central nervous system: brain and spinal cord (p. 611)

cephalization (sehf uh lih ZAY shuhn): evolution of a definite head end; from the Greek *kephalikos,* meaning "head" (p. 442)

cerebellum (sehr uh BEHL uhm): small structure in the rear part of the brain that controls balance, posture, and coordination; from the Latin *cerebellum,* diminutive for "brain" (p. 612)

cerebrum (suh REE bruhm): large rounded area of the brain divided into left and right hemispheres; from the Latin *cerebrum,* meaning "brain" (p. 612)

cervix: ring of strong, flexible muscles forming the lower part of the uterus; from the Latin *cervix,* meaning "neck" (p. 724)

chelicera (kuh LIHS uh ruh): one of the first pair of appendages of arachnids and their relatives; used for feeding, and in some species as poison-delivering fangs; from the Greek *chēlē,* meaning "claw" (p. 501)

chemical bond: attractive force that holds two atoms together in a molecule; from the Latin *alchymia,* meaning "alchemy" (p. 27)

chemical reaction: the forming and breaking of covalent bonds (p. 29)

chemosynthesis: production of food using the energy contained in inorganic molecules; from the Greek *syn-* + *tithenai,* meaning "together" + "to place" (p. 335)

chitin (KY tihn): polysaccharide that makes up the tough cell walls of fungi and the exoskeletons of insects and other arthropods (p. 374)

chlamydia (KLU mih dee uh): common sexually-transmitted disease in the United States; caused by the *Chlamydia* bacterium and with symptoms similar to gonorrhea (p. 740)

chlorofluorocarbon (KLAWR oh FLOH roh KAHR buhn): any of a group of compounds that contain carbon, chlorine, and fluorine; often used as coolants, propellants, or foaming agents (p. 292)

chlorophyll (KLAWR uh fihl): the major pigment in plants that helps gather energy for photosynthesis; from the Greek *chlōros* and *phyllon,* meaning "pale green" and "a leaf" (p. 87)

chloroplast (KLAWR uh plast): carbohydrate-producing organelles found in green plants and protists; from the Greek *chlōros,* meaning "pale green" (p. 52)

chorion (KAWR ee ahn): watertight protective membrane surrounding the embryos of reptiles, birds, and mammals; from the Greek *chorion,* meaning "fetal membrane" (p. 473)

chromatin (KROH muh tihn): a threadlike complex of DNA and proteins in the nucleus of a nondividing cell (p. 101)

chromosome (KROH muh sohm): compact structure of tightly coiled DNA that forms prior to mitosis; from the Greek *chrōma,* meaning "color" (pp. 33, 101)

cilia (SIL ee uh): hairlike projections on certain cells and mobile organisms; from the Latin *celare,* meaning "to conceal" (p. 358)

circulatory system: network of blood-carrying vessels that transports nutrients, oxygen, wastes, and carbon dioxide; from the Latin *circulari,* meaning "to form a circle" (p. 446)

cladogram (KLAD uh gram): branching diagram showing evolutionary relationships among a set of organisms; from the Greek *klados,* meaning "branch" (p. 321)

class: collection of similar orders; from the Latin *classis,* meaning "class" or "division" (p. 318)

cloning (KLOHN ihng): growing a large number of genetically identical cells from one cell; from the Greek *klōn,* meaning "twig" (p. 154)

codon (KOH dahn): three-nucleotide sequence in mRNA specifying a particular amino acid; basic unit of genetic code; from the Latin *code,* meaning "book" (p. 143)

coelacanth (SEE luh kanth): any member of a group of lobe-finned bony fishes; from the Greek *koilos* and *akantha,* meaning "hollow" and "thorn" (p. 531)

coelom (SEE luhm): fluid-filled body cavity that separates the muscles of the body wall from the muscles that surround the gut (p. 446)

coevolution (KOH ev uh LOO shuhn): situation in

which two or more species evolve in response to each other; from the Latin *co-* + *evolvere,* meaning "with" + "to roll" (p. 273)

commensalism (kuh MEN suhl ihz uhm): ecological relationship between two organisms in which one benefits and the other neither benefits nor is harmed; from the Latin *com-* + *mensa,* meaning "with" + "table" (p. 277)

community: all the organisms that live in an ecosystem (p. 254)

competition: situation in which two or more organisms attempt to use the same scarce resource (p. 279)

competitive exclusion: process in which one species is outcompeted and dies out within an ecosystem (p. 279)

complete metamorphosis: series of abrupt changes in the form of young insects as they develop into adults; stages include larva, nymph, pupa (or chrysalis), adult (p. 509)

composting (KOHM pohst ihng): the process of mixing organic debris and allowing it to begin decomposition to make an organic fertilizer (p. 423)

cone: 1. in plants, clusters of scales or modified leaves that produce gametophytes (p. 385); 2. receptor cell in the retina of the eye that detects color, produces sharp images, and detects bright light; from the Greek *kōnos,* meaning "cone" (p. 624)

conjugation: the transfer of genetic material from one bacterium to another; from the Latin *conjugatus,* meaning "to join together" (p. 332)

connective tissue: bones, blood, ligaments and tendons; from the Latin *texere,* meaning "to weave" (p. 576)

consumer: organism that obtains its energy by eating other organisms; from the Latin *consumere,* meaning "to take up" (p. 256)

continental drift: movement of the continents over geologic time (p. 202)

contour feather: feathers that cover the body of a bird, giving shape to its wings and tail (p. 553)

control experiment: a comparison of an experimental group with a control group (p. 12)

control group: the group not exposed to the test condition (p. 12)

convergent evolution: process whereby organisms that are not related to each other evolve similar features (p. 320)

copepod (KOH puh pahd): very small, sometimes parasitic crustacean living in either salt water or fresh water; from the Greek *kōpē,* meaning "oar" (p. 518)

cotyledon (kaht uh LEED uhn): leaflike structure of a plant embryo that stores food or helps absorb food stored elsewhere; from the Greek *kotylēdon,* meaning "cavity" (p. 384)

covalent bond: chemical bond formed by sharing of electrons; from the Latin *co-* + *valere,* meaning "with" + "to be strong" (p. 27)

crop rotation: alternating cultivation of two or more crops in the same field (p. 422)

crossing-over: exchange of genetic material between paired homologous chromosomes during meiosis (p. 112)

cross-pollination: transfer of pollen from the anther of a flower to the stigma of a flower on another plant of the same species (p. 402)

cuticle (KYOOT ih kuhl): waxy, waterproof layer that covers the surface of most plants; from the Latin *cuticula,* meaning "skin" (p. 381)

cyanobacteria (SY ahn oh bak TIHR ee ah): group of photosynthetic bacteria thought to be the first organisms to release oxygen; from the Greek *cyano-* + *bactērion,* meaning "blue" + "small staff" (p. 196)

cytokinesis (syt oh kuh NEE sihs): division of the cytoplasm following mitosis; from the Greek *kytos* and *kinēsis,* meaning "a hollow" and "motion" (p. 105)

cytoplasm (SYT uh plaz uhm): material inside the cell membrane containing the necessary components for cell life; from the Greek *plassein,* meaning "to form" (p. 50)

cytosol (SYT uh sohl): liquid portion of the cytoplasm (p. 51)

decapod (DEHK uh pahd): a crustacean with 10 legs, such as a lobster or a shrimp; from the Greek *deka* and *pous,* meaning "ten" and "foot" (p. 517)

decomposer: organism that obtains its energy by consuming organic wastes; from the Latin *de-* + *componere,* meaning "from" + "to put together" (p. 256)

dendrite (DEHN dryt): short branch of a nerve cell that carries impulses toward the cell body; from the Greek *dendron,* meaning "tree" (p. 601)

derived character: a unique characteristic, such as hair, that defines a group of organisms (p. 321)

dermis: layer of skin below the epidermis; from the Greek *derma,* meaning "skin" (p. 578)

deuterostome (DOOT uhr oh stohm): animal in which the first opening in the embryo does not form the mouth, but often becomes the anus; from the Greek *deuteros* and *stoma,* meaning "second" and "mouth" (p. 451)

development: process of growth and cell specialization (p. 17)

diabetes mellitus (deye uh BEET eez muh LYT uhs): a chronic disease in which cells are unable to absorb glucose from blood and tissue fluids; from the Latin *diabetes* and *mellitus,* meaning "siphon" and "honeyed" (p. 645)

diffusion (dif FYOO zhuhn): movement of molecules from a region of higher concentration to one of lower concentration; from the Latin *diffusus,* meaning "poured in different directions" (p. 67)

diploid (DIH ployd): describes cells that have two sets of chromosomes; from the Greek *diploos,* meaning "double" (p. 103)

direct development: developmental pattern in arthropods in which the young are miniature copies of adults (p. 509)

directional selection: situation in which selection for or against an allele in a population is unopposed (p. 183)

divergence: accumulation of differences between two or more species or populations; from the Latin *divergere,* meaning "to turn apart" (p. 185)

diversity: measure of the number of species living in an ecosystem (p. 254)

DNA fingerprint: a pattern of dark bands that reflects the composition of an organism's DNA (p. 166)

dominant: describes the allele that is expressed as a trait (p. 119)

double fertilization: fertilization in angiosperms in which one sperm fuses with an egg and a second sperm fuses with two nuclei (p. 403)

double helix: spiral of two strands, as in DNA structure; from the Greek *helissein,* meaning "to turn around" (p. 139)

down feather: fine feathers that grow underneath contour feathers and that are specialized for insulation (p. 553)

ecological race: a population of a species that differs genetically from other populations of the same species because they have adapted to different environments (p. 184)

ecology: study of relationships of organisms to their environment; from the Greek *oikos* and *logos,* meaning "house" and "study of" (p. 254)

ecosystem: self-sustaining collection of organisms and their physical environment (p. 254)

ectoderm (EHK tuh durm): outer layer of cells of an embryo; from the Greek *ektos* and *derma,* meaning "outside" and "skin" (p. 441)

ectopic pregnancy: pregnancy in which the embryo grows somewhere other than in the uterus; from the Greek *ektopos,* meaning "out of place" (p. 740)

ectotherm (EHK tuh thurm): animal that uses heat from the environment to control its body temperature; from the Greek *thermē,* meaning "heat" (p. 208)

ejaculation (ee jak yoo LAY shuhn): expulsion of sperm from the male's body; from the Latin *ejaculari,* meaning "to throw out" (p. 721)

electron: elementary particle with negative electric charge; from the Greek *elektron,* meaning "amber" (p. 26)

element: matter that is made of only one kind of atom; from the Latin *elementum,* meaning "first principle" (p. 26)

embryo: fertilized egg that has begun to divide; from the Greek *en-* + *bryein,* meaning "in" + "to swell" (p. 726)

emphysema (em fuh SEE muh): condition of the lung in which it has lost its elasticity, greatly reducing the efficiency of gas exchange, often caused by smoking; from the Greek *emphysēma,* meaning "inflation" (p. 674)

endocrine (EN duh krihn) **gland**: an organ that is part of the endocrine system, which produces most of the body's hormones and which releases them without ducts directly into the bloodstream; from the Greek *krinein,* meaning "to separate" (p. 631)

endocytosis (ehn doh seye TOH sihs): process in which a cell surrounds and engulfs a particle; from the Greek *kytos,* meaning "hollow" (p. 72)

endoderm (EHN doh durm): inner layer of cells of the embryo; from the Greek *endon* and *derma,* meaning "within" and "skin" (p. 441)

endoplasmic reticulum (ehn duh PLAZ mihk rih TIK yuh luhm): organelle that consists of a network of membranes that transports substances made by the cell; from the Greek *plassein* and Latin *reticulum,* meaning "to form" and "network" (p. 52)

endoskeleton: internal supporting structure, such as bones; from the Greek *skeletos,* meaning "dried up" (p. 452)

endotherm (EHN doh thurm): animal that uses heat produced by metabolism to control its body temperature; from the Greek *thermē,* meaning "heat" (p. 208)

enzyme: protein that acts as a catalyst for a biochemical reaction; from the Greek *en-* + *zyme,* meaning "in" + "leaven" (p. 79)

epidermis: 1. in plants, the outermost layer, or layers, of cells that cover the plant body (p. 394); 2. outermost layer of the skin, consisting of several layers of cells; from the Greek *epi-* + *derma,* meaning "upon" + "skin" (p. 578)

epididymis (ehp uh DIHD ih mihs): duct on top of and behind each testis that temporarily stores sperm; from the Greek *didymos,* meaning "testicle" (p. 720)

epithelial (ehp uh THEE lee uhl) **tissue**: tightly connected cells, arranged in flat sheets, such as the skin (p. 575)

esophagus (ih SAHF uh guhs): muscular tube from the pharynx that permits food to descend to the stomach; from the Greek *oisein* and *phagein,* meaning "to be going to carry" and "to eat" (p. 706)

essential amino acid: the nine amino acids needed by, but not produced by, the animal body and supplied by a mixed diet; in humans, one of nine specific compounds containing nitrogen (p. 415)

eubacteria (YOO bak TIHR ee ah): group of bacteria thought to have evolved more recently than the archaebacteria; from the Greek *eu-* + *baktērion,* meaning "true" + "small staff" (p. 195)

eukaryote (yoo KAR ee oht): organisms whose cells contain a membrane-bound nucleus and membrane-bound organelles, such as plants, animals, fungi, and protists; from the Greek *karyon* and *-otes,* meaning "nut (kernel)" and "inhabitant" (p. 50)

evolution (ehv uh LOO shuhn): genetic change in a species over time (p. 17)

exhalation: exit of air from the lungs during breathing, equalizing air pressure (p. 672)

exocytosis (ehk soh sye TOH sihs): process in which a cell discharges substances or waste; from the Greek *kytos,* meaning "hollow" (p. 72)

exon: portions of DNA that code for proteins (p. 148)

exoskeleton: supporting structure that surrounds the body; from the Greek *skeletos,* meaning "dried up" (p. 451)

experiment: the controlled test of a hypothesis; from the Latin *experimentum,* meaning "a trial, test" (p. 12)

external fertilization: fertilization that takes place outside the body of either parent (p. 472)

facilitated diffusion: transport of substances through channels in either direction through the cell membrane; from the Latin *facere* and *diffusus,* meaning "to do or make" and "poured in different directions" (p. 69)

fallopian tube: tube that carries eggs released from the ovaries to the uterus; named for Gabriel Fallopius, Italian anatomist (p. 723)

family: group of closely related genera (p. 318)

feedback: a system of self-regulation (p. 634)

feedback inhibition: slowing or stopping of an early reaction in a series of biochemical reactions when levels of the end product become high (p. 94)

fermentation: the incomplete breakdown of organic compounds in the absence of oxygen; from the Latin *fermentum,* meaning "leavening, yeast" (p. 91)

fertility: a soil's ability to supply plants with nutrients (p. 422)

fertilization: the fusion of gametes (p. 728)

fertilizer: substance that adds nutrients to soil (p. 423)

fetus: the embryo from the ninth week of development until birth (p. 729)

"fight or flight" reaction: reaction produced by hormones of the adrenal medulla in response to certain stresses such as fear, anger, pain, excitement, or physical exertion, enabling the body to respond to emergencies; long-term reactivity is associated with stress-related disorders and disease (p. 639)

filtration: first stage of urine formation, in which smaller solutes are forced out of the blood (p. 711)

flagellum (fluh JEHL uhm): whiplike structure of some cells, used for locomotion; from the Latin *flagellum,* meaning "whip" (p. 333)

follicle (FAHL ih kuhl): in reproductive biology, a fluid-filled chamber within the ovary in which an egg matures; from the Latin *folliculus,* meaning "small bag" (p. 726)

fracture: broken bone; from the Latin *fractus,* meaning "broken" (p. 586)

fruit: enlarged ovary of a flowering plant that contains seeds; from the Latin *fructus,* meaning "enjoyment" (p. 403)

fundamental niche: total niche that an organism could potentially use within an ecosystem (p. 278)

gamete (GAM eet): a haploid reproductive cell, such as an egg or sperm cell; from the Greek *gametē,* meaning "a wife" (p. 110)

gametophyte (guh MEET uh fyt): in a plant's life cycle, the haploid plant (stage) that produces gametes and grows from a spore; from the Greek *gamein* and *phyein,* meaning "to marry" and "to grow" (p. 381)

gastrulation (gas troo LAY shuhn): process of forming layers of cells in the embryo; from the Greek *gastēr* meaning "stomach" (p. 441)

gene: segment of DNA that codes for a trait (p. 7)

gene expression: use of genetic information in DNA to make proteins (p. 141)

genetic code: the correspondence between nucleotide triplets in DNA and the amino acids in proteins; from the Greek *gignesthai* and the Latin *code,* meaning "to be born" and "book" (p. 143)

genetic disorder: the harmful effects produced by certain mutations (p. 128)

genetic engineering: the process of moving genes from the chromosomes of one organism to those of another (p. 153)

genetics (juh NEHT ihks): science of biological inheritance (p. 117)

genotype (JEE nuh typ): constitution of an organism in terms of its genes; from the Greek *genos,* meaning "race, kind" (p. 119)

genus (JEE nuhs): category in biological classification in which closely related species are grouped; from the Latin *genus,* meaning "race, kind" (p. 314)

germination: to sprout or start to grow; from the Latin *germinare,* meaning "to sprout" (p. 405)

gill: one of the structures used by aquatic animals for gas exchange (p. 464)

global warming: the increase in global temperatures (p. 293)

glycolysis (gly KAHL uh sihs): process that breaks down glucose in a cell's cytoplasm and releases a small amount of energy; from the Greek *glykys* and *lysis,* meaning "sweet" and "dissolving" (p. 90)

goiter (GOY tuhr): swollen thyroid gland caused by lack of iodine; from the Latin *guttur,* meaning "throat" (p. 642)

Golgi (GOHL jee) **apparatus**: organelle that delivers substances released by the endoplasmic reticulum to specific places in a cell; named after Camillo Golgi, Italian neurologist (p. 52)

gradualism: hypothesis that evolution occurs at a slow, constant rate (p. 186)

grain: small seedlike fruit of cereal grasses; from the Latin *granum,* meaning "seed, kernel" (p. 411)

Gram staining: use of dyes to stain bacterial cell walls to aid in identification and diagnosis; named after Hans Christian Gram, Danish physician (p. 332)

greenhouse effect: warming of Earth's atmosphere resulting from heat trapped by carbon dioxide and other gases (p. 262)

green manure: crop that is grown to be plowed back into the soil to add organic matter (p. 422)

ground tissue: tissue that makes up the bulk of a plant's body, some parts of which are for storage and the rest for strength (p. 394)

guard hair: long, thick outer hair responsible for the color of an animal's coat (p. 564)

gut: internal passage through which food passes while being digested (p. 440)

gymnosperm (JIHM noh spurm): seed plant that does not produce its seeds within a fruit; from the Greek *gymnos* and *speirein,* meaning "naked" and "to sow" (p. 385)

habitat: physical location where an organism lives in an ecosystem; from the Latin *habitare,* meaning "to inhabit" (p. 254)

half-life: the time it takes for one half of the radioactive atoms in a sample to decay (p. 176)

haploid (HAH ployd): describes cells that have only one set of chromosomes; from the Greek *haploos,* meaning "single" (p. 103)

hemodialysis (hee moh deye AL uh sis): process of removing blood from an artery, purifying it by dialysis, and returning it to a vein; from the Greek *haima* and *dialyein,* meaning "flowing blood" and "to dissolve" (p. 712)

hemoglobin (HEE moh gloh bihn): an iron-containing molecule in red blood cells that gives blood its color and which carries oxygen and carbon dioxide (p. 652)

herbivore (HUR buh vawr): consumer that eats only plants; from the Latin *herba* and *vorare,* meaning "grass" and "to devour" (p. 257)

heredity (huh RED ih tee): transmission of genetic information from parent to offspring (p. 17)

hermaphrodite (huhr MAHF roh deyet): animal with both female and male sexual organs; from the Greek myth in which Hermaphroditus shared a body with a nymph (p. 483)

heterotroph (HET uhr oh trohf): organism that cannot make its own food and feeds on other organisms or organic wastes; from the Greek *hetero-* + *trophē,* meaning "different" + "food" (p. 257)

heterozygous (heht uh roh ZY guhs): having two different alleles for a trait; from the Greek *zygon,* meaning "yoke" (p. 119)

histamine (HIHS tuh meen): chemical messenger that can increase mucus production and stimulate capillaries to swell and release fluids in response to the presence of an antigen (p. 688)

HIV: Human Immunodeficiency Virus, the virus that causes AIDS and which attacks and destroys helper T-cells (p. 345)

homeostasis (hoh mee oh STAY sihs): tendency to maintain stability in an organism amid environmental change; from the Greek *homo-* + *histanai,* meaning "similar" + "to stand" (p. 16)

hominid (HAHM ih nihd): two-legged primates such as humans and australopithecines; from the Latin *homo,* meaning "man" (p. 218)

homologous (hoh MAHL uh guhs) **structures**: structures that have a common ancestry; from the Greek *homologos,* meaning "agreeing" (p. 178)

homozygous (hoh muh ZY guhs): having identical alleles for a trait; from the Greek *zygon,* meaning "yoke" (p. 119)

hormone: chemical signal, secreted by specialized glands, that travels through the bloodstream and regulates the body's activities; from the Greek *hormeinō,* meaning "to stimulate" (p. 631)

human genome: entire collection of genes within human cells (p. 166)

humus (HYOO muhs): organic component of soil (p. 421)

hybrid: offspring that results from interbreeding between members of different species (p. 323)

hydrogen bond: linkage formed when a hydrogen atom in one molecule makes an additional bond to an atom in another molecule; from the Greek *hydōr* + *genēs,* meaning "water" + "born" (p. 44)

hydroponics: cultivation of plants in nutrient solutions instead of in soil; from the Greek *ponein,* meaning "to toil" (p. 423)

hypertension: condition in which the blood pressure is continually higher than normal; from the Greek *hyper* + the Latin *tensus,* meaning "above, over" + "stretched" (p. 669)

hyperthyroidism: overproduction of thyroxine with overstimulation of the thyroid gland (p. 641)

hypha (HY fuh): one of the slender filaments that make up the body of a fungus; from the Greek *hyphē,* meaning "web" (p. 374)

hypothalamus (heye poh THAL uh muhs): part of the upper end of the brain stem that regulates activities in the nervous and endocrine systems; from the Greek *hypo-* + *thalamos,* meaning "under" + "inner chamber" (p. 613)

hypothesis: a testable explanation for an observation; from the Greek *hypotithenai,* meaning "to put under" (p. 12)

hypothyroidism: production of too little thyroxine (p. 642)

immune response: immune system's defensive action to protect itself against a specific pathogen or antigen; from the Latin *in* + *munus,* meaning "in" + "service" (p. 683)

immunity: resistance to disease either natural or acquired (p. 687)

immunological surveillance: the constant patrol of the immune system, watching for surface changes on the body's cells (p. 690)

implantation: attachment of the embryo to the uterine wall within a few days after fertilization (p. 729)

imprinting: a type of learning that enables an animal to recognize members of its own species (p. 236)

incomplete metamorphosis: in arthropods, the gradual transformation into adult form (p. 509)

inducer: molecule that binds to the repressor protein, thereby enabling DNA to be transcribed; from the Latin *inducere,* meaning "to lead into" (p. 146)

inflammatory response: the body's response to injury, indicated by redness, swelling, and fever; from the Latin *flammare,* meaning "to flame" (p. 680)

inhalation: entry of air into the lungs during breathing, equalizing air pressure (p. 672)

insulin: a peptide hormone that interacts with a receptor at the cell membrane and helps glucose cross the membrane and enter the cytoplasm of a cell (p. 644)

internal fertilization: fertilization that occurs inside the female's body (p. 473)

introns: noncoding portions of the DNA (p. 148)

ion: an atom that has lost or gained one or more electrons and therefore has a net electric charge; from the Greek *ienai,* meaning "to go" (p. 27)

ionic bond: force of attraction between oppositely charged ions (p. 27)

joint: place where two or more bones connect; from the Latin *junctus,* meaning "joined" (p. 589)

karyotype (KAH ree oh typ): array of the chromosomes found in an individual's cells arranged in order of size and shape; from the Greek *karyon,* meaning "a nut, kernel" (p. 103)

keystone species: species whose niche affects many other species in the ecosystem so that it cannot be readily replaced if lost (p. 281)

kidney: one of two organs located toward the back of the abdominal cavity that maintains homeostasis, removes wastes, regulates water in the blood, and adjusts concentrations of other substances in the blood (p. 711)

kin selection: a behavioral mechanism in which an individual helps relatives (kin) reproduce rather than producing its own offspring (p. 245)

kingdom: collection of similar phyla (p. 318)

labor: second stage of pregnancy, when the uterus contracts strongly in order to push the baby out of the mother's body (p. 736)

large intestine: organ of the digestive system that stores, compacts, and then eliminates indigestible material; from the Latin *intestinus,* meaning "inward" (p. 709)

lateral line: row of pressure-sensitive cells within a fluid-filled canal on each side of a fish; from the Latin *latus,* meaning "side" (p. 462)

learning: the development of behaviors through experience (p. 236)

legume (LEHG yoom): member of a family of plants that includes peas and beans; from the Latin *legumen,* meaning "that which can be gathered" (p. 415)

leukocyte (LOO koh syt): white blood cell, called the "soldier" of the immune system (p. 680)

lichen (LY kuhn): mutualistic relationship between a fungus and an alga; from the Greek *leichein,* meaning "to lick" (p. 378)

ligament: strong, elastic bands of connective tissue joining bones together; from the Latin *ligare,* meaning "to bind" (p. 589)

lipid (LIHP ihd): class of organic compounds that includes fats and waxes that organisms may use to store energy; from the Greek *lipos,* meaning "fat" (p. 31)

lipid bilayer: the double layer of phospholipids found in cell membranes and organelles (p. 46)

liver: large glandular organ that secretes bile and modifies substances contained in the blood; metabolizes carbohydrates, fat, and proteins, and removes toxins from the blood (p. 708)

loam: soil that is a mixture of sand, silt, and clay (p. 421)

lobe-finned fish: a group of fishes having fleshy fins supported by a series of bones (p. 531)

lumber: wood from trees that has been cut down and sawed into boards and planks (p. 417)

lung cancer: cancer in the lung (p. 674)

lymphatic system: vessels that return to the circulatory system excess fluids that bathe the cells; from the Latin *lympha,* meaning "spring water" (p. 657)

lymphocyte (LIM foh syt): specialized white blood cell of the immune system formed in the lymphatic system; from the Latin *lympho* + *kytos* meaning "spring water" + "hollow" (p. 680)

macromolecule (MAHK roh MAWL uh kyool): large molecule, such as DNA or proteins, composed of several simple structural units; from the Greek *makros* and the Latin *moles,* meaning "long" and "mass" (p. 29)

macronutrient: nutrient that plants and animals need in relatively large amounts (p. 420)

macrophage (MAHK roh fayj): white blood cell which recognizes and destroys foreign invaders of the animal body; from the Greek *makros* + *phagein,* meaning "long" + "to eat" (p. 681)

magnification: a microscope's ability to make an object appear larger (p. 56)

Malpighian (mal PIHG ee uhn) **tubule**: in insects and spiders, organ responsible for filtering blood; named after Italian physiologist Marcello Malpighi (p. 471)

mammary gland: gland of female mammals that secretes milk for suckling the young; from the Latin *mamma,* meaning "breast" (p. 559)

mandible: jaw of certain animals; from the Latin *mandibulum,* meaning "jaw" (p. 506)

marrow: spongelike center of long bones where blood cells are produced (p. 585)

marsupial (mahr SOO pee uhl): mammal that lacks a placenta and has an external pouch for young; from the Greek *marsypos,* meaning "pouch" (p. 560)

mass extinction: an episode during which large numbers of species became extinct (p. 197)

medusa: one of the two life stages of cnidarians, the jellyfish stage; from Greek mythology, a snake-haired beast-woman called Medusa (p. 481)

meiosis (my OH sihs): a form of cell division that reduces the number of chromosomes in a cell by half; from the Greek *meioun,* meaning "to make smaller" (p. 110)

melanin (MEHL uh nihn): brownish-colored pigment of skin cells; from the Greek *melas,* meaning "black" (p. 581)

melanoma (mehl uh NOH muh): cancer arising from the melanin-producing cells, which are located in the lowest epidermal layer (p. 583)

memory cell: B cell or T-cell that remains after infection and continues to patrol the body's tissues (p. 686)

menopause: shutdown of menstrual and ovarian cycles naturally in middle age or earlier by surgery; from the Latin *mens,* meaning "month" (p. 727)

menstrual (MEHN struhl) cycle: the series of changes occurring in the uterus and ovaries each month (p. 725)

meristem (MEHR uh stehm): region of actively dividing cells that give rise to all of the other plant tissues; from the Greek *meristos,* meaning "divided" (p. 394)

mesoderm: layer of tissue between the endoderm and ectoderm that develops into muscle, reproductive organs, and circulatory vessels; from the Greek *mesos* and *derma,* meaning "middle" and "skin" (p. 443)

metabolism (muh TAB uh liz uhm): the sum of all the chemical processes that use of release energy in an organism; from the Greek *metabolē,* meaning "change" (p. 78)

metamorphosis (meht uh MAWR fuh sihs): change in form, structure, or function of an animal as a result of development; from the Greek *metamorphē,* meaning "change" (p. 509)

micronutrient: nutrient that plants and animals need in very small amounts (p. 420)

microvillus: any of the hairlike projections coating the villi of the small intestine; from the Greek *mikros* and the Latin *villus,* meaning "small" and "tuft of hair" (p. 708)

mineral: inorganic compound, one not made from living things; from the Latin *minera,* meaning "ore" (p. 702)

mitochondrion (myt uh KAHN dree uhn): energy-producing organelles in eukaryotes; from the Greek *mitos* and *chondros,* meaning "thread" and "grain" (p. 52)

mitosis (my TOH sihs): nuclear division; involves copying and separating chromosomes into two new nuclei; from the Greek *mitos,* meaning "thread" (p. 105)

molecule: group of atoms that form the smallest unit of a substance that can retain its chemical properties; from the Latin *moles,* meaning "mass" (p. 28)

monotreme (MAHN oh treem): mammal that lays eggs and has a single opening for the digestive and urinary tracts and for reproduction; from the Greek *monos* and *trēma,* meaning "single" and "hole" (p. 560)

mucous (MYOO kuhs) membrane: epithelial layer that releases mucus and is impenetrable to most pathogens; from the Latin *mucus,* meaning "slimy secretion" (p. 679)

mulch: bark chips or other organic matter placed on top of soil and used to prevent evaporation of water from the soil around a plant (p. 426)

multicellular (MUHL tee SEHL yoo lar) **organism**: a living thing made of more than one cell (pp. 63, 196)

muscle tissue: tissue made of cells that contract and return to original length; from the Latin *musculum* and *texere,* meaning "little mouse (muscle)" and "to weave" (p. 576)

mutagen (MYOOT uh juhn): environmental agent that can alter the structure of DNA; from the Latin *mutare,* meaning "to change" (p. 140)

mutation (myoo TAY shuhn): abrupt change in genotype of an organism (pp. 17, 109, 127)

mutualism (MYOO choo uhl ihz uhm): ecological

relationship which benefits all organisms involved (pp. 199, 276)

mycorrhizae (MY koh REYE zee): mutualistic association of certain fungi with root cells of some vascular plants; from the Greek *mykes* and *rhiza,* meaning "fungus" and "root" (p. 199)

mycelium (my SEE lee uhm): tangled mass of fungal hyphae that makes up the body of a fungus (p. 374)

myelin (MY uh lihn) **sheath**: a fatty, segmented covering wrapped around axons, which increases the speed of nerve impulse transmissions (p. 604)

natural selection: process by which organisms best suited to their environmental conditions are most likely to survive and reproduce (pp. 17, 175)

negative feedback: a process in which an end product inhibits the first step of a system (p. 634)

nematocyst (NEHM uh toh sihst): stinging, harpoon-like structure on the tentacles and outer body surface of cnidarians; from the Greek *nēma* and *kystis,* meaning "thread" and "sac" (p. 481)

nephridium (nee FRIHD ee uhm): tubule that collects wastes from the coelom and discharges them from the body; from the Greek *nephros,* meaning "kidney" (p. 487)

nephron (NEHF rahn): blood-cleaning unit of the kidney (p. 711)

nerve: bundles of neurons that extend throughout the body (p. 601)

nerve tissue: tissue made up of cells that generate electrical impulses and transfer them to other cells; from the Latin *nervus* and *texere,* meaning "sinew, nerve, string" and "to weave" (p. 576)

neuroendocrine system: term covering the closely-linked nervous system and endocrine system; from the Greek *neuron* and *endon* + *krinein,* meaning "nerve" and "within" + "to separate" (p. 632)

neuron (NOO rahn): basic cell of the nervous system, specialized to carry nerve impulses; from the Greek *neuron,* meaning "nerve" (p. 601)

neurotransmitter: chemical released by a neuron into a synapse that may stimulate or inhibit other neurons (p. 605)

niche (NIHCH): an organism's "profession," the sum of its interactions with its environment (p. 278)

nitrogen fixation: conversion of atmospheric nitrogen to ammonia by bacteria (pp. 159, 260)

nonpolar molecule: molecule with an even charge distribution (p. 45)

nonrenewable resource: natural resource, such as iron ore, that is not regenerated by natural processes (p. 295)

notochord (NOHT uh kawrd): rod of cartilage that extends along the back in chordates; from the Greek *nōton* and Latin *chorda,* meaning "back" and "cord" (p. 204)

nucleic (noo CLAY ihk) acid: DNA or RNA; large chainlike molecule containing phosphoric acid, sugar, and purine and pyrimidine bases; from the Latin *nucleus,* meaning "nut or kernel" (p. 33)

nucleotide (NOO klee oh tyd): structural unit of a nucleic acid (subunit of DNA or RNA) consisting of three parts: a sugar, a phosphate group, and a nitrogenous base (pp. 33, 139)

nucleus (NOO klee uhs): organelle that directs cell activities and stores DNA; from the Latin *nucleus,* meaning "nut or kernel" (p. 50)

nutrition: science that studies how food affects the functions of living organisms (p. 699)

nymph (NIHMF): stage in the development of an insect, which is a small but incomplete version of the adult; from the Greek *nymphe,* meaning "a young goddess" (p. 509)

obesity (oh BEES ih tee): condition in which a high proportion of the body weight is stored fat; from the Latin *obesus,* meaning "devoured" (p. 697)

omnivore (AHM nih vawr): organism that eats both plants and animals; from the Latin *omni-* + *vorare,* meaning "all" + "to devour" (p. 257)

opposable thumb: digit that can be bent toward other digits (p. 213)

order: collection of similar families of organisms (p. 318)

organ: collection of different kinds of tissue that are dedicated to one or a few functions; from the Latin *organum,* meaning "instrument" or "tool" (pp. 443, 577)

organelle (awr guh NEHL): specialized compartments that carry out specific functions in eukaryotic cells (p. 51)

organic molecule: molecule with carbon-carbon bonds (p. 29)

organ system: group of interrelated organs that carry out one or a few essential body functions (pp. 443, 577)

osmosis (ahz MOH sihs): movement of water through a membrane from an area of high concentration to an area of low concentration; from the Greek *othein,* meaning "to push" (p. 68)

osmotic (ahz MAH tihk) pressure: pressure created by force of water pushing against a cell membrane (p. 68)

osteoporosis (ahs tee oh puh ROH sihs): condition of bone resulting from loss of bone minerals that make bone dense, causing brittleness and fractures; from the Greek *osteon* and Latin *porosis,* meaning "bone" and "porous condition" (p. 586)

ovary: one of the two female organs that produce eggs (p. 723)

overconsumption: eating too much food, a major factor leading to nutritional problems in the United States (p. 699)

ovulation (ahv yoo LAY shun): release of an egg from the ovary; from the Latin *ovum,* meaning "egg" (p. 725)

ovule (AHV yool): structure within a seed plant that contains the female gametophyte and becomes a seed when mature (p. 400)

ovum: female gamete, or egg (plural, ova) (p. 723)

oxidative (AHKS ih day tihv) **respiration**: series of chemical reactions that occur in mitochondria, and the process by which cells get most of their energy; from the Latin *respirare,* meaning "to breathe" (p. 90)

ozone: unstable pale-blue gas that absorbs ultraviolet radiation; from the Greek *ozein,* meaning "to smell" (p. 198)

pacemaker: small bundle of cells near the entrance to the right atrium of the heart that emits an electrical signal to start the heartbeat (p. 659)

pancreas (PAN kree uhs): large gland situated behind the stomach that secretes digestive juices into the small intestine and that produces insulin and glucagon; from the Greek *pan-* + *kreas,* meaning "all" + "flesh" (p. 708)

parasitism: ecological relationship in which one organism lives on or in another organism and absorbs nutrients from it (p. 276)

pathogen: disease-causing agent; from the Greek *pathos,* meaning "suffering" (p. 338)

pedigree (PEHD uh gree): record that tracks the inheritance of a trait through several generations of a family (p. 130)

pedipalp: second pair of appendages on spiders and other arachnids; used to capture and manipulate prey and for courtship displays and mating; from the Latin *pes* and *palpus,* meaning "foot" and "soft palm of the hand" (p. 501)

penicillin (pen ih SIHL ihn): antibiotic compound obtained from the mold *Penicillium* (p. 343)

pepsin: enzyme present in the stomach that, with hydrochloric acid and other enzymes, breaks proteins into small chains of amino acids; from the Greek *pepsis,* meaning "digestion" (p. 707)

peptide hormone: hormone made of amino acids that send messages from outside the cell; from the Greek *peptein* and *hormeinō* meaning "to digest" and "to stimulate" (p. 638)

peripheral nervous system: the part of the nervous system that carries messages between the central nervous system and the rest of the body (p. 611)

peristalsis (pehr uh STAHL sihs): wavelike muscle contractions that propel food through the digestive tract; from the Greek *peristaltikos,* meaning "surrounding" (p. 707)

pH: term used to describe the acidity of a solution (p. 291)

phagocyte (FAG oh syt): type of white blood cell that consumes pathogens; from the Greek *phagein,* meaning "to eat" (p. 680)

phenotype (FEE nuh typ): physical characteristics of an organism; from the Greek *phainein,* meaning "to show" (p. 119)

phloem (FLOH ehm): vascular tissue of plants that transports sugars and other dissolved nutrients; from the Greek *phloos,* meaning "bark" (p. 394)

phospholipid (fahs foh LIHP ihd): a molecule made of two nonpolar fatty acid chains joined to a short polar phosphorus-containing head (p. 46)

photosynthesis (foh toh SIHN thuh sihs): process by which organisms use the energy of light to convert inorganic molecules into organic molecules; from the Greek *synthenai,* meaning "to put together" (p. 84)

phylum (FEYE luhm): any of the broad, principal divisions of a kingdom; from the Greek *phylon,* meaning "tribe" (pp. 197, 318)

phytoplankton (feye tuh PLANK tuhn): photosynthetic plankton, usually found near the surface of a body of water; from the Greek *phyein* and *planktos,* meaning "to grow" and "wandering" (p. 362)

pigment: molecule that absorbs and reflects different colors of light; from the Latin *pingere,* meaning "to paint" (p. 86)

pistil: female part of a flower that produces ovules, which contain the female gametophyte; from the Latin *pistillum,* meaning "pestle" (p. 400)

pituitary (puh TOO uh tehr ee) **gland**: small endocrine gland that produces at least six different hormones and is attached by a stalk to the base of the brain; from the Latin *pituita,* meaning "phlegm" (p. 633)

placenta (pluh SEHN tuh): structure through which nutrients and oxygen are transferred to the fetus and which blocks some harmful agents from entering the fetal blood supply; from the Greek *plakous,* meaning "flat cake" (p. 474)

placental (pluh SENT uhl) **mammal**: mammal with a placenta, through which offspring are nourished until birth (p. 560)

plankton: passively floating or weakly mobile aquatic algae, bacteria, and animals; from the Greek *planktos,* meaning "wandering" (p. 263)

plant breeding: practice of producing new varieties of plants through artificial selection (p. 430)

plasma (PLAZ muh): protein-rich fluid that makes up 55 percent of the blood; from the Greek *plassein,* meaning "to form" (p. 652)

plasmid (PLAZ mihd): small, circular DNA molecules found in cytoplasm of bacteria (p. 155)

platelet: fragment of a cell that is formed within bone marrow and helps form blood clots; from the Greek *platys,* meaning "broad, flat" (p. 653)

polar molecule: molecule in which charge is unevenly distributed (p. 44)

pollen grain: male gametophyte of a seed plant; from the Greek *palē,* meaning "dust" (p. 385)

pollination: the transfer of pollen from male to female flower structures (p. 385)

pollution: addition of harmful substances to the environment; from the Latin *polluere,* meaning "to soil" (p. 289)

polyp: one of the two life stages of cnidarians, generally attached to a hard surface; from the Greek *poly-* + *pous,* meaning "many" + "foot" (p. 481)

prediction: expected outcome if a hypothesis is proved to be accurate; from the Latin *praedicere,* meaning "to foretell" (p. 12)

primary growth: growth that results in the lengthening of a plant's roots and shoots (p. 394)

primary immune response: immune response that occurs after the first exposure to a pathogen (p. 686)

primary succession: regular pattern of species replacement occurring where nothing has lived before (p. 280)

primate: order of mammals to which people, apes, monkeys, and prosimians belong; from the Latin *primus,* meaning "first" (p. 213)

probability: the likelihood that something will happen (p. 121)

producer: organism that is able to make its own food (p. 256)

prokaryote (pro KAR ee oht): single-celled organism lacking a true nucleus; from the Greek *pro-* and *karyon* and *otes,* meaning "before" and "nut or kernel" and "inhabitant" (p. 50)

prosimian (proh SIHM ee uhn): member of a group of tree-dwelling primates; from the Latin *pro* + *simia,* meaning "before" + "ape" (p. 214)

prostaglandins (pras tuh GLAN dihnz): a group of lipids produced in small quantities by specialized cells that function in the body as hormones; from the Greek *proístanai,* meaning "to put in front of" (p. 646)

protein (PROH teen): any of the polymers of amino acids; from the Greek *prōtos,* meaning "first" (p. 32)

proton pump: mechanism in cells that expels protons, which drives cell metabolism (p. 70)

protostome (PROHT uh stohm): animal in which the initial depression that starts gastrulation becomes the mouth of the adult organism; from the Greek *prōtos* and *stoma,* meaning "first" and "mouth" (p. 451)

pseudocoelom (SOO doh see luhm): body cavity between the mesoderm and endoderm of roundworms (nematodes); from the Greek *koilia,* meaning "hollow" (p. 445)

pseudopod (SOO doh pahd): projection of the cell body used for movement or feeding; from the Greek *podos,* meaning "foot" (p. 358)

punctuated equilibria: hypothesis that evolution occurs at an irregular rate; from the Latin *punctus, aequus,* and *libra,* meaning "point," "equal," and "balance" (p. 186)

purine (PYUR een): one of the two larger nucleotide bases, namely adenine and guanine (p. 139)

pyrimidine (py RIHM uh deen): one of the two smaller nucleotide bases, namely cytosine, thymine (in DNA), and uracil (in RNA) (p. 139)

pyruvic (py ROO vihk) **acid**: three-carbon organic compound; two molecules result from the splitting of a molecule of glucose during glycolysis (p. 91)

radial symmetry: wheel-like symmetry in which body parts radiate from a central axis; from the Latin *radius* and *symmetria,* meaning "spoke, ray" and "measured together" (p. 441)

radioactive dating: technique for measuring the age of an object by determining the ratio of the concentrations of a radioisotope to that of a stable isotope in it (p. 176)

radula (RAJ oo lah): flexible toothy structure in the mouth of mollusks that is used for feeding; from the Latin *radere,* meaning "to scrape" (p. 488)

ray-finned fish: bony fish whose fins are fan-shaped and supported by bony rays (p. 531)

reabsorption: second process of urine formation, in which filtered fluid passes into the nephron, returning important substances to the blood (p. 711)

realized niche: that part of a fundamental niche that an organism actually occupies as a result of competition or predation (p. 279)

receptor protein: cell membrane protein that senses chemical signals and receives other chemicals that "fit" sites on its surface (p. 64)

recessive: describes an allele that is not expressed in heterozygous individuals (p. 119)

recombinant (ree KAHM buh nuhnt) **DNA**: molecule formed when fragments of DNA from two or more different organisms are spliced together (p. 153)

red blood cell: cell containing hemoglobin that carries oxygen from the lungs to all cells of the body and carries carbon dioxide back to the lungs to be exhaled (p. 652)

reflex: sudden, involuntary response to a stimulus; from the Latin *reflexus,* meaning "reflected" (p. 619)

renewable resource: natural resource, such as water or timber, that can be replaced by a natural process (p. 295)

replication: copying of DNA before cell division (p. 140)

repressor protein: protein that blocks transcription by preventing RNA polymerase from moving along the gene; from the Latin *repressus,* meaning "held back" (p. 146)

resistance exercise: exercise that matches the force of a muscle against the force generated by a movable or an immovable object (p. 595)

resolution: the capacity of a microscope to distinguish different objects that are close to one another (p. 56)

restriction enzyme: enzyme that recognizes and binds to specific short sequences of DNA, then cuts the DNA at a specific site within that sequence; from the Greek *en-* + *zyme,* meaning "in" + "leaven" (p. 154)

retina: layer of rods and cones that covers the surface of the back of the eye (p. 624)

ribonucleic acid (RNA): a single-stranded nucleic acid containing the bases adenine, guanine, cytosine, and uracil (p. 141)

ribosome (RY buh sohm): RNA-containing organelle that is the site for protein synthesis (p. 50)

rod: receptor cell in the retina of the eye that is extremely sensitive to light and can detect shades of gray (p. 624)

root cap: thimble-like cluster of cells that covers and protects the tip of a root (p. 391)

root hair: projection from a cell in the epidermis near the end of a root of a plant, greatly increasing the surface area of a root (p. 391)

saturated fat: a type of fat in which the carbon atoms are bonded to as many hydrogen atoms as possible; usually in a solid state at room temperature (p. 31)

science: a way of investigating the world and observing nature in order to form general rules about what causes things to happen; from the Latin *scire,* meaning "to know" (p. 5)

scientific name: the unique two-word name assigned to each kind of organism on earth (p. 313)

secondary growth: growth that results in the thickening of the plant body; in woody plants, growth that produces wood through the production of rings of new xylem and phloem (p. 394)

secondary immune response: immune response to a previously encountered pathogen; usually occurs much faster and produces more antibodies than the first response (p. 686)

secondary succession: regular pattern of species replacement occurring where there has been previous growth (p. 280)

second messenger: molecule that carries information from a hormone outside the cell membrane into the cytoplasm (p. 638)

sedentary lifestyle: too little physical activity, a major factor leading to nutritional problems in the United States (p. 699)

seed: plant embryo that is surrounded by a protective coat (p. 381)

seedling: the young plant that develops from the embryo of a germinated seed (p. 384)

segmentation: characteristic of animals that are composed of repeated body units; from the Latin *segmentum,* meaning "part" (p. 448)

selective transport: process by which only one type of molecule is transported through membrane protein channels (p. 69)

self-pollination: pollination that occurs when pollen grains fall from the anther of a flower to the stigma of the same flower or another flower on the same plant (p. 402)

semen: fluid consisting of sperm mixed with the secretions of several glands, expelled from the penis during ejaculation; from the Latin *semen,* meaning "seed" (p. 721)

sensitivity: the responsiveness of an organ or an organism (p. 23)

sensory neuron: nerve cell that carries nerve impulses from sense organs to the central nervous system (p. 618)

sensory organ: a specialized organ of the body that responds to stimuli by producing nerve impulses in a sensory neuron (p. 621)

sex chromosome: either member of a pair of chromosomes responsible for sex determination (p. 103)

sex hormone: hormone responsible for the development and maintenance of sexual characteristics; from the Greek *hormeinō*, meaning "to stimulate" (p. 643)

sex-linked inheritance: a pattern of heredity in which traits are transmitted by genes located on the X chromosome (p. 129)

sexual selection: an evolutionary mechanism in which traits that enhance the ability of individuals to acquire mates increase in frequency (p. 244)

shoot: the aboveground part of a plant (p. 392)

skeletal muscle: muscle attached to the bones and responsible for movement; from the Latin *musculus*, meaning "little mouse (muscle)" (p. 590)

small intestine: digestive organ that receives partially digested food from the stomach, where it is further broken down and where nutrients are transferred to the blood; from the Latin *intestinus*, meaning "inward" (p. 708)

smooth muscle: muscle of the internal organs that is not under conscious control; from the Latin *musculus*, meaning "little mouse (muscle)" (p. 590)

sodium-potassium pump: active-transport mechanism that transports sodium out of and potassium into cells (p. 70)

species (SPEE sheez): a group of organisms that can interbreed with each other but not with members of any other group; from the Latin *species*, meaning "appearance, kind" (pp. 17, 314)

spicule (SPIHK yool): needlelike, bony structure that forms the skeleton of a sponge; from the Latin *spica*, meaning "point" (p. 479)

spinneret: nozzle-like structure on the abdomen of spiders that directs the flow of silk from glands in the abdomen (p. 502)

spore: reproductive cell capable of developing into a new organism; from the Greek *speirein*, meaning "to sow" (p. 375)

sporophyte (SPOHR uh fyt): in the life cycle of a plant, the diploid plant (stage) that produces spores and results from the fusing of two gametes; from the Greek *phyein*, meaning "to grow" (p. 381)

sprain: injury resulting from a ligament that is overstretched (p. 589)

stability: ability of an ecosystem to resist change in the face of disturbance (p. 281)

stamen (STAY mehn): male part of a flower that produces pollen grains (the male gametophytes); from the Latin *stamen*, meaning "thread" (p. 400)

steroid hormone: hormone synthesized from cholesterol, such as the sex hormones and anterior pituitary hormones testosterone or estrogen; from the Greek *hormeinō*, meaning "to stimulate" (p. 637)

stoma (STOH muh): tiny pore between two guard cells in a plant leaf; from the Greek *stoma*, meaning "mouth" (p. 392)

substrate (SUHB strayt): molecule on which an enzyme acts during a chemical reaction; from the Latin *substernere*, meaning "to strew about" (p. 80)

succession: regular progression of species replacement in a developing ecosystem; from the Latin *succedere*, meaning "to follow" (p. 280)

swim bladder: gas-filled sac in bony fishes that provides buoyancy (p. 529)

symbiosis (sihm beye OH sihs): close, long-term association between two or more species; from the Greek *syn* + *bios*, meaning "together" + "life" (p. 276)

synapse (SIHN ahps): junction of a neuron with another cell; from the Greek *synaptein*, meaning "to fasten together" (p. 605)

target: in the endocrine system, the specific destination (tissue or organ) of a hormone's action (p. 631)

taxonomy: science of classifying living things (p. 317)

T-cell lymphocyte: white blood cell that controls the immune response or attacks infected cells (p. 684)

tendinitis: the painful inflammation of a tendon caused by too much friction or stress (p. 596)

tendon: inelastic, strong, fibrous connective tissue attaching muscles to bones; from the Latin *tendere*, meaning "to stretch" (p. 592)

teratogens (tuh RAT oh jinz): chemicals, diseases, or other agents that cause malformation to a fetus (p. 731)

testes (TEHS tees): oval-shaped organs of the male that produce sperm; from the Latin *testis*, meaning "witness" (p. 720)

theory: explanation of several phenomena based on hypotheses that have been tested many times; from the Greek *thēorein*, meaning "to look at" (p. 13)

tissue: group of similar cells that are organized into a functional unit; from the Latin *texere*, meaning "to weave" (pp. 440, 575)

trachea (TRAY kee uh): the cartilaginous, membranous tube by which air passes to and from the lungs in humans and many vertebrates, also one of the tubes leading into an insect's body cavity from pores (spiracles) in the surface of the exoskeleton (p. 467)

transcription: first stage of gene expression, in which an RNA copy of DNA is made (p. 141)

transformation (trans fawr MAY shuhn): the transfer of genetic material from one organism to another, as when a bacterium takes up foreign DNA; first observed by Griffith (p. 137)

translation: second stage of gene expression, in which three kinds of RNA work together to assemble amino acids into a protein molecule (p. 141)

translocation: transport of organic molecules made by photosynthesis from the leaves or storage tissues throughout the body of a plant (p. 397)

transpiration: loss of water by the leaves and stem of a plant; from the Latin *trans-* + *spirare*, meaning "across + to breathe" (p. 397)

transposon (tranz POH zahn): "jumping gene," a gene with the ability to move from one location to another in a chromosome; from the Latin *transponere*, meaning "to transfer" (p. 148)

trimester: one of the three 3-month periods of the 9-month pregnancy; from the Latin *trimestris*, meaning "of three months" (p. 733)

trophic (TROHF ihk) **level**: group of organisms whose energy source is the same number of steps away from the sun; from the Greek *trophē*, meaning "food" (p. 257)

tropism (TROH pihz uhm): growth response of a plant that occurs toward or away from a stimulus (p. 398)

tuber: enlarged end of a rhizome that grows underground and can be used for food; from the Latin *tuber*, meaning "swelling" (p. 416)

tumor: formation of a mass of cells as a result of abnormally rapid cell division (pp. 108, 690)

underhair: shorter, thinner hairs under the guard hair coat of animals, which reduce the amount of body-heat loss (p. 564)

unsaturated fat: a type of fat in which some carbon atoms are linked by double bonds; usually in a liquid state at room temperature (p. 31)

urea: nitrogenous waste that is excreted by mammals, and requires some water for elimination; from the Greek *ouron*, meaning "urine" (p. 470)

uric acid: nontoxic solid waste eliminated by reptiles and birds (p. 470)

uterus: hollow muscular organ of the female where the fetus develops during pregnancy; from the Latin *uterus*, meaning "womb" (p. 723)

vaccine (vak SEEN): a preparation of killed or weakened pathogens introduced into the body to produce immunity; from the Latin *vacca*, meaning "cow" (pp. 163, 342)

vagina (vuh JEYE nuh): muscular tube in the female reproductive system that receives sperm ejaculated by a male during intercourse and through which the infant passes during birth; from the Latin *vagina*, meaning "sheath" (p. 724)

vascular tissue: tubes and vessels within a plant and consisting of cells that transport water and nutrients; from the Latin *vasculum*, meaning "small vessel" (p. 380)

vas deferens (vas DEHF uh REHNZ): duct that connects each epididymis to the urethra and carries sperm during ejaculation; from the Latin *vas deferentia*, meaning "deferent vessel" (p. 720)

vector: agent used to transfer genes in genetic engineering; from the Latin *vectus*, meaning "carried" (p. 156)

vegetative propagation: growing of new plants from nonreproductive plant parts, such as roots, stems, and leaves (p. 428)

vein: blood vessel that returns blood to the heart (p. 654)

ventricle (VEHN trih kuhl): chamber of the heart that pumps blood out; from the Latin *venter*, meaning "belly" (p. 658)

vertebrate: animal with a backbone; from the Latin *vertebratus*, meaning "jointed" (p. 204)

vestigial (ves TIHJ ee uhl) **structure**: structure that is a remnant of an organism's evolutionary past and has no function; from the Latin *vestigium*, meaning "footprint" (p. 178)

villus: any of the projections lining the small intestine; from the Latin *villus*, meaning "tuft of hair" (p. 708)

virulent (VEER yoo luhnt): causing disease (p. 137)

virus: any of the microscopic particles that invade cells of plants, animals, fungi, and bacteria, often destroying the cells; from the Latin *virus*, meaning "poison" (p. 345)

vitamin: organic substance needed by the body in relatively small amounts (p. 703)

voltage-sensitive channel: protein channel found in membranes of cells that respond to electrical signals (p. 65)

water vascular system: complex arrangement of water-filled tubes that function in the locomotion, feeding, and gas exchange in echinoderms; from the Latin *vasculum*, meaning "small vessel" (p. 493)

white blood cell: colorless blood cell that protects the body against invading bacteria, viruses or other foreign cells (p. 653)

wood: tissue that consists primarily of secondary xylem produced in plants that live more than one growing season (p. 417)

xylem (ZY luhm): vascular tissue of plants that transports water and minerals; from the Greek *xylon*, meaning "wood" (p. 394)

zygote (ZY goht): fertilized egg; cell that results from the fusing of an egg and a sperm (pp. 382, 729)

Index

Page numbers in boldface type indicate introductions of key terms. When "*illus.*" or "*table*" appears without page numbers following, the illustrations and tables are found, along with text references, on pages listed prior to "*illus.*" or "*table.*"

F

H

M

R

Credits

List of Abbreviations

All photographs attributed to SP/Foca were taken by Sergio Purtell/Foca Co.

AA = Animals Animals; AP = Archive Photos; AP/WW = AP/Wide World; AT = Alexander & Turner; BPA = Biophoto Associates; BPS = BPS & Biological Photo Service; CBMI = Claire Booth Medical Illustration; CM = Custom Medical; CP = Culver Pictures; CS = Comstock; HS = Highlight Studios; ES = Earth Scenes; FPG = FPG International, Corp.; KBI = Karapelou BioMedical Illustrations; KC= Kip Carter, M.S. certified medical illustrator; LPI = Lynne Prentice Illustration; MA = rep. Mattleson Associates, Ltd.; MC = MediChrome; MMA = Margulies Medical Art; MT = rep. Melissa Turk; MHI = Mark Hallett Illustrations; NSI = Natural Science Illustration; OSF = Oxford Scientific Films; PA = Peter Arnold, Inc.; PR = Photo Researchers; PT= Phototake; RWS = rep. Richard W. Salzman S.F., CA; SPL = Science Photo Library; SS = Studio Sloth; TBA = The Bettmann Archive; TM = Teri J. McDermott, M.A., certified medical illustrator; TMMK = Teri J. McDermott, M.A., certified medical illustrator/Michael Kress-Russick, M.A., M.S.; TSA = Tom Stack & Associates; TSM = The Stock Market; TWC = The Wildlife Collection; VU = Visuals Unlimited; WCA = Woodfin Camp & Assoc.

All Safety Symbols were designed by Michael Helfenbein/Evolution Design.

Cover Frans Lanting/Minden Pictures; **ii** SP/Foca; **iii** (b) SP/Foca; (t) Claire Booth/CBMI; **iv** Courtesy of George Johnson; **v** European Space Agency/SPL/PR**;** **vi** (tr) R.Margulies/MMA; (b) E. Alexander/ AT; (tl) European Space Agency/SPL/PR**;** **vii** (t) T. Narashima/SS; (b) Daniel Kirk/BA; **viii** (tl, tr) Mark Hallet/MHI; (b) SP/ Foca; **ix** (tr, bl) SP/Foca; (tl) Laurie O'Keefe/HS; **x** (tl) Deborah Daugherty/Laurie O'Keefe/HS; (heart) Peg Gerrity/HS; (b) Wendy Smith-Griswold; (frog) SP/Foca; **xi** SP/Foca; **xii** (t) Walter Stuart/RWS; (c) E. Alexander/AT; (bl) T. McDermott/Michael Kress-Russick/TMMK; **xiii** (t) Lennart Nilsson/Bonnier Alba AB; (c) SP/Foca; calculator Permission of Texas Instruments; **xiv** (frog, vessel) Walter Stuart/RWS; (tl) Lynne Prentice/LPI; (cr) David Beck/RMF; (cell) Robert Margulies/ MMA; (turtle lung) Kip Carter/KC; (leaf, turtle) SP/Foca; **xv** (l) Kerry Klayman; (cl) Michael J. Balick/PA; (cr) Carlos Goldin/SPL/PR; (bl) Biophoto Associates/Science Source/PR; (bc) SP/Foca; **xvi** (tr) Ted Horowitz/TSM; (tl) Michel Viard/PA; (petri dishes) SP/Foca; **xvii** (tl) Ralph A. Reinhold/AA; (scientist) ©1992/CM; (br) SP/Foca; **2-3** Prof. P. Motta/Dept. of Anatomy/University "La Sapienza", Rome/SPL/PR; **3** (t) Yoav Levy/PT; (l) Simon Fraser/RVI, Newcastle-Upon-Tyne/SPL/ PR; (b) SIU/VU; **4** Kevin Schafer/PA; **5** (t) Courtesy Smithsonian Institution Photo; (b) © 1996 Discover Magazine; **06** (tl) SP/Foca; **6-7** (c) John W. Karapelou/KBI; **7** (b) Michael Minardi/PA; (tr) Tom Raymond/MediChrome; **8** (c,r) Courtesy Oil Spill Public Information Center; (tl) SP/Foca; **8-9** (c) European Space Agency/SPL/PR; **9** (l) Peter May/PA; (bc) Mark Moffett/MP; (tc) John Colwell/Grant Heilman Photography; (tr) Bruce Ando/Index Stock; **10** SP/Foca; **11** Randall Zwingler/LM; **12** (l) Corbis-Bettmann; (c) UPI/Corbis-Bettmann; (r) Oxford Scientific Films/AA; **13** (l) Manfred Kage/PA; (c) UPI/Corbis-Bettmann; (r) Courtesy New York Academy of Medicine; **14** THE FAR SIDE cartoon by Gary Larson is reprinted by permission of Chronicle Features, San Francisco, CA. All rights reserved.; **15** Carlyn Iverson; **16** Jay Wasserman/VU; **17** (bl) James L. Amos/PR; (br) Frans Lanting/MP; (tr) Mark Newman/BC; **18** Gerard Lacz/PA; **19** (tl) Courtesy Oil Spill Public Information Center; (tr) Carlton Ray/PR; (b) Frans Lanting/MP; (cl) Oxford Scientific Films/AA; **20** (t) Kevin Schafer/PA; **21** Matt Meadows/PA; **22** Max Gibbs/Scientific Oxford Films/AA; **23** (t) Nigel Westwood/Survival Oxford Scientific Films/ES; (b) Cabisco/VU; **24** Cabisco/VU; **25** Cabisco/VU; **26** (t) Shelby Thorner/David Madison; (b) Robert Margulies/MMA; **27** (tl) SP/Foca; (cl) Dennis Kunkel/PT; (cr) Robert Margulies/MMA; (tr) Claire Booth/CBMI; **28** (tl,bl,br) Claire Booth/ CBMI; (tr) Ralph A. Reinhold/ES; **29** (br) Jeff Rotman/PA; (bl) SP/Foca; **30** SP/Foca; **31** Claire Booth/CBMI; **32** Claire Booth/CBMI; **33** Robert Margulies/ MMA; **34** (l) Claire Booth/CBMI; (hair) SP/Foca; (starch) Omikron/Science Source/PR; (fat) Manfred Kage/PA; (chromo) BPA/Science Source/PR; **35** (tl) Cabisco/VU; (b) Jeff Rotman/PA; (tr) SP/Foca; (c) Robert Margulies/MMA; **36** Max Gibbs/Scientific Oxford Films/AA; **37** Robert Margulies/MMA; **38** (bl) Sinclair Stammers/SPL/PR; (tr) Carlos Goldin/SPL/PR; (c,l) Omikron-Science Source/PR; **39** Robert Margulies/MMA; **40** © 1993 B.S.I.P./CM; **41** (tl) Nancy Kedersha/SPL/CM; (bl) M. Abbey Photo/PR; (cl,br) SP/Foca; (tr) © W. Gregory Brown/AA; **42** (tl,bl) SP/Foca; (br,tr) Claire Booth/CBMI; **43** Claire Booth/CBMI; **44** (tl,bc,br) Claire Booth/CBMI; (bl) SP/Foca; **45** (t-br) SP/Foca; (bl) Claire Booth/CBMI; **46** Catherine Twomey/Craven Design Studios, Inc.; **47** Robert Margulies/MMA; **48** Robert Margulies/MMA; **49** Robert Margulies/MMA; **50** (l) CNRI/SPL/PR; (r) M. I. Walker/PR; **51** Robert Margulies/MMA; **53** (c) Robert Margulies/MMA; (tr) D. W. Fawcett, S. Ito/PR; (cl) BPA/PR; (br) B. Bowers/PR; (tl) David M. Phillips/The Population Council/Science Source/PR; **54** Robert Margulies/MMA; **55** (tr) Bruce Berman/TSM; (tl) Howard Sochurek/TSM; Biophoto Associates/Science Source; (bl) Gervase Pevanik/WARD'S; (bkgd) SP/Foca; **56** (r) M. Abbey Photo/PR; (l) SP/Foca; **57** (r) Electra/PT; (l) Sinclair Stammers/SPL/PR; **58** (r) Dennis Kunkel/CNRI/PT; (l) Philippe Plailly/Eurelios/SPL/PR; **59** (tl) Claire Booth/CBMI; (c,bl) Robert Margulies/MMA; (tr) SP/Foca; **60** (t) © 1993 B. S.I.P./CM; **61** Catherine Twomey/Craven Design Studios, Inc.; **62** Dr. Dennis Kunkel/PT; **63** (tl) Institut Pasteur/CNRI/PT; (r) Ed Reschke/PA; (bl) © Zig Leszczynski/AA; **64** (c) SP/Foca; (t,b) Robert Margulies/MMA; **65** (cl,tl) Claire Booth/CBMI; (bc, br) Robert Margulies/MMA; **66** (t) © 1991 Kevin Beebe/CM; (bl) T. Narashima/SS; (bc) Carolina Biological Supp/PT; (br) Robert Margulies/MMA; **67** SP/Foca; **68** (tl,tr) SP/Foca; (tc) T. Narashima/SS; **69** (l) SP/Foca; (r) Robert Margulies/MMA; **70-71** Robert Margulies/MMA; **72** (t,r) T. Narashima/SS; (bl) Secchi-Lecaque/Roussel-UCLAF/CNRI/SPL/PR; **73** (tr) HMS Images/IM; (c) Robert Margulies/MMA; (bl) SP/Foca; (br) T. Narashima/SS; **74** Dr. Dennis Kunkel/PT; **75** Robert Margulies/MMA; **76** Frans Lanting/MP; **77** (br) SP/Foca; (tl) Wayne H. Chasan/IM; **78** SP/Foca; **79** Walter H. Hodge/PA; **80** Robert Margulies/MMA; **81** (l) SP/Foca; (r) Fred Breummer/PA; **82** Dr. Arnold Brody/SPL/PR; **83** (t) Foca; (b) Claire Booth/CBMI; **84** (l) NASA; (cl) © 1991 CS; (cr) David C. Fritts/AA; (r) Mark Stouffer; **85** SP/Foca; (inset) © Colin Milkins/Oxford Scientific Films/ES; **86** SP/Foca; **86-87** SP/Foca; **87** (tl) John D. Cunningham/VU; (tr) Robert Margulies/MMA; (br) E. Alexander/AT; **88** E. Alexander/AT; **89** Claire Booth/CBMI; **90** © 1992 Bruce Fitz; **91** (tl,tr) SP/Foca; (c,bl) Claire Booth/CBMI; **92** (moclecules) Claire Booth/CMI; (grapes, boy) SP/Foca; **93** Claire Booth/CBMI; **94** Robert Margulies/MMA; **95** (tr) OSF/AA; (tl) Walter H. Hodge/PA; (tc) David C. Fritts/AA; (bc) E. Alexander/AT; (b) SP/Foca; **96** Frans Lanting/MP; **98** Omikron/Science Source/PR; **99** (t) Richard Nowitz/PT/PNI; (b) Tony Craddock/SPL/PR; **100** Ken Wagner/PT; **101** (tr) © 1992 CM; (b) Biophoto Associates/PR; **102** (c) Robert Margulies/MMA; (br) ©1995 SPL/CM; **103** (l) © 1993 Robert Becker/CM; (r) © 1990 CM; **104** (r) T. Narashima/SS; (l) Institut Pasteur/CNRI/PT; **105** Biology Media/PR; **106** T. Narashima/SS; **107** Claire Booth/CBMI; **108** (l) Dr. Tony Brain/SPL/PR; (r) Robert Margulies/MMA; **109** (l) Robert Margulies/ MMA; **110** (tl,tr,tc) SP/Foca; (bc,cl) © 1995 SPL/CM; (cr) © 1992 SPL/CM; (br) Jason Burns/Ace/PT; **111** (tl,tr) T. Narashima/SS; (bl) CNRI/PT; (br) L. Zamboni-D. Fawcett/VU; **112** T. Narashima/SS; **113** (bl) © 1995 SPL/CM; (br) © 1992 SPL/CM; (tr) Maria Taglienti/IM; (t) Biophoto Associates/PR; (c) Biology Media/PR; **114** Ken Wagner/PT; **115** (tl, tc, tr) SP/Foca; (cl,bc) © 1995 SPL/CM; (cr) © 1992 SPL/CM; (br) Jason Burns/Ace/PT; **116** David Madison; **117** (c) CP; (br) Karen Kluglein/MA; (tl) SP/Foca; **118** Karen Kluglein/MA; **119** Karen Kluglein/MA; **120** Karen Kluglein/MA; **121** Karen Kluglein/MA; **123** (c) Karen Kluglein/MA; **124** Karen Kluglein/MA; **125** (tl) SP/Foca; (bl,br) Larry Lefever/Grant Heilman Photography; **126** (t) SP/Foca; (l) Michio Hoshino/MP; (b) Jim Brandenburg/MP; **127** © David York/Medichrome; **128** (b) Al Lamke/LEN/PT; (c) © Simon Fraser/SPL/CM;**129** (l) Courtesy The Rice Family; (cr) Richard Hutchings/PR; (br) T. McDermott/Michael Kress-Russick/ TMMK; **130** Courtesy The Davidson Family; **131** (l) SP/Foca; (bkgd) Biophoto Associates/Science Source/PR; (t) Dr. Jeremy Burgess/ SPL/PR; **132** Ted Thai/Time Magazine; **133** (tr,c) SP/Foca; (tl) Karen Kluglein/MA; (b) © David York/Medichrome; **134** (b) Jim Brandenburg/MP; (t) David Madison; **136** Jean Claude Revy/PT; **137** (tl) SP/Foca; (bacteria) Claire Booth/CBMI; (mice) Susan Johnston Carlson; **138** (t) Claire Booth/CBMI; (b) Lee. D. Simon/PR; **139** (t) PA; (b) Robert Margulies/Virginia Ferrante; **140** (b) Robert Margulies/MMA; (t) SP/Foca; **141** Robert Margulies/MMA; **142** Robert Margulies/MMA; **143** SP/Foca; **144** E. Alexander/ C. Turner/AT; **145** E. Alexander/ C. Turner/AT; **146-147** Robert Margulies/Christine Schaar/MMA; **148** (l) SP/Foca; (r) D. Goldberg/Sygma; **149** (tr) PA; (tl, b) Robert Margulies/MMA; (c) E. Alexander/AT; **150** (b) Foca; (t) Jean Claude Revy/PT; **151** E. Alexander/AT; **152** Ted Horowitz/TSM; **153** (tl) Agricultural Research Service, USDA; (br) R. Kessel & G. Shih/VU; (bl) SP/Foca; (bc) Robert Margulies/MMA; **154** (bl) SP/Foca; (tl, tr, tc) Robert Margulies/MMA; (br) E. Alexander/AT; **155** (tl) David M. Phillips/VU; (c) E. Alexander/AT; (tr) R. Kessel & G. Shih/VU; **156** E. Alexander/AT; **157** Ted Horowitz/TSM; **158** E. Alexander/AT;

159 (tl, br) SP/Foca; (bc) © E. H. Newcomb & S. R. Tandon/Univ. of Wisconsin/BPS;**160** B. Hansen/NSI; **161** Monsanto; **162** (c) R. Kessel & G. Shih/VU; (t) E. Alexander/AT; (b) SP/Foca; **163** (illus.) E. Alexander/AT; (tr) SP/Foca; **164** Dr. A. Liepins/SPL/SS/PR; **165** (cl) Kerry Klayman; (bkgd) Roland Seitre/PA; (tl) Karl & Kay Amman/BC; (tr) James P. Blair/National Geographic Image Collection; (bc) Michael J. Balick/PA; **166** Manfred Kage/PA; **167** (c) Monsanto; (tr) @1992 SPL/CM; (tl) E. Alexander/AT; (b) SP/Foca; **168** (t) Ted Horowitz/TSM; **169** (l) E. Alexander/AT; (tr) David M. Phillips/VU; (br) Robert Margulies/MMA; **170** (c) Monsanto; (bl) Robert Margulies/MMA; **170-171** (r) SP/Foca; **171** Robert Margulies/MMA; **172** Frans Lanting/MP; **173** (tl) Michio Hoshino/MP; (b) Lazlo Kubinyi/Rep. Gerald & Cullen Rapps, Inc.; (r) Corbis-Bettmann; **174** (tc) John S. Dunning/PR; (bl, bc) M. A. Chappell/AA; (br) Leonard Lee Rue III/AA; **175** (tl) Corbis-Bettmann; (tr) Courtesy Department of Library Services, American Museum of Natural History; (bl) Corbis-Bettmann; **176** SP/Foca; **177** (r) Chris Forsey; (l) David Beck/RMF; **178** T. McDermott/Michael Kress-Russick/TMMK; **179** T. McDermott/Michael Kress-Russick/TMMK; **180** Marius Muresanu/Foca; **181** Daniel Kirk/BA; **182** Breck P. Kent/AA; **183** C. Turner/AT; **184-185** (sparrows) Daniel Kirk/BA; (bkgd) Randall Zwingler/LM; **186** Carlyn Iverson; **187** (tr, bl) Breck P. Kent/AA; (tl) Corbis-Bettmann; (c) Chris Forsey; **188** (t) Frans Lanting/MP; **190** 1993 Louis Psihoyos/Matrix; **191** (b) T. Narashima/SS; (t) SP/Foca; **192** UPI/Corbis-Bettmann; **193** Yemi; **194** (l, c) SP/Foca; (r) David M. Phillips/VU;**195** (t) Corale Brierley/VU; (b) Richard Thom/VU; **196** (tl) CNRI/SPL/PR; (bl) Ralph Robinson/VU; (bc) D. Gottshall/VU; (tr, br) SP/Foca; (cr) Richard Thom/VU; **197** Lynn Prentice; **198** (l) Claire Booth/CBMI; (r) NASA/Science Source/PR; **199** (l) SP/Foca; (r) R. L. Peterson, Univ. of Guelph/BPS; **200** (tl, tc, tr) Carlyn Iverson; (bl, cc) David Beck/RMF; (bc) Pat Ortega; (cl) Lynn Prentice/LPI; (cr) Courtesy Peabody Mus. Nat. History/Yale University/painted by Rudolph F. Zallinger; (br) Robert Frank; (cr-inset) Lynn Prentice/LPI; **201** (tc) Lynn Prentice/LPI; (eukaryote) Carlyn Iverson; (cc) Courtesy Peabody Mus. Nat. History/Yale University/painted by Rudolph F. Zallinger; (tr, bl, T. Rex) Pat Ortega; (cr) John Burgoyne; (cl) David Ashby; (br) Robert Frank; (human) John Burgoyne; **202** Carlyn Iverson; **203** (bkgd) Courtesy Peabody Mus. Nat. History/Yale University/painted by Rudolph F. Zallinger; (inset) Lynn Prentice/LPI; **204** (t) David Beck/RMF; (b) © Breck P. Kent/AA; **205** (t) Robert Frank; (b) Victoria McCormick/AA; **206** (l) Stephen Dalton/AA; (r) Robert Frank; **207** (bkgd) Robert Frank; (c) Mark Hallet/MHI; (sillhouette) Kevin A. Somerville/MIS; **208** (b) AA; (r) Robert A. Tyrell/OSF/AA; (c) Steve Kauffman/TS; **209** (tc) Corale Brierley/VU; (tr) Dr. Fred Espenak/SPL/PR; (tl) David M. Phillips/VU; (bc) NASA/Science Source/ PR; (b) Steve Kauffman/TS; **210** 1993 Louis Psihoyos/Matrix; **211** Yemi; **212** Gamma Press Images/ Gamma Liaison; **213** (b) Todd Buck/HS; (tl) Kenneth Garrett/ National Geographic Image Collection; **214** E. H. Pao/PR; **215** (bl) Patti Murray/AA; (tr) Leonard Lee Rue III/TS; **216** S. Turner/OSS/ AA; **217** (gibbons) E. R. Degginger/AA; (orangutan) Christopher Arnesen/TS; (gorilla) Pat Crowe/AA; (bonobo) Michael Dick/AA; (chimpanzee, humans) SP/Foca; **218** (tl, br) Bianca Lavies/National Geographic Image Collection; (bl) Gilbert M. Grosvenor/National Geographic Image Collection; (tr) Volkmar Kurt Wentzel/National Geographic Image Collection; **219** (tl, tr) SP/Foca; (bl, br) Dana Geraths; **220** Mark Hallet/MHI; **221** (tl, br) Institute of Human Origins; (b) John Reader/SPL/PR; (jaws) Dana Geraths; **222** (l) John Reader/SPL/PR; (r) Courtesy Department of Library Services, American Museum of Natural History; **223** Mark Hallet/MHI; **224-225** (skulls) Mark Hallet/MHI; (bkgd) Randall Zwingler/LM; **226** (t) SP/Foca; (b) Randall Zwingler/LM; **227** John Burgoyne; **228** (t) Courtesy Department of Library Services, American Museum of Natural History; (b) AR, NYC; **229** (tr) Reprinted courtesy OMNI Magazine © 1991; (tl) Leonard Lee Rue III/TS; (c) Institute of Human Origins; (b) John Burgoyne; (jaw) Dana Geraths; **230** Gamma Press Images/Gamma Liaison; **231** Mark Hallet/MHI; **232** Phil Farnes/PR; **233** (tr) Klaus Uhlenhut/AA; (b) David J. Boyle/AA; **234** (t) Robin Brandt/Natural Selection; (b) George B. Schaller; **235** Susan Johnston Carlson; **236** (tl) Susan Johnston Carlson; (br) Thomas McAvoy/Life Magazine; **237** (b) Omikron/Science Source/ PR; (t) SP/Foca; **238** (bkgd) SP/Foca; (l) Courtesy of Jennifer Dusek; (tr) J. Bucher/PR; (tl) J. M. Labat/Jacana/PR; **239** (t) S. Feld/H. Armstrong Roberts; **240** (t) Gary W. Griffen/AA; (c) Peter Weimann/ AA; (b) Joe McDonald/AA; **241** (b) Michael Fogden/BC; (t, bc) Ken Cole/AA; (tc) Patti Murray/AA; **242** (tl) © Bill Silker Jr./AA; (bl) IFA Bilderteam/BC; (br) Michael Fogden/Oxford Scientific Films/AA; **243** (t) J. C. Carton/BC; (b) Hans Reinhard/BC; **244** (tr) Max Gibbs/ Oxford Scientific Films/AA; (br) Martin Harvey/TWC; (bl) Adrienne T. Gibson/AA; **245** (l) Patti Murray/AA; (r) B. Heinrich/University of Vermont; **246** Susan Johnston Carlson; **247** (c) Robin Brandt/ Natural Selection; (b) Max Gibbs/Oxford Scientific Films/AA; (tr) feature THE FAR SIDE © 1988 FARWORKS, INC. Dist. by UNIVERSAL PRESS SYNDICATE. Reprinted with permission. All rights reserved.; **248** Phil Farnes/PR; **249** Susan Johnston Carlson; **250-251** James Randklev/TS; **251** (t) Eric Kamp/Index Stock; (c) H. Armstrong Roberts; (b) Tim Brown/Index Stock; **252** Richard Day/Panoramic Images; **253** (t) Lee Rentz/BC; (b) © 1989 Kevin Schafer/PA; **254** Lynn Prentice/LPI; **255** Carlyn Iverson; **256** (l) David Beck/RMF; (r) © Ed Reschke/PA; **257** David Beck/RMF; **258** Carlyn Iverson; **259** Carlyn Iverson; **260** (bkgd, plants, soil bacteria) Carlyn Iverson; (molecules) Claire Booth/CBMI; (roots, deer) Lynn Prentice/LPI; **261** B. Hansen/NSI; **262** (plant, jeep, bkgd) Claire Booth/CBMI; (deer, well) Lynn Prentice/LPI;**263** SP/Foca; **264** (t) Klaus Uhlenhut/ES; (b) Gregory Dimijian/PR; (maps) Foca; **265** (t) Bill Bachman/PR; (c) Mickey Gibson/ES; (b) Stephen J. Krasemann/PR; (maps) Foca; (t, b) Stephen J. Krasemann/PR; **266** (maps) Foca; **267** (l, b) Courtesy Carmen R. Cid-Benevento; (bkgd) AA/ES; (tl) Stephen Dalton/AA; (tr) Zig Leszczynski/AA; **268** David Beck/RMF; **269** (tl) © Ed Reschke/PA; (c) B. Hansen/NSI; (b) Stephen J. Krasemann/PR; (tr) Raymond Mendez/AA; **270** Richard Day/Panoramic Images; **271** David Beck/RMF; **272** Gregory K. Scott/PR; **273** (t) Arthur M. Greene/BC; (b) Bill Wood/BC; **274** (b) Fletcher W. K./PR; (flies) SP/Foca; **275** SP/Foca; **276** (b) L. West/PR; (inset) V. Ahmadjian/VU; **277** (t) Fred Bavendam; (b) F. Gohier/PR; **278** Peter Weimann/AA; **279** Wendy Smith-Griswold/MT; **280** (t) Michael P. Gadomski/BC; (c) L.L. Rue III/BC; (b) Michael P. Gadomski/ES; **281** (tl) Adrienne T. Gibson/AA; (br) SP/Foca; **282** (l) Randall Zwingler/LM; (r) Alan Blank/BC; **283** (tl) Tom McHugh/PR; (cl) Milton H. Tierney, Jr./VU; (bl) Patrick Montagne/PR; (r) Earth Imaging/Tony Stone Images; **284** (l) Bob Gossington/BC; (r) Provided by Dr. S. Bradleigh Vinson, Texas A&M University; **285** (c) Fred Bavendam; (b) Adrienne T. Gibson/AA; (t) AP/WW Photo; **286** Gregory K. Scott/PR; **288** Rafael Macia/PR; **289** (t) SP/Foca; (b) Earl Roberge/PR; **290** (t) AP/WW Photo; (b) Courtesy Oil Spill Public Information Center; **291** (l) © 1989 D. A. Glawe, Univ. of Ill./BPS; (bkgd) Randall Zwingler/LM; (tr) Master Communications; (br) 1991 Ray Pfortner/PA; **292** (l) NASA/Mark Marten/PR; **292-293** NASA/Goddard Institute for Space Studies//SPL/PR; **294** Donna Binder/Impact Visuals/PNI; **295** SP/Foca; **296** (l,r) Gene Alexander/Soil Conservation Service; (c) Tim McCabe/Soil Conservation Service; **297** (c, r) Michael J. Balick/PA; (l) R. Cramm/Okapia/PR; **299** Foca; **300** Mickey Gibson/AA; **301** SP/Foca; **302** (t) SP/Foca; magazines Courtesy Mother Earth News; Wildlife Conservation; (egret on cover) Art Wolfe; National Audubon Society; **303** SP/Foca; **304** SP/Foca; **305** (b) SP/Foca; (tr) James H. Karales/PA; (c) Michael J. Balick/PA; (tl) Courtesy Oil Spill Public Information Center; **306** Rafael Macia/PR; **308** (bl) Richard Gorbun/Leo de Wys Inc.; (t) NOAA/SPL/PR; **309** (b) Michael Newman/PhotoEdit; **310-311** Andris Apse/Panoramic Images; **311** (c) Martin Bond/SPL/PR; (b) Jim Richardson/Richardson Photography; (t) SP/Foca; **312** SP/Foca; **313** (t) SP/Foca; (b) Charles W. Mann/PR; **314** SP/Foca; **315** (tl) Ray Coleman/PR; (tc) Randall Zwingler/LM; (tr) Dan Sudia/PR; (bc) M. Fogden/OSF/AA; (br) Bridgeman, London/AR, NYC; (bl) Mark Hallet/MHI; **316** (tl, br) SP/Foca; (bl) Courtesy Paul Melancon/Univ. of Colorado, Boulder; (tc) Courtesy Alma Solis/SEL/USDA; (bc) Courtesy Lucy Bunkley-Williams, Ernest H. Williams, Univ. of Puerto Rico at Mayaguez; **317** (t) SP/Foca; (b) 1992 The Pierpont Morgan Library/AR; **319** David Beck/RMF; **320** (l, r) E. Alexander/AT; (tc) G. Soury/Jacana/PR; (bc) Chuck Davis Photography; **321** (moss) K. G. Vock/Okapia/PR; (fern, flowers, seeds) SP/Foca; (tree)1991 Tom Bean/TS; (vascular system) P. Dayanandan/PR; **322** Susan Johnston Carlson/MT; **323** Mark Boulrow/PR; **324** (b) Neil G. McDaniel/PR; (t) Eric Grave/PT; **325** (l) T. J. Beveridge/VU; (c) Oliver Meckes/PR; (r) Manfred Kage/PA; **326** (tl) Robert Lee/PR; (bl) SP/Foca; (cr) Jeremy Stafford-Deitsch/Ellis nature Photography; **327** (tr) Bob Orsillo; (tl) SP/Foca; (c) 1992 The Pierpont Morgan Library/AR; (b) Neil G. McDaniel/PR; **328** (t, cr) SP/Foca; (cl) Dan Sudia/PR; (b) Neil G. McDaniel/PR; **329** (moss) K. G. Vock/Okapia/PR; (fern, flowers, seeds) SP/Foca; (tree)1991 Tom Bean/TS; (vascular system) P. Dayanandan/PR; **331** (tl) Stanley Flegler/VU; (bl) Biophoto Associates/Science Source/PR; (cl) Oliver Meckes/Ottawa/PR; (cr) CNRI/SPL/PR; (br) CNRI/SPL/PR; **332** (tl) Manfred Kage/PA; (insets) CNRI/SPL/PR; (tr, br) T. Narashima/SS; (bl) Manfred Kage/PA; **333** (bkgd) John Coulter/VU; (c) Robert Margulies/MMA; (b) Foca; **334** (l) Michael S. Renner/BC; (r) Biophoto Associates/PR; **335** (l) Fred McConnaughey/Science Source/PR; (r) Carl O. Wirsen/VU; **336** (t) Larry Jensen/VU; (b) C. Bradley Simmons/BC; **337** (cl) SP/Foca; (t) Cabisco/VU; (cr) Martin Rotker/PT; (b) AFP/Corbis-Bettmann; **338** (t) John D. Cunningham /VU; (c) SPL/PR; (b) Ed Reschke/PA; **339** (t) SP/Foca; **340** (inset) Manfred Kage/PA; (b) Cabisco/VU; (tl) Kent Wood/PR; **341** AP/WW Photo; **342** (t) 1989 Cytographics/VU; (tc) David M. Phillips/VU; (bc) Genentech/VU; (b) © L. O'Shaughnessy/Medichrome; **343** (tl) National Library of Medicine/SPL/PR; (c) S. Flegler/VU; (r) SP/Foca; (l) D. Newman/VU; **344** Karen Kluglein/MA; **345** (t) National Cancer Inst./Science Source/PR; (b) © 1986 A. J. Olsen/The Scripps Research Institute; **346** (c) John W. Karapelou/KBI; (b) Foca; **347** T. Narashima/SS; **348** Lowell Georgia Science Source/PR; **349** (b) AP/WW Photo; (t) Courtesy J. Wickett/World Health Organization; **350** (viruses) Karen Kluglein/MA; (AIDS) John W. Karapelou/KBI;

351 (c) National Library of Medicine/SPL/PR; (tl) Robert Margulies/MMA; (b) John W. Karapelou/KBI; (tr) Lennart Nilsson/Boehringer Ingelheim, International, GmbH; **352** (t) CDC/LL/PA; (bl) John W. Karapelou/KBI; (br) Robert Margulies/MMA; **353** T. Narashima/SS; **354** (t) NIBSC/Science Source/PR; (c) John Madere/TSM; (bl) NIBSC/SPL/PR; **355** Hank Morgan/Rainbow/PNI; **356** Andrew Syred/SPL/PR; **357** (t) Manfred Kage/PA; (b) Eric Grave/PT; **358** (c-inset) Dennis Kunkel/BPS; (b) Claire Booth/CBMI; (t, inset) Eric Grave/PT; **359** Robert & Linda Mitchell; **360** (tl) Walter H. Hodge/PA; (bl) SP/Foca; (bc) © Phil Gates, Univ. of Durham, England/BPS; (br) Biophoto/PR; **361** (c) Laurie O'Keefe; (bkgd) Michael P. Gadomski/PR; (b) Foca; **362** (bkgd) Deborah Daugherty/ Peg Gerrity/HS); (l) Manfred Kage/PA; (c) SP/Foca; (r) Robert & Linda Mitchell; **363** (l) David M. Phillips/VU; (r) Kevin Schafer/PA; **364** (l) B. Beatty/VU; (inset) Cabisco/VU; (tr) Eric Grave/PR; (br) G. Büttner/ Naturbild/Okapia/PR; **365** (bl) Biophoto Associates/Science Source/PR; (tl) AA/ES; (inset) Eric Grave/Science Source/PR; (br) Biophoto Associates/Science Source/PR; **366** (b) Hans Pfeltschinger/PA; (t) Kim Taylor/BC/PNI; **367** (full) C. Turner/AT; (life cycle) Carlyn Iverson; **368** Randall Zwingler/LM; **369** (tr) THE FAR SIDE cartoon by Gary Larson is reprinted by permission of Chronicle Features, SF., CA. All rights reserved.; (tl) Eric Grave/PT; (c) Eric Grave/PR; (b) Hans Pfeltschinger/PA; **370** Andrew Syred/ SPL/PR; **371** Carlyn Iverson; **372** A. Gurmaskin/VU; **373** (bl) SP/Foca; (t) Bill Beatty/VU; (bc) Ed Reschke/PA; (br) © David Scharf/ PA; **374** (c) M. Fogden/BC; (l) E. Gueho-CNRI/SPL/PR; (r) Wendy Smith Griswold; **375** Deborah Daugherty/HS; **376** (t) SP/Foca; (bc) Scott Camazine/PR; (tc) David Newman/VU; (b) Bob Gossington/BC; **377** (r) Katherine LoBuglio/The Field Museum; (l) Gregory M. Mueller/The Field Museum; **378** Bill Ruth/BC; **379** (l) Dr. Jeremy Burgess/SPL/PR; (r) SP/Foca; (c) St. Mary's Hospital Medical School/ Science Source Photo Library/PR; **380** David Beck/RMF; **381** Science VU/VU; **382** (r) CBC/PT/PNI; (l) Jeff Lepore/PR; **383** Susan Johnston Carlson/MT; **384** SP/Foca; **385** (l) Glenn Oliver/VU; (tl) Mickey Gibson/ES; (bl) Ed Reschke/PA; (tc) Karen Kluglein/MA; (br) Robert & Linda Mitchell; (r) SP/Foca; **386** SP/Foca; **387** (tl) Scott Camazine/PR; (tr) Dan Franck/Foca; (c) CBC/PT/PNI; (bl) SP/Foca; **388** (cl) SP/Foca; (t) A. Gurmaskin/VU; (cr) David Newman/VU; (bl) Bob Gossington/BC; (br) Scott Camazine/PR; **390** Robert P. Carr/BC; **391** (c) Robert & Linda Mitchell; (inset) R. F. Evert; (tl) Stouffer Productions/ES; **392** SP/Foca; **393** (tl) Dr. Jeremy Burgess/SPL/PR; (cr) SP/Foca; (tr) C. Gerald Van Dyke/VU; (cl) Robert Margulies/ MMA; (bl, br) E. Alexander/AT; **394** (l) Deborah Daugherty/HS; (r) M. I. Walker/PR; (c) SP/Foca; **395** (flowers, leaves) SP/Foca; (embryos) Carlyn Iverson; (bl) Cabisco/VU; (br) J. R. Waaland, Univ. of Wash, Seattle/BPS; **396** (l) © J. R. Waaland, Univ. of Wash., Seattle/BPS; **396-397** (zucchini plant) SP/Foca; (t) © E. Zeiger, Stanford Univ./BPS; (b) © J. R. Waaland, Univ. of Wash., Seattle/BPS; **398** Sylvan Wittwer/VU; **399** (tc) Alvin E. Staffan/PR; (br) BC; (bc) Dan Suzio/PR; (tr) SP/Foca; **400** (t) SP/Foca; (b) B. Hansen/NSI; **401** (bkgd) J. Shaw/BC; (c) Susan Johnston Carlson; (b) Kevin A. Somerville/MIS; **402-3** SP/Foca; **403** (cl, cr) Deborah Daugherty/HS; (r) D. Cavagnaro; **404** (tl, br) SP/Foca; (bl) Robert Maier/AA; (tr) Joyce Photographics/PR; **405** (cl) Jane Burton/BC; (cr) R. J. Erwin/PR; (tr) C. C. Lockwood/ES; (tl, bl) SP/Foca; **406** Deborah Daugherty/Pat Gerrity/HS; **407** (tr, tl, c) SP/Foca; (b) B. Hansen/NSI; **408** Robert P. Carr/BC; **409** SP/Foca; **410** John Kieffer/PA; **411** (b) Randall Zwingler/LM; (tl) SP/Foca; **412** SP/Foca; (inset) Courtesy Ray F. Evert, Univ. of Wisconsin; **413** (l) © W. H. Hodge/PA; (r) Bruno P. Zehnder/PA; **414** SP/Foca; (l) © Marshall D. Sunberg/ Louisiana State Univ.; **415** (t) SP/Foca; (b) Siegfried Tauqueur/Leo de Wys Inc.; **416** SP/Foca; **417** (l) Bios/M. Gunther/PA; (tc) Tommaso Guicciardini/SPL/PR; (r) Bruce Ando/Index Stock; (bc) SP/Foca; **418** (t) Ed Reschke/PA; (b) Wendy Smith Griswold; **419** (l) SP/Foca; (c) Jacques Jangoux/PA; (r) David R. Frazier Photolibrary; **420** (b) Wendy Smith Griswold; (t) Eric Lessing/AR; **421** SP/Foca; **422** (l) David Cavagnaro/PA; (r) P. Murray/ES; **423** SP/Foca; **424** (c) Randall Zwingler/LM; (br, bl) Robert & Linda Mitchell; (t) John Gerlach/ES; **425** (l) Courtesy of Bob Murase; (bkgd, tl) SP/Foca; (tr) John Elk/BC; **426** Geo F. Godfrey/ES; **427** Patti Murray/AA; **428** (r) Mary Lou Gripshover; (l) Jane Burton/BC; **429** (tr) Dr. Nigel Smith/ES; (l) African Violet Society of America Inc.; (br) Rutgers Cooperative Extension; **430** Holt Studios International/Nigel Cattlin/PR; **431** (t) SP/Foca; (c) Bios/M. Gunther/PA; (b) Geo. F. Godfrey/ES; **432** (b) SP/Foca; (t) John Kieffer/PA; **434-435** Natalie Fobes/TS; **435** (b) Joe McDonald/AA; (t) Robert Perron/PR; **436** AA/ES; **437** (tl) David Scharf/PA; (bl) Marty Snyderman/VU; (tc) Calvin Larsen/PR; (bc) A. Kerstitch/VU; (br) Zig Leszczynski/AA; (tr) SP/Foca; **438** Laurie O'Keefe; **439** (t) Nancy Sefton/PR; (b) Laurie O'Keefe/HS; **440** (t) T. McDermott/Michael Kress-Russick/TMMK; (b) Carolina Biological Supp/PT; **441** C. Turner/AT; **442** Michael Abbey/PR; **443** (t) C. Turner/AT; (b) Michael Abbey/PR; **444** (t) Claire Booth/CBMI; (b) John D. Cunningham/VU; **445** (t) Courtesy Dr. Paul Goldstein; (b) Laurie O'Keefe; **446** C. Turner/AT; **447** (t) Laurie O'Keefe; (b) SP/Foca; **448** Marty Snyderman/VU; **449** (t) Laurie O'Keefe/HS; (b) © Stephen J. Krasemann/PA; **450** (t) Deborah Daugherty/Laurie O'Keefe/HS; (bl) H Pfletschinger/PA; **451** (tr) Zig Leszczynski/AA; (cr) G. C. Kelley/PR; (l, br) SP/Foca; **452** (t,c) T. McDermott/Michael Kress-Russick/TMMK; (b) G. Dimijian, M. D./PR; **453** (t) Zig Leszczynski/AA; (b) T. McDermott/Michael Kress-Russick/TMMK; **454** (fluke) Claire Booth/CBMI; (wasp) Deborah Daugherty/Laurie O'Keefe/HS; (sponge, worm, nematode, snail) Laurie O'Keefe/HS; (hydra, lancelet, sea star) T. McDermott/HS; **455** (tl) Nancy Sefton/PR; (c) Courtesy Dr. Paul Goldstein; (b) Deborah Daugherty/Laurie O'Keefe/HS; (tr) SP/Foca; **456** (t) AA/ES; (br, bl) SP/Foca; **457** C. Turner/AT; **458** Johnny Johnson/AA; **459** (t) Robert Maier/AA; (b) E. R. Degginger/AA; **460** Peg Gerrity/HS; **461** (t) SP/Foca; (b) Peg Gerrity/HS; **462** (tl) © Wildlife Conservation Society; (tr) Art Wolfe/TS; (br) SP/Foca; **463** Hans Pfletschinger/PA; **464** (br) Peg Gerrity/HS; (t, bl) SP/Foca; **465** (c, b) Peg Gerrity/HS; (t) SP/Foca; **466** (c) Peg Gerrity/HS; (tl) Tom McHugh/Steinhart Aquarium/PR; (br) SP/Foca; **467** (illus.) Peg Gerrity/HS; (photos) SP/Foca; **468** (illus.) Kip Carter/KC; (photo) SP/Foca; **469** (illus.) Kip Carter/KC; (photo) SP/Foca; **470** Claire Booth/CBMI; **471** Peg Gerrity/HS; **472** (bl) PA; (inset) Kim Taylor/BC; (tl) J&B Photo/AA; **473** (l) E. R. Degginger/AA; (r) David Beck/RMF; **474** CNRI/PT; **475** (c) Peg Gerrity/HS; (b) David Beck/RMF; (tr) Drawing by Mort Gerberg © 1992, The New Yorker Magazine, Inc.; (tl) SP/Foca; **476** Johnny Johnson/AA; **477** David Beck/RMF; **478** W. Gregory Brown/AA; **479** (tl) Joyce Burek/AA; (b) Andrew J. Martinez/PR; **480** Laurie O'Keefe; **481** Lynn Prentice/LPI; **482** (c) Lynn Prentice/LPI; (tr) Kevin A. Somerville/MIS; **483** Walter Stuart/RWS; **484** (tl) CNRI/SPL/PR; (c) Lynn Prentice/LPI; (l) SP/Foca; (br) C. Prescott-Allen/AA; **485** (t) BPA/PR; (b) R. Calentine/VU; **486** (l) SP/Foca; (r) Courtesy Mayo Foundation; **487** (l) A. Kerstitch/VU; (r) Glenn M. Oliver/VU; **488** (b) Tom McHugh/Steinhart Aquarium/PR; (c) David M. Phillips/VU; (tl) SP/Foca; **489** (bkgd) Carl Roessler/AA; (c) Lynn Prentice/LPI; (t) Kevin A. Somerville/MIS; **490** (tl) Robert Dunne/PR; (b) Laurie O'Keefe; **491** (tl) Glenn M. Oliver/VU; (bl) M. A Chappell/ AA; (bc) Allan Morgan/PA; (br) SP/Foca; **492** (tl) Alvin E. Staffan/PR; (cr) OSF/AA; (bl) Ed Reschke/PA; (br) Patti Murray/AA; (cl) SP/Foca; **493** (l) Fred McConnaughey/PR; (r) SP/Foca; **494** (l) Fred McConnaughey/PR; (r) Zig Leszczynski/AA; **495** (tl) CNRI/SPL/PR; (b) Fred McConnaughey/PR; (tr, c) SP/Foca; **496** W. Gregory Brown/AA; **498** (t) John Shaw/BC; (bl) M. Fogden/AA; (c) M. Lockwood/VIREO; **499** F.Whitehead/AA; **500** Hans Reinhard/BC; **501** (tl) Mantis Wildlife Films/Oxford Scientific Films/AA; (b) SP/Foca; **502** Lynn Prentice/LPI; **503** (l) Patti Murray/AA; (tr) Leonard Lee Rue III/VU; (br) Larry Miller/PR; **504** (t) SP/Foca; (bl) David York/The Stock Shop; (br) ©1982 Dwight Kuhn; **505** (br) Breck P. Kent/AA/ES; (l) SP/Foca; **506** SP/Foca; **507** (flea) John D. Cunningham/VU; SP/Foca; **508** (c) Wendy Smith Griswold; (bkgd) Patti Murray/ES; (br) Foca; **509** (tl) Dick Poe/VU; (cl) William J. Weber/VU; (bl, c) Dan Kline/VU; (r) E. R. Degginger/AA; **510** (t) Kjell B. Sandved/BC; (l) BC; (b) Patti Murray/AA; (r) Raymond A. Mendez/AA; **511** (bkgd, tr) SP/Foca; (tl) Kim Taylor/BC; (l) © J. Arturo Lozaro; **512** (l) E. R. Degginger/BC; (c) Oxford Scientific Films/AA; (r) Bob Gossington/BC; **513** (l) Donald Specker/AA; (c) AA/ES; (r) J. H. Robinson/AA; **514** (b) A. Kerstitch/VU; (t) SP/Foca; **515** SP/Foca; **516** (c) Lynn Prentice/LPI; (bkgd) E. R. Degginger/BC; (b) Kevin A. Somerville/MIS; **517** SP/Foca; (bl) AA/ES; **518** (l) Sinclair Stammers/SPL/PR; (r) Peter Parks/OSF/AA; **519** (tr) Rainbow/PNI; (tl, c, b) SP/Foca; **520** Hans Reinhard/BC; **521** SP/Foca; **522** Zig Leszczynski/AA; **523** (tl) Amos Nachoum/PT; (b) Robert Frank/MT; **524** (t) Tom Stack/TSA; (b, c) Tom McHugh/ Steinhart Aquarium/PR; **525** (b) David Beck/RMF; (t) Laurie O'Keefe; **526** (l) Fred Bavendam/PA; (br) SP/Foca; (bl) Jeffrey L. Rotman/PA; **526-527** J. Stafford-Deitsch/Ellis nature Photography; **528** SP/Foca; **529** (t) Laurie O'Keefe; (b) Mickey Gibson/AA; **530** (c) David Beck/RMF; (bkgd) Hans Reinhard/BC; (b) Kevin A. Somerville/MIS; **531** (t) Tom McHugh/PR; (b) David Beck/RMF; **532** Peter Scoones/Planet Earth Pictures; **533** (l) Zig Leszczynski/AA; (r) SP/Foca; **534** (tl) Hans Pfletschinger/PA; (cl, bl) Stephen Dalton/PR; (br) SP/Foca; **535** (c) Walter Stuart/RWS; (bkgd) Schafer & Hill/PA; (b) Kevin A. Somerville/MIS; **536** (tl) Juan M. Renjifo/AA; (tr) Tom McHugh/PR; (cr, br) SP/Foca; **537** (tl) Robert Frank/MT; (c) Tom McHugh/PR; (tr) Michael Fogden/AA; (b) SP/Foca; **538** Zig Leszczynski/AA; **539** SP/Foca; **540** Hans Reinhard/BC; **541** SP/Foca; **542** (t) Kip Carter/KC; **542-543** (b) SP/Foca; **543** (tl) Jeff Lepore/PR; **544** (branch) David Beck/RMF; (lizard) Lynn Prentice/LPI; (bkgd) BC; (hand) Kevin A. Somerville/MIS; **545** (t) Pond & Giles/PS; (b) Pat Ortega; **546-547** Pat Ortega; **548** (t) M. H. Sharp/PR; (c, b) SP/Foca; **549** (t) John Cancolosi; (cr, br) E. R. Degginger/PR; **550** (t) Zig Leszczynski/AA; (b) Pat Ortega; **551** (t to b) Pat Ortega; SP/Foca; SP/Foca; M. H. Sharp/PR; John Cancalosi; **552** James L. Amos/PR; **553** (l, r, cr) SP/Foca; (cl) Cabisco/VU; **554** Dan Guravich/PR; **555** (c) Daniel Kirk/BA; (bkgd) Erwin and Peggy Bauer/BC; (bird) Virgil Wong; **555** (hand) Kevin A. Somerville/MIS; **556-557** Lynn Prentice/LPI; **558** David Ashby; **559** BC; **560** (t) F. Prenzel/AA; (b) Jean Philippe Varin/JACANA/PR; **561**

(l) © 1991 John Cancalosi/TSA; (r) E. R. Degginger/AA; **562-563** Lynn Prentice/Craig Tyler/LPI; **564** SP/Foca; **565** (tr) John Daugherty/Lynn Prentice/LPI; (antlers) SP/Foca; (b) Ana Laura Gonzalez/AA; (br) Pat & Tom Leeson/PR; **566** SP/Foca; **567** (c) B. Hansen/NSI; (bkgd) Chris Faust/Panoramic Images; (tl) Kevin A. Somerville/MIS; **568** (l) Merlin D. Tuttle/Bat Conservation International/PR; (r) Gilbert Grant/PR; **569** (tl, c, b,) SP/Foca; (tr) THE FAR SIDE cartoon by Gary Larson is reprinted by permission of Chronicle Features, San Francisco, CA. All rights reserved; (bc) BC; **570** Hans Reinhard/BC; **571** (t, c) SP/Foca; (b) Lynn Prentice/LPI; **572** David M. Phillips/PR; **573** (b) Srulik Haramaty/PT; (t) Oliver Meckes/Ottawa/PR; **574** Darrell Wong/TS; **575** (cl) S. Elems/VU; (tl) CNRI/SPL/PR; (bl, cr) BPA/PR; (br) Ed Reschke/PA; (c) Peg Gerrity/HS; **576** (l) Robert Becker/CM; (r) J. Gross, Biozentrum/SPL/PR; (bc) CMRI/SPL/PR; **577** (t) SP/Foca; (bl, br) Peg Gerrity/HS; **578** Robert Margulies/MMA; **579** SP/Foca; **580** (tr) David Scharf/PA; (l, b) SP/Foca; (cr) CNRI/SPL/PR; **581** SP/Foca; **582** Andre Jenny/Stock South/PNI; **583** (t) SPL/PR; (b) James Stevenson/PR; **584** SP/Foca; **585** Robert Margulies/MMA; **586** (l) SP/Foca; (inset) Doug Sizemore/VU; **586-587** (chart) Foca; **587** (l) SP/Foca; (tr, br) Michael Klein/PA; **588** SP/Foca; **589** (illus.) Peg Gerrity/HS; (photos) SP/Foca; **590** Peg Gerrity/HS; **591** (illus.) T. Narashima/SS; (c) SP/Foca; (tl) M. Abbey/PR; (cl) John Burbridge/SPL/PR; (bl) BPA/PR; **592** (c) SP/Foca; (l) Don W. Fawcett/VU; (r) Robert Caughey/VU; **593** Robert Margulies/MMA; **594** (l) SP/Foca; (r) Eric Sanford/TSA; (bkgd) Tom Hollyman/PR; **595** (l) SP/Foca; (r) Rafael Macia/PR; **596** (t) SPL/CM; (cr) Walter Stuart/RWS; (bl) Todd Buck/HS; (cl) Robert Margulies; (br) Peg Gerrity/HS; **597** (tl) Peg Gerrity/HS; (tr) S. C. Reuman/VU; (tc) David Scharf/PA; (bc, bl) SP/Foca; **598** Darrell Wong/TS; **599** SP/Foca; **600** Bruce Curtis/PA; **601** (t) Don Fawcett/Science Source/PR; (b) Biophoto Associates/PR; **602** Kip Carter/KC; **603** Foca; **604** Kip Carter/KC; **605** Kip Carter/KC; **606** Kip Carter/KC; **607** Kip Carter/KC; **608** Kip Carter/KC; **610** Bob Daemmrich/Boston Stock, Inc./PNI; **611** Kip Carter; **612** SPL/CM; **613** (b) Todd Buck/HS; (t, tc) SP/Foca; (bc) Bruce Curtis/PA; **614** Fred Hossler/VU; **615** (l) Pete Saloutos/TSM; (r) Howard Sochurek/TSM; **616** (t) G. Soury/Jacana/PR; (bl) Michael Fogden/OSF/AA; (br) Dan Sudia/PR; (illus.) Todd Buck/HS; **617** (illus.) Todd Buck/HS; (photo) SP/Foca; **618** (r) Alfred Pasieka/PR; (l) John W. Karapelou/KBI; **619** (l) Ed Reschke/PA; (r) J. L. Weber/PA; **620** (b) SIU/VU; (t) SP/Foca; **622** (photo) SP/Foca; (illus.) T. McDermott/Michael Kress Russick/TMMK; **623** (illus.) T. McDermott/Michael Kress-Russick/TMMK; (photo) SP/Foca; **624** (tl, br) SP/Foca; (tc) John W. Karapelou/KBI; (tr) Robert Margulies/MMA; **625** (r) SP/Foca; (l) John W. Karapelou/KBI; **626** (r) T. McDermott/Michael Kress-Russick/TMMK; (l, c) SP/Foca; **627** (tl) Biophoto Associates/PR; (c) Todd Buck/HS; (b) John W. Karapelou/KBI; (tr) NIH/Science Source/PR; **628** Bruce Curtis/PA; **629** Kip Carter; **630** Secchi-Lecaque/Roussel-UCLAF/CNRI/SPL/PR; **631** (b) Peg Gerrity/HS; (t) Tim Brown/Index Stock; **632** SP/Foca; **633** (t) Peg Gerrity/HS; (r) Kevin A. Somerville/MIS; (photos) SP/Foca; **634** SP/Foca; **635** (tr, bl) Robert Margulies/MMA; (tl, br) Peg Gerrity/HS; **636** Robert Margulies/MMA; **637** Walter Stuart/RWS; **638** Walter Stuart/RWS; **639** Everett Collection; **640** Kevin A. Somerville/MIS; **641** John McDonough/Sports Illustrated; **642** (t) John Paul Kay/PA; (b) Tom Stack/TSA; (c) SP/Foca; **643** Peg Gerrity/HS; **644** SP/Foca; **645** SP/Foca; **646** (tr) @1992/ CM; (l) David Madison; (bc, bl, br) SP/Foca; **647** (tr) Corbis-Bettmann; (c) Robert Margulies/MMA; (b) John McDonough/Sports Illustrated; (tl) SP/Foca; **648** Secchi-Lecaque/Roussel-UCLAF/CNRI/SPL/PR; **650** CMRI/SPL/PR; **651** (b) Corbis-Bettmann; (t) © 1990 CM Stock Photo; **652** (r) David M. Phillips/VU; (l) Todd Buck/HS; **653** (cl) David M. Phillips/VU; (r) Robert Caughey/VU; **654** T. McDermott/Michael Kress-Russick/TMMK; **655** Walter Stuart/RWS; **656** (l) Kevin Beebe/CM; (r) Bob Daemmrich/Boston Stock, Inc./PNI; **657** Todd Buck/HS; **658** SP/Foca; **659** E. Alexander/AT; **660** C. Turner/AT; **661** C. Turner/AT; **662** (heart cycle) Virgil Wong; (tl) Treat Davidson/Nat'l Audobon Society/PR; (bl) SP/Foca; (tr, br) Kip Carter; **663** (heart cycle) Virgil Wong; (tr) SP/Foca; (b) Miriam Austerman/AA; (tl, bl) Kip Carter; **664** (b) Todd Buck/HS; (t) SP/Foca; **665** (bkgd, b) SP/Foca; (l) Courtesy of Dora Duncan; (t) Ted Horowitz/TSM; **667** (c) Todd Buck/HS; (bl) OMIKRON/Science Source; (r) Robert Becker/CM; (b) Biophoto Associates/Science Source/PR; **668** Cabisco/VU; **669** SP/Foca; **670** Jean Marc Barey/PR; **671** (tr) C. Turner/AT; (br) David M. Phillips/VU; **672** Todd Buck/HS; **673** Todd Buck/HS; **674** (l) Art by Jude Dominique, reprinted from Kids Say Don't Smoke by permission of Workman Publishing, © Workman Press; (r) © 1991 Art by Melissa Antonow/reprinted from Kids Say Don't Smoke by permission of Workman Press; **675** (tr, tl) David M. Phillips/VU; (c) SP/Foca; (b) C. Turner/AT; **676** CMRI/SPL/PR; **677** Cabisco/VU; **678** Lennart Nilsson/Bonnier Alba AB; **679** (b) Kevin A. Somerville/MIS; (t) Dennis Kunkel/CNRI/PT; **680** Claire Booth/CBMI; **681** Robert Margulies/MMA; **682** George W. Wilder/VU; **683** Lennart Nilsson/THE BODY VICTORIOUS/National Geographiic Society/Boehringer Ingelheim Int'l GmbH; **684** Walter Stuart/RWS; **685** (cells) Walter Stuart/RWS; (plasma) Bob Margulies/MMA; **687** Guy Gillette/Science Source/PR; **688** Dr. Larry Arlian/Wright State Univ.; **689** Claire Booth/CBMI; **690** Dr. A. Liepins/SPL/PR; **691** Lennart Nilsson/Boehringer Ingelheim Int'l GmbH; **692** PEOPLE Weekly © 1990 Kim Komenich; **693** (br) SP/Foca; (tl) Michael Grecco/Outline Press/Sygma; (tr) Photofest; **695** (tr) Lennart Nilsson/THE BODY VICTORIOUS/National Geographiic Society/Boehringer Ingelheim Int'l GmbH; (bl) Dr. Larry Arlian/Wright State Univ.; (tl) Robert Margulies/MMA; (c) Walter Stuart/RWS; **696** Lennart Nilsson/Bonnier Alba AB; **697** Walter Stuart/RWS; **698** CNRI/SPL/PR; **699** SP/Foca; **700** Courtesy USDA; **701** SP/Foca; **702** SP/Foca; **703** SP/Foca; **704** SP/Foca; **705** (l) © Danna Byrom; (t) © CS Inc.; (bkgd, b) SP/Foca; **706** SP/Foca; **707** T. McDermott; **708** (b) Kevin A. Somerville/MIS; (tl) T. McDermott/Michael Kress-Russick/TMMK; **709** (tl) T. McDermott/Michael Kress-Russick/TMMK; (b) Robert Margulies/MMA; **710** Walter Stuart/RWS; **711** Robert Margulies/MMA; **712** (r) Yemi; (l) 1987 SIU/CM; **713** (tr) SIU/CM; (tl) Courtesy USDA; (b) Walter Stuart/RWS; (c) SP/Foca; **714** CNRI/SPL/PR; **715** T. McDermott/Michael Kress-Russick/TMMK; **716** (t) CTK/Sovfoto/Eastfoto/PNI; (b) Srulik Haramaty/PT; **717** Pete Saloutos/TSM; **718** Lennart Nilsson/Bonnier Alba AB; **719** (t) David M . Phillips/VU; (c) Peg Gerrity/HS; (b) Lennart Nilsson/Bonnier Alba AB; **720** Robert Margulies/MMA; **721** D. Fawcett/PR; **722** Peg Gerrity/HS; **723** (r) E. Alexander/AT; (l) Kevin A. Somerville/MIS; **724** Robert Margulies/MMA; **725** E. Alexander/AT; **727** Peg Gerrity/HS; **728** Lennart Nilsson/Bonnier Alba AB; **729** (t) E. Alexander/AT; (b) Lennart Nilsson/A CHILD IS BORN/Dell Publishing Co.; **730** (r) C. Turner/AT; (l) Robert Margulies/MMA; **731** SP/Foca; **732** (t) Courtesy Dr. A. P. Streissguth/School of Medicine, Univ. of Wash., Seattle; (b) SP/Foca; **733** Kevin A. Somerville/MIS; **734** (t) Biophoto Associates/Science Source/PR; (cl, cr, b) Lennart Nilsson/Bonnier Alba AB; **735** Lennart Nilsson/Bonnier Alba AB; **736** C. Turner/AT; **737** Lawrence Migdale/PR; **738** EM unit, CVL Weybridge/SPL/PR; **740** Dr. R. Dourmashkin/SPL/PR; **741** (tr) Courtesy George Johnson; (tl) Peg Gerrity/HS; (c) E. Alexander/AT; (bc) Biophoto Associates/Science Source/PR; (b) EM unit, CVL Weybridge/SPL/PR; **742** Lennart Nilsson/Bonnier Alba AB; **744** (tr) PA; **744-745** SP/Foca; **747** SP/Foca; **749** (t) Ted Horowitz/TSM; (r) SP/Foca; **750** SP/Foca; **753** SP/Foca; **754** SP/Foca; **755** SP/Foca; **756** PA; **757** (c) SP/Foca; (b) Ted Horowitz/TSM; **758** (l) Laurence Gould/Oxford Scientific Films/AA; (c) Hal Beral/Photo Network/PNI; (bar) Herb Schmitz/TS; **759** AA/ES; **760** SP/Foca; **761** Matt Meadows/PA; **762** SP/Foca; **763** (l) Hank Morgan/PR; (t) SP/Foca; **764** Renee Lynn/TS; **765** (t) SP/Foca; (b) Kaluzny/Thatcher/TS; **766** Michael Rosenfeld/TS; **767** SP/Foca; **770** Biology Media/PR; **771** (t) SP/Foca; (b) Mark Gibson/Index Stock; **774** SP/Foca; **775** (t) Ed Reschke/PA; (b) Peter Menzel/Boston Stock, Inc./PNI; **776** Tom Galliher/Sharpshooters; **777** J. Nettis/H. Armstrong Roberts; **780** (r) Manfred Kage/PA; (l) Ed Reschke/PA; **781** (tl, tr) SP/Foca; (b) Will & Deni McIntyre/PR; **783** SP/Foca; **784** SP/Foca; **785** (t) SP/Foca; (b) James Holmes/Cellmark Diagnostics/SPL/PR; **786** (l) SP/Foca; Calculator, Permission of Texas Instruments; (r) SP/Foca; **787** Jose L. Pelaez/TSM; **788** Carlyn Iverson; **789** (t) Carlyn Iverson; (b) Johnny Johnson/AA; **790** (t) Nathan Benn/WCA/PNI; (b) SP/Foca; **791** (b) Eugene Mopsik/H. Armstrong Roberts; (t) SP/Foca; **792** SP/Foca; **793** SP/Foca; **794** Mark Sherman/BC; **795** (b) © CS Inc./Michael Thompson; (t) SP/Foca; **796** (bl) Robert & Linda Mitchell; (tl) Hans Reinhard/BC; (r) SP/Foca; **797** Michel Viard/PA; **798** SP/Foca; **799** (t) SP/Foca; (b) Michel Viard/PA; **800** John Paling/Oxford Scientific Films/ES; **801** Stephen Frisch/Boston Stock, Inc./PNI; **802** (t) Nancy Honey/TS; (b) SP/Foca; **803** SP/Foca; **804** (l) Ralph A. Reinhold/AA; (r) SP/Foca; **805** Eric Kamp/Index Stock; **806** SP/Foca; **807** SP/Foca; **808** (bl) Martha Cooper/PA; (br) Jim Tuten/AA; **809** Werner H. Muller/PA; **812** Sinclair Stammers/SPL/PR; **813** Estate of Sybil Shelton/PA; **816** Diane Nelson/CM; **817** Kevin Beebe/CM; **818** Bob Daemmrich/Boston Stock, Inc./PNI; **819** (t) D. R. Stoecklein/TSM; (b) SIU/PR; **820** SP/Foca; **821** Matthew Borkoski/Index Stock; **823** @1992/CM; **826** (tl) SP/Foca; **826-827** SP/Foca; **828** (tl) M. Abbey Photo/PR; (l, r) SP/Foca; **829** (tl) Dr. Arnold Brody/SPL/PR; (tc) Electra/PT; (tr) Dennis Kunkel/CNRI/PT; (bl, r) SP/Foca; **830** SP/Foca; **831** (t) SP/Foca; (cl) Carlyn Iverson; **843** SP/Foca; **844** SP/Foca; **845** (t) David Cavagnaro/PA; (b) SP/Foca; **846** (t, cr) Courtesy USDA; (b) SP/Foca; **848** Lynne Prentice/LPI; (human) Kevin Somerville; **849** (methanococcus) Wendy Smith-Griswold; (other illus.) Lynne Prentice/LPI; **850** Lynne Prentice/LPI; **851** Lynne Prentice/LPI; **852** Lynne Prentice/LPI; **853** Lynne Prentice/LPI; **854** Lynne Prentice/LPI; **855** Lynne Prentice/LPI